Advances in Modal Logic

Volume 15

Advances in Modal Logic
Volume 15

Edited by

Agata Ciabattoni
David Gabelaia

and

Igor Sedlár

ISBN 978-1-84890-467-5

College Publications
Scientific Director: Dov Gabbay
Managing Director: Jane Spurr

http://www.collegepublications.co.uk

Contents

Preface ix

Abstracts of Invited Talks 1

Alexandru Baltag
Knowability as Continuity: The Modal Logic of Continuous and Uniform Dependence 3

Marta Bílková
Group Epistemics, (Co-)algebraically 5

David Fernández-Duque
Modal Logics in Dynamical Systems 9

Cláudia Nalon
Efficient Theorem-Proving for Modal Logics 13

Contributed papers 17

Bahareh Afshari, Lide Grotenhuis, Graham E. Leigh and Lukas Zenger
Intuitionistic Master Modality 19

Juan P. Aguilera and Grigorii Stepanov
Strong Completeness of the Closed Fragment of GLP 41

Rodrigo Nicolau Almeida and Nick Bezhanishvili
Coalgebraic Semantics for Intuitionistic Modal Logic 59

Rodrigo Nicolau Almeida and Silvio Ghilardi
Unification With Simple Variable Restrictions and Admissibility of Π_2-Rules 79

Melissa Antonelli and Jan von Plato
On the Proof Theory of Apodictic Syllogistic 101

PHILIPPE BALBIANI AND HANS VAN DITMARSCH
Towards Dynamic Distributed Knowledge 125

ALEXANDRU BALTAG AND SONJA SMETS
Logics for Data Exchange and Communication 147

LINE VAN DEN BERG, MANUELA BUSANICHE, MIGUEL MARCOS AND GEORGE METCALFE
Towards an Algebraic Theory of KD45-Like Logics 171

NICK BEZHANISHVILI, LAURA BUSSI, VINCENZO CIANCIA, DAVID FERNÁNDEZ-DUQUE AND DAVID GABELAIA
Logics of Polyhedral Reachability 187

NICK BEZHANISHVILI, DAVID FERNÁNDEZ-DUQUE AND REIHANE ZOGHIFARD
The Goldblatt-Thomason Theorem for Derivative Spaces 205

MARTA BÍLKOVÁ, HANS VAN DITMARSCH, ROMAN KUZNETS AND ROJO RANDRIANOMENTSOA
Bisimulation for Impure Simplicial Complexes 225

THOMAS BOLANDER AND ALESSANDRO BURIGANA
Better Bounded Bisimulation Contractions 249

JESSE COMER
Lovász Theorems for Modal Languages 269

YIFENG DING AND YIPU LI
Some General Completeness Results for Propositionally Quantified Modal Logics 293

YIFENG DING, KRISHNA MANOORKAR, MATTIA PANETTIERE AND RUODING WANG
Toward the van Benthem Characterization Theorem for Non-Distributive Modal Logic 315

DAMIANO FORNASIERE, JOHANNES MARTI AND GIOVANNI VARRICCHIONE
Frame Definability in Conditional Logic 335

DANIEL GAINA
Birkhoff Style Proof Systems for Hybrid-Dynamic Quantum Logic ... 357

Rustam Galimullin and Louwe B. Kuijer
Varieties of Distributed Knowledge 379

Vitor Greati and Revantha Ramanayake
Deducibility in the Full Lambek Calculus with Weakening Is HAck-Complete 401

Wesley H. Holliday
Modal Logic, Fundamentally 423

Ahmee Christensen
Logics of Knowability 447

Agi Kurucz, Frank Wolter and Michael Zakharyaschev
The Interpolant Existence Problem for Weak K4 and Difference Logic 465

Xinghan Liu and Emiliano Lorini
The Complexity of Reasoning about Classifiers 485

Guillaume Massas
Goldblatt-Thomason Theorems for Fundamental (Modal) Logic 505

Borja Sierra Miranda, Thomas Studer and Lukas Zenger
Coalgebraic Proof Translations for Non-Wellfounded Proofs 527

Mojtaba Mojtahedi and Konstantinos Papafilippou
Projectivity Meets Uniform Post-Interpolant: Classical and Intuitionistic Logic 549

Valentin Müller
Natural Deduction, Normalization and Subformula Property for Kreisel-Putnam Logic 565

Wolfgang Poiger
Positive Modal Logic Over Finite MV-Chains 587

Vít Punčochář and Ivo Pezlar
Informative Presupposition in Inquisitive Logic 609

Dmitry Rozplokhas
LEGO-Like Small Model Constructions for Åqvist's Logics 631

MIKHAIL RYBAKOV, DMITRY SHKATOV AND DMITRIJ SKVORTSOV

On the System of Positive Slices in the Structure of Superintuitionistic Predicate Logics 653

SOFÍA SANTIAGO-FERNÁNDEZ, JOOST J. JOOSTEN AND DAVID FERNÁNDEZ-DUQUE

A Tree Rewriting System for the Reflection Calculus 675

YANJING WANG AND JUNHUA YU

Point-Set Neighborhood Logic 697

XUEFENG WEN

A Modal Logic for Reasoning in Contexts 719

YUANZHE YANG

A First-order Modal Logic of Strict Implication on Varying-Domain Models 741

Preface

Advances in Modal Logic (AiML) is an initiative founded in 1995 and aimed at presenting an up-to-date picture of the state of the art in modal logic and its many applications. It consists of a conference series together with volumes based on the conferences. The conference series is the main international forum at which research on all aspects of modal logic is presented. The first installment was held in 1996 in Berlin, Germany, and since then it has been organized biennially, with meetings in 1998 in Uppsala, Sweden; in 2000 in Leipzig, Germany (jointly with ICTL-2000); in 2002 in Toulouse, France; in 2004 in Manchester, UK; in 2006 in Noosa, Australia; in 2008 in Nancy, France; in 2010 in Moscow, Russia; in 2012 in Copenhagen, Denmark; in 2014 in Groningen, the Netherlands; in 2016 in Budapest, Hungary; in 2018 in Bern, Switzerland (jointly with LATD 2018); in 2020 in Helsinki, Finland and in 2022 in Rennes, France (co-located with LAMAS&SR 2022). Information about AiML and related events, including conference proceedings, is available at the website www.aiml.net.

The fifteenth conference in the AiML series was organized by Igor Sedlár (the Institute of Computer Science of the Czech Academy of Sciences, ICS CAS) with the assistance of Marta Bílková, Petr Cintula, Wesley Fussner, Zuzana Haniková, Raheleh Jalali, Filip Jankovec, Chun-Yu Lin, Hana Stříhavková (ICS CAS), Olga Bažantová, Ondrej Majer, Vít Punčochář (the Institute of Philosophy of the Czech Academy of Sciences), Vojtěch Kolman (Faculty of Arts of Charles University). The conference was co-located with the 21st International Conference on Relational and Algebraic Methods in Computer Science (RAMiCS 2024). The website of AiML 2024 can be found at www.cs.cas.cz/aiml2024/.

This volume contains abstracts of invited talks and contributed papers from the conference. The invited talks were given by

- Alexandru Baltag (University of Amsterdam)
- Marta Bílková (Czech Academy of Sciences)
- David Fernández-Duque (University of Barcelona)
- Claudia Nalon (University of Brasilia)

The Programme Committee received 64 regular paper submissions. Of these, 34 were selected for this volume by a reviewing process where every paper received three independent expert reviews.

The volume includes papers on propositional modal logics, their products and fusions, predicate modal logics, dynamic epistemic logics, spatial logics, deontic logics, many-valued modal logics, provability logics, graded modal logic, non-distributive modal logic, apodictic syllogistic, inquisitive, intuitionistic, substructural, conditional, fundamental, probabilistic, hybrid, quantum logics and other related logics. The topics of the papers at AiML 2024 include history of modal reasoning, proof theory, model theory, interpolation, unification, canonicity and canonical extensions, propositional quantifiers, bisimulations, neighbourhood semantics, polyhedral and simplicial semantics, coalgebraic semantics, decidability and complexity results, Kripke and topological completeness, as well as results in the style of van Benthem characterization theorem, Goldblatt-Thomason theorem, Hennessy-Milner theorem and Lovász theorem.

In addition, there were 17 submissions for short presentations at the conference, and 12 were accepted after a reviewing process whereby each submission received at least 3 independent expert reviews.

The members of the AiML 2024 Programme Committee were:

- Philippe Balbiani (Institut de Recherche en Informatique de Toulouse)
- Lev Beklemishev (Steklov Institute of Mathematics)
- Nick Bezhanishvili (University of Amsterdam)
- Willem Conradie (University of the Witwatersrand)
- Tommaso Flaminio (IIIA–CSIC)
- Silvio Ghilardi (University of Milan)
- Nina Gierasimczuk (Technical University of Denmark)
- Valentin Goranko (Stockholm University)
- Rajeev Gore
- Wesley Holliday (University of California, Berkeley)
- John Horty (University of Maryland)
- Emil Jeřábek (Institute of Mathematics, Czech Academy of Sciences)
- Clemens Kupke (University of Strathclyde)
- Agi Kurucz (King's College London)
- Roman Kuznets (TU Wien)
- Emiliano Lorini (Institut de Recherche en Informatique de Toulouse)
- Tim Lyon (Technische Universität Dresden)
- Tommaso Moraschini (University of Barcelona)
- Larry Moss (Indiana University Bloomington)
- Sara Negri (University of Genova)
- Nicola Olivetti (Aix-Marseille Université)
- Alessandra Palmigiano (VU Amsterdam)
- Xavier Parent (TU Wien)
- Elaine Pimentel (UCL)
- Vít Punčochář (Institute of Philosophy, Czech Academy of Sciences)
- Revantha Ramanayake (University of Groningen)
- Renate A. Schmidt (The University of Manchester)
- Igor Sedlár (Czech Academy of Sciences)
- Ilya Shapirovsky (New Mexico State University)
- Viorica Sofronie-Stokkermans (University of Koblenz)

- Balder Ten Cate (ILLC, University of Amsterdam)
- Hans van Ditmarsch (University of Toulouse, CNRS, IRIT)
- Yde Venema (University of Amsterdam)
- Rineke Verbrugge (University of Groningen)
- Heinrich Wansing (Ruhr University Bochum)
- Frank Wolter (University of Liverpool)

The Programme Committee was chaired by:

- Agata Ciabattoni (TU Wien)
- David Gabelaia (TSU Razmadze Mathematical Institute)

The Steering Committee of AiML for 2024–2026 consists of:

- Guram Bezhanishvili (New Mexico State University)
- Agata Ciabattoni (TU Wien)
- Giovanna D'Agostino (University of Udine)
- David Fernández-Duque (University of Barcelona)
- David Gabelaia (TSU Razmadze Mathematical Institute)
- Nicola Olivetti (Aix-Marseille Université)
- Alessandra Palmigiano (VU Amsterdam)
- Igor Sedlár (Czech Academy of Sciences)
- Rineke Verbrugge (University of Groningen)

Many other people assisted with the reviewing process, including:

Jean-Marie Le Bars, Bartosz Bednarczyk, Hanoch Ben-Yami, Katalin Bimbo, Adam Bjorndahl, Luca Carai, Ivano Ciardelli, Giorgio Cignarale, Abhishek De, Yiwen Ding, Andrea De Domenico, Sergey Drobyshevich, Lev Dvorkin, Sebastian Enqvist, Raül Espejo-Boix, Jie Fan, Nicholas Ferenz, Camillo Fiorentini, Damiano Fornasiere, Tim French, Peter Fritz, Rustam Galimullin, Didier Galmiche, Han Gao, Francesco Antonio Genco, Iris van der Giessen, Marianna Girlando, Roberto Giuntini, Quentin Gougeon, Giuseppe Greco, Gianluca Grilletti, Jim de Groot, Ronald de Haan, Ullrich Hustadt, Joost Joosten, Bruce Kapron, Roman Kniazev, Andrey Kudinov, Louwe B. Kuijer, Serafina Lapenta, Tadeusz Litak, Xinghan Liu, Minghui Ma, Ondrej Majer, Matteo Manighetti, Krishna Balajirao Manoorkar, Miguel Martins, Andrea Masini, Daniel Mery, Marianela Morales, Cláudia Nalon, Victor Barroso Nascimento, Pavel Naumov, Karl Nygren, Sergei Odintsov, Mattia Panettiere, Fabio Papacchini, Konstantinos Papafilippou, Edi Pavlović, Mati Pentus, Luiz Carlos Pereira, Ivo Pezlar, Adam Přenosil, Luca Reggio, Mikhail Rybakov, Simon Santschi, Thomas Schlögl, Ezra Schoen, Ian Shillito, Dmitry Shkatov, Sergey Slavnov, Vladislav Sliusarev, John Stell, Grigorii Stepanov, Apostolos Tzimoulis, Amanda Vidal, Cheng-Syuan Wan, Yì Nicholas Wáng, Kentaro Yamamoto, Lukas Zenger, Margherita Zorzi.

The organisers of the conference are grateful to the following sponsors for their support: The Association for Logic, Language and Information, Czech

Science Foundation (projects 22-01137S, 22-23022L and 22-16111S), Czech Society for Cybernetics and Informatics, Institute of Computer Science of the Czech Academy of Sciences, Institute of Philosophy of the Czech Academy of Sciences, Faculty of Arts of Charles University, the MOSAIC project EU H2020-MSCA-RISE-2020, Grant Agreement number 101007627, and the Strategy AV21 program of the Czech Academy of Sciences.

The editors of this volume would like to thank all the members of the Programme Committee and the subreviewers for their time and effort, all the authors of contributed papers and invited speakers for their cooperation in the production of these proceedings, and Filip Jankovec for his help with the typesetting.

24 July 2024

Agata Ciabattoni,
David Gabelaia,
Igor Sedlár

Abstracts of Invited Talks

Knowability as Continuity: The Modal Logic of Continuous and Uniform Dependence

Alexandru Baltag

Institute of Logic, Language and Computation
University of Amsterdam
Amsterdam, The Netherlands

I investigate modal logics for various forms of functional dependence between variables in topological spaces. The motivation is epistemic, the aim being to model and study knowable informational dependence between empirical questions (for which only approximate answers are empirically observable). In this context, knowability of the dependence corresponds to continuity of the underlying function. I also investigate epistemic independence in topological terms and show that it is compatible with functional (but non-continuous) dependence. I then proceed to study a stronger notion of knowability based on uniformly continuous dependence. On the technical logical side, I determine the complete logics of languages that combine general functional dependence, continuous dependence, and uniformly continuous dependence. This talk is based on recent joint work with Johan van Benthem.

Group Epistemics, (Co-)algebraically

Marta Bílková [1]

The Czech Academy of Sciences, Institute of Computer Science, Prague

Abstract

In the vast majority of contributions to multi-agent epistemic, doxastic, and coalition logic, a group is reduced to its extension, i.e., the set of its members. Membership in groups is common knowledge to all agents, with the counterintuitive consequence that groups change identity when their membership changes, and it precludes uncertainty about who is a member of a given group. This idealisation does not reflect the structure of groups or the structured way in which collective epistemic attitudes emerge in the intended application of logical models. Epistemic logics of intensional groups lift the extensionality assumptions above by seeing groups as given to us intensionally by a common property that can change their extension from world to world. We will outline an abstract algebraic and coalgebraic framework that replaces agent or group labels of epistemic modalities with names, and gives them an algebraic structure relevant to the types of collective epistemic attitudes in question. (The talk is based on ongoing collaborative work with Zoé Christoff, Wesley Fussner, Olivier Roy, and Igor Sedlár).

Keywords: Epistemic logic, coalgebraic logic, neighborhood semantics, common knowledge, distributed knowledge.

One of the common assumptions of multi-agent epistemic logic is that groups of agents are given *extensionally* as sets of agents. As a consequence, group membership is common knowledge to all agents, and any change in membership implies a corresponding change in the identity of a group. This is not how we usually think of groups: we routinely reason in various contexts without knowing the extensions of groups - we might, for example, refer to groups such as "bot accounts", "democrats", or "correct processes" - and we are not satisfied with reducing groups to their extensions either, since they can clearly change over the state space of a system or over possible states of the world. Epistemic logics of intensional groups lift the above assumptions by seeing groups as given to us *intensionally* by a common property that can change its extension from world to world—as, e.g., being a "consistent subgroup" does. In their seminal work [6,5], Grove and Halpern introduced a multi-agent epistemic logic in which groups are labelled by abstract *names* and the naming

[1] Work on this paper was supported by the grant no. 22-23022L CELIA of the Czech Science Foundation.

relationship can vary from world to world. The language originally contains two types of modalities: $E_n\varphi$ means "everyone named n knows that φ", and $S_n\varphi$ means "someone named n knows that φ"; they further consider a natural extension of the basic framework where names are replaced by formulas expressing *structured* group-defining concepts. Motivated mainly by applications such as dynamic networks of processes, another framework where the set of agents can vary from state to state has been developed in a form of term-modal logic. Introduced by [4], it builds on first-order logic by indexing modalities with terms that can be quantified over. Epistemic logic with names of [6] was in a sense seminal to the development of term-modal logic, and can be seen as its simple decidable fragment (a closely related language of implicitly quantified modal logic was studied in [11]). Grove and Halpern's work is enjoying a recent resurgence of interest in the epistemic logic community: for instance, in [1], we considered expansions of epistemic logic with names based on with non-rigid versions of common and distributed knowledge, derived using $E_n\varphi$ and $S_n\varphi$ modalities respectively. On the other hand, Humml and Schröder [9] generalize Grove and Halpern's approach to structured names represented by formulas defining group membership, including, for example, formulas of the description logic ALC. Their abstract group epistemic logic AGEL contains a common knowledge modality as the only modality and, unlike in [1,6], their group names are rigid.

In [2], we adopted the perspective that (i) both "everyone labeled a knows" and "someone labeled a knows" modalities form a minimal epistemic language for group knowledge where groups are understood intensionally, and (ii) that their labels reflect their structured nature. We used languages built on top of classical propositional language containing modalities $[a], \langle a]$ indexed by elements of an algebra of a given signature of interest, we set up a general relational semantics involving an algebra of group labels to index (sets of) relations in each world, and shown how some related logics can be modelled in such a way. The minimal logic (for the empty signature) coincides with multimodal expansion of a logic known as bi-modal monotone logic [10], and traces back to Brown's Logic of ability [3].

A fully abstract account of modal logics based on this minimal logic is clearly desirable, and although monotone modal logic is well known and investigated coalgebraically, the combination of monotone and normal operators involved in our case has not received the same kind of attention. Following previous work by Hansen et al. [7,8], we give a coalgebraic account of the logics considered in [2]. We concentrate on linking two-sorted algebras (involving propositions and group labels/types of knowledge) with appropriate neighborhood frame semantics, in terms of an appropriate (topological or discrete) duality. This can hopefully be further applied, e.g. to obtain a definability theorem or to design a multi-type proof theory for basic logic. We also discuss some particular examples of algebraic signatures that give rise to interesting and useful variants of group knowledge, such as distributed or common knowledge.

References

[1] Bílková, M., Z. Christoff and O. Roy, *Revisiting epistemic logic with names*, in: J. Halpern and A. Perea, editors, *TARK 2021*, 2021, pp. 39–54.

[2] Bílková, M. and I. Sedlár, *Epistemic logics of structured intensional groups*, in: R. Verbrugge, editor, Proceedings Nineteenth conference on *Theoretical Aspects of Rationality and Knowledge,* Oxford, United Kingdom, 28-30th June 2023, Electronic Proceedings in Theoretical Computer Science **379** (2023), pp. 113–130.

[3] Brown, M. A., *On the logic of ability*, Journal of Philosophical Logic **17** (1988), pp. 1–26.

[4] Fitting, M., L. Thalmann and A. Voronkov, *Term-modal logics*, Studia Logica **69** (2001), pp. 133–169.

[5] Grove, A. J., *Naming and identity in epistemic logic part ii: a first-order logic for naming*, Artificial Intelligence **74** (1995), pp. 311–350.

[6] Grove, A. J. and J. Y. Halpern, *Naming and identity in epistemic logics. Part I: The propositional case*, Journal of Logic and Computation **3** (1993), pp. 345–378.

[7] Hansen, H. H. and C. Kupke, *A coalgebraic perspective on monotone modal logic*, in: J. Adámek and S. Milius, editors, *Proceedings of the Workshop on Coalgebraic Methods in Computer Science, CMCS 2004, Barcelona, Spain, March 27-29, 2004*, Electronic Notes in Theoretical Computer Science **106** (2004), pp. 121–143.
URL `https://doi.org/10.1016/j.entcs.2004.02.028`

[8] Hansen, H. H., C. Kupke and R. A. Leal, *Strong completeness for iteration-free coalgebraic dynamic logics*, in: J. Díaz, I. Lanese and D. Sangiorgi, editors, *Theoretical Computer Science - 8th IFIP TC 1/WG 2.2 International Conference, TCS 2014, Rome, Italy, September 1-3, 2014. Proceedings*, Lecture Notes in Computer Science **8705** (2014), pp. 281–295.
URL `https://doi.org/10.1007/978-3-662-44602-7_22`

[9] Humml, M. and L. Schröder, *Common knowledge of abstract groups*, in: *AAAI '23*, 2023.

[10] Lellmann, B., *Combining monotone and normal modal logic in nested sequents - with countermodels*, in: S. Cerrito and A. Popescu, editors, *TABLEAUX 2019, Proceedings*, Lecture Notes in Computer Science **11714** (2019), pp. 203–220.

[11] Padmanabha, A. and R. Ramanujam, *Propositional modal logic with implicit modal quantification*, in: *ICLA 2019*, Springer, 2019, pp. 6–17.

Modal Logics in Dynamical Systems

David Fernández-Duque [1] [2]

Department of Philosophy
University of Barcelona
Barcelona, Spain

Abstract

Dynamic topological logic and its variant, intuitionistic temporal logic, combine topological semantics with linear temporal logic in order to reason about topological dynamics. I will provide a historical overview of the field, from its inception to recent results and open questions.

Keywords: Dymamical systems, topological semantics, linear temporal logic, intuitionistic logic

---

A common theme cutting across the natural and social sciences is the need to model the behaviour over time of a system whose evolution follow fixed laws. Mathematically, such phenomena may be represented as the continuous action of a semigroup Θ (e.g. the real or natural numbers) on a topological space X. The case where $\Theta = \mathbb{N}$ is particularly compelling from a modal perspective, as it lends itself to formalisation using the well-understood linear temporal logic (LTL) [8]. Such actions may be represented simply by a function S sending each point to its temporal successor, and for our purposes, a *dynamical system* consists of a pair (X, S), where X is a topological space and $S\colon X \to X$ is continuous.

The function S yields semantics for LTL. Meanwhile, it was already known by McKinsey and Tarski in the 1940s [12] that the topological structure of X may be used to interpret the modal logic S4. These observations led Artemov et al. [1] to introduce the logic S4C, combining the topological interior modality ($\blacksquare$) with the 'next' modality ($\circ$) from LTL. Here C stands for 'continuous', and Artemov et al. showed that the resulting logic is sound and complete for the class of dynamical systems, moreover enjoying the finite model property and hence decidability.

Kremer and Mints [11] established analogous results for the variant S4H, interpreted over dynamical systems where S is a homeomorphism. They moreover noted that by adding the 'henceforth' modality ($\Box$), non-trivial recurrence

---

[1] `fernandez-duque@ub.edu`

[2] Partially funded by the SNSF-FWO Lead Agency Grant 200021L_196176 (SNSF)/ G0E2121N (FWO).

phenomena such as the Poincaré recurrence theorem could be formalised in a modal setting. Kremer and Mints dubbed the resulting tri-modal logic *dynamic topological logic* (DTL). They also showed that, unlike the logic S4C, DTL can distinguish between its topological and Kripke semantics, from which in particular it follows that it cannot have the finite model property.

Unfortunately, DTL was soon shown to be undecidable by Konev et al. [9], and even non-axiomatisable when interpreted over systems with a homeomorphism [10]. This spurred a search for decidable variants of DTL, with Gabelaia et al. [7] proposing DTL with finite iterations and myself [4], following a suggestion of Mints, interpretations over *minimal systems,* characterised by the orbit of every point being dense. Nevertheless, in both of these cases, the complexity is non-primitive recursive. I also showed that DTL does enjoy a natural axiomatisation, albeit over an extended language [5]. In contrast, the Kripke variant of DTL has not been axiomatised or even proven to be computably enumerable.

Despite these advances, the quest for a decidable variant of dynamical topological logic interpretable over arbitrary dynamical systems with unbounded time remained elusive. One proposal that seemed to slip under the radar was Kremer's unpublished note (but see [3]), where he proposed an intuitionistic variant of DTL. Kremer showed that this variant does not validate key LTL axioms with $\Box$. Over a decade later, I realised that it *does* validate all axioms involving only 'eventually' ($\Diamond$) and indeed this variant of the logic, now dubbed intuitionistic (linear) temporal logic (ITL), is indeed decidable in primitive recursive time [6] (with the precise complexity not yet known). Moreover, both minimality and Poincaré recurrence are representable in this framework, thus finally obtaining a decidable framework for reasoning about arbitrary dynamic topological spaces as envisioned by Kremer and Mints.

Since then, Boudou, Diéguez and I have axiomatised $\mathsf{ITL}_{\Diamond\forall}$ [2] and, in forthcoming work, McLean, Zenger and I showed that by incorporating *coimplication,* $\mathsf{ITL}_{\Diamond\Box}$ becomes naturally axiomatisable, the first axiomatisation for any DTL-style Kripke-based logic where that does not coincide with its topological variant. The techniques developed here may lead to axiomatising Kripke DTL, as well as expanding products of modal logics in a more general setting [7].

In this talk we will overview these and other related results, highlighting notable proof ideas which are characteristic of the field and are likely to find greater applicability in modal logics in general and, specifically, other nonclassical dynamic logics. I will also discuss future directions and open questions, both in the field and in potential applications to other non-classical dynamic logics.

References

[1] Artëmov, S. N., J. M. Davoren and A. Nerode, *Topological semantics for hybrid systems,* in: S. I. Adian and A. Nerode, editors, *Logical Foundations of Computer Science, 4th*

International Symposium, LFCS'97, Yaroslavl, Russia, July 6-12, 1997, Proceedings, Lecture Notes in Computer Science **1234** (1997), pp. 1–8.

[2] Boudou, J., M. Diéguez and D. Fernández-Duque, *Complete intuitionistic temporal logics for topological dynamics*, J. Symb. Log. **87** (2022), pp. 995–1022.

[3] Boudou, J., M. Diéguez, D. Fernández-Duque and P. Kremer, *Exploring the jungle of intuitionistic temporal logics*, Theory Pract. Log. Program. **21** (2021), pp. 459–492.

[4] Fernández-Duque, D., *Dynamic topological logic interpreted over minimal systems*, Journal of Philosophical Logic **40** (2011), pp. 767–804.

[5] Fernández-Duque, D., *A sound and complete axiomatization for dynamic topological logic*, Journal of Symbolic Logic **77** (2012), pp. 947–969.

[6] Fernández-Duque, D., *The intuitionistic temporal logic of dynamical systems*, Logical Methods in Computer Science **14** (2018), pp. 1–35.

[7] Gabelaia, D., A. Kurucz, F. Wolter and M. Zakharyaschev, *Non-primitive recursive decidability of products of modal logics with expanding domains*, Annals of Pure and Applied Logic **142** (2006), pp. 245–268.

[8] Goldblatt, R., "Logics of Time and Computation," Number 7 in CSLI Lecture Notes, Center for the Study of Language and Information, 1992, 2 edition, second edition.

[9] Konev, B., R. Kontchakov, F. Wolter and M. Zakharyaschev, *Dynamic topological logics over spaces with continuous functions*, in: G. Governatori, I. M. Hodkinson and Y. Venema, editors, *Advances in Modal Logic 6*, (2006), pp. 299–318.

[10] Konev, B., R. Kontchakov, F. Wolter and M. Zakharyaschev, *On dynamic topological and metric logics*, Stud. Logica **84** (2006), pp. 129–160.

[11] Kremer, P. and G. Mints, *Dynamic topological logic*, Ann. Pure Appl. Log. **131** (2005), pp. 133–158.

[12] McKinsey, J. C. C. and A. Tarski, *The algebra of topology*, Ann. of Math. **45** (1944), pp. 141–191.

Efficient Theorem-Proving for Modal Logics

Cláudia Nalon [1]

Department of Computer Science, University of Brasília, Brazil

Abstract

K_SP is a resolution-based theorem prover for multimodal normal logics that implements decision procedures for the local and global satisfiability problems for the multimodal logic K_n, all logics in the cube and logics of confluence (with parameters 1 or 0). In this talk, we discuss two different resolution-based calculi for those logics that are implemented in the prover with particular focus on the characteristics that have impact on the efficiency of the implementation

Keywords: Modal Logics, Resolution, Automated Reasoning.

Modal logics have long been used in Computer Science for describing and reasoning about complex systems, with applications to programming languages, knowledge representation and reasoning, terminological reasoning, and verification. The basic normal multimodal modal logic is K_n, which extends the classical language with new operators, $[a]$ and $\langle a \rangle$, with $a \in \mathcal{A}_n = \{1, \ldots, n\}$, a fixed finite set of indexes. The local satisfiability problem for the multimodal propositional case is PSPACE-complete [2]. The global satisfiability and the local satisfiability under global constraints problems for K_n are EXPTIME-complete [13]. Some applications might require specific extensions of modal logics that better represent constraints for a particular given domain, which can be achieved by adding axioms to K_n. Here, we consider the logics in the modal cube (the ones obtained by adding one or more of the axioms $\mathbf{B}_a$, $\mathbf{D}_a$, $\mathbf{T}_a$, $\mathbf{4}_a$, and $\mathbf{5}_a$) and a subset of logics of confluence (the ones obtained by adding one or more of the instances of the parameterised axiom $\mathbf{G}_a^{p,q,r,s}$: $\langle a \rangle^p [a]^q \varphi \rightarrow [a]^r \langle a \rangle^s \varphi$, with $p, q, r, s \in \{0, 1\}$). Those axioms and their first-order characterisation are given in Fig. 1.

Given the wide range of applications for which those logics can be used for both representation and reasoning, and considering the complexity of their corresponding satisfiability problems, the development of automatic, efficient tools is desirable. Several proof methods and tools for modal reasoning exist, either in the form of methods applied direct to the modal language or obtained by translation into more expressive languages (first-order or high-order logic,

[1] Email: *nalon@unb.br*. Joint work with Ullrich Hustadt (University of Liverpool), Clare Dixon (University of Manchester) and Fabio Papacchini (Lancaster University in Leipzig).

(p,q,r,s)	Name	Axioms	Property	Condition on Frames
$(0,0,1,1)$ $(1,1,0,0)$	$\mathbf{B}_a$	$\varphi \to [a]\langle a\rangle\varphi$ $\langle a\rangle[a]\varphi \to \varphi$	symmetric	$wR_aw_1 \to w_1R_aw$
$(0,0,1,0)$ $(1,0,0,0)$	$\mathbf{Ban}_a$	$\varphi \to [a]\varphi$ $\langle a\rangle\varphi \to \varphi$	modally banal	$wR_aw_1 \to w = w_1$
$(0,1,0,1)$	$\mathbf{D}_a$	$[a]\varphi \to \langle a\rangle\varphi$	serial	$\exists w_1(wR_aw_1)$
$(1,0,1,0)$	$\mathbf{F}_a$	$\langle a\rangle\varphi \to [a]\varphi$	functional	$wR_aw_1 \wedge wR_aw_2 \to w_1 = w_2$
$(0,0,0,1)$ $(0,1,0,0)$	$\mathbf{T}_a$	$\varphi \to \langle a\rangle\varphi$ $[a]\varphi \to \varphi$	reflexive	wR_aw
$(1,0,1,1)$ $(1,1,1,0)$	$\mathbf{5}_a$	$\langle a\rangle\varphi \to [a]\langle a\rangle\varphi$ $\langle a\rangle[a]\varphi \to [a]\varphi$	Euclidean	$wR_aw_1 \wedge wR_aw_2 \to w_1R_aw_2$
$(1,1,1,1)$	$\mathbf{G1}_a$	$\langle a\rangle[a]\varphi \to [a]\langle a\rangle\varphi$	convergent	$wR_aw_1 \wedge wR_aw_2 \to$ $\exists w_3(w_1R_aw_3 \wedge w_2R_aw_3)$
$(0,1,1,1)$ $(1,1,0,1)$	$\mathbf{G}_a^{0,1,1,1}$	$[a]\varphi \to [a]\langle a\rangle\varphi$ $\langle a\rangle[a]\varphi \to \langle a\rangle\varphi$	0,1,1,1 convergent	$wR_aw_1 \to$ $\exists w_2(wR_aw_2 \wedge w_1R_aw_2)$
$(0,1,2,0)$ $(2,0,1,0)$	$\mathbf{4}_a$	$[a]\varphi \to [a][a]\varphi$ $\langle a\rangle\langle a\rangle\varphi \to \langle a\rangle\varphi$	transitive	$wR_aw_1 \wedge w_1R_aw_2 \to w_1R_aw_2$

Fig. 1. Axioms and conditions on frames for logics of confluence, where $\mathbf{G}_a^{p,q,r,s}$: $\langle a\rangle^p[a]^q\varphi \to [a]^r\langle a\rangle^s\varphi$; the free variables (w, w_1, w_2) are universally quantified.

for instance). Here we briefly present two different proof methods for K_n, namely the clausal modal resolution procedures described in [3,4].

The calculus given in [3] (GMR) operates on a very simple normal form, called SNF_K, which allows for total separation of the propositional and modal parts of a problem. However, the clausal language requires clauses to be in the scope of the universal operator [*], which might lead to the production of clauses that cannot ever be used in a refutation. The calculus given in [4] (MLR) minimises this problem by taking advantage of well-known properties of the satisfiability problem for K_n, namely: a formula φ is satisfiable if, and only if, it is satisfiable in a finite tree-like model; and also that checking the satisfiability of a subformula associated with a level ml of the tree depends only on the subformulae occurring at levels greater or equal to ml.The clausal normal form, called SNF_{ml}, introduces labels in $\mathbb{N} \cup \{*\}$ to clauses corresponding to those levels, and resolution inferences can only be applied whenever the labels unify. The special label $*$ is used if a clause is true at all depths/worlds in a model and it normally only occurs in the normal form if we want to check a formula for global satisfiability. This restriction also allows for a very efficient implementation. The inference rules for MLR are given in Figure 2 and can only be applied if the unification on labels is defined, where the function σ on sets of labels is given by $\sigma(\{ml, *\}) = ml$; and $\sigma(\{ml\}) = ml$; otherwise, σ is undefined. For the GMR calculus [3], the inference rules are similar to those in Figure 2, but inference rules are applied within the scope of [*] instead of using the labels. The additional inference rules are given in Figure 3, where constant **start** is meant to be true only at the root of a model.

K$_\mathsf{S}$P is a theorem prover for the basic multimodal logic K_n, first presented in [5] (see also [7] for an extended presentation of its architecture and features). It is written in C and includes both calculi mentioned above. The main loop is

$$\text{[LRES]}\ \frac{\begin{array}{l} ml_1 : D \vee l \\ ml_2 : D' \vee \neg l \end{array}}{ml : D \vee D'} \qquad \text{[GEN2]}\ \frac{\begin{array}{l} ml_1 : l'_1 \rightarrow [a] l_1 \\ ml_2 : l'_2 \rightarrow [a] \neg l_1 \\ ml_3 : l'_2 \rightarrow \langle a \rangle l_2 \end{array}}{ml : \neg l'_1 \vee \neg l'_2 \vee \neg l'_3} \qquad \text{[MRES]}\ \frac{\begin{array}{l} ml_1 : l_1 \rightarrow [a] l \\ ml_2 : l_2 \rightarrow \langle a \rangle \neg l \end{array}}{ml : \neg l_1 \vee \neg l_2}$$

$$\text{[GEN1]}\ \frac{\begin{array}{l} ml_1 : l'_1 \rightarrow [a] \neg l_1 \\ \quad \vdots \\ ml_m : l'_m \rightarrow [a] \neg l_m \\ ml_{m+1} : l' \rightarrow \langle a \rangle \neg l \\ ml_{m+2} : l_1 \vee \ldots \vee l_m \vee l \end{array}}{ml : \neg l'_1 \vee \ldots \vee \neg l'_m \vee \neg l'} \qquad \text{[GEN3]}\ \frac{\begin{array}{l} ml_1 : l'_1 \rightarrow [a] \neg l_1 \\ \quad \vdots \\ ml_m : l'_m \rightarrow [a] \neg l_m \\ ml_{m+1} : l' \rightarrow \langle a \rangle l \\ ml_{m+2} : l_1 \vee \ldots \vee l_m \end{array}}{ml : \neg l'_1 \vee \ldots \vee \neg l'_m \vee \neg l'}$$

Fig. 2. Rules in MLR, where $ml = \sigma(\{ml_1, \ldots, ml_{m+1}, ml_{m+2} - 1\})$ in GEN1, GEN3; $ml = \sigma(\{ml_1, ml_2\})$ in LRES, MRES; and $ml = \sigma(\{ml_1, ml_2, ml_3\})$ in GEN2.

based on the given-clause algorithm, a variation of the set of support strategy [14], a refinement which restricts the set of choices of clauses participating in a derivation step. For the modal calculus, the set of clauses is partitioned according to the modal layer at which clauses are true, noting that for the GMR calculus, except for the initial clauses, all clauses are in the same level.

Previous experimental evaluations [1,5,6,7,11,12] indicate that K_SP with MLR works well on problems with high nesting of modal operators where the separation of modal levels can be exploited to improve the efficiency of reasoning [5,6,11,8,12]. The evaluations also indicate that K_SP's performance is comparable to other state-of-the-art provers with built-in support for K_1.

For the remaining logics in the cube, the implementation of the MLR calculus can deal with the satisfiability problem via the reductions given in [11,8,12]. The reductions are functions that take a formula in the extended logic and produce an equisatisfiable set of SNF_{ml} clauses in K_n. Some of those functions might return infinite sets. However, bounds for each of the logics and their combinations were determined so only a finite number of clauses is required. The GMR calculus can deal with any combination of the multimodal logics with axioms presented in Fig. 1, as given in [3,10], where the inference rules for each logic are added to the calculus. Evaluations with K_SP and GMR show good performance for logics including the axioms $\mathbf{4}_1$ and $\mathbf{5}_1$ [8,9].

We finish by noting that all evaluations ever performed, as indicated above, considered reasoning for monomodal logics (either K_1 or its extensions in the modal cube). However, K_SP is truly multimodal, in the sense that not only multiple modalities can be used, but also more than one logic can be specified in order to represent (and reason about) a problem. With GMR, K_SP implements

$$\text{[IRES1]}\ \frac{\begin{array}{l} [*](\ \mathbf{start} \rightarrow D \vee l) \\ [*](\ \mathbf{start} \rightarrow D' \vee \neg l) \end{array}}{[*](\ \mathbf{start} \rightarrow D \vee D')} \qquad \text{[IRES2]}\ \frac{\begin{array}{l} [*](\ \mathbf{start} \rightarrow D \vee l) \\ [*](\ \mathbf{true} \rightarrow D' \vee \neg l) \end{array}}{[*](\ \mathbf{start} \rightarrow D \vee D')}$$

Fig. 3. GMR rules for initial clauses.

decision procedures for logics which are interesting for some applications (e.g. including seriality and functionality for temporal modalities), but these features have never been extensively tested. There is a lack of provers that deal with all those logics and also a need for more challenging benchmarks. The combination of the global and local calculus is currently under implementation. We believe that with this new feature, we can further improve on the number of problems solved using the reductions given in [9] and previous works.

References

[1] Goré, R. and C. Kikkert, *Cegar-tableaux: Improved modal satisfiability via modal clause-learning and sat*, in: A. Das and S. Negri, editors, *TABLEAUX 2021* (2021), pp. 74–91.

[2] Halpern, J. Y. and Y. Moses, *A guide to completeness and complexity for modal logics of knowledge and belief*, Artificial Intelligence **54** (1992), pp. 319–379.

[3] Nalon, C. and C. Dixon, *Clausal resolution for normal modal logics*, J. Algorithms **62** (2007), pp. 117–134.

[4] Nalon, C., C. Dixon and U. Hustadt, *Modal resolution: Proofs, layers, and refinements*, ACM Trans. Comput. Log. **20** (2019), pp. 23:1–23:38.

[5] Nalon, C., U. Hustadt and C. Dixon, *K_SP: A resolution-based prover for multimodal K*, in: N. Olivetti and A. Tiwari, editors, *IJCAR 2016*, LNCS **9706** (2016), pp. 406–415.

[6] Nalon, C., U. Hustadt and C. Dixon, *K_SP: A resolution-based prover for multimodal K, abridged report*, in: C. Sierra, editor, *Proceedings of IJCAI 2017* (2017), pp. 4919–4923.

[7] Nalon, C., U. Hustadt and C. Dixon, *K_SP: Architecture, refinements, strategies and experiments*, J. Autom. Reason. **64** (2020), pp. 461–484.

[8] Nalon, C., U. Hustadt, F. Papacchini and C. Dixon, *Local reductions for the modal cube*, in: *IJCAR 2022*, LNCS **13385** (2022), pp. 486–505.

[9] Nalon, C., U. Hustadt, F. Papacchini and C. Dixon, *Buy one get 14 free: Evaluating local reductions for modal logic*, in: B. Pientka and C. Tinelli, editors, *Proceedings of CADE-29*, Lecture Notes in Artificial Intelligence **14132** (2023), pp. 382–400.

[10] Nalon, C., J. Marcos and C. Dixon, *Clausal resolution for modal logics of confluence*, in: S. Demri, D. Kapur and C. Weidenbach, editors, *Proceedings of IJCAR 2014*, LNCS **8562** (2014), pp. 322–336.

[11] Papacchini, F., C. Nalon, U. Hustadt and C. Dixon, *Efficient local reductions to basic modal logic*, in: A. Platzer and G. Sutcliffe, editors, *CADE 2021*, LNCS **12699** (2021), pp. 76–92.

[12] Papacchini, F., C. Nalon, U. Hustadt and C. Dixon, *Local is best: Efficient reductions to modal logic K*, J. Autom. Reason. **66** (2022), pp. 639–666.

[13] Spaan, E., "Complexity of Modal Logics," Ph.D. thesis, University of Amsterdam (1993).

[14] Wos, L., G. A. Robinson and D. F. Carson, *Efficiency and Completeness of the Set of Support Strategy in Theorem Proving*, Journal of the ACM **12** (1965), pp. 536–541.

Contributed Papers

Intuitionistic Master Modality

Bahareh Afshari[a 1] Lide Grotenhuis[b 1] Graham E. Leigh[a 2]
Lukas Zenger[c 3]

[a] {bahareh.afshari,graham.leigh}@gu.se, *University of Gothenburg, Sweden*

[b] l.m.grotenhuis@uva.nl, *University of Amsterdam, The Netherlands*

[c] lukas.zenger@unibe.ch, *University of Bern, Switzerland*

Abstract

We present a cyclic sequent calculus for intuitionistic modal logic with the master modality. Formulas of the logic are evaluated over bi-relational Kripke models with three different frame conditions: functional frames, 'triangle' confluent frames, and arbitrary frames. It is shown that the calculus is sound and complete for all three classes of models. This, in particular, proves that intuitionistic modal logic with the master modality cannot distinguish between arbitrary models and functional models. Soundness is established by a standard argument while completeness is proven via a detour to non-wellfounded proofs, using a proof-search argument that draws on analyticity of the calculus. The framework is robust in the sense that it can be naturally adapted to account for various frame conditions, such as serial models, reflexive models or S4-models, as well as for a polymodal extension that can be interpreted as intuitionistic common knowledge.

Keywords: Modal logic, Intuitionistic logic, Sequent calculus, Cyclic proofs.

---

1 Introduction

Intuitionistic modal logic has a long history with contributions from various fields, ranging from proof theory and philosophical logic to type theory and programming language theory. The logics studied can be roughly divided into two camps: *intuitionistic modal logics*, aimed at capturing an intuitionistic meta-reading of possible world semantics [23], and *constructive modal logics*, built for modelling particular computational properties such as staged or contextual computation [13]. More recently, extensions of these logics with fixed point operators, referred to as Intuitionistic Fixed Point Modal Logics (IFPML), have

---

[1] Supported by Dutch Research Council [OCENW.M20.048].

[2] Supported by Knut and Alice Wallenberg Foundation [2020.0199].

[3] Supported by Swiss National Science Foundation [200021L_196176] and the European Union's Horizon 2020 research and innovation program under the Marie Skłodowska-Curie grant agreement No 101007627.

gained increasing attention. Examples include intuitionistic linear-time temporal logic [8,2,3,4,1], intuitionistic common knowledge logic [12], intuitionistic Gödel-Löb logic [6,22,11] and intuitionistic modal μ-calculus [19].

The mathematical theory underpinning IFPML is little explored compared to its classical counterpart. In the classical realm, games, automata and, more recently, cyclic proofs have shown to be particularly suitable for the study of fixed point modal logics [10,7,21]. In contrast to the more traditional finitary proof systems with induction rules, cyclic proof systems are often analytic, and therefore better suited for proof search. However, the proof-theoretic techniques used for classical FPML do not readily apply to the intuitionistic case, as the Kripke semantics of the latter involves frame conditions that often require more complex proof calculi. In the realm of IFPML, non-wellfounded and cyclic proof systems have so far only been developed for intuitionistic linear-time temporal logic [1,16].

This work is part of a larger programme to establish frameworks and techniques for studying IFPML ranging from intuitionistic versions of basic modal logics to intuitionistic modal μ-calculus. Here, we study the language $\mathcal{L}_{\mathrm{IM}}$ which extends the language of IPC with the basic modality $\Box$ and the master modality $\boxplus$. A formula $\boxplus\varphi$ is characterised as the greatest fixed point of the propositional function $p \mapsto \varphi \wedge \Box p$. Formulas are evaluated over bi-relational Kripke models $(W, \leq, R, V)$, where $\leq$ is the intuitionistic partial order and R the modal accessibility relation. The monotonicity property, that $w \leq v$ and $w \models \varphi$ implies $v \models \varphi$, can be built directly into the semantics, as we will initially do. An alternative approach is to impose frame conditions on $\leq$ and R, such as *triangle confluence*: if $w \leq v$ and vRu, then wRu.

For triangle confluent models, the truth conditions for the modalities reduce to the classical ones. We consider the class of all bi-relational models, the class of models with a functional modal relation (called *functional models*), and the class of triangle confluent models, inducing the logics IM_{K}, IM_{f}, IM_{t}, respectively. The logic IM_{f} can be viewed as a weak version of intuitionistic linear-time temporal logic.

We introduce a cyclic proof system cIM, and establish soundness with respect to IM_{K} and completeness with respect to both IM_{f} and IM_{t}. This implies that the three logics are equivalent, thereby showing that our language cannot distinguish between arbitrary bi-relational models, functional models, and triangle models. While the result for arbitrary bi-relational models and triangle models was already known for $\mathcal{L}_{\mathrm{IM}}$ without the master modality (see e.g., [14]), the fact that $\mathcal{L}_{\mathrm{IM}}$ cannot distinguish functional models from arbitrary ones, is, to the best of our knowledge, a new result. It should be noted that the equivalence of the three classes of models holds in particular in the absence of the fixed point modality $\boxplus$; one can simply restrict the syntax and carry out the completeness proofs as done in this paper.

The calculus cIM is a natural modal extension of the standard multi-conclusion calculus for intuitionistic propositional logic (see e.g., [17]). To ensure soundness, the calculus uses a focus annotation that keeps track of good traces. As cIM is cut-free, it is analytic and hence suitable for effective proof search. A similar calculus for classical modal logic with the master modality over S5-frames is presented in [20]; that calculus differs from cIM in that it requires analytic cuts due to the S5 frame conditions.

Completeness of cIM proceeds via a detour into a non-wellfounded proof calculus specifically designed for proof-search. Inspired by the game-theoretic arguments in [18], we present a modular framework for proof-search as a two-player infinite game between Prover and Refuter, such that every unprovable sequent induces a countermodel of a particular form. The form of Prover's turn can be adapted as to obtain a particular frame condition. In this way, we obtain completeness of the non-wellfounded calculus for triangle and functional models. Completeness of cIM is obtained by showing that a non-wellfounded proof induces a cyclic proof. In addition, we also show completeness of a single-conclusion version of cIM.

Due to the modular approach, the calculus cIM and the proof methods are robust in the sense that they can easily be adapted to account for various frame conditions, such as serial frames, reflexive frames, and S4-frames. Furthermore, cIM can be adapted to a polymodal version of $\mathcal{L}_{\mathrm{IM}}$ to obtain an analytic calculus for the intuitionistic common knowledge logic considered in [12].

2 Syntax and semantics

The *language* of $\mathcal{L}_{\mathrm{IM}}$ consists of a countable set of atomic propositions Prop, the constant $\bot$, logical connectives $\wedge, \vee, \to$ and modal operators $\Box$ and $\boxast$. The operator $\boxast$ is called the *master modality*. *Formulas* of $\mathcal{L}_{\mathrm{IM}}$ are given by the grammar:

$$\varphi ::= \bot \mid p \mid \varphi \wedge \varphi \mid \varphi \vee \varphi \mid \varphi \to \varphi \mid \Box\varphi \mid \boxast\varphi$$

where $p \in \mathsf{Prop}$. Define $\top := \bot \to \bot$ and $\Box^k\varphi$ by $\Box^0\varphi := \varphi$ and $\Box^{k+1}\varphi := \Box\Box^k\varphi$. The set of formulas of $\mathcal{L}_{\mathrm{IM}}$ is denoted Fm. Greek letters $\varphi, \psi, \ldots$ etc., possibly with subscript, are meta-variables for formulas.

Definition 2.1 The *closure* of a formula φ is the smallest set $\mathsf{Cl}(\varphi)$ which contains φ, is closed under the subformula relation, and contains $\Box\boxast\psi$ whenever $\boxast\psi \in \mathsf{Cl}(\varphi)$. Given a set Γ of formulas, we define its closure $\mathsf{Cl}(\Gamma) := \cup_{\phi\in\Gamma}\mathsf{Cl}(\phi)$.

The following lemma is proven by a straightforward induction on the structure of the formula φ.

Lemma 2.2 *For any formula φ, the closure $\mathsf{Cl}(\varphi)$ is finite.*

Formulas are evaluated in bi-relational (Kripke) models.

Definition 2.3 A *(bi-relational) model* is a tuple $M = (W, \leq, R, V)$ where

(i) $W \neq \emptyset$ is a set;

(ii) $(W, \leq)$ is a partial order;

(iii) $V \colon W \to \mathcal{P}(\mathsf{Prop})$ is monotone in $\leq$: if $w \leq v$ then $V(w) \subseteq V(v)$;

(iv) $R \subseteq W \times W$ is a binary relation.

Fig. 1. Forth-down confluence (left) and triangle confluence (right). Dashed lines represent the relations each confluence condition stipulates the existence of.

Elements of W are called *worlds*, and given some world $w \in W$, we call the tuple (M, w) a *pointed model.* The function V is called a *valuation*, the relation $\leq$ is called the *intuitionistic order* and R is called the *modal accessibility relation.* If $w \leq v$, then v is called an *intuitionistic successor* of w, and if wRv, then we call v a *modal successor* of w. A model is called *functional* if the modal accessibility relation R is functional, i.e., if wRv and wRu, then $v = u$.

For any binary relation S, we let S^* denote the reflexive and transitive closure of S. Given a model $M = (W, \leq, R, V)$, we let $\tilde{R}$ denote the composition $\leq; R$. Note that, since $\leq$ is reflexive, $w\tilde{R}^*v$ holds if and only if there exist a natural number n and worlds $u_0, \ldots, u_n$ such that $u_0 = w$, $u_n = v$ and for all $0 \leq i < n$ we have $u_i R u_{i+1}$ or $u_i \leq u_{i+1}$. The *truth relation* $\models$ is defined inductively by the following clauses, where $p \in \mathsf{Prop}$ and $w \in W$.

$$
\begin{array}{lll}
M, w \not\models \bot, & & \\
M, w \models p & \text{iff} & p \in V(w) \\
M, w \models \varphi \wedge \psi & \text{iff} & M, w \models \varphi \text{ and } M, w \models \psi, \\
M, w \models \varphi \vee \psi & \text{iff} & M, w \models \varphi \text{ or } M, w \models \psi, \\
M, w \models \varphi \rightarrow \psi & \text{iff} & \text{for all } v \geq w \text{ if } M, v \models \varphi, \text{ then } M, v \models \psi, \\
M, w \models \Box\varphi & \text{iff} & \text{for all } v \in W \text{ if } w\tilde{R}v, \text{ then } M, v \models \varphi, \\
M, w \models \boxast\varphi & \text{iff} & \text{for all } v \in W \text{ if } w\tilde{R}^*v, \text{ then } M, v \models \varphi.
\end{array}
$$

Validity and satisfiability over a class of models are defined as expected.

As remarked, monotonicity is built-in to the semantics:

Lemma 2.4 (Monotonicity of $\models$) *Let $\varphi \in \mathsf{Fm}$ and let $M = (W, \leq, R, V)$ be a model with $w, v \in W$. If $w \leq v$ and $M, w \models \varphi$, then $M, v \models \varphi$.*

2.1 Triangle models

We introduce a subclass of models in which the intuitionistic order and the modal accessibility relation satisfy a particular confluence property. For this subclass of models, the classical truth conditions for the modalities suffice to obtain the monotonicity lemma (cf. Lemma 2.4).

Definition 2.5 A *triangle model* is a model $M = (W, \leq, R, V)$ where $\leq$ and R are *triangle confluent*: if $w \leq v$ and vRu, then wRu (see Figure 1).

Given a triangle model $M = (W, \leq, R, V)$, a second truth relation $\models_t$ can be given which differs from $\models$ only in the modal clauses:

$$
\begin{array}{lll}
M, w \models_t \Box\varphi & \text{iff} & \text{for all } v \in W \text{ if } wRv, \text{ then } M, v \models_t \varphi, \\
M, w \models_t \boxast\varphi & \text{iff} & \text{for all } v \in W \text{ if } wR^*v, \text{ then } M, v \models_t \varphi.
\end{array}
$$

Triangle confluence implies $R = \tilde{R}$, so we obtain the next two lemmas.

Lemma 2.6 (Monotonicity of $\models_t$) *Let $\varphi \in \mathsf{Fm}$ and let $M = (W, \leq, R, V)$ be a triangle model with $w, v \in W$. If $w \leq v$ and $M, w \models_t \varphi$, then $M, v \models_t \varphi$.*

Lemma 2.7 *Let $\varphi \in \mathsf{Fm}$ and let (M, w) be a pointed triangle model. Then $M, w \models_t \varphi$ if and only if $M, w \models \varphi$.*

Triangle confluence is a special case of *forth-down confluence*: if $w \leq v$ and vRu, then there exists $s \in W$ with wRs and $s \leq u$ (illustrated in Figure 1). Forth-down confluence is sufficient for monotonicity. However, every model $M = (W, \leq, R, V)$, and so, in particular, every forth-down confluent model, induces a triangle model $M' = (W, \leq, (\leq; R), V)$ in a truth-preserving way. So the logic over forth-down models is identical to the logic over triangle models.

Denote by $\mathrm{IM_K}$, $\mathrm{IM_f}$ and $\mathrm{IM_t}$ the set of valid formulas over the class of bi-relational models, the class of functional models and the class of triangle models, respectively. By definition, $\mathrm{IM_K} \subseteq \mathrm{IM_f} \cap \mathrm{IM_t}$. In the next section we present a non-wellfounded and a cyclic calculus which are each sound and analytically complete for these logics. As a corollary, the three notions of validity coincide: $\mathrm{IM_K} = \mathrm{IM_f} = \mathrm{IM_t}$.

3 Proof systems

An *annotated formula* is a pair (φ, a) where φ is a formula and $a \in \{\mathsf{f}, \mathsf{u}\}$, where f designates that the formula is *in focus* and u that the formula is *unfocused.* Annotated formulas are written as φ^a. We display annotated formulas without using brackets, e.g. the formula $\varphi \to \psi^{\mathsf{u}}$ should be read as $(\varphi \to \psi)^{\mathsf{u}}$. Finite sets of annotated formulas are denoted by Γ, Δ, Σ and Π with or without subscripts. For a set of annotated formulas Γ define

$$\Gamma^- = \{\varphi \mid \varphi^a \in \Gamma\} \text{ and } \Gamma^{\mathsf{u}} = \{\varphi^{\mathsf{u}} \mid \varphi^a \in \Gamma\}.$$

A *sequent* is an ordered pair $\Gamma \Rightarrow \Delta$ where Γ and Δ are finite sets of annotated formulas, such that the following conditions hold.

(i) Every formula in Γ is unfocused.

(ii) At most one formula in Δ is in focus.

(iii) If a formula φ is in focus, then $\varphi = \boxtimes\psi$ or $\varphi = \Box\boxtimes\psi$ for some formula ψ.

We use σ to denote sequents and write Γ_σ and Δ_σ for the left and right side of σ respectively. The *interpretation* of σ is the formula $\sigma^I := \bigwedge \Gamma_\sigma^- \to \bigvee \Delta_\sigma^-$ where $\bigwedge \emptyset = \top$ and $\bigvee \emptyset = \bot$. Note that annotations convey no semantic meaning. Given a pointed model (M, w) we write $M, w \models \sigma$ iff $M, w \models \sigma^I$. The *closure* of σ is the set $\mathsf{Cl}(\sigma) := \mathsf{Cl}(\Gamma_\sigma^-) \cup \mathsf{Cl}(\Delta_\sigma^-)$.

Our calculi employ multi-conclusion sequents, i.e., sequents $\Gamma \Rightarrow \Delta$ where Δ may contain more than one formula. This streamlines the proof-search argument for completeness as it allows writing the disjunction and left-implication rules in invertible form. But it is not an essential restriction; Section 5.4 demonstrates how a single-conclusion proof can be obtained from any multi-conclusion one.

$$\frac{}{\Gamma, \varphi^{\mathsf{u}} \Rightarrow \varphi^{a}, \Delta}\ \mathsf{id} \qquad \frac{}{\Gamma, \bot^{\mathsf{u}} \Rightarrow \Delta}\ \bot$$

$$\frac{\Gamma, \varphi^{\mathsf{u}}, \psi^{\mathsf{u}} \Rightarrow \Delta}{\Gamma, \varphi \wedge \psi^{\mathsf{u}} \Rightarrow \Delta}\ \wedge\mathsf{L} \qquad \frac{\Gamma \Rightarrow \varphi^{\mathsf{u}}, \Delta \quad \Gamma \Rightarrow \psi^{\mathsf{u}}, \Delta}{\Gamma \Rightarrow \varphi \wedge \psi^{\mathsf{u}}, \Delta}\ \wedge\mathsf{R}$$

$$\frac{\Gamma, \varphi^{\mathsf{u}} \Rightarrow \Delta \quad \Gamma, \psi^{\mathsf{u}} \Rightarrow \Delta}{\Gamma, \varphi \vee \psi^{\mathsf{u}} \Rightarrow \Delta}\ \vee\mathsf{L} \qquad \frac{\Gamma \Rightarrow \varphi^{\mathsf{u}}, \psi^{\mathsf{u}}, \Delta}{\Gamma \Rightarrow \varphi \vee \psi^{\mathsf{u}}, \Delta}\ \vee\mathsf{R}$$

$$\frac{\Gamma, \varphi \to \psi^{\mathsf{u}} \Rightarrow \varphi^{\mathsf{u}}, \Delta \quad \Gamma, \psi^{\mathsf{u}} \Rightarrow \Delta}{\Gamma, \varphi \to \psi^{\mathsf{u}} \Rightarrow \Delta}\ {\to}\mathsf{L} \qquad \frac{\Gamma, \varphi^{\mathsf{u}} \Rightarrow \psi^{\mathsf{u}}}{\Gamma \Rightarrow \varphi \to \psi^{\mathsf{u}}, \Delta}\ {\to}\mathsf{R}$$

$$\frac{\Gamma, \varphi^{\mathsf{u}}, \Box\boxast\varphi^{\mathsf{u}} \Rightarrow \Delta}{\Gamma, \boxast\varphi^{\mathsf{u}} \Rightarrow \Delta}\ \boxast\mathsf{L} \qquad \frac{\Gamma \Rightarrow \varphi^{\mathsf{u}}, \Delta \quad \Gamma \Rightarrow \Box\boxast\varphi^{a}, \Delta}{\Gamma \Rightarrow \boxast\varphi^{a}, \Delta}\ \boxast\mathsf{R}$$

$$\frac{\Gamma \Rightarrow \varphi^{\mathsf{u}}, \Delta}{\Gamma \Rightarrow \varphi^{\mathsf{f}}, \Delta}\ \mathsf{u} \qquad \frac{\Gamma \Rightarrow \varphi^{\mathsf{f}}, \Delta}{\Gamma \Rightarrow \varphi^{\mathsf{u}}, \Delta}\ \mathsf{f}$$

$$\frac{\Gamma \Rightarrow \varphi^{a}}{\Pi, \Box\Gamma \Rightarrow \Box\varphi^{a}, \Sigma}\ \Box$$

Table 1
The rules of the calculus IM

Definition 3.1 The sequent calculus IM consists of the rules depicted in Table 1 for all values of $a \in \{\mathsf{u}, \mathsf{f}\}$.

All rules except u and f are called *logical rules*, of which the rules id and $\bot$ are called *axioms*. The logical rules $\boxast\mathsf{L}$ and $\boxast\mathsf{R}$ reflect the equivalence $\boxast\varphi \leftrightarrow \varphi \wedge \Box\boxast\varphi$. The rules u and f are the *focus rules*: the rule u takes a sequent with no formula in focus and puts one formula in focus. The rule f does the opposite: it takes a sequent with a formula in focus and changes its annotation to unfocused. The names of these rules are motivated by the fact that they will later be read bottom-up.

Note that the rules ${\to}\mathsf{R}$ and $\Box$ have single-conclusion premises and all other rules are invertible in the sense that the conclusion is valid if and only if all premises are. We therefore refer to $\Box$ and ${\to}\mathsf{R}$ as the *non-invertible* rules and to the other rules as *invertible*. For each rule except $\Box$, the distinguished formula in the conclusion is called *principal* and the distinguished formula(s) in the premises are called its *residual(s)*. For example, for ${\to}\mathsf{L}$ the principal formula is $\varphi \to \psi^{\mathsf{u}}$ and its residuals are $\varphi \to \psi^{\mathsf{u}}$, φ^{u} and ψ^{u}. For the rule $\Box$, all formulas in the conclusion are called principal and each formula in the premise is the residual of its corresponding boxed formula in the conclusion (formulas in Σ and Π have no residuals); for example, the formula φ^{a} is the residual of $\Box\varphi^{a}$. In every rule application, any formula that is neither principal nor residual is called a *side formula*.

Observe that the calculus IM does not contain a cut-rule. In fact, IM is *an-*

$$\frac{}{\Gamma, \varphi^{\mathsf{u}} \Rightarrow \varphi^{a}}\ \mathsf{id} \qquad\qquad \frac{}{\Gamma, \bot^{\mathsf{u}} \Rightarrow \Delta}\ \bot$$

$$\frac{\Gamma, \varphi^{\mathsf{u}}, \psi^{\mathsf{u}} \Rightarrow \Delta}{\Gamma, \varphi \wedge \psi^{\mathsf{u}} \Rightarrow \Delta}\ \wedge\mathsf{L} \qquad\qquad \frac{\Gamma \Rightarrow \varphi^{\mathsf{u}} \quad \Gamma \Rightarrow \psi^{\mathsf{u}}}{\Gamma \Rightarrow \varphi \wedge \psi^{\mathsf{u}}}\ \wedge\mathsf{R}$$

$$\frac{\Gamma, \varphi^{\mathsf{u}} \Rightarrow \Delta \quad \Gamma, \psi^{\mathsf{u}} \Rightarrow \Delta}{\Gamma, \varphi \vee \psi^{\mathsf{u}} \Rightarrow \Delta}\ \vee\mathsf{L} \qquad\qquad \frac{\Gamma \Rightarrow \varphi_i^{\mathsf{u}}}{\Gamma \Rightarrow \varphi_0 \vee \varphi_1^{\mathsf{u}}}\ \vee_i\mathsf{R}$$

$$\frac{\Gamma, \varphi \to \psi^{\mathsf{u}} \Rightarrow \varphi^{\mathsf{u}} \quad \Gamma, \psi^{\mathsf{u}} \Rightarrow \Delta}{\Gamma, \varphi \to \psi^{\mathsf{u}} \Rightarrow \Delta}\ {\to}\mathsf{L} \qquad\qquad \frac{\Gamma, \varphi^{\mathsf{u}} \Rightarrow \psi^{\mathsf{u}}}{\Gamma \Rightarrow \varphi \to \psi^{\mathsf{u}}}\ {\to}\mathsf{R}$$

$$\frac{\Gamma, \varphi^{\mathsf{u}}, \Box\boxtimes\varphi^{\mathsf{u}} \Rightarrow \Delta}{\Gamma, \boxtimes\varphi^{\mathsf{u}} \Rightarrow \Delta}\ \boxtimes\mathsf{L} \qquad\qquad \frac{\Gamma \Rightarrow \varphi^{\mathsf{u}} \quad \Gamma \Rightarrow \Box\boxtimes\varphi^{a}}{\Gamma \Rightarrow \boxtimes\varphi^{a}}\ \boxtimes\mathsf{R}$$

$$\frac{\Gamma \Rightarrow \varphi^{\mathsf{u}}}{\Gamma \Rightarrow \varphi^{\mathsf{f}}}\ \mathsf{u} \qquad\qquad \frac{\Gamma \Rightarrow \varphi^{\mathsf{f}}}{\Gamma \Rightarrow \varphi^{\mathsf{u}}}\ \mathsf{f}$$

$$\frac{\Gamma \Rightarrow \varphi^{a}}{\Pi, \Box\Gamma \Rightarrow \Box\varphi^{a}}\ \Box$$

Table 2
The single-conclusion version of IM, where $|\Delta| \leq 1$.

alytic in the sense that, for every rule instance r of IM, every formula occurring in a premise of r belongs to the closure of some formula in the conclusion of r. Analyticity of IM will play a crucial role in the completeness proof of the cyclic calculus defined below.

We also introduce a single-conclusion version of IM, where sequents $\Gamma \Rightarrow \Delta$ satisfy the condition that $|\Delta| \leq 1$. The rules of the single-conclusion version of IM are depicted in Table 2.

The condition that sequents have at most one formula in focus imposes restrictions on rule applications, as is illustrated by the following lemma. The proof is analogous to the proof of [20, Lemma 4.1].

Lemma 3.2 *If in an instance of* $\boxtimes\mathsf{R}$ *the principal formula is in focus, then the left premise has no formula in focus.*

Proof. Suppose towards contradiction that there is an application of $\boxtimes\mathsf{R}$ in which the principal formula is in focus and the left premise has a formula in focus, too.

$$\frac{\Gamma \Rightarrow \varphi^{\mathsf{u}}, \Delta \qquad \Gamma \Rightarrow \Box\boxtimes\varphi^{\mathsf{f}}, \Delta}{\Gamma \Rightarrow \boxtimes\varphi^{\mathsf{f}}, \Delta}\ \boxtimes\mathsf{R}$$

Since φ^{u} is by definition not in focus, the formula in focus must occur in Δ. Since the right premise cannot contain two formulas in focus, the formula in focus in Δ must be $\Box\boxtimes\varphi^{\mathsf{f}}$. But then the conclusion sequent contains two formulas in focus, a contradiction. □

In the following we introduce a non-wellfounded and a cyclic proof system based on the rules of IM. We remark that the annotations are not vital for the non-wellfounded system; it is possible to define a sound and complete non-wellfounded proof system based on the rules of IM without annotations.[4] However, the annotations are essential in our soundness proof of the cyclic system.

3.1 Non-wellfounded calculus nIM

A *derivation* in nIM of a sequent σ is a finite or countably infinite tree whose nodes are labelled by sequents according to the rules of IM and whose root is labelled by σ. We read derivations 'upwards', so that the premise of a rule is considered to be a successor of the conclusion. Given a derivation π, a *path* through π is a finite or infinite sequence of nodes $\rho = \rho_0, \rho_1, \rho_2, \ldots$ of π such that for each index i the node ρ_{i+1} (if it exists) is a direct successor of ρ_i. A *branch* is path that starts at the root and is either infinite or ending in a leaf. We will often tacitly identify a node in a derivation with the sequent labelling it and thereby paths with sequences of sequents.

Definition 3.3 A nIM-*proof* of a sequent σ is a derivation in nIM of σ, such that every leaf is labelled by an axiom and every infinite branch ρ has a *good* suffix ρ': every sequent in ρ' has a formula in focus and ρ' contains infinitely many applications of $\boxast$R where the principal formula is in focus.

Lemma 3.4 *Every good suffix contains infinitely many applications of* $\Box$.

3.2 Cyclic calculus cIM

A *derivation* in cIM of a sequent σ is a nIM-derivation of σ that is finite.

Definition 3.5 A path ρ in a cIM-derivation is *successful* if the following hold.

(i) Every sequent in ρ has a formula in focus.

(ii) The path ρ passes through at least one instance of $\boxast$R where the principal formula is in focus.

Given a cIM-derivation π, a pair of nodes (u, v) of π is called a *repetition* if there exists a path from u to v and both nodes are labelled by the same sequent. A repetition (u, v) is *successful* if the path from u to v is successful.

Definition 3.6 A cIM-*proof* of a sequent σ is a cIM-derivation π of σ such that every leaf l of π is either labelled by an axiom or there exists a node $c(l)$ in π such that $(c(l), l)$ is a successful repetition.

We analogously define *single-conclusion* derivations and proofs in nIM and cIM, where instead of the multi-conclusion rules of IM we use their single-conclusion version depicted in Table 2. Until Section 5.4, we will only study the multi-conclusion systems. In the following, 'sequent', 'proof', etc. will therefore refer to 'multi-conclusion sequent', 'multi-conclusion proof', and so

[4] For such a system the global soundness condition on infinite branches is formulated in a different way than presented here using formula traces.

on. When referring to single-conclusion proofs, we will explicitly do so. We conclude this section by giving an example of a proof in nIM and cIM.

Example 3.7 We illustrate that the induction axiom $(\varphi \wedge \boxast(\varphi \to \Box\varphi)) \to \boxast\varphi$ for the master modality is provable in nIM and cIM. We first present a cyclic proof, where we write γ for the formula $\varphi \to \Box\varphi$.

$$
\dfrac{\dfrac{\dfrac{\dfrac{\dfrac{}{\varphi^{\mathsf{u}}, \boxast\gamma^{\mathsf{u}} \Rightarrow \varphi^{\mathsf{u}}}\,\mathsf{id} \qquad \dfrac{\dfrac{\dfrac{}{\varphi^{\mathsf{u}}, \gamma^{\mathsf{u}}, \Box\boxast\gamma^{\mathsf{u}} \Rightarrow \varphi^{\mathsf{u}}, \Box\boxast\varphi^{\mathsf{f}}}\,\mathsf{id} \qquad \dfrac{\varphi^{\mathsf{u}}, \boxast\gamma^{\mathsf{u}} \Rightarrow \boxast\varphi^{\mathsf{f}}}{\varphi^{\mathsf{u}}, \Box\varphi^{\mathsf{u}}, \Box\boxast\gamma^{\mathsf{u}} \Rightarrow \Box\boxast\varphi^{\mathsf{f}}}\,\Box}{\varphi^{\mathsf{u}}, \gamma^{\mathsf{u}}, \Box\boxast\gamma^{\mathsf{u}} \Rightarrow \Box\boxast\varphi^{\mathsf{f}}}\,{\to}\mathsf{L}}{\varphi^{\mathsf{u}}, \boxast\gamma^{\mathsf{u}} \Rightarrow \Box\boxast\varphi^{\mathsf{f}}}\,\boxast\mathsf{L}}{\varphi^{\mathsf{u}}, \boxast\gamma^{\mathsf{u}} \Rightarrow \boxast\varphi^{\mathsf{f}}}\,\boxast\mathsf{R}}{\varphi^{\mathsf{u}}, \boxast\gamma^{\mathsf{u}} \Rightarrow \boxast\varphi^{\mathsf{u}}}\,\mathsf{f}}{\varphi \wedge \boxast\gamma^{\mathsf{u}} \Rightarrow \boxast\varphi^{\mathsf{u}}}\,\wedge\mathsf{L}}{\Rightarrow (\varphi \wedge \boxast\gamma) \to \boxast\varphi^{\mathsf{u}}}\,{\to}\mathsf{R}
$$

Denote the right-most leaf by v and the node that is the premise of the rule instance of f by u. Observe that v and u are labelled by the exact same sequent. Moreover, the path from u to v always has a formula in focus and passes through an instance of $\boxast\mathsf{R}$ where the principal formula is in focus. Therefore the pair (u, v) forms a successful repetition. As every other leaf is labelled by an axiom, the depicted derivation is in fact a cyclic proof. From this cyclic proof one may obtain a non-wellfounded proof of the induction axiom by unfolding the tree infinitely often over the repetition (u, v).

4 Soundness

This section establishes soundness of cIM with respect to bi-relational models. [5] The proof closely follows the soundness proof of [20]. Since cyclic proofs may contain non-axiomatic leafs (i.e. those leafs that belong to a successful repetition), a more complex argument than an induction on the height of a proof to establish soundness is required. Intuitively, the cyclic calculus is sound because success of repetitions ensures that along every repetition, *progress* is made in the form of a modal step: as a result, when proving $\boxast\phi$ we are essentially proving $\Box^n\phi$ for every $n < \omega$.

To prove this formally, we argue contrapositively. We first assign to each invalid sequent with a focused formula a *measure* in the form of a natural number. Then, assuming there is a cyclic proof π of an invalid sequent, we show there must be a successful repetition (u, v) in π of invalid sequents whose measure strictly decreases along the path (u, v). As u and v are labelled by the same sequent, this contradicts the fact that measures are well-defined.

Lemma 4.1 *If the conclusion of a rule instance* r *of* IM *is invalid, then there exists a premise of* r *that is invalid.*

[5] Although the calculus nIM is only used to show completeness of cIM, let us note here that nIM is also sound: this follows from soundness of cIM and Theorem 5.14.

Proof. Straightforward by inspection of the rules. □

Let σ be a sequent that has a formula in focus, i.e., Δ_σ contains a formula of the form $\Box^j \boxplus \varphi^{\mathsf{f}}$ for $j \in \{0,1\}$. Denote by $\sigma(n)$ the sequent $\Gamma_\sigma \Rightarrow \Delta_\sigma, \Box^j \Box^n \varphi^{\mathsf{u}}$, i.e., the sequent expanding the right side of σ by the formula $\Box^j \Box^n \varphi^{\mathsf{u}}$.

Lemma 4.2 *If σ has a formula in focus and is invalid, then there exists a natural number n such that $\sigma(n)$ is invalid.*

Proof. Let σ be an invalid sequent with a formula in focus. Then there exists a formula $\Box^j \boxplus \varphi^{\mathsf{f}} \in \Delta_\sigma$ for $j \in \{0,1\}$, and a pointed bi-relational model (M, w) with $M, w \not\models \sigma$. So in particular $M, w \not\models \Box^j \boxplus \varphi$. If $j = 0$, then there exists a world v with $w\tilde{R}^* v$ and $M, v \not\models \varphi$. Since $\leq$ is reflexive there are worlds $u_0, \ldots, u_{2n}$ such that $u_0 = w$, $u_{2n} = v$ and for all $0 \leq i < 2n$ it holds that if i is even, then $u_i \leq u_{i+1}$ and if i i odd, then $u_i R u_{i+1}$. Therefore $w\tilde{R}^n v$, implying that $M, w \not\models \Box^n \varphi$. If $j = 1$, then there is a world v with $w\tilde{R}v$ and $M, v \not\models \boxplus\varphi$. By the previous case we have that $M, v \not\models \Box^n \varphi$ for some n. Hence $M, w \not\models \Box\Box^n \varphi$. Therefore $M, w \not\models \sigma(n)$ for some natural number n. □

As a consequence, every invalid sequent σ with a formula in focus can be associated a measure:

$$\mu(\sigma) := \min\{n \in \omega \mid \sigma(n) \text{ is invalid}\}.$$

We may now prove a strenghtening of Lemma 4.1.

Lemma 4.3 *Suppose*

$$\frac{\sigma_1 \quad \cdots \quad \sigma_n}{\sigma}\ \mathsf{r}$$

is a rule instance of IM. *If σ is invalid, then there is an i such that σ_i is invalid. If both σ and σ_i have a formula in focus, then, moreover,*

$$\mu(\sigma_i) \leq \mu(\sigma),$$

where the inequality is strict if $\mathsf{r} = \boxplus\mathsf{R}$ *and the principal formula is in focus.*

Proof. By Lemma 4.1 it suffices to only consider the case where both the conclusion and at least one premise have a formula in focus. We first treat the case that the formula in focus is not principal. Then $\mathsf{r} \notin \{\to\mathsf{R}, \Box\}$, as this would contradict the existence of a premise with a focused formula. By inspection of the rules, note that then *every* premise must have a formula in focus, and so the following is a correct rule instance of r.

$$\frac{\sigma_1(\mu(\sigma)) \quad \cdots \quad \sigma_n(\mu(\sigma))}{\sigma(\mu(\sigma))}\ \mathsf{r}$$

By Lemma 4.1, since $\sigma(\mu(\sigma))$ is invalid, there exists a premise $\sigma_i(\mu(\sigma))$ that is invalid. Hence σ_i is invalid and $\mu(\sigma_i) \leq \mu(\sigma)$.

Now suppose that the formula in focus is principal in r. Then $\mathsf{r} = \boxplus\mathsf{R}$ or $\mathsf{r} = \Box$. In the first case, σ is of the form $\Gamma \Rightarrow \boxplus\varphi^{\mathsf{f}}, \Delta$ with premises σ_1 and σ_2

given by $\Gamma \Rightarrow \varphi^{\mathsf{u}}, \Delta$ and $\Gamma \Rightarrow \Box\boxtimes\varphi^{\mathsf{f}}, \Delta$, respectively. As there exists a pointed model (M, w) that falsifies $\sigma(\mu(\sigma))$, w has an intuitionistic successor v such that $M, v \models \Gamma$ and $M, v \not\models \boxtimes\varphi \vee \Box^{\mu(\sigma)}\varphi \vee \bigvee \Delta^-$. If $\mu(\sigma) = 0$, then $M, v \not\models \varphi$, so (M, v) falsifies the left premise σ_1. By Lemma 3.2, σ_1 does not have a formula in focus, and so the statement of the lemma holds. If $\mu(\sigma) > 0$, then $M, v \not\models \Box\Box^{\mu(\sigma)-1}\varphi$. Hence (M, v) falsifies $\sigma_2(\mu(\sigma) - 1)$. So σ_2 is invalid and we have $\mu(\sigma_2) < \mu(\sigma)$.

In the second case, the conclusion σ is of the form $\Pi, \Box\Gamma \Rightarrow \Box\boxtimes\varphi^{\mathsf{f}}, \Sigma$ and the single premise σ_1 is of the form $\Gamma \Rightarrow \boxtimes\varphi^{\mathsf{f}}$. Note that invalidity of $\sigma(\mu(\sigma))$ implies the invalidity of $\bigwedge \Gamma^- \to \Box^{\mu(\sigma)}\varphi$, which in turn implies invalidity of $\sigma_1(\mu(\sigma))$. So σ_1 is invalid and we have $\mu(\sigma_1) \leq \mu(\sigma)$. □

Theorem 4.4 *If there is a* cIM*-proof of a sequent σ, then σ is valid over the class of bi-relational models.*

Proof. Let π be a cIM-proof of σ and suppose for contradiction that σ is invalid. By repeatedly applying Lemma 4.3 we obtain a path of invalid sequents

$$\rho = \sigma_1, \sigma_2 \ldots, \sigma_n$$

through π such that $\sigma = \sigma_1$ and σ_n is a leaf. As σ_n cannot be an axiom and π is a proof, there exists some σ_i such that (σ_i, σ_n) is a successful repetition. Then the path from σ_i to σ_n always has a formula in focus and passes through at least one instance of $\boxtimes$R in which the formula in focus is principal. Hence, by construction, we have $\mu(\sigma_n) < \mu(\sigma_i)$, contradicting that $\sigma_n = \sigma_i$. □

The above result also implies that cIM is sound for the class of functional models and the class of triangle models. In addition, soundness of the single-conclusion version of cIM follows, as any single-conclusion proof induces a multi-conclusion proof via weakening.

5 Completeness

We now turn our attention to completeness of the cyclic calculus with respect to triangle and functional models. The argument proceeds in two steps. First, we set up a general framework for proving completeness via proof-search games, from which completeness of the non-wellfounded calculus nIM is deduced. We then show how to transform an arbitrary nIM-proof into a single-conclusion nIM-proof, and lastly how to transform a (single-conclusion) nIM-proof into a (single-conclusion) cIM-proof.

5.1 Proof-search games

Each sequent σ will be associated with a *proof-search tree* which will form the arena of a two-player game between *Prover*, whose winning strategies establish proofs of σ, and *Refuter*, whose winning strategies describe countermodels for σ. Completeness then becomes a corollary of determinacy of the game.

A proof-search tree for σ is built by applying rules bottom-up to σ. The invertible rules are applied first until a *saturated* sequent is obtained.

Definition 5.1 A sequent $\Gamma \Rightarrow \Delta$ is *saturated* if the following hold.

(i) If $\varphi \wedge \psi^{\mathsf{u}} \in \Gamma$, then $\varphi^{\mathsf{u}} \in \Gamma$ and $\psi^{\mathsf{u}} \in \Gamma$.

(ii) If $\varphi \vee \psi^{\mathsf{u}} \in \Gamma$, then $\varphi^{\mathsf{u}} \in \Gamma$ or $\psi^{\mathsf{u}} \in \Gamma$.

(iii) If $\varphi \rightarrow \psi^{\mathsf{u}} \in \Gamma$, then $\varphi^{\mathsf{u}} \in \Delta$ or $\psi^{\mathsf{u}} \in \Gamma$.

(iv) If $\boxast\varphi^{\mathsf{u}} \in \Gamma$, then $\varphi^{\mathsf{u}} \in \Gamma$ and $\Box\boxast\varphi^{\mathsf{u}} \in \Gamma$.

(v) If $\varphi \wedge \psi^{\mathsf{u}} \in \Delta$, then $\varphi^{\mathsf{u}} \in \Delta$ or $\psi^{\mathsf{u}} \in \Delta$.

(vi) If $\varphi \vee \psi^{\mathsf{u}} \in \Delta$, then $\varphi^{\mathsf{u}} \in \Delta$ and $\psi^{\mathsf{u}} \in \Delta$.

(vii) If $\boxast\varphi^{a} \in \Delta$, then $\varphi^{\mathsf{u}} \in \Delta$ or $\Box\boxast\varphi^{a} \in \Delta$.

Given a sequent σ, a formula occurring in σ is said to be *saturated* if σ satisfies the corresponding clause above for that formula.

As we are working with set sequents, formulas can simultaneously function as principal and as side formulas. We call an application of a rule *preserving* if the principal formula(s) also occurs as a side formula. For example, an application of $\boxast$L as depicted in Table 1 is preserving if $\boxast\varphi^{\mathsf{u}} \in \Gamma$. Intuitively, when a rule is applied preservingly to a sequent σ, we ensure that the premise(s) preserve the information contained in σ by preserving all formulas of σ. Note that applications of the non-invertible rules $\Box$ and $\rightarrow$R are never preserving.

The particular form of the proof-search tree depends on the kind of countermodel one wants to obtain from a winning strategy of Refuter. The general form of the proof-search trees employed here can be defined as follows.

Definition 5.2 Fix some inference rule C and a sequent σ. A *proof-search tree (with choice rule* C*)* for σ is a finite or countably infinite tree T whose nodes are labelled by sequents according to C and the invertible logical rules of IM such that:

(i) The root is labelled by $\Gamma_\sigma \Rightarrow \Delta_\sigma$;

(ii) Every invertible rule is applied preservingly;

(iii) No invertible rule is applied to a sequent in which the principal formula is already saturated;

(iv) A node is a leaf if and only if it is labelled by an axiom or by a saturated sequent to which the C-rule cannot be applied;

(v) The C-rule is only applied to saturated sequents.

The C-rule replaces both the non-invertible and the focus rules. Each completeness proof we present will be relative to a suitable 'choice' rule C.

Note that every sequent σ has a proof-search tree. Due to property (ii), every sequent can be saturated by finitely many invertible rule applications. By property (iii), we then obtain the following result.

Lemma 5.3 *Every infinite branch of a proof-search tree contains infinitely many applications of* C.

Given a proof-search tree T with choice rule C for a sequent σ, a game $G(T, \mathsf{C})$

can be defined between the players Prover and Refuter where a play corresponds to a branch in T: reading upwards, invertible rules represent a choice of admissible moves for Refuter and the C-rule represents a choice of admissible moves for Prover. Prover wins a play ρ if and only if ρ is finite and ends in an axiom, or ρ is infinite and has a good suffix. All other plays are won by Refuter.

A winning strategy for Refuter corresponds to a *refutation* of σ.

Definition 5.4 A *refutation* of a sequent σ is a subtree S of a proof-search tree T for σ satisfying the following properties.

(i) S contains the root of T.

(ii) No leaf is an axiom.

(iii) No infinite branch of S has a good suffix.

(iv) If S contains a node u that is labelled by the conclusion of a C-application, then S contains all direct successors of u in T.

(v) If S contains a node u that is labelled by the conclusion of any other rule than C, then S contains exactly one direct successor of u in T.

Note that property (iv) reflects that a refutation S considers all possible moves by Prover, while property (v) reflects that S prescribes a particular move for Refuter given some partial initial play. Property (ii) and (iii) then ensure that any play where Refuter plays according to S, is won by Refuter.

Whereas a winning strategy for Refuter corresponds to a refutation of σ, a winning strategy for Prover should correspond to a proof of σ; it must be checked that this is indeed the case for a particular choice for C.

It is routine to check that the set of winning plays in $G(T, \mathsf{C})$ for each player is Borel, and so it follows from Martin's determinacy theorem [15] that every sequent has a refutation or a proof. Thus, in order to prove completeness, the key result to obtain is that a refutation induces a countermodel. To show this, we will make use of the following 'canonical' model construction. For this construction, we assume that the premises of the choice rule C have been partitioned into two groups: the *intuitionistic premises* and the *modal premises.*

Definition 5.5 Let σ be a sequent and let S be a refutation of σ. The *canonical model* based on S is the model $M_S = (W, \leq, R, V)$ defined as follows.

(i) $W = S/{\sim}$, where $s \sim t$ iff there exists a path between s and t in which no instance of the C-rule occurs.

(ii) $\leq$ is the reflexive, transitive closure of the relation $\leq_0 \subseteq W \times W$ given by

$$w \leq_0 v \text{ iff there exist } s \in w \text{ and } t \in v \text{ such that } s \text{ is the conclusion and } t \text{ an intuitionistic premise of the same C-rule instance.}$$

(iii) $R \subseteq W \times W$ is such that

$$wRv \text{ iff there exist } s \in w \text{ and } t \in v \text{ such that } s \text{ is the conclusion and } t \text{ is a modal premise of the same C-rule instance.}$$

(iv) $V : w \mapsto \Gamma_w \cap \mathsf{Prop}$ where Γ_w is the left side of the sequent labelling the unique node in w that is the conclusion of a C-rule application.

The construction of M_S reflects the idea that applying invertible rules to sequents in S provides more information about the current 'world', while applying non-invertible rules (captured by the C-rule) corresponds to taking either a modal or an intuitionistic step.

5.2 Completeness of nIM with respect to triangle models

To show completeness of nIM with respect to triangle models, we consider the choice rule $\mathsf{C_t}$ given by

$$\frac{\Pi, \Box\Gamma, \varphi_0^{\mathsf{u}} \Rightarrow \psi_0^{\mathsf{u}} \quad \cdots \quad \Pi, \Box\Gamma, \varphi_l^{\mathsf{u}} \Rightarrow \psi_l^{\mathsf{u}} \quad \Gamma \Rightarrow \chi_0^{b_0} \quad \cdots \quad \Gamma \Rightarrow \chi_m^{b_m}}{\Pi, \Box\Gamma \Rightarrow \{(\varphi_i \to \psi_i)^{\mathsf{u}}\}_{i=0}^{l}, \{\Box\chi_i^{a_i}\}_{i=0}^{m}, \Sigma} \ \mathsf{C_t}$$

where the annotations b_i are equal to f whenever the underlying formula χ_i is a ⊞-formula, and equal to u otherwise. Moreover, we require that $\Pi \cup \Sigma$ contains no $\Box$-formulas and that Σ contains no $\to$-formulas. The modal premises are those of the form $\Gamma \Rightarrow \chi_i^{b_i}$, and the others are the intuitionistic premises.

Lemma 5.6 *If T is a proof-search tree for σ with choice rule $\mathsf{C_t}$, then a winning strategy for Prover in $G(T, \mathsf{C_t})$ corresponds to a nIM-proof of σ.*

Proof. Just like a winning strategy for Refuter corresponds to a subtree of T that is a refutation, a winning strategy for Prover corresponds to a subtree S of T that contains the root, all direct successors of nodes labelled by an invertible rule and exactly one direct successor of nodes labelled by the $\mathsf{C_t}$-rule. Note that the latter can be viewed as an application of $\to$R, or an application of $\Box$ with a subsequent application of f in case the principal formula is an unfocused ⊞-formula. Moreover, due to the winning conditions of Prover, every leaf of S must be an axiom and every infinite branch must have a good suffix. Thus we can view S as a proof of σ. □

Proposition 5.7 *If a sequent has a refutation with choice rule $\mathsf{C_t}$, then it is falsified in a triangle model.*

Proof. Let T be a proof-search tree for σ with choice rule $\mathsf{C_t}$ and let S be a subtree of T that is a refutation of σ. Let $M_S = (W, \leq, R, V)$ be the canonical model based on S, and let $M = (W, \leq, (\leq; R), V)$ be the induced triangle model. For each $w \in W$, let $\Gamma_w \Rightarrow \Delta_w$ be the sequent labelling the unique node in w that is the conclusion of a $\mathsf{C_t}$-rule application.

Let φ be a formula. By induction on the logical complexity of φ, we simultaneously prove that for any $w \in W$ we have (a) $M, w \models \varphi$ if $\varphi \in \Gamma_w^-$ and (b) $M, w \not\models \varphi$ if $\varphi \in \Delta_w^-$. The proof relies on the fact that the sequent $\Gamma_w \Rightarrow \Delta_w$ is saturated, since it is the conclusion of a $\mathsf{C_t}$-application. We only treat the connectives $\to$ and ⊞. Recall that, for triangle models, we can simply use the classical truth conditions for the modalities.

The case of $\to$. (a). If $\varphi \to \psi \in \Gamma_w^-$ and $v \geq w$, then by definition of $\mathsf{C_t}$ and the fact that invertible rules are applied preservingly, we have $\varphi \to \psi \in \Gamma_v^-$.

So by saturation and the induction hypothesis (IH), we have $M, v \models \psi$ or $M, v \not\models \varphi$, so we obtain $M, w \models \varphi \to \psi$. (b). If $\varphi \to \psi \in \Delta_w^-$, then by construction of $\leq$ there exists a $v \geq_0 w$ such that $\varphi \in \Gamma_v^-$ and $\psi \in \Delta_v^-$. So by the IH, we obtain $M, v \models \varphi$ and $M, v \not\models \psi$, so $M, w \not\models \varphi \to \psi$.

The case of $\boxtimes$. (a). Let $\boxtimes\varphi \in \Gamma_w^-$ and wR^*v. Saturation implies that $\Box\boxtimes\varphi \in \Gamma_w^-$, so by definition of $\mathsf{C_t}$ and the fact that invertible rules are applied preservingly, we have $\Box\boxtimes\varphi \in \Gamma_u^-$ for all $u \geq w$. This means that $\boxtimes\varphi \in \Gamma_s^-$ if wRs. Iterating the argument, we find that $\boxtimes\varphi \in \Gamma_v^-$. Saturation then gives $\varphi \in \Gamma_v^-$, so $M, v \models \varphi$ by the IH. (b). If $\boxtimes\varphi \in \Delta_w^-$, then saturation implies $\varphi \in \Delta_w^-$ or $\Box\boxtimes\varphi \in \Delta_w^-$. Suppose, for contradiction, that for all wR^*v we have $\varphi \notin \Delta_v^-$. Let $s \in w$ be the last node in w, i.e., s is the conclusion of a $\mathsf{C_t}$-application. Then we can define an infinite path ρ in S starting from s as follows: at each $\mathsf{C_t}$-application, we pick the modal premise that has $\boxtimes\varphi^{\mathsf{f}}$ as consequent. Note that saturation and the fact that no wR^*v satisfies $\varphi \in \Delta_v^-$ implies that this is always possible. The path ρ then forms a good suffix of the infinite branch of S in which it is contained, contradicting that S is a refutation. So there must be some wR^*v with $\varphi \in \Delta_v^-$, and thus $M, v \not\models \varphi$ by the IH. We conclude that $M, w \not\models \boxtimes\varphi$.

We conclude that the root of M falsifies the sequent σ. □

Theorem 5.8 *The calculus* nIM *is complete for the logic* $\mathrm{IM_t}$.

Proof. Suppose a sequent σ is valid over the class of triangle models. Let T be a proof-search tree with choice rule $\mathsf{C_t}$ for σ and consider the two-player game $G(T, \mathsf{C_t})$. By determinacy, exactly one of the two players Prover and Refuter has a winning strategy. By validity of σ and Proposition 5.7, Refuter cannot have a winning strategy. Hence Prover has a winning strategy. Therefore, by Lemma 5.6, σ is provable in nIM. □

5.3 Completeness of nIM with respect to functional models

When constructing the proof-search tree for $\mathrm{IM_f}$, we have to ensure that the induced countermodel will be functional. This means that, when we reach a saturated sequent of the form $\Gamma \Rightarrow \Box\chi_1, \ldots, \Box\chi_m, \Delta$ we can only pick *one* χ_i that will be falsified in the (unique) modal successor. This problem can be solved by adding in extra intuitionistic successors, so that the remaining χ_i can be falsified at *their* modal successor. To keep track of which right $\Box$-formula has to be 'taken care of' at a particular step, the proof-search tree will be labelled by *indexed sequents* $\Gamma \Rightarrow_k \Delta$, that is, sequents equipped with a natural number k that we call the *index* of the sequent.

Definition 5.9 An *indexed proof-search tree* for a sequent σ consists of an enumeration $\chi_0, \chi_1, \ldots, \chi_n$ of formulas in $\Box^{-1}\mathsf{Cl}(\sigma) := \{\chi \mid \Box\chi \in \mathsf{Cl}(\sigma)\}$ and a finite or infinite tree T whose nodes are labelled by indexed sequents such that:

(i) T is a proof-search tree for σ with choice rule [6]

$$\frac{\{\Pi, \Box\Gamma, \varphi_i^{\mathsf{u}} \Rightarrow_0 \psi_i^{\mathsf{u}}\}_{i=0}^{l} \quad \Gamma_\tau \Rightarrow_{(k+1)_m} \Delta_\tau \quad \Gamma \Rightarrow_0 \chi_{i_k}^{a}}{\Pi, \Box\Gamma \Rightarrow_k \{\varphi_i \to \psi_i^{\mathsf{u}}\}_{i=0}^{l}, \{\Box\chi_{i_j}^{a_i}\}_{j=0}^{m}, \Sigma}\ \mathsf{C_f}$$

where τ is the sequent labelling the conclusion (so Γ_τ and Δ_τ denote the left and right side of the conclusion), and a equals f if χ_{i_k} is a $\boxast$-formula and equals u otherwise. We require that $i_0 < i_1 < \cdots < i_m$, $k < m$ and $(k+1)_m$ denotes $k+1$ modulo m. Moreover, $\Pi \cup \Sigma$ contains no $\Box$-formulas and Σ contains no $\to$-formulas. Note that the premise $\Gamma_\tau \Rightarrow_{(k+1)_m} \Delta_\tau$ differs from the conclusion only in the index. The rightmost premise is a modal premise and the others are intuitionistic premises.

(ii) Invertible rule applications leave the index of a sequent unchanged.

Lemma 5.10 *If T is a proof-search tree for σ with choice rule $\mathsf{C_f}$, then a winning strategy for Prover in $G(T, \mathsf{C_f})$ corresponds to a nIM-proof of σ.*

Proof. As in the proof of Lemma 5.6, a winning strategy for Prover is a subtree S of T that contains exactly one direct successor of each node labelled by the $\mathsf{C_f}$-rule. In case this direct successor is not labelled by the premise of the form $\Gamma_\tau \Rightarrow_{(k+1)_m} \Delta_\tau$, we can simply view this as an application of $\to\mathsf{R}$ or $\Box$ (possibly with a subsequent application of f) by forgetting the indices. In case the direct successor *is* labelled by the premise of the form $\Gamma_\tau \Rightarrow_{(k+1)_m} \Delta_\tau$, we simply view this as no rule application at all, as the left and right side of the sequent have not changed. Due to the winning conditions of Prover, every leaf of S must be an axiom and every infinite branch must have a good suffix. Thus S corresponds to a nIM-proof. □

Proposition 5.11 *If a sequent has a refutation with C-rule $\mathsf{C_f}$, then it has a functional countermodel.*

Proof. Let σ be a sequent and T be an indexed proof-search tree based on some enumeration $\chi_1, \ldots, \chi_n$ of $\Box^{-1}\mathsf{Cl}(\sigma)$. Let S be a subtree of T that is a refutation of σ and let $M = (W, \leq, R, V)$ be the canonical model based on S. Note, R is functional, as every $\mathsf{C_f}$-rule application has only one right premise.

We follow the proof of Proposition 5.7. For any formula ϕ, by induction we simultaneously prove that for any $w \in W$ we have (a) $M, w \models \varphi$ if $\varphi \in \Gamma_w^-$ and (b) $M, w \not\models \varphi$ if $\varphi \in \Delta_w^-$. Here we only treat the connective $\Box$.

(a). If $\Box\varphi \in \Gamma_w^-$ and $w \leq vRu$ then, by definition of the $\mathsf{C_f}$-rule and the fact that invertible rules are applied preservingly, $\varphi \in \Gamma_u^-$. The IH then implies $M, u \models \varphi$, so $M, w \models \Box\varphi$. (b). Let $\Box\varphi \in \Delta_w^-$. As $\Gamma_w \Rightarrow \Delta_w$ is the conclusion of the $\mathsf{C_f}$-rule, it must be of the form $\Pi, \Box\Gamma \Rightarrow_k \{\varphi_i \to \psi_i\}_{i=0}^{l}, \{\Box\chi_{i_j}\}_{j=0}^{m}, \Sigma$ with $\varphi = \chi_{i_p}$ for some p. Now, by construction of $\leq$ and the rule $\mathsf{C_f}$, it follows

[6] Strictly speaking, we only defined proof-search trees for 'plain' sequents, that is, sequents without an index. However, if we extend the syntax by allowing formulas of the form k with $k \in \omega$, then we can simply define an indexed sequent $\Gamma \Rightarrow_k \Delta$ as the plain sequents $k, \Gamma \Rightarrow \Delta$. We prefer the former notation as it highlights the specific role of the index k.

that there exists a $v \geq w$ such that $\Gamma_v \Rightarrow \Delta_v$ is equal to $\Pi, \Box\Gamma \Rightarrow_p \{\varphi_i \rightarrow \psi_i\}_{i=0}^{l}, \{\Box\chi_{i_j}\}_{j=0}^{m}, \Sigma$. So, by construction of R, there exists a u with vRu and $\chi_{i_p} \in \Delta_u^-$. The IH then implies $M, u \not\models \varphi$, so $M, w \not\models \Box\varphi$.

We conclude that the root of M falsifies the sequent σ. □

Combining these results yields completeness of nIM with respect to functional models. The proof of the following theorem is analogous to the proof of Theorem 5.8.

Theorem 5.12 *The calculus* nIM *is complete for the logic* $\mathrm{IM_f}$.

5.4 Completeness of the cyclic calculus

With completeness of the non-wellfounded calculus at hand, we are ready to prove completeness of the cyclic calculus. In addition, we can show that its single-conclusion version is also complete. To this end, we first show that our use of multi-conclusion sequents is not a real restriction.

Lemma 5.13 *If a single-conclusion sequent σ has a* nIM*-proof, then it has a single-conclusion* nIM*-proof.*

Proof. Given any nIM-proof π, let its *trunk* π_{tr} be the derivation obtained from π by cutting off each branch right above the premise of the lowest application of →R or □. As π is a proof, note that π_{tr} must be finite due to Lemma 3.4. Working top-down, each sequent in π_{tr} will be replaced by a single-conclusion one. The first rule instances will be of type →R, □, ⊥ or id, which are straightforward to treat; the premises are single-conclusion by definition, and if, for example, the conclusion is $\Pi, \Box\Gamma \Rightarrow \Box\varphi^a, \Delta$ with $\Box\varphi^a$ principal, replace it by $\Pi, \Box\Gamma \Rightarrow \Box\varphi^a$. Now consider a highest sequent $\Gamma \Rightarrow \Delta$ in π_{tr} with $|\Delta| > 1$. This will occur as the conclusion of a (possibly incorrect) rule instance

$$\frac{\sigma_1 \quad \cdots \quad \sigma_n}{\Gamma \Rightarrow \Delta}\ \mathsf{r}$$

with single-conclusion premises. Then, either (1) there exists a $\delta \in \Delta$ such that

$$\frac{\sigma_1 \quad \cdots \quad \sigma_n}{\Gamma \Rightarrow \delta}\ \mathsf{r}$$

is a correct instance of r, or (2) there is a premise σ_i with $\Gamma_{\sigma_i} = \Gamma$ and $\Delta_{\sigma_i} \subseteq \Delta$. We treat r = →L as an exemplary case. If the conclusion is $\Gamma', \varphi \rightarrow \psi^{\mathsf{u}} \Rightarrow \Delta$ with $\varphi \rightarrow \psi^{\mathsf{u}}$ principal, then the premises are of the form $\Gamma', \varphi \rightarrow \psi^{\mathsf{u}} \Rightarrow \chi^a$ and $\Gamma', \psi^{\mathsf{u}} \Rightarrow \zeta^b$. If $\chi^a = \varphi^{\mathsf{u}}$, then $\delta := \zeta^b$ satisfies property (1). Otherwise, we must have $\chi^a \in \Delta$, so the premise $\Gamma', \varphi \rightarrow \psi^{\mathsf{u}} \Rightarrow \chi^a$ satisfies property (2). Note that rule instances of □ only satisfy property (1).

Property (1) means that $\Gamma \Rightarrow \Delta$ can be replaced by $\Gamma \Rightarrow \delta$, whereas (2) means that the node labelled by $\Gamma \Rightarrow \Delta$ can simply be deleted. Iterating this, we then obtain a single-conclusion derivation π_{tr}^{sc} such that replacing the trunk π_{tr} by π_{tr}^{sc} in π yields a nIM-proof π'. By construction, if π proves the sequent $\Gamma \Rightarrow \Delta$ then π' proves $\Gamma \Rightarrow \delta$ for some $\delta \in \Delta$.

Now let π be an nIM-proof of σ. Given a node s in π, we let $\uparrow s$ denote the nIM-proof induced by the upset of s in π. We define a sequence $(\pi_i)_{i<\omega}$ of finite, single-conclusion derivations as follows. Let π_0 be π^{sc}_{tr}, and given π_i, let π_{i+1} be the result of replacing each leaf s in π_i by the derivation $(\uparrow s)^{sc}_{tr}$. It is then easy to see that the limit π' of this construction gives a single-conclusion nIM-derivation of σ. To see that π' is in fact a proof, consider an infinite branch β' in π'. By construction of π', the branch β' is a single-conclusion and possibly condensed version of an infinite branch β in π. As π is a proof, β has a good suffix ρ, which in turn corresponds to a suffix ρ' of β'. By Lemma 3.4 and the fact that rule instances of $\Box$ always satisfy property (1), it follows that ρ' contains infinitely many $\Box$-applications. Thus, since ρ always has a formula in focus and $\Box$ has a single-conclusion premise, ρ' always has a formula in focus as well. These two facts together then imply that ρ' must contain infinitely many $\boxplus$R-applications in which, due to the sequent being single-conclusion, the principal formula is in focus. Hence ρ' is a good suffix of β' and π' is a single-conclusion proof of σ. ☐

Theorem 5.14 *If a sequent σ has a (single-conclusion)* nIM*-proof, then it has a (single-conclusion)* cIM*-proof.*

Proof. Let π be a (single-conclusion) nIM-proof of σ. First note that π contains only finitely many sequents, as analyticity of IM ensures that each such sequent only contains formulas in the finite set $\mathsf{Cl}(\sigma)$. Now let π' be the derivation obtained from π by cutting off each branch after the first successful repeat (if it exists). We prove that π' is finite. Suppose, for contradiction, that it is not. By König's lemma, π' then has an infinite branch ρ. Then ρ is also an infinite branch of the proof π, and thus it must have a good suffix ρ'. However, as ρ' contains only finitely many sequents, it follows that ρ' must contain a successful repeat. This contradicts that ρ is an infinite branch of π'. Thus π' is a (single-conclusion) cIM-proof. ☐

From soundness of cIM (Theorem 4.4) and the two completeness results (Theorem 5.8 and 5.12) for nIM, we then obtain the following result.

Theorem 5.15 *The calculus* cIM *and its single-conclusion version are sound and complete for* $\mathrm{IM_K}$*,* $\mathrm{IM_t}$ *and* $\mathrm{IM_f}$*. In particular, we have* $\mathrm{IM_K} = \mathrm{IM_f} = \mathrm{IM_t}$.

6 Discussion

We close the paper by summarising some natural adaptions of the system cIM and possible interpretations of the language $\mathcal{L}_{\mathrm{IM}}$.

6.1 Intuitionistic temporal logic

The system cIM can be adapted to a sound and complete system cIM^s with respect to serial models and total functional models.[7] The modal rule $\Box$ is

[7] We call a model $(W, \leq, R, V)$ *serial* if the relation R is serial and *total functional* if R is both serial and functional.

replaced by the rule:

$$\frac{\Gamma \Rightarrow \Delta_0}{\Pi, \Box\Gamma \Rightarrow \Box\Delta, \Sigma} \; \Box_s$$

where $\Delta_0 \subseteq \Delta$ and $|\Delta_0| \leq 1$. As in Section 5.3, completeness of the non-wellfounded calculus with respect to total functional frames is shown by proof-search on indexed sequents. The choice rule $\mathsf{C_f}$ is adapted so as to allow a right premise with an empty consequent in case the conclusion contains no $\Box$-formula. As a result, each world in the induced canonical model necessarily has a modal successor, so the obtained countermodel will be total functional.

Completeness for total functional models induces an interpretation of the language $\mathcal{L}_{\mathrm{IM}}$ as an intuitionistic version of linear-time temporal logic (LTL). For each world w, the unique modal successor $R(w)$ may be interpreted as its temporal successor. The modal operator $\Box$ is interpreted as the 'next' operator X and the master modality $\boxast$ as the 'henceforth' operator. In contrast to classical LTL, the evaluation of a formula $\mathsf{X}\varphi$ at world w does not depend solely on $R(w)$, but also on $R(v)$ for all worlds $v \geq w$. As we have no confluence condition on $\leq$ and R, classical temporal tautologies such as $\mathsf{X}(\varphi \vee \psi) \to (\mathsf{X}\varphi \vee \mathsf{X}\psi)$ do not hold in this setting. The obtained temporal logic is therefore weaker than those considered in [4,1].

6.2 Intuitionistic common knowledge

Jäger and Marti introduce an intuitionistic version of common knowledge logic in [12] employing a polymodal extension of the language $\mathcal{L}_{\mathrm{IM}}$ with finitely many box operators $\Box_0, \ldots, \Box_n$. The formula $\Box_i\varphi$ is read as *agent i knows φ* and $\boxast\varphi$ as *φ is common knowledge.* This language is interpreted over triangle models with a modal relation for each $i \leq n$. Jäger and Marti present a finitary calculus for this logic based on an induction rule, which is complete for the class of triangle models and can be extended to complete calculi for reflexive models and S4-models. The proof of completeness, however, makes essential use of the cut rule, and a cut-elimination theorem is not given.

The calculus cIM can be adapted to the polymodal language by incorporating rules

$$\frac{\Gamma \Rightarrow \varphi^a}{\Pi, \Box_i\Gamma \Rightarrow \Box_i\varphi^a, \Delta} \; \Box_i$$

for each $i \leq n$, and appropriate modification of the rules for $\boxast$. The resulting system cIM_p is easily shown to be sound and complete with respect to (polymodal) triangle models using the presented methods and an appropriate adaption of the choice rule $\mathsf{C_t}$. Moreover, cIM_p (and so, in particular, cIM) can be extended to account for reflexive and for S4-models. For reflexive models we add for each i the rule $\Box_i^T$ below to cIM_p, and for S4-models we additionally replace the rules $\Box_i$ by $\Box_i^{S4}$:

$$\frac{\Gamma, \varphi^{\mathsf{u}} \Rightarrow \Delta}{\Gamma, \Box_i\varphi^{\mathsf{u}} \Rightarrow \Delta} \; \Box_i^T \qquad\qquad \frac{\Box_i\Gamma \Rightarrow \varphi^a}{\Pi, \Box_i\Gamma \Rightarrow \Box_i\varphi^a, \Sigma} \; \Box_i^{S4}$$

To establish completeness, the choice rule $\mathsf{C_t}$ needs only be adapted for S4-models, which is given by simply replacing Γ by $\Box_i\Gamma$ in the right premises of (the polymodal) $\mathsf{C_t}$. As cIM_p and its extensions are cut-free and analytic, these systems are suitable for proof-search and may therefore be considered an improvement of Jäger and Marti's work. Whether cIM_p can be adapted to account for S5-models is unknown to us.

6.3 Future work

We have presented cyclic calculi for intuitionistic modal logic with $\Box$ and the master modality $\boxplus$. Two natural directions for further research are to extend the language by diamonds, or to allow for more fixed point operators. Concerning the former, note that $\Box$ and $\Diamond$ are not interdefinable in the intuitionistic setting. As a result, obtaining monotonicity in the presence of diamond operators requires other confluence conditions that are less robust than triangle confluence with respect to proof-search. It seems that more complex calculi are needed in this case, such as a nested or labelled calculi [23,9,5]. With respect to adding more fixed points, the current work seems to generalise more readily. A natural candidate in this regard is intuitionistic modal logic with $\Box$ and arbitrary least and fixed points. As triangle confluence still suffices in this general case, proof-search can be carried out in a similar fashion as done here.

Another open question is the complexity of the validity-checking problem for $\mathrm{IM_K}$. We conjecture that the validity problem has an EXPTIME upper bound and suspect that a similar approach as taken in [20] works: translate the calculus cIM into a parity game with a constant number of priorities. As such a game can be decided in polynomial time in the size of the arena (due to the fact that the number of priorities is constant [10]), and the size of the arena is exponential in the size of the formula (by analiticity of cIM), an exponential upper bound follows. Whether the lower bound is also exponential is unclear.

References

[1] Afshari, B., L. Grotenhuis, G. E. Leigh and L. Zenger, *Ill-founded proof systems for intuitionistic linear-time temporal logic*, Automated Reasoning with Analytic Tableaux and Related Methods **14278** (2023), pp. 223–241.

[2] Balbiani, P., J. Boudou, M. Diéguez and D. Fernández-Duque, *Intuitionistic linear temporal logics*, ACM Trans. Comput. Logic **21** (2019).

[3] Boudou, J., M. Diéguez and D. Fernández-Duque, *A decidable intuitionistic temporal logic*, in: V. Goranko and M. Dam, editors, *26th EACSL Annual Conference on Computer Science Logic (CSL 2017)*, Leibniz International Proceedings in Informatics (LIPIcs) **82** (2017), pp. 14:1–14:17.

[4] Boudou, J., M. Diéguez and D. Fernández-Duque, *Complete intuitionistic temporal logics for topological dynamics*, Journal of Symbolic Logic **87** (2022), pp. 995–1022.

[5] Das, A. and S. Marin, *On intuitionistic diamonds (and lack thereof)*, in: *International Conference on Automated Reasoning with Analytic Tableaux and Related Methods*, Springer, 2023, pp. 283–301.

[6] Das, A., I. van der Giessen and S. Marin, *Intuitionistic Gödel-Löb logic, à la Simpson: Labelled systems and birelational semantics*, in: A. Murano and A. Silva, editors,

32nd EACSL Annual Conference on Computer Science Logic (CSL 2024), Leibniz International Proceedings in Informatics, LIPIcs (2024), pp. 22:1–22:18.

[7] Demri, S., V. Goranko and M. Lange, "Temporal Logics in Computer Science: Finite-State Systems," Cambridge Tracts in Theoretical Computer Science, Cambridge University Press, 2016.

[8] Fernández-Duque, D., *The intuitionistic temporal logic of dynamical systems*, Logical Methods in Computer Science **14** (2018).

[9] Girlando, M., R. Kuznets, S. Marin, M. Morales and L. Straßburger, *Intuitionistic S4 is decidable*, in: *2023 38th Annual ACM/IEEE Symposium on Logic in Computer Science (LICS)*, IEEE, 2023, pp. 1–13.

[10] Grädel, E., W. Thomas and T. Wilke, editors, "Automata, Logics, and Infinite Games: A Guide to Current Research," Lecture Notes in Computer Science, Springer Berlin, Heidelberg, 2002.

[11] Iemhoff, R., *Reasoning in circles*, in: J. van Eijck, R. Iemhoff and J. J. Joosten, editors, *Liber Amicorum Alberti: A Tribute to Albert Visser*, College Publications, 2016 .

[12] Jäger, G. and M. Marti, *Intuitionistic common knowledge or belief*, Journal of applied logic **18** (2016), pp. 150–163.

[13] Kavvos, G. A., *The many worlds of modal λ-calculi: I. Curry–Howard for necessity, possibility and time*, arXiv preprint arXiv:1605.08106 (2016).

[14] Litak, T. and A. Visser, *Lewis meets Brouwer: constructive strict implication*, Indagationes Mathematicae **29** (2018), pp. 36–90.

[15] Martin, D. A., *Borel determinacy*, Annals of Mathematics **102** (1975), pp. 363–371.

[16] Menéndez Turata, G., "Cyclic proof systems for modal fixpoint logics," Ph.D. thesis, Universiteit van Amsterdam (2024).

[17] Negri, S. and J. von Plato, "Structural Proof Theory," New York: Cambridge University Press, 2001.

[18] Niwinski, D. and I. Walukiewicz, *Games for the mu-calculus*, Theoretical Computer Science **163** (1996), pp. 99–116.

[19] Pacheco, L., *Game semantics for the constructive μ-calculus*, arXiv preprint arXiv:2308.16697 (2024).

[20] Rooduijn, J. M. W. and L. Zenger, *An analytic proof system for common knowledge logic over S5*, in: *David Fernández-Duque, Alessandra Palmigiano and Sophie Pinchinat (eds.) Advances in Modal Logic*, 2022, pp. 659–680.

[21] Rowe, R., *Non-well-founded and cyclic proof theory: A bibliography*, `https://reubenrowe.github.io/cyclic-proof-bibliography/`.

[22] Sierra-Miranda, B., *Cyclic proofs for iGL via corecursion*, arXiv preprint arXiv:2310.10785 (2023).

[23] Simpson, A., "The Proof Theory and Semantics of Intuitionistic Modal Logic," Ph.D. thesis, University of Edinburgh (1994).

Strong Completeness of the Closed Fragment of GLP

Juan P. Aguilera

Institute of Discrete Mathematics and Geometry, Vienna University of Technology. Wiedner Hauptstraße 8–10, 1040 Vienna, Austria.

Grigorii Stepanov [1]

Institute of Discrete Mathematics and Geometry, Vienna University of Technology. Wiedner Hauptstraße 8–10, 1040 Vienna, Austria.

Abstract

The polymodal provability logic GLP is known to be Kripke incomplete w.r.t. Kripke semantics. However, it was shown by Ignatiev that the closed (i.e. variable free) fragment of GLP is Kripke complete w.r.t. a certain Kripke frame, which is now known as the Ignatiev frame, moreover he showed that it is complete w.r.t. to a specific set of its points, which are called the main axis. We show that the closed fragment of GLP is strongly complete w.r.t. to an extended variant of the Ignatiev frame, as well is that it is not strongly complete w.r.t. to the main diagonal and moreover w.r.t. to neither Icard not Beklemishev-Gabelaia topological spaces, which are known to provide topological completeness for the closed fragment of GLP and GLP itself correspondingly.

Keywords: Provability logic, strong completeness, Ignatiev frame, GLP.

1 Introduction

The interest in the provability logic stems from the investigations of Gödel's incompleteness theorems. Löb [24] formulated three conditions on the provability predicate of Peano Arithmetic that form a useful modification of the conditions that Hilbert and Bernays [19] introduced for their proof of Gödel's second incompleteness theorem. Friedman [18] posed the problem of axiomatizing the set of valid arithmetical formulæ built from expressions of the form "φ is provable" by means of Boolean connectives and provability assertions. Boolos [13] (and independently Bernardi, Montagna, and van Benthem) proved that Löb's axiomatization was complete when restricting to closed (i.e., variable-free) formulæ, building on work of Segerberg [25] on the Kripke semantics of Löb's logic

[1] Corresponding author. Email: `grigorii.stepanov@tuwien.ac.at`. This work was partially supported by FWF grant P36837

GL. Solovay [27] later extended Boolos' theorem to a completeness theorem of GL for its arithmetical interpretation.

We consider a modal logic with infinitely many modalities ($[n]$ and $\langle n\rangle$ for $n < \omega$). The intended interpretation of $[n]\varphi$ is "φ is provable in the formal system T_n," where T_n is an arithmetical or set-theoretic system such that T_{n+1} is stronger than T_n for each n. We shall not concern ourselves with the arithmetical interpretation of this polymodal provability logic here, however, and instead focus mainly on its relational semantics. This polymodal extension of GL – called GLP – was introduced by Japaridze [23] and is complete with respect to various choices of T_n; see e.g., Japaridze [23] or Fernandez-Duque and Joosten [17]. Beklemishev [7] and Beklemishev and Pakhomov [15] show how GLP can be applied for the purposes of ordinal analysis and other proof-theoretic results. The first author and Pakhomov [5] have recently proved the completeness of the extension GLP.3 of GLP for a set-theoretic interpretation, extending a result of Solovay [27] for the unimodal case.

Much work has been carried out on the models of GL and GLP, in part due to the fact that these are generally complicated, as we shall recall below. For instance, it is not difficult to show that GLP has no nontrivial Kripke frames and in particular is not Kripke-complete. Nonetheless, Ignatiev [22] proved a relational completeness theorem for the closed fragment of GLP with respect to what is nowadays called the *Ignatiev frame* $\mathfrak{I}$. The purpose of this article is to investigate the question of whether the closed fragment of GLP is strongly complete with respect to this semantics.

Definition 1.1 A modal logic L is *strongly Kripke-complete* with respect to a class of Kripke frames $\mathcal{C}$ if every consistent set of L-formulæ has a model in $\mathcal{C}$.

Here recall that a set of formulæ Γ is consistent relative to a logic L if $\mathsf{L} \nvdash \neg \bigwedge \Delta$ for any finite $\Delta \subset \Gamma$.

We shall prove:

Theorem 1.2 *The closed fragment of the provability logic* GLP *is strongly Kripke-complete with respect to* $\{\mathfrak{I}\}$*, where* $\mathfrak{I}$ *is the Ignatiev frame.*

Strictly speaking, the frame we use is a slight extension of Ignatiev's original frame, and this is necessary, as we shall see below. The idea to establish the result is to use maximal consistent sets of formulæ to inductively extract coordinates for the unique point in the Ignatiev frame which satisfies the given set of formulæ. Indeed, in addition to Theorem 1.2, we mention various counterexamples to strong completeness with respect to other semantics. In particular, we give a counterexample to the strong completeness of the closed fragment of GLP with respect to so-called *Icard spaces*. These were studied by Icard [20,21] as a variant of the blow-up constructions of Beklemishev [8] and provide a simple semantics for the closed fragment of GLP. The same argument shows that GLP is not strongly complete with respect to the topological spaces constructed by Beklemishev-Gabelaia [10]. The first author [4] had previously observed that GLP is also not strongly complete with respect to its so-called "standard topological models" (see Beklemishev and Gabelaia [11] and Bagaria [6] for more on these) and so at the current stage it is not clear whether there are any natural candidates for strongly complete models. Nonetheless, Shamkanov [26] has proved a nice *global completeness* result making use of an illfounded proof system for GLP and the global consequence relation, from which strong completeness for topological models follows, however it is open whether strong completeness holds for any nice collection of ordinal or otherwise easily described spaces, and we shall see below that it fails for what might otherwise be the first natural candidates. A strong completeness theorem of GL for a slight variation of its relational semantics was obtained by the first author and Fernández-Duque [3]. This result extends to GLP – a proof of this shall appear in a forthcoming article.

2 Preliminaries

2.1 The logic GLP

We begin with some preliminary notions, definitions, and recall some relevant results. For general background on provability logic, we refer the reader to Boolos [14].

Definition 2.1 We consider the modal language $\mathcal{L}$ generated according to the following rules:

$$\mathcal{L} = \bot \mid p \mid \varphi \wedge \psi \mid \neg\varphi \mid [n]\varphi$$

where p is a propositional variable and $n < \omega$. The closed (variable free) fragment of $\mathcal{L}$ is defined by

$$\mathcal{L}_0 = \bot \mid \varphi \wedge \psi \mid \neg\varphi \mid [n]\varphi$$

Finally, the language $\mathcal{L}_{[k;l)}$ is defined by restricting the modalities in $\mathcal{L}$ to the interval $[k, l)$. We conventionally use $\langle n\rangle\varphi = \neg[n]\neg\varphi$.

Definition 2.2 The logic I is the minimal set of $\mathcal{L}$-formulæ closed under modus ponens and necessitation and containing the following axioms:

(i) All Boolean tautologies

(ii) $[n](\varphi \to \psi) \to ([n]\varphi \to [n]\psi)$;

(iii) $[n]([n]\varphi \to \varphi) \to [n]\varphi$ for all $n < \omega$;

(iv) $[m]\varphi \to [n][m]\varphi$ for all $m \leq n < \omega$;

(v) $\langle m\rangle\varphi \to [n]\langle m\rangle\varphi$ for all $m < n < \omega$;

The logic GLP is obtained from I by adding the monotonicity axiom:

(vi) $[m]\varphi \to [n]\varphi$ for all $m < n < \omega$.

Definition 2.3 A *Kripke frame* is a tuple $F = (W, R_0, R_1, \dots)$, where W is a set and $R_i \subset W \times W$ for each $i < \omega$. Given a Kripke frame F and a function $v : \mathsf{Vars} \to \mathcal{P}(W)$, we say that $M = (F, v)$ is a *Kripke model*, which yields the following interpretation $[\![\cdot]\!]$ of modal formulæ:

- $[\![\bot]\!] = \emptyset$;
- $[\![p]\!] = v(p)$, where $p \in \mathsf{Vars}$;
- $[\![\varphi \wedge \psi]\!] = [\![\varphi]\!] \cap [\![\psi]\!]$;
- $[\![\neg\varphi]\!] = W \setminus [\![\varphi]\!]$;
- $[\![\langle n\rangle\varphi]\!] = \{x : \exists y \in [\![\varphi]\!]\, xR_ny\}$;

We say that a formula φ holds at a point x in a model M if $x \in [\![\varphi]\!]$, in which case we write $M, x \Vdash \varphi$. We write $M \Vdash \varphi$ to mean $M, x \Vdash \varphi$ for some $x \in M$ and $M \models \varphi$ to mean $M, x \Vdash \varphi$ for all $x \in F$ and $F \models \varphi$ to mean $M \models \varphi$ for all models of the form $M = \langle F, v\rangle$.

We say that a Kripke frame is an I-*frame*, if each R_i is converse well-founded and transitive and the following holds:

$$\forall x, y\big(xR_ny \to \forall z(xR_mz \leftrightarrow yR_mz)\big)$$

if $m < n$.

For each point $x \in W$ we assign an *i-degree* $\deg_i(x)$, which is 0 for each element that has no i-successors, and otherwise it is

$$\deg_i(x) = \sup\{\deg_i(y) + 1 : y \text{ is an immediate } R_i\text{-successor of } x\}.$$

The study of models of GLP is complicated, as these are hard to produce.

Example 2.4 Let $(W, R_0, R_1, R_2, \ldots)$ be a Kripke frame such that $W \models \mathsf{GLP}$. From the validity of axiom (v) it follows that if wR_0v and wR_1u, then uR_0v.

Suppose thus that wR_1v. From the validity of axiom (vi) it follows that wR_0v and thus that vR_0v, by the comment above. However, it is well known that the validity of Löb's axiom requires that frames be wellfounded and in particular irreflexive. Thus we conclude all GLP frames satisfy $R_1 = R_2 = \cdots = \varnothing$. In particular, GLP is not Kripke-complete.

This example in fact shows that in I-frames there cannot be more than one arrow between any two points.

The problem of Kripke-incompleteness is sometimes circumvented by considering other kinds of models, such as topological models; we shall return to these later. It also motivates the study of its closed fragment, which is sufficient for many of its applications in particular to proof theory and ordinal analysis. As was shown by Ignatiev, the closed fragment of GLP coincides with the closed fragment of I:

Theorem 2.5 ([22]) *Let* $\mathsf{GLP}_0 = \mathsf{GLP} \cap \mathcal{L}_0$ *and* $\mathsf{I}_0 = \mathsf{I} \cap \mathcal{L}_0$. *Then* $\mathsf{GLP}_0 = \mathsf{I}_0$.

Ignatiev showed that I is complete with respect to the class of I-frames. In particular $\mathsf{I}_0 = \mathsf{GLP}_0$ is Kripke-complete. In addition, it is complete with respect to a single certain frame, called the *Ignatiev frame*, which we shall recall below. Let us first remind ourselves some basic facts about ordinal numbers.

2.2 On ordinals $\leq \varepsilon_0$

Definition 2.6 ε_0 is the least ordinal α that satisfies the equation $\omega^\alpha = \alpha$.

The ordinal ε_0 satisfies

$$\varepsilon_0 = \sup\Big\{ \underbrace{\omega^{\omega^{\cdot^{\cdot^{\cdot^{\omega}}}}}}_{n \text{ times}} : n < \omega \Big\} = \omega^{\omega^{\cdot^{\cdot^{\cdot^{\omega^{\cdot^{\cdot^{\cdot}}}}}}}}.$$

Definition 2.7 Given an ordinal $\alpha > 0$, it can be uniquely represented in its *Cantor normal form* as

$$\alpha = \omega^{\lambda_0} + \omega^{\lambda_1} + \cdots + \omega^{\lambda_n}$$

where $\lambda_0 \geq \lambda_1 \geq \cdots \geq \lambda_n$. Assuming $\alpha > 0$ we define $\log(\alpha) = \lambda_n$ and its iteration, $\log^0(\alpha) = \alpha$ and $\log^{n+1}(\alpha) = \log(\log^n(\alpha))$. Similarly, we put $\log(0) = 0$.

Note that for any ordinal $\alpha < \varepsilon_0$, there is large enough $m < \omega$, such that $\log^n(\alpha) = 0$ for any $n > m$, however $\log^n(\varepsilon_0) = \varepsilon_0$ for any $n < \omega$.

2.3 The frame $\mathfrak{I}$

Definition 2.8 Let $\iota \leq \varepsilon_0$. The Ignatiev frame $\mathfrak{I}_{\leq\iota}$ consists of functions

$$\boldsymbol{\alpha} : \omega \to \iota + 1$$

with the property that $\alpha_{i+1} \leq \log(\alpha_i)$. Here and below, we write α_i to mean $\alpha(i)$ in order to regard $\boldsymbol{\alpha}$ as a sequence. For $\boldsymbol{\alpha}, \boldsymbol{\beta} \in \mathfrak{I}_{\leq\iota}$, we define $\boldsymbol{\alpha} R_k \boldsymbol{\beta}$ if and only if the following hold:

- $\forall i < k \ \alpha_i = \beta_i$;
- $\alpha_k > \beta_k$.

Definition 2.9 Let ι be an ordinal. We define $\mathfrak{I}_{<\iota}$ to be the union of $\{\mathfrak{I}_{\leq\eta} : \eta < \iota\}$. We also use $\mathfrak{I}$ to denote the union of $\{\mathfrak{I}_{<\eta} : \eta \in \mathsf{Ord}\}$, though this shall only be needed occasionally for notational convenience.

Note that the structures $\mathfrak{I}_{\leq\iota}$ strictly extend each other as ι increases, so $\mathfrak{I}_{<\iota}$ makes sense model-theoretically. In the past, typically only the model $\mathfrak{I}_{<\varepsilon_0}$ has been considered, but here we shall find use for the slightly larger model $\mathfrak{I}_{\leq\varepsilon_0}$. For sequences $\boldsymbol{\alpha} \in \mathfrak{I}_{<\varepsilon_0}$, we always find some $n < \omega$ such that $\alpha_k = 0$ for all $k > n$, i.e., elements of $\mathfrak{I}_{<\varepsilon_0}$ are finitely supported. We may occasionally identify elements of $\mathfrak{I}$ with their restriction to their support. In $\mathfrak{I}_{\leq\varepsilon_0}$ the only sequence with infinite support is the constant-ε_0 sequence. For all finitely supported sequences we write $|\boldsymbol{\alpha}|$ to denote the least $i > 0$ such that $\alpha_{i-1} = 0$.

Let us introduce some notational conventions. First, given an ordinal α, we define a sequence $\hat{\alpha}$ by

$$\hat{\alpha}_i = \log^i(\alpha).$$

For all $k < \omega$ we always denote $\boldsymbol{\alpha}_k = \langle \alpha_0, \ldots, \alpha_k \rangle$, $\langle \alpha, \boldsymbol{\beta} \rangle = \langle \alpha, \beta_i \rangle_{i<|\boldsymbol{\beta}|}$, $\langle \boldsymbol{\alpha}, \beta \rangle = \langle \alpha_i, \beta \rangle_{i<|\boldsymbol{\alpha}|}$ and $\langle \boldsymbol{\alpha}, \boldsymbol{\beta} \rangle = \langle \alpha_i, \beta_j \rangle_{i<|\boldsymbol{\alpha}|, j<|\boldsymbol{\beta}|}$. We also define pointwise addition, that is $\boldsymbol{\alpha} + \boldsymbol{\beta}$ is a sequence given by $\alpha_i + \beta_i$ for all $i < \omega$, as well as pointwise comparison which is denoted by $\preceq$. Finally, we define a useful operator that glues the first n elements of a sequence with the rest of another sequence, we define a sequence $(\boldsymbol{\alpha} :_n \boldsymbol{\beta})$ by:

$$(\boldsymbol{\alpha} :_n \boldsymbol{\beta})_i = \begin{cases} \alpha_i, & i < n; \\ \beta_i, & i \geq n; \end{cases}$$

The motivation behind it is the following, for any $\boldsymbol{\alpha}, \boldsymbol{\beta} \in \mathfrak{I}$, $\boldsymbol{\alpha} \succeq \boldsymbol{\beta}$ implies $(\boldsymbol{\alpha} :_n \boldsymbol{\beta}) \in \mathfrak{I}$, since $\log(\alpha_{n-1}) \geq \alpha_n \geq \beta_n$.

It can be verified that $\mathfrak{I}$ is an I-frame. The following theorem shows that considering this very frame is enough for establishing completeness.

Theorem 2.10 ([22]) $\mathsf{GLP}_0 \vdash \varphi$ *if and only if* $\mathfrak{I}_{<\varepsilon_0} \models \varphi$.

In fact, Ignatiev showed a slightly stronger fact:

Definition 2.11 Let $\boldsymbol{\alpha} = \langle \alpha_0, \ldots, \alpha_n \rangle \in \mathfrak{I}$. We say that $\boldsymbol{\alpha}$ belongs to the *main axis* of $\mathfrak{I}$ if it satisfies $\log(\alpha_i) = \alpha_{i+1}$ for all $i < n$, i.e., if it is of the form $\hat{\gamma}$ for some γ. We denote by $\mathrm{ma}(\mathfrak{I})$ the main axis of $\mathfrak{I}$.

For each $\alpha_0 < \varepsilon_0$, there is precisely one point in the main axis of $\mathfrak{I}$ whose first coordinate is α_0, namely $\hat{\alpha}_0$. Ignatiev proved that every closed formula consistent with GLP is satisfied somewhere along the main axis of $\mathfrak{I}_{<\varepsilon_0}$.

The goal of this article is to study the question of whether $\mathfrak{I}$ is strongly complete for the closed fragment of GLP. We shall show:

Theorem 2.12 (Strong completeness)

(A) *The closed fragment of* GLP *is strongly complete with respect to* $\mathfrak{I}_{\leq\varepsilon_0}$. *More precisely: let* Γ *be a set of closed* $\mathcal{L}$*-formulæ. Then the following are equivalent:*
  - (i) Γ *is consistent with* GLP*; and*
  - (ii) $\mathfrak{I}_{\leq\varepsilon_0}, \boldsymbol{\alpha} \Vdash \Gamma$ *for some* $\boldsymbol{\alpha} \in \mathfrak{I}_{\leq\varepsilon_0}$.

(B) *Moreover,*
  - (i) *The closed fragment of* GLP *is not strongly complete with respect to* $\mathfrak{I}_{<\varepsilon_0}$*; and*
  - (ii) *The closed fragment of* GLP *is not strongly complete with respect to the main axis of* $\mathfrak{I}_{\leq\varepsilon_0}$.

3 Proof of Theorem 2.12

3.1 Proof of strong completeness

In this section, we prove Theorem 2.12. We will find it convenient to introduce some further notation. Given a finite tuple of ordinals $\boldsymbol{\alpha}$, we denote

$$\mathfrak{I}_{\boldsymbol{\alpha}} = \{\boldsymbol{\gamma} \in \mathfrak{I} : \forall i < |\boldsymbol{\alpha}|\, \gamma_i = \alpha_i\}.$$

The intuition behind the notation is that when $\boldsymbol{\alpha}$ is a singleton α, then $\mathfrak{I}_\alpha$ consists of precisely the points in $\mathfrak{I}$ of "level" α, and $\mathfrak{I}_{\leq\iota}$ gathers precisely all the levels $\mathfrak{I}_\alpha$ with $\alpha \leq \iota$.

Definition 3.1 Let $\varphi \in \mathcal{L}$. We define a formula $\uparrow\varphi$ by induction on the syntactical complexity of φ:

$$\uparrow\varphi = \begin{cases} \bot, & \varphi = \bot; \\ p, & \varphi = p; \\ \uparrow\psi \wedge \uparrow\pi, & \varphi = \psi \wedge \pi; \\ \neg\uparrow\psi, & \varphi = \neg\psi; \\ [k+1]\psi, & \varphi = [k]\varphi,\ k < \omega; \\ \langle k+1\rangle\psi, & \varphi = \langle k\rangle\varphi,\ k < \omega. \end{cases}$$

Iterations of $\uparrow$ are defined by $\uparrow^0\varphi = \varphi$ and $\uparrow^{(k+1)}\varphi = \uparrow(\uparrow^k\varphi)$. For any $\Gamma \subset \mathcal{L}$ we write $\uparrow^k\Gamma = \{\uparrow^k\varphi : \varphi \in \Gamma\}$.

The next lemma shows how formulæ reflect the similarity of modal relations.

Claim 3.2 *For all $\varphi \in \mathcal{L}_0$, $\mathfrak{I}, \boldsymbol{\gamma} \Vdash \varphi$ if and only if for all $\boldsymbol{\alpha}$ and $k := |\boldsymbol{\alpha}|$, we have $\mathfrak{I}, \langle\boldsymbol{\alpha}, \boldsymbol{\gamma}\rangle \Vdash \uparrow^k\varphi$ whenever $\langle\boldsymbol{\alpha}, \boldsymbol{\gamma}\rangle \in \mathfrak{I}$.*

Proof. Fix $\boldsymbol{\alpha}$. We reason by induction on the complexity of φ:

- Booleans and $\top$, $\bot$ are clear;
- Suppose $\varphi = \langle i\rangle\psi$. First, we make the observation that if $\boldsymbol{\gamma} R_i \boldsymbol{\beta}$, then $\langle\boldsymbol{\alpha}, \boldsymbol{\beta}\rangle \in \mathfrak{I}$ and, moreover $\langle\boldsymbol{\alpha}, \boldsymbol{\gamma}\rangle R_{i+k} \langle\boldsymbol{\alpha}, \boldsymbol{\beta}\rangle$. Now, we see that $\mathfrak{I}, \boldsymbol{\gamma} \Vdash \varphi$ if and only if there exists $\boldsymbol{\beta} \in \mathfrak{I}$ such that $\boldsymbol{\gamma} R_i \boldsymbol{\beta}$ and $\boldsymbol{\beta} \Vdash \psi$. By the observation above and the induction hypothesis, for such a $\boldsymbol{\beta}$ we have $\langle\boldsymbol{\alpha}, \boldsymbol{\beta}\rangle \Vdash \uparrow^k\psi$. By the second part of the observation above, we see that $\mathfrak{I}, \boldsymbol{\gamma} \Vdash \varphi$ if and only if there is $\boldsymbol{\beta}$ such that $\langle\boldsymbol{\alpha}, \boldsymbol{\gamma}\rangle R_{i+k} \langle\boldsymbol{\alpha}, \boldsymbol{\beta}\rangle$ and $\langle\boldsymbol{\alpha}, \boldsymbol{\beta}\rangle \Vdash \uparrow^k\psi$. Using the definition of R_{i+k} we see that this is equivalent to $\langle\boldsymbol{\alpha}, \boldsymbol{\gamma}\rangle \Vdash \uparrow^k\varphi$, as desired;
- $\varphi = [i]\psi$ follows from duality using the fact that $\uparrow$ commutes with negation.

This proves the claim. □

Definition 3.3 We say that a formula $\varphi \in \mathcal{L}$ is a *worm* if it has the form $\langle k_0\rangle \ldots \langle k_n\rangle\top$.

For any worm $\varphi = \langle k_0\rangle \ldots \langle k_n\rangle\top$ and any $\boldsymbol{\alpha} \in \mathfrak{I}$ such that $\boldsymbol{\alpha} \Vdash \varphi$, we can always find a sequence $(\boldsymbol{\alpha}^1, \ldots, \boldsymbol{\alpha}^n)$ witnessing that $\boldsymbol{\alpha} \Vdash \varphi$, i.e., such that $\boldsymbol{\alpha}^0 R_{k_1} \boldsymbol{\alpha}^1 \ldots R_{k_n} \boldsymbol{\alpha}^n$. The next lemma says that if a worm is satisfied at some $\boldsymbol{\alpha} \in \mathfrak{I}$, then it is also satisfied at $\langle\beta\rangle + \boldsymbol{\alpha}$ for each $\beta \leq \varepsilon_0$ (recall that $\langle\beta\rangle = \langle\beta, 0, 0, \ldots\rangle$ and "+" is a poinwise addition, so $\langle\beta\rangle + \boldsymbol{\alpha} = \langle\beta + \alpha_0, \alpha_1, \ldots\rangle$).

Claim 3.4 *For any sequence $(\boldsymbol{\alpha}^1, \ldots, \boldsymbol{\alpha}^n) \subset \mathfrak{I}$, $\boldsymbol{\alpha}^0 R_{k_1} \boldsymbol{\alpha}^1 \ldots R_{k_n} \boldsymbol{\alpha}^n$ implies that for any $\beta \leq \varepsilon_0$ it is true that $(\langle\beta\rangle + \boldsymbol{\alpha}^0) R_{k_1} (\langle\beta\rangle + \boldsymbol{\alpha}^1) \ldots R_{k_n} (\langle\beta\rangle + \boldsymbol{\alpha}^n)$.*

Proof. One can see $\alpha_1^i \leq \log(\beta + \alpha_0^i)$ for each $1 \leq i \leq n$, since $\log(\beta + \alpha_0^i) = \log \alpha_0^i$. Hence, $\langle\beta\rangle + \boldsymbol{\alpha}^i \in \mathfrak{I}$ for each $i \leq n$. The fact that $(\langle\beta\rangle + \boldsymbol{\alpha}^i)R_{k_{i+1}}(\langle\beta\rangle + \boldsymbol{\alpha}^{i+1})$ for each $i < n$ is straightforward. □

The next lemma relies upon a construction which is very similar to one introduced and analyzed in [7], however we aim to introduce an explicit proof particular to the Ignatiev frame. Indeed, [7, Lemma 12] relies on assigning ordinals to worms in order to compare them, while we will need the opposite procedure of assigning worms to ordinals. It is possible that the lemma is well known but we were unable to locate a reference. We also refer the reader to Fernández-Duque and Joosten [16] or to Borges and Joosten [1] for some similar arguments.

Lemma 3.5 *For each $k < \omega$ and $\alpha < \varepsilon_0$ there exists a worm φ_k^α, such that whenever $\boldsymbol{\beta} \in \mathfrak{I}_{\leq\varepsilon_0}$ we have $\boldsymbol{\beta} \Vdash \langle k\rangle\varphi_k^\alpha$ if and only if $\beta_k > \alpha$. Moreover, for each $\alpha \leq \varepsilon_0$ there exists a set of formulæ T_k^α such that for each $\boldsymbol{\gamma}$, $\boldsymbol{\gamma} \Vdash T_k^\alpha$ if and only if $\gamma_k = \alpha$.*

Proof. For any two worms $\varphi = \langle k_0\rangle \ldots \langle k_n\rangle\top, \psi = \langle l_0\rangle \ldots \langle l_m\rangle\top$, we write $\varphi\psi$ for $\langle k_0\rangle \ldots \langle k_n\rangle\langle l_0\rangle \ldots \langle l_m\rangle\top$. We build φ_0^α inductively:

- $\alpha = 0$ then $\varphi_0^\alpha = \top$;
- $\alpha = \beta + 1 + \gamma$ for some $\beta \geq \gamma$, then $\varphi_0^\alpha = \varphi_0^\gamma\langle 0\rangle\varphi_0^\beta$;
- $\alpha = \omega^\beta$ then $\varphi_0^\alpha = \uparrow\varphi_0^\beta$;

First, note that this definition of φ_0^α is exhaustive, in the sense that every $\alpha < \varepsilon_0$ is indeed assigned a formula φ_0^α. However, the assignment is not unique; the following claim shows that this is not a problem.

Claim 3.6 *For each α, φ_0^α is well defined in the sense that if $\alpha = \beta + 1 + \gamma = \delta + 1 + \zeta$, then $\mathsf{GLP} \vdash \varphi_0^\gamma\langle 0\rangle\varphi_0^\beta \leftrightarrow \varphi_0^\zeta\langle 0\rangle\varphi_0^\delta$. Moreover, for each $\alpha > \beta$, $\mathsf{GLP} \vdash \varphi_0^\alpha \to \langle 0\rangle\varphi_0^\beta$.*

The proof of Claim 3.6 can be found in the appendix. Now for $\alpha < \varepsilon_0$ we set $T_0^\alpha = \{\langle 0\rangle\varphi^{\alpha'} : \alpha' < \alpha\} \cup \{\neg\langle 0\rangle\varphi^\alpha\}$ and $T_0^{\varepsilon_0} = \{\langle 0\rangle\varphi^{\alpha'} : \alpha' < \varepsilon_0\}$. Let us first assume that the first part of the lemma is true for $k = 0$. Then, the "moreover" part follows immediately from the definition, for $k = 0$. In addition, if we inductively let $\varphi_{k+1}^\alpha = \uparrow\varphi_k^\alpha$ and $T_{k+1}^\alpha = \uparrow T_k^\alpha$, then we see that the lemma holds for all k by Claim 3.2.

It remains to verify that the formulæ work as desired for $k = 0$. We start by showing that φ_0^α holds at $\hat{\alpha}$. Afterwards we will show that φ_0^α fails at all $\boldsymbol{\beta} \in \mathfrak{I}_{\leq\varepsilon_0}$ with $\beta_0 < \alpha$. We proceed by induction:

- Suppose $\alpha = 0$. Then $\top$ definitely holds at $\hat{0}$.
- Suppose $\alpha = \beta+1+\gamma$ with $\beta \geq \gamma$. Write $\varphi_0^\gamma = \langle n_1\rangle \ldots \langle n_m\rangle\top$ and choose a sequence $\hat{\gamma}R_{n_1}\boldsymbol{\delta}^1 \ldots R_{n_m}\boldsymbol{\delta}^m$ witnessing $\hat{\gamma} \Vdash \varphi_0^\gamma$. Noting that $\langle\beta+1\rangle + \hat{\gamma} = \widehat{\beta+1+\gamma}$, we get $\widehat{\beta+1+\gamma}R_{n_1}(\langle\beta+1\rangle + \boldsymbol{\delta}^1) \ldots R_{n_m}(\langle\beta+1\rangle + \boldsymbol{\delta}^m)$ by Claim 3.4. This sequence witnesses $\widehat{\beta+1+\gamma} \Vdash \langle n_1\rangle \ldots \langle n_m\rangle\psi$ for any ψ that holds in $\langle\beta+1\rangle + \boldsymbol{\delta}^m$; in particular $\langle\beta+1\rangle + \boldsymbol{\delta}^m \Vdash \langle 0\rangle\varphi_0^\beta$, since

$(\langle\beta+1\rangle+\boldsymbol{\delta}^m)R_0\hat{\beta}$.

- Suppose $\alpha=\omega^\beta$. We have $\hat{\alpha}=\langle\alpha,\beta,\log\beta,\dots\rangle=\langle\alpha,\hat{\beta}\rangle$. By induction hypothesis we have $\hat{\beta}\Vdash\varphi_0^\beta$, so $\hat{\alpha}\Vdash\varphi_1^\beta$ by Claim 3.2. Since $\varphi_1^\beta=\varphi_0^\alpha$, the claim follows.

We have shown that φ_0^α holds at $\hat{\alpha}$. Before moving on, we state a claim.

Claim 3.7 *For any worm* $\psi=\langle n_1\rangle\dots\langle n_m\rangle\top$ *and for any* $\boldsymbol{\gamma}\in\mathfrak{I}_{\le\varepsilon_0}$, $\boldsymbol{\gamma}\Vdash\psi$ *implies* $\hat{\gamma}_0\Vdash\psi$.

Proof. Let $(\boldsymbol{\alpha}^1,\dots,\boldsymbol{\alpha}^m)\subset\mathfrak{I}_{\le\varepsilon_0}$ be such that $\boldsymbol{\gamma}R_{n_1}\boldsymbol{\alpha}^1\dots R_{n_m}\boldsymbol{\alpha}^m$. Take $\boldsymbol{\delta}^1=(\hat{\gamma}_0:_{n_1}\boldsymbol{\alpha}^1)$ and $\boldsymbol{\delta}^{i+1}=(\boldsymbol{\delta}^i:_{n_{i+1}}\boldsymbol{\alpha}^{i+1})$ (recall the definition on p. 46). One can see that $\boldsymbol{\delta}^i\succeq\boldsymbol{\alpha}^i$ for each $1\le i\le m$, hence $\boldsymbol{\delta}^i\in\mathfrak{I}_{\le\varepsilon_0}$ for each $1\le i\le n$ (recall that $\succeq$ denotes pointwise comparison), hence $\hat{\gamma}_0\Vdash\psi$ is witnessed by $\hat{\gamma}_0R_{n_1}\boldsymbol{\delta}^1\dots R_{n_m}\boldsymbol{\delta}^m$. □

Now, assume towards a contradiction that for some $\gamma<\alpha$ there exists $\boldsymbol{\beta}$ such that $\gamma=\beta_0$ and $\boldsymbol{\beta}\Vdash\varphi_0^\alpha$. By Claim 3.7, we have $\hat{\gamma}\Vdash\varphi_0^\alpha$. Assuming without loss of generality that γ is least, it follows from the induction hypothesis on α that $\hat{\gamma}\Vdash\varphi_0^\alpha\wedge\neg\langle 0\rangle\varphi_0^\gamma$. However, $\mathsf{GLP}\vdash\varphi_0^\alpha\to\langle 0\rangle\varphi_0^\gamma$ by Claim 3.6, which is a contradiction. This finishes the proof of Lemma 3.5. □

Below, we use the formulæ φ_k^α and the sets T_k^α defined in the proof of Lemma 3.5.

Claim 3.8 *Let* $\beta<\gamma<\varepsilon_0$ *and* $k<\omega$. *Then, we have* $\mathsf{I}\vdash\uparrow^k\langle 0\rangle\varphi_0^\gamma\to\uparrow^k\langle 0\rangle\varphi_0^\beta$.

Proof. Instead of arguing syntactically, we appeal to Ignatiev's completeness theorem (Theorem 2.10). Let $\boldsymbol{\alpha}\in\mathfrak{I}_{<\varepsilon_0}$ be such that $\mathfrak{I}_{<\varepsilon_0},\boldsymbol{\alpha}\Vdash\uparrow^k\langle 0\rangle\varphi_0^\gamma$. By Lemma 3.5, we must have $\alpha_k\ge\gamma$ and thus $\alpha_k\ge\beta$, so that by Lemma 3.5 once more we obtain $\mathfrak{I}_{<\varepsilon_0},\boldsymbol{\alpha}\Vdash\uparrow^k\langle 0\rangle\varphi_0^\beta$. By the completeness of $\mathfrak{I}_{<\varepsilon_0}$, we thus have $\mathsf{I}\vdash\uparrow^k\langle 0\rangle\varphi_0^\gamma\to\uparrow^k\langle 0\rangle\varphi_0^\beta$, as desired. □

Claim 3.9 *Let* Γ *be a maximal set of closed* $\mathcal{L}$*-formulæ consistent with* GLP. *Then for each* $k<\omega$, *there exists a unique* $\alpha\le\varepsilon_0$ *such that* $T_k^\alpha\subset\Gamma$.

Proof. We let

$$\alpha=\sup\Big\{\beta+1:\beta<\varepsilon_0\wedge\uparrow^k\langle 0\rangle\varphi_0^\beta\in\Gamma\Big\}.$$

By maximality, we must then have $\neg\uparrow^k\langle 0\rangle\varphi_0^\alpha\in\Gamma$ or $\alpha=\varepsilon_0$. Similarly, by maximality, Γ contains all consequences of I, thus, by Claim 3.8 the set of β such that $\uparrow^k\langle 0\rangle\varphi_0^\beta\in\Gamma$ forms an initial segment of the ordinals. Since

$$\begin{aligned}T_k^\alpha&=\{\uparrow^k\varphi:\varphi\in T_0^\alpha\}\\&=\{\uparrow^k\langle 0\rangle\varphi_0^{\alpha'}:\alpha'<\alpha\}\cup\{\neg\uparrow^k\langle 0\rangle\varphi_0^\alpha\},\end{aligned}$$

it follows that $T_k^\alpha\subset\Gamma$. □

As a consequence of this claim, it follows that, for such a Γ, there is at most one $w\in\mathfrak{I}_{\le\varepsilon_0}$ such that $\mathfrak{I}_{\le\varepsilon_0},w\Vdash\Gamma$. We are now ready to prove the first part of the strong completeness theorem, which we restate for convenience.

Theorem 2.12(A) *Let $\Gamma \subset L_0$ be a maximal set of closed $\mathcal{L}$-formulæ consistent with* GLP*. Then there is $w \in \mathfrak{I}_{\leq\varepsilon_0}$ such that $w \Vdash \Gamma$.*

Proof. In this proof we assume that all formulæ of Γ are in the language:

$$\mathcal{L}' = \bot \mid \top \mid \varphi \wedge \psi \mid \varphi \vee \psi \mid [n]\varphi \mid \langle n \rangle \varphi$$

That is, we only consider formulas without negation, which we can do without loss of generality. Nonetheless, we may use the notation $\neg\eta$ for some η as shorthand for the negation normal form of $\neg\eta$. For each $i < \omega$, we let

$$\Gamma_i = \Big\{[i]\varphi : [i]\varphi \in \Gamma\Big\} \cup \Big\{\langle i \rangle \psi : \langle i \rangle \psi \in \Gamma\Big\}.$$

Our plan is to treat these strata consecutively.

We show inductively that for each $k < \omega$ there exists α_k such that:

(i) if $k \neq 0$, then $\alpha_i \leq \log(\alpha_{i-1})$ for each $0 < i < k$, and

(ii) $\mathfrak{I}_{\leq\varepsilon_0}, \boldsymbol{\gamma} \Vdash \bigcup_{i \leq k} \Gamma_i$ whenever $\gamma_i = \alpha_i$ for each $i \leq k$.

We denote by $(\#)_k$ the sentence asserting the existence of such an α_k.

Inductively, suppose that $(\#)_i$ has been proved for all $i < k$. We prove $(\#)_k$. Let $\boldsymbol{\alpha}_{k-1} = \langle \alpha_0, \ldots, \alpha_{k-1} \rangle$. The next claim asserts that it is enough to consider only formulæ that contain only boxes and diamonds with index $\geq k$.

Lemma 3.10 *Let $\Gamma'_k = \Gamma_k \cap \mathcal{L}_{[k;\omega)}$ and suppose $\beta \leq \log(\alpha_{k-1})$. Then $\langle \boldsymbol{\alpha}_{k-1}, \beta \rangle \Vdash \Gamma'_k$ if and only if $\langle \boldsymbol{\alpha}_{k-1}, \beta \rangle \Vdash \Gamma_k$.*

Proof. For all $i < k$ and any $\mathcal{L}'$-formula $\varphi \in \Gamma_k$ we modify it as follows: substitute each subformula of φ of the form $\langle i \rangle \psi$ or $[i]\psi$ by $\top$ if it belongs to Γ or by $\bot$ if it does not belong to Γ. We call the result of such a substitution $\varphi^{(k)}$. Formally,

$$\begin{array}{rll}
\bot^{(k)} = \bot; & \top^{(k)} = \top; & \\
(\psi \wedge \varphi)^{(k)} = \psi^{(k)} \wedge \varphi^{(k)}; & (\psi \vee \varphi)^{(k)} = \psi^{(k)} \vee \varphi^{(k)}; & \\
(\langle i \rangle \varphi)^{(k)} = \langle i \rangle (\varphi^{(k)}); & ([i]\varphi)^{(k)} = [i](\varphi^{(k)}) & \text{for } j \geq k; \\
(\langle i \rangle \varphi)^{(k)} = \top \text{ if } \langle i \rangle \varphi \in \Gamma; & (\langle i \rangle \varphi)^{(k)} = \bot \text{ otherwise} & \text{for } i < k; \\
([i]\varphi)^{(k)} = \top \text{ if } [i]\varphi \in \Gamma; & ([i]\varphi)^{(k)} = \bot \text{ otherwise} & \text{for } i < k;
\end{array}$$

Let $\Gamma_k^{(k)} = \{\varphi^{(k)} : \varphi \in \Gamma_k\}$. Then, $\Gamma_k^{(k)}$ is consistent provided Γ_k is; indeed:

Claim 3.11 *For any $\psi \in \mathcal{L}_0$ and $k \leq n < \omega$, $\Gamma \vdash [n](\psi \leftrightarrow \psi^{(k)})$.*

Here, entailment refers to provability from the axioms of GLP.

Proof. It is sufficient to show that $\Gamma \vdash [k](\psi \leftrightarrow \psi^{(k)})$. We prove this by induction on the complexity of ψ.

- Suppose $\psi = \eta \wedge \zeta$. By induction hypothesis, $\Gamma \vdash [k](\eta \leftrightarrow \eta^{(k)})$ and $\Gamma \vdash [k](\zeta \leftrightarrow \zeta^{(k)})$; hence $\Gamma \vdash [k](\eta \wedge \zeta \leftrightarrow \eta^{(k)} \wedge \zeta^{(k)})$ by normality. The case $\psi = \eta \vee \zeta$ is similar.

- Suppose $\xi = \langle i \rangle \eta$ with $i < k$. If $\langle i \rangle \eta$ belongs to Γ, then $\Gamma \vdash [k]\langle i \rangle \eta$, so $\Gamma \vdash [k](\langle i \rangle \eta \leftrightarrow \top)$; otherwise if $\langle i \rangle \eta$ does not belong to Γ, then $[i]\neg\eta \in \Gamma$ and thus by GLP we have $\Gamma \vdash [k][i]\neg\eta$, so that $\Gamma \vdash [k](\langle i \rangle \eta \leftrightarrow \bot)$. The case $\xi = [i]\eta$ with $i < k$ is analogous.
- Finally, suppose $\xi = \langle i \rangle \eta$ with $i \geq k$. By induction hypothesis, $\Gamma \vdash [k](\eta \leftrightarrow \eta^{(k)})$. This implies $\Gamma \vdash [k][k](\eta \leftrightarrow \eta^{(k)})$, and hence $\Gamma \vdash [k][i](\eta \leftrightarrow \eta^{(k)})$ by GLP. It is easy to deduce from this, using normality, that $\Gamma \vdash [k](\langle i \rangle \eta \leftrightarrow \langle i \rangle \eta^{(k)})$. The case $\xi = [i]\eta$ with $i \geq k$ is similar.

This proves the claim. □

Claim 3.12 $\Gamma \vdash \Gamma_k^{(k)}$.

Proof. We show that $\Gamma \vdash \varphi$ implies $\Gamma \vdash \varphi^{(k)}$ for each $\varphi \in \Gamma_k$. This is proved by induction. The Boolean cases, as well as the base cases $\top, \bot$ are straightforward. We consider the case where $\varphi = \langle n \rangle \psi$; the boxes can be treated similarly. There are two subcases:

- Suppose $n \geq k$. By the previous claim we have $\Gamma \vdash [n](\psi \to \psi^{(k)})$. By normality, $\Gamma \vdash [n](\psi \to \psi^{(k)}) \to (\langle n \rangle \psi \to \langle n \rangle \psi^{(k)})$, so that $\Gamma \vdash \langle n \rangle \psi$ implies $\Gamma \vdash \langle n \rangle \psi^{(k)}$.
- Suppose $n < k$. Then $\varphi^{(k)} = \top$, so the conclusion follows immediately.

This proves the claim. □

Claim 3.13 *For any $w \in \mathfrak{I}_{\leq \varepsilon_0}$ such that $w = \langle \boldsymbol{\alpha}_{k-1}, \boldsymbol{\beta} \rangle$, $\mathfrak{I}_{\leq \varepsilon_0}, w \Vdash \Gamma_k^{(k)}$ if and only if $\mathfrak{I}_{\leq \varepsilon_0}, w \Vdash \Gamma_k$.*

Proof. Let $\varphi \in \mathcal{L}'$. We reason by induction on the syntactic construction of φ. We have the following three cases:

(i) Booleans are clear, as are the constants $\bot$ and $\top$.

(ii) Suppose $\varphi = \langle i \rangle \psi$ or $\varphi = [i]\psi$ with $i < k$. By the induction hypothesis $(\#)_i$ for $i < k$, we have that $\mathfrak{I}_{\leq \varepsilon_0}, w \Vdash \varphi$ if and only if $\varphi \in \Gamma_i$. By definition of $\Gamma^{(k)}$, we thus have

$$\varphi \in \Gamma_i \leftrightarrow \varphi^{(k)} = \top \leftrightarrow \varphi^{(k)} \neq \bot \leftrightarrow \mathfrak{I}_{\leq \varepsilon_0}, w \Vdash \varphi^{(k)}.$$

(iii) Suppose $\varphi = \langle i \rangle \psi$ with $i \geq k$. Then, we have $\mathfrak{I}_{\leq \varepsilon_0}, w \Vdash \varphi$ if and only if there $u \in \mathfrak{I}_{\leq \varepsilon_0}$ such that $w R_i u$ and $\mathfrak{I}_{\leq \varepsilon_0}, u \Vdash \psi$. By the inductive hypothesis, this occurs if and only if $\mathfrak{I}_{\leq \varepsilon_0}, u \Vdash \psi^{(k)}$ and thus if and only if $\mathfrak{I}_{\leq \varepsilon_0}, w \Vdash \varphi^{(k)}$.

(iv) Suppose $\varphi = [i]\psi$ with $i \geq k$. This is analogous to the previous case.

This completes the proof of Claim 3.13. □

Now, one can also see that $\Gamma_k^{(k)} = \Gamma'_k$. Obviously, $\Gamma'_k \subset \Gamma_k^{(k)}$, moreover from the previous claim we have that $\Gamma_k^{(k)} \subset \Gamma$, thus $\Gamma_k^{(k)} \subset \Gamma'_k$ follows. This completes the proof of Lemma 3.10. □

One can see that having fixed α_i for all $i < k$ the set of worlds accessible by R_k depends only on the α_k in $\boldsymbol{\alpha} = \langle \alpha_0, \ldots, \alpha_k \ldots \rangle$. We must choose an ordinal α_k such that $\langle \boldsymbol{\alpha}_k, \boldsymbol{\beta} \rangle \Vdash \Gamma'_k$ whenever $\langle \boldsymbol{\alpha}_k, \boldsymbol{\beta} \rangle \in \mathfrak{I}_{\leq \varepsilon_0}$. The following lemma asserts the existence of such an ordinal.

Lemma 3.14 *There exists a unique ordinal α such that $\langle \boldsymbol{\alpha}_{k-1}, \alpha \rangle \Vdash \Gamma'_k$.*

Proof. Fix an enumeration $\{\varphi'_i : i < \omega\}$ of all formulæ φ such that $[k]\varphi \in \Gamma'_k$ and an enumeration $\{\psi'_i : i < \omega\}$ of all formulæ ψ such that $\langle k \rangle \psi \in \Gamma'_k$. Let

$$\Gamma'_k(i) = \{\langle k \rangle \varphi'_j : j < i\}.$$

Since $\Gamma'_k(i) \subset \mathcal{L}_{[k;\omega)}$ there is $\Sigma_i \subset \mathcal{L}$ for each $i < \omega$ such that $\Gamma'_k(i) = \uparrow^k \Sigma_i$ (so that $\Gamma'_k = \uparrow^k \Sigma$ where $\Sigma = \bigcup_i \Sigma_i$). By Ignatiev's Completeness Theorem 2.10, there is a point $\langle \gamma \rangle \in \mathfrak{I}$ that $\langle \gamma \rangle \Vdash \Sigma_i$ (note that each formula in Σ is of the form $\langle 0 \rangle \psi$ or $[0]\varphi$); let α^i_k be the least such γ and let $\alpha = \sup\{\alpha^i_k : i < \omega\}$. Evidently, $\langle \alpha \rangle \Vdash \langle 0 \rangle \psi$ for each $\langle 0 \rangle \psi \in \Sigma$. Suppose towards a contradiction that $\langle \alpha \rangle \Vdash \langle 0 \rangle \neg \varphi$ for some φ with $[0]\varphi \in \Sigma$. That is, there exists $\boldsymbol{\gamma} \in \mathfrak{I}_{<\alpha}$ such that $\boldsymbol{\gamma} \Vdash \neg\varphi$. Let $m < \omega$ be large enough so that $\alpha^n_k \Vdash \langle 0 \rangle \neg \varphi$ for every $n > m$. Since each α^i_k was chosen minimal, we have

$$\mathfrak{I} \models \bigwedge_{i<m+1} \Sigma_i \to \langle 0 \rangle \neg \varphi.$$

By Theorem 2.10, we have

$$\Vdash \bigwedge_{i<m+1} \Sigma_i \to \langle 0 \rangle \neg \varphi,$$

contradicting the consistency of Σ and thus of Γ. This shows that $\mathfrak{I}_{\leq \varepsilon_0}, \langle \alpha \rangle \Vdash \Sigma$. Now we are to show that $\alpha < \log(\alpha_{k-1})$.

Claim 3.15 $\alpha \leq \log(\alpha_{k-1})$.

Proof. Let $\gamma < \alpha$. Thus, $\langle \alpha \rangle \Vdash \varphi^\gamma_0$. we want to show $\alpha < \gamma$. Since $T^{\alpha_{k-1}}_{k-1} \subset \Gamma$, we have $\langle k-1 \rangle \varphi^{\omega^\gamma}_{k-1} \in \Gamma$.

Suppose towards a contradiction that $\alpha_{k-1} = \xi + \omega^\gamma$, where $\gamma < \alpha$. Since $T^{\alpha_{k-1}}_{k-1} \subset \Gamma$, we have in particular that $\neg\langle k-1 \rangle \varphi^{\omega^\gamma}_{k-1} \langle 0 \rangle \varphi^\xi_{k-1} \in \Gamma$, following the construction in the proof of Lemma 3.5. According to this construction, we have $\varphi^{\omega^\gamma}_{k-1} = \uparrow \varphi^\gamma_{k-1} = \varphi^\gamma_k$. From this, we get $\neg\langle k \rangle (\uparrow \varphi^\gamma_{k-1}) \langle 0 \rangle \varphi^\xi \in \Gamma$. It follows that $\neg\langle k \rangle \varphi^\gamma_k \in \Gamma$, since $\varphi^\xi \in \Gamma$, hence $\alpha < \gamma$. □

Thus, by Claim 3.2 we have $\langle \alpha \rangle \Vdash \Sigma$, hence $\langle \boldsymbol{\alpha}_{k-1}, \alpha \rangle \Vdash \Gamma'_k$. This proves the lemma. □

With it, we finish the proof of # for arbitrary k. Therefore, setting $\boldsymbol{\alpha} = \langle \alpha_k \rangle_{k<\omega}$ we get $\boldsymbol{\alpha} \Vdash \Gamma$.

□

3.2 Counterexamples

In this subsection, we include the counterexamples for the two incompleteness results in the statement of Theorem 2.12.

Lemma 3.16 *Let* $\Gamma = \{\langle i \rangle \top : i < \omega\}$. *Then,* Γ *is consistent with* GLP, *but for all* $\boldsymbol{\alpha} \in \mathfrak{I}_{<\varepsilon_0}$, *we have* $\mathfrak{I}_{<\varepsilon_0}, \boldsymbol{\alpha} \nVdash \Gamma$.

Proof. It is easy to verify directly that $\mathfrak{I}_{\leq\varepsilon_0}, \langle \varepsilon_0, \varepsilon_0, \varepsilon_0, \ldots \rangle \Vdash \Gamma$, so that indeed Γ is consistent with GLP. However, each ordinal $\alpha < \varepsilon_0$ satisfies $\log^k \alpha = 0$ for some $k < \omega$ and thus each point in $\mathfrak{I}_{<\varepsilon_0}$ satisfies $[k]\bot$ for some k. □

Lemma 3.17 *Let* $\Gamma = \{\langle 0 \rangle^k \top : k < \omega\} \cup \{[1]\bot\} \cup \{[0][1]\bot\}$. *Then,* Γ *is consistent with* GLP, *but for all* $\boldsymbol{\alpha} \in \mathrm{ma}(\mathfrak{I}_{\leq\varepsilon_0})$, *we have* $\mathfrak{I}_{<\varepsilon_0}, \boldsymbol{\alpha} \nVdash \Gamma$.

Proof. It is easy to verify directly that $\mathfrak{I}_{\leq\varepsilon_0}, \langle \omega, 0 \rangle \Vdash \Gamma$, so that indeed Γ is consistent with GLP. Suppose $\mathfrak{I}_{<\varepsilon_0}, \boldsymbol{\alpha} \Vdash \Gamma$. Then by $\langle 0 \rangle^k \top$, we must have $\alpha_0 > k$. By $[1]\bot$, we must have $\alpha_1 = 0$ and by $[0][1]\bot$ we must have $\alpha_0 \leq \omega$, so the only point in $\mathfrak{I}_{\leq\varepsilon_0}$ which satisfies Γ is $\langle \omega, 0 \rangle$, which is not on the main axis of $\mathfrak{I}_{\leq\varepsilon_0}$. □

4 Counterexamples to strong completeness for topological spaces

In this section we are to show that strong completeness of GLP_0 (and therefore of GLP) fails for the hitherto studied topological (ordinal) models.

4.1 Topological semantics

We start with the definition of topological models for GL.

Definition 4.1 Given a topological space (X, τ) for each $A \subset X$ we denote $d_\tau A = \{x : \forall U \in \tau \exists y \neq x (y \in U \cap A)\}$. We call d_τ the derivative operator. We omit the index if there's no risk of confusion.

Definition 4.2 A *topological model* is a tuple (X, τ, v), where (X, τ) is a topological space and $v : \mathsf{Vars} \to \mathcal{P}(X)$ is an interpretation of variables. Such an interpretation is extended to arbitrarily formulæ by the following clauses:

- $[\![p]\!] = v(p)$;
- $[\![\neg\varphi]\!] = X \setminus [\![\varphi]\!]$;
- $[\![\varphi \wedge \psi]\!] = [\![\varphi]\!] \cap [\![\psi]\!]$;
- $[\![\Diamond\varphi]\!] = d_\tau [\![\varphi]\!]$;

One can show that GL is valid in a space (X, τ) if and only if it is *scattered*, i.e., if every $A \subset X$ has an isolated point. A natural example of such space is an ordinal with its order topology (also called interval topology). In fact, GL is complete for such semantics and indeed it suffices to restrict to a single ordinal:

Theorem 4.3 (Abashidze [2], Blass [12]) GL *is sound and complete with respect to every ordinal* $\Omega \geq \omega^\omega$ *when equipped with the order topology.*

Similarly, we define the semantics for the polymodal case:

Definition 4.4 A *topological model* is a tuple $(X, \tau_i, v)_{i<\omega}$, where $(X, \tau_i)_{i<\omega}$ is a polytopological space and $v : \mathsf{Vars} \to \mathcal{P}(X)$ is an interpretation. As before, v is readily extended to arbitrary formulæ:

- $[\![p]\!] = v(p)$;
- $[\![\neg\varphi]\!] = X \setminus [\![\varphi]\!]$;
- $[\![\varphi \wedge \psi]\!] = [\![\varphi]\!] \cap [\![\psi]\!]$;
- $[\![\langle i\rangle\varphi]\!] = d_{\tau_i}[\![\varphi]\!]$;

We say that topological space is a *GLP-space* if it is a model of GLP. Generally, GLP can be shown to be strongly complete with respect to GLP-spaces (cf. [26]). However, it is open whether we could obtain strong completeness to some more specific and tame spaces, in particular ordinal spaces. In particular, if we consider certain *canonical* topologies, then regular completeness is already non-trivial, namely independent from ZFC. For instance Beklemishev in [9] has shown consistency of completeness for the canonical topologies for GLB (GLP restricted to only two modalities) are studied, whereas from the results by Blass [12] it follows that it is consistent that GLP is incomplete to any canonical topologies.

4.2 Icard spaces

By Ignatiev's theorem, given a closed formula consistent with GLP, there is a point on the *main axis* of $\mathfrak{I}$ that satisfies it, i.e., for any closed formula φ consistent with GLP there is an ordinal $\alpha < \varepsilon_0$ such that $\hat{\alpha} \Vdash \varphi$. Since the main axis of $\mathfrak{I}$ is wellordered in length ε_0, this motivates the question of whether a natural topology on ε_0 yields completeness for GLP or its closed fragment and leads to the notion of *Icard spaces*:

Definition 4.5 Let Θ_λ denote the space $\langle \lambda, \tau_n : n < \omega\rangle$, where for each $n < \omega$, τ_n is generated by:

$$\{\alpha : \log^m(\alpha) < \beta\} \quad \{\alpha : \log^k \alpha > \beta\}$$

where $m \leq n$, $k < n$.

These spaces provide a complete semantics for the closed fragment of GLP.

Theorem 4.6 ([20]) $\mathsf{GLP}_0 \vdash \varphi$ *if and only if* $\Theta_{\varepsilon_0} \models \varphi$.

This result gives completeness of GLP_0 with respect to ordinal spaces. Icard spaces do not validate all the axioms of GLP, however. The first complete semantics for GLP introduced was given by what are nowadays called *Beklemishev-Gabelaia spaces.* We briefly recall their definition although it will not be used in an essential way in what follows. In order to introduce BG spaces we define certain operations on topologies:

Definition 4.7 Given a scattered space (X, τ) we call τ^+ the topology generated by $\tau \cup \{d_\tau A : A \subset X\}$.

The derivative operation $d = d_\tau$ can be iterated transfinitely by putting $d^{\alpha+1}(A) = d(d^\alpha(A))$ and $d^\lambda(A) = \bigcap_{\alpha<\lambda}(A)$ at limit stages.

Definition 4.8 Let (X, τ) be a scattered space. We define the rank function $\rho_\tau : X \to \mathsf{Ord}$ by $\rho_\tau(x) = \inf\{\alpha : x \notin d^{\alpha+1}X\}$.

Definition 4.9 Let (X, τ) be a scattered space. We say that σ is an ℓ-extension of τ if $\tau \subset \sigma$, σ and τ have the same rank function, but σ adds no new neighborhoods at points of successor rank.

Below, we speak of maximal ℓ-extensions of some τ: these are ℓ-extensions of τ which have no proper ℓ-extension.

Definition 4.10 Let $\lambda \in \mathsf{Ord}$. We call $(\lambda, \sigma_n : n < \omega)$ a Beklemishev-Gabelaia space (or BG-space) if σ_0 is a maximal ℓ-extension of the order topology on λ and, for each n, σ_{n+1} is a maximal ℓ-extension of σ_n^+.

It can be inductively checked for any BG-space, the rank functions are given by $\rho_{\sigma_n} = \log^{n+1}$. Hence, the ranks of points in BG-spaces are the same as in Icard spaces.

4.3 Counterexamples to strong completeness

Theorem 4.11 GLP_0 *is not strongly complete with respect to any Icard space.* GLP *is not strongly complete with respect to any Beklemishev-Gabelaia space.*

Proof. The counterexample is the same as before. We let

$$\Gamma = \{\langle 0 \rangle^k \top : k < \omega\} \cup \{[1]\bot\} \cup \{[0][1]\bot\}.$$

We had seen in the proof of Theorem 2.12 that this set is consistent with GLP. Suppose towards a contradiction that $\alpha \Vdash \Gamma$ for some ordinal α in some Icard space (the proof for Beklemishev-Gabelaia spaces is the same). Write α in the form

$$\alpha = \beta + \omega^\gamma.$$

Since $\alpha \Vdash [0][1]\bot$, it follows that all ordinals α' sufficiently close to α satisfy $\log^2(\alpha') = 0$, which implies $\gamma = \log(\alpha) \leq \omega$. Since $\alpha \Vdash [1]\bot$, we must have $\log^2(\alpha) = 0$, so γ is finite. Hence, α is of the form $\beta + \omega^k$ for some k and thus satisfies $[0]^{k+1}\bot$, which is a contradiction. □

References

[1] A.A. Borges and J.J. Joosten. The Worm Calculus. *Advances in Modal Logic*, 12:13–27, 2018.

[2] M. Abashidze. Ordinal completeness of the Gödel-Löb modal system. *Intensional Logics and the Logical Structure of Theories*, pages 49–73, 1985. in Russian.

[3] J. P. Aguilera and D. Fernández-Duque. Strong Completeness of Provability Logic for Ordinal Spaces. *The Journal of Symbolic Logic*, 82(2):608–628, June 2017.

[4] J. P. Aguilera. A topological completeness theorem for transfinite provability logic. *Arch. Math. Log.*, 62:751–788, 2022.

[5] J. P. Aguilera and F. Pakhomov. The Logic of Correct Models. Preprint.

[6] J. Bagaria. Derived topologies on ordinals and stationary reflection. *Trans. Amer. Math. Soc.*, 2019.

[7] L.D. Beklemishev. Provability algebras and proof-theoretic ordinals, I. *Annals of Pure and Applied Logic*, 128(1-3):103–123, 2004.

[8] L. D. Beklemishev. Kripke semantics for provability logic GLP. *Annals of Pure and Applied Logic*, 161(6):756–774, 2010.

[9] L. D. Beklemishev. Ordinal completeness of bimodal provability logic glb. In Nick Bezhanishvili, Sebastian Löbner, Kerstin Schwabe, and Luca Spada, editors, *Logic, Language, and Computation*, pages 1–15, Berlin, Heidelberg, 2011. Springer Berlin Heidelberg.
[10] L. D. Beklemishev and D. Gabelaia. Topological completeness of the provability logic GLP. *Annals of Pure and Applied Logic*, 164(12):1201–1223, December 2013.
[11] L. D. Beklemishev and D. Gabelaia. Topological interpretations of provability logic. *Leo Esakia on Duality in Modal and Intuitionistic Logics*, pages 257–290, 2014.
[12] A. Blass. Infinitary combinatorics and modal logic. *J. Symbolic Logic*, 55(2):761–778, 1990.
[13] G. S. Boolos. Friedman's 35th problem has an affirmative solution. *Notices Amer. Math. Soc.*, 22:A–646, 1975.
[14] G. S. Boolos. *The Logic of Provability*. Cambridge University Press, 1993.
[15] L. D. Beklemishev and F. Pakhomov. Reflection algebras and conservation results for theories of iterated truth. *Annals of Pure and Applied Logic*, 173(5), 2022.
[16] D. Fernández-Duque and J. J. Joosten. Models of transfinite provability logic. *J. Symbolic Logic*, 78:543–561, 2013.
[17] D. Fernández-Duque and J. J. Joosten. The omega-rule interpretation of transfinite provability logic. *Ann. Pure Appl. Logic*, 169:333–371, 2018.
[18] H. M. Friedman. One hundred and two problems in mathematical logic. *J. Symbolic Logic*, 1975.
[19] D. Hilbert and P. Bernays. *Grundlagen der Mathematik. Bd. II.*, volume 50 of *Grundlehren Math. Wiss.* Springer, Cham, 1939.
[20] T. F. Icard. *Models of Provability Logic.* 2008. M.Sc. Thesis. University of Amsterdam.
[21] T. F. Icard. A topological study of the closed fragment of GLP. *J. Log. and Comp.*, 21(4):683–696, 2011.
[22] K. N. Ignatiev. On strong provability predicates and the associated modal logics. *The Journal of Symbolic Logic*, 58(1):249–290, 1993.
[23] G. Japaridze. The polymodal provability logic. In *Intensional logics and the logical structure of theories: material from the Fourth Soviet-Finnish Symposium on Logic.* Telavi, 1988.
[24] M. H. Löb. Solution of a Problem of Leon Henkin. *The Journal of Symbolic Logic*, 20(2):115–118, 1955.
[25] K. Segerberg. An essay in classical modal logic. *Filosofiska Föreningen och Filosofiska Institutionen vid Uppsala Universitet*, 1971.
[26] D. S. Shamkanov. Global Neighbourhood Completeness of the Provability Logic GLP. In *Advances in Modal Logic*, volume 13, page 581–596, 2020.
[27] R. M. Solovay. Provability interpretations of modal logic. *Israel Journal of Mathematics*, 25(3-4):287–304, September 1976.

A Proof of Claim 3.6

Towards proving Claim 3.6 we recall a certain construction that was introduced in [7]. Given a worm φ we one can assign an ordinal $o_n(\varphi) < \varepsilon_0$ with the property that for any pair of worms $\varphi, \psi \in \mathcal{L}_{[n;\omega)}$, $\mathsf{GLP} \vdash \varphi \to \langle n \rangle \psi$ implies $o_n(\varphi) > o_n(\psi)$. The definition is the following:

- $o_n(\langle n \rangle^k \top) = k$;
- $o_n(\varphi_0 \langle n \rangle \ldots \langle n \rangle \varphi_n) = \omega^{o_{n+1}(\varphi_n)} + \cdots + \omega^{o_{n+1}(\varphi_0)}$, where $\varphi_i \in \mathcal{L}_{[n+1;\omega)}$ for each $i \leq n$;

Note that $o_n(\varphi) = o_{n+1}(\uparrow\varphi)$.

Claim A.1 (Beklemishev) *For any pair of worms φ, ψ in the language $\mathcal{L}_{[n;\omega)}$, exactly one of the following holds:*

(i) $\mathsf{GLP} \vdash \varphi \to \langle n \rangle \psi$,

(ii) $\mathsf{GLP} \vdash \psi \to \langle n \rangle \varphi$,

(iii) $\mathsf{GLP} \vdash \varphi \leftrightarrow \psi$.

Moreover, which of the three alternatives occurs is determined by comparing the ordinals $o_n(\varphi)$ and $o_n(\psi)$.

Proof. In [7], Beklemishev identifies a particular collection of worms called *normal forms.* It is shown that each worm is equivalent to a unique normal form and that the set of normal forms is isomorphic to $(\varepsilon_0, <)$ under the ordering $\varphi <_0 \psi :\leftrightarrow \mathsf{GLP} \vdash \psi \to \langle 0 \rangle \varphi$. Moreover, the transformation to normal forms preserves the ordinal $o_0(\varphi)$. More generally, normal forms in the language $\mathcal{L}_{[n;\omega)}$ are isomorphic to $(\varepsilon_0, <)$ under the ordering $\varphi <_n \psi :\leftrightarrow \mathsf{GLP} \vdash \psi \to \langle n \rangle \varphi$ and moreover, the transformation of worms in $\mathcal{L}_{[n;\omega)}$ to normal forms preserves the ordinal $o_n(\varphi)$. The claim follows from this easily. □

The proof of Claim A.1 shows that if $o_n(\varphi) < o_n(\psi)$, then $\mathsf{GLP} \vdash \psi \to \langle n \rangle \varphi$, provided that φ and ψ are of the form $\uparrow^n \chi$ for some χ, i.e., that φ and ψ belong to $\mathcal{L}_{[n;\omega)}$. We now prove Claim 3.6, which we restate for convenience:

Claim 3.6 *For each $\alpha < \varepsilon_0$, φ_n^α is well-defined in the sense that for any $\beta, \gamma, \delta, \zeta$ such that $\alpha = \beta + 1 + \gamma = \delta + 1 + \zeta$ we have $\mathsf{GLP} \vdash \varphi_n^\gamma \langle n \rangle \varphi_n^\beta \leftrightarrow \varphi_n^\zeta \langle n \rangle \varphi_n^\delta$. Moreover, for each $\alpha > \beta$, $\mathsf{GLP} \vdash \varphi_n^\alpha \to \langle n \rangle \varphi_n^\beta$.*

Proof. It is sufficient to show that $o_n(\varphi_n^\alpha) = \alpha$ for each $\alpha < \varepsilon_0$. If so, then the claim immediately from Claim A.1. We do this by induction.

(i) When $\alpha < \omega$ or when $\alpha = \omega^\beta$, the identity follows immediately from the fact that $o_n(\varphi) = o_{n+1}(\uparrow\varphi)$ for all φ.

(ii) Suppose $\alpha = \beta + 1 + \gamma$ with $\beta \geq \gamma$. Write $\beta = \beta_1 + 1 + \cdots + 1 + \beta_k$ and $\gamma = \gamma_1 + 1 + \cdots + 1 + \gamma_m$ where each term is either infinite and additively indecomposable or else equal to zero. Then,

$$
\begin{aligned}
o_n\left(\varphi_n^{\alpha}\right) &= o_n\left(\varphi_n^{\beta+1+\gamma}\right) \\
&= o_n\left(\uparrow^n \varphi_0^{\beta+1+\gamma}\right) && \text{definition of } \varphi_n \\
&= o_0\left(\varphi_0^{\beta+1+\gamma}\right) && o_n(\varphi) = o_{n+1}(\uparrow\varphi \\
&= o_0\left(\varphi_0^{\beta_1+1+\cdots+1+\beta_k+1+\gamma_1+\cdots+1+\gamma_m}\right) \\
&= o_0\left(\varphi_0^{\gamma_m}\langle 0\rangle \ldots \langle 0\rangle \varphi_0^{\gamma_1}\langle 0\rangle \varphi_0^{\beta_k}\langle 0\rangle \ldots \langle 0\rangle \varphi_0^{\beta_1}\right) && \text{definition of } \varphi \\
&= \omega^{o_1(\varphi_0^{\beta_1})} + \cdots + \omega^{o_1(\varphi_0^{\beta_k})} + \omega^{o_1(\varphi_0^{\gamma_1})} + \cdots + \omega^{o_1(\varphi_0^{\gamma_m})} && \text{definition of } o
\end{aligned}
$$

Now, each term $\xi = \beta_i$ or $\xi = \gamma_i$ is either equal to 0, in which case $o_1(\varphi_0^{\xi}) = 0$ and thus $\omega^{o_1(\varphi_0^{\xi})} = 1$; or otherwise is infinite and additively indecomposable ordinal and thus of the form ω^{ζ} for some ζ. If so, then

$$\omega^{o_1(\varphi_0^{\xi})} = \omega^{o_1(\varphi_0^{\omega^{\zeta}})} = \omega^{o_1(\uparrow\varphi_0^{\zeta})} = \omega^{o_0(\varphi_0^{\zeta})} = \omega^{\zeta} = \xi = 1 + \xi$$

where the third-to-last equality makes use of the induction hypothesis. Applying this to each term, we have:

$$
\begin{aligned}
o_n(\varphi_n^{\alpha}) &= \omega^{o_1(\varphi_0^{\beta_1})} + \cdots + \omega^{o_1(\varphi_0^{\beta_k})} + \omega^{o_1(\varphi_0^{\gamma_1})} + \cdots + \omega^{o_1(\varphi_0^{\gamma_m})} \\
&= \beta_1 + 1 + \cdots + 1 + \beta_k + 1 + \gamma_1 + 1 + \cdots + 1 + \gamma_n \\
&= \beta + 1 + \gamma = \alpha,
\end{aligned}
$$

as desired.

This proves the claim. $\square$

Coalgebraic Semantics for Intuitionistic Modal Logic

Rodrigo Nicolau Almeida [1]

Institute for Logic Language and Computation (ILLC) - University of Amsterdam
Science Park 107
1098 XG Amsterdam

Nick Bezhanishvili

Institute for Logic Language and Computation (ILLC) - University of Amsterdam
Science Park 107
1098 XG Amsterdam

Abstract

We give a new coalgebraic semantics for intuitionistic modal logic with $\Box$. In particular, we provide a colagebraic representation of intuitionistic descriptive modal frames and of intuitonistic modal Kripke frames based on image-finite posets. This gives a solution to a problem in the area of coalgebaic logic for these classes of frames, raised explicitly by Litak (2014) and de Groot and Pattinson (2020). Our key technical tool is a recent generalization of a construction by Ghilardi, in the form of a right adjoint to the inclusion of the category of Esakia spaces in the category of Priestley spaces. As an application of these results, we study bisimulations of intuitionistic modal frames, describe dual spaces of free modal Heyting algebras, and provide a path towards a theory of coalgebraic intuitionistic logics.

Keywords: Coalgebra, Intuitionistic Modal Logic, Modal Heyting Algebras.

1 Introduction

Coalgebraic semantics of classical modal and positive modal logics have been thoroughly investigated [26,21,25,19]. In the case of classical modal logic this is done via the powerset coalgebras in the context of Kripke frames and by the Vietoris coalgebras for descriptive frames (see [19,26]). For positive modal logic this is obtained via the convex set functor and its topological analogue [3,6,21,4].

Intuitonistic modal logics, on the other hand, have so far escaped this kind of analysis. Such logics are quite varied, stemming from several distinct proposals over what is the appropriate notion of a "constructive" version of modal logic,

[1] r.dacruzsilvapinadealmeida@uva.nl.

see, e.g., [23,17,22,15,27]. They all have well-known algebraic semantics in the form of modal Heyting algebras, a Kripke-style semantics [15,20] via so-called "Intuitionistic modal frames" [27] and an order-topological semantics in terms of intuitionistic descriptive modal frames (see e.g., [27], where these are called □-frames, ◇-frames, etc, depending on the specific signature used). However, the question of how to represent such frames as coalgebras for an appropriate functor on the category of posets with p-morphisms or respecitively on the category of Esakia spaces and continuous p-morphisms (which is dual to the category of Heyting algebras) was open in the research community for a long time. It was raised explicitly by Litak [20] and by de Groot and Pattinson [2] [13], where the existence of such a representation was left as an open question.

In this paper we resolve the problem of defining coalgebraic semantics for intuitionistic modal logic, albeit for particular classes of intuitionistic modal frames. More concretely, we will represent intuitionistic descriptive modal frames and intuitionistic modal Kripke frames based on image-finite posets as particular coalgebras. Intuitionistic descriptive modal frames can be seen as Esakia spaces equipped with a modal relation. We will represent these as particular coalgebras on the category of Esakia spaces. For this we will be relying on a recent generalization [2] of a construction by Ghilardi [18]. As our main results show, descriptive intuitionistic modal frames can be seen as coalgebras for an appropriate composition of a variant of the classical upwards-Vietoris set functor and a right adjoint to the inclusion of the category of Esakia spaces with continuous p-morphisms in the category of Priestley spaces. Similar results are likewise obtained for the categories of image-finite posets with p-morphisms and posets with monotone maps, with appropriate modifications, giving us a coalgebraic representation of all image-finite intuitionistic modal frames. We observe, however, that this does not provide a representation of all intuitionistic modal frames, since the usage of image-finiteness is quite crucial.

We provide three applications of these constructions: (1) a colagebraic notion of a bisimulation for intuitionistic descrptive and image-finite Kripke frames, (2) a concrete description of the dual space of the free modal Heyting algebra on finitely many generators, and (3) a new definition for coalgebraic intuitionistic logic laying a path towards a theory of such logics following the line of research suggested at the end of [13].

The outline of the paper is as follows: in Section 2 we recall the necessary preliminaries from the theory of coalgebras over **Set** and **Stone**, as well as the relevant versions of intuitionistic modal logic and its semantics. In Section 3 we review the key aspects of the construction from [2] we will need for our purpose. We provide our main results showcasing the coalgebraic semantics, and its equivalence with the classical descriptive frame semantics, in Section 4. We present our applications in Section 5. We conclude in Section 6 by pointing to further research directions in the study of these representations.

---

[2] We point out that de Groot and Pattinson [13] also use the so-called dialgebraic representations in their work.

2 Preliminaries

2.1 Duality Theory and Coalgebra

We assume throughout that the reader is familiar with Stone and Priestley duality (as presented, e.g. in [11]).

Definition 2.1 Let $(X, \leq, \tau)$ be an ordered-topological space. We say that X is a *Priestley space* if (X, τ) is compact, and it satisfies the *Priestley Separatiom Axiom*: if $x \not\leq y$, there is a clopen upset U such that $x \in U$ and $y \notin U$.

Definition 2.2 Let $(X, \leq), (Y, \leq)$ be two posets, and $f : X \to Y$ be a map between them. The map f is said to be *monotone* if whenever $x, y \in X$, and $x \leq y$ then $f(x) \leq f(y)$. We say that f is a *p-morphism* if it is monotone and in addition, whenever $x \in X$, $y \in Y$ and $f(x) \leq y$, then there is some x' such that $x \leq x'$ and $f(x') = y$.

We say that a map $f : X \to Y$ between Priestley spaces is a *Priestley morphism* if it is continuous and monotone. We denote by **Pries** the category of Priestley spaces, with Priestley morphisms. We will in particular need the restriction of such a duality to *Esakia duality*, which we now recall:

Definition 2.3 Let $(X, \leq)$ be a Priestley space. We say that X is an *Esakia space* if whenever U is a clopen set in X, then $\downarrow U$ is clopen as well.

Given Esakia spaces X, Y, a morphism $f : X \to Y$ is said to be an *Esakia morphism* if it is a Priestley morphism and a p-morphism between the underlying posets. We denote by **Esa** the category of Esakia spaces with Esakia morphisms.

It is well-known (see e.g. [16]) that the category **Esa** is dual to the category of Heyting algebras and Heyting algebra homomorphisms. We also recall Jónnson-Tarski duality (see e.g. [7,9]):

Definition 2.4 Let (X, R) be a Stone space where $R \subseteq X \times X$. We say that X is a *modal space* if:

(i) For each $x \in X$, the set $R[x]$ is closed;

(ii) For each clopen set U, $R^{-1}[U]$ is clopen.

Given a modal space (X, R), let $\mathfrak{X} = (X, R, \mathcal{A})$ be the triple where $\mathcal{A} = \mathsf{Clop}(X)$ is the set of clopen subsets. We call $\mathfrak{X}$ a *descriptive general frame over* X. Given $f : X \to Y$ a map between descriptive general frames, we say that it is a descriptive morphism if f is continuous, and whenever $x \in X$ and $y \in Y$, and $f(x)Ry$, then there is some $x' \in X$ such that xRx' and $f(x') = y$. We denote by **DG** the category of descriptive general frames with descriptive morphisms. [3]

In parallel with these categories, we will work throughout with some categories of posets. Recall that a poset $(P, \leq)$ is said to be *image-finite* if for each $x \in P$, $\uparrow x = \{y : x \leq y\}$ is finite. We will work with the following:

[3] We note that this definition of descriptive general frames is equivalent to the usual one that can be found in e.g. [7,9].

(i) **Pos**, the category of posets with monotone maps;

(ii) The (non-full) subcategory $\mathbf{Pos}_p$ of **Pos** where we restrict maps to p-morphisms;

(iii) The subcategory $\mathbf{ImFinPos}_p$ of *image-finite* posets with p-morphisms.

As we will note below (see Section 3.2 for further discussion), the categories **Pos** and $\mathbf{ImFinPos}_p$ play a similar role to the categories **Pries** and **Esa**, respectively, in a discrete setting.

Definition 2.5 Given a category $\mathbb{C}$, and an endofunctor $F : \mathbb{C} \to \mathbb{C}$, a pair (A, f) of an object A and a morphism $f : A \to F(A)$ is called an F-*coalgebra* (or just a coalgebra, if the relevant F is clear from context).

Given two F-coalgebras (A, f_A) and (B, f_B), we say that a morphism $h : A \to B$ is a *coalgebra morphism* between (A, f_A) and (B, f_B) if it makes the following diagram commute:

$$\begin{array}{ccc} A & \xrightarrow{\;h\;} & B \\ {\scriptstyle f_A}\downarrow & & \downarrow{\scriptstyle f_B} \\ F(A) & \xrightarrow[F(h)]{} & F(B) \end{array}$$

Fig. 1. Coalgebra morphism compatibility

Such morphisms compose in the obvious way. We write $\mathbf{CoAlg}(F)$ for the category of F-coalgebras and coalgebra morphisms.

We refer the reader to [25] for all of the facts about coalgebra we will assume here. If one considers in particular the category **Set** of sets and functions, the powerset functor $\mathcal{P}$ admits a particularly transparent description of its coalgebras: they are precisely the Kripke frames. Tracing its origins in the work of Esakia [14] (see also [26] for an in-depth discussion), it has been realised that also *descriptive general frames* can be represented as coalgebras of a specific endofunctor on the category **Stone** of Stone spaces and continuous functions:

Definition 2.6 Let X be a Stone space. Let $V(X)$ be the set of closed subset of X. We give this set a topology consisting of the "hit-and-miss" topology, i.e., by giving it the topology determined by the subbasis consisting of

$$[U] = \{C \in V(X) : C \subseteq U\} \text{ and } \langle V \rangle = \{C \in V(X) : C \cap V \neq \emptyset\}$$

where U, V range over clopen subsets of X. We call this space the *Vietoris hyperspace* of X. Moreover, given a continuous function $f : X \to Y$, we define $V(f)$ to be the direct image of f.

Then we have the following (see e.g. [19,26]):

Proposition 2.7 *The assignment V as above defines an endofunctor on* **Stone**. *Moreover, for each X a Stone space, the categories* $\mathbf{CoAlg}(V)$ *and* **DG** *are equivalent.*

2.2 Intuitionistic Modal Logic

We consider the language of $\mathsf{IPC}_\Box$ of intuitionistic logic with an additional unary operation $\Box$. Throughout this section we refer to *intuitionistic modal logic*, $\mathsf{IPC}_\Box$, as the logic axiomatised by:

(i) The axioms from IPC;

(ii) $\Box(\phi \wedge \psi) \leftrightarrow \Box\phi \wedge \Box\psi$;

(iii) $\Box\top \leftrightarrow \top$.

Using the usual completeness methods, one can show that this logic is complete with respect to some Kripke-style semantics. The relevant semantics for such a logic is given over modal intuitionistic frames (sometimes called $\Box$-frames, to distinguish them from the semantics of richer intuitionistic modal logics):

Definition 2.8 Let $(X, \leq, R)$ be a triple where $(X, \leq)$ is a partial order, $R \subseteq X \times X$. We say this is a *modal intuitionistic frame* if it satisfies

$$R = \leq \circ R \circ \leq.$$

Given two modal intuitionistic frames $(X, \leq, R)$ and $(Y, \leq, R)$ we say that a map $f : X \to Y$ is a *modal p-morphism* if f is a p-morphism with respect to $\leq$ and R. Let **ImFinK** be the category of image-finite modal intuitionistic frames with modal p-morphisms.

Models are constructed by taking valuations in $\mathsf{Up}(X)$, the set of upsets of the poset. The semantics of the intuitionistic connectives is kept the same, whilst the clause for the $\Box$-operator is similar to classical modal logic: a model $\mathfrak{M}, x \Vdash \Box\phi$ if and only if whenever $y \in R[x]$ then $\mathfrak{M}, y \Vdash \phi$.

Note that if X is a poset, then $\mathsf{Up}(X)$ carries a richer structure: ordering upwards closed subsets using *reverse inclusion*, we obtain that $(\mathsf{Up}(X), \supseteq)$ is a poset. And such an assignment is in fact an endofunctor on the category **Pos**, sending monotone maps to their direct image.

Now, if $f : X \to \mathsf{Up}(X)$ is a coalgebra for this functor, then we can think of this map as picking for each $x \in X$ a set of modal successors; and dually, given a modal intuitionistic frame $(X, \leq, R)$, the map $R_X : X \to \mathsf{Up}(X)$ will provide a coalgebra. Hence we have:

Proposition 2.9 *$\Box$-frames $\mathfrak{X} = (X, \leq, R)$, are in 1-1 correspondence with monotone maps $R_{\mathfrak{X}} : X \to \mathsf{Up}(X)$ defined by*

$$x \mapsto R[x].$$

This suggests that the category of coalgebras for the endofunctor $\mathsf{Up}(-)$ should be equivalent to the category of modal intuitionistic frames with modal p-morphisms. However, following the discussion in [20], note that we are working within the category **Pos**, where maps are only required to be monotone, which means that there may be coalgebra morphisms which are not p-morphisms for the $\leq$-relation. If one tentatively restricts to $\mathbf{Pos}_p$, then the problem of coalgebra morphisms failing to be p-morphisms can be avoided. Indeed, $\mathsf{Up}(-)$ restricts to an endofunctor on this (non-full) subcategory. However, this means that the above assignment $R_X : X \to \mathsf{Up}(X)$ might no longer be a morphism in this category, since there is no guarantee that it will be a p-morphism.

Hence, despite the intuitive connection between the functor $\mathsf{Up}(-)$ and the semantics of intuitionistic modal logic, something seems to be missing for a coalgebraic representation. We will return to this in the next section. For now we will need to also discuss how this plays out in the context of general frames.

2.3 Descriptive general frames for IML

Just like in classical modal logic, in order to address the phenomenon of Kripke incompleteness, more general structures are required, in the form of *intuitionistic general frames.* We recall here this semantics [27].

Definition 2.10 Let $(X, \leq, R)$ be a triple where $(X, \leq)$ is an Esakia space and $R \subseteq X \times X$. We say that $(X, \leq, R)$ is a *modal Esakia space* if:

(i) Whenever U is a clopen upset, then $\Box_R U$ is a clopen upset, where $\Box_R U = \{x \in X : R[x] \subseteq U\}$.

(ii) For each $x \in X$, $R[x]$ is a closed upset.

We call the quadruple $(X, \leq, R, \mathcal{A})$ where $(X, \leq, R)$ is a modal Esakia space and $\mathcal{A} = \mathsf{ClopUp}(X)$ a *descriptive intuitionistic modal frame* (or $\Box$-general frame, for short).

Given a map $f : X \to Y$ between $\Box$-general frames, we say that this is a modal p-morphism if it is a continuous p-morphism with respect to both relations. We denote by **DiG** the category of $\Box$-general frames with modal p-morphisms.

We will need the following fact, which is derived essentially from the persistence condition on valuations:

Lemma 2.11 *If $(X, \leq, R, \mathcal{A})$ is a $\Box$-general frame, then R and $\leq$ satisfy the following mix law:*

$$R {=} {\leq} \circ R \circ {\leq}$$

Consequently, we can think of a $\Box$-general frame as a modal intuitionistic frame equipped with a compatible topology, or as an Esakia space equipped with a compatible relation R. In the latter case we say that $(X, R, \leq, \mathcal{A})$ is a $\Box$-general frame over X.

By combining the classical Jónnson-Tarski duality and Esakia duality, $\Box$-general frames have been shown to provide a general completeness result for $\mathsf{IPC}_\Box$ with respect to $\Box$-general frames (see e.g. [21,27,8]).

We consider the following variation of the previously outlined Vietoris endofunctor: given an Esakia space X, we write

$$\mathsf{V}^{\uparrow}(X) := \{C \subseteq X : C \text{ is a closed upset}\}$$

with a topology given by a subbasis of sets of the form

$$[U], \langle X - V\rangle$$

where U, V range over clopen upsets. Then we have the following fact, a proof of which can be found in the Appendix [4].

Proposition 2.12 *Given a Priestley space X, the order-topological space $(V^{\uparrow}(X), \supseteq)$ is again a Priestley space.*

Proof. Note that this satisfies the Priestley separation axiom, since the original space does: if $C \not\leq D$, then $C \not\supseteq D$. Hence there is clopen upset such that $D \subseteq U$ and $C \nsubseteq U$; this means that $C \in [U]$ an $D \notin [U]$, which was to show.

Now we show compactness. Assume that

$$\mathsf{V}^{\uparrow}(X) = \bigcup_{i\in I}[U_i] \cup \bigcup_{j\in J}\langle X - V_j\rangle$$

where U_i, V_j are clopen upsets. Look at $C = X - \bigcup_{j\in J} X - V_j$. Now, if $C = \emptyset$, then X is covered by $X - V_j$, so we can extract a finite subcover, say $X = X - V_0 \cup ... \cup X - V_n$. Now if $A \in \mathsf{V}^{\uparrow}(X)$, then A must intersect one of the above subsets, since it is non-empty, and so

$$\mathsf{V}^{\uparrow}(X) = \langle X - V_0\rangle \cup ... \cup \langle X - V_n\rangle.$$

Otherwise $C \neq \emptyset$, so $C \in \mathsf{V}^{\uparrow}(X)$. By construction, $C \in [U_i]$ for some $i \in I$, so because it is closed, using compactness we have obtain a finite subcover of $X - U_i$,

$$X - U_i \subseteq X - V_0 \cup ... \cup X - V_n.$$

And then we can show that

$$\mathsf{V}^{\uparrow}(X) = [U_i] \cup \langle X - V_0\rangle \cup ... \cup \langle X - V_n\rangle,$$

which again shows compactness. □

In analogy with what we noted for intuitionistic modal frames, we can at this point note the following (see e.g. [13]):

Proposition 2.13 *$\Box$-general frames $\mathfrak{X} = (X, \leq, R, \mathcal{A})$ are in 1-1 correspondence with Priestley morphisms $R_{\mathfrak{X}} : X \to \mathsf{V}^{\uparrow}(X)$ defined by*

$$x \mapsto R[x].$$

Proof. First note that if $x \leq y$ then from yRz, by Lemma 2.11, we have xRz. So we have that $R[x] \supseteq R[y]$, which means that the map is monotone. To see that it is continuous, note that if $U \subseteq X$ is a clopen upset, then

$$R_{\mathfrak{X}}^{-1}[[U]] = \{x : R[x] \in [U]\} = \{x : R[x] \subseteq U\} = \Box_R U$$

---

[4] We include this proof for completeness. It actually follows from the results mentioned, for e.g. in [4], where $(V^{\uparrow}(X), \subseteq)$ is noted to be a Priestley space, by realising that the order-dual of a Priestley space is again a Priestley space.

which is a clopen upset by assumption.

Conversely, assume that $f : X \to \mathsf{V}^{\uparrow}(X)$ is a Priestley morphism. Then define a relation R as follows:

$$xRy \iff y \in f(x).$$

Note that by construction $R[x] = f(x)$ will be a closed upset. Moreover, if U is a clopen upset, then we have

$$f^{-1}[[U]] = \{x : f(x) \in [U]\} = \Box_R U$$

is clopen, since f is continuous. It is clear that these two assignments are each other's inverses. □

Just like before, one can ask whether $\mathsf{CoAlg}(V^{\uparrow})$ is equivalent to the category of □-frames, and the exact same pattern repeats here: not all Priestley morphisms will be p-morpshims, so one may wish to restrict to **Esa**; and whilst $\mathsf{V}^{\uparrow}(-)$ is an endofunctor on Esakia spaces, the assignment $R_X : X \to \mathsf{V}^{\uparrow}(X)$ may not be a p-morphism.

Having gotten to this point, it is natural to wonder if indeed one needs different kinds of frames, possibly induced by a different kind of functor. As we will show, however, □-frames are enough, if one considers the right endofunctor on the category of Esakia spaces. This will be the subject of the next section.

3 Generalizing Ghilardi

In this section we recover the key technical tools which will be needed in the sequel. These results can be found in [2] and generalize the ideas from Ghilardi's classical construction of the free Heyting algebra through a step-by-step method; see also [5,24] for more details on this perspective. Here, reversing the order of previous sections, we start by handling the topological case:

3.1 The functor V_G

Definition 3.1 Let X, Y, Z be Priestley spaces, and $g : X \to Y$ and $f : Y \to Z$ be Priestley morphisms. We say that f is *open relative to* g [5] (g-open for short) if it satisfies the following:

$$\forall a \in X, \forall b \in Y, (f(a) \leq b \implies \exists a' \in X, (a \leq a' \ \&\ g(f(a')) = g(b)). \qquad (*)$$

Given $S \subseteq X$, we say that S is *rooted* if there is a point $x \in S$ such that for each $y \in S$, we have $x \leq y$. We say that $S \subseteq X$, a closed subset, is *g-open* (understood as a poset with the restricted partial order relation) if the inclusion is itself g-open. Equivalently:

$$\forall s \in S, \forall b \in X(s \leq b \implies \exists s' \in S(s \leq s' \ \&\ g(s') = g(b)).$$

[5] We recall that p-morphisms between posets X and Y correspond precisely to the open maps between the underlying topological spaces of X, Y when given the Alexandroff topology. The terminology here, standard in the literature, derives from this analogy.

Definition 3.2 Let $g : X \to Y$ be a map between Priestley spaces. Then consider

$$V_g(X) := \{C \subseteq X : C \text{ is closed, rooted and } g\text{-open }\},$$

with the topology given by a subbasis consisting of sets of the form

$$[U], \langle V \rangle$$

where U, V are clopen subsets of X.

The following is proven in [2, Lemmas 10 and 11]:

Proposition 3.3 *Given $g : X \to Y$ a Priestley morphism, the order-topological space $(V_g(X), \supseteq)$ is a Priestley space, equipped with a Priestley surjection $r_g : V_g(X) \to X$ sending each rooted subset to its root.*

We refer to r_g as the *root map* with respect to $V_g(X)$. The key property which this construction enjoys which we will need is the following:

Lemma 3.4 *Given a Priestley morphism $g : X \to Y$, and given a Priestley space Z with a g-open Priestley morphism $h : Z \to X$, there exists a unique r_g-open, continuous and monotone map h' such that the triangle in Figure 2 commutes.*

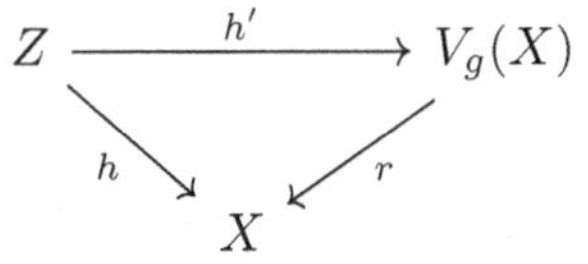

Fig. 2. Commuting Triangle of Priestley spaces

Definition 3.5 Let $g : X \to Y$ be a Priestley morphism. The *g-Vietoris complex* $(V^g_\bullet(X), \leq_\bullet)$ over X, is a sequence

$$(V_0(X), V_1(X), ..., V_n(X), ...)$$

connected by morphisms $r_i : V_{i+1}(X) \to V_i(X)$ such that:

(i) $V_0(X) = Y$ and $V_1(X) = X$;

(ii) $r_0 = g$;

(iii) For $i > 1$, $V_{i+1}(X) := V_{r_i}(V_i(X))$;

(iv) For $i > 0$ $r_{i+1} = r_{r_i} : V_{i+1}(X) \to V_i(X)$ is the root map.

We denote the projective limit of this family (in the category **Pries**) by $V^g_G(X)$ (by duality, and the fact that the category of distributive lattices is cocomplete, e.g. [1]). When g is the terminal map to the one element poset, we often omit it.

The proof of the following proposition, which will be instrumental in our work, can be found in the Appendix:

Proposition 3.6 *Let X be an Esakia space, Y a Priestley space, and assume that $f : X \to Y$ is a Priestley morphism. Then there is a unique Esakia morphism $\overline{f} : X \to V_G(Y)$, extending f. This is given as follows: the family $f_n : X \to V_n(Y)$, given by*

(i) $f_0 = f$;

(ii) $f_{n+1}(x) = f_n[\uparrow x]$;

consisting of continuous functions, and $\overline{f} : X \to V_G(Y)$ is given by

$$\overline{f}(x) = (f_0(x), f_1(x), ...).$$

Proof. Using Lemma 3.4 repeatedly, starting with the terminal map and proceeding along the roots, we get, a sequence defined by

(i) For every $x \in X$, $f_0(x) = f(x)$;

(ii) For every $x \in X$, $f_{n+1}(x) = f_n[\uparrow x]$.

By uniqueness of inverse limits, an extension $f^{\infty} : X \to V_G(X)$ given by

$$x \mapsto (f_0(x), f_1(x), f_2(x), ...).$$

such that this map commutes with all the root maps. Such a map is certainly unique, so if we can show that it is an Esakia morphism, we are done. So assume that $f^{\infty}(x) \leq y$. Consider the following:

$$S = \uparrow x \cup \{f_n^{-1}[y(n)] : n \in \omega\}.$$

This is a family of closed subsets in X, which is an Esakia space. Moreover, it has the finite intersection property: if we consider only finitely many elements, note that since $y(n) \subseteq f_n(x)$, then there is some k such that $x \leq k$, and $f^{\infty}(k)$ agrees with y up to the level n, i.e., the finite intersection $\uparrow x \cap f_0^{-1}[y(0)] \cap ... \cap f_n^{-1}[y(n)]$ is non-empty. By compactness, there exists some $x' \in \bigcap S$, which means precisely that there is some $x \leq x'$ such that $f^{\infty}(x') = y$. This shows that f^{∞} is a p-morphism, as desired. □

Using this one obtains:

Theorem 3.7 *The assignment V_G is an endofunctor on the category* **Pries** *of Priestley spaces and Priestley morphisms; indeed it is the right adjoint to the inclusion of the category* **Esa** *of Esakia spaces and Esakia morphisms into* **Pries**.

3.2 The functor P_G

Having the analysis of the previous section, we can ask whether something similar can be done for posets in general. Indeed this is the case, if one restricts to *image-finite* posets; we will present the construction, and then comment a bit on this restriction.

Definition 3.8 Let $g : X \to Y$ be a monotone map between posets. Then consider:

$$P_g(X) = \{C \subseteq X : C \text{ is finite, rooted and } g\text{-open}\}.$$

Definition 3.9 Let $g : X \to Y$ be a monotone map between image-finite posets. The g-discrete complex $(P^g_\bullet(X), \leq_\bullet)$ over X, is a sequence

$$(P_0(X), P_1(X), ..., P_n(X), ...)$$

connected by morphisms $r_i : P_{i+1}(X) \to P_i(X)$ such that

(i) $P_0(X) = Y$ and $P_1(X) = X$;

(ii) $r_0 = g$;

(iii) For $i > 1$, $P_{i+1}(X) := P_{r_i}(P_i(X))$;

(iv) For $i > 1$, $r_{i+1} := r_{r_i} : P_{i+1}(X) \to P_i(X)$ is the root map.

We denote the *image-finite part* of the projective limit of this family (in **Pos**) by $P^g_G(X)$. When g is the terminal map to the one element poset we often omit it.

The following propositions are the analogues of Propositions 3.6 and Theorem 3.7.

Proposition 3.10 *Let X be an image-finite poset, Y a poset, and assume that $f : X \to Y$ is a monotone map. Then there is a unique p-morphism $\overline{f} : X \to P_G(Y)$, extending f. This is given as follows: the family $f_n : X \to P_n(Y)$, given by*

(i) $f_0 = f$;

(ii) $f_{n+1}(x) = f_n[\uparrow x]$;

consists of monotone maps, and $\overline{f} : X \to V_G(Y)$ is given by

$$\overline{f}(x) = (f_0(x), f_1(x), ...)$$

Theorem 3.11 *The assignment P_G is an endofunctor on the category* **Pos** *of posets and monotone maps; indeed it is the right adjoint to the inclusion of the category* $\mathbf{ImFinPos}_p$ *of Image-finite posets spaces and p-morphisms into* **Pos**.

Remark 3.12 The reader may wonder why the restriction is done to the image-finite case. For the purpose of the results of this section, the issue lies in proving Proposition 3.10: without the restriction to image-finiteness, it is not clear that the lifting $\overline{f}(x)$ will indeed be a p-morphism. But there are more general considerations which make this a natural restriction: as discussed in [2, Section 7], there are several category-theoretic facts which make such a category much better behaved – for instance, it is monadic over the category of posets, precisely through the above construction, and it corresponds to the profinite completion of the category of finite posets with p-morphisms. By contrast, the category $\mathbf{Pos}_p$ does not immediately seem to enjoy such properties.

In light of this remark, the restriction to $\mathbf{ImFinPos}_p$ will be assumed throughout this paper, and the question of how to provide coalgebraic representations for arbitrary posets is left open.

4 Main Results

In this section we show that the category $\mathbf{CoAlg}(\mathsf{V}_G(\mathsf{V}^{\uparrow}(-)))$ is equivalent to the category **DiG** of descriptive □-frames with modal p-morphisms. In particular, we will prove the following:

Theorem 4.1 *Let $(X, \leq)$ be an Esakia space. Then the following are in one-to-one correspondence:*

(i) *□-frames over X;*

(ii) *Priestley morphisms $f : X \to \mathsf{V}^{\uparrow}(X)$;*

(iii) *Esakia morphisms $f' : X \to \mathsf{V}_G(\mathsf{V}^{\uparrow}(X))$.*

Proof. (i) is equivalent to (ii) by Lemma 2.13. For (iii) to (ii), if $f' : X \to \mathsf{V}_G(\mathsf{V}^{\uparrow}(X))$ is an Esakia morphism, then certainly the map

$$f := f' \circ \pi_0$$

which projects everything to the first coordinate is a Priestley morphism. So we focus on the implication from (ii) to (iii). Given the map $f : X \to \mathsf{V}^{\uparrow}(X)$, by Proposition 3.6 there is a unique Esakia morphism $\overline{f} : X \to \mathsf{V}_G(\mathsf{V}^{\uparrow}(X))$ extending it. It is moreover clear, by definition, that these two assignments are each other's inverses. □

We moreover have the following:

Theorem 4.2 *The category* **DiG** *is equivalent to the category* $\mathbf{CoAlg}(\mathsf{V}_G(\mathsf{V}^{\uparrow}(-)))$.

Proof. Similarly to above, we show that if $f : X \to Y$ is a modal p-morphism, then we can lift this to a coalgebra morphism making the right diagram commute. Let $i_X : X \to \mathsf{V}_G(\mathsf{V}^{\uparrow}(X))$ and $i_Y : Y \to \mathsf{V}_G(\mathsf{V}^{\uparrow}(Y))$ be the coalgebra maps. It is clear, since $\mathsf{V}_G(\mathsf{V}^{\uparrow}(-))$ map is a functor, that this lifts to a map, which essentially depends on the map

$$f^* : \mathsf{V}^{\uparrow}(X) \to \mathsf{V}^{\uparrow}(Y).$$

Indeed, this amounts, for $x \in X$, to have that $f^*(\pi_0[i_X(x)]) = f[R[x]]$ equal to $R[f(x)]$, since the liftings will commute with these maps; but it is clear that such an equality means precisely that f is a modal p-morphism. □

Exactly the same arguments, using Proposition 2.9, and using the appropriate discrete versions of the results shown in Section 3, provide the following:

Theorem 4.3 *The category* **ImFinK** *of image-finite modal Kripke frames is equivalent to the category* $\mathbf{CoAlg}(P_G(\mathsf{Up}(-)))$.

Remark 4.4 The above result can now explain the phenomenon we alluded to in Section 2. The noted correspondence between □-frames and coalgebras for $\mathsf{V}^{\uparrow}$, witnesses an equivalence between the category $\mathbf{DiG}^m$ of □-general frames with Priestley morphisms satisfying the p-morphism conditions for R, and the category of coalgebras for that functor. This provides a coalgebraic representation for positive modal logic over $\Box$. But this should not suffice to model the implication. One way to do so, is to introduce a construction which freely adds the implications. This is exactly the role played by our functor $\mathsf{V}_G(\mathsf{V}^{\uparrow}(-))$.

5 Applications

In this section we provide a few applications of our characterizations, exploiting and expanding the coalgebraic ideas exposed above.

5.1 Bisimulations of $\Box$-frames

As an illustration of the correctness of this coalgebraic representation, we show that the notion of bisimulation one obtains corresponds to the ones we would naturally want for $\Box$-frames and $\Box$-general frames. We show this for the functor $P_G(\mathsf{Up}(-))$, though similar results could easily be derived for V_G, since the former is slightly simpler to handle.

Definition 5.1 Let $\mathfrak{X} = (X, \leq, R)$ and $\mathfrak{Y} = (Y, \leq', R')$ be two $\Box$-frames. We say that a relation $\sim \subseteq X \times Y$ is a $\Box$-bisimulation if:

(i) Whenever $x, y \in X$ and xSy and $x \sim x'$ where $x' \in Y$, then there is some $y' \in Y$ such that ySy' and $y \sim y'$;

(ii) Whenever $x', y' \in Y$ and $x \in X$ and $x'Sy'$ and $x \sim x'$ then there is some $y \in X$ such that $y \sim y'$.

Where S ranges over $\leq$ and R.

The following theorem, proven in the Appendix, then follows by similar techniques to those used above:

Theorem 5.2 *Let $(X, \leq, R)$ and $(Y, \leq, R)$ be two image-finite modal intuitionistic frames. Then the following are in one-to-one correspondence:*

(i) *$\Box$-bisimulations between X and Y;*

(ii) *Bisimulations for the endofunctor $\mathsf{P}_G(\mathsf{Up}(-))$.*

Proof. Recall that a bisimulation in the category of image-finite posets with p-morphisms, for the functor $\mathsf{P}_G(\mathsf{Up}(-))$ amounts to a relation $B \subseteq X \times Y$ endowed with a coalgebra structure making the following diagram commute:

$$\begin{array}{ccccc} X & \xleftarrow{\;\pi_X\;} & B & \xrightarrow{\;\pi_Y\;} & Y \\ \downarrow & & \downarrow & & \downarrow \\ \mathsf{P}_G(\mathsf{Up}(X)) & \longleftarrow & \mathsf{P}_G(\mathsf{Up}(B)) & \longrightarrow & \mathsf{P}_G(\mathsf{Up}(Y)) \end{array}$$

Fig. 3. Bisimulations for the P_G

Now given the bisimulation B, first define a structure on $X \times Y$ by sending (x, y) to $R[x] \times R[y]$; then lift this to a map to $P_G(\mathsf{Up}(X \times Y))$ using Proposition 3.10; the uniqueness ensures that the desired map commutes.

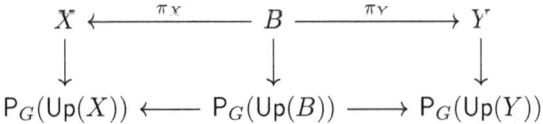

Conversely, assume that we have $B \subseteq X \times Y$ yielding a commuting diagram. Then we claim that B is a $\Box$-bisimulation. Because the projetions are required to be p-morphisms, the bisimulation satisfies the clauses for the $\leq$-relation, and the fact that it does so for the R-relation amounts to projecting onto the first coordinate. $\Box$

5.2 Constructing the Free Intuitionistic Modal algebra

The step-by-step construction of free algebras has long been employed in constructing modal algebras. This amounts to giving access to these algebras in a way that exploits the simplicity of working with simpler algebras, like Boolean algebras. In addition to that, these methods typically depend on the finitarity of some underlying algebra, and it is not always clear how to extend them to the infinite case (see [5] for some discussion of this). Algebraically, the mechanism at play is the existence of a locally finite reduct over which one can "layer" a non-locally finite operation. However, when in the face of two non-locally finite operations – such as when having Heyting implications and modalities – the situation can quickly become difficult. This is what we will be concerned with in this section: the construction of free modal Heyting algebras.

Definition 5.3 Let X be an Esakia space. Define the following sequence:

$$(M_0(X), M_1(X), ..., M_n(X), ...)$$

and a sequence of morphisms $\pi_k : M_k(X) \to M_{k-1}(X)$, for $k > 0$ and $\pi_0 : M_0(X) \to M_0(X)$ as follows:

(i) $M_0(X) = X$;

(ii) $M_{n+1}(X) := X \times \mathsf{V}_G(\mathsf{V}^{\uparrow}(M_n(X)))$;

(iii) $\pi_0 = id_{M_0}$;

(iv) $\pi_{n+1}(x, C) = (x, \pi_n[C])$.

For each k, define the relation $R_k \subseteq (X \times \mathsf{V}_G(\mathsf{V}^{\uparrow}(M_k(X)) \times M_k(X)$, given by

$$(x, C)R_k y \iff y \in \pi_0[C].$$

Let $M_\infty(X)$ be the inverse limit (in the category of Priestley spaces) of these spaces. Define a relation R_ω as follows: if $x, y \in M_\infty(X)$

$$xR_\omega y \iff \forall k \in \omega, x(k+1)R_k y(k).$$

First we note the following:

Proposition 5.4 *Given any Esakia space X, $M_\infty(X)$ is a modal Esakia space.*

Proof. The fact that the inverse limit is an Esakia space follows straightforwardly from duality. Moreover, given any point $x \in M_\infty(X)$ $R_\omega[x] = \bigcap\{y : x(k+1)R_k y(k)\}$, so to show this is point-closed it suffices to show that R_k is point-closed. But this amounts to the fact that $x(k+1) = (z, C)$ where $C \in \mathsf{V}_G(\mathsf{V}^{\uparrow}(M_k(X)))$, and so $y \in R_k[x(k+1)]$ if and only if $y \in \pi_0[C]$, i.e., $R_k[x(k+1)] = \pi_0[C]$, which is closed by construction. Similarly, if $U \subseteq M_\infty(X)$ is a clopen downset, then note that by construction, for some n, $\pi_n[U] \subseteq M_n(X)$ is a clopen downset. Then consider

$$\Box_{R_{n+1}}[\pi_n[U]] = \{x \in M_{n+1}(X) : \forall y \in M_n(xR_{n+1}y \to y \in \pi_n[U])\}.$$

Since $xR_{n+1}y$ holds if $y \in \pi_0[C]$, this amounts, after unfolding the definitions, to

$$[\pi_n[U]] = \{C \subseteq \mathsf{V}^\uparrow(M_n(X)) : C \subseteq \pi_n[U]\},$$

which is clopen upset by construction. By the topology of inverse limits, this implies that $\Box_{R_\omega} U$ is clopen upset as well. □

Now suppose that Y is a modal Esakia space, and assume that $p : Y \to X$ is a monotone map. Then define the map $p_1 : Y \to \mathsf{V}^\uparrow(X)$ as follows:

$$\overline{p}(y) = p[R[y]];$$

this is well defined, since if $z \in R[y]$, and $p(z) \leq w$, then by the p-morphism condition, there is some z' such that $p(z') = w$, and $z \leq z'$; then $z' \in R[y]$, as well, so $p[R[y]]$ is an upwards closed set, and closed as well. To see that this is continuous, note that if U is a clopen upset, then

$$\overline{p}^{-1}[[U]] = \{y \in Y : p[R[y]] \subseteq U\} = \{y \in Y : R[y] \subseteq p^{-1}[U]\} = \Box p^{-1}[U]$$

and since $p^{-1}[U]$ is a clopen upset, the latter is clopen because Y is a modal Esakia space. Using Proposition 3.6, we obtain a p-morphism $\overline{\overline{p}} : Y \to V_G(\mathsf{V}^\uparrow(X))$, and hence a map

$$\begin{aligned} p_1 : Y &\to X \times \mathsf{V}_G(\mathsf{V}^\uparrow(X)) \\ y &\mapsto (p(y), \overline{\overline{p}}(y)). \end{aligned}$$

We thus define a sequence of p-morphisms $p_n : Y \to M_n(X)$, which in turn induces a unique map $p_\infty : Y \to M_\infty(X)$, which is likewise a p-morphism. Then we can show the following:

Proposition 5.5 *Given X an Esakia space and Y a modal Esakia space, and $p : Y \to X$ a p-morphism, the unique lifting $p_\infty : Y \to M_\infty(X)$ is a modal Esakia morphism.*

Proof. It suffices to show that if xRy, then $p_{k+1}(x) R_k p_k(y)$. To see this, in turn it suffices to show that $p_k(y) \in \pi_0[p_{k+1}(x)]$, i.e., $p_k(y) \in \overline{p_k}(x)$. But this is given by definition, since $\overline{p_k}(x) = p_k[R[x]]$. Hence we have the result. □

Theorem 5.6 *Given X a finite set*[6]*, consider $\mathcal{P}(X)$ as a poset with reverse inclusion. Then $M_\infty(\mathcal{P}(X))$ is the dual to the free modal Heyting algebra on X many generators.*

[6] The restriction to finiteness is insubstantial; if one wishes to consider arbitrary sets, replace $\mathcal{P}(X)$ by the Priestley space dual to the free distributive lattice generated by X generators.

5.3 Coalgebraic Intuitionistic Logic

The results of the previous sections point to a possible avenue for coalgebraic intuitionistic logic. To illustrate this we will provide here a coalgebraic semantics for intuitionistic neighbourhood frames. This is done in the case of image-finite posets, to maintain consistency with the available results in the literature, and for simplicity.

Recall that the logic $\mathsf{IPC}_{\Box,N}$ is defined over the same language as IPC, but omitting the two normality axioms. An *intuitionistic neighbourhood frame* is a triple $(X, \leq, N)$ of a poset together with a monotone map $N : X \to \mathcal{P}(\mathsf{Up}(X))$, where $\mathcal{P}(-)$ is ordered by inclusion (see e.g. [12] for a related approach, see also [10]). The morphisms between such frames are functions $f : X \to X'$ satisfying

$$a' \in N'(f(x)) \iff f^{-1}(a') \in N(x)$$

for all $x \in X$ and $a' \subseteq X'$. We denote by **ImFinN** the category of image-finite neighbourhood $\Box$-frames.

Consider the assignment:

$$\mathcal{P}(\mathsf{Up}(-)) : \mathbf{Pos} \to \mathbf{Pos},$$

which sends a poset P to the powerset (with inclusion) of its upset (with reverse inclusion), and sends monotone maps to their direct image. Then this is in an endofunctor on **Pos**. Just like before, we can consider the composite functor $\mathsf{P}_G(\mathcal{P}(\mathsf{Up}(-)))$. Then we have:

Theorem 5.7 *There is an equivalence between* $\mathbf{CoAlg}(\mathsf{P}_G(\mathcal{P}(-)))$ *and the category* **ImFinN**.

The proof of this theorem follows by using exactly the same tools as before, and would go through just as well if we replaced **ImFinN** by the corresponding category of neighbourhood $\Box$-general frames, and the functors by their appropriate Vietoris-style variations. In fact, it suggests a more general phenomenon which we now point out:

Definition 5.8 Let $F : \mathbf{Pries} \to \mathbf{Pries}$ be an endofunctor on the category of Priestley spaces; we define $F^* : \mathbf{Esa} \to \mathbf{Esa}$: the *intuitionistic lifting* of F to be the functor obtained by composition in the following diagram:

$$\begin{array}{ccc} \mathbf{Pries} & \xrightarrow{F} & \mathbf{Pries} \\ {\scriptstyle I}\uparrow & & \downarrow{\scriptstyle P_G} \\ \mathbf{Esa} & \xrightarrow[F^*]{} & \mathbf{Esa} \end{array}$$

Fig. 4. Intuitionistic Lifting of functor F

Similarly, given an endofunctor $F : \mathbf{Pos} \to \mathbf{Pos}$ on the category of posets with monotone maps, we define $F^* : \mathbf{ImFinPos_p} \to \mathbf{ImFinPos_p}$ the *intuitionistic reflection* of F to be the functor obtained by composing the obvious diagram.

The results presented so far indicate a way to move from *positive distributive logics* to *intuitionistic logics*, using the mechanism of intuitionistic lifting. Hence it opens an avenue into investigations of intuitionistic coalgebraic logic. We leave a systematic study of the properties of intuitionistic lifting of functors, in a coalgebraic setting, for further work.

6 Conclusions and Future Work

In this paper we established a coalgebraic representation for descriptive intuitionistic modal frames, and for image-finite modal Kripke frames. Using this we provided a construction of the free modal Heyting algebra generated by an Esakia space, which has as a special case the free modal Heyting algebra generated by a set of generators. The work presented leaves several questions open, of a technical and conceptual nature.

Of a more technical nature, there are several facts to be clarified about P_g and V_g which would be of interest; for instance, it is not clear, assuming that X is an Esakia space, that $V_g(X)$ remains an Esakia space. Such technical facts would be relevant for the study of normal forms in intuitionistic modal logic.

In terms of the scope of the approach, there are naturally several possible lines of development. Like intuitionistic logic, S4 modal logic is axiomatised by axioms of rank greater than 1, which has so far kept it from being represented coalgebraically; we expect that similar techniques to the ones exposed here should account for these and similar cases.

Finally, related to our final remarks, the question of how to develop coalgebraic intuitionistic logic remains open, though the approach exposed here shows one recipe: first develop positive distributive logic, and then lift it to an intuitionistic setting. We leave a full study of this situation for future research.

7 Acknowledgements

The authors would like to thank the two anonymous referees for suggestions that improved the presentation of this work.

References

[1] Adámek, J. and J. Rosický, *Which categories are varieties? ((co)algebraic pearls)* (2021). URL `https://drops.dagstuhl.de/entities/document/10.4230/LIPIcs.CALCO.2021.6`

[2] Almeida, R. N., *Colimits of heyting algebras through esakia duality* (2024). URL `https://arxiv.org/abs/2402.08058`

[3] Balan, A., A. Kurz and J. Velebil, *Positive fragments of coalgebraic logics*, Logical Methods in Computer Science **Volume 11, Issue 3** (2015). URL `http://dx.doi.org/10.2168/LMCS-11(3:18)2015`

[4] Bezhanishvili, G., J. Harding and P. Morandi, *Remarks on hyperspaces for priestley spaces*, Theoretical Computer Science **943** (2023), p. 187–202. URL `http://dx.doi.org/10.1016/j.tcs.2022.12.001`

[5] Bezhanishvili, N. and M. Gehrke, *Finitely generated free heyting algebras via birkhoff duality and coalgebra*, Logical Methods in Computer Science **Volume 7, Issue 2** (2011). URL `https://doi.org/10.2168/lmcs-7(2:9)2011`

[6] Bezhanishvili, N. and A. Kurz, *Free modal algebras: A coalgebraic perspective*, in: T. Mossakowski, U. Montanari and M. Haveraaen, editors, *Algebra and Coalgebra in Computer Science, Second International Conference, CALCO 2007, Bergen, Norway, August 20-24, 2007, Proceedings*, Lecture Notes in Computer Science **4624** (2007), pp. 143–157.
URL `https://doi.org/10.1007/978-3-540-73859-6_10`

[7] Blackburn, P., M. de Rijke and Y. Venema, **Cambridge tracts in theoretical computer science**, Cambridge University Press, Cambridge, England, 2002.

[8] Celani, S. and R. Jansana, *A new semantics for positive modal logic*, Notre Dame Journal of Formal Logic **38** (1997).
URL `http://dx.doi.org/10.1305/ndjfl/1039700693`

[9] Chagrov, A. and M. Zakharyaschev, "Modal Logic," Oxford Logic Guides, Clarendon Press, Oxford, England, 1997.

[10] Dalmonte, T., C. Grellois and N. Olivetti, *Intuitionistic non-normal modal logics: A general framework*, Journal of Philosophical Logic **49** (2020), p. 833–882.
URL `http://dx.doi.org/10.1007/s10992-019-09539-3`

[11] Davey, B. A. and H. A. Priestley, "Introduction to Lattices and Order," Cambridge University Press, Cambridge, 2002, 175–200 pp.

[12] de Groot, J., *Goldblatt-thomason theorems for modal intuitionistic logics*, ArXiv **abs/2011.10221** (2020).
URL `https://api.semanticscholar.org/CorpusID:227118992`

[13] de Groot, J. and D. Pattinson, *Modal intuitionistic logics as dialgebraic logics*, in: *Proceedings of the 35th Annual ACM/IEEE Symposium on Logic in Computer Science*, LICS '20 (2020), pp. 355–369.
URL `http://dx.doi.org/10.1145/3373718.3394807`

[14] Esakia, L., *Topological kripke models*, Dokl. Akad. Nauk SSSR **214** (1974), pp. 298–301.

[15] Esakia, L., *The modalized heyting calculus: a conservative modal extension of the intuitionistic logic*, Journal of Applied Non-Classical Logics **16** (2006), p. 349–366.
URL `http://dx.doi.org/10.3166/jancl.16.349-366`

[16] Esakia, L., "Heyting Algebras: Duality Theory," Springer, 2019, english translation of the original 1985 book.

[17] Fischer Servi, G., *On modal logic with an intuitionistic base*, Studia Logica **36** (1977), p. 141–149.
URL `http://dx.doi.org/10.1007/BF02121259`

[18] Ghilardi, S., *Free heyting algebras as bi-heyting algebras*, C. R. Math. Rep. Acad. Sci. Canada (1992), pp. 240–244.

[19] Kupke, C., A. Kurz and Y. Venema, *Stone coalgebras*, Electronic Notes in Theoretical Computer Science **82** (2003), p. 170–190.
URL `http://dx.doi.org/10.1016/S1571-0661(04)80638-8`

[20] Litak, T., "Constructive Modalities with Provability Smack," Springer Netherlands, 2014 p. 187–216.
URL `http://dx.doi.org/10.1007/978-94-017-8860-1_8`

[21] Palmigiano, A., *A coalgebraic view on positive modal logic*, Theoretical Computer Science **327** (2004), p. 175–195.
URL `http://dx.doi.org/10.1016/j.tcs.2004.07.026`

[22] Servi, G. F., "Semantics for a Class of Intuitionistic Modal Calculi," Springer Netherlands, 1980 p. 59–72.
URL `http://dx.doi.org/10.1007/978-94-009-8937-5_5`

[23] Simpson, A. K., "The proof theory and semantics of intuitionistic modal logic," Ph.D. thesis, University of Edinburgh (1994).

[24] van Gool, S., *Constructing the lindenbaum algebra for a logic step-by-step using duality (extended version)*, in: *Future Directions of Logic, Proceedings of PhDs in Logic III*, 2011, pp. 1–15.
URL `https://api.semanticscholar.org/CorpusID:14809947`

[25] Venema, Y., "6 Algebras and coalgebras," Elsevier, 2007 p. 331–426.
URL `http://dx.doi.org/10.1016/S1570-2464(07)80009-7`

[26] Venema, Y. and J. Vosmaer, "Modal Logic and the Vietoris Functor," Springer Netherlands, 2014 p. 119–153.
URL http://dx.doi.org/10.1007/978-94-017-8860-1_6

[27] Wolter, F. and M. Zakharyaschev, "Intuitionistic Modal Logic," Springer Netherlands, 1999 p. 227–238.
URL http://dx.doi.org/10.1007/978-94-017-2109-7_17

Unification With Simple Variable Restrictions and Admissibility of Π_2-Rules

Rodrigo Nicolau Almeida [1]

Institute for Logic Language and Computation (ILLC) - University of Amsterdam
Science Park 107
1098 XG Amsterdam

Silvio Ghilardi [2]

Dipartimento di Matematica
Via C. Saldini 50
20133 Milano

Abstract

We develop a method to recognize admissibility of Π_2-rules, relating this problem to a specific instance of the unification problem with linear constants restriction [3], called here "unification with simple variable restriction". It is shown that for logical systems enjoying an appropriate algebraic semantics and a finite approximation of left uniform interpolation, this unification with simple variable restriction can be reduced to standard unification. As a corollary, we obtain the decidability of admissibility of Π_2-rules for many logical systems.

Keywords: Unification, Admissibility, Π_2-rules.

1 Introduction

Non-standard rules have often been used in the context of logical systems to axiomatise specific classes of models. Their use traces its origin to the work of Gabbay [11] as well as Takeuti and Titani [23], and has been the subject of some attention, especially with a focus on axiomatisation and admissibility of such rules [25,7,8,1]. Nevertheless, in most of these contexts specific assumptions have been made on what counts as such a non-standard rule, which make it difficult to provide a unified account of what these rules should be, and which make the current results available in the literature difficult to transfer: for instance, whilst in [8], some connections were made between the solution of the admissibility problem for some modal logic systems, the existence of uniform

[1] r.dacruzsilvapinadealmeida@uva.nl.

[2] silvio.ghilardi@unimi.it

interpolants, and some problems of unification, it is not clear how to generalize this to settings such as the Takeuti and Titani rule.

In this paper we start from the simple observation that, when left uniform interpolants are available, admissibility of Π_2 rules can be reduced to admissibility of standard rules just by eliminating bound context variables via such interpolants. It is less obvious that one can get the same result when assuming only that left uniform interpolants are 'finitely approximable': in fact, in this case one needs to show that such finite approximations are stable under substitutions. We obtain the result by employing techniques from two different sources: on one side, we reduce our task to subobject manipulations in the opposite category of finitely presented algebras (in the style of [17]) and on the other side we connect admissibility problems for Π_2 rules to a dedicated version of E-unification theory obtained by specializing the "Unification with Linear Constant Restrictions" employed in [3] to handle combined E-unificaton problems and general E-unification problems.

The structure of the paper is as follows: in Section 2 we define formally the problem of admissibility of Π_2-rules, and state our main result. In Section 3 we introduce the problem of unification with simple variable restriction, and provide an equivalent algebraic presentation of it. In Section 4 we recall the correspondence between some logical properties we need and their reformulations in the opposite of the category of finitely presented algebras. In Section 5 we prove our main theorem, showing that under suitable assumptions, the unification type for simple variable restriction is finitary and reduces to the standard unification type. In Section 6 we provide some applications; in Section 8 we conclude and highlight some limitations of our method. In Section 7, we analyze the prominent example of nuclear implicative semilattices.

2 Admissibility of Π_2-rules in Logical Systems

Throughout we will assume that we are working in a functional signature $\mathcal{L}$ comprising at least a constant symbol; the set of *terms* (aka propositional formulas, or just *formulas*) is denoted by $Fm_{\mathcal{L}}$. A *logic* $\vdash$ in this language is a relation $\vdash \subseteq \wp(Fm_{\mathcal{L}}) \times Fm_{\mathcal{L}}$ satisfying the usual identity, monotonicity, transitivity, structurality (i.e. invariance under substitutions) and finitarity conditions (see [10, Definitions 1.5-1.6]). We use the letters $\phi, \psi, \ldots$ for $\mathcal{L}$-formulas and letters $p, q, \ldots$ or $x, y, \ldots$ for variables; we compactly represent a tuple of distinct variables as $\overline{p}$. The notation $\phi(\overline{p})$ means that the formula ϕ has free variables included in the tuple $\overline{p}$. Since our *tuples of variables* are assumed to be formed by *distinct* elements, we emphasize that when we write e.g. $\phi(\overline{p}, \overline{q})$, we mean that the tuples $\overline{p}, \overline{q}$ are made of distinct variables and are also disjoint from each other. Notations like $\phi(\overline{\psi}/\overline{p})$ (or just $\phi(\overline{\psi})$) denote the result of substituting $\overline{p}$ by $\overline{\psi}$ inside of ϕ. If Γ, Δ are sets of formulas, $\Gamma \vdash \Delta$ means $(\Gamma, \phi) \in \vdash$ for all $\phi \in \Delta$.

When a logic is algebraizable, most of the definitions we shall introduce can be transferred back and forth from the corresponding class of algebras. Recall that $\vdash$ is algebraizable iff there is a quasivariety of $\mathcal{L}$-algebras $\mathbf{K}$ and there are essentially inverse structural transformers between $\mathcal{L}$-formulas and $\mathcal{L}$-equations, mapping elements of $\vdash$ to **K**-valid quasi-equations and vice versa (see [10, Definitions 3.11]). We say that our logic $\vdash$ (which we assumed to be finitary) is *strongly algebraizable* iff **K** is actually a variety.

Assumption 2.1 *For the whole paper, we fix a language $\mathcal{L}$ and a strongly algebraizable logic $\vdash$ in it; we call* **K** *the equivalent algebraic semantics of $\mathcal{L}$-algebras and E an equational theory axiomatizing* **K**.

We shall also fix a Hilbert-style derivation system $\vdash_S$ associated to the logic $\vdash$; vacuously one always exists (simply considering the set of rules $\Gamma \vdash \phi$ whenever Γ is finite and $\Gamma \vdash \phi$, see [10] for further details). We will begin by outlining in general what a Π_2-rule is in this context; the definitions here are analogous to the ones presented in [1].

Definition 2.2 Let $\Gamma = \{\phi_i(\overline{p}, \overline{q}) : i \leq n\}$ and $\psi(\overline{q})$ be formulas in the language $\mathcal{L}$. The Π_2-*rule* associated with this sequence of formulas is denoted $\forall \overline{p}\Gamma/^2\psi$ (sometimes without the universal quantifier, when the variables are clear from context) [3] and usually displayed as:

$$\frac{\forall \overline{p}\ (\phi_0(\overline{p}, \overline{q}), ..., \phi_n(\overline{p}, \overline{q}))}{\psi(\overline{q}).}$$

Given such a collection of formulas $\Gamma = \{\phi_i\}_i$, we refer to $\overline{p}$ as the *bound context of* Γ, or generally, the *bound context associated to* Γ, and sometimes denote it as $F_c(\phi_i)$ or $F_c(\Gamma)$; we refer to propositional variables not ocurring in F_c as the *free context.* Whenever the bound context is empty, the rule is referred to as a *standard* rule.

Example 2.3 Let $\mathcal{L}$ be the language of modal logic. Gabbay's *irreflexivity rule* is the rule

$$\frac{\forall p\, ((\Box p \to p) \vee \phi)}{\phi}$$

where ϕ is any formula, such that p does not occur in ϕ. This was used in [11] to obtain completeness with respect to a class of irreflexive frames.

Example 2.4 Let $\mathcal{L}$ be the language of modal algebras with a binary modality $\rightsquigarrow$, called the signature of *contact algebras.* Consider the following rule:

$$\frac{\forall p\, ((p \rightsquigarrow p) \wedge (\phi \rightsquigarrow p) \wedge (p \rightsquigarrow \psi)) \to \chi}{(\phi \rightsquigarrow \psi) \to \chi.}$$

This rule was discussed in [7] and [8], and used to axiomatise the strict implication calculus. Notice however that our notion of Π_2-rule is more general than the Π_2-rules introduced in [7] and [8] (for instance, the Π_2-rules introduced there do not have all standard rules as special cases).

[3] This notation serves to emphasise both the fact that these are distinct from usual rules, and the second-order nature of these rules, but we point out that it is purely formal.

Example 2.5 Let $\mathcal{L}$ be the language of Heyting algebras. Consider the following rule, often called the *Takeuti-Titani rule* or the *density rule*:

$$\frac{\forall r(g \to ((p \to r) \vee (r \to q) \vee c))}{g \to (p \to q) \vee c}$$

Such a rule has been fruitfully used to axiomatise classes of Gödel algebras and other MV-algebras (see e.g. [21,5]). We will return to it as an example later.

We now explain how Π_2-rules can be used within the derivation system $\vdash_S$:

Definition 2.6 Let Σ be a set of Π_2-rules. Given a formula ϕ we say that ϕ is *derivable* using the Π_2-rules in Σ, and write $\vdash_{S\oplus\Sigma} \phi$, provided there is a sequence $\psi_0, ..., \psi_n$ of formulas such that:

- $\psi_n = \phi$;
- For each ψ_i we have that either:

(i) ψ_i is an instance of an axiom of $\vdash_S$ or,
(ii) ψ_i is obtained using a rule from $\vdash_S$, from some previous $\psi_{j_0}, ..., \psi_{j_k}$ or,
(iii) $\psi_i = \chi(\overline{\xi}/\overline{q})$ and $\psi_{j_k} = \mu_k(\overline{r}/\overline{p}, \overline{\xi}/\overline{q})$ for $0 \leq j_k < i \leq n$, where
 · $\overline{r}$ is a renaming of $\overline{p}$, away from $\overline{\xi}$, i.e., a set of fresh variables not ocurring in $\overline{\xi}$;
 · $\Delta = \{\mu_k(\overline{p}, \overline{q}) : k \in \{0, ..., m\}\}$;
 · $\forall\overline{p}\Delta/^2\chi \in \Sigma$;
 · $\chi = \chi(\overline{q})$.

An *extended calculus* is a calculus of the kind $\vdash_{S\oplus\Sigma}$. We can now write what it means for a rule to be *admissible*:

Definition 2.7 Let $\Gamma/^2\phi$ be a Π_2-rule, and $\vdash_{S\oplus\Sigma}$ some extended calculus. We say that the rule $\Gamma/^2\phi$ is *admissible* in $\vdash_{S\oplus\Sigma}$ if for all ψ:

$$\vdash_{S\oplus\Sigma\oplus\Gamma/^2\phi} \psi \implies \vdash_{S\oplus\Sigma} \psi.$$

In light of the definition of derivation using non-standard rules, we want to "internalize" the notion of admissibility for ordinary (i.e. non extended) calculi. A standard rule is admissible over $\vdash$ iff in every substitution making the premises into theorems of $\vdash$ also makes the conclusion into a theorem of $\vdash$.[4] In order to obtain a similar characterization here, we need the notion of C-invariant substitution. Given a finite set $C = \{p_1, ..., p_n\}$ of propositional variables, a *C-invariant substitution* is a substitution σ mapping the p_i into themselves and the other propositional variables q into formulas $\sigma(q)$ not containing the variables in C.

Lemma 2.8 *Let $\forall\overline{p}\Gamma/^2\phi$ be a Π_2-rule and let $C := \{p_1, ..., p_n\}$. Then $\forall\overline{p}\Gamma/^2\phi$ is admissible over $\vdash_S$ if and only if whenever σ is a C-invariant substitution and we have $\vdash_S \sigma(\Gamma)$, then we have also $\vdash_S \sigma(\phi)$.*

---

4 It should be noticed however that this characterization does not always hold, for instance it fails for multiple-conclusion rules, see [20] for a thorough discussion.

Proof. Follows by induction on the structure of the derivation and by the fact that $\vdash_S$ is invariant under variable renamings. □

Given a logic $\vdash$, by the *$\vdash$-admissibility problem for Π_2-rules* we mean the problem of determining, given a triple (Γ, ϕ, C) (where $C = \overline{p}$ is a set of propositional letters ocurring in Γ but not in ϕ), whether the Π_2-rule $\forall \overline{p}\Gamma/^2\phi$ is admissible over $\vdash$. It is well-known that the $\vdash$-admissibility problem for standard rules need not be decidable (see for example [2, Theorem 4]), and that it is decidable for a substantial number of logical systems encountered in practice, such as S4, S5, IPC and lax logic [15] amongst others. In this paper we will obtain a general result concerning the decidability of admissibility for Π_2-rules, under specific assumptions which we now proceed to review.

First, we need to recall some notions concerning different interpolation properties. The following definition is modelled after [9, Section 2]:

Definition 2.9 We say that a logic $\vdash$ has the *Maehara interpolation property* if for any finite sets $\overline{p}, \overline{q}, \overline{r}$ of propositional variables and for any finite sets of formulas $\Sigma(\overline{p}, \overline{q}), \Delta(\overline{q}, \overline{r}), \Sigma'(\overline{p}, \overline{q})$ in the language $\mathcal{L}$ the following holds: if $\Sigma(\overline{p}, \overline{q}) \cup \Delta(\overline{q}, \overline{r}) \vdash \Sigma'(\overline{p}, \overline{q})$, then there exists a set of formulas $\Pi(\overline{q})$, such that $\Delta(\overline{q}, \overline{r}) \vdash \Pi(\overline{q})$ and $\Sigma(\overline{p}, \overline{q}) \cup \Pi(\overline{q}) \vdash \Sigma'(\overline{p}, \overline{q})$.

Definition 2.10 We say that a logic $\vdash$ has *right uniform deductive interpolation* if for any finite sets $\overline{p}, \overline{q}$ of propositional variables, and finite set of formulas $\Sigma(\overline{p}, \overline{q})$ there exists a finite set of formulas $\Pi(\overline{q})$ such that for any finite set of propositional variables $\overline{r}$ and for any finite set of formulas $\Delta(\overline{q}, \overline{r})$

$$\Sigma(\overline{p}, \overline{q}) \vdash \Delta(\overline{q}, \overline{r}) \iff \Pi(\overline{q}) \vdash \Delta(\overline{q}, \overline{r}).$$

There is a specular *left uniform deductive interpolation property* saying that for any finite sets $\overline{q}, \overline{r}$ of propositional variables, and finite set of formulas $\Delta(\overline{q}, \overline{r})$ there exists a finite set of formulas $\Theta(\overline{q})$ such that for any finite set of propositional variables $\overline{p}$ and for any set of formulas $\Sigma(\overline{p}, \overline{q})$

$$\Sigma(\overline{p}, \overline{q}) \vdash \Delta(\overline{q}, \overline{r}) \iff \Sigma(\overline{p}, \overline{q}) \vdash \Theta(\overline{q}).$$

The two properties are not equivalent [5]; the latter property is sometimes denoted by saying that the logic $\vdash$ has *global post-interpolants.* In this paper, we consider the following strictly weaker version of left uniform deductive interpolation:

Definition 2.11 We say that $\vdash$ has *left-finitary uniform deductive interpolation* if for any finite sets $\overline{q}, \overline{r}$ of propositional variables, and finite set of formulas $\Delta(\overline{q}, \overline{r})$ there is a finite collection of finite sets of formulas $\Theta_1(\overline{q}), \ldots, \Theta_n(\overline{q})$ such

[5] See Example 6.2 below for a counterexample. Equivalence can hold in very special contexts, typically when compact congruences are Boolean (this is the case of [8, Section 4] for the presence of a universal modality).

that:

$$(i)\ \Theta_i(\overline{q}) \vdash \Delta(\overline{q}, \overline{r}) \text{ for each } i \leq n \text{ and}$$
$$(ii) \text{ for any } \overline{p} \text{ and } \Sigma(\overline{p}, \overline{q}),$$
$$\Sigma(\overline{p}, \overline{q}) \vdash \Delta(\overline{q}, \overline{r}) \implies \Sigma(\overline{p}, \overline{q}) \vdash \Theta_i(\overline{q})$$
$$\text{for some } i \leq n.$$

When we move to concrete decidability problems and say that $\vdash$ has left-finitary uniform deductive interpolation, we also assume that the above finite sets $\Theta_1(\overline{q}), \ldots, \Theta_n(\overline{q})$ are *effectively computable* from $\Delta(\overline{q}, \overline{r})$. The main result of the paper is as follows:

Theorem 2.12 *Suppose that* $\vdash$ *has*

(i) *the Maehara Interpolation Property;*

(ii) *Right-Uniform Deductive Interpolation;*

(iii) *Left-Finitary Deductive Uniform Interpolation.*

Then if both $\vdash$ *itself and the* $\vdash$*-admissibility problem for standard rules are decidable, so is the* $\vdash$*-admissibility problem for* Π_2*-rules.*

After establishing the algebraic analogues of these syntactic properties (in Section 4), the proof of this Theorem will be given as Corollary 5.5. As a consequence, we obtain that several well-studied logical systems have such a decidable problem: among them we have S5, GL, Grz, IPC, LC and lax logic [18], all of which satisfy the hypotheses of the Theorem. As for logics without left uniform deductive interpolation property, we mention the $\{\wedge, \top, \rightarrow\}$ -fragment of IPC and the $\{\ell, \wedge, \top, \rightarrow\}$-fragment of lax logic: these systems have Maehara interpolation, are locally finite (by Diego theorem and extensions [6]) and have decidable admissibility problems for standard rules. In Section 6, we illustrate with examples how the algorithm from Theorem 2.12 works.

3 Unification with Simple Variable Restrictions

In this section we recall some essential concepts from unification theory which will be needed in our work; the reader can find more general information on unification and E-unification theory in [4] and the references contained therein. Our aim is to introduce *unification with simple variable restrictions*: this is a special case of unification with linear constant restrictions as introduced in [3], where unification with linear constant restrictions is an essential ingredient for building combined unification algorithms.

We let $Fm_{\mathcal{L}}(\overline{p})$ be the set of $\mathcal{L}$-formulas containing at most the variables $\overline{p}$; in algebraic terms, $Fm_{\mathcal{L}}(\overline{p})$ is the absolutely free algebra over $\overline{p}$ (also called the *term algebra* over $\overline{p}$). A *substitution* is an $\mathcal{L}$-morphism of term algebras $\sigma : Fm_{\mathcal{L}}(\overline{p}) \to Fm_{\mathcal{L}}(\overline{q})$; hence σ can be represented as a finite set of variable-term pairs,

$$\sigma = \{p_1 \leftarrow \phi_1(\overline{q}), ..., p_n \leftarrow \phi_n(\overline{q})\}.$$

We say that $\overline{p}$ is the domain and $\overline{q}$ the co-domain of the substitution σ. Recall that a substitution σ is said to be *C-invariant* (for a finite set C of propositional variables included into the domain and in the codomain of σ) iff it maps the $p \in C$ into themselves and the $q \notin C$ into terms $\sigma(q)$ not containing the variables in C.

Given two terms ϕ, ψ, and an equational theory E, we write $\phi =_E \psi$ to mean that $E \models \phi = \psi$.[6] We say that two substitutions σ and τ having the same domain and codomain are E-equivalent, briefly $\sigma =_E \tau$, if and only if $\sigma(p) =_E \tau(p)$ holds for every variable p in their domain. We say that $\tau : Fm_{\mathcal{L}}(\overline{p}) \to Fm_{\mathcal{L}}(\overline{q})$ is *less general than* $\sigma : Fm_{\mathcal{L}}(\overline{p}) \to Fm_{\mathcal{L}}(\overline{q}')$ (with respect to E) if there is a substitution $\theta : Fm_{\mathcal{L}}(\overline{q}) \to Fm_{\mathcal{L}}(\overline{q}')$ such that

$$\tau =_E \theta \circ \sigma.$$

Definition 3.1 Given a finite set of propositional variables C, an *E-unification problem with simple variable restriction* (briefly a *C-unification problem*) is a finite set of pairs of terms in the variables $\overline{p}$ (with $C \subseteq \overline{p}$)[7]

$$(P_C) \qquad (\phi_1(\overline{p}), \psi_1(\overline{p})), ..., (\phi_k(\overline{p}), \psi_k(\overline{p}));$$

a solution to such a problem or a *C-unifier* is a C-invariant substitution σ of domain $Fm_{\mathcal{L}}(\overline{p})$ such that

$$\sigma(\phi_1) =_E \sigma(\psi_1), ..., \sigma(\phi_k) =_E \sigma(\psi_k).$$

When $C = \emptyset$, we speak of *standard* unification problems, or just unification problems. We also write $U_E^{svr}(P_C)$ for the set of C-unifiers for the problem P_C.

Once C and (P_C) (a unification problem as above) are fixed, given two C-unifiers τ and σ, we say that τ is *less general* than σ, and write $\tau \leq_C \sigma$ if it is less general than σ as a substitution. Hence, given a unification problem (P_C) with simple variable restrictions, this definition of $\leq_C$ induces a preorder on $U_E^{svr}(P_C)$. A *C-unification basis* from this set is a subset $B \subseteq U_E^{svr}(P_C)$ such that for every $\sigma' \in U_E^{svr}(P_C)$ there is $\sigma \in B$ such that $\sigma' \leq_C \sigma$ holds; a *most general C-unifier* (C-mgu) of (P_C) is a $\sigma \in U_E^{svr}(P_C)$ such that $\{\sigma\}$ is a C-unification basis.

Definition 3.2 We say that E has *finitary simple-variable-restriction (svr) unification type* iff every C-unification problem (P_C) has a finite C-unification basis; E has *unitary scr-unification type* iff every C-unification problem (P_C) has a C-mgu.

When C is empty we have as special case the (standard) notion of *unifier*, *mgu* and *finitary/unitary unification type*; in such case, we indicate unification

---

[6] Of course, since we assume algebraizability, all definitions in this section could be equivalent stated inside the logical context of $\vdash$ by applying the appropriate transformers.

[7] In this paper, we do not consider free constants – i.e., fixed propositional variables which are interpreted freely in our algebras – in unification problems; considering them, would lead to consider *parameters* in inference rules.

problems $(P_\emptyset)$ with (P) and the corresponding set of unifiers as $U_E(P)$ instead of $U_E^{svr}(P_\emptyset)$. The next Proposition (which is an immediate consequence of the above definition and Lemma 2.8) shows the connection between finitary C-unification type and admissibility of Π_2-rules:

Proposition 3.3 *Assume that $\vdash_S$ is decidable. Then if E has finitary svr-unification type (with computable finite C-unification bases), then the $\vdash$-admissibility problem for Π_2-rules is decidable too.*

We mention that there is a connection in the reverse direction that works in case our logic $\vdash$ is decidable, consistent and has a $\bot$-proposition.[8] In fact, in this case C-unifiability of (P_C) is equivalent to the non-admissibility of the rule $\forall \overline{p}\Gamma/^2\bot$, where $C = \overline{p}$, and where Γ is the finite set of formulas obtained by appying the transformers to the equations in $\phi_i = \psi_i$ $(i = 1, \ldots, k)$.

3.1 Algebraic Characterization of C-Unification

It will be convenient for our purposes to see unification problems with simple variable restriction from the point of view of finitely presented algebras (following the approach of [13] for standard unification problems). For that purpose, given our equational theory E, we will work in $\mathsf{Alg}_{fp}^{op}(E)$, the opposite of the category of finitely presented E-algebras (recall that an algebra is finitely presented iff it is isomorphic to a finitely generated free algebra divided by a finitely generated congruence). Given a finitely presented E-algebra $\mathcal{A}$, we write it as $\mathcal{A}^*$ when we see it in the opposite category $\mathsf{Alg}_{fp}^{op}(E)$ (thus, $\mathcal{A}^*$ is just a formal dual of $\mathcal{A}$); in particular $\mathbf{F}(X)^*$ is the formal dual of $\mathbf{F}(X)$, the free algebra on the finitely many generators X. A similar notation is used for morphisms (notice that we have $(\sigma \circ \tau)^* = \tau^* \circ \sigma^*$ for contravariancy). Given an object $\mathcal{B}^*$ in $\mathsf{Alg}_{fp}^{op}(E)$, we write $\mathsf{Sub}_r(\mathcal{B}^*)$ for the set of regular subobjects[9] of $\mathcal{B}^*$ in the category $\mathsf{Alg}_{fp}^{op}(E)$; we recall that such regular subobjects correspond, dually, to the finitely presented quotients of $\mathcal{B}$ (see e.g. [17] for details).

In this context, an E-unification problem with simple variable restrictions is a pair $(\mathcal{A}, C)$, where C is a finite set of free constants, and $\mathcal{A}^* \in \mathsf{Sub}_r(\mathbf{F}(X)^* \times \mathbf{F}(C)^*)$. A solution to this problem, which we call suggestively a *C-unifier*, is a homomorphism $\sigma : \mathbf{F}(X) \to \mathbf{F}(Z)$, such that $\sigma^* \times 1$ factors in such a way that the diagram of Figure 1 commutes:

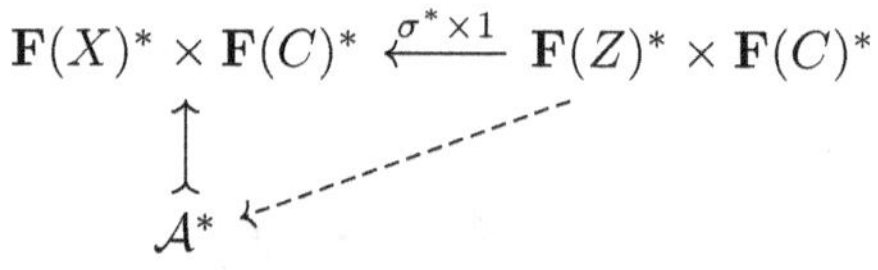

Fig. 1. Solution to Unification Problem with Simple Variable Restrictions

---

[8] This means that there is a constant $\bot \in \mathcal{L}$ such that $(\{\bot\}, \phi)$ belongs to $\vdash$ for all ϕ.

[9] Recall that a regular subobject of X is an equivalence class (wrt to iso) of monomorphisms (with codomain X) which happens to be equalizers of a pair of parallel arrows.

Given such an E-unification problem with simple variable restrictions, $(\mathcal{A}, C)$, and two C-unifiers $\sigma : \mathbf{F}(X) \to \mathbf{F}(Z)$ and $\gamma : \mathbf{F}(X) \to \mathbf{F}(W)$, we say that σ is more general than γ, and write $\sigma \leq \gamma$ if there is a homomorphism $k : \mathbf{F}(Z) \to \mathbf{F}(W)$ such that $k \circ \sigma = \gamma$; as a consequence, the outer triangle of the following diagram commutes:

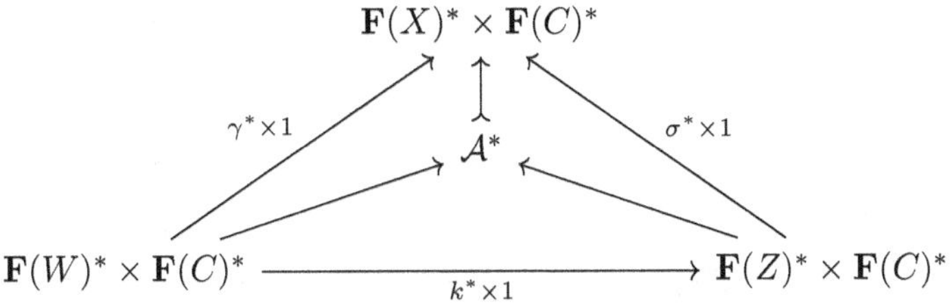

Fig. 2.

(the commutativity of the inner triangle follows as a consequence from the fact that $\mathcal{A}^* \hookrightarrow \mathbf{F}(X)^* \times \mathbf{F}(C)^*$ is mono).

Remark 3.4 Using the fact that our language $\mathcal{L}$ contains at least a constant symbol, *we have that* $\sigma \leq \gamma$ *iff there is* $l : \mathbf{F}(Z) + \mathbf{F}(C) \to \mathbf{F}(W) + \mathbf{F}(C)$ *such that* $(\sigma^* \times 1) \circ l^* = \gamma^* \times 1$ [10]. In fact, homsets among free algebras are not empty, so if we have $(\sigma^* \times 1) \circ l^* = \gamma^* \times 1$, then letting $l^* = \langle l_1^*, l_2^* \rangle$, we can put $k^* := l_1^* \circ \langle 1, \alpha^* \rangle$ (where α is any morphism $\mathbf{F}(C) \longrightarrow \mathbf{F}(W)$) and then prove $\sigma^* \circ k^* = \gamma^*$ via elementary properties of products. The latter is seen as follows: from $(\sigma^* \times 1) \circ l^* = \gamma^* \times 1$, taking first components of pairs, we get $\sigma^* \circ l_1^* = \gamma^* \circ \pi_{\mathbf{F}(X)}$, so that

$$\sigma^* \circ k^* = \sigma^* \circ l_1^* \circ \langle 1, \alpha^* \rangle = \gamma^* \circ \pi_{\mathbf{F}(X)} \circ \langle 1, \alpha^* \rangle = \gamma^* \quad .$$

It is easy to see that the above definition of comparison for C-unifiers gives a preorder; we can write $U_E^{svr}(\mathcal{A}, C)$ for the preordered set of C-unifiers for $\mathcal{A}$. This 'algebraic' approach to C-unification is equivalent to the 'symbolic' approach of Definition 3.2, as the following proposition, proved in the Appendix shows:

Proposition 3.5 *Let* (P_C) *a* E*-unification problem with simple variable restriction. If* $\mathcal{A}$ *is a finitely presented algebra with presentation* (P_C)*, then the antisymmetric quotients of the preordered sets* $U_E^{svr}(\mathcal{A}, C)$ *and* $U_E^{svr}(P_C)$ *are isomorphic.*

4 Interpolation and Finitely Presented Algebras

We analyzed C-unification inside the opposite of the category of finitely presented algebras; we now do the same for the interpolation properties mentioned

[10] Here $\mathbf{F}(Z) + \mathbf{F}(C)$ denotes the coproduct of $\mathbf{F}(Z)$ and $\mathbf{F}(C)$ in the category of finitely presented algebras.

in Theorem 2.12. We first recall some well-known results from universal algebra.

Definition 4.1 We say that a class of $\mathcal{L}$-algebras **K** enjoys the property that "Injections are Transferable" (IT) if whenever $f : \mathcal{A} \to \mathcal{B}$ is a homomorphism, and $g : \mathcal{A} \to \mathcal{C}$ is a monomorphism, then there are a morphism $h : \mathcal{C} \to \mathcal{E}$ and a monomorphism $h' : \mathcal{B} \to \mathcal{E}$ such that $h' \circ f = h \circ g$ (see Figure 3).

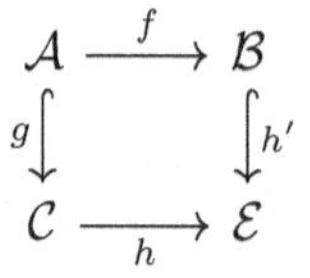

Fig. 3. Injections are Transferable

The following result is proved in [22, Lemma 26], see also [9]:

Theorem 4.2 *The following are equivalent:*

(i) $\vdash$ *has the Maehara Interpolation Property;*

(ii) **K** *has the property (IT).*

We note that it can be shown that (IT) is also equivalent to the conjunction of the congruence extension property and the amalgamation property (this is likewise proven in [22, Theorem 29]); we shall make brief use of this further characterization in some examples below. Also notice that if (IT) holds in **K**, then it holds also in the full subcategory of **K** formed by the finitely presented algebras: this is because such a subcategory is closed under pushouts, and because a homomorphism g which fits into a factoring of $h = f \circ g$, where h is a monomorphism, is itself a monomorphism (so that (IT) for finitely presented algebras comes from the universal property of pushouts).

Right uniform deductive interpolation can likewise be given concrete meaning, in a way which is directly inside finitely presented algebras:

Definition 4.3 We say that **K** is *coherent* if finitely generated subalgebras of finitely presented algebras are again finitely presented.

The following is shown in [19, Theorem 2.3]

Theorem 4.4 $\vdash$ *has right uniform deductive interpolation if and only if* **K** *is coherent.*

The notion of an r-regular category [17] can be useful to summarize the properties of the opposite of the category of finitely presented algebras coming from the above facts.

Definition 4.5 Let **C** be a category. We say that **C** is *r-regular* if it satisfies the following:

(i) it has all finite limits;

(ii) epimorphisms are stable under pullback;

(iii) every arrow has an epi/regular mono factorization.

The next proposition is folklore; it shows the importance of our base assumptions on E (we provide a proof in the appendix, for completeness):

Proposition 4.6 *If* $\mathsf{Alg}(E)$ *satisfies (IT) and coherence, then* $\mathsf{Alg}^{op}_{fp}(E)$ *is an r-regular category.*

The special property of left-finitary uniform deductive interpolation can likewise be given a straightforward algebraic interpretation inside the category of finitely presented algebras. Our notation for r-regular categories is mostly consistent with [17]. In particular, for an object X and an arrow $f : Y \longrightarrow X$, we indicate with $\mathsf{Sub}_r(X)$ the poset of the regular subobjects of X and by f^{-1} : $\mathsf{Sub}_r(X) \longrightarrow \mathsf{Sub}_r(Y)$ the operation of taking pullback along f. Projections like $X \times Y \longrightarrow Y$ are indicated as π_Y; the identity arrow for an object X (= the maximum regular subobject of X) is indicated both with 1_X or just with 1 for simplicity.

Definition 4.7 Let $\mathbf{C}$ be a category with finite limits. Given $T \in \mathsf{Sub}_r(X \times Z)$, we say that a finite collection $B_1, ..., B_n \in \mathsf{Sub}_r(Z)$ is a $\forall_X$*-factorization* of T (or just a $\forall$-factorization of T) if:

(i) $\pi_Z^{-1}(B_i) \leq T$ for each $i \leq n$;

(ii) for every $C \in \mathsf{Sub}_r(Z)$, such that $\pi_Z^{-1}(C) \leq T$, there is some $i \leq n$ such that $C \leq B_i$.

If $n = 1$ we say that this is a *singular* $\forall$-factorization. We say that $\mathbf{C}$ has the $\forall$*-factorization property* (singular $\forall$-factorization property) if for all objects X, Z and any $S \in \mathsf{Sub}_r(X \times Z)$, there is a $\forall_X$-factorization (resp. singular $\forall_X$-factorization) of S. The equational theory E (or the equivalent algebraic semantics $\mathbf{K}$) has the $\forall$*-factorization property* (singular $\forall$-factorization property) iff so does $\mathsf{Alg}^{op}_{fp}(E)$.

The following proposition is immediate from the definitions:

Proposition 4.8 *The logic* $\vdash$ *has left-finitary deductive interpolation if and only if* $\mathsf{Alg}^{op}_{fp}(E)$ *has the* $\forall$*-factorization property.*

The $\forall$*-factorization property* may follow from some natural assumptions: for instance, if E is locally finite it trivially holds, since there are only finitely many subobjects. In addition, if E is an equational theory axiomatising a logical calculus L, and L has global post-interpolants, then the $\forall$-factorization property holds and moreover $\forall$-factorizations are singular.

5 Finitary svr-Unification Types

In this section we present a proof of Theorem 2.12. Throughout this section, *we assume that* $\mathbf{K}$ *has (IT), Coherence properties and the* $\forall$*-factorization property.* We recall the following fact, proved in [17, Proposition 3.1, pp.51]:

Proposition 5.1 *Let* $\mathbf{C}$ *be an r-regular category. Then the pullback functors on regular subobjects have left adjoints satisfying the Beck-Chevalley condition:*

for every arrow $f : Y \to X$ *in* **C**, *and every regular subobject* $S \in \mathsf{Sub}_r(Y)$, *there is a regular subobject* $\exists_f(S) \in \mathsf{Sub}_r(X)$ *such that:*

$$\exists_f(S) \leq T \text{ iff } S \leq f^{-1}(T)$$

holds for every regular subobject $T \in \mathsf{Sub}_r(X)$; *in addition, for every pullback square as in Figure 4, and every regular subobject* $S \in \mathsf{Sub}_r(Y_1)$, *the following condition holds:*

$$f_2^{-1}(\exists_{f_1}(S)) = \exists_{p_2}(p_1^{-1}(S)).$$

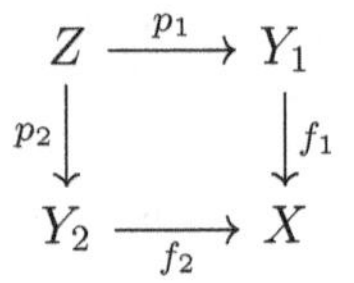

Fig. 4. Beck-Chevalley Pullback Square

We now proceed to show that in r-regular categories, $\forall$-factorizations are stable under pullbacks.

Lemma 5.2 *Let* **C** *be an r-regular category; consider the pullback of Figure 5. If* $S \in \mathsf{Sub}_r(X \times Z)$ *and* $B_1, ..., B_n$ *is a* $\forall_Z$*-factorization of* S, *then*

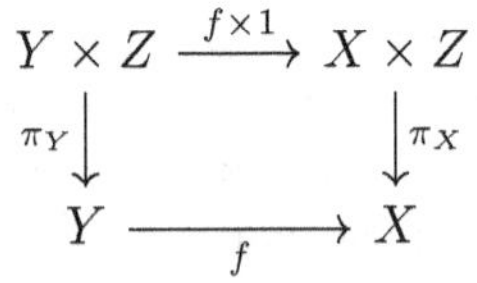

Fig. 5. Pullback Stability

$f^{-1}(B_1), ..., f^{-1}(B_n)$ *is a* $\forall_Z$*-factorization of* $(f \times 1)^{-1}(S)$.

Proof. Let $i \leq n$ be arbitrary, and assume that $B_1, ..., B_n \in \mathsf{Sub}_r(X)$ are a $\forall_Z$-factorization of S. First we want to show that

$$\pi_Y^{-1}(f^{-1}(B_i)) \leq (f \times 1)^{-1}(S).$$

Note that since $\pi_X^{-1}(B_i) \leq S$, then we have $(f \times 1)^{-1}(\pi_X^{-1}(B_i)) \leq (f \times 1)^{-1}(S)$ by usual facts on pullbacks. Moreover, since $f \circ \pi_Y = \pi_X \circ (f \times 1)$ we have that $\pi_Y^{-1}(f^{-1}(B_i)) \leq (f \times 1)^{-1}(S)$.

For the second property, consider an arbitrary $C \in \mathsf{Sub}_r(Y)$ such that $\pi_Y^{-1}(C) \leq (f \times 1)^{-1}(S)$. Using Proposition 5.1, we can apply the functor $\exists_{f \times 1}$, to obtain

$$\exists_{f \times 1}(\pi_Y^{-1}(C)) \leq \exists_{f \times 1}(f \times 1)^{-1}(S) \leq S.$$

Using Beck-Chevalley, this implies that $\exists_{f \times 1}(\pi_Y^{-1}(C)) = \pi_X^{-1}(\exists_f(C)) \leq S$. So by the $\forall_Z$-factorization property, there is some B_i such that $\exists_f(C) \leq B_i$. By adjunction, $C \leq f^{-1}(B_i)$ follows, which shows the property. □

Lemma 5.3 *Given* $\mathcal{A}^* \in \mathsf{Sub}_r(\mathbf{F}(X)^* \times \mathbf{F}(C)^*)$ *and* $\overline{\sigma} : \mathbf{F}(X) \to \mathbf{F}(X')$, *we have that* $\overline{\sigma}$ *is a scr-unifier of* $(\mathcal{A}, C)$ *iff* $1 \leq (\overline{\sigma}^* \times 1)^{-1}(\mathcal{A}^*)$.

Proof. Notice that if $1 \leq (\overline{\sigma}^* \times 1)^{-1}(\mathcal{A}^*)$, then there is an arrow making the triangle of Figure 1 commute. Conversely, if there is such an arrow, then by the universal property of pullbacks if follows that the monic $(\overline{\sigma}^* \times 1)^{-1}(\mathcal{A}^*) \hookrightarrow \mathbf{F}(X')^* \times \mathbf{F}(C)^*$ has a left inverse, so it is an isomorphism. □

We now prove the key technical result of this section: we reduce E-unification with simple variable restriction to standard E-unification:

Theorem 5.4 *Let* $\mathcal{A}^* \in \mathsf{Sub}_r(\mathbf{F}(X)^* \times \mathbf{F}(C)^*)$, *and* $\mathcal{B}_1^*, ..., \mathcal{B}_n^*$ *be a* $\forall_{\mathbf{F}(C)^*}$*-factorization of* $\mathcal{A}^*$. *We have*

$$U_E^{svr}(\mathcal{A}, C) \ \simeq \ U_E(\mathcal{B}_1) \cup \cdots \cup U_E(\mathcal{B}_n)$$

(the isomorphisms being a preordered sets isomorphism).

Proof. Let $\overline{\sigma} \in U_E^{svr}(\mathcal{A}, C)$ be a scr-unifier; then, by Lemma 5.3, we have $\pi^{-1}_{\mathbf{F}(X')^*}(1) = 1 \leq (\overline{\sigma}^* \times 1)^{-1}(\mathcal{A}^*)$. Using Lemma 5.2 we have that $(\overline{\sigma}^*)^{-1}(\mathcal{B}_1^*), ..., (\overline{\sigma}^*)^{-1}(\mathcal{B}_n^*)$ is a $\forall$-factorization of $(\overline{\sigma}^* \times 1)^{-1}(\mathcal{A}^*)$, hence we have

$$1 \leq (\overline{\sigma}^*)^{-1}(\mathcal{B}_i^*)$$

for some i, which means (again by Lemma 5.3 applied to the case $C = \emptyset$) that $\overline{\sigma}$ is a unifier of $\mathcal{B}_i$. Thus we have $\overline{\sigma} \in \bigcup_i U_E(\mathcal{B}_i)$.

Conversely, if $\overline{\sigma} : \mathbf{F}(X) \to \mathbf{F}(X')$ belongs to $U_E(\mathcal{B}_i)$, then by Lemma 5.3 (case $C = \emptyset$) we have $1_{\mathbf{F}(X')^*} \leq (\overline{\sigma}^*)^{-1}(\mathcal{B}_i^*)$. Since $\pi^{-1}_{\mathbf{F}(X)^*}(\mathcal{B}_i^*) \leq \mathcal{A}^*$ we get $\pi^{-1}_{\mathbf{F}(X')^*}((\overline{\sigma}^*)^{-1}(\mathcal{B}_i^*)) = (\overline{\sigma}^* \times 1)^{-1}(\pi^{-1}_{\mathbf{F}(X)^*}(\mathcal{B}_i^*)) \leq (\overline{\sigma}^* \times 1)^{-1}(\mathcal{A}^*)$. Then by transitivity we obtain

$$1 \cong \pi^{-1}_{\mathbf{F}(X')^*}(1_{\mathbf{F}(X')^*}) \leq \pi^{-1}_{\mathbf{F}(X')^*}((\overline{\sigma}^*)^{-1}(\mathcal{B}_i^*)) \leq (\overline{\sigma}^* \times 1)^{-1}(\mathcal{A}^*),$$

which means that $\overline{\sigma} \in U_E^{svr}(\mathcal{A})$ by Lemma 5.3.

Finally, note that the preorder relation is defined in the same way in $U_E^{svr}(\mathcal{A}, C)$ and in $U_E(\mathcal{B}_1) \cup ... \cup U_E(\mathcal{B}_n)$. □

As a consequence, we can supply the proof of Theorem 2.12:

Proof. By the results from Section 4, Maehara Interpolation, Right-Uniform Interpolation and Left-Finitary Interpolation correspond, respectively, to (IT), Coherence and the $\forall$-factorization Property. Consider now the Π_2-rule $\forall\overline{r}\Delta(\overline{q}, \overline{r})/^2\psi(\overline{q})$ and let $\Theta_1(\overline{q}), \ldots, \Theta_n(\overline{q})$ be the finite sets of formulas mentioned in Definition 2.11 for $\Delta(\overline{q}, \overline{r})$. What Theorem 5.4 says (applying the relevant transformers from formulas to equations and back) is that for every $\overline{r}$-invariant substitution σ, we have $\vdash \sigma(\Delta)$ iff we have $\vdash \sigma(\Theta_i)$ for some i. Thus, in view of Lemma 2.8, the rule $\forall\overline{r}\Delta/^2\psi$ is admissible iff one of the standard rules Θ_i/ψ is admissible. □

Theorem 5.4 has also some important consequences regarding the decidability of the unification problem with simple variable restrictions:

Corollary 5.5 *Suppose that* **K** *satisfies (IT), Coherence and* $\forall$*-factorization property. Then:*

(i) *If E has finitary unification type, then it has finitary unification type for the problem with simple variable restrictions.*

(ii) *If E has unitary unification type and $\forall$-factorizations are singular, then it has unitary unification type for the problem with simple constant restrictions.*

(iii) *If E-unification is decidable and $\forall$-factorizations are computable, then E-unification with simple constant restrictions is decidable as well.*

6 Applications

In this section we supply some example applications.

Admissibility of Π_2-rules via Unification

We can use Corollary 5.5 and Proposition 3.3 in order to directly obtain a decision procedure for admissibility of Π_2-rules via unification. This goes as follows (let us call τ the structural transformer from formulas to equations which is granted from the algebraizability hypothesis, see [10, Definitions 3.11]): given a Π_2-rule $\forall\overline{p}\Gamma/^2\phi$, we first compute a basis of $\overline{p}$-unifiers $\sigma_0, ..., \sigma_n$ for the svr-unification problem given by $\{\tau(\psi) \mid \psi \in \Gamma\}$, and for each of these unifiers – using the decidability of our logic – we check whether $\vdash \rho(\sigma_i(\tau(\phi)))$ holds or not, where ρ is the inverse transformer of τ. The procedure for computing the basis of $\overline{p}$-unifiers amounts to the following: using left-finitary deductive uniform interpolation, we compute the finitely many "approximants" of Γ with respect to $\overline{p}$; using decidability of unification, we compute for the transformed equations of each such approximant a finite basis of *standard* unifiers. As shown by Theorem 5.4 above, these correspond to a basis of $\overline{p}$-unifiers of $\{\tau(\psi) \mid \psi \in \Gamma\}$. In practical cases, there is no need to apply structural transformers (from formulas to equations and back) because many standard unification algorithms in the literature oriented to propositional logics [2] takes as input directly formulas (not their transformed equations). Thus, below we shall directly speak of 'unifiers' and of 'C-unifiers' of a set of formulas Γ (meaning with that the 'unifiers' and the 'C-unifiers' of the transformed set of equations $\{\tau(\psi) \mid \psi \in \Gamma\}$).

As an illustration of how to use our techniques to study admissibility, we turn to the Takeuti-Titani rule, mentioned in Example 2.5. Such a rule was proven to be admissible over a large class of algebraic signatures, through syntactic methods, by Metcalfe and Montagna [21], generalising a proof of Baaz and Veith [5].

Example 6.1 Let E be the theory of Gödel algebras, i.e., Heyting algebras $\mathcal{H}$ satisfying the additional axiom

$$(p \to q) \vee (q \to p) \ = \ \top.$$

Their associated logical system is often denoted by LC (for 'linear calculus'). We will show that the TT rule

$$\frac{\forall r\,(g \to ((p \to r) \vee (r \to q) \vee c))}{g \to (p \to q) \vee c}$$

is admissible. Following our remarks above, it suffices to show that all of the C-unifiers (where $C = \{r\}$) of the the formula $g \to ((p \to r) \vee (r \to q) \vee c)$ are C-unifiers for $g \to ((p \to q) \vee c)$. Since standard E-unification is unitary [26] and the conditions of the previous section are satisfied – we are in the locally finite case and, indeed, uniform post-interpolants exist – it suffices to show that the most general standard unifier for the formula

$$\forall_r\,(g \to (p \to r) \vee (r \to q) \vee c)$$

(namely for the the uniform post-interpolant of $g \to ((p \to r) \vee (r \to q) \vee c)$ wrt r) is also a standard unifier for the consequent $g \to (p \to q) \vee c$.

In LC and in intuitionistic logic systems, uniform interpolants admit a bisimulation semantics which works for Kripke frames, as we proceed to explain. Such semantics can be used to check that a certain formula is really the uniform interpolant of another given one. The finite Kripke frames corresponding to finite Gödel algebras are precisely the finite frames $\mathfrak{F} = (W, R)$ which are prelinear, i.e. such that for each $x \in W$, $R[x] = \{y \mid xRy\}$ is a linear order. By the results from [17], for any prelinear finite Kripke frame, and formula $\phi(\overline{p}, \overline{q})$, for each Kripke model V over $\mathfrak{F}$ and over the propositional letters $\overline{p}$, we have for each $x \in W$

$$(\mathfrak{F}, V), x \Vdash \forall_{\overline{q}}\phi(\overline{p}, \overline{q}) \iff \text{for any } (\mathfrak{F}', V')\ \overline{p}\text{-bisimilar model } (\mathfrak{F}', V'), x \Vdash \phi(\overline{q}, \overline{p}).$$

Using this semantics, we can then show that:

$$\forall_r\,(g \to (p \to r) \vee (r \to q) \vee c) \equiv g \to (p \to q) \vee c.$$

The right to left side follows from second order intuitionistic propositional logic with the help of the LC-valid formula $(r \to q) \vee (q \to r)$. For the other side we need bisimulation semantics. Suppose that we have points $x \geq y \in \mathfrak{F}$ such that $(\mathfrak{F}, V), x \Vdash g$, $(\mathfrak{F}, V), x \nVdash c$, $(\mathfrak{F}, V), y \Vdash p$, $(\mathfrak{F}, V), y \nVdash q$. Form the bisimulation expansion containing a duplicate y' as an immediate successor of y where y refutes r, and y' forces r. This expansion provides a bisimilar model such that $g \to (p \to r) \vee (r \to q) \vee c$ fails at x. Having such an equivalence, the result immediately follows, since the uniform interpolant we obtained is precisely the formula in the consequent of the Takeuti-Titani rule. △

Finitarity and Unitarity of Unification Type

In Section 5, we noted that there is a clear connection between svr-unification types and classical unification. It is natural to ask whether in fact the type is always preserved. The next example shows that it may happen that E-unification is unitary and svr-unification type is only finitary.

Example 6.2 Consider the equational theory of *implicative semilattices*, denoted ISL; this corresponds to the $(\top, \wedge, \to)$-fragment of IPC. Such an equational theory is locally finite, has the amalgamation property and the congruence extension property (hence it fulfills the hypotheses of Theorem 2.12, by the remark we made after Theorem 4.2). It is known that ISL has unitary elementary unification type [12]. We show that in the setting of unification with simple variable restrictions the unification type becomes finitary. For consider the following svr-unification problem, where we put $C = \{z\}$ (the formula we use is taken from [24, Example 4.5]):

$$((x \to z) \land (y \to z) \to z, \top).$$

Note that this problem has two incomparable C-unifiers, namely $\sigma = \{x \mapsto \top, y \mapsto y\}$ and $\tau = \{y \mapsto \top, x \mapsto x\}$. But we claim that there can be no unifier more general than both of them. For suppose that there was one, say μ. We must have $\mu(x) \neq_E \top$ and $\mu(y) \neq_E \top$ (otherwise μ would be less general than σ or τ). This implies that there are rooted Kriple models (with respective roots r_1, r_2) such that $(\mathfrak{F}_1, V_1), r_1 \nVdash \mu(x)$ and $(\mathfrak{F}_2, V_2), r_2 \nVdash \mu(y)$. Since z does not occur in $\mu(x), \mu(y)$, we can freely suppose that $(\mathfrak{F}_1, V_1), r_1 \Vdash z$ and $(\mathfrak{F}_2, V_2), r_2 \Vdash z$. Now build another rooted Kripke model $(\mathfrak{F}, V)$ by taking the disjoint union of $(\mathfrak{F}_1, V_1)$ and $(\mathfrak{F}_2, V_2)$ and by attaching it a new root r; we also stipulate that $(\mathfrak{F}, V), r \nVdash z$. Now then we have that,

$$(\mathfrak{F}, V), r \nVdash (\mu(x) \to z) \land (\mu(y) \to z) \to z \ ,$$

so μ cannot be a C-unifier[11]. $\triangle$

7 svr-Unification in Nuclear Implicative Semilattices

As a further nontrivial example, we show that the $\{\ell, \land, \top, \to\}$-fragment of lax logic satisfies the hypotheses of Theorem 2.12. The Maehara Interpolation Property follows by the deduction theorem and by inspecting the proof of the interpolation property for lax logic in [18]; Right-Uniform Interpolation and Left-Finitary Uniform Interpolation follow from local finiteness of this variety, shown in [6]. The decidability of the admissibility problems for standard rules comes from *finitarity of unification and computability of finite unification bases*: we will show such properties below (using methods different methods from those adopted for lax logic in [15]).

First we need the following folklore fact (implicit in [14]):

Proposition 7.1 *Let* $\mathbf{K}$ *be a locally finite variety such that subalgebras of finite projective* $\mathbf{K}$*-algebras are projective. Then unification in* $\mathbf{K}$ *is finitary. Moreover, if* (P) *is a unification problem, then the unifiers from a finite unification basis for this problem can be chosen so as not to contain more variables than those already occurring in* (P).

Proof. Let $\mathcal{A}$ be a finitely presented $\mathbf{K}$-algebra, which is finite by local finiteness; let it be a quotient $q : \mathbf{F}(X) \longrightarrow \mathcal{A}$ of the finitely generated free alge-

[11] Incidentally, we notice that the above argument independently proves that ISL does not have left uniform interpolation (if it had, by Theorem 5.4, it would also have unitary svr-unification type as it has unitary unification type).

bra $\mathbf{F}(X)$. Let $\sigma : \mathbf{F}(X) \longrightarrow \mathbf{F}(Y)$ be a unifier for $\mathcal{A}$;[12] as such, σ factors through q as $\overline{\sigma} \circ q$. Taking the image factorization, we can further factorize $\overline{\sigma}$ as $\mathcal{A} \xrightarrow{\overline{\sigma}_0} P \xrightarrow{\iota} \mathbf{F}(Y)$, where P is projective as a subalgebra of a projective algebra. By projectivity, the surjective map $\overline{\sigma}_0 \circ q : \mathbf{F}(X) \longrightarrow P$ has a section s (i.e. $\overline{\sigma}_0 \circ q \circ s = 1_P$). Now $s \circ \overline{\sigma}_0 \circ q : \mathbf{F}(X) \longrightarrow \mathbf{F}(X)$ is another unifier for $\mathcal{A}$ (because it factors though q) and is more general than σ because

$$\sigma \circ (s \circ \overline{\sigma}_0 \circ q) = (\iota \circ \overline{\sigma}_0 \circ q) \circ (s \circ \overline{\sigma}_0 \circ q) = \iota \circ \overline{\sigma}_0 \circ q = \sigma$$

Thus unifiers of $\mathcal{A}$ having domain and codomain $\mathbf{F}(X)$ form a unification basis. Since $\mathbf{F}(X)$ is finite, and hence there can be only finitely many unifiers with this domain and codomain, this unification basis is finite. □

The variety algebraizing the $\{\ell, \wedge, \top, \rightarrow\}$-fragment of lax logic is formed by *nuclear implicative semilattices*, namely by the algebras $\mathcal{A} = (A, \top, \wedge, \rightarrow, \ell)$, where $(A, \top, \wedge, \rightarrow)$ is an implicative semilattice and $\ell : A \longrightarrow A$ is a *nucleus*, i.e., a unary operator satisfying the conditions

$$x \leq \ell x, \qquad \ell(x \wedge y) = \ell x \wedge \ell y, \qquad \ell\ell x \leq \ell x \quad .$$

The category of finite nuclear semilattices $\mathbf{NIS}_{fin}$ (which by local finiteness, coincides with the category of finitely presented such algebras) is dual to the category $\mathbf{SF}_{fin}$ of finite S-posets and morphisms [6]. An S-poset is a triple $(X, \leq, S)$, where $(X, \leq)$ is a poset and $S \subseteq X$ is a subset; a morphism

$$f : (X, \leq, S) \longrightarrow (Y, \leq, T)$$

between S-posets is a partial map f satisfying the following conditions (we let $dom(f)$ be the domain of f, and $x < y$ mean $x \leq y$ and $x \neq y$):

(i) if $x < y$ and $x, y \in dom(f)$ then $f(x) < f(y)$;

(ii) if $x \in dom(f)$ and $f(x) < y$ there there is x' such that $x < x', x' \in dom(f)$ and $f(x') = y$;

(iii) $f^{-1}(T) = dom(f) \cap S$;

(iv) if $s \in S, s \leq x$ and $x \in dom(f)$, then there are $s', x' \in dom(f)$ such that $s \leq s' \leq x'$, $s' \in S$ and $f(x') = f(x)$.

We also need the following fact [6, Proposition 5.1]:

Lemma 7.2 *The dual of $f : (X, \leq, S) \longrightarrow (Y, \leq, T)$ is injective iff f is surjective and the dual of f is surjective iff f is injective and totally defined.*

Let α be an antichain in an S-poset $(X, \leq, S)$, i.e., a set of mutually $\leq$-incomparable elements; a *cover* of α is some $s \in S$ such that α is the set of the immediate successors of s.

[12] We adopt notation and definitions consistent with those introduced in Subsection 3.1. Adopting the approach of [12,14] (which views a unifier of $\mathcal{A}$ directly as a morphism with domain $\mathcal{A}$ and codomain a finitely presented projective algebra) would simplify the arguments.

Lemma 7.3 *The dual of a finite S-poset* $(X, \leq, S)$ *is a projective nuclear implicative semilattice iff every antichain* $\alpha \subseteq X$ *such that* $\alpha \not\subseteq S$ *has a cover.*

Proof. If the dual of $(X, \leq, S)$ is projective and $\alpha \subseteq X$ is such that $\alpha \not\subseteq S$ in order to find a cover of α it is sufficient to embed $(X, \leq, S)$ into the S-poset $(X \cup \{*\}, \leq, S \cup \{*\})$ obtained from $(X, \leq, S)$ by adding an extra element $*$ covering α: we show that the retract $r : (X \cup \{*\}, \leq, S \cup \{*\}) \longrightarrow (X, \leq, S)$ must map $*$ to a cover $r(*)$ of α. First, since r is a retract, we must have $r(x) = x$ for $x \in X$. Then, $*$ must be in the domain of r by (i)-(iv): in fact, we have $* \leq d$ for $d \in \alpha \setminus S$ and so there must be s' such that $s' \in dom(r) \cap (S \cup \{*\})$ and $* \leq s' \leq d$. This s' must coincide with $*$ (because d is an immediate successor of $*$, and $d \notin S$), so $* \in dom(r)$; also $r(*) \in S$ by (iii) and $r(*)$ covers α by (i)-(ii).

Conversely, suppose that $(X, \leq, S)$ satisfies the condition of the lemma and take a total embedding into a finite S-poset $(X, \leq, S) \hookrightarrow (Y, \leq, T)$. We find a retract $r : (Y, \leq, T) \longrightarrow (X, \leq, S)$ by defining $r(y)$ by induction on the height of $y \in Y$. If $y \in X$, we put $r(y) = y$; otherwise let α be the antichain of the minimal elements of $\{r(y') \mid y < y' \ \&\ y' \in dom(r)\}$. If $y \notin T$ or $\alpha \subseteq S$, then $dom(r)$ will not include y, otherwise we take $r(y)$ to be a cover of α. It is easy to check that r is a retract of the inclusion $(X, \leq, S) \hookrightarrow (Y, \leq, T)$. To prove that r satisfies the above conditions (i)-(iv), one shows by induction on the height of y that such conditions are satisfied by the restriction of r to the cone $\{y' \in Y \mid y \leq y'\}$. □

Lemma 7.4 *Subalgebras of projective finite nuclear implicative semilattices are projective.*

Proof. Let $(X, \leq, S)$ be dual to a finite projective nuclear implicative semilattice and let $f : (X, \leq, S) \longrightarrow (Y, \leq, T)$ be surjective. Take an antichain $\alpha \subseteq Y$ such that $\alpha \not\subseteq T$; let β be the antichain formed by the minimal elements of $f^{-1}(\alpha)$; we have that $f(\beta) = \alpha$ and $\beta \not\subseteq S$ by condition (iii) above. Thus there is a cover $s \in S$ for β and by (iv) we must have that $s \in dom(f)$ and that $f(s)$ is a cover of α. □

Theorem 7.5 *Admissibility of standard rules (and consequently also of* Π_2*-rules) is decidable in the* $\{\ell, \wedge, \top, \rightarrow\}$*-fragment of lax logic.*

Proof. By Proposition 7.1 and Lemma 7.4, standard unification for this logic has finitary unification type and finite bases of unifiers are computable. This guarantees decidability of admissibility of standard rules and also of Π_2-rules, as explained at the beginning of Section 6. □

8 Conclusions and Further Work

In this work we analyzed a new type of unification problems which are properly situated in the literature between elementary unification and the so-called unification with linear constant restrictions. Their interest here lies primarily in the connection with admissibility of non-standard Π_2-rules over logics. We supplied some first results and we discuss here several natural open questions.

One natural question that applies to logics, is whether the admissibility of more complex logical rules could be related to unification with linear constant restrictions. Such a question can be motivated also from a model theoretic point of view, since it is connected to decision problems for the positive theory of free algebraic structures.

Finally, and most importantly, it would be interesting to explore unification with simple variable restrictions for systems not covered by the results of the present paper. A natural example in this sense is the modal logic S4, which is well-known not to enjoy uniform interpolation [16]. We note that the obvious approach to attack this problem – a generalization of the *projective approximations* from [13] – does not work in the obvious way, since the key technique of Lowenheim Substitutions seems not to be available.

Appendix

In this appendix we collect some missing (mostly folklore) technical proofs.

(A) Proof of Proposition 3.5

We recall that, given finite sets of variables Z, Z' and a substitution σ with domain Z and codomain Z', we can canonically associate with it the homomorphism $\eta(\sigma) : \mathbf{F}(Z) \longrightarrow \mathbf{F}(Z')$ mapping the equivalence class of a term $t \in Fm_{\mathcal{L}}(Z)$ to the equivalence class of the term $\sigma(t) \in Fm_{\mathcal{L}}(Z')$. This correspondence is *bijective*, if we identify substitutions up to $=_E$: for that reason we always used the same letters $\sigma, \tau, \ldots$ for substitutions and free algebra homomorphisms. In this subsection, however, we conveniently distinguish between σ and the associated homomorphism $\eta(\sigma)$. Also note that the correspondence $\sigma \mapsto \eta(\sigma)$ commutes with compositions, in the sense that it maps the composition of substitutions into the compositions of homomorphisms.

In addition, notice that if X, X', C are finite disjoint sets, and $\eta(\sigma) : \mathbf{F}(X \cup C) \longrightarrow \mathbf{F}(X' \cup C)$ is the free algebras homomorphism induced by a C-invariant substitution σ, then there is a substitution $\overline{\sigma} : \mathbf{F}(X) \to \mathbf{F}(X')$ such that $\eta(\sigma) \simeq \eta(\overline{\sigma}) + 1$, up to the isomorphisms $\mathbf{F}(X) + \mathbf{F}(C) \simeq \mathbf{F}(X \cup C)$ and $\mathbf{F}(X') + \mathbf{F}(C) \simeq \mathbf{F}(X' \cup C)$.

Proposition 3.5 *Let (P_C) a E-unification problem with simple variable restriction. If $\mathcal{A}$ is a finitely presented algebra with presentation (P_C), then the antisymmetric quotients of the preordered sets $U_E^{svr}(\mathcal{A}, C)$ and $U_E^{svr}(P_C)$ are isomorphic.*

Proof. Let (P_C) be the unification problem with simple variable restriction

$$(P_C) \; = \; (s_1, t_1), ..., (s_k, t_k)$$

where $X = \{x_1, ..., x_n\} \cup C = \{c_1, ..., c_m\}$ are the disjoint sets of variables occuring in these terms. The corresponding finitely presented algebra algebra $\mathcal{A}$ is $\mathbf{F}(X \cup C)/S$ where S is the smallest congruence generated by the set of pairs $\{(s_1, t_1), ..., (s_k, t_k)\}$.

Define now a map $e : U_E^{svr}(P_C) \to U_E^{svr}(\mathcal{A}, C)$ as follows: let $\sigma \in U_E^{svr}(P_C)$, and suppose that $Y \cup C$ is its codomain, where Y are the variables occurring in the terms $\sigma(x_1), ..., \sigma(x_n)$ (note that this set is disjoint from C). Thus σ can be restricted to a substitution $\overline{\sigma}$ with domain X and codomain Y. Such σ and $\overline{\sigma}$ induce homomorphisms

$$\eta(\overline{\sigma}) : \mathbf{F}(X) \longrightarrow \mathbf{F}(Y), \qquad \eta(\sigma) : \mathbf{F}(X \cup C) \longrightarrow \mathbf{F}(Y \cup C)$$

such that $\eta(\overline{\sigma}) + 1 = \eta(\sigma)$. We put $e(\sigma) := \eta(\overline{\sigma})$. Recalling the definitions from Section 3 and Subsection 3.1, it is clear that $\sigma \in U_E^{svr}(P_C)$ iff $e(\sigma) \in U_E^{svr}(\mathcal{A}, C)$ (in fact $e(\sigma)^* \times 1 = \eta(\overline{\sigma})^* \times 1 = \eta(\sigma)^*$ factors through $\mathcal{A}$ iff the kernel of $\eta(\sigma)$ contains the equivalence classes in $\mathbf{F}(X \cup C)$ of the pairs of terms $(s_1, t_1), ..., (s_k, t_k)$, which precisely means that σ unifies them).

The map $e : U_E^{svr}(P_C) \to U_E^{svr}(\mathcal{A})$ is bijective up to $=_E$ equivalence of substitutions, thus it becomes a real bijection when we identify substitutions up to the comparison order. This order is preserved and reflected by e if we compare the preordered sets $U_E^{svr}(\mathcal{A}, C)$ and $U_E^{svr}(P_C)$ using the equivalent definition for $U_E^{svr}(\mathcal{A}, C)$ given by the Remark 3.4 of Subsection 3.1. In fact, for two substitutions $\sigma_1 : Fm_{\mathcal{L}}(Z) \longrightarrow Fm_{\mathcal{L}}(Z_1)$ and $\sigma_2 : Fm_{\mathcal{L}}(Z) \longrightarrow Fm_{\mathcal{L}}(Z_2)$, we have that $\sigma_1 \leq \sigma_2$ iff there is a substitution $\theta : Fm_{\mathcal{L}}(Z_2) \longrightarrow Fm_{\mathcal{L}}(Z_1)$ such that $\eta(\theta \circ \sigma_2) = \eta(\theta) \circ \eta(\sigma_2) = \eta(\sigma_1)$. □

(B) r-Regularity of $\mathsf{Alg}_{fp}^{op}(E)$

Proposition A.1 *Let E be an equational theory enjoying coherence and (IT). Then $\mathsf{Alg}_{fp}^{op}(E)$ is an r-regular category.*

Proof. The fact that $\mathsf{Alg}_{fp}^{op}(E)$ has all finite limits is immediate, given the standard fact that $\mathsf{Alg}_{fp}^{op}(E)$ has all finite colimits. To see the factorization properties, assume that $f^* : \mathcal{B}^* \to \mathcal{A}^*$ is a map of $\mathsf{Alg}_{fp}^{op}(E)$; then $f : \mathcal{A} \to \mathcal{B}$ is a homomorphism, which as usual has an image factorization

$$\mathcal{A} \xrightarrow{f} Im(f) \xrightarrow{i} \mathcal{B}.$$

Note that $Im(f)$ is a finitely generated subalgebra of $\mathcal{B}$, since it is a quotient of the finitely presented algebra $\mathcal{A}$; hence by coherence, $Im(f)$ is itself finitely presented; this means that the image factorization lives inside of the category $\mathsf{Alg}_{fp}^{op}(E)$, and hence by duality, $\mathsf{Alg}_{fp}^{op}(E)$ has the desired factorization properties.

Finally, assume that we have a pullback square as in the Figure below, and that g^* is an epimorphism. By (IT) and duality, there exist an epimorphism $p : \mathcal{E}^* \to \mathcal{B}^*$ and a homomorphism $p' : \mathcal{E}^* \to \mathcal{C}^*$ commuting the outer square below. Since $\mathcal{D}^*$ is a pullback, there is a connecting morphism $k : \mathcal{E}^* \to \mathcal{D}^*$; but this means that h_2 is an epimorphism as well, since it is the second component of an epimorphism. □

References

[1] Almeida, R. N., *π_2-rule systems and inductive classes of Gödel algebras* (2023). URL `https://arxiv.org/abs/2311.07189`

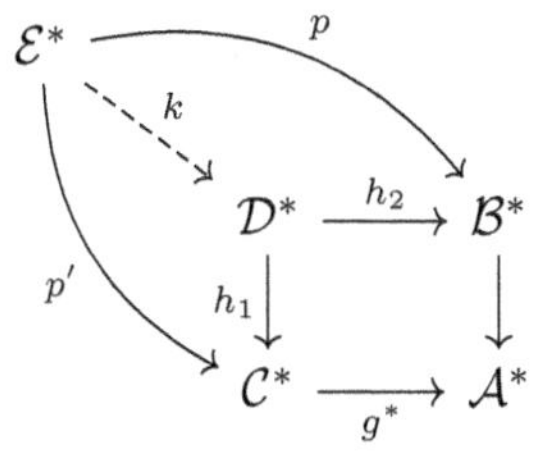

[2] Baader, F. and S. Ghilardi, *Unification in modal and description logics*, Logic Journal of IGPL **19** (2010), pp. 705–730.
URL `https://doi.org/10.1093/jigpal/jzq008`

[3] Baader, F. and K. U. Schulz, *Unification in the union of disjoint equational theories: Combining decision procedures*, Journal of Symbolic Computation **21** (1996), pp. 211–243.
URL `https://www.sciencedirect.com/science/article/pii/S0747717196900097`

[4] Baader, F. and J. Siekmann, *Unification theory*, in: D. M. Gabbay, C. J. Hogger and J. A. Robinson, editors, *Handbook of logic in artificial intelligence and logic programming: Volume 2: Deduction methodologies*, Handbook of Logic in Artificial Intelligence and Logic Programming, Clarendon Press, Oxford, England, 1994 pp. 41–101.

[5] Baaz, M. and H. Veith, *An axiomatization of quantified propositional Gödel logic using the takeuti-titani rule*, in: *Logic Colloquium '98*, Cambridge University Press, 2017 pp. 91–104.
URL `https://doi.org/10.1017/9781316756140.008`

[6] Bezhanishvili, G., N. Bezhanishvili, L. Carai, D. Gabelaia, S. Ghilardi and M. Jibladze, *Diego's theorem for nuclear implicative semilattices*, Indag. Math. (N.S.) **32** (2021), pp. 498–535.
URL `https://doi.org/10.1016/j.indag.2020.12.005`

[7] Bezhanishvili, G., N. Bezhanishvili, T. Santoli and Y. Venema, *A strict implication calculus for compact hausdorff spaces*, Annals of Pure and Applied Logic **170** (2019), p. 102714.

[8] Bezhanishvili, N., L. Carai, S. Ghilardi and L. Landi, *Admissibility of pi2-inference rules: interpolation, model completion, and contact algebras*, Annals of Pure and Applied Logic **174** (2023), p. 103169.
URL `https://www.sciencedirect.com/science/article/pii/S0168007222000847`

[9] Czelakowski, J., *Sentential logics and maehara interpolation property*, Studia Logica **44** (1985), pp. 265–283.
URL `https://doi.org/10.1007/bf00394446`

[10] Font, J. M., "Abstract algebraic logic – An introductory textbook," College Publications, London, 2016.

[11] Gabbay, D. M., *An irreflexivity lemma with applications to axiomatizations of conditions on tense frames*, in: *Aspects of Philosophical Logic*, Springer Netherlands, 1981 pp. 67–89.
URL `https://doi.org/10.1007/978-94-009-8384-7_3`

[12] Ghilardi, S., *Unification through Projectivity*, Journal of Logic and Computation **7** (1997), pp. 733–752.
URL `https://doi.org/10.1093/logcom/7.6.733`

[13] Ghilardi, S., *Best solving modal equations*, Annals of Pure and Applied Logic **102** (2000), pp. 183–198.
URL `https://doi.org/10.1016/s0168-0072(99)00032-9`

[14] Ghilardi, S., *Unification, finite duality and projectivity in varieties of Heyting algebras*, Ann. Pure Appl. Log. **127** (2004), pp. 99–115.
URL `https://doi.org/10.1016/j.apal.2003.11.010`

[15] Ghilardi, S. and G. Lenzi, *Unification in lax logic*, J. Algebr. Hyperstruct. Log. Algebras **3** (2022), pp. 61–75.
URL `https://doi.org/10.1086/286397`

[16] Ghilardi, S. and M. Zawadowski, *Undefinability of propositional quantifiers in the modal system s4*, Studia Logica **55** (1995), pp. 259–271.
URL https://doi.org/10.1007/bf01061237

[17] Ghilardi, S. and M. Zawadowski, "Sheaves, Games, and Model Completions," Springer Netherlands, 2002.
URL https://doi.org/10.1007/978-94-015-9936-8

[18] Iemhoff, R., *Proof theory for lax logic* (2022).
URL https://arxiv.org/abs/2209.08976v1

[19] Kowalski, T. and G. Metcalfe, *Uniform interpolation and coherence*, Annals of Pure and Applied Logic **170** (2019), pp. 825–841.
URL https://doi.org/10.1016/j.apal.2019.02.004

[20] Metcalfe, G., *Admissible rules: From characterizations to applications*, in: C. L. Ong and R. J. G. B. de Queiroz, editors, *Logic, Language, Information and Computation - 19th International Workshop, WoLLIC 2012, Buenos Aires, Argentina, September 3-6, 2012. Proceedings*, Lecture Notes in Computer Science **7456** (2012), pp. 56–69.
URL https://doi.org/10.1007/978-3-642-32621-9_4

[21] Metcalfe, G. and F. Montagna, *Substructural fuzzy logics*, The Journal of Symbolic Logic **72** (2007), pp. 834–864.
URL http://www.jstor.org/stable/27588573

[22] Metcalfe, G., F. Montagna and C. Tsinakis, *Amalgamation and interpolation in ordered algebras*, Journal of Algebra **402** (2014), pp. 21–82.
URL https://doi.org/10.1016/j.jalgebra.2013.11.019

[23] Takeuti, G. and S. Titani, *Intuitionistic fuzzy logic and intuitionistic fuzzy set theory*, Journal of Symbolic Logic **49** (1984), pp. 851–866.

[24] van Gool, S. J., G. Metcalfe and C. Tsinakis, *Uniform interpolation and compact congruences*, Annals of Pure and Applied Logic **168** (2017), pp. 1927–1948.
URL https://www.sciencedirect.com/science/article/pii/S0168007217300684

[25] Venema, Y., *Derivation rules as anti-axioms in modal logic*, The Journal of Symbolic Logic **58** (1993), pp. 1003–1034.

[26] Wronski, A., *Transparent unification problems*, Reports on Mathematical Logic **29** (1995), pp. 105–107.

On the Proof Theory of Apodictic Syllogistic

Melissa Antonelli [1]

HIIT & University of Helsinki
Pietari Kalmin katu 5
00560 Helsinki, Finland

Jan von Plato

University of Helsinki
Unioninkatu 40
00014 Helsinki, Finland

Abstract

The success of syllogistic has been such that many of the features originally characterizing this logic have changed over decades. The present study aims at offering a faithful, possibly comprehensive, account of Aristotle's deductive logic, by extending the inherently proof-theoretical approach introduced by von Plato in 2009/16. Concretely, the main novelty consists in treating syllogistic as a natural deduction system of rules and 'perfecting' arguments (for imperfect moods) as derivability proofs in tree form. Nothing is added to the original source but, in this way, assertoric and apodictic syllogistic are transparently reconstructed and Aristotle's perfecting proofs are proved to be always correct. This would also be a first step to make the whole modal fragment fully comprehensible and to rehabilitate syllogistic as a fertile theory in the context of natural language reasoning.

Keywords: Proof Theory, Apodictic Syllogistic, Derivation Trees, Aristotle.

1 Introduction

It is *communis opinio* that, in contrast to the assertoric theory, Aristotle's modal syllogistic is obscure or even inconsistent. Following the proof-theoretical approach introduced in 2009/16 by the second author [47], which has already led to a transparent study of the assertoric fragment and to the remarkable normal form theorem, we provide a plain and perfectly intelligible reconstruction of apodictic syllogistic.

One of the primary goals of the *Analytica* was to present a well-defined deductive system and to study its properties. [2] Coherently, our approach is

---

[1] The author thanks Helsinki Institute for Information Technology (HIIT) for the support.

[2] Aristotle's syllogistic may be seen as a system of deductive logic (or as a prototypical logical theory [11]). In this respect, *An.Pr.* represents the historically-first attempt to develop a

essentially proof-theoretical. To meticulously inspect the original source, we treat syllogistic as a natural deduction system of rules and present perfecting (i.e. derivability) proofs in tree form. Indeed, in *An.Pr.* I, Aristotle himself systematically 'perfects' all imperfect syllogisms. Yet, the linearity of the text partially hides the deductive structure of these proofs, which is rather made explicit by means of our two-dimensional trees. Advantages offered by the arboreal form are evident, as it transparently shows the skeleton of the argument by keeping "track of what depends on what within the proof" [47, p. 16]. This brings great benefits for both exegesis and logical comprehension.

Our study aims at extending the approach by [47] and at offering a *comprehensive* and *faithful* account on Aristotle's deductive logic, in order to show that his modal syllogistic is not only comprehensible but also potentially relevant for contemporary logic. Methodologically, our analysis is loyal to the original source, to which nothing is added. Nonetheless, by presenting perfecting proofs in tree form, we have obtained a clear reconstruction not only for the plain assertoric system, but also for (part of) the controversial modal one. The readers themselves can compare our reconstructive proposal with the corresponding quotation and verify that Aristotle's proofs are always correct. In this way, purely- and mixed-apodictic syllogistic are fully vindicated.

1.1 The Structure of the Paper

Before effectively reconstructing assertoric and apodictic syllogistic, we try to clarify the *original* meaning of the term 'syllogism' in Aristotle's *corpus.* [3] This is done in Section 2. Then, in Section 3, we cursorily, but systematically, deal with the presentation of assertoric syllogistic. The reconstruction of this fragment is not problematic, but is relevant as its language and rules are at the basis of the mixed-apodictic system subsequently considered. In addition, already this investigation shows how the proof-theoretical approach offers a straightforward way to analyze the correctness of perfecting proofs. In Section 4, such an approach is extended to the apodictic fragment. In particular, we consider purely-apodictic syllogistic first (Section 4.2), then moving to the mixed-apodictic system, the language of which includes assertoric and apodictic propositions (Section 4.3). We conclude by comparing our work with related studies, in Section 5, and by pointing at possible directions for future research, in Section 6.

2 Rehabilitating Syllogistic

Aristotle's *Analytica Priora* marks the beginning of deductive logic. Later on, syllogistic gains an enormous success, becoming the hegemonic paradigm of Western logic until its crisis at the end of the XIX century. After the advent of mathematical logic, syllogistic has become more and more marginal, being

comprehensive and self-contained deductive system and to study its properties.

[3] Exegetical problems stem both from the 'double-nature' of the term in Aristotelian texts, where the author uses it with connected but different meanings, and from theoretical changes it has undertaken over decades. For further details, see Section 2.

considered as nothing but a monadic fragment of first-order logic, and nowadays it is commonly seen as a limited and quite sterile theory, the interest of which is merely historical. Actually, when looking at the paucity of its language, it appears striking that syllogistic has been the dominant deductive system for about two millennia.

In fact, historical inspection, both of the original source and of its subsequent interpretations, proves a wide scope of this logic, which cannot be confined to the assertoric fragment. Many occurrences of the term in ancient and medieval sources draw an image of syllogistic which massively diverges from today's logic handbooks and, when delving deeper into the Aristotelian *corpus* and the development of the disciplines, perplexities arise on what exactly a 'syllogism' is. We will show that, against what is commonly thought, syllogistic holds valuable properties related to its affinity with natural language syntax and to ordinary reasoning, which have also been recently revived by studies in Natural Logic [45,36,26].

2.1 Aristotle's 'syllogism'

The term 'syllogism' is not a neologism introduced by Aristotle. In earlier Greek texts, this word is witnessed as signifying 'to reason', 'to calculate' or 'to conclude an argument', see [8,5,38]. These definitions are not far from the ones appearing in the *Organon*, where syllogisms are presented as deductions or valid arguments [4, 24b19-22], [33, 100a25-29], and not as *specific kinds* of inference, namely the assertoric ones [4, I, 1-6]. In addition, Aristotle's effective use of the term is quite liberal, not excluding 'syllogistic-style' (non-assertoric) arguments, e.g. modal syllogisms [4, I, 3,8ff.] or syllogisms with more than two premisses or with singular subjects [4, 70a27]. [4]

As some scholars noticed, the original meaning of the word 'syllogism' does not correspond to the one of today's handbooks, which roughly overlaps with the assertoric theory presented in [4, 1-6]. This latter seems actually connected to the "secondary sense" that Aristotle attributes to the word, the "primary" one being that of deduction in general [5, pp. 23-25]. So, strikingly, the best translation for the word 'syllogism' is not its English calque, but a term with different linguistic roots:

> Logicians normally use "syllogism" to *mean* one of the specific forms of valid argument Aristotle discussed in *An.Pr.* I 1-6 but Aristotle's definition of *syllogismos* comprehend a much wider class: pretty much any valid argument, or at least any argument with a conclusion different from any of its premisses. If we translate *sullogismos* as "syllogism", the broad scope of this definition is obscured and with it the nature of Aristotle's logical theorizing. [39, p. 30]

Also the reception of Aristotle's work in early commentators and in the late ancient tradition suggests a 'wide scope' interpretation of the notion. [5]

[4] This polysemous, sometimes ambiguous use of the term is coherent with Aristotle's use of the language in general [31].

[5] In Late Antiquity, hypothetical and assertoric syllogistic coexist, and the term is equally used for both, see [40, p. 2]. For further details, see for instance [6,7,13,40].

On the other hand, the subsequent success of syllogistic has been such that many original features of this theory have been modified and misread several times and in a variety of ways, both by restricting and by broadening it. In fact, even assuming that syllogistic had to be originally identified with assertoric syllogistic, it clearly emerges that this core theory has been variously (re-)interpreted over centuries. For instance, during the Middle Ages, both Arabic and Latin traditions tended to extend Aristotelian definitions, by considering relational or three-premisses syllogisms or syllogisms with singular terms. [6]

For all these reasons, it seems reasonable to distinguish (at least) between syllogism *lato sensu*, the meaning of which corresponds to 'deduction' or 'valid argument', and *stricto sensu*, referring to specific assertoric inferences constituted by two categorical premisses and by three terms. As anticipated, it is likely that Aristotle's original use of the word is closer to that recently developed by Natural Logic, i.e. denoting a (correct) inference expressed in natural language, than to the (toy) definition attributed to it from the Nineteenth century on in contraposition to mathematical logic. This simple consideration would shed new light on the real scope of syllogistic and on its potential interest for contemporary logic. When syllogistic is seen for what it is – namely, a (first) logical system for human reasoning – the study of its fragments, including the (non-assertoric) modal and relational ones, acquires new importance.

3 On Assertoric Syllogistic

In this Section, we briefly, but systematically, reconstruct the fragment introduced by Aristotle in *An.Pr.* I, 1-7. [7] This is crucial as, although it is not controversial, [8] "assertoric syllogistic provides the foundation for the entire syllogistic system" [35, p. 11]. In Section 3.1, we cursorily introduce core notions at the basis of the whole Aristotelian syllogistic. Then, in Section 3.2, we present a formalization for the assertoric language and primitive rules that constitute assertoric syllogistic. Finally, in Section 3.3, we consider each perfecting proof by Aristotle and show them correct. In this way we provide a *plain* and *faithful* presentation of these reduction proofs, departing from the original source only by moving from linear to tree form and by using the nowadays standard (Medieval) names for syllogisms. [9]

---

[6] For an overview of Medieval theories of syllogisms, see [43].

[7] In *An.Pr.* I, 1 Aristotle introduced basic notions related to syllogistic in general, here summarized in Section 3.1. Assertoric primitive syllogisms are presented in *An.Pr.* I, 2. Then, *An.Pr.* I, 4, 5 and 6 deal with the first, second and third figures, respectively. In *An.Pr.* I, 7, Aristotle shows that perfect `Darii` and `Ferio` (and related imperfect moods) can actually be perfected due to `Barbari` and `Celarent`, introducing what we define as the minimal system $\mathbf{SYL}^*$, see Appendix A.

[8] In fact, controversies concern its semantical interpretation.

[9] For space reasons, we do not deal with the historical development of assertoric syllogistic and related mnemonics. For further details on the topic, see [43].

3.1 Aristotle's Assertoric Syllogisms

As seen, the definition of assertoric syllogistic has changed over two millennia. According to Aristotle, there are fourteen assertoric syllogisms, i.e. concluding arguments defined by two categorical premisses. They were originally divided into three *figures*, based on the relation linking the middle term and the extremes. He doesn't consider a fourth figure, or the notion of mood. As we shall see, Aristotle's theory is perfectly intelligible, and incoherence appears only in later versions of it (see, for instance, [17]).

Assertoric syllogisms are usually defined as valid arguments constituted by two assertoric premisses and an assertoric conclusion. In fact, when inspecting the original source, it emerges that a syllogism should probably be identified with its *pair of productive premisses* [34].[10] Aristotle establishes syllogistic validity either by 'semantic justification'[11] or by one 'method of proof'. Accordingly, syllogisms are divided into *perfect* or *complete* ones, which are somehow self-evident, and *imperfect* or *incomplete* ones, which are reducible to the first class [4, 24b23-27]. The core of *An.Pr.* I is the 'perfection' or 'reduction' of *all* imperfect syllogisms to perfect ones: by systematically considering every combination of premisses for each figure, Aristotle proves that either a given conclusion follows from them or that no conclusion can. In the latter case, the aforementioned premisses are said not to 'syllogize' and a counterexample is (usually) provided.

Three main 'kinds' of perfecting proofs have been recognized in the literature: ostensive, indirect and by *ecthesis.*[12] In fact, our reconstruction deals with the first two 'methods' only [4, 19a30-33]. Whenever possible, Aristotle constructs ostensive proofs, that is, after assuming the premisses of the syllogism he wants to 'perfect', he applies first-figure and conversion 'rules'. In this way, imperfect syllogisms are reduced to perfect (self-evident) ones [4, 29a33-35]. Indirect or "through impossibility" proofs are needed for syllogisms that are not directly provable, i.e., `Bocardo` and `Baroco`: the conclusion is assumed not to follow from the premisses, and given the premisses and the contradictory of the conclusion, falsity is derived.[13] We will show that, given the primitive syllogistic system, 'perfecting' proofs for imperfect syllogisms are nothing but

[10] Otherwise, assertoric syllogisms would be more than fourteen; in particular, subaltern moods should also be counted (see Appendix A).

[11] It is usually done by invoking the meaning of the given logical expression and appealing to the so-called *dictum de omni et nullo*, see for example [23,46,24].

[12] The intuition behind *ecthesis* seems not far from the one at the basis of semantical justification for perfect syllogisms, see [47, pp. 332-333]. However, its nature remains ambiguous [34, p. 13], lying somewhere in between a semantic explanation and an informal proof device. Being it a redundant or an essential part of the system, *ecthesis* appears in several passages of *An.Pr.*, e.g. it is mentioned as an alternative proof to complete indirect moods [4, 28a22-26, 28b13-15] and to justify perfect (non-mixed) apodictic [`Baro`] and [`Boc`] (several commentators also consider Π^-- and Σ^--conversion rules as obtained through it [4, 25a14-26]).

[13] Indirect proofs are also used to obtain (more complicated) alternatives to direct ones, see, for instance, [4, 27a14-15] (for `Cesare`) or [4, 28a28-30] (for `Felapton`) or [4, 28b13-15] (for `Disamis`) (see also Appendix A). For further details see [1].

derivability proofs. Indeed, in each of the aforementioned cases, Aristotle assumes the premisses of the imperfect syllogisms that he wants to perfect and, by applying primitive rules, obtains the conclusion.

3.2 The Primitive System SYL

We introduce the assertoric language and two primitive proof systems, **SYL** and **SYL***, the rules of which are justified by original quotation reference. Actually, although the two systems are equivalent (see Appendix A), only the less economical **SYL** is used by Aristotle in his perfecting proofs.

3.2.1 The Assertoric Language

Assertoric propositions are either universal or particular and either affirmative or negative [4, 24a16-20].[14] Subjects and predicates are called *terms*. All propositions express the belonging of a term, the predicate, to another term, the subject. We abbreviate these four forms following the notation by [47, p. 327]:

- $\Pi^+(A, B)$ for affirmative universal propositions: *Every A is B*
- $\Pi^-(A, B)$ for negative universal propositions: *No A is B*
- $\Sigma^+(A, B)$ for affirmative particular propositions: *Some A is B*
- $\Sigma^-(A, B)$ for negative particular propositions: *Some A is not B.*[15]

We also introduce a syntactical negation symbol so that, for any given assertoric proposition (P), also its negation ($P^\neg$) is part of our formal language.[16]

Definition 3.1 Let A, B be terms and Q be a proposition in the assertoric language. Then, the language of assertoric syllogistic $\mathcal{L}_{\mathbf{SYL}}$ is defined as follows:

$$P ::= \Pi^+(A,B) \mid \Pi^-(A,B) \mid \ \Sigma^+(A,B) \mid \Sigma^-(A,B) \mid \ Q^\neg .$$

3.2.2 Introducing the Rules of SYL

Here we present the primitive rules of the assertoric syllogistic system **SYL**. Most of them are introduced by quoting the corresponding source text. As we shall see, in a few cases, no explicit justification is given in *An.Pr.*, but can be found in other works of Aristotle's *corpus*.

The Rule of Contradictory Pairs. In *De Int.* 7, Aristotle introduces the contradictory pairs, $\Pi^+(A, B)$-$\Sigma^-(A, B)$ and $\Pi^-(A, B)$-$\Sigma^+(A, B)$ [2, 17a33-b19]. The corresponding rules are defined accordingly, by using the negation symbol $(\cdot)^\neg$: as seen, given an assertoric proposition P, its contradictory opposite is denoted by $P^\neg$. We follow Whitaker's terminology calling them "rules of contradictory pairs" [49].

---

[14] Aristotle introduces indeterminate propositions as well, but these have no role in assertoric syllogistic.

[15] Observe that the order of subject and predicate is the reverse of the original one.

[16] Notice that different choices are possible. For instance, in [47], the assertoric language is 'minimal' since negation is defined as a meta-logical symbol. Following this approach, our RCPs and RAA would "collapse" into a unique rule.

Definition 3.2 The *rules of contradictory pairs* (RCPs) are defined below:[17]

$$\frac{\Pi^-(A,B)}{\Sigma^+(A,B)^\neg}\ \mathtt{RCP}_1 \qquad \frac{\Pi^+(A,B)}{\Sigma^-(A,B)^\neg}\ \mathtt{RCP}_2$$

$$\frac{\Sigma^+(A,B)^\neg}{\Pi^-(A,B)}\ \mathtt{RCP}_3 \qquad \frac{\Sigma^-(A,B)^\neg}{\Pi^+(A,B)}\ \mathtt{RCP}_4$$

Intuitively, these are justified by the fact that "of every contradictory pair, one member is true and the other false" [49, p. 79].

The rule of non-contradiction. According to Aristotle's principle of non-contradiction, contradictory propositions cannot be both true at the same time [3, 1005b19-20, 1005b25-30, 1011b13-14].

Definition 3.3 The *rule of non-contradiction* is the schematic rule below:

$$\frac{P \qquad P^\neg}{\bot}\ \bot\mathtt{I}$$

Reductio ad Absurdum. As said, together with probative or direct deduction, Aristotle considers indirect proofs [4, 41a23-32, 29a35-29b1].[18] In fact, he variously uses *reductio ad absurdum* (RAA) or 'proofs through impossibility' in syllogistic derivations by absurdly assuming that the conclusion to be proved does not hold, so that (RCPs given) its contradictory opposite does. But then absurdity follows, and the desired conclusion comes out to hold.

Definition 3.4 The *schematic rule of reductio ad absurdum* (RAA) is defined as follows:

$$\begin{array}{c}[P^\neg]^1\\ \vdots\\ \dfrac{\bot}{P}\ \mathtt{RAA},1\end{array}$$

Since Aristotle does not give a unique and formal definition of RAA, the given schematic representation is not the only possible one. In [47], the second author introduces a more economical rule, encapsulating both our RAA and $\mathtt{RCP}_{1\text{-}4}$.

Rules of Conversion. Aristotle presents three one-premiss rules of conversion: universal negative conversion [4, 25a6–7], universal affirmative conversion [4, 25a7–8], and particular affirmative conversion [4, 25a9–10].[19]

Definition 3.5 The *rule of negative, universal* ($\Pi^-\mathtt{C}$), of *affirmative, universal* ($\Pi^+\mathtt{C}$), and of *affirmative, particular conversion* ($\Sigma^+\mathtt{C}$) are defined as follows:

[17] Actually, we should consider eight RCPs, but, in the presence of RAA, $\mathtt{RCP}_1$-$\mathtt{RCP}_4$ are enough to derive all imperfect syllogisms as done in *An.Pr.* I, 1-7, see [2, 17b19-26]. RCPs are also at the basis of the development of the so-called square of the opposition, see e.g. [41].

[18] Remarkably, when reading [4, 41a23-32] through the lenses of modern logic, it looks striking that the example of 'proof through impossible' introduced by Aristotle is the paradigmatic example of a proof of negation. Namely, the given proof of the incommensurability of the diagonal and the square is not a genuine indirect proof, but a direct proof deriving falsity.

[19] For their (semantical) justification, see [47, pp. 329–331].

$$\frac{\Pi^-(P,S)}{\Pi^-(S,P)}\ \Pi^-\mathtt{C} \qquad \frac{\Pi^+(P,S)}{\Sigma^+(S,P)}\ \Pi^+\mathtt{C} \qquad \frac{\Sigma^+(P,S)}{\Sigma^+(S,P)}\ \Sigma^+\mathtt{C}$$

Rules $\Pi^-\mathtt{C}$ and $\Sigma^+\mathtt{C}$ are commonly called (negative and positive) *conversio simplex*, whereas the Latin name for $\Pi^+\mathtt{C}$ is *conversio per accidens.* Observe also that $\Sigma^+\mathtt{C}$ is derivable in a 'more economical' system **SYL***, where $\Sigma^+\mathtt{C}$ is not primitive, see Appendix A.

Syllogistic Rules. Finally, Aristotle introduces *complete* or *perfect* syllogisms, namely four first-figure, two-premisses rules, such that the middle term is the subject of one of the premisses, later labelled as major premiss, and the predicate of the other, the so-called minor premiss [4, 25b31-38]. As said, differently from the notion of mood, that of figure is original to Aristotle [4, 26b25-34]. He starts with perfect `Barbara` in [4, 25b37–40], and `Celarent` [4, 26a1–2]. Then, after considering a couple of premisses which do not syllogize, he introduces `Darii` [4, 26a23–25] and `Ferio` [4, 26a25–26].

Definition 3.6 *First-figure syllogisms* are defined as follows:

$$\frac{\Pi^+(B,A) \qquad \Pi^+(C,B)}{\Pi^+(C,A)}\ \mathtt{Barbara} \qquad \frac{\Pi^-(B,A) \qquad \Pi^+(C,B)}{\Pi^-(C,A)}\ \mathtt{Celarent}$$

$$\frac{\Pi^+(B,A) \qquad \Sigma^+(C,B)}{\Sigma^+(C,A)}\ \mathtt{Darii} \qquad \frac{\Pi^-(B,A) \qquad \Sigma^+(C,B)}{\Sigma^-(C,A)}\ \mathtt{Ferio}$$

Aristotle also shows that `Darii` and `Ferio` are not 'really primitive', as they can be derived using (minimal) first-figure `Barbara` and `Celarent` together with `RAA`, `RCP`s and conversion rules.[20]

3.2.3 The System SYL (and SYL*)

In [4], Aristotle presents two syllogistic systems: **SYL** corresponds to the system which is effectively used in *An.Pr.* to 'perfect' imperfect moods, whereas **SYL*** is more economical as it does not include `Darii`, `Ferio` and $\Sigma^+\mathtt{C}$.

Definition 3.7 The minimal system **SYL*** is made of the four rules of contradictory pairs $\mathtt{RCP}_1$-$\mathtt{RCP}_4$, the two conversion rules $\Pi^-\mathtt{C}, \Pi^+\mathtt{C}$, the rule of *reductio* `RAA`, the rule of non-contradiction $\bot\mathtt{I}$, and the two syllogistic rules `Barbara` and `Celarent`. The system **SYL** is made of all rules defining **SYL*** plus the rule of positive *conversio simplex* $\Sigma^+\mathtt{C}$, and first-figure `Darii` and `Ferio`.

The two systems are equivalent in the sense that exactly the same set of (imperfect) syllogisms is derivable in them, see Appendix A.

3.3 Imperfect Syllogisms

Aristotle considers syllogisms in the second and third figures as *incomplete.* Most of *An.Pr.* I, 1-6 is devoted to "perfecting" *each of them.*[21] In modern

[20] For a brief presentation of the mentioned derivability proofs, see Appendix A. Full details can be found in [1].

[21] Actually, in some cases proofs are only sketched, for example the ones of second-figure `Camestres` or third-figure `Felapton` and `Datisi`.

terms, after presenting the primitive rule system(s) **SYL** (and **SYL***), Aristotle proves the derivability of imperfect syllogisms in it. He also claims that these 'perfectable' moods represent the totality of the two-premisses rules which are derivable in **SYL**, so providing a proof of syntactical completeness.

3.3.1 The Second Figure

Second-figure syllogisms are syllogisms in which both terms in the conclusion appear in subject position in the premisses [4, 26a4-27a3]. In presenting this figure, variables A, B, C, used for perfect syllogisms, are substituted by N, M, X, respectively. There are four second-figure syllogisms: for each of them we quote Aristotle's original proof and show its correctness – namely, that any rule applied is a proper instance of one in **SYL** – by simply presenting it in tree form.

Cesare. Aristotle considers `Cesare` first and shows it derivable in [4, 27a5–8]:

$$\dfrac{\dfrac{\Pi^-(N,M)}{\Pi^-(M,N)}\,\Pi^-\mathtt{C} \qquad \Pi^+(X,M)}{\Pi^-(X,N)}\,\mathtt{Celarent}$$

Festino. Then, `Festino` is presented and derived in [4, 27a32–37]:

$$\dfrac{\dfrac{\Pi^-(N,M)}{\Pi^-(M,N)}\,\Pi^-\mathtt{C} \qquad \Sigma^+(X,M)}{\Sigma^-(X,N)}\,\mathtt{Ferio}$$

Baroco. The subsequent derivability proof is that of `Baroco` [4, 27a37–b2], and it is not ostensive:

$$\dfrac{\dfrac{\Sigma^-(X,M) \qquad \dfrac{\dfrac{\Pi^+(N,M) \qquad \dfrac{[(\Sigma^-(X,N))^\neg]^1}{\Pi^+(X,N)}\,\mathtt{RCP}_4}{\Pi^+(X,M)}\,\mathtt{Barbara}}{(\Sigma^-(X,M))^\neg}\,\mathtt{RCP}_2}{\bot}\,\bot\mathtt{I}}{\Sigma^-(X,N)}\,\mathtt{RAA,1}$$

Camestres. Imperfect `Camestres` is also presented in [4, 27b–4], but its proof is omitted as close to `Cesare`'s one:

$$\dfrac{\dfrac{\dfrac{\Pi^-(X,M)}{\Pi^-(M,X)}\,\Pi^-\mathtt{C} \qquad \Pi^+(N,M)}{\Pi^-(N,X)}\,\mathtt{Celarent}}{\Pi^-(X,N)}\,\Pi^-\mathtt{C}$$

3.3.2 The Third Figure

Finally, Aristotle moves to the third figure, in which the subject and predicate of the conclusion are the predicates of the premisses [4, 28a10–14]. Here, he uses the variables S, R, P for middle term, conclusions' subject and predicate, respectively.

Darapti. First, he introduces `Darapti` and shows it derivable in [4, 28a22–26]:

$$\dfrac{\Pi^+(S,P) \qquad \dfrac{\Pi^+(S,R)}{\Sigma^+(R,S)}\,\Pi^+\mathtt{C}}{\Sigma^+(R,P)}\,\mathtt{Darii}$$

Felapton. Then, `Felapton` is considered and ostensively derived [4, 28a26-30]:

$$\dfrac{\Pi^-(S,P) \qquad \dfrac{\Pi^+(S,R)}{\Sigma^+(R,S)}\,\Pi^+\mathtt{C}}{\Sigma^-(R,P)}\,\mathtt{Ferio}$$

Disamis. Next, `Disamis` is proved derivable in [4, 29b8–11]:

$$\dfrac{\dfrac{\Pi^+(S,R) \qquad \dfrac{\Sigma^+(S,P)}{\Sigma^+(P,S)}\,\Sigma^+\mathtt{C}}{\Sigma^+(P,R)}\,\mathtt{Darii}}{\Sigma^+(R,P)}\,\Sigma^+\mathtt{C}$$

Datisi. `Datisi` is presented as perfectable, but no explicit derivability proof is given. Yet, this can be easily established by applying $\Sigma^+\mathtt{C}$ and perfect `Darii`:

$$\dfrac{\Pi^+(S,P) \qquad \dfrac{\Sigma^+(S,R)}{\Sigma^+(R,S)}\,\Sigma^+\mathtt{C}}{\Sigma^+(R,P)}\,\mathtt{Darii}$$

Bocardo. The proof of third-figure `Bocardo` is indirect [4, 28b17–20]:

$$\dfrac{\dfrac{\Sigma^-(S,P) \qquad \dfrac{\dfrac{\dfrac{[(\Sigma^-(R,P))^\neg]^1}{\Pi^+(R,P)}\,\mathtt{RCP}_4 \qquad \Pi^+(S,R)}{\Pi^+(S,P)}\,\mathtt{Barbara}}{(\Sigma^-(S,P))^\neg}\,\mathtt{RCP}_2}{\bot}\,\bot\mathtt{I}}{\Sigma^-(R,P)}\,\mathtt{RAA},1$$

Ferison. Finally, `Ferison` is defined and shown derivable in [4, 28b31–36]:

$$\dfrac{\Pi^-(S,P) \qquad \dfrac{\Sigma^+(S,R)}{\Sigma^+(R,S)}\,\Sigma^+\mathtt{C}}{\Sigma^-(R,P)}\,\mathtt{Ferio}$$

4 On Apodictic Syllogistic

The reconstruction of Aristotle's modal syllogistic is controversial and a variety of attempts for a consistent interpretation have appeared in the literature. Most of them focus on the reconstruction of semantical interpretations. On the contrary, as for the assertoric fragment, we treat apodictic syllogistic as a deductive system (that is, as it is actually presented in the source) without suggesting any interpretation. Again, our rule system is obtained directly from *An.Pr.*, and proofs are presented in tree form. Also in this case, we show that Aristotle always applies his rules coherently. Concretely, we start by introducing the apodictic language (Section 4.1). Then, we briefly consider the purely-apodictic system (Section 4.2). Finally, we deal with the mixed-apodictic fragment (Section 4.3).

4.1 The Apodictic Language

Syllogistic propositions are characterized not only by quality and quantity, but also by *modality*: assertoric, necessary, and possible [4, 25a1–2].[22] We abbreviate apodictic predications of the forms *All/No S is/is not necessarily P* as: $\Pi^+[S,P]$, $\Pi^-[S,P]$, $\Sigma^+[S,P]$ and $\Sigma^-[S,P]$. Observe that modalities are not treated as logical operators, but as part of the structure of atomic formulas to which the four quantifiers are applied. In addition, our notation for apodictic predication is modeled upon the standard box operator coming from modal logics.

4.2 Purely Apodictic Syllogistic

We start by presenting the language of purely apodictic syllogistic in Section 4.2.1. Then, in Section 4.2.2, its rule system $\mathbf{AP}_1$ is introduced. Finally, in Section 4.2.3, we briefly consider imperfect purely-apodictic moods.

4.2.1 The Language of Purely Apodictic Syllogistic

The language of purely apodictic syllogistic is defined by universal and particular, positive and negative apodictic propositions.

Definition 4.1 Let A, B be terms. Then, the language of purely apodictic syllogistic $\mathcal{L}_{\mathbf{AP}_1}$ is defined as follows:

$$P ::= \Pi^+[A,B] \mid \Pi^-[A,B] \mid \ \Sigma^+[A,B] \mid \Sigma^-[A,B].$$

4.2.2 The Purely Apodictic System $\mathbf{AP}_1$

Aristotle's treatment of purely apodictic syllogistic is extremely coincise. Apodictic conversions [4, 25a26–36] and perfect moods [4, 29b35–30a3] are defined analogously to the corresponding assertoric ones.

Definition 4.2 *Purely apodictic conversion rules* are defined below:

$$\frac{\Pi^-[P,S]}{\Pi^-[S,P]}\ \Pi^-[\mathtt{C}] \qquad \frac{\Pi^+[P,S]}{\Sigma^+[S,P]}\ \Pi^+[\mathtt{C}] \qquad \frac{\Sigma^+[P,S]}{\Sigma^+[S,P]}\ \Sigma^+[\mathtt{C}]$$

Definition 4.3 *Perfect purely apodictic syllogisms* are the following six ones:

$$\frac{\Pi^+[B,A] \qquad \Pi^+[C,B]}{\Pi^+[C,A]}\ [\mathtt{Bar}] \qquad \frac{\Pi^-[B,A] \qquad \Pi^+[C,B]}{\Pi^-[C,A]}\ [\mathtt{Cel}]$$

$$\frac{\Pi^+[B,A] \qquad \Sigma^+[C,B]}{\Sigma^+[C,A]}\ [\mathtt{Dar}] \qquad \frac{\Pi^-[B,A] \qquad \Sigma^+[C,B]}{\Sigma^-[C,A]}\ [\mathtt{Fer}]$$

$$\frac{\Pi^+[A,B] \qquad \Sigma^-[C,B]}{\Sigma^-[C,A]}\ [\mathtt{Baro}] \qquad \frac{\Sigma^-[B,A] \qquad \Pi^+[B,C]}{\Sigma^-[C,A]}\ [\mathtt{Boc}]$$

Indeed, differently from **SYL**, in $\mathbf{AP}_1$ [Baro] and [Boc] are primitive.[23]

Definition 4.4 The *purely-apodictic system* $\mathbf{AP}_1$ is made of the three purely apodictic rules of conversion and six perfect purely apodictic syllogisms above.

[22] Actually, Aristotle seems to use two different notions of possibility, but we leave the systematic analysis of the possible fragments to future work.

[23] Actually, they are presented using the second- and third-figure term variables, namely (N, M, X) and (S, P, R), respectively, and are justified by *echtesis*.

4.2.3 Imperfect Purely Apodictic Syllogisms

Derivability proofs in $\mathbf{AP}_1$ are obtained precisely as the corresponding assertoric (ostensive) ones, see Section 3.3. There are four first-figure, four second- and six third-figure purely apodictic syllogisms. For space reasons, we consider just two examples of derivability in $\mathbf{AP}_1$. Second-figure [Festino] can be perfected by Π^-[C] and [Fer] as follows,

$$\dfrac{\dfrac{\Pi^-[N,M]}{\Pi^-[M,N]}\,\Pi^-[\mathtt{C}] \qquad \Sigma^+[X,M]}{\Sigma^-[X,N]}\,[\mathtt{Fer}]$$

Third-figure [Disamis] is shown derivable in $\mathbf{AP}_1$ by Σ^+[C] and [Dar],

$$\dfrac{\dfrac{\Pi^+[S,R] \qquad \dfrac{\Sigma^+[S,P]}{\Sigma^+[P,S]}\,\Sigma^+[\mathtt{C}]}{\Sigma^+[P,R]}\,[\mathtt{Dar}]}{\Sigma^+[R,P]}\,\Sigma^+[\mathtt{C}]$$

First Figure	Second Figure	Third Figure
[Bar]	[Cesare]	[Darapti]
[Cel]	[Festino]	[Felapto]
[Dar]	[Baro]	[Disamis]
[Fer]	[Camestres]	[Datisi]
		[Boc]
		[Ferison]

A proof of syntactical completeness can also be provided by a systematic inspection that shows these 14 syllogisms to represent the *totality* of the purely-apodictic ones.[24] In fact, this is true only when defining a syllogism as a *pair* of productive premisses [34]. Otherwise, three further imperfect moods have to be added, namely, first-figure [Barbari] and [Celaront] and second-figure [Cesaro]. On the other hand, subaltern [Camestros] is not derivable (at least ostensively) in $\mathbf{AP}_1$.[25]

4.3 Mixed Apodictic Syllogistic AP

Then, Aristotle introduces *mixed apodictic* syllogistic. Mixed apodictic syllogisms are made of an assertoric and an apodictic premiss and from them an apodictic conclusion is inferred.

4.3.1 The Mixed Apodictic Language

The mixed apodictic language is obtained by enriching that for **SYL** with apodictic predications.

[24] For further details, see Appendix B.

[25] Indeed, differently from the other subaltern moods, its derivation would require negative subalternation conversion, which is not, in turn, derivable in $\mathbf{AP}_1$. The investigation of (possible) 'apodictic RAA' is left to future study.

Definition 4.5 Let A, B be terms, and Q be a proposition in the assertoric language. Then, the language of mixed apodictic syllogistic $\mathcal{L}_{\mathbf{AP}}$ is as below:

$$P ::= \Pi^{+/-}(A,B) \mid \Pi^{+/-}[A,B] \mid \Sigma^{+/-}(A,B) \mid \Sigma^{+/-}[A,B] \mid\ Q^{\perp}.$$

4.3.2 The Mixed Apodictic System AP

Aristotle presents the first-figure mixed moods accompanying them with explanatory remarks in [4, 30a16–b2].

Definition 4.6 The four *perfect mixed-apodictic syllogisms* are as follows:

$$\frac{\Pi^{+}[B,A] \qquad \Pi^{+}(C,B)}{\Pi^{+}[C,A]}\ \texttt{B[a]rbar[a]} \qquad \frac{\Pi^{-}[B,A] \qquad \Pi^{+}(C,B)}{\Pi^{-}[C,A]}\ \texttt{C[e]lar[e]nt}$$

$$\frac{\Pi^{+}[B,A] \qquad \Sigma^{+}(C,B)}{\Sigma^{+}[C,A]}\ \texttt{D[a]ri[i]} \qquad \frac{\Pi^{-}[B,A] \qquad \Sigma^{+}(C,B)}{\Sigma^{-}[C,A]}\ \texttt{F[e]ri[o]}$$

By adding these rules to the ones of **SYL** and $\mathbf{AP}_1$, the mixed apodictic system **AP** is obtained.

Definition 4.7 The *mixed apodictic system* **AP** is made of the rules of **SYL** plus those of $\mathbf{AP}_1$.

4.3.3 Imperfect Mixed Apodictic Syllogisms

Aristotle shows that there are nine imperfect mixed-apodictic syllogisms: three in the second figure and six in the third.

Mixed-Apodictic Second Figure. The three mixed-apodictic, second-figure moods are `C[e]sar[e]`, `Cam[e]str[e]s` and `F[e]stin[o]`. `C[e]sar[e]` is considered first and shown derivable by Π^{-}`[C]` and mixed-apodictic `C[e]lar[e]nt`:

> For first let the privative[26] be necessary and let it not be possible for A to belong to any B, but let A merely belong to C. Then, since the privative converts, neither is it possible for B to belong to any A. But A belongs to every C; consequently, it is not possible for B to belong to any C, for C is below A. [4, 30b6–14]

$$\cfrac{\cfrac{\Pi^{-}[B,A]}{\Pi^{-}[A,B]}\ \Pi^{-}\texttt{[C]} \qquad \Pi^{+}(C,A)}{\Pi^{-}[C,B]}\ \texttt{C[e]lar[e]nt}$$

Then, `Cam[e]str[e]s` is derived in a similar way:

> And likewise also if the privative is put in relation to C. For if it is not possible for A to belong to any C, then neither will it be possible for C to belong to any A. But A belongs to every B; consequently, it will not be possible for C to belong to any of the Bs, for it becomes the first figure again. Therefore, neither will it be possible for B to belong to C, for it converts similarly. [4, 30b14–9]

$$\cfrac{\cfrac{\cfrac{\Pi^{-}[C,A]}{\Pi^{-}[A,C]}\ \Pi^{-}\texttt{[C]} \qquad \Pi^{+}(B,A)}{\Pi^{-}[B,C]}\ \texttt{C[e]lar[e]nt}}{\Pi^{-}[C,B]}\ \Pi^{-}\texttt{[C]}$$

[26] This corresponds to our notion of negative proposition.

Finally, F[e]stin[o] is introduced and shown perfectable by purely apodictic $\Pi^-[\mathtt{C}]$ and mixed apodictic F[e]r[i]o:

> The situation will also be similar in the case of the particular decisions. For when the privative premise is both universal and necessary. First, then, let the privative premise be both universal and necessary, and let it not be possible for A to belong to any B, but let A belong to some C. Then, since the privative converts, neither would it be possible for B to belong to any A. But A belongs to some one of the Cs, so that B will of necessity not belong to some one of the Cs. [4, 31a1–11]

$$\dfrac{\dfrac{\Pi^-[B,A]}{\Pi^-[A,B]}\,\Pi^-[\mathtt{C}] \qquad \Sigma^+(C,A)}{\Sigma^-[C,B]}\,\mathtt{F[e]ri[o]}$$

There is no mixed-apodictic syllogism corresponding to Baroco [4, 31a11–16].

Mixed-Apodictic Third Figure. Aristotle presents completion proofs for third-figure D[a]rapt[i], Dar[a]pt[i], F[e]lapt[o]n, D[a]tis[i], Dis[a]m[i]s and F[e]ris[o]n. He starts with D[a]rapt[i], which is derived due to $\Pi^-\mathtt{C}$ and perfect D[a]ri[i]:

> For let both premises first be positive, and let both A and B belongs to every C, but let AC be necessary. Then, since B belongs to every C, C will also belong to some B because the universal converts into a particular. Consequently, if A belongs to every C of necessity, and C belongs to some B, then it is also necessary for A to belong to some B (for B is under C). This first figure therefore comes about. [4, 31a24-31]

$$\dfrac{\Pi^+[C,A] \qquad \dfrac{\Pi^+(C,B)}{\Sigma^+(B,C)}\,\Pi^+\mathtt{C}}{\Sigma^+[B,A]}\,\mathtt{D[a]ri[i]}$$

Dar[a]pt[i] is shown derivable in a similar way, but also using apodictic $\Sigma^+[\mathtt{C}]$:

> And it will also be proved in the same way if BC is necessary. For C converts to some A; consequently, if B belongs to every C of necessity, then it will also belong to some A of necessity. [4, 31a31–35]

$$\dfrac{\dfrac{\Pi^+[C,B] \qquad \dfrac{\Pi^+(C,A)}{\Sigma^+(A,C)}\,\Pi^+\mathtt{C}}{\Sigma^+[A,B]}\,\mathtt{D[a]ri[i]}}{\Sigma^+[B,A]}\,\Sigma^+[\mathtt{C}]$$

Then, F[e]lapt[o]n is perfected relying on $\Pi^+\mathtt{C}$ and mixed-apodictic F[e]ri[o]:

> Next, let AC be privative, BC affirmative, and the privative necessary. Then, since C converts to some B and A belongs of necessity to no C, A will also of necessity not belong to some B: for B is below C. [4, 31a35–38]

$$\dfrac{\Pi^-[C,A] \qquad \dfrac{\Pi^+(C,B)}{\Sigma^+(B,C)}\,\Pi^+\mathtt{C}}{\Sigma^-[B,A]}\,\mathtt{F[e]ri[o]}$$

Subsequently, mixed-apodictic D[a]tis[i] is proved derivable in $\mathbf{AP}_1$ due to $\Sigma^+\mathtt{C}$ and mixed-apodictic D[a]ri[i]:

> But if one term is universal, the other is particular, and both are positive, then the conclusion will be necessary whenever the universal is necessary. The demonstration is the same as the previous one, for the positive particular also converts: Thus if it is necessary for B to belong to every C and A is below C, then it is necessary for B to belong to some A. [4, 31b12–17]

$$\dfrac{\Pi^+[C,B] \qquad \dfrac{\Sigma^+(C,A)}{\Sigma^+(A,C)}\,\Sigma^+\mathtt{C}}{\Sigma^+[A,B]}\,\mathtt{D[a]ri[i]}$$

Next, `Dis[a]m[i]s` is perfected in a similar way, by using also $\Sigma^+[\mathtt{C}]$:

> But if it is necessary for B to belong to some A, then it is also necessary for A to belong to some B (for it converts). [4, 31b17-20]

$$\dfrac{\dfrac{\Pi^+[C,B] \qquad \dfrac{\Sigma^+(C,A)}{\Sigma^+(A,C)}\,\Sigma^+\mathtt{C}}{\Sigma^+[A,B]}\,\mathtt{D[a]ri[i]}}{\Sigma^+[B,A]}\,\Sigma^+[\mathtt{C}]$$

Finally, `F[e]ris[o]n` is proved perfectable via $\Sigma^+\mathtt{C}$ and perfect `F[e]ri[o]`:

> But if one of the terms is positive and the other privative, then when the universal is both privative and necessary, the conclusion will also be necessary. For if it is not possible for A to belong to any C and B belongs to some C, then it is necessary for A not to belong to some B. [4, 31b33–36]

$$\dfrac{\Pi^-[C,A] \qquad \dfrac{\Sigma^+(C,B)}{\Sigma^+(B,C)}\,\Sigma^+\mathtt{C}}{\Sigma^-[B,A]}\,\mathtt{F[e]ri[o]}$$

First Figure	Second Figure	Third Figure
B[a]rbar[a]	C[e]sar[e]	D[a]rapt[i]
		Dar[a]pt[i]
C[e]lar[e]nt	Cam[e]str[e]s	F[e]lapt[o]n
D[a]ri[i]	F[e]stin[o]	Dis[a]m[i]s
F[e]ri[o]		D[a]tis[i]
		F[e]ris[o]n

5 Related Works

Reconstructions of Aristotle's logic via modern tools date back at least to the 1950s [19], whereas first interpretations of syllogistic as a natural deduction (not axiomatic) systems are due to Corcoran's [9,10] and Smiley's [37] 1973 works. [27] Other proposals following this line were then presented, for instance, in [21] and in [41]. Yet, none of these proof systems is made of rules in arboreal form. A turning point is instead represented by [47,48], in which syllogistic derivations

[27] In particular, Corcoran introduced a perfecting system where *reductio* is not an inference rule and, as for **SYL**, positive *conversio simplex* and perfect *Darii* and *Ferio* correspond to primitive rules. On the contrary, as for **SYL***, Smiley does not consider these three rules (but the contradictory symbol is treated as a meta-linguistic one).

are arranged as trees.[28] Another related, original reconstruction was developed by Dyckhoff in [12], as based on the notion of inverses of syllogisms.[29] The present work is strongly inspired by [47]. In fact, Section 3 can be seen as a systematic application of the second author's approach to *An.Pr.* I, 1-6, and Section 4 as its generalization to the apodictic fragment.

On the other hand, investigations of Aristotle's modal syllogistic started to appear already in the 1960s [22]. Later on, different reconstructive proposals for this system have been presented in the literature, sometimes even in a formal setting or based on contemporary logical tools [15,44,32,30,42,20,35]. Differently from our study, most of these analyses are somehow 'semantically-oriented' and do not provide a plain reconstruction, to be directly checked by the reader.

Notably, all the mentioned works are historically-driven, that is, aim at reconstructing Aristotle's original system. In addition to that, we hope that our study could point at new (modal) research directions in the framework of Natural Logic. Indeed, one of the goals of this discipline is to define systems which, differently from FOL, are tailored on human reasoning. From the 1970s on, it has led to the revival of syllogistic logic,[30] and, more recently, implementations and applications of (extended) syllogistic systems in AI and NLP have also been provided [28,29,27]. Yet, to the best of our knowledge, no apodictic natural logic system has been developed so far.

6 Conclusion

This paper aims at 'rehabilitating' syllogistic logic by challenging the widespread idea of its "lack of sophistication" [18, p. ix] and at showing its potential appeal for contemporary logic. In particular, we have systematically applied to assertoric and extended to apodictic syllogistic the proof-theoretical approach by [47]. When treating syllogistic as a natural-deduction rule system and presenting perfecting derivations in tree form, it is strikingly straightforward to see that they are all correct, and that the apodictic syllogistic is not a "realm of darkness" [35, p. 1] anymore. In the future, we wish to complete our investigation of Aristotle's modal syllogistic by taking the contingent and mixed apodictic-contingent fragments into account. We hope that, as a byproduct, our reconstruction would also shed new light on the semantics associated with Aristotle's modal logic. In addition, we plan to start a comprehensive study of the proof-theoretical properties of these new syllogistic rule systems.

---

[28] In fact, a pioneering use of tree form to investigate Aristotle's perfection proofs already appears in two examples by [17].

[29] Slightly different reconstructions are those focussing on *ecthesis*, see e.g. in [38,16].

[30] The history of Natural Logic can be divided into three main phases: in the 1970s, a renewed interest emerges for term logic; in the 1980s-90s, the discipline acquires a defined status, and first foundational works are written [45,36]; more recently (non-)monotonic aspects of extended syllogistic and related implementations or applications in AI have also been investigated [25,14].

References

[1] Antonelli, M. and J. von Plato, *Nested indirect proofs in Aristotle's deductive logic* (sumbitted).

[2] Aristotle, "De Interpretatione," Princeton University Press, 1984.

[3] Aristotle, *Mataphysics*, in: P. U. Press, editor, *The complete work of Aristotle: The revised Oxford translation*, W.D. Ross and J. Barnes, 1984 pp. 1552–1728.

[4] Aristotle, "Prior Analytics," Hackett Publishing Company, 1989.

[5] Barnes, J., *Proof and the syllogism*, in: *Aristotle on Sciences: The Posterior Analytics*, Berti, E., 1981 pp. 97–139.

[6] Bobzien, S., "Stoic syllogistic," Clarendon Press, 1996.

[7] Bobzien, S., *Pre-Stoic hypothetical syllogistic in Galen's* Institution Logica, Bulletin of the Institute of Classical Studies **77** (2002), pp. 57–72.

[8] Cavini, W., *Sillogismo aristotelico e teoria logica*, Nuova Civiltá delle Macchine **9** (1991), pp. 29–36.

[9] Corcoran, J., *Completeness of an ancient logic*, The Journal of Symbolic Logic **37** (1973), pp. 696–702.

[10] Corcoran, J., *A mathematical model of Aristotle's syllogistic*, Archiv für Geschichte der Philosophie **55** (1973), pp. 191–219.

[11] Corcoran, J., *Aristotle's prototype rule-based underlying logic*, Logical Universalis **12** (2018), pp. 9–35.

[12] Dyckhoff, R., *Indirect proof and inversions of syllogisms*, Bulletin of Symbolic Logic **25** (2019), pp. 196–207.

[13] Ebbesen, S., *The traditions of Ancient Logic-cum-grammar in the Middle Ages - What's the problem?*, Vivarium **45** (2007), pp. 136–152.

[14] Hu, H., Q. Chen and L. Moss, *Natural language inference with monotonicity*, in: S. Dobnik, S. Chatzikyriakidis and V. Demberg, editors, *Proc. 13th International Conference of Computational Semantics*, 2019, pp. 8–15.

[15] Johnson, F., *Models for modal syllogism*, Notre Dame Journal of Formal Logic **30** (1989), pp. 271–255.

[16] Joray, P., *The principle of contradiction and* echtetis *in Aristotle's syllogistic*, History and Philosophy of Logic **35** (2014), pp. 219–236.

[17] Kneale, W. and M. Kneale, "The Development of Logic," Clarendon Press, 1962.

[18] Lear, J., "Aristotle and Logical Theory," Cambridge University Press, 1980.

[19] Lukasiewicz, J., "Aristotle's Syllogistic Form the Standpoint of Modern Formal Logic," Clarendon Press, 1957.

[20] Malink, M., *A reconstruction of Aristotle's modal logic*, History and Philosophy of Logic **27** (2006), pp. 95–141.

[21] Martin, J. M., *Aristotle's natural deduction reconsidered*, History and Philosophy of Logic **18** (1997), pp. 1–15.

[22] McCall, S., "Aristotle's Modal Syllogisms," North-Holland, 1963.

[23] Mignucci, M., *Parts, quantification, and Aristotle*, The Monist **83** (2000), pp. 3–21.

[24] Morison, B., *Aristotle, etc.*, Phronesis **53** (2008), pp. 209–222.

[25] Moss, L., *Completeness theorems for syllogistic fragments*, in: F. Hamm and S. Kepser, editors, *Logics for Linguistic Structures*, de Gruyter, 2008 pp. 143–173.

[26] Moss, L., *Natural logic and semantics*, in: B. H. J. T. Aloni, M. and K. Schulz, editors, *Proc. Logic, Language and Meaning, 17th Amsterdam Colloqium*, 2009, pp. 84–93.

[27] Moss, L., *Implementations of natural logics*, in: C. Benzmüller and J. Otten, editors, *ARQNL*, 2018, pp. 1–10.

[28] Moss, L., J. Hemann and C. Swords, *Two advances in the implementation of syllogistic logic*, in: Balducci et al., editors, *Proc. NLPAR '15*, 2015.

[29] Moss, L. and M. Wollowski, *Natural logic in AI and cognitive science*, in: *Proc. Midwest Artificial Intelligence and Cognitive Science (MAICS '17)*, 2017, pp. 41–46.

[30] Nortmann, U., *The logic of necessity in Aristotle – an outline of approaches to the modal syllogistic, together with a general account of* de dicto *and* de re *necessity*, History and Philosophy of Logic **23** (2002), pp. 253–265.

[31] Owen, G., *Aristotle on time*, in: P. Machamer and R. Turnbull, editors, *Motion and time, space and matter: Interrelations in the history of philosphy of science*, University Press, 1976 pp. 3–27.

[32] Patterson, R., *Aristotle's modal logic. essence and entailment in the Organon*, Cambridge University Press (1995).

[33] Pickard-Cambridge, W., "Topoi," Princeton University Press, 1984.

[34] Read, S., *Aristotle's theory of the assertoric syllogism* (2017), `https://www.st-andrews.ac.uk/~slr/The_Syllogism.pdf`.

[35] Rini, A., "Aristotle's Modal Logic: *Prior Analytics* A8-22 in Predicate Logic," Springer, 2011.

[36] Sánchez Valencia, V., "Studies on Natural Logic and Categorial Grammar," Ph.D. thesis, University of Amsterdam (1991).

[37] Smiley, T., *What is a syllogism?*, Journal of Philosophical Logic **2** (1973), pp. 136–154.

[38] Smith, R., *What is Aristotelian ecthesis?*, History and Philosophy of Logic **3** (1982), pp. 113–127.

[39] Smith, R., *Logic*, in: *The Cambridge Companion to Aristotle*, 1995 pp. 27–65.

[40] Speca, A., "Hypotetical Syllogistic and Stoic Logic," Brill, 2001.

[41] Thom, P., "The Syllogism," Philosophia, 1981.

[42] Thom, P., "The logic of essentialism. An interpretation of Aristotle's modal syllogistic," Kluwer, 1996.

[43] Thom, P., *The syllogism and its transformations*, in: C. Novaes and S. Read, editors, *The Cambridge Companion to Medieval Logic*, Cambridge University Press, 2016 .

[44] Thomason, S., *Semantic analysis of the modal syllogistic*, Journal of Philosophical Logic **22** (1993), pp. 111–128.

[45] van Eijck, J., *Generalized quantifiers in traditional logic*, in: A. Ter Maulen and J. van Benthem, editors, *Generalized Quantifiers in Natural Language*, De Gruyter, 1985 .

[46] Vlastis, J., *Mereology in Aristotle's assertoric syllogistic*, History and Philosophy of Logic **40** (2018), pp. 1–11.

[47] von Plato, J., *Aristotle's deductive logic: a proof-theoretical study*, in: *Concepts of Proof in Mathematics, Philosophy, and Computer Science*, de Gruyter, 2016 .

[48] von Plato, J., "The Great Formal Machinery Works," Princeton University Press, 2017.

[49] Whitaker, C., "Aristotle's *De Interpretatione*: Contradiction and Dialectic," Oxford Scholarship, 2002.

Appendix

A The system SYL*

The language of **SYL*** is the same as that of **SYL** (Definition 3.1).
The derivability of $\Sigma^+\texttt{C}$ can be obtained by $\texttt{RCP}_{1,3}$ together with $\bot\texttt{I}$ and $\texttt{RAA}$:

$$\dfrac{\dfrac{\Sigma^+(A,B) \qquad \dfrac{\dfrac{\dfrac{[(\Sigma^+(B,A))^\neg]^1}{\Pi^-(B,A)}\,\texttt{RCP}_3}{\Pi^-(A,B)}\,\Pi^-\texttt{C}}{(\Sigma^+(A,B))^\neg}\,\texttt{RCP}_1}{\bot}\,\bot\texttt{I}}{\Sigma^+(B,A)}\,\texttt{RAA},1$$

First-figure `Darii` and `Ferio` are also derivable in **SYL***. Aristotle himself proves this after concluding his presentation of assertoric syllogistic [4, 29b6–9]. Then, in [4, 29b9-11], `Darii` is shown to be derivable by `RAA`, second-figure `Camestres`, and $\bot\texttt{I}$:

$$\dfrac{\dfrac{\Sigma^+(C,B) \qquad \dfrac{\dfrac{\Pi^+(B,A) \qquad \dfrac{[(\Sigma^+(C,A))^\neg]^1}{\Pi^-(C,A)}\,\texttt{RCP}_3}{\Pi^-(C,B)}\,\texttt{Camestres}}{(\Sigma^+(C,B))^\neg}\,\texttt{RCP}_1}{\bot}\,\bot\texttt{I}}{\Sigma^+(C,A)}\,\texttt{RAA},1$$

Since `Camestres` is derivable from `Celarent`, `Darii` comes out to be derivable in primitive **SYL*** as well:

$$\dfrac{\dfrac{\Sigma^+(C,B) \qquad \dfrac{\dfrac{\dfrac{\dfrac{\dfrac{[(\Sigma^+(C,A))^\neg]^1}{\Pi^-(C,A)}\,\texttt{RCP}_3}{\Pi^-(A,C)}\,\Pi^-\texttt{C} \qquad \Pi^+(B,A)}{\Pi^-(B,C)}\,\texttt{Celarent}}{\Pi^-(C,B)}\,\Pi^-\texttt{C}}{(\Sigma^+(C,B))^\neg}\,\texttt{RCP}_1}{\bot}\,\bot\texttt{I}}{\Sigma^+(C,A)}\,\texttt{RAA},1$$

On the other hand, `Ferio` is proved to be perfectable relying on second-figure `Cesare` [4, 29b11-15]:[31]

$$\dfrac{\dfrac{\Sigma^+(C,B) \qquad \dfrac{\dfrac{\Pi^-(B,A) \qquad \dfrac{[(\Sigma^-(C,A))^\neg]^1}{\Pi^+(C,A)}\,\texttt{RCP}_4}{\Pi^-(C,B)}\,\texttt{Cesare}}{(\Sigma^+(C,B))^\neg}\,\texttt{RCP}_1}{\bot}\,\bot\texttt{I}}{\Sigma^-(C,A)}\,\texttt{RAA},1$$

[31] Actually, in the original formulation, this proof is quite elliptical.

Given that `Cesare` is derivable in $\mathbf{SYL}^*$ via `Celarent`, also `Ferio` comes out as provable in $\mathbf{SYL}^*$.

Therefore, as noticed by Aristotle in [4, 29b15-20], the basic assertoric system can alternatively (but equally) be defined using primitive `Barbara` and `Celarent` only.

B Syntactical Completeness (Proof Sketch)

The proof of syntactical completeness for $\mathbf{AP}_1$ and $\mathbf{AP}$ is established by systematically inspecting all possible cases. For space reasons and since proofs are tedious but not complicated, only a few cases are considered here. Overall, it emerges that, except for the subaltern moods, the set of syllogisms presented by Aristotle represents the totality of syllogisms derivable in $\mathbf{AP}_1$ and $\mathbf{AP}$ systems (which, as seen, are in turn supposed to actually represent Aristotle's primitive rule systems). Remarkably, subaltern moods are proper moods only when syllogisms are defined by their sets of premisess *together with their conclusion.* On the other hand, when, following [34], syllogisms are defined as pairs of *productive* premisses, a subaltern syllogism is nothing but a (somehow weaker) version of its superaltern, corresponding 'mood'; indeed, for instance, [`Barbara`] and its subaltern [`Barbari`] are actually the same syllogism, i.e. a (perfect) first-figure, purely apodictic syllogism defined by productive premisses of the form $\Pi^+[B, A]$ and $\Pi^+[C, B]$.

B.1 Syntactical Completeness of $\mathbf{AP}_1$

We start by considering the only detour conversions involved, namely that for $\Pi^-[\mathtt{C}]$:

$$\dfrac{\dfrac{\begin{matrix}\vdots\\ \Pi^-[P,S]\end{matrix}}{\Pi^-[S,P]}\;\Pi^-[\mathtt{C}]}{\Pi^-[P,S]}\;\Pi^-[\mathtt{C}] \qquad \rightsquigarrow \qquad \begin{matrix}\vdots\\ \Pi^-[S,P]\end{matrix}$$

and the one for $\Sigma^+[\mathtt{C}]$:

$$\dfrac{\dfrac{\begin{matrix}\vdots\\ \Sigma^+[P,S]\end{matrix}}{\Sigma^+[S,P]}\;\Sigma^+[\mathtt{C}]}{\Sigma^+[P,S]}\;\Sigma^+[\mathtt{C}] \qquad \rightsquigarrow \qquad \begin{matrix}\vdots\\ \Sigma^+[S,P]\end{matrix}$$

Then, syntactical completeness of $\mathbf{AP}_1$ is established by tedious, but systematical, inspection of all possible combinations of premisses. The whole proof crucially relies on the following three propositions.

Proposition B.1 *In* $\mathbf{AP}_1$*, no syllogistic conclusion follows from two particular premisses.*

Proof Sketch. $\mathbf{AP}_1$ contains no primitive rule with two particular premisses nor a conversion rule to derive a universal proposition from a primitive one. □

Proposition B.2 *In* $\mathbf{AP}_1$*, no syllogistic conclusion follows from two negative premisses.*

Proof Sketch. $\mathbf{AP}_1$ contains no primitive rule with two negative premisses, nor a conversion rule to derive a positive proposition from a negative one. □

Proposition B.3 *If the major premiss of a first-figure pair is particular, then the couple does not syllogize.*

Proof Sketch. By straightforward inspection. □

Theorem B.4 *The totality of purely apodictic syllogisms (primitive or) derivable in* $\mathbf{AP}_1$ *is presented in Figure B.1.*

Proof Sketch. For space reason, we explicitly consider the first-figure case only. We start noticing that by Proposition B.1, the pairs $(\Sigma^+[B,A],\Sigma^+[C,B])$, $(\Sigma^+[B,A],\Sigma^-[C,B])$, $(\Sigma^-[B,A],\Sigma^+[C,B])$ and $(\Sigma^-[B,A],\Sigma^-[C,B])$ do not syllogize. Similarly, by Proposition B.2, the pairs $(\Pi^-[B,A],\Pi^-[C,B])$, $(\Pi^-[B,A],\Sigma^-[C,B])$, $(\Sigma^-[B,A],\Pi^-[C,B])$ and, as seen, $(\Sigma^-[B,A],\Sigma^-[C,B])$ are not syllogisms.

Let us consider positive premisses. First of all, we have already seen that the pairs $(\Pi^+[B,A],\Pi^+[C,B])$ and $(\Pi^+[B,A],\Sigma^+[C,B])$ do syllogize.[32] On the other hand, no syllogism is derivable from the pair $(\Sigma^+[B,A],\Pi^+[C,B])$ by Proposition B.3. As said, $(\Sigma^+[B,A],\Sigma^+[C,B])$ cannot syllogize (Prop. B.1).

We can now consider positive and negative premisses together. Again, the pairs $(\Pi^-[B,A],\Pi^+[C,B])$ and $(\Pi^-[B,A],\Sigma^+[C,B])$ do syllogize as, respectively, [Cel] and [Fer] are primitive rule.[33] On the other hand, by Prop. B.3, $(\Sigma^-[B,A],\Pi^+[C,B])$ and $(\Sigma^-[B,A],\Sigma^-[C,B])$ do not conclude syllogistically. For the same reason, the pairs $(\Sigma^+[B,A],\Pi^-[C,B])$ and $(\Sigma^+[B,A],\Sigma^-[C,B])$ do not lead to syllogistic conclusions. That the pairs $(\Pi^+[B,A],\Pi^-[C,B])$ and $(\Pi^+[B,A],\Sigma^-[C,B])$ do not syllogize is proved by inspection.[34] Since, as seen, a negative pair of premisses never syllogizes (Prop. B.2), this concludes our proof for the first figure.

---

[32] Indeed, [Bar] and [Fer] are primitive rule. Also subaltern [Barbari] can be shown derivable in $\mathbf{AP}_1$. On the contrary, it can be easily proved that no negative conclusion follows from positive premisses so pseudo-moods [Barbare], [Barbaro], [Darie] and [Dario] are not syllogisms. Also [Daria] is not a syllogism as it can be shown by inspection that from $(\Pi^+[B,A],\Sigma^+[C,B])$ no universal conclusion follows.

[33] The former pair would possibly lead also to a particular negative conclusion as subaltern [Celaront] is derivable in $\mathbf{AP}_1$. It can also be easily proved by inspection that from $\Pi^-[B,A]$ and $\Sigma^+[C,B]$ no universal conclusion follows.

[34] About the former, (i) there is no primitive, first-figure syllogism with the given premisses, (ii) converting the minor premiss only (via Π^-[C]) would lead to $(\Pi^+[B,A],\Pi^-[B,C])$, but there is no third-figure syllogism in this form, (iii) converting the major premiss only (via Π^+[C]) would lead to $(\Sigma^+[A,B],\Pi^-[C,B])$, and there is no second-figure syllogism in this form; if converting it again via Σ^+[C], no conclusion can follow by Prop. B.1 (no other non-detour conversion is possible), (iv) by converting both premisses, the pair $(\Pi^-[B,C],\Sigma^+[A,B])$ is obtained, which can be easily shown not to give any conclusion with subject C and predicate A. Concerning the latter, (i) it is clear that there is no primitive syllogism with these premisses, (ii) the major premiss converts only via Π^+[C], leading to a particular proposition and, (iii) by Prop. B.1, no syllogistic conclusion follows from two particular premisses.

First Figure	Second Figure	Third Figure
[Barbara]		
[Barbari]		[Darapti]
	[Camestres]	
		[Disamis]
[Celarent]	[Cesare]	
[Celaront]	[Cesaro]	[Felapton]
		[Bocardo]
[Darii]		[Datisi]
	[Baroco]	
[Ferio]	[Festino]	[Ferison]

Fig. B.1. Primitive and Derivable Purely Apodictic Syllogisms

From the systematic analysis of all the three syllogistic figures, it emerges that precisely the syllogisms in Figure B.1 are those derivable in $\mathbf{AP}_1$. In particular, the derivability of subaltern [Barbari] is established by perfect [Bar] and apodictic conversions Π^+[C] and Σ^+[C]:

$$\dfrac{\dfrac{\dfrac{\Pi^+[B,A] \qquad \Pi^+[C,B]}{\Pi^+[C,A]}\,[\mathtt{Bar}]}{\Sigma^+[A,C]}\,\Pi^+[\mathtt{C}]}{\Sigma^+[C,A]}\,\Sigma^+[\mathtt{C}]$$

First-figure subaltern [Celaront] is derived in $\mathbf{AP}_1$ via perfect [Fer] and apodictic conversions Π^+[C] and Σ^+[C]:

$$\dfrac{\Pi^-[B,A] \qquad \dfrac{\dfrac{\Pi^+[C,B]}{\Sigma^+[B,C]}\,\Pi^+[\mathtt{C}]}{\Sigma^+[C,B]}\,\Sigma^+[\mathtt{C}]}{\Sigma^-[C,A]}\,[\mathtt{Fer}]$$

Finally, second-figure [Cesaro] can be perfected by means of perfect [Fer] and all apodictic conversion rules:

$$\dfrac{\dfrac{\Pi^-[A,B]}{\Pi^-[B,A]}\,\Pi^-[\mathtt{C}] \qquad \dfrac{\dfrac{\Pi^+[C,B]}{\Sigma^+[B,C]}\,\Pi^+[\mathtt{C}]}{\Sigma^+[C,B]}\,\Sigma^+[\mathtt{C}]}{\Sigma^-[C,A]}\,[\mathtt{Fer}]$$

□

B.2 Syntactical Completeness of AP

By a similar systematical inspection it emerges that, as sketched by Aristotle himself, there are 4 (perfect) first-figure 3 (imperfect) second-figure and 6 (imperfect) third-figure, mixed apodictic syllogisms. When defining syllogisms in terms of a pair of premisses *together with* their conclusion, two subaltern first-figure syllogisms and one subaltern second-figure syllogism have to be added.

First Figure	Second Figure	Third Figure
B[a]rbar[a]		
B[a]rbar[i]		D[a]rapt[i]
		Dar[a]pt[i]
	Cam[e]str[e]s	
C[e]lar[e]nt	C[e]sar[e]	F[e]lapt[o]n
C[e]lar[o]nt	C[e]sar[o]	
D[a]ri[i]		D[a]tis[i]
		Dis[a]m[i]s
F[e]ri[o]	F[e]stin[o]	
		F[e]ris[o]n

Fig. B.2. Primitive and Derivable Mixed Apodictic Syllogisms

In particular, subaltern B[a]rbar[i] is derived in **AP** as follows:

$$\dfrac{\dfrac{\dfrac{\Pi^+[B,A] \qquad \Pi^+(C,B)}{\Pi^+[C,A]}\ \texttt{B[a]rbar[a]}}{\Sigma^+[A,C]}\ \Pi^+[\texttt{C}]}{\Sigma^+[C,A]}\ \Sigma^+[\texttt{C}]$$

Perfection proof for subaltern C[e]lar[e]nt is below:

$$\dfrac{\Pi^-[B,A] \qquad \dfrac{\dfrac{\Pi^+(C,B)}{\Sigma^+(B,C)}\ \Pi^+\texttt{C}}{\Sigma^+(C,B)}\ \Sigma^+\texttt{C}}{\Sigma^-[C,A]}\ \texttt{F[e]ri[o]}$$

Finally, second-figure, subaltern C[e]sar[o] is shown to be derivable in **AP** as:

$$\dfrac{\dfrac{\dfrac{\Pi^-[C,B]}{\Pi^-[B,C]}\ \Pi^-[\texttt{C}] \qquad \Pi^+(A,B)}{\Pi^-[A,C]}\ \texttt{C[e]lar[e]nt}}{\Pi^-[C,A]}\ \Pi^-[\texttt{C}]$$

Towards Dynamic Distributed Knowledge

Philippe Balbiani and Hans van Ditmarsch

Institut de Recherche en Informatique de Toulouse
CNRS - INPT - UT3, France

Abstract

We propose a novel notion of distributed knowledge called *dynamic distributed knowledge* that corresponds to what a group of agents know after they share their knowledge. Its interpretation in Kripke models therefore combines static (modal accessibility) with dynamic (update) aspects. In prior work the static and dynamic aspects were also investigated, but separately, where the latter was called *resolution*. Unlike the usual distributed knowledge, in this work called *static distributed knowledge*, if a group of agents has dynamic distributed knowledge of a proposition, then it has dynamic distributed knowledge that the proposition is common knowledge. We report on its expressivity, axiomatization, and bisimulation characterization.

Keywords: modal logic, distributed knowledge, axiomatization, dynamics

1 Introduction and survey of related work

Distributed knowledge. Distributed knowledge is a well-known group notion of knowledge [21,22,27,19,29,20,14,31,34]. If agent a knows that p implies q and agent b knows that p, then agents a and b have distributed knowledge of q. Neither agent knows q individually. They have to share their knowledge.

> Distributed knowledge of a proposition for a group of agents is true in a world of a given Kripke model, if *in that model* it is true in all worlds accessible by the relation that is the intersection of the equivalence relations for all agents in the group.

Observe that the proposition is then interpreted in the given model. What if we define distributed knowledge dynamically?

> Distributed knowledge of a proposition for a group of agents is true in a world of a given Kripke model, if *in the model wherein the relation for each agent in the group is replaced by the intersection of the equivalence relations for all agents in the group*, it is true in all worlds accessible by that relation.

So we now interpret the proposition in the updated model. Let us call this novel notion *dynamic distributed knowledge*, whereas the standard notion is henceforth called *static distributed knowledge*. Dynamic distributed knowledge for a singleton group of agents is, as for static distributed knowledge, the same as individual knowledge for that agent. But we will see that for two or more agents it is different from static distributed knowledge.

With the novel notion we avoid some standard issues when sharing static distributed knowledge. For example, when a and b have static distributed knowledge that b does not know that a knows p, after sharing their information b knows that a knows p. The act of sharing (the update) may change the truth value of propositions, such as is here the case. With the novel notion we can say that in this case a and b have dynamic distributed knowledge that b knows that a knows p. This is then even common knowledge between them.

We lose some of the properties of static distributed knowledge. Static distributed knowledge of a proposition implies that it is true. But dynamic distributed knowledge does not imply that it is true. The above is an example, a and b have dynamic distributed knowledge that b knows that a knows p, but where ('right now') b does not know that a knows p. Other properties are as for static distributed knowledge, such as positive and negative introspection.

Our results. Consider a logical language where apart from the boolean connnectives the only modalities are those for dynamic distributed knowledge. For this logic of dynamic distributed knowledge we provide a modal logical semantics, determine various validities of interest and translations relating it to static distributed knowledge, we show its decidability, and propose a notion of shared bisimulation for which we show the Hennessy-Milner property. We give a complete axiomatization for an extension of the language. The complete axiomatization of the logic of dynamic distributed knowledge and the expressivity with respect to static distributed knowledge are left for further research.

Related work. Roots of the notion of distributed knowledge are found in sociology, economics, and philosophy [21,22,29]. In the epistemic logical literature an early source for static distributed knowledge is [19], wherein the notion was called *implicit knowledge*. A formal semantics or proof theory was not given. The later journal version [20] gives a standard Kripke model semantics but not an axiomatization, and now calls the notion *distributed knowledge*. Intriguingly, the slightly earlier publication [27] gives an axiomatization however without any claim of completeness. Complete axiomatizations are then given in [14,31], where the latter continue to call the notion implicit knowledge. The axiomatization proposed in [27] is the one proved to be complete in [14,31]. For a review, see also [34].

Different dynamic epistemic logics have been proposed for sharing distributed knowledge [1,8,11]. These works have in common that static dynamic knowledge operators are distinguished from dynamic modalities for sharing distributed knowledge. The dynamic distributed knowledge $\mathbb{D}_B\varphi$ that we propose in this work is definable in (all of) [1,8,11], for example, as $R_B D_B \varphi$ in [1], where *resolution* R_B means sharing distributed knowledge and where D_B is (static) distributed knowledge of [14] and related. (Resolution as resolving distributed knowledge is unrelated to resolution in logic programming.)

Instead of distributed knowledge $\mathbb{D}_B$ interpreted by the relation $\sim_B$ that is the intersection of the relations $\sim_a$ for $a \in B$, one can also consider a notion of distributed knowedge where $\sim_B$ may be contained in that intersection; in general, where $\sim_B \subseteq \sim_C$ for any $C \subseteq B \subseteq A$. In completeness proofs [14,1] this features for technical convenience. However, it is by now also an intuitive notion of group epistemics, where the group knows more than the sum of its individual parts, for example in [7], as *correlated knowledge*, and in [17].

Notions of distributed knowledge for weaker frame classes than S5 are proposed in [13,2]. Dynamic distributed knowledge that we propose is based on ideas in [13], where another considered base logic is K4, as in [2]. In [17] distributed knowledge is investigated for KB4.

The proof theory of distributed knowledge has been investigated in [18].

Recent publications involving distributed knowledge and its dynamics are [15,9,37,25].

Overview Section 2 presents the logical language and semantics and Section 3 focusses on its validities. Section 4 proposes shared bisimulation and compares it to other notions of bisimulation. Section 5 gives two different embeddings of dynamic distributed knowledge in static distributed knowledge. Section 6 and Section 7 prepare the ground for a complete axiomatization of the language with dynamic distributed knowledge and resolution, and then show its completeness. Elementary proofs are omitted.

2 The logic of dynamic distributed knowledge

Let a finite nonempty set A of agents and a countable set P of propositional variables (atoms) be given.

Language. The language $\mathcal{L}_{\mathbb{D}DR}$ is defined by the BNF

$$\varphi ::= p \mid \top \mid (\varphi \wedge \varphi) \mid \neg\varphi \mid \mathbb{D}_B\varphi \mid D_B\varphi \mid R_B\varphi$$

where $p \in P$ and $B \subseteq A$. We follow the standard rules for omission of the parentheses. The sublanguage $\mathcal{L}_{\mathbb{D}}$ with only modalities $\mathbb{D}_B\varphi$ is the *language of dynamic distributed knowledge.* The sublanguage $\mathcal{L}_D$ with only modalities $D_B\varphi$ is the *language of static distributed knowledge.* Notice that $\mathbb{D}_\emptyset$, $D_\emptyset$ and $R_\emptyset$ are modalities of $\mathcal{L}_{\mathbb{D}DR}$. For all $B \subseteq A$, let $|B|$ be the cardinality of B.

Other propositional and modal connectives are defined by abbreviation. In particular, the dual $\widehat{\mathbb{D}}_B\varphi$ is defined as $\neg\mathbb{D}_B\neg\varphi$, and the dual $\widehat{D}_B\varphi$ is defined as $\neg D_B\neg\varphi$. For $\mathbb{D}_{\{a_1,\dots,a_n\}}$ we often write $\mathbb{D}_{a_1\dots a_n}$, and similarly for D. Modality $\mathbb{D}_B$ is *dynamic distributed knowledge* for B, and formula $\mathbb{D}_B\varphi$ is read as 'the agents in group B have dynamic distributed knowledge of φ'. This notion is novel. The typography of $\mathbb{D}$ is chosen to evoke two (here superimposed) letters D, as in *D*ynamic *D*istributed, which may help the reader to distinguish it from the notation for static distributed knowledge. Modality D_B is *static distributed knowledge* for B. This is the notion of distributed knowledge known from the literature. Modality R_B is known as *resolution*, explained later. For $\mathbb{D}_a\varphi$ (i.e., $\mathbb{D}_{\{a\}}\varphi$) or $D_a\varphi$ we write $K_a\varphi$, and for $\widehat{\mathbb{D}}_a\varphi$ and $\widehat{D}_a\varphi$ we write $\widehat{K}_a\varphi$. The fragment of $\mathcal{L}_{\mathbb{D}}$ with only modalities $\mathbb{D}_\emptyset$ (of which the semantics are that of the universal modality) and K_a for $a \in A$ is called $\mathcal{L}_{KU}$, and without $\mathbb{D}_\emptyset$

we get $\mathcal{L}_K$, the *language of individual knowledge.* If $\vec{B} = B_1 \dots B_n$ then for $\mathbb{D}_{B_1} \dots \mathbb{D}_{B_n}\varphi$ we may write $\mathbb{D}_{\vec{B}}\varphi$; where $\mathbb{D}_\epsilon\varphi := \varphi$.

Frames and models. The structures are multi-agent epistemic *frames* $(W, \sim)$ and *models* $(W, \sim, V)$ where W is a domain of *worlds*, $\sim$ is a function from the set of agents to equivalence relations (indistinguishability relations, knowledge relations) on W (where we write $\sim_a$ for the knowledge relation for agent $a \in A$) and V is function from the set of atoms to the powerset of W, namely mapping each atom to the subset of worlds where it is true. We write $\sim_B$ for $\cap_{b\in B} \sim_b$ (thus, $\sim_\emptyset = W \times W$).

If $M = (W, \sim, V)$ and $B \subseteq A$ then $M^B = (W, \sim^B, V)$ where for $a \in B$, $\sim_a^B := \bigcap_{b\in B} \sim_b$ and for $a \notin B$, $\sim_a^B := \sim_a$. In *updated* model M^B the relations for the agents $a \in B$ have been *updated* from $\sim_a$ to $\sim_B$. For $M^{\{a_1,\dots,a_n\}}$ we write $M^{a_1\dots a_n}$, whereas for $(M^B)^C$ we write M^{BC}. In M^B we have that $\sim_B^B = \sim_b^B$ for all $b \in B$, unlike in M. Note that $M^a = M$ for all $a \in A$ and that $M^\emptyset = M$. Given $\vec{B} = B_1 \dots B_n \in \mathcal{P}(A)^*$ and $M = (W, \sim, V)$, for $M^{B_1\dots B_n}$ we write $M^{\vec{B}}$ (where $M^\epsilon = M$) and we write $\sim^{\vec{B}}$ for its knowledge function (where $\sim^\epsilon = \sim$). Given $\vec{B}, \vec{C} \in \mathcal{P}(A)^*$, $\vec{B}$ is a prefix of $\vec{C}$, notation $\vec{B} \sqsubseteq \vec{C}$, if there is a $\vec{E} \in \mathcal{P}(A)^*$ such that $\vec{C} = \vec{B}\vec{E}$.

Consider an operation $\mathsf{see} : \mathcal{P}(A) \to \mathcal{P}(A)^* \to \mathcal{P}(A)$ that associates to each $C \subseteq A$ and to each sequence $\vec{B}$ of subsets of A the set ($\mathsf{see}_C(\vec{B})$ denoted) $\mathsf{see}_C(\vec{B})$ such that $\sim_{\mathsf{see}_C(\vec{B})}$ is the knowledge relation for all agents in C after the $\vec{B}$ update.

$$\begin{array}{lll} \mathsf{see}_C(\epsilon) & := C & \\ \mathsf{see}_C(\vec{B}E) & := \mathsf{see}_C(\vec{B}) & \text{if } C \cap E = \emptyset \\ \mathsf{see}_C(\vec{B}E) & := \mathsf{see}_{C\cup E}(\vec{B}) & \text{if } C \cap E \neq \emptyset \end{array}$$

Intuitively, $\mathsf{see}_C(\vec{B})$ determines what each agent c in group C can 'see' after the $\vec{B}$ update, that is, how its knowledge relation has been restricted after the successive sharings of knowledge with the different groups of agents in $\vec{B}$. For example, if $C = \{a\}$ and $\vec{B} = \{a, b\}\{a, c\}$, then after the first $\{a, b\}$ update agent a already sees what a and b see (know), her relation is now ($\sim_{\{a,b\}}$ denoted) $\sim_{ab}$, and after the second update she sees what all three agents a, b, c can see (know), her relation is now $\sim_{abc}$. Etcetera.

Lemma 2.1 *Let $M = (W, \sim, V)$ be a model. For all $B, C \subseteq A$: (i) $\sim_B^\emptyset = \sim_B$, (ii) $\sim_B^B = \sim_B$, (iii) if $C \subseteq B$ then $\sim^{BC} = \sim^B$, (iv) if $B \cap C = \emptyset$ then $\sim_C^B = \sim_C$, and (v) if $B \cap C \neq \emptyset$ then $\sim_C^B = \sim_{B\cup C}$.*

Lemma 2.2 *For all $C \subseteq A$ and $\vec{B} \in \mathcal{P}(A)^*$, $C \subseteq \mathsf{see}_C(\vec{B})$.*

Lemma 2.3 *For all $(W, \sim)$ and $\vec{B} \in \mathcal{P}(A)^*$ and $C \subseteq A$, $\sim_C^{\vec{B}} = \sim_{\mathsf{see}_C(\vec{B})}$.*

Proof. Induction on the length of $\vec{B}$. Clearly, $\sim_C^\epsilon = \sim_C = \sim_{\mathsf{see}_C(\epsilon)}$. Then, for $E \subseteq A$, if $C \cap E = \emptyset$: $\sim_C^{\vec{B}E} = \sim_C^{\vec{B}} =$ (ind.) $\sim_{\mathsf{see}_C(\vec{B})} = \sim_{\mathsf{see}_C(\vec{B}E)}$. Otherwise, if $C \cap E \neq \emptyset$: $\sim_C^{\vec{B}E} = \sim_E^{\vec{B}} \cap \sim_C^{\vec{B}} = \sim_{C\cup E}^{\vec{B}} =$ (ind.) $\sim_{\mathsf{see}_{C\cup E}(\vec{B})} = \sim_{\mathsf{see}_C(\vec{B}E)}$. □

Lemma 2.4 *For all* $(W, \sim)$ *and* $\vec{B}, \vec{C} \in \mathcal{P}(A)^*$, *if* $\sim_a^{\vec{B}} = \sim_a^{\vec{C}}$ *for all* $a \in A$, *then* $\sim^{\vec{B}} = \sim^{\vec{C}}$.

Update lattice. The function see plays a role in the axiomatization and in the canonical frame (Section 7). With the function see we can compare sequences $\vec{B} \in \mathcal{P}(A)^*$, and therefore, updates. We define $\vec{B} \preceq \vec{C}$ if for all $a \in A$, $\mathsf{see}_a(\vec{B}) \subseteq \mathsf{see}_a(\vec{C})$, and $\vec{B} \approx \vec{C}$ if $\vec{B} \preceq \vec{C}$ and $\vec{C} \preceq \vec{B}$ as well as $\vec{B} \prec \vec{C}$ if $\vec{B} \preceq \vec{C}$ but not $\vec{C} \preceq \vec{B}$.

Lemma 2.5 *For all* $B, C \subseteq A$, *if* $C \subseteq B$ *then* $BC \approx B$.

The quotient of $\mathcal{P}(A)^*$ with respect to the equivalence relation $\approx$ defines an *update lattice*. The maximum element of this lattice is denoted ω. It is the equivalence class of $\vec{B}$ such that $\mathsf{see}_a^{\vec{B}} = A$ for all $a \in A$. The minimum element is the equivalence class of ϵ. Note that, if $\vec{B} \approx \vec{C}$, then for all epistemic models $M = (W, \sim, V)$, $M^{\vec{B}} = M^{\vec{C}}$ (as $\sim^{\vec{B}} = \sim^{\vec{C}}$, cf. Lemma 2.3).

Gossip. There is a strong relation between a sequence $\vec{B} = B_1 \dots B_n$ of two-element sets, representing a sequence of resolution updates [1], and a sequence of calls between two agents in *gossip protocols* [26,23,4,3]. In that case the tuple $(\mathsf{see}_{a_1}(\vec{B}), \dots, \mathsf{see}_{a_{|A|}}(\vec{B}))$ is known as the *secret distribution* resulting from the call sequence $\vec{B}$, given an initial distribution where agents only know their own secrets, and the depth of the update lattice is the maximum number $\binom{|A|}{2}$ of informative calls [5].

Semantics. We define the satisfaction relation $\models$ by induction on $\varphi \in \mathcal{L}_{\mathbb{D}DR}$ (where $p \in P$ and $B \subseteq A$).

$$
\begin{array}{lll}
M, w \models p & \text{iff} & w \in V(p) \\
M, w \models \top & \text{iff} & \text{always} \\
M, w \models \neg\varphi & \text{iff} & M, w \not\models \varphi \\
M, w \models \varphi \wedge \psi & \text{iff} & M, w \models \varphi \text{ and } M, w \models \psi \\
M, w \models \mathbb{D}_B\varphi & \text{iff} & M^B, v \models \varphi \text{ for all } v \in W \text{ such that } w \sim_B v \\
M, w \models D_B\varphi & \text{iff} & M, v \models \varphi \text{ for all } v \in W \text{ such that } w \sim_B v \\
M, w \models R_B\varphi & \text{iff} & M^B, w \models \varphi
\end{array}
$$

A formula $\varphi \in \mathcal{L}_{\mathbb{D}DR}$ is *valid on a frame* $(W, \sim)$ if for all models $M = (W, \sim, V)$ and for all $w \in W$, $M, w \models \varphi$. A formula $\varphi \in \mathcal{L}_{\mathbb{D}DR}$ is *valid* if it is valid on all epistemic frames. The set $L_{\mathbb{D}}$ of validities of $\mathcal{L}_{\mathbb{D}}$ is *the logic of dynamic distributed knowledge*. The set L_D of the validities of $\mathcal{L}_D$ is *the logic of static distributed knowledge*.

The interpretation of $K_a\varphi$ as $\mathbb{D}_a\varphi$ and $D_a\varphi$ is standard, as $\bigcap_{b\in\{a\}} \sim_b = \sim_a$, so that $M^a = M$, and that of resolution R_B is as in [1]. It is elementary that:

Lemma 2.6 *For all* $B \subseteq A$ *and* $\varphi \in \mathcal{L}_{\mathbb{D}DR}$, $\mathbb{D}_B\varphi \leftrightarrow D_B R_B\varphi$ *and* $\mathbb{D}_B\varphi \leftrightarrow R_B D_B\varphi$ *are valid.*

Therefore, the modality $\mathbb{D}_B$ is definable from D_B and R_B in the language $\mathcal{L}_{\mathbb{D}DR}$. We recall that our research interest is the language $\mathcal{L}_{\mathbb{D}}$ *without* D_B and R_B operators, and how it relates to $\mathcal{L}_D$.

3 Validities

We first list a number of invalidities for $\mathcal{L}_{\mathbb{D}}$, to defeat the expectation of the reader versed in static distributed knowledge. We then list $\mathcal{L}_{\mathbb{D}}$ validities describing the properties of a single modality $\mathbb{D}_B$, and show that it has the properties of KD45. We finally list $\mathcal{L}_{\mathbb{D}}$ validities describing interaction between different modalities $\mathbb{D}_B$ and $\mathbb{D}_C$.

Invalidities Invalid (that is, not valid for all B, C and for all φ) for these semantics are the following formula schemata. We contrast them with well-known validities for static distributed knowledge and resolving distributed knowledge.

$\not\models \mathbb{D}_B\varphi \to \varphi$	$\models D_B\varphi \to \varphi$
$\not\models \mathbb{D}_B\varphi \to \mathbb{D}_C\varphi$ when $B \subseteq C$	$\models D_B\varphi \to D_C\varphi$ when $B \subseteq C$
$\not\models \mathbb{D}_B\neg\varphi \leftrightarrow \neg\mathbb{D}_B\varphi$	$\models R_B\neg\varphi \leftrightarrow \neg R_B\varphi$

Example 3.1 Simple counterexamples suffice to show that the above are invalid. The model on the left is for two agents a and b and a single atom p such that a knows but b is uncertain about p. The model on the right results after a and b share their knowledge.

$$M: \quad w(p) \overset{b}{\text{———}} v(\neg p) \qquad\qquad M^{ab}: \quad w(p) \qquad v(\neg p)$$

We can observe that

$M, w \models \mathbb{D}_{ab}K_b p$	whereas $M, w \not\models K_b p$	so	$\not\models \mathbb{D}_{ab}\varphi \to \varphi$
$M, w \models K_b\neg K_b p$	whereas $M, w \not\models \mathbb{D}_{ab}\neg K_b p$	so	$\not\models \mathbb{D}_b\varphi \to \mathbb{D}_{ab}\varphi$
$M, w \models \neg K_b p$	whereas $M, w \not\models K_b\neg p$	so	$\not\models \neg\mathbb{D}_b\varphi \to \mathbb{D}_b\neg\varphi$

Although $R_B\neg\varphi \leftrightarrow \neg R_B\varphi$ is valid [36], as the update is a total function, our $\mathbb{D}_B\varphi$ corresponds to $R_B D_B\varphi$, and just as $\mathbb{D}_B\neg\varphi$ is not equivalent to $\neg\mathbb{D}_B\varphi$, also $R_B D_B\neg\varphi$ is not equivalent to $\neg R_B D_B\varphi$ in [36]. Although $\mathbb{D}_B\varphi \to \varphi$ is invalid, that is, not valid for all B and φ, it is easy to see that $\mathbb{D}_B p \to p$ is valid, as atoms do not change their value after update, and as the actual world is indistinguishable for B. As a consequence, uniform substitution is not validity preserving for these semantics ($\models \varphi$ does not imply $\models \varphi[p/\psi]$); so the logic of dynamic distributed knowledge is not a normal modal logic.

Mono-modal validities We continue with validities involving a single $\mathbb{D}_B$.

Proposition 3.2 *Valid / validity preserving for these semantics are:*

$\models \mathbb{D}_B(\varphi \to \psi) \to (\mathbb{D}_B\varphi \to \mathbb{D}_B\psi)$	$\models \widehat{\mathbb{D}}_B\varphi \to \mathbb{D}_B\widehat{\mathbb{D}}_B\varphi$
$\models \mathbb{D}_B\varphi \to \mathbb{D}_B\mathbb{D}_B\varphi$	$\models \varphi$ *implies* $\models \mathbb{D}_B\varphi$
$\models \mathbb{D}_B\varphi \to \widehat{\mathbb{D}}_B\varphi$	

Proof. We use that $M^{BB} = M^B$ and that $\sim_B^B = \sim_B$. More details omitted.□

It follows that the modalities $\mathbb{D}_B$ have the properties of KD45 (we also have Modus Ponens), and thus represent consistent group belief. Still, the $\mathbb{D}_B$ also have other properties, such as the mentioned $\models \mathbb{D}_B p \to p$, which makes it clear that the logic of dynamic distributed knowledge is not (multi-)KD45.

Multi-modal validities Finally, validities relating $\mathbb{D}_B$ and $\mathbb{D}_C$ for $C \neq B$.

Proposition 3.3 $\models \mathbb{D}_B\mathbb{D}_C\varphi \leftrightarrow \mathbb{D}_B\varphi$, *where* $\emptyset \neq C \subseteq B \subseteq A$.

Proof. Let (M, w) be given. Then: $M, w \models \mathbb{D}_B\mathbb{D}_C\varphi$, iff $M^B, v \models \mathbb{D}_C\varphi$ for all $v \sim_B^\epsilon w$, iff $M^{BC}, u \models \varphi$ for all $u \sim_C^B v$ and for all $v \sim_B^\epsilon w$, iff $(*)$ $M^{BC}, u \models \varphi$ for all $u \sim_B^\epsilon v$ and for all $v \sim_B^\epsilon w$, iff (Lemma 2.1) $M^B, u \models \varphi$ for all $u \sim_B^\epsilon v$ and for all $v \sim_B^\epsilon w$, iff $M^B, u \models \varphi$ for all $u \sim_B^\epsilon w$, iff $M, u \models \mathbb{D}_B\varphi$.

$(*)$: As $C \subseteq B$ for all $c \in C$, $\sim_c^B = \bigcap_{b \in B} \sim_b$. Furthermore, for all $b \in B$, $\sim_b^B = \bigcap_{b \in B} \sim_b$. Therefore, $u \sim_C^B v$ iff $u \sim_B^B v$. □

The instantiation of Prop. 3.3 for $B = C$ is that $\mathbb{D}_B\mathbb{D}_B\varphi \leftrightarrow \mathbb{D}_B\varphi$ is valid, from which we also get $\mathbb{D}_B\varphi \rightarrow \mathbb{D}_B\mathbb{D}_B\varphi$.

We continue with multi-modal validities that resemble the validities $D_B\varphi \rightarrow \varphi$ and $D_B\varphi \rightarrow D_C\varphi$ for static distributed knowledge, however, subject to restrictions. Let us define $\mathcal{L}_{\mathbb{D}}\uparrow C$ as the fragment of $\mathcal{L}_{\mathbb{D}}$ with only $\mathbb{D}_B$ modalities for $C \subseteq B \subseteq A$.

Lemma 3.4 *Let* $B, C \subseteq A$ *with* $B \subseteq C$, *and let* $\varphi \in \mathcal{L}_{\mathbb{D}}\uparrow C$. *Then for all models* $M = (W, \sim, V)$ *and* $w \in W$, $M^B, w \models \varphi$ *iff* $M^C, w \models \varphi$.

Proof. The proof is by induction on φ. □

Proposition 3.5 $\models \mathbb{D}_B\varphi \rightarrow \varphi$ *whenever* $\varphi \in \mathcal{L}_{\mathbb{D}}\uparrow B$.

Proof. The proof is by induction on $\varphi \in \mathcal{L}_{\mathbb{D}}$ in negation normal form. Let $M = (W, \sim, V)$ be given and $w \in W$. The cases of p and $\neg p$, conjunction and disjunction (using Lemma 3.4) have been omitted.

Case $\mathbb{D}_C\varphi$. $M, w \models \mathbb{D}_B\mathbb{D}_C\varphi$, iff (semantics) $M^B, v \models \mathbb{D}_C\varphi$ for all $v \sim_B w$, iff $M^{BC}, u \models \varphi$ for all $v \sim_B w$ and for all $u \sim_C^B v$, iff ($B \subseteq C$ implies that $M^{BC} = M^C$ and $\sim_C^B = \sim_C$) $M^C, u \models \varphi$ for all $v \sim_B w$ and for all $u \sim_C v$, which implies (as $w \sim_B w$) $M^C, u \models \varphi$ for all $u \sim_C w$, iff $M, w \models \mathbb{D}_C\varphi$.

Case $\widehat{\mathbb{D}}_C\varphi$. The proof is similar to the case $\mathbb{D}_C\varphi$. □

Proposition 3.6 $\models \mathbb{D}_B\varphi \rightarrow \mathbb{D}_C\varphi$ *whenever* $B \subseteq C$ *and* $\varphi \in \mathcal{L}_{\mathbb{D}}\uparrow C$.

Proof. Suppose $B \subseteq C$. Let $\varphi \in \mathcal{L}_{\mathbb{D}}\uparrow C$. Suppose $M, w \models \mathbb{D}_B\varphi$. Hence, $M^B, v \models \varphi$ for all $v \sim_B w$. Thus, by Lemma 3.4, $M^C, v \models \varphi$ for all $v \sim_B w$. Since $B \subseteq C$, therefore $\sim_C \subseteq \sim_B$. Since $M^C, v \models \varphi$ for all $v \sim_B w$, therefore $M^C, v \models \varphi$ for all $v \sim_C w$. Consequently, $M, w \models \mathbb{D}_C\varphi$. □

All above seems not enough for a complete axiomatization, see Section 7.

4 Bisimulation

We propose a novel notion of *shared bisimulation* and we compare it to the notions of individual bisimulation and collective bisimulation from the literature. For binary relations Z we write $(x, y) \in Z$ as well as Zxy.

Individual and collective bisimulation. Let $M = (W, \sim, V)$ and $M' = (W', \sim', V')$ be given. A non-empty relation $Z \subseteq W \times W'$ is a *collective bisimulation* [28] if for all $(w, w') \in Z$, $p \in P$, and $B \subseteq A$:

- **atoms**: $w \in V(p)$ iff $w' \in V'(p)$
- **forth**: if $w \sim_B v$ there is $v' \in W'$ such that $w' \sim'_B v'$ and $(v, v') \in Z$
- **back**: if $w' \sim'_B v'$ there is $v \in W$ such that $w \sim_B v$ and $(v, v') \in Z$

Pointed models (M, w) and (M', w') are *collective bisimilar*, notation $(M, w) \leftrightarroweq^c (M', w')$, if there exists a collective bisimulation Z between M and M' with $(w, w') \in Z$. Relation Z is an *individual bisimulation* [10] if **forth** and **back** are satisfied for all singleton sets $\{a\}$ where $a \in A$; (M, w) and (M', w') are *individually bisimilar*, notation $(M, w) \leftrightarroweq^i (M', w')$, if there exists an individual bisimulation Z between M and M' with $(w, w') \in Z$.

Shared bisimulation. The novel notion of *shared bisimulation* is given by defining a set of $\vec{B}$-shared bisimulations, for all $\vec{B} \in \mathcal{P}(A)^*$.

Definition 4.1 Let $M = (W, \sim, V)$ and $M' = (W', \sim', V')$ be given. A *shared bisimulation* Z is a collection $\{Z^{\vec{B}} \mid \vec{B} \in \mathcal{P}(A)^*\}$ of non-empty relations $Z^{\vec{B}} \subseteq W \times W'$ such that for all $(w, w') \in Z^{\vec{B}}$, $p \in P$, and $B \subseteq A$:

- **atoms**: $w \in V(p)$ iff $w' \in V'(p)$
- **forth**: if $w \sim^{\vec{B}}_B v$ there is $v' \in W'$ such that $w' \sim'^{\vec{B}}_B v'$ and $(v, v') \in Z^{\vec{B}B}$
- **back**: if $w' \sim'^{\vec{B}}_B v'$ there is $v \in W$ such that $w \sim^{\vec{B}}_B v$ and $(v, v') \in Z^{\vec{B}B}$

When $\{Z^{\vec{B}} \mid \vec{B} \in \mathcal{P}(A)^*\}$ is a shared bisimulation between M and M', each $Z^{\vec{B}}$ is called a *$\vec{B}$-shared bisimulation.* Pointed models (M, w) and (M', w') are *shared bisimilar*, notation $(M, w) \leftrightarroweq^s (M', w')$, if there exists a shared bisimulation Z between M and M' with $(w, w') \in Z^\epsilon$.

We recall the update lattice from Section 2. We can similarly view the $\vec{B}$-shared bisimulations $Z^{\vec{B}}$ for all $\vec{B} \in \mathcal{P}(A)^*$ of a shared bisimulation Z as such a lattice, with top Z^ω and bottom Z^ϵ. There are therefore only finitely many $\vec{B}$-shared bisimulations to consider, modulo $\approx$-equivalence. Note that, given equivalence classes $[w]_A$ in W and $[w']_A$ in W' containing the same valuations, relating (for all w, w') states in $[w]_A$ to states in $[w']_A$ with the same valuation, is an ω-shared bisimulation.

Proposition 4.2 ([28]) *Collective bisimilar is (implies) individual bisimilar.*

Proposition 4.3 *Collective bisimilar is shared bisimilar.*

Proof. Let $Z \subseteq W \times W'$ be a collective bisimulation. Then $Z_2 := \{Z_2^{\vec{B}} \mid \vec{B} \in \mathcal{P}(A)^*\}$ where $Z_2^{\vec{B}} = Z$ for all $\vec{B}$ is a shared bisimulation. Details omitted. □

Proposition 4.4 *Shared bisimilar is individual bisimilar.*

Proof. Let $Z = \{Z^{\vec{B}} \mid \vec{B} \in \mathcal{P}(A)^*\}$ be a shared bisimulation between $M = (W, \sim, V)$ and $M' = (W', \sim', V')$. We show that $Z^{A*} := \bigcup\{Z^{\vec{a}} \mid \vec{a} \in A^*\}$ is an individual bisimulation between M and M'. (Each $Z^{a_1 \cdots a_n}$ is considered $Z^{\{a_1\} \dots \{a_n\}}$ with $\{a_1\} \dots \{a_n\} \in \mathcal{P}(A)^*$.) **Atoms** is obvious. Consider **forth**. Let be given $Z^{A*}ww'$ and $w \sim_b v$. As $Z^{A*}ww'$, there is $\vec{a} \in A^*$ with $Z^{\vec{a}}ww'$.

As $M^{\vec{a}} = M$, $w \sim_b v$ iff $w \sim_b^{\vec{a}} v$. As $Z^{\vec{a}}$ is a $\vec{a}$-shared bisimulation, there is a $v' \sim_b^{\vec{a}} w'$ with $Z^{\vec{a}b}vv'$. As $Z^{\vec{a}b} \subseteq Z^{A*}$, therefore $Z^{A*}vv'$. Furthermore, $v' \sim_b^{\vec{a}} w'$ iff $v' \sim_b^{\vec{a}} w'$ because $M'^{\vec{a}} = M'$. **Back** is similar. □

Example 4.5 It is well-known that individual bisimilar may not be collective bisimilar. A typical counterexample involves the models M and M' below.

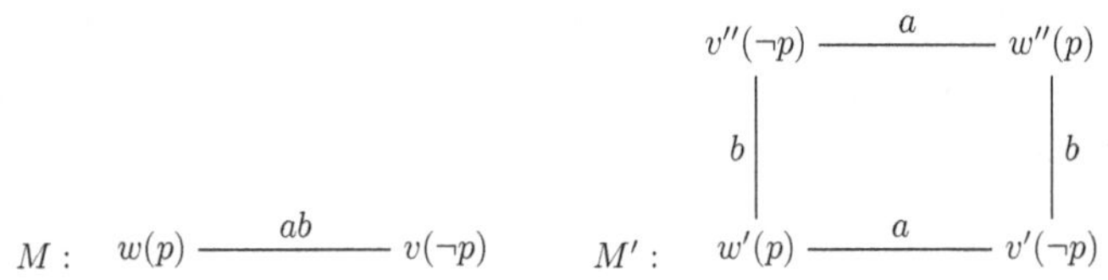

An individual bisimulation between M and M' consists of linking the p worlds and linking the $\neg p$ worlds. For example, pair (w, w') in this bisimulation. However, it is not a collective bisimulation: although $w \sim_{ab} v$ in M and where we note that $v \in V(p)$, there is no world s in M' such that $w' \sim'_{ab} s$ and that has the same value of p as v in M. Only $s = w'$ fulfills that role and $w' \notin V(p)$. So, we cannot satisfy **atoms**. Indeed, $M, w \not\models D_{ab}p$, whereas $M', w' \models D_{ab}p$.

This also demonstrates that individual bisimilar may not be shared bisimilar, as **atoms** has also to be satisfied for a shared bisimulation, and indeed, similarly, $M, w \not\models \mathbb{D}_{ab}p$, whereas $M', w' \models \mathbb{D}_{ab}p$.

It remains a, rather annoyingly, open question whether collective bisimilar and shared bisimilar are the same or are different, but we conjecture that collective bisimilarity is a more refined notion to distinguish structures than shared bisimilarity. An answer to this question might help to compare the expressivity of dynamic distributed knowledge and static distributed knowledge.

Hennessy-Milner characterization We continue by characterizing shared bisimilarity. For that we need some additional terminology.

Given $M = (W, \sim, V)$, $M' = (W', \sim', V')$, $w \in W$, and $w' \in W'$, pointed models (M, w) and (M', w') are *modally equivalent*, notation $(M, w) \equiv (M', w')$, if for all $\varphi \in \mathcal{L}_{\mathbb{D}}$, $M, w \models \varphi$ iff $M', w' \models \varphi$.

A model $M = (W, \sim, V)$ is *image-finite* if for all agents $a \in A$ and worlds $v \in W$ the equivalence class $[v]_a = \{u \in W \mid v \sim_a u\}$ is finite.

Proposition 4.6 *Shared bisimilar implies modally equivalent.*

Proof. We need to show that for arbitrary pointed models (M, w) and (M', w'), $(M, w) \leftrightarrows^s (M', w')$ implies $(M, w) \equiv (M', w')$. In order to prove that, we will prove the following statement.

> Let be given $\varphi \in \mathcal{L}_{\mathbb{D}}$, models $M = (W, \sim, V)$ and $M' = (W', \sim', V')$, and a shared bisimulation Z between M and M'. Then for all $w \in W$, $w' \in W'$, and $\vec{B} \in \mathcal{P}(A)^*$: $Z^{\vec{B}}ww'$ implies $(M^{\vec{B}}, w) \equiv (M'^{\vec{B}}, w')$. (∗)

The proof is by induction on φ. Elementary cases are omitted.

Case $\mathbb{D}_B\varphi$: Assume $M'^{\vec{B}}, w' \models \mathbb{D}_B\varphi$. To prove $M^{\vec{B}}, w \models \mathbb{D}_B\varphi$, assume $v \in W$ with $v \sim_B^{\vec{B}} w$. From initial assumption $Z^{\vec{B}}ww'$, $v \sim_B^{\vec{B}} w$, and **forth** it follows that there is $v' \in W'$ such that $v' \sim_B^{\vec{B}} w'$ and $Z^{\vec{B}B}vv'$. Also, from assumption $M'^{\vec{B}}, w' \models \mathbb{D}_B\varphi$ and $v' \sim_B^{\vec{B}} w'$ it follows that $M'^{\vec{B}B}, v' \models \varphi$. From $M'^{\vec{B}B}, v' \models \varphi$, $Z^{\vec{B}B}vv'$, and induction it now follows that $M^{\vec{B}B}, v \models \varphi$. As v was arbitrary, from the semantics of $\mathbb{D}_B$ it now follows that $M^{\vec{B}}, w \models \mathbb{D}_B\varphi$. The other direction is proved similarly.

The case $\vec{B} = \epsilon$ of $(*)$ gets us that $Z^\epsilon ww'$ implies $(M, w) \equiv (M', w')$. □

Proposition 4.7 *Modally equivalent implies shared bisimilar on the class of image-finite models.*

Proof. We need to show that for arbitrary image-finite pointed models (M, w) and (M', w'), $(M, w) \equiv (M', w')$ implies $(M, w) \leftrightarrows^s (M', w')$. In order to prove that, we will prove the following statement.

> Let be given $M = (W, \sim, V)$ and $M' = (W', \sim', V')$ with $w \in W$ and $w' \in W'$, and such that $(M, w) \equiv (M', w')$. Consider the set of relations $Z = \{Z^{\vec{B}} \mid \vec{B} \in \mathcal{P}(A)^*\}$ such that for all $v \in W$, $v' \in W'$, $Z^{\vec{B}}vv'$ iff $(M^{\vec{B}}, v) \equiv (M'^{\vec{B}}, v')$. Then Z is a shared bisimulation between M and M'. $(**)$

Suppose towards a contradiction that Z is not a shared bisimulation. Then there is $\vec{B} \in \mathcal{P}(A)^*$ such that $Z^{\vec{B}}$ is not a $\vec{B}$-shared bisimulation. Then there are $w, v \in W$ and $w' \in W'$ with $Z^{\vec{B}}ww'$ and $w \sim_B^{\vec{B}} v$ but for all $v' \in W'$ with $w' {\sim'}_B^{\vec{B}} v'$, not $Z^{\vec{B}B}vv'$. As M' is imagine-finite, we can enumerate such v' as $v'_1, \ldots, v'_n$. By the definition of Z, this means that $(M^{\vec{B}B}, v) \not\equiv (M'^{\vec{B}B}, v'_1)$, $\ldots$, and $(M^{\vec{B}B}, v) \not\equiv (M'^{\vec{B}B}, v'_n)$. Let $\psi_1, \ldots, \psi_n$ be distinguishing formulas such that $M^{\vec{B}B}, v \models \psi_1$ but $M'^{\vec{B}B}, v'_1 \not\models \psi_1$, $\ldots$, and $M^{\vec{B}B}, v \models \psi_n$ but $M'^{\vec{B}B}, v'_n \not\models \psi_n$. Note that $M^{\vec{B}B}, v \models \bigwedge_{i=1}^n \psi_i$. By the semantics of dynamic distributed knowledge we therefore have that $M^{\vec{B}}, w \models \widehat{\mathbb{D}}_B \bigwedge_{i=1}^n \psi_i$. On the other hand, $M'^{\vec{B}}, w' \not\models \widehat{\mathbb{D}}_B \bigwedge_{i=1}^n \psi_i$, as in any of the $\mathsf{see}_B(\vec{B})$-accessible $v'_i = v'_1, \ldots, v'_n$ the conjunction $\bigwedge_{i=1}^n \psi_i$ is false: ψ_1 is false in v'_1, and therefore $\bigwedge_{i=1}^n \psi_i$ is false in v'_1.

This contradicts the assumption that $Z^{\vec{B}}$ is not a $\vec{B}$-shared bisimulation, and therefore the initial assumption that Z is not a shared bisimulation.

As Z is a shared bisimulation, Z^ϵ is an ϵ-shared bisimulation. The original assumption $(M, w) \equiv (M', w')$ means that $Z^\epsilon ww'$, by the definition of Z^ϵ. By the definition of shared bisimilarity $Z^\epsilon ww'$ means that $(M, w) \leftrightarrows^s (M', w')$. □

Example 4.8 Not all $Z^{\vec{B}}$ are collective bisimulations. Consider the model below for agents a, b, c and variables p and q (valuations label worlds, $\overline{p} = \neg p$).

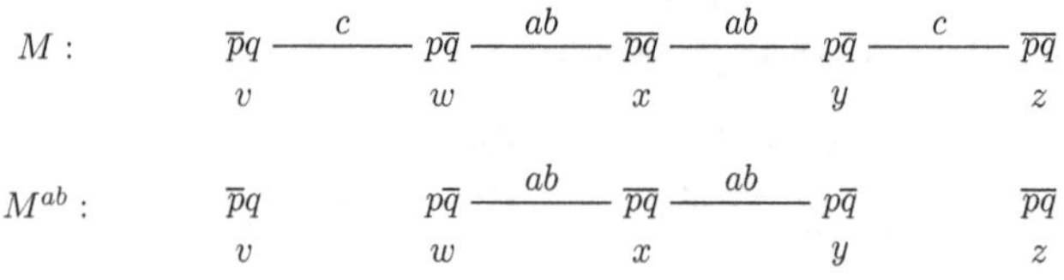

A maximal shared (auto)bisimulation on this model consists of:

— $Z^\epsilon = Z^a = Z^b = \{(v,v),(w,w),(x,x),(y,y),(z,z)\}$

— $Z^{ab} = Z^\omega = \{(v,v),(w,w),(x,x),(y,y),(z,z)\} \cup \{(w,y),(y,w)\}$

In this case we have that $Z^\epsilon \subset Z^\omega$ and that Z^ω is not a collective bisimulation, or even an individual bisimulation, as worlds w and y are not (even individual) bisimilar: $M, w \models K_c q$ whereas $M, y \models K_c q$. However, in the M^{ab} update they are. So, $Z^{ab}st$ does not imply $Z^\epsilon st$ for arbitrary worlds s and t. However, given $s = w$, there always is *some* t such that $Z^{ab}st$ and $Z^\epsilon st$, namely $t = w$.

5 Translations

We present two translations. We show that every formula in $\mathcal{L}_{I\!D}$ is equivalent to one in $\mathcal{L}_D$, using a translation τ, and we show that on the class of frames *with local agreement* (for any two agents one always knows more than the other) every formula in $\mathcal{L}_{I\!D}$ is equivalent to one in $\mathcal{L}_{KU}$, using a translation δ.

5.1 Dynamic into static distributed knowledge

Lemma 2.6 showed that $I\!D_B\varphi$ is definable as either $D_B R_B \varphi$ or as $R_B D_B \varphi$. In [1] it is shown that every formula in $\mathcal{L}_{DR}$ is equivalent to a formula in $\mathcal{L}_D$. Combining the two proves that every $\varphi \in \mathcal{L}_{I\!D}$ is equivalent to a $\psi \in \mathcal{L}_D$. Instead of this indirect proof, we define a translation τ, indexed by groups of agents $B \subseteq A$, such that any $\varphi \in \mathcal{L}_{I\!D}$ is equivalent to $\tau_\emptyset(\varphi) \in \mathcal{L}_D$.

Definition 5.1 Let translation $(\tau_B)_{B\subseteq A}$ be the family of functions from $\mathcal{L}_{I\!DDR}$ to $\mathcal{L}_D$ inductively defined below. For $\tau_{B_1}(\dots\tau_{B_m}(\varphi)\dots)$ we write $\tau_{\vec{B}}(\varphi)$.

$$
\begin{array}{ll|lll}
\tau_B(p) & = p & \tau_B(I\!D_C\varphi) = & D_C\tau_B(\tau_C(\varphi)) & \text{if } B\cap C = \emptyset \\
\tau_B(\top) & = \top & \tau_B(I\!D_C\varphi) = & D_{B\cup C}\tau_B(\tau_C(\varphi)) & \text{if } B\cap C \neq \emptyset \\
\tau_B(\neg\varphi) & = \neg\tau_B(\varphi) & \tau_B(D_C\varphi) = & D_C\tau_B(\varphi) & \text{if } B\cap C = \emptyset \\
\tau_B(\varphi\wedge\psi) & = \tau_B(\varphi)\wedge\tau_B(\psi) & \tau_B(D_C\varphi) = & D_{B\cup C}\tau_B(\varphi) & \text{if } B\cap C \neq \emptyset \\
 & & \tau_B(R_C\varphi) = & \tau_B(\tau_C(\varphi)) &
\end{array}
$$

Lemma 5.2 *Let $\varphi \in \mathcal{L}_{I\!DDR}$. For all $B, C \subseteq A$, $\tau_B(I\!D_C\varphi) = \tau_B(D_C R_C\varphi)$.*

Lemma 5.3 *Let $\vec{B} \in \mathcal{P}(A)^*$. For all $C \subseteq A$, there exists $C' \subseteq A$ such that $C \subseteq C'$ and for all $\varphi \in \mathcal{L}_{I\!DDR}$, $\tau_{\vec{B}}(I\!D_C\varphi) = D_{C'}\tau_{\vec{B}}(\varphi)$.*

Proof. By induction on $\vec{B}$. □

Lemma 5.4 *Let $\varphi \in \mathcal{L}_{I\!DDR}$. For all models $M = (W, \sim, V)$, for all $u \in W$ and for all $B \subseteq A$, $M^B, u \models \varphi$ if and only if $M, u \models \tau_B(\varphi)$.*

Proof. The proof is done by induction on φ. Let $M = (W, \sim, V)$ be a model, $u \in W$ and $B \subseteq A$. We only show the case $I\!D_C\psi$.

Case $B \cap C = \emptyset$. Then: (i) $M^B, u \models I\!D_C\psi$, iff (ii) for all $v \in W$, if $u \sim^B_C v$ then $M^{BC}, v \models \psi$, iff (iii) for all $v \in W$, if $u \sim_C v$ then $M^{BC}, v \models \psi$, iff (iv) for all $v \in W$, if $u \sim_C v$ then $M^B, v \models \tau_C(\psi)$, iff (v) for all $v \in W$, if $u \sim_C v$ then $M, v \models \tau_B(\tau_C(\psi))$, iff (vi) $M, u \models D_C\tau_B(\tau_C(\psi))$.

The equivalence between (ii) and (iii) is a consequence of $B \cap C = \emptyset$, and that between (iii), (iv) and (v) is a consequence of the induction hypothesis.

Case $B \cap C \neq \emptyset$. Then: (i) $M^B, u \models \mathbb{D}_C \psi$, iff (ii) for all $v \in W$, if $u \sim^B_C v$ then $M^{BC}, v \models \psi$, iff (iii) for all $v \in W$, if $u \sim_{B \cup C} v$ then $M^{BC}, v \models \psi$, iff (iv) for all $v \in W$, if $u \sim_{B \cup C} v$ then $M^B, v \models \tau_C(\psi)$, iff (v) for all $v \in W$, if $u \sim_{B \cup C} v$ then $M, v \models \tau_B(\tau_C(\psi))$, iff (iv) $M, u \models D_{B \cup C} \tau_B(\tau_C(\psi))$.

The equivalence between (ii) and (iii) is a consequence of $B \cap C \neq \emptyset$, and that between (iii), (iv) and (v) is a consequence of the induction hypothesis.□

Corollary 5.5 *Any $\varphi \in \mathcal{L}_{\mathbb{D}}$ is equivalent to $\tau_{\emptyset}(\varphi) \in \mathcal{L}_D$.*

As a further consequence, satisfiability in the logic $L_{\mathbb{D}}$ is *decidable*, because the logic L_D is decidable [14,34].

5.2 Agents with local agreement

A *frame (model) with local agreement*, first considered in [16], is a multi-agent epistemic frame $(W, \sim)$ (resp. model $(W, \sim, V)$) such that for all $a, b \in A$ and for all $s \in W$, either $[s]_a \subseteq [s]_b$, or $[s]_b \subseteq [s]_a$. Validity of formulas φ on frames with local agreement is denoted $\models^{\mathsf{agree}} \varphi$. We show that on the class of frames with local agreement every formula in $\mathcal{L}_{\mathbb{D}}$ is equivalent to a formula in $\mathcal{L}_K$. Intuitively, local agreement formalizes that given any two agents, one always knows more than the other. A consequence easily shown is that in any state of a model there is a hierarchy between the agents (a total order) from one who knows most to one who knows least:

Lemma 5.6 *Let $(W, \sim)$ be a frame with local agreement. For all $B \subseteq A$, if $B \neq \emptyset$ then for all $s \in W$ there exists $a \in B$ such that for all $b \in B$, $[s]_a \subseteq [s]_b$.*

Lemma 5.7 *Preservation after update: Let $(W, \sim)$ be a frame with local agreement. For all $B \subseteq A$, the frame $(W, \sim^B)$ is a frame with local agreement.*

Definition 5.8 Let translation $(\delta_B)_{B \subseteq A}$ be the family of functions from $\mathcal{L}_{\mathbb{D}}$ to $\mathcal{L}_{\mathbb{D}}$ inductively defined below. For $\delta_{B_1}(\dots \delta_{B_m}(\varphi) \dots)$ we write $\delta_{\vec{B}}(\varphi))$.

$$\begin{array}{llll}
\delta_B(\varphi) = \varphi \text{ if } |B| \leq 1 & \delta_B(\neg\varphi) & = \neg\delta_B(\varphi) & \text{if } |B| \geq 2 \\
\delta_B(p) = p \text{ if } |B| \geq 2 & \delta_B(\varphi \wedge \psi) & = \delta_B(\varphi) \wedge \tau_B(\psi) & \text{if } |B| \geq 2 \\
\delta_B(\top) = \top \text{ if } |B| \geq 2 & \delta_B(\mathbb{D}_\emptyset \varphi) & = \mathbb{D}_\emptyset \delta_B(\varphi) & \text{if } |B| \geq 2
\end{array}$$

$$\begin{array}{ll}
\delta_B(\mathbb{D}_C \varphi) = \bigvee\{K_c \delta_B(\delta_C(\varphi)) \mid c \in C\} & \text{if } |B| \geq 2,\ C \neq \emptyset \text{ and } B \cap C = \emptyset \\
\delta_B(\mathbb{D}_C \varphi) = \bigvee\{K_c \delta_B(\delta_C(\varphi)) \mid c \in B \cup C\} & \text{if } |B| \geq 2,\ C \neq \emptyset \text{ and } B \cap C \neq \emptyset
\end{array}$$

Define $\delta : \mathcal{L}_{\mathbb{D}} \to \mathcal{L}_{\mathbb{D}}$ replacing every subformula of shape $\mathbb{D}_B \varphi$ by $\delta_B(\mathbb{D}_B \varphi)$.[1]

Lemma 5.9 *For all $B \subseteq A$ with $|B| \geq 2$ and for all $\varphi \in \mathcal{L}_D$, $\delta_B(\varphi) \in \mathcal{L}_{KU}$.*

Proof. By induction on φ. □

Lemma 5.10 *For all models $M = (W, \sim, V)$ with local agreement, formulas $\varphi \in \mathcal{L}_{\mathbb{D}}$, $s \in W$ and $B \subseteq A$: $M^B, s \models \varphi$ iff $M, s \models \delta_B(\varphi)$.*

---

[1] $\delta(p) = p$, $\delta(\top) = \top$, $\delta(\varphi \wedge \psi) = \delta(\varphi) \wedge \delta(\psi)$, $\delta(\neg\varphi) = \neg\delta(\varphi)$, and $\delta(\mathbb{D}_B \varphi) = \delta_B(\mathbb{D}_B \varphi)$

Proof. By induction on φ. We only show the case $\varphi = \mathbb{D}_C\psi$ where $C \neq \emptyset$. We assume $|B| \geq 2$, for otherwise the result trivially holds.

From left to right, suppose $M^B, s \models \mathbb{D}_C\psi$ and $M, s \not\models \delta_B(\mathbb{D}_C\psi)$. Now, there are two cases: either $B \cap C = \emptyset$, or $B \cap C \neq \emptyset$. In the former case, $M, s \not\models \bigvee\{K_c\delta_B(\delta_C(\psi)) \mid c \in C\}$. Hence, for all $c \in C$, $M, s \not\models K_c\delta_B(\delta_C(\psi))$. By Lemma 5.6, let $c \in C$ be such that for all $d \in C$, $[s]_c \subseteq [s]_d$. Thus, $M, s \not\models K_c\delta_B(\delta_C(\psi))$. Consequently, there exists $t \in W$ such that $s \sim_c t$ and $M, t \not\models \delta_B(\delta_C(\psi))$. Since for all $d \in C$, $[s]_c \subseteq [s]_d$, therefore $s \sim_C t$. Moreover, by induction hypothesis, $M^{BC}, t \not\models \psi$. Since $B \cap C = \emptyset$ and $s \sim_C t$, therefore $s \sim_C^B t$. Since $M^B, s \models \mathbb{D}_C\psi$, therefore $M^{BC}, t \models \psi$: a contradiction. In the latter case, $M, s \not\models \bigvee\{K_c\delta_B(\delta_C(\psi)) \mid c \in B \cup C\}$. Hence, for all $c \in B \cup C$, $M, s \not\models K_c\delta_B(\delta_C(\psi))$. By Lemma 5.6, let $c \in B \cup C$ be such that for all $d \in B \cup C$, $[s]_c \subseteq [s]_d$. Thus, $M, s \not\models K_c\delta_B(\delta_C(\psi))$. Consequently, there exists $t \in W$ such that $s \sim_c t$ and $M, t \not\models \delta_B(\delta_C(\psi))$. Since for all $d \in B \cup C$, $[s]_c \subseteq [s]_d$, therefore $s \sim_{B \cup C} t$. Moreover, by induction hypothesis, $M^{BC}, t \not\models \psi$. Since $B \cap C \neq \emptyset$ and $s \sim_{B \cup C} t$, therefore $s \sim_C^B t$. Since $M^B, s \models \mathbb{D}_C\psi$, therefore $M^{BC}, t \models \psi$: a contradiction.

From right to left, suppose $M^B, s \not\models \mathbb{D}_C\psi$ and $M, s \models \delta_B(\mathbb{D}_C\psi)$. Now, there are two cases: either $B \cap C = \emptyset$, or $B \cap C \neq \emptyset$. In the former case, $M, s \models \bigvee\{K_c\delta_B(\delta_C(\psi)) \mid c \in C\}$. Hence, there exists $c \in C$ such that $M, s \models K_c\delta_B(\delta_C(\psi))$. Since $M^B, s \not\models \mathbb{D}_C\psi$, therefore there exists $t \in W$ such that $s \sim_C^B t$ and $M^{BC}, t \not\models \psi$. Thus, by induction hypothesis, $M, t \not\models \delta_B(\delta_C(\psi))$. Since $B \cap C = \emptyset$ and $s \sim_C^B t$, therefore $s \sim_C t$. Since $c \in C$, therefore $s \sim_c t$. Since $M, s \models K_c\delta_B(\delta_C(\psi))$, therefore $M, t \models \delta_B(\delta_C(\psi))$: a contradiction. In the latter case, $M, s \models \bigvee\{K_c\delta_B(\delta_C(\psi)) \mid c \in B \cup C\}$. Hence, there exists $c \in B \cup C$ such that $M, s \models K_c\delta_B(\delta_C(\psi))$. Since $M^B, s \not\models \mathbb{D}_C\psi$, therefore there exists $t \in W$ such that $s \sim_C^B t$ and $M^{BC}, t \not\models \psi$. Thus, by induction hypothesis, $M, t \not\models \delta_B(\delta_C(\psi))$. Since $B \cap C \neq \emptyset$ and $s \sim_C^B t$, therefore $s \sim_{B \cup C} t$. Since $c \in B \cup C$, therefore $s \sim_c t$. Since $M, s \models K_c\delta_B(\delta_C(\psi))$, therefore $M, t \models \delta_B(\delta_C(\psi))$: a contradiction. □

Lemma 5.11 *For all* $\varphi \in \mathcal{L}_{\mathbb{D}}$, $|B| \geq 2$: $\models^{\mathsf{agree}} \mathbb{D}_B\varphi \leftrightarrow \bigvee\{K_b\delta_B(\varphi) \mid b \in B\}$.

Proof. Let $\varphi \in \mathcal{L}_{\mathbb{D}}$. For the sake of the contradiction, suppose $|B| \geq 2$ and $\not\models^{\mathsf{agree}} \mathbb{D}_B\varphi \leftrightarrow \bigvee\{K_b\delta_B(\varphi) \mid b \in B\}$. Hence, there exists a (M, s) with local agreement such that either $M, s \models \mathbb{D}_B\varphi$ and $M, s \not\models \bigvee\{K_b\delta_B(\varphi) \mid b \in B\}$, or $M, s \not\models \mathbb{D}_B\varphi$ and $M, s \models \bigvee\{K_b\delta_B(\varphi) \mid b \in B\}$, where $M = (W, \sim, V)$. In the former case, since $(W, \sim)$ is a frame with local agreement and $B \neq \emptyset$, therefore there exists $a \in B$ such that for all $b \in B$, $[s]_a \subseteq [s]_b$. Since $M, s \not\models \bigvee\{K_b\delta_B(\varphi) \mid b \in B\}$, therefore $M, s \not\models K_a\delta_B(\varphi)$. Thus, there exists $t \in W$ such that $s \sim_a t$ and $M, t \not\models \delta_B(\varphi)$. Consequently, by Lemma 5.10, $M^B, t \not\models \varphi$. Since $s \sim_a t$ and for all $b \in B$, $[s]_a \subseteq [s]_b$, therefore $s \sim_B t$. Since $M^B, t \not\models \varphi$, therefore $M, s \not\models \mathbb{D}_B\varphi$: a contradiction. In the latter case, there exists $t \in W$ such that $s \sim_B t$ and $M^B, t \not\models \varphi$. Hence, by Lemma 5.10, $M, t \not\models \delta_B(\varphi)$. Moreover, there exists $a \in B$ such that $M, s \models K_a\delta_B(\varphi)$. Since $s \sim_B t$, therefore $s \sim_a t$. Since $M, s \models K_a\delta_B(\varphi)$, therefore $M, t \models \delta_B(\varphi)$: a contradiction. □

Proposition 5.12 *Let* φ *in* $\mathcal{L}_{\mathbb{D}}$. *Then* $\models^{\mathsf{agree}} \varphi \leftrightarrow \delta(\psi)$.

Proof. By Lemmas 5.9 and 5.11. □

If $\varphi \in \mathcal{L}_{\mathbb{D}}$, then $\delta(\varphi) \in \mathcal{L}_{KU}$, and if $\varphi \in \mathcal{L}_{\mathbb{D}}$ without modality $\mathbb{D}_{\emptyset}$ (the universal modality), then $\delta(\varphi) \in \mathcal{L}_K$. Consequently, every φ in the language of dynamic distributed knowledge without the universal modality is equivalent to a formula in the multi-agent language of individual knowledge.

We initially considered frames with local agreement in order to obtain a complete axiomatization of dynamic distributed knowledge 'for an interesting fragment'. We ended up proving that this fragment can be translated into multi-agent individual knowledge. This seems to make it less interesting. However, local agreement might have worthwhile applications in modelling hierarchical knowledge relations between groups of agents.

6 On standard frames and semi-standard frames

Preparing the ground for Section 7 on the axiomatization, we generalize the epistemic frames introduced in Section 2, that we name *standard frames*, to the *semi-standard frames* wherein an equivalence relation $\sim_B$ is associated with each $B \subseteq A$ and that satisfy the requirement that $\sim_C \subseteq \sim_B$ whenever $B \subseteq C$. It may then be that $\sim_B$ is a proper subset of $\bigcap_{b \in B} \sim_b$ (the group knows more than the sum of its individuals). The completeness proofs involving distributed knowledge often involve 'unravelling' a canonical model based on a semi-standard frame into one that is based on a standard frame with the same information content [14,1]. We prove this result explicitly, as our proof applies to other frame classes (than those with equivalence relations), which may be of interest.

Definition 6.1 A frame $(W, \sim)$ is *semi-standard* if for all $B, C \subseteq A$, if $B \subseteq C$ then $\sim_C \subseteq \sim_B$. It is *standard* if, also, for all $B, C \subseteq A$, $\sim_B \cap \sim_C \subseteq \sim_{B \cup C}$.

We note that the results of Section 5.1 translating $\mathcal{L}_{\mathbb{D}}$ into $\mathcal{L}_D$ without loss of generality also hold for semi-standard frames: all clauses in the proof of crucial Lemma 5.4 showing that $M^B, u \models \varphi$ iff $M, u \models \tau_B(\varphi)$ equally apply to models based on semi-standard frames.

Definition 6.2 A *bounded morphism from a frame* $(W', \sim')$ *to a frame* $(W, \sim)$ is a function $\pi : W' \longrightarrow W$ such that

(**FC**) for all $B \subseteq A$ and for all $s', t' \in W'$, if $s' \sim'_B t'$ then $\pi(s') \sim_B \pi(t')$,

(**BC**) for all $B \subseteq A$, for all $s' \in W'$ and for all $t \in W$, if $\pi(s') \sim_B t$ then there exists $t' \in W'$ such that $s' \sim'_B t'$ and $\pi(t') = t$.

A frame $(W, \sim)$ is a *bounded morphic image of a frame* $(W', \sim')$ if there exists a bounded morphism from $(W', \sim')$ to $(W, \sim)$.

Since, in any frame, the accessibility relation associated to the empty set of agents is the universal relation on the frame, therefore in our setting every bounded morphism is a surjective function.

Lemma 6.3 *If the frame* $(W, \sim)$ *is a bounded morphic image of the frame* $(W', \sim')$ *then for all* $\varphi \in \mathcal{L}_{\mathbb{D}DR}$*, if* $(W', \sim') \models \varphi$ *then* $(W, \sim) \models \varphi$.

Proof. Suppose the frame $(W, \sim)$ is a bounded morphic image of the frame $(W', \sim')$. Let $\varphi \in \mathcal{L}_{\mathbb{D}DR}$. Suppose $(W', \sim') \models \varphi$. Suppose $(W, \sim) \not\models \varphi$. Hence, by Cor. 5.5, $(W, \sim) \not\models \tau_\emptyset(\varphi)$. Since the frame $(W, \sim)$ is a bounded morphic image of the frame $(W', \sim')$, therefore $(W', \sim') \not\models \tau_\emptyset(\varphi)$. See [10, Theorem 3.14] and [12, Cor. 3.16]. Thus, by Cor. 5.5, $(W', \sim') \not\models \varphi$: a contradiction. □

Lemma 6.4 *Let* $(W, \sim)$ *be a frame. Then* $(W, \sim)$ *is semi-standard, iff for all* $B, C \subseteq A$, $\sim_{B \cup C} \subseteq \sim_B \cap \sim_C$.

Lemma 6.5 *Every semi-standard frame is the bounded morphic image of a standard frame.*

Proof. Let $(W, \sim)$ be a semi-standard frame. For all $B \subseteq A$ and for all $u \in W$, let $[u]_B$ be the equivalence class of u modulo $\sim_B$. For all $X, Y \in \mathcal{P}(W)$, let $X + Y = (X \setminus Y) \cup (Y \setminus X)$. Notice that $(\mathcal{P}(W), \emptyset, W, +, \cap)$ is a Boolean ring. Let $\mathcal{F}$ be the set (with typical elements denoted α, β, etc) of all functions of type $\mathcal{P}(A) \times A \longrightarrow \mathcal{P}(W)$. Let $W' = W \times \mathcal{F}$. Let $\sim' : \mathcal{P}(A) \longrightarrow \mathcal{P}(W' \times W')$ be the function such that for all $B \subseteq A$, $\sim'_B$ is the binary relation on W' such that for all $(s, \alpha), (t, \beta) \in W'$, $(s, \alpha) \sim'_B (t, \beta)$ if and only if

(C₁) for all $E \subseteq A$, $[s]_E + \Sigma_{a \in E} \alpha(E, a) = [t]_E + \Sigma_{a \in E} \beta(E, a)$,

(C₂) for all $E \subseteq A$ and $a \in A$, if $a \in B$ and $a \in E$ then $\alpha(E, a) = \beta(E, a)$.

The proof of the lemma is now established in Facts 6.6, 6.7, 6.8, and 6.9. □

Fact 6.6 *The couple* $(W', \sim')$ *is a frame.*

Fact 6.7 *The frame* $(W', \sim')$ *is semi-standard.*

Proof. Let $B, B' \subseteq A$. Suppose $B \subseteq B'$. Suppose $\sim'_{B'} \not\subseteq \sim'_B$. Hence, there exist $(s, \alpha), (t, \beta) \in W'$ such that $(s, \alpha) \sim'_{B'} (t, \beta)$ and $(s, \alpha) \not\sim'_B (t, \beta)$. Thus, either there exists $E \subseteq A$ such that $[s]_E + \Sigma_{a \in E} \alpha(E, a) \neq [t]_E + \Sigma_{a \in E} \beta(E, a)$, or there exists $E \subseteq A$ and there exists $a \in A$ such that $a \in B$, $a \in E$ and $\alpha(E, a) \neq \beta(E, a)$. In the former case, since $(s, \alpha) \sim'_{B'} (t, \beta)$, therefore $[s]_E + \Sigma_{a \in E} \alpha(E, a) = [t]_E + \Sigma_{a \in E} \beta(E, a)$: a contradiction. In the latter case, since $B \subseteq B'$, therefore $a \in B'$. Since $(s, \alpha) \sim'_{B'} (t, \beta)$ and $a \in E$, therefore $\alpha(E, a) = \beta(E, a)$: a contradiction. □

Fact 6.8 *The semi-standard frame* $(W', \sim')$ *is standard.*

Proof. Let $B, B' \subseteq A$. Suppose $\sim'_{B \cup B'} \not\supseteq \sim'_B \cap \sim'_{B'}$. Hence, there exist $(s, \alpha), (t, \beta) \in W'$ such that $(s, \alpha) \not\sim'_{B \cup B'} (t, \beta)$, $(s, \alpha) \sim'_B (t, \beta)$ and $(s, \alpha) \sim'_{B'} (t, \beta)$. Thus, for all $F \subseteq A$, $[s]_F + \Sigma_{b \in F} \alpha(F, b) = [t]_F + \Sigma_{b \in F} \beta(F, b)$. Moreover, for all $F \subseteq A$ and for all $b \in A$, if $b \in B$ and $b \in F$ then $\alpha(F, b) = \beta(F, b)$ and for all $F \subseteq A$ and for all $b \in A$, if $b \in B'$ and $b \in F$ then $\alpha(F, b) = \beta(F, b)$. Since $(s, \alpha) \not\sim'_{B \cup B'} (t, \beta)$, therefore there exists $E \subseteq A$ and there exists $a \in A$ such that $a \in B \cup B'$, $a \in E$ and $\alpha(E, a) \neq \beta(E, a)$. Consequently, either $a \in B$, or $a \in B'$. In the former case, since for all $F \subseteq A$ and for all $b \in A$, if $b \in B$ and $b \in F$ then $\alpha(F, b) = \beta(F, b)$, therefore

$\alpha(E, a) = \beta(E, a)$: a contradiction. In the latter case, since for all $F \subseteq A$ and for all $b \in A$, if $b \in B'$ and $b \in F$ then $\alpha(F, b) = \beta(F, b)$, therefore $\alpha(E, a) = \beta(E, a)$: a contradiction. □

Let $\pi : W' \longrightarrow W$ be the function such that for all $(s, \alpha) \in W'$, $\pi(s, \alpha) = s$.

Fact 6.9 *π is a bounded morphism from $(W', \sim')$ to $(W, \sim)$.*

Proof. Let $B \subseteq A$ and $(s, \alpha), (t, \beta) \in W'$ be such that $(s, \alpha)\sim'_B(t, \beta)$. Hence, for all $E \subseteq A$, $[s]_E + \Sigma_{a \in E}\alpha(E, a) = [t]_E + \Sigma_{a \in E}\beta(E, a)$. Moreover, for all $E \subseteq A$ and for all $a \in A$, if $a \in B$ and $a \in E$ then $\alpha(E, a) = \beta(E, a)$. Thus, $[s]_B + \Sigma_{a \in B}\alpha(B, a) = [t]_B + \Sigma_{a \in B}\beta(B, a)$. Moreover, for all $a \in A$, if $a \in B$ then $\alpha(B, a) = \beta(B, a)$. Consequently, $[s]_B = [t]_B$. Hence, $s\sim_B t$.

Let $B \subseteq A$, $(s, \alpha) \in W'$ and $t \in W$ be such that $\pi(s, \alpha)\sim_B t$. Thus, $s\sim_B t$. Consequently, for all $E \subseteq A$, if $E \subseteq B$ then $s\sim_E t$. Hence, for all $E \subseteq A$, if $E \subseteq B$ then $[s]_E = [t]_E$. For all $E \subseteq A$, if $E \not\subseteq B$ then let $\zeta(E) \in A$ be such that $\zeta(E) \in E$ and $\zeta(E) \notin B$. Let β be the function of type $\mathcal{P}(A) \times A \longrightarrow \mathcal{P}(W)$ such that for all $E \subseteq A$ and for all $a \in A$,

- if $a \in E$ and $a \in B$ then $\beta(E, a) = \alpha(E, a)$,
- elsif $a \in E$, $a \notin B$ and $a = \zeta(E)$ then $\beta(E, a) = \Sigma_{b \in E \setminus B}\alpha(E, b)$,
- else $\beta(E, a) = \emptyset$.

The reader may easily verify that $(s, \alpha)\sim'_B(t, \beta)$. □

Proposition 6.10 *For all $\varphi \in \mathcal{L}_{\mathbb{D}R}$, if φ is valid on standard frames then φ is valid on semi-standard frames.*

Proof. By Lemmas 6.3 and 6.5. □

The same proof technique can be used for frames with other than equivalence relations $\sim_a$, and thus used for extensions with distributed knowledge of other logics than S5, such as KD45 (consistent belief) and KB4 (partial equivalence relations).

7 Axiomatization

7.1 Introduction

There are multiple roads towards determining whether a formula in $\psi \in \mathcal{L}_{\mathbb{D}}$ is semantically equivalent to a formula ψ' in some other language, and that is a theorem in an axiomatization for the logic of that language. All these inspire the quest to find an axiomatization of $L_{\mathbb{D}}$. (i) Replace in ψ all $\mathbb{D}_B$ by $R_B D_B$ and determine whether the resulting ψ'' is a theorem in the axiomatization of the logic of [1], which proceeds by rewriting $\psi'' \in \mathcal{L}_{DR}$ into a $\psi' \in \mathcal{L}_D$, and then determine whether ψ' is a theorem of the logic of distributed knowledge [14]. We considered axiomatizing the fragment of [1] with packed modalities $R_B D_B \varphi$ [36]. (ii) Use the translation $\tau_\emptyset : \mathcal{L}_{\mathbb{D}} \to \mathcal{L}_D$ of Section 5 (Def. 5.1) and determine whether $\psi' = \tau_\emptyset(\psi)$ is a theorem in L_D [14]. We considered axiomatizing the clauses τ_B of that translation. (iii) Axiomatize the logic for special frame classes: use the translations δ_B of Section 5 and Prop. 5.12 for frames with local agreement. Thus we obtain a $\psi' \in \mathcal{L}_{KU}$, or even in $\mathcal{L}_K$ if ψ did not contain $\mathbb{D}_\emptyset$ (the universal modality), which can be determined a theorem in the logic (multi-)S5. This frame restriction trivializes $L_{\mathbb{D}}$.

None of these roads, despite looking hard, provided us with an axiomatization of $L_{\mathbb{D}}$. The 'hybrid' nature of $\mathbb{D}_B$ that is simultaneously statically and dynamically interpreted is a complication. A similar case seems the *logic of knowability* [35] with quantifiers over announcements but no announcement modalities, for which no axiomatization is known, unlike for APAL [6].

Instead of a complete axiomatization for the logic $L_{\mathbb{D}}$ of dynamic distributed knowledge we therefore present a complete axiomatization for the logic $L_{\mathbb{D}R}$ of dynamic distributed knowledge *and resolution.* Our results may compare to the axiomatizations in [1] for the logic L_{DR}, and to a variant of that in [8].

7.2 The logic of dynamic distributed knowledge and resolution

Definition 7.1 Let $\boldsymbol{L}_{\mathbb{D}R}$ be the least set of $\mathcal{L}_{\mathbb{D}R}$-formulas produced by the below, where in (IR2): $\mathsf{see}_{E_1}(\vec{B}_1) \subseteq \mathsf{see}_C(\vec{B})$, ..., $\mathsf{see}_{E_k}(\vec{B}_k) \subseteq \mathsf{see}_C(\vec{B})$.

$(F0)$	all propositional tautologies	$(F5)$	$R_B p \leftrightarrow p$
$(F1)$	$R_B(\varphi \to \psi) \to (R_B\varphi \to R_B\psi)$	$(F6)$	$\neg R_B \bot$
$(F2)$	$\mathbb{D}_B\varphi \to R_B\varphi$	$(F7)$	$R_B\neg\varphi \leftrightarrow \neg R_B\varphi$
$(F3)$	$\mathbb{D}_B\varphi \to R_B\mathbb{D}_B\varphi$	$(IR0)$	from $\varphi \to \psi$ and φ infer ψ
$(F4)$	$\neg\mathbb{D}_B\varphi \to \mathbb{D}_B\neg\mathbb{D}_B\varphi$	$(IR1)$	from φ infer $R_B\varphi$

$$(IR2) \text{ from } R_{\vec{B}_1} R_{E_1}\varphi_1 \wedge \ldots \wedge R_{\vec{B}_k} R_{E_k}\varphi_k \to R_{\vec{B}} R_C\psi \text{ infer}$$
$$R_{\vec{B}_1} \mathbb{D}_{E_1}\varphi_1 \wedge \ldots \wedge R_{\vec{B}_k} \mathbb{D}_{E_k}\varphi_k \to R_{\vec{B}} \mathbb{D}_C\psi$$

Note that rather few axioms and rules of $\boldsymbol{L}_{\mathbb{D}R}$ involve dynamic distributed knowledge $\mathbb{D}_B$, namely $(F2)$, $(F3)$, $(F4)$, and $(IR2)$, and that rule $(IR2)$ is very involved: there is an instantiation of $(IR2)$ for every $k \in \mathbb{N}$. This suggests a complete axiomatization for $L_{\mathbb{D}}$ might also need a complex derivation rule.

Lemma 7.2 *Derivable in* $\boldsymbol{L}_{\mathbb{D}R}$ *are* $\mathbb{D}_B(\varphi \to \psi) \to (\mathbb{D}_B\varphi \to \mathbb{D}_B\psi)$, $R_B(\varphi \vee \psi) \leftrightarrow (R_B\varphi \vee R_B\psi)$, $R_B(\varphi \wedge \psi) \leftrightarrow (R_B\varphi \wedge R_B\psi)$, $\neg\mathbb{D}_B\bot$, *and '*φ *impl.* $\mathbb{D}_B\varphi$*'.*

Proof. This is shown by using $(F1)$, $(F6)$, $(F7)$, $(IR1)$ and $(IR2)$. □

Interestingly, we were not able to find a derivation of the validity $\mathbb{D}_B\varphi \to \mathbb{D}_B\mathbb{D}_B\varphi$ in this axiomatization $\boldsymbol{L}_{\mathbb{D}R}$. It is derivable as the axiomatization will be shown complete.

Lemma 7.3 *For all* $\varphi \in \mathcal{L}_{\mathbb{D}R}$*, if* $\varphi \in \boldsymbol{L}_{\mathbb{D}R}$ *then* φ *is valid on standard frames.*

Proof. It suffices to demonstrate that the $\mathcal{L}_{\mathbb{D}R}$-formulas $(F1)$–$(F7)$ are valid on standard frames and the inference rules $(IR1)$ and $(IR2)$ preserve validity on standard frames. We only show the case of the inference rule $(IR2)$.

Suppose the inference rule $(IR2)$ does not preserve validity on standard frames. Hence, there exists $k \in \mathbb{N}$, there exist $\vec{B}_1, \ldots, \vec{B}_k \in \mathcal{P}(A)^*$, there exist $E_1, \ldots, E_k \subseteq A$, there exists $\vec{B} \in \mathcal{P}(A)^*$ and there exists $C \subseteq A$ such that

$\mathsf{see}_{E_1}(\vec{B}_1) \subseteq \mathsf{see}_C(\vec{B})$, ..., $\mathsf{see}_{E_k}(\vec{B}_k) \subseteq \mathsf{see}_C(\vec{B})$ and there exists $\varphi_1, \ldots, \varphi_k \in \mathcal{L}_{\mathbb{D}R}$ and there exists $\psi \in \mathcal{L}_{\mathbb{D}R}$ such that $R_{\vec{B}_1} R_{E_1}\varphi_1 \wedge \ldots \wedge R_{\vec{B}_k} R_{E_k}\varphi_k \rightarrow R_{\vec{B}} R_C\psi$ is valid on standard frames and $R_{\vec{B}_1} \mathbb{D}_{E_1}\varphi_1 \wedge \ldots \wedge R_{\vec{B}_k} \mathbb{D}_{E_k}\varphi_k \rightarrow R_{\vec{B}} \mathbb{D}_C\psi$ is not valid on standard frames. Thus, there exists a standard model $M = (W, \sim, V)$ and $u \in W$ such that $M, u \models R_{\vec{B}_1} \mathbb{D}_{E_1}\varphi_1$, ..., $M, u \models R_{\vec{B}_k} \mathbb{D}_{E_k}\varphi_k$ and $M, u \not\models R_{\vec{B}} \mathbb{D}_C\psi$. Consequently, there exists $v \in W$ such that $u \sim_C^{\vec{B}} v$ and $M^{\vec{B}C}, v \not\models \psi$. Hence, by Lemma 2.3, $u \sim_{\mathsf{see}(\vec{B},C)} v$. Moreover, $M, v \not\models R_{\vec{B}} R_C\psi$. Since $R_{\vec{B}_1} R_{E_1}\varphi_1 \wedge \ldots \wedge R_{\vec{B}_k} R_{E_k}\varphi_k \rightarrow R_{\vec{B}} R_C\psi$ is valid on standard frames, therefore there exists $i \in (k)$ such that $M, v \not\models R_{\vec{B}_i} R_{E_i}\varphi_i$. Thus, $M^{\vec{B}_i E_i}, v \not\models \varphi_i$. Since $\mathsf{see}_{E_i}(\vec{B}_i) \subseteq \mathsf{see}_C(\vec{B})$, from $u \sim_{\mathsf{see}_C(\vec{B})} v$ we obtain $u \sim_{\mathsf{see}_{E_i}(\vec{B}_i)} v$. Consequently, by Lemma 2.3, $u \sim_{E_i}^{\vec{B}_i} v$. Since $M^{\vec{B}_i E_i}, v \not\models \varphi_i$, therefore $M, u \not\models R_{\vec{B}_i} \mathbb{D}_{E_i}\varphi_i$: a contradiction. □

7.3 Completeness

We now proceed to show the completeness of the logic $\boldsymbol{L}_{\mathbb{D}R}$.

A set s of $\mathcal{L}_{\mathbb{D}R}$-formulas is $\boldsymbol{L}_{\mathbb{D}R}$-*consistent* if for all $n \in \mathbb{N}$ and for all $\varphi_1, \ldots, \varphi_n \in s$, $\neg(\varphi_1 \wedge \ldots \wedge \varphi_n) \notin \boldsymbol{L}_{\mathbb{D}R}$.

Lemma 7.4 *$\boldsymbol{L}_{\mathbb{D}R}$ is an $\boldsymbol{L}_{\mathbb{D}R}$-consistent set of $\mathcal{L}_{\mathbb{D}R}$-formulas.*

Proof. By Lemma 7.3. □

An $\boldsymbol{L}_{\mathbb{D}R}$-consistent set s of $\mathcal{L}_{\mathbb{D}R}$-formulas is *maximal* if for all $\boldsymbol{L}_{\mathbb{D}R}$-consistent sets t of $\mathcal{L}_{\mathbb{D}R}$-formulas, if $s \subseteq t$ then $s = t$.

Lemma 7.5 *For all $\boldsymbol{L}_{\mathbb{D}R}$-consistent sets s of $\mathcal{L}_{\mathbb{D}R}$-formulas, there exists a maximal $\boldsymbol{L}_{\mathbb{D}R}$-consistent set t of $\mathcal{L}_{\mathbb{D}R}$-formulas such that $s \subseteq t$.*

Proof. See [10, Lemma 4.17] and [12, Lemma 5.1] for details. □

For convenience of the exposition, from here on we write $\boldsymbol{L}$ instead of $\boldsymbol{L}_{\mathbb{D}R}$ in this section.

Let $W_{\boldsymbol{L}}$ be the set of all maximal $\boldsymbol{L}$-consistent sets of all $\mathcal{L}_{\mathbb{D}R}$-formulas. Let $\sim_{\boldsymbol{L}}$: $\mathcal{P}(A) \longrightarrow \mathcal{P}(W_{\boldsymbol{L}} \times W_{\boldsymbol{L}})$ be the function such that for all $C \subseteq A$, $\sim_{\boldsymbol{L},C}$ is the binary relation on $W_{\boldsymbol{L}}$ such that for all $s, t \in W_{\boldsymbol{L}}$, $s \sim_{\boldsymbol{L},C} t$ if and only if for all $\vec{B} \in \mathcal{P}(A)^*$ and for all $E \subseteq A$, if $\mathsf{see}_E(\vec{B}) \subseteq C$ then for all $\varphi \in \mathcal{L}_{\mathbb{D}R}$, if $R_{\vec{B}} \mathbb{D}_E\varphi \in s$ then $R_{\vec{B}} R_E\varphi \in t$.

Lemma 7.6 *The pair $(W_{\boldsymbol{L}}, \sim_{\boldsymbol{L}})$ is a frame.*

Proof. By Lemmas 7.4 and 7.5, $W_{\boldsymbol{L}}$ is nonempty. Hence, it suffices to demonstrate that for all $C \subseteq A$, $\sim_{\boldsymbol{L},C}$ is reflexive and Euclidean (and therefore an equivalence relation). Let $C \subseteq A$.

Suppose $\sim_{\boldsymbol{L},C}$ is not reflexive. Thus, there exists $s \in W_{\boldsymbol{L}}$ such that $s \not\sim_{\boldsymbol{L},C} s$. Consequently, there are $\vec{B} \in \mathcal{P}(A)^*$ and $E \subseteq A$ such that $\mathsf{see}_E(\vec{B}) \subseteq C$ and there is $\varphi \in \mathcal{L}_{\mathbb{D}R}$ such that $R_{\vec{B}} \mathbb{D}_E\varphi \in s$ and $R_{\vec{B}} R_E\varphi \notin s$. Hence, by using $(F1)$, $(F2)$ and $(IR1)$, $R_{\vec{B}} R_E\varphi \in s$: a contradiction.

Suppose $\sim_{\boldsymbol{L},C}$ is not Euclidean. Thus, there exist $s, t, u \in W_{\boldsymbol{L}}$ such that $s \sim_{\boldsymbol{L},C} t$, $s \sim_{\boldsymbol{L},C} u$ and $t \not\sim_{\boldsymbol{L},C} u$. Consequently, there are $\vec{B} \in \mathcal{P}(A)^*$

and $E \subseteq A$ such that $\mathsf{see}_E(\vec{B}) \subseteq C$ and there exists $\varphi \in \mathcal{L}_{\mathbb{D}R}$ such that $R_{\vec{B}}\mathbb{D}_E\varphi \in t$ and $R_{\vec{B}}R_E\varphi \notin u$. Since $s \sim_{\boldsymbol{L},C} u$ and $\mathsf{see}_E(\vec{B}) \subseteq C$, therefore $R_{\vec{B}}\mathbb{D}_E\varphi \notin s$. Hence, by using $(F1)$, $(F7)$ and $(IR1)$, $R_{\vec{B}}\neg\mathbb{D}_E\varphi \in s$. Thus, by using $(F1)$, $(F4)$ and $(IR1)$, $R_{\vec{B}}\mathbb{D}_E\neg\mathbb{D}_E\varphi \in s$. Since $s \sim_{\boldsymbol{L},C} t$ and $\mathsf{see}_E(\vec{B}) \subseteq C$, therefore $R_{\vec{B}}R_E\neg\mathbb{D}_E\varphi \in t$. Consequently, by using $(F1)$, $(F7)$ and $(IR1)$, $R_{\vec{B}}R_E\mathbb{D}_E\varphi \notin t$. Hence, by using $(F1)$, $(F3)$ and $(IR1)$, $R_{\vec{B}}\mathbb{D}_E\varphi \notin t$: a contradiction. □

Lemma 7.7 *The frame* $(W_{\boldsymbol{L}}, \sim_{\boldsymbol{L}})$ *is semi-standard.*

Proof. Let $C, C' \subseteq A$. Suppose $C \subseteq C'$. Suppose $\sim_{\boldsymbol{L},C'} \not\subseteq \sim_{\boldsymbol{L},C}$. Hence, there exists $s, t \in W_{\boldsymbol{L}}$ such that $s \sim_{\boldsymbol{L},C'} t$ and $s \not\sim_{\boldsymbol{L},C} t$. Thus, there exists $\vec{B} \in \mathcal{P}(A)^*$ and there exists $E \subseteq A$ such that $\mathsf{see}_E(\vec{B}) \subseteq C$ and there exists $\varphi \in \mathcal{L}_{\mathbb{D}R}$ such that $R_{\vec{B}}\mathbb{D}_E\varphi \in s$ and $R_{\vec{B}}R_E\varphi \notin t$. Since $C \subseteq C'$, therefore $\mathsf{see}_E(\vec{B}) \subseteq C'$. Since $s \sim_{\boldsymbol{L},C'} t$ and $R_{\vec{B}}\mathbb{D}_E\varphi \in s$, therefore $R_{\vec{B}}R_E\varphi \in t$: a contradiction. □

Definition 7.8 The semi-standard frame $(W_{\boldsymbol{L}}, \sim_{\boldsymbol{L}})$ is called *canonical frame of* $\boldsymbol{L}$. The *canonical valuation of* $\boldsymbol{L}$ is the function $V_{\boldsymbol{L}}$ associating to each atom p the set $V_{\boldsymbol{L}}(p)$ of all $s \in W_{\boldsymbol{L}}$ such that $p \in s$. The triple $M_{\boldsymbol{L}} = (W_{\boldsymbol{L}}, \sim_{\boldsymbol{L}}, V_{\boldsymbol{L}})$ is called *canonical model of* $\boldsymbol{L}$.

Lemma 7.9 *Let* $\varphi \in \mathcal{L}_{\mathbb{D}R}$. *For all* $s \in W_{\boldsymbol{L}}$ *and for all* $\vec{B} \in \mathcal{P}(A)^*$, $M_{\boldsymbol{L}}^{\vec{B}}, s \models \varphi$ *if and only if* $R_{\vec{B}}\varphi \in s$.

Proof. By induction on φ. We only show the case $\mathbb{D}_C\psi$.

From right to left, suppose $M_{\boldsymbol{L}}^{\vec{B}}, s \not\models \mathbb{D}_C\psi$ and $R_{\vec{B}}\mathbb{D}_C\psi \in s$. Hence, there exists $t \in W_{\boldsymbol{L}}$ such that $s \sim_{\boldsymbol{L},C}^{\vec{B}} t$ and $M_{\boldsymbol{L}}^{\vec{B}C}, t \not\models \psi$. Thus, by Lemma 2.3, $s \sim_{\boldsymbol{L},\mathsf{see}_C(\vec{B})} t$. Moreover, by induction hypothesis, $R_{\vec{B}}R_C\psi \notin t$. Since $R_{\vec{B}}\mathbb{D}_C\psi \in s$, therefore $R_{\vec{B}}R_C\psi \in t$: a contradiction.

From left to right, suppose $M_{\boldsymbol{L}}^{\vec{B}}, s \models \mathbb{D}_C\psi$ and $R_{\vec{B}}\mathbb{D}_C\psi \notin s$. Let $t_0 = \{R_{\vec{B'}}R_{E'}\chi \mid \vec{B'} \in \mathcal{P}(A)^*$ and $E' \subseteq A$ are such that $\mathsf{see}_{E'}(\vec{B'}) \subseteq \mathsf{see}_C(\vec{B})$ and $\chi \in \mathcal{L}_{\mathbb{D}R}$ is such that $R_{\vec{B'}}\mathbb{D}_{E'}\chi \in s\} \cup \{\neg R_{\vec{B}}R_C\psi\}$.

We first observe that t_0 is an $\boldsymbol{L}$-consistent set of $\mathcal{L}_{\mathbb{D}R}$-formulas. This can be shown as follows. Suppose t_0 is not an $\boldsymbol{L}$-consistent set of $\mathcal{L}_{\mathbb{D}R}$-formulas. Hence, there exist $k \in \mathbb{N}$, $\vec{B'_1}, \ldots, \vec{B'_k} \in \mathcal{P}(A)^*$ and $E'_1, \ldots, E'_k \subseteq A$ such that $\mathsf{see}_{E'_1}(\vec{B'_1}) \subseteq \mathsf{see}_E(\vec{B})$, …, $\mathsf{see}_{E'_k}(\vec{B'_k}) \subseteq \mathsf{see}_E(\vec{B})$, and there exist $\chi_1, \ldots, \chi_k \in \mathcal{L}_{\mathbb{D}R}$ such that $R_{\vec{B'_1}}\mathbb{D}_{E'_1}\chi_1, \ldots, R_{\vec{B'_k}}\mathbb{D}_{E'_k}\chi_k \in s$ and $R_{\vec{B'_1}}R_{E'_1}\chi_1 \wedge \ldots \wedge R_{\vec{B'_k}}R_{E'_k}\chi_k \to R_{\vec{B}}R_C\psi \in \boldsymbol{L}$. Thus, by using $(IR2)$, $R_{\vec{B'_1}}\mathbb{D}_{E'_1}\chi_1 \wedge \ldots \wedge R_{\vec{B'_k}}\mathbb{D}_{E'_k}\chi_k \to R_{\vec{B}}\mathbb{D}_C\psi \in \boldsymbol{L}$. Since $R_{\vec{B'_1}}\mathbb{D}_{E'_1}\chi_1, \ldots, R_{\vec{B'_k}}\mathbb{D}_{E'_k}\chi_k \in s$, therefore $R_{\vec{B}}\mathbb{D}_C\psi \in s$: a contradiction.

Now using that t_0 is an $\boldsymbol{L}$-consistent set of $\mathcal{L}_{\mathbb{D}R}$-formulas, by Lemma 7.5, there exists $t \in W_{\boldsymbol{L}}$ such that $t_0 \subseteq t$. Hence, for all $\vec{B'} \in \mathcal{P}(A)^*$ and $E' \subseteq A$, if $\mathsf{see}_{E'}(\vec{B'}) \subseteq \mathsf{see}_C(\vec{B})$ then for all $\chi \in \mathcal{L}_{\mathbb{D}R}$, if $R_{\vec{B'}}\mathbb{D}_{E'}\chi \in s$ then $R_{\vec{B'}}R_{E'}\chi \in t$. Moreover, $\neg R_{\vec{B}}R_C\psi \in t$. Thus, $s \sim_{\mathsf{see}_C(\vec{B})} t$. Also, $R_{\vec{B}}R_C\psi \notin t$. Consequently, by Lemma 2.3, $s \sim_C^{\vec{B}} t$. Since $M_{\boldsymbol{L}}^{\vec{B}}, s \models \mathbb{D}_C\psi$, therefore $M_{\boldsymbol{L}}^{\vec{B}C}, t \models \psi$. Hence, by induction hypothesis, $R_{\overrightarrow{B}} R_C \psi \in t$: a contradiction. □

Lemma 7.10 *For all $\varphi \in \mathcal{L}_{\mathbb{D}R}$, φ is true in $M_{\boldsymbol{L}}$ if and only if $\varphi \in \boldsymbol{L}$.*

Proof. By Lemma 7.9. □

Lemma 7.11 *For all $\varphi \in \mathcal{L}_{\mathbb{D}R}$, if φ is valid in the class of all semi-standard frames then $\varphi \in \boldsymbol{L}$.*

Proof. By Lemmas 7.7 and 7.10. □

Theorem 7.12 *For all $\varphi \in \mathcal{L}_{\mathbb{D}R}$: (i) $\varphi \in \boldsymbol{L}$, iff (ii) φ is valid on standard frames, iff (iii) φ is valid on semi-standard frames.*

Proof. By Lemmas 6.10, 7.3 and 7.11. □

8 Conclusions and further research

Conclusions. We proposed a logic of *dynamic distributed knowledge* with modalities combining static and dynamic features. We considered its validities, gave translations to static distributed knowledge, showed its decidability, proposed and characterized a notion of shared bisimulation, and gave a complete axiomatization for the extension with resolution.

Further research Interpreting dynamic distributed knowledge on Kripke models with other frame properties than equivalence relations, extending S5, is worth pursuing. Dynamic distributed knowledge based on K4 (*transitive* models) is investigated in [13]. Static distributed knowledge based on KB4 (models with *partial equivalence relations*) is investigated in distributed computing [17] to model *crashed* (or *dead*) agents, or *faulty* agents [32]. A crashed or faulty agent satisfies $K_a\bot$. We also consider extending the language $\mathcal{L}_{\mathbb{D}}$ with *common knowledge*. It is easy to see that $\mathbb{D}_B\varphi \leftrightarrow \mathbb{D}_B C_B \varphi$ is valid. We conjecture this is the only interaction axiom needed in order to get a complete axiomatization, extending a presumed one for $\mathbb{D}$ and the standard one for C [34]. Finally we consider incorporating *relativized common knowledge* [24,30], (dynamic) *common distributed knowledge* [8,33], semi-standard Kripke models ($\sim_C \subseteq \sim_B$ for $B \subseteq C$) [7,17], and applications in game theory and economics [13].

Acknowledgements At the origin of these ideas are Roman Kniazev and Michele Crescenzi. We are greatful for their interaction and for their support. We thank the AiML reviewers for their comments.

References

[1] Ågotnes, T. and Y. Wáng, *Resolving distributed knowledge*, Artif. Intell. **252** (2017), pp. 1–21.

[2] Alechina, N., P. Balbiani and D. Shkatov, *Modal logics for reasoning about infinite unions and intersections of binary relations*, J. Appl. Non Class. Logics **22** (2012), pp. 275–294.

[3] Apt, K. and D. Wojtczak, *Verification of distributed epistemic gossip protocols*, J. Artif. Intell. Res. **62** (2018), pp. 101–132.

[4] Attamah, M., H. van Ditmarsch, D. Grossi and W. van der Hoek, *Knowledge and gossip*, in: *Proc. of 21st ECAI* (2014), pp. 21–26.

[5] Attamah, M., H. van Ditmarsch, D. Grossi and W. van der Hoek, *The pleasure of gossip*, in: C. Başkent, L. Moss and R. Ramanujam, editors, *Rohit Parikh on Logic, Language and Society* (2017), pp. 145–163.

[6] Balbiani, P., A. Baltag, H. van Ditmarsch, A. Herzig, T. Hoshi and T. D. Lima, *'Knowable' as 'known after an announcement'*, Review of Symbolic Logic **1(3)** (2008), pp. 305–334.

[7] Baltag, A. and S. Smets, *Correlated knowledge: an epistemic-logic view on quantum entanglement*, International Journal of Theoretical Physics **49** (2010), pp. 3005–3021.

[8] Baltag, A. and S. Smets, *Learning what others know*, in: *Proc. of 23rd LPAR*, EPiC Series in Computing **73**, 2020, pp. 90–119.

[9] Baltag, A. and S. Smets, *Logics for data exchange and communication* (2024), to appear in the proceedings of the 15th AiML Prague.

[10] Blackburn, P., M. de Rijke and Y. Venema, "Modal Logic," Cambridge University Press, 2001.

[11] Castañeda, A., H. van Ditmarsch, D. Rosenblueth and D. Velázquez, *Communication pattern logic: Epistemic and topological views*, Journal of Philosophical Logic (2023).

[12] Chagrov, A. and M. Zakharyaschev, "Modal Logic," Oxford University Press, 1997.

[13] Crescenzi, M., *Group knowledge and individual introspection*, CoRR **abs/2305.08729** (2023), manuscript.

[14] Fagin, R., J. Halpern and M. Vardi, *What can machines know? on the properties of knowledge in distributed systems*, J. ACM **39** (1992), pp. 328–376.

[15] Galimullin, R. and L. Kuijer, *Varieties of distributed knowledge* (2024), to appear in the proceedings of the 15th AiML Prague.

[16] Gargov, G., *Two completeness theorems in the logic for data analysis*, Technical report, Institute of Computer Science, Polish Academy of Sciences (1986), Tech. Report 581.

[17] Goubault, E., R. Kniazev, J. Ledent and S. Rajsbaum, *Semi-simplicial set models for distributed knowledge*, in: *LICS*, 2023, pp. 1–13.

[18] Hakli, R. and S. Negri, *Proof theory for distributed knowledge*, in: F. Sadri and K. Satoh, editors, *Proc. of the 8th CLIMA*, Lecture Notes in Computer Science **5056**, 2007, pp. 100–116.

[19] Halpern, J. and Y. Moses, *Knowledge and common knowledge in a distributed environment*, in: *Proc. of the 3rd PODC*, 1984, pp. 50–61.

[20] Halpern, J. and Y. Moses, *Knowledge and common knowledge in a distributed environment*, Journal of the ACM **37(3)** (1990), pp. 549–587.

[21] Hayek, F., *The use of knowledge in society*, American Economic Review **35** (1945), pp. 519–530.

[22] Hilpinen, R., *Remarks on personal and impersonal knowledge*, Canadian Journal of Philosophy **7** (1977), pp. 1–9.

[23] Kermarrec, A.-M. and M. van Steen, *Gossiping in distributed systems*, SIGOPS Oper. Syst. Rev. **41** (2007), pp. 2–7.

[24] Kooi, B. and J. van Benthem, *Reduction axioms for epistemic actions*, in: *Proc. of Advances in Modal Logic*, University of Manchester, 2004, pp. 197–211.

[25] Lindqvist, J., F. Velázquez-Quesada and T. Ågotnes, *Cautious distributed belief*, in: C. Areces and D. Costa, editors, *Proc. of the 4th DaLí*, Lecture Notes in Computer Science **13780**, 2022, pp. 106–124.

[26] Mukund, M. and M. Sohoni, *Keeping track of the latest gossip in a distributed system*, Distributed Computing **10** (1997), pp. 137–148.

[27] Parikh, R. and R. Ramanujam, *Distributed processes and the logic of knowledge*, in: *Proc. of Logics of Programs*, Lecture Notes in Computer Science **193**, 1985, pp. 256–268.

[28] Roelofsen, F., *Distributed knowledge*, Journal of Applied Non-Classical Logics **17** (2007), pp. 255–273.

[29] Swanson, D., *Undiscovered public knowledge*, The Library Quarterly: Information, Community, Policy **56** (1986), pp. 103–118.

[30] van Benthem, J., J. van Eijck and B. Kooi, *Logics of communication and change*, Information and Computation **204(11)** (2006), pp. 1620–1662.

[31] van der Hoek, W. and J.-J. Meyer, *Making some issues of implicit knowledge explicit*, Int. J. Found. Comput. Sci. **3** (1992), pp. 193–223.

[32] van Ditmarsch, H., K. Fruzsa and R. Kuznets, *A new hope*, in: D. Fernández-Duque and A. Palmigiano, editors, *Proc. of the 14th AiML* (2022), pp. 349–369.

[33] van Ditmarsch, H., E. Goubault, J. Ledent and S. Rajsbaum, *Knowledge and simplicial complexes*, in: B. Lundgren and N. Nuñez Hernández, editors, *Philosophy of Computing, Philosophical Studies Series 143* (2022), pp. 1–50.

[34] van Ditmarsch, H., J. Halpern, W. van der Hoek and B. Kooi, *An introduction to logics of knowledge and belief*, in: H. van Ditmarsch, J. Halpern, W. van der Hoek and B. Kooi, editors, *Handbook of epistemic logic*, 2015, pp. 1–51.

[35] van Ditmarsch, H., W. van der Hoek and P. Iliev, *Everything is knowable – how to get to know whether a proposition is true*, Theoria **78(2)** (2012), pp. 93–114.

[36] Wang, Y., *Beyond knowing that: A new generation of epistemic logics*, in: H. van Ditmarsch and G. Sandu, editors, *Jaakko Hintikka on knowledge and game-theoretical semantics*, Outstanding contributions to logic 12 (2018), pp. 499–533.

[37] Wáng, Y. and T. Ågotnes, *Simpler completeness proofs for modal logics with intersection*, in: M. Martins and I. Sedlár, editors, *Proc. of the 3rd DaLí*, Lecture Notes in Computer Science **12569**, 2020, pp. 259–276.

Logics for Data Exchange and Communication

Alexandru Baltag[1] Sonja Smets[2]

University of Amsterdam
Institute for Logic, Language and Computation, NL

Abstract

We present a new family of dynamic logics that can model complex scenarios of data exchange and communication. In the context of multi-agent systems, these are acts by which individual agents or groups can publicly or privately access all the information stored at specific locations. In addition to having the full power of standard dynamic epistemic logics (DEL) to model propositional communication, our logics can handle the type of information exchanges in which the data that is being communicated is non-propositional. To axiomatize these logics, in the presence of common and distributed knowledge modalities, we first have to enrich them with polyadic modalities, representing various complex levels of conditional group knowledge. While expressible in the extension of PDL with intersection (which is decidable but does not have FMP), these logics are better behaved: we use an innovative method to show that they do have FMP. We axiomatize these static logics using PDL-type systems, then axiomatize the corresponding dynamic data-exchange logics and prove their decidability, by reducing them to their static base via Recursion/Reduction Axioms.

Keywords: Dynamic Epistemic Logic; Communication protocols; Common Knowledge; Distributed Knowledge; Distributed computing.

1 Introduction

In this paper, we provide a formal setting for reasoning about interactions in which agents gain (or are given) access to information sources, thus being potentially able to 'read' all the data available at that source. More specifically, we assume that at any given time information is locally stored (in both propositional and non-propositional form, e.g. numerical data) at specific locations, databases or 'sites' (e.g. files, folders, data sets, websites, etc.). Each such database can itself be thought of as being associated with an 'agent': it may actually represent the knowledge base of a (natural or artificial) agent, or else we can think of the source itself as an abstract 'agent' ('possessing' all data that is stored at that site, as well as all the information that is logically entailed by these data). The same piece of data can be stored at multiple locations,

[1] thealexandrubaltag@gmail.com.
[2] s.j.l.smets@uva.nl.

and each location can store multiple pieces of data. A 'reading' agent may gain access to another agent's data and 'copy' them in her own database, either because she is granted such access by the source-agent, or because she illegally gain such access ('hacking'). Several agents may simultaneously access multiple sources. So a data-exchange event can be *public* (when it is common knowledge that the information stored at a given location has become visible to everybody, cf. the WikiLeaks case), or *semi-public* (when it is accessible only to some agents, but this fact is common knowledge), or *semi-private* (when some databases are publicly accessed, others are privately accessed, and these facts are known to some but not to all of the agents), or even *fully private* (when both the information and the access are unknown to outsiders).

Standard temporal-epistemic logics [20,23], and dynamic approaches e.g. Public Announcement Logic (PAL) [24] and Dynamic Epistemic Logic (DEL) [7,18,15,9], deal with complex communication scenarios only by *always making explicit the specific sentences* that are being communicated. So all data is assumed to be propositional, which is not always convenient. Moreover, sometimes there might be *no sentences* in our language that capture all relevant data. And, even if such sentences exist, they may be huge, or may be state-dependent: there may be no single formula that uniformly captures (across all the states of the model) the information that is being learnt. [3]

In [12], we introduced a general setting for representing a large class of such data-exchange events (called 'reading events' in that paper), together with a number of dynamic-epistemic logics to reason about them. However, the setting and the results in [12] suffered from a number of limitations. First, the setting did *not* constitute a proper generalization of classical DEL, since it did not fully include it: unlike the DEL epistemic events in [7], the 'reading events' in [12] did not come with (propositional) *preconditions.* As such, these events could happen in *any* state, and thus they *did not carry any specific propositional information*! So, for instance, public announcements $!\varphi$ (of some proposition φ) were not representable as reading events: hence, the dynamic logics in [12] did not formally include the public announcement logic PAL [4], and could not embed the logic of public announcements and common knowledge PAC.

Second, completeness and decidability were proved in [12] only for an even more restricted class of events (-the *'semi-public' ones*). Admittedly, a more general axiomatization was proposed (with a Dynamic Induction Rule in the style of [7]), but its completeness was left as a conjecture. [5]

Third, though decidability was obtained for the logics in [12], this was done via establishing only a Finite 'Pseudo-Model' Property (and then unravelling

[3] E.g. when an agent a learns another agent's secret password x_b, the information that is being gained can be expressed by a formula of the form $x_b = n$, where n is the natural number corresponding to the actual password, but... this number varies across possible states.

[4] Though PAL is not formally included in any of the logics in [12], it is known to be co-expressive with basic modal logic, and it is thus embeddable in those logics

[5] While we believe this conjecture is correct, we chose a different path in this paper, by enlarging the static base to obtain a more transparent axiomatization.

the pseudo-model into an infinite model, to obtain completeness). The Finite Model Property remained open for these logics, and this was *not* a trivial issue: the only known logic that embeds all the logics in [12] is PDL with Intersection, which is known *not* to have FMP (though it is decidable).

In this paper, we address all these limitations. The logics considered here combine classical DEL and the setting in [12]: by endowing data-exchange events with preconditions, they can represent the acquisition of *both* propositional *and* non-propositional information. To obtain simple axiomatizations, we extend the static base logics, in a spirit similar to the work on 'Epistemic PDL' (E-PDL), used in [16] as a static base for classical DEL [6], as well as to the work on 'Group Epistemic PDL' (GE-PDL) [25].[7] But we achieve this by adapting and extending a very recent approach to classical DEL ([5], unpublished manuscript, under review): the idea is to introduce static n-ary modalities for "event-conditional common knowledge" $C_A^e\overline{\varphi}$ (where $\overline{\varphi}$ is an n-tuple of formulas of the same length as the size $n = |E|$ of the corresponding event model), that generalize both the notion of 'relativized common knowledge' $C_A^\theta\varphi$ from [16] and the common distributed knowledge $Cd_{A_1,\ldots,A_n}\varphi$ from [12]. These operators can represent infinite (regular) iterations of distributed knowledge and conditionalization, expressing *regular levels of conditional group knowledge*, that generalize the levels in [22]. Though expressible in GE-PDL (and thus also in PDL with Intersection, but *not* in E-PDL), we directly axiomatize them (as generalizations of common knowledge), and use them to reduce the dynamic logics to their static base, via recursion/reduction laws. In contrast to both E-PDL and GE-PDL [8], the recursion axioms obtained in this way are of great transparency and simplicity.

Finally, we prove FMP for the resulting logics, by using an innovative method, which to our knowledge has never been used before: while based on on the standard technique of making copies of the same worlds, we avoid infinite unraveling by using elementary notions of modular arithmetic (-congruences modulo a natural number) to directly convert every finite pseudo-model into a finite model (in which distributed knowledge is 'standard').

2 Data-exchange logics

We introduce a family of logics for data-exchange and communication. Following the approach to DEL syntax introduced in [6], and then adopted in a number of papers [10,8,5], our languages will be parametrized by data-exchange *signatures*, corresponding to naturally closed classes of data-exchange events.

---

[6] E-PDL is just PDL with an epistemic interpretation: the basic modalities $[a]\varphi$ are interpreted as "knowledge by agent a", based on some indistinguishability relation $\sim_a$, $[\bigcup_{a\in A} a]^*\varphi$ captures "common knowledge of φ in the group A, $[\bigcup_{a\in A}(a;?\theta)]^*\varphi$ captures conditional (or "relativized") common knowledge of φ given θ, etc.

[7] GE-PDL is an extension of E-PDL, based on relations $\sim_A$ for groups A of agents (obtained by taking intersections $\bigcap_{a\in A}\sim_a$), then closing them under the PDL operations.

[8] Though of great theoretical importance, the reduction axioms in [16] and [25] are extremely complex, taking several pages to state.

2.1 Data-exchange frames and signatures

Throughout this paper, we fix a finite set $\mathcal{A}$ of *agents* $a, b, \ldots$, and a finite set P of *atomic sentences* $p, q, \ldots$. We use capital letters $A, B, \ldots$ for *sets of agents.*

Epistemic frames and epistemic models An *epistemic frame* is a structure $\mathbf{S} = (S, \sim)$, consisting of a set S of 'states' $s \in S$, and a map $\sim$ assigning to each agent $a \in \mathcal{A}$ some equivalence relation $\sim_a \subseteq S \times S$, called *(epistemic) indistinguishability relation.* An *epistemic (state) model* $\mathbf{M} = (\mathbf{S}, \|\bullet\|)$ consists of an epistemic frame $\mathbf{S} = (S, \sim)$ and a *valuation map* $\| \bullet \| : P \to \mathcal{P}(S)$, associating to each atom p some set of states $\|p\| \subseteq S$.

Data-exchange frames A *data-exchange frame* is a structure $\mathbf{\Sigma} = (\Sigma, \sim, \bullet_\bullet)$, consisting of a *finite* epistemic frame $(\Sigma, \sim)$ (whose 'states' $\sigma \in \Sigma$ are called *action types*) and an *access map* $\bullet(\bullet)$, associating to each agent $a \in \mathcal{A}$ and action type $\sigma \in \Sigma$ some set of agents $\sigma(a) \subseteq \mathcal{A}$, satisfying two constraints:

(i) *agents access their own data:* $a \in \sigma(a)$ for all $a \in A$;

(ii) *agents know what databases they access:* $\sigma \sim_a \lambda$ implies $\sigma(a) = \lambda(a)$.

The notation for access maps can be naturally extended to *sets of agents*, by putting $\sigma(A) := \bigcup_{a \in A} \sigma(a)$, for any $A \subseteq \mathcal{A}$.

We follow here the approach in [6], by distinguishing between action 'types' and full-fledged 'events': action-types are *not* yet fully specified events (since the propositional information carried by the event is not yet specified). An action-type $\sigma \in \Sigma$ represents a 'type' of data-exchange, in which every agent a gains access to all data stored in the locations (or possessed by the agents) in $\sigma(a)$. The epistemic relation $\sim_a$ represents agent a's knowledge about the type of the on-going exchange: a cannot distinguish type σ from types $\sigma' \sim_a \sigma$.

We now proceed to give examples of data-exchange frames.

Example: Public Announcements The *public announcement frame* $\mathbf{\Sigma}_! = (\{!\}, \sim, \bullet_\bullet)$ is a frame with *only one action-type*, typically denoted by !, with *loops for all agents* (i.e. every $\sim_a$ is just the identity relation), and with the access map given by singletons $!(a) = \{a\}$ for all $a \in \mathcal{A}$.

Example: Public and Semi-Public Sharing (*"Tell Us All You Know"*) For sets $A, B \subseteq \mathcal{A}$, the *semi-public sharing frame* $\mathbf{\Sigma}_{!(A:B)}$ has only one action-type, denoted by $!(A : B)$, with loops for all agents and with access map given by: $!(A : B)(a) = B \cup \{a\}$ for all $a \in A$, and $!(A : B)(a) = \{a\}$ for $a \notin A$. It is common knowledge that all agents in A gain access to the databases of all agents in B. A special case is B*-public sharing* frame $\mathbf{\Sigma}_{!B} := \mathbf{\Sigma}_{!(\mathcal{A}:B)}$, obtained by taking $A = \mathcal{A}$ for the unique action-type $!B :=!(\mathcal{A} : B)$: all agents in B publicly share their information (with everybody).

Example: Semi-public Data-Exchange Frames We can generalize the previous examples to a wider class of frames: for any map $\alpha : \mathcal{A} \to \mathcal{P}(\mathcal{A})$, the associated *semi-public exchange frame* $\mathbf{\Sigma}_{!\alpha}$ consists of *only one-action type* $!\alpha$, with *loops for all agents* (i.e. every $\sim_a$ is just the identity relation), and access map given by putting $(!\alpha)(a) = \alpha(a) \cup \{a\}$ for all $a \in \mathcal{A}$. So, in semi-public actions, it is common knowledge who can access/read whose database, but the actual data may not be necessarily seen by all agents: the *type* of data exchange is public, but data themselves may be only privately available. A more compact description of the semi-public frames $!\alpha$ is obtained by grouping together in separate groups all the agents a having the same $\alpha(a)$. For each $n \geq 0$ and each n-sized family $\{(A_i, B_i) : i \leq n\} = \{(A_1, B_1), \ldots, (A_n, B_n)\} \subseteq \mathcal{P}(\mathcal{A}) \times \mathcal{P}(\mathcal{A})$ consisting of pairs (A_i, B_i) of groups $A_i, B_i \subseteq \mathcal{A}$, the frame $\mathbf{\Sigma}_{!\{A_1:B_1,\ldots,A_n:B_n\}}$ is just the semi-public frame whose unique action-type, denoted by $!\{(A_i : B_i) : i \leq n\} = !\{A_1 : B_1, \ldots, A_n : B_n\}$, has the access map given by $!\{A_1 : B_1, \ldots, A_n : B_n\}(a) = \{a\} \cup \bigcup\{B_i : i \leq n, a \in A_i\}$ for all $a \in \mathcal{A}$. This is semi-public sharing *between multiple groups*: it is common knowledge that all agents in each group A_i gain access to all data of the agents in the corresponding B_i.

Special Case: Sharing *within* Groups For sets $A_1, \ldots, A_n \subseteq \mathcal{A}$, the *multiple in-group sharing frame* is the special case of the previous type of frame in which we take $B_i = A_i$ for all i. Its action-type, denoted by $!\{A_1, \ldots, A_n\} = !\{A_1 : A_1, \ldots, A_n : A_n\}$, describes a type of data exchange in which it is common knowledge that every agent in each group A_i shares all her knowledge with the agents in the same group A_i. The special case $n = 1$ is the so-called *resolution* action $!\{A\}$ (in which it is common knowledge that all agents in A share their information with each other), first introduced in [2].[9]

Example: Private Announcement with Common Knowledge of Alternatives. Given $A \subseteq \mathcal{A}$ and $n \geq 1$, the *A-private announcement* frame $\mathbf{\Sigma}_{!_{A,n}} = (\{!^1_{A,n}, \ldots, !^n_{A,n}\}, \sim, \bullet_\bullet)$ (with *common knowledge of n alternatives* has a domain consisting of n action-types; the access map is $!(a) = \{a\}$ for all a; and $\sim_a$ is the identity relation for $a \in A$, and the universal relation for $a \notin A$.

Example: Secret Hacking Suppose that it is common knowledge that the only possible data-collecting action at the moment is the one by which an agent a (the "hacker") tries to gain access to another agent b's information base (the "source"). Only the hacker knows whether or not he actually succeeds to gain access to b's data (though everybody else is aware of this possibility). We can represent this scenario in a data-exchange frame with two action-types $SH_{a:b}$ (successful, secret hacking) and $UH_{a:b}$ (unsuccessful hacking attempt). The access maps are: $SH_{a:b}(a) = \{a, b\}$, $SH_{a:b}(b) = \{b\}$, and $SH_{a:b}(c) = \{c\}$ for all other agents $c \neq a, b$; while $UH_{a:b}$ is just the identity map on agents. Besides loops for all agents, we have $SH_{a:b} \sim_b UH_{a:b}$, and $SH_{a:b} \sim_c UH_{a:b}$ for all 'outsiders' $c \neq a, b$ (i.e. only a can distinguish between the two possibilities).

Secret Detection of Hacking If we modify the previous scenario to allow the possibility that the source-agent b might detect the hacking attack (so that she might come to know that she is being hacked by a); to have an S5 frame, we assume that everybody is aware of this possibility, but only b knows whether she actually detects an attack or not. The data-exchange frame will now contain three action-types $DH_{a:b}$ (detected hacking), $SH_{a:b}$ (successful, i.e.

[9] Resolution was previously considered under different names in [13] and [11].

undetected, hacking) and $UH_{a:b}$ (unsuccessful hacking attempt). The access maps for $SH_{a:b}$ and $UH_{a:b}$ are the same as in the previous example, and the access map for $DH_{a:b}$ is the same as for $SH_{a:b}$. Besides loops for all agents, we also have $DH_{a:b} \sim_a SH_{a:b}$ (i.e. a doesn't know whether his hacking is detected or not) and $SH_{a:b} \sim_b UH_{a:b}$ (b can't distinguish between undetected hacking and unsuccessful hacking attempt), while $\sim_c$ is the universal relation for all outsiders $c \neq a, b$ (they can't distinguish between any of the three cases).

Secret Mutual Hacking Suppose that agents a and b simultaneously attempt to hack each other's database, though their attempts might be unsuccessful. We assume now that the attempts cannot be detected, so only the hacker knows whether a hacking attack is successful or not. The data-exchange frame contains four action-types $SH_{a:b,b:a}$ (both hacking attacks are successful), $SH_{a:b}$ (only a succeeds to hack b's database), $SH_{b:a}$ (only b succeeds in her hacking attack on a) and UH (all hacking is unsuccessful). The access map for $SH_{a:b}$ is the same as in previous examples, and by switching the roles of a and b in it we obtain the access map for $SH_{b:a}$; the access map for UH is the same as the one for $UH_{a:b}$ in the previous examples; finally, the access map for $SH_{a:b,b:a}$ is $SH_{a:b,b:a}(a) = SH_{a:b,b:a}(b) = \{a, b\}$, and $SH_{a:b,b:a}(c) = \{c\}$ in rest. Besides loops for all agents, we have $SH_{a:b,b:a} \sim_a SH_{a:b} \sim_b UH$ and $SH_{a:b,b:a} \sim_b SH_{b:a} \sim_a UH$, while $\sim_c$ is again the universal relation for $c \neq a, b$.

In order to introduce our key notion of data-exchange signature, we need to first define an operation (*sequential product*) on data-exchange frames.

Product of data-exchange frames Given two data-exchange frames $\mathbf{\Sigma} = (\Sigma, \sim_a, \bullet_\bullet)$ and $\mathbf{\Sigma}' = (\Sigma', \sim'_a, \bullet'_\bullet)$, their *product* is the frame $\mathbf{\Sigma} \times \mathbf{\Sigma}' = (\Sigma \times \Sigma', \sim''_a, \bullet''_\bullet)$, where $\Sigma \times \Sigma'$ is the *Cartesian product* of the two sets of action-types; the epistemic relations are the *product of the underlying relations*: $(\sigma, \sigma') \sim_a (\lambda, \lambda')$ iff $\sigma \sim_a \lambda$ and $\sigma' \sim_a \lambda'$; and the access map is the *composition of the underlying access maps*: $(\sigma, \sigma')(a) := \sigma(\sigma'(a)) = \bigcup_{b \in \sigma'(a)} \sigma(b)$.

Distinguished sets of frames A *distinguished set* is a countable set $\mathcal{F}$ of data-exchange frames s.t. every two frames in $\mathcal{F}$ are mutually disjoint (i.e. have no action-types in common).

Frames up-to-isomorphism Given a class $\mathcal{F}$ of data-exchanges frames, *the class* $\mathcal{F}^\dagger$ *of all* $\mathcal{F}$*-frames up-to-isomorphism* is defined by first choosing a representative $\mathbf{\Sigma} = (\Sigma, \sim_a, \bullet_\bullet) \in \mathcal{F}$ from each isomorphism class in the quotient $\mathcal{F}/\cong$ (of $\mathcal{F}$ modulo isomorphism), then replacing it by an isomorphic copy $\mathbf{\Sigma}^\dagger = (\Sigma^\dagger, \sim_a, \bullet_\bullet)$, obtained by taking $\Sigma^\dagger := \{(\sigma, \mathbf{\Sigma}) : \sigma \in \Sigma\}$, and in rest isomorphically copying the relational and access-map structures.

Observation *For every class* $\mathcal{F}$ *of data-exchanges frames, the class* $\mathcal{F}^\dagger$ *of all* $\mathcal{F}$*-frames up-to-isomorphism is a distinguished set.*

Signatures A *data-exchange signature* is a distinguished set $\mathcal{S}$ of data-exchange frames, that contains the public announcement frame $\mathbf{\Sigma}_!$ and is closed under products of frames up-to-isomorphism (i.e., if $\mathbf{\Sigma}, \mathbf{\Sigma}' \in \mathcal{S}$ then there exists $\mathbf{\Sigma}'' \in \mathcal{S}$ s.t. $\mathbf{\Sigma} \times \mathbf{\Sigma}' \cong \mathbf{\Sigma}''$).

Examples are: the *universal signature* $\mathcal{S}_U$, consisting of all data-exchange frames up-to-isomorphism; the *public announcement signature* $\mathcal{S}_!$, which is just the singleton $\{\mathbf{\Sigma}_!\}$; the *semi-public signature* $\mathcal{S}_{SP} = \{\mathbf{\Sigma}_{!\alpha} | \alpha : \mathcal{A} \to \mathcal{P}(\mathcal{A})\}$, consisting of all semi-public frames $\mathbf{\Sigma}_{!\alpha}$. These classes are already closed under products, because of the isomorphisms $\mathbf{\Sigma}_! \times \mathbf{\Sigma}_! \cong \mathbf{\Sigma}_!$ and $\mathbf{\Sigma}_{!\alpha} \times \mathbf{\Sigma}_{!\beta} \cong \mathbf{\Sigma}_{!(\alpha\circ\beta)}$, where $\alpha, \beta : \mathcal{A} \to \mathcal{P}(\mathcal{A})$ are access maps and $\alpha \circ \beta$ is their functional composition, given by putting $(\alpha \circ \beta)(a) = \alpha(\beta(a))$ for all $a \in \mathcal{A}$.

In contrast, the *signature* $\mathcal{S}_{PriCom}$ *of private announcements with common knowledge of alternatives* requires an additional closure step: $\mathcal{S}_{PriCom}$ consists of all *finite products* of frames of the form $!_{A,n}$ (for all $n \in N$ and all $A \subseteq \mathcal{A}$). [10]

2.2 Syntax of the logics $LDC^e\langle e\rangle$

Given a signature $\mathcal{S}$, the *set of formulas* φ of the language $LDC^e\langle e\rangle_{\mathcal{S}}$ is defined by the following recursive syntax

$$\varphi \ ::= \ p \,|\, \neg\varphi \,|\, \varphi \wedge \varphi \,|\, D_A\varphi \,|\, C_A^{\sigma\overline{\varphi}}\overline{\varphi} \,|\, \langle \sigma\overline{\varphi}\rangle\varphi$$

where p are atomic sentences, a are agents, $A \subseteq \mathcal{A}$ are non-empty sets of agents, σ are action-type symbols coming from any data-exchange frame $\mathbf{\Sigma} = (\Sigma, \sim, \bullet_\bullet) \in \mathcal{S}$ in the given signature $\mathcal{S}$, φ are formulas and $\overline{\varphi} = (\varphi_\sigma)_{\sigma\in\Sigma}$ are Σ-indexed tuples of formulas.

The set of *event expressions* (or 'events', for short) of the language $LDC^e\langle e\rangle_{\mathcal{S}}$ consists of all the expressions of the form $e ::= \sigma\overline{\varphi}$, occurring in any of the above formulas. Using letters e for events in $LDC^e\langle e\rangle_{\mathcal{S}}$, the above syntax can be restated in the more familiar form:

$$\varphi \ ::= \ p \,|\, \neg\varphi \,|\, \varphi \wedge \varphi \,|\, D_A\varphi \,|\, C_A^{e}\overline{\varphi} \,|\, \langle e\rangle\varphi$$

We read D_A as *distributed knowledge* (among the group A), and read C_A^e as *event-conditional common knowledge* (among the group A, conditional on event e), while $\langle e\rangle$ are dynamic modalities for events. Note that the operators C_A^e are *polyadic* modalities, which can be applied to corresponding tuples of formulas $\overline{\varphi}$, to produce new formulas $C_A^e\overline{\varphi}$, representing complex "levels of conditional group knowledge" (about the propositions in the tuple $\overline{\varphi}$). These levels can be thought of in terms of regular patterns of infinite nestings of knowledge and conditional operators, that generalize the 'regular levels of knowledge" investigated in [22]. Their *conditional* nature becomes more easily apparent when we think of event expressions $e := \sigma\overline{\theta}$ as being the result of plugging a tuple of formulas $\overline{\theta}$ into an action-type σ (coming from a data-exchange frame $\mathbf{\Sigma} \in \mathcal{S}$). The formulas $C_A^{\sigma\overline{\theta}}\overline{\varphi}$ could thus be seen as *(group epistemic) conditionals* $\overline{\theta} \,\Box\!\!\to_A^\sigma \overline{\varphi}$, in which the antecedent formulas $\overline{\theta}$ play the role of 'conditions' (or hypothesis) for knowledge, while the formulas in the tuple $\overline{\varphi}$ play the role of epistemic 'conclusions' (that are conditionally known). [11]

---

[10] More generally, every distinguished set $\mathcal{C}$ generates a signature $\mathcal{S}_{\mathcal{C}}$, obtained by adding the frame $\mathbf{\Sigma}_!$ to $\mathcal{C}$ and closing under finite products.

[11] The alternative conditional notation $C_A^{\sigma\overline{\theta}}\overline{\varphi}$ fits well the above-mentioned interpretation in terms of "levels of conditional group knowledge" within group A.

The *static language* $LDC^{e}_{\mathcal{S}}$ is the fragment obtained by dropping dynamic modalities $\langle e\rangle\varphi$ from the above recursive grammar of $LDC^{e}\langle e\rangle_{\mathcal{S}}$. Note that the static fragment is indeed 'static', despite its use of syntactic event expressions: its semantics will only use the evaluation of formulas in the *current model.* In contrast, when occurring in dynamic modalities $\langle e\rangle\theta$, the event expressions $e = \sigma\overline{\varphi}$ will really denote *epistemic events*, living in a DEL-style *event model* (and the semantics of $\langle e\rangle\theta$ will use the evaluation of θ in an *updated model*).

Example: the language of public announcements This is just the special case $LDC^{e}\langle e\rangle_{\mathcal{S}_!}$, where $\mathcal{S}_! = \{\Sigma_!\}$ is the public-announcement signature. Unfolding our syntax in this case, we obtain the format:

$$\varphi \ ::= \ p \,|\, \neg\varphi \,|\, \varphi\wedge\varphi \,|\, D_A\varphi \,|\, C_A^{!\varphi}\varphi \,|\, \langle !\varphi\rangle\varphi$$

The construct $C_A^{!\varphi}\psi$ is traditionally written $C_A^{\varphi}\psi$, and it is a 'factive' version of van Benthem's *relativized common knowledge* [16] (represented a form of 'conditional' common knowledge).

Example: the language of semi-public data-exchange events Another special case is obtained by taking the semi-public signature $\mathcal{S}_{SP} = \{!\alpha|\alpha : \mathcal{A} \to \mathcal{P}(\mathcal{A})\}$. In a more explicit form, this language is given by the recursive format:

$$\varphi \ ::= \ p \,|\, \neg\varphi \,|\, \varphi\wedge\varphi \,|\, D_A\varphi \,|\, C_A^{!\{A_1:B_1,\ldots,A_n:B_n\}\varphi}\varphi \,|\, \langle !\{A_1 : B_1, \ldots, A_n : B_n\}\varphi\rangle\varphi$$

for all sets of pairs $\{(A_1, B_1), \ldots, (A_n, B_n)\} \subseteq \mathcal{P}(\mathcal{A}) \times \mathcal{P}(\mathcal{A})$, of any finite size $n \geq 0$.[12] Here, $!\{A_1 : B_1, \ldots, A_n : B_n\}\varphi$ represents a semi-public exchange event, having a precondition φ (so that this event can only happen when φ is true), and in which it is common knowledge that all agents in each group A_i gain access to all data of the agents in the corresponding group B_i.

Other examples include the *language of private announcements with common knowledge of alternatives* (based on the signature $\mathcal{S}_{PriCom}$), and the universal data-exchange language (based on the universal signature $\mathcal{S}_U$).

Event Models An $\mathcal{S}$-*event model* $\mathbf{E} = (E, \sim_a, \bullet(\bullet), pre)$ is a data-exchange frame $(E, \sim_a, \bullet(\bullet)) \in \mathcal{S}$ together with a 'precondition' map *pre*, associating to each action-type $e \in E$ some formula pre_e in $LDC^{e}\langle e\rangle_{\mathcal{S}}$, called the *precondition* of e. When considered within an event model, we refer to the action types $e \in E$ as *events.* Intuitively, pre_e gives the *condition of possibility* for event e.

Event models can be considered as semantic objects in their own respect, see e.g. [7,6,18]. However, in this paper, we will only consider them as *meta-syntactic notations*, that refer to event expressions $\sigma\overline{\varphi}$, and encode their relations to other such event expressions in a conveniently succinct way. We refer to these as *syntactic event models* (though often dropping the 'syntactic' qualification for brevity):

---

[12] The case $n = 0$ allows us to embed the logic public announcements: for $n = 0$, we get the event $!\,\emptyset\varphi$, which is the same as the public announcement $!\varphi$.

Syntactic Event Models Given any data-exchange frame $\mathbf{\Sigma} = (\Sigma, \sim, \bullet_\bullet) \in \mathcal{S}$ in our signature, and any Σ-indexed tuple of formulas $\overline{\varphi} = (\varphi_\sigma)_{\sigma\in\Sigma}$, we consider the *corresponding (syntactic) event model* $\mathbf{\Sigma}\overline{\varphi} = (\Sigma\overline{\varphi}, \sim_a, \bullet(\bullet), pre)$ defined as follows: the set of events is $\Sigma\overline{\varphi} := \{\sigma\overline{\varphi} : \sigma \in \Sigma\}$ consists of all event expressions $\sigma\overline{\varphi}$, with $\sigma \in \Sigma$; the accessibility relations and access maps on events are inherited from the underlying data-exchange frame: $\sigma\overline{\varphi} \sim_a \lambda\overline{\varphi}$ iff $\sigma \sim_a \lambda$, and $(\sigma\overline{\varphi})(a) := \sigma(a)$; while the precondition function *pre* maps each event $\sigma\overline{\varphi} \in \Sigma\overline{\varphi}$ into the corresponding σ-formula $pre_{\sigma\overline{\varphi}} := \varphi_\sigma$ in the Σ-indexed tuple $\overline{\varphi}$.

Notation We will denote by $\mathcal{E}(\mathcal{S})$ the *set of syntactic event models associated to the language* $LDC^{e}\langle e\rangle_{\mathcal{S}}$, consisting of all event models of the form $E = \mathbf{\Sigma}\overline{\varphi}$ with $\mathbf{\Sigma} \in \mathcal{S}$ and $\overline{\varphi} = (\varphi_\sigma)_{\sigma\in\Sigma}$ with all formulas φ_σ in $LDC^{e}\langle e\rangle_{\mathcal{S}}$.

Event-indexed tuples Given the natural bijection $\sigma \mapsto \sigma\overline{\varphi}$ between the action-types in the frame $\mathbf{\Sigma}$ and the corresponding events in the syntactic event model $\mathbf{E} = \mathbf{\Sigma}\overline{\varphi}$, the Σ-indexed tuple $\overline{\varphi} = (\varphi_\sigma)_{\sigma\in\Sigma}$ can also be thought of as being *indexed by events* $e \in E = \Sigma\overline{\varphi}$, i.e., written as $\overline{\varphi} = (\varphi_e)_{e\in E}$.

Sequential composition of event models. Given two (syntactic) event models $\mathbf{E} = (E, \sim_a, \bullet(\bullet), pre)$ and $\mathbf{F} = (F, \sim_a, \bullet(\bullet), pre)$, based respectively on data-exchange frames $\mathbf{\Sigma}, \mathbf{\Lambda} \in \mathcal{S}$ and on tuples of formulas $\overline{\varphi} = (\varphi_\sigma)_{\sigma\in\Sigma}$ and $\overline{\theta} = (\theta_\lambda)_{\lambda\in\Lambda}$ we form their *sequential product* $\mathbf{E};\mathbf{F}$, by taking as the underlying frame (the frame in $\mathcal{S}$ that is isomorphic to) the *product* $\mathbf{\Sigma} \times \mathbf{\Lambda}$ of the two frames, and as underlying tuple the $\Sigma \times \Lambda$-indexed tuple

$$\langle e\rangle\overline{\theta} \;=\; (\langle\sigma\overline{\varphi}\rangle\theta_\lambda)_{(\sigma,\lambda)\in\Sigma\times\Lambda}$$

Using the natural bijection between $\mathbf{E} \times \mathbf{F}$ and $\mathbf{\Sigma} \times \mathbf{\Lambda}$, we can identify the events in $\mathbf{E};\mathbf{F}$ with pairs $(e, f) \in \mathbf{E}\times\mathbf{F}$. We denote by $e; f$ the event expression corresponding to the pair (e, f) (via the bijection), and note that our choice of the above $\Sigma \times \Lambda$-indexed tuple of formulas gives us:

$$pre_{e;f} \;=\; \langle e\rangle pre_f$$

Observation. *The set $\mathcal{E}(\mathcal{S})$ of syntactic event models associated to each language $LDC^{e}\langle e\rangle_{\mathcal{S}}$ is closed under sequential composition.*

Subformulas of a formula or event expression. The set of *proper subformulas* of a given formula φ or event e is recursively defined by putting: $Sub(p) := \emptyset$; $Sub(\neg\varphi) = Sub(D_A\varphi) := \{\varphi\} \cup Sub(\varphi)$; $Sub(\varphi \wedge \psi) := \{\varphi, \psi\} \cup Sub(\varphi) \cup Sub(\psi)$; $Sub(C^{e}_{A}\varphi) = Sub(\langle e\rangle\varphi) := Sub(e) \cup \{\varphi\} \cup Sub(\varphi)$; and finally, for events $e = \sigma\overline{\varphi}$, where $\sigma \in \Sigma \in \mathcal{S}$ is an action type and $\overline{\varphi} = (\varphi_1, \ldots, \varphi_n)$ is an n-tuple of formulas, we put $Sub(e) := \{\varphi_i : 1 \le i \le n\} \cup \bigcup_{1\le i\le n} Sub(\varphi_i)$.

Note that this definition has the consequence that: for all events $e, f \in E$ and $|E|$-tuples of formulas $\overline{\varphi}$, *pre_f and φ_f are proper subformulas of $C^{e}_{A}\overline{\varphi}$*.

Some abbreviations. We have the usual Boolean notions $\varphi \vee \psi$, $\varphi \to \psi$, $\varphi \leftrightarrow \psi$, the Diamond-type duals of our static epistemic modalities $\langle D_A\rangle\varphi$, as well as the Box-type dual $[e]\varphi$ of our dynamic operator $\langle e\rangle\varphi$. In addition, we adopt the following abbreviations:

$K_a\varphi$	$:=$	$D_{\{a\}}\varphi$	(*individual knowledge*)
$D^e_A\varphi$	$:=$	$D_{e(A)}(pre_e \to \varphi)$	(*event-conditional distributed knowledge*)
$K^e_a\varphi$	$:=$	$D^e_{\{a\}}\varphi$	(*event-conditional knowledge*)
$C_A\varphi$	$:=$	$C^{!\top}_A\varphi$	(*common knowledge*)
$C^\theta_A\varphi$	$:=$	$C^{!\theta}_A\varphi$	(*relativized common knowledge*)

where ! is the public announcement action-type (living in the public announcement frame $\mathbf{\Sigma}_!$), and $\top$ is any propositional tautology. Also, for two $\mathbf{E}$-indexed tuples of formulas $\overline{\varphi} = (\varphi_e)_{e\in\mathbf{E}}$ and $\overline{\psi} = (\psi_e)_{e\in\mathbf{E}}$, we put

$$\overline{\varphi \to \psi} := (\varphi_e \to \psi_e)_{e\in\mathbf{E}}$$

for the tuple of implications between the corresponding formulas.

Action of a data-exchange model over indexed tuples Finally, here is another useful piece of meta-syntactic notation: given event models $\mathbf{E}$ and $\mathbf{F}$, and an $\mathbf{F}$-indexed tuple $\overline{\varphi} = (\varphi_f)_{f\in\mathbf{F}}$, we put

$$\langle\mathbf{E}\rangle\overline{\varphi} := (\langle e\rangle\varphi_f)_{e\in\mathbf{E},f\in\mathbf{F}}$$

Note that we can think of $\langle\mathbf{E}\rangle\overline{\varphi}$ as *an* $\mathbf{E};\mathbf{F}$*-indexed tuple* $(\langle e\rangle\varphi_f)_{(e;f)\in\mathbf{E};\mathbf{F}}$.

2.3 Semantics

We interpret formulas on epistemic state models $\mathbf{M}$, and we interpret event expressions in syntactic $\mathcal{S}$-event models $\mathbf{E}$. As usual in DEL, we simultaneously define (i) a *satisfaction relation* $s \models_{\mathbf{M}} \varphi$ between states in a model and formulas, and (ii) an *update product operation* taking any state model $\mathbf{M}$ and any $\mathcal{S}$-based event model $\mathbf{E}$ into an updated state model $\mathbf{M} \bigotimes \mathbf{E}$ (their update product). The definition is by mutual recursion. For (i), the truth clauses for propositional logic are as usual, while for the modal operators, we put:

$$s \models_{\mathbf{M}} D_A\varphi \;\text{ iff }\; w \models \varphi \text{ for all states } w \sim_A s,$$

where $\sim_A := \bigcap_{a\in A} \sim_a$ is the intersection of all the relations $\sim_a$ with $a \in A$;

$$s \models_{\mathbf{M}} C^e_A\overline{\varphi} \;\text{ iff }\; s \models_{\mathbf{M}} \varphi_e,\, K^f_a\varphi_f,\, K^f_aK^h_b\varphi_h, \ldots,$$

for all finite chains $e \sim_a f \sim_b h \ldots$ with $a, b \ldots \in A$;[13] and

$$s \models_{\mathbf{M}} \langle e\rangle\varphi \quad\text{iff}\quad (s,e) \models_{\mathbf{M}\bigotimes\mathbf{E}} \varphi.$$ [14]

[13] Note the analogy with the semantic definition of common knowledge, in terms of a nested iteration of knowledge operators. But also note the differences: event-conditional common knowledge is in fact a *nested iteration of distributed knowledge operators and conditionals.*

[14] The statement on the right-hand side presupposes that $(s,e) \in M \bigotimes E$, hence this definition gives us $s \not\models \langle e\rangle\varphi$ whenever $(s,e) \notin M \bigotimes E$: this is the *existential* version of the dynamic modality.

For (ii): the update product $\mathbf{M} \bigotimes \mathbf{E} = (M \bigotimes E, \sim_a, \| \bullet \|)$ of a state model $\mathbf{M} = (M, \sim_a, \| \bullet \|^M)$ and an event model $\mathbf{E} = (E, \sim_a, \bullet(\bullet), pre)$ is given by:

$$M \bigotimes E = \{(s,e) \in M \times E : s \models_{\mathbf{M}} pre_e\}$$

$$(s,e) \sim_a (s',e') \text{ iff } s \sim_{e(a)} s' \text{ and } e \sim_a e'$$

$$\|p\|^{\mathbf{M} \otimes \mathbf{E}} = \{(s,e) \in M \bigotimes E : s \in \|p\|^{\mathbf{M}}\}$$

Intuitively, this definition can be justified as follows. The pair (s,e) denotes the *output-state* produced by performing data-exchange action e on input-state s: *this output-state exists only if action e can be executed at s*, i.e. if s satisfies the precondition of e. In that case, the output state (s,e) is *unique*: our data-exchange events are *deterministic*, changing the state in a fully-determined way. The *new valuation is inherited from the old one*: these are *pure informational events*, which do not change the ontic facts p.

The crucial difference with standard DEL lies in the definition of the new epistemic relations $\sim_a$ on output-states: note the key role played in this definition by the relation $\sim_{e(a)}$ on input-states. This says that agent a's new knowledge after a data-exchange event e is the result of putting together the knowledge gained by accessing the prior information of all agents in $e(a)$ and her knowledge about the event e itself.

As noted in [6], we can think of event models $\mathbf{E}$ as "acting" on state models $\mathbf{M}$ to produce the product models $\mathbf{M} \bigotimes \mathbf{E}$, and we can then see that the 'sequential composition' operation $\mathbf{E}; \mathbf{F}$ defined above behaves indeed like a sequential composition of actions: the map $((s,e), f) \mapsto (s, (e; f))$ is a natural isomorphism between $(\mathbf{M} \bigotimes \mathbf{E}) \bigotimes \mathbf{F}$ and $\mathbf{M} \bigotimes (\mathbf{E}; \mathbf{F})$.

3 Axiomatization, completeness and FMP

Let $\mathbf{LDC^e}\langle \mathbf{e} \rangle_{\mathcal{S}}$ be the proof system in Table 1, and let $\mathbf{LDC^e}_{\mathcal{S}}$ be the system for the static logic $LDC^{e}_{\mathcal{S}}$ obtained by removing the recursion axioms (group IV) from Table 1 (and restricting all instances of remaining axioms to $LDC^{e}_{\mathcal{S}}$).

Theorem 3.1 *Given the above notations, we have the following:*

(i) *the system* $\mathbf{LDC^e}_{\mathcal{S}}$ *is sound and complete for the static fragment* $LDC^{e}_{\mathcal{S}}$;

(ii) *the logic* $LDC^{e}_{\mathcal{S}}$ *is decidable and has the FMP;*

(iii) *the full dynamic logic* $LDC^e\langle e \rangle_{\mathcal{S}}$ *has the same expressivity as its static fragment* $LDC^{e}_{\mathcal{S}}$ *(and hence it is also decidable and has the FMP);*

(iv) *the system* $\mathbf{LDC^e}\langle \mathbf{e} \rangle_{\mathcal{S}}$ *is sound and complete for the logic* $LDC^e\langle e \rangle_{\mathcal{S}}$.

The meaning of the axioms We now give some explanations of the axioms and rules of $\mathbf{LDC^e}\langle \mathbf{e} \rangle_{\mathcal{S}}$. The ones in Group (II) are just the standard laws of distributed knowledge (having as a special case the standard $S5$ axioms for individual knowledge). As for the axioms and rules for C^{e}_{A} (Group III), note the analogy with the standard axiomatization for common knowledge.

(I) Axioms and rules of classical propositional logic	
(II) Axioms and rules for (distributed) knowledge:	
(D-Necessitation)	From φ, infer $D_A\varphi$
(D-Distribution)	$D_A(\varphi \to \psi) \to (D_A\varphi \to D_A\psi)$
(Veracity)	$D_A\varphi \to \varphi$
(Pos. Introspection)	$D_A\varphi \to D_A D_A\varphi$
(Neg. Introspection)	$\neg D_A\varphi \to D_A\neg D_A\varphi$
(Monotonicity)	$D_A\varphi \to D_B\varphi$, for sets $A \subseteq B$
(III) Axioms and rules for event-conditional common knowledge	
Given $\mathbf{E} = (E, \sim_a, \bullet(\bullet), pre) \in \mathcal{E}(\mathcal{S})$, $e \in E$ and $\overline{\varphi} = (\varphi_{e'})_{e' \in E}$:	
(C^{e}-Necessitation)	From $\{\varphi_e\}_{e \in E}$, infer $C^{\mathrm{e}}_A\overline{\varphi}$
(C^{e}-Distribution)	$C^{\mathrm{e}}_A\overline{\varphi \to \psi} \to (C^{\mathrm{e}}_A\overline{\varphi} \to C^{\mathrm{e}}_A\overline{\psi})$
(C^{e}-Mix)	$C^{\mathrm{e}}_A\overline{\varphi} \to \left(\varphi_e \wedge K^{e'}_a C^{e'}_A\overline{\varphi}\right)$, for $e' \in E$, $a \in A$ s.t. $e \sim_a e'$
(C^{e}-Induction Rule)	From $\{\varphi_f \to K^{f'}_a\varphi_{f'} : f, f' \in E, a \in A \text{ s.t. } f \sim_a f'\}$, infer $\varphi_e \to C^{\mathrm{e}}_A\overline{\varphi}$
(IV) Recursion/reduction axioms and rules	
Given $\mathbf{E} = (E, \sim_a, \bullet(\bullet), pre), \mathbf{F} = (F, \ldots) \in \mathcal{E}(\mathcal{S})$, $e \in E$, $f \in F$, $\overline{\varphi} = (\varphi_f)_{f \in F}$:	
($[\mathrm{e}]$-Necessitation)	From φ, infer $[\mathrm{e}]\varphi$
($[\mathrm{e}]$-Distribution)	$[\mathrm{e}](\varphi \to \psi) \to ([\mathrm{e}]\varphi \to [\mathrm{e}]\psi)$
(Atomic Permanence)	$\langle e\rangle p \leftrightarrow (pre_e \wedge p)$
(Partial Functionality)	$\langle e\rangle\neg\varphi \leftrightarrow (pre_e \wedge \neg\langle e\rangle\varphi)$
(D-Action)	$\langle e\rangle D_A\varphi \leftrightarrow \left(pre_e \wedge \bigwedge_{e' \sim_A e} D^{e'}_A\langle e'\rangle\varphi\right)$
(C^f-Action)	$\langle e\rangle C^f_A\overline{\varphi} \leftrightarrow C^{e;f}_A\langle \mathbf{E}\rangle\overline{\varphi}$

Table 1

The proof system $\mathbf{LDC^e}\langle\mathbf{e}\rangle_{\mathcal{S}}$, where we used the abbreviations from Section 2.2: in particular, $\langle\mathbf{E}\rangle\overline{\varphi} := (\langle e\rangle\varphi_f)_{e \in E, f \in F}$, for event models $\mathbf{E}, \mathbf{F}$ and F-indexed tuples $\overline{\varphi} = (\varphi_f)_{f \in F}$. The system $\mathbf{LDC^e}_{\mathcal{S}}$ is obtained by removing the last group (IV).

The increased complexity is due to the fact that C^{e}_A is a polyadic modality, so its Induction Rule requires multiple implications as premises to prove its conclusion. Another dis-analogy with standard common knowledge is 'buried' in the notation $K^f_a\psi$ for $D_{f(a)}(pre_f \to \psi)$: the role of individual knowledge K_a in the usual definition of C_A is now played by conditional distributed knowledge!

The first four axioms and rules in group (IV) are standard. The recursion axiom (D-Action) can be restated more explicitly as $\langle e\rangle D_A\varphi \leftrightarrow (pre_e \wedge \bigwedge_{e' \sim_A e} D_{e(A)}[e']\varphi)$, where $[e']\psi$ is the universal dual of $\langle e'\rangle\psi$ (and where we used the fact that $e' \sim_A e$ implies $e'(A) = e(A)$). This says that group A's distributed knowledge after an exchange e is pre-encoded by the prior distributed knowledge among the *larger* group $e(A)$ (of *agents whose data are being accessed by anybody in* A). And, as a special case of this, *individual knowledge after an exchange is pre-encoded by distributed knowledge*: $\langle e\rangle K_a\varphi \leftrightarrow \left(pre_e \wedge \bigwedge_{f \sim_a e} D_{e(a)}[f]\varphi\right)$

Finally, the simplicity of the recursion axiom for C^f_A illustrates the usefulness of our new polyadic modalities: *their recursion law simply amounts to event-composition*! Indeed, the law says that f-conditional common knowledge after an event e is pre-encoded by $e; f$-conditional common knowledge.[15]

3.1 Sketch of the proof of Theorem 3.1

While relegating many details to the Appendix, we briefly explain here the plan and main steps of the proof (and spelling out the most innovative parts).

Step 1: Pseudo-model completeness and FMP via filtration.

We need a detour through a more general type of models, called *pseudo-models*, in which distributed knowledge relations $\sim_A$ are treated as basic, undefined relations (rather than as intersections of individual knowledge relations).

Pseudo-model semantics. A *pseudo-model* is a structure $\mathbf{S} = (S, \sim_A, \|\bullet\|)_A$, where: S is a set of states; $\sim_A \subseteq S \times S$ are equivalence relations (one for each non-empty set $A \subseteq \mathcal{A}$) with the property that: $A \subseteq B$ and $s \sim_A w$ imply $s \sim_B w$; and $\| \bullet \| : Prop \to \mathcal{P}(S)$ is a valuation function, mapping atomic propositions p into sets of states $\|p\| \subseteq S$. Given a pseudo-model $\mathbf{S}$, we can define the *satisfaction* relation $s \models \varphi$, by using the standard clauses for atomic formulas and propositional connectives, defining D_A as the Kripke modality for $\sim_A$, and defining $C^e_A\varphi$ as in standard models.

Models are *standard* pseudo-models. A pre-model $\mathbf{S}$ is *standard* if $\sim_{A\cup B}$ is the intersection of $\sim_A$ and $\sim_B$, for all sets of agents A and B. It should be obvious that *every model comes with a standard premodel structure*, by defining $\sim_A$ (for each non-empty $A \subseteq \mathcal{A}$) to be the *intersection* $\bigcap_{a\in A} \sim_a$ of all relations $\sim_a$ with $a \in A$. Vice-versa, *every standard pre-model can be seen as a model*, by defining $\sim_a$ to be the same as $\sim_{\{a\}}$, for every $a \in \mathcal{A}$. Moreover, *pseudo-model semantics is indeed a generalization of our intended semantics*: when restricted to standard pseudo-models, the pseudo-model semantics amounts to the same as model semantics. So *models are essentially the same as standard pre-models*: from now on we will *identify* the two.

Proposition 3.2 *The system* $\mathbf{LDC^e}_{\mathcal{S}}$ *is sound and complete wrt pseudo-model semantics, and has the Finite Pseudo-Model Property.*

The *proof* uses a variant of the standard method of *filtration*, adapted to our polyadic modalities. We relegate the details to Appendix A.

Step 2: From finite pseudo-models to finite models.

First, note that pseudo-models *are* relational $S5$-models, in the usual sense, except that their basic relations are labelled by sets of agents instead of single agents. Hence, we can apply to them all the standard Modal Logic concepts and theory [17]. In particular, we get for free the notions of *bisimulation* and *p-morphism* (=functional bisimulation) between pseudo-models.

Moreover, we can easily check invariance under p-morphisms:

---

[15] Contrast this with the complexity of the recursion axiom for iteration π^* in E-PDL.

Lemma 3.3 *All formulas of our static logic LDC^e are preserved under pseudo-model p-morphisms.*

To prove this, we need a non-standard notion of complexity:

Lemma 3.4 *There exists $\gamma : LDC^e \to On$, associating to every formula φ in LDC^e some (countable) ordinal $\gamma(\varphi) \in On$, s.t. the associated well-founded strict partial order $<$ on formulas in LDC^e (called γ-complexity order, and defined by putting $\varphi < \psi$ iff $\gamma(\varphi) < \gamma(\psi)$) has the following properties:*

- *if φ is a proper subformula of ψ, then $\varphi < \psi$ (and so in particular, $pre_f < C^e_A\overline{\varphi}$ for all e and f living in the same event model E);*
- *$C^e_A\overline{\varphi}$ has higher complexity than all the formulas involved in its semantics:*

$$\varphi_e, K^f_a\varphi_f, K^f_a K^h_b\varphi_h, \ldots < C^e_A\overline{\varphi}$$

for all events $e, f, h, \ldots$ in a given event model, and all agents $a, b, \ldots \in \mathcal{A}$.

Proof. We define γ by setting: $\gamma(p) = 0$; $\gamma(\neg\varphi) = \gamma(D_A\varphi) = \gamma(\varphi) + 1$; $\gamma(\varphi \wedge \psi) = max\{\gamma(\varphi), \gamma(\psi)\} + 1$; and $\gamma(C^e_A\overline{\varphi}) = max\{\gamma(pre_f) : f \in E\} \cup \{\gamma(\varphi_f) : f \in E\} + \omega$ (where E is the event model in which e lives).[16] □

Proof of Lemma 3.3. Let $\pi : \mathbf{S} \to \mathbf{S}'$ be a p-morphism between pseudo-models. We show that for all formulas φ in LDC^e, we have:

$$s \models_{\mathbf{S}} \varphi \text{ iff } \pi(s) \models_{\mathbf{S}'} \varphi,$$

for all $s \in S$. The proof is by *induction on the order* $<$. The atomic case and the inductive steps for Boolean connectives and modalities $D_A\varphi$, go as usual, using the Atomic Preservation and the back and forth clauses in the definition of p-morphism, as well as the induction hypothesis (where we use the fact that every formula has higher γ-complexity than its subformulas). Similarly, the inductive step for $C^e_A\overline{\varphi}$ uses the fact that $C^e_A\overline{\varphi}$ has higher γ-complexity than all the formulas involved in its semantics ($\varphi_e, K^f_a\varphi_f, K^f_a K^h_b\varphi_h, \ldots$); so the induction hypothesis, those formulas are preserved by π, and hence $C^e_A\varphi$ is also preserved by π.

Given Proposition 3.2 and Lemma 3.3, it is clear that in order to prove completeness and FMP for our intended models, it is enough to show:

Lemma 3.5 *(**Intersection Lemma.**) Every finite pseudo-model $\mathbf{S} = (S, \sim_A, \| \bullet \|)$ is a p-morphic image of some finite model $\overline{\mathbf{S}} = (\overline{S}, \sim_a, \| \bullet \|)$.*

Proof. For every non-empty set of agents $A \subseteq \mathcal{A}$, let $[s]_A := \{s' \in S : s \sim_A s'\}$ be the equivalence class of any state $s \in S$ modulo $\sim_A$. We denote by $S/\sim_A := \{[s]_A : s \in S\}$ the quotient of S modulo $\sim_A$ (consisting of all equivalence classes). Clearly $S/\sim_A$ is *finite* (since S is finite), and so let N_A be its *cardinality* (=the *number of equivalence classes* modulo $\sim_A$).

[16] The map γ is well-defined, since the relevant event models E are finite, so at each step we are only taking the maximum of finitely many ordinals.

For each non-empty $A \subseteq \mathcal{A}$, consider some *fixed enumeration* (without repetitions) of the equivalence classes in $S/\sim_A$. For every state $s \in S$, denote by $|s|_A$ the *index* of its equivalence class $[s]_A$ in this enumeration. We obviously have that $|s|_A \in \{0, 1, \dots, N_A - 1\}$, for all $s \in S$. Note that $|s|_A = |s'|_A$ implies $[s]_A = [s']_A$, i.e. $s \sim_A s'$ (since our enumeration has no repetitions).

We now construct our new model $\overline{\mathbf{S}}$. The set of states will be given by

$$\overline{S} \;:=\; S \times \prod_{A \subseteq \mathcal{A}, A \neq \emptyset} \bigcup \{0, 1, \dots, N_A - 1\}^{\mathcal{A}}.$$

In other words, the new states are of the form $(s, \overline{i}) := (s, i_A)_{A \subseteq \mathcal{A}, A \neq \emptyset}$, consisting of an old state $s \in S$ and a tuple $\overline{i} = (i_A)_{A \subseteq \mathcal{A}}$ of assignments $i_A : \mathcal{A} \to \{0, 1, \dots, N_A - 1\}$ (indexed by non-empty sets $A \subseteq \mathcal{A}$), that map agents $a \in \mathcal{A}$ to indeces $i_A(a) \in \{0, 1, \dots, N_A - 1\}$ in the enumeration of $S/\sim_A$.

Restriction, Sum and Accessibility. We use the notation $i_A|B$ to denote the *restriction of the map i_A to the agents in B*. We also put

$$\overline{i}(A) \;:=\; \sum_{a \in A} i_A(a)$$

for the *sum of all the values of i_A over agents in A*. Finally, we define a *basic accessibility relation* $\sim$ on $\overline{S}$, by

$$(s, \overline{i}) \sim (s', \overline{i}') \;\text{ iff }\; |s|_A - \overline{i}(A) \equiv |s'|_A - \overline{i}'(A) \,(mod\, N_A) \text{ for all } A \subseteq \mathcal{A} \text{ s.t. } A \neq \emptyset,$$

where $\equiv$ denotes *congruence modulo N_A*.

Obviously, accessibility $\sim$ is an equivalence relation on $\overline{S}$. In our construction, $\sim$ will constitute a precondition for epistemic equivalence $\sim_a$.

The model. We now structure the set $\overline{S}$ into a *model* $\overline{\mathbf{S}}$, as follows: the *valuation* is inherited from the pseudo-model $\mathbf{S}$, i.e. $\|p\|_{\overline{\mathbf{S}}} := \{(s, \overline{i}) \in S : s \in \|p\|_{\mathbf{S}}\}$; and we define the basic *indistinguishability relations* $\sim_a$ on $\overline{\mathbf{S}}$ by putting

$$(s, \overline{i}) \sim_a (s', \overline{i}') \;\text{ iff }\; (s, \overline{i}) \sim (s', \overline{i}') \text{ and } i_A(a) = i'_A(a) \text{ for all } A \neq \emptyset.$$

It should be clear that all $\sim_a$ are equivalence relations, so the structure $\overline{\mathbf{S}} = (\overline{S}, \sim_a, \| \bullet \|)$ is a *model*. Hence, if for every non-empty set $B \subseteq \mathcal{A}$ we take $\sim_B$ to be the *intersection* $\bigcap_{b \in B} \sim_b$, then $\overline{\mathbf{S}}$ is a standard pseudo-model.

It is obvious that, for every non-empty such set B of agents, we have:

$$(s, \overline{i}) \sim_B (s', \overline{i}') \;\text{ iff }\; (s, \overline{i}) \sim (s', \overline{i}') \text{ and } i_A|B = i'_A|B \text{ for all } A \neq \emptyset.$$

All is left to show is that $\mathbf{S}$ *is a p-morphic image of* $\overline{\mathbf{S}}$. For this, we take the *first-component projection* $\pi : \overline{S} \to S$, given by $\pi(s, \overline{i}) := s$, and we will prove that π *is a surjective p-morphism of pseudo-models.*

Surjectivity and Atomic Preservation are obvious.

To check the 'forth' clause, we assume that $(s, \overline{i}) \sim_B (s', \overline{i}')$, and we need to show that $s \sim_B^{\mathbf{S}} s'$. For this, we apply the above characterization of the relation

$\sim_B$ to the set $A := B$: our assumption implies both that $i_B|B = i'_B|B$, and hence that $\overline{i}(B) = \overline{i}'(B)$, and that $|s|_B - \overline{i}(B) \equiv |s'|_B - \overline{i}'(B)\,(mod\, N_B)$. Adding the last two equalities, we obtain that $|s|_B \equiv |s'|_B\,(mod\, N_B)$, i.e. $|s|_B = |s'|_B$, and thus $s \sim_B^{\mathbf{S}} s'$, as desired.

Finally, to check the 'back' clause, let $(s, \overline{i}) \in \overline{S}$ be such that $s \sim_B s'$ for some $s' \in S$. We need to find some $\overline{i}' = (i'_A)_{A \subseteq \mathcal{A}}$ satisfying $(s, \overline{i}) \sim_B (s', \overline{i}')$. For this, we simply put $i'_A := i_A$ for every $A \subseteq B$; while, for $A \not\subseteq B$, we choose some $a \in A - B$, then put $i'_A(b) := i_A(b)$ for all $b \in \mathcal{A} - \{a\}$, and $i'_A(a) := i_A(a) + |s'|_A - |s|_A\,(mod\, N_A)$. Using again the above characterization of the relations $\sim_B^{\overline{\mathbf{S}}}$, we can verify that we have $(s, i) \sim_B (s', i')$, as desired. □

This finishes the proof of parts 1 and 2 of Theorem 3.1 (completeness and decidability for the static fragments LDC^{e}_{S} of our signature-based logics).

Step 3: Reduction and completeness for the dynamic logics $LDC^{e}\langle e\rangle_S$. Checking the *soundness* of the axioms and rules of $\mathbf{LDC^{e}\langle e\rangle}_S$ is a tedious, but easy verification, so we skip the details. For *completeness*, we use the reduction axioms to recursively show that each formula in $LDC^{e}\langle e\rangle_S$ is provably equivalent to a 'static' formula in LDC^{e}_{S}. This establishes parts 3 and 4 of Theorem 3.1. Once again, we relegate the details to Appendix B.

4 Conclusions and future work

The purpose of this paper was to axiomatize the logic of all data-exchange events (with preconditions), in the presence of both common and distributed knowledge. In order to have elegant and transparent recursion/reduction axioms, we extended the static base with new polyadic modalities, denoting complex levels of conditional group knowledge. This in itself is an original technical contribution to the field of Dynamic Epistemic Logic.

Furthermore, we proved decidability by directly establishing FMP, a strategy that was not previously available for logics with distributed knowledge (or intersection of programs). For this, we employed a highly innovative method, that uses notions of elementary number theory (congruences modulo a natural number) to control the proliferation of world-copies: this is the second original technical contribution of our paper. [17]

We should note that our congruence-based method can be extended to *prove FMP for GE-PDL, and even for 'Group Epistemic Mu-Calculus' GE_μ* (i.e. Mu-Calculus based on modalities $[A]\varphi = D_A\varphi$ for distributed knowledge). This is in sharp contrast to PDL with Intersection (which does *not* have FMP). We are planning to write up these proofs in a follow-up paper.

Other on-going work pertains investigating *non-S5 versions* of our data-exchange events. This would allow us to capture notions of *belief* and treat

[17] In addition, we adapted a number of notions (pseudo-models, filtration, Fischer-Ladner closure, bisimulations and p-morphisms etc), to yield variants that are appropriate for our polyadic generalizations of the common knowledge operator.

hidden, unsuspected, or deceiving actions: e.g., *fully secret hacking* (in which the hacked agent does not even suspect the hacking), *fully private announcements* a la Gerbrandy [21] (where the outsiders are unaware that the announcement may be happening), etc. For this, one may need to extend the static base with conditional beliefs and other ingredients borrowed from Belief Revision Theory, as it has already been done for classical DEL [14,10].

Yet another line of on-going research involves adding *explicit non-propositional data*, in the form of *variables* x (taking e.g. different numerical values $x(s)$ in different possible states s), and express the agents' possession of such data via epistemic operators for *(distributed) knowledge of the value of a variable* $K_A x$, as it was done in [26,19,3]. This allows us to explicitly formalize the data-acquisition effects of our data-exchange events: e.g., after accessing the database of some other agent b who already knew the value of a variable x, agent a comes to know the value of x as well. Proving completeness and decidability for these logics poses interesting technical challenges, but we are in the process of finishing a manuscript dealing with these challenges.

Appendix

A Completeness wrt finite pseudo-models

In this section, we prove Proposition 3.2. Soundness of $\mathbf{LDC^e}$ on pseudo-models is an easy verification. But completeness require a bit more work.

Fisher-Ladner Closure. Given a formula φ_0 in LDC^e_S, its Fisher-Ladner closure is the smallest set Φ that contains φ_0, and is closed under subformulas and single negations[18] $\sim\varphi$, as well as under the following rule:

if $C^e_A\phi \in \Phi$ for some event model E, event $e \in E$ and set of agents A, then $K^f_a C^f_A \phi \in \Phi$ for all $a \in A$ and all $f \in E$ s.t. $e \sim_a f$,

One can easily check that *the Fisher-Ladner closure of any formula is finite.*

Finite Canonical Pseudo-Model. For a given formula φ_0, let $\Phi = \Phi(\varphi_0)$ be its Fisher-Ladner closure. The *finite canonical pseudo-model for* φ_0 is the structure $\mathbf{S}^c = (S^c, \sim_B, \| \bullet \|)$, where: S^c is the set of all maximally consistent theories $T \subseteq \Phi$ (over the finite sublanguage given by the Fisher-Ladner closure of φ_0); the group epistemic relations $\sim_A$ are given by putting, for all $T, W \in S^c$:

$$T \sim_A W \quad \text{iff} \quad \forall (D_A\varphi) \in \Phi \; (D_A\varphi \in T \leftrightarrow D_A\varphi \in W)\,;$$

and the valuation is given by putting, for all $p \in Prop$: $\|p\| = \{T \in S^c : p \in T\}$.

It is easy to check that S^c *is indeed a pseudo-model*: $\sim_A$ are obviously equivalence relations, and the other conditions are ensured by the axioms. It is also clear that S^c *is finite*, of size $|S^c| \leq 2^{|\Phi|}$.

For our next results, we need another abbreviation:

Hat Notation For every theory $T \in S^c$, we put $\widehat{T} := \bigwedge T$.

---

[18] The single negation $\sim\varphi$ is defined as: $\sim\varphi := \psi$ if φ is of the form $\neg\psi$; and $\sim\varphi := \neg\varphi$ if φ is *not* of the form $\neg\psi$ (for any ψ).

Lemma A.1 *("Diamond Lemma") For all theories $T, W \in S^c$, we have:*

(i) *If $\widehat{T} \wedge \langle D_A \rangle \widehat{W}$ is consistent, then $T \sim_A W$;*

(ii) *If $D_A\varphi \in \Phi - T$, then there exists $T' \in S^c$ s.t. $(\sim \varphi) \in T'$ and $T \sim_A T'$.*

The proof of this lemma is standard, so we skip the details.

Lemma A.2 *("Truth Lemma") Given a finite canonical pseudo-model S^c over some Fisher-Ladner closure Φ, we have for all $\varphi \in \Phi$:*

$$T \models_{\mathbf{S}^c} \varphi \;\; \textit{iff} \;\; \varphi \in T, \; \textit{for every } T \in S^c$$

Proof. Induction on subformulas. The atomic case is taken care by the canonical valuation, whole the Boolean cases are trivial.

Inductive step for $D_A\varphi$. *Left-to-right*: assume that $T \models D_A\varphi$ for some $(D_A\varphi) \in \Phi$, and suppose towards a contradiction that $(D_A\varphi) \notin T$, hence $(D_A\varphi) \in \Phi - T$. By the Diamond Lemma, there exists some $T' \in S^c$ with $(\sim \varphi) \in T'$ and $T \sim_A T'$. But $T \sim_A T'$ and $T \models D_A\varphi$ imply that $T' \models \varphi$, which by the induction hypothesis gives us $\varphi \in T'$, thus contradicting the fact that $(\sim \varphi) \in T'$ (given the consistency of T').

Right-to-left: Assume that $(D_A\varphi) \in T$. To prove that $T \models D_A\varphi$, let $W \in S^c$ be s.t. $T \sim_A W$; it is enough to show that $\varphi \in W$. For this, note that, by the definition of $T \sim_A W$ in our canonical pseudo-model, $(D_A\varphi) \in T$ implies $(D_A\varphi) \in W$, which in its turn implies that $\varphi \in W$ (by the Veracity axiom).

Inductive step for $C^e_A\overline{\varphi}$, with $e \in E$, where $\mathbf{E}$ is an event model (involving only preconditions in Φ of lower complexity than $C^e_A\overline{\varphi}$). To prove this case, we need some useful notation and two preliminary claims. For all $e \in E$, put

$$\eta_e := \bigvee\{\widehat{W} : W \in S^e_{\overline{\varphi}}\}, \quad \text{where } S^e_{\overline{\varphi}} := \{W \in S^c : W \models C^e_A\overline{\varphi}\},$$

and denote by $\overline{\eta} := (\eta_e)_{e \in E}$ the resulting E-tuple of formulas.

Claim 1: For all $e, f \in E$, $a \in A$ with $e \sim_a f$, we have $\vdash \eta_e \to K^f_a \eta_f$.

Proof of Claim 1: Suppose not. Using the definition of K^f_a as an abbreviation, we must have that $\eta_e \wedge \langle D_{f(a)} \rangle (pre_f \wedge \neg \eta_f)$ is consistent (where $\langle D_{f(a)} \rangle \theta := \neg D_{f(a)} \neg \theta$ is the existential dual of $D_{f(a)}$). Given the definition of $\overline{\eta}$, and the easily proven theorem $\vdash \bigvee\{\widehat{V} : V \in S^c\}$, this means there exist $W \in S_c$, $V \in S^c$ such that $W \models C^e_A\overline{\varphi}$, $V \not\models C^f_A\overline{\varphi}$ and $\widehat{W} \wedge \langle D_{f(a)} \rangle (pre_f \wedge \widehat{V})$ is consistent. But this implies that $W \sim_{f(a)} V$ (using the first part of the Diamond Lemma) and that $pre_f \in V$ (since $pre_f \in \Phi$ is consistent with $\widehat{V}$, and V is a maximally consistent subset of Φ), and thus $V \models pre_f$ (by the induction hypothesis). But by (C^e)-Mix Axiom, $W \models C^e_A\overline{\varphi}$ implies $W \models K^f_a C^f_A\overline{\varphi}$, i.e. $W \models D_{f(a)}(pre_f \to C^f_A\overline{\varphi})$. From this together with $W \sim_{f(a)} V$ and $V \models pre_f$, we infer $V \models C^f_A\overline{\varphi}$, contradicting the previously established fact $V \not\models C^f_A\overline{\varphi}$.

Claim 2: For all $e \in E$, we have $\vdash \eta_e \to C^e_A\overline{\varphi}$.

Proof of Claim 2: First, by (C^e)-Mix, we have $\vdash C^e_A\overline{\varphi} \to \varphi_e$, hence $W \models \varphi_e$ (for all $e \in E$ and $W \in S^e_{\overline{\varphi}}$). By the induction hypothesis, we get that $\varphi_e \in W$, and hence that $\vdash \widehat{W} \to \varphi_e$, for all $e \in E$ and $W \in S^{e}_{\overline{\varphi}}$. By the definition of η_e, we then obtain that $\vdash \eta_e \to \varphi_e$ is a theorem for all $e \in E$. By applying C^e-Necessitation, we derive $\vdash C^{e}_{A}\overline{\eta \to \varphi}$, and by C^e-Distribution we obtain $\vdash C^{e}_{A}\overline{\eta} \to C^{e}_{A}\overline{\varphi}$. On the other hand, we can also prove the theorem $\vdash \eta_e \to C^{e}_{A}\overline{\eta}$ for all $e \in E$ (by applying the C^e-Induction Rule to Claim 1 above). Putting these together, we obtain that $\vdash \eta_e \to C^{e}_{A}\overline{\varphi}$ for all $e \in E$, as desired.

We now proceed to prove the inductive case for $C^{e}_{A}\overline{\varphi}$:

Left-to-right implication: assume that $T \models C^{e}_{A}\overline{\varphi}$. Then (by the very definition of η_e), we have that $\vdash \widehat{T} \to \eta_e$. Together with Claim 2, this gives us that $\vdash \widehat{T} \to C^{e}_{A}\overline{\varphi}$, which implies that $C^{e}_{A}\overline{\varphi} \in T$ (since $C^{e}_{A}\overline{\varphi} \in \Phi$ and T is a maximally consistent subset of Φ), as desired.

Right-to-left: Assume that $C^{e}_{A}\overline{\varphi} \in T$. To prove that $T \models C^{e}_{A}\overline{\varphi}$, let $e = e_0 \sim_{a^1} e_1 \sim_{a^1} \ldots e_{n-1} \sim_{a^n} e_n$ be any finite chain with all $a^i \in A$, and we need to show that $T \models K^{e_1}_{a^1} \ldots K^{e_n}_{a^n}\varphi_{e_n}$, i.e. that $T \models D_{e_1(a^1)}(pre_{e_1} \to \ldots D_{e_n(a^n)}(pre_{e_n} \to \varphi_{e_n})\ldots)$. For this, let $T = T^0 \sim_{e_1(a^1)} T^1 \ldots T^{n-1} \sim_{e_n(a^n)} T^n$, with $T^k \models pre_{e_k}$ for all $k = 1, n$, and we need to prove that $T^n \models \varphi_{e_n}$. For that, we first need to establish a more general statement:

Claim 3: $C^{e_k}_{A}\overline{\varphi} \in T^k$ for all $1 \leq k \leq n$.

Proof of Claim 3: Induction on k. For $k = 1$, the claim is true by the assumption that $(Cd_{\mathcal{B}}\varphi) \in T$. For the inductive step: assume that $C^{e_{k-1}}_{A}\overline{\varphi} \in T^{k-1}$, and use the theorem $\vdash C^{e_{k-1}}_{A}\overline{\varphi} \to D_{e_k(a^k)}(pre_{e_k} \to C^{e_k}_{A}\overline{\varphi})$ (which is just the explicit unfolding of $\vdash C^{e_{k-1}}_{A}\overline{\varphi} \to K^{e_k}_{a^k}C^{e_k}_{A}\overline{\varphi}$, itself an instance of C^e-Mix), to obtain $D_{e_k(a^k)}(pre_{e_k} \to C^{e_k}_{A}\overline{\varphi}) \in T^{k-1}$. From this and $T^{k-1} \sim_{e_k(a^k)} T^k$ (and the canonical definition of $\sim_A$), we get that $D_{e_k(a^k)}(pre_{e_k} \to C^{e_k}_{A}\overline{\varphi}) \in T^k$. Using Veracity and the fact that $pre_{e_k} \in T^k$ (which follows from $T^k \models pre_{e_k}$ by the induction hypothesis), we conclude that $C^{e_k}_{A}\overline{\varphi} \in T^k$.

Applying Claim 3 to $k := n$, we obtain $C^{e_n}_{A}\overline{\varphi} \in T^n$. Using C^e-Mix, we infer $\varphi_{e_n} \in T^n$, which by the induction hypothesis yields $T^n \models \varphi_{e_n}$, as desired. □

Proof of Proposition 3.2 (*Completeness of* $\mathbf{LDC^e}_{\mathcal{S}}$ *wrt finite pseudo-models*) Given a consistent formula φ_0, the Lindenbaum Lemma gives us some $T_0 \in S^c$ s.t. $\varphi_0 \in T_0$. By the Truth Lemma, T_0 satisfies φ_0 in $\mathbf{S}^c$.

B Completeness and Reduction of Dynamic Logics

Lemma B.1 *The following are standard consequences of* $[e]$*-Necessitation and* $[e]$*-Distributivity (holding in general for all normal modalities* $[e]$*):*

(i) (Replacement of Equivalent Formulas) *If* $\vdash \varphi \leftrightarrow \psi$ *is a theorem in* $\mathbf{LDC^e\langle e\rangle}$, *then* $\vdash \langle e\rangle\varphi \leftrightarrow \langle e\rangle\psi$ *is also a theorem.*

(ii) (Distributivity over Conjunction) *The following is a theorem in* $\mathbf{LDC^e\langle e\rangle}$*:* $\vdash [e](\varphi \wedge \psi) \leftrightarrow ([e]\varphi \wedge [e]\psi)$

The second item, in combination with the reduction axiom for negation (i.e. the Partial Functionality axiom), gives us:

Lemma B.2 ($\wedge$-Reduction Law) *The following is a theorem in* $\mathbf{LDC^e}\langle\mathbf{e}\rangle$*:*

$$\vdash\ \langle e\rangle(\varphi\wedge\psi)\leftrightarrow(\langle e\rangle\varphi\wedge\langle e\rangle\psi)$$

Lemma B.3 (Replacement of Equivalent Event Models) *Let* $\mathbf{\Sigma}\in\mathcal{S}$ *be a data-exchange frame, of size* $n=|\Sigma|$*, and let* $\overline{\theta},\overline{\theta'}$ *be* n*-tuples of formulas, such that we have* $\vdash\ \theta_i\leftrightarrow\theta'_i$ *for all* $i=1,n$*. Consider the event models* $E=\Sigma\overline{\theta}$ *and* $E'=\Sigma\overline{\theta'}$*, and the map* $':E\to E'$*, given by* $e=\sigma\overline{\theta}\mapsto e'=\sigma\overline{\theta'}$*. Then, for all* $e\in E$*, agents* a*, sets of agents* A*, formulas* φ *and tuples* $\overline{\varphi}$*, we have:*

(i) $':E\to E'$ *is an isomorphism of data-exchange frames, with the property that* $\vdash\ pre_e\leftrightarrow pre_{e'}$*;*

(ii) $\vdash\ K^e_a\varphi\leftrightarrow K^{e'}_a\varphi$*;*

(iii) $\vdash\ C^e_A\overline{\varphi}\leftrightarrow C^{e'}_A\overline{\varphi}$*.*

Proof. The first item is a trivial verification, while the second follows from the normality of D_A, together with the definition of $K^e_a\varphi:=D_{e(a)}(pre_e\to\varphi)$ and the fact that events e and e' have the same access maps and equivalent preconditions. For the left-to-right implication in item 3: by the Mix Axiom, we have $\vdash\ C^e_A\overline{\varphi}\to K^f_aC^e_A\overline{\varphi}$, for all agents $a\in A$ and events $e,f\in E$ s.t. $e\sim_a f$; putting this together with $\vdash\ K^f_aC^e_A\overline{\varphi}\to K^{f'}_aC^e_A\overline{\varphi}$ (by item 2), we obtain $\vdash\ C^e_A\overline{\varphi}\to K^{f'}_aC^e_A\overline{\varphi}$ for all such and a; then apply the C^e_A-Induction Rule to conclude that $\vdash\ C^e_A\overline{\varphi}\to C^{e'}_A\overline{\varphi}$. The right-to-left implication is similar. □

We can now prove a preliminary "one-step reduction" result:

Lemma B.4 *Let* θ *be any 'static' formula in* LDC^e*, and* $E=\Sigma\overline{\theta}$ *be any event model over the static fragment (i.e. with all* θ_i *in* LDC^e*). Then, for every event* $e\in E$*, there exists some formula* θ_e *in the static fragment* LDC^e*, s.t.*

$$\vdash\ \langle e\rangle\theta\leftrightarrow\theta_e\qquad\textit{is provable in }\mathbf{LDC^e}\langle\mathbf{e}\rangle.$$

Proof. Induction on the subformula complexity of the static formula θ:

For $\theta:=p$, Atomic Permanence gives us the appropriate formula $\theta_e:=p$.

For $\theta:=\neg\psi$: by the induction hypothesis, there exists some 'static' formula ψ_e, s.t. $\vdash\ \langle e\rangle\psi\leftrightarrow\psi_e$. By the reduction axiom for negation (i.e. Partial Functionality), we get $\vdash\ \theta\leftrightarrow(pre_e\wedge\neg\psi_e)$. So we can take $\theta_e:=pre_e\wedge\neg\psi_e$.

For $\theta:=\phi\wedge\psi$: put $\theta_e:=\phi_e\wedge\psi_e$ (using induction and Lemma B.2).

For $\theta:=D_A\psi$, we apply the induction hypothesis to ψ; hence for every event $f\sim_A e$, there exists some static formula ψ_f s.t. $\vdash\ [f]\psi\leftrightarrow\psi_f$. Putting this together with the Reduction Axiom for D_A, we get $\vdash\ \theta\leftrightarrow(pre_e\wedge\bigwedge_{f\sim_B e}D^f_A\psi_f)$. So we can take $\theta_e:=pre_e\wedge\bigwedge_{f\sim_B e}D^f_A\psi_f$.

For $\theta:=C^f_A\overline{\varphi}$, where $f=\sigma\overline{\psi}$ comes from some event model $F=\Sigma\overline{\psi}$ over LDC^e: note that, since $\theta=C^f_A\overline{\varphi}=C^{\sigma\overline{\psi}}_A\overline{\varphi}$, all the formulas φ_i and ψ_j are subformulas of θ. So, by the induction hypothesis, for each $e_i\in E$ and each $f_j\in F$ with precondition $pre_{f_j}=\psi_j$, there exist static formulas $\varphi'_{ij},\psi'_{ij}\in LDC^e$, s.t. we have $\vdash\ \langle e_i\rangle\varphi_j\leftrightarrow\varphi'_{ij}$ and $\vdash\ \langle e_i\rangle\psi_j\leftrightarrow\psi'_{ij}$. By the

Reduction Axiom for $C_A^f\overline{\varphi}$, we also have $\vdash \langle e_i\rangle C_A^{f_j}\overline{\varphi} \leftrightarrow C_A^{e_i;f_j}\langle E\rangle\overline{\varphi}$. Since each precondition $pre_{e_i;f_j} = \langle e_i\rangle\psi_j$ is equivalent to the corresponding static formula ψ'_{ij}, we can use Lemma B.3 to replace the event model $E;F = \{e_i;f_j\}_{i,j}$ by another event model $G = \{g_{i,j}\}_{i,j}$ having the same data-exchange structure but with the static formulas $\{\psi'_{ij}\}$ as preconditions of the corresponding $g_{i;j}$; so we obtain that $\vdash C_A^{e_i;f_j}\langle E\rangle\overline{\varphi} \leftrightarrow C_A^{g_{i,j}}\langle E\rangle\overline{\varphi}$. Using this, the Normality of C^e and the fact that each $\langle e_i\rangle\varphi_j$ is equivalent to the corresponding φ'_{ij}, we obtain that $\vdash C_A^{e_i;f_j}\langle E\rangle\overline{\varphi} \leftrightarrow C_A^{g_{i,j}}\overline{\varphi'}$, where $\overline{\varphi'} := \{\varphi_{i,j}\}_{i,j}$. Since G is an event model with preconditions ψ'_{ij} in the static fragment LDC^e, and also all $\varphi'_{ij} \in LDC^e$, the formulas $C_A^{g_{i,j}}\overline{\varphi'}$ are all in LDC^e as well. Applying this to the specific i and j for which $e_i := e \in E$ and $f_j := f \in F$ and recalling that $\theta = C_A^f\overline{\varphi}$, we obtain that $\vdash \langle e\rangle\theta \leftrightarrow \theta_e$, for $\theta_e := C_A^{g_{i,j}}\overline{\varphi'}$ belonging to LDC^e, as desired. □

Now we can establish our full reduction result:

Lemma B.5 *For every formula θ of the dynamic logic $LDC^e\langle e\rangle$, there exists some formula θ' of the static language LDC^e, s.t.*

$$\vdash \theta \leftrightarrow \theta' \qquad \textit{is provable in } \mathbf{LDC^e\langle e\rangle}.$$

Proof. Induction on the subformula complexity of the dynamic formula θ:

For $\theta := p$, we can take $\theta' = \theta$ (since this is already in LDC^e).

The Boolean cases $\theta := \neg\psi$ and $\theta := \phi \wedge \psi$ are also trivial.

For $\theta := D_A\psi$, apply the induction hypothesis to ψ; so there exists some 'static' formula ψ', such that $\vdash \psi \leftrightarrow \psi'$. By D_A-Necessitation and D_A-Distribution, we get that $\vdash D_A\psi \leftrightarrow D_A\psi'$ (so we can take $\theta' := D_A\psi'$).

For $\theta := C_A^e\overline{\psi}$, with $e \in E$, where E is an event model of size $|E| = n$, and $\overline{\psi} = \{\psi_i\}_{i\leq n}$ is an n-tuple of formulas: apply the induction hypothesis to obtain an n-tuple $\overline{\psi'} = \{\psi'_i\}_{i\leq n}$ of static formulas $\psi'_i \in LDC^e$, s.t. we have $\vdash \psi_i \leftrightarrow \psi'_i$ for all i. Using the Normality of C^e, we obtain $\vdash C_A^e\overline{\psi} \leftrightarrow C_A^e\overline{\psi'}$. Since the preconditions of the events in E are subformulas of θ, we can use the induction hypothesis to obtain 'static' formulas in LDC^e that are provably equivalent to each precondition of an event in E; then use Lemma B.3 to replace the event model E by another event model $E' = \{e' : e \in E\}$, having the same data-exchange structure, but with static preconditions $pre_{e'} \in LDC^e$ s.t. $\vdash pre_e \leftrightarrow pre_{e'}$ for all e, obtaining $\vdash C_A^e\overline{\psi'} \leftrightarrow C_A^{e'}\overline{\psi'}$ for all $e \in E$. Putting all these together, we conclude that $\vdash C_A^e\overline{\psi} \leftrightarrow \theta'$, where $\theta' := C_A^{e'}\overline{\psi'}$.

For $\theta := \langle e\rangle\psi$: by the induction hypothesis, there exists some 'static' formula ψ', such that $\vdash \psi \leftrightarrow \psi'$. By Lemma B.1, we get that $\vdash \langle e\rangle\psi \leftrightarrow \langle e\rangle\psi'$, and by Lemma B.4 we obtain another static formula $\psi'_e \in LDC^e$, s.t. $\vdash \langle e\rangle\psi' \leftrightarrow \psi'_e$. Putting these together, we get $\vdash \langle e\rangle\psi \leftrightarrow \psi'_e$ (so we can take $\theta' := \psi'_e$). □

Finally, we can prove the last part of Theorem 3.1:

Proof of completeness of $\mathbf{LDC^e\langle e\rangle}$: Let θ be a consistent formula of $LDC^e\langle e\rangle$. By Lemma B.5, there exists some θ' in LDC^e s.t. $\vdash \theta \leftrightarrow \theta'$ is a theorem in $\mathbf{LDC^e\langle e\rangle}$. So θ' is consistent. By Theorem 3.1(1) (completeness for $\mathbf{LDC^e}$), θ' is satisfiable. Since $\vdash \theta \leftrightarrow \theta'$, θ is also satisfiable.

References

[1] Adriaans, P. and J. van Benthem, *Handbook of Philosophy of Information*, (2008), Elsevier.

[2] Agotnes, T. and Y.N. Wang, *Resolving Distributed Knowledge*, in: *Artificial Intelligence*, (2017), 252, pp. 1-21.

[3] Baltag, A., *To Know is to Know the Value of a Variable*, in: *Adv. in Modal Logic 2016*, (2016), pp. 135-155.

[4] Baltag, A. and J. van Benthem, *A Simple Logic of Functional Dependence*, in: *Journal of Philosophical logic*, (2021), vol. 50, pp. 939-1005.

[5] Baltag, A. and J. van Benthem, *Updates, Generalized p-Morphisms, and (Co-)Recursive Equations*, To appear in van Benthem, J. and F. Liu (eds.), *Graph Games and Logic Design - Recent developments and further directions*, Springer, (2024).

[6] Baltag, A. and L. S. Moss, *Logics for Epistemic Programs*, in: Symons, J. and J. Hintikka. (eds.), W. van der Hoek (special section editor), Synthese, (2004), 139(2), pp. 165-224.

[7] Baltag, A.; L. Moss and S. Solecki, *The Logic of Public Announcements, Common Knowledge, and Private Suspicions*, in: *Proceedings TARK 98*, (1998), pp. 43-56.

[8] Baltag, A.; L. Moss and S. Solecki, *Logics for epistemic actions: completeness, decidability, expressivity*, in: *Logics* (2023), 1 (2), pp. 97-147.

[9] Baltag, A. and B. Renne, *Dynamic Epistemic Logic*, in: *Stanford Encyclopedia of Philosophy*, (2016).

[10] Baltag, A. and S Smets, *A qualitative theory of dynamic interactive belief revision*, in *Texts in logic and games* (2008), vol. 3, pp. 9–58), Amsterdam University Press.

[11] Baltag, A. and S. Smets, *Protocols for Belief Merge: Reaching Agreement via Communication*, in: *Logic Journal of the IGPL*, (2013), 21(3), pp. 468-487.

[12] Baltag, A. and S. Smets, *Learning what Others Know*, in: Kovacs, L. and E. Albert (eds.), *LPAR23 proceedings of the International Conference on Logic for Programming AI and Reasoning, EPiC Series in Computing*, (2020), Volume 73, pp. 90-110.

[13] van Benthem, J., *One is a lonely number*, in: Koepke, P.; Z. Chatzidakis and W. Pohlers, (eds.) *Logic Colloquium 2002*, (2002), pp. 96-129, ASL and A.K. Peters, Wellesley MA.

[14] van Benthem, J., *Dynamic logic of belief revision*, in: *Journal for Applied Non-Classical Logics* (2007), 17(2), pp. 129–155.

[15] van Benthem, J., *Logical Dynamics of Information and Interaction*, (2011), Cambridge University Press, Cambridge UK.

[16] van Benthem, J.; J. van Eijck and B. Kooi, *Logics of Communication and Change*, in: *Information and Communication*, (2006), 204:11, pp. 1620-1662.

[17] Blackburn, P.; M. de Rijke and Y. Venema, *Modal Logic*, (2000), Cambridge University Press, Cambridge.

[18] van Ditmarsch, H.; W. van der Hoek and B. Kooi, *Dynamic Epistemic Logic*, (2007), Springer, Dordrecht.

[19] van Eijck, J; M. Gattinger and Y. Wang, *Knowing Values and Public Inspection*, in: *Proceedings 7th Indian Conference on Logic and its Applications* (2007), Kanpur, 77–90.

[20] Fagin, R.; J. Halpern, Y. Moses and M. Vardi, *Reasoning About Knowledge*, (1995), The MIT Press, Cambridge MA.

[21] Gerbrandy, J., *Bisimulations on Planet Kripke*, (1998), Doctoral Dissertation, ILLC, UvA. ILLCS DS Series DS-1999-01. ISBN: 9057760193, 9789057760198

[22] Parikh, R., *Levels of Knowledge, Games and Group Action*, in: *Research in Economics*, (2003), 57, pp. 267-281.

[23] Parikh, R. and R. Ramanujam, *A Knowledge-Based Semantics of Messages*, in: *Journal of Logic, Language and Information*, (2003), 12, pp. 453-467.

[24] Plaza, J., *Logics of Public Communication*, in: *Proceedings 4th International Symposium on Methodologies for Intelligent Systems*, (1989), pp. 201-216.

[25] van Wijk, S., *Coalitions in Epistemic Planning.MSc Thesis*, (2015), ILLC, UvA. *https://www.illc.uva.nl/Research/Publications/Reports/MoL/*

[26] Wang, Y., *Beyond Knowing That: A New Generation of Epistemic Logics*, in *Hintikka Volume* (2018), Springer, Dordrecht, 499-533.

Towards an Algebraic Theory of KD45-Like Logics

Line van den Berg

Mathematical Institute, University of Bern, Switzerland

Manuela Busaniche Miguel Marcos

Universidad Nacional del Litoral and CONICET, Santa Fe, Argentina
{mbusaniche,mmarcos}@santafe-conicet.gov.ar

George Metcalfe [1]

Mathematical Institute, University of Bern, Switzerland
george.metcalfe@unibe.ch

Abstract

Algebraic semantics are introduced for a family of 'KD45-like' modal substructural logics as a generalization of Bezhanishivili's pseudomonadic algebras for the modal logic KD45. It is shown that these structures correspond to ordered pairs consisting of an $\mathsf{FL_e}$-algebra (or commutative pointed residuated lattice) and a subalgebra with a suitable lattice filter, extending a similar result for 'S5-like' logics. It is then shown that if the $\mathsf{FL_e}$-algebra reduct belongs to a variety that has the superamalgamation property, then the structure equipped with an additional constant is representable as an algebra of functions from a set of worlds to an $\mathsf{FL_e}$-algebra of the same variety.

Keywords: Modal Logic, Substructural Logic, $\mathsf{FL_e}$-algebra, Residuated Lattice.

1 Introduction

Non-classical modal logics — extensions of intermediate, substructural, and many-valued logics with various modal operators — have been motivated and studied in a wide range of contexts (see, e.g., [4,7–9,13,14,20,22]). These logics are introduced either semantically via a relational semantics, or syntactically, via an axiomatization or class of algebraic structures, but general systematic accounts relating these two perspectives are mostly lacking in the literature. Such an account has been given in [12] for 'S5-like' modal logics corresponding to one-variable fragments of first-order logics, and the main aim of the current

[1] This project has received funding from the European Union's Horizon 2020 research and innovation programme under the Marie Skłodowska-Curie grant agreement No 101007627.

paper is to explore how far this approach extends to 'KD45-like' modal logics. In particular, we introduce, and provide representation theorems for, algebraic structures that generalize both Bezhanishivili's pseudomonadic algebras for the modal logic KD45 [3] and algebras for the 'KD45-like' many-valued modal logics with a possibilistic semantics considered in [5,6,10,21].

In Section 2, we consider, as a general starting point, expansions with unary operations $\Box$ and $\Diamond$ of *$\mathcal{L}$-lattices*: algebraic structures for a signature $\mathcal{L}$ with a lattice reduct that provide semantics for a broad range of non-classical logics. As shown in [12], if the expansion satisfies equations corresponding to the modal axioms for S5, then its image under $\Box$ (equivalently, $\Diamond$) forms a subalgebra of its $\mathcal{L}$-lattice reduct. We show here that this is also the case for expansions that fail the 'reflexivity' equations $\Box x \leq x$ and $x \leq \Diamond x$, corresponding to the 'T' axiom distinguishing S5 from KD45. However, whereas the expansions satisfying the reflexivity equations are in one-to-one correspondence with $\mathcal{L}$-lattices equipped with a 'relatively complete' subalgebra [12], this is not the case in general.

In Section 3, we address this issue in the setting of $\mathsf{FL_e}$-algebras (also known as commutative pointed residuated lattices): algebraic structures that serve as semantics for a broad family of substructural logics (see [16,18]). We introduce *pm-$\mathsf{FL_e}$-algebras* as expansions of $\mathsf{FL_e}$-algebras with unary operations $\Box$ and $\Diamond$ that include the aforementioned pseudomonadic algebras [3] and algebras for many-valued modal logics [5,6,10,21]. We then prove that these structures are in one-to-one correspondence with ordered pairs consisting of an $\mathsf{FL_e}$-algebra and a subalgebra equipped with a suitable lattice filter (Theorem 3.9).

In Section 4, we establish a functional representation theorem for 'pointed' pm-$\mathsf{FL_e}$-algebras expanded with a constant that, as prescribed by two additional equations, generates the corresponding lattice filter. More precisely, following similar proofs in [2,12] for one-variable fragments of first-order logics, we prove that if the $\mathsf{FL_e}$-reduct of a pointed pm-$\mathsf{FL_e}$-algebra belongs to a variety $\mathcal{V}$ that has the superamalgamation property, then the latter can be represented as an algebra of functions from a set of worlds to a member of $\mathcal{V}$ (Theorem 4.3). In the case where $\mathcal{V}$ is also closed under regular completions, this yields completeness results for a corresponding 'KD45-like' logic with respect to complex algebras based on frames consisting of a set of worlds W and a map $\pi \colon W \to A$ for some $\mathbf{A} \in \mathcal{V}$, generalizing the 'simplified semantics' for KD45 described in [19] and 'possibilistic semantics' for many-valued modal logics considered in [5,21].

2 Modal Operators via Subalgebras

In this section, we consider a general class of lattice-ordered algebraic structures with unary operations $\Box$ and $\Diamond$ that satisfy equations corresponding to axioms common to 'KD45-like' logics. In particular, we show that this class includes exactly the algebraic structures in the general framework whose images under $\Box$ and $\Diamond$ coincide and form a common subalgebra.

Let us call an algebraic structure $\langle L, \wedge, \vee, \Box, \Diamond \rangle$ with a lattice reduct $\langle L, \wedge, \vee \rangle$ and unary operations $\Box$ and $\Diamond$ an *e-lattice* if it satisfies the following equations satisfied by Bezhanishivili's pseudomonadic algebras[2] for KD45 [3]:

$$\begin{array}{llll} (\mathrm{L1}_\Box) & \Box(x \wedge y) \approx \Box x \wedge \Box y & (\mathrm{L1}_\Diamond) & \Diamond(x \vee y) \approx \Diamond x \vee \Diamond y \\ (\mathrm{L2}_\Box) & \Box\Box x \approx \Box x & (\mathrm{L2}_\Diamond) & \Diamond\Diamond x \approx \Diamond x \\ (\mathrm{L3}_\Box) & \Box\Diamond x \approx \Diamond x & (\mathrm{L3}_\Diamond) & \Diamond\Box x \approx \Box x. \end{array}$$

It follows directly, using $(\mathrm{L1}_\Box)$ and $(\mathrm{L1}_\Diamond)$, that the unary operations $\Box$ and $\Diamond$ of any e-lattice are order-preserving; that is, letting $\alpha \leq \beta$ stand for $\alpha \wedge \beta \approx \alpha$, every e-lattice satisfies the quasi-equations

$$(\mathrm{L4}_\Box)\ x \leq y \implies \Box x \leq \Box y \qquad (\mathrm{L4}_\Diamond)\ x \leq y \implies \Diamond x \leq \Diamond y.$$

Also $\Box A := \{\Box a \mid a \in A\} = \{\Diamond a \mid a \in A\}$, by $(\mathrm{L3}_\Box)$ and $(\mathrm{L3}_\Diamond)$, and $\Box$ and $\Diamond$ restrict to identity maps on this set, by $(\mathrm{L2}_\Box)$ and $(\mathrm{L2}_\Diamond)$.

Note that the equations $(\mathrm{L1}_\Box)$-$(\mathrm{L3}_\Box)$ and $(\mathrm{L1}_\Diamond)$-$(\mathrm{L3}_\Diamond)$ are associated with the 'K', '4', and '5' axioms of KD45, but the following equation corresponding to the 'D' axiom $\Box p \to \Diamond p$ is *not* satisfied by all e-lattices:

$$(\mathrm{LD})\ \Box x \leq \Diamond x.$$

Now let $\mathcal{L}$ be any algebraic signature containing binary operation symbols $\wedge$ and $\vee$, and denote by $\mathcal{L}_\Box$ the extension of $\mathcal{L}$ with unary operation symbols $\Box$ and $\Diamond$. We call an $\mathcal{L}$-structure $\mathbf{A}$ with a lattice reduct $\langle A, \wedge, \vee\rangle$ an *$\mathcal{L}$-lattice*, and call an $\mathcal{L}_\Box$-structure $\langle \mathbf{A}, \Box, \Diamond\rangle$ with an e-lattice reduct $\langle A, \wedge, \vee, \Box, \Diamond\rangle$ an *e-$\mathcal{L}$-lattice* if it satisfies for each n-ary operation symbol $\star$ of $\mathcal{L}$ the equation

$$(\star_\Box)\ \Box(\star(\Box x_1, \ldots, \Box x_n)) \approx \star(\Box x_1, \ldots, \Box x_n).$$

It follows easily, using $(\star_\Box)$, $(\mathrm{L3}_\Box)$, and $(\mathrm{L3}_\Diamond)$, that every e-$\mathcal{L}$-lattice $\langle \mathbf{A}, \Box, \Diamond\rangle$ also satisfies for each n-ary operation symbol $\star$ of $\mathcal{L}$ the equation

$$(\star_\Diamond)\ \Diamond(\star(\Diamond x_1, \ldots, \Diamond x_n)) \approx \star(\Diamond x_1, \ldots, \Diamond x_n).$$

Moreover, as the next result demonstrates, these additional equations express precisely what is required of an $\mathcal{L}_\Box$-structure $\langle \mathbf{A}, \Box, \Diamond\rangle$ with an e-lattice reduct $\langle A, \wedge, \vee, \Box, \Diamond\rangle$ to ensure that $\Box A$ forms a subalgebra of the $\mathcal{L}$-structure $\mathbf{A}$.

Proposition 2.1 *Let $\langle \mathbf{A}, \Box, \Diamond\rangle$ be any $\mathcal{L}_\Box$-structure such that $\mathbf{A}$ is an $\mathcal{L}$-lattice and $\langle A, \wedge, \vee, \Box, \Diamond\rangle$ is an e-lattice. Then $\langle \mathbf{A}, \Box, \Diamond\rangle$ is an e-$\mathcal{L}$-lattice if and only if $\Box A$ forms a subalgebra $\Box\mathbf{A}$ of $\mathbf{A}$.*

---

[2] A *pseudomonadic algebra* is an algebraic structure $\langle \mathbf{B}, \exists\rangle$ such that $\mathbf{B} = \langle B, \wedge, \vee, \neg, 0, 1\rangle$ is a Boolean algebra and $\exists$ is a unary operation satisfying the following equations:

$$\begin{array}{llll} (\mathrm{B1}) & \exists 0 \approx 0 & (\mathrm{B3}) & \exists(x \wedge \exists y) \approx \exists x \wedge \exists y \\ (\mathrm{B2}) & \exists(x \vee y) \approx \exists x \vee \exists y & (\mathrm{B4}) & \neg\exists x \leq \exists\neg x. \end{array}$$

Proof. If $\langle \mathbf{A}, \Box, \Diamond \rangle$ is an e-$\mathcal{L}$-lattice, then $\Box A$ is closed under each operation $\star^{\mathbf{A}}$ of $\mathbf{A}$, by ($\star_\Box$), and hence forms a subalgebra. For the converse, suppose that $\Box\mathbf{A}$ is a subalgebra of $\mathbf{A}$ and consider any n-ary operation symbol $\star$ of $\mathcal{L}$ and $a_1, \ldots, a_n \in A$. By assumption, $\Box a_1, \ldots, \Box a_n \in \Box A$, so $\star(\Box a_1, \ldots, \Box a_n) \in \Box A$. Hence $\Box(\star(\Box a_1, \ldots, \Box a_n)) = \star(\Box a_1, \ldots, \Box a_n)$, by (L2$_\Box$). $\Box$

It is shown in [12] that every e-$\mathcal{L}$-lattice $\langle \mathbf{A}, \Box, \Diamond \rangle$ satisfying two additional 'reflexivity' axioms can be uniquely identified with the ordered pair $\langle \mathbf{A}, \Box\mathbf{A} \rangle$. Following [12], let us call an e-$\mathcal{L}$-lattice an *m-$\mathcal{L}$-lattice* if it satisfies the equations

$$(\mathrm{L5}_\Box)\ \Box x \leq x \qquad (\mathrm{L5}_\Diamond)\ x \leq \Diamond x.$$

It is easily checked that for any m-$\mathcal{L}$-lattice $\langle \mathbf{A}, \Box, \Diamond \rangle$ and $a \in A$,

$$\Box a = \max\{b \in \Box A \mid b \leq a\} \quad \text{and} \quad \Diamond a = \min\{b \in \Box A \mid a \leq b\}.$$

Conversely, call a subalgebra $\mathbf{A}_0$ of an $\mathcal{L}$-lattice $\mathbf{A}$ *relatively complete* if for any $a \in L$, the set $\{b \in L_0 \mid b \leq a\}$ has a maximum and the set $\{b \in L_0 \mid a \leq b\}$ has a minimum; in this case, defining $\Box_0 a := \max\{b \in A_0 \mid b \leq a\}$ and $\Diamond_0 a := \min\{b \in A_0 \mid a \leq b\}$ for each $a \in A$, yields an m-$\mathcal{L}$-lattice $\langle \mathbf{A}, \Box_0, \Diamond_0 \rangle$. Combining these two observations yields:

Theorem 2.2 ([12, Theorem 3.5]) *There exists a one-to-one correspondence between m-$\mathcal{L}$-lattices and ordered pairs $\langle \mathbf{A}, \mathbf{A}_0 \rangle$ such that $\mathbf{A}$ is an $\mathcal{L}$-lattice and $\mathbf{A}_0$ is a relatively complete subalgebra of $\mathbf{A}$, implemented by the maps $\langle \mathbf{A}, \Box, \Diamond \rangle \mapsto \langle \mathbf{A}, \Box\mathbf{A} \rangle$ and $\langle \mathbf{A}, \mathbf{A}_0 \rangle \mapsto \langle \mathbf{A}, \Box_0, \Diamond_0 \rangle$.*

A further 'functional' representation theorem is obtained in [12] for certain classes of m-$\mathcal{L}$-lattices. Let $\mathcal{V}$ be any variety of $\mathcal{L}$-lattices. A *V-formation* $\langle \mathbf{A}, \mathbf{B}_1, \mathbf{B}_2, f_1, f_2 \rangle$ in $\mathcal{V}$ consists of algebras $\mathbf{A}, \mathbf{B}_1, \mathbf{B}_2 \in \mathcal{V}$ and embeddings $f_1 \colon \mathbf{A} \to \mathbf{B}_1$, $f_2 \colon \mathbf{A} \to \mathbf{B}_2$. A *superamalgam* $\langle \mathbf{C}, g_1, g_2 \rangle$ in $\mathcal{V}$ of this V-formation consists of an algebra $\mathbf{C} \in \mathcal{V}$ and embeddings $g_1 \colon \mathbf{B}_1 \to \mathbf{C}$, $g_2 \colon \mathbf{B}_2 \to \mathbf{C}$ such that $g_1 \circ f_1 = g_2 \circ f_2$ and for any $b_1 \in B_1$, $b_2 \in B_2$ and distinct $i, j \in \{1, 2\}$,

$$g_i(b_i) \leq g_j(b_j) \implies g_i(b_i) \leq g_i \circ f_i(a) = g_j \circ f_j(a) \leq g_j(b_j) \text{ for some } a \in A.$$

$\mathcal{V}$ is said to have the *superamalgamation property* if every V-formation in $\mathcal{V}$ has a superamalgam in $\mathcal{V}$.

Theorem 2.3 ([12, Theorem 4.1]) *Let $\mathcal{V}$ be a variety of $\mathcal{L}$-lattices that has the superamalgamation property. Then every m-$\mathcal{L}$-lattice with an $\mathcal{L}$-lattice reduct in $\mathcal{V}$ is isomorphic to an m-$\mathcal{L}$-lattice $\langle \mathbf{B}, \Box, \Diamond \rangle$ such that $\mathbf{B}$ is a subalgebra of $\mathbf{A}^W$ for some $\mathbf{A} \in \mathcal{V}$ and set W, and contains for each $f \in B$ the constant maps*

$$\Box f \colon W \to A;\ u \mapsto \bigwedge_{v \in W} f(v) \quad \textit{and} \quad \Diamond f \colon W \to A;\ u \mapsto \bigvee_{v \in W} f(v).$$

The main goal of this paper is to obtain versions of Theorems 2.2 and 2.3 for a broad family of e-$\mathcal{L}$-lattices serving as algebraic semantics for 'KD45-like' logics.

3 A Correspondence Theorem

The correspondence established in Theorem 2.2 does not apply to e-$\mathcal{L}$-lattices that fail to satisfy either (L5$_\Box$) or (L5$_\Diamond$). In this section, we establish a refined version of this correspondence in the setting of $\mathsf{FL_e}$-algebras.

Let us fix $\mathcal{L}$ to be the signature with binary operation symbols $\wedge$, $\vee$, $\cdot$, and $\to$ and constant symbols e and f. An $\mathsf{FL_e}$-*algebra* — also known as a *commutative pointed residuated lattice* — is an $\mathcal{L}$-lattice $\langle A, \wedge, \vee, \cdot, \to, \mathrm{e}, \mathrm{f}\rangle$ such that $\langle A, \cdot, \mathrm{e}\rangle$ is a commutative monoid and $\to$ is the residuum of $\cdot$, that is, $a \cdot b \leq c \iff a \leq b \to c$, for all $a, b, c \in A$.

Remark 3.1 Varieties of $\mathsf{FL_e}$-algebras provide algebraic semantics for a broad family of non-classical logics (see, e.g., [16, 18]); in particular:

- $\mathsf{FL_e}$-algebras serve as algebraic semantics for the full Lambek calculus with exchange, and $\mathsf{FL_{ew}}$-algebras — $\mathsf{FL_e}$-algebras satisfying $\mathrm{f} \leq x \leq \mathrm{e}$ — play the same role for the extension of this calculus with weakening rules;
- Heyting algebras, serving as algebraic semantics for intuitionistic logic, are term-equivalent to $\mathsf{FL_{ew}}$-algebras satisfying $x \approx x \cdot x$ (just identify $\cdot$ and $\wedge$), and Boolean algebras, serving as algebraic semantics for classical logic, are term-equivalent to Heyting algebras satisfying $\neg\neg x \approx x$, where $\neg x := x \to \mathrm{f}$.
- BL-algebras, serving as algebraic semantics for Hájek's basic fuzzy logic, are term-equivalent to $\mathsf{FL_{ew}}$-algebras satisfying $(x \to y) \vee (y \to x) \approx \mathrm{e}$ and $x \wedge y \approx x \cdot (x \to y)$, while MV-algebras and Gödel algebras, serving as algebraic semantics for Łukasiewicz logic and Gödel-Dummett logic, are term-equivalent to BL-algebras satisfying $\neg\neg x \approx x$ and $x \approx x \cdot x$, respectively.

We call an e-$\mathcal{L}$-lattice $\langle \mathbf{A}, \Box, \Diamond\rangle$ with $\mathsf{FL_e}$-algebra reduct $\mathbf{A}$ a *pm*-$\mathsf{FL_e}$-*algebra* if it satisfies the additional equations

$$\text{(P1)} \quad \Box(x \to \Box y) \approx \Diamond x \to \Box y \qquad \text{(P2)} \quad \Box(\Box x \to y) \approx \Box x \to \Box y.$$

Every m-$\mathcal{L}$-lattice satisfies (P1) and (P2) (see [12, Example 3.1]) and will hence also be referred to as an *m*-$\mathsf{FL_e}$-*algebra.*

It follows easily using (L3$_\Box$), (P1), and (P2) that every pm-$\mathsf{FL_e}$-algebra $\langle \mathbf{A}, \Box, \Diamond\rangle$ also satisfies the equations

$$\text{(P3)} \quad \Box(x \to \Diamond y) \approx \Diamond x \to \Diamond y \qquad \text{(P4)} \quad \Box(\Diamond x \to y) \approx \Diamond x \to \Box y.$$

The following lemma establishes an additional property of pm-$\mathsf{FL_e}$-algebras that is useful in comparing them with algebras for other modal logics.

Lemma 3.2 *Every pm*-$\mathsf{FL_e}$-*algebra* $\langle \mathbf{A}, \Box, \Diamond\rangle$ *satisfies*

$$\text{(P5)} \quad \Diamond(x \cdot \Diamond y) \approx \Diamond x \cdot \Diamond y.$$

Proof. Consider any $a, b \in A$. Using $(\cdot_\Diamond)$, $(\to_\Diamond)$, (P3), and standard properties of $\mathsf{FL_e}$-algebras, we obtain

$$\begin{aligned}\Diamond(a \cdot \Diamond b) \to (\Diamond a \cdot \Diamond b) &= \Diamond(a \cdot \Diamond b) \to \Diamond(\Diamond a \cdot \Diamond b)\\ &= \Box((a \cdot \Diamond b) \to \Diamond(\Diamond a \cdot \Diamond b))\\ &= \Box(a \to (\Diamond b \to \Diamond(\Diamond a \cdot \Diamond b)))\\ &= \Box(a \to \Diamond(\Diamond b \to \Diamond(\Diamond a \cdot \Diamond b)))\\ &= \Diamond a \to \Diamond(\Diamond b \to \Diamond(\Diamond a \cdot \Diamond b))\\ &= \Diamond a \to (\Diamond b \to (\Diamond a \cdot \Diamond b))\\ &\geq \mathrm{e},\end{aligned}$$

and therefore, by residuation, $\Diamond(a \cdot \Diamond b) \leq \Diamond a \cdot \Diamond b$. Similarly,

$$\begin{aligned}(\Diamond a \cdot \Diamond b) \to \Diamond(a \cdot \Diamond b) &= \Diamond a \to (\Diamond b \to \Diamond(a \cdot \Diamond b))\\ &= \Diamond a \to \Diamond(\Diamond b \to \Diamond(a \cdot \Diamond b))\\ &= \Box(a \to \Diamond(\Diamond b \to \Diamond(a \cdot \Diamond b)))\\ &= \Box(a \to (\Diamond b \to \Diamond(a \cdot \Diamond b)))\\ &= \Box((a \cdot \Diamond b) \to \Diamond(a \cdot \Diamond b))\\ &= \Diamond(a \cdot \Diamond b) \to \Diamond(a \cdot \Diamond b)\\ &\geq \mathrm{e},\end{aligned}$$

and therefore, by residuation, $\Diamond a \cdot \Diamond b \leq \Diamond(a \cdot \Diamond b)$. □

Note that if $\langle \mathbf{A}, \Box, \Diamond \rangle$ is a pm-$\mathsf{FL_e}$-algebra such that $\mathbf{A}$ satisfies $x \leq \mathrm{e}$ and $x \wedge (x \to y) \leq y$ (in particular, if $\mathbf{A}$ is a Heyting algebra), then for any $a \in A$,

$$\Box a = \Box a \wedge \mathrm{e} = \Box a \wedge \Box(a \to \Diamond a) = \Box(a \wedge (a \to \Diamond a)) \leq \Box\Diamond a = \Diamond a,$$

that is, $\langle \mathbf{A}, \Box, \Diamond \rangle$ satisfies (LD).

Remark 3.3 Let us call a pm-$\mathsf{FL_e}$-algebra $\langle \mathbf{A}, \Box, \Diamond \rangle$ *Boolean* if $\mathbf{A}$ is a Boolean algebra according to the term-equivalence described in Remark 3.1. In this case, defining $\neg a := a \to \mathrm{f}$ produces a pseudomonadic algebra $\langle A, \wedge, \vee, \neg, \mathrm{f}, \mathrm{e}, \Diamond \rangle$. Conversely, given any pseudomonadic algebra $\langle \mathbf{B}, \exists \rangle$, defining $a \to b := \neg a \vee b$ and $\forall a := \neg\exists\neg a$ produces a Boolean pm-$\mathsf{FL_e}$-algebra $\langle A, \wedge, \vee, \wedge, \to, \mathrm{e}, \mathrm{f}, \forall, \exists \rangle$. So pseudomonadic algebras are term-equivalent to Boolean pm-$\mathsf{FL_e}$-algebras. Similarly, pm-$\mathsf{FL_e}$-algebras with a Gödel algebra reduct are term-equivalent to algebraic semantics for the modal Gödel-Dummett logic defined over a class of many-valued serial, transitive, and Euclidean frames in [10] and studied in [21] as a possibilistic logic.

Remark 3.4 It is straightforward to show that pm-$\mathsf{FL_e}$-algebras satisfying (LD) with a BL-algebra reduct are term-equivalent to the pseudomonadic BL-algebras proposed in [6] as algebraic semantics for '$\mathsf{KD45}$-like' extensions of Hájek's basic fuzzy logic. In particular, a class of pseudomonadic BL-algebras with an MV-algebra reduct is studied in [5] as algebraic semantics for a logic of possibilistic Kripke frames defined over finite totally ordered MV-algebras. To see that not all pm-$\mathsf{FL_e}$-algebras with a BL-algebra (or even MV-algebra) reduct satisfy (LD), consider the MV-algebra $\mathbf{Ł}_3 = \langle\{0, \frac{1}{2}, 1\}, \min, \max, \cdot, \to, 0, 1\rangle$ (in the language of $\mathsf{FL_e}$-algebras), where $a \cdot b := \max(0, a + b - 1)$ and $a \to b := \min(1, 1 - a + b)$, and define

$$\Box a := \begin{cases} 1 & \text{if } a = 1 \text{ or } a = \frac{1}{2} \\ 0 & \text{if } a = 0, \end{cases} \qquad \Diamond a = \begin{cases} 1 & \text{if } a = 1 \\ 0 & \text{if } a = 0 \text{ or } a = \frac{1}{2}. \end{cases}$$

Then $\langle \mathbf{Ł}_3, \Box, \Diamond\rangle$ does not satisfy (LD), since $\Box\frac{1}{2} = 1 > 0 = \Diamond\frac{1}{2}$.

Next, we introduce a generalization of the concepts of focal elements and c-relatively complete subalgebras defined in [6] for pseudomonadic BL-algebras (see Remark 3.4). Given any pm-$\mathsf{FL_e}$-algebra $\langle \mathbf{A}, \Box, \Diamond\rangle$, let

$$\mathcal{U}_{\mathbf{A}} := \{a \in A \mid \mathrm{e} \leq \Box a\}.$$

Recall that a *lattice filter* of an $\mathsf{FL_e}$-algebra $\mathbf{A}$ is a subset $L \subseteq A$ such that $L = \uparrow L := \{b \in A \mid a \leq b \text{ for some } a \in L\}$ and $a, b \in L$ implies $a \wedge b \in L$; if also $\mathrm{e} \in L$, then we call L an e-*lattice filter* of $\mathbf{A}$.

Lemma 3.5 *Let $\langle \mathbf{A}, \Box, \Diamond\rangle$ be any pm-$\mathsf{FL_e}$-algebra. Then $\mathcal{U}_{\mathbf{A}}$ is an* e*-lattice filter of $\mathbf{A}$ and $\mathcal{U}_{\mathbf{A}} = \uparrow\{\Box a \to a \mid a \in A\} \supseteq \{a \to \Diamond a \mid a \in A\}$. In particular, if $\mathbf{A}$ satisfies $x \leq \mathrm{e}$, then $\mathcal{U}_{\mathbf{A}} = \{a \in A \mid \mathrm{e} = \Box a\} = \{\Box a \to a \mid a \in A\}$.*

Proof. Observe first that $\mathrm{e} \in \mathcal{U}_{\mathbf{A}}$, since $\Box\mathrm{e} = \mathrm{e}$ by ($\mathrm{e}_\Box$). Also, if $a \in \mathcal{U}_{\mathbf{A}}$ and $b \in A$ satisfies $a \leq b$, then $\mathrm{e} \leq \Box a \leq \Box b$, by ($\mathrm{L4}_\Box$), so $b \in \mathcal{U}_{\mathbf{A}}$. Finally, if $a, b \in \mathcal{U}_{\mathbf{A}}$, then $\mathrm{e} \leq \Box a \wedge \Box b = \Box(a \wedge b)$, by ($\mathrm{L1}_\Box$), so $a \wedge b \in \mathcal{U}_{\mathbf{A}}$. Hence $\mathcal{U}_{\mathbf{A}}$ is an e-lattice filter of $\mathbf{A}$.

Next, for convenience, let $T := \uparrow\{\Box a \to a \mid a \in A\}$. If $a \in \mathcal{U}_{\mathbf{A}}$, then $a = \mathrm{e} \to a \geq \Box a \to a$, so $a \in T$. That is, $\mathcal{U}_{\mathbf{A}} \subseteq T$. But also $\Box a \to a \in \mathcal{U}_{\mathbf{A}}$ for each $a \in A$, since $\Box(\Box a \to a) = \Box a \to \Box a \geq \mathrm{e}$, by (P2), and, since $\mathcal{U}_{\mathbf{A}}$ is a lattice filter, $T = \mathcal{U}_{\mathbf{A}}$. Finally, $a \to \Diamond a \in \mathcal{U}_{\mathbf{A}}$ for each $a \in A$, since $\Box(a \to \Diamond a) = \Diamond a \to \Diamond a \geq \mathrm{e}$, by (P3). □

The next lemma shows that the operations $\Box$ and $\Diamond$ of any pm-$\mathsf{FL_e}$-algebra $\langle \mathbf{A}, \Box, \Diamond\rangle$ can be recovered from the subalgebra $\Box\mathbf{A}$ and e-lattice filter $\mathcal{U}_{\mathbf{A}}$.

Lemma 3.6 *Let $\langle \mathbf{A}, \Box, \Diamond\rangle$ be any pm-$\mathsf{FL_e}$-algebra. Then for any $a \in A$,*

$$\Box a = \max\{b \in \Box A \mid b \leq c \to a \textit{ for some } c \in \mathcal{U}_{\mathbf{A}}\};$$
$$\Diamond a = \min\{b \in \Box A \mid c \cdot a \leq b \textit{ for some } c \in \mathcal{U}_{\mathbf{A}}\}.$$

Proof. Fix $a \in A$, noting that $\Box a \in \{b \in \Box A \mid b \leq c \to a \text{ for some } c \in \mathcal{U}_{\mathbf{A}}\}$, since $\Box a \to a \in \mathcal{U}_{\mathbf{A}}$ and $\Box a \leq (\Box a \to a) \to a$. Now suppose that $b \in \Box A$ is such that $b \leq c \to a$ for some $c \in \mathcal{U}_{\mathbf{A}}$. Then, using residuation twice, $c \leq b \to a$. But $\Box b = b$, by ($\mathrm{L2}_\Box$), so, using ($\mathrm{L4}_\Box$) and (P2),

$$\mathrm{e} \leq \Box c \leq \Box(b \to a) = \Box(\Box b \to a) = \Box b \to \Box a = b \to \Box a,$$

and, by residuation, $b \leq \Box a$.

Similarly, $\Diamond a \in \{b \in \Box A \mid c \cdot a \leq b \text{ for some } c \in \mathcal{U}_{\mathbf{A}}\}$, using the fact that $a \to \Diamond a \in \mathcal{U}_{\mathbf{A}}$. Now suppose that $b \in \Box A$ is such that $c \cdot a \leq b$ for some $c \in \mathcal{U}_{\mathbf{A}}$. Then $c \leq a \to b = a \to \Box b$, since $b = \Box b$, and, using (L4$_\Box$) and (P1),

$$\mathrm{e} \leq \Box c \leq \Box(a \to \Box b) = \Diamond a \to \Box b = \Diamond a \to b,$$

and, by residuation, $\Diamond a \leq b$. □

We now provide a characterization of the subalgebra and e-lattice filter of an $\mathsf{FL_e}$-algebra required to obtain a corresponding pm-$\mathsf{FL_e}$-algebra. For any subalgebra $\mathbf{B}$ and e-lattice filter $\mathcal{U}$ of an $\mathsf{FL_e}$-algebra $\mathbf{A}$, let us call the ordered pair $\langle \mathbf{B}, \mathcal{U} \rangle$ a *$\mathcal{U}$-relatively complete subalgebra of* $\mathbf{A}$ if

(S1) for each $a \in A$, the set $\{b \in B \mid b \leq c \to a \text{ for some } c \in \mathcal{U}\}$ has a maximum and the set $\{b \in B \mid c \cdot a \leq b \text{ for some } c \in \mathcal{U}\}$ has a minimum;

(S2) $B \cap \mathcal{U} \subseteq \uparrow\mathrm{e}$.

Lemma 3.7 *Let $\langle \mathbf{A}, \Box, \Diamond \rangle$ be any pm-$\mathsf{FL_e}$-algebra. Then $\langle \Box\mathbf{A}, \mathcal{U}_{\mathbf{A}} \rangle$ is a $\mathcal{U}_{\mathbf{A}}$-relatively complete subalgebra of* $\mathbf{A}$.

Proof. Recall that $\Box\mathbf{A}$ is a subalgebra of $\mathbf{A}$, by Proposition 2.1, and $\mathcal{U}_{\mathbf{A}}$ is an e-lattice filter of $\mathbf{A}$, by Lemma 3.5. Moreover, (S1) follows immediately from Lemma 3.6, so it remains to observe for (S2) that if $b \in \Box A \cap \mathcal{U}_{\mathbf{A}}$, then $\mathrm{e} \leq \Box b = b$, and therefore $\Box A \cap \mathcal{U}_{\mathbf{A}} \subseteq \uparrow\mathrm{e}$. □

Lemma 3.8 *Let $\langle \mathbf{B}, \mathcal{U} \rangle$ be any $\mathcal{U}$-relatively complete subalgebra of an $\mathsf{FL_e}$-algebra $\mathbf{A}$, and define*

$$\Box_{\langle \mathbf{B}, \mathcal{U} \rangle} a := \max\{b \in B \mid b \leq c \to a \textit{ for some } c \in \mathcal{U}\};$$
$$\Diamond_{\langle \mathbf{B}, \mathcal{U} \rangle} a := \min\{b \in B \mid c \cdot a \leq b \textit{ for some } c \in \mathcal{U}\}.$$

Then $\langle \mathbf{A}, \Box_{\langle \mathbf{B}, \mathcal{U} \rangle}, \Diamond_{\langle \mathbf{B}, \mathcal{U} \rangle} \rangle$ is a pm-$\mathsf{FL_e}$-algebra, $B = \Box_{\langle \mathbf{B}, \mathcal{U} \rangle} A$, and $\mathcal{U} = \mathcal{U}_{\mathbf{A}}$.

Proof. For convenience, let us abbreviate $\Box_{\langle \mathbf{B}, \mathcal{U} \rangle}$ and $\Diamond_{\langle \mathbf{B}, \mathcal{U} \rangle}$ by $\Box$ and $\Diamond$, respectively. We observe first that condition (S1) implies that the required maximum and minimum exist for each $a \in A$. Next we show that $\Box$ and $\Diamond$ restrict to identity maps on B.

Claim 1. $\Box a = a$, for each $a \in B$.

Proof. Note first that $a \leq \mathrm{e} \to a$, so, since $a \in B$ and $\mathrm{e} \in \mathcal{U}$, the definition of $\Box a$ yields $a \leq \Box a$. But also, by definition, $\Box a \leq c \to a$ for some $c \in \mathcal{U}$. Hence, by residuation twice, $c \leq \Box a \to a$, and, since $\mathbf{B}$ is a subalgebra of $\mathbf{A}$ and $\mathcal{U}$ is an e-lattice filter, $\Box a \to a \in B \cap \mathcal{U}$. So $\Box a \to a \in \uparrow\mathrm{e}$, by (S2), and, by residuation, $\Box a \leq a$. That is, $\Box a = a$. □

Claim 2. $\Diamond a = a$, for each $a \in B$.

Proof. Note first that $\mathrm{e} \cdot a \leq a$, so, since $a \in B$ and $\mathrm{e} \in \mathcal{U}$, the definition of $\Diamond a$ yields $\Diamond a \leq a$. But also, by definition, $c \cdot a \leq \Diamond a$ for some $c \in \mathcal{U}$. So, by residuation, $c \leq a \to \Diamond a$, and, since $\mathbf{B}$ is a subalgebra of $\mathbf{A}$ and $\mathcal{U}$ is an e-lattice filter, $a \to \Diamond a \in B \cap \mathcal{U}$. Hence $a \to \Diamond a \in \uparrow \mathrm{e}$, by (S2), and, by residuation, $a \leq \Diamond a$. That is, $\Diamond a = a$. $\square$

Claim 1 yields $B \subseteq \Box A$; hence, since the other inclusion is immediate, $B = \Box A$. Observe also that for any $c \in \mathcal{U}$, since $\mathrm{e} \leq c \to c$, it follows from the definition of $\Box c$ that $\mathrm{e} \leq \Box c$. So $\mathcal{U} \subseteq \mathcal{U}_{\mathbf{A}}$. For the converse, if $\mathrm{e} \leq \Box c$ for some $c \in A$, then, again by the definition of $\Box c$, there exists a $c' \in \mathcal{U}$ such that $\mathrm{e} \leq c' \to c$. So $c' \leq c$ and, since $\mathcal{U}$ is an e-lattice filter, also $c \in \mathcal{U}$. Hence $\mathcal{U}_{\mathbf{A}} = \mathcal{U}$.

We now check that $\langle \mathbf{A}, \Box, \Diamond \rangle$ satisfies the equations to be a pm-$\mathsf{FL_e}$-algebra. First consider (L1$_\Box$). Let $a, b \in A$. It is clear from the definition of $\Box$ that $\Box(a \wedge b) \leq \Box a \wedge \Box b$. For the converse inequality, let $c_a, c_b \in \mathcal{U}$ be such that $\Box a \leq c_a \to a$ and $\Box b \leq c_b \to b$. Since $c_a \wedge c_b \leq c_a$ and $c_a \wedge c_b \leq c_b$,

$$\Box a \wedge \Box b \leq \Box a \leq c_a \to a \leq (c_a \wedge c_b) \to a$$
$$\Box a \wedge \Box b \leq \Box b \leq c_b \to b \leq (c_a \wedge c_b) \to b,$$

and hence

$$\Box a \wedge \Box b \leq ((c_a \wedge c_b) \to a) \wedge ((c_a \wedge c_b) \to b) = (c_a \wedge c_b) \to (a \wedge b).$$

So, since $\mathcal{U}$ is a lattice filter, $c_a \wedge c_b \in \mathcal{U}$ and $\Box a \wedge \Box b \leq \Box(a \wedge b)$.

Now consider (L1$_\Diamond$). Let $a, b \in A$. It is clear from the definition of $\Diamond$ that $\Diamond(a \vee b) \geq \Diamond a \vee \Diamond b$. For the converse inequality, let $c_a, c_b \in \mathcal{U}$ be such that $\Diamond a \geq c_a \cdot a$ and $\Diamond b \geq c_b \cdot b$. Then

$$\Diamond a \vee \Diamond b \geq \Diamond a \geq c_a \cdot a \geq (c_a \wedge c_b) \cdot a$$
$$\Diamond a \vee \Diamond b \geq \Diamond b \geq c_b \cdot b \geq (c_a \wedge c_b) \cdot b,$$

and hence

$$\Diamond a \vee \Diamond b \geq ((c_a \wedge c_b) \cdot a) \vee ((c_a \wedge c_b) \cdot b) = (c_a \wedge c_b) \cdot (a \vee b).$$

So, since $\mathcal{U}$ is a lattice filter, $c_a \wedge c_b \in \mathcal{U}$ and $\Diamond a \vee \Diamond b \geq \Diamond(a \vee b)$.

The cases of (L2$_\Box$), (L2$_\Diamond$), (L3$_\Box$), (L3$_\Diamond$), and ($\star_\Box$) for $\star \in \{\cdot, \to, \wedge, \vee\}$ all follow easily from Claims 1 and 2 and the fact that $\mathbf{B}$ is a subalgebra of $\mathbf{A}$.

Next, consider (P1). For any $a, b \in A$,

$$\Diamond a \to \Box b \leq (a \cdot (a \to \Diamond a)) \to \Box b = (a \to \Diamond a) \to (a \to \Box b).$$

But $\Diamond a \to \Box b \in \Box A = B$ and $a \to \Diamond a \in \mathcal{U}_{\mathbf{A}} = \mathcal{U}$, by Lemma 3.5, so, using Claim 1 and the definition of $\Box$,

$$\Diamond a \to \Box b \leq \Box(a \to \Box b).$$

On the other hand, there is a $c \in \mathcal{U}$ such that $\Box(a \to \Box b) \leq c \to (a \to \Box b) = (c \cdot a) \to \Box b$. By residuation twice, $c \cdot a \leq \Box(a \to \Box b) \to \Box b$. From the definition of $\Diamond$ and using residuation once more, it follows that $\Box(a \to \Box b) \leq \Diamond a \to \Box b$.

Finally, consider (P2). Let $c \in \mathcal{U}$ be such that

$$\Box(\Box a \to b) \leq c \to (\Box a \to b) = \Box a \to (c \to b).$$

Then $\Box(\Box a \to b) \cdot \Box a \leq c \to b$. Recalling that $\mathbf{B} = \Box\mathbf{A}$ is a subalgebra of $\mathbf{A}$, $\Box(\Box a \to b) \cdot \Box a \in B$, so $\Box(\Box a \to b) \cdot \Box a \leq \Box b$ and $\Box(\Box a \to b) \leq \Box a \to \Box b$. For the converse inequality, if $c \in \mathcal{U}$ is such that $\Box b \leq c \to b$, then

$$\Box a \to \Box b \leq \Box a \to (c \to b) = c \to (\Box a \to b).$$

Using once more the fact that $\mathbf{B} = \Box\mathbf{A}$ is a subalgebra of $\mathbf{A}$, $\Box a \to \Box b \in B$ and $\Box a \to \Box b \leq \Box(\Box a \to b)$. □

Combining Lemmas 3.7 and 3.8, we obtain:

Theorem 3.9 *There is a one-to-one correspondence between pm-$\mathsf{FL_e}$-algebras and ordered pairs $\langle \mathbf{A}, \langle \mathbf{B}, \mathcal{U} \rangle \rangle$ such that $\mathbf{A}$ is an $\mathsf{FL_e}$-algebra and $\langle \mathbf{B}, \mathcal{U} \rangle$ is a $\mathcal{U}$-relatively complete subalgebra of $\mathbf{A}$, implemented by the maps $\langle \mathbf{A}, \Box, \Diamond \rangle \mapsto \langle \mathbf{A}, \langle \Box\mathbf{A}, \mathcal{U}_\mathbf{A} \rangle \rangle$ and $\langle \mathbf{A}, \langle \mathbf{B}, \mathcal{U} \rangle \rangle \mapsto \langle \mathbf{A}, \Box_{\langle \mathbf{B}, \mathcal{U} \rangle}, \Diamond_{\langle \mathbf{B}, \mathcal{U} \rangle} \rangle$.*

Remark 3.10 Theorem 3.9 is a proper generalization of Theorem 2.2 in the setting of $\mathsf{FL_e}$-algebras. Just observe that if $\langle \mathbf{A}, \Box, \Diamond \rangle$ is an m-$\mathsf{FL_e}$-algebra, then $\mathcal{U}_\mathbf{A} = \uparrow \mathrm{e}$ and $\langle \mathbf{A}, \langle \Box\mathbf{A}, \mathcal{U}_\mathbf{A} \rangle \rangle$ can be identified with $\langle \mathbf{A}, \Box\mathbf{A} \rangle$, where $\Box\mathbf{A}$ is a relatively complete subalgebra of $\mathbf{A}$.

This result can also be specialized to pm-$\mathsf{FL_e}$-algebras that satisfy (LD), including (up to term-equivalence) the pseudomonadic BL-algebras discussed in Remark 3.4.

Lemma 3.11 *A pm-$\mathsf{FL_e}$-algebra $\langle \mathbf{A}, \Box, \Diamond \rangle$ satisfies* (LD) *if and only if $\{b \in \Box A \mid b \geq c \cdot d$ for some $c, d \in \mathcal{U}_\mathbf{A}\} \subseteq \uparrow \mathrm{e}$.*

Proof. Assume first that $\langle \mathbf{A}, \Box, \Diamond \rangle$ satisfies (LD). If $b \geq c \cdot d$ for some $b \in \Box A$ and $c, d \in \mathcal{U}_\mathbf{A}$, then $b \geq \Diamond c \geq \Box c \geq \mathrm{e}$. Now suppose that $\{b \in \Box A \mid b \geq c \cdot d$ for some $c, d \in \mathcal{U}_\mathbf{A}\} \subseteq \uparrow \mathrm{e}$. For any $a \in A$, since $\Box a \to a, a \to \Diamond a \in \mathcal{U}_\mathbf{A}$, $\Box a \to \Diamond a \in \Box A$, and $(\Box a \to a) \cdot (a \to \Diamond a) \leq \Box a \to \Diamond a$, it follows by assumption that $\mathrm{e} \leq \Box a \to \Diamond a$ which implies $\Box a \leq \Diamond a$. □

Corollary 3.12 *There is a one-to-one correspondence between pm-$\mathsf{FL_e}$-algebras that satisfy* (LD) *and ordered pairs $\langle \mathbf{A}, \langle \mathbf{B}, \mathcal{U} \rangle \rangle$ such that $\mathbf{A}$ is an $\mathsf{FL_e}$-algebra, $\langle \mathbf{B}, \mathcal{U} \rangle$ is a $\mathcal{U}$-relatively complete subalgebra of $\mathbf{A}$, and $\{b \in B \mid b \geq c \cdot d$ for some $c, d \in \mathcal{U}\} \subseteq \uparrow \mathrm{e}$.*

We conclude this section by introducing an expansion of pm-$\mathsf{FL_e}$-algebra $\langle \mathbf{A}, \Box, \Diamond \rangle$ with a constant u that generates the e-lattice filter $\mathcal{U}_A$. To this end, let us call an algebraic structure $\langle \mathbf{A}, \Box, \Diamond, u \rangle$ with a pm-$\mathsf{FL_e}$-algebra reduct $\langle \mathbf{A}, \Box, \Diamond \rangle$ and constant u a *pointed pm-$\mathsf{FL_e}$-algebra* if it satisfies the equations

$$\text{(U1)}\ \Box u \approx \mathrm{e} \qquad \text{(U2)}\ u \leq \Box x \to x.$$

Observe that $\langle \mathbf{A}, \Box, \Diamond, u \rangle$ satisfies $\Box u \approx u$ if and only if it satisfies $\Box x \leq x$ and $x \leq \Diamond x$, i.e., if and only if $\langle \mathbf{A}, \Box, \Diamond \rangle$ is an m-$\mathsf{FL_e}$-algebra.

If $\langle \mathbf{A}, \Box, \Diamond, u\rangle$ is a pointed pm-$\mathsf{FL_e}$-algebra, then $\mathcal{U}_A = \uparrow u$. Just observe that if $a \in \uparrow u$, then $\mathrm{e} = \Box u \leq \Box a$, by (U1) and (L4$_\Box$), so $a \in \mathcal{U}_A$, and, conversely, if $a \in \mathcal{U}_A$, then $a \in \uparrow\{\Box b \to b \mid b \in A\} \subseteq \uparrow u$, by Lemma 3.5 and (U2). On the other hand, if $\langle \mathbf{B}, \mathcal{U}\rangle$ is a $\mathcal{U}$-relatively complete subalgebra of an $\mathsf{FL_e}$-algebra $\mathbf{A}$ such that $\mathcal{U} = \uparrow u$ for some $u \in \mathcal{U}$, then $\langle \mathbf{A}, \Box, \Diamond, u\rangle$ is a pointed pm-$\mathsf{FL_e}$-algebra, where for each $a \in A$,

$$\Box a := \max\{b \in B \mid b \leq u \to a\} = \Box_{\langle B,\mathcal{U}\rangle} a;$$
$$\Diamond a := \min\{b \in B \mid a \cdot u \leq b\} = \Diamond_{\langle B,\mathcal{U}\rangle} a.$$

A direct application of Theorem 3.9 therefore yields the following:

Corollary 3.13 *Pointed pm-$\mathsf{FL_e}$-algebras are in one-to-one correspondence with ordered pairs $\langle \mathbf{A}, \langle \mathbf{B}, u\rangle\rangle$ such that $\mathbf{B}$ is a subalgebra of $\mathbf{A}$, $B \cap \uparrow u \subseteq \uparrow \mathrm{e}$, and the sets $\{b \in B \mid b \leq u \to a\}$ and $\{b \in B \mid au \leq b\}$ have a maximum and minimum, respectively, for each $a \in A$.*

4 A Functional Representation Theorem

Let $\mathcal{V}$ be any variety of $\mathsf{FL_e}$-algebras and let $pm\mathcal{V}_u$ denote the variety of pointed pm-$\mathsf{FL_e}$-algebras with an $\mathcal{L}$-lattice reduct in $\mathcal{V}$. In this section, we prove — adapting a strategy introduced for Heyting algebras in [2] and generalized to a universal algebraic setting in [12] — that if $\mathcal{V}$ has the superamalgamation property, then every $\langle \mathbf{A}, \Box, \Diamond, u\rangle \in pm\mathcal{V}_u$ can be represented as an algebraic structure consisting of functions from a set of worlds W to some $\mathbf{L} \in \mathcal{V}$.

Let $\mathbf{A}$ be any $\mathsf{FL_e}$-algebra. A *possibilistic* $\mathbf{A}$*-frame* is an ordered pair $\langle W, \pi\rangle$ consisting of a non-empty set W and a map $\pi\colon W \to A$ satisfying $\bigvee \pi[W] = \mathrm{e}$. A *universal* $\mathbf{A}$*-frame* is a possibilistic $\mathbf{A}$-frame $\langle W, \pi\rangle$ satisfying $\pi[W] = \{\mathrm{e}\}$. The following lemma describes the construction of pointed pm-$\mathsf{FL_e}$-algebras for a possibilistic $\mathbf{A}$-frame $\langle W, \pi\rangle$ consisting of maps $f\colon W \to A$, where the operations of $\mathbf{A}$ are defined pointwise and $\Box f$ and $\Diamond f$ are constant functions mapping each $w \in W$ to some fixed $c \in A$.

Lemma 4.1 *Let $\mathbf{A}$ be any $\mathsf{FL_e}$-algebra and let $\langle W, \pi\rangle$ be a possibilistic $\mathbf{A}$-frame. Suppose that $\mathbf{B}$ is a subalgebra of $\mathbf{A}^W$ such that $\pi \in B$ and for each $f \in B$, the following constant functions exist and belong to B:*

$$\Box f := \bigwedge\{\pi(w) \to f(w) \mid w \in W\}, \qquad \Diamond f := \bigvee\{\pi(w) \cdot f(w) \mid w \in W\}.$$

Then $\langle \mathbf{B}, \Box, \Diamond, \pi\rangle$ is a pointed pm-$\mathsf{FL_e}$-algebra.

Proof. To see that $\langle \mathbf{B}, \Box, \Diamond\rangle$ satisfies (L1$_\Box$), observe that for any $f, g \in B$,

$$\begin{aligned}
\Box(f \wedge g) &= \bigwedge\{\pi(w) \to (f \wedge g)(w) \mid w \in W\} \\
&= \bigwedge\{(\pi(w) \to f(w)) \wedge (\pi(w) \to g(w)) \mid w \in W\} \\
&= \bigwedge\{\pi(w) \to f(w) \mid w \in W\} \wedge \bigwedge\{\pi(w) \to g(w) \mid w \in W\} \\
&= \Box f \wedge \Box g.
\end{aligned}$$

The condition ($L1_\Diamond$) is analogous. To check the remaining conditions for an e-$\mathcal{L}$-lattice, it suffices to observe that $\Box f$ and $\Diamond f$ are constant functions for any $f \in B$, and for any constant function $c\colon W \to A$ (writing c also for the constant value taken by the function c),

$$\Box c = \bigwedge\{\pi(w) \to c \mid w \in W\} = \bigvee \pi[W] \to c = \mathrm{e} \to c = c$$
$$\Diamond c = \bigvee\{\pi(w) \cdot c \mid w \in W\} = \bigvee \pi[W] \cdot c = \mathrm{e} \cdot c = c.$$

In particular, for $\star \in \{\wedge, \vee, \cdot, \to\}$ and any $f, g \in B$, the functions $\Box f$, $\Box g$, and $\Box f \star \Box g$ are all constant and hence $\Box(\Box f \star \Box g) = \Box f \star \Box g$.

Observe next for (P1) that for any $f, g \in B$,

$$\begin{aligned}\Box(f \to \Box g) &= \bigwedge\{\pi(w) \to (f \to \Box g)(w) \mid w \in W\}\\ &= \bigwedge\{(\pi(w) \cdot f(w)) \to \Box g \mid w \in W\}\\ &= \bigvee\{\pi(w) \cdot f(w) \mid w \in W\} \to \Box g\\ &= \Diamond f \to \Box g,\end{aligned}$$

and, similarly, $\Box(\Box f \to g) = \Box f \to \Box g$, for (P2).

For (U1), observe that $\pi(w) \leq \mathrm{e}$ for each $w \in W$, since $\bigvee \pi[W] = \mathrm{e}$, so

$$\mathrm{e} \leq \bigwedge\{\pi(w) \to \pi(w) \mid w \in W\} \leq \bigwedge\{\pi(w) \to \mathrm{e} \mid w \in W\} = \bigvee \pi[W] \to \mathrm{e} = \mathrm{e},$$

that is, $\Box\pi = \mathrm{e}$. Finally, for (U2), given any $f \in B$, clearly $\pi(w) \leq (\pi(w) \to f(w)) \to f(w)$ for each $w \in W$, so $\pi \leq \Box f \to f$. □

Observe that if $\mathbf{A}$ is a *complete* $\mathsf{FL_e}$-algebra (i.e., $\bigwedge X$ and $\bigvee X$ exist in A for all $X \subseteq A$), then every possibilistic $\mathbf{A}$-frame $\langle W, \pi\rangle$ is associated with the *complex* $\mathbf{A}$*-algebra* $\langle \mathbf{A}^W, \Box, \Diamond, \pi\rangle$, where for each $f \in A^W$,

$$\Box f := \bigwedge\{\pi(w) \to f(w) \mid w \in W\}, \qquad \Diamond f := \bigvee\{\pi(w) \cdot f(w) \mid w \in W\}.$$

For any class $\mathcal{K}$ of $\mathsf{FL_e}$-algebras, we call a (pointed) pm-$\mathsf{FL_e}$-algebra $\mathcal{K}$-*functional* if it is isomorphic to one constructed as in Lemma 4.1 with $\mathbf{A} \in \mathcal{K}$.

Remark 4.2 If $\langle W, \pi\rangle$ is a universal $\mathbf{A}$-frame, then $\langle \mathbf{B}, \Box, \Diamond\rangle$, as constructed in Lemma 4.1, satisfies $\Box f = \bigwedge\{f(w) \mid w \in W\}$ and $\Diamond f = \bigvee\{f(w) \mid w \in W\}$ for all $f \in B$, and is an m-$\mathsf{FL_e}$-algebra. Conversely, if $\langle \mathbf{B}, \Box, \Diamond\rangle$, as constructed in Lemma 4.1, is an m-$\mathsf{FL_e}$-algebra, then $\mathrm{e} \leq \Box\pi \to \pi = \mathrm{e} \to \pi = \pi \leq \mathrm{e}$, so $\pi = \mathrm{e}$ and $\langle W, \pi\rangle$ is a universal $\mathbf{A}$-frame. That is, an m-$\mathsf{FL_e}$-algebra is $\mathcal{V}$-functional if and only if it is isomorphic to an algebraic structure constructed as described in Lemma 4.1 with $\mathbf{A} \in \mathcal{V}$, where $\langle W, \pi\rangle$ is a universal $\mathbf{A}$-frame.

Theorem 4.3 *Let $\mathcal{V}$ be a variety of $\mathsf{FL_e}$-algebras that has the superamalgamation property. Then every member of pm$\mathcal{V}_u$ is $\mathcal{V}$-functional.*

Proof. Consider any $\langle \mathbf{A}, \Box, \Diamond, u\rangle \in pm\mathcal{V}_u$. We define for $W := \mathbb{N}^{>0}$ a sequence of algebras $\langle \mathbf{A}_i\rangle_{i\in W}$ in $\mathcal{V}$ and sequences of embeddings $\langle f_i \colon \Box\mathbf{A} \to \mathbf{A}_i\rangle_{i\in W}$, $\langle g_i \colon \mathbf{A} \to \mathbf{A}_i\rangle_{i\in W}$, and $\langle s_i \colon \mathbf{A}_{i-1} \to \mathbf{A}_i\rangle_{i\in W}$ by induction as follows. Let $\mathbf{A}_0 := \mathbf{A}$ and let $f_0 \colon \Box\mathbf{A} \to \mathbf{A}$ be the inclusion embedding. Inductively, for each $i \in W$, there exists a superamalgam $\langle \mathbf{A}_i, s_i, g_i\rangle$ of the V-formation $\langle \Box\mathbf{A}, \mathbf{A}_{i-1}, \mathbf{A}, f_{i-1}, f_0\rangle$, and we define also $f_i := s_i \circ f_{i-1} = g_i \circ f_0 = g_i|_{\Box A}$.

Now let $\mathbf{L}$ be the direct limit of the system $\langle\langle \mathbf{A}_i, s_i\rangle\rangle_{i\in W}$ with associated sequence of embeddings $\langle l_i \colon \mathbf{A}_i \to \mathbf{L}\rangle_{i\in W}$. The first two superamalgamation steps of this construction are depicted in the following diagram:

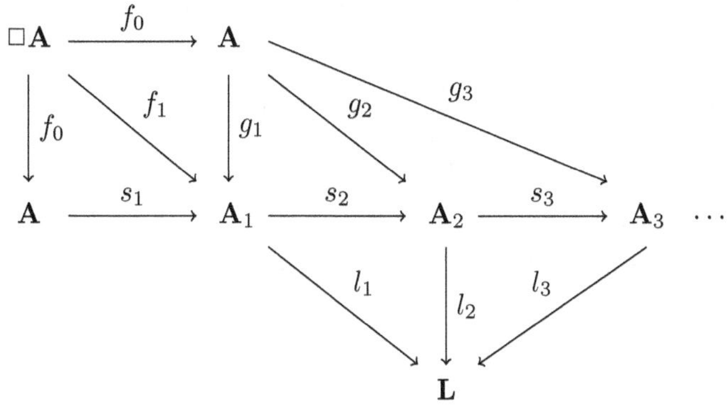

The operations of $\mathbf{L}^W$ are defined pointwise, so $B := \{\langle l_i \circ g_i(a)\rangle_{i\in W} \mid a \in A\}$ is the universe of a subalgebra $\mathbf{B}$ of $\mathbf{L}^W$ that contains $\pi := \langle l_i \circ g_i(u)\rangle_{i\in W}$. Now consider any $a \in A$ and let

$$S_a := \{\pi(i) \to (l_i \circ g_i(a)) \mid i \in W\} \quad \text{and} \quad T_a := \{\pi(i) \cdot (l_i \circ g_i(a)) \mid i \in W\}.$$

Fix $i \in W$. We show that $l_i \circ g_i(\Box a) = \bigwedge S_a$, the proof that $l_i \circ g_i(\Diamond a) = \bigvee T_a$ being very similar. In particular, it then follows that for any $i \in W$,

$$\mathrm{e} = \Diamond(\mathrm{e}) = l_i \circ g_i(\Diamond \mathrm{e}) = \bigvee T_{\mathrm{e}} = \bigvee \pi[W].$$

Note first that for any $k \in W$, using the definition of f_k and the fact that $\mathbf{L}$ is a direct limit,

$$l_k \circ g_k(\Box a) = l_k \circ f_k(\Box a) = l_{k+1} \circ s_{k+1} \circ f_k(\Box a) = l_{k+1} \circ g_{k+1}(\Box a).$$

Note also that $\Box a \leq u \to a$. Hence for each $j \in W$,

$$l_i \circ g_i(\Box a) = l_j \circ g_j(\Box a) \leq l_j \circ g_j(u \to a) = \pi(j) \to (l_j \circ g_j(a)).$$

So $l_i \circ g_i(\Box a)$ is a lower bound of S_a. Now suppose that $c \in L$ is another lower bound of S_a. Then there exist $k \in W$ and $d \in A_k$ such that $c = l_k(d)$ and

$$l_{k+1} \circ s_{k+1}(d) = l_k(d) = c \leq \pi(k+1) \to (l_{k+1} \circ g_{k+1}(a)) = l_{k+1} \circ g_{k+1}(u \to a).$$

But l_{k+1} is an embedding, so $s_{k+1}(d) \leq g_{k+1}(u \to a)$. Hence there exists, since $\langle \mathbf{A}_{k+1}, s_{k+1}, g_{k+1} \rangle$ is a superamalgam of $\langle \Box\mathbf{A}, \mathbf{A}_k, \mathbf{A}, f_k, f_0 \rangle$, an element $b \in \Box A$ such that

$$s_{k+1}(d) \leq s_{k+1} \circ f_k(b) = g_{k+1} \circ f_0(b) \leq g_{k+1}(u \to a).$$

But s_{k+1} and g_{k+1} are embeddings and f_0 is the inclusion map, so $d \leq f_k(b)$ and $b \leq u \to a$. Hence $b \leq \Box a$, since $b \in \Box A$ and $\Box a = \max\{b' \in \Box A \mid b' \leq u \to a\}$. So $f_k(b) \leq f_k(\Box a) = g_k(\Box a)$, and, using the first inequality,

$$c = l_k(d) \leq l_k \circ f_k(b) \leq l_k \circ g_k(\Box a) = l_i \circ g_i(\Box a).$$

So $l_i \circ g_i(\Box a)$ is the greatest lower bound of S_a.

It remains to prove that the following map is an isomorphism:

$$f \colon \langle \mathbf{A}, \Box, \Diamond, u \rangle \to \langle \mathbf{B}, \Box, \Diamond, \pi \rangle; \quad a \mapsto \langle l_i \circ g_i(a) \rangle_{i \in W}.$$

Since the operations of $\mathbf{L}^W$ are defined pointwise, it is easily checked that f is an isomorphism of $\mathsf{FL_e}$-algebras. Moreover, for any $a \in L$ and $i \in W$,

$$\begin{aligned} f(\Box a)(i) &= l_i \circ g_i(\Box a) \\ &= \bigwedge \{\pi(j) \to l_j \circ g_j(a) \mid j \in W\} \\ &= \Box(l_i \circ g_i(a)) \\ &= \Box f(a)(i), \end{aligned}$$

and, similarly, $f(\Diamond a)(i) = \Diamond f(a)(i)$. □

Theorem 4.3 can also be understood as a general completeness theorem for a family of 'KD45-like' logics. Let $\mathbf{Fm}^u_\Box$ denote the formula algebra of the language of pointed pm-$\mathsf{FL_e}$-algebras over a countably infinite set of variables. For any class $\mathcal{K}$ of pointed pm-$\mathsf{FL_e}$-algebras, we obtain a consequence relation (logic) over $\mathrm{Fm}^u_\Box$ by defining for $\Gamma \cup \{\varphi\} \subseteq \mathrm{Fm}^u_\Box$,

$$\begin{aligned} \Gamma \models_{\mathcal{K}} \varphi :\Longleftrightarrow {} & \text{for every } \mathbf{A} \in \mathcal{K} \text{ and homomorphism } h \colon \mathbf{Fm}^u_\Box \to \mathbf{A}, \\ & \mathrm{e} \leq h(\psi) \text{ for all } \psi \in \Gamma \implies \mathrm{e} \leq h(\varphi). \end{aligned}$$

Then the theorem states that for any variety $\mathcal{V}$ of $\mathsf{FL_e}$-algebras that has the superamalgamation property, the class of $\mathcal{V}$-functional pointed pm-$\mathsf{FL_e}$-algebras provides a semantics for the logic $\models_{pm\mathcal{V}_u}$.

Remark 4.4 It is well-known that exactly eight varieties of Heyting algebras have the superamalgamation property [17]. Also, it has been proved recently that uncountably many varieties of $\mathsf{FL_e}$-algebras — including the varieties of $\mathsf{FL_e}$-algebras and $\mathsf{FL_{ew}}$-algebras — have the superamalgamation property [15], equivalent in this setting to the Craig interpolation property for the associated substructural logic (see, e.g., [18]). Hence, for such a variety $\mathcal{V}$, every member of $pm\mathcal{V}_u$ is $\mathcal{V}$-functional, and $\mathcal{V}$-functional pointed pm-$\mathsf{FL_e}$-algebras provide a semantics for a 'KD45-like' modal intermediate or substructural logic.

In certain cases, we obtain a more refined result. Let us say that a variety $\mathcal{V}$ of $\mathsf{FL_e}$-algebras *admits regular completions* if, for any $\mathbf{A} \in \mathcal{V}$, there exists an embedding of $\mathbf{A}$ into a complete member $\mathbf{B}$ of $\mathcal{V}$ that preserves all existing meets and joins. For such a variety, every member of $pm\mathcal{V}_u$ embeds into a complex $\mathbf{A}$-algebra. Just observe that for any $\langle \mathbf{A}, \Box, \Diamond, u \rangle \in pm\mathcal{V}_u$, the direct limit $\mathbf{L} \in \mathcal{V}$ constructed in the proof of Theorem 4.3 embeds into a complete member $\bar{\mathbf{L}}$ of $\mathcal{V}$ and hence $\langle \mathbf{A}, \Box, \Diamond, u \rangle$ is isomorphic to a subalgebra of the complex algebra $\langle \bar{\mathbf{L}}^W, \Box, \Diamond, \pi \rangle$.

Corollary 4.5 *Let $\mathcal{V}$ be a variety of $\mathsf{FL_e}$-algebras that is closed under regular completions and has the superamalgamation property, and let $\mathcal{P}_\mathcal{V}$ denote the class of complex algebras $\langle \mathbf{A}^W, \Box, \Diamond, \pi \rangle$ where $\mathbf{A}$ is a complete member of $\mathcal{V}$ and $\langle W, \pi \rangle$ is a possibilistic $\mathbf{A}$-frame. Then for any $\Gamma \cup \{\varphi\} \subseteq \mathrm{Fm}^u_\Box$,*

$$\Gamma \models_{pm\mathcal{V}_u} \varphi \iff \Gamma \models_{\mathcal{P}_\mathcal{V}} \varphi.$$

Remark 4.6 A sufficient condition for a class of $\mathsf{FL_e}$-algebras to admit regular completions is closure under MacNeille completions. In particular, this is the case for $\mathcal{BA}$ and $\mathcal{HA}$; indeed, they are the only non-trivial varieties of Heyting algebras that have this property [1]. Infinitely many varieties of $\mathsf{FL_e}$-algebras — in particular, the varieties of $\mathsf{FL_e}$-algebras and $\mathsf{FL_{ew}}$-algebras — are closed under MacNeille completions, which is equivalent in this setting to the existence of a certain form of analytic sequent calculus [11].

In this paper, we have taken the first steps towards an algebraic theory of a family of '$\mathsf{KD45}$-logics' that are complete with respect to semantics defined by possibilistic frames over a variety $\mathcal{V}$ of $\mathsf{FL_e}$-algebras, obtaining a representation theorem for $pm\mathcal{V}_u$ (Theorem 4.3) when $\mathcal{V}$ has the superamalgamation property and a completeness theorem with respect to complex algebras when $\mathcal{V}$ also admits regular completions (Corollary 4.5). However, there are several issues to address in future work. First, it would be preferable to obtain analogous results for $pm\mathcal{V}$, i.e., without assuming the constant u in the language. Second, we aim to extend our approach to varieties that lack the superamalgamation property such as the varieties of MV-algebras and BL-algebras that serve as algebraic semantics for Łukasiewicz logic and Hájek's basic logic, respectively.

References

[1] G. Bezhanishvili and J. Harding, *MacNeille completions of Heyting algebras*, Houston J. Math. **30** (2004), 937–952.

[2] ———, *Functional monadic Heyting algebras*, Algebra Universalis **48** (2002), 1–10.

[3] N. Bezhanishvili, *Pseudomonadic algebras as algebraic models of doxastic modal logic*, Math. Log. Q. **48** (2002), 624–646.

[4] F. Bou, F. Esteva, L. Godo, and R. Rodríguez, *On the minimum many-valued logic over a finite residuated lattice*, J. Logic Comput. **21** (2011), no. 5, 739–790.

[5] M. Busaniche, P. Cordero, M. Marcos, and R.O. Rodriguez, *An algebraic semantics for possibilistic finite-valued Łukasiewicz logic*, Internat. J. Approx. Reason. **159** (2023), 108924.

[6] M. Busaniche, P. Cordero, and R.O. Rodriguez, *Pseudomonadic BL-algebras: an algebraic approach to possibilistic BL-logic*, Soft Computing **23** (2019), no. 7, 2199–2212.

[7] X. Caicedo, G. Metcalfe, R. Rodríguez, and J. Rogger, *Decidability in order-based modal logics*, J. Comput. System Sci. **88** (2017), 53–74.

[8] X. Caicedo, G. Metcalfe, R. Rodríguez, and O. Tuyt, *One-variable fragments of intermediate logics over linear frames.*, Inform. and Comput. **287** (2022).

[9] X. Caicedo and R. Rodríguez, *Standard Gödel modal logics*, Studia Logica **94** (2010), no. 2, 189–214.

[10] ———, *Bi-modal Gödel logic over* $[0,1]$*-valued Kripke frames*, J. Logic Comput. **25** (2015), no. 1, 37–55.

[11] A. Ciabattoni, N. Galatos, and K. Terui, *MacNeille completions of FL-algebras*, Algebra Universalis **66** (2011), no. 4, 405–420.

[12] P. Cintula, G. Metcalfe, and N. Tokuda, *One-variable fragments of first-order logics*, 2024. *Bull. Symb. Log.*, to appear.

[13] M.C. Fitting, *Many-valued modal logics*, Fundamenta Informaticae **15** (1991), no. 3–4, 235–254.

[14] ———, *Many-valued modal logics II*, Fundamenta Informaticae **17** (1992), 55–73.

[15] W. Fussner and S. Santschi, *Interpolation in linear logic and related systems*, 2023. Manuscript. Available at https://arxiv.org/abs/2305.05051.

[16] N. Galatos, P. Jipsen, T. Kowalski, and H. Ono, *Residuated lattices: An algebraic glimpse at substructural logics*, Elsevier, 2007.

[17] L.L. Maksimova, *Craig's theorem in superintuitionistic logics and amalgamable varieties of pseudo-Boolean algebras*, Algebra Logika **16** (1977), 643–681.

[18] G. Metcalfe, F. Paoli, and C. Tsinakis, *Residuated structures in algebra and logic*, Mathematical Surveys and Monographs, vol. 277, American Mathematical Society, 2023.

[19] A. Pietruszczak, *Simplified Kripke style semantics for modal logics K45, KB4 and KD45*, Bull. Sect. Logic Univ. Łódź **38** (2009), no. 3–4, 163–171.

[20] G. Restall, *Modalities in substructural logics*, Logique et Analyse **35** (1992), 303–321.

[21] R. Rodríguez, O. Tuyt, F. Esteva, and L. Godo, *Simplified Kripke semantics for K45-like Gödel modal logics and its axiomatic extensions*, Studia Logica **110** (2022), 1081–1114.

[22] A. Simpson, *The proof theory and semantics of intuitionistic modal logic*, PhD Dissertation, 1994.

Logics of Polyhedral Reachability

Nick Bezhanishvili

University of Amsterdam, Amsterdam, The Netherlands

Laura Bussi Vincenzo Ciancia

National Research Council, Pisa, Italy

David Fernández-Duque

University of Barcelona, Barcelona, Spain

David Gabelaia

TSU Razmadze Mathematical Institute, Tbilisi, Georgia

Abstract

Polyhedral semantics is a recently introduced branch of spatial modal logic, in which modal formulas are interpreted as piecewise linear subsets of an Euclidean space. Polyhedral semantics for the basic modal language has already been well investigated. However, for many practical applications of polyhedral semantics, it is advantageous to enrich the basic modal language with a reachability modality. Recently, a language with an Until-like spatial modality has been introduced, with demonstrated applicability to the analysis of 3D meshes via model checking. In this paper, we exhibit an axiom system for this logic, and show that it is complete with respect to polyhedral semantics. The proof consists of two major steps: First, we show that this logic, which is built over Grzegorczyk's system Grz, has the finite model property. Subsequently, we show that every formula satisfied in a finite poset is also satisfied in a polyhedral model, thereby establishing polyhedral completeness.

Keywords: Spatial logic, topological semantics, polyhedral semantics, completeness.

1 Introduction

Spatial modal logic is a well-established subdiscipline of modal logic, see e.g., [3]. Its primary focus lies in reasoning about spatial entities and their interrelations. In the topological semantics of modal logic, the modal operators $\Diamond$ and $\Box$ are interpreted as the topological operators of closure and interior, respectively. The classic result of McKinsey and Tarski states that the modal logic of all topological spaces is S4. Moreover, S4 is the logic of any dense-in-itself metric space [19]. For modern proofs of this and related topological completeness results, we refer to [5].

Recently, a variant of topological semantics was introduced for polyhedra. Polyhedra can be seen as piecewise linear subsets of an n-dimensional Euclidean space. For each polyhedron P, one interprets formulas into a Boolean algebra of its subpolyhedra. It is easy to see that the closure of a polyhedron is again a polyhedron, thereby providing a *polyhedral semantics for modal logic.* From the standpoint of domain modeling/language expressiveness, polyhedral semantics is easily seen to encompass 3D meshes (which is the natural application domain). As a matter of expressiveness, digital images can be considered as polyhedral models, especially in contexts such as medical imaging, where pixels and voxels ('volumetric picture elements') have a dimensionality and are considered hyperrectangles.

Polyhedral semantics has been introduced and studied in a sequence of papers [9,17,1,2] encompassing both intuitionistic and modal frameworks. These realms are interconnected through the Gödel translation and the theory of modal companions [11]. In this paper, we focus specifically on modal logics. It follows from [9] and [1] that the modal logic of all polyhedra is Grzegorczyk's modal logic Grz. This is the modal logic of finite posets [11]. In [1], a general criterion for a modal or intermediate logic to be complete for polyhedra, the so-called "nerve criterion", has been established, which enabled showing that many well-known modal logics such as Grz.2 and Grz.3 are not polyhedrally complete. On the other hand, Scott's logic and logics axiomatized by Jankov formulas of particular, star-like trees are polyhedrally complete. The logic of convex polyhedra was studied in [2], and the full characterization of polyhedral logics of flat polygons was announced in [18].

However, for many applications, it is important to enrich the modal language with an Until-like *spatial reachability modality.* A prominent example is the research line on *spatial model checking* (see [12] and the references therein for a lightweight introduction), where the basic modal language is enhanced with reachability [15], and interpreted on finitely representable spaces, such as images, graphs, or polyhedra. Further examples of such applications are shown in [13,14,8], where authors define bisimilarity modulo reachability. So far, the methodology has been applied in a variety of application domains, among which we mention medical imaging (see e.g. [4]) and analysis of video streams [10].

Formally, the reachability modality, which we denote by γ, is interpreted as follows: $\gamma(\varphi, \psi)$ is true at a point x if *there is a path starting at point x and ending at some point y satisfying ψ, and every intermediate point along the path satisfies φ.* Polyhedral semantics with this modality was investigated in [7]. We will give one illustrative example of the use of this modality. In Figure 1, there is a maze represented as triangulated polyhedron (a finite union of points, segments, triangles and tetrahedra). We can think of the red cube as the actual state, green cubes as safe exits, darker cubes as unsafe passage rooms, and white cubes as safe passage rooms. All these rooms are being connected by corridors. Then, the formula $\texttt{red} \wedge \gamma(\texttt{red} \vee \texttt{corridor} \vee \texttt{white}, \texttt{green})$ is true at the red

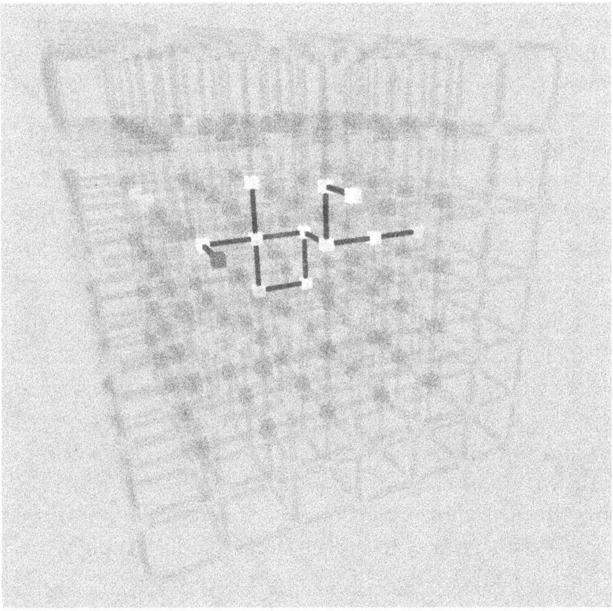

Fig. 1. Example of a polyhedral model depicting a maze, a starting area (in red), exit areas (in green), safe places (white), and a 'path' witnessing that from the starting area, an exit can be safely reached. All the points in the red area make the formula $\texttt{red} \wedge \gamma(\texttt{red} \vee \texttt{corridor} \vee \texttt{white}, \texttt{green})$ true.

state if and only if there is a safe exit out of the current state of the maze. For more examples and details, we refer to [7].

In this paper, we introduce the logic PLR, the *polyhedral logic of reachability*. Although in polyhedral semantics, as well as in the posets, the modality $\Diamond$ can be expressed via γ with the formula $\gamma(\varphi, \top)$, we still keep $\Diamond$ in the language for convenience. The reachability-free modal fragment of PLR is the modal logic Grz. We provide axioms for reachability. Our main result states that PLR is sound and complete with respect to the class of all polyhedra. One of the key insights of polyhedral logics is the fact that all the logical information concerning a polyhedron is encoded in the face posets of its triangulations [1,7].

We prove our main result in stages. To start with, we show that the formula of our language with reachability is satisfiable in a polyhedral model iff it is satisfiable on a finite poset. This is shown using the results of [9] and [1] to construct from each finite poset F a nerve poset $N(F)$ consisting of all non-empty chains of F ordered by inclusion. Utilizing the results from [7], we demonstrate that if F refutes a formula in our language, then so does $N(F)$. Again relying on the results from [9] and [1], we construct a triangulated polyhedron P whose face poset is $N(F)$. We then show that a polyhedral model and its face poset model satisfy precisely the same formulas. This implies that the polyhedron P also refutes the formula refuted on F. The converse direction

– that a formula refutable on a polyhedron is refutable on a finite poset – also follows from the correspondence between polyhedral models and their face poset models, so the reachability logic of polyhedra is shown to be the same as the reachability logic of finite posets. Next we syntactically define a logic PLR. To show that it is sound and complete with respect to finite posets we first prove completeness for an intermediary logic ALR (which is based on S4) employing a variant of filtration, and then extend this to PLR (based on Grz), by the method of cutting clusters in a filtrated model. Combining the above results we obtain that PLR is sound and complete with respect to polyhedral semantics.

The paper is organized as follows: in Section 2 we provide the reader with background information about topological spaces and polyhedra. Section 3 introduces the reachability language and establishes its topological, Kripke and polyhedral semantics. Section 4 recalls the key construction of the nerve of a poset, which links polyhedral and Kripke semantics. In Section 5 we provide an axiomatic definition of the two target logics, ALR and PLR. In Section 6 we present our first result, the finite model property of the S4-reachability logic ALR. In Section 7 we extend this to the finite model property with respect to posets of the Grz-reachability logic PLR and prove the polyhedral completeness of PLR. In Section 8 we draw conclusions and highlight some directions for future work.

2 Alexandroff spaces and polyhedra

In this section we briefly review some notions from topology, particularly polyhedra, setting up the stage for our semantics. We begin by reviewing Alexandroff spaces, which provide a link between topological semantics and the familiar Kripke semantics for logics above S4.

2.1 Topological and Alexandroff spaces

We assume the familiarity with the basic concepts of topology such as topological spaces, closed and open sets, closure and interior, basis, etc. We refer to [16] for all these notions. Recall that each preordered set (i.e., a set with a reflexive and transitive relation) can be viewed as a special topological space, in which an arbitrary intersection of open sets is open. Such spaces are known as *Alexandroff spaces.* For a preordered set (X, R) and $x \in X$ we let $R[x] = \{y \in X \mid xRy\}$. Call $U \subseteq X$ an R-upset if $R(U) = U$ where $R(U) = \{y \in X \mid \exists x \in U(xRy)\}$. An R-downset is defined dually, and for a partially ordered set we simply say an upset or downset. The collection τ_R of all R-upsets of (X, R) is an Alexandroff topology on X such that the closure of a set U is the set $R^{-1}(U) = \{y \in X : R[y] \cap U \neq \varnothing\}$, and $\{R[x] : x \in X\}$ is a basis for τ_R. Conversely, given a topological space (X, τ) one can define a *specialization order* R_τ on X by saying that $xR_\tau y$ iff every open set containing x also contains y. It is well known that if (X, τ) is an Alexandroff space, then $\tau = \tau_{R_\tau}$, and for every preordered set (X, R) we have $R = R_{\tau_R}$. Because of this we will not distinguish Alexandroff spaces and preordered sets. Note that any finite topological space is clearly Alexandroff.

2.2 Simplicial complexes and polyhedra

We recall here some definitions about polyhedra. Most of the definitions are drawn from [1,7].

Definition 2.1 A *d-simplex* σ is the convex hull of a finite set $V = \{v_0, \ldots, v_d\} \subseteq \mathbb{R}^m$ of $d+1$ affinely independent points.

We recall that $v_0, \ldots, v_d$ are affinely independent if $v_1 - v_0, \ldots, v_d - v_0$ are linearly independent. The number d is said to be the *dimension* of σ, while the points $v_0, \ldots, v_d$ are said to be its *vertices*. Note that simplices are bounded, convex and compact subspaces of $\mathbb{R}^n$.

A *face* of σ is the convex hull τ of a set $T \subseteq \{v_0, \ldots, v_d\}$, with $T \neq \emptyset$: it is straightforward to see that τ is also a simplex, and we thus have a partial order on simplices given by $\tau \preccurlyeq \sigma$ if τ is a face of σ.

Definition 2.2 The *relative interior* of a simplex σ is the set

$$\sigma^\circ = \{\Sigma_{i=0}^d \lambda_i v_i \mid \forall i.\ \lambda_i \in (0,1] \textit{ and } \Sigma_{i=0}^d \lambda_i = 1\}.$$

The relative interior of a simplex coincides with its topological interior in the subspace defined by the affine span of the simplex. Any simplex can be partitioned into the relative interiors of its faces. We write $\sigma^\circ \preccurlyeq^\circ \tau^\circ$ whenever $\sigma \preccurlyeq \tau$, and it is easy to see that this is also a partial order, isomorphic to $\preccurlyeq$.

More complex spaces, aptly known as *simplicial complexes,* are obtained by taking finite unions of simplices.

Definition 2.3 A *simplicial complex* K is a finite set of simplices of $\mathbb{R}^n$ such that:

1. If $\sigma \in K$ and $\tau \preccurlyeq \sigma$, then $\tau \in K$;
2. If $\sigma, \tau \in K$ with $\sigma \cap \tau \neq \varnothing$, then $\sigma \cap \tau \preccurlyeq \sigma$.

Simplicial complexes inherit the relations $\preccurlyeq^\circ$ and $\preccurlyeq$. The dimension of a simplicial complex K is the maximum of the dimensions of its simplices, while the face relation $\preccurlyeq$ is given by the union of the face relations of the simplices composing K. Given a simplicial complex K, the *polyhedron* of K (denoted by $|K|$) is the set-theoretic union of its simplices. Most importantly, given a polyhedron $|K|$, each of its points belongs to one and only one of the relative interiors of its simplices, so that the elements of K° induce a partition of $|K|$:

Lemma 2.4 *Each point of $|K|$ belongs to the relative interior of exactly one simplex in K. That is, $K^\circ = \{\sigma^\circ \mid \sigma \in K\}$ is a partition of $|K|$.*

K° is called a *simplicial partition* of $|K|$, and elements of K° are called *cells.* Note that the simplicial complex decomposition of a polyhedron is not unique.

We next recall the notion of a *topological path.*

Definition 2.5 A *topological path* in a topological space X is a continuous function $\pi : [0,1] \to X$, where $[0,1]$ is equipped with the subspace topology of $\mathbb{R}$. We say that π is a path *from a to b* if $\pi(0) = a$ and $\pi(1) = b$.

If $P = |K|$ for a given simplicial complex K and $\sigma \in K$, we say that π *traverses* σ if there is $x \in (0,1)$ such that $\pi(x) \in \sigma^\circ$.

Topological paths have long been at the core of methods in fields such as algebraic topology and complex analysis. As we will see, they also give rise to an 'until-like' operator in modal logics of space – the main feature of our logics.

3 Language and semantics

In this section, we introduce our logic and its semantics. The semantics will have three variants, based on Kripke frames, polyhedra, and topological spaces, with the first two being special cases of the third.

3.1 Language with reachability modality

Our logics are based on the language generated by the following grammar:

$$\varphi \equiv \mathbb{P} \mid \neg\varphi \mid \varphi \wedge \psi \mid \Box\varphi \mid \gamma(\varphi, \psi)$$

Here, $\mathbb{P}$ denotes a set of propositional variables typically assumed to be countably infinite. We let $\mathcal{L}_\gamma$ denote the set of all formulas of this language. In this paper, 'formula' refers exclusively to elements of $\mathcal{L}_\gamma$.

3.2 Topological and polyhedral semantics

Next we introduce topological semantics for $\mathcal{L}_\gamma$. As special cases, we obtain polyhedral and Kripke semantics.

Definition 3.1 A *topological model* is a triple $\mathcal{X} = (X, \tau, [\![\cdot]\!])$, where (X, τ) is a topological space and $[\![\cdot]\!] : \mathbb{P} \to 2^X$ is a valuation map. The interior operator for τ is denoted $\mathcal{I}$.

Given a topological model $\mathcal{X} = (X, \tau, [\![\cdot]\!])$ we extend $[\![\cdot]\!]$ to all formulas $\varphi \in \mathcal{L}_\gamma$ via the satisfaction relation $\models$ by stipulating $[\![\varphi]\!] = \{x \in X \mid \mathcal{X}, x \models \varphi\}$ and using the following recursive definition:

- $\mathcal{X}, x \models p \iff x \in [\![p]\!]$
- $\mathcal{X}, x \models \neg\varphi \iff \mathcal{X}, x \not\models \varphi$
- $\mathcal{X}, x \models \varphi_1 \wedge \varphi_2 \iff \mathcal{X}, x \models \varphi_1$ and $\mathcal{X}, x \models \varphi_2$
- $\mathcal{X}, x \models \Box\varphi \iff x \in \mathcal{I}([\![\varphi]\!])$
- $\mathcal{X}, x \models \gamma(\varphi, \psi) \iff$ there exists a topological path π such that $\pi(0) = x$, $\pi(1) \in [\![\psi]\!]$ and $\pi[(0,1)] \subseteq [\![\varphi]\!]$

On occasion we may write $x \models \varphi$ instead of $\mathcal{X}, x \models \varphi$ when $\mathcal{X}$ is clear from context. With this, we may define polyhedral models as a special case of topological models.

Definition 3.2 A *polyhedral model* is a pair $\mathcal{X} = (K, [\![\cdot]\!])$, where K is a simplicial complex and $[\![\cdot]\!] : \mathbb{P} \to 2^K$. We identify $\mathcal{X}$ with the topological model $\mathcal{X}_{\text{top}} = (|K|, \tau_K, [\![\cdot]\!]_{\text{top}})$, where τ_K is the subspace topology on $|K|$ inherited from $\mathbb{R}^n$ and $[\![p]\!]_{\text{top}} = \bigcup_{\sigma \in [\![p]\!]} \sigma^\circ$ for every $p \in \mathbb{P}$.

It is easily observed that if $\mathcal{X} = (K, [\![\cdot]\!])$ and $\mathcal{X}' = (K', [\![\cdot]\!])$ are two models with $|K| = |K'|$ (and sharing their propositional valuation), then for each $x \in |K|$ and $\varphi \in \mathcal{L}_\gamma$ we have that $\mathcal{X}, x \models \varphi \iff \mathcal{X}', x \models \varphi$.

Definition 3.3 An *Alexandroff model* is a triple $\mathcal{M} = (W, \preccurlyeq, [\![\cdot]\!])$, where W is any set, $\preccurlyeq$ is a preorder on W, and $[\![\cdot]\!] : \mathbb{P} \to 2^W$.

An Alexandroff model $\mathcal{M} = (W, \preccurlyeq, [\![\cdot]\!])$ is a *finite poset model* if $\preccurlyeq$ is a partial order and W is finite.

The valuation $[\![\varphi]\!]$ is extended to arbitrary formulas φ by identifying $\mathcal{M}$ with the topological model obtained by equipping W with the upset topology.

As usual, we write $w \prec v$ if $w \preccurlyeq v$ and $v \not\preccurlyeq w$. Our Alexandroff models are simply S4 models, but we use this terminology to stress that S4 frames coincide with Alexandroff spaces equipped with their specialization preorder, allowing us to apply Definition 3.1 to Alexandroff models. However, in this setting it will also be convenient to characterize the semantics of $\gamma(\cdot, \cdot)$ without an explicit reference to continuous paths.

Definition 3.4 Given an Alexandroff model $\mathcal{M} = (W, \preccurlyeq, [\![\cdot]\!])$ and a formula φ, we define the *reachability relation* R^φ to be the least relation so that $w \; R^\varphi \; v$ if there is $u \in [\![\varphi]\!]$ such that either

(i) $w \preccurlyeq u \succcurlyeq v$, or

(ii) $w \; R^\varphi \; u$ and $u \; R^\varphi \; v$.

This relation can equivalently be presented in terms of *up-down paths.*

Definition 3.5 Let $\mathcal{M} = (W, \preccurlyeq, V)$ be an Alexandroff model. A sequence $(w_0, \ldots, w_k) \subseteq W$ is said to be an *up-down path* if $k = 2j$ for some $j > 0$, $w_0 \preccurlyeq w_1$, $w_{k-1} \succcurlyeq w_k$, and whenever $0 < i < j$, we have that $w_{2i-1} \succ w_{2i} \prec w_{2i+1}$.

Thus, an up-down path is a path $w_0 \preccurlyeq w_1 \succ w_2 \prec w_3 \succ \ldots \prec w_{k-1} \succcurlyeq w_k$.

Lemma 3.6 *Given a finite Alexandroff model $\mathcal{M} = (W, \preccurlyeq, [\![\cdot]\!])$, $w, v \in W$, and a formula φ, the following are equivalent:*

(i) $w \; R^\varphi \; v$.

(ii) *There is an up-down path $(w_0, \ldots, w_k) \subseteq W$ such that $w_0 = w$, $w_k = v$, and for all $i \in (0, k)$, $w_i \in [\![\varphi]\!]$.*

(iii) *There is a topological path $\pi \colon [0, 1] \to W$ such that $\pi(0) = w$, $\pi(1) = v$, and for all $t \in (0, 1)$, $\pi(t) \in [\![\varphi]\!]$.*

Proof. We prove that the second item implies the first; the other direction is similar. Suppose that $(w_0, \ldots, w_k) \subseteq W$ is an up-down path such that for all $i \in (0, k)$, $w_i \in [\![\varphi]\!]$. We prove by induction on k that $w_0 \; R^\varphi \; w_k$.

We know that $w_1 \succcurlyeq w_2$, so that $w_0 \preccurlyeq w_1 \succcurlyeq w_2$ witnesses $w_0 \; R^\varphi \; w_2$ by the 'base case' of the definition of R^φ. If $k = 2$ we are done, otherwise $k > 2$ and the induction hypothesis yields $w_2 \; R^\varphi \; w_k$. Since in this case $w_2 \in [\![\varphi]\!]$, this witnesses the inductive clause for R^φ, and once again $w_0 \; R^\varphi \; w_k$.

To show that (ii) implies (iii) suppose that $(w_0, \ldots, w_{2k})$ is an up-down path. To build a corresponding topological path, consider points $\frac{i}{k} \in [0, 1]$ with $0 \leq i \leq k$ and let $\pi(x) = w_{2i}$ if $x = \frac{i}{k}$ and $\pi(x) = w_{2i+1}$ if $x \in \left(\frac{i}{k}, \frac{i+1}{k}\right)$. Then π is continuous and satisfies the requirements of (iii).

For the other direction suppose $\pi : [0,1] \to W$ is a topological path with $\pi(0) = w$, $\pi(1) = v$ and $\pi[(0,1)] \subseteq [\![\varphi]\!]$. Since $(0,1)$ is a connected subspace of $[0,1]$ and π is continuous, $\pi[(0,1)]$ is a connected subspace of the Alexandroff space W. Moreover, it follows from the continuity of π that $w = \pi(0)$ and $v = \pi(1)$ are in the closure of $\pi[(0,1)]$. Hence there exist $w_1, v_1 \in \pi[(0,1)]$ with $w \preccurlyeq w_1$ and $v \preccurlyeq v_1$. It is not hard to check that if A is a connected subspace of an Alexandroff space, then any two elements of A are connected via a zigzag path [6, Lemma 3.4]. In particular, the connectedness of $\pi[(0,1)] \subseteq [\![\varphi]\!]$ ensures now the existence of an up-down path as required in (ii). □

Given this context, up-down paths are closely related to topological paths. To further utilize this connection, first we define a simplicial model which is a Kripke companion to a given polyhedral model.

Definition 3.7 For a polyhedral model $\mathcal{X} = (K, [\![\cdot]\!])$, the *Kripke companion of* $\mathcal{X}$ is $\mathcal{M}(\mathcal{X}) = (K, \preccurlyeq, [\![\cdot]\!])$, where simplices of K are regarded as points, $\preccurlyeq$ is the face relation of Section 2.2, and $[\![p]\!]$ is unchanged.

We define $\varsigma \colon |K| \to K$ by letting $\varsigma(x)$ be the unique $\sigma \in K$ such that $x \in \sigma^\circ$.

Example 3.8 *Let us illustrate with a simple example of a polyhedron consisting of a single triangle with vertices a, b, c, including its interior. Then, $|K|$ consists of the set of points in the closed triangle, while $K = \{\{a\}, \{b\}, \{c\}, \overline{bc}, \overline{ac}, \overline{ab}, |K|\}$, where $\overline{xy}$ denotes the closed line segment from x to y. For $x \in |K|$, $\varsigma(x)$ is the smallest face to which x belongs. Thus for example, $\varsigma(a) = \{a\}$ since vertices are 'singleton faces', $\varsigma(x) = \overline{ab}$ if x is* strictly *between a and b, and $\varsigma(x) = |K|$ only when x is in the topological interior of $|K|$, i.e. if it does not lie on any of the edges of our triangle.*

Lemma 3.9 *Let $\mathcal{X} = (K, [\![\cdot]\!])$ be a polyhedral model. Then, the map $\varsigma \colon |K| \to K$ is continuous with respect to the Alexandroff topology on K.*

Moreover, if $x, y \in |K|$ and $N \subseteq K$, the following are equivalent:

(i) *There is a topological path from x to y traversing all the elements of N.*

(ii) *There is an up-down path $(\sigma_0, \ldots, \sigma_n)$ from $\varsigma(x)$ to $\varsigma(y)$ such that $N = \{\sigma_1, \ldots, \sigma_n\}$.*

Proof. First we check that ς is continuous. For this it suffices to show that if $C \subseteq K$ is closed, then $\varsigma^{-1}[C] = \{x \in |K| : x \in \bigcup C^\circ\}$ is closed. We have that C is closed iff it is downward-closed under $\preccurlyeq$, i.e. if $\tau \preccurlyeq \sigma \in C$ then $\tau \in C$. From this it follows that if $\sigma \in C$ then $\sigma \subseteq \bigcup C^\circ$, as if $x \in \sigma$ it follows that $x \in \tau^\circ$ for some $\tau \preccurlyeq \sigma$ and, as observed, $\tau \in C$. Since C is closed, we thus have that $\varsigma^{-1}[C] = \bigcup C^\circ = \bigcup C$, and since the latter is closed, so is the former.

From this it is immediate that (i) implies (ii), since if $\pi \colon [0,1] \to |K|$ is a topological path then $\varsigma \circ \pi \colon [0,1] \to K$ is also a topological path (being continuous) and clearly the two traverse the same faces. For the converse, it suffices to consider the case where N has three (possibly repeating) elements $\sigma_0 \preccurlyeq \sigma_1 \succcurlyeq \sigma_2$ and to find a path from $x \in \sigma_0^\circ$ to $y \in \sigma_2^\circ$ traversing only σ_1, as such paths can then be strung together inductively. Choose $z \in \sigma_1^\circ$; then, it can easily be checked that the piecewise linear path π with $\pi(0) = x$, $\pi(1/2) = z$ and $\pi(1) = y$ has the required properties. □

Combining lemmas 3.6 and 3.9, we see that the semantics of $\gamma(\cdot, \cdot)$ for a polyhedral model coincides with that of its Kripke companion, in the following sense.

Lemma 3.10 *Let $\mathcal{X} = (K, [\![\cdot]\!])$ be a polyhedral model and $x \in |K|$. Then, for every formula φ, we have that $\mathcal{X}, x \models \varphi \iff \mathcal{M}(\mathcal{X}), \varsigma(x) \models \varphi$.*

It follows readily from Lemma 3.10 that any formula satisfiable on a polyhedral model is also satisfiable on a finite poset model. The converse is also true, but the transformation from Kripke models to polyhedral ones requires an extra step, as we see in the next section.

4 The nerve of a Kripke model

Not every finite poset model $\mathcal{M}$ is of the form $\mathcal{M}(\mathcal{X})$ for some polyhedral model $\mathcal{X}$, but it is possible to transform $\mathcal{M}$ into a new model which *does* have this property. This new model is the *nerve* of $\mathcal{M}$.

Definition 4.1 Given a poset W, its *nerve*, $N(W)$, is the collection of finite non-empty chains in W ordered by set-theoretic inclusion. We define a function $\max : N(W) \to W$ sending each element of $N(W)$ to its maximal element and use this map to 'pull back' the valuation from W to $N(W)$; to be precise, we set $[\![p]\!]^N = \{c \mid \max(c) \in [\![p]\!]\}$. The *nerve model* of $\mathcal{M}$ is then the model $N(\mathcal{M}) = (N(W), \subseteq, [\![\cdot]\!]^N)$.

The map $\max : N(W) \to W$ mapping a chain to its maximal element is a p-morphism [1]. We can always find a polyhedral model $\mathcal{X}$ such that $N(\mathcal{M}) = \mathcal{M}(\mathcal{X})$, as shown, e.g. in [1].

Theorem 4.2 *If $\mathcal{M}$ is any finite poset model then there exists a polyhedral model $\mathcal{X}$ such that $N(\mathcal{M}) = \mathcal{M}(\mathcal{X})$.*

The function max will preserve the truth of formulas in our language. To make this precise, we must introduce the notion of up-down morphism.

Definition 4.3 Let $\mathcal{M} = (W, \preccurlyeq, [\![\cdot]\!])$, $\mathcal{M}' = (W', \preccurlyeq', [\![\cdot]\!]')$ be finite poset models. A function $f : W \to W'$ is an *up-down morphism* if:

- (atom) if $w \in W$, $w \in [\![p]\!]$, then $f(w) \in [\![p]\!]'$;
- (forth) if $w \preccurlyeq u \succcurlyeq v$, then $f(w) \preccurlyeq f(u) \succcurlyeq f(v)$, and
- (back) if $f(w) \preccurlyeq' u' \succcurlyeq' v'$, then there is an up-down path $(v_0, \ldots, v_k) \subseteq W$ such that $v_0 = w$, $f(v_k) = v'$, and for all $i \in (0, k)$, $f(v_i) = u'$.

It is easy to see that an up-down morphism is also a p-morphism. Indeed, if $w \in W$ and $w \preccurlyeq v$, then we have an up-down path $w \preccurlyeq v \succcurlyeq v$, and this implies $f(w) \preccurlyeq' f(v) \succcurlyeq' f(v)$, so in particular $f(w) \preccurlyeq' f(v)$. On the other hand, if $f(w) \preccurlyeq' v'$, then we consider the up-down path $f(w) \preccurlyeq' v' \succcurlyeq' v'$ and we get that there is an up-down path $(v_0, \ldots, v_n) \subseteq W$ such that $v_0 = w$ and $f(v_1) = v'$, which yields the back condition for $\preccurlyeq$ since $v_0 \preccurlyeq v_1$.

Lemma 4.4 *Let $\mathcal{M} = (W, \preccurlyeq, [\![\cdot]\!])$, $\mathcal{M}' = (W', \preccurlyeq', [\![\cdot]\!]')$ be finite poset models and suppose that $f : W \to W'$ is an up-down morphism. Then, for every formula $\varphi \in \mathcal{L}_\gamma$ we have $[\![\varphi]\!] = f^{-1}[[\![\varphi]\!]']$.*

Proof. The proof proceeds by a standard structural induction on φ. The case for $\Box\varphi$ follows from up-down morphisms being also p-morphisms. The case for $\gamma(\varphi, \psi)$ requires observing that if $(w_0, \ldots, w_{2n})$ is an up-down path in W witnessing that $\gamma(\varphi, \psi)$ holds on $w = w_0$, then the 'forth' clause can be applied n times to obtain a path $(w'_0, \ldots, w'_{2n'})$ witnessing $\gamma(\varphi, \psi)$ on $f(w_0)$. If instead $\gamma(\varphi, \psi)$ holds on $f(w)$, a symmetric argument shows that $\gamma(\varphi, \psi)$ holds on w.□

To see that the nerve of a Kripke model satisfies the formulas satisfied in the original model, it suffices to see that max is an up-down morphism.

Lemma 4.5 *If $\mathcal{M} = (W, \preccurlyeq, [\![\cdot]\!])$ is a finite poset model then the map* max: $N(W) \to W$ *is an up-down morphism.*

Proof. The atomic clause is taken care of by the definition of $[\![\cdot]\!]^N$. For the (forth) condition, if $c, c_1, c_2 \in N(W)$ with $c \subseteq c_1 \supseteq c_2$, it readily follows that $\max(c) \preccurlyeq \max(c_1) \succcurlyeq \max(c_2)$.

Let us now take $w, u, v \in W$ such that $w \preccurlyeq u \succcurlyeq v$. We note that $N(W)$ contains the chains $\{w\} \subseteq \{w, u\} \supseteq \{u\} \subseteq \{v, u\} \supseteq \{v\}$, and that this sequence forms an up-down path. Moreover, we have $\max(\{w\}) = w$, $\max(\{v\}) = v$ and $\max(\{w, u\}) = \max(\{u\}) = \max(\{v, u\}) = u$. Hence the (back) condition is also satisfied. □

It follows that the satisfiability in finite poset models is the same as the satisfiability in polyhedral models.

Theorem 4.6 *A formula φ is satisfiable on the class of finite poset models if and only if it is satisfiable on the class of polyhedral models.*

Proof. That any formula satisfied on a polyhedral model is satisfied on a finite poset model follows from Lemma 3.10.

For the converse, if φ is satisfied on a finite poset model $\mathcal{M}$, then φ is also satisfied on $N(\mathcal{M})$ by lemmas 4.4 and 4.5. Theorem 4.2 and Lemma 3.10 then imply that φ is satisfied on a polyhedral model. □

We conclude that the logic of polyhedral models coincides with the logic of finite poset models. With this in mind, we may restrict our attention to the latter for the remainder of the text.

5 Logics of reachability

In this section, we introduce axiomatically two key logics of this paper – ALR and PLR. The former will be based on S4 and the latter on Grz, both familiar modal logics related to transitive frames and topological spaces. However, it is necessary to introduce new axioms and rules for the reachability operator. In particular, we define two rules: the first one is quite intuitive, and it basically states that implication preserves reachability, or that reachability is monotone in both coordinates.

The second rule, or *induction rule*, is a bit more subtle. Intuitively, it works as a back propagation rule for the ψ formula, where the first premise is needed to propagate through a down step of an up-down path, while the second one is needed to propagate through an up step.

Definition 5.1 Axioms of the Alexandroff reachability logic ALR are given by all the propositional tautologies and Modus Ponens, S4 axioms and rules for $\Box$, plus the following:

Axiom 1. $\psi \vee (\varphi \wedge \gamma(\varphi,\psi)) \to \Box(\varphi \to \gamma(\varphi,\psi))$

Axiom 2. $\Diamond(\varphi \wedge \gamma(\varphi,\psi)) \to \gamma(\varphi,\psi)$

Rule 1. $\dfrac{\varphi \to \varphi' \quad \psi \to \psi'}{\gamma(\varphi,\psi) \to \gamma(\varphi',\psi')}$

Rule 2. $\dfrac{\psi \to \Box(\varphi \to \psi) \quad \varphi \wedge \Diamond(\varphi \wedge \psi) \to \psi}{\gamma(\varphi,\psi) \to \Diamond(\varphi \wedge \psi)}$.

The polyhedral reachability logic PLR is obtained by adding the Grz axiom $\Box(\Box(p \to \Box p) \to p) \to \Box p$ to ALR.

We note in passing that the above two rules can also be formulated as axioms using a defined operator $[\pi]\varphi \equiv \neg\gamma(\top,\neg\varphi)$ which semantically acts as a 'global' box modality inside the path-connected component of the point of evaluation. Indeed, it is not hard to see that $[\pi]\varphi$ will be true at a point w in an Alexandroff model iff φ is true at each v which connects with w by a zigzag path, that is φ is universally true inside the connected component containing w. With this in mind, Rule 1 can be expressed as:

$$[\pi](\varphi \to \varphi') \wedge [\pi](\psi \to \psi') \wedge \gamma(\varphi,\psi) \to \gamma(\varphi',\psi')$$

The next proposition shows soundness of our logics.

Proposition 5.2 *The logic* ALR *is sound for the class of Alexandroff spaces and the logic* PLR *is sound for the class of finite poset models.*

Proof. Let $\mathcal{M} = (W, \preccurlyeq, [\![\cdot]\!])$ be an Alexandroff model.

Axiom 1: $\psi \vee (\varphi \wedge \gamma(\varphi,\psi)) \to \Box(\varphi \to \gamma(\varphi,\psi))$

Take $w \not\models \Box(\varphi \to \gamma(\varphi,\psi))$: then $\exists v \succcurlyeq w.\ v \not\models (\varphi \to \gamma(\varphi,\psi))$, namely $v \models \varphi$, $v \not\models \gamma(\varphi,\psi)$. Suppose $w \models \psi \vee (\varphi \wedge \gamma(\varphi,\psi))$. We then have two cases:

- $w \models \psi$: we have $v \preccurlyeq v \succcurlyeq w$. Hence $v\ R^{\varphi}\ w$ and $v \models \gamma(\varphi,\psi)$, a contradiction.
- $w \models (\varphi \wedge \gamma(\varphi,\psi))$: again we have $v \preccurlyeq v \succcurlyeq w$, hence $v\ R^{\varphi}\ w$. From $w \models \gamma(\varphi,\psi)$ we also get $w\ R^{\varphi}\ u$ for some $u \models \psi$. But then, since $w \models \varphi$, $v\ R^{\varphi}\ w$ and $w\ R^{\varphi}\ u$ we obtain $v\ R^{\varphi}\ u$. Together with $u \models \psi$ this implies $v \models \gamma(\varphi,\psi)$, again a contradiction.

Axiom 2: $\Diamond(\varphi \wedge \gamma(\varphi,\psi)) \to \gamma(\varphi,\psi)$

Take $w \models \Diamond(\varphi \wedge \gamma(\varphi,\psi))$, $w \not\models \gamma(\varphi,\psi)$. Then there is a point v such that $w \preccurlyeq v$ and $v \models \varphi$, $v \models \gamma(\varphi,\psi)$. Hence $w\ R^{\varphi}\ v$ and $v\ R^{\varphi}\ u$ for some $u \models \psi$.

This implies $w\ R^\varphi\ u$ and thus, $w \models \gamma(\varphi, \psi)$ – a contradiction.

Rule 1:

$$\frac{\varphi \to \varphi' \quad \psi \to \psi'}{\gamma(\varphi,\psi) \to \gamma(\varphi',\psi')}$$

Take a model $(W, R, [\![.]\!])$ and suppose that $\forall w \in W$, $w \models \varphi \implies w \models \varphi'$, $w \models \psi \implies w \models \psi'$. Take points u, v such that $u\ R^\varphi\ v$ and $v \models \psi$. Then it easily follows that $u\ R^{\varphi'}\ v$ and $v \models \psi'$. Hence $\gamma(\varphi,\psi) \to \gamma(\varphi',\psi')$ is valid in the model.

Rule 2 (inductive rule):

$$\frac{\psi \to \Box(\varphi \to \psi) \quad \varphi \land \Diamond(\varphi \land \psi) \to \psi}{\gamma(\varphi,\psi) \to \Diamond(\varphi \land \psi)}.$$

Take a model $\mathcal{M}$ such that for all $w \in W$ we have $w \models \psi \to \Box(\varphi \to \psi)$ and $w \models \varphi \land \Diamond(\varphi \land \psi) \to \psi$. Suppose $u \models \gamma(\varphi, \psi)$. Then there is an up-down path $w_0 \preccurlyeq w_1 \succ w_2 \prec w_3 \succ \ldots \prec w_{2k-1} \succcurlyeq w_{2k}$ with $w_0 = u$, $w_{2k} \models \psi$ and $w_i \models \varphi$ for all i with $0 < i < 2k$. We aim to show that $w_1 \models \psi$ thereby demonstrating $u \models \Diamond(\varphi \land \psi)$. We will in fact show by a backwards induction that $w_i \models \psi$ for all i with $0 < i < 2k$. Indeed, given the aforementioned up-down path, we start by considering the last point (where ψ holds) and we build the proof by proceeding backward on the path. To this end, consider first w_{2k-1}. From $w_{2k} \models \psi \to \Box(\varphi \to \psi)$ and $w_{2k} \models \psi$ we get $w_{2k} \models \Box(\varphi \to \psi)$. Since $w_{2k} \preccurlyeq w_{2k-1}$ and $w_{2k-1} \models \varphi$, we have $w_{2k-1} \models \psi$. If $k = 1$ we are done, otherwise $w_{2k-2} \models \varphi$ and we continue. From $w_{2k-2} \preccurlyeq w_{2k-1} \models \varphi \land \psi$ we obtain $w_{2k-2} \models \Diamond(\varphi \land \psi)$. Utilizing $w_{2k-2} \models \varphi \land \Diamond(\varphi \land \psi) \to \psi$ we conclude $w_{2k-2} \models \psi$. Continuing in this fashion, we will end up with $w_1 \models \psi$ as desired.

If moreover $\mathcal{M}$ is a finite poset model, thus based on a finite poset, then it is well known that the Grz axiom is valid on $\mathcal{M}$. □

Lemma 5.3 *$\gamma(\varphi,\psi) \to \Diamond\varphi$ is a theorem of* ALR.

Proof. Since $\psi \to \top$ is a tautology, applying Rule 1 gives $\gamma(\varphi,\psi) \to \gamma(\varphi,\top)$. Taking easily derivable $\top \to \Box(\varphi \to \top)$ and $\varphi \land \Diamond(\varphi \land \top) \to \top$ as premises of Rule 2, we conclude $\gamma(\varphi,\top) \to \Diamond(\varphi \land \top)$. Modus Ponens now yields $\Diamond\varphi$. □

6 Alexandroff completeness

In order to prove completeness of the logic, we will use a widely known technique in the realm of modal logics. We will go through the construction of a *canonical model*, which will be filtered modulo a specifically crafted set of formulas. Finiteness of such a set will give us a finite model, which is essential in showing that the axiom system also enjoys the *finite model property*. We will then prove a filtration lemma, to achieve completeness.

Definition 6.1 Let $\Lambda \in \{\mathsf{ALR}, \mathsf{PLR}\}$. The *canonical model for* Λ is the structure $\mathcal{M}_c = (W_c, \preccurlyeq_c, [\![\cdot]\!]_c)$, where W_c is the set of all Λ-*theories* (maximal Λ-consistent sets), $T \preccurlyeq_c S$ if whenever $\Box\varphi \in T$, it follows that $\Box\varphi \in S$, and $[\![p]\!]_c = \{T \in W_c : p \in T\}$.

The results in this section apply to both the canonical model for ALR and for PLR, beginning with the following standard lemma.

Lemma 6.2 *For any formula* φ, $\Diamond\varphi \in T \in W_c$ *iff there is* $S \succcurlyeq_c T$ *with* $\varphi \in S$.

6.1 Filtration

Say that a set Σ of formulas is *adequate* if it is closed under subformulas and single negations, and whenever $\gamma(\varphi, \psi) \in \Sigma$, then $\Box(\varphi \to \gamma(\varphi, \psi)) \in \Sigma$ and $\Diamond(\varphi \wedge \gamma(\varphi, \psi)) \in \Sigma$. For a set Γ of formulas let $\mathrm{sub}(\Gamma)$ denote the set of all subformulas of formulas in Γ.

Lemma 6.3 *Let* Γ *be a finite set of formulas. Then the smallest adequate set containing* Γ *is also finite.*

Proof. Let $\Gamma_1 = \Gamma \cup \{\Box(\varphi \to \gamma(\varphi, \psi)), \Diamond(\varphi \wedge \gamma(\varphi, \psi)) \mid \gamma(\varphi, \psi) \in \mathrm{sub}(\Gamma)\}$. Clearly Γ_1 is finite. Moreover, Γ and Γ_1 have the same set of subformulas of the form $\gamma(\varphi, \psi)$.

Let now $\Sigma = \mathrm{sub}(\Gamma_1) \cup \{\neg\varphi \mid \varphi \in \mathrm{sub}(\Gamma_1), \varphi \neq \neg\psi\}$. It is straightforward to check that Σ is finite and adequate. □

Definition 6.4 Let Σ be adequate. Define $T \sim_\Sigma S$ if $T \cap \Sigma = S \cap \Sigma$. We define $\mathcal{M}_\Sigma = (W_\Sigma, \preccurlyeq_\Sigma, [\![\cdot]\!]_\Sigma)$ as follows:

(i) $W_\Sigma = \{T_\Sigma : T \in W_c\}$, where T_Σ is the equivalence class of T under $\sim_\Sigma$.

(ii) $T_\Sigma \preccurlyeq_\Sigma S_\Sigma$ if whenever $\Box\varphi \in T \cap \Sigma$, it follows that $\Box\varphi \in S$.

(iii) $[\![p]\!]_\Sigma = \{T_\Sigma : p \in T \cap \Sigma\}$.

It easily follows from the definitions that $S \preccurlyeq_c T$ implies $S_\Sigma \preccurlyeq_\Sigma T_\Sigma$.

Given $\varphi \in \Sigma$, we define a φ-reachability relation on $\mathcal{M}_\Sigma$ inductively by letting $T_\Sigma\ R^\varphi_\Sigma\ S_\Sigma$ if there is $U_\Sigma \in W_\Sigma$ such that $\varphi \in U$ and either

(i) $T_\Sigma \preccurlyeq_\Sigma U_\Sigma \succcurlyeq_\Sigma S_\Sigma$, or

(ii) $T_\Sigma\ R^\varphi_\Sigma\ U_\Sigma$ and $U_\Sigma\ R^\varphi_\Sigma\ S_\Sigma$.

It easily follows from this definition that R^φ_Σ is a symmetric relation.

Lemma 6.5 *If* $\gamma(\varphi, \psi) \in \Sigma$, $T_\Sigma\ R^\varphi_\Sigma\ S_\Sigma$, *and either* $\psi \in S$ *or* $\varphi \wedge \gamma(\varphi, \psi) \in S$, *then* $\gamma(\varphi, \psi) \in T$.

Proof. By induction on the structure of R^φ_Σ from T_Σ to S_Σ.
In the base case, there is U_Σ such that $\varphi \in U$, $T_\Sigma \preccurlyeq_\Sigma U_\Sigma$ and $S_\Sigma \preccurlyeq_\Sigma U_\Sigma$. Since $\gamma(\varphi, \psi) \in \Sigma$ and Σ is adequate, we have $\Box(\varphi \to \gamma(\varphi, \psi)) \in \Sigma$. Suppose $\psi \in S$ or $\varphi \wedge \gamma(\varphi, \psi) \in S$. By Axiom 1 and Modus Ponens, $\Box(\varphi \to \gamma(\varphi, \psi)) \in S$. From $S_\Sigma \preccurlyeq_\Sigma U_\Sigma$ it follows that $\Box(\varphi \to \gamma(\varphi, \psi)) \in U$ and by the reflexivity axiom of S4 we get $\varphi \to \gamma(\varphi, \psi) \in U$. Since $\varphi \in U$, we may conclude $\gamma(\varphi, \psi) \in U$. By adequacy of Σ we also have $\Diamond(\varphi \wedge \gamma(\varphi, \psi)) \in \Sigma$ and since $T_\Sigma \preccurlyeq_\Sigma U_\Sigma$, $\varphi \wedge \gamma(\varphi, \psi) \in U$, we obtain $\Diamond(\varphi \wedge \gamma(\varphi, \psi)) \in T$. By Axiom 2, $\gamma(\varphi, \psi) \in T_\Sigma$, as required.

For the inductive case we have that there is U with $\varphi \in U$, $T_\Sigma\ R^\varphi_\Sigma\ U_\Sigma$, and $U_\Sigma\ R^\varphi_\Sigma\ S_\Sigma$. By the induction hypothesis, since $\psi \in S$ and $U_\Sigma\ R^\varphi_\Sigma\ S_\Sigma$ we obtain

$\gamma(\varphi,\psi) \in U$. It follows that $\varphi \wedge \gamma(\varphi,\psi) \in U$ and since $T_\Sigma\ R^\varphi_\Sigma\ U_\Sigma$, once again by the induction hypothesis, we have $\gamma(\varphi,\psi) \in T$. □

Definition 6.6 Let Σ be an adequate and finite set. For $T_\Sigma \in W_\Sigma$ we define $\chi(T_\Sigma) = \bigwedge(T \cap \Sigma)$ (note that this is well defined by the definition of $\sim_\Sigma$).

Lemma 6.7 *Let Σ be an adequate and finite set. Suppose $U_\Sigma \in W_\Sigma$ is a world of the filtrated model, $\varphi \in \Sigma$ and let $\chi = \bigvee\{\chi(s) : s \in R^\varphi(U_\Sigma)\}$. Then,*

(i) $\mathsf{ALR} \vdash \Diamond(\varphi \wedge \chi) \to \chi$

(ii) $\mathsf{ALR} \vdash \varphi \wedge \chi \to \Box(\varphi \to \chi)$.

Proof.
We will make use of the easily seen observation that for any $S \in W_c$ we have $\chi \in S$ iff $S_\Sigma \in R^\varphi(U_\Sigma)$.

To prove the first item, suppose that $\Diamond(\varphi\wedge\chi) \to \chi$ is not provable, i.e. $\neg\chi \wedge \Diamond(\varphi \wedge \chi)$ is consistent. Then there is $T \in W_c$ with $\Diamond(\varphi \wedge \chi) \wedge \neg\chi \in T$. By Lemma 6.2, there exists $S \succcurlyeq_c T$ such that $\varphi \wedge \chi \in S$. Since $\chi \in S$, we have that $S_\Sigma \in R^\varphi(U_\Sigma)$. On the other hand, since $\varphi \in S$ and $T_\Sigma \preccurlyeq_\Sigma S_\Sigma \succcurlyeq_\Sigma S_\Sigma$, we also have $T_\Sigma\ R^\varphi_\Sigma\ S_\Sigma$ and hence $T_\Sigma \in R^\varphi(U_\Sigma)$, implying $\chi \in T$, a contradiction.

Suppose now that $\varphi\wedge\chi \to \Box(\varphi \to \chi)$ is not provable, i.e. $(\varphi\wedge\chi)\wedge\Diamond\neg(\varphi \to \chi)$ is consistent. Then there is $T \in W_c$ such that $(\varphi \wedge \chi) \wedge \Diamond(\varphi \wedge \neg\chi) \in T$, and again by Lemma 6.2 we get $S \succcurlyeq_c T$ such that $\varphi \wedge \neg\chi \in S$. Since $\chi \in T$, we have that $T_\Sigma \in R^\varphi(U_\Sigma)$. We now note that $\varphi \in S$, and thus $T_\Sigma\ R^\varphi\ S_\Sigma$, as $T_\Sigma \preccurlyeq_\Sigma S_\Sigma \succcurlyeq_\Sigma S_\Sigma$. Then $U_\Sigma\ R^\varphi\ S_\Sigma$ also holds, implying $\chi \in S$, again a contradiction. □

Lemma 6.8 *If $\gamma(\varphi,\psi) \in T\cap\Sigma$, then there exists S with $T_\Sigma\ R^\varphi_\Sigma\ S_\Sigma$ and $\psi \in S$.*

Proof. Toward a contradiction, assume that $\gamma(\varphi,\psi) \in T$ and for all $S_\Sigma \in R^\varphi(T_\Sigma)$ we have that $\psi \notin S$. Let χ be as in the previous lemma, for $U_\Sigma = T_\Sigma$. Taking contrapositives we obtain:

(i) $\mathsf{ALR} \vdash \neg\chi \to \Box(\varphi \to \neg\chi)$

(ii) $\mathsf{ALR} \vdash \varphi \wedge \Diamond(\varphi \wedge \neg\chi) \to \neg\chi$.

By Rule 2, $\mathsf{ALR} \vdash \gamma(\varphi,\neg\chi) \to \Diamond(\varphi \wedge \neg\chi)$.

Further, since $\psi \notin S$ for each $S_\Sigma \in R^\varphi(T_\Sigma)$, we get $\mathsf{ALR} \vdash \chi \to \neg\psi$, hence $\mathsf{ALR} \vdash \psi \to \neg\chi$. By Rule 1, $\mathsf{ALR} \vdash \gamma(\varphi,\psi) \to \gamma(\varphi,\neg\chi)$. Together with $\mathsf{ALR} \vdash \gamma(\varphi,\neg\chi) \to \Diamond(\varphi \wedge \neg\chi)$ this yields $\mathsf{ALR} \vdash \gamma(\varphi,\psi) \to \Diamond(\varphi \wedge \neg\chi)$.

Since we assumed $\gamma(\varphi,\psi) \in T$ at the outset, $\Diamond(\varphi \wedge \neg\chi) \in T$ follows, and hence there is $S \succcurlyeq_c T$ such that $\varphi \wedge \neg\chi \in S$. Since $S_\Sigma \succcurlyeq_\Sigma T_\Sigma$ and $\varphi \in S$, we obtain $T_\Sigma\ R^\varphi_\Sigma\ S_\Sigma$, thus by our definition of χ, $\chi \in S$, contradicting $\neg\chi \in S$. □

With this we are able to prove a key filtration lemma.

Lemma 6.9 *Let Σ be a finite adequate set of formulas and let $\varphi \in \Sigma$. Then for each $T \in W_c$ we have $\varphi \in T$ iff $\mathcal{M}_\Sigma, T_\Sigma \models \varphi$.*

Proof. Standard induction on formulas using lemmas 6.2, 6.5 and 6.8. □

Theorem 6.10 ALR *is complete for the class of Alexandroff models and has the finite model property.*

Proof. Suppose ALR $\nvdash \neg\varphi$. Then there is $T \in W_c$ with $\varphi \in T$. By Lemma 6.3 there exists a finite adequate set Σ containing sub(φ). It follows from Lemma 6.9 that $\mathcal{M}_\Sigma, T_\Sigma \models \varphi$. Since $\mathcal{M}_\Sigma$ is finite, the proof is finished.□

7 Polyhedral completeness

Now that we have obtained completeness for Alexandroff spaces, it remains to show that PLR is complete for the class of finite poset models, hence by Theorem 4.6, also for the class of polyhedral logics. The proof builds on that for ALR, but requires some additional steps.

For a given adequate set Σ we define $\Sigma_1 = \Sigma \cup \{\Diamond(\neg\varphi \wedge \Diamond\varphi) \mid \Diamond\varphi \in \Sigma\} \cup \{\Diamond(\varphi \wedge \neg\psi) \mid \gamma(\varphi, \psi) \in \Sigma\}$ and let $\widehat{\Sigma}$ be the smallest adequate set containing Σ_1. It is a straightforward consequence of Lemma 6.3 that if Σ is finite, $\widehat{\Sigma}$ is also finite.

Given a finite adequate set Σ we now take the filtration of the canonical model for PLR via $\widehat{\Sigma}$ to obtain a finite model $\mathcal{M}_{\widehat{\Sigma}}$. We know from Lemma 6.9 that for $\varphi \in \widehat{\Sigma}$, $\mathcal{M}_{\widehat{\Sigma}}, T_{\widehat{\Sigma}} \models \varphi \iff \varphi \in T$. The model $\mathcal{M}_{\widehat{\Sigma}}$ may however fail to be a poset. The next definition allows for a necessary transformation of $\mathcal{M}_{\widehat{\Sigma}}$.

Definition 7.1 Let $\mathcal{M} = (W, \preccurlyeq, [\![\cdot]\!])$ be an Alexandroff model. We define $\text{cut}(\mathcal{M}) = (W', \preccurlyeq', [\![\cdot]\!]')$ where $W' = W$, $[\![p]\!]' = [\![p]\!]$ for all atomic propositions and for all $x, y \in W'$, $x \preccurlyeq' y \iff x = y$ or $x \prec y$[1].

Note that if $\preccurlyeq$ is a preorder, then $\preccurlyeq'$ is a partial order, so cut($\mathcal{M}$) is a finite poset model.

Let $\mathcal{M}^{\text{cut}}_{\widehat{\Sigma}} = \text{cut}(\mathcal{M}_{\widehat{\Sigma}})$, and let $\preccurlyeq^{\text{cut}}$ be the accessibility relation in $\mathcal{M}^{\text{cut}}_{\widehat{\Sigma}}$.

Lemma 7.2 *Let $\Diamond\varphi \in \Sigma$ and suppose for some $T \in W_c$ we have $\Diamond\varphi \in T$, $\varphi \notin T$. Then there is $S \in W_c$ such that $T_{\widehat{\Sigma}} \prec_{\widehat{\Sigma}} S_{\widehat{\Sigma}}$ and $\mathcal{M}_{\widehat{\Sigma}}, S_{\widehat{\Sigma}} \models \varphi$.*

Proof. From $\Diamond\varphi \in T$, by axiom Grz, we obtain $\Diamond(\varphi \wedge \neg\Diamond(\neg\varphi \wedge \Diamond\varphi)) \in T$. Then by Lemma 6.2 there is $S \succcurlyeq_c T$ with $\varphi \wedge \neg\Diamond(\neg\varphi \wedge \Diamond\varphi) \in S$. By the construction of $\widehat{\Sigma}$, we have $\neg\Diamond(\neg\varphi \wedge \Diamond\varphi) \in \widehat{\Sigma}$. Hence $T_{\widehat{\Sigma}} \preccurlyeq_{\widehat{\Sigma}} S_{\widehat{\Sigma}}$, $\mathcal{M}_{\widehat{\Sigma}}, S_{\widehat{\Sigma}} \models \varphi$ and $\mathcal{M}_{\widehat{\Sigma}}, S_{\widehat{\Sigma}} \models \neg\Diamond(\neg\varphi \wedge \Diamond\varphi)$. To see that in fact $T_{\widehat{\Sigma}} \prec_{\widehat{\Sigma}} S_{\widehat{\Sigma}}$, suppose the contrary, that $S_{\widehat{\Sigma}} \preccurlyeq_{\widehat{\Sigma}} T_{\widehat{\Sigma}}$, and note that $\neg\Diamond(\neg\varphi \wedge \Diamond\varphi)$ can be rewritten as $\Box(\Diamond\varphi \to \varphi)$. Then since $\mathcal{M}_{\widehat{\Sigma}}, T_{\widehat{\Sigma}} \models \Diamond\varphi$, we get $\mathcal{M}_{\widehat{\Sigma}}, T_{\widehat{\Sigma}} \models \varphi$. But this implies $\varphi \in T$, which contradicts $\varphi \notin T$. □

Lemma 7.3 *For any $\gamma(\varphi, \psi) \in \Sigma$, if $\mathcal{M}_{\widehat{\Sigma}}, T_{\widehat{\Sigma}} \models \gamma(\varphi, \psi)$, then there exists an up-down path $(v_0, v_1, \ldots, v_k)$ witnessing this such that for all $i < k$, either $v_i = v_{i+1}$, $v_i \prec_{\widehat{\Sigma}} v_{i+1}$, or $v_i \succ_{\widehat{\Sigma}} v_{i+1}$.*

Proof. If $\mathcal{M}_{\widehat{\Sigma}}, T_{\widehat{\Sigma}} \models \gamma(\varphi, \psi)$ then by Lemma 3.6 we have a path $(v_0, v_1, \ldots, v_k)$ satisfying all of the above properties except possibly when $i = 0$ or $i = k - 1$. If $v_0 \prec_{\widehat{\Sigma}} v_1$ fails and $v_0 \neq v_1$ then $v_0 \preccurlyeq_{\widehat{\Sigma}} v_1$ and $v_0 \succcurlyeq_{\widehat{\Sigma}} v_1$. By lemmas 5.3

[1] Recall that $x \prec y$ is a shorthand for $x \preccurlyeq y$ and $y \not\preccurlyeq x$.

and 7.2 there is $S \in W_c$ such that $\mathcal{M}_{\widehat{\Sigma}}, S_{\widehat{\Sigma}} \models \varphi$ and $T_{\widehat{\Sigma}} \prec_{\widehat{\Sigma}} S_{\widehat{\Sigma}}$. We thus obtain a new up-down path $(\tilde{v}_0, \ldots, \tilde{v}_k) := (v_0, S_{\widehat{\Sigma}}, v_2, \ldots, v_k)$ where $\tilde{v}_0 \prec_{\widehat{\Sigma}} \tilde{v}_1$. If $\tilde{v}_{k-1} \succ_{\widehat{\Sigma}} \tilde{v}_k$ fails and $\tilde{v}_{k-1} \neq \tilde{v}_k$, we likewise choose U such that $\mathcal{M}_{\widehat{\Sigma}}, U_{\widehat{\Sigma}} \models \varphi$ and $\tilde{v}_k \prec_{\widehat{\Sigma}} U_{\widehat{\Sigma}}$, and replace $\tilde{v}_{k-1}$ by $U_{\widehat{\Sigma}}$. The resulting path has all required properties. □

Lemma 7.4 *If $\varphi \in \Sigma$ and $T \in W_c$ is arbitrary, we have:*

$$\mathcal{M}_{\widehat{\Sigma}}, T_{\widehat{\Sigma}} \models \varphi \iff \mathcal{M}^{\mathrm{cut}}_{\widehat{\Sigma}}, T_{\widehat{\Sigma}} \models \varphi.$$

Proof. Proceed by induction on φ. Atomic proposition and Boolean cases are immediate and thus omitted.

Consider the formula $\Diamond\varphi$.

We first show the right-to-left implication. Suppose $\mathcal{M}^{\mathrm{cut}}_{\widehat{\Sigma}}, T_{\widehat{\Sigma}} \models \Diamond\varphi$. Then there is $S_{\widehat{\Sigma}} \succcurlyeq^{\mathrm{cut}} T_{\widehat{\Sigma}}$ such that $\mathcal{M}^{\mathrm{cut}}_{\widehat{\Sigma}}, S_{\widehat{\Sigma}} \models \varphi$. By the definition of $\preccurlyeq^{\mathrm{cut}}$ we obtain $T_{\widehat{\Sigma}} \preccurlyeq S_{\widehat{\Sigma}}$ and by the inductive hypothesis we have $\mathcal{M}_{\widehat{\Sigma}}, S_{\widehat{\Sigma}} \models \varphi$. Then $\mathcal{M}_{\widehat{\Sigma}}, T_{\widehat{\Sigma}} \models \Diamond\varphi$.

Now for the left-to-right implication. We have that $\mathcal{M}_{\widehat{\Sigma}}, T_{\widehat{\Sigma}} \models \Diamond\varphi$. Then we have two cases:

- $\mathcal{M}_{\widehat{\Sigma}}, T_{\widehat{\Sigma}} \models \varphi$: by the IH, $\mathcal{M}^{\mathrm{cut}}_{\widehat{\Sigma}}, T_{\widehat{\Sigma}} \models \varphi$ and hence $\mathcal{M}^{\mathrm{cut}}_{\widehat{\Sigma}}, T_{\widehat{\Sigma}} \models \Diamond\varphi$.
- $\mathcal{M}_{\widehat{\Sigma}}, T_{\widehat{\Sigma}} \not\models \varphi$: by Lemma 7.2 there is $S_{\widehat{\Sigma}} \succ_{\widehat{\Sigma}} T_{\widehat{\Sigma}}$ with $\mathcal{M}_{\widehat{\Sigma}}, S_{\widehat{\Sigma}} \models \varphi$. Again by the IH, $\mathcal{M}^{\mathrm{cut}}_{\widehat{\Sigma}}, S_{\widehat{\Sigma}} \models \varphi$ and it follows that $\mathcal{M}^{\mathrm{cut}}_{\widehat{\Sigma}}, T_{\widehat{\Sigma}} \models \Diamond\varphi$.

Consider now the case of a formula $\gamma(\varphi, \psi)$.

The right-to-left implication is trivial, as $T_{\widehat{\Sigma}} \preccurlyeq^{\mathrm{cut}} S_{\widehat{\Sigma}} \implies T_{\widehat{\Sigma}} \preccurlyeq_{\widehat{\Sigma}} S_{\widehat{\Sigma}}$, and hence any up-down path in $\mathcal{M}^{\mathrm{cut}}_{\widehat{\Sigma}}$ witnessing $\gamma(\varphi, \psi)$ is also an up-down path in $\mathcal{M}_{\widehat{\Sigma}}$ witnessing, by IH, $\gamma(\varphi, \psi)$.

As for the left-to-right implication, if $\mathcal{M}_{\widehat{\Sigma}}, T_{\widehat{\Sigma}} \models \gamma(\varphi, \psi)$, by Lemma 7.3 there exists an up-down path $(v_0, v_1, \ldots, v_k)$ in $\mathcal{M}_{\widehat{\Sigma}}$ witnessing $\gamma(\varphi, \psi)$ such that for all $i < k$, either $v_i = v_{i+1}$, $v_i \prec_{\widehat{\Sigma}} v_{i+1}$, or $v_i \succ_{\widehat{\Sigma}} v_{i+1}$. It is immediate from the latter that this is also an up-down path in $\mathcal{M}^{\mathrm{cut}}_{\widehat{\Sigma}}$. Using the IH, we have $\mathcal{M}^{\mathrm{cut}}_{\widehat{\Sigma}}, T_{\widehat{\Sigma}} \models \gamma(\varphi, \psi)$. □

With this, we obtain completeness for finite poset models.

Proposition 7.5 *If φ is valid over the class of all finite poset models, then* $\mathsf{PLR} \vdash \varphi$.

Proof. Arguing by contraposition, if $\neg\varphi$ is consistent with PLR then there is a theory T with $\neg\varphi \in T$. Let Σ be the smallest adequate set containing $\neg\varphi$, which is finite by Lemma 6.3. Then Lemma 6.9 together with Lemma 7.4 yield $\mathcal{M}^{\mathrm{cut}}_{\widehat{\Sigma}}, T_{\widehat{\Sigma}} \models \neg\varphi$, in other words $\mathcal{M}^{\mathrm{cut}}_{\widehat{\Sigma}}, T_{\widehat{\Sigma}} \not\models \varphi$. Since $\mathcal{M}^{\mathrm{cut}}_{\widehat{\Sigma}}$ is a finite poset, the proof is finished. □

Putting together Proposition 7.5, Proposition 5.2 and Theorem 4.6, we obtain our main result.

Theorem 7.6 *Given a formula φ, the following are equivalent:*

(i) $\mathsf{PLR} \vdash \varphi$;

(ii) *φ is valid over the class of all finite poset models;*

(iii) *φ is valid over the class of all polyhedral models.*

8 Conclusions

Polyhedral semantics of modal logics with reachability operators enables reasoning about many interesting real-world scenarios. In this work, we provided sound and complete axiomatization of the polyhedral reachability logic.

There are many different directions for future research from both theoretical and practical perspectives. Here we focus on some theoretical questions. Having obtained axiomatization and polyhedral completeness for modal logic with reachability, as the next step one could study more general topological completeness for such logics. On general topological models some of the theorems of PLR are no longer valid. Namely, $\Diamond$ is no longer definable through γ. This makes the axiomatization more challenging. Natural follow up problems are axiomatizing the reachability logics for the class of all topological spaces and for other interesting classes of spaces such as Euclidean spaces, hereditarily irresolvable spaces, scattered spaces, locally connected spaces, etc.

The proposed axiomatisation may also have an impact on applications. Indeed, in light of the results obtained in [13,14,8], the proposed axioms can be used both for automated reasoning and for model checking, in order to perform minimisation of formulas: a way to do this is to mix the existing model checking approach with equality saturation tools (see e.g. [20]).

Acknowledgements We thank Mamuka Jibladze and Evgeny Kuznetsov for fruitful discussions about the polyhedral semantics of reachability modality. The research was supported by the bilateral project between CNR (Italy) and SRNSF (Georgia) "Model Checking for Polyhedral Logics" (grant #CNR-22-010), by the Shota Rustaveli National Science Foundation of Georgia grant #FR-22-6700, by European Union - Next GenerationEU - National Recovery and Resilience Plan (NRRP), Investment 1.5 Ecosystems of Innovation, Project "Tuscany Health Ecosystem" (THE), CUP: B83C22003930001 and by European Union - Next-GenerationEU - National Recovery and Resilience Plan (NRRP) – Mission 4 Component 2, Investment N. 1.1, CALL PRIN 2022 D.D. 104 02-02-2022 – (Stendhal) CUP N. B53D23012850006.

References

[1] Adam-Day, S., N. Bezhanishvili, D. Gabelaia and V. Marra, *Polyhedral completeness of intermediate logics: the nerve criterion*, The Journal of Symbolic Logic (2022), pp. 1–41.

[2] Adam-Day, S., N. Bezhanishvili, D. Gabelaia and V. Marra, *The intermediate logic of convex polyhedra* (2023), arXiv:2307.16600 [math.LO].

[3] Aiello, M., I. Pratt-Hartmann and J. Benthem, van, editors, "Handbook of Spatial Logics," Springer, 2007.

[4] Belmonte, G., V. Ciancia, D. Latella and M. Massink, *Voxlogica: A spatial model checker for declarative image analysis*, in: *Tools and Algorithms for the Construction and Analysis of Systems, TACAS*, LNCS **11427** (2019), pp. 281–298.

[5] Benthem, J. v. and G. Bezhanishvili, *Modal logics of space*, in: M. Aiello, I. Pratt-Hartmann and J. v. Benthem, editors, *Handbook of Spatial Logics*, Springer, 2007 pp. 217–298.

[6] Bezhanishvili, G. and D. Gabelaia, *Connected modal logics*, Archive for Mathematical Logic **50** (2011), pp. 287–317.

[7] Bezhanishvili, N., V. Ciancia, D. Gabelaia, G. Grilletti, D. Latella and M. Massink, *Geometric model checking of continuous space*, Logical Methods in Computer Science **Volume 18, Issue 4** (2022).

[8] Bezhanishvili, N., V. Ciancia, D. Gabelaia, M. Jibladze, D. Latella, M. Massink and E. P. de Vink, *Weak simplicial bisimilarity for polyhedral models and slcs-eta*, in: V. Castiglioni and A. Francalanza, editors, *Formal Techniques for Distributed Objects, Components, and Systems* (2024), pp. 20–38.

[9] Bezhanishvili, N., V. Marra, D. McNeill and A. Pedrini, *Tarski's theorem on intuitionistic logic, for polyhedra*, Annals of Pure and Applied Logic **169** (2018), p. ii.

[10] Bussi, L., V. Ciancia, F. Gadducci, D. Latella and M. Massink, *Towards model checking video streams using VoxLogicA on GPUs*, in: J. Bowles, G. Broccia and R. Pellungrini, editors, *From Data to Models and Back*, LNCS **13268** (2022), pp. 78–90.

[11] Chagrov, A. and M. Zakharyaschev, "Modal logic," Number 35 in Oxford Logic Guides, The Clarendon Press, Oxford University Press, New York, 1997.

[12] Ciancia, V., G. Belmonte, D. Latella and M. Massink, *A hands-on introduction to spatial model checking using voxlogica*, in: *Model Checking Software* (2021), pp. 22–41.

[13] Ciancia, V., D. Gabelaia, D. Latella, M. Massink and E. P. de Vink, *On bisimilarity for polyhedral models and slcs*, in: M. Huisman and A. Ravara, editors, *Formal Techniques for Distributed Objects, Components, and Systems* (2023), pp. 132–151.

[14] Ciancia, V., J. F. Groote, D. Latella, M. Massink and E. P. de Vink, *Minimisation of spatial models using branching bisimilarity*, in: M. Chechik, J.-P. Katoen and M. Leucker, editors, *Formal Methods* (2023), pp. 263–281.

[15] Ciancia, V., D. Latella, M. Loreti and M. Massink, *Model Checking Spatial Logics for Closure Spaces*, Logical Methods in Computer Science **Volume 12, Issue 4** (2016).

[16] Engelking, R., "General Topology," Sigma series in pure mathematics, Heldermann, 1989.

[17] Gabelaia, D., K. Gogoladze, M. Jibladze, E. Kuznetsov and M. Marx, *Modal logic of planar polygons* (2018), arXiv:1807.02868 [math.LO].

[18] Gabelaia, D., M. Jibladze, E. Kuznetsov and L. Uridia, *Characterization of flat polygonal logics* (2019), abstract of the talk given at the conference *Topology, Algebra, and Categories in Logic*, Nice. https://math.unice.fr/tacl/assets/2019/abstracts.pdf.

[19] McKinsey, J. C. C. and A. Tarski, *The algebra of topology*, Annals of Mathematics **45** (1944), pp. 141–191.

[20] Willsey, M., C. Nandi, Y. R. Wang, O. Flatt, Z. Tatlock and P. Panchekha, *egg: Fast and extensible equality saturation* **5** (2021).
URL `https://doi.org/10.1145/3434304`

The Goldblatt-Thomason Theorem for Derivative Spaces

Nick Bezhanishvili

ILLC, University of Amsterdam, The Netherlands

David Fernández-Duque [1]

Department of Philosophy University of Barcelona, Spain

Reihane Zoghifard [2]

School of Mathematics, Institute for Research in Fundamental Sciences (IPM), Iran

Abstract

The Goldblatt-Thomason theorem is a classic result of modal definability of Kripke frames. Its topological analogue for the closure semantics has been proved by ten Cate et al. (2009). In this paper we prove a version of the Goldblatt-Thomason theorem for topological semantics via the Cantor derivative. We work with derivative spaces which provide a natural generalisation of topological spaces on the one hand and of weakly transitive frames on the other.

Keywords: Modal definability, topological semantics, Cantor derivative, Alexandroff extensions.

1 Introduction

The Goldblatt-Thomason theorem [9] is one of the classic results in modal logic. It states that an elementary class of Kripke frames is modally definable iff it is closed under generated subframes, bounded morphic images, disjoint unions and reflects ultrafilter extensions. In [15] van Benthem gave a model theoretic proof of this theorem. Since then a number of Goldblatt-Thomason theorems have been proved in different contexts. Van Benthem provided a

[1] David Fernández-Duque was supported by the FWO-FWF Lead Agency grant G030620N (FWO)/I4513N (FWF) and by the SNSF–FWO Lead Agency Grant 200021L_196176/G0E2121N.

[2] Reihane Zoghifard acknowledges the receipt of a grant from the CIMPA-ICTP Research in Pairs Program. She was in part supported by a grant from IPM (No.1401030021).

version of this result for finite Kripke frames [14]. Ten Cate [13] investigated Goldblatt-Thomason theorems for hybrid languages. Gabelaia [8] gave a topological version of this theorem for c-semantics (and [12] proved it for extended languages). The Goldblatt-Thomason theorem for coalgebraic modal logic was established in [10].

In this paper we prove the Goldblatt-Thomason theorem for the topological derived set semantics (d-semantics). While in the topological c-semantics the modal diamond $\Diamond$ is interpreted as the closure c, in the derived set semantics the diamond $\Diamond$ is interpreted as the Cantor derivative d. Recall that in a topological space a derived set $d(A)$ of a set A consists of the points x such that for every open neighbourhood U_x of x the intersection $A \cap (U_x \setminus \{x\}) \neq \varnothing$. It is well known that the logic of all topological spaces for c-semantics is $\mathsf{S4}$ [11] and that the logic of all topological spaces for the d-semantics is the logic $\mathsf{wK4}$ of all weakly transitive frames (see, e.g., [16]), where a relation is weakly transitive if

$$\forall x \forall y \forall z((Rxy \wedge Ryz) \rightarrow (Rxz \vee x = z)).$$

Recall that the *ultrafilter extension* of a Kripke frame (X, R) is the ultrafilter frame of the modal algebra $(\mathcal{P}(X), \Diamond_R)$. In the case of topological c-semantics the role of ultrafilter extensions is played by the Alexandroff extensions [8,12]. The *Alexandroff extension* of a topological space (X, τ) is the Alexandroff space associated with the ultrafilter frame of the $\mathsf{S4}$-algebra $(\mathcal{P}(X), \mathsf{c})$. The topological version of the Goldblatt-Thomason theorem states that an elementary class (i.e., an L_t-definable class, see below) of topological spaces is modally definable in the c-semantics iff it is closed under open subspaces, interior images (i.e., images under continuous and open maps), topological sums and reflects Alexandroff extensions [8,12].

We define the *d-Alexandroff extension* of a topological space (X, τ) as the ultrafilter frame of the $\mathsf{wK4}$-algebra $(\mathcal{P}(X), d)$. However, while the Alexandroff extension of a topological space is an Alexandroff topological space, the d-Alexandroff extension of a topological space may not be a topological space, which complicates the matter. In order to overcome this difficulty, instead of topological spaces we work with *derivative spaces* introduced in [6,1]. Derivative spaces generalize topological spaces on the one hand and weakly transitive Kripke frames on the other. We show that the d-Alexandroff extension of a derivative space is a derivative space (in fact, it is always a weakly transitive frame).

The original Goldblatt-Thomason theorem, as well as its topological variant, gives a characterization of modally definable classes that are elementary. We introduce an appropriate first-order language for studying derivative spaces. Similarly to the language $\mathcal{L}^2$ and its fragment $\mathcal{L}_t$ used to study model theory of topological spaces [7,12], the language $\mathcal{L}_2$ is a two sorted first-order language with two sorts of variables, where one ranges over points of the space and the other ranges over basic subsets.

We also work with d-analogues of generated subframes, bounded morphic images and disjoint unions—d-subspaces, d-morphic images and d-sums. Our main result (Theorem 5.2) states that an $\mathcal{L}_2$-definable class of derivative spaces is modally definable iff it is closed under d-subspaces, d-morphic images and d-sums and reflects d-Alexandroff extensions.

For proving the result we use the model-theoretic approach of [14,12]. In order to do this we develop some model theory of derivative spaces. We introduce an equivalent presentation of derivative spaces, which we call *based spaces* as they resemble the presentation of topological spaces in terms of their bases. Based space presentation enables us to apply model-theoretic techniques to derivative spaces more easily. We define ultraproducts and saturation of based spaces. Using based spaces as an equivalent presentation of derivative spaces and also seeing them as structures of $\mathcal{L}_2$ enable us to apply model-theoretic results, such as compactness, to our models. Utilizing this we show that any derivative space has a saturated ultrapower with a d-map from this ultrapower to the d-Alexandroff extension of the space. This is one of our main technical lemmas for proving the Goldblatt-Thomason theorem.

As a special case we obtain the Goldblatt-Thomason theorem for topological derivative spaces (Corollary 5.6), i.e., for derivative spaces that are topological spaces. Since the d-Alexandroff extension of a topological derivative space is not necessarily a topological derivative space, we consider a reflection from a wider class which contains topological spaces and weakly transitive Kripke frames. Then, we show that an $\mathcal{L}_2$-elementary class of topological spaces is modally definable iff it is closed under disjoint unions, open subspaces, d-morphic images and reflects weak transitive extensions.

Since topological spaces with c-semantics satisfy all conditions of the Goldblatt-Thomason theorem, these spaces are definable over the class of all derivative spaces. Indeed, they are definable by the formula $p \to \Diamond p$. On the other hand, the class of all derivative spaces associated with weakly transitive Kripke frames is not modally definable over the class of all derivative spaces, since it does not reflect d-Alexandroff extensions. It also follows from our Goldblatt-Thomason theorem that the class of T_1-spaces is not modally definable over the class of all topological derivative spaces, while the class of T_d-spaces is definable.

2 Preliminaries: Derivative spaces and based spaces

In this section we review the main semantical structures used in the text: derivative spaces, and, specifically, their presentation as based spaces. Derivative spaces were introduced by Fernández-Duque and Iliev [6] as 'convergence spaces', in order to unify topological and Kripke semantics for the logic of the Cantor derivative. They were renamed *derivative spaces* by Baltag et al. [1]. These are a special case of the more general *derivative algebras* of Esakia [5]. Derivative spaces may moreover be presented as *based spaces,* similarly to how topological spaces may be presented in terms of a basis. The latter presentation will be the most convenient for us, since many classic results for first order logic can be readily applied to based spaces.

Derivative spaces are sets of points equipped with an operator satisfying the basic properties of the Cantor derivative, although this operator need not coincide with the 'true' derivative of the induced topological space.

Definition 2.1 A *derivative space* is a pair (X, d) where $X \neq \varnothing$ is a set of points and $d : \mathcal{P}(X) \to \mathcal{P}(X)$, called a derivative operator, is a map that satisfies the following conditions for all $A, B \subseteq X$:

(i) $d(\varnothing) = \varnothing$;

(ii) $d(A \cup B) = d(A) \cup d(B)$;

(iii) $d(d(A)) \subseteq A \cup d(A)$.

The dual of d, called the *co-derivative*, is defined as $\tilde{d}(A) = X - d(X - A)$, for each $A \subseteq X$.

The following two examples show that we can define two different derivative spaces over a given topological space. In other words, a derivative space is a generalized notion of a topological space.

Example 2.2 Let $\mathcal{X} = (X, \tau)$ be a topological space, then (X, c) is a derivative space, where c is the topological closure operator of τ. Any derivative space of this form is called a *topological closure space*. Note that in addition to the conditions of the derivative operator in Definition 2.1, the topological closure operator also satisfies $A \subseteq \mathsf{c}(A)$ and $\mathsf{c}(\mathsf{c}(A)) \subseteq \mathsf{c}(A)$ for all $A \subseteq X$.

In 1944, by interpreting the modal operator $\Diamond$ as the topological closure operator, McKinsey and Tarski introduced a topological semantics of modal logic [11], nowadays named c-semantics.

Example 2.3 Let $\mathcal{X} = (X, \tau)$ be a topological space. The Cantor derivative operator of $\mathcal{X}$, denoted by d_τ, is the operator that assigns to each subset of X the set of its limit points, i.e., for any $A \subseteq X$

$$d_\tau(A) = \{a \in X \mid \forall O \in \tau \; (a \in O \Rightarrow A \cap O - \{a\} \neq \varnothing)\}.$$

Then (X, d_τ) is a derivative space.

In [11], McKinsey and Tarski also noted that the modal operator $\Diamond$ interpreted as the Cantor derivative operator gives another topological semantics for modal logic. It is known as the d-semantics for modal logic. Since $\mathsf{c}(A) = A \cup d_\tau(A)$, the d-semantics is more expressive than c-semantics. For example, the class of T_0 spaces is not definable in c-semantics ([12, Cor. 37]), but it is definable in d-semantics ([2, Cor. 1]).

Examples 2.2 and 2.3 show that the class of derivative spaces contains the class of topological spaces, equipped with either the closure or Cantor derivative operators. On the other hand, any derivative space (X, d) can be seen as a topological space by defining the closure operator c_d as $\mathsf{c}_d(A) = A \cup d(A)$. We denote this induced topology by τ_d. Note that given an arbitrary derivative space (X, d), its derivative operator d does not necessarily coincide with the Cantor derivative of its induced topology τ_d. Precisely, for any $A \subseteq X$ we have

$$d(A) = d_{\tau_d}(A) \cup \mathrm{ref}(A), \tag{1}$$

where d_{τ_d} is the Cantor operator of τ_d and $\text{ref}(A) = \{a \in A \mid a \in d(\{a\})\}$. The elements of ref$(A)$ are called *reflexive points.*

To see that Equation 1 holds, first assume that $a \notin d_{\tau_d}(A) \cup \text{ref}(A)$. Since $a \notin d_{\tau_d}(A)$, there is an open $O \in \tau_d$ such that $a \in O$ and $O \cap A - \{a\} = \varnothing$. By the second condition of d, we have $d(A) = d(A \cap O) \cup d(A \cap O^c)$. Since $a \in O = Int_d(O)$, we have $a \notin d(O^c) \subseteq O^c$, thus $a \notin d(A \cap O^c)$. Furthermore, if $a \in A$, then $A \cap O = \{a\}$ and $a \notin \text{ref}(A)$ implies that $a \notin d(A \cap O)$. If $a \notin A$, then $A \cap O = \varnothing$, and thus $a \notin d(A \cap O)$. Therefore, $a \notin d(A)$.

For the other direction, first assume that $a \in \text{ref}(A)$. Then $a \in A$ and $a \in d(\{a\})$ which implies that $a \in d(A)$. Now assume that $a \notin \text{ref}(A)$ and $a \notin d(A)$. If $a \notin A$, then $a \in A^c \cap d^c(A) = Int_d(A^c) = O \in \tau_d$. Thus, $a \notin d_{\tau_d}(A)$, because $O \cap A - \{a\} = \varnothing$. If $a \in A$, then $a \notin d(\{a\})$. Let $O = Int_d(A^c \cup \{a\})$, this implies that $a \notin d_{\tau_d}(A)$.

We call a subset $A \subseteq X$, *d-closed* if it is closed in τ_d, i.e., $d(A) \subseteq A$. So, $A \subseteq X$ is *d-open* if $A \subseteq \tilde{d}(A) = X - d(X - A)$.

Another special case of derivative spaces comes from weakly transitive Kripke frames. A Kripke frame (W, R) is weakly transitive if $(wRv \wedge vRz \Rightarrow w = z \vee wRz)$, for all $w, v, z \in W$.

Example 2.4 Let (W, R) be a weakly transitive Kripke frame. We obtain a derivative space (W, d_R) by defining, for any $A \subseteq W$,

$$d_R(A) = \{w \in W \mid wRs \text{ for some } s \in A\}.$$

Then, τ_{d_R} is the upset topology over (W, R), i.e., $O \subseteq W$ is open if for any $a \in O$ we have $R(a) = \{s \mid aRs\} \subseteq O$. Also, $\text{ref}(A) = \{a \in A \mid aRa\}$. Therefore, if R is a weakly transitive and irreflexive relation, then d_R is the Cantor derivative operator of τ_R. Also, if R is a transitive and reflexive relation, then d_R is the topological closure operator of τ_R.

As mentioned above, the class of derivative spaces contains some interesting classes of structures, e.g., topological spaces and weakly transitive Kripke frames. On the other hand, the class of derivative spaces is a subclass of the class of monotonic neighbourhood structures.

Definition 2.5 A *neighbourhood derivative space* is a pair $(X, \mathcal{N})$, where $\mathcal{N} : X \to 2^{2^X}$ is a derivative neighbourhood assignment that satisfies the following conditions for any $a \in X$:

(i) $X \in \mathcal{N}(a)$,

(ii) if $A \in \mathcal{N}(a)$ and $A \subseteq B$, then $B \in \mathcal{N}(a)$,

(iii) if $A, B \in \mathcal{N}(a)$, then $A \cap B \in \mathcal{N}(a)$,

(iv) if $a \in A \in \mathcal{N}(a)$, then $\{b \in X \mid A \in \mathcal{N}(b)\} \in \mathcal{N}(a)$.

In [1], it is mentioned that there is an equivalent presentation of derivative spaces as neighbourhood derivative spaces. For a given derivative space (X, d) and $a \in X$, let

$$\mathcal{N}_d(a) = \{A \subseteq X \mid a \notin d(X - A)\} = \{A \subseteq X \mid a \in \tilde{d}(A)\}.$$

We call the members of $\mathcal{N}_d(a)$ *d-neighbourhoods of* a. Note that every open neighbourhood of a with respect to the topology τ_d is a d-neighbourhood of a, but the converse is not true in general. For example, when d is the Cantor derivative of a topological space, then $A \in \mathcal{N}_d(a)$ if there is an open neighbourhood O of a such that $O - \{a\} \subseteq A$. In other words, A is d-open if for any $a \in A$ we have $A \in \mathcal{N}_d(a)$.

Conversely, for a given derivative neighbourhood space $(X, \mathcal{N})$, we can define an operator d over $\mathcal{P}(X)$ as follows:

$$d(A) = \{a \in X \mid \forall O \in \mathcal{N}(a)\ O \cap A \neq \varnothing\}.$$

Then (X, d) is a derivative space.

In addition to this equivalent presentation of derivative spaces, we can also consider another one which plays a similar role as bases for topological spaces relative to derivative spaces.

Definition 2.6 Let $\mathcal{B} : X \to 2^{2^X}$ be a function over $X \neq \varnothing$. Then $\mathcal{B}$ is called a *basic neighbourhood assignment* if for each $a \in X$ the following conditions hold:

(i) $\mathcal{B}(a) \neq \varnothing$,

(ii) if $A, B \in \mathcal{B}(a)$, then there is $C \in \mathcal{B}(a)$ such that $C \subseteq A \cap B$,

(iii) if $A \in \mathcal{B}(a)$ and $b \in A$, then there is $B \in \mathcal{B}(b)$ such that $B \subseteq A \cup \{a\}$.

Any set $A \in \mathcal{B}(a)$ is called a *basic d-neighbourhood* of a. Also, $\bigcup_{a \in X} \mathcal{B}(a)$ is called a *derivative base* (or set of basic d-neighbourhoods) of X, and $(X, \mathcal{B})$ is called a *based space.*

Note that basic neighbourhood assignments need not be monotone (i.e., closed under supersets), but this does not affect their modal logic, as one may obtain an equivalent monotone structure from them.

Lemma 2.7 *Let $\mathcal{B}$ be a basic neighbourhood assignment over X. For each $a \in X$, let $\mathcal{N}_{\mathcal{B}}(a)$ be the closure of $\mathcal{B}(a)$ under supersets. Then $\mathcal{N}_{\mathcal{B}}$ is a derivative neighbourhood assignment.*

Moreover, the function $d_{\mathcal{B}} : \mathcal{P}(X) \to \mathcal{P}(X)$ defined as

$$d_{\mathcal{B}}(A) = \{a \in X \mid \forall O \in \mathcal{B}(a)\ O \cap A \neq \varnothing\}$$

is a derivative operator, and $\mathcal{N}_{d_{\mathcal{B}}} = \mathcal{N}_{\mathcal{B}}$.

Proof. By definition it is clear that $\mathcal{N}_{\mathcal{B}}$ is a derivative neighbourhood assignment. It is not hard to see that $d_{\mathcal{B}}$ satisfies (i-ii) of Definition 2.1. To see (iii), assume that $a \notin A$ and $a \notin d_{\mathcal{B}}(A)$. Thus there is $O \in \mathcal{B}(a)$ such that $O \cap A = \varnothing$. We show that $O \cap d_{\mathcal{B}}(A) = \varnothing$. If $O = \varnothing$, then clearly $a \notin d_{\mathcal{B}} d_{\mathcal{B}}(A)$. Otherwise, assume that $b \in O \cap d_{\mathcal{B}}(A)$. Then by the third condition of Definition 2.6, there is $B \in \mathcal{B}(b)$ such that $B \subseteq O \cup \{a\}$. This implies that $B \cap A = \varnothing$, which contradict with $b \in d_{\mathcal{B}}(A)$. □

Likewise, we may readily obtain a based space from a given derivative space by assigning to each point its set of 'punctured neighbourhoods'.

Lemma 2.8 *Let (X, d) be a derivative space and $\mathcal{B}_d : X \to 2^{2^X}$ be an operator which assigns to each $a \in X$ the set of its punctured d-neighbourhoods, i.e.,*

$$\mathcal{B}_d(a) = \{O \subseteq X \mid a \in \tilde{d}(O) \ \&\ O \subseteq \tilde{d}(O \cup \{a\})\}.$$

Then $\mathcal{B}_d$ is a basic neighbourhood assignment with $d_B = d$.

Proof. $\mathcal{B}_d$ satisfies (i) and (ii) of Definition 2.6, since $\tilde{d}(X) = X$ and $\tilde{d}(A \cap B) = \tilde{d}(A) \cap \tilde{d}(B)$. For (iii), assume that $A \in \mathcal{B}_d(a)$ and $b \in A$. We have to show that there is $B \in \mathcal{B}(b)$ such that $B \subseteq A \cup \{a\}$. Since $A \in \mathcal{B}_d(a)$, we have $a \in \tilde{d}(A)$ and $A \subseteq \tilde{d}(A \cup \{a\})$. This implies that $b \in \tilde{d}(A \cup \{a\})$ and $A \cup \{a\} \subseteq \tilde{d}(A \cup \{a\}) = \tilde{d}(A \cup \{a\} \cup \{b\})$. Thus $A \cup \{a\} \in \mathcal{B}_d(b)$. □

So, we can consider based spaces as an equivalent presentation of derivative spaces. In other words, we can identify a class K of derivative spaces as a class K' of based spaces $(X, \mathcal{B})$ such that $(X, d_B) \in K$. There are various technical advantages to working with based spaces.

Example 2.9 Assume that (X, τ) is a topological space and σ is a basis for τ. Then the function $\mathcal{B}_1$ defined as $\mathcal{B}_1(a) = \{O \in \sigma \mid a \in O\}$ for each $a \in X$, is a basic neighbourhood assignment of X and $d_{B_1} = \mathsf{c}$. Also the function $\mathcal{B}_2$ defined as $\mathcal{B}_2(a) = \{O - \{a\} \mid a \in O \in \sigma\}$ is a basic neighbourhood assignment with $d_{B_2} = d_\tau$.

3 Model theory of derivative spaces

Our main focus in this paper is on the basic modal language. Let $\mathbb{P}$ be a countable set of propositional variables. Modal formulas are constructed recursively from $\mathbb{P}$ using Boolean connectives and modal operator $\Diamond$.

Definition 3.1 A based model is a triple $\mathfrak{M} = (X, \mathcal{B}, [\![\cdot]\!])$ where $\mathcal{X} = (X, \mathcal{B})$ is a based space and $[\![\cdot]\!] : \mathbb{P} \to \mathcal{P}(X)$ is a valuation function.

The satisfaction of formulas is defined by structural induction. For propositional variables and Boolean connectives we have the standard definitions, and for modal operator we have

$$\mathfrak{M}, a \models \Diamond\varphi \text{ iff } A \cap [\![\varphi]\!] \neq \varnothing \text{ for all } A \in \mathcal{B}(a).$$

Then we have $\mathfrak{M}, a \models \Box\varphi$ iff there exists $A \in \mathcal{B}(a)$ such that $A \subseteq [\![\varphi]\!]$.

In other words, for any derivative model $\mathfrak{M} = (X, d, [\![\cdot]\!])$, which is obtained by adding a valuation function to a derivative space (X, d), we have

$$\mathfrak{M}, a \models \Diamond\varphi \text{ iff } a \in d([\![\varphi]\!]),$$

and thus, $\mathfrak{M}, a \models \Box\varphi$ iff $a \in \tilde{d}([\![\varphi]\!])$.

A formula φ is *valid* in the based (derivative) model $\mathfrak{M}$, denoted by $\mathfrak{M} \models \varphi$, if $\mathfrak{M}, a \models \varphi$ for all $a \in X$. Also, φ is valid on the based (resp. derivative) space $\mathcal{X}$, denoted by $\mathcal{X} \models \varphi$, if it is valid in any based (resp. derivative) model over $\mathcal{X}$. Furthermore, for any class of based space K by $K \models \varphi$, we mean that $\mathcal{X} \models \varphi$ for all $\mathcal{X} \in K$. The notion of validity can be extended to the set of modal formulas in the usual way, for instant a set of modal formulas Γ is valid on $\mathcal{X}$, denoted by $\mathcal{X} \models \Gamma$, whenever $\mathcal{X} \models \varphi$ for all $\varphi \in \Gamma$.

3.1 Corresponding language

The original Goldblatt-Thomason theorem [9] provides a characterization of modally definable conditions for classes of Kripke frame that are elementary, i.e., definable in the corresponding first-order language. Ten Cate et al. [12] also prove a version of the Goldblatt-Thomason theorem for those topological closure spaces that are definable in a suitable corresponding language $\mathcal{L}_t$. In this section, we introduce a first-order language which is appropriate for studying derivative spaces. This language is interpreted over based spaces as an equivalent presentation of derivative spaces.

Let $\mathcal{L}_2$ be a two-sorted first-order language. The first sort is reserved to range over the set of points of a space, and we denote its variables by $x, y, \ldots$. The second sort is used to range over basic subsets of the space, and we denote its variables by $U, V, \ldots$. The language $\mathcal{L}_2$ contains two binary relations ε and ν, where ε relates point variables with basic subset variables and ν relates basic subset variables with point variables. Also, $\mathcal{L}_2$ contains a unary predicate P_p for each proposition p. The formulas of $\mathcal{L}_2$ are defined as follows: [3]

$$\varphi ::= x = y \mid x \,\varepsilon\, U \mid U \,\nu\, x \mid U = V \mid P_p(x) \mid \neg\varphi \mid \varphi \wedge \varphi \mid \exists x\varphi \mid \exists U\varphi.$$

Any based model $\mathfrak{M} = (X, \mathcal{B}, [\![\cdot]\!])$ can be seen as an $\mathcal{L}_2$-structure

$$\mathcal{M}_{\mathfrak{M}} = (X, \bigcup_{a \in X} \mathcal{B}(a), \varepsilon^{\mathfrak{M}}, \nu^{\mathfrak{M}}, \{P_p \mid p \in \mathbb{P}\}),$$

where $a \,\varepsilon^{\mathfrak{M}}\, O$ means that $a \in O$ and $O \,\nu^{\mathfrak{M}}\, a$ means that $O \in \mathcal{B}(a)$.

Clearly, not every $\mathcal{L}_2$-structure is a based model. However, the class of based models is characterized by an $\mathcal{L}_2$-theory.

Assume that $\mathcal{M} = (D_1, D_2, \varepsilon^{\mathcal{M}}, \nu^{\mathcal{M}}, \{P_p \mid p \in \mathbb{P}\})$ is an $\mathcal{L}_2$-model. For any $A \in D_2$, let $\|A\| = \{a \in D_1 \mid a \,\varepsilon^{\mathcal{M}}\, A\}$.

Lemma 3.2 *There exists a finite set of $\mathcal{L}_2$-formulas Γ_{basic} such that if $\mathcal{M}$ is any $\mathcal{L}_2$-model such that $\mathcal{M} \models \Gamma_{basic}$, then $\mathfrak{M} = (D_1, \mathcal{B}, [\![\cdot]\!])$ is a based model with a basic neighbourhood assignment $\mathcal{B}(a) = \{\|A\| \mid A \in D_2 \;\&\; A \,\nu^{\mathcal{M}}\, a\}$. Moreover, $\mathcal{M}_{\mathfrak{M}} \cong \mathcal{M}$.*

Proof. We define Γ_{basic} to be the following $\mathcal{L}_2$-formulas, which express the axioms of basic d-neighbourhoods:

- $\forall x \exists U (U \,\nu\, x)$,
- $\forall x \forall U, V (U \,\nu\, x \wedge V \,\nu\, x \to \exists W (W \,\nu\, x \wedge \forall y (y \,\varepsilon\, W \to y \,\varepsilon\, U \wedge y \,\varepsilon\, V)))$,
- $\forall x \forall U (U \,\nu\, x \to \forall z (z \,\varepsilon\, U \to \exists W (W \,\nu\, z \wedge \forall y (y \,\varepsilon\, W \to y = x \vee y \,\varepsilon\, U)))))$,
- $\forall U, V (\forall x (x \varepsilon U \leftrightarrow x \varepsilon V) \to U = V)$.

□

[3] The language $\mathcal{L}_2$ is similar to the language $\mathcal{L}^2$, introduced for studying topological spaces [7], except that it has an extra predicate ν.

Then the function d defined over $\mathcal{P}(D_1)$ as

$$d(A) = \{a \in D_1 \mid \forall O \in D_2\ (O\ \nu^{\mathcal{M}}\ a \Rightarrow \|O\| \cap A \neq \varnothing)\},$$

is a derivative operator with $d_{\mathcal{B}} = d$.

Just as the standard translation interprets modal formulas with their Kripke semantics as first order formulas, we may interpret the modal language in $\mathcal{L}_2$ with respect to its derivational semantics, i.e., with respect to those $\mathcal{L}_2$-models that arising from a based model.

Definition 3.3 Given a designated first order variable x, we recursively define a translation Tr_x from modal formulas to $\mathcal{L}_2$-formulas as follows:

$$\begin{aligned} Tr_x(p) &= P_p(x) \\ Tr_x(\neg\varphi) &= \neg Tr_x(\varphi) \\ Tr_x(\varphi \wedge \psi) &= Tr_x(\varphi) \wedge Tr_x(\psi) \\ Tr_x(\Diamond\varphi) &= \forall U\ (U\ \nu\ x \rightarrow \exists y(y\ \varepsilon\ U \wedge Tr_y(\varphi))) \end{aligned}$$

Proposition 3.4 *For any based model* $\mathfrak{M}$ *and any modal formula* φ *we have*

$$\mathfrak{M}, a \models \varphi \text{ iff } \mathfrak{M} \models Tr_x(\varphi)[a].$$

Remark 3.5 In the translation of modal formulas, the quantifiers over basic subset variables appear in restricted forms. An $\mathcal{L}_2$-formula α is called *negative (positive)* in set variable U if all free occurrences of U in α are under odd (even) number of negation signs. Let $\mathcal{L}_d$ be a language obtained by restricting the use of quantifiers over set variables of $\mathcal{L}_2$ as follows:

- $\exists U(U\ \nu\ x \wedge \alpha)$, where α is negative in U,
- $\forall U(U\ \nu\ x \rightarrow \alpha)$, where α is positive in U.

Then one can easily see that $\mathcal{L}_d$ is invariant under basic d-neighbourhoods, i.e., the satisfaction of $\mathcal{L}_d$-formulas is independent of interpreting $U\ \nu\ x$ as 'U is a d-neighbourhood of x' or as 'U is a *basic* d-neighbourhood of x'. [4]

3.2 Ultraproducts

Ultraproducts are an essential tool in estabilshing the original Goldblatt-Thomason theorem, as well as its topological variant; they will also be used throughout our own proof. In this section, we review their definitions and basic properties.

Let I be a non-empty set. A set $\mathcal{D} \subseteq \mathcal{P}(I)$ is a *filter* over I, if $I \in \mathcal{D}$ and $\mathcal{D}$ is closed under finite intersections and supersets. A filter $\mathcal{D}$ is called an *ultrafilter* if for all $A \subseteq I$, either $A \in \mathcal{D}$ or $I - A \in \mathcal{D}$. The ultrafilter theorem states that any subset of $\mathcal{P}(I)$ with the finite intersection property can be extended to an ultrafilter over I; in other words, if $\mathcal{D}$ is a filter such

[4] Similarly, over the class of topological spaces, the language $\mathcal{L}_t$ is introduced as a fragment of $\mathcal{L}^2$ which is invariant under topological bases, see [7,12].

that whenever $X_1, \ldots, X_n \in \mathcal{D}$, it follows that $X_1 \cap \ldots \cap X_n \neq \varnothing$, then there is an ultrafilter $\mathcal{U} \supseteq \mathcal{D}$.

Let $(X_i : i \in I)$ be a family of non-empty sets and $\prod_{i \in I} X_i$ be the Cartesian product of this family, i.e., $\prod_{i \in I} X_i = \{(a_i)_{i \in I} \mid a_i \in X_i\}$. Two elements $(a_i)_{i \in I}, (b_i)_{i \in I} \in \prod_{i \in I} X_i$ are $\mathcal{D}$-equivalent, denoted by $(a_i)_{i \in I} \sim_{\mathcal{D}} (b_i)_{i \in I}$, if $\{i \in I \mid a_i = b_i\} \in \mathcal{D}$. It is clear that $\sim_{\mathcal{D}}$ is an equivalence relation. We denote the equivalence class of $(a_i)_{i \in I}$ by $[(a_i)]$. Let $\prod_{\mathcal{D}} X_i$ be the set of all equivalence classes.

Assume $(\mathcal{X}_i : i \in I)$ is a family of based spaces, and $\mathcal{D}$ is an ultrafilter over I. Define the function $\mathcal{B}_{\mathcal{D}}$ over $\prod_{\mathcal{D}} X_i$ as

$$\mathcal{B}_{\mathcal{D}}([(a_i)]) = \{\prod_{\mathcal{D}} U_i \mid \{i \in I \mid U_i \in \mathcal{B}_i(a_i)\} \in \mathcal{D}\},$$

where $U_i \subseteq X_i$ and $\mathcal{B}_i$ is a basic neighbourhood assignment of X_i. Then one can easily see that $(\prod_{\mathcal{D}} X_i, \mathcal{B}_{\mathcal{D}})$ is a based space.

Definition 3.6 The *d-ultraproduct* of a family of based models $(\mathfrak{M}_i : i \in I)$ is a model $\prod_{\mathcal{D}} \mathfrak{M}_i = (\prod_{\mathcal{D}} X_i, \mathcal{B}_{\mathcal{D}}, [\![\cdot]\!]_{\mathcal{D}})$, where $[\![p]\!]_{\mathcal{D}} = \prod_{\mathcal{D}} [\![p]\!]_i$ for any proposition $p \in \mathbb{P}$.

If $\mathfrak{M}_i = \mathfrak{M}$ for each $i \in I$, then $\prod_{\mathcal{D}} \mathfrak{M}_i$ is called an *ultrapower* of $\mathfrak{M}$. We denote by $\hat{a}$ the class $[(a_i)]$ where $a_i = a$ for each $i \in I$.

Example 3.7 Let $(\mathcal{X}_i : i \in I)$ be a family of topological spaces (X_i, σ_i) where σ_i is a topological base. For each $i \in I$, consider its basic neighbourhood assignment $\mathcal{B}_1$ defined in Example 2.9. Then $\mathcal{B}_{\mathcal{D}}$ is a topological base over $\prod_{\mathcal{D}} X_i$ (see Definition 15 in [12]).

The following proposition can be proved by induction on the complexity of modal formulas. Also, it can be obtained by the Łoś theorem for first-order $\mathcal{L}_2$-formulas ([4, Thm. 4.1.9]) and Proposition 3.4.

Proposition 3.8 *For any family of based models $(\mathfrak{M}_i : i \in I)$ and for any modal formula φ, we have*

$$\prod_{\mathcal{D}} \mathfrak{M}_i, [(a_i)] \models \varphi \ \textit{ iff } \ \{i \in I \mid \mathfrak{M}_i, a_i \models \varphi\} \in \mathcal{D}.$$

Since the ultraproduct of based spaces is a based space, the Łoś Theorem applied to the two-sorted first order language $\mathcal{L}_2$-formulas implies that $\mathcal{L}_2$ has the compactness property over that class of based models.

3.3 d-Saturation

The next key ingredient in a proof for the Goldblatt-Thomason theorem is saturation; essentially, a structure is saturated if any set $\Gamma(x)$ of formulas with one free variable whose finite subsets are satisfied on some point of the structure is uniformly satisfied on a single point. By modifying the notion of $\mathcal{L}_t$-saturation introduced in [12] for topological closure spaces, we provide an appropriate notion for derivative spaces.

Definition 3.9 Let $\mathfrak{M} = (X, \mathcal{B}, [\![\cdot]\!])$ be a based model. A subset $A \subseteq X$ is *point-saturated* if for any $\mathcal{L}_2$-type $\Gamma(x)$, i.e., a set of $\mathcal{L}_2$-formulas with one point free variable x and without any basic subset variable, we have $\Gamma(x)$ is satisfiable in A provided that it is finitely satisfiable in A. A based model $(X, \mathcal{B}, [\![\cdot]\!])$ is *derivative saturated* (or *d-saturated* for short) if it satisfies the following conditions:

(i) X is point-saturated.

(ii) For any $a \in X$, any $O \in \mathcal{B}(a)$ is point-saturated.

(iii) For any $a \in X$ there exists $O_a \in \mathcal{B}(a)$ such that for any formula $\varphi(x)$ which is true in all members of some neighbourhood of a, is true over all members of O_a.

Proposition 3.10 *For any based model* $\mathfrak{M}$, *there is an ultrafilter* $\mathcal{D}$ *such that* $\prod_{\mathcal{D}} \mathfrak{M}$ *is d-saturated.*

Proof. From model theory, we know that for any $\mathcal{L}_2$-model, and thus for any based model, $\mathfrak{M}$ there is an ultrafilter $\mathcal{D}$ such that $\prod_{\mathcal{D}} \mathfrak{M}$ is ω-saturated (see [4], Theorem 6.1.8). Note that Lemma 3.2 and the Łoś Theorem of first-order formulas imply that $\prod_{\mathcal{D}} \mathfrak{M}$ is a based model. We show that $\prod_{\mathcal{D}} \mathfrak{M}$ is also d-saturated.

(i) It immediately follows by ω-saturation of $\mathfrak{M}$.

(ii) Let a be any point of $\prod_{\mathcal{D}} X$, and $O \in \mathcal{B}_{\mathcal{D}}(a)$ be one of its basic neighbourhoods in $\prod_{\mathcal{D}} \mathfrak{M}$. By the definition, $O = \prod_{\mathcal{D}} O_i$ which is point-saturated, since for any set of $\mathcal{L}_2$-type $\Gamma(x)$ of O we can consider the $\mathcal{L}_2$-type $\Gamma'(x) = \{x \varepsilon U\} \cup \Gamma(x)$. Since $\prod_{\mathcal{D}} \mathfrak{M}$ is ω-saturated, there is $b \in \prod_{\mathcal{D}} X$ such that $\prod_{\mathcal{D}} \mathfrak{M} \models \Gamma'(b)$. This means that Γ is satisfiable in O.

(iii) Assume that $a \in \prod_{\mathcal{D}} X$ is given and $\Gamma(x)$ is the set of all $\mathcal{L}_2$-formulas $\varphi(x)$ such that $\varphi(x)$ holds throughout some basic neighbourhood of a. Now consider the following set of $\mathcal{L}_2$-formulas:

$$\Gamma(U) = \{U \,\nu\, a\} \cup \{\forall y (y \,\varepsilon\, U \to \varphi(y)) \mid \varphi \in \Gamma\}.$$

Since $\prod_{\mathcal{D}} \mathfrak{M}$ is an ω-saturated, there is $\prod_{\mathcal{D}} O_i \in \mathcal{B}(a)$ such that $\prod_{\mathcal{D}} \mathfrak{M} \models \Gamma(\prod_{\mathcal{D}} O_i)$.

□

4 Definability

Our main goal in this paper is to study modal definability conditions for classes of derivative spaces. As mentioned above, based spaces provide an equivalent presentation of derivative spaces, and we investigate the definability conditions of based spaces. This will allow us to have more flexibility on the corresponding language, as well as grant us access to standard techniques for first order logic. More precisely, for any given class $\mathcal{C}$ of derivative spaces, we identify $\mathcal{C}$ with the class $\mathcal{C}'$ of based spaces $(X, \mathcal{B})$ such that (X, d_B) belongs to $\mathcal{C}$.

Definition 4.1 Let $\mathcal{C}$ be a class of based spaces. A class $K \subseteq \mathcal{C}$ is *modally definable over* $\mathcal{C}$ if there is a set of modal formulas Σ such that for any based space $\mathcal{X} \in \mathcal{C}$ we have $\mathcal{X} \in K$ iff $\mathcal{X} \models \Sigma$.

A class K is *modally definable* if it is modally definable over the class of all based spaces.

Example 4.2 For a based space $\mathcal{X} = (X, \mathcal{B})$, let $\sigma = \bigcup_{a \in X} \mathcal{B}(a) \cup \{\varnothing\}$. Then $\mathcal{X} \models p \to \Diamond p$ iff σ is a topological base and $d_{\mathcal{B}}$ is the closure operator over the topology generated by σ. Note that σ is a topological base iff for any $a \in X$ and $A \in \mathcal{B}(a)$ we have $a \in A$. In other words, the class of all topological closure spaces, i.e., derivative spaces (X, d) for which $d = \mathsf{c}$, is defined by $p \to \Diamond p$.

As mentioned in Example 2.3, the d-semantics is more expressive than the c-semantics. The following is another example witnessing the more expressive power of d-semantics.

Example 4.3 A topological space (X, τ) is a T_d-space if every $x \in X$ is an intersection of an open and a closed set; equivalently if $d_\tau(A)$ is closed for all $A \subseteq X$. The class of T_d-spaces is not definable in c-semantics (see [12, Cor. 37]). But, it is definable over the class of all topological derivative spaces by $\Box p \to \Box\Box p$ [2]. More generally, this formula defines the class of all derivative spaces with the property that $d(d(A)) \subseteq d(A)$ for each $A \subseteq X$.

Example 4.4 It is easy to see that the Löb formula $\Box(\Box p \to p) \to \Box p$ is valid on derivative space $\mathcal{X} = (X, d)$ iff $d(A) = d(A - d(A))$, for any $A \subseteq X$. Also, for any topological space $\mathcal{X}$ we have $\mathcal{X}$ is scattered iff $d_\tau(A) = d_\tau(A - d_\tau(A))$ for any $A \subseteq X$. Thus, for any derivative space $\mathcal{X} = (X, d)$, if (X, τ_d) is scattered and $d_{\tau_d} = d$, then the Löb formula is valid on $\mathcal{X}$ (see, e.g., [16]).

We can also show the converse of that, i.e., if the Löb formula is valid on the derivative space $\mathcal{X}$, then (X, τ_d) is scattered and $d_{\tau_d} = d$. So the class of all scattered spaces is definable by the Löb formula over the class of topological derivative spaces.

4.1 Invariance Results

Any modally definable class must be invariant under any operation that preserves modal formulas. We identify three constructions over based spaces (resp. derivative spaces) that have this property; our Goldblatt-Thomason theorem will then state that these conditions precisely characterize modal definability.

Definition 4.5 Let $(\mathcal{X}_i : i \in I)$ be a family of disjoint based spaces. The *d-sum* of this family is a space $\biguplus_{i \in I} \mathcal{X}_i = (X, \mathcal{B})$, where $X = \biguplus_{i \in I} X_i$ and $\mathcal{B} : X \to 2^{2^X}$ is defined by $\mathcal{B}(a) = \mathcal{B}_i(a)$, for $a \in X_i$.

Proposition 4.6 *Let $(\mathcal{X}_i : i \in I)$ be a family of disjoint based spaces. Then for any formula φ, we have $\biguplus_{i \in I} \mathcal{X}_i \models \varphi$ iff $\mathcal{X}_i \models \varphi$ for all $i \in I$.*

Example 4.7 The class of finite based spaces is not modally definable, since an infinite sum of finite spaces will validate any formula valid on finite spaces. Similarly, the class of all derivative spaces with a finite derivative operator, i.e., with a derivative operator which assigns to each subset a finite subset, is not modally definable.

Definition 4.8 Let $\mathcal{X}$ be a based space. A *d-open subspace* of $\mathcal{X}$ is a space $\mathcal{O} = (O, \mathcal{B}_O)$, where O is a d-open subset of X (i.e. for any $a \in O$ there is $O' \in \mathcal{B}(a)$ such that $O' \subseteq O$) and $\mathcal{B}_O(a) = \{A \cap O \mid A \in \mathcal{B}(a)\}$, for any $a \in O$.

For a weakly transitive Kripke frame (X, R) with $\mathcal{B}_R(a) = \{R(a)\}$, a d-open subspace is a generated subframe. Also, for a based space $(X, \mathcal{B})$, if $\sigma = \bigcup_{a \in X} \mathcal{B}(a) \cup \{\varnothing\}$ is a topological base, then a d-open subspace is a topologically open subspace.

Proposition 4.9 *Let $\mathcal{O} = (O, \mathcal{B}_O)$ be a d-open subset of a based space $\mathcal{X}$. Then $\mathcal{X} \models \varphi$ implies that $\mathcal{O} \models \varphi$, for any formula φ.*

Proof. Assume that $\mathfrak{M}_O$ is an arbitrary model based on $\mathcal{O}$. Let $\mathfrak{M}$ be a model based on $\mathcal{X}$ with $[\![p]\!]_{\mathfrak{M}} \cap O = [\![p]\!]_{\mathfrak{M}_O}$, for any proposition p. By induction on the complexity of modal formulas, one can show that for any formula φ and any $a \in O$, we have

$$\mathfrak{M}, a \models \varphi \text{ iff } \mathfrak{M}_O, a \models \varphi.$$

The base case and the ones for Boolean connectives are clear. Let $\varphi = \Diamond\psi$. First assume that $\mathfrak{M}_O, a \models \Diamond\psi$. For any $U \in \mathcal{B}(a)$, we have $U \cap O \in \mathcal{B}_O(a)$. Thus, $U \cap O \cap [\![\psi]\!]_{\mathfrak{M}_O} \neq \varnothing$. Then induction hypothesis implies that $U \cap [\![\psi]\!]_{\mathfrak{M}_O} \neq \varnothing$. Now assume that $\mathfrak{M}, a \models \Diamond\psi$. Let $U \in \mathcal{B}_O(a)$, then there is $U' \in \mathcal{B}(a)$ such that $U = U' \cap O$. Since $a \in O$, there is $U_a \in \mathcal{B}(a)$ such that $U_a \subseteq O$. By the assumption we have $U' \cap U_a \in \mathcal{B}(a)$ and $U' \cap U_a \cap [\![\psi]\!]_{\mathfrak{M}} \neq \varnothing$. Then $U \cap U_a \in \mathcal{B}_O(a)$ and $U \cap U_a \cap [\![\psi]\!]_{\mathfrak{M}_O} \neq \varnothing$. □

Example 4.10 The class of all based spaces with some point $a \in X$ such that $\varnothing \in \mathcal{B}(a)$ is not modally definable. To see this consider a based space $\mathcal{X} = (\{w_0, w_1, w_2\}, \mathcal{B}_R)$ with $R = \{(w_1, w_2), (w_2, w_2)\}$ and its d-open subspace $\mathcal{Y}$ over $\{w_1, w_2\}$.

Definition 4.11 Let $(X, \mathcal{B})$ and $(X', \mathcal{B}')$ be two based spaces. The function $f : X \to X'$ is a *d-morphism* if

(i) for any $A \in \mathcal{B}(a)$, there exists $A' \in \mathcal{B}'(f(a))$ such that for any $b' \in A'$ there is $b \in A$ with $f(b) = b'$,

(ii) for any $A' \in \mathcal{B}'(f(a))$, there exists $A \in \mathcal{B}(a)$ such that for any $b \in A$ there is $b' \in A'$ with $f(b) = b'$.

If f is a surjective d-morphism, then we say that $\mathcal{X}'$ is a *d-morphic image* of $\mathcal{X}$.

It can easily be checked that a function $f : X \to X'$ is a d-morphism between two derivative spaces (X, d) and (X', d') if $f^{-1}(d'(A')) = d(f^{-1}(A'))$, for any $A' \subseteq X'$.

As usual, validity of formulas is preserved under d-morphic images.

Proposition 4.12 *Assume that $\mathcal{X}$ and $\mathcal{X}'$ are two based (or derivative) spaces. If $\mathcal{X}'$ is a d-morphic image of $\mathcal{X}$, then $\mathcal{X} \models \varphi$ implies $\mathcal{X}' \models \varphi$.*

We may also use d-morphisms to identify open subspaces.

Proposition 4.13 *$\mathcal{O} = (O, d_O)$ is a d-open subspace of $\mathcal{X} = (X, d)$, iff $O \subseteq X$ and the inclusion function $i : O \to X$ is a d-morphism.*

Example 4.14 The classes of T_1 and T_2-spaces are not definable over the class of all topological derivative spaces. To see this, let $\mathcal{X}$ be the topological derivative space over the ordinal ω^2 equipped with the interval topology. Then, for any $A \subseteq \omega^2$, $d(A)$ is the set of all limit ordinals in A, i.e, $\omega.k = \omega + \cdots + \omega$, k-times, for $0 < k < \omega$. Thus, $\mathcal{X}$ is both T_1 and T_2-space. Now, consider the function f from $\mathcal{X}$ to the Sierpiński space, i.e., the space $(\{0,1\}, \{\varnothing, \{1\}, \{0,1\}\})$, defined as $f(\omega.k) = 0$ for each $0 < k < \omega$ and $f(x) = 1$ for others. Let d' be the Cantor derivative operator over the Sierpiński space. Then, f is a d-morphism since $f^{-1}(d'(A')) = d(f^{-1}(A'))$, for any $A' \subseteq \{0,1\}$. So, since the Sierpiński space is not T_1 neither T_2 the result is obtained by Proposition 4.12.

In the Goldblatt-Thomason theorem for Kripke semantics, there is another important construction, namely, the ultrafilter extension, which reflects the validity of modal formulas. In [12] a similar construction is introduced for c-semantics of topological spaces, named Alexandroff extension. Now, we define an analogous notion for derivative spaces.

Let $\mathcal{X} = (X, \mathcal{B})$ be a based space and X^* be the set of all ultrafilters over X. Recall that for each $A \subseteq X$, we have $d_{\mathcal{B}}(A) = \{a \mid \forall B \in \mathcal{B}(a)\ B \cap A \neq \varnothing\}$. Define a binary relation R^* over X^* as follows:

$$\mathfrak{u} R^* \mathfrak{u}' \text{ iff } A \in \mathfrak{u}' \text{ implies } d_{\mathcal{B}}(A) \in \mathfrak{u} \text{ for any } A \subseteq X.$$

Lemma 4.15 *R^* is a weakly transitive relation over X^*.*

Proof. Suppose $\mathfrak{u}, \mathfrak{u}', \mathfrak{u}'' \in X^*$ and $\mathfrak{u} R^* \mathfrak{u}'$ and $\mathfrak{u}' R^* \mathfrak{u}''$. We show that if $\mathfrak{u} \neq \mathfrak{u}''$, then $\mathfrak{u} R^* \mathfrak{u}''$. So, we have to show that $d_B(B) \in \mathfrak{u}$ whenever $B \in \mathfrak{u}''$ for any $B \subseteq X$. Take an arbitrary $B \in \mathfrak{u}''$. Since $\mathfrak{u} \neq \mathfrak{u}''$, there is $A \subseteq X$ such that $A \in \mathfrak{u}''$ and $A^c \in \mathfrak{u}$. Then $A \cap B \in \mathfrak{u}''$, and thus $d_B d_B(A \cap B) \in \mathfrak{u}$. This implies that $d_B(A \cap B) \cup (A \cap B) \in \mathfrak{u}$. Hence, $(d_B(A \cap B) \cap A^c) \cup (A \cap B \cap A^c) \in \mathfrak{u}$. So, $d_B(A \cap B) \cap A^c \in \mathfrak{u}$. Therefore, $d_B(A \cap B) \subseteq d_B(B) \in \mathfrak{u}$. □

Note that, in general, (X^*, R^*) is not irreflexive. To see this, consider the topological derivative space of $(\mathbb{N}, \tau)$ where $\tau = \{\varnothing, \text{ all co-finite sets}\}$. Then for any $A \subseteq \mathbb{N}$, we have

$$d(A) = \begin{cases} \varnothing & \text{if } A \text{ is finite} \\ \mathbb{N} & \text{if } A \text{ is infinite} \end{cases}$$

Thus, any non-principal ultrafilter in $\mathbb{N}^*$ is reflexive since each of its members is infinite. Indeed, any non-principal ultrafilter is an R^*-successor of all members of $\mathbb{N}^*$.

Definition 4.16 Let $\mathcal{X} = (X, \mathcal{B})$ be a based space. The *d-Alexandroff extension* of $\mathcal{X}$ is the space $\mathcal{X}^* = (X^*, \mathcal{B}^*)$, where $\mathcal{B}^*(\mathfrak{u}) = \{R^*(\mathfrak{u})\}$, for each $\mathfrak{u} \in X^*$.

In other words, for any derivative space $\mathcal{X} = (X, d)$, the *d-Alexandroff extension* of $\mathcal{X}$ is the space $\mathcal{X}^* = (X^*, d^*)$ where

$$d^*(A) = \{\mathfrak{u} \mid \exists \mathfrak{u}' \in A \ s.t \ \mathfrak{u}R^*\mathfrak{u}'\},$$

for any $A \subseteq X^*$.

Example 4.17 For any weakly transitive Kripke frame (W, R), its d-Alexandroff extension is equal to its ultrafilter extension (W^*, R^{ue}).

Note that the d-Alexandroff extension of a derivative space (X, d) is the derivative space corresponding to the ultrafilter frame of the wK4-algebra $(\mathcal{P}(X), d_\tau)$ (see [3]).

Example 4.18 For any topological closure space (X, τ), its d-Alexandroff extension is equal to its Alexandroff extension defined in [12].

But for any topological derivative space, its d-Alexandroff extension is not necessarily a topological derivative space. In other words, the class of all topological derivative spaces is not closed under d-Alexandroff extensions.

Proposition 4.19 *For any based space $\mathcal{X}$ we have, $\mathcal{X}^* \models \varphi$ implies $\mathcal{X} \models \varphi$.*

Proof. If $\mathcal{X} \not\models \varphi$. Then there is a model $\mathfrak{M} = (\mathcal{X}, [\![\cdot]\!])$ and $a \in X$ such that $\mathfrak{M}, a \not\models \varphi$. Let $\mathfrak{M}^* = (\mathcal{X}^*, [\![\cdot]\!]^*)$, where $[\![p]\!]^* = \{\mathfrak{u} \in X^* \mid [\![p]\!] \in \mathfrak{u}\}$, for each $p \in \mathbb{P}$.

Claim. $\mathfrak{M}^*, \mathfrak{u} \models \theta$ iff $[\![\theta]\!] \in \mathfrak{u}$, for each modal formula θ.

This claim implies that $\mathfrak{M}^*, \mathfrak{u}_a \not\models \varphi$, where $\mathfrak{u}_a$ is the principal ultrafilter generated by a, and thus $\mathcal{X}^* \not\models \varphi$.

Proof of the Claim: By induction on the complexity of formula θ. By definition of $[\![\cdot]\!]^*$ and by the ultrafilter properties it is easy to see that the claim holds for atomic formulas and Boolean connectives. Now let $\theta = \Diamond\psi$. If $\mathfrak{u} \models \Diamond\psi$, then $[\![\psi]\!]^* \cap B^* \neq \varnothing$ for all $B^* \in \mathcal{B}^*(\mathfrak{u})$. Specifically, $R^*(\mathfrak{u}) \cap [\![\psi]\!]^* \neq \varnothing$. So, there is $\mathfrak{u}'$ such that $\mathfrak{u}R^*\mathfrak{u}'$ and $\mathfrak{u}' \in [\![\psi]\!]^*$. By induction hypothesis we have $[\![\psi]\!] \in \mathfrak{u}'$, which implies that $[\![\Diamond\psi]\!] \in \mathfrak{u}$. For the other direction, suppose that $[\![\Diamond\psi]\!] \in \mathfrak{u}$. Let $\mathfrak{u}_0 = \{A \subseteq X \mid d_{\mathcal{B}}(A) \in \mathfrak{u}\}$. Then $\mathfrak{u}_0 \cup \{[\![\psi]\!]\}$ has the finite intersection property, and thus it can be extended to an ultrafilter $\mathfrak{u}'$. Then $[\![\psi]\!] \in \mathfrak{u}'$ and $\mathfrak{u}R^*\mathfrak{u}'$. □

Recall that a topological space (X, τ) is Alexandroff if every point has a minimal open neighbourhood. The Alexandroff extension of any topological closure space, is an Alexandroff space, and this implies that the class of Alexandroff spaces is not definable in the c-semantics (see Corollary 41 in [12]). We can extend this notion to derivative spaces. A derivative space (X, d) is called *Alexandroff* if every point has a minimal basic d-neighbourhood.

Example 4.20 The class of all Alexandroff derivative spaces is not modally definable. For any arbitrary derivative space $\mathcal{X}$, even for non-Alexandroff one, $\mathcal{X}^*$ is Alexandroff, since $R^*(\mathfrak{u})$ is the only element of $\mathcal{B}^*(\mathfrak{u})$, for all $\mathfrak{u} \in X^*$.

5 The Goldblatt-Thomason Theorem

We are now ready to prove our main result. We begin with the following useful fact.

Proposition 5.1 *For any based space $\mathcal{X}$, there exists a based ultrapower $\prod_{\mathcal{D}} \mathcal{X}$ with a surjective d-morphism $f : \prod_{\mathcal{D}} X \to X^*$.*

Proof. For any $A \subseteq X$, add a new unary predicate P_A to the language $\mathcal{L}_2$ and let $\mathfrak{M}$ be an $\mathcal{L}_2$-model of $\mathcal{X}$ with the natural interpretation of new predicates, i.e., P_A is interpreted as A. Let $\mathfrak{M}_{\mathcal{D}}$ be the d-saturated ultrapower of $\mathfrak{M}$ as in Proposition 3.10.

Let T be the set of $\mathcal{L}_2$-sentences of the following forms:

(i) $\exists x\ P_A(x)$, for any non-empty $A \subseteq X$.

(ii) $\forall x(P_{A\cap B}(x) \leftrightarrow P_A(x) \wedge P_B(x))$.

(iii) $\forall x(\neg P_A(x) \leftrightarrow P_{A^c}(x))$.

(iv) $\forall x(P_{d(A)}(x) \leftrightarrow \forall U(U\ \nu\ x \to \exists y(y \in U \wedge P_A(y)))$.

(v) $\forall x(P_{\tilde{d}(A)}(x) \leftrightarrow \exists U(U\ \nu\ x \wedge \forall y(y \in U \to P_A(y))))$

Then we have $\mathfrak{M}_{\mathcal{D}} \models T$, since $\mathfrak{M} \models T$.

Now define a function $f : \prod_{\mathcal{D}} X \to X^*$ as follows:

$$f(a) = \{A \subseteq X \mid a \in (P_A)^{\mathfrak{M}_{\mathcal{D}}}\}.$$

We have to show that f is a d-morphism.

- f is well-defined, since (ii) and (iii) imply that $f(a)$ is an ultrafilter.
- To see that f is surjective, for any given $\mathfrak{u} \in X^*$, let $\Gamma_{\mathfrak{u}}(x) = \{P_A(x) \mid A \in \mathfrak{u}\}$. By (ii), $\Gamma_{\mathfrak{u}}$ is finitely satisfiable in $\mathfrak{M}_{\mathcal{D}}$. Since $\mathfrak{M}_{\mathcal{D}}$ is d-saturated, there is a point $a \in \prod_{\mathcal{D}} X$ such that $\mathfrak{M}_{\mathcal{D}} \models \Gamma_{\mathfrak{u}}(a)$.
- To show f is a d-morphism, first assume that $O \in \mathcal{B}_{\mathcal{D}}(a)$. We have to show there exists $O^* \in \mathcal{B}^*(f(a))$ such that for any $\mathfrak{u} \in O^*$ there is $b \in O$ with $f(b) = \mathfrak{u}$. Let $O^* = R^*(f(a))$. For a given $\mathfrak{u} \in O^*$, we have $f(a)R^*\mathfrak{u}$. Let $\Gamma(x) = \{P_A(x) \mid d(A) \in \mathfrak{u}\}$. Then, by (iv), $\Gamma(x)$ is finitely satisfiable in any $B \in \mathcal{B}_{\mathcal{D}}(a)$. Specifically, Γ is finitely satisfiable in O. Now d-saturation of $\mathfrak{M}_{\mathcal{D}}$ implies that O is point saturated, and thus there is $b \in O$ such that $\Gamma(b)$ is true and $f(b) = \mathfrak{u}$.

  For the other direction, assume that $O^* \in \mathcal{B}^*(f(a))$, we have to show that there is $O \in \mathcal{B}_{\mathcal{D}}(a)$ such that $f(O) \subseteq O^*$. Let O_a be the d-neighbourhood of a as defined in part 3 of the definition of d-saturation (cf. Definition 3.9). Let $\Gamma = \{P_A(x) \mid \tilde{d}(A) \in f(a)\}$. Then, by (v), for any finite subset $\Gamma' \subseteq \Gamma$, we have $\bigwedge \Gamma'$ is true throughout some $O \in \mathcal{B}_{\mathcal{D}}(a)$. So, Γ is true throughout O_a and $O_a \cap f^{-1}(A^*) \neq \varnothing$. Thus, there is $b \in O_a$ such that $f(b) \in A^*$ and $f(a)R^*f(b)$.

□

With the above proposition, we are ready to prove our main result.

Theorem 5.2 *Let K be an $\mathcal{L}_2$-elementary class of based spaces. Then, K is modally definable iff K is closed under d-sums, d-open subspaces, d-morphisms and reflects d-Alexandroff extensions.*

Proof. The left-to-right direction follows from the results in the previous section. For the other direction, let $Log(K) = \{\varphi \mid K \models \varphi\}$. We show that for any based space $\mathcal{X}$ we have $\mathcal{X} \in K$ iff $\mathcal{X} \models Log(K)$. Clearly, $Log(K)$ is valid on any space in K. Now suppose that $\mathcal{X} \models Log(K)$. Consider a language containing a propositional variable p_A for any $A \subseteq X$, and let $\mathfrak{M} = (\mathcal{X}, [\![\cdot]\!])$ where $[\![\cdot]\!]$ is a natural valuation in this language, i.e., $[\![p_A]\!] = A$ for any $A \subseteq X$. Now take Δ to be the theory containing all the formulas in the following form for any $A, B \subseteq X$:

$$\begin{aligned} p_{A^c} &\leftrightarrow \neg p_A \\ p_{A\cap B} &\leftrightarrow p_A \wedge p_B \\ p_{d(A)} &\leftrightarrow \Diamond p_A \\ p_{\tilde{d}(A)} &\leftrightarrow \Box p_A \end{aligned}$$

Then $\mathfrak{M} \models \Delta$.

Claim. For any $a \in X$ there exists a model $\mathfrak{N}_a$ based on some $\mathcal{Y}_a \in K$ and $b \in Y_a$ such that $\mathfrak{N}_a \models \Delta$ and $\mathfrak{N}_a, b \models p_{\{a\}}$.

Proof of the claim: Suppose that $a \in X$. Let

$$\Delta_a = \{\Box\varphi \wedge \varphi \mid \varphi \in \Delta\} \cup \{p_a\}.$$

Then Δ_a is finitely satisfiable in K, since otherwise $\neg\delta \in Log(K)$ for some finite subset δ of Δ_a, which is a contradiction with $\mathfrak{M} \models Log(K)$.

Since K is an elementary class, it is closed under ultraproducts, so we can assume that there is a d-saturated model $\mathfrak{N}$ based on some $\mathcal{Y} \in K$ and $b \in Y$ such that $\mathfrak{N}, b \models \Delta_a$. By d-saturation, b has a d-neighbourhood $U_b \in \mathcal{B}(b)$ such that φ holds throughout U_b for all $\varphi \in \Delta$. Let $\mathfrak{N}_a$ be a d-open subspace of $\mathfrak{N}$ generated by $O = U_b \cup \{b\}$, and this completes the proof of the claim.

Let $\mathfrak{N}$ be the d-saturated ultrapower of $\biguplus_{a\in X} \mathfrak{N}_a$ (see Proposition 3.10). By the closure of K under d-sums and d-ultraproducts, $\mathcal{Y} = \prod_{\mathcal{D}} \uplus_{a\in X} \mathcal{Y}_a \in K$.

Now, by Proposition 5.1, we know that there is a d-morphism from $\mathcal{Y}$ to $\mathcal{X}^*$. The closure of K under d-morphisms implies that $\mathcal{X}^* \in K$. So $\mathcal{X} \in K$ since K reflects d-Alexandroff extensions. □

As mentioned above, for any derivative space (X, d), the frame (X^*, R^*) is not necessarily irreflexive. Specially, for any topological derivative space, (X^*, R^*) is not necessary irreflexive, so (X^*, d^*) is not a topological derivative space, i.e., d^* is not necessarily equal to $d_{\tau_{d^*}}$. Thus, for giving a version of the Goldblatt-Thomason theorem for such classes of structures, i.e., for those that are not closed under the d-Alexandroff extension, we consider the following definition.

Definition 5.3 Let K be a class of derivative spaces. We say that K *reflects d-Alexandroff images* whenever for some derivative space $\mathcal{Y} \in K$, if $\mathcal{X}^*$ is a d-morphic image of $\mathcal{Y}$, then $\mathcal{X}$ is in K.

Let $\mathcal{C}$ be a class of derivative spaces that is closed under d-ultraproducts, d-sums and d-open subspaces, but it is not necessarily closed under d-morphisms and d-Alexandroff extensions. We say that a subclass K of $\mathcal{C}$ is closed under d-morphisms over $\mathcal{C}$ whenever if $\mathcal{X} \in K$ and $\mathcal{Y} \in \mathcal{C}$ and there is a d-morphism from $\mathcal{X}$ to $\mathcal{Y}$, then $\mathcal{Y} \in K$.

Theorem 5.4 *Let $\mathcal{C}$ be a class of derivative spaces with the above closure properties and $K \subseteq \mathcal{C}$ be an $\mathcal{L}_2$-elementary class of derivative spaces. Then K is modally definable over $\mathcal{C}$ iff K is closed under d-sums, d-open subspaces, d-morphisms over $\mathcal{C}$ and reflects d-Alexandroff images.*

Note that in Theorem 5.4 and its proof we do not need $\mathcal{X}^*$ be in K or even in $\mathcal{C}$.

Specifically for the class of topological derivative spaces, we can also express the above theorem in another way based on the following definition.

Definition 5.5 Suppose that K is a class of topological spaces. Let K^+ be the class of all Kripke frames that are d-morphic images of some elements of K. We say that a class K of topological spaces *reflects the weak transitive extensions* from K^+, whenever for any topological space (X, τ) if (X^*, R^*) is in K^+, then (X, τ) is in K.

Corollary 5.6 *Let K be an $\mathcal{L}_2$-elementary class of topological spaces. K is modally definable iff it is closed under disjoint unions, open subspaces, and d-morphic images and reflects weak transitive extensions from K^+.*[5]

6 Concluding Remarks

In this paper we give a version of the Goldblatt-Thomason theorem for derivative spaces. There are some lines of research that can be considered for future work.

One of the natural further directions is to investigate definability for extended languages which provide the ability to define more properties. For example one can show that the class of all topological derivative spaces is definable by the hybrid formula $@_i \neg \Diamond i$ while it is not definable in the basic language. Utilising the method used for topological c-semantics in [12], one can extend our results to extended languages such as modal logic with universal modality and hybrid logics.

First-order modal logic (FML) is another extension of modal logic. There is some work on definability in the context of FML, e.g. [17,18]. For example, [18] gives a version of the Goldblatt-Thomason FML for Kripke frames. It uses a well-known technique from classical model theory called Morleyization (or atomization). Providing a version of the Goldblatt-Thomason theorem for FML with respect to topological semantics is an interesting future work.

The other challenging extension of modal logic is the modal μ-calculus (μML)—modal logic enriched with fixed point operators. There are many ob-

---

5 Note that over the class of topological derivative spaces, any $\mathcal{L}_2$-formula can be rephrased as an $\mathcal{L}_t$-formula. Thus we can assume that K is $\mathcal{L}_t$-elementary (see footnote 4).

stacles for this logic, for example, μML does not enjoy the compactness property, which plays a significant role in proving the Goldblatt-Thomason theorem. To tackle this problem, one might need to consider simpler extensions such as modal languages with a tangled operator.

This also suggests a line of study of general definability characterization theorems, for classes of structures that are not necessarily elementary. There are many important examples of classes of Kripke frames or topological spaces that are definable, but not elementary. For example, the class of scattered spaces (Example 4.4) is definable by Löb's formula, but is not elementary.

References

[1] Baltag, A., N. Bezhanishvili and D. Fernández-Duque, *The topological mu-calculus: completeness and decidability*, in: *2021 36th Annual ACM/IEEE Symposium on Logic in Computer Science (LICS)*, IEEE, 2021, pp. 1–13.

[2] Bezhanishvili, G., L. Esakia and D. Gabelaia, *Spectral and T_0-spaces in d-semantics*, in: *International Tbilisi Symposium on Logic, Language, and Computation*, Springer, 2009, pp. 16–29.

[3] Blackburn, P., M. De Rijke and Y. Venema, "Modal Logic,", Cambridge University Press, 2001.

[4] Chang, C. and H. Keisler, "Model Theory," 3rd edn, North Holland, Amsterdam, 1989.

[5] Esakia, L., *Intuitionistic logic and modality via topology*, Annals of Pure and Applied Logic **127** (2004), pp. 155–170, provinces of logic determined.

[6] Fernández-Duque, D. and P. Iliev, *Succinctness in subsystems of the spatial μ-calculus*, FLAP **5** (2018), pp. 827–874.

[7] Flum, J. and M. Ziegler, "Topological Model Theory," Springer Berlin Heidelberg, 1980.

[8] Gabelaia, D., "Modal definability in topology," Master's Thesis, University of Amsterdam (2001).

[9] Goldblatt, R. I. and S. K. Thomason, *Axiomatic classes in propositional modal logic*, in: J. N. Crossley, editor, *Algebra and Logic* (1975), pp. 163–173.

[10] Kurz, A. and J. Rosický, *The Goldblatt-Thomason theorem for coalgebras*, in: *Algebra and Coalgebra in Computer Science: Second International Conference, CALCO 2007, Bergen, Norway, August 20-24, 2007. Proceedings 2*, Springer, 2007, pp. 342–355.

[11] McKinsey, J. C. C. and A. Tarski, *The algebra of topology*, Annals of mathematics (1944), pp. 141–191.

[12] ten Cate, B., D. Gabelaia and D. Sustretov, *Modal languages for topology: Expressivity and definability*, Annals of Pure and Applied Logic **159** (2009), pp. 146–170.

[13] ten Cate, B. D., "Model theory for extended modal languages," Ph.D. thesis, University of Amsterdam (2004).

[14] van Benthem, J., *Notes on modal definability.*, Notre Dame Journal of Formal Logic **30** (1988), pp. 20–35.

[15] van Benthem, J., *Modal frame classes revisited*, Fundamenta Informaticae **18** (1993), pp. 307–317.

[16] van Benthem, J. and G. Bezhanishvili, *Modal logics of space*, in: *Handbook of spatial logics*, Springer, 2007 pp. 217–298.

[17] van Benthem, J. F., *Frame correspondences in modal predicate logic*, Proofs, categories and computations: Essays in honor of Grigori Mints (2010), pp. 1–14.

[18] Zoghifard, R. and M. Pourmahdian, *First-order modal logic: frame definability and a Lindström theorem*, Studia Logica **106** (2018), pp. 699–720.

Bisimulation for Impure Simplicial Complexes

Marta Bílková [1] and Hans van Ditmarsch

Czech Acad Sci, Inst Comp Sci *Un. of Toulouse, CNRS, IRIT, France*

Roman Kuznets [2] and Rojo Randrianomentsoa [2]

TU Wien, Austria

Abstract

As an alternative to Kripke models, simplicial complexes are a versatile semantic primitive on which to interpret epistemic logic. Given a set of vertices, a simplicial complex is a downward closed set of subsets, called simplexes, of the vertex set. A maximal simplex is called a facet. Impure simplicial complexes represent that some agents (processes) are dead. It is known that impure simplicial complexes categorically correspond to so-called partial epistemic (Kripke) models. In this contribution, we define a notion of bisimulation to compare impure simplicial complexes and show that it has the Hennessy–Milner property. These results are for a logical language including atoms that express whether agents are alive or dead. Without these atoms no reasonable standard notion of bisimulation exists, as we amply justify by counterexamples, because such a restricted language is insufficiently expressive.

Keywords: Epistemic logic, simplicial complexes, crashing agents, bisimulation.

1 Introduction

Impure simplicial complexes provide semantics for multiagent epistemic logic for distributed systems with crash failures. In this paper, we consider a three-valued semantics with the third value 'undefined' used, e.g., for propositional atoms and knowledge statements pertaining to crashed agents. We show that the Hennessy–Milner property fails for the standard notion of bisimulation and standard epistemic language. To ameliorate the situation, we extend the language with global atoms representing whether a given agent is alive or crashed and prove the Hennessy–Milner property for this extended language. The standard proof of this property, which relies on the symmetry between the two boolean truth values, does not work for our three-valued logic. To adapt it,

[1] Work of Marta Bílková on this paper was supported by the grant no. 22-23022L CELIA of the Czech Science Foundation.

[2] This research was funded in whole or in part by the Austrian Science Fund (FWF) project ByzDEL [10.55776/P33600].

we employ a tailor-made notion of a life tree, which encodes definability and enables us to define a localized translation from undefined to true formulas.

Survey of the literature Simplicial complexes are well-known in combinatorial topology. There have been promising and exciting recent connections between combinatorial topology and epistemic logic [6–8,11,12,14,15,18,19,21].

Combinatorial topology has been used in distributed computing to model concurrency and asynchrony since [3, 17, 29]. Higher-dimensional topological properties [24,25] allow for an epistemic representation. The basic structure in combinatorial topology is the *simplicial complex*, a downward closed collection of subsets called *simplexes* of a set of *vertices*. Geometric manipulations such as subdivision have natural combinatorial counterparts.

Epistemic logic investigates knowledge and belief, and change thereof, in multiagent systems. A foundational study is [26]. Knowledge change was extensively modeled in temporal epistemic logics [16, 23, 33] and in dynamic epistemic logics [1, 13].

An epistemic logic interpreted on *pure* simplicial complexes was proposed in [11]. It shows a categorical correspondence between Kripke models and simplicial complexes, and based on that, the resulting logic is multiagent S5 augmented with the locality axiom $K_a p_a \vee K_a \neg p_a$ stating that all agents know their local state. Action models [1] are used to model distributed computing tasks.

In *pure complexes* and their temporal developments, all processes remain active (are alive). They describe *asynchronous* message passing. In impure complexes some processes may have crashed (are dead). They can be used to describe *synchronous* message passing (with timeouts) [24]. *Impure complexes* correspond to Kripke models with partial equivalence relations (symmetric and transitive relations). Epistemic logics interpreted on impure simplicial complexes were proposed in [14, 21]. In [21], a two-valued semantics is proposed and the authors axiomatize the logic as multiagent KB4 where, if process a is dead, then $K_a\bot$ is true. In [14], a three-valued modal logical semantics is proposed for the same language where the third value stands for 'undefined', e.g., dead processes cannot know or be ignorant of any proposition, nor can live processes know or be ignorant of factual propositions involving processes they know to be dead. This logic was axiomatized in [34] in a version of multiagent S5 called $\mathsf{S5}^{\bowtie}$. Its notion of knowledge relates to "belief as defeasible knowledge" of [31]. For example, $K_a\varphi \rightarrow \varphi$ is valid in the sense that, if $K_a\varphi$ is true, then φ is not false (but may be undefined). The three- and two-valued approaches are compared in [15].

Subsequent developments of simplicial epistemics include generalizations from individual knowledge to distributed knowledge [19] and from simplicial complexes to (semi-)simplicial sets [6, 18, 19]. Dynamics of complexes were investigated in [8].

The propositional base of our three-valued semantics is known as Paraconsistent Weak Kleene logic (PWK) [5, 9, 22, 27, 35]. Bisimulations and limits of expressivity (Hennessy–Milner property) for many-valued modal logics based on algebra-valued Kripke frames, or more generally coalgebras, were considered in [2, 30], restricting to algebras in question being residuated lattices or even finite MTL chains (of which the PWK three-valued matrix is neither).

Motivating example Consider three agents a, b, and c with *local atoms* p_a, p_b, and p_c respectively describing their local state. The value for a is 1 when p_a is true or 0 when false, and similarly for b and c. In combinatorial topology, such information can be represented in a simplicial complex. In simplicial model $\mathcal{C}'$ of Fig. 1, every agent knows (i.e., there is mutual knowledge) that the values of a and c are 1 and that of b is 0. In simplicial model $\mathcal{C}$, on the other hand, agent a is uncertain whether agent c is still alive. This uncertainty is represented by edge X and triangle Y intersecting in the vertex (labeled with) 1_a.

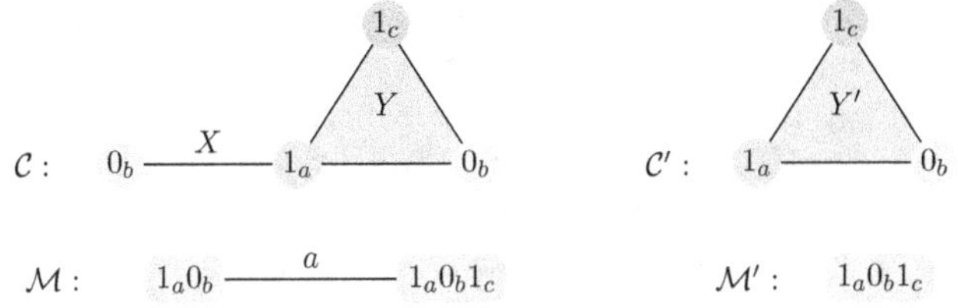

Fig. 1. Simplicial models and corresponding partial epistemic models

Model $\mathcal{C}$ encodes that a is uncertain whether agent c is alive but knows, e.g., that the value of b is 0 as this is the case in both edge X and triangle Y, which intersect in the a-vertex. Model $\mathcal{C}'$ encodes that a knows that agent c is alive, and also that a knows that the value of b is 0. In our logical semantics, one cannot evaluate formulas referring to the knowledge or local atoms of agents that are dead. These are then undefined. The semantics is, therefore, three-valued: a formula can be true, false, or undefined. For example, in edge X of $\mathcal{C}$, formulas p_c (the atom) and $K_c p_a$ are undefined. However, also in X, a formula like $\widehat{K}_a p_c$ is defined (and true): a considers it possible (namely, in the indistinguishable triangle Y) that the value of c is 1 (that p_c is true).

A so-called *global atom* like a, b, or c represents that agents a, b, and c respectively are alive. It is elementary to distinguish structures $(\mathcal{C}, Y)$ and $(\mathcal{C}', Y')$ this way: in $(\mathcal{C}, Y)$ it is true that a considers it possible that c is dead, formalized as $\widehat{K}_a \neg c$, whereas in $(\mathcal{C}', Y')$ this is false: agent a there knows that c is alive, $K_a c$. A corresponding notion of bisimilarity between structures should, therefore, also distinguish them. In this contribution, we will define such a notion and show its elementary properties.

In the language without such global atoms, we cannot say that a considers it possible that c is dead. For instance, a naive attempt to use $K_c\top$ to represent global atom c fails for two separate reasons. Firstly, $\neg K_c\top$ is undefined rather than true when c is actually dead; secondly, the boolean constant $\top$, which is true in all structures, is itself not expressible without global atoms. Indeed, for every formula without global atoms, there is a singleton structure where this formula is undefined [34].

More generally, there is no way to distinguish $(\mathcal{C}, Y)$ from $(\mathcal{C}', Y')$ without expanding the language. We show (Prop. 3.3) that in the logical language with atoms p_a, p_b, and p_c and modalities K_a, K_b, and K_c only, $(\mathcal{C}, Y)$ and $(\mathcal{C}', Y')$ make the same formulas true, false, and undefined: they are *modally equivalent*. Therefore, $(\mathcal{C}, Y) \leftrightarrows^* (\mathcal{C}', Y')$ should hold for any reasonable notion $\leftrightarrows^*$ of *bisimilarity*. But now we have a problem: assuming such a notion, as X and Y intersect in a, when making a bisimilar **forth** step from Y to X, we should be able to emulate that in $\mathcal{C}'$ by moving from Y' somewhere. But no modally equivalent match exists, making it impossible to define $\leftrightarrows^*$.

All these issues similarly exist in the corresponding local epistemic (Kripke) models $\mathcal{M}$ and $\mathcal{M}'$ pictured below simplicial models $\mathcal{C}$ and $\mathcal{C}'$ in Fig. 1. In place of equivalence relations of epistemic models, *partial epistemic models* have partial equivalence relations (symmetric and transitive relations). Model $\mathcal{C}'$ corresponds to the singleton model $\mathcal{M}'$, with reflexive access for a, b, and c and making p_a and p_b false and p_c true there, whereas model $\mathcal{C}$ corresponds to a two-world Kripke model $\mathcal{M}$ where only a cannot distinguish the two worlds and wherein the relation for c is restricted to the world where it is alive. In fact, $\mathcal{M}$ is an equivalence class of two Kripke models, one where p_c is assigned value true in the left world and another one where it is assigned false. Since c is dead in that world, however, the assigned value of p_c there is moot: p_c is considered undefined, allowing us to conflate the two models.

Results in this contribution We distinguish two epistemic languages to describe impure simplicial complexes: one without and one with global atoms. These languages are interpreted on simplicial models, where we also extend a known correspondence between those and partial epistemic models to the language with global atoms.

For the language without global atoms, we prove that even simple structures like $(\mathcal{C}, Y)$ and $(\mathcal{C}', Y')$ in Fig. 1 may have the same information content (be modally equivalent), but lack a corresponding notion of bisimulation where **back** and **forth** steps can be locally checked.[1] One can only employ "brute force" here, taking properties of the whole model into account. Hence, the language without global atoms is insufficiently expressive.

For the language with global atoms, we define bisimulation on simplicial models. By way of an intermediate structure called a *life tree*, which provides an efficient way to check definability, we then establish the Hennessy–Milner characterization. In our three-valued semantics, with values true, false, and undefined, it is non-standard to show that modal equivalence implies bisimilarity on finitary structures. In the usual method for the two-valued semantics, one can, w.l.o.g., assume that a formula distinguishing a point in one structure from a (finite) set of points in the other structure is true in the former and not true (false) in the latter. But in the three-valued semantics, instead of being true, it can be undefined. In that case, life trees and the local geometry of the former

[1] Note that our analysis here and in Example 3.5 pertains only to the standard notion of bisimulation. The existence of weaker notions, e.g., of ρ-bisimulations [10], which do not depend on expressivity, remains open.

structure are used to construct another distinguishing formula that is true.

Finally, we translate our results to a three-valued Kripke semantics interpreted on partial epistemic models. We define a *life bisimulation* between partial epistemic models, which can be easily seen to be weaker than standard bisimulation. The Hennessy–Milner property follows from that for complexes via a categorical equivalence.

Overview Section 2 defines the two logical languages and simplicial models and provides the simplicial semantics to interpret these languages on them. Section 3 explains the difficulties of defining bisimulation for the language with only local atoms. Section 4 then defines bisimulation for the language with global atoms and establishes the Hennessy–Milner characterization. Section 5 presents the results for bisimulation in the corresponding setting of Kripke models.

2 Epistemic Simplicial Semantics

We consider a finite set A of *agents* (or *processes*) $a, b, \dots$ and a set $P = A \sqcup \bigsqcup_{a \in A} P_a$ of *propositional atoms* where sets P_a are countable and mutually disjoint sets of *local atoms for agent* a, denoted $p_a, q_a, p'_a, q'_a, \dots$ We overload the meaning of A as it also stands for the set of *global atoms*. Atoms $a \in A$ represent 'agent a is alive.' We define two languages: the *local language* with only local atoms and the *glocal language* with both local and global atoms.

Definition 2.1 Let $a \in A$ and $p_a \in P_a$. *Language* $\mathcal{L}^+$ is defined by

$$\varphi ::= a \mid p_a \mid \neg\varphi \mid (\varphi \wedge \varphi) \mid \widehat{K}_a\varphi.$$

Language $\mathcal{L}^-$ is the fragment of $\mathcal{L}^+$ without global atoms a. Boolean connectives are defined in a standard way; $K_a\varphi := \neg\widehat{K}_a\neg\varphi$. In $\mathcal{L}^+$, we can additionally define boolean constants $\top := a \vee \neg a$ for some fixed $a \in A$ and $\bot := \neg\top$.

Definition 2.2 A *simplicial model* $\mathcal{C}$ is a triple (C, χ, ℓ) such that:

- A (*simplicial*) *complex* $C \neq \varnothing$ is a collection of *simplexes* such that
  (i) every $X \in C$ is a non-empty finite subset of a given set $\mathcal{V}$ of *vertices*;
  (ii) for each simplex $X \in C$, if $\varnothing \neq Y \subseteq X$, then $Y \in C$;
  (iii) $\{v\} \in C$ for each vertex $v \in \mathcal{V}$.
- A *chromatic function* $\chi\colon \mathcal{V} \to A$ is a map from vertices to agents such that for each $X \in C$ and $v, u \in X$, if $\chi(v) = \chi(u)$, then $v = u$.
- A *valuation function* $\ell\colon \mathcal{V} \to 2^{P \setminus A}$ is a map from vertices to sets of local atoms such that $\ell(v) \subseteq P_a$ for each $v \in \mathcal{V}$ with $\chi(v) = a$.

For each simplex $X \in C$, we define $\chi(X) := \{\chi(v) \mid v \in X\}$ and $\ell(X) := \bigsqcup_{v \in X} \ell(v)$. For arbitrary simplexes $X, Y \in C$, if $Y \subseteq X$, we say that Y is a *face* of X. Since each simplex is a face of itself, we use 'simplex' and 'face' interchangeably. Faces Y and Z are *a-adjacent* iff $a \in \chi(Y \cap Z)$. A face X is a *facet* iff for any simplex $Y \in C$, if $Y \supseteq X$, then $Y = X$. The set of all facets is denoted $\mathcal{F}(C)$. The *dimension of a simplex* X is $|X| - 1$. The *dimension of a simplicial model* $\mathcal{C}$ is the largest dimension of its facets. A simplicial model $\mathcal{C}$ is *pure* iff all its facets have dimension $|A| - 1$. Otherwise, $\mathcal{C}$ is *impure*. A *pointed simplicial model* is a pair $(\mathcal{C}, X)$ where $X \in \mathcal{F}(C)$. We often omit 'pointed'.

Definition 2.3 Let $\mathcal{C} = (C, \chi, \ell)$ be a simplicial model. *Definability relation* $\bowtie$ and *satisfaction relation* $\models$ for $X \in \mathcal{F}(C)$ are defined recursively on $\varphi \in \mathcal{L}^+$:

$\mathcal{C}, X \bowtie a$ always;
$\mathcal{C}, X \bowtie p_a$ iff $a \in \chi(X)$;
$\mathcal{C}, X \bowtie \neg\varphi$ iff $\mathcal{C}, X \bowtie \varphi$;
$\mathcal{C}, X \bowtie \varphi \land \psi$ iff $\mathcal{C}, X \bowtie \varphi$ and $\mathcal{C}, X \bowtie \psi$;
$\mathcal{C}, X \bowtie \widehat{K}_a\varphi$ iff $\mathcal{C}, Y \bowtie \varphi$ for some $Y \in \mathcal{F}(C)$ with $a \in \chi(X \cap Y)$.

$\mathcal{C}, X \models a$ iff $a \in \chi(X)$;
$\mathcal{C}, X \models p_a$ iff $p_a \in \ell(X)$;
$\mathcal{C}, X \models \neg\varphi$ iff $\mathcal{C}, X \bowtie \varphi$ and $\mathcal{C}, X \not\models \varphi$;
$\mathcal{C}, X \models \varphi \land \psi$ iff $\mathcal{C}, X \models \varphi$ and $\mathcal{C}, X \models \psi$;
$\mathcal{C}, X \models \widehat{K}_a\varphi$ iff $C, Y \models \varphi$ for some $Y \in \mathcal{F}(C)$ with $a \in \chi(X \cap Y)$.

Example 2.4 For simplicial model $\mathcal{C}$ from Fig. 1:

- *Atoms and knowledge of dead agents.* Illustrating the novel aspects of the semantics, $\mathcal{C}, X \not\bowtie p_c$ since $c \notin \chi(X) = \{a, b\}$. Consequently, $\mathcal{C}, X \not\bowtie \neg p_c$, $\mathcal{C}, X \not\models p_c$, and $\mathcal{C}, X \not\models \neg p_c$. For the same reason, $\mathcal{C}, X \not\bowtie \widehat{K}_c p_a$. Thus, $\mathcal{C}, X \not\bowtie \neg\widehat{K}_c p_a$, $\mathcal{C}, X \not\models \widehat{K}_c p_a$, and $\mathcal{C}, X \not\models \neg\widehat{K}_c p_a$.
- *Knowledge of a live agent concerning dead agents.* Although $\mathcal{C}, X \not\bowtie p_c$, still $\mathcal{C}, X \models \widehat{K}_a p_c$ because $a \in \chi(X \cap Y) = \{a\}$ and $\mathcal{C}, Y \models p_c$. More surprisingly, also $\mathcal{C}, X \models K_a p_c$ because, given the two facets X and Y that agent a considers possible, as far as a knows, p_c is true. This knowledge is defeasible because a may learn that the actual facet is X and not Y, which she also considers possible.

3 Bisimulation for Simplicial Models with Local Atoms?

Before we define bisimulation for simplicial models in the language with global atoms, we explain the difficulties in finding such a notion for the language without. The introductory section mentioned that pointed simplicial models $(\mathcal{C}, Y)$ and $(\mathcal{C}', Y')$ from Fig. 1 are modally equivalent in language $\mathcal{L}^-$. We now establish it formally.

Definition 3.1 Pointed simplicial models $(\mathcal{C}, Y)$ and $(\mathcal{C}', Y')$ are *modally equivalent in a language* $\mathcal{L}$, written $(\mathcal{C}, Y) \equiv_{\mathcal{L}} (\mathcal{C}', Y')$, iff for each $\varphi \in \mathcal{L}$:

$$\mathcal{C}, Y \bowtie \varphi \iff \mathcal{C}', Y' \bowtie \varphi, \tag{1}$$
$$\mathcal{C}, Y \models \varphi \iff \mathcal{C}', Y' \models \varphi, \tag{2}$$
$$\mathcal{C}, Y \models \neg\varphi \iff \mathcal{C}', Y' \models \neg\varphi. \tag{3}$$

It is easy to show that, in fact, (2) alone is sufficient:

Lemma 3.2 (Criterion of modal equivalence) *If* (2) *holds for all* $\varphi \in \mathcal{L}$ *for pointed simplicial models* $(\mathcal{C}, Y)$ *and* $(\mathcal{C}', Y')$, *then* $(\mathcal{C}, Y) \equiv_{\mathcal{L}} (\mathcal{C}', Y')$.

Proof. We need to establish (1) and (3) for all $\varphi \in \mathcal{L}$. (3) for φ follows from (2) for $\neg\varphi$. To show (1) for φ, assume first that $\mathcal{C}, Y \bowtie \varphi$. Then either $\mathcal{C}, Y \models \varphi$ or $\mathcal{C}, Y \models \neg\varphi$. By (2) for φ and $\neg\varphi$, either $\mathcal{C}', Y' \models \varphi$ or $\mathcal{C}', Y' \models \neg\varphi$. In either case,

$\mathcal{C}', Y' \bowtie \varphi$. We have proved the left-to-right direction of (1). The right-to-left direction is symmetric. □

Lemma 3.2 seems to considerably simplify the definition of three-valued modal equivalence, as it now appears to be the same as in the two-valued semantics. In practice, it is not much help in proofs, as $\mathcal{C}, Y \not\models \varphi$ may imply falsity as well as being undefined. It is more practical to check both definability (1) and truth (2). (See, e.g., the proof of Theorem 4.2, where both (1) and (2) are needed in the induction step for (2) for $\neg$.)

Proposition 3.3 *$(\mathcal{C}, Y) \equiv_{\mathcal{L}^-} (\mathcal{C}', Y')$ for $(\mathcal{C}, Y)$ and $(\mathcal{C}', Y')$ from Fig. 1.*

Proof. Let us first prove by induction on the construction of $\varphi \in \mathcal{L}^-$ an auxiliary statement:

$$\mathcal{C}, X \bowtie \varphi \quad \Longrightarrow \quad (\mathcal{C}, X \models \varphi \Longleftrightarrow \mathcal{C}, Y \models \varphi). \tag{4}$$

For local atoms, this is obvious from the construction of $\mathcal{C}$. The cases for $\neg$ and $\wedge$ are straightforward. For $\varphi = \widehat{K}_a\psi$, the truth value is determined by the vertex 1_a, which is shared between X and Y. No formula $\widehat{K}_c\psi$ is defined in X. It remains to consider $\varphi = \widehat{K}_b\psi$ such that $\mathcal{C}, X \bowtie \widehat{K}_b\psi$:

$$\mathcal{C}, X \models \widehat{K}_b\psi \quad \Longleftrightarrow \quad \mathcal{C}, X \models \psi \quad \stackrel{\text{IH}}{\Longleftrightarrow} \quad \mathcal{C}, Y \models \psi \quad \Longleftrightarrow \quad \mathcal{C}, Y \models \widehat{K}_b\psi$$

where the IH can be applied to ψ because $\mathcal{C}, X \bowtie \widehat{K}_b\psi$ implies $\mathcal{C}, X \bowtie \psi$.

By Lemma 3.2, it is sufficient to prove (2). (Note that (1) is trivial here since all formulas are defined in both Y and Y'.) We use induction on $\varphi \in \mathcal{L}^-$. For local atoms, (2) is obvious from the construction of $\mathcal{C}$ and $\mathcal{C}'$. The cases for $\neg$ and $\wedge$ are straightforward. For $\varphi = \widehat{K}_i\psi$ with $i \in \{b, c\}$, property (2) follows from the IH and the fact that, for each of Y and Y', the only facet b-/c-adjacent to it is itself. Finally, for $\varphi = \widehat{K}_a\psi$,

$$\mathcal{C}, Y \models \widehat{K}_a\psi \quad \Longleftrightarrow \quad \mathcal{C}, X \models \psi \text{ or } \mathcal{C}, Y \models \psi \quad \stackrel{(4)}{\Longleftrightarrow} \quad \mathcal{C}, Y \models \psi \quad \stackrel{\text{IH}}{\Longleftrightarrow}$$
$$\stackrel{\text{IH}}{\Longleftrightarrow} \quad \mathcal{C}', Y' \models \psi \quad \Longleftrightarrow \quad \mathcal{C}', Y' \models \widehat{K}_a\psi$$

where (4) is only used when $\mathcal{C}, X \models \psi$ and, hence, $\mathcal{C}, X \bowtie \psi$. □

A natural notion of bisimulation for simplicial models, as adapted from Kripke models, is the following:

Definition 3.4 A *bisimulation* between simplicial models $\mathcal{C} = (C, \chi, \ell)$ and $\mathcal{C}' = (C', \chi', \ell')$ is a non-empty binary relation $\mathcal{B} \subseteq \mathcal{F}(C) \times \mathcal{F}(C')$ such that, whenever $X\mathcal{B}X'$, the following conditions are fulfilled:

Atoms: $\chi(X) = \chi'(X')$ and $\ell(X) = \ell'(X')$.

Forth: For each agent $a \in A$, if $Y \in \mathcal{F}(C)$ and $a \in \chi(X \cap Y)$, then there exists $Y' \in \mathcal{F}(C')$ with $a \in \chi'(X' \cap Y')$ and $Y\mathcal{B}Y'$.

Back: For each agent $a \in A$, if $Y' \in \mathcal{F}(C')$ and $a \in \chi'(X' \cap Y')$, then there exists $Y \in \mathcal{F}(C)$ with $a \in \chi(X \cap Y)$ and $Y\mathcal{B}Y'$.

Simplicial models are called *bisimilar* iff there exists a bisimulation $\mathcal{B}$ between them. Pointed simplicial models $(\mathcal{C}, X)$ and $(\mathcal{C}', X')$ are *bisimilar*, written $(\mathcal{C}, X) \leftrightarrows (\mathcal{C}', X')$, iff there exists a bisimulation $\mathcal{B}$ between $\mathcal{C}$ and $\mathcal{C}'$ with $X\mathcal{B}X'$.

A proper notion of bisimulation should enjoy the Hennessy–Milner property: two pointed simplicial models are modally equivalent iff they are bisimilar. Since $(C, Y) \equiv_{\mathcal{L}^-} (C', Y')$ by Prop. 3.3, they are supposed to be bisimilar. The **forth** requirement of Def. 3.4 demands to find a facet of $\mathcal{C}'$ bisimilar to facet X of $\mathcal{C}$. But no facet of $\mathcal{C}'$ can be bisimilar to X because none is modally equivalent to X.[2] Thus, the only possibility to define bisimulation coinciding with modal equivalence would be to rule out such troublesome adjacent facets from the scope of **forth/back**, similar to how a disconnected component does not prevent establishing a bisimulation. The difficulty here is that X is not disconnected. Morally, X need not be considered here in **forth** step from Y because strictly fewer formulas are defined in X than in Y (while for all φ, $\mathcal{C}, X \bowtie \varphi$ implies $\mathcal{C}, Y \bowtie \varphi$ and $\mathcal{C}, X \vDash \varphi$ implies $\mathcal{C}, Y \vDash \varphi$). Unfortunately, as we demonstrate in Example 3.5, no local condition exists that would enable us to make a determination whether to consider an adjacent facet in **forth/back** or not. More precisely, such a determination cannot be made based exclusively on the values of χ and ℓ in all facets reachable in (at most) k consecutive steps from one facet to an adjacent one, for any fixed k.

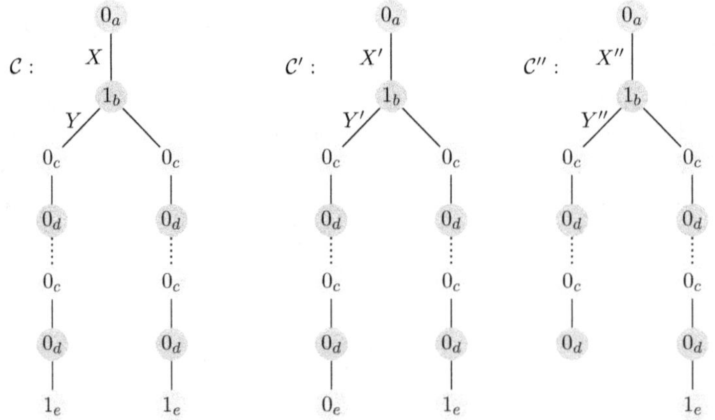

Fig. 2. Differences between $(\mathcal{C}, X)$, $(\mathcal{C}', X')$, and $(\mathcal{C}'', X'')$ are not local.

Example 3.5 Consider simplicial models $\mathcal{C}$, $\mathcal{C}'$, and $\mathcal{C}''$ in Fig. 2, where the number of yellow 0_c vertices (purple 0_d vertices) on each vertical branch is m. The three models differ only in the e-labeled vertices at the bottom, which can be reached from facets X, X', and X'' respectively in no fewer than $2m + 1$ consecutive steps. Despite all facets within at most

---

[2] In fact, even if the semantics is defined on all simplexes rather than only on facets (see [15]), even then there are no simplexes in $\mathcal{C}'$ that are modally equivalent to X because $\mathcal{C}, X \vDash \neg p_b$ but $\mathcal{C}, X \nvDash \widehat{K}_b p_c$.

$2m$ steps from X, X', and X'' coinciding, we have $(\mathcal{C}, X) \not\equiv_{\mathcal{L}^-} (\mathcal{C}', X')$ because $\mathcal{C}, X \vDash K_b(K_cK_d)^m p_e$ whereas $\mathcal{C}', X' \nvDash K_b(K_cK_d)^m p_e$.

Assume towards a contradiction that there exists some notion $\leftrightarrows^*$ of bisimulation that allows to ignore some of the adjacent facets in **forth**/**back** steps based on some local condition and that this notion satisfies the Hennessy–Milner property. In particular, $(\mathcal{C}, X) \not\leftrightarrows^* (\mathcal{C}', X')$. Since the only difference between $\mathcal{C}$ and $\mathcal{C}'$ can be reached through Y and Y' respectively, $\leftrightarrows^*$ cannot avoid the **back** step from X' to Y'. At the same time, Y'' in $\mathcal{C}''$ is the same as Y' in $\mathcal{C}'$, and so are all facets reachable by at most $2m-1$ consecutive steps from Y'' respectively Y'. Given that m can be made arbitrarily large, no local condition can differentiate Y'' from Y', meaning that $\leftrightarrows^*$, in testing $(\mathcal{C}, X)$ and $(\mathcal{C}'', X'')$, must also consider Y'' as a **back** step from X''. It is easy to see that Y'' is not modally equivalent to any facet of $\mathcal{C}$. Indeed, $\mathcal{C}'', Y'' \vDash p_b \wedge \neg p_c$, yet $\mathcal{C}'', Y'' \nvDash (K_cK_d)^m p_e$. Thus, $(\mathcal{C}, X) \not\leftrightarrows^* (\mathcal{C}'', X'')$ must hold due to the failure of **back** to Y''. At the same time, one can show, as in Prop. 3.3, that $(\mathcal{C}, X) \equiv_{\mathcal{L}^-} (\mathcal{C}'', X'')$. This contradiction with the Hennessy–Milner property shows that no reasonable standard bisimulation $\leftrightarrows^*$ exists for language $\mathcal{L}^-$.

What is the reason for the failure of Hennessy–Milner property for language $\mathcal{L}^-$? Kupke and Pattinson, in the context of coalgebraic semantics of modal logics [28], call the Hennessy–Milner property the "gold standard" regarding the question of "expressivity of modal logic," or, more specifically, "whether the logic [...] provide[s] enough power to describe particular properties of a system." Similarly, and specifically in the context of many-valued bisimulations, Marti and Metcalfe [30] equate the formal question of "whether analogues of the Hennessy–Milner property [...] hold for image-finite models of many-valued modal logics" with the informal attempt "to determine whether the language is expressive enough to distinguish image-finite models of many-valued modal logics." Combining these two separate areas, in studying many-valued modal logics coalgebraically, Bílková and Dostál [2] ask whether "finitary modal languages [are] expressive for bisimilarity" by checking the Hennessy–Milner property. Thus, the proper question is not why Hennessy–Milner fails, but what properties language $\mathcal{L}^-$ is not able to express. Put this way, the answer is not hard to guess. What is missing is the ability to talk about agents being dead/alive in the object language. In the next section, we show that adding global atoms for agents being alive indeed remedies the situation.

4 Bisimulation for Simplicial Models with Glocal Atoms

In this section, we switch to language $\mathcal{L}^+$ with global atoms. Hence, $\equiv$ from now on is understood as $\equiv_{\mathcal{L}^+}$. Note that the **atoms** clause of Def. 3.4 requires that not only local atoms have the same truth values, i.e., $\ell(X) = \ell'(X')$, but that global atoms do too, i.e., $\chi(X) = \chi'(X')$.

Example 4.1 The three simplicial models below are pairwise bisimilar. Moreover, there exist bisimulations that relate each facet of one model to some facet of another in either direction (we call such bisimulations *total*). In particular,

$\{(X,X'),(X,Z'),(Y,Y')\}$ is a bisimulation between the left and middle models. Similarly, $\{(X,X''),(Y,Y''),(Y,Z'')\}$ is a bisimulation between the left and right models. The atomic harmony for global atoms is a consequence of only relating facets with the same set of agents.

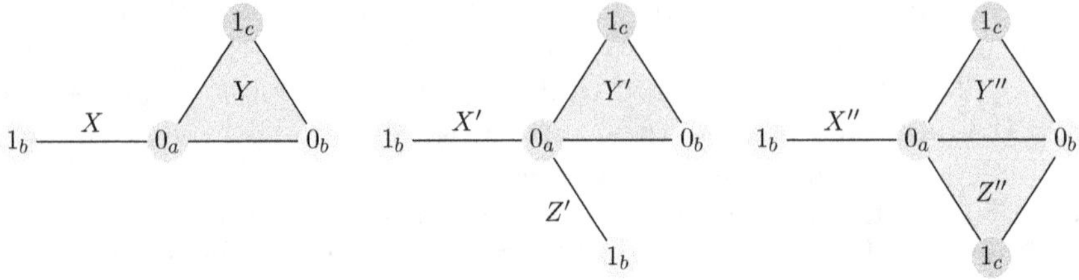

The following theorem is the easier direction of the Hennessy–Milner property: bisimilarity implies modal equivalence.

Theorem 4.2 (Bisimilarity implies modal equivalence) *For arbitrary pointed simplicial models* $(\mathcal{C},X)$ *and* $(\mathcal{C}',X')$,

$$(\mathcal{C},X) \leftrightarrows (\mathcal{C}',X') \quad\Longrightarrow\quad (\mathcal{C},X) \equiv_{\mathcal{L}^+} (\mathcal{C}',X').$$

Proof. Let $(\mathcal{C},X) \leftrightarrows (\mathcal{C}',X')$ for $\mathcal{C} = (C,\chi,\ell)$ and $\mathcal{C}' = (C',\chi',\ell')$. Let $X\mathcal{B}X'$ for a bisimulation $\mathcal{B}$ between $\mathcal{C}$ and $\mathcal{C}'$. By Lemma 3.2, to prove $(\mathcal{C},X) \equiv (\mathcal{C}',X')$, it is sufficient to prove (2) for $Y = X$ and $Y' = X'$ for all $\varphi \in \mathcal{L}^+$. We prove both (2) and (1) for all facets $Y \in \mathcal{F}(C)$ and $Y' \in \mathcal{F}(C')$ such that $Y\mathcal{B}Y'$ by mutual induction on the construction of $\varphi \in \mathcal{L}^+$.

- For global atom $\varphi = a$: (1) is trivial because a is defined on all facets. (2) follows from $\chi(Y) = \chi'(Y')$ part of **atoms**.
- For local atom $\varphi = p_a$: (1) and (2) follow from $\chi(Y) = \chi'(Y')$ and $\ell(Y) = \ell'(Y')$ parts of **atoms** respectively.
- For negation $\varphi = \neg\psi$:

  (1) $\mathcal{C},Y \bowtie \neg\psi \iff \mathcal{C},Y \bowtie \psi \overset{\text{IH(1)}}{\iff} \mathcal{C}',Y' \bowtie \psi \iff \mathcal{C}',Y' \bowtie \neg\psi.$

  (2) $\mathcal{C},Y \vDash \neg\psi \iff (\mathcal{C},Y \bowtie \psi \text{ and } \mathcal{C},Y \nvDash \psi) \overset{\text{IH(1),(2)}}{\iff}$

  $(\mathcal{C}',Y' \bowtie \psi \text{ and } \mathcal{C}',Y' \nvDash \psi) \iff \mathcal{C}',Y' \vDash \neg\psi.$
- For conjunction $\varphi = \psi \wedge \chi$, the argument is similar.
- For modality $\varphi = \widehat{K}_a\psi$, the argument for (1) and (2) is similar, thus, we only prove (1). From left to right, assume $\mathcal{C},Y \bowtie \widehat{K}_a\psi$. Then there is $Z \in \mathcal{F}(C)$ such that $a \in \chi(Y \cap Z)$ and $\mathcal{C},Z \bowtie \psi$. By **forth** for $Y\mathcal{B}Y'$, there is $Z' \in \mathcal{F}(C')$ such that $a \in \chi'(Y' \cap Z')$ and $Z\mathcal{B}Z'$. Then $\mathcal{C}',Z' \bowtie \psi$ by the IH(1) for ψ and $Z\mathcal{B}Z'$. Thus, $\mathcal{C}',Y' \bowtie \widehat{K}_a\psi$. The other direction is symmetric using **back** instead of **forth**. □

Before proving the other, harder direction of the Hennessy–Milner property, we first explain where the standard proof fails. In order to show the converse, i.e., that modal equivalence implies bisimilarity, one typically defines a binary relation of modal equivalence between states of two given models and shows that it is a bisimulation, under a suitable finitary assumption. As mentioned earlier, the standard assumption for Kripke models is that of image-finiteness [30]. The translation of this assumption into simplicial models yields the property of *star-finiteness*: a simplicial model is *star-finite* iff, for each agent a, the number of facets a-adjacent to any given facet is finite. Assuming star-finite $(\mathcal{C}, X)$ and $(\mathcal{C}', X')$ such that $(\mathcal{C}, X) \equiv (\mathcal{C}', X')$, consider **forth**. Let Y be a-adjacent to X in $\mathcal{C}$, and assume, towards a contradiction, that there is no modally equivalent Y' in $\mathcal{C}'$ that is a-adjacent to X'. Then, for each facet a-adjacent to X' in $\mathcal{C}'$, there must be a formula distinguishing it from Y. By the finitary assumption, there are only finitely many such facets $Y_1', \ldots, Y_k'$ and formulas $\psi_1, \ldots, \psi_k$ such that ψ_i distinguishes Y from Y_i'. Since a formula distinguishes facets iff its negation does, in the two-valued semantics, w.l.o.g., we may assume that each ψ_i is true in Y and false in Y_i', so that $\widehat{K}_a(\psi_1 \wedge \cdots \wedge \psi_k)$ is true in X and false in X', contradicting the assumption of their modal equivalence. By contrast, in the three-valued semantics some of ψ_i's may be undefined in Y, which cannot be sidestepped by switching to the negation.

Thus, to implement the same method, instead of negation, one needs a more complex way of transforming a distinguishing formula ψ_i that is undefined in Y but defined in Y_i' into another distinguishing formula $f(\psi_i)$ that is true in Y and false or undefined in Y_i'. Unfortunately, such transformation cannot be done independently of $(\mathcal{C}, X)$, as follows from the following proposition:

Proposition 4.3 *For formula $\varphi = p_a \wedge K_b p_c$, no formula $\varphi^{\not\bowtie} \in \mathcal{L}^+$ satisfies*

$$\mathcal{C}, X \not\bowtie \varphi \qquad \Longleftrightarrow \qquad \mathcal{C}, X \vDash \varphi^{\not\bowtie} \tag{5}$$

for all pointed simplicial models $(\mathcal{C}, X)$.

Proof. For any $\varphi^{\not\bowtie} \in \mathcal{L}^+$, there exists a unique purely propositional formula $\xi(x_1, \ldots, x_m)$, with $m \geq 0$, such that $\varphi^{\not\bowtie} = \xi\big(\widehat{K}_{i_1}\psi_1, \ldots, \widehat{K}_{i_m}\psi_m\big)$ for some agents $i_1, \ldots, i_m \in A$ and formulas $\psi_1, \ldots, \psi_m \in \mathcal{L}^+$. We consider three cases: (i) $i_j = a$ for some $1 \leq j \leq m$, (ii) $i_j \neq a$ for some $1 \leq j \leq m$, and (iii) $m = 0$, i.e., $\varphi^{\not\bowtie} = \xi$ is purely propositional. For each case, we construct $\mathcal{C} = (C, \chi, \ell)$ and $X \in \mathcal{F}(C)$ to violate (5).

(i) $\mathcal{C}, X \not\bowtie p_a \wedge K_b p_c$ if $a \notin \chi(X)$, but $\mathcal{C}, X \not\bowtie \varphi^{\not\bowtie}$ because of $\widehat{K}_a\psi_j$.

(ii) $\mathcal{C}, X \not\bowtie p_a \wedge K_b p_c$ if $\chi(X) = \{a\}$, but $\mathcal{C}, X \not\bowtie \varphi^{\not\bowtie}$ because of $\widehat{K}_{i_j}\psi_j$.

(iii) Consider $(\mathcal{C}, X)$ and $(\mathcal{C}', X')$ depicted below with the same evaluations of all $p_a \in P_a$ on both a-colored vertices and of all $p_b \in P_b$ on both b-colored ones. Then $\mathcal{C}, X \not\bowtie p_a \wedge K_b p_c$ while $\mathcal{C}', X' \bowtie p_a \wedge K_b p_c$. Thus, according to (5), any $\varphi^{\not\bowtie}$ should distinguish $(\mathcal{C}, X)$ from $(\mathcal{C}', X')$. But it is easy to prove by construction that any purely propositional formula ξ has the same truth value (true, false, or undefined) in $(\mathcal{C}, X)$ as in $(\mathcal{C}', X')$.

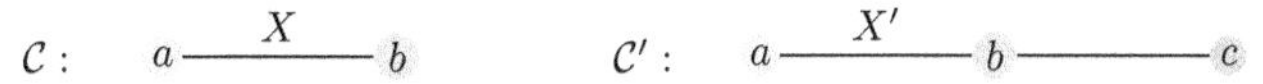

□

Since a purely syntactical transformation from undefined to true is impossible, we will present such a transformation modulo a given $(\mathcal{C}, X)$ (Lemma 4.11). The construction relies on the modal structure of formula φ, which is encoded in a form of what we call a *life tree*. With this localized transformation in place, we will be able to complete the proof of the remaining direction of Hennessy–Milner along the lines of the standard proof (Prop. 4.13).

Definition 4.4 A *life tree* $\mathfrak{T} = (\mathcal{T}, \mathfrak{X})$ is a directed rooted tree $\mathcal{T} = (V, E)$ supplied with a function $\mathfrak{X}$ that labels each node $\sigma \in V$ with a set $\mathfrak{X}(\sigma) \subseteq A$ of agents and each edge $(\sigma, \tau) \in E$ with an agent $\mathfrak{X}(\sigma, \tau) \in A$ such that $\mathfrak{X}$ satisfies the property $\mathfrak{X}(\sigma, \tau) \in \mathfrak{X}(\sigma) \cap \mathfrak{X}(\tau)$ for each $(\sigma, \tau) \in E$. We call (σ, τ) an *a-edge* iff $\mathfrak{X}(\sigma, \tau) = a$. For the root of a life tree, we typically use ρ, possibly with a subscript, for instance, ρ_1 could be used for the root of $\mathcal{T}_1$. The *a-grafting* $\mathfrak{T}^a$ of a life tree $\mathfrak{T}$ is obtained by adding a to the label of its root.

Definition 4.5 The *life tree* $\mathfrak{T}_\varphi$ *of a formula* $\varphi \in \mathcal{L}^+$ is a labeled tree $(\mathcal{T}, \mathfrak{X})$ defined by recursion on the construction of φ:

- $\mathfrak{T}_a$ is the tree consisting of a single root ρ with $\mathfrak{X}(\rho) = \varnothing$;
- $\mathfrak{T}_{p_a}$ is the tree consisting of a single root ρ with $\mathfrak{X}(\rho) = \{a\}$;
- $\mathfrak{T}_{\neg\varphi} := \mathfrak{T}_\varphi$;
- $\mathfrak{T}_{\varphi\wedge\psi}$ is obtained by merging the roots ρ_φ of $\mathfrak{T}_\varphi = (\mathcal{T}_\varphi, \mathfrak{X}_\varphi)$ and ρ_ψ of $\mathfrak{T}_\psi = (\mathcal{T}_\psi, \mathfrak{X}_\psi)$ and their labels, i.e., by first taking the disjoint union of $\mathfrak{T}_\varphi$ and $\mathfrak{T}_\psi$ with all the labeling preserved and then merging both roots ρ_φ and ρ_ψ into a new root ρ with the label $\mathfrak{X}_\varphi(\rho_\varphi) \cup \mathfrak{X}_\psi(\rho_\psi)$ without changing labels of any edges or of any other nodes;
- $\mathfrak{T}_{\widehat{K}_a\varphi}$ is obtained by adding a new root ρ to the a-grafting $\mathfrak{T}^a_\varphi$ of $\mathfrak{T}_\varphi$, labeling this new root ρ with $\{a\}$, and adding a new a-edge from ρ to the root ρ_φ of $\mathfrak{T}^a_\varphi$, with $\mathfrak{X}$ taking over all labels of $\mathfrak{T}^a_\varphi$ without change.

It is easy to show by induction on the construction of formula life trees that

Proposition 4.6 (Correctness for formula life trees) *The life tree $\mathfrak{T}_\varphi$ of any formula $\varphi \in \mathcal{L}^+$ is a life tree. In addition, for any child σ of the root of $\mathcal{T}$, the subtree of $\mathfrak{T}_\varphi$ rooted in σ is the a-grafting $\mathfrak{T}^a_\psi$ of the life tree $\mathfrak{T}_\psi$ of some subformula ψ of φ for some agent a.*

Example 4.7 The life trees of three formulas are depicted in Fig. 3. Note that $\mathfrak{T}^a_{\widehat{K}_b\neg p_d\wedge\widehat{K}_c p_d}$ is a subtree of $\mathfrak{T}_{\widehat{K}_a(\widehat{K}_b\neg p_d\wedge\widehat{K}_c p_d)}$.

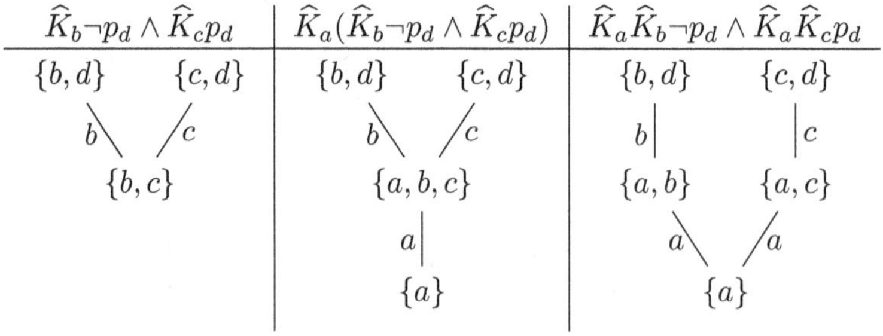

Fig. 3. Life trees of three sample formulas

Definition 4.8 An *embedding* of a life tree $\mathfrak{T} = (\mathcal{T}, \mathfrak{X})$ with $\mathcal{T} = (V, E)$ into $(\mathcal{C}, X)$ with $\mathcal{C} = (C, \chi, \ell)$ is a function $e \colon V \to \mathcal{F}(C)$ such that

(i) the root ρ of $\mathcal{T}$ is mapped to X, i.e., $e(\rho) = X$;
(ii) $\mathfrak{X}(\sigma) \subseteq \chi\big(e(\sigma)\big)$ for all tree nodes $\sigma \in V$;
(iii) $\mathfrak{X}(\sigma, \tau) \in \chi\big(e(\sigma) \cap e(\tau)\big)$ for all tree edges $(\sigma, \tau) \in E$.
Then we write $\mathcal{C}, X \bowtie_e \mathfrak{T}$, or simply $\mathcal{C}, X \bowtie \mathfrak{T}$, if such an embedding exists.

The following two lemmas show that formula life trees provide a practical and compact syntactical description of definability.

Lemma 4.9 (Equivalence of embeddability and definability) *For any formula $\varphi \in \mathcal{L}^+$ and pointed simplicial model $(\mathcal{C}, X)$,*

$$\mathcal{C}, X \bowtie \varphi \quad \Longleftrightarrow \quad \mathcal{C}, X \bowtie \mathfrak{T}_\varphi. \tag{6}$$

Proof. Let $\mathcal{C} = (C, \chi, \ell)$ and $\mathfrak{T}_\varphi = \big((V, E), \mathfrak{X}\big)$ with root $\rho \in V$. We prove (6) for all $X \in \mathcal{F}(C)$ by induction on a formula $\varphi \in \mathcal{L}^+$.

- For $\varphi = a$: Since $\mathcal{C}, X \bowtie a$ always holds, we need to show that $\mathcal{C}, X \bowtie \mathfrak{T}_a$. By Def. 4.8(i), no function other than $e = \{(\rho, X)\}$ can be an embedding of $\mathfrak{T}_a$ into $(\mathcal{C}, X)$. To show that it is, given that $\mathfrak{T}_a$ has no edges, it suffices to check Def. 4.8(ii), which holds since $\mathfrak{X}(\rho) = \varnothing \subseteq \chi(X)$.
- For $\varphi = p_a$: Again, Def. 4.8(i) means that $e(\rho) = X$ and there is at most one embedding of $\mathfrak{T}_{p_a}$ into $(\mathcal{C}, X)$. Since in this case $\mathfrak{X}(\rho) = \{a\}$, it remains to note that $\mathcal{C}, X \bowtie p_a$ iff $a \in \chi(X)$ iff $\mathfrak{X}(\rho) \subseteq \chi(X)$ iff $\mathcal{C}, X \bowtie \mathfrak{T}_{p_a}$.
- For $\varphi = \neg\psi$: $\mathcal{C}, X \bowtie \neg\psi$ iff $\mathcal{C}, X \bowtie \psi$ iff by the IH $\mathcal{C}, X \bowtie \mathfrak{T}_\psi$ iff $\mathcal{C}, X \bowtie \mathfrak{T}_{\neg\psi}$.
- For $\varphi = \psi \wedge \xi$: By Def. 2.3, $\mathcal{C}, X \bowtie \psi \wedge \xi$ iff $\mathcal{C}, X \bowtie \psi$ and $\mathcal{C}, X \bowtie \xi$. By the IH, this is equivalent to $\mathcal{C}, X \bowtie \mathfrak{T}_\psi$ and $\mathcal{C}, X \bowtie \mathfrak{T}_\xi$. It remains to prove that this is equivalent to $\mathcal{C}, X \bowtie \mathfrak{T}_{\psi\wedge\xi}$. Let the life trees $\mathfrak{T}_\psi = \big((V_\psi, E_\psi), \mathfrak{X}_\psi\big)$ and $\mathfrak{T}_\xi = \big((V_\xi, E_\xi), \mathfrak{X}_\xi\big)$ have roots ρ_ψ and ρ_ξ respectively.

  Assume first that $\mathcal{C}, X \bowtie_{e_\psi} \mathfrak{T}_\psi$ and $\mathcal{C}, X \bowtie_{e_\xi} \mathfrak{T}_\xi$ for $e_\psi\colon V_\psi \to \mathcal{F}(C)$ and $e_\xi\colon V_\xi \to \mathcal{F}(C)$. Note that $e_\psi(\rho_\psi) = e_\xi(\rho_\xi) = X$ by Def. 4.8(i). It is easy to see that $\mathcal{C}, X \bowtie_e \mathfrak{T}_{\psi\wedge\xi}$ for the function $e\colon V \to \mathcal{F}(C)$ defined by

$$e(\theta) := \begin{cases} X & \text{if } \theta = \rho, \\ e_\psi(\theta) & \text{if } \theta \in V_\psi \setminus \{\rho_\psi\}, \\ e_\xi(\theta) & \text{if } \theta \in V_\xi \setminus \{\rho_\xi\}. \end{cases}$$

  Assume now, conversely, that $\mathcal{C}, X \bowtie_e \mathfrak{T}_{\psi\wedge\xi}$ for $e\colon V \to \mathcal{F}(C)$. It is easy to see that $\mathcal{C}, X \bowtie_{e_\psi} \mathfrak{T}_\psi$ and $\mathcal{C}, X \bowtie_{e_\xi} \mathfrak{T}_\xi$ for $e_\psi\colon V_\psi \to \mathcal{F}(C)$ and $e_\xi\colon V_\xi \to \mathcal{F}(C)$ defined by

$$e_\psi(\sigma) := \begin{cases} X & \text{if } \sigma = r_\psi, \\ e(\sigma) & \text{if } \sigma \in V_\psi \setminus \{r_\psi\}, \end{cases} \qquad e_\xi(\tau) := \begin{cases} X & \text{if } \tau = \rho_\xi, \\ e(\tau) & \text{if } \tau \in V_\xi \setminus \{r_\xi\}. \end{cases}$$

- For $\varphi = \widehat{K}_a\psi$: By Def. 2.3, $\mathcal{C}, X \bowtie \widehat{K}_a\psi$ iff $\mathcal{C}, Y \bowtie \psi$ for some $Y \in \mathcal{F}(C)$ with $a \in \chi(X \cap Y)$. By the IH, the latter is equivalent to $\mathcal{C}, Y \bowtie \mathfrak{T}_\psi$ for some $Y \in \mathcal{F}(C)$ with $a \in \chi(X \cap Y)$. It remains to prove that this is equivalent to $\mathcal{C}, X \bowtie \mathfrak{T}_{\widehat{K}_a\psi}$. Let the life tree $\mathfrak{T}_\psi = \big((V_\psi, E_\psi), \mathfrak{X}_\psi\big)$ have root ρ_ψ, which is the only child of ρ in $\mathfrak{T}_{\widehat{K}_a\psi}$, where $V = V_\psi \sqcup \{\rho\}$.

Assume first that $\mathcal{C}, X \bowtie_e \mathfrak{T}_{\widehat{K}_a\psi}$, for some $e\colon V \to \mathcal{F}(C)$. By Def. 4.8(i), $e(\rho) = X$. For $Y = e(\rho_\psi)$, it is easy to see that $\mathcal{C}, Y \bowtie_{e\restriction V_\psi} \mathfrak{T}^a_\psi$ and, hence, $\mathcal{C}, Y \bowtie_{e\restriction V_\psi} \mathfrak{T}_\psi$. Finally, we have $a = \mathfrak{X}(\rho, \rho_\psi) \in \chi\big(e(\rho) \cap e(\rho_\psi)\big)$ by Def. 4.8(iii), i.e., $a \in \chi(X \cap Y)$.

Assume now, conversely, that $\mathcal{C}, Y \bowtie_{e_\psi} \mathfrak{T}_\psi$ for some $Y \in \mathcal{F}(C)$ with $a \in \chi(X \cap Y)$ and $e_\psi\colon V_\psi \to \mathcal{F}(C)$. Let us show that $\mathcal{C}, X \bowtie_e \mathfrak{T}_{\widehat{K}_a\psi}$ for $e := e_\psi \sqcup \{(\rho, X)\}$. Indeed, $e(\rho) = X$ fulfilling Def. 4.8(i). To show Def. 4.8(ii) for ρ, note that $\mathfrak{X}(\rho) = \{a\} \subseteq \chi(X \cap Y) \subseteq \chi(X) = \chi\big(e(\rho)\big)$. $\mathfrak{X}(\rho, \rho_\psi) = a \in \chi(X \cap Y) = \chi\big(e(\rho) \cap e(\rho_\psi)\big)$ proves Def. 4.8(iii) for (ρ, ρ_ψ).

$$\mathfrak{X}(\rho_\psi) = \{a\} \cup \mathfrak{X}_\psi(\rho_\psi) \subseteq \chi(X \cap Y) \cup \chi\big(e_\psi(\rho_\psi)\big) \subseteq \chi(Y) \cup \chi(Y) = \chi\big(e(\rho_\psi)\big)$$

shows Def. 4.8(ii) for ρ_ψ. Def. 4.8(ii)–(iii) for the remaining nodes and edges easily follow from the same properties of e_ψ. Thus, $\mathcal{C}, X \bowtie \mathfrak{T}_{\widehat{K}_a\psi}$. □

Lemma 4.10 (Criterion of non-embeddability) *Let $(\mathcal{C}, X)$ be a pointed simplicial model with $\mathcal{C} = (C, \chi, \ell)$ and $\mathfrak{T}_\varphi = \big((V, E), \mathfrak{X}\big)$ be the life tree of a formula $\varphi \in \mathcal{L}^+$ with root ρ. Then $\mathcal{C}, X \not\bowtie \mathfrak{T}_\varphi$ iff at least one of the following two criteria is satisfied:*

(i) *either $\mathfrak{X}(\rho) \not\subseteq \chi(X)$ or*

(ii) *there is a subformula ψ of φ and an a-edge $(\rho, \rho_\psi) \in E$ to the root ρ_ψ of a-grafting $\mathfrak{T}^a_\psi$ such that $\mathcal{C}, Y \not\bowtie \mathfrak{T}_\psi$ for any $Y \in \mathcal{F}(C)$ with $a \in \chi(X \cap Y)$.*

Proof. Let $(\rho, \sigma_1), \dots, (\rho, \sigma_k) \in E$ be all edges from ρ in $\mathfrak{T}_\varphi$, where $k \geq 0$ and let $\mathfrak{X}(\rho, \sigma_i) = a_i$. By Prop. 4.6, each σ_i is the root of $\mathfrak{T}^{a_i}_{\psi_i} = \big((V_i, E_i), \mathfrak{X}_i\big)$ for some subformula ψ_i of φ. In particular, $V = \{\rho\} \sqcup \bigsqcup_{i=1}^k V_i$.

First, let $\mathcal{C}, X \not\bowtie \mathfrak{T}_\varphi$, and assume towards a contradiction that both criteria (i) and (ii) fail. Then for each ψ_i, there is $Y_i \in \mathcal{F}(C)$ with $a_i \in \chi(X \cap Y_i)$ such that $\mathcal{C}, Y_i \bowtie_{e_i} \mathfrak{T}_{\psi_i}$ for some $e_i\colon V_i \to \mathcal{F}(C)$ with $e_i(\sigma_i) = Y_i$. It is easy to show that $\mathcal{C}, X \bowtie_e \mathfrak{T}_\varphi$ for

$$e(\tau) := \begin{cases} X & \text{if } \tau = \rho, \\ e_i(\tau) & \text{if } \tau \in V_i. \end{cases}$$

Assume now that $\mathcal{C}, X \bowtie_e \mathfrak{T}_\varphi$ for some $e\colon V \to \mathcal{F}(C)$, where $e(\rho) = X$ by Def. 4.8(i) and $\mathfrak{X}(\rho) \subseteq \chi(X)$ by Def. 4.8(ii). Hence, (i) fails. Let $Y_i := e(\sigma_i)$. It is easy to verify that $\mathcal{C}, Y_i \bowtie_{e\restriction V_i} \mathfrak{T}_{\psi_i}$. It remains to note that, by Def. 4.8(iii), $a_i = \mathfrak{X}(\rho, \sigma_i) \in \chi\big(e(\rho) \cap e(\sigma_i)\big)$, i.e., $a_i \in \chi(X \cap Y_i)$. Thus, (ii) also fails. □

Using life trees, we can now provide the localized transformation from undefined to true formulas:

Lemma 4.11 *Let $\mathcal{C}, X \not\bowtie \varphi$ for some formula $\varphi \in \mathcal{L}^+$ and some star-finite pointed simplicial model $(\mathcal{C}, X)$. Then there is a formula $\varphi^{\bowtie}_X \in \mathcal{L}^+$ such that*

$$\mathcal{C}, X \vDash \neg\varphi^{\bowtie}_X, \tag{7}$$

$$\mathcal{C}', X' \bowtie \varphi \implies \mathcal{C}', X' \vDash \varphi^{\bowtie}_X \quad \textit{for all } (\mathcal{C}', X'). \tag{8}$$

Proof. Let $\mathcal{C} = (C, \chi, \ell)$. By Lemma 4.9, $\mathcal{C}, X \not\bowtie \mathfrak{T}_\varphi$. We construct $\varphi_X^{\bowtie}$ by recursion on the $\mathsf{Depth}(\mathfrak{T}_\varphi)$ of the life tree $\mathfrak{T}_\varphi = \big((V, E), \mathfrak{X}\big)$ with root $\rho \in V$.

If $\mathfrak{X}(\rho) \nsubseteq \chi(X)$ meaning that $a \notin \chi(X)$ for some $a \in \mathfrak{X}(\rho)$, we define $\varphi_X^{\bowtie} := \bigwedge_{c \in \mathfrak{X}(\rho)} c$. This $\varphi_X^{\bowtie}$ satisfies (7) because a is false in X due to $a \notin \chi(X)$. In addition, (8) holds because $\mathcal{C}', X' \bowtie \varphi$ for $\mathcal{C}' = (C', \chi', \ell')$ is equivalent to $\mathcal{C}', X' \bowtie \mathfrak{T}_\varphi$ by Lemma 4.9, which implies that $\mathfrak{X}(\rho) \subseteq \chi'(X')$ and, hence, $\mathcal{C}', X' \vDash \varphi_X^{\bowtie}$. This case requires no recursive calls.

If $\mathfrak{X}(\rho) \subseteq \chi(X)$, then, by Lemma 4.10, criterion (ii) must hold, i.e., there is a subformula ψ of φ and an a-edge $(\rho, \rho_\psi) \in E$ to the root ρ_ψ of the a-grafting $\mathfrak{T}_\psi^a$ where $\mathfrak{T}_\psi = \big((V_\psi, E_\psi), \mathfrak{X}_\psi\big)$ such that $\mathcal{C}, Y \not\bowtie \mathfrak{T}_\psi$ for any Y from $\mathsf{star}_a(X) := \{Y \in \mathcal{F}(\mathcal{C}) \mid a \in \chi(X \cap Y)\}$, which is finite by star-finiteness and $X \in \mathsf{star}_a(X)$. Indeed, $a = \mathfrak{X}(\rho, \rho_\psi) \in \mathfrak{X}(\rho) \subseteq \chi(X) = \chi(X \cap X)$. Let $\mathsf{star}_a(X) = \{Y_1, \dots, Y_r\}$. By Lemma 4.9, $\mathcal{C}, Y_i \not\bowtie \psi$ for $1 \leq i \leq r$, so we can recursively construct $\psi_{Y_i}^{\bowtie}$ such that

$$\mathcal{C}, Y_i \vDash \neg\psi_{Y_i}^{\bowtie}, \tag{9}$$
$$\mathcal{C}', X' \bowtie \psi \quad \Longrightarrow \quad \mathcal{C}', X' \vDash \psi_{Y_i}^{\bowtie} \quad \text{for all } (\mathcal{C}', X'). \tag{10}$$

for each $1 \leq i \leq r$ because $\mathsf{Depth}(\mathfrak{T}_\psi) < \mathsf{Depth}(\mathfrak{T}_\varphi)$. We define a sequence of formulas $\xi_1, \dots, \xi_r$ as follows:

$$\xi_1 := \psi_{Y_1}^{\bowtie}$$
$$\xi_{k+1} := \begin{cases} \xi_k & \text{if } \mathcal{C}, Y_{k+1} \nvDash \xi_k; \\ \xi_k \wedge \psi_{Y_{k+1}}^{\bowtie} & \text{otherwise, i.e., if } \mathcal{C}, Y_{k+1} \vDash \xi_k. \end{cases}$$

By induction on $1 \leq k \leq r$ we prove that

$$\mathcal{C}, Y_{j_k} \vDash \neg\xi_k \text{ for some } 1 \leq j_k \leq k; \tag{11}$$
$$\mathcal{C}, Y_l \nvDash \xi_k \text{ for any } 1 \leq l \leq k. \tag{12}$$

For the *base case* $k = 1$ we must choose $j_1 := 1$, which works because $\xi_1 = \psi_{Y_1}^{\bowtie}$, yielding both $\mathcal{C}, Y_1 \vDash \neg\psi_{Y_1}^{\bowtie}$ by (9) for (11) and $\mathcal{C}, Y_1 \nvDash \psi_{Y_1}^{\bowtie}$ for (12). For the *induction step*, assume (11)–(12) hold for k and consider ξ_{k+1}:

- If $\xi_{k+1} = \xi_k$ because $\mathcal{C}, Y_{k+1} \nvDash \xi_k$, then we set $j_{k+1} := j_k$. By the IH(11), $\mathcal{C}, Y_{j_k} \vDash \neg\xi_k$, yielding $\mathcal{C}, Y_{j_{k+1}} \vDash \neg\xi_{k+1}$, fulfilling (11). Further, by the IH(12), $\mathcal{C}, Y_l \nvDash \xi_k$ for any $1 \leq l \leq k$ and $\mathcal{C}, Y_{k+1} \nvDash \xi_k$ by construction. Hence, $\mathcal{C}, Y_l \nvDash \xi_{k+1}$ for any $1 \leq l \leq k+1$ fulfilling (12).
- If, on the other hand, $\xi_{k+1} = \xi_k \wedge \psi_{Y_{k+1}}^{\bowtie}$ because $\mathcal{C}, Y_{k+1} \vDash \xi_k$, then we set $j_{k+1} := k+1$. By (9), $\mathcal{C}, Y_{k+1} \vDash \neg\psi_{Y_{k+1}}^{\bowtie}$, so $\mathcal{C}, Y_{k+1} \vDash \neg(\xi_k \wedge \psi_{Y_{k+1}}^{\bowtie})$ fulfilling (11). Further, by the IH(12), $\mathcal{C}, Y_l \nvDash \xi_k$ for any $1 \leq l \leq k$ and, as already discussed, $\mathcal{C}, Y_{k+1} \nvDash \psi_{Y_{k+1}}^{\bowtie}$. Hence, $\mathcal{C}, Y_l \nvDash \xi_k \wedge \psi_{Y_{k+1}}^{\bowtie}$ for any $1 \leq l \leq k+1$ fulfilling (12).

This completes the induction proof of (11)–(12). In particular, for $k = r$, we have that ξ_r is not true in any facet a-adjacent to X and is false, hence, defined in at least one such facet. Thus, $\mathcal{C}, X \vDash \neg\widehat{K}_a\xi_r$, and (7) holds for $\varphi_X^{\bowtie} := \widehat{K}_a\xi_r$.

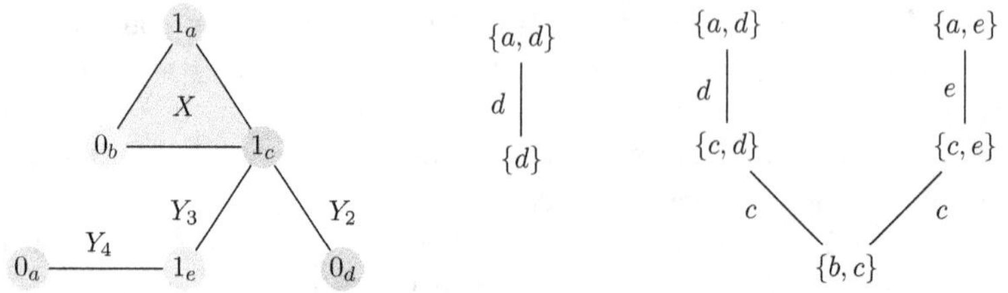

Fig. 4. From right to left: life tree $\mathfrak{T}_\varphi$ of formula $\varphi = p_b \land \widehat{K}_c\widehat{K}_d p_a \land \widehat{K}_c\widehat{K}_e\neg p_a$; life tree $\mathfrak{T}_\psi$ of its subformula $\psi = \widehat{K}_d p_a$; and simplicial model $(\mathcal{C}, X)$ such that $\mathcal{C}, X \not\bowtie \varphi$.

It remains to show that (8) also holds. Consider any $(\mathcal{C}', X')$ with $\mathcal{C}' = (C', \chi', \ell')$ such that $\mathcal{C}', X' \bowtie \varphi$. By Lemma 4.9, $\mathcal{C}', X' \bowtie_{e'} \mathfrak{T}_\varphi$ for some $e' \colon V \to \mathcal{F}(C')$ with $e'(\rho) = X'$. Let $e'(\rho_\psi) = Z'$. Since $(\rho, \rho_\psi) \in E$ is an a-edge, $a \in \chi'(X' \cap Z')$ by Def. 4.8(iii). It is easy to see that $\mathcal{C}', Z' \bowtie_{e' \restriction V_\psi} \mathfrak{T}_\psi$, thus, $\mathcal{C}', Z' \bowtie \psi$ by Lemma 4.9. By (10), $\mathcal{C}', Z' \vDash \psi^{\bowtie}_{Y_i}$ for all $1 \leq i \leq r$. Given that ξ_r is the conjunction of some of $\psi^{\bowtie}_{Y_i}$'s, we conclude that $\mathcal{C}', Z' \vDash \xi_r$. Consequently, $\mathcal{C}', X' \vDash \widehat{K}_a \xi_r$, i.e., $\mathcal{C}', X' \vDash \varphi^{\bowtie}_X$, which completes the proof of (8). □

Example 4.12 Consider the pointed simplicial model $(\mathcal{C}, X)$ in Fig. 4 (left). Let $\varphi = p_b \land \widehat{K}_c\widehat{K}_d p_a \land \widehat{K}_c\widehat{K}_e\neg p_a$. Its life tree $\mathfrak{T}_\varphi = \big((V, E), \mathfrak{X}\big)$ with root ρ is depicted in Fig. 4 (right). We will show that $\mathcal{C}, X \not\bowtie \varphi$ by using Lemmas 4.9 and 4.10, while simultaneously constructing $\varphi^{\bowtie}_X$ in two different ways. All agents from $\mathfrak{X}(\rho) = \{b, c\}$ of its root ρ are present in $\chi(X) = \{a, b, c\}$. Thus, criterion (i) of Lemma 4.10 fails. But consider the left subtree $\mathfrak{T}^c_\psi$ of $\mathfrak{T}_\varphi$, which corresponds to subformula $\psi = \widehat{K}_d p_a$ of φ. Life tree $\mathfrak{T}_\psi$, see Fig. 4 (middle), cannot be embedded into any of the facets c-adjacent to X: indeed, label $\{d\}$ of its root ρ_ψ is included in neither $\chi(X) = \{a, b, c\}$ nor $\chi(Y_3) = \{c, e\}$, resulting in $\psi^{\bowtie}_X = \psi^{\bowtie}_{Y_3} = d$, according to the construction in the proof of Lemma 4.11. For the remaining, third c-adjacent facet Y_2 with $\chi(Y_2) = \{c, d\}$, the label $\{d\}$ of ρ_ψ is included, however, $\mathfrak{T}_\psi$ has a d-edge from its root to a node labeled $\{a, d\}$, which corresponds to subformula p_a, and a is absent in the only facet d-adjacent to Y_2, which is Y_2 itself. Therefore, the same construction yields $\psi^{\bowtie}_{Y_2} = \widehat{K}_d a$.

How $\varphi^{\bowtie}_X$ is constructed from $\psi^{\bowtie}_X$, $\psi^{\bowtie}_{Y_2}$, and $\psi^{\bowtie}_{Y_3}$ depends on the chosen ordering among these three c-adjacent facets:

Ordering Y_2, X, Y_3	Ordering X, Y_2, Y_3
$\xi_1 = \psi^{\bowtie}_{Y_2} = \widehat{K}_d a$	$\xi'_1 = \psi^{\bowtie}_X = d$
$\xi_2 = \xi_1 = \widehat{K}_d a$ as $\mathcal{C}, X \nvDash \widehat{K}_d a$	$\xi'_2 = \xi'_1 \land \psi^{\bowtie}_{Y_2} = d \land \widehat{K}_d a$ as $\mathcal{C}, Y_2 \vDash d$
$\xi_3 = \xi_2 = \widehat{K}_d a$ as $\mathcal{C}, Y_3 \nvDash \widehat{K}_d a$	$\xi'_3 = \xi'_2 = d \land \widehat{K}_d a$ as $\mathcal{C}, Y_3 \nvDash d \land \widehat{K}_d a$

Accordingly, the two orderings provide two alternative formulas $\widehat{K}_c\widehat{K}_d a$ or $\widehat{K}_c(d \land \widehat{K}_d a)$ for the role of $\varphi^{\bowtie}_X$. Either of them satisfies (7)–(8).

Theorem 4.13 (Modal equivalence implies bisimilarity) *For arbitrary star-finite pointed simplicial models* $(\mathcal{C}, X)$ *and* $(\mathcal{C}', X')$,

$$(\mathcal{C}, X) \equiv_{\mathcal{L}^+} (\mathcal{C}', X') \quad \Longrightarrow \quad (\mathcal{C}, X) \leftrightarrows (\mathcal{C}', X').$$

Proof. Let $(\mathcal{C}, X) \equiv_{\mathcal{L}^+} (\mathcal{C}', X')$ for $\mathcal{C} = (C, \chi, \ell)$ and $\mathcal{C}' = (C', \chi', \ell')$. We define a binary relation $\mathcal{B} \subseteq \mathcal{F}(C) \times \mathcal{F}(C')$ so that $Y\mathcal{B}Y'$ iff $(\mathcal{C}, Y) \equiv_{\mathcal{L}^+} (\mathcal{C}', Y')$. In particular, $X\mathcal{B}X'$. Towards a contradiction, assume that $\mathcal{B}$ is not a bisimulation. We can prove without difficulty that **atoms** is satisfied. Thus, either **forth** or **back** fails. The two cases are symmetric, and we consider only the former. Assume that $Z\mathcal{B}Z'$, but for some $a \in A$ and some $Y \in \mathcal{F}(C)$ with $a \in \chi(Z \cap Y)$, there is no $Y' \in \mathcal{F}(C')$ such that $a \in \chi'(Z' \cap Y')$ and $Y\mathcal{B}Y'$. Let $\mathsf{star}_a(Z') = \{Y_1', \dots, Y_n'\}$ in $\mathcal{C}'$ (this set is finite because $\mathcal{C}'$ is star-finite). By construction of $\mathcal{B}$, we have $(\mathcal{C}, Y) \not\equiv_{\mathcal{L}^+} (\mathcal{C}', Y_i')$ for any $1 \leq i \leq n$. In other words, there exist formulas $\varphi_1, \dots, \varphi_n$ such that for each $1 \leq i \leq n$ one of the following three statements holds:

$$\mathcal{C}, Y \vDash \varphi_i \qquad \text{but} \qquad \mathcal{C}', Y_i' \nvDash \varphi_i, \tag{13}$$
$$\mathcal{C}, Y \vDash \neg\varphi_i \qquad \text{but} \qquad \mathcal{C}', Y_i' \nvDash \neg\varphi_i, \tag{14}$$
$$\mathcal{C}, Y \not\bowtie \varphi_i \qquad \text{but} \qquad \mathcal{C}', Y_i' \bowtie \varphi_i. \tag{15}$$

We transform these distinguishing formulas into $\psi_1, \dots, \psi_n$ as follows:

$$\psi_i := \begin{cases} \varphi_i & \text{if (13) holds for } \varphi_i, \\ \neg\varphi_i & \text{if (14) holds for } \varphi_i, \\ \neg(\varphi_i)_Y^{\bowtie} & \text{if (15) holds for } \varphi_i. \end{cases}$$

Note that now $\mathcal{C}, Y \vDash \psi_i$ and $\mathcal{C}', Y_i' \nvDash \psi_i$ for all $1 \leq i \leq n$. Indeed, for the first two clauses, it follows directly from (13)–(14), while for the last clause this is a consequence of Lemma 4.11. In other words, we have replaced the initial distinguishing formulas φ_i with the distinguishing formulas ψ_i that are all true in $(\mathcal{C}, Y)$. We will show that $\widehat{K}_a \bigwedge_{i=1}^n \psi_i$ distinguishes Z from Z'. This contradicts our assumption that $(\mathcal{C}, Z) \equiv_{\mathcal{L}^+} (\mathcal{C}', Z')$ and, thus, proves that $\mathcal{B}$ is a bisimulation.

Since $\mathcal{C}, Y \vDash \psi_i$ for all $1 \leq i \leq n$, also $\mathcal{C}, Y \vDash \bigwedge_{i=1}^n \psi_i$. Given $a \in \chi(Z \cap Y)$, we obtain $\mathcal{C}, Z \vDash \widehat{K}_a \bigwedge_{i=1}^n \psi_i$. Since $\mathcal{C}', Y_i' \nvDash \psi_i$, it follows that $\mathcal{C}', Y_i' \nvDash \bigwedge_{i=1}^n \psi_i$ for any $1 \leq i \leq n$. Thus, $\mathcal{C}', Z' \nvDash \widehat{K}_a \bigwedge_{i=1}^n \psi_i$. □

Theorem 4.14 (Hennessy–Milner property) *For arbitrary star-finite pointed simplicial models* $(\mathcal{C}, X)$ *and* $(\mathcal{C}', X')$*,*

$$(\mathcal{C}, X) \equiv_{\mathcal{L}^+} (\mathcal{C}', X') \quad \Longleftrightarrow \quad (\mathcal{C}, X) \leftrightarrows (\mathcal{C}', X').$$

Proof. This follows from Theorems 4.2 and 4.13. □

5 Life Bisimulation for Partial Epistemic Models

The main purpose of this section is to provide a bridge to a more familiar formalism for those who are not yet fluent in simplicial semantics. In particular, we compare simplicial models to Kripke models, including our results for bisimulation. Accordingly, we omit proofs in this section because they are obtained via the categorical equivalence with simplicial models. For the model correspondence in language $\mathcal{L}^-$ we recall [14], which only requires a minor modification to be extended to language $\mathcal{L}^+$. We continue by defining life bisimulation and will contrast it with standard bisimulation. Note that languages $\mathcal{L}^+$ and $\mathcal{L}^-$ remain the same throughout, we merely interpret them on Kripke models now.

Definition 5.1 A binary relation $\sim$ on a set S is called a *partial equivalence relation* iff $\sim$ is an equivalence relation on some subset $S' \subseteq S$.

In particular, partial equivalence relations are transitive and symmetric and induce a partition of S'. Note also that S' is uniquely determined by $\sim$: $S' = \{s \in S \mid s \sim s\}$.

Definition 5.2 Consider a pair $(S, \sim)$ where S is the *domain* of (*global*) *states* and $\sim: A \to 2^{S \times S}$ maps each agent $a \in A$ to a partial equivalence relation $\sim_a$ on S. Let S_a be the subset of S such that $\sim_a$ is an equivalence relation on S_a, i.e., the set of states where a is alive. Let $[s]_a := \{t \in S \mid s \sim_a t\}$ denote equivalence classes of $\sim_a$ on S_a. $(S, \sim)$ is *image-finite* iff $[s]_a$ is finite for all $a \in A$ and $s \in S_a$. Given $s \in S$, the set $A_s := \{a \in A \mid s \in S_a\}$ contains the agents that are alive in state s. Relation $\sim_a$ is *proper* iff for all distinct $s, t \in S$ there is an agent $b \in A_s$ such that $s \not\sim_b t$.

For such a pair $(S, \sim)$, a triple $\mathcal{M} = (S, \sim, L)$ is a *partial epistemic model* iff all $\sim_a$ are proper and a *valuation function* $L: S \to 2^P$ satisfies: for all $a \in A$, $p_a \in P_a$, $s, t \in S_a$, and $u \in S$,

- if $s \sim_a t$, then $p_a \in L(s)$ iff $p_a \in L(t)$, and
- $a \in L(u)$ iff $a \in A_u$.

$(\mathcal{M}, s)$ for $s \in S$ is a *pointed partial epistemic model* (we often omit 'pointed').

As in Def. 2.3, to interpret $\varphi \in \mathcal{L}^+$ in a global state s of a partial epistemic model $\mathcal{M}$, by induction on the structure of φ, we define $\bowtie$ to determine whether φ is defined and $\models$ to determine its truth value when defined.

Definition 5.3 Given a partial epistemic model $\mathcal{M} = (S, \sim, L)$, for all $s \in S$ we define $\bowtie$ and $\models$ by induction on $\varphi \in \mathcal{L}^+$:

$\mathcal{M}, s \bowtie a$ always;
$\mathcal{M}, s \bowtie p_a$ iff $s \in S_a$;
$\mathcal{M}, s \bowtie \neg\varphi$ iff $\mathcal{M}, s \bowtie \varphi$;
$\mathcal{M}, s \bowtie \varphi \wedge \psi$ iff $\mathcal{M}, s \bowtie \varphi$ and $\mathcal{M}, s \bowtie \psi$;
$\mathcal{M}, s \bowtie \widehat{K}_a\varphi$ iff $\mathcal{M}, t \bowtie \varphi$ for some $t \in S$ such that $t \sim_a s$.

$\mathcal{M}, s \models a$ iff $s \in S_a$;
$\mathcal{M}, s \models p_a$ iff $s \in S_a$ and $p_a \in L(s)$;
$\mathcal{M}, s \models \neg\varphi$ iff $\mathcal{M}, s \bowtie \varphi$ and $\mathcal{M}, s \not\models \varphi$;
$\mathcal{M}, s \models \varphi \wedge \psi$ iff $\mathcal{M}, s \models \varphi$ and $\mathcal{M}, s \models \psi$;
$\mathcal{M}, s \models \widehat{K}_a\varphi$ iff $\mathcal{M}, t \models \varphi$ for some $t \in S$ such that $t \sim_a s$.

Formula φ is *valid* iff $\mathcal{M}, s \bowtie \varphi$ implies $\mathcal{M}, s \models \varphi$ for all $(\mathcal{M}, s)$. The *denotation* of φ in $\mathcal{M}$ is defined as the set $[\![\varphi]\!]_{\mathcal{M}} := \{s \in S \mid \mathcal{M}, s \models \varphi\}$.

Partial epistemic models $(\mathcal{M}, s)$ and $(\mathcal{M}', s')$ are *modally equivalent*, denoted $(\mathcal{M}, s) \equiv^{\mathsf{par}} (\mathcal{M}', s')$, iff for all $\varphi \in \mathcal{L}^+$, $\mathcal{M}, s \bowtie \varphi \iff \mathcal{M}', s' \bowtie \varphi$,

$\mathcal{M}, s \models \varphi \iff \mathcal{M}', s' \models \varphi$, and $\mathcal{M}, s \models \neg\varphi \iff \mathcal{M}', s' \models \neg\varphi$.

A consequence of our semantics is that local atoms p_a for agent a may be assigned to states s outside of S_a, i.e., where a is dead. Such atoms p_a are undefined in s (and atom a is false) whether $p_a \in L(s)$ or $p_a \notin L(s)$. Partial epistemic models, therefore, contain superfluous information. An alternative knowledge representation would make them truly partial, albeit at the expense of comparing life bisimulation with standard bisimulation.

We now recall the correspondence between impure simplicial models and partial epistemic models from [14], which generalizes the equivalence of categories from [20] between pure simplicial models and local proper Kripke models where all relations are equivalence relations.

Definition 5.4 Operation σ (for *S*implicial) maps each partial epistemic model $\mathcal{M} = (S, \sim, L)$ to a simplicial model $\sigma(\mathcal{M}) = (C, \chi, \ell)$ as follows:

- vertices are pairs $\big([s]_a, a\big)$ for all $s \in S$ and $a \in A_s$;
- C consists of simplexes $\big\{\big([s]_a, a\big) \mid a \in B\big\}$ for all $s \in S$ and $\emptyset \neq B \subseteq A_s$;
- $\chi\Big(\big([s]_a, a\big)\Big) := a$ for each vertex $\big([s]_a, a\big)$;
- $\ell\Big(\big([s]_a, a\big)\Big) := P_a \cap L(s)$ for each vertex $\big([s]_a, a\big)$.

We let $\sigma(s)$ denote the facet $\big\{\big([s]_a, a\big) \mid a \in A_s\big\}$.

Operation κ (for *K*ripke) maps each simplicial model $\mathcal{C} = (C, \chi, \ell)$ to a partial epistemic model $\kappa(\mathcal{C}) = (S, \sim, L)$ as follows:

- $S := \mathcal{F}(C)$ consists of facets $X \in \mathcal{F}(C)$;
- $X \sim_a Y$ iff $a \in \chi(X \cap Y)$ for any agent a and global states X and Y;
- $L(X) := \ell(X) \cup \chi(X)$ for any global state X.

As σ maps each state s in $\mathcal{M}$ to a facet $\sigma(s)$ in $\sigma(\mathcal{M})$ and κ maps each facet X in $\mathcal{C}$ to a state X in $\kappa(\mathcal{C})$, these maps are also between structures $(\mathcal{M}, s)$ respectively $(\mathcal{C}, X)$: we let $\sigma(\mathcal{M}, s) := \big(\sigma(\mathcal{M}), \sigma(s)\big)$ and $\kappa(\mathcal{C}, X) := \big(\kappa(\mathcal{C}), X\big)$.

We recall from [14] that for all $\varphi \in \mathcal{L}^-$, $\mathcal{M}, s \bowtie \varphi \iff \sigma(\mathcal{M}, s) \bowtie \varphi$ and $\mathcal{C}, X \bowtie \varphi \iff \kappa(\mathcal{C}, X) \bowtie \varphi$. It is straightforward to extend this to $\mathcal{L}^+$.

Proposition 5.5 *Let* $\varphi \in \mathcal{L}^+$.

- *For all pointed partial epistemic models* $(\mathcal{M}, s)$*:*
  $\mathcal{M}, s \bowtie \varphi \iff \sigma(\mathcal{M}, s) \bowtie \varphi$ *and* $\mathcal{M}, s \models \varphi \iff \sigma(\mathcal{M}, s) \models \varphi$.
- *For all pointed simplicial models* $(\mathcal{C}, X)$*:*
  $\mathcal{C}, X \bowtie \varphi \iff \kappa(\mathcal{C}, X) \bowtie \varphi$ *and* $\mathcal{C}, X \models \varphi \iff \kappa(\mathcal{C}, X) \models \varphi$.

Figure 1 contains examples of corresponding simplicial models and partial epistemic models. Similarly, simplicial models from Example 4.1 correspond to partial epistemic models from Example 5.7 below.

We now define *life bisimulation* for partial epistemic models, show how it corresponds to impure simplicial models, and how it is different from the standard notion of bisimulation for Kripke models.

Definition 5.6 A *life bisimulation* between partial epistemic models $\mathcal{M} = (S, \sim, L)$ and $\mathcal{M}' = (S', \sim', L')$, notation $\mathcal{Z} : \mathcal{M} \leftrightarroweq \mathcal{M}'$, or $\mathcal{Z} : (\mathcal{M}, s) \leftrightarroweq (\mathcal{M}', s')$ given $s\mathcal{Z}s'$, is a non-empty binary relation $\mathcal{Z} \subseteq S \times S'$

such that for all $s \in S$ and $s' \in S'$ with $s\mathcal{Z}s'$ the following three conditions are satisfied:

- **Atoms**: $L(s) \cap A = L'(s') \cap A$ and, additionally, $L(s) \cap P_a = L'(s') \cap P_a$ for each $a \in A_s$ (note that here $A_s = L(s) \cap A = L'(s') \cap A = A_{s'}$).
- **Forth**: for all $a \in A_s$, for all $t \sim_a s$, there is a $t' \sim'_a s'$ such that $t\mathcal{Z}t'$.
- **Back**: for all $a \in A_{s'}$, for all $t' \sim'_a s'$, there is a $t \sim_a s$ such that $t\mathcal{Z}t'$.

Life bisimulation $\mathcal{Z}$ is *total* iff the domain and codomain of $\mathcal{Z}$ are S respectively S'. When $\mathcal{Z}$ is omitted, a life bisimulation must exist.

A *standard bisimulation* [4], notation $\underline{\leftrightarrow}^{\mathsf{st}}$, can be obtained from a life bisimulation by replacing the requirements $a \in A_s$ (twice) and $a \in A_{s'}$ in the above definition with $a \in A$. Standard bisimilarity, therefore, implies life bisimilarity: if $(\mathcal{M}, s) \underline{\leftrightarrow}^{\mathsf{st}} (\mathcal{M}', s')$, then $(\mathcal{M}, s) \underline{\leftrightarrow} (\mathcal{M}', s')$. On the class of multiagent S5 models, where all partial equivalence relations are equivalence relations, $(\mathcal{M}, s) \underline{\leftrightarrow}^{\mathsf{st}} (\mathcal{M}', s')$ iff $(\mathcal{M}, s) \underline{\leftrightarrow} (\mathcal{M}', s')$. In the states of S5 models, all formulas are defined, and the semantics becomes two-valued. Other than that, comparing life and standard bisimulations is a bit like comparing apples to onions, as our semantics is three-valued. See also the example below.

Example 5.7 Partial epistemic models $\mathcal{M}$, $\mathcal{M}'$, and $\mathcal{M}''$ below correspond to the simplicial models of Example 4.1 and are life bisimilar. The states are named with the values of the local atoms of the live agents. As in Example 4.1, a total bisimulation $\mathcal{Z}$ between $\mathcal{M}$ and $\mathcal{M}'$ requires that $X\mathcal{Z}X'$ and $X\mathcal{Z}Z'$, while $\mathcal{Z}$ between $\mathcal{M}$ and $\mathcal{M}''$ must have $Y\mathcal{Z}Y''$ and $Y\mathcal{Z}Z''$.

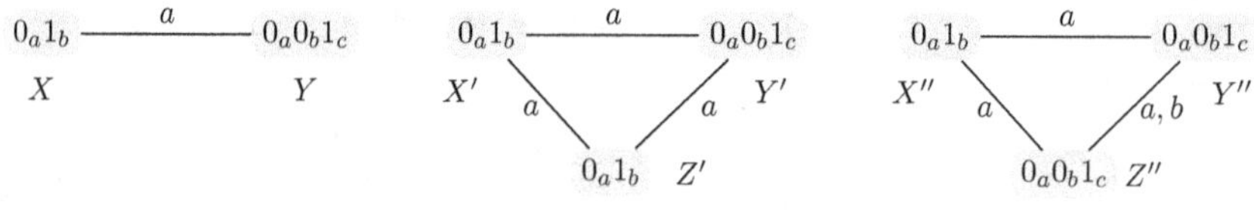

Let p_c be false in X, false in X', true in Z', and true in X''. Then no two of these models are standard bisimilar. (*i*) Neither $\mathcal{M}$ and $\mathcal{M}'$, nor $\mathcal{M}$ and $\mathcal{M}''$ are standard bisimilar because $\mathcal{M}'$ and $\mathcal{M}''$ contain a state s with $L(s) = \{p_b, p_c\}$ whereas $\mathcal{M}$ does not contain such a state. (*ii*) Whereas $\mathcal{M}'$ and $\mathcal{M}''$ are not standard bisimilar because $p_c \in L(s)$ for all states in $\mathcal{M}''$ but not in $\mathcal{M}'$.

We will now show that the σ and κ transformations preserve bisimilarity, and also the Hennessy–Milner property for life bisimulation between partial epistemic models. There are actually two ways to go about this: a direct proof using life trees and embeddings in partial epistemic models and an indirect proof using the results already obtained for simplicial models. We will do the latter as that proof is easy.

Proposition 5.8 (i) *$\mathcal{M} \underline{\leftrightarrow} \mathcal{M}'$ implies $\sigma(\mathcal{M}) \underline{\leftrightarrow} \sigma(\mathcal{M}')$ for any partial epistemic models $\mathcal{M}$ and $\mathcal{M}'$.*
(ii) *$\mathcal{C} \underline{\leftrightarrow} \mathcal{C}'$ implies $\kappa(\mathcal{C}) \underline{\leftrightarrow} \kappa(\mathcal{C}')$ for any simplicial models $\mathcal{C}$ and $\mathcal{C}'$.*

Corollary 5.9 *$\mathcal{M} \underline{\leftrightarrow} \kappa(\sigma(\mathcal{M}))$ and $\mathcal{C} \underline{\leftrightarrow} \sigma(\kappa(\mathcal{C}))$.*

Consequently, $\mathcal{M} \underline{\leftrightarrow} \mathcal{M}'$ iff $\kappa(\sigma(\mathcal{M})) \underline{\leftrightarrow} \kappa(\sigma(\mathcal{M}'))$ and, by the same token,

$\mathcal{C} \leftrightarroweq \mathcal{C}'$ iff $\sigma(\kappa(\mathcal{C})) \leftrightarroweq \sigma(\kappa(\mathcal{C}'))$. Same for the pointed versions.

Theorem 5.10 (Hennessy–Milner property) *For any image-finite partial epistemic models* $(\mathcal{M}, s)$ *and* $(\mathcal{M}', s')$

$$(\mathcal{M}, s) \leftrightarroweq (\mathcal{M}', s') \iff (\mathcal{M}, s) \equiv^{\mathsf{par}} (\mathcal{M}', s').$$

6 Conclusions and Future Work

In this paper, we concentrate on the question of what is a natural notion of bisimulation for impure simplicial complexes. The notion of bisimulation we have defined is indeed natural for the following two reasons: (I) On the categorically equivalent structures, namely partial Kripke frames, this is the abstract notion of bisimulation coming from coalgebraic many-valued logic (cf. [2]) because partial Kripke frames can be represented by a well-behaved set endofunctor. (II) It has the expected structural **forth**-and-**back** conditions, which are easy to check on finite structures, and it captures our intuition on process or behavioral equivalence in this case.

For this structurally natural notion of bisimulation, we have shown that the local language is invariant but fails to be sufficiently expressive (the Hennessy–Milner property fails). Similar situation has been encountered, e.g., in [30] or in [2], for example, in case of most Gödel modal logics. We have demonstrated how to enhance the language with global atoms to ensure sufficient expressivity and proved the Hennessy–Milner property for the extended language. Unlike the two-valued case, the latter turned out to be quite non-trivial. The reason for that is the lack of symmetry among our three values. If two boolean-valued structures are not modally equivalent, the distinguishing formula is either true or false in the first structure, but, if the other distinguishing truth value is desired, it is sufficient to take the negation. In our three-valued logic, there is an additional case of the distinguishing formula being undefined in the first structure. The transformation of it into a defined (true) formula is so non-trivial that it cannot even be done in a structure-independent way.

Future work For the local language, one can ask for a matching notion of model equivalence, for example, in the spirit of logic-induced bisimulations of [10]. For the fully abstract coalgebraic treatment of the logics in this paper as three-valued coalgebraic logics, as well as for investigating which three-valued first-order fragment corresponds to modal formulas in a sense of Van Benthem Theorem, essential insights need to be developed first. While the propositional logic PWK does not behave well in terms of abstract algebraic logic and only relatively recently was investigated in this context [5,32], the first-order expansion of PWK matching our interpretation of modalities has not been explored. The three-element weak Kleene algebra underlying the semantics is not a lattice, which impedes the use of the usual methods, e.g., those employed in [2].

Other directions for further research were proposed by the anonymous reviewers. Among them was the question of the complexity of checking whether two structures are bisimilar. We thank them for their comments and suggestions.

References

[1] A. Baltag, L. S. Moss, and S. Solecki. The logic of public announcements, common knowledge, and private suspicions. In I. Gilboa, editor, *Theoretical Aspects of Rationality and Knowledge: Proceedings of the Seventh Conference (TARK 1998)*, pages 43–56. Morgan Kaufmann, 1998. Available from: http://tark.org/proceedings/tark_jul22_98/p43-baltag.pdf.

[2] M. Bílková and M. Dostál. Expressivity of many-valued modal logics, coalgebraically. In J. Väänänen, Å. Hirvonen, and R. de Queiroz, editors, *Logic, Language, Information, and Computation: 23rd International Workshop, WoLLIC 2016, Puebla, Mexico, August 16–19th, 2016, Proceedings*, volume 9803 of *Lecture Notes in Computer Science*, pages 109–124. Springer, 2016. doi:10.1007/978-3-662-52921-8_8.

[3] O. Biran, S. Moran, and S. Zaks. A combinatorial characterization of the distributed 1-solvable tasks. *Journal of Algorithms*, 11(3):420–440, September 1990. doi:10.1016/0196-6774(90)90020-F.

[4] P. Blackburn, M. de Rijke, and Y. Venema. *Modal Logic*, volume 53 of *Cambridge Tracts in Theoretical Computer Science*. Cambridge University Press, 2001. doi:10.1017/CBO9781107050884.

[5] S. Bonzio, J. Gil-Férez, F. Paoli, and L. Peruzzi. On Paraconsistent Weak Kleene logic: Axiomatisation and algebraic analysis. *Studia Logica*, 105(2):253–297, April 2017. doi:10.1007/s11225-016-9689-5.

[6] C. Cachin, D. Lehnherr, and T. Studer. Synergistic knowledge. In S. Dolev and B. Schieber, editors, *Stabilization, Safety, and Security of Distributed Systems: 25th International Symposium, SSS 2023, Jersey City, NJ, USA, October 2–4, 2023, Proceedings*, volume 14310 of *Lecture Notes in Computer Science*, pages 552–567. Springer, 2023. doi:10.1007/978-3-031-44274-2_41.

[7] A. Castañeda, H. van Ditmarsch, R. Kuznets, Y. Moses, and U. Schmid. Epistemic and topological reasoning in distributed systems (Dagstuhl Seminar 23272). *Dagstuhl Reports*, 13(7):34–65, March 2024. doi:10.4230/DagRep.13.7.34.

[8] A. Castañeda, H. van Ditmarsch, D. A. Rosenblueth, and D. A. Velázquez. Communication pattern logic: Epistemic and topological views. *Journal of Philosophical Logic*, 52(5):1445–1473, October 2023. doi:10.1007/s10992-023-09713-8.

[9] R. Ciuni and M. Carrara. Semantic analysis of weak Kleene logics. *Journal of Applied Non-Classical Logics*, 29(1):1–36, 2019. doi:10.1080/11663081.2018.1547514.

[10] J. de Groot, H. H. Hansen, and A. Kurz. Logic-induced bisimulations. In N. Olivetti, R. Verbrugge, S. Negri, and G. Sandu, editors, *Advances in Modal Logic*, volume 13, pages 289–308. College Publications, 2020. Available from: http://www.aiml.net/volumes/volume13/deGroot-Hansen-Kurz.pdf.

[11] H. van Ditmarsch, É. Goubault, M. Lazić, J. Ledent, and S. Rajsbaum. A dynamic epistemic logic analysis of equality negation and other epistemic covering tasks. *Journal of Logical and Algebraic Methods in Programming*, 121:100662, June 2021. doi:10.1016/j.jlamp.2021.100662.

[12] H. van Ditmarsch, É. Goubault, J. Ledent, and S. Rajsbaum. Knowledge and simplicial complexes. In B. Lundgren and N. A. Nuñez Hernández, editors, *Philosophy of Computing: Themes from IACAP 2019*, volume 143 of *Philosophical Studies Series*, pages 1–50. Springer, 2022. doi:10.1007/978-3-030-75267-5_1.

[13] H. van Ditmarsch, W. van der Hoek, and B. Kooi. *Dynamic Epistemic Logic*, volume 337 of *Synthese Library*. Springer, 2007. doi:10.1007/978-1-4020-5839-4.

[14] H. van Ditmarsch and R. Kuznets. Wanted dead or alive: Epistemic logic for impure simplicial complexes. Eprint 2103.03032, arXiv, 2023. Accepted to *Journal of Logic and Computation*. doi:10.48550/arXiv.2103.03032.

[15] H. van Ditmarsch, R. Kuznets, and R. Randrianomentsoa. On two- and three-valued semantics for impure simplicial complexes. In A. Achilleos and D. Della Monica, editors, *Proceedings of the Fourteenth International Symposium on Games, Automata, Logics, and Formal Verification, Udine, Italy, 18–20th September 2023*, volume 390 of *Electronic*

Proceedings in Theoretical Computer Science, pages 50–66. Open Publishing Association, 2023. doi:10.4204/EPTCS.390.4.

[16] C. Dixon, C. Nalon, and R. Ramanujam. Knowledge and time. In H. van Ditmarsch, J. Y. Halpern, W. van der Hoek, and B. Kooi, editors, *Handbook of Epistemic Logic*, pages 205–259. College Publications, 2015.

[17] M. J. Fischer, N. A. Lynch, and M. S. Paterson. Impossibility of distributed consensus with one faulty process. *Journal of the ACM*, 32(2):374–382, April 1985. doi:10.1145/3149.214121.

[18] É. Goubault, R. Kniazev, and J. Ledent. A many-sorted epistemic logic for chromatic hypergraphs. In A. Murano and A. Silva, editors, *32nd EACSL Annual Conference on Computer Science Logic: CSL 2024, February 19–23, 2024, Naples, Italy*, volume 288 of *Leibniz International Proceedings in Informatics (LIPIcs)*, pages 30:1–30:18. Schloss Dagstuhl – Leibniz-Zentrum für Informatik, 2024. doi:10.4230/LIPIcs.CSL.2024.30.

[19] É. Goubault, R. Kniazev, J. Ledent, and S. Rajsbaum. Semi-simplicial set models for distributed knowledge. In *2023 38th Annual ACM/IEEE Symposium on Logic in Computer Science (LICS), 26–29 June 2023, Boston, USA*. IEEE, 2023. doi:10.1109/LICS56636.2023.10175737.

[20] É. Goubault, J. Ledent, and S. Rajsbaum. A simplicial complex model for dynamic epistemic logic to study distributed task computability. *Information and Computation*, 278:104597, June 2021. doi:10.1016/j.ic.2020.104597.

[21] É. Goubault, J. Ledent, and S. Rajsbaum. A simplicial model for $KB4_n$: Epistemic logic with agents that may die. In P. Berenbrink and B. Monmege, editors, *39th International Symposium on Theoretical Aspects of Computer Science: STACS 2022, March 15–18, 2022, Marseille, France (Virtual Conference)*, volume 219 of *Leibniz International Proceedings in Informatics (LIPIcs)*, pages 33:1–33:20. Schloss Dagstuhl – Leibniz-Zentrum für Informatik, 2022. doi:10.4230/LIPIcs.STACS.2022.33.

[22] S. Halldén. *The Logic of Nonsense*. Lundequist, 1949.

[23] J. Y. Halpern and Y. Moses. Knowledge and common knowledge in a distributed environment. *Journal of the ACM*, 37(3):549–587, July 1990. doi:10.1145/79147.79161.

[24] M. Herlihy, D. Kozlov, and S. Rajsbaum. *Distributed Computing through Combinatorial Topology*. Morgan Kaufmann, 2014. doi:10.1016/C2011-0-07032-1.

[25] M. Herlihy and N. Shavit. The topological structure of asynchronous computability. *Journal of the ACM*, 46(6):858–923, November 1999. doi:10.1145/331524.331529.

[26] J. Hintikka. *Knowledge and Belief: An Introduction to the Logic of the Two Notions*. Cornell University Press, 1962.

[27] S. C. Kleene. *Introduction to Metamathematics*. D. van Nostrand Company, 1952.

[28] C. Kupke and D. Pattinson. Coalgebraic semantics of modal logics: An overview. *Theoretical Computer Science*, 412(38):5070–5094, September 2011. doi:10.1016/j.tcs.2011.04.023.

[29] M. C. Loui and H. H. Abu-Amara. Memory requirements for agreement among unreliable asynchronous processes. In F. P. Preparata, editor, *Parallel and Distributed Computing*, volume 4 of *Advances in Computing Research: A Research Annual*, pages 163–183. JAI Press, 1987.

[30] M. Marti and G. Metcalfe. A Hennessy-Milner property for many-valued modal logics. In R. Goré, B. Kooi, and A. Kurucz, editors, *Advances in Modal Logic*, volume 10, pages 407–420. College Publications, 2014. Available from: http://www.aiml.net/volumes/volume10/Marti-Metcalfe.pdf.

[31] Y. Moses and Y. Shoham. Belief as defeasible knowledge. *Artificial Intelligence*, 64(2):299–321, December 1993. doi:https://doi.org/10.1016/0004-3702(93)90107-M.

[32] F. Paoli and M. Pra Baldi. Extensions of paraconsistent weak Kleene logic. *Logic Journal of the IGPL*, 29(5):798–822, October 2021. doi:10.1093/jigpal/jzaa024.

[33] A. Pnueli. The temporal logic of programs. In *18th Annual Symposium on Foundations of Computer Science*, pages 46–57. IEEE, 1977. doi:10.1109/SFCS.1977.32.

[34] R. Randrianomentsoa, H. van Ditmarsch, and R. Kuznets. Impure simplicial complexes: Complete axiomatization. *Logical Methods in Computer Science*, 19(4):3:1–3:35, October 2023. doi:10.46298/lmcs-19(4:3)2023.

[35] D. E. Szmuc. An epistemic interpretation of Paraconsistent Weak Kleene logic. *Logic and Logical Philosophy*, 28(2):277–330, June 2019. doi:10.12775/LLP.2019.014.

Better Bounded Bisimulation Contractions

Thomas Bolander [1]

Technical University of Denmark, Denmark

Alessandro Burigana [2]

Free University of Bozen-Bolzano, Italy

Abstract

Bisimulations are standard in modal logic and, more generally, in the theory of state-transition systems. The quotient structure of a Kripke model with respect to the bisimulation relation is called a bisimulation contraction. The bisimulation contraction is a minimal model bisimilar to the original model, and hence, for (image-)finite models, a minimal model modally equivalent to the original. Similar definitions exist for bounded bisimulations (k-bisimulations) and bounded bisimulation contractions. Two finite models are k-bisimilar if and only if they are modally equivalent up to modal depth k. However, the quotient structure with respect to the k-bisimulation relation does not guarantee a minimal model preserving modal equivalence to depth k. In this paper, we remedy this asymmetry to standard bisimulations and provide a novel definition of bounded contractions called rooted k-contractions. We prove that rooted k-contractions preserve k-bisimilarity and are minimal with this property. Finally, we show that rooted k-contractions can be exponentially more succinct than standard k-contractions.

Keywords: Modal logic, Kripke models, Bounded bisimulations, Bisimulation contractions, Exponential succinctness.

1 Introduction

Bisimulation plays a central role in countless fields, such as modal logic, set theory, formal verification, concurrency theory, process calculus, and others. Two structures are bisimilar if they are indistinguishable with respect to some behavioral property. In the case of Kripke models, bisimilarity between two models mean that their accessibility relations are structurally equivalent, and it then follows that two (image-)finite models are bisimilar if and only if they are modally equivalent (modal equivalence means they satisfy the same formulas) [3]. When using Kripke models for computational purposes, it is often

[1] E-mail: `tobo@dtu.dk`.

[2] E-mail: `burigana@inf.unibz.it`.

desirable to keep the models as small as possible while preserving their logical properties, *e.g.*, preserving modal equivalence. The quotient structure of a Kripke model with respect to the bisimulation relation is called a *bisimulation contraction*, giving us a minimal model modally equivalent to the original [4].

Bounded bisimulations only preserve structural equivalence up to some depth k. In the case of Kripke models, two models are k-bisimilar if their accessibility relations are structurally equivalent up to depth k. Two finite models are then k-bisimilar if and only if they are modally equivalent to modal depth k [3]. When the modalities are used to represent knowledge in an epistemic logic, k-bisimilarity means that higher-order reasoning is preserved to depth k. If we are only interested in reasoning to depth k, say in a setting with agents having bounded rationality, it seems intuitive to consider the quotient structure with respect to k-bisimilarity in the attempt to find a minimal model preserving modal equivalence to depth k. We call the quotient structure with respect to k-bisimilarity the *standard k-contraction* [17]. However, as we will show, the standard k-contraction does not guarantee a minimal model preserving modal equivalence to depth k. We provide an alternative notion of k-bisimulation contraction, called a *rooted k-contraction*, that indeed guarantees minimality.

The inspiration for this paper came from bounded rationality in epistemic planning [2]. Epistemic planning is concerned with computing plans for agents having incomplete information about the world and each other's knowledge, using epistemic logic as the underlying formalism [5]. Unfortunately, epistemic planning is in general undecidable, *i.e.*, it has an undecidable plan existence problem [5,7]. In the search for interesting decidable fragment of epistemic planning, we decided to limit the reasoning capabilities of agents to some fixed depth k (e.g. limiting to depth 2 would mean that agents can reason about what they know about the knowledge of others, but not what they know about what others know about them). In order to define such *depth-limited epistemic planning* formally, we needed a notion of contraction that would preserve modal equivalence to depth k. We originally started out using standard k-contractions, but soon discovered that they didn't guarantee minimality of the contracted models, in many cases actually quite far from it. We then set out on a quest to try to find a better notion of k-contraction that would guarantee minimality among k-bisimilarity preserving models, hence also reestablishing the symmetry to the corresponding existing results for standard bisimulations. In this paper we report on the results of that quest. We will report on the results of applying these notions to epistemic planning in a separate paper.

2 Preliminaries

In this section, we recall some basic notions in modal logic, *i.e.*, pointed Kripke models, bisimulation and bounded bisimulation [3]. Let $\mathcal{P}$ be a countable set of atomic propositions and $\mathcal{I}$ a finite set of modality indices. The language $\mathcal{L}$ of *multi-modal logic* is defined by the following BNF (where $p \in \mathcal{P}$ and $i \in \mathcal{I}$):

$$\varphi ::= p \mid \neg\varphi \mid \varphi \wedge \varphi \mid \Box_i \varphi.$$

Symbols $\top$, $\bot$, $\vee$ and $\Diamond_i$ are defined as usual. *Modal depth* is defined inductively on the structure of formulas: $md(p) = 0$ (for all $p \in \mathcal{P}$), $md(\neg\varphi) = md(\varphi)$, $md(\varphi_1 \wedge \varphi_2) = \max\{md(\varphi_1), md(\varphi_2)\}$ and $md(\Box_i\varphi) = 1 + md(\varphi)$.

Definition 2.1 A *model* of $\mathcal{L}$ is a triple $M = (W, R, V)$ where:

- $W \neq \varnothing$ is a finite set of *(possible) worlds*;
- $R : \mathcal{I} \to 2^{W \times W}$ assigns to each $i \in \mathcal{I}$ an *accessibility relation* R_i;
- $V : \mathcal{P} \to 2^W$ is a *valuation function* assigning to each atom a set of worlds.

A *pointed model* $\mathcal{M}$ is a pair (M, w_d), where $w_d \in W$ is the *designated world.*

We also use wR_iv for $(w, v) \in R_i$. We call an *i-edge*, or simply an *edge*, such a pair of worlds. A *path* is a sequence of worlds connected by edges. Note that we have restricted our attention to finite models. All of our results generalize to infinite models, but for many of the results we then have to make additional assumptions such as the models being image-finite, the underlying set of propositional atoms being finite or the number of modalities being finite [3]. To avoid this additional layer of complexity, and since all of our intended applications are within finite models, we restrict to those throughout the paper.

Definition 2.2 Let $M = (W, R, V)$ be a model of $\mathcal{L}$ and let $w \in W$.

$$\begin{array}{lll} (M,w) \models p & \text{iff} & w \in V(p) \\ (M,w) \models \neg\varphi & \text{iff} & (M,w) \not\models \varphi \\ (M,w) \models \varphi \wedge \psi & \text{iff} & (M,w) \models \varphi \text{ and } (M,w) \models \psi \\ (M,w) \models \Box_i\varphi & \text{iff} & \text{for all } v \text{ if } wR_iv \text{ then } (M,v) \models \varphi \end{array}$$

We say that two pointed models $\mathcal{M}$ and $\mathcal{M}'$ *agree on* (the formulas of) a set $\Phi \subseteq \mathcal{L}$ if, for all $\phi \in \Phi$, $\mathcal{M} \models \phi$ iff $\mathcal{M}' \models \phi$. We recall below the notions of bisimulation and bounded bisimulation (k-bisimulations) [3,15].

Definition 2.3 Let (M, w_d) and (M', w'_d) be two pointed models, with $M = (W, R, V)$ and $M' = (W', R', V')$. A *bisimulation* between (M, w_d) and (M', w'_d) is a non-empty binary relation $Z \subseteq W \times W'$ with $(w_d, w'_d) \in Z$ and satisfying:

- [atom] If $(w, w') \in Z$, then for all $p \in \mathcal{P}$, $w \in V(p)$ iff $w' \in V'(p)$.
- [forth] If $(w, w') \in Z$ and wR_iv, then there exists $v' \in W'$ such that $w'R'_iv'$ and $(v, v') \in Z$.
- [back] If $(w, w') \in Z$ and $w'R'_iv'$, then there exists $v \in W$ such that wR_iv and $(v, v') \in Z$.

If a bisimulation between (M, w_d) and (M', w'_d) exists, we say that (M, w_d) and (M', w'_d) are *bisimilar*, denoted $(M, w_d) \leftrightarrows (M', w'_d)$. When $(M, w) \leftrightarrows (M, w')$ for some worlds w, w' of the same model M, we simply write $w \leftrightarrows w'$, and say that w and w' are *bisimilar*. Finally, we denote the *bisimulation (equivalence) class* of a world $w \in W$ as $[w]_{\leftrightarrows} = \{v \in W \mid w \leftrightarrows v\}$.

Proposition 2.4 ([3]) *Two pointed models are bisimilar iff they agree on* $\mathcal{L}$.

Definition 2.5 Let $k \geq 0$ and let (M, w_d) and (M', w'_d) be two pointed models, with $M = (W, R, V)$ and $M' = (W', R', V')$. A *k-bisimulation* between (M, w_d) and (M', w'_d) is a sequence of non-empty binary relations $Z_k \subseteq \cdots \subseteq Z_0 \subseteq W \times W'$ with $(w_d, w'_d) \in Z_k$ and satisfying, for all $h < k$:

- [atom] If $(w, w') \in Z_0$, then for all $p \in \mathcal{P}$, $w \in V(p)$ iff $w' \in V'(p)$.
- [forth$_h$] If $(w, w') \in Z_{h+1}$ and wR_iv, then there exists $v' \in W'$ such that $w'R'_iv'$ and $(v, v') \in Z_h$.
- [back$_h$] If $(w, w') \in Z_{h+1}$ and $w'R'_iv'$, then there exists $v \in W$ such that wR_iv and $(v, v') \in Z_h$.

If a k-bisimulation between (M, w_d) and (M', w'_d) exists, we say that (M, w_d) and (M', w'_d) are *k-bisimilar*, denoted $(M, w_d) \leftrightarroweq_k (M', w'_d)$. When $(M, w) \leftrightarroweq_k (M', w')$, we often simply write $w \leftrightarroweq_k w'$, and say that w and w' are *k-bisimilar* (when M and M' are clear from the context). Finally, we denote the *k-bisimulation (equivalence) class* of a world $w \in W$ as $[w]_k = \{v \in W \mid w \leftrightarroweq_k v\}$.

Note that a k-bisimulation between pointed models is also an h-bisimulation for all $h \leq k$, and hence that k-bisimilar worlds are also h-bisimilar for all $h \leq k$.

Proposition 2.6 ([3]) *Two pointed models are k-bisimilar iff they agree on* $\{\phi \in \mathcal{L} \mid md(\phi) \leq k\}$, i.e., *on all of formulas up to modal depth k.*

Definition 2.7 Let (M, w_d) be a pointed model. The *depth* $d(w)$ of a world w is the length of the shortest path from w_d to w (∞ if no such path exists). Given $k \geq 0$, the *restriction* $M \restriction k$ of M to k is the sub-model containing all worlds with depth at most k (and preserving all edges between them).

Lemma 2.8 ([3]) *Let M and k be as above. Then, for every world w of $M \restriction k$, we have* $(M \restriction k, w) \leftrightarroweq_{k-d(w)} (M, w)$.

3 Defining Rooted k-Contractions

The notion of bisimulation contraction is well-known in modal logic. The *(bisimulation) contraction* of a pointed model $\mathcal{M} = ((M, R, V), w_d)$, that we denote with $\lfloor \mathcal{M} \rfloor$, is defined as the *quotient structure* of $\mathcal{M}$ with respect to $\leftrightarroweq$, *i.e.*, $\lfloor \mathcal{M} \rfloor = ((W', R', V'), [w_d]_{\leftrightarroweq})$, where $W' = \{[w]_{\leftrightarroweq} \mid w \in W\}$, $R'_i = \{([w]_{\leftrightarroweq}, [v]_{\leftrightarroweq}) \mid wR_iv\}$, and $V'(p) = \{[w]_{\leftrightarroweq} \in W' \mid w \in V(p)\}$ [15]. It is relatively straightforward to prove that: (i) $\lfloor \mathcal{M} \rfloor$ is bisimilar to $\mathcal{M}$; and (ii) $\lfloor \mathcal{M} \rfloor$ is a minimal model bisimilar to $\mathcal{M}$. A similar definition exists for *k-(bisimulation) contractions.* Namely, the k-contraction of $\mathcal{M}$, that we denote with $\lfloor \mathcal{M} \rfloor_k$, has been defined as the quotient structure of $\mathcal{M}$ with respect to $\leftrightarroweq_k$ [8,17]. We call this the *standard k-contraction* of $\mathcal{M}$. However, although such a contracted model is k-bisimilar to the original one [8,17], in general it is not minimal, as the following example shows.

Example 3.1 Consider the chain model $\mathcal{M}$ in Figure 1 (left). Since p is true in all worlds, and the length of the chain is k, a minimal model k-bisimilar to $\mathcal{M}$ is a singleton pointed model with a loop (Figure 1, right). This is because the loop model preserves all formulas up to depth k, cf. Proposition 2.6. However,

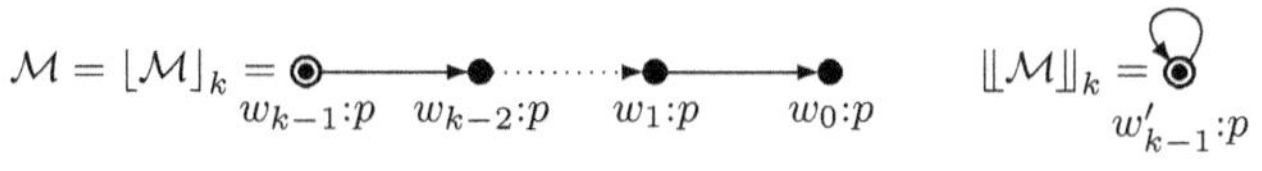

Fig. 1. Standard ($\lfloor \mathcal{M} \rfloor_k$) and rooted ($\llbracket \mathcal{M} \rrbracket_k$) k-contractions of chain $\mathcal{M}$ (symbol $\llbracket \mathcal{M} \rrbracket_k$ is borrowed from Definition 3.10). Each world w is denoted by a bullet labeled by its name, followed by the atomic propositions that hold in w. An arrow labeled with i from w to v means that wR_iv. We omit the labels on arrows whenever $|\mathcal{I}| = 1$. The designated world is represented by a circled bullet.

d	b (k=1)	b (k=2)	b (k=3)
0	1	2	3
1	0	1	2
2	−1	0	1

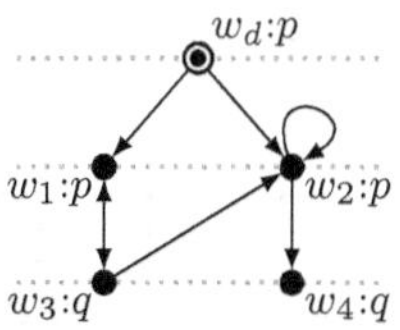

Fig. 2. Depth (d) and bound (b) of worlds for $k = 1, 2$ and 3.

the standard k-contraction of $\mathcal{M}$ is simply $\mathcal{M}$ itself: First note that for all $h \leq k-1$, the formula $\Diamond^h \Box \bot$ is true only in world w_h of $\mathcal{M}$ (it expresses the existence of a path of length h to a world from which no world is accessible). Hence, any two worlds of $\mathcal{M}$ can be distinguished by a formula $\Diamond^h \Box \bot$ of depth $h \leq k-1$. This implies that no two distinct worlds of $\mathcal{M}$ are modally equivalent to modal depth k, and hence cannot be part of the same k-bisimulation class. Thus, the standard k-contraction of $\mathcal{M}$ is $\mathcal{M}$ itself (or, more precisely, the k-contraction of $\mathcal{M}$ is isomorphic to $\mathcal{M}$).

We now move to introduce our *rooted k-contractions*. First, in this section, we show how to define a notion of a rooted k-contraction that guarantees the resulting model to have the smallest number of worlds among any model k-bisimilar to the original one (we call this property *world minimality*). Later we then define a stronger notion of rooted k-contraction that additionally guarantees the set of edges of the contracted model to be minimal (called *edge minimality*). In what follows, we fix a constant $k \geq 0$ and a pointed model $\mathcal{M} = (M, w_d)$ with $M = (W, V, R)$. Recall the notion of the depth $d(w)$ of a world w (Definition 2.7). We now introduce the notion of *bound* of a world.

Definition 3.2 The *bound* of a world w is $b(w) = k - d(w)$.

Lemma 3.3 *If xR_iy, then $b(y) \geq b(x) - 1$.*

Proof. $d(y)$ is the length of the shortest path from the designated world to y. Such a path either goes through the edge $(x, y) \in R_i$, or there is a shorter path to y. Hence, $d(y) \leq d(x) + 1 \Leftrightarrow k - b(y) \leq k - b(x) + 1 \Leftrightarrow b(y) \geq b(x) - 1$. $\square$

Example 3.4 The notion of *bound* of a world will play a key role in the definition of rooted k-contractions. Figure 2 shows an example of bound of worlds for $k = 1, 2$ and 3. Now consider the pointed models $\mathcal{N}_1$ and $\mathcal{N}_2$ of Figure 3.

Fig. 3. Two 2-bisimilar pointed models: $\mathcal{N}_1$ (left) and $\mathcal{N}_2$ (right).

Taking the standard 2-contraction of $\mathcal{N}_1$ would result in the model $\mathcal{N}_1$ itself, by a similar argument as in Example 3.1 (no two worlds of $\mathcal{N}_1$ are 2-bisimilar, actually not even 1-bisimilar). However, $\mathcal{N}_1$ is not world minimal among models 2-bisimilar to $\mathcal{N}_1$. That is true for $\mathcal{N}_2$, however. Any model 2-bisimilar to $\mathcal{N}_1$ must have at least three worlds (one for each atomic proposition), which is exactly what $\mathcal{N}_2$ has. The 2-bisimulation between $\mathcal{N}_1$ and $\mathcal{N}_2$ is defined by: $Z_2 = \{(w_d, w'_d)\}$; $Z_1 = Z_2 \cup \{(w_1, w'_1), (w_2, w'_2)\}$; and $Z_0 = Z_1 \cup \{(w_3, w'_2)\}$. Notice that $\mathcal{N}_2$ has been obtained by $\mathcal{N}_1$ by redirecting all incoming edges of w_3 to w_2 and deleting the worlds that are no longer reachable from the designated world. In standard bisimulation contractions, we simply identify worlds that are bisimilar, meaning that we can merge them into one world. In the case of $\mathcal{N}_1$ and $\mathcal{N}_2$, we cannot just trivially merge w_2 and w_3 into one world, as they don't have the same successor worlds (we might be left with w_4 as a successor to w_2, which destroys 2-bisimilarity). The idea of redirecting edges rather than merging worlds forms a crucial part of the intuition behind our rooted k-contractions. The point is that w_2 can be used as a "representative" for w_3 when we perform the contraction, and hence we can get rid of w_3. The reason that w_2 works as a representative for w_3 is that w_3 is at depth 2, so has bound 0. Intuitively this means that w_3 can be represented by any world that it is 0-bisimilar to, as to maintain k-bisimilarity at the designated worlds, we only need to require $(k-d)$-bisimilarity of the worlds at depth d, *i.e.*, worlds of bound b only need to be b-bisimilar. These intuitions are made formally precise in the following.

Lemma 3.5 *Let $k \geq 0$, let (M, w_d) be a pointed model, with $M = (W, R, V)$ and let $x, y \in W \setminus \{w_d\}$ be two distinct worlds such that $b(x) \geq b(y) \geq 0$ and $x \leftrightarroweq_{b(y)} y$. Let (M', w_d), with $M' = (W', R', V')$, be the pointed model obtained by deleting y from (M, w_d) and redirecting its incoming edges to x. More precisely:*

- $W' = W \setminus \{y\}$;
- $R'_i = (R_i \cap (W' \times W')) \cup \{(w, x) \mid w R_i y\}$;
- $V'(p) = V(p) \cap W'$, *for all* $p \in \mathcal{P}$.

Then $(M, w_d) \leftrightarroweq_k (M', w_d)$.

Proof. To avoid confusion, we remark that, for all w, $b(w)$ refers to the bound that w has in $\mathcal{M}$ (and not in $\mathcal{M}'$). Similarly, $w \leftrightarroweq_h w'$ means $(M, w) \leftrightarroweq_h$

(M, w'). For all $0 \leq h \leq k$, let $Z_h \subseteq W \times W'$ be the following binary relation:

$$Z_h = \{(w, w') \in W \times W' \mid w \leftrightarrows_h w', b(w) \geq h \text{ and } b(w') \geq h\}$$

We now show that $Z_k, \ldots, Z_0$ is a k-bisimulation between $\mathcal{M}$ and $\mathcal{M}'$. Clearly, $Z_k \subseteq \cdots \subseteq Z_0$. Also, trivially, $(w_d, w_d) \in Z_k$ (recalling that $w \leftrightarrows_h w'$ means $(M, w) \leftrightarrows_h (M, w')$). We get [atom] since $(w, w') \in Z_0$ implies $w \leftrightarrows_0 w'$ and thus $w \in V(p)$ iff $w' \in V'(p)$ (by [atom] of Definition 2.5).

We now show [forth$_h$]. Let $h < k$, $(w, w') \in Z_{h+1}$ and wR_iv. We need to find a $v' \in W'$ such that $w'R'_iv'$ and $(v, v') \in Z_h$. Since $(w, w') \in Z_{h+1}$, we have $w \leftrightarrows_{h+1} w'$, $b(w) \geq h+1$ and $b(w') \geq h+1$. From $w \leftrightarrows_{h+1} w'$, there exists $u \in W$ such that $w'R_iu$ and $v \leftrightarrows_h u$. Since wR_iv, $w'R_iu$, $b(w) \geq h+1$ and $b(w') \geq h+1$, by Lemma 3.3, we get $b(v) \geq h$ and $b(u) \geq h$. We have two cases. (i) If $u \neq y$, then by construction of R'_i, we get $w'R'_iu$. Since $v \leftrightarrows_h u$, $b(v) \geq h$ and $b(u) \geq h$, letting $v' = u$ we get $(v, v') \in Z_h$. (ii) If $u = y$, then by construction of R'_i, we get $w'R'_ix$. From $b(y) = b(u) \geq h$ and $x \leftrightarrows_{b(y)} y$, we get $x \leftrightarrows_h y$. Since $v \leftrightarrows_h y \leftrightarrows_h x$, $b(v) \geq h$ and $b(y) \geq h$, letting $v' = x$ we get $(v, v') \in Z_h$. This concludes [forth$_h$].

Now for [back$_h$]. Let $h < k$, $(w, w') \in Z_{h+1}$ and $w'R'_iv'$. We need to find v such that wR_iv and $(v, v') \in Z_h$. Since $(w, w') \in Z_{h+1}$, we have $w \leftrightarrows_{h+1} w'$, $b(w) \geq h+1$ and $b(w') \geq h+1$. We have two cases. (i) If $v' \neq x$, then by construction of R'_i, we get $w'R_iv'$. Since $w \leftrightarrows_{h+1} w'$, there exists $v \in W$ such that wR_iv and $v \leftrightarrows_h v'$. As in [forth$_h$], Lemma 3.3 gives $b(v) \geq h$ and $b(v') \geq h$. Thus, $(v, v') \in Z_h$. (ii) If $v' = x$, then by construction of R'_i, we get $w'R_ix$ or $w'R_iy$. If $w'R_ix$, we can reason as in (i). If $w'R_iy$, pick v with wR_iv and $v \leftrightarrows_h y$. As before, $b(v) \geq h$ and $b(y) \geq h$. Since $b(x) \geq b(y) \geq h$ and $x \leftrightarrows_{b(y)} y$, we get $x \leftrightarrows_h y$, and thus $v \leftrightarrows_h x$. Hence $(v, x) \in Z_h$, as required. □

The lemma tells us that if we're only interested in preserving k-bisimilarity, a world y can be deleted from a model if there exists a distinct world x such that $b(x) \geq b(y)$ and $x \leftrightarrows_{b(y)} y$. This leads us to the following definition.

Definition 3.6 Let x, y be two worlds with non-negative bound. We say that x *represents* y, denoted by $x \succeq y$, iff $b(x) \geq b(y)$ and $x \leftrightarrows_{b(y)} y$. If furthermore $b(x) > b(y)$, we say that x *strictly represents* y, denoted by $x \succ y$. The set of *maximal representatives* of W is the set of worlds $W^{\max} = \{x \in W \mid b(x) \geq 0 \text{ and } \neg\exists y \in W (y \succ x)\}$. We say that a world x is a *maximal representative* of y if $x \in W^{\max}$ and $x \succeq y$.

Note that every world $w \in W$ with $b(w) \geq 0$ has at least one maximal representative: Any chain $w \prec w' \prec w'' \prec \cdots$ is finite (since W is finite) and must hence end in a maximal representative of w. We are going to build our rooted k-contractions on the maximal representatives, the intuition being that all other worlds can be represented by one of these and hence be deleted, cf. Lemma 3.5.

Proposition 3.7 *For any pointed model $((W, R, V), w_d)$, we have:*

1) $w_d \in W^{\max}$;

2) if $w \succ v$, then $v \notin W^{\max}$;

3) if $b(w) < 0$, then $w \notin W^{\max}$;

4) if $w, v \in W^{\max}$ and $w \leftrightarrows_{b(w)} v$ then $b(w) = b(v)$.

Proof. Item 1 follows since w_d has bound $b(w_d) = k$ and there can not be any world with a greater bound. Items 2 and 3 immediately follow by definition of $W^{\max}$. Item 4 is by contradiction: Suppose that $w \leftrightarrows_{b(w)} v$ and $b(w) > b(v)$ (the case $b(v) > b(w)$ being symmetric). Since $w \leftrightarrows_{b(w)} v$, it follows that $w \leftrightarrows_{b(v)} v$, which implies that $w \succ v$, contradicting the fact that $v \in W^{\max}$. □

Example 3.8 Let $\mathcal{M}$ the pointed model in Figure 2, let W be its set of worlds and let $k = 2$. Using Proposition 2.6, we can show that for all worlds $w, v \in W$, if $b(w) > b(v)$ then $w \not\leftrightarrows_{b(v)} v$: w_3 and w_4 of bound 0 are not propositionally equivalent to any world of greater bound (they are the only ones satisfying q); w_1 and w_2 of bound 1 both satisfy $\Diamond q$ of modal depth 1, which is not satisfied by the only world of greater bound, w_d. Hence, all worlds are maximal representatives, *i.e.*, $W^{\max} = W$. We immediately get $W^{\max} = W$ for $k = 3$ as well, since if worlds w and v are not n-bisimilar, they are also not $(n+1)$-bisimilar. For $k = 1$, Proposition 3.7(3) gives $w_3, w_4 \notin W^{\max}$, and since $w_d \succ w_1, w_2$ (they satisfy the same atomic propositions), we get $W^{\max} = \{w_d\}$.

Definition 3.9 The *representative class* of a world w is the class $[w]_{b(w)}$, which we denote with the compact notation $[\![w]\!]$.

Definition 3.10 Let $\mathcal{M} = ((W, R, V), w_d)$ and let $k \geq 0$. The *rooted k-contraction* of $\mathcal{M}$ is the pointed model $[\![\mathcal{M}]\!]_k = ((W', R', V'), [\![w_d]\!])$, where:

- $W' = \{[\![x]\!] \mid x \in W^{\max}\}$;
- $R'_i = \{([\![x]\!], [\![y]\!]) \mid x, y \in W^{\max}, \exists z(xR_iz \text{ and } y \leftrightarrows_{b(x)-1} z) \text{ and } b(x) > 0\}$;[3]
- $V'(p) = \{[\![x]\!] \mid x \in W^{\max} \text{ and } x \in V(p)\}$.[4]

The definition of R'_i requires some explanation. At first, one might think that defining the set of i-edges as $\{([\![x]\!], [\![y]\!]) \mid x, y \in W^{\max} \text{ and } xR_iy\}$ would be sufficient. However, this is not the case. To show this, consider the pointed model $\mathcal{N}_1$ in Figure 3 and let $k = 2$. One can easily show that $W^{\max} = \{w_d, w_1, w_2\}$, hence the rooted 2-contraction has worlds $W' = \{[\![w_d]\!], [\![w_1]\!], [\![w_2]\!]\}$. Since $(w_1, w_2) \notin R$, defining the accessibility relation

[3] The definition of R'_i doesn't depend on the choice of maximal representatives: If $[\![x]\!] = [\![x']\!]$ and $[\![y]\!] = [\![y']\!]$ and $x, x', y, y' \in W^{\max}$, then $b(x) > 0$ iff $b(x') > 0$ and $\exists z(xR_iz$ and $y \leftrightarrows_{b(x)-1} z)$ iff $\exists z'(x'R_iz'$ and $y' \leftrightarrows_{b(x')-1} z')$. To prove this, first note that since $[\![x]\!] = [\![x']\!]$, we get $x \in [\![x']\!]$ and $x' \in [\![x]\!]$. Similarly for y and y'. Since $x \in [\![x']\!] = [x']_{b(x')}$, we then get $x \leftrightarrows_{b(x')} x'$, and hence $b(x) = b(x')$, by Proposition 3.7(4). Hence, $b(x) > 0$ iff $b(x') > 0$. Now suppose $\exists z(xR_iz$ and $y \leftrightarrows_{b(x)-1} z)$. We need to show that $\exists z'(x'R_iz'$ and $y' \leftrightarrows_{b(x')-1} z')$. Since $x \leftrightarrows_{b(x')} x'$ and xR_iy, there exists a z' such that $x'R_iz'$ and $y \leftrightarrows_{b(x')-1} z'$. Since xR_iy then $b(y) \geq b(x) - 1$, by Lemma 3.3. As $b(x) = b(x')$, we get $b(y) \geq b(x') - 1$. Since $y' \in [\![y]\!] = [y]_{b(y)}$, we get $y' \leftrightarrows_{b(y)} y$ and hence $y' \leftrightarrows_{b(x')-1} y$. Combining $y \leftrightarrows_{b(x')-1} z'$ and $y' \leftrightarrows_{b(x')-1} y$, we get $y' \leftrightarrows_{b(x')-1} z'$, as required.

[4] The definition of $V'(p)$ is well-defined since from $x' \in [\![x]\!]$ we get that $x' \leftrightarrows_{b(x)} x$ and, thus, $x \in V(p)$ iff $x' \in V'(p)$, by [atom] of Definition 2.5.

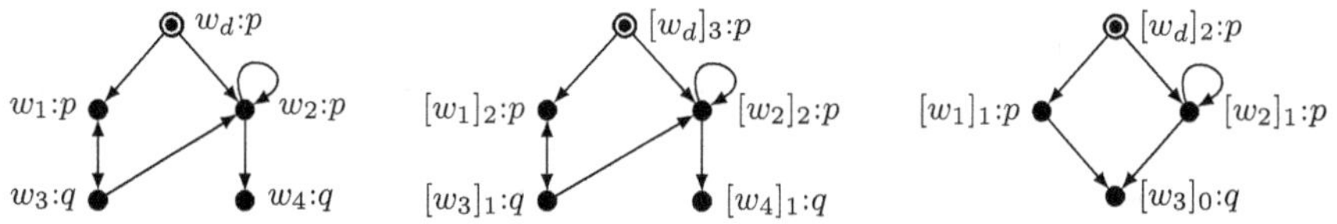

Fig. 4. Pointed model $\mathcal{M}$ (left) of Figure 2, $\lfloor\!\lfloor \mathcal{M} \rfloor\!\rfloor_3$ (center) and $\lfloor\!\lfloor \mathcal{M} \rfloor\!\rfloor_2$ (right).

of the contracted model by $R' = \{([\![x]\!], [\![y]\!]) \mid x, y \in W^{\max} \text{ and } xRy\}$ would imply $([\![w_1]\!], [\![w_2]\!]) \notin R'$. But then the formula $\Diamond r$ would be true in w_1 and not in $[\![w_1]\!]$, *i.e.*, $w_1 \not\leftrightarrows_1 [\![w_1]\!]$, implying that $\mathcal{N}_1$ is not even 2-bisimilar to its own rooted 2-contraction! With our current definition of R', we actually get that the rooted 2-contraction of $\mathcal{N}_1$ is exactly the model $\mathcal{N}_2$ of Figure 3 that we in Example 3.4 showed to be a world-minimal model 2-bisimilar to $\mathcal{N}_1$.

Lemma 3.11 *Let $\mathcal{M} = ((W, R, V), w_d)$ and $\lfloor\!\lfloor \mathcal{M} \rfloor\!\rfloor_k = ((W', R', V'), w'_d)$. For all $i \in \mathcal{I}$ we have $R'_i \supseteq \{([\![x]\!], [\![y]\!]) \mid x, y \in W^{\max}, xR_iy \text{ and } b(x) > 0\}$.*

Proof. Let $x, y \in W^{\max}$ be such that xR_iy and $b(x) > 0$. We need to show that $([\![x]\!], [\![y]\!]) \in R'_i$. By definition of R'_i, this is true if there exists a $z \in W$ such that xR_iz and $y \leftrightarrows_{b(x)-1} z$. Since xR_iy, we can take $z = y$. □

Example 3.12 Let $\mathcal{M} = (M, w_d)$ with $M = (W, R, V)$ be the pointed model of Figure 2, shown again in Figure 4 (left). Let $\lfloor\!\lfloor \mathcal{M} \rfloor\!\rfloor_3 = (M', w'_d)$, with $M' = (W', R', V')$, shown in Figure 4 (center). In Example 3.8, we showed that, when $k = 3$, we have $W^{\max} = W$. From Definition 3.10, we then have $W' = \{[w_d]_3, [w_1]_2, [w_2]_2, [w_3]_1, [w_4]_1\}$ and $w'_d = [w_d]_3$, where $[w_d]_3 = \{w_d\}$, $[w_1]_2 = \{w_1\}$, $[w_2]_2 = \{w_2\}$, $[w_3]_1 = \{w_3\}$ and $[w_4]_1 = \{w_4\}$. Since $W^{\max} = W$ and $[\![w]\!] = w$ for all $w \in W$, it is easy to check that we also get $R' = \{[\![x]\!], [\![y]\!] \mid xRy\}$. Let now $\lfloor\!\lfloor \mathcal{M} \rfloor\!\rfloor_2 = (M', w'_d)$, with $M' = (W', R', V')$, shown in Figure 4 (right). Again, from Example 3.8, we have $W^{\max} = W$. From Definition 3.10, we then have $W' = \{[w_d]_2, [w_1]_1, [w_2]_1, [w_3]_0\}$ and $w'_d = [w_d]_2$, where $[w_d]_2 = \{w_d\}$, $[w_1]_1 = \{w_1\}$, $[w_2]_1 = \{w_2\}$ and $[w_3]_0 = \{w_3, w_4\}$. In this case we get $R' = \{([\![x]\!], [\![y]\!]) \mid xRy \text{ and } b(x) > 0\}$.

4 Properties of Rooted k-Contractions

The first crucial property to show is that rooted k-contractions are k-bisimilar to their original models.

Theorem 4.1 *Let $\mathcal{M}$ be a pointed model and let $k \geq 0$. Then, $\mathcal{M} \leftrightarrows_k \lfloor\!\lfloor \mathcal{M} \rfloor\!\rfloor_k$.*

Proof. Let $\mathcal{M} = ((W, R, V), w_d)$ and $\lfloor\!\lfloor \mathcal{M} \rfloor\!\rfloor_k = ((W', R', V'), w'_d)$. For all $0 \leq h \leq k$, let $Z_h \subseteq W \times W'$ be the following binary relation:

$$Z_h = \{(x, [\![x']\!]) \mid x' \in W^{\max}, x' \leftrightarrows_h x \text{ and } b(x) \geq h\}.$$

We now show that the sequence $Z_k, \ldots, Z_0$ is a k-bisimulation between $\mathcal{M}$ and $\lfloor\!\lfloor \mathcal{M} \rfloor\!\rfloor_k$. Clearly, $Z_k \subseteq \cdots \subseteq Z_0$. Moreover, since $w_d \in W^{\max}$ (by Proposition 3.7(1)) and $b(w_d) = k$, it follows that $(w_d, [\![w_d]\!]) \in Z_k$.

We first show [atom]. Let $(x, x'') \in Z_0$. Then, by definition of Z_0, we have that $x'' = [\![x']\!]$ for some $x' \in W^{\max}$ such that $x' \leftrightarrows_0 x$ and $b(x) \geq 0$. From $x'' = [\![x']\!]$, we get $x'' \leftrightarrows_{b(x')} x'$ and hence $x'' \leftrightarrows_0 x'$. Thus, $x'' \leftrightarrows_0 x' \leftrightarrows_0 x$ and, by [atom] of Definition 2.5, we then get $x'' \in V'(p)$ iff $x \in V(p)$, as required.

Before moving to [forth$_h$] and [back$_h$], we show the following claim.

Claim 1. Let $h \leq k$, $x \in W$ and $x' \in W^{\max}$ be such that $x \leftrightarrows_h x'$ and $b(x) \geq h$. Then $b(x') \geq h$.

Proof of claim. Since $x' \in W^{\max}$, we have $x \not\succ x'$, hence either $b(x) \not> b(x')$ or $x \not\leftrightarrows_{b(x')} x'$. If $b(x) \not> b(x')$, we get $b(x') \geq b(x) \geq h$, as required. If $x \not\leftrightarrows_{b(x')} x'$, then since $x \leftrightarrows_h x'$, we must have $b(x') > h$. This proves the claim.

We now show [forth$_h$]. Let $h < k$, $(x, x'') \in Z_{h+1}$, and xR_iy. Then $x'' = [\![x']\!]$, where $x' \in W^{\max}$, $x' \leftrightarrows_{h+1} x$ and $b(x) \geq h + 1$. From Claim 1, we get $b(x') \geq h + 1$. We need to find a world $y'' \in W'$ such that $x''R'_iy''$ and $(y, y'') \in Z_h$. Since xR_iy and $x \leftrightarrows_{h+1} x'$, there exists $z \in W$ such that $x'R_iz$ and $y \leftrightarrows_h z$. Let y' be a maximal representative of z. Then, $y' \in W^{\max}$, $b(y') \geq b(z)$ and $y' \leftrightarrows_{b(z)} z$. Since $x'R_iz$, by Lemma 3.3 we have $b(z) \geq b(x') - 1$. Since $y' \leftrightarrows_{b(z)} z$, then we also get $y' \leftrightarrows_{b(x')-1} z$. We now have $x', y' \in W^{\max}$, $x'R_iz$, $y' \leftrightarrows_{b(x')-1} z$ and $b(x') \geq h+1 > 0$, which by Definition 3.10 means $[\![x']\!]R'_i[\![y']\!]$. Letting $y'' = [\![y']\!]$, we have hence found a y'' such that $x''R'_iy''$. The only thing left to prove now is that $(y, y'') \in Z_h$. By definition of Z_h, it suffices to prove $y' \in W^{\max}$, $y' \leftrightarrows_h y$ and $b(y) \geq h$. We already have $y' \in W^{\max}$. Since $y' \leftrightarrows_{b(x')-1} z$ and $b(x') \geq h+1$, we get $y' \leftrightarrows_h z$. As we also have $y \leftrightarrows_h z$, we get $y' \leftrightarrows_h y$, as required. Finally, since xR_iy and $b(x) \geq h + 1$, Lemma 3.3 gives $b(y) \geq h$. This concludes [forth$_h$].

We now show [back$_h$]. Let $h < k$, $(x, x'') \in Z_{h+1}$, and $x''R'_iy''$. We need to find a world $y \in W$ such that xR_iy and $(y, y'') \in Z_h$. Since $x''R'_iy''$, by Definition 3.10 there exist $x', y' \in W^{\max}$ and $z \in W$ such that $x'' = [\![x']\!]$, $y'' = [\![y']\!]$, $x'R_iz$, $y' \leftrightarrows_{b(x')-1} z$ and $b(x') > 0$. Since $(x, x'') \in Z_{h+1}$, then $x'' = [\![\hat{x}]\!]$, where $\hat{x} \in W^{\max}$, $\hat{x} \leftrightarrows_{h+1} x$ and $b(x) \geq h + 1$. By Claim 1, we get $b(\hat{x}) \geq h + 1$. Since $x', \hat{x} \in W^{\max}$ and $[\![x']\!] = [\![\hat{x}]\!]$, by Proposition 3.7(4) we get $b(x') = b(\hat{x})$. From $x' \leftrightarrows_{b(x')} \hat{x}$ and $b(x') \geq h + 1$ we get $x' \leftrightarrows_{h+1} \hat{x}$ and, hence, $x' \leftrightarrows_{h+1} x$. Since $x' \leftrightarrows_{h+1} x$ and $x'R_iz$, there exists $y \in W$ such that xR_iy and $y \leftrightarrows_h z$. Only left to show is that $(y, y'') \in Z_h$. From $y' \leftrightarrows_{b(x')-1} z$ and $b(x') \geq h + 1$, we get $y' \leftrightarrows_h z$ and, thus, $y' \leftrightarrows_h y$. Since xR_iy and $b(x) \geq h + 1$, Lemma 3.3 gives $b(y) \geq h$. We now have $y' \in W^{\max}$, $y' \leftrightarrows_h y$ and $b(y) \geq h$. Thus, $(y, y'') \in Z_h$, as required. This concludes [back$_h$]. □

We now prove world minimality. To show this property, it is useful to group the worlds of a rooted k-contraction wrt. to their bound and analyze them separately. Specifically, we prove that each such group of worlds is minimal. To this end, we first show an intermediate result, namely that a maximal representative x of $\mathcal{M}$ and its representative class $[\![x]\!]$ have the same bound (wrt. $\mathcal{M}$ and $\lfloor\!\lfloor \mathcal{M} \rfloor\!\rfloor_k$, respectively). This result highlights the link between the notions of maximal representatives and representative classes, since a representative class $[\![x]\!]$ maintains the same bound as the maximal representative x.

Lemma 4.2 *Let $\mathcal{M} = ((W, R, V), w_d)$ be a pointed model with rooted k-contraction $\llbracket \mathcal{M} \rrbracket_k = ((W', R', V'), w'_d)$ and let $x \in W^{\max}$. Then $b(x) = b(\llbracket x \rrbracket)$.*

Proof. The proof is by induction on $h = b(x)$. For the base case, we consider $h = k$ (we do induction from $h = k$ down to $h = 0$). By definition, only the designated world of a model has bound k, so we immediately get $b(w_d) = b(\llbracket w_d \rrbracket) = k$, concluding the base case. Assume now by induction hypothesis (I.H.) that for all $x \in W^{\max}$ with $b(x) = h > 0$ we have $b(x) = b(\llbracket x \rrbracket)$. Let $y \in W^{\max}$ with $b(y) = h-1$. We need to show $b(\llbracket y \rrbracket) = h-1$. Since $b(y) = h-1$, there must exist an x with xR_iy and $b(x) = h$. We now prove $x \in W^{\max}$ by contradiction: Assuming $x \notin W^{\max}$, there exists $x' \in W^{\max}$ such that $x' \succ x$, *i.e.*, $b(x') > b(x)$ and $x' \leftrightarrows_{b(x)} x$. Since xR_iy, there exists y' such that $x'R_iy'$ and $y \leftrightarrows_{b(x)-1} y'$. As $b(x) = h$ and $b(y) = h - 1$ we get $y \leftrightarrows_{b(y)} y'$. Since $x'R_iy'$ and $b(x') > b(x) = h$, Lemma 3.3 gives $b(y') > h - 1$ and thus $b(y') > b(y)$. We now have $b(y') > b(y)$ and $y' \leftrightarrows_{b(y)} y$, which means $y' \succ y$, contradicting $y \in W^{\max}$. Thus, $x \in W^{\max}$. Since $x, y \in W^{\max}$, xR_iy and $b(x) = h > 0$, Lemma 3.11 gives $\llbracket x \rrbracket R'_i \llbracket y \rrbracket$. Since $b(\llbracket x \rrbracket) = h$ (by I.H.), Lemma 3.3 then gives $b(\llbracket y \rrbracket) \geq h-1$. We also have $b(\llbracket y \rrbracket) \leq h-1$, since if $b(\llbracket y \rrbracket) \geq h$, then I.H. would give $b(y) \geq h$, contradicting $b(y) = h - 1$. Thus $b(\llbracket y \rrbracket) = h - 1$, as required. □

Corollary 4.3 *If $x \neq y$ are worlds of $\llbracket \mathcal{M} \rrbracket_k$ and $b(x) = b(y) = h$ then $x \not\leftrightarrows_h y$.*

Proof. Let $\llbracket \mathcal{M} \rrbracket_k = (M, w_d)$ and $\mathcal{M} = (M', w'_d)$. By Definition 3.10, we have $x = \llbracket x' \rrbracket$ and $y = \llbracket y' \rrbracket$ for some worlds $x', y' \in W^{\max}$. By Lemma 4.2 we get $b(x) = b(x')$ and $b(y) = b(y')$, and hence $b(x') = b(y') = h$. Since $x \neq y$ and $b(x') = b(y') = h$, we get $x' \not\leftrightarrows_h y'$. From the proof of Theorem 4.1 we get $(M', x') \leftrightarrows_h (M, x)$ and $(M', y') \leftrightarrows_h (M, y)$, and hence $x \not\leftrightarrows_h y$. □

For a model $M = (W, R, V)$ and $h \geq 0$, let W_h denote the subset of worlds with bound h, *i.e.*, $W_h = \{w \in W \mid b(w) = h\}$ (given a $k \geq 0$).

Lemma 4.4 *Let $k \geq 0$, let $\mathcal{M} = ((W, V, R), w_d)$ be a rooted k-contraction, and $\mathcal{M}' = ((W', V', R'), w'_d)$ be a world-minimal pointed model k-bisimilar to $\mathcal{M}$. Then, for any $0 \leq h \leq k$, the relation $\leftrightarrows_h$ is a bijection between W_h and W'_h.*

Proof. Since $\mathcal{M} = ((W, R, V), w_d)$ is a rooted k-contraction, we have $\mathcal{M} = \llbracket \mathcal{M}'' \rrbracket_k$ for some $\mathcal{M}'' = ((W'', R'', V''), w''_d)$. We first show that for each $x \in W_h$, there is a unique $x' \in W'_h$ such that $x \leftrightarrows_h x'$. Given $x \in W_h$, we have $b(x) = h$, which implies the existence of a path of length $k - h$ from w_d to x. Since $w_d \leftrightarrows_k w'_d$, by repeated application of [forth], we get a path of length $k-h$ from w'_d to a world x' with $x \leftrightarrows_h x'$. Since x' is reachable by a path of length $k - h$ from w'_d, we have $b(x') \geq k - (k - h) = h$. We now show that $b(x') \leq h$ by contradiction, which together with $b(x') \geq h$ gives $b(x') = h$, thus obtaining $x' \in W'_h$. Assume $b(x') > h$, *i.e.*, $d(x') < k - h$. Then, there is a path of length $< k - h$ from w'_d to x' and, since $w_d \leftrightarrows_k w'_d$, by repeated applications of [back], we get a path of length $< k - h$ from w_d to a world y with $y \leftrightarrows_h x'$. As above, since y is reachable by such a path, we have $b(y) > h$. From Definition 3.10, there exists $x'', y'' \in W^{\max}$ such that $x = \llbracket x'' \rrbracket$ and $y = \llbracket y'' \rrbracket$. Lemma 4.2 now gives $b(x'') = b(x) = h$ and $b(y'') = b(y) > h$. Since $x'', y'' \in W^{\max}$ and since both have bound $\geq h$, we have that $(x'', [\![x'']\!]), (y'', [\![y'']\!]) \in Z_h$, where Z_h is the binary relation defined in the proof of Theorem 4.1. The proof of Theorem 4.1 shows that $Z_0, \ldots, Z_h$ is an h-bisimulation, so $x'' \leftrightarroweq_h [\![x'']\!]$ and $y'' \leftrightarroweq_h [\![y'']\!]$. We then get $y'' \leftrightarroweq_h [\![y'']\!] = y \leftrightarroweq_h x' \leftrightarroweq_h x = [\![x'']\!] \leftrightarroweq_h x''$, showing that $y'' \leftrightarroweq_h x''$. We now have $b(y'') > h = b(x'')$ and $y'' \leftrightarroweq_h x''$, which implies $y'' \succ x''$, contradicting $x'' \in W^{\max}$. This gives the required proof by contradiction that $x' \in W'_h$. So far, we showed that for any $x \in W_h$ there exists a world $x' \in W'_h$ such that $x' \leftrightarroweq_h x'$. Since $\mathcal{M}'$ is world minimal, there is no $y' \in W'_h$ such that $x' \leftrightarroweq_h y'$. Therefore, such an x' is unique in W'_h, as required.

We now show that for each $x' \in W'_h$, there exists a unique $x \in W_h$ with $x \leftrightarroweq_h x'$. Letting $x' \in W'_h$, we can first reason symmetrically as above to show that there exists a world $x \in W_h$ such that $x \leftrightarroweq_h x'$ and $b(x) \geq h$ (using [back] instead of [forth]). Symmetrically to before, we now show that $b(x) \leq h$ by contradiction, from which we can conclude $x \in W_h$. Assume $b(x) > h$, *i.e.*, $d(x) < k - h$. Then there is a path of length $< k - h$ from w_d to x and, since $w_d \leftrightarroweq_k w'_d$, by repeated applications of [forth], we get a path of length $< k - h$ from w'_d to a world y' with $x \leftrightarroweq_h y'$. As above, since y' is reachable by such a path, we have $b(y') > h$. We now have $b(y') > h = b(x')$ and $y' \leftrightarroweq_h x \leftrightarroweq_h x'$. By Lemma 3.5, this implies that there exists a pointed model $\mathcal{N}$ with world set $W' \backslash \{x'\}$ such that $\mathcal{N} \leftrightarroweq_k \mathcal{M}'$, contradicting the fact that $\mathcal{M}'$ is a world-minimal pointed model k-bisimilar to $\mathcal{M}$. This complete the proof by contradiction that $x \in W_h$. The only thing left to prove is uniqueness of x. Suppose x' was h-bisimilar to another world $y \in W_h$. Then we would have $x \neq y$, $b(x) = b(y) = h$, and $x \leftrightarroweq_h y$, contradicting Corollary 4.3. □

Theorem 4.5 *Let $\mathcal{M}$ be a pointed model and let $k \geq 0$. Then $\lfloor\!\lfloor \mathcal{M} \rfloor\!\rfloor_k$ is a world-minimal model k-bisimilar to $\mathcal{M}$,* i.e., *it has the least number of worlds among all models k-bisimilar to $\mathcal{M}$.*

Proof. Let $\mathcal{M}' = \lfloor\!\lfloor \mathcal{M} \rfloor\!\rfloor_k$, let $\mathcal{M}''$ be a world-minimal pointed model with $\mathcal{M}' \leftrightarroweq_k \mathcal{M}''$ and let W' and W'' be the world sets of $\mathcal{M}'$ and $\mathcal{M}''$, respectively. We need to show that $|W'| = |W''|$. From Lemma 4.4, we immediately get that $|W'_h| = |W''_h|$ for all $0 \leq h \leq k$, and hence $|W'| = |W''|$, as required. □

5 Minimal Contractions

We have defined rooted k-contractions and shown them to be world minimal. However, Definition 3.10 does not guarantee that the resulting k-contraction is also edge minimal, as we now exemplify.

Example 5.1 Let $\mathcal{M} = ((W, R, V), w_d)$ be the pointed model in Figure 4 left and let $\lfloor\!\lfloor \mathcal{M} \rfloor\!\rfloor_3 = ((W', R', V'), w'_d)$ be its rooted 3-contraction (Figure 4 center). Recall from Example 3.12 that $W^{\max} = W$ and $b(w_3) = 1$. Since $w_3 R w_1$ and $w_3 R w_2$, Lemma 3.11 hence gives us $[\![w_3]\!] R' [\![w_1]\!]$ and $[\![w_3]\!] R' [\![w_2]\!]$. However, including only one of those edges in R' is sufficient to guarantee 3-bisimilarity to $\mathcal{M}$: $b(w_3) = 1$ and thus $[\![w_3]\!]$ only needs to preserve 1-bisimilarity to w_3.

When not all edges are required, we need to decide which to preserve. To this end, we introduce the notion of *least h-representative* of a world.

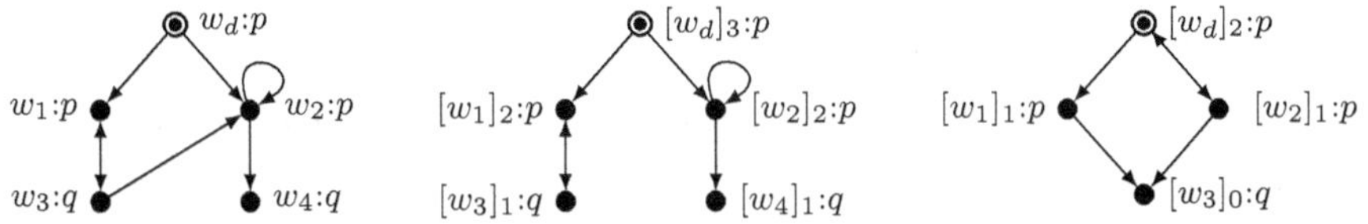

Fig. 5. Pointed model $\mathcal{M}$ (left) of Example 5.5, $\llbracket\mathcal{M}\rrbracket_3^<$ (center) and $\llbracket\mathcal{M}\rrbracket_2^<$ (right).

Definition 5.2 Let $<$ be a total order on W and let $0 \leq h \leq k$. The *least h-representative* of $w \in W$ is the world $\min_h(w) = \min_<\{v \in W^{\max} \mid v \leftrightarrows_h w\}$.

The least h-representative of a world w is the minimal (wrt. $<$) maximal representative v such that $v \leftrightarrows_h w$. We now present the revised definition of rooted k-contraction guaranteeing minimality among k-bisimilar models both in number of worlds and edges (and hence minimality in terms of overall size).

Definition 5.3 Let $\mathcal{M} = ((W, R, V), w_d)$, let $k \geq 0$ and let $<$ be a total order on W. The *rooted k-contraction* of $\mathcal{M}$ wrt. $<$ is the pointed model $\llbracket\mathcal{M}\rrbracket_k^< = ((W', R', V'), \llbracket w_d \rrbracket)$, where:

- $W' = \{\llbracket x \rrbracket \mid x \in W^{\max}\}$;
- $R'_i = \{(\llbracket x \rrbracket, \llbracket \min_{b(x)-1}(y) \rrbracket) \mid x \in W^{\max}, xR_iy \text{ and } b(x) > 0\}$;
- $V'(p) = \{\llbracket x \rrbracket \mid x \in W^{\max} \text{ and } x \in V(p)\}$.

Well-definedness of the definition (independence of choice of representatives) is guaranteed in the same way as for Definition 3.10.

Lemma 5.4 *Let $\mathcal{M} = ((W, R, V), w_d)$, $\llbracket\mathcal{M}\rrbracket_k^< = ((W', R', V'), w'_d)$ and $\llbracket\mathcal{M}\rrbracket_k = ((W'', R'', V''), w''_d)$. Then, $R'_i \subseteq R''_i$.*

Proof. Let $x'R'_iy'$. Then by Definition 5.3, there exists $x, y \in W$ such that $x' = \llbracket x \rrbracket$, $y' = \llbracket \min_{b(x)-1}(y) \rrbracket$, $x \in W^{\max}$, xR_iy and $b(x) > 0$. From Definition 5.2, we immediately get $\min_{b(x)-1}(y) \leftrightarrows_{b(x)-1} y$ and $\min_{b(x)-1}(y) \in W^{\max}$. Since $x, \min_{b(x)-1}(y) \in W^{\max}$, xR_iy, $\min_{b(x)-1}(y) \leftrightarrows_{b(x)-1} y$, and $b(x) > 0$, choosing $z = y$ in Definition 3.10, we get $(\llbracket x \rrbracket, \llbracket \min_{b(x)-1}(y) \rrbracket) \in R''_i$, *i.e.*, $x'R''_iy'$. □

Example 5.5 Let $\mathcal{M} = ((W, R, V), w_d)$ be the pointed model of Figure 2 (also shown in Figure 5 left). Let $<$ be a total order on W such that $w_d < w_1 < w_2 < w_3 < w_4$. The pointed models $\llbracket\mathcal{M}\rrbracket_3^<$ and $\llbracket\mathcal{M}\rrbracket_2^<$, also shown in Figure 5, will now be analyzed. First let $\llbracket\mathcal{M}\rrbracket_3^< = ((W', R', V'), w'_d)$. From Example 3.12, we have $W^{\max} = W$, $W' = \{[w_d]_3, [w_1]_2, [w_2]_2, [w_3]_1, [w_4]_1\}$ and $w'_d = [w_d]_3$. Now for the edges of $\llbracket\mathcal{M}\rrbracket_3^<$. Notice that for all xRy of $\mathcal{M}$ such that $b(x) \geq 2$, we have that $\min_{b(x)-1}(y) = y$. Thus, we have $[w_d]_3R'[w_1]_2$, $[w_d]_3R'[w_2]_2$, $[w_1]_2R'[w_3]_1$, $[w_2]_2R'[w_2]_2$ and $[w_2]_2R'[w_4]_1$. Finally, consider the edges w_3Rw_1 and w_3Rw_2 as discussed in Example 5.1. Since $\min_{b(w_3)-1}(w_1) = \min_{b(w_3)-1}(w_2) = w_1$, we have $[w_3]_1R'[w_1]_2$ (and not $[w_3]_1R'[w_2]_2$).

Now let $\llbracket\mathcal{M}\rrbracket_2^< = ((W', V', R'), w'_d)$. From Example 3.12, we have $W^{\max} = W$, $W' = \{[w_d]_2, [w_1]_1, [w_2]_1, [w_3]_0\}$ and $w'_d = [w_d]_2$. From $\min_{b(w_d)-1}(w_1) = w_1$ and $\min_{b(w_d)-1}(w_2) = w_2$ we get $[w_d]_2R'[w_1]_1$ and $[w_d]_2R'[w_2]_1$. From

$\min_{b(w_1)-1}(w_3) = \min_{b(w_2)-1}(w_4) = w_3$ we get $[w_1]_1 R'[w_3]_0$ and $[w_2]_1 R'[w_3]_0$. Finally, since $\min_{b(w_2)-1}(w_2) = w_d$, we have $[w_2]_1 R'[w_d]_2$.

Theorem 5.6 *Let $\mathcal{M}$ be a pointed model and let $k \geq 0$. Then $\mathcal{M} \leftrightarroweq_k \llbracket \mathcal{M} \rrbracket_k^<$.*

Proof. Let $\mathcal{M} = ((W, R, V), w_d)$ and $\llbracket \mathcal{M} \rrbracket_k^< = ((W', R', V'), w'_d)$. For all $0 \leq h \leq k$, let $Z_h \subseteq W \times W'$ be as in the proof of Theorem 4.1:

$$Z_h = \{(x, \llbracket x' \rrbracket) \mid x' \in W^{\max}, x' \leftrightarroweq_h x \text{ and } b(x) \geq h\}.$$

We now show that $Z_k, \ldots, Z_0$ is a k-bisimulation between $\mathcal{M}$ and $\llbracket \mathcal{M} \rrbracket_k^<$. From the proof of Theorem 4.1, we immediately get $Z_k \subseteq \cdots \subseteq Z_0$, $(w_d, \llbracket w_d \rrbracket) \in Z_k$ and [atom]. Moreover, from the same proof, we also get [back$_h$], since by Lemma 5.4 we have $R'_i \subseteq R''_i$, where $((W'', R'', V''), w''_d)$ is the k-contraction of $\mathcal{M}$ of Definition 3.10.

To show [forth$_h$], let $h < k$, $(x, x'') \in Z_{h+1}$, and xR_iy. Then $x'' = \llbracket x' \rrbracket$, where $x' \in W^{\max}$, $x' \leftrightarroweq_{h+1} x$ and $b(x) \geq h+1$. We need to find $y'' \in W'$ such that $x''R'_iy''$ and $(y, y'') \in Z_h$. Since xR_iy and $x \leftrightarroweq_{h+1} x'$, there exists $z \in W$ such that $x'R_iz$ and $y \leftrightarroweq_h z$. Let $y' = \min_{b(x')-1}(z)$. Then $y' \in W^{\max}$ and $y' \leftrightarroweq_{b(x')-1} z$. By Claim 1 of Theorem 4.1, we get $b(x') \geq h+1$, and thus $y' \leftrightarroweq_h z \leftrightarroweq_h y$. Since xR_iy and $b(x) \geq h+1$, Lemma 3.3 gives $b(y) \geq h$. Let $y'' = \llbracket y' \rrbracket$. We now have $y' \in W^{\max}$, $y' \leftrightarroweq_h y$ and $b(y) \geq h$, which by definition of Z_h means that $(y, y'') \in Z_h$. By Definition 5.3, from $x' \in W^{\max}$, $x'R_iz$, $y' = \min_{b(x')-1}(z)$ and $b(x') \geq h+1 > 0$, we get $x''R'_iy''$, as required. □

Lemma 5.7 *Let $\mathcal{M} = ((W, R, V), w_d)$ be a pointed model with rooted k-contraction $\llbracket \mathcal{M} \rrbracket_k^< = ((W', R', V'), w'_d)$ and let $x \in W^{\max}$. Then $b(x) = b(\llbracket x \rrbracket)$.*

Proof. The proof mimics the proof of Lemma 4.2, except we have fewer edges in $\llbracket \mathcal{M} \rrbracket_k^<$ than in $\llbracket \mathcal{M} \rrbracket_k$, so we cannot rely on Lemma 3.11. The proof is again by induction on $h = b(x)$ (from $h = k$ down to $h = 0$), and the base case is identical to the proof of Lemma 4.2. Assume now by induction hypothesis (I.H.) that for all $x \in W^{\max}$ with $b(x) = h > 0$ we have $b(x) = b(\llbracket x \rrbracket)$. Let $y \in W^{\max}$ with $b(y) = h-1$. We need to show $b(\llbracket y \rrbracket) = h-1$. Since $b(y) = h-1$, there must exist an x with xR_iy and $b(x) = h$. From this it follows that $x \in W^{\max}$ exactly as in the proof of Lemma 4.2 (we are here reasoning about $\mathcal{M}$ only). Since $x \in W^{\max}$, xR_iy and $b(x) = h > 0$, Definition 5.3 gives $\llbracket x \rrbracket R'_i \llbracket \min_{b(x)-1}(y) \rrbracket$, *i.e.*, $\llbracket x \rrbracket R'_i \llbracket \min_{b(y)}(y) \rrbracket$. Since $b(\llbracket x \rrbracket) = h$, Lemma 3.3 then gives $b(\llbracket \min_{b(y)}(y) \rrbracket) \geq h-1$. Since $\min_{b(y)}(y) \leftrightarroweq_{b(y)} y$, we get that $\min_{b(y)}(y) \succeq y$. Since $y \in W^{\max}$, we must then have $b(\min_{b(y)}(y)) \leq b(y)$, since otherwise we would have $\min_{b(y)}(y) \succ y$, contradicting the maximality of y. Since $\min_{b(y)}(y) \in W^{\max}$, we must also have $b(\min_{b(y)}(y)) \geq b(y)$, since otherwise we would have $y \succ \min_{b(y)}(y)$, contradicting the maximality of $\min_{b(y)}(y)$. We can thus conclude $b(\min_{b(y)}(y)) = b(y)$. We also have $b(\llbracket \min_{b(y)}(y) \rrbracket) \leq h-1$, since if $b(\llbracket \min_{b(y)}(y) \rrbracket) \geq h$, then I.H. would give $b(\min_{b(y)}(y)) \geq h > h-1 = b(y)$, contradicting what we just concluded. We can thus conclude $b(\llbracket \min_{b(y)}(y) \rrbracket) = h-1$. Now note that since $\min_{b(y)}(y) \leftrightarroweq_{b(y)} y$ and $b(\min_{b(y)}(y)) = b(y)$, we get $\llbracket \min_{b(y)}(y) \rrbracket = \llbracket y \rrbracket$, and hence $b(\llbracket y \rrbracket) = b(\llbracket \min_{b(y)}(y) \rrbracket) = h-1$, as required. □

Corollary 5.8 *If $x \neq y$ are worlds of $\llbracket \mathcal{M} \rrbracket_k^<$ and $b(x) = b(y) = h$ then $x \not\leftrightarroweq_h y$.*

Proof. The proof is identical to that of Corollary 4.3 by using Lemma 5.7 instead of Lemma 4.2. □

Lemma 5.9 *Let $k \geq 0$, let $\mathcal{M}$ be a rooted k-contraction wrt. $<$, $\mathcal{M}'$ be a world-minimal pointed model k-bisimilar to $\mathcal{M}$ and let W and W' be their world sets. Then, for any $0 \leq h \leq k$, the relation $\leftrightarroweq_h$ is a bijection between W_h and W'_h.*

Proof. The proof is identical to that of Lemma 4.4 by using Definition 5.3, Lemma 5.7 and Theorem 5.6 instead of Definition 3.10, Lemma 4.2 and Theorem 4.1, respectively. □

Theorem 5.10 *Let $\mathcal{M}$ be a pointed model and $k \geq 0$. Then $\llbracket \mathcal{M} \rrbracket_k^<$ is a minimal pointed model k-bisimilar to $\mathcal{M}$ (i.e., it is both world and edge minimal).*

Proof. For any model $M = (W, R, V)$, let $(R_i)_h$ denote the set of i-edges outgoing from worlds in W_h, *i.e.*, $(R_i)_h = R_i \cap (W_h \times W)$. Let $\mathcal{M} = ((W, R, V), w_d)$, $\mathcal{M}' = \llbracket \mathcal{M} \rrbracket_k^< = ((W', R', V'), w'_d)$, and let $\mathcal{M}'' = ((W'', R'', V''), w''_d)$ be a minimal pointed model with $\mathcal{M}' \leftrightarroweq_k \mathcal{M}''$. We can prove that $|W''| = |W'|$ as in Theorem 4.5 by using Lemma 5.9 and Corollary 5.8 instead of Lemma 4.4 and Corollary 4.3, respectively. We then only need to show that $|R'_i| \leq |R''_i|$ for all $i \in \mathcal{I}$. To achieve a contradiction, assume $|(R'_i)_h| > |(R''_i)_h|$ for some i and h. Then $(x', y') \in (R'_i)_h$ for some x', y'. This implies $x' \in W'_h$ and hence $b(x') = h$. By Lemma 5.9, there then exists $x'' \in W''_h$ such that $x' \leftrightarroweq_h x''$. By [back] and [forth], this implies that each i-successor of x'' is $(h-1)$-bisimilar to an i-successor of x' and vice versa. Since $|(R'_i)_h| > |(R''_i)_h|$, this is only possible if there exist two distinct i-successors y'_1 and y'_2 of x' that are $(h-1)$-bisimilar. We can now reason as follows for $n = 1, 2$. Since $b(x') = h$ and $(x', y'_n) \in (R'_i)_h$, Lemma 3.3 gives us $b(y'_n) \geq h - 1$; and Definition 5.3 further gives us the existence of $x \in W^{\max}$ and $y_n \in W$ such that $x R_i y_n$, $x' = \llbracket x \rrbracket$ and $y'_n = \llbracket \min_{h-1}(y_n) \rrbracket$. Since, by definition, $\min_{h-1}(y_n) \in W^{\max}$, and since $b(y'_n) \geq h - 1$, Lemma 5.7 gives $b(\min_{h-1}(y_n)) \geq h - 1$. From $\min_{h-1}(y_n) \in W^{\max}$ and $b(\min_{h-1}(y_n)) \geq h - 1$, we then get $(\min_{h-1}(y_n), \llbracket \min_{h-1}(y_n) \rrbracket) \in Z_{h-1}$, where Z_{h-1} is the binary relation defined in the proof of Theorem 5.6. The proof of that theorem shows that $Z_0, \ldots, Z_{h-1}$ is an $(h-1)$-bisimulation, so $\min_{h-1}(y_n) \leftrightarroweq_{h-1} \llbracket \min_{h-1}(y_n) \rrbracket$. We now get $y_1 \leftrightarroweq_{h-1} \min_{h-1}(y_1) \leftrightarroweq_{h-1} \llbracket \min_{h-1}(y_1) \rrbracket = y'_1 \leftrightarroweq_{h-1} y'_2 = \llbracket \min_{h-1}(y_2) \rrbracket \leftrightarroweq_{h-1} \min_{h-1}(y_2) \leftrightarroweq_{h-1} y_2$. This shows that $y_1 \leftrightarroweq_{h-1} y_2$, and hence $\min_{h-1}(y_1) = \min_{h-2}(y_2)$ and thus $y'_1 = \llbracket \min_{h-1}(y_1) \rrbracket = \llbracket \min_{h-1}(y_2) \rrbracket = y'_2$, contradicting $y'_1 \neq y'_2$. □

6 Exponential Succinctness

In this section, we show that, for any $k \geq 0$, rooted k-contractions can be exponentially more succinct than standard k-contractions. This means that we can create models of arbitrary size for which the rooted k-contraction is exponentially smaller than the corresponding standard k-contraction.

A binary tree of height k has $2^{k+1} - 1$ nodes. We will build a model $\mathcal{M}_k = ((W_k, R_k, V_k), \varepsilon)$ on a binary tree of height k such that the standard

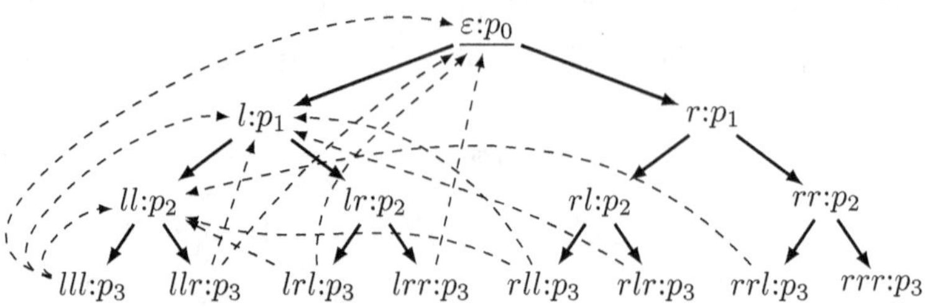

Fig. 6. The pointed model $\mathcal{M}_k = ((W_k, R_k, V_k), \varepsilon)$ with designated world ε, for $k = 3$. The solid and dashed edges are the accessibility edges of $\Box_s$ and $\Box_d$, respectively.

k-contraction of $\mathcal{M}_k$ still has all $2^{k+1} - 1$ nodes as worlds, but the rooted k-contraction is just a path in the tree, so a model with $k+1$ worlds. Hence the rooted k-contraction is exponentially smaller.

The nodes of a binary tree of height k can be represented as strings of length at most k over the alphabet $\{l, r\}$, where l and r represent the left and right child of a node, respectively. This is illustrated in Figure 6 for $k = 3$. The root is named ε (the empty string), and the left and right children of a node σ are σl and σr, respectively. More precisely, we let the set of worlds W_k of $\mathcal{M}_k$ be the set of strings of length at most k over the alphabet $\{l, r\}$, *i.e.*, $W_k = \{\sigma \in \{l, r\}^* \mid |\sigma| \leq k\}$. The tree edges (solid edges in Figure 6) are the edges from σ to σl (left child) and to σr (right child). We let $\mathcal{I} = \{s, d\}$, where the accessibility relation $(R_k)_s$ of the modality $\Box_s$ represents the solid edges. Hence we let $(R_k)_s = \{(\sigma, \sigma\alpha) \in W_k \times W_k \mid \alpha \in \{l, r\}\}$. The accessibility relation $(R_k)_d$ of the modality $\Box_d$ represents the dashed edges, described later.

As shown in Figure 6, the valuation function V_k of $\mathcal{M}_k$ is such that each world at depth n of the tree makes p_n true and all other propositions false. More precisely, the set of atomic propositions of $\mathcal{M}_k$ is $\mathcal{P} = \{p_0, \ldots, p_k\}$, and we let $V_k(p_n) = \{\sigma \in W_k \mid |\sigma| = n\}$. Suppose we decided to let $(R_k)_d = \emptyset$, *i.e.*, ignore the dashed edges. Then, the model is simply a binary tree where each world is labelled by an atomic proposition denoting the depth of the world. Clearly, any two worlds at the same depth are then bisimilar. Hence, the bisimulation contraction of $\mathcal{M}_k$ will simply be a chain model with $k+1$ worlds where the first world is labelled p_0, the second p_1, etc. We add the dashed edges to ensure that $\mathcal{M}_k$ cannot be contracted further when considering standard k-contractions, but where the rooted k-contraction is still the simple chain model.

We now describe the dashed edges of the model. The dashed edges are from the leaf nodes of the binary tree to nodes of the leftmost branch of the tree. The crucial property of these edges is that each leaf node has edges to a different subset of the nodes of the leftmost branch. Hence no two leaf nodes will satisfy the same formulas of modal depth 1. Our specific choice is to put an edge from a leaf node σ to the leftmost node at depth n iff the $(n+1)$st letter of σ is l, see Figure 6. More precisely, we let $(R_k)_d = \{(\alpha_1 \cdots \alpha_k, l^n) \in W_k \times W_k \mid \alpha_{n+1} = l\}$ (where l^n as usual denotes the string with n occurrences of l).

Lemma 6.1 *Let σ, τ be distinct worlds of $\mathcal{M}_k$ at depth n. Then $\sigma \leftrightarrows_{k-n} \tau$ but not $\sigma \leftrightarrows_{k-n+1} \tau$.*

Proof. Since σ and τ are worlds at depth n of the tree, they satisfy the same formulas up to modal depth $k - n$, since none of the dashed edges can be reached by formulas of modal depth $\leq k - n$ from worlds at depth n. From Proposition 2.6 we can then conclude that $\sigma \leftrightarrows_{k-n} \tau$, as required.

Since $\sigma \neq \tau$, they differ in at least one of their positions, say position $m+1$ (*i.e.*, they differ in the $(m+1)$st letter). Suppose without loss of generality that σ has l in position $m+1$ and τ has r. Then by construction of $\mathcal{M}_k$ we have that the formula $\Diamond_s^{n-k}\Diamond_d\, p_m$ is true in σ but not in τ: There exists a solid path of length $n - k$ from σ to a leaf node that can see a p_m world via a dashed edge, but there is no such path from τ. Hence σ and τ are not $(n-k+1)$-bisimilar.□

We can now reason as follows. Since the standard k-contraction only identifies worlds that are k-bisimilar, it will never be able to identify worlds at the same depth of the tree since, according to the lemma, any such two distinct worlds σ and τ are not $(k - n + 1)$-bisimilar and hence not k-bisimilar. It will not be able to identify worlds at different levels either, as they have distinct valuations. Hence the standard k-contraction of $\mathcal{M}_k$ will contain all worlds of the original tree, so $2^{k+1} - 1$ worlds. Compare this to the rooted k-contraction. The worlds of the rooted k-contraction are of the form $[\![\sigma]\!]$ where $\sigma \in W_k^{\max}$. Note that if σ is a world of $\mathcal{M}_k$ at depth n, then $b(\sigma) = k - n$, and hence $[\![\sigma]\!] = [\sigma]_{k-n}$. By the lemma, we then get that any two worlds σ, τ at the same depth of $\mathcal{M}_k$ belong to the same representative class $[\![\sigma]\!]$ (since they are $(k-n)$-bisimilar). Hence the rooted k-contraction can have at most $k + 1$ worlds, one per level of the tree. We have hence proved the following succinctness result.

Theorem 6.2 (Exponential succinctness) *There exist models $\mathcal{M}_k$, $k \geq 0$, for which the rooted k-contraction has $\Theta(k)$ worlds whereas the standard k-contraction has $\Theta(2^k)$ worlds.*

7 Related and Future Work

Implementation and computational complexity The most commonly used technique to calculate bisimulation contractions is *partition refinement* [1,16]. In these algorithms, the worlds are initially partitioned into equivalence classes called *blocks*, for instance the blocks of the initial partition might consist of all worlds having the same valuation. The algorithms then refine the partition iteratively until some stopping condition is met, and the worlds of the contracted model are then (usually) the blocks of that final partition. Many partition refinement algorithms exist, the most famous probably being the one by Paige and Tarjan [16]. The Paige and Tarjan algorithm doesn't immediately lend itself to be adapted to do k-bisimulation contractions, as it is not partitioning blocks in a stratified manner. Other simpler partition refinement algorithms lend themselves more directly to be adapted to do k-bisimulation contractions. One such partition refinement algorithm for k-bisimulation contraction has been presented by Bolander and Lequen [9]. In that algorithm, the first h refinement steps lead to a partition consisting exactly of the h-bisimulation equivalence classes. Hence, by running the algorithm for k refinement steps, we compute all classes $[w]_h$ for all $w \in W$ and $h \leq k$. As shown in that paper, this can be done in polynomial time in the size of the model.

Having computed all classes $[w]_h$ for all $w \in W$ and $h \leq k$, we can easily compute the k-contraction (Definition 5.3). First we compute W', which can also be done in polynomial time. [5] We then compute the accessibility relations R'_i, also in polynomial time. [6] Computing V' is trivial, and we hence have a PTIME algorithm for computing rooted k-contractions. A version of this algorithm has been implemented and is further discussed in Bolander, Burigana and Montali [6] (under submission at the time of writing), as well as being applied to doing depth-bounded epistemic planning.

Rooted k-Contractions and Modal Structures A series of works by Fagin, Geanakoplos, Halpern and Vardi [10,11,12,13,14] explored an alternative semantics to modal logic, with a particular emphasis on epistemic logic. In these papers, they introduce and analyze objects called *modal structures*, which are essentially infinite sequences of functions $\mathbf{f} = \langle f_0, f_1, \dots \rangle$ where a function f_k describes the knowledge of a set of agents up to modal depth k. Among the many results they provide, the authors analyze the relation between pointed Kripke models and modal structures. Interestingly, it turns out that different k-bisimilar pointed models can be associated to the same subsequence $\mathbf{f}_{\leq k} = \langle f_0, \dots, f_k \rangle$ of functions, called the *k-prefix* of the modal structure $\mathbf{f}$. Hence, $\mathbf{f}_{\leq k}$ can be seen as representing a class of k-bisimilar pointed models.

In their works, the authors consider α-sequences of functions, where α is a generic ordinal. Moreover, they also consider an infinitary language of modal logic, where conjunctions of infinite formulas are allowed. Conversely, in this paper we worked under the assumption that k is a finite ordinal, and we only consider finite conjunctions. As a future avenue of research, we would like to investigate the possibility of closing the gap between rooted k-contractions and modal structures. We would like to generalize our notion of rooted k-contraction to account for generic ordinals and infinite models. This, in turn, would allow us to investigate the interplay between rooted k-contractions and modal structures.

---

[5] A naive algorithm to do this runs as follows. First we compute $W^{\max}$. For each world x with $b(x) \geq 0$, we can iterate through all worlds y with $b(y) > b(x)$ and check whether $[y]_{b(x)} = [x]_{b(x)}$. If this is not true for any y, we put x in $W^{\max}$, since then no y strictly represents x, cf. Definition 3.6. From $W^{\max}$, we can then easily compute W', since it is simply the set of $[x]_{b(x)}$ for $x \in W^{\max}$. This algorithm runs in polynomial time since we already computed all the relevant equivalence classes $[w]_h$.

[6] A naive algorithm to compute R'_i is as follows. For each $x \in W^{\max}$ with $b(x) > 0$ we can iterate through all worlds $y \in W$ such that xR_iy. For each such y, we compute the least $(b(x)-1)$-representative z of y by taking the $<$-minimal world in the set $W^{\max} \cap [y]_{b(x)-1}$ (see Definition 5.2), and we add the pair $([\![x]\!], [\![z]\!])$ to R'_i. Assuming we can compute whether $x < y$ in constant time, this also runs in polynomial time.

References

[1] Aceto, L., A. Ingólfsdóttir and J. Srba, *The algorithmics of bisimilarity*, in: *Advanced Topics in Bisimulation and Coinduction*, Cambridge tracts in theoretical computer science **52**, Cambridge University Press, Cambridge, England, 2012 pp. 100–172.

[2] Belle, V., T. Bolander, A. Herzig and B. Nebel, *Epistemic planning: Perspectives on the special issue*, Artificial Intelligence (2022), p. 103842.

[3] Blackburn, P., M. d. Rijke and Y. Venema, "Modal Logic," Cambridge Tracts in Theoretical Computer Science, Cambridge University Press, 2001.

[4] Blackburn, P. and J. van Benthem, *Modal logic: A semantic perspective*, in: *Handbook of Modal Logic* (2006).

[5] Bolander, T. and M. B. Andersen, *Epistemic planning for single- and multi-agent systems*, Journal of Applied Non-Classical Logics **21** (2011), pp. 9–34.

[6] Bolander, T., A. Burigana and M. Montali, *Depth-bounded epistemic planning* (2024), under review. Preprint URL: `https://arxiv.org/abs/2406.01139`.

[7] Bolander, T., T. Charrier, S. Pinchinat and F. Schwarzentruber, *DEL-based epistemic planning: Decidability and complexity*, Artificial Intelligence **287** (2020), pp. 1–34.

[8] Bolander, T. and A. Lequen, *Parameterized complexity of dynamic belief updates: A complete map*, J. Log. Comput. **33** (2023), pp. 1270–1300.

[9] Bolander, T. and A. Lequen, *Parameterized complexity of dynamic belief updates: A complete map*, Journal of Logic and Computation **33** (2023), pp. 1270–1300.

[10] Fagin, R., *A Quantitative Analysis of Modal Logic*, Journal of Symbolic Logic **59** (1994), pp. 209–252.

[11] Fagin, R., J. Geanakoplos, J. Y. Halpern and M. Y. Vardi, *The Expressive Power of the Hierarchical Approach to Modeling Knowledge and Common Knowledge*, in: *Proceedings of the 4th Conference on Theoretical Aspects of Reasoning about Knowledge*, Tark '92 (1992), pp. 229–244.

[12] Fagin, R., J. Geanakoplos, J. Y. Halpern and M. Y. Vardi, *The Hierarchical Approach to Modeling Knowledge and Common Knowledge*, International Journal of Game Theory **28** (1999), pp. 331–365.

[13] Fagin, R., J. Y. Halpern and M. Y. Vardi, *A Model-Theoretic Analysis of Knowledge*, Journal of the ACM **38** (1991), pp. 382–428.

[14] Fagin, R. and M. Y. Vardi, *An Internal Semantics for Modal Logic: Preliminary Report*, in: *Proceedings of the Seventeenth Annual ACM Symposium on Theory of Computing - STOC '85* (1985), pp. 305–315.

[15] Goranko, V. and M. Otto, *Model theory of modal logic*, in: P. Blackburn, J. Van Benthem and F. Wolter, editors, *Handbook of Modal Logic*, Studies in Logic and Practical Reasoning **3**, Elsevier, 2007 pp. 249–329.

[16] Paige, R. and R. E. Tarjan, *Three partition refinement algorithms*, SIAM J. Comput. **16** (1987), pp. 973–989.
URL `https://doi.org/10.1137/0216062`

[17] Yu, Q., X. Wen and Y. Liu, *Multi-agent epistemic explanatory diagnosis via reasoning about actions*, in: F. Rossi, editor, *IJCAI 2013, Proceedings of the 23rd International Joint Conference on Artificial Intelligence, Beijing, China, August 3-9, 2013* (2013), pp. 1183–1190.

Lovász Theorems for Modal Languages

Jesse Comer

University of Pennsylvania

Abstract

A famous result due to Lovász states that two finite relational structures M and N are isomorphic if, and only if, for all finite relational structures T, the number of homomorphisms from T to M is equal to the number of homomorphisms from T to N. Since first-order logic (FO) can describe finite structures up to isomorphism, this can be interpreted as a characterization of FO-equivalence via *homomorphism-count indistinguishability* with respect to the class of finite structures. We identify classes of labeled transition systems (LTSs) such that homomorphism-count indistinguishability with respect to these classes, where "counting" is done within an appropriate semiring structure, captures equivalence with respect to positive-existential modal logic, graded modal logic, and hybrid logic, as well as the extensions of these logics with either backward or global modalities. Our positive results apply not only to finite structures, but also to certain well-behaved infinite structures. We also show that equivalence with respect to positive modal logic and equivalence with respect to the basic modal language are not captured by homomorphism-count indistinguishability with respect to any class of LTSs, regardless of which semiring is used for counting.

Keywords: Homomorphism, Semiring, Graded Modal Logic, Hybrid Logic.

1 Introduction

Lovász's theorem [24] grew out of the study of a fundamental computational problem in graph theory and complexity theory: the graph isomorphism problem. This problem is significant because it is not known to be solvable in polynomial time, but is also not known to be **NP**-complete. In fact, recent work has shown that the problem can be resolved in quasipolynomial time [6], and it is considered to be a potential member of the conjectured class of **NP**-intermediate problems, which exist if and only if $\mathbf{P} \neq \mathbf{NP}$ [22]. Due to the high running time of known exact algorithms for the problem, and the difficulty in determining a lower bound on its complexity, researchers have turned toward the study of heuristic algorithms, such as the color-refinement algorithm, which can distinguish many (but not all) non-isomorphic graphs [7].

The Lovász theorem relates homomorphisms to isomorphisms; while originally stated for structures with a single relation of arbitrary finite arity, it will be convenient for our purposes to consider its generalization to arbitrary finite relational structures. A map between two finite relational structures is a *homomorphism* if, whenever a tuple of elements in the first structure occurs in some relation, then the image of that tuple must also occur in the corresponding relation in the second structure. Given finite structures M and N, we write $\hom_{\mathbb{N}}(N, M)$ to denote the number of homomorphisms from N to M. Given a class $\mathcal{C}$ of finite structures and a fixed finite structure M, we can form the *homomorphism count vector* of M with respect to the class $\mathcal{C}$: the sequence $\hom_{\mathbb{N}}(\mathcal{C}, M) = \langle \hom_{\mathbb{N}}(A, M)\rangle_{A\in\mathcal{C}}$. Using this notation, Lovász's result in [24] can be stated as follows: two finite relational structures M and N are isomorphic if and only if $\hom_{\mathbb{N}}(\mathcal{M}, M) = \hom_{\mathbb{N}}(\mathcal{M}, N)$, where $\mathcal{M}$ is the class of all finite structures. Informally, this says that homomorphism count indistinguishability with respect to $\mathcal{M}$ *captures* isomorphism between finite structures.

Every class of finite structures $\mathcal{C}$ induces an equivalence relation $\sim_{\mathcal{C}}$ on finite structures defined by $M \sim_{\mathcal{C}} N$ if and only if $\hom_{\mathbb{N}}(\mathcal{C}, M) = \hom_{\mathbb{N}}(\mathcal{C}, N)$. Dvořák initiated the study of such equivalence relations for proper subclasses $\mathcal{C}$ of $\mathcal{M}$, showing that two undirected graphs are homomorphism count indistinguishable with respect to the class of trees if and only if they are indistinguishable by the color-refinement algorithm [16]. This was later proven independently by Dell et. al. [15]. In fact, Dvořák and Dell et. al. proved a more general result: homomorphism count indistinguishability with respect to graphs of tree-width at most k captures indistinguishability by the k-dimensional Weisfeiler-Leman (WL) method, where the color-refinement algorithm is the special case for $k = 1$.

Given two graphs with adjacency matrices A and B, an isomorphism between them can be interpreted as a permutation matrix X such that $AX = B$. If we drop the requirement that X contain only binary values, allowing instead positive rational number entries such that each column and row sums to 1, then X is a *fractional isomorphism* [27]. The existence of a fractional isomorphism between two graphs is strictly weaker than the existence of an isomorphism, and so induces a less-refined equivalence relation on the class of all graphs. Fractional isomorphisms are an inherently *linear algebraic* notion, and yet it has also been shown that two graphs are indistinguishable by the color-refinement algorithm if and only if a fractional isomorphism exists between them [31,32].

The *two-variable fragment* (FO^2) is the fragment of first-order logic in which only two variables are allowed. An important extension of FO^2 is the *two-variable fragment with counting quantifiers* (C^2), which contains quantifiers of the form $\exists^{\geq k}$, where $\exists^{\geq k} x\varphi(x)$ asserts the existence of at least k elements satisfying $\varphi(x)$. C^2 is an expressive, but decidable, fragment of FO [19]. A theorem of Cai et. al. shows that two graphs are C^2-equivalent if and only if they are indistinguishable by the the color-refinement algorithm [11]. In fact, they show that two graphs are invariant under the k-variable fragment with counting quantifiers (C^k), which naturally generalizes C^2, if and only if they are indistinguishable by the $(k-1)$-dimensional WL method (for $k \geq 2$).

In artificial intelligence, *graph neural networks* (GNNs) are a type of machine learning architecture which have found numerous applications in the social and physical sciences [33,34]. In [25], Morris et. al. showed that GNNs can distinguish precisely those graphs distinguishable by the color-refinement algorithm. Inspired by the observation that C^2 and the color-refinement algorithm can be generalized to C^k and the k-dimensional WL method, respectively, the authors proposed *k-dimensional GNNs.* They showed that these k-dimensional GNNs can distinguish non-isomorphic graphs with the same expressive power as the k-dimensional WL method.

We have now seen that several seemingly distinct notions – the color-refinement algorithm from graph theory, fractional isomorphism from linear algebra, the two-variable fragment with counting quantifiers from logic, and graph neural networks from machine learning – all induce the same equivalence class on the class of undirected graphs. We have also seen that similar equivalences also hold for the natural generalizations of these notions. Furthermore, they are all undergirded by the same phenomenon: the expressive power of homomorphism count vectors restricted to particular classes of structures.

Due to these connections, Atserias et. al. set out to study which equivalence relations on graphs can be expressed by restricting homomorphism vectors to some fixed class of graphs [5]. In particular, they provide negative results showing that chromatic equivalence and FO^k-equivalence cannot be captured by homomorphism count indistinguishability with respect to any class of graphs. They also introduce a more general perspective, which we also take, in which "counting" can be performed in an arbitrary semiring.

Main Contributions. This paper aims to characterize logical equivalence with respect to various modal languages via homomorphism count indistinguishability with respect to appropriate classes of labeled transition systems (LTSs). The main results are as follows.

(i) Positive-existential modal equivalence is captured by homomorphism count indistinguishability over the Boolean semiring with respect to the class of *trees.* The extended languages with backward and global modalities are captured by the classes of *connected, acyclic LTSs* and *forests,* respectively, over the Boolean semiring.

(ii) Graded modal equivalence is captured by homomorphism count indistinguishability over the natural semiring with respect to the class of trees. The extended languages with backward and global modalities are captured by the classes of *connected, acyclic* LTSs and *forests*, respectively, over the natural semiring.

(iii) Equivalence with respect to hybrid logic is captured by homomorphism count indistinguishability over the natural semiring with respect to the class of *point-generated* LTSs. The extended language with backward modalities is captured by the class of *connected* LTSs.

(iv) Equivalence of LTSs with respect to positive modal logic and the basic modal language cannot be captured by restricting the left homomorphism count vector over any semiring to any class of LTSs.

These results capture equivalence relations even over certain infinite structures, which we specify in the respective sections. The negative result (iv) is similar in spirit to the negative results from [5] mentioned above, but is more general in that it rules out homomorphism count indistinguishability characterizations for arbitrary semirings. Some of these results were obtained in the author's MSc thesis [13].

2 Preliminaries

We assume familiarity with the syntax and semantics of first-order logic (FO). We use σ and τ to denote first-order signatures, and we work primarily over *modal signatures* of the form $\sigma = \text{Prop} \cup \mathbb{A}$, where Prop is a finite set of unary predicate symbols (called *proposition letters*) and $\mathbb{A}$ is a finite set of binary predicate symbols (called *actions* or *transitions*). All of the modal languages discussed in this paper will be variants of the basic (multi)modal language ML, which is defined by the following recursive syntax:

$$\varphi := p \mid \varphi \wedge \varphi \mid \varphi \vee \varphi \mid \neg\varphi \mid \Diamond_i \varphi \mid \Box_i \varphi,$$

where $p \in \text{Prop}$ and $\Diamond_i, \Box_i$ are *modalities* for the action $R_i \in \mathbb{A}$. We define the semantics of ML by the well-known *standard translation* of ML to FO:

$$\begin{aligned} ST_x(p) &:= P(x), & ST_x(\neg\varphi) &:= \neg ST_x(\varphi), \\ ST_x(\varphi \wedge \psi) &:= ST_x(\varphi) \wedge ST_x(\psi), & ST_x(\Diamond_i\varphi) &:= \exists y(R_i(x,y) \wedge ST_y(\varphi)), \\ ST_x(\varphi \vee \psi) &:= ST_x(\varphi) \vee ST_x(\psi), & ST_x(\Box_i\varphi) &:= \forall y(R_i(x,y) \to ST_y(\varphi)). \end{aligned}$$

We write M, N, S, T to denote (possibly infinite) first-order structures and a, b, c, d, m, n, s, t to denote elements of structures. Given a relation symbol F, we write F^M to denote the interpretation of F in the structure M. For a k-ary relation symbol F, we say that $F^M(m_1, \ldots, m_k)$ *holds* if the tuple $\langle m_1, \ldots, m_k \rangle$ is in F^M, in which case we say that $F^M(m_1, \ldots, m_k)$ is a *fact* of M. Given a fact f, we write $\text{el}(f)$ for the set of elements occurring in f. A *pointed* structure, denoted $(M, a_1, \ldots, a_n)$, is a first-order structure M together with a tuple of *distinguished elements* $a_1, \ldots, a_n \in \text{dom}(M)$. A *labeled transition system* (LTS) is a pointed structure $M_a = (M, a)$ over a modal signature with exactly one distinguished element. We write that M_a is a σ-LTS to emphasize that it is defined over the modal signature σ. We refer to elements of $\text{dom}(M)$ as *states*. Given a σ-LTS M and a state $m \in \text{dom}(M)$, we define

$$\begin{aligned} \text{Succ}^M_\sigma[m] &:= \{n \in M \mid R^M(m,n) \text{ holds for some } R \in \sigma\}, \text{ and} \\ \text{Pred}^M_\sigma[m] &:= \{n \in M \mid R^M(n,m) \text{ holds for some } R \in \sigma\} \end{aligned}$$

to be the sets of *σ-successors* and *σ-predecessors*, respectively, of m in M. For $R \in \sigma$, we also write $\text{Succ}^M_R[m]$ and $\text{Pred}^M_R[m]$ for the successors (resp. predecessors) of m in M along an R transition. A σ-LTS M_a is *image-finite* if $\text{Succ}^M_\sigma[m]$ is finite for each $m \in \text{dom}(M)$, and *degree-finite* if both $\text{Succ}^M_\sigma[m]$ and $\text{Pred}^M_\sigma[m]$ are finite for each $m \in \text{dom}(M)$. We also write $\lambda^M_\sigma(m)$ to denote the set of proposition letters $p \in \sigma$ such that $M_m \models p$. Each modal language $\mathcal{L}$ discussed in this paper has an associated *satisfaction relation* $\models$ between LTSs and formulas of $\mathcal{L}$. If two σ-LTSs M_a and N_b satisfy the same formulas of $\mathcal{L}$ over signature σ, then we write $M_a \equiv^{\sigma}_{\mathcal{L}} N_b$.

Homomorphism Count Vectors. Let $(M,\overline{a})$ and $(N,\overline{b})$ be pointed σ-structures, where $\overline{a} = a_1,\dots,a_n \in \text{dom}(M)$ and $\overline{b} = b_1,\dots,b_n \in \text{dom}(N)$. A map $h : dom(M) \to dom(N)$ is a *homomorphism* from $(M,\overline{a})$ to $(N,\overline{b})$ if $a_i \mapsto b_i$ for each $i \leq n$ and, for each k-ary relation symbol $R \in \sigma$, we have that $R^N(h(s_1),\dots,h(s_k))$ holds whenever $R^M(s_1,\dots,s_k)$ holds. An *isomorphism* is a bijective homomorphism whose inverse is also a homomorphism; if an isomorphism from $(M,\overline{a})$ to $(N,\overline{b})$ exists, we write $(M,\overline{a}) \cong (N,\overline{b})$. A homomorphism $h : (M,\overline{a}) \to (N,\overline{b})$ is *fully surjective* if it is surjective and if, for all k-ary relation symbols $R \in \sigma$, whenever $\langle t_1,\dots,t_k\rangle \in R^N$, there also exists a tuple $\langle s_1,\dots,s_k\rangle \in R^M$ such that $\langle h(s_1),\dots,h(s_k)\rangle = \langle t_1,\dots,t_k\rangle$. We say that $(M,\overline{a})$ and $(N,\overline{b})$ are *homomorphically equivalent* if there exist homomorphisms $h : (M,\overline{a}) \to (N,\overline{b})$ and $g : (N,\overline{b}) \to (M,\overline{a})$.

Borrowing from database-theoretic terminology, an FO formula of the form

$$\varphi(x_1,\dots,x_n) := \exists y_1,\dots,y_m \left(\bigwedge_{j\in J} \alpha_j \right),$$

where J is a finite index set and each α_j is an atomic formula, is called a *conjunctive query* (CQ). Each CQ φ corresponds to a finite pointed structure whose domain is the variables of the formula, where each free variable is a distinguished element, and whose facts are the atomic formulas occurring in the formula [12]. This structure is the *canonical instance* of φ (notation: $\texttt{inst}(\varphi)$). Any FO formula containing only atomic formulas, existential quantifiers, and conjunction can be converted to a CQ by pulling all quantifiers to the front and renaming variables as necessary, so we will use the notation $\texttt{inst}$ for arbitrary formulas of this form. The following useful fact equates satisfying assignments for a conjunctive query with homomorphisms out of its canonical instance.

Fact 2.1 *Let $\varphi(x_1,\dots,x_n)$ be a CQ and $(M,a_1,\dots,a_n)$ a structure over the same signature. A homomorphism $h : \texttt{inst}(\varphi) \to (M,a_1,\dots,a_n)$ is a satisfying assignment for φ in $(M,a_1,\dots,a_n)$ such that $x_i \mapsto a_i$ for each $i \leq n$.*

We write $\text{Hom}((M,\overline{a}),(N,\overline{a}))$ to denote the collection of all homomorphisms from $(M,\overline{a})$ to $(N,\overline{b})$. A *semiring* is an algebraic structure $\mathcal{S} = \langle S,+,\cdot,0,1\rangle$, where $\langle S,+,0\rangle$ is a commutative monoid, $\langle S,\cdot,1\rangle$ is a monoid, $\cdot$ distributes over $+$, and $a\cdot 0 = 0\cdot a = 0$ for all $a \in A$. We define the *homomorphism count* from $(M,\overline{a})$ to $(N,\overline{b})$ over $\mathcal{S}$ to be

$$\text{hom}_{\mathcal{S}}((M,\overline{a}),(N,\overline{a})) := \text{count}_{\mathcal{S}}(|\text{Hom}((M,\overline{a}),(N,\overline{a}))|),$$

where $\text{count}_{\mathcal{S}} : \mathbb{N} \to S$ is defined by

$$\text{count}_{\mathcal{S}}(n) := \begin{cases} 0_{\mathcal{S}} & \text{if } n = 0 \\ \sum_{1\leq i\leq n} 1_{\mathcal{S}}, & \text{otherwise,} \end{cases}$$

where the summation is defined by iterated addition in $\mathcal{S}$. Note that $\hom_{\mathcal{S}}((M,\overline{a}),(N,\overline{a}))$ is only defined when $|\mathrm{Hom}((M,\overline{a}),(N,\overline{a}))|$ is finite. Our notion of counting is essentially just iterated addition, within some semiring, of the multiplicative unit of that semiring with itself. The decision to use semirings is not a canonical choice, but is general enough to cover all known results on homomorphism count indistinguishability.

Definition 2.2 Let $(M,\overline{a})$ be a τ-structure with n distinguished elements, and let $\mathcal{C}$ be a class of finite τ-structures, each with n distinguished elements, such that $\mathrm{Hom}((N,\overline{b}),(M,\overline{a}))$ is finite for each $(N,\overline{b})$ in $\mathcal{C}$. The *left homomorphism vector* (or *left profile*) of $(M,\overline{a})$ over $\mathcal{S}$ restricted to $\mathcal{C}$ is the $\mathcal{C}$-indexed sequence

$$\hom_{\mathcal{S}}(\mathcal{C},(M,\overline{a})) := \langle \hom_{\mathcal{S}}((N,\overline{b}),(M,\overline{a}))\rangle_{(N,\overline{b})\in\mathcal{C}}.$$

The term *left* is used here because the sequence includes homomorphism counts *from* structures in $\mathcal{C}$ *to* the structure $(M,\overline{a})$.

We work mostly with the Boolean semiring $\mathbb{B} = \langle\{0,1\},\vee,\wedge,\top,\bot\rangle$ and the natural number semiring $\mathbb{N} = \langle\omega,+,\cdot,0,1\rangle$. We write $\mathcal{M}^n_\tau$ for the class of all finite τ-structures with n distinguished elements. Note that $\hom_{\mathbb{B}}((M,\overline{a}),(N,\overline{b})) = 1$ when a homomorphism from $(M,\overline{a})$ to $(N,\overline{b})$ exists, and $\hom_{\mathbb{B}}((M,\overline{a}),(N,\overline{b})) = 0$ otherwise. It follows easily that $\hom_{\mathbb{B}}(\mathcal{M}^n_\tau,(M,\overline{a})) = \hom_{\mathbb{B}}(\mathcal{M}^n_\tau,(N,\overline{b}))$ if and only if $(M,\overline{a})$ and $(N,\overline{b})$ are homomorphically-equivalent. Using the notation of Definition 2.2, Lovász's theorem can be stated as follows.

Theorem 2.3 (Lovász's Theorem, [24]) *Let $(M,\overline{a})$ and $(N,\overline{b})$ be finite τ-structures with n distinguished elements, where τ is a finite relational signature. Then $hom_{\mathbb{N}}(\mathcal{M}^n_\tau,(M,\overline{a})) = hom_{\mathbb{N}}(\mathcal{M}^n_\tau,(N,\overline{b}))$ if and only if $(M,\overline{a}) \cong (N,\overline{b})$.*

The following definition was introduced in [5] to generalize Lovász's result.

Definition 2.4 If $\mathcal{C}$ is a class of τ-structures, we write $\mathrm{Inj}(\mathcal{C})$ to denote the class of τ-structures $(N,\overline{b})$ such that there exists some injective homomorphism $h:(N,\overline{b})\to(M,\overline{a})$ for some $(M,\overline{a})\in\mathcal{C}$. Similarly, we write $\mathrm{Sur}(\mathcal{C})$ to denote the class of τ-structures $(N,\overline{b})$ such that there exists some fully-surjective homomorphism $h:(M,\overline{a})\to(N,\overline{b})$ for some $(M,\overline{a})\in\mathcal{C}$. We define the *extension class* of $\mathcal{C}$ to be $\mathrm{Ext}(\mathcal{C}) := \mathrm{Inj}(\mathcal{C})\cap\mathrm{Sur}(\mathcal{C})$.

Theorem 2.5 ([5]) *Let $\mathcal{C}$ be a non-empty class of finite pointed τ-structures, each with the same number of distinguished elements. Then for all $(M,\overline{a}),(N,\overline{b})\in\mathcal{C}$, we have $hom_{\mathbb{N}}(Ext(\mathcal{C}),(M,\overline{a})) = hom_{\mathbb{N}}(Ext(\mathcal{C}),(N,\overline{b}))$ if and only if $(M,\overline{a})\cong(N,\overline{b})$.*

Important Classes of Structures. We now define the classes of structures relevant to our results (examples of each can be found in Figure 1). Let M_a be a σ-LTS. Given states $m,n\in\mathrm{dom}(M)$, a *σ-path of length k from m to n* is a sequence $\pi = \langle f_1,\ldots,f_k\rangle$ of binary facts such that $m\in\mathrm{el}(f_1)$, $n\in\mathrm{el}(f_k)$, and $\mathrm{el}(f_i)\cap\mathrm{el}(f_{i+1})\neq\emptyset$ for each $i<k$. A σ-path is *simple* if it contains no duplicate facts. A *connected component* of M_a is a maximal set $S\subseteq\mathrm{dom}(M)$

such that, for each distinct pair of states $m, n \in S$, there exists a σ-path π from m to n. We say that M_a is *connected* if $\text{dom}(M)$ is a connected component of M_a, and we say that M_a is *acyclic* if there are no simple σ-paths from some $m \in \text{dom}(M)$ to itself. A *directed σ-path of length k* from m to n is a length-k tuple $\langle (b_0, b_1), (b_1, b_2), \ldots, (b_{k-1}, b_k) \rangle$ such that for each $j < k$, there is some $R \in \sigma$ such that $R^M(b_j, b_{j+1})$ holds. Note that all directed σ-paths can be seen as a special case of σ-paths.

A σ-LTS M_a is *point-generated* if, for each $m \in \text{dom}(M)$, there's a directed σ-path from a to m. If there is a unique directed σ-path from a to each $m \in \text{dom}(M)$, then M_a is a *σ-tree.* The *depth* of a state m in a point-generated σ-LTS M_a is the length $\text{depth}(m)$ of the shortest directed σ-path from a to m; we set $\text{depth}(a) = 0$. The *depth* of a point-generated σ-LTS is the supremum of the depths of its elements. Given a point-generated σ-LTS M_a, we define M_a^k to be the pointed substructure of M_a containing all elements in $\text{dom}(M)$ of depth at most k. Depth for connected structures is defined analogously via σ-paths. If (M_j, a_j) is a σ-tree for each $j \in J$, where J is some finite index set, and (M, a) is a σ-LTS obtained by taking the disjoint union $M = \biguplus_{j \in J} M_j$ and setting $a = a_j$ for some $j \in J$, then (M, a) is a *σ-forest.*

Definition 2.6 We use the following notation for these classes of structures.

(i) $\mathcal{T}_\sigma^k$ is the class of finite σ-trees of depth at most k.

(ii) $\mathcal{A}_\sigma^k$ is the class of finite connected, acyclic σ-LTSs of depth at most k.

(iii) $\mathcal{F}_\sigma$ is the class of finite σ-forests.

(iv) $\mathcal{PG}^k$ is the class of finite point-generated σ-LTSs of depth at most k.

(v) $\mathcal{C}_\sigma^k$ is the class of finite connected σ-LTSs of depth at most k.

(vi) We set $\mathcal{T}_\sigma := \bigcup_{k \in \omega} \mathcal{T}_\sigma^k$; we define $\mathcal{A}_\sigma$, $\mathcal{PG}_\sigma$, and $\mathcal{C}_\sigma$ similarly.

The following class inclusions are clear from the definitions:

$$\mathcal{T}_\sigma \subseteq \mathcal{PG}_\sigma \subseteq \mathcal{C}_\sigma, \quad \mathcal{T}_\sigma \subseteq \mathcal{A}_\sigma \subseteq \mathcal{C}_\sigma, \quad \text{and} \quad \mathcal{T}_\sigma \subseteq \mathcal{F}_\sigma.$$

The following two facts are easily verified (cf. Definition 2.4).

Fact 2.7 $\mathcal{T}_\sigma^k = \mathit{Ext}(\mathcal{T}_\sigma^k)$.

Fact 2.8 $\mathcal{PG}_\sigma^k - \mathit{Ext}(\mathcal{PG}_\sigma^k)$.

The next lemma, used in Sections 4 and 5, is proven by constructing an ascending chain of local isomorphisms whose union is a full isomorphism.

Lemma 2.9 *If M_a and N_b are point-generated σ-LTSs such that M_a^k and N_b^k are finite and isomorphic for all $k \in \mathbb{N}$, then $M_a \cong N_b$.*

For the remainder of the paper, we fix a modal signature $\sigma = \text{Prop} \cup \mathbb{A}$, where Prop is a finite set of (unary) proposition letters and $\mathbb{A} = \{R_i \mid i \in I\}$ is a set of (binary) *actions* (or *transitions*) indexed by some finite set I.

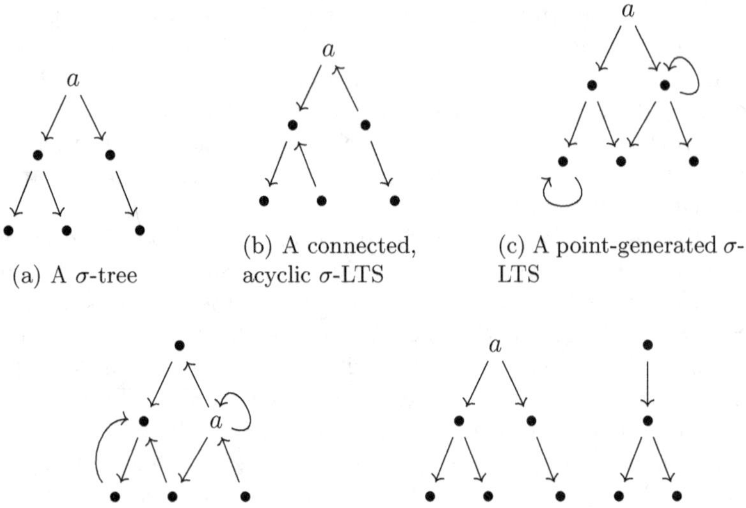

(a) A σ-tree

(b) A connected, acyclic σ-LTS

(c) A point-generated σ-LTS

(d) A connected σ-LTS

(e) A σ-forest

Fig. 1. Examples of σ-LTSs.

3 Positive-Existential Modal Logic

We begin with a characterization of equivalence with respect to positive-existential modal logic (notation: $\mathrm{ML}_{\Diamond}^{+}$) by restricting the left homomorphism vector over the Boolean semiring to the class of σ-trees. $\mathrm{ML}_{\Diamond}^{+}$ is the fragment of ML lacking both negation and the $\Box$ modality, and we write $\mathrm{ML}_{\Diamond}^{+,k}$ for the collection of $\mathrm{ML}_{\Diamond}^{+}$ formulas of modal depth at most k. The key observation leading to this theorem is the following proposition.

Proposition 3.1 *A σ-LTS T_c is in $\mathcal{T}_\sigma^k$ if and only if $T_c \cong$ $\mathtt{inst}(ST_x(\varphi))$ for some disjunction-free $\varphi \in \mathrm{ML}_{\Diamond}^{+,k}$, where ST_x denotes the standard translation.*

Proof. For the forward direction, we show by induction on the depth of σ-trees T_c that T_c is isomorphic to the canonical instance of some $\mathrm{ML}_{\Diamond}^{+,k}$ formula. For each element $s \in T$, define $\mathrm{mark}_s^{+,T} := \bigwedge_{p \in \lambda_\sigma^T(s)} p$. For the base case, if $\mathrm{depth}(T_c) = 0$, then $\mathrm{dom}(T) = \{c\}$. Then clearly $\mathtt{inst}(ST_x(\mathrm{mark}_c^{+,T})) \cong T_c$. Now suppose that every σ-tree of depth $j < k$ is isomorphic to the canonical instance of some $\mathrm{ML}_{\Diamond}^{+,k}$ formula, and let T_c be an arbitrary σ-tree of depth k. Let $\mathrm{Succ}_\sigma^T[c] = \{s_1, \ldots, s_n\}$, and let $T_{s_1}^1, \ldots, T_{s_n}^n$ denote the corresponding rooted subtrees of T_c. By the inductive hypothesis, there exist formulas $\varphi_1, \ldots, \varphi_n$ such that $\mathtt{inst}(ST_x(\varphi_i)) \cong T_{s_i}^i$ for each $i \leq n$. For each $i \leq n$, let j_i be the unique index in I such that $R_{j_i}^T(c, s_i)$ holds. Then $\mathtt{inst}(ST_x(\mathrm{mark}_s^{+,T} \wedge \bigwedge_{i \leq n} \Diamond_{j_i} \varphi_i))$ is easily seen to be isomorphic to T_c.

For the reverse direction, we show by induction on the complexity of $\mathrm{ML}_{\Diamond}^{+}$ formulas φ that $\mathtt{inst}(ST_x(\varphi))$ is a σ-tree. For the base case, if $\varphi = p$ for some $p \in \mathrm{Prop}$, then $\mathtt{inst}(ST_x(\varphi))$ is a single state at which the proposition letter p is true, which is a σ-tree. For the inductive step, either $ST_x(\varphi) = ST_x(\psi_1) \wedge ST_x(\psi_2)$ for some formulas ψ_1, ψ_2, or $ST_x(\varphi) = \exists y(R_i(x, y) \wedge \psi)$ for some formula ψ. In the first case, $\texttt{inst}(ST_x(\varphi))$ is the σ-tree obtained by equating the roots of $\texttt{inst}(ST_x(\psi_1))$ and $\texttt{inst}(ST_x(\psi_2))$. In the second case, $\texttt{inst}(ST_x(\varphi))$ is the σ-tree obtained by adding a new root to $\texttt{inst}(ST_x(\psi))$, with the old root as its unique R_i-successor. □

Note that if T_c is a finite σ-tree and M_a is an image-finite σ-LTS, then there are only finitely many homomorphisms from T_c to M_a. It follows that if M_a is image-finite, then $\hom_{\mathcal{S}}(\mathcal{T}^k, M_a)$ is well-defined for all semirings $\mathcal{S}$.

Theorem 3.2 *If M_a and N_b are image-finite σ-LTSs, then $M_a \equiv_{\mathrm{ML}_\diamond^{+,k}} N_b$ if and only if $\hom_{\mathbb{B}}(\mathcal{T}_\sigma^k, M_a) = \hom_{\mathbb{B}}(\mathcal{T}_\sigma^k, N_b)$.*

Proof. For the left-to-right direction, suppose that $M_a \equiv_{\mathrm{ML}_\diamond^{+,k}} N_b$, and let T_c be an arbitrary finite σ-tree of depth at most k. By Proposition 3.1, let φ be a disjunction-free $\mathrm{ML}_\diamond^{+,k}$ formula such that $T_c \cong \texttt{inst}(ST_x(\varphi))$. Then

$$\begin{aligned} \hom_{\mathbb{B}}(T_c, M_a) = 1 &\iff M_a \models \varphi && \text{(Fact 2.1)} \\ &\iff N_b \models \varphi && \text{(Assumption)} \\ &\iff \hom_{\mathbb{B}}(T_c, N_b) = 1 && \text{(Fact 2.1).} \end{aligned}$$

Hence $\hom_{\mathbb{B}}(\mathcal{T}_\sigma^k, M_a) = \hom_{\mathbb{B}}(\mathcal{T}_\sigma^k, N_b)$. The other direction is symmetric. □

Corollary 3.3 *If M_a and N_b are image-finite σ-LTSs, then $M_a \equiv_{\mathrm{ML}_\diamond^{+}} N_b$ if and only if $\hom_{\mathbb{B}}(\mathcal{T}_\sigma, M_a) = \hom_{\mathbb{B}}(\mathcal{T}_\sigma, N_b)$.*

$\mathrm{ML}_\diamond^+$ with backward and global modalities. We now state two results for $\mathrm{ML}_\diamond^+$ extended with backward and global modalities. The proofs are similar to, yet simpler than, those for $\mathrm{ML}_\#$ with the backward and global modalities given in Section 4, and so we omit them.

Definition 3.4 Given a pointed σ-LTS M_a and a formula φ, we define

$M, a \models \blacklozenge_i^{\geq k}\varphi$ if there exist at least k many elements $b \in \mathrm{Pred}_{R_i}^M[a]$ such that $M, b \models \varphi$.

We call $\blacklozenge_i^{\geq k}$ a *backward modality* for the action R_i, where $i \in I$. We write $\mathrm{ML}_\diamond^{+,B}$ for the extension of $\mathrm{ML}_\diamond^+$ with the modalities $\blacklozenge_i^{\geq 1}$, and $\mathrm{ML}_\diamond^{+,B,k}$ for the fragment of $\mathrm{ML}_\diamond^{+,B}$ containing formulas of modal depth at most k.

Theorem 3.5 *If M_a and N_b are degree-finite σ-LTSs, then $M_a \equiv^\sigma_{\mathrm{ML}_\diamond^{+,B,k}} N_b$ if and only if $\hom_{\mathbb{B}}(\mathcal{A}_\sigma^k, M_a) = \hom_{\mathbb{B}}(\mathcal{A}_\sigma^k, N_b)$.*

Definition 3.6 Given a pointed σ-LTS M_a and a formula φ, we define

$M_a \models \mathrm{E}^{\geq k}\varphi$ if there exist at least k many elements $b \in M$ such that $M_b \models \varphi$.

We refer to $\mathrm{E}^{\geq k}$ as a *global modality*. Let $\mathrm{ML}_\diamond^{+,G}$ denote the extension of $\mathrm{ML}_\diamond^+$ with the global modality for $k = 1$.

Theorem 3.7 *If M_a and N_b are finite σ-LTSs, then $M_a \equiv^{\sigma}_{\mathrm{ML}^{+}_{\Diamond},G} N_b$ if and only if $hom_{\mathbb{B}}(\mathcal{F}_\sigma, M_a) = hom_{\mathbb{B}}(\mathcal{F}_\sigma, N_b)$.*

Note that Theorems 3.2, 3.5, and 3.7 are stated for image-finite, degree-finite, and finite LTSs, respectively, *only* due to the fact that our notion of counting requires a finite number of homomorphisms in order to be well-defined. However, this is an artificial constraint: if we were to treat the Boolean homomorphism count only as an indicator that $\mathrm{Hom}(T_c, M_a)$ is non-empty, then the Boolean left profiles with respect to $\mathcal{T}_\sigma$, $\mathcal{A}_\sigma$, and $\mathcal{F}_\sigma$ would be well-defined for, and hence all of these results would apply to, *arbitrary* σ-LTSs.

4 Graded Modal Logic

We now turn to graded modal logic (notation: $\mathrm{ML}_{\#}$), which is the extension of the basic modal language with *graded modalities* $\Diamond_i^{\geq k}$ for each $k \in \mathbb{Z}^+$ and $i \in I$, where $\Diamond_i^{\geq k}\varphi$ asserts that there are at least k many R_i-successors of the current state at which φ is true [17,18]. Recall the notion of a tree-unraveling.

Definition 4.1 The *unraveling* of a σ-LTS M_a is the σ-LTS $\mathtt{unr}(M_a)$ where

(i) $\mathrm{dom}(\mathtt{unr}(M_a))$ is the set of strings $w = \langle w_1, \ldots, w_n\rangle$ over $\mathrm{dom}(M)$ with $a = w_1$ and for each $i < n$, there is $R \in \mathbb{A}$ such that $R^M(w_i, w_{i+1})$ holds,

(ii) $R^{\mathtt{unr}(M_a)} = \{(w, w^\frown\langle u\rangle) \mid (\mathtt{last}(w), u) \in R^M\}$ for each $R \in \mathbb{A}$, and

(iii) $p^{\mathtt{unr}(M_a)} = \{w \in \mathrm{dom}(\mathtt{unr}(M_a)) \mid \mathtt{last}(w) \in p^M\}$ for $p \in \mathrm{Prop}$,

where $\langle a\rangle$ is the unique distinguished element of the model, $\mathtt{last}$ is the function mapping strings to their last element, and $w^\frown w'$ denotes string concatenation.

If M_a is a σ-LTS, then $\mathtt{unr}(M_a)$ is a (possibly infinite) σ-tree. Furthermore, if M_a is image-finite, then $\mathtt{unr}^k(M_a)$, the substructure of $\mathtt{unr}(M_a)$ containing only states of depth at most k, is a finite σ-tree of depth k. The unraveling construction is known to preserve the truth of $\mathrm{ML}_{\#}$ formulas.

Theorem 4.2 (Unraveling Invariance, [10]) *If M_a is a σ-LTS, then for each $\varphi \in \mathrm{ML}_{\#}$, we have that $M_a \models \varphi$ if and only if $\mathtt{unr}(M_a) \models \varphi$.*

In fact, $\mathrm{ML}^k_{\#}$ formulas can describe finite σ-trees of depth k up to isomorphism. Recall that if M_a is a σ-tree, then M_a^k denotes the substructure of M_a containing only elements of depth at most k. We write $\mathtt{unr}^k(M_a)$ for the substructure of $\mathtt{unr}(M_a)$ containing only elements of depth at most k.

Proposition 4.3 *For each $T_c \in \mathcal{T}^k_\sigma$, there is a formula $\varphi \in \mathrm{ML}^k_{\#}$ such that, if M_a is a σ-tree, then $M_a \models \varphi$ if and only if $M^k_a \cong T_c$.*

Proof. For any state $s \in T$, define $\mathrm{mark}^T_s := \left(\bigwedge_{p\in\lambda^T_\sigma(s)} p\right) \wedge \left(\bigwedge_{p\notin\lambda^T_\sigma(s)} \neg p\right)$. We proceed by strong induction on k. For $T_c \in \mathcal{T}^0_\sigma$, clearly mark^T_s meets the requirements of the claim. Now suppose the claim holds for $\mathcal{T}^j_\sigma$ for all $j < k$, and let $T_c \in \mathcal{T}^k_\sigma$. Let $\mathrm{Succ}^T_\sigma[c] = \{s_1, \ldots, s_n\}$, and let $T^i_{s_i}$ denote the subtree of T_c rooted at s_i for each $i \leq n$. By the inductive hypothesis, there is a formula φ_i satisfying the claim for each $T^i_{s_i}$. Some of these may be equivalent, so let $\varphi'_1, \ldots, \varphi'_m$ be the sequence of formulas obtained by removing duplicates. For each $i \le m$, let n_i denote the number of elements $s_j \in \mathrm{Succ}^T_\sigma[c]$ such that the formula associated with $T^j_{s_j}$ is φ'_i. The formula $\varphi := \left(\mathrm{mark}^T_c \wedge \Diamond^n_i \top \wedge \bigwedge_{i \le m} \Diamond^{=n_i}_i \varphi'_i\right)$ satisfies the requirements of the claim.□

The next lemma shows that homomorphism counts from finite σ-trees are preserved between a σ-LTS and its unraveling up to depth k.

Lemma 4.4 *If M_a is an image-finite LTS, then for each $k \in \mathbb{N}$, we have that $hom_{\mathbb{N}}(\mathcal{T}^k, M_a) = hom_{\mathbb{N}}(\mathcal{T}^k, \boldsymbol{unr}^k(M_a))$.*

Proof. Given a directed tree-shaped LTS T_c of depth at most k, we construct injections between $\mathrm{Hom}(T_c, M_a)$ and $\mathrm{Hom}(T_c, \mathtt{unr}^k(M_a))$.

($\le$) For each $h \in \mathrm{Hom}(T_c, M_a)$, we define partial maps $\hat{h}_i : T_c \to \mathtt{unr}^k(M_a)$ for $i \le k$ by recursion on the depth of elements of T_c, where $\hat{h}_0(c) = \langle a \rangle$ and $\hat{h}_{i+1}(m) = \hat{h}_i(\mathrm{parent}(m))^\frown h(m)$. We claim that $\hat{h} = \bigcup_{i \le k} \hat{h}_i$ is a homomorphism. To see that $\hat{h}$ preserves proposition letters, observe that, for any proposition letter $p \in P$, we have that if $p^T(m)$ holds, then $p^M(h(m))$ holds since h is a homomorphism, and so $p^{\mathtt{unr}^k(M_a)}(\hat{h}(m))$ holds since $h(m) = \mathtt{last}(\hat{h}(m))$.

To show that $\hat{h}$ preserves actions, it suffices to show that, for all $R \in \mathbb{A}$, all states $m \in T_c$, and all $s \in \mathrm{Succ}^M_R[m]$, we have that $R^{\mathtt{unr}^k(M_a)}(\hat{h}(m), s)$ holds. This follows from the observations that (1) $\hat{h}(s)$ extends $\hat{h}(m)$ (by the definition of $\hat{h}$), and (2) if $R^M(\mathtt{last}(\hat{h}(x)), \mathtt{last}(\hat{h}(s)))$ holds, then $R^M(h(x), h(s_j))$ (since h_i is a homomorphism), in which case $R^{\mathtt{unr}^k(M_a)}(\hat{h}(m), s)$ holds by the definition of unravelings. Thus $\hat{h}$ is a homomorphism. Furthermore, it's clear that the map $h \mapsto \hat{h}$ is an injection from $\mathrm{Hom}(T_c, M_a)$ to $\mathrm{Hom}(T_c, \mathtt{unr}^k(M_a))$.

($\ge$) For each $g \in \mathrm{Hom}(T_c, \mathtt{unr}^k(M_a))$, define $\hat{g} : T_c \to M_a$ to be the map $m \mapsto \mathtt{last}(g(m))$. By the definition of unravelings, $\hat{g}$ is a homomorphism. We claim that $g \mapsto \hat{g}$ is an injective map from $\mathrm{Hom}(T_c, \mathtt{unr}^k(M_a))$ to $\mathrm{Hom}(T_c, M_a)$. To see this, let $g, g' : T_c \to unr(M_a)$ be homomorphisms, and let $\hat{g}, \hat{g}'$ be the corresponding maps in $\mathrm{Hom}(T_c, M_a)$. Suppose that $\hat{g} = \hat{g}'$. We now show by induction on depth of the elements of T_c that $g = g'$.

The base case is immediate, since $g(c) = g'(c) = \langle a \rangle$. Now suppose inductively that g and g' agree on all elements of depth less than k, and let $m \in T_c$ be some element of depth k. By assumption, we have that $\hat{g}(m) = \hat{g}'(m)$, and so $last(g(m)) = last(g'(m))$. Let n denote the unique predecessor of m (i.e., its parent). Clearly n has depth less than k, and so $g(n) = g'(n)$. Since g and g' are homomorphisms and $R_i(n, m)$ holds for some $i \in I$, we have that $R_i^{unr(M_a)}(g(n), g(m))$ and $R_i^{unr(M_a)}(g'(n), g'(m))$ hold. Then by the definition of the actions for unravelings, we have that $g(m) = g(n)^\frown \hat{g}(m) = g'(m)$. □

We are now ready to prove our characterization result for $\mathrm{ML}_{\#}$.

Theorem 4.5 *For image-finite LTSs M_a and N_b, the following are equivalent:*

(i) *$hom_{\mathbb{N}}(\mathcal{T}^k, M_a) = hom_{\mathbb{N}}(\mathcal{T}^k, N_b)$,*

(ii) *$\boldsymbol{unr}^k(M_a) \cong \boldsymbol{unr}^k(N_b)$,*

(iii) $M_a \equiv^{\sigma}_{\mathrm{ML}^k_\#} N_b$.

Proof. The equivalence of *(ii)* and *(iii)* is a consequence of Proposition 4.3, Theorem 4.2 and the observation that satisfaction of $\mathrm{ML}^k_\#$ formulas in σ-trees depends only on the elements up to depth k. For *(i)* to *(ii)*, suppose that $\mathrm{hom}_{\mathbb{N}}(\mathcal{T}^k, M_a) = \mathrm{hom}_{\mathbb{N}}(\mathcal{T}^k, N_b)$. Then by Lemma 4.4, we have that $\mathrm{hom}_{\mathbb{N}}(\mathcal{T}^k, \mathbf{unr}^k(M_a)) = \mathrm{hom}_{\mathbb{N}}(\mathcal{T}^k, \mathbf{unr}^k(N_b))$. Since $\mathbf{unr}^k(M_a), \mathbf{unr}^k(N_b) \in \mathcal{T}^k$, this implies, by Fact 2.7 and Theorem 2.5, that $\mathbf{unr}^k(M_a) \cong \mathbf{unr}^k(N_b)$. For *(ii)* to *(i)*, suppose that $\mathbf{unr}^k(M_a) \cong \mathbf{unr}^k(N_b)$. Then $\mathrm{hom}_{\mathbb{N}}(\mathcal{T}^k, \mathbf{unr}^k(M_a)) = \mathrm{hom}_{\mathbb{N}}(\mathcal{T}^k, \mathbf{unr}^k(N_b))$, and so $\mathrm{hom}_{\mathbb{N}}(\mathcal{T}^k, M_a) = \mathrm{hom}_{\mathbb{N}}(\mathcal{T}^k, N_b)$ by Lemma 4.4.□

Using Lemma 2.9, we easily obtain the following corollary.

Corollary 4.6 *For image-finite LTSs M_a and N_b, the following are equivalent:*

(i) $\mathit{hom}_{\mathbb{N}}(\mathcal{T}, M_a) = \mathit{hom}_{\mathbb{N}}(\mathcal{T}, N_b)$,

(ii) $\mathbf{unr}(M_a) \cong \mathbf{unr}(N_b)$,

(iii) $M_a \equiv^{\sigma}_{\mathrm{ML}_\#} N_b$.

$\mathrm{ML}_\#$ with backward modalities. Let $\mathrm{ML}^B_\#$ denote the extension of $\mathrm{ML}_\#$ with backward modalities for each $k \in \mathbb{N}$ (cf. Definition 3.4). We write $\mathrm{ML}^{B,k}_\#$ for the fragment of $\mathrm{ML}^B_\#$ formulas of modal depth at most k. Fix an expansion $\sigma_B = \mathrm{Prop} \cup \mathbb{A} \cup \mathbb{A}_B$ of σ, where $\mathbb{A}_B = \{B_i \mid i \in I\}$ is disjoint from $\mathbb{A}$.

Definition 4.7 The *backward expansion* of a σ-LTS M_a is the σ_B-expansion M^B_a of M_a given by setting $B_i^{M^B} = \{\langle n, m\rangle \mid R_i^M(m,n) \text{ holds}\}$ for each $i \in I$.

Recall that $\mathcal{A}^k_\sigma$ denotes the class of connected acyclic σ-LTSs of depth at most k, and that $\mathcal{A}_\sigma = \bigcup_{k\in\omega} \mathcal{A}^k_\sigma$ (cf. Definition 2.6).

Definition 4.8 Given some $T_c \in \mathcal{A}_\sigma$, we define a σ_B-LTS $T^{\downarrow}_c := (T^{\downarrow}, c)$ with $\mathrm{dom}(T^{\downarrow}) := \mathrm{dom}(T)$ and $\lambda^{T^{\downarrow}}_\sigma(m) := \lambda^T_\sigma(m)$ for all $m \in \mathrm{dom}(T)$, where

(i) If $R^T_i(m,n)$ holds where $\mathit{depth}(m) < \mathit{depth}(n)$, then $R^{T^{\downarrow}}_i(m,n)$ holds.

(ii) If $R^T_i(m,n)$ holds where $\mathit{depth}(n) < \mathit{depth}(m)$, then $B^{T^{\downarrow}}_i(m,n)$ holds.

Intuitively, $(\cdot)^{\downarrow}$ replaces all R_i transitions "pointing toward" the root c with B_i transitions in the opposite direction. For all $T_c \in \mathcal{A}_\sigma$, clearly $T^{\downarrow}_c$ is a σ_B-tree.

Definition 4.9 Let S_d be a σ_B-LTS. Define a σ-LTS $\mathtt{flip}(S_d) := (\mathtt{flip}(S), d)$ with $\mathrm{dom}(\mathtt{flip}(S)) = \mathrm{dom}(S)$ and $\lambda^{\mathtt{flip}(S)}_\sigma(m) := \lambda^S_\sigma(m)$ for all $m \in \mathrm{dom}(S)$, where $R^{\mathtt{flip}(S)}_i = R^S_i \cup (B^S_i)^{-1}$ for each $i \in I$.

Intuitively, $\mathtt{flip}$ forms a σ-LTS from a σ_B-LTS S_d by replacing B_i transitions in S_d by the corresponding R_i transition in the opposite direction. If S_d is a σ_B-tree, then $\mathtt{flip}(S_d)$ is a connected acyclic σ-LTS. The $(\cdot)^{\downarrow}$ transformation on connected, acyclic σ-LTSs and the $\mathtt{flip}$ transformation on σ_B-trees are exact inverses of one another: $T_c = \mathtt{flip}(T^{\downarrow}_c)$ for all connected acyclic σ-LTSs T_c, and $S_d = (\mathtt{flip}(S_d))^{\downarrow}$ for all σ_B-trees S_d. These operations also clearly preserve the depth of the structures to which they are applied.

When we consider homomorphisms from finite connected acyclic σ-LTSs T_c, image-finiteness is not enough to guarantee that $\mathrm{Hom}(T_c, M_a)$ is finite. However, if M_a is degree-finite, then $\bigcup_{h \in Hom(T_c,M_a)} Im(h)$ is finite, and hence $\mathrm{hom}_{\mathbb{N}}(\mathcal{A}_\sigma, M_a)$ is well-defined. Furthermore, note that if M_a is a degree-finite σ-LTS, then M_a^B is also a degree-finite σ^B-LTS.

Proposition 4.10 *Let M_a be a degree-finite σ-LTS. Then*

(i) *If T_c is in $\mathcal{A}_\sigma$, then $hom_{\mathbb{N}}(T_c, M_a) = hom_{\mathbb{N}}(T_c^{\downarrow}, M_a^B)$.*

(ii) *If T_c is in $\mathcal{T}_{\sigma_B}$, then $hom_{\mathbb{N}}(T_c, M_a^B) = hom_{\mathbb{N}}(\texttt{flip}(T_c), M_a)$.*

Proof. [Sketch] For part (i), we show that a map $h : \mathrm{dom}(T) \to \mathrm{dom}(M)$ is a homomorphism from T_c to M_a if and only if it is also a homomorphism from $T_c^{\downarrow}$ to M_a^B, which is straightforward from the definitions of $(\cdot)^{\downarrow}$ and M_a^B. The proof of part (ii) is analogous. □

Lemma 4.11 *Let M_a and N_b be degree-finite σ-LTSs. Then*

$$hom_{\mathbb{N}}(\mathcal{A}_\sigma^k, M_a) = hom_{\mathbb{N}}(\mathcal{A}_\sigma^k, N_b) \iff hom_{\mathbb{N}}(\mathcal{T}_{\sigma_B}^k, M_a^B) = hom_{\mathbb{N}}(\mathcal{T}_{\sigma_B}^k, N_b^B).$$

Proof. [Sketch] Observe that $(\cdot)^{\downarrow}$ is a bijective map from $\mathcal{A}_\sigma^k$ to $\mathcal{T}_{\sigma_B}^k$, while `flip` is its inverse. The forward direction is by contraposition. If we have $\mathrm{hom}_{\mathbb{N}}(T_c, M_a^B) \neq \mathrm{hom}_{\mathbb{N}}(T_c, N_b^B)$ for some σ_B-tree T_c of depth at most k, then we have $\mathrm{hom}_{\mathbb{N}}(T_c^{\downarrow}, M_a) \neq \mathrm{hom}_{\mathbb{N}}(T_c^{\downarrow}, N_b)$ by Proposition 4.10. The reverse direction is proven by contraposition in a similar fashion. □

Lemma 4.12 *Let M_a and N_b be degree-finite σ-LTSs. Then $M_a \equiv^{\sigma}_{\mathrm{ML}_\#^{B,k}} N_b$ if and only if $M_a^B \equiv^{\sigma_B}_{\mathrm{ML}_\#^k} N_b^B$.*

Proof. [Sketch] Consider the translation tr from $\mathrm{ML}_\#^{B,k}$ to $\mathrm{ML}_\#^k$ which replaces backward modalities with the corresponding forward modalities in $\mathbb{A}_{\sigma_B}$. It is a straightforward induction to show that $M_a \models \varphi$ if and only if $M_a^B \models \mathrm{tr}(\varphi)$. Since this translation is bijective, the result follows immediately. □

We now prove our $\mathrm{ML}_\#^B$ characterization result.

Theorem 4.13 *Let M_a and N_b be degree-finite σ-LTSs. Then*

$$\mathrm{hom}(\mathcal{A}_\sigma^k, M_a) = \mathrm{hom}(\mathcal{A}_\sigma^k, N_b) \iff M_a \equiv^{\sigma}_{\mathrm{ML}_\#^{B,k}} N_b.$$

Proof. Let M_a and N_b be degree-finite σ-LTSs. Then we have that

$$\begin{aligned}
\mathrm{hom}(\mathcal{A}_\sigma^k, M_a) &= \mathrm{hom}(\mathcal{A}_\sigma^k, N_b) \\
&\iff \mathrm{hom}(\mathcal{T}_{\sigma_B}^k, M_a^B) = \mathrm{hom}(\mathcal{T}_{\sigma_B}^k, N_b^B) && \text{(Lemma 4.11)} \\
&\iff M_a^B \equiv^{\sigma_B}_{\mathrm{ML}_\#^k} N_b^B && \text{(Theorem 4.5)} \\
&\iff M_a \equiv^{\sigma}_{\mathrm{ML}_\#^{B,k}} N_b. && \text{(Lemma 4.12)}
\end{aligned}$$

This completes the proof. □

$\mathrm{ML}_\#$ with the global modality. Let $\mathrm{ML}_\#^G$ denote the extension of $\mathrm{ML}_\#$ with the global modalities for each $k \in \mathbb{N}$ (cf. Definition 3.6). The proof of our characterization result for $\mathrm{ML}_\#^G$ mirrors that of Theorem 4.13. Let $\sigma_G = \mathrm{Prop} \cup \mathbb{A} \cup \{R_G\}$, where R_G is a fresh action not in $\mathbb{A}$. We write $\Diamond_G^{\geq k}$ for the graded modalities associated with the action R_G. The following definition is analogous to the "backwards expansion" (cf. Definition 4.7).

Definition 4.14 Given a pointed σ-LTS M_a, the *global expansion* of M_a is the σ_G-expansion M_a^G of M_a given by setting $R_G^{M^G} = \mathrm{dom}(M) \times \mathrm{dom}(M)$.

Recall that $\mathcal{F}_\sigma^k$ denotes the class of connected acyclic σ-LTSs of depth at most k, and that $\mathcal{F}_\sigma = \bigcup_{k \in \omega} \mathcal{F}_\sigma^k$ (cf. Definition 2.6). The next definition defines a relation between structures in $\mathcal{F}_\sigma$ and those in $\mathcal{T}_\sigma$. Its role is analogous to that of `flip` (cf. Definition 4.9) in the proof of Theorem 4.13.

Definition 4.15 Let $T_c \in \mathcal{F}_\sigma$. We say that a σ_G-expansion T_c' of T_c is an *R_G-connection of* T_c if T_c' is a σ_G-tree and $T_c = T_c' \restriction \sigma$.

It's easy to see that every $T_c' \in \mathcal{T}_{\sigma_G}$ is an R_G-connection of some $T_c \in \mathcal{F}_\sigma$. Similarly, for all $T_c \in \mathcal{F}_\sigma$, we have that $T_c = T_c' \restriction \sigma$ (the reduct of T_c' to the signature σ) for some $T_c' \in \mathcal{T}_{\sigma_G}$. From these definitions, it is straightforward to prove the following analogues of Lemma 4.11 and Lemma 4.12.

Lemma 4.16 *Let M_a and N_b be finite σ-LTSs. Then*

$$hom_{\mathbb{N}}(\mathcal{F}_\sigma, M_a) = hom_{\mathbb{N}}(\mathcal{F}_\sigma, N_b) \iff hom_{\mathbb{N}}(\mathcal{T}_{\sigma_G}, M_a^G) = hom_{\mathbb{N}}(\mathcal{T}_{\sigma_G}, N_b^G).$$

Lemma 4.17 *Let M_a and N_b be finite σ-LTSs. Then $M_a \equiv^\sigma_{\mathrm{ML}_\#^G} N_b$ if and only if $M_a^G \equiv^{\sigma_G}_{\mathrm{ML}_\#} N_b^G$.*

In the case of the global modality, we state our result only for finite σ-LTSs M_a. This is necessary, since homomorphisms out of σ-forests could map connected components which do not contain c to any connected component in M_a, and so $\mathrm{Hom}(T_c, M_a)$ may be infinite even if M_a is degree-finite.

Theorem 4.18 *If M_a and N_b are finite σ-LTSs, then $M_a \equiv^\sigma_{\mathrm{ML}_\#^G} N_b$ if and only if $hom_{\mathbb{N}}(\mathcal{F}_\sigma, M_a) = hom_{\mathbb{N}}(\mathcal{F}_\sigma, N_b)$.*

Proof. By Lemma 4.16, Theorem 4.5, and Lemma 4.17. □

5 Hybrid Logic

The hybrid logic $\mathrm{HL}(\downarrow, @)$ is the extension of the basic modal language with the $\downarrow$-binder, the @-operator, and a countably infinite collection WVAR of *world variables* [3]. Formulas of $\mathrm{HL}(\downarrow, @)$ are generated by the following grammar:

$$\varphi := p \mid x \mid \varphi \wedge \varphi \mid \varphi \vee \varphi \mid \neg\varphi \mid \Diamond_i\varphi \mid \Box_i\varphi \mid \downarrow x.\varphi \mid @_x\varphi,$$

where $i \in I$, $p \in \mathrm{Prop}$, and $x \in \mathrm{WVAR}$ [1]. A world variable x occurs *free* in a formula φ if it does not occur in a subformula of φ of the form $\downarrow x.\psi$, and *bound* otherwise. A formula is a *sentence* if it contains no free (world) variables.

[1] Readers familiar with $\mathrm{HL}(\downarrow, @)$ should note that we omit *nominals* from our presentation.

An *assignment* for a σ-LTS M_a is a map $g : \text{WVAR} \to \text{dom}(M)$. Given an assignment g, a world variable x_i, and a state $m \in \text{dom}(M)$, we let $g[x_i \mapsto m]$ denote the assignment which is the same as g, except that it maps x_i to m. The semantics (omitting the propositional, Boolean, and modal clauses, which are defined as usual) for HL($\downarrow$, @) are given as follows

$$\begin{array}{ll} M_a, g \models x & \text{if } g(x) = a \text{ for } x \in \text{WVAR}, \\ M_a, g \models \downarrow x_i.\varphi & \text{if } M_a, g[x_i \mapsto a] \models \varphi, \text{ and} \\ M_a, g \models @_x\varphi & \text{if } M_b, g \models \varphi, \text{ where } g(x) = b, \end{array}$$

For HL($\downarrow$, @) sentences φ, the assignment chosen does not matter, and so we write $M_a \models \varphi$ instead of $M_a, g \models \varphi$. Given a σ-LTS M_a, the *submodel of M generated by a* is the structure $\mathtt{gsub}(M_a)$, defined to be the smallest substructure M'_a of M_a containing a and such that, whenever $b \in \text{dom}(M')$ and $R^M(b, c)$ holds, then $c \in \text{dom}(M')$. Clearly $\mathtt{gsub}(M_a)$ is a point-generated σ-LTS. The following known result relates HL($\downarrow$, @) to the generated submodel-invariant fragment of FO, where a formula φ is *invariant for generated submodels* if, for any σ-LTS M_a, we have $M_a \models \varphi$ if and only if $\mathtt{gsub}(M_a) \models \varphi$.

Theorem 5.1 (Generated Submodel Invariance, [3]) *If $\varphi(x)$ is a first-order formula in a modal signature, then $\varphi(x)$ is equivalent to a (nominal-free)* HL($\downarrow$, @) *sentence if and only if $\varphi(x)$ is invariant for generated submodels.*

We write $\mathtt{gsub}^k(M_a)$ to denote the substructure of $\mathtt{gsub}(M_a)$ containing only elements of depth at most k. If M_a is image-finite, then $\mathtt{gsub}^k(M_a)$ is finite for all $k \in \mathbb{N}$. The next proposition follows easily from Theorem 5.1.

Proposition 5.2 *For each $N_b \in \mathcal{PG}^k_\sigma$, there is a formula $\varphi \in$ HL($\downarrow$, @) such that, if M_a is an image-finite point-generated σ-LTS, then $M_a \models \varphi$ if and only if $\mathtt{gsub}^k(M_a) \cong N_b$.*

Proof. Fix some N_b in $\mathcal{PG}^k$ with $\text{dom}(N) = \{b_1 \dots b_n\}$, where $b = b_1$. Let $\delta(x_1, \dots, x_n)$ be the FO formula expressing that the x_i are distinct, and that for all y, y is reachable from x_1 by a directed σ-path of length at most n if and only if $y = x_j$ for some $1 \leq j \leq n$. Consider the FO formula

$$\psi(x_1) := \exists x_2 \dots \exists x_n \left(\delta(x_1, \dots, x_n) \wedge \left(\bigwedge_{i,j \leq n : R^N(b_i, b_j)} R(x_i, x_j) \right) \right).$$

If M_a is an image-finite point-generated LTS, then $M \models \psi(a)$ if and only if $\mathtt{gsub}^k(M_a) \cong N_b$. Clearly $\psi(x_1)$ is a first-order formula in a modal signature which is invariant for generated submodels, and so there exists a nominal-free HL($\downarrow$, @) sentence equivalent to $\psi(x_1)$, which is what we wanted to show. □

The next lemma is obvious, since homomorphisms preserve path lengths and map distinguished elements to distinguished elements.

Lemma 5.3 *If M_a is an image-finite LTS, then for each $k \in \mathbb{N}$, we have that $hom_{\mathbb{N}}(\mathcal{PG}^k_\sigma, M_a) = hom_{\mathbb{N}}(\mathcal{PG}^k_\sigma, \mathtt{gsub}^k(M_a))$.*

We now prove our characterization result for HL($\downarrow$, @).

Theorem 5.4 *For image-finite LTSs M_a and N_b, the following are equivalent:*

(i) $hom_{\mathbb{N}}(\mathcal{PG}_\sigma, M_a) = hom_{\mathbb{N}}(\mathcal{PG}_\sigma, N_b)$,

(ii) $\mathtt{gsub}(M_a) \cong \mathtt{gsub}(N_b)$,

(iii) $M_a \equiv_{\mathrm{HL}(\downarrow, @)} N_b$.

Proof. For (i) to (ii), suppose that $\hom_{\mathbb{N}}(\mathcal{PG}_\sigma, M_a) = \hom_{\mathbb{N}}(\mathcal{PG}_\sigma, N_b)$. Then clearly $\hom_{\mathbb{N}}(\mathcal{PG}^k_\sigma, M_a) = \hom_{\mathbb{N}}(\mathcal{PG}^k_\sigma, N_b)$, and so by Lemma 5.3, we have that $\hom_{\mathbb{N}}(\mathcal{PG}^k_\sigma, \mathtt{gsub}^k(M_a)) = \hom_{\mathbb{N}}(\mathcal{PG}^k_\sigma, \mathtt{gsub}^k(N_b))$. Hence by Fact 2.8 and Theorem 2.5, $\mathtt{gsub}^k(M_a) \cong \mathtt{gsub}^k(N_b)$ for each $k \in \mathbb{Z}^+$, and so by Lemma 2.9, we have that $\mathtt{gsub}(M_a) \cong \mathtt{gsub}(N_b)$. For (ii) to (i), suppose that we have $\mathtt{gsub}(M_a) \cong \mathtt{gsub}(N_b)$. Since the range of a homomorphism from a point-generated σ-LTS to M_a (resp. N_b) is contained within $\mathtt{gsub}(M_a)$ (resp. $\mathtt{gsub}(N_b)$), we have that $\hom_{\mathbb{N}}(\mathcal{PG}_\sigma, M_a) = \hom_{\mathbb{N}}(\mathcal{PG}_\sigma, N_b)$. The direction (ii) to (iii) is immediate from Theorem 5.1. For (iii) to (ii), Proposition 5.2 gives us formulas $\varphi_k \in \mathrm{HL}(\downarrow, @)$ such that $N_b \models \varphi$ if and only if $\mathtt{gsub}^k(N_b) \cong \mathtt{gsub}^k(M_a)$ for all $k \in \mathbb{N}$. Since $\mathtt{gsub}^k(M_a) \models \varphi$, we have by Theorem 5.1 that $M_a \models \varphi$, and so by the assumption that $M_a \equiv^k_{\mathrm{HL}(\downarrow, @)} N_b$, we have $N_b \models \varphi$. Hence $\mathtt{gsub}^k(M_a) \cong \mathtt{gsub}^k(N_b)$ for all $k \in \mathbb{N}$. Then by Lemma 2.9, $\mathtt{gsub}(M_a) \cong \mathtt{gsub}(N_b)$. □

We do not provide a version of this theorem which is parametrized by modal depth, as we did for $\mathrm{ML}_\#$ (cf. Theorem 4.5), because Proposition 5.2 does not offer a bound (as a function of k) on the modal depth of the HL($\downarrow$, @) formula describing a point-generated submodel of depth at most k up to isomorphism.

Backward and Global Modalities. HL(E, $\downarrow$, @), the extension of HL($\downarrow$, @) with the global modality, is known to have the expressive power of full first-order logic [4], which we noted previously is captured by the left profile over the natural number semiring with respect to the class of all structures. This implies that HL(E, $\downarrow$, @) equivalence is captured by restricting the left profile over the natural semiring to the class of all σ-LTSs. We now provide a characterization result for $\mathrm{HL}^B(\downarrow, @)$, the extension of HL($\downarrow$, @) with the backward modalities for $k = 1$ (cf. Definition 3.4). As in Section 4, we fix an expanded signature $\sigma_B = \mathrm{Prop} \cup \mathbb{A} \cup \mathbb{A}_B$, where $\mathbb{A}_B = \{B_i \mid i \in I\}$ is disjoint from $\mathbb{A}$.

Definition 5.5 Let $T_c \in \mathcal{C}_\sigma$ and $T'_c \in \mathcal{PG}_{\sigma_B}$ with $\mathrm{dom}(T) = \mathrm{dom}(T')$. We say that T'_c is a *$\mathcal{PG}$-augmentation* of M_a if, for each $i \in I$, there exists some $X_i \subseteq R^T_i$ such that $R^{T'}_i = R^T_i \setminus X_i$ and $B^{T'}_i = B^T_i \cup X^{-1}_i$.

Thus T'_c is a $\mathcal{PG}$-augmentation of T_c if it can be obtained by replacing R_i transitions in T_c with B_i transitions in the opposite direction. Recall the `flip` operation (cf. Definition 4.9). If T'_c is a point-generated σ_B-LTS, then clearly $\mathtt{flip}(T'_c)$ is a connected σ-LTS.

Definition 5.6 Let M_a be a connected σ_B-LTS. We write $\mathcal{R}^{M_a}$ to denote the set of elements in $\mathrm{dom}(M)$ reachable by a directed σ_B-path from a, and we set $\mathcal{U}^{M_a} = \mathrm{dom}(M) \setminus \mathcal{R}^{M_a}$. If all transitions from elements of $\mathcal{U}^{M_a}$ to elements of

$\mathcal{R}^{M_a}$ are actions in $\mathbb{A}$, then we write $\mathcal{P}(M_a)$. Furthermore, if $\mathcal{P}(M_a)$ is satisfied, then we define $\texttt{exp}(M_a)$ to be the σ_B-LTS with $\text{dom}(\texttt{exp}(M_a)) = \text{dom}(M)$ and $p^{\texttt{exp}(M_a)} = p^M$ for all $p \in \text{Prop}$, such that for all $i \in I$,

$$R_i^{\texttt{exp}(M_a)} = R_i^M \setminus \{\langle m, n\rangle \mid \langle m, n\rangle \in R_i^M, m \in \mathcal{U}^{M_a}, n \in \mathcal{R}^{M_a}\}, \text{ and}$$
$$B_i^{\texttt{exp}(M_a)} = B_i^M \cup \{\langle n, m\rangle \mid \langle m, n\rangle \in R_i^M, m \in \mathcal{U}^{M_a}, n \in \mathcal{R}^{M_a}\}.$$

Intuitively, $\mathcal{P}(M_a)$ asserts that all transitions out of elements of $\mathcal{U}^{M_a}$ to elements of $\mathcal{R}^{M_a}$ are actions in $\mathbb{A}$. The $\texttt{exp}$ operation replaces R_i transitions from elements of $\mathcal{U}^{M_a}$ to elements of $\mathcal{R}^{M_a}$ with the corresponding B_i actions in the opposite direction. The next proposition shows that $\texttt{exp}$ is an operation on the class of connected σ_B-LTSs M_a satisfying $\mathcal{P}(M_a)$ which grows the set of elements reachable by a σ_B-path from a. The proof is straightforward.

Proposition 5.7 *For all $M_a \in \mathcal{C}_{\sigma_B}$ satisfying $\mathcal{P}(M_a)$, we have that $\mathcal{R}^{M_a} \subseteq \mathcal{R}^{\texttt{exp}(M_a)}$ and $\mathcal{U}_a^{\texttt{exp}(M_a)} \subseteq \mathcal{U}^{M_a}$, and these inclusions are proper if $\mathcal{U}^{M_a} \neq \emptyset$. Furthermore, $\texttt{exp}(M_a)$ is a connected σ_B-LTS satisfying $\mathcal{P}(M_a)$.*

Proposition 5.8 *If T_c is in $\mathcal{C}_\sigma$, then there is a $\mathcal{PG}$-augmentation T_c' of T_c.*

Proof. Suppose T_c is a finite connected σ-LTS. Since T_c contains no σ_B transitions, it clearly satisfies $\mathcal{P}(T_c)$. Recall from Proposition 5.7 that $\mathcal{U}^{\texttt{exp}(T_c)}$ is a proper subset of $\mathcal{U}^{T_c}$ whenever $\mathcal{U}^{T_c} \neq \emptyset$, and so there exists some $k \in \mathbb{N}$ such that $\mathcal{U}^{\texttt{exp}^k(T_c)} = \emptyset$, where the exponent k indicates iterated application of the $\texttt{exp}$ operation (which we can do by Proposition 5.7). Hence $\mathcal{R}^{\texttt{exp}^k(T_c)} = \text{dom}(\texttt{exp}^k(T_c))$, and so $\texttt{exp}^k(T_c)$ is a point-generated σ_B-LTS. Furthermore, $T_c' := \texttt{exp}^k(T_c)$ is clearly a $\mathcal{PG}$-augmentation of T_c. □

The following proposition is straightforward to prove from the definitions.

Proposition 5.9 *Let M_a be a degree-finite σ-LTS. Then*

(i) *If T_c is in $\mathcal{C}_\sigma$, then $hom_{\mathbb{N}}(T_c, M_a) = hom_{\mathbb{N}}(T_c', M_a^B)$ for any $\mathcal{PG}$-augmentation T_c' of T_c.*

(ii) *If T_c' is in $\mathcal{PG}_{\sigma_B}$, then $hom_{\mathbb{N}}(\texttt{flip}(T_c'), M_a) = hom_{\mathbb{N}}(T_c', M_a^B)$.*

Lemma 5.10 *Let M_a and N_b be degree-finite σ-LTSs. Then $hom_{\mathbb{N}}(\mathcal{C}_\sigma, M_a) = hom_{\mathbb{N}}(\mathcal{C}_\sigma, N_b)$ if and only if $hom_{\mathbb{N}}(\mathcal{PG}_{\sigma_B}, M_a^B) = hom_{\mathbb{N}}(\mathcal{PG}_{\sigma_B}, N_b^B)$.*

Proof. Both directions are by contraposition. For the reverse direction, if $\text{hom}_{\mathbb{N}}(\mathcal{C}_\sigma, M_a) \neq \text{hom}_{\mathbb{N}}(\mathcal{C}_\sigma, N_b)$, then there is a finite connected σ-LTS T_c such that $\text{hom}_{\mathbb{N}}(T_c, M_a) \neq \text{hom}_{\mathbb{N}}(T_c, N_b)$. Then by Proposition 5.8, there exists a $\mathcal{PG}$-augmentation T_c' of T_c. Then by Lemma 5.9, we have that $\text{hom}_{\mathbb{N}}(T_c', M_a) \neq \text{hom}_{\mathbb{N}}(T_c', N_b)$, and hence $\text{hom}_{\mathbb{N}}(\mathcal{PG}_{\sigma_B}, M_a^B) \neq \text{hom}_{\mathbb{N}}(\mathcal{PG}_{\sigma_B}, N_b^B)$. The forward direction is similar, using the $\texttt{flip}$ function and Lemma 5.9. □

The proof of the next lemma is analogous to that of Lemma 4.12.

Lemma 5.11 *Let M_a and N_b be degree-finite σ-LTSs. Then $M_a \equiv^\sigma_{\text{HL}^B(\downarrow, @)} N_b$ if and only if $M_a^B \equiv^{\sigma_B}_{\text{HL}(\downarrow, @)} N_b^B$.*

Theorem 5.12 *If M_a and N_b are degree-finite σ-LTSs, then $M_a \equiv^{\sigma}_{\mathrm{HL}^B(\downarrow, @)} N_b$ if and only if* $\hom(\mathcal{C}_\sigma, M_a) = \hom(\mathcal{C}_\sigma, N_b)$.

Proof. By Lemma 5.10, Theorem 5.4, and Lemma 5.11. □

6 Negative Results

Recall that ML denotes the basic (multi)modal language. *Positive modal logic* (notation: ML^+) is the fragment of ML without negation. We now show that ML^+-equivalence and ML-equivalence do not admit homomorphism count indistinguishability characterizations. For this, it will be convenient to work with the modal equivalence relations corresponding to these languages.

Definition 6.1 Let M_a and N_b denote σ-LTSs. A *directed simulation* from M_a to N_b is a relation $Z \subseteq \mathrm{dom}(M) \times \mathrm{dom}(N)$ with $(a, b) \in Z$ such that

(prop$^-$) If $(m, n) \in Z$, then $\lambda^M_\sigma(m) \subseteq \lambda^N_\sigma(n)$;

(forth) For each $i \in I$, if $(m, n) \in Z$ and there's $s \in M$ such that $R^M_i(m, s)$, then there's some $t \in N$ such that $R^N_i(n, t)$ and $(s, t) \in Z$; and

(back) For each $i \in I$, if $(m, n) \in Z$ and there's $t \in N$ such that $R^N_i(n, t)$, then there's some $s \in M$ such that $R^M_i(m, s)$ and $(s, t) \in Z$.

If directed simulations from M_a to N_b and N_b to M_a exist, then we say that they are *directed simulation equivalent* (notation: $M_a \leftrightarroweq_d N_b$). Z is a *bisimulation* between M_a and N_b if it also satisfies the stronger condition (prop) asserting that $\lambda^M_\sigma(m) = \lambda^N_\sigma(n)$ whenever $(m, n) \in Z$. If a bisimulation between σ-LTSs M_a and N_b exists, then they are *bisimilar* (notation: $M_a \leftrightarroweq N_b$).

Directed simulation equivalence and bisimulation capture ML^+-equivalence and ML-equivalence, respectively, over image-finite σ-LTSs.

Theorem 6.2 *(Directed Simulation Equivalence Invariance, [21]) For image-finite σ-LTSs M_a and N_b, we have that $M_a \leftrightarroweq_d N_b$ if and only if $M_a \equiv_{\mathrm{ML}^+} N_b$.*

Theorem 6.3 *(Bisimulation Invariance, [9]) For image-finite σ-LTSs M_a and N_b, we have that $M_a \leftrightarroweq N_b$ if and only if $M_a \equiv_{ML} N_b$.*

For equivalence relations $\sim$ and $\approx$ on σ-LTSs, if $M_a \sim N_b$ implies $M_a \approx N_b$, then we say that $\sim$ is *finer* than $\approx$, and $\approx$ is *coarser* than $\sim$. A function f with $\mathrm{dom}(f) = \mathbb{N}$ is *ultimately periodic* if there exist $P \in \mathbb{Z}^+$ and $L \in \mathbb{N}$ such that $f(n) = f(n + P)$ for all $n \geq L$. If L and P are the least integers such that the ultimate periodicity condition is satisfied, then we refer to the sequence $\langle f(0), \ldots, f(L-1)\rangle$ as the *preperiod* of f, and we refer to the sequence $\langle f(L), \ldots, f(L + P - 1)\rangle$ as the *periodic segment* of f.

Proposition 6.4 *Let $\mathcal{S} = \langle S, +_S, \cdot_S, 0_S, 1_s\rangle$ be a semiring such that $\mathrm{count}_\mathcal{S}$ is not injective. Then $\mathrm{count}_\mathcal{S}$ is ultimately periodic. Furthermore, the preperiod and the periodic segment are disjoint, and there are no elements which occur more than once in either the preperiod or the periodic segment.*

Proof. If $\mathrm{rng}(\mathrm{count}_\mathcal{S})$ is not injective, then there exists some least m such that $\mathrm{count}_\mathcal{S}(m) = \mathrm{count}_\mathcal{S}(L)$ for some $L < m$. Let $P = m - L$. Observe that $\text{count}_{\mathcal{S}}(a+b) = \text{count}_{\mathcal{S}}(a) +_{S} \text{count}_{\mathcal{S}}(b)$ for all $a, b \in \mathbb{N}$, by associativity of addition in $\mathcal{S}$. We show by induction on $n \in \mathbb{N}$ that $\text{count}_{\mathcal{S}}(n) = \text{count}_{\mathcal{S}}(n+P)$ if $n \geq L$. If $n = L$, then $\text{count}_{\mathcal{S}}(n) = \text{count}_{\mathcal{S}}(m) = \text{count}_{\mathcal{S}}(n+P)$. Now suppose inductively that $\text{count}_{\mathcal{S}}(n) = \text{count}_{\mathcal{S}}(n+P)$. Then

$$\begin{aligned} \text{count}_{\mathcal{S}}(n+1) &= \text{count}_{\mathcal{S}}(n) \ +_S \ \text{count}_{\mathcal{S}}(1) && \text{(Assoc. of } +_S) \\ &= \text{count}_{\mathcal{S}}(n+P) \ +_S \ \text{count}_{\mathcal{S}}(1) && \text{(Inductive Hypothesis)} \\ &= \text{count}_{\mathcal{S}}(n+1+P). && \text{(Assoc. of } +_S) \end{aligned}$$

Hence $\text{count}_{\mathcal{S}}(n) = \text{count}_{\mathcal{S}}(n+P)$ for all $n \geq L$. By the above argument, the periodic segment begins with the first appearance of an element of $\mathcal{S}$ which occurs twice in $\text{rng}(\text{count}_{\mathcal{S}})$, and so the preperiod does not contain repeated elements. This also implies that the preperiod and period are disjoint. Finally, the fact that the periodic segment must also not contain any duplicate elements is clear, since successive elements are obtained by adding 1_S, and so a duplicate element must mark the start of another repetition of the periodic segment. □

Theorem 6.5 *Let $\mathcal{S} = \langle S, +_S, \cdot_S, 0_S, 1_S \rangle$ be an arbitrary semiring, and let $\sim$ denote any relation finer than directed simulation and coarser than bisimulation. There does not exist a class $\mathcal{C}$ of σ-LTSs such that, for all finite σ-LTSs M_a and N_b, we have $hom_{\mathcal{S}}(\mathcal{C}, M_a) = hom_{\mathcal{S}}(\mathcal{C}, N_b)$ if and only if $M_a \sim N_b$.*

Proof. Suppose toward a contradiction that some such class $\mathcal{C}$ exists. For $n \in \mathbb{Z}^+$, let K^n_a denote the σ-LTS with n states, distinguished element a, $p^{K^n} = \text{dom}(K^n)$, and $R^{K^n} = \text{dom}(K^n) \times \text{dom}(K^n)$. Clearly $K^n_a \leftrightarroweq K^{n'}_{a'}$ for all $n, n' \in \mathbb{Z}^+$. Furthermore, for all σ-LTSs T_c with $|\text{dom}(T)| = k$, every map $h : T_c \to K^n_a$ with $h(c) = a$ is a homomorphism, so $|\text{Hom}(T_c, K^n_a)| = n^{k-1}$.

We first rule out that $\mathcal{C}$ contains only σ-LTSs T_c with $|\text{dom}(T)| = 1$. If it did, then for all σ-LTSs S_d, we have $\text{hom}_{\mathcal{S}}(T_c, S_d) = 1$ if and only if $\text{Hom}(T_c, S_d) \neq \emptyset$. Consider the σ-LTSs in Figure 2. By homomorphic equivalence, $\text{Hom}(T_c, M_a) \neq \emptyset$ if and only if $\text{Hom}(T_c, N_b) \neq \emptyset$. Hence $\text{hom}_{\mathcal{S}}(\mathcal{C}, M_a) = \text{hom}_{\mathcal{S}}(\mathcal{C}, N_b)$. However, since $M_a \not\simeq_d N_b$, this implies that $M_a \not\sim N_b$, contradicting our assumption about $\mathcal{C}$. Thus $\mathcal{C}$ must contain a structure T_c with $|\text{dom}(T)| = k+1$ for some $k \in \mathbb{Z}^+$.

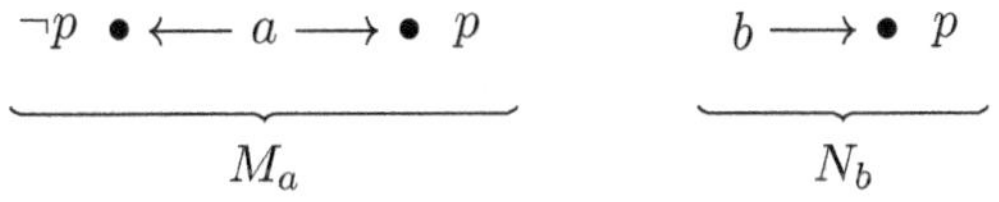

Fig. 2. Homomorphically-equivalent σ-LTSs such that $M_a \not\simeq N_b$ and $M_a \not\simeq_d N_b$.

We now claim that $\text{count}_{\mathcal{S}}$ is non-injective. If $\text{count}_{\mathcal{S}}$ were injective, then $\text{hom}_{\mathcal{S}}(T_c, K^1_a) = \text{count}_{\mathcal{S}}(1) \neq \text{count}_{\mathcal{S}}(2^k) = \text{hom}_{\mathcal{S}}(T_c, K^2_a)$. Since $K^1_a \leftrightarroweq K^2_a$ (and hence $K^1_a \sim K^2_a$), this contradicts our assumption about $\mathcal{C}$. Thus we may assume that $\text{count}_{\mathcal{S}}$ is non-injective, and so by Proposition 6.4, it is ultimately periodic: there exist $P \in \mathbb{Z}^+$ and $L \in \mathbb{N}$ such that $\text{count}_{\mathcal{S}}(n) = \text{count}_{\mathcal{S}}(n+P)$ for all $n \geq L$. Let $\pi = \pi_0 \ldots \pi_{P-1}$ denote the periodic segment of $\text{count}_{\mathcal{S}}$, and assume that L and P are minimal, so that, by Proposition 6.4, π contains no duplicate elements. Figure 3 depicts the range of $\text{count}_{\mathcal{S}}$.

$$0_{\mathcal{S}} \quad 1_{\mathcal{S}} \quad \cdots \quad \text{count}_{\mathcal{S}}(L-1) \quad \pi_0 \quad \cdots \quad \pi_{P-1} \quad \pi_0 \quad \cdots \quad \pi_{P-1} \quad \cdots$$

Fig. 3. The counting sequence in $\mathcal{S}$.

We now distinguish several cases, deriving a contradiction in each.

(i) If $0_{\mathcal{S}}$ occurs in π, then $L = 0$ (i.e., $\text{count}_{\mathcal{S}}$ is *purely* periodic), and $\pi_0 = 0_{\mathcal{S}}$. Then $\text{count}_{\mathcal{S}}(n) = \text{count}_{\mathcal{S}}(n \mod P)$ for all $n \in \mathbb{N}$. Hence we have that $\text{hom}_{\mathcal{S}}(T_c, K_a^P) = \text{count}_{\mathcal{S}}(P^k) = \text{count}_{\mathcal{S}}(P^k \mod P) = 0_{\mathcal{S}}$, while $\text{hom}(T_c, K_a^1) = \text{count}_{\mathcal{S}}(1) = 1_{\mathcal{S}}$. This implies that $\text{hom}_{\mathcal{S}}(T_c, K_a^1) \neq \text{hom}_{\mathcal{S}}(T_c, K_a^P)$, which is a contradiction since $K_a^1 \leftrightarrows K_a^P$.

(ii) If $1_{\mathcal{S}}$ occurs in π but $0_{\mathcal{S}}$ does not, then $\pi_0 = 1_{\mathcal{S}}$. Distinguish cases.

(a) If $P = 1$, then for all σ-LTSs S_d, we have $\text{hom}_{\mathcal{S}}(T_c, S_d) = 1$ if and only if $\text{Hom}(T_c, S_d) \neq \emptyset$. Consider the example in Figure 2: by homomorphic equivalence, we have $\text{hom}_{\mathcal{S}}(\mathcal{C}, M_a) = \text{hom}_{\mathcal{S}}(\mathcal{C}, N_b)$. This is again a contradiction, since $M_a \not\leftrightarrows_d N_b$.

(b) If $P > 1$, then $\text{count}_{\mathcal{S}}(0) = 0_{\mathcal{S}}$, and $\text{count}_{\mathcal{S}}(n) = \pi_{((n-1) \mod P)}$ for $n > 0$. Then since $P^k - 1 \mod P = P - 1$, we have that $\text{hom}_{\mathcal{S}}(T_c, K_a^P) = \pi_{((P^K-1) \mod P)} = \pi_{P-1}$. Furthermore, since $\pi_0 = 1$, $P - 1 \neq 0$, and the periodic segment contains no repeated elements, $\pi_{P-1} \neq 1_{\mathcal{S}}$. Hence $\text{hom}_{\mathcal{S}}(T_c, K_a^1) \neq \text{hom}_{\mathcal{S}}(T_c, K_a^P)$.

(iii) If $1_{\mathcal{S}}$ does not occur in π, then $\text{hom}_{\mathcal{S}}(T_c, K_a^n) = n^k \neq 1_{\mathcal{S}}$ for n sufficiently large. Hence we have $\text{hom}_{\mathcal{S}}(T_c, K_a^1) \neq \text{hom}_{\mathcal{S}}(T_c, K_a^n)$.

Since we reach a contradiction in each case, no such class $\mathcal{C}$ can exist. □

7 Discussion

Our positive characterization results, summarized in Figure 4, could also be seen as characterizations of certain modal equivalence relations, just as our negative result (Theorem 6.5) was. For example, image-finite LTSs are $\text{ML}_{\#}$-equivalent if and only if there exists a *graded bisimulation* between them [29]. Similarly, two LTSs are equivalent with respect to nominal-free $\text{HL}(\downarrow, @)$ formulas if and only if there is an ω-bisimulation between them [3].

Language	Captured by
$\text{ML}_{\diamond}^{+}$	$\text{hom}_{\mathbb{B}}(\mathcal{T}_\sigma, M_a)$
$\text{ML}_{\diamond}^{+,B}$	$\text{hom}_{\mathbb{B}}(\mathcal{A}_\sigma, M_a)$
$\text{ML}_{\diamond}^{+,G}$	$\text{hom}_{\mathbb{B}}(\mathcal{F}_\sigma, M_a)$
$\text{ML}_{\#}$	$\text{hom}_{\mathbb{N}}(\mathcal{T}_\sigma, M_a)$
$\text{ML}_{\#}^{B}$	$\text{hom}_{\mathbb{N}}(\mathcal{A}_\sigma, M_a)$

Language	Captured by
$\text{ML}_{\#}^{G}$	$\text{hom}_{\mathbb{N}}(\mathcal{F}_\sigma, M_a)$
$\text{HL}(\downarrow, @)$	$\text{hom}_{\mathbb{N}}(\mathcal{PG}_\sigma, M_a)$
$\text{HL}^{B}(\downarrow, @)$	$\text{hom}_{\mathbb{N}}(\mathcal{C}_\sigma, M_a)$
ML^{+}	None
ML	None

Fig. 4. Summary of Characterization Results.

Related work. An initial catalyst for investigating a left-profile characterization for $ML_{\#}$ was recent work by Barcelo et. al. showing that nodes of undirected graphs are indistinguishable by a special case of GNNs (aggregate-combine GNNs) if and only if they are graded modal equivalent [8]. Given that $ML_{\#}$ is a syntactic fragment of C^2, and that both C^2-equivalence and indistinguishability by GNNs can be captured by the restriction of the left homomorphism vector to the class of undirected trees, this result naturally suggested that a similar restriction to appropriate classes of trees should capture graded modal logic, as we have shown (cf. Theorem 4.5).

Sections 3, 4, and 5 provide homomorphism count indistinguishability characterizations using model-theoretic methods. An important line of related work studies categorical generalizations of Lovász's original result; early work in this direction includes [20] and [26]. More recent work on *game comonads* formalizes model-comparison games (such as the bisimulation game) in category-theoretic terms [1,2]. These game comands can be used to derive homomorphism count indistinguishability results from general categorical results. For example, Theorem 3.2 is a consequence of a general categorical result proven in [2]. Similarly, a weaker version of Theorem 4.5, applying to finite structures, was obtained in [14] using these methods. Categorical and topological arguments were used in [28] to provide the first Lovász-style results for classes of infinite structures.

Early negative results pertaining to characterizations of logical equivalences via homomorphism count indistinguishability begin with [5], in this case limited to negative results with respect to counting done in the Boolean and natural number semirings. In [23], the authors show that equivalence with respect to *linear-algebraic logic* cannot be captured by homomorphism count indistinguishability with respect to any class of graphs, both when counting is done in the natural numbers, and when counting is done in an arbitrary finite prime field. The present paper goes a step further, using the more general algebraic structure of semirings as the basis of counting for its negative results.

Future work. Our combinatorial model-theoretic arguments for Theorems 3.2, 4.5, and 5.4 are analogous to earlier results for Lovász-style theorems. However, the method of lifting these results to extensions of the languages with backward or global modalities is, to the author's knowledge, a novel approach. One future avenue of research would be to generalize these methods to a categorical setting. Furthermore, while the aforementioned categorical work has provided interesting sufficient conditions for Lovász-style theorems, there is not yet a concise *necessary* condition for a logic to admit such a result. Another interesting avenue of research would be to use the insight gained from our broad negative result in Theorem 6.5 to identify such a condition. A last direction for future work is to identify modal relations captured by homomorphism indistinguishability with respect to *finite* classes of LTSs; these are naturally related to the notion of *homomorphism query algorithms* [30].

Acknowledgements. I sincerely thank Balder ten Cate for his invaluable advising during the development of much of this work, and I thank Scott Weinstein and Val Tannen for helpful comments on an earlier draft.

References

[1] Abramsky, S., A. Dawar and P. Wang, *The pebbling comonad in finite model theory*, in: *2017 32nd Annual ACM/IEEE Symposium on Logic in Computer Science (LICS)*, 2017, pp. 1–12.

[2] Abramsky, S. and N. Shah, *Relating structure and power: Comonadic semantics for computational resources*, Journal of Logic and Computation **31** (2021), pp. 1390–1428.

[3] Areces, C., P. Blackburn and M. Marx, *Hybrid logics: characterization, interpolation and complexity*, Journal of Symbolic Logic **66** (2001), p. 977–1010.

[4] Areces, C. and B. ten Cate, *14 hybrid logics*, in: P. Blackburn, J. Van Benthem and F. Wolter, editors, *Handbook of Modal Logic*, Studies in Logic and Practical Reasoning **3**, Elsevier, 2007 pp. 821–868.

[5] Atserias, A., P. G. Kolaitis and W.-L. Wu, *On the expressive power of homomorphism counts*, in: *2021 36th Annual ACM/IEEE Symposium on Logic in Computer Science (LICS)*, IEEE, 2021, pp. 1–13.

[6] Babai, L., *Graph isomorphism in quasipolynomial time*, in: *Proceedings of the forty-eighth annual ACM symposium on Theory of Computing*, 2016, pp. 684–697.

[7] Babai, L., P. Erdős and S. M. Selkow, *Random graph isomorphism*, SIAM Journal on Computing **9** (1980), pp. 628–635.

[8] Barceló, P., E. V. Kostylev, M. Monet, J. Pérez, J. Reutter and J.-P. Silva, *The Logical Expressiveness of Graph Neural Networks*, in: *8th International Conference on Learning Representations (ICLR 2020)*, Virtual conference, Ethiopia, 2020.

[9] Benthem, J. v., "Modal correspondence theory," Ph.D. thesis, University of Amsterdam (1976).

[10] Benthem, J. v., B. t. Cate and J. Vaananen, *Lindstrom theorems for fragments of first-order logic*, Logical Methods in Computer Science **5** (2009).

[11] Cai, J.-Y., M. Fürer and N. Immerman, *An optimal lower bound on the number of variables for graph identifications*, Combinatorica **12** (1992), pp. 389–410.

[12] Chandra, A. K. and P. M. Merlin, *Optimal implementation of conjunctive queries in relational data bases*, in: *Proceedings of the ninth annual ACM symposium on Theory of computing*, 1977, pp. 77–90.

[13] Comer, J., *Homomorphism counts, database queries, and modal logics* (2023).

[14] Dawar, A., T. Jakl and L. Reggio, *Lovász-type theorems and game comonads*, in: *2021 36th Annual ACM/IEEE Symposium on Logic in Computer Science (LICS)*, 2021, pp. 1–13.

[15] Dell, H., M. Grohe and G. Rattan, *Lovász Meets Weisfeiler and Leman*, in: I. Chatzigiannakis, C. Kaklamanis, D. Marx and D. Sannella, editors, *45th International Colloquium on Automata, Languages, and Programming (ICALP 2018)*, Leibniz International Proceedings in Informatics (LIPIcs) **107** (2018), pp. 40:1–40:14.

[16] Dvořák, Z., *On recognizing graphs by numbers of homomorphisms*, Journal of Graph Theory **64** (2010), pp. 330–342.

[17] Fine, K., *In so many possible worlds.*, Notre Dame Journal of Formal Logic **13** (1972), pp. 516 – 520.

[18] Goble, L. F., *Grades of modality*, Logique et Analyse **13** (1970), pp. 323–334.

[19] Gradel, E., M. Otto and E. Rosen, *Two-variable logic with counting is decidable*, in: *Proceedings of Twelfth Annual IEEE Symposium on Logic in Computer Science*, 1997, pp. 306–317.

[20] Isbell, J., *Some inequalities in hom sets*, Journal of Pure and Applied Algebra **76** (1991), pp. 87–110.

[21] Kurtonina, N. and M. d. Rijke, *Simulating without negation*, Journal of Logic and Computation **7** (1997), pp. 501–522.

[22] Ladner, R. E., *On the structure of polynomial time reducibility*, J. ACM **22** (1975), p. 155–171.

[23] Lichter, M., B. Pago and T. Seppelt, *Limitations of Game Comonads for Invertible-Map Equivalence via Homomorphism Indistinguishability*, in: A. Murano and A. Silva,

editors, *32nd EACSL Annual Conference on Computer Science Logic (CSL 2024)*, Leibniz International Proceedings in Informatics (LIPIcs) **288** (2024), pp. 36:1–36:19.

[24] Lovász, L., *Operations with structures*, Acta Math. Acad. Sci. Hungar **18** (1967), pp. 321–328.

[25] Morris, C., M. Ritzert, M. Fey, W. L. Hamilton, J. E. Lenssen, G. Rattan and M. Grohe, *Weisfeiler and leman go neural: Higher-order graph neural networks*, Proceedings of the AAAI Conference on Artificial Intelligence **33** (2019), pp. 4602–4609.

[26] Pultr, A., *Isomorphism types of objects in categories determined by numbers of morphisms*, Acta Sci. Math. Szeged **35** (1973), pp. 155–160.

[27] Ramana, M. V., E. R. Scheinerman and D. Ullman, *Fractional isomorphism of graphs*, Discrete Mathematics **132** (1994), pp. 247–265.

[28] Reggio, L., *Polyadic sets and homomorphism counting*, Advances in Mathematics **410** (2022), p. 108712.

[29] Rijke, M. d., *A note on graded modal logic*, Studia Logica **64** (2000), pp. 271–283.

[30] ten Cate, B., V. Dalmau, P. G. Kolaitis and W.-L. Wu, *When Do Homomorphism Counts Help in Query Algorithms?*, in: G. Cormode and M. Shekelyan, editors, *27th International Conference on Database Theory (ICDT 2024)*, Leibniz International Proceedings in Informatics (LIPIcs) **290** (2024), pp. 8:1–8:20.

[31] Tinhofer, G., *Graph isomorphism and theorems of birkhoff type*, Computing (Wien. Print) **36** (1986), pp. 285–300.

[32] Tinhofer, G., *A note on compact graphs*, Discrete Applied Mathematics **30** (1991), pp. 253–264.

[33] Wu, Z., S. Pan, F. Chen, G. Long, C. Zhang and P. S. Yu, *A comprehensive survey on graph neural networks*, IEEE Transactions on Neural Networks and Learning Systems **32** (2021), pp. 4–24.

[34] Zhou, J., G. Cui, S. Hu, Z. Zhang, C. Yang, Z. Liu, L. Wang, C. Li and M. Sun, *Graph neural networks: A review of methods and applications*, AI Open **1** (2020), pp. 57–81.

Some General Completeness Results for Propositionally Quantified Modal Logics

Yifeng Ding, Yipu Li

Department of Philosophy and Religious Studies, Peking University

Abstract

We study the completeness problem for propositionally quantified modal logics on quantifiable general frames, where the admissible sets are the propositions the quantifiers can range over and expressible sets of worlds are admissible, and Kripke frames, where the quantifiers range over all sets of worlds. We show that any normal propositionally quantified modal logic containing all instances of the Barcan scheme is strongly complete with respect to the class of quantifiable general frames validating it. We also provide a sufficient condition for the truth of all formulas, possibly with quantifiers, to be preserved under passing from a quantifiable general frame to its underlying Kripke frame. This is reminiscent of both the idea of elementary submodel in model theory and the persistence concepts in propositional modal logic. The key to this condition is the concept of finite diversity (Fritz 2023), and with it, we show that if Θ is a set of Sahlqvist formulas whose class of Kripke frames has finite diversity, then the smallest normal propositionally quantified modal logic containing Θ, Barcan, a formula stating the existence of world propositions, and a formula stating the definability of successor sets, is Kripke complete. As a special case, we have a simple finite axiomatization of the logic of Euclidean Kripke frames.

Keywords: Propositional quantifier, Sahlqvist formula, canonical frame, diversity, completeness

1 Introduction

Propositionally quantified modal logics (PQMLs henceforth) are modal logics augmented with propositional quantifiers, a special kind of quantifiers that can be intuitively understood as capturing the quantification implicit in English sentences such as "Everything Jane believes is false" and "It's likely that there's something I will never know". One can also understand the expression "For all the police knows, John is dead already" as "That John is dead already is compatible with everything the police knows" and see that propositional quantification is involved.

Normal propositional modal logic has been studied fruitfully in relation to the possible world-based Kripke frames. For propositional quantifiers, it is natural to consider them as quantifying over sets of possible worlds, since in Kripke frames, propositional variables are interpreted as sets of worlds, and propositional quantifiers bind these variables. [1] This immediately gives PQMLs a second-order flavor, and indeed they are also known as second-order propositional modal logics when interpreted on Kripke frames. We may also impose a distinction between sets of worlds that count as propositions (possible value of propositional variables) and sets of worlds that do not count, thereby adding a *propositional domain* (also called the set of admissible sets) to Kripke frames and obtaining general frames. General frames in which every formula under every variable assignment expresses a proposition are called *quantifiable* frames, and they have the desirable logical property of validating instantiation reasonings such as from "Everything I believe is true" and "I believe that the Moon is made of cheese" to "The Moon is made of cheese".

The theory of PQMLs based on Kripke frames and quantifiable frames has been studied since the early days of modal logic (see [22] for a nice survey). The early landmark paper is Fine's [15] in which, among many other things, it is shown that the PQML consisting of formulas valid on the class of reflexive and transitive Kripke frames is not recursively axiomatizable while the PQML of the class of Kripke frames with a universal relation is decidable and can be axiomatized by simply adding to the modal logic S5 the standard quantificational logic and an atomicity principle stating that there is always a world proposition that is itself true and entails every truth. Related questions of decidability, axiomatizability, and expressivity under various kinds of frames were under sustained investigation [29,28,1,31,8,32,4,30,18,42,12,3]. A recent breakthrough is by Fritz in [19] where he established many new results on the decidability and axiomatizability of PQMLs in a very general fashion. Prior to this work, it was not even known whether the PQML of Euclidean Kripke frames is decidable, and the decidability of the PQML of the Kripke frames validating KD45 was only established in [12].

In this paper, we focus on the question of when an axiomatically defined PQML is complete w.r.t. the quantifiable/Kripke frames it defines. In particular, we aim for a counterpart of the celebrated Sahlqvist Completeness Theorem for PQMLs. For example, the 5 axiom $\Diamond p \to \Box\Diamond p$ defines the class of Euclidean Kripke frames, and we know by Sahlqvist Completeness Theorem that 5 is also sufficient to axiomatize the normal modal logic of Euclidean Kripke frames. But is 5 sufficient also for the PQML of Euclidean Kripke frames, even with the help of some atomicity principles? A full generalization of the Sahlqvist Completeness Theorem was already declared impossible by Fine's results, as the PQML of the class of all Kripke frames is not recursively axiomatizable but

[1] This is obviously not the only way: another well known approach to propositionally quantified formulas is to view them as uniform interpolants [34,41,9,5], and semantically we also have the related *bisimulation quantifiers* [23,16,17,14,37,10]. A detailed comparison is well beyond the scope of this paper, and we only note that understanding propositional quantifiers as bisimulation quantifiers would immediately lead to the *atomless principle* $\forall p(\Diamond p \to \exists q(\Diamond(p \wedge q) \wedge \Diamond(p \wedge \neg q)))$, assuming a reasonably rich model class. As we will see, one of the hallmarks of Kripke frames is that they validate the opposite *atomicity principle*, and quantifiable frames in general are silent on this issue.

the class is trivially definable. Thus, we must build on some general condition for axiomatizability, and [19] provided an elegant one: the finiteness of the *diversity* of Kripke frames. Given a Kripke frame $\mathbb{F}$, two worlds in it are called duplicates if the permutation switching them is an automorphism of $\mathbb{F}$. Being duplicates is an equivalence relation, and the diversity of $\mathbb{F}$ is the cardinality of the set of equivalence classes of this relation, while the *diversity* of a class C of Kripke frames is the supremum of the diversity of all *point-generated* subframes of frames in C. Fritz [19] shows that if a class C of Kripke frames is defined by a finite set Θ of formulas and has finite diversity, then the PQML of C is decidable. Decidability itself does not provide much information on whether there can be a simple and intuitive axiomatization or if a given axiomatically defined logic is complete, but as we mentioned above, Fine gave a simple axiomatization of the PQML of Kripke frames with a universal relation (equivalently, Kripke frames validating $\mathtt{S5}$), and more recently, for Kripke frames validating $\mathtt{KD45}$, it is shown [12] that we only need to add the quantificational axioms and rules, the Barcan scheme $\mathtt{Bc}$, and the atomicity principle for completeness. Can this result be generalized? Our finding is that if Θ is a set of Sahlqvist formulas and the class $\mathsf{KFr}(\Theta)$ of Kripke frames it defines has diversity n, then the normal PQML axiomatized by Θ, the Barcan scheme $\mathtt{Bc}$, an atomicity principle $\mathtt{At}^n$, and an axiom $\mathtt{R}^n$ stating that the successor sets of propositions are propositions, is sound and complete w.r.t. $\mathsf{KFr}(\Theta)$. The last two axioms are parametrized by n as they use iterated modalities to simulate the reflexive and transitive closure of the primitive modality. While we can replace the condition of Θ being Sahlqvist by suitable technical conditions, we cannot drop it completely; and while for $\mathtt{KD45}$ $\mathtt{R}^n$ is redundant, for $\mathtt{K5}$ it is not.

Our method is based on saturated (witnessed) maximally consistent sets and the saturated canonical general frame built from them. Fine [15] claims that this method can be used to show the completeness of $\mathtt{S5}$ with atomicity principle w.r.t. Kripke frames with a universal relation but provided only an extremely terse sketch. We will first show that if Λ is a normal PQML with the Barcan scheme $\mathtt{Bc}$, then its saturated canonical general frame is quantifiable and indeed validates Λ. A corollary is that any normal PQML containing $\mathtt{Bc}$ is sound and complete w.r.t. the class of quantifiable frames it defines. Note that since quantifiable frames are essentially multi-sorted first-order structures, PQMLs on them are no longer second-order and we are not bound by the non-axiomatizability issues. This basic general completeness result may have been realized by multiple scholars before, but a formal proof for the fully general statement appears to be missing.

For the main theorem, we start with the saturated canonical general frame and gradually turn it into a Kripke frame that validates the original logic and satisfies a given consistent set of formulas. The strategy is as follows: first (1) take a point-generated general frame, then (2) keep only the worlds that are named by world propositions and obtain what we call the *atomic general frame*, and (3) finally show that we can expand the propositional domain of the atomic general frame to the full powerset without affecting the semantic value of any formula. Step (2) also happens in some completeness proofs of hybrid logics (see e.g. [39], Definition 5.2.6). The resulting atomic general frame has the special property of being discrete and tense, and Sahlqvist formulas are shown to be persistent over these general frames [40]. This is essential in showing that the final frame validates the original logic. Step (3) can also be utilized to show completeness for the monadic second-order theory of ω [35].

In Section 2 we review the basic definitions and provide a Tarski-Vaught-style test for the expansion of the propositional domain to the full powerset to preserve the value of all formulas with propositional quantifiers. In Section 3, we show that the saturated canonical general frame of any PQML containing Bc validates the logic. In Section 4, we show how finite diversity allows a quantifiable frame to pass the Tarski-Vaught test. In Section 5 we prove our main result. Finally, we conclude in Section 6.

2 Preliminaries

Definition 2.1 We fix a countably infinite set Prop of propositional variables and define the language $\mathcal{L}$ of PQMLs by the following grammar:

$$\varphi ::= p \mid \neg\varphi \mid (\varphi \vee \varphi) \mid \Diamond\varphi \mid \exists p\varphi$$

where $p \in \mathsf{Prop}$ and $\Diamond$ is the sole modality in this language. The other common connectives $\wedge, \rightarrow, \leftrightarrow, \Box$ and the universal quantifier $\forall p$ are defined as abbreviations as usual. We also define iterated modalities recursively: $\Diamond^0\varphi = \varphi$, $\Diamond^{n+1}\varphi = \Diamond\Diamond^n\varphi$, $\Diamond^{\leqslant 0}\varphi = \varphi$, and $\Diamond^{\leqslant n+1}\varphi = \Diamond^{\leqslant n}\varphi \vee \Diamond^{n+1}\varphi$. $\Box^n$ and $\Box^{\leqslant n}$ are defined dually. Let $\mathsf{Fv}(\varphi)$ be the set of free variables of φ and $\mathcal{L}_{qf}$ the quantifier-free fragment of $\mathcal{L}$.

Definition 2.2 A *Kripke frame* is a pair $\mathbb{F} = (W, R)$ where W is non-empty and $R \subseteq W^2$. A *general frame* is a triple $\mathbf{F} = (W, R, B)$ where (W, R) is a Kripke frame and $B \subseteq \wp(W)$ is non-empty and closed under Boolean operations and $m_\Diamond^{\mathbf{F}}$, defined by $m_\Diamond^{\mathbf{F}}(X) = \{w \in W \mid \exists u \in X, wRu\}$. $R[w] = \{v \in W \mid wRv\}$ and $R[X] = \bigcup_{w\in X} R[w]$. When $B = \wp(W)$ we call $\mathbf{F}$ *full*.

A valuation v for $\mathbf{F}$ (or B) is a function from Prop to B. We define the semantic value $[\![\varphi]\!]^{\mathbf{F}}(v)$ of formulas φ relative to valuation v in the general frame $\mathbf{F} = (W, R, B)$ inductively by

$$[\![p]\!]^{\mathbf{F}}(v) = v(p) \quad [\![\neg\varphi]\!]^{\mathbf{F}}(v) = W \setminus [\![\varphi]\!]^{\mathbf{F}}(v) \quad [\![\varphi \vee \psi]\!]^{\mathbf{F}}(v) = [\![\varphi]\!]^{\mathbf{F}}(v) \cup [\![\psi]\!]^{\mathbf{F}}(v)$$
$$[\![\Diamond\varphi]\!]^{\mathbf{F}}(v) = m_\Diamond^{\mathbf{F}}([\![\varphi]\!]^{\mathbf{F}}(v)) \quad [\![\exists p\varphi]\!]^{\mathbf{F}}(v) = \bigcup\{[\![\varphi]\!]^{\mathbf{F}}(v[X/p]) \mid X \in B\}.$$

Here $v[X/p]$ is the function that is identical to v except that $v[X/p](p) = X$. The intended meaning of $[\![\varphi]\!]^{\mathbf{F}}(v)$ is that it is the set of worlds where φ is true. A formula φ is *valid* on $\mathbf{F}$ when $[\![\varphi]\!]^{\mathbf{F}}(v) = W$ for all valuations v for $\mathbf{F}$.

Valuation and semantics for any Kripke frame $\mathbb{F} = (W, R)$ is defined the same as for $(W, R, \wp(W))$. That is, semantically a Kripke frame is equivalent to the full general frame based on it.

A *quantifiable frame* is a general frame $\mathbf{F} = (W, R, B)$ such that for any $v \in B^{\mathsf{Prop}}$ and any $\varphi \in \mathcal{L}$, $[\![\varphi]\!]^{\mathbf{F}}(v) \in B$. That is, B is 'closed under semantics'.

Notation: Given a Kripke frame (W, R) and two general frames $\mathbf{F} = (W, R, A)$ and $\mathbf{G} = (W, R, B)$ based on it, for any valuation $v \in (A \cap B)^{\mathsf{Prop}}$, it is clear that if $[\![\varphi]\!]^{\mathbf{F}}(v) \neq [\![\varphi]\!]^{\mathbf{G}}(v)$, it is only because of the difference between A and B. Hence, when it is clear which Kripke frame (W, R) is in discussion, we write $[\![\varphi]\!]^{(W,R,A)}$ simply as $[\![\varphi]\!]^{A}$ or even $[\![\varphi]\!]$ when φ is quantifier-free. Also, it is routine to show that $[\![\varphi]\!]^{\mathbf{F}}(v)$ only depends on $v|_{\mathsf{Fv}(\varphi)}$, the restriction of v to $\mathsf{Fv}(\varphi)$. Thus, for any partial function v from Prop to B such that $dom(v) \supseteq \mathsf{Fv}(\varphi)$, we take $[\![\varphi]\!]^{\mathbf{F}}(v)$ to be the unique element in $\{[\![\varphi]\!]^{\mathbf{F}}(v') \mid v \subseteq v' \in B^{\mathsf{Prop}}\}$.

Now we present the Tarski-Vaught-style test for expanding the propositional domain safely.

Definition 2.3 Given a Kripke frame (W, R) and $\varnothing \neq A, B \subseteq \wp(W)$, A is a PQML-*invariant subdomain* of B if $A \subseteq B$ and for any $\varphi \in \mathcal{L}$ and $v \in A^{\mathsf{Prop}}$, $[\![\varphi]\!]^{A}(v) = [\![\varphi]\!]^{B}(v)$.

Lemma 2.4 *Given a Kripke frame (W, R) and $\varnothing \neq B \subseteq \wp(W)$, B is a* PQML-*invariant subdomain of $\wp(W)$ iff for any $\varphi \in \mathcal{L}$, $p \in \mathsf{Fv}(\varphi)$, $v \in B^{\mathsf{Prop}}$, $w \in W$, and $X \in \wp(W)$, if $w \in [\![\varphi]\!]^{\wp(W)}(v[X/p])$ then there is $Y \in B$ with $w \in [\![\varphi]\!]^{\wp(W)}(v[Y/p])$.*

Proof. Left-to-Right: suppose B is a PQML-invariant subdomain of $\wp(W)$ and $w \in [\![\varphi]\!]^{\wp(W)}(v[X/p])$. Then $w \in [\![\exists p\varphi]\!]^{\wp(W)}(v)$. Then by assumption, $w \in [\![\exists p\varphi]\!]^{B}(v)$ and hence $w \in \bigcup\{[\![\varphi]\!]^{B}(v[Y/p]) \mid Y \in B\}$. So there is $Y \in B$ such that $w \in [\![\varphi]\!]^{B}(v[Y/p])$. Finally since B is a PQML-invariant subdomain of $\wp(W)$, and $v[Y/p] \in B^{\mathsf{Prop}}$, $[\![\varphi]\!]^{B}(v[Y/p]) = [\![\varphi]\!]^{\wp(W)}(v[Y/p])$. Consequently there is $Y \in B$ such that $w \in [\![\varphi]\!]^{\wp(W)}(v[Y/p])$.

Right-to-Left: assume the stated criteria and use induction on φ. Only the step for $\exists$ is non-trivial, where we need to show that $\bigcup\{[\![\varphi]\!]^{\wp(W)}(v[X/p]) \mid X \in \wp(W)\} = \bigcup\{[\![\varphi]\!]^{B}(v[Y/p]) \mid Y \in B\}$ with $v \in B^{\mathsf{Prop}}$. By IH, we only need to show that $\bigcup\{[\![\varphi]\!]^{\wp(W)}(v[X/p]) \mid X \in \wp(W)\} = \bigcup\{[\![\varphi]\!]^{\wp}(W)(v[Y/p]) \mid Y \in B\}$, and the right-to-left inclusion is trivial. For the other inclusion, take any $w \in \bigcup\{[\![\varphi]\!]^{\wp(W)}(v[X/p]) \mid X \in \wp(W)\}$ and use the criteria. □

Substitutions play an important role in our later proofs. Many authors define substitution only when it is free to do so, but we need the version that renames the bound variables when conflicts arise.

Definition 2.5 A *substitution* is a function $\sigma : \mathsf{Prop} \to \mathcal{L}$, and σ_p^{ψ} for any $p \in \mathsf{Prop}$ and $\psi \in \mathcal{L}$ is the substitution that is identical with σ except that $\sigma_p^{\psi}(p) = \psi$. Given a substitution σ, we extend it to $\mathcal{L}$ recursively so that that $\sigma(\neg\varphi) = \neg\sigma(\varphi)$, $\sigma(\Diamond\varphi) = \Diamond\sigma(\varphi)$, $\sigma(\varphi \vee \psi) = \sigma(\varphi) \vee \sigma(\psi)$, and $\sigma(\exists p\varphi) = \exists q\sigma_p^q(\varphi)$ where $q = p$ if $p \notin \bigcup_{r \in \mathsf{Fv}(\exists p\varphi)} \mathsf{Fv}(\sigma(r))$ and otherwise q is the first variable not used in $\exists p\varphi$ and any $\sigma(r)$ for $r \in \mathsf{Fv}(\exists p\varphi)$.

Let ι be the identity substitution. Then, using the above definition, $\iota_p^{\psi}(\varphi)$ is the result of substituting ψ for p in φ with the necessary renamings of bound variables. In particular, $\iota(\varphi) = \varphi$. Then, the standard substitution lemma connecting syntactic and semantic substitution is:

Lemma 2.6 *On any general frame* $\mathbf{F} = (W, R, B)$, *valuation* v *for* $\mathbf{F}$, *and substitution* σ, *define valuation* $\sigma \star v : p \mapsto [\![\sigma(p)]\!]^{\mathbf{F}}(v)$. *Then for any* $\varphi \in \mathcal{L}$, $[\![\varphi]\!]^{\mathbf{F}}(\sigma \star v) = [\![\sigma(\varphi)]\!]^{\mathbf{F}}(v)$.

Finally, we introduce logic. For the convenience of certain proofs, we opted for $\Diamond$ as the primitive modality. For this reason, the `Dual` axiom is necessary.

Definition 2.7 A *normal propositionally quantified modal logic* (NPQML) is a set $\Lambda \subseteq \mathcal{L}$ satisfying the following conditions:

- (Taut) all propositional tautologies are in Λ;
- axiom $\mathtt{K} = \Box(p \to q) \to (\Box p \to \Box q)$ and $\mathtt{Dual} = \Diamond p \leftrightarrow \neg\Box\neg p$ are in Λ;
- (EI) all instances of $\iota_p^\psi(\varphi) \to \exists p\varphi$ are in Λ;
- (Nec) if $\varphi \in \Lambda$, then $\Box\varphi \in \Lambda$;
- (MP) if $\varphi, (\varphi \to \psi) \in \Lambda$, then $\psi \in \Lambda$;
- (EE) if $\varphi \to \psi \in \Lambda$ with $p \notin \mathsf{Fv}(\psi)$, then $\exists p\varphi \to \psi \in \Lambda$.

For any axioms or axiom schemes $A_1, A_2, \ldots, A_n$, we write $\mathsf{K}_\Pi A_1 A_2 \ldots A_n$ for the smallest NPQML containing all (instances) of all A_i's.

Fact 2.8 *For any* φ *and* ψ *obtained by renaming some bound variables in* φ, $(\varphi \leftrightarrow \psi) \in \Lambda$ *for any* NPQML Λ.

Recall that the famous *Barcan scheme* Bc is $\Diamond\exists p\varphi \to \exists p\Diamond\varphi$.

Fact 2.9 *For any class* C *of quantifiable frames, the set* Λ *of formulas valid on each member of* C *is a* NPQML *containing all instances of* Bc.

Definition 2.10 For any set $\Theta \subseteq \mathcal{L}$, let $\mathsf{KFr}(\Theta)$ be the class of Kripke frames validating all formulas in Θ and let $\Theta\pi+$ be the set of formulas valid on all members of $\mathsf{KFr}(\Theta)$.

3 General completeness for quantifiable frames

This section introduces saturated canonical general frames for NPQMLs containing all instances of the Barcan scheme and shows that they validate the original logic and are automatically quantifiable. A consequence is the following:

Theorem 3.1 *Any* NPQML $\Lambda \supseteq \mathsf{K}_\Pi\mathsf{Bc}$ *is strongly complete w.r.t. the class of quantifiable frames validating* Λ.

Fix a NPQML $\Lambda \supseteq \mathsf{K}_\Pi\mathsf{Bc}$. To construct the saturated canonical general frame for Λ, we extend Prop to Prop^+ with countably infinitely many new variables and obtain the extended language $\mathcal{L}^+$. Semantics and logics for $\mathcal{L}^+$ are defined completely analogously. Let Λ^+ be the smallest NPQML in $\mathcal{L}^+$ extending Λ.

Definition 3.2 A set $\Gamma \subseteq \mathcal{L}$ is Λ-*consistent* if there is no finite $A \subseteq \Gamma$ s.t. $\neg(\bigwedge A) \in \Lambda$. A *maximally* Λ-*consistent set* (Λ-MCS) is a Λ-consistent set s.t. all of its proper extensions in $\mathcal{L}$ are not Λ-consistent. Λ^+-consistency and Λ^+-MCSs are defined in the same way using Λ^+ and $\mathcal{L}^+$.

A Λ^+-MCS Γ is *saturated* if for any $\exists p\varphi \in \Gamma$, there is $q \in \mathsf{Prop}^+$ not occurring in $\exists p\varphi$ s.t. $\iota^q_p(\varphi) \in \Gamma$.

Now define the *saturated canonical general frame* $\mathbf{F}_\Lambda = (W, R, B)$ where

- W is the set of all saturated Λ^+-MCSs,
- wRv iff for all $\varphi \in v$, $\Diamond\varphi \in w$,
- $B = \{[\varphi] \mid \varphi \in \mathcal{L}^+\}$ where $[\varphi] = \{w \in W \mid \varphi \in w\}$.

The following two lemmas are completely analogous to their first-order modal logic counterparts. The proof of the first extension lemma is almost identical for example to the proof of Theorem 14.1 of [33]. The full power of (EI) is not used, and all we need is (EE) and that NPQMLs can prove equivalences between formulas that differ only by renaming of bound variables. The second existence lemma can also be proved by repeating the steps in the proof of the existence lemma for first-order modal logic with Bc. For an example, see the proof of Theorem 14.2 of [33].

Lemma 3.3 *Any Λ-consistent set of $\mathcal{L}$ formulas $\Gamma \subseteq \mathcal{L}$ can be extended to a saturated Λ^+-MCS Γ^+.*

Lemma 3.4 *For any $w \in W$, if $\Diamond\varphi \in w$, then there is $u \in R[w]$ s.t. $\varphi \in u$.*

The truth lemma is replaced by a more general statement for all valuations arising from substitutions. This is a common idea in algebraic semantics.

Lemma 3.5 *For any substitution σ for $\mathcal{L}^+$, define its associated valuation $[\sigma] : p \mapsto [\sigma(p)]$. Then for any $\varphi \in \mathcal{L}^+$ and all σ, $[\![\varphi]\!]^{\mathbf{F}_\Lambda}([\sigma]) = [\sigma(\varphi)]$.*

Proof. By induction on φ. The cases for variables and negation go by

$$[\![p]\!]^{\mathbf{F}_\Lambda}([\sigma]) = [\sigma](p) = [\sigma(p)].$$
$$[\![\neg\varphi]\!]^{\mathbf{F}_\Lambda}([\sigma]) = W \setminus [\![\varphi]\!]^{\mathbf{F}_\Lambda}([\sigma]) = W \setminus [\sigma(\varphi)] = [\neg\sigma(\varphi)] = [\sigma(\neg\varphi)].$$

The case for disjunction is similar. For the modal case, note that Lemma 3.4 implies that for any φ, $[\Diamond\varphi] = m^{\mathbf{F}}_{\Diamond}([\varphi])$, so this is again easy.

For the quantifier case, recall that $\sigma(\exists p\varphi) = \exists q\sigma^q_p(\varphi)$ for a suitable q. Now observe that due to saturation and (EI),

$$[\exists q\sigma^q_p(\varphi)] \subseteq \bigcup\{[\iota^r_q\sigma^q_p(\varphi)] \mid r \in \mathsf{Prop}^+\} \subseteq \bigcup\{[\iota^\psi_q\sigma^q_p(\varphi)] \mid \psi \in \mathcal{L}^+\} \subseteq [\exists q\sigma^q_p(\varphi)].$$

Next, observe that given how q is chosen when performing $\sigma(\exists p\varphi) = \exists q\sigma^q_p(\varphi)$, $\iota^\psi_q\sigma^q_p(\varphi)$ and $\sigma^\psi_p(\varphi)$ differ only by renaming of bound variables and thus are logically equivalent. Hence, $[\iota^\psi_q\sigma^q_p(\varphi)] = [\sigma^\psi_p(\varphi)]$. Also, $[\sigma^\psi_p] = [\sigma][[\psi]/p]$. Thus, with IH and recalling that $B = \{[\psi] \mid \psi \in \mathcal{L}^+\}$,

$$\begin{aligned}[\sigma(\exists p\varphi)] &= \bigcup\{[\sigma^\psi_p(\varphi)] \mid \psi \in \mathcal{L}^+\} = \bigcup\{[\![\varphi]\!]^{\mathbf{F}_\Lambda}([\sigma^\psi_p]) \mid \psi \in \mathcal{L}^+\}\\ &= \bigcup\{[\![\varphi]\!]^{\mathbf{F}_\Lambda}([\sigma][[\psi]/p]) \mid \psi \in \mathcal{L}^+\} = \bigcup\{[\![\varphi]\!]^{\mathbf{F}_\Lambda}([\sigma][X/p]) \mid X \in B\}\\ &= [\![\exists p\varphi]\!]^{\mathbf{F}_\Lambda}([\sigma]).\end{aligned}$$

□

Now we are ready to prove Theorem 3.1. First we show that $\mathbf{F}_\Lambda$ is indeed a quantifiable frame, for any valuation $v \in B^{\mathsf{Prop}}$ for $\mathbf{F}_\Lambda$, given how B is defined, consider the substitution $\sigma_v : p \mapsto \varphi_p$ where $v(p) = [\varphi_p]$. Then $v = [\sigma_v]$. It follows that $\mathbf{F}_\Lambda$ is quantifiable by Lemma 3.5. Next, $\mathbf{F}_\Lambda \vDash \Lambda$ as for $\varphi \in \Lambda \subseteq \Lambda^+$ and arbitrary $v \in B^{\mathsf{Prop}}$, $[\varphi]^{\mathbf{F}_\Lambda}(v) = [\varphi]^{\mathbf{F}_\Lambda}(\sigma_v) = [\sigma_v(\varphi)] = W$ since $\sigma(\varphi)$ is also in Λ^+. Finally, taking σ as ι in Lemma 3.5, $[\![\varphi]\!]^{\mathbf{F}_\Lambda}([\iota]) = [\varphi]$, which means under valuation $[\iota]$, each Λ^+-MCS is satisfied by itself on a general frame $\mathbf{F}_\Lambda$ that validates Λ. Hence Theorem 3.1 follows.

4 From finite diversity to PQML-invariant subdomain

We first introduce the concepts of duplicates and diversity.

Definition 4.1 Given a Kripke frame $\mathbb{F} = (W, R)$, we say that worlds $w, u \in W$ are *duplicates* if the permutation (wu) of W that exchanges w and u is an automorphism of $\mathbb{F}$. Let Δ be this relation of being duplicates ($\mathbb{F}$'s duplication relation), which clearly is an equivalence relation on W, and then let W/Δ be the set of Δ's equivalence classes. The *diversity* of $\mathbb{F}$ is the cardinality of W/Δ. The *diversity* of a Kripke frame class is the supremum of the diversity of all point-generated subframes of the frames in that class (if exists).

Intuitively, duplicate classes are 'positions' a world could be in, and the diversity of a Kripke frame counts the number of positions in that frame. We use some examples to illustrate the concept of diversity.

Example 4.2 While cyclic frames of the form $(\{0, 1, \ldots, n-1\}, \{(i, i+1 \textit{ mod } n) \mid i = 0 \ldots n-1\})$ are highly symmetric, no two distinct worlds are duplicates of each other, as switching them and only them is not an automorphism.

Example 4.3 The diversity of $\mathsf{KFr}(\mathsf{D45})$ is 2, and the diversity of $\mathsf{KFr}(\mathsf{5})$ is 3. It is well known that a point-generated Kripke frame validating $\mathsf{D45}$ is either a clique $(W, W \times W)$ or a point looking at a clique $(\{r\} \cup W, (\{r\} \cup W) \times W)$. Clearly, Kripke frames of the later form has exactly two duplicate classes as every $x, y \in W$ are duplicates of each other.

It is also well known that the most non-trivial kind of point-generated Kripke frames validating $\mathsf{5}$ are of the form $(\{r\} \cup W, \{r\} \times U \cup W \times W)$ where $r \notin W$ and $\varnothing \neq U \subseteq W$. Then, there are three duplicate classes: $\{r\}, U, W \setminus U$.

Example 4.4 We give a Sahlqvist definable frame class of diversity 4. Consider Sahlqvist formulas $\varphi_1 = \Box(\Diamond p \to \Box\Diamond p)$ and $\varphi_2 = \Diamond\Diamond p \to \Box\Diamond p$. Let $\mathbb{F} = (W, R)$ be a Kripke frame that validates the two formulas and is point-generated from $r \in W$. By φ_2, for any $x, y \in R[r]$, $R[x] = R[y]$. By φ_1, all worlds $w \in W \setminus \{r\}$ locally validates $\mathsf{5}$: $(W, R), w \vDash \forall p(\Diamond p \to \Box\Diamond p)$. Now we discuss several cases:

- If $R[r] = \varnothing$, then $(W, R) = (\{r\}, \varnothing)$ with diversity 1.
- If $r \in R[r]$, then (W, R) is a universally connected clique, again with diversity 1.
- If we are not in the above two cases, and there is $u \in R[r]$ that is reflexive,

let A be the set of reflexive worlds in $R[r]$, B be the irreflexive worlds in $R[r]$, and C be $R[u]$. Then observe that W is $\{r\} \cup B \cup C$ where $r \notin B \cup C$, $B \cap C = \varnothing$, and $A \subseteq C$, and R is $\{r\} \times (B \cup A) \cup (B \cup C) \times C$. Clearly, such a frame has at most 4 duplicate classes: $\{r\}$, B (could be empty), A, and $C \setminus A$ (could be empty).

- If we are not in the above three cases, then $R[r]$ is non-empty (call it A and let u be a member of it), and every world in $R[r]$ is irreflexive. If $R[u]$ is empty, then (W, R) is $(\{r\} \cup A, \{r\} \times A)$. If $R[u]$ is non-empty, call it B, and let C be $R[v]$ for any $v \in B$. The choice of v is irrelevant due to the 5 axiom. Thus, (W, R) is $(\{r\} \cup A \cup C, \{r\} \times A \cup A \times B \cup C \times C)$ where $r \notin A \cup C$, $A \cap C = \varnothing$, and $B \subseteq C$. Again, (W, R) has at most 4 duplicate classes: $\{r\}$, A, B and $C \setminus B$.

Example 4.5 Finite frames have only finite diversity. Thus, the axioms $\mathtt{Alt}^n = \bigwedge_{i<n+1} p_i \to \bigvee_{i<j<n+1} \Diamond(p_i \wedge p_j)$ and $\mathtt{Trs}^m = \Diamond^{\leqslant m} p \to \Diamond^{\leqslant m+1} p$ together define Kripke frame classes of diversity at most n^{m+1}.

The following lemma collects some easy but very useful properties of the duplicate classes and how they interact with R and $m_\Diamond$.

Lemma 4.6 *For any Kripke frame $\mathbb{F} = (W, R)$ and its duplication relation Δ,*

- *for any $D_1 \neq D_2 \in W/\Delta$, there is $w \in D_1$ and $u \in D_2$ s.t. wRu iff for all $w \in D_1$ and $u \in D_2$, wRu;*
- *for any $D \in W/\Delta$, the only possible configurations for $R|_D$ are: D^2, $\varnothing$, and when $|D| \geqslant 2$, $D^2 \setminus id_D$ and id_D (id_D is the identity relation on D).*

For convenience we define the binary relation R_Δ on W/Δ s.t. $D_1 R_\Delta D_2$ iff there is $w \in D_1$ and $u \in D_2$ s.t. wRu. Here we allow $D_1 = D_2$.

Now we discuss the possible ways $m_\Diamond^{\mathbf{F}}(X) \cap D$ is determined.

- *In case $R|_D = D^2$ or $R|_D = \varnothing$, clearly $m_\Diamond^{\mathbf{F}}(X) \cap D$ is either D or $\varnothing$, and it is D iff there is $D' \in R_\Delta[D]$ s.t. $|X \cap D'| \geqslant 1$.*
- *In case $R|_D = D \setminus id_D$ with $|D| \geqslant 2$, if there is $D' \in R_\Delta[D] \setminus \{D\}$ s.t. $|X \cap D'| \geqslant 1$, then $m_\Diamond^{\mathbf{F}}(X) \cap D = D$, otherwise,*
  - · *if $|X \cap D| \geqslant 2$, then also $m_\Diamond^{\mathbf{F}}(X) \cap D = D$,*
  - · *if $|X \cap D| = 1$, then $m_\Diamond^{\mathbf{F}}(X) \cap D = D \setminus X$, and*
  - · *if $|X \cap D| = 0$, then $m_\Diamond^{\mathbf{F}}(X) \cap D = \varnothing$.*
- *In case $R|_D = id_D$ with $|D| \geqslant 2$, if there is $D' \in R_\Delta[D] \setminus \{D\}$ s.t. $|X \cap D'| \geqslant 1$, then $m_\Diamond^{\mathbf{F}}(X) \cap D = D$, and otherwise, $m_\Diamond^{\mathbf{F}}(X) \cap D = X$.*

What follows is the core of the proof of our main result. We want to show that if the underlying Kripke frame of the quantifiable frame (W, R, B) has finite diversity and B contains all singletons and duplicate classes, then B is a PQML-invariant subdomain of $\wp(W)$. The key idea is that whenever φ is true at w where at most one $p \in \mathsf{Fv}(\varphi)$ is evaluated to a set $X \subseteq W$ that is not necessarily in B, we can always swap the valuation of p to a $Y \in B$ while keeping φ true at w. For this to be true, we must establish that φ's truth at w is insensitive to certain changes in the valuation of p. For monadic second-order logic, this can be done with EF-game, but for modal logic, we cannot only focus on how φ's truth at w is insensitive to change since modality requires us to also consider the truth of φ at other worlds. We must take a more global perspective and strive to show that φ's 'meaning' is insensitive to certain changes. In the end, we arrive at a qualified quantifier-elimination: when restricted to a duplicate class D and relative to a valuation v, the 'meaning' of φ can be written as a Boolean formula $f_\varphi(v, D)$ using variables in $\mathsf{Fv}(\varphi)$, and when valuations u and v are close enough, $f_\varphi(v, D) = f_\varphi(u, D)$.

Definition 4.7 For any finite $\mathsf{p} \subseteq \mathsf{Prop}$, we write $\langle \mathsf{p} \rangle$ for the Boolean language generated from p. $at(\langle \mathsf{p} \rangle)$ is the finite set of all formulas $l_1 \wedge \cdots \wedge l_k$ where each l_i is either p_i or $\neg p_i$ and $p_1, \ldots, p_k$ list all elements in p. These formulas correspond to the atoms in the Lindenbaum algebra of $\langle \mathsf{p} \rangle$.

Definition 4.8 Given a Kripke frame (W, R) with duplicate relation Δ, for any finite $\mathsf{p} \subseteq \mathsf{Prop}$, $u, v \in \wp(W)^\mathsf{p}$, and $n \in \mathbb{N}$, $u \approx_n v$ if for all $D \in W/\Delta$ and $\zeta \in at(\langle \mathsf{p} \rangle)$, $|[\![\zeta]\!](u) \cap D| = |[\![\zeta]\!](v) \cap D|$ or both $|[\![\zeta]\!](u) \cap D|, |[\![\zeta]\!](v) \cap D| \geqslant 2^n$.

Lemma 4.9 *Let $u, v \in \wp(W)^\mathsf{p}$ for some finite $\mathsf{p} \subseteq \mathsf{Prop}$.*

(i) *If $u \approx_n v$, then not only for $\zeta \in at(\langle \mathsf{p} \rangle)$, for any $\beta \in \langle \mathsf{p} \rangle$ and $D \in W/\Delta$, $|[\![\beta]\!](u) \cap D| = |[\![\beta]\!](v) \cap D|$ or both $|[\![\beta]\!](u) \cap D|, |[\![\beta]\!](v) \cap D| \geqslant 2^n$.*

(ii) *If $u \approx_n v$ and $p \notin \mathsf{p}$, for any $X \in \wp(W)$ there is $Y \in \wp(W)$ s.t. $u[X/p] \approx_{n-1} v[Y/p]$ (since $p \notin dom(u)$, $u[X/p] = u \cup \{(p, X)\}$).*

Proof. The first part is easy since every $[\![\beta]\!](u)$ is the union of some $[\![\zeta]\!](u)$'s where $\zeta \in at(\langle \mathsf{p} \rangle)$. The second part: for each $\zeta \in at(\langle \mathsf{p} \rangle)$ and $D \in W/\Delta$, choose a set $Y_{\zeta, D} \subseteq [\![\zeta]\!](v) \cap D$ such that:

- if $|([\![\zeta]\!](u) \cap D) \cap X| < 2^{n-1}$, then $|Y_{\zeta, D}| = |([\![\zeta]\!](u) \cap D) \cap X|$;
- if $|([\![\zeta]\!](u) \cap D) \setminus X| < 2^{n-1}$, then $|([\![\zeta]\!](v) \cap D) \setminus Y_{\zeta, D}| = |([\![\zeta]\!](u) \cap D) \setminus X|$;
- if both $|([\![\zeta]\!](u) \cap D) \cap X|, |([\![\zeta]\!](u) \cap D) \setminus X| \geqslant 2^{n-1}$, then $|Y_{\zeta, D}| = 2^{n-1}$.

Given that $u \approx_n v$, the above conditions can be satisfied. Then either $|([\![\zeta]\!](u) \cap D) \cap X| = |Y_{\zeta, D}|$ or both $|([\![\zeta]\!](u) \cap D) \cap X|, |Y_{\zeta, D}| \geqslant 2^{n-1}$, and the same goes for $|([\![\zeta]\!](v) \cap D) \setminus Y_{\zeta, D}|$ and $|([\![\zeta]\!](u) \cap D) \setminus X|$. Then, with $Y = \bigcup_{\zeta \in at(\langle \mathsf{p} \rangle), D \in W/\Delta} Y_{\zeta, D}$, $u[X/p] \approx_{n-1} v[Y/p]$. □

Let $qd(\varphi)$ be the quantifier depth of φ.

Lemma 4.10 *Given a Kripke frame (W, R) with duplicate relation Δ, for each $\varphi \in \mathcal{L}$, there is a function $f_\varphi : (\wp(W)^{\mathsf{Fv}(\varphi)} \times W/\Delta) \to \langle \mathsf{Fv}(\varphi) \rangle$ such that*

- *for any $v \in \wp(W)^{\mathsf{Fv}(\varphi)}$ and $D \in W/\Delta$, $[\![\varphi]\!](v) \cap D = [\![f_\varphi(v, D)]\!](v) \cap D$;*
- *for any $u \approx_{qd(\varphi)+1} v \in \wp(W)^{\mathsf{Fv}(\varphi)}$, $f_\varphi(u, D) = f_\varphi(v, D)$ for all $D \in W/\Delta$.*

In particular, recursively define f as follows and it will witness the lemma: for the base and Boolean cases: $f_p(v, D) = p$, $f_{\neg \varphi}(v, D) = \neg f_\varphi(v, D)$, $f_{\varphi \vee \psi}(v, D) = f_\varphi(v|_{\mathsf{Fv}(\varphi)}, D) \vee f_\psi(v|_{\mathsf{Fv}(\psi)}, D)$. For the modal case, we copy the analysis in Lemma 4.6 with $X = [\![f_\varphi(v, D)]\!](v)$:

- *In case* $R|_D = D^2$ *or* $\varnothing$, *if there is* $D' \in R_\Delta[D]$ *s.t.* $|X \cap D'| \geqslant 1$ *then* $f_{\Diamond\varphi}(v, D) = \top$, *otherwise* $f_{\Diamond\varphi}(v, D) = \bot$.
- *In case* $R|_D = D \setminus id_D$ *with* $|D| \geqslant 2$, *if there is* $D' \in R_\Delta[D] \setminus \{D\}$ *s.t.* $|X \cap D'| \geqslant 1$, *then* $f_{\Diamond\varphi}(v, D) = \top$, *otherwise,*
  - *if* $|X \cap D| \geqslant 2$, *then also* $f_{\Diamond\varphi}(v, D) = \top$,
  - *if* $|X \cap D| = 1$, *then* $f_{\Diamond\varphi}(v, D) = \neg f_\varphi(v, D)$, *and*
  - *if* $|X \cap D| = 0$, *then* $f_{\Diamond\varphi}(v, D) = \bot$.
- *In case* $R|_D = id_D$ *with* $|D| \geqslant 2$, *if there is* $D' \in R_\Delta[D] \setminus \{D\}$ *s.t.* $|X \cap D'| \geqslant 1$, *then* $f_{\Diamond\varphi}(v, D) = \top$, *and otherwise,* $f_{\Diamond\varphi}(v, D) = f_\varphi(v, D)$.

For the quantifier case let $f_{\exists p\varphi}(v, D)$ *be the following:*

$$\bigvee\{\zeta \in at(\langle \mathsf{Fv}(\exists p\varphi)\rangle) \mid [\![\zeta]\!](v) \cap \bigcup_{X \in \wp(W)} [\![f_\varphi(v[X/p], D)]\!](v[X/p]) \cap D \neq \varnothing\}.$$

Proof. We first show by induction that whenever $u \approx_{qd(\varphi)+1} v$, $f_\varphi(u, D) = f_\varphi(v, D)$. The base case and the inductive steps for Boolean connectives are trivial. For the modal case, suppose $u \approx_{qd(\Diamond\varphi)+1} v$. Then $u \approx_{qd(\varphi)+1} v$. Using IH, let $\beta = f_\varphi(u, D) = f_\varphi(v, D)$ and let $X = [\![\beta]\!](u), X' = [\![\beta]\!](v)$. Now at least $u \approx_1 v$, so for any $E \in W/\Delta$, either $|X \cap E| = |X' \cap E|$ or both $|X \cap E|$ and $|X' \cap E| \geqslant 2$. This means in the case analysis defining $f_{\Diamond\varphi}(u, D)$ and $f_{\Diamond\varphi}(v, D)$, the same case must be active, and $f_{\Diamond\varphi}(u, D) = f_{\Diamond\varphi}(v, D)$.

For the quantifier case, suppose $u \approx_{qd(\exists p\varphi)+1} v$ with $\mathsf{p} = \mathsf{Fv}(\exists p\varphi)$ and $u, v \in \wp(W)^{\mathsf{p}}$. Then $u \approx_{qd(\varphi)+2} v$. Now pick any $\zeta \in at(\langle \mathsf{p}\rangle)$ and suppose ζ is a disjunct of $f_{\exists p\varphi}(u, D)$. Then there is $X \in \wp(W)$ s.t. $[\![\zeta]\!](u) \cap [\![f_\varphi(u[X/p], D)]\!](u[X/p]) \cap D \neq \varnothing$. Since $p \notin \mathsf{p}$, $[\![\zeta]\!](u) = [\![\zeta]\!](u[X/p])$. So we have $[\![\zeta \wedge f_\varphi(u[X/p], D)]\!](u[X/p]) \cap D \neq \varnothing$. Now by Lemma 4.9, there is $Y \in \wp(W)$ s.t. $u[X/p] \approx_{qd(\varphi)+1} v[Y/p]$. So with IH, we can let $\beta = \zeta \wedge f_\varphi(u[X/p], D) = \zeta \wedge f_\varphi(v[Y/p], D)$ and now $[\![\beta]\!](u[X/p]) \cap D \neq \varnothing$. Then $[\![\beta]\!](v[Y/p]) \cap D \neq \varnothing$. This means ζ is also a disjunct of $f_{\exists p\varphi}(v, D)$. The above argument can be reversed, so $f_{\exists p\varphi}(u, D)$ and $f_{\exists p\varphi}(v, D)$ have the same disjuncts and thus are the same formula.

Now we show that $[\![\varphi]\!](v) \cap D = [\![f_\varphi(v, D)]\!](v) \cap D$. Again this is by induction and the non-quantifier cases are easy. For easy notation, let $\mathsf{p} = \mathsf{Fv}(\exists p\varphi)$ and $\beta_X = f_\varphi(v[X/p], D)$. By IH, $[\![\exists p\varphi]\!](v) \cap D = \bigcup_{X \in \wp(W)} [\![\beta_X]\!](v[X/p]) \cap D$. Given the definition of $f_{\exists p\varphi}(v, D)$ and that $\mathcal{C} := \{[\![\zeta]\!](v) \cap D \mid \zeta \in at(\langle \mathsf{p}\rangle)\}$ forms a partition of D, all we need to show is that $[\![\exists p\varphi]\!](v) \cap D$ is a union of cells in $\mathcal{C}$. For this, it suffices to show that for any $\zeta \in at(\langle \exists p\varphi\rangle)$ and $w_1, w_2 \in [\![\zeta]\!](v) \cap D$, if $w_1 \in \bigcup_{X \in \wp(W)} [\![\beta_X]\!](v[X/p]) \cap D$ then w_2 is also in $\bigcup_{X \in \wp(W)} [\![\beta_X]\!](v[X/p]) \cap D$. So suppose that $w_1, w_2 \in [\![\zeta]\!](v) \cap D$ for some $\zeta \in at(\langle p\rangle)$ and there is $X \in \wp(W)$ s.t. $w_1 \in [\![\beta_X]\!](v[X/p]) \cap D$. Recall that $(w_1 w_2)$ is the permutation of W that exchanges w_1 and w_2. Let $Y = (w_1 w_2)[X]$. Since w_1, w_2 are both in $[\![\zeta]\!](v)$ and ζ is in $at(\langle \mathsf{p}\rangle)$, for any $q \in \mathsf{p}$, $v(q) = (w_1 w_2)[v(q)]$ as $w_1 \in v(q)$ iff $w_2 \in v(q)$. Recall also that w_1, w_2 are in the same duplication class D, so for any $D' \in W/\Delta$, $w_1 \in D'$ iff $w_2 \in D'$. From all these, it is clear that $v[X/p] \approx_{qd(\varphi)+1} v[Y/p]$, since $(w_1 w_2)[\cdot]$ commutes with all Boolean operations and thus for any $\gamma \in \langle \mathsf{p} \cup \{p\} \rangle$ and $D' \in W/\Delta$, $[\![\gamma]\!](v[Y/p]) \cap D' = (w_1 w_2)[[\![\gamma]\!](v[X/p]) \cap D']$, meaning also that they are of the same cardinality. Then $\beta_X = \beta_Y$. Recall that $w_1 \in [\![\beta_X]\!](v[X/p]) \cap D$. Apply $(w_1 w_2)$ to both sides and we have $w_2 \in [\![\beta_Y]\!](v[Y/p]) \cap D$. □

Theorem 4.11 *For any general frame (W, R, B) with (W, R) having finite diversity, B is a* PQML*-invariant subdomain of $\wp(W)$ if for any $w \in W$ and $D \in W/\Delta$, $\{w\}, D \in B$. (Only B's closure under Boolean operations is used.)*

Proof. It suffices to show that for any $w \in W$, $\varphi \in \mathcal{L}$ with $\mathsf{p} = \mathsf{Fv}(\exists p\varphi)$ and $n = qd(\varphi) + 1$, $v \in B^{\mathsf{p}}$, and $X \in \wp(W)$, there is $Y \in B$ s.t. $v[X/p] \approx_n v[Y/p]$ and $w \in X$ iff $w \in Y$, since if so, then by Lemma 4.10, $w \in [\![\varphi]\!](v[X/p]) \cap D$ iff $w \in [\![f_\varphi(v[X/p], D)]\!](v[X/p]) \cap D$ iff $w \in [\![f_\varphi(v[Y/p], D)]\!](v[X/p]) \cap D$ iff $w \in [\![f_\varphi(v[Y/p], D)]\!](v[Y/p]) \cap D$ (recall that $f_\varphi(v[Y/p], D)$ is a Boolean formula) iff $w \in [\![\varphi]\!](v[Y/p]) \cap D$ and thereby by Lemma 2.4 we are done.

Notice that by assumption we have a finite partition $\mathcal{C} = \{[\![\zeta]\!](v) \cap D \mid \zeta \in at(\langle \mathsf{p} \rangle), D \in W/\Delta\}$ of W and each $C \in \mathcal{C}$ is in B by assumption. For $Y \in B$, it is enough to make sure that for all $C \in \mathcal{C}$, $C \cap Y$ or $C \setminus Y$ is finite. For $v[X/p] \approx_n v[Y/p]$, it is enough to make sure that for all $C \in \mathcal{C}$, $|C \cap Y| = |C \cap X|$ or $|C \cap Y|, |C \cap X| \geqslant 2^n$, and $|C \setminus Y| = |C \setminus X|$ or $|C \setminus Y|, |C \setminus X| \geqslant 2^n$. Thus, for each $C \in \mathcal{C}$, let $Y_C = C \cap X$ if either $C \cap X$ or $C \setminus X$ is finite, and otherwise when both are infinite, if $w \in C \cap X$, let Y_C be $C \setminus Z$ for some $Z \subseteq C \setminus X$ with $|Z| = 2^n$, and otherwise let Y_C be some subset of $C \cap X$ with $|Y_C| = 2^n$. Clearly $Y = \bigcup_{C \in \mathcal{C}} Y_C$ satisfies the requirements. □

5 General completeness with finite-diversity

We first define the extra axioms needed.

Definition 5.1
- $\mathtt{Q}^n(\varphi) = \Diamond^{\leqslant n}\varphi \wedge \forall p(\Box^{\leqslant n}(\varphi \to p) \vee \Box^{\leqslant n}(\varphi \to \neg p))$ where p is the first variable not in $\mathsf{Fv}(\varphi)$;
- $\mathtt{At}^n = \forall q(\Diamond^{\leqslant n} q \to \exists p(\mathtt{Q}^{\leqslant n}(p) \wedge \Box^{\leqslant n}(p \to q)))$;
- $\mathtt{R}^n = \forall p \exists q(\Box^{\leqslant n}(p \to \Box q) \wedge \forall r(\Box^{\leqslant n}(p \to \Box r) \to \Box^{\leqslant n}(q \to r))$.

The iterated modalities $\Diamond^{\leqslant n}$ and $\Box^{\leqslant n}$ used here are meant to capture the global modality (at least at the root of a point-generated model). Then, $\mathtt{Q}^n(\varphi)$ states that the proposition φ expresses is a maximally specific possible proposition, a non-bottom proposition that settles the truth or falsity of all propositions. Semantically, assuming that enough propositions exist to distinguish possible worlds, such maximally specific possible propositions must be singletons, true at exactly one possible world. Even without assuming that worlds can be distinguished by propositions, maximally specific possible propositions are atoms in the algebra of all propositions, and can themselves serve as worlds. Thus, they are often called world propositions. Using $\mathtt{Q}^n$, $\mathtt{At}^n$ is the atomicity principle stating that whatever is possible is entailed by a maximally specific possible proposition. In other words, the algebra of all propositions is atomic. Such an atomicity principle features prominently in many works [15,8,25,11,20] with both technical and philosophical significance. The use of the nested modality $\Box^{\leqslant n}$ is needed as we are not assuming strong modal axioms and we need to simulate a global modality. The formula $\mathtt{R}^n$ is meant to capture the fact that $\wp(W)$ is trivially closed under taking successor sets: if $X \in \wp(W)$, then $R[X] \in \wp(W)$. Again, using $\Box^{\leqslant n}$ as a substitute for the global modality, $\mathtt{R}^n$ says that for any proposition p, there is a proposition q such that whenever p is true, 'necessarily q' is true, and q is the strongest proposition with this property. The validity of these two formulas over Kripke frames is easy to see using singleton sets and successor sets.

Using the two axioms $\mathtt{At}^n$ and $\mathtt{R}^n$, our main theorem is the following:

Theorem 5.2 *Let $\Theta \subseteq \mathcal{L}_{qf}$ be a set of Sahlqvist formulas s.t. the class $\mathsf{KFr}(\Theta)$ of Kripke frames validating Θ has diversity n. Then $\mathsf{K}_\Pi\Theta\mathtt{BcAt}^n\mathtt{R}^n$ is sound and strongly complete for $\mathsf{KFr}(\Theta)$.*

An outline of the proof is in order. First, we show that finite diversity means finite depth, and the logic $\mathsf{K}_\Pi\Theta\mathtt{BcAt}^n\mathtt{R}^n$ recognizes this. This means that $\Box^{\leqslant n}$ simulates global modality well enough. Second, to satisfy a consistent set of formulas, we take a point-generated general frame $\mathbf{F}_a$ in the canonical saturated general frame where a extends this consistent set of formulas. Third, we keep only the 'isolated' worlds in $\mathbf{F}_a$, i.e., those worlds whose singleton is a proposition, and arrive at $\mathbf{F}_a^{at}$. This step does not disturb the truth of formulas at the remaining worlds. Finally, we expand the propositional domain of $\mathbf{F}_a^{at}$ to the full powerset. We have to show that this step again keeps the truth value of all formulas in $\mathcal{L}$ unchanged, and also the validity of Θ. These two points only depend on that the propositional domain of $\mathbf{F}_a^{at}$ contains all singletons, is closed under taking successor sets, and contains all the duplicate classes of the underlying Kripke frame. Lemma 5.7 establishes these three properties.

Now we begin the proof. Fix a set Θ of Sahlqvist formulas with the diversity of $\mathsf{KFr}(\Theta)$ being n and let $\Lambda = \mathsf{K}_\Pi\Theta\mathtt{BcAt}^n\mathtt{R}^n$. We first establish a logical point:

Lemma 5.3 *$\Diamond^{n+1}p \to \Diamond^{\leqslant n}p$ is a theorem of Λ. Thus, denoting $\Diamond^{\leqslant n}$ by E and the dual $\Box^{\leqslant n}$ by A, Λ proves that A is an* S4 *modality that commutes with $\forall$, and A works like the reflexive and transitive closure of $\Box$ in that for example (1) $\mathsf{E}\Diamond\varphi \to \mathsf{E}\varphi \in \Lambda$ and (2) for any $m \in \mathbb{N}$, $\Diamond^m\varphi \to \mathsf{E}\varphi \in \Lambda$.*

Proof. First we show that any Kripke frame $\mathbf{G} \in \mathsf{KFr}(\Theta)$ must also validate $\Diamond^{n+1}p \to \Diamond^{\leqslant n}p$. Suppose not, then we have some $wRx_1Rx_2\dots Rx_nRu$, which is also a shortest path from w to u. This path is also present and shortest in the subframe $\mathbf{G}_w$ of $\mathbf{G}$ generated from w. Now note that $w, x_1, \dots, x_n$ are pairwise non-duplicates within $\mathbf{G}_w$, since if there were a duplicate pair, then the path can be shortened. This contradicts that $\mathsf{KFr}(\Theta)$ has diversity n. Hence any Kripke frame $\mathbf{G} \in \mathsf{KFr}(\Theta)$ validates $\Diamond^{n+1}p \to \Diamond^{\leqslant n}p$. Since the normal propositional modal logic axiomatized by Θ is Kripke complete, $\Diamond^{n+1}p \to \Diamond^{\leqslant n}p$ is in Λ. The remaining claims follow easily from basic normal modal reasoning and $\mathtt{Bc}$. □

In the following we continue using A for $\Box^{\leqslant n}$ and E for $\Diamond^{\leqslant n}$ and drop the superscripts on $\mathtt{Q}^n$, $\mathtt{At}^n$, and $\mathtt{R}^n$.

Now we start with the canonical saturated general frame $\mathbf{F}_\Lambda = (W, R, B)$. Recall that this involves expanding the language to $\mathcal{L}^+$ built from variables in Prop^+ and extending Λ conservatively to Λ^+. Lemma 5.3 transfer to Λ^+ without problems. For Theorem 5.2, clearly it is enough to show that every $w \in W$ can be satisfied in a Kripke frame validating Θ. Thus fix an arbitrary $a \in W$ and consider the general frame $\mathbf{F}_a = (W_a, R_a, B_a)$ generated from a, defined as follows:

- W_a is $R^*[a]$ where R^* is the reflexive and transitive closure of R;
- $R_a = R \cap (W_a \times W_a)$;
- $B_a = \{X \cap W_a \mid X \in B\}$; we write $[\varphi]_a$ for $[\varphi] \cap W_a$, and with this notation, $B_a = \{[\varphi]_a \mid \varphi \in \mathcal{L}^+\}$.

We show that A and E work as universal and existential modalities at a and $\mathtt{Q}$ works as intended.

Lemma 5.4 *For any $\varphi \in \mathcal{L}^+$, $\mathsf{A}\varphi \in a$ iff $[\varphi]_a = W_a$, and similarly $\mathsf{E}\varphi \in a$ iff $[\varphi]_a$ is non-empty. Also, $\mathtt{Q}(\varphi) \in a$ iff $[\varphi]_a$ is a singleton.*

Proof. For the first part, we just prove $\mathsf{E}\varphi \in a$ iff $[\varphi]_a \neq \varnothing$. Since a is an MCS and together with Lemma 5.3, $\mathsf{E}\varphi \in a$ iff there is m s.t. $\Diamond^m\varphi \in a$. By standard reasoning in canonical models, i.e., repeated use of Lemma 3.4, this is true iff there is $u \in W_a$ s.t. $\varphi \in u$.

Now for $\mathtt{Q}(\varphi)$, again since a is a saturated MCS, $\mathtt{Q}(\varphi) \in a$ iff (1) $\mathsf{E}\varphi \in a$ and (2) for any $\psi \in \mathcal{L}^+$, $\mathsf{A}(\varphi \to \psi) \vee \mathsf{A}(\varphi \to \neg\psi) \in a$. (The second point uses saturation.) Using the first part of this lemma, (1) translates to $[\varphi]_a \neq \varnothing$, and (2) translates to that for any $\psi \in \mathcal{L}^+$, $[\varphi]_a \subseteq [\psi]_a$ or $[\varphi]_a \subseteq (W_a \setminus [\psi]_a)$. If $[\varphi]_a$ is a singleton, these two points are clearly true. Conversely, if $[\varphi]_a$ is empty, (1) is clearly false. If instead there are distinct $x, y \in [\varphi]_a$, then there must be a formula $\psi \in \mathcal{L}^+$ s.t. $\psi \in x$ but $\psi \notin y$, making (2) false. □

This means the atoms of B_a (as a Boolean algebra under set inclusion) are precisely $\{[\varphi]_a \mid \mathtt{Q}(\varphi) \in a\}$. We focus on these atoms and define the *atomic subframe* $\mathbf{F}_a^{at}$ of $\mathbf{F}_a$ as $(W_a^{at}, R_a^{at}, B_a^{at})$ where

- $W_a^{at} = \{w \in W_a \mid \{w\} \in B_a\}$;
- $R_a^{at} = R_a \cap (W_a^{at} \times W_a^{at})$;
- $B_a^{at} = \{X \cap W_a^{at} \mid X \in B_a\}$.

We write $[\varphi]_a^{at} = [\varphi]_a \cap W_a^{at}$. Then $B_a^{at} = \{[\varphi]_a^{at} \mid \varphi \in \mathcal{L}^+\}$. A key property of $\mathbf{F}_a^{at}$ is that every world in it is named by a formula given how W_a^{at} is defined. For each $w \in W_a^{at}$, we fix a formula χ_w s.t. $\{w\} = [\chi_w]_a$.

We want to immediately make sure that $a \in W_a^{at}$.

Lemma 5.5 *The singleton $\{a\}$ is in B_a and thus $a \in W_a^{at}$.*

Proof. By a formal derivation, $\exists p(p \wedge \mathtt{Q}(p))$ is in Λ. Indeed, with $\mathtt{S4}$ normal modal reasoning and $\mathtt{Bc}$, we can derive $\mathtt{Q}(p) \to \mathsf{A}(p \to \mathtt{Q}(p))$ in Λ^+. The main steps include

- $(\mathsf{A}(p \to q) \vee \mathsf{A}(p \to \neg q)) \to \mathsf{A}(\mathsf{A}(p \to q) \vee \mathsf{A}(p \to \neg q))$
- $\forall q(\mathsf{A}(p \to q) \vee \mathsf{A}(p \to \neg q)) \to \mathsf{A}\forall q(\mathsf{A}(p \to q) \vee \mathsf{A}(p \to \neg q))$
- $\mathsf{A}(p \to \mathsf{E}p)$ and then $\forall q(\mathsf{A}(p \to q) \vee \mathsf{A}(p \to \neg q)) \to \mathsf{A}(p \to \mathtt{Q}(p))$.

Now suppose $\neg\exists p(p \wedge \mathtt{Q}(p))$. Then we derive $\mathsf{E}\forall p(p \to \neg\mathtt{Q}(p))$. With $\mathtt{At}$, we derive $\exists p(\mathtt{Q}(p) \wedge \mathsf{A}(p \to \forall p(p \to \neg\mathtt{Q}(p))))$. A contradiction follows. Using (EE), we only need to derive a contradiction from $\mathtt{Q}(p) \wedge \mathsf{A}(p \to \forall p(p \to \neg\mathtt{Q}(p)))$. From $\mathsf{A}(p \to \forall p(\mathtt{Q}(p) \to \neg p))$ we have $\mathsf{A}(p \to \neg\mathtt{Q}(p))$. And recall $\mathtt{Q}(p) \to \mathsf{A}(p \to \mathtt{Q}(p))$ is in Λ^+. This means we derive $\mathsf{A}(p \to \mathtt{Q}(p))$ and then $\mathsf{A}\neg p$, which contradicts the $\mathsf{E}p$ part in $\mathtt{Q}(p)$.

Thus $\exists p(p \wedge \mathtt{Q}(p)) \in a$, and since a is saturated, there is $r \in \mathsf{Prop}^+$ s.t. $r \wedge \mathtt{Q}(r) \in a$. Then $[r]_a$ must be $\{a\}$ and $\{a\} \in B_a$. □

From this, we show that $\mathbf{F}_a^{at}$ behaves as a canonical general frame and thus is quantifiable:

Lemma 5.6
- *For any $w \in W_a^{at}$ and $\varphi \in \mathcal{L}^+$, $\varphi \in w$ iff $\mathsf{A}(\chi_w \to \varphi) \in a$ iff $\mathsf{E}(\chi_w \wedge \varphi) \in a$. Also, for any $w, u \in W_a^{at}$, $wR_a^{at}u$ iff $\mathsf{E}(\chi_w \wedge \Diamond\chi_u) \in a$.*
- *For any $\Diamond\varphi \in w \in W_a^{at}$, there is $u \in R_a^{at}[w]$ s.t. $\varphi \in u$.*
- *For any substitution σ, define the associated valuation $[\sigma]_a^{at}$ for $\mathbf{F}_a^{at}$ by $[\sigma]_a^{at}(p) = [\sigma(p)]_a^{at}$. Then $[\![\varphi]\!]^{\mathbf{F}_a^{at}}([\sigma]_a^{at}) = [\sigma(\varphi)]_a^{at}$.*

Proof. By definition, $[\chi_w]_a = \{w\}$ and $[\chi_u]_a = \{u\}$. As we have shown, $\mathsf{E}(\chi_w \wedge \Diamond\chi_u) \in a$ iff $[\chi_w \wedge \Diamond\chi_u]_a$ is non-empty, iff $\Diamond\chi_u \in w$, and iff $wR_a^{at}u$. The universal case is dual to the existential case.

Now suppose $\Diamond\varphi \in w \in W_a^{at}$. First, apply $\mathtt{R}$ to χ_w. Using saturation at a, we have a formula (indeed a variable) which we denote by $\chi_{R[w]}$ s.t. $\mathsf{A}(\chi_w \to \Box\chi_{R[w]}) \wedge \forall r(\mathsf{A}(\chi_w \to \Box r) \to \mathsf{A}(\chi_{R[w]} \to r))$ is in a. By plugging in $\neg r$ for r and contraposing, we have $\forall r(\mathsf{E}(\chi_{R[w]} \wedge r) \to \mathsf{E}(\chi_w \wedge \Diamond r)) \in a$. Also, from $\mathsf{A}(\chi_w \to \Box\chi_{R[w]}) \in a$ and $\Diamond\varphi \in w$, $\mathsf{E}(\chi_w \wedge \Diamond(\chi_{R[w]} \wedge \varphi)) \in a$. This means $\mathsf{E}\Diamond(\chi_{R[w]} \wedge \varphi)$ and thus $\mathsf{E}(\chi_{R[w]} \wedge \varphi) \in a$. Then by $\mathtt{At}$ (or the atomicity of B_a), there is $u \in W_a^{at}$ s.t. φ and $\chi_{R[w]} \in u$. Then $\mathsf{E}(\chi_{R[w]} \wedge \chi_u) \in a$, and thus $\mathsf{E}(\chi_w \wedge \Diamond\chi_u) \in a$, which means $u \in R_a^{at}[w]$.

For the last part, recall that for the original canonical saturated frame $\mathbf{F}_\Lambda$, Lemma 3.5 applies: for any φ and substitution σ for $\mathcal{L}^+$, $[\![\varphi]\!]^{\mathbf{F}_\Lambda}([\sigma]) = [\sigma(\varphi)]$. $\mathbf{F}_a^{at}$ is obtained by restricting $\mathbf{F}_\Lambda$ to W_a^{at}. Thus, it is enough to show that for any valuation v for $\mathbf{F}_\Lambda$, writing $v|_a^{at}$ for the restricted valuation defined by $v|_a^{at}(p) = v(p) \cap W_a^{at}$, $[\![\varphi]\!]^{\mathbf{F}_a^{at}}(v|_a^{at}) = [\![\varphi]\!]^{\mathbf{F}_\Lambda}(v) \cap W_a^{at}$. Using induction on φ, the base and the Boolean cases are trivial, as the operation of relative negation and intersection commutes intersecting with W_a^{at}. For the modal case, we need $m_\Diamond^{\mathbf{F}_a^{at}}(X \cap W_a^{at}) = m_\Diamond^{\mathbf{F}_\Lambda}(X) \cap W_a^{at}$ where X is assumed to be in B due to IH. The left-to-right inclusion is trivial since $\mathbf{F}_a^{at}$ is a restriction of $\mathbf{F}_\Lambda$. For the right-to-left inclusion, first write X as $[\psi]$ for some $\psi \in \mathcal{L}^+$ and use the second bullet point of this lemma. The quantifier case is not much different from the case for disjunction, using only the distribution of intersection over arbitrary union and that $B_a^{at} = \{X \cap W_a^{at} \mid X \in B\}$. □

Now we start to show that B_a^{at} has what it takes to be a PQML-invariant subdomain of $\wp(W_a^{at})$ over the underlying Kripke frame $\mathbb{F}_a^{at} = (W_a^{at}, R_a^{at})$.

Lemma 5.7 • *For any $w \in W_a^{at}$, $\{w\} \in B_a^{at}$.*

- *For any $X \in B_a^{at}$, $R_a^{at}[X] \in B_a^{at}$.*
- *$\mathbf{F}_a^{at}$ is point-generated from a and $\mathbb{F}_a^{at} \models \Theta$. Thus it has diversity $k \leqslant n$.*
- *Let Δ be the duplicate relation of $\mathbb{F}_a^{at}$. Then each $D \in W_a^{at}/\Delta$ is in B_a^{at}.*

Proof. The first bullet point is trivial. For the second bullet point, pick any $X \in B_a^{at}$. Then we have some φ s.t. $X = [\varphi]_a^{at}$. Reasoning in the saturated MCS a and apply R to φ, we obtain a ψ s.t. $\mathsf{A}(\varphi \to \Box\psi) \in a$ and for any $\gamma \in \mathcal{L}^+$, $\mathsf{E}(\psi \wedge \gamma) \to \mathsf{E}(\varphi \wedge \Diamond\gamma) \in a$. Now for any $w \in X$, $\varphi \in w \in W_a^{at}$. Thus $\mathsf{E}(\chi_w \wedge \varphi) \in a$. Together with $\mathsf{A}(\varphi \to \Box\psi)$, $\mathsf{E}(\chi_w \wedge \Box\psi) \in a$, meaning $\Box\psi \in w$. Thus $R[w] \subseteq [\psi]$ and hence $R_a^{at}[w] \subseteq [\psi]_a^{at}$. Since w is chosen arbitrarily from X, $R_a^{at}[X] \subseteq [\psi]_a^{at}$. On the other hand, suppose $u \in [\psi]_a^{at}$. Then $\mathsf{E}(\psi \wedge \chi_u) \in a$. Then $\mathsf{E}(\varphi \wedge \Diamond\chi_u) \in a$ and thus there is $w \in W_a^{at}$ s.t. φ and $\Diamond\chi_u$ are in w. This means $w \in X$ and $wR_a^{at}u$. So in sum, $[\psi]_a^{at} \subseteq R_a^{at}[X]$.

For the third bullet point, we need to first show that every $w \in W_a^{at}$ is reachable from a within $\mathbf{F}_a^{at}$. If $w \in W_a^{at}$, then at least $\mathsf{E}\chi_w \in a$. This means for some $m \leqslant n$, $\Diamond^m\chi_w \in a$. By repeated use of the second bullet point of Lemma 5.6, there is indeed a path from a to w inside $\mathbf{F}_a^{at}$. Now, by the third bullet point of Lemma 5.6, $\mathbf{F}_a^{at} \models \Theta$. For this to transfer to $\mathbb{F}_a^{at}$, we rely on the assumption that Θ consists of Sahlqvist formulas, which are $\mathcal{AT}$-persistent in the sense that if they are valid on a general frame whose set of admissible sets contains all singletons (the atomic/$\mathcal{A}$ part) and is closed under taking successor set (the tense/$\mathcal{T}$ part), then they are also valid on the underlying Kripke frame. That Sahlqvist formulas are $\mathcal{AT}$-persistent has been observed for example in [40]. The idea is that for any $\varphi \in \Theta$, if $\mathbb{F}_a^{at} \not\models \varphi$, then there is a falsifying valuation that only uses sets in B_a^{at} so that it is also a valuation for $\mathbf{F}_a^{at}$, contradicting that $\mathbf{F}_a^{at} \models \Theta$. This special valuation is obtained by the standard minimal valuation technique for Sahlqvist formulas and note that in minimal valuations, only finite unions of sets of the form $(R_a^{at})^m[\{w\}]$ are used, which are in B_a^{at} by the assumed closure properties.

Now that $\mathbb{F}_a^{at}$ has diversity $k \leqslant n$, there are also $b_1, b_2, \ldots, b_k \in W_a^{at}$ each representing a duplicate class. For each i, we show that the duplicate class D that b_i is in is in B_a^{at}. Now if $D = \{b_i\}$ then we have shown that it is in B_a^{at}. So assume that there is $c \neq b_i$ in D. Observe that for any $w \in W_a^{at} \setminus \{b_1, \ldots, b_k\}$, $w \in D$ iff w and c are duplicates, and iff the following are true:

- for any j, $wR_a^{at}b_j$ iff $cR_a^{at}b_j$;
- for any j, $b_jR_a^{at}w$ iff $b_jR_a^{at}c$;
- w is reflexive iff c is reflexive;
- $wR_a^{at}c$ iff $cR_a^{at}w$.

The above conditions are all expressible in B_a^{at} using singletons, the $R_a^{at}[\cdot]$ operation, the $m_\Diamond^{\mathbb{F}_a^{at}}$ operation, and also the set of reflexive points in $\mathbb{F}_a^{at}$ defined by sentence $\forall p(\Box p \to p)$. In fact, in the original canonical saturated general frame $\mathbb{F}_\Lambda$, $[\forall p(\Box p \to p)]$ is already the set of all reflexive worlds since any two worlds are separated by a proposition in B. Indeed, the set

$$\begin{aligned} \{m_\Diamond^{\mathbb{F}_a^{at}}(\{b_j\}) \mid cR_a^{at}b_j\} \cup \{W_a^{at} \setminus m_\Diamond^{\mathbb{F}_a^{at}}(\{b_j\}) \mid \text{not } cR_a^{at}b_j\} \cup \\ \{R_a^{at}[b_j] \mid b_jR_a^{at}c\} \cup \{W_a^{at} \setminus R_a^{at}(b_j) \mid \text{not } b_jR_a^{at}c\} \cup \\ \{[\forall p(\Box p \to p)]_a^{at} \mid cR_a^{at}c\} \cup \{W_a^{at} \setminus [\forall p(\Box p \to p)]_a^{at} \mid \text{not } cR_a^{at}c\} \cup \\ \{(m_\Diamond^{\mathbb{F}_a^{at}}(\{c\}) \cap R_a^{at}[c]) \cup ((W_a^{at} \setminus m_\Diamond^{\mathbb{F}_a^{at}}(\{c\})) \cap (W_a^{at} \setminus R_a^{at}[c]))\} \end{aligned}$$

contains all the required conditions, the intersection of which we denote by X. Then $(X \setminus \{b_1, \dots, b_k\}) \cup \{b_i\}$ is the duplicate class D that b_i is in. □

Putting pieces together, for any Λ^+-MCS Σ, it is satisfied on $\mathbf{F}_a^{at}$ by itself under valuation $[\iota]_a^{at}$ by Lemma 5.6. By Lemma 5.7 and Theorem 4.11, B_a^{at} is a PQML-invariant subdomain of $\wp(W_a^{at})$. By definition of PQML-invariant subdomain and $[\iota]_a^{at} \in (B_a^{at})^{\mathsf{Prop}}$, Σ is satisfied on $\mathbb{F}_a^{at}$ by itself under valuation $[\iota]_a^{at}$. Finally, by the third bullet point of Lemma 5.7, $\mathbb{F}_a^{at}$ is a frame validating Θ. This completes the proof of theorem 5.2.

By observing where we used the Sahlqvist condition on Θ, we note a different way of stating our main result.

Definition 5.8 For any formula $\varphi \in \mathcal{L}$, it is *$\mathcal{ATQ}$-persistent* if for any quantifiable general frame (W, R, B) such that all singleton subsets of W is in B and B is closed under $R[\cdot]$, if φ is valid on (W, R, B), then φ is valid on the underlying (W, R).

Definition 5.9 For any NPQML $\Lambda \subseteq \mathcal{L}$, it is Kripke $\mathcal{L}_{qf}$-complete if for any $\varphi \in \mathcal{L}_{qf}$ that is valid on all Kripke frames in $\mathsf{KFr}(\Lambda)$, φ is in Λ.

Corollary 5.10 *Let $\Theta \subseteq \mathcal{L}$ be a set of $\mathcal{ATQ}$-persistent formulas such that $\mathsf{KFr}(\Theta)$ has diversity n and $\mathsf{K}_\Pi\Theta\mathtt{BcAt}^n\mathtt{R}^n$ is Kripke $\mathcal{L}_{qf}$-complete. Then $\mathsf{K}_\Pi\Theta\mathtt{BcAt}^n\mathtt{R}^n$ is the NPQML of $\mathsf{KFr}(\Theta)$.*

Now we consider a special case, the class $\mathsf{KFr}(\mathtt{5})$ of Euclidean frames. We have observed that $\mathsf{KFr}(\mathtt{5})$ has diversity 3. Note that it also validates $\Diamond^3 p \to \Diamond^2 p$. Thus, the NPQML of Euclidean Kripke frames can be axiomatized as $\mathsf{K}_\Pi\mathtt{5BcAt}^2\mathtt{R}^2$. In other words, $\mathtt{5}\pi+$ is $\mathsf{K}_\Pi\mathtt{5BcAt}^2\mathtt{R}^2$. However, unlike $\mathtt{D45}\pi+ = \mathsf{K}_\Pi\mathtt{D45BcAt}^1$, $\mathtt{R}^2$ is indispensable for the axiomatization of the logic of Euclidean frames. For this, we construct a quantifiable frame $\mathbf{F}$ such that $\mathbf{F}$ validates $\mathsf{K}_\Pi\mathtt{5BcAt}^2$ but not $\exists p(\Box p \wedge \forall q(\Box q \to \Box^2(p \to q)))$, a formula that is valid on any Kripke frame using the successor sets.

Let $\mathbf{F} = (W, R, B)$ where

- $W = \mathbb{N} \cup \{w\}$, where $w \notin \mathbb{N}$;
- $R = \mathbb{N}^2 \cup \{(w, 2n) \mid n \in \mathbb{N}\}$; i.e. R is total in $\mathbb{N}$ and w sees all even numbers;
- $B = \{X \subseteq W \mid X \text{ is finite or cofinite with respect to } W\}$.

Proposition 5.11 **F** *is a quantifiable frame validating* $\mathsf{K}_\Pi\mathtt{5BcAt}^2$ *but not* $\exists p(\Box p \wedge \forall q(\Box q \to \Box^2(p \to q)))$.

Proof. First, let us check that **F** is a quantifiable frame. Consider first the subframes $\mathbf{G} = (\mathbb{N}, \mathbb{N}^2, B')$ generated from (any) $n \in \mathbb{N}$, where $B' = \{X \cap \mathbb{N} \mid X \in B\}$. Observe that B' is also $\{X \subseteq \mathbb{N} \mid X$ is finite or $\mathbb{N} \setminus X$ is finite$\}$. We first show that **G** is quantifiable. There is only one duplicate class $\mathbb{N}$ of $(\mathbb{N}, \mathbb{N}^2)$, and B' is a discrete field of sets, so Theorem 4.11 applies: for any formula φ and valuation $v : \mathsf{Prop} \to B'$, $[\![\varphi]\!]^{B'}(v) = [\![\varphi]\!]^{\wp(\mathbb{N})}(v)$. By Lemma 4.10, there is then a Boolean formula α such that $[\![\varphi]\!]^{\wp(\mathbb{N})}(v) = [\![\alpha]\!](v)$. So $[\![\varphi]\!]^{B'}(v)$ is a Boolean combination of propositions in B', which is in B' as B' is a field of sets. This shows that **G** is quantifiable.

Now, since for any $n \in \mathbb{N}$, its point-generated subframe is **G**, by an easy induction, for any valuation v (with $v|_\mathbb{N}$ being its point-wise restriction to $\mathbb{N}$) and any formula φ, $n \in [\![\varphi]\!]^\mathbf{F}(v)$ iff $n \in [\![\varphi]\!]^\mathbf{G}(v|_\mathbb{N})$. In other words, $[\![\varphi]\!]^\mathbf{F}(v) \cap \mathbb{N} = [\![\varphi]\!]^\mathbf{G}(v|_\mathbb{N})$. As **G** is quantifiable, $[\![\varphi]\!]^\mathbf{F}(v) \cap \mathbb{N}$ is finite or cofinite in $\mathbb{N}$. But $[\![\varphi]\!]^\mathbf{F}(v)$ and $[\![\varphi]\!]^\mathbf{F}(v) \cap \mathbb{N}$ differ by at most one point w. So $[\![\varphi]\!]^\mathbf{F}(v)$ is finite or cofinite in W, and is in B. This shows that **F** is quantifiable.

As **F** is quantifiable, as soon as we verify that **F** validates $\mathtt{5}$ and $\mathtt{At}^3$, **F** then automatically validates $\mathsf{K}_\Pi\mathtt{5BcAt}^3$. But the underlying Kripke frame of **F** already validates $\mathtt{5}$, and B contains all singleton subsets of W.

To see that $\exists p(\Box p \wedge \forall q(\Box q \to \Box^2(p \to q)))$ is invalid, we only need to note that for any $X \in B$ such that $X \supseteq R[w]$, there is always a $Y \in B$ such that $Y \subsetneq X$ and $Y \supseteq R[w]$. This is because X must contain infinitely many odd numbers, and we can delete one odd number to obtain Y. This shows that at w, the formula $\forall p(\Box p \to \exists q(\Box q \wedge \Diamond^2(p \wedge \neg q)))$ is true at w. Thus, its negation $\exists p(\Box p \wedge \forall q(\Box q \to \Box^2(p \to q)))$ is false at w. □

Thus, $\mathsf{K}_\Pi\mathtt{5BcAt}^2$ is not the logic of Euclidean frames.

We can also show that the requirement of Θ consisting of Sahlqvist formulas cannot be dispensed with altogether either. Consider the axioms $\mathtt{T}$: $p \to \Diamond p$, $\mathtt{M}$: $\Diamond\Box\neg p \vee \Diamond\Box p$, $\mathtt{E}$: $\Diamond(\Diamond p \wedge \Box q) \to \Box(\Diamond p \vee \Box q)$, and $\mathtt{Q}$: $(\Diamond p \wedge \Box(p \to \Box p)) \to p$ (we reuse the letter $\mathtt{Q}$) used in [38]. It is not hard to show that the only Kripke frames validating $\mathtt{TMEQ}$ are those with the identity accessibility relation, and thus $\mathsf{KFr}(\mathtt{TMEQ})$ has diversity 1. However, $\mathsf{K}_\Pi\mathtt{TMEQBcAt}^1\mathtt{R}^1$ is not the NPQML of $\mathsf{KFr}(\mathtt{TMEQ})$, as it does not derive $p \leftrightarrow \Box p$ that is valid on Kripke frames with the identity accessibility relation. The idea is to observe that the veild recession general frame $\mathbf{F} = (\mathbb{Z}, R, B)$ where

- nRm iff $m \geqslant n - 1$, and
- $X \in B$ iff there is n such that for all $m \geqslant n$, $m \in X$ iff $n \in X$ (call this property *settled after* n),

is quantifiable and validates $\mathtt{TMEQAt}^1\mathtt{R}^1$, but not $p \to \Box p$. Thus, the set of validities of **F** separates $\mathsf{K}_\Pi\mathtt{TMEQAt}^1\mathtt{R}^1$ from $p \to \Box p$ and shows that $\mathsf{K}_\Pi\mathtt{TMEQAt}^1\mathtt{R}^1$ is not $\mathtt{TMEQ}\pi+$. It is easy to check that **F** validates the special axioms, but it takes more work to show that **F** is quantifiable. Here we need to use truncated point-generated submodels and the key step is to show that for all sufficiently large m and n, if $\exists p\varphi$ is true at m using set X as the valuation for p and d is the modal depth of $\exists p\varphi$, then $\exists p\varphi$ is also true at n as we can shift X by $n-m$ and then the depth-d truncated submodel generated from n is isomorphic to the depth-d truncated submodel generated from m, making this shifted X a witness to $\exists p\varphi$ for n. Due to the space limit, we refer the readers to a later online version for detailed proofs.

6 Conclusion

We conclude with some ideas for possible future research. First, the result that all NPQMLs containing Bc is complete for the class of quantifiable frames it defines is fairly standard and expected, but can still be generalized, for example, to *neighborhood* frames. Of course, Bc will be dropped from the logic, and similar work has been done for first-order modal logic [2]. One may also consider the alternative semantics in [24] where $\forall p\varphi$ is true at w iff there is a proposition X that contains w and entails all propositions expressible by φ as we vary the proposition denoted by p. We also believe that it would be instructive to rewrite the proof of Theorem 3.1 in terms of Lindenbaum algebras and duality theory, as this may help us generalize the result.

For our second result, there are two natural ideas to generalize. The first is dropping $\mathtt{At}^n$ and consider completeness w.r.t. algebraic semantics based on complete and completely multiplicative modal algebras. It should be noted that, as is clear in our proof, axiom $\mathtt{At}^n$ corresponds to the existence of 'world propositions' that can later be interpreted as possible worlds, and the world propositions serve as the names of the possible worlds. Hybrid logics use world propositions in a much more direct way by taking them as a primitive syntactical category, namely the nominals, and a recent work [6] has considered propositionally quantified hybrid modal logic. As is mentioned in that paper, Arthur Prior is a strong proponent of both. However, the idea of there being no maximally specified possible worlds but only partial states [27,36,26,13] is also worth investigating in this context (though PQML together with plural quantifiers are used to argue for there being world propositions [20]), and algebraic semantics allowing atomless elements in the algebras is a natural way to model this. Previous works in this line include [25,11,12].

The other direction for generalization is dropping the finite diversity condition in some way. The condition that $\mathsf{KFr}(\Theta)$ has finite diversity is admittedly a very restrictive one, and it is worth investigating the exact scope of this condition, especially together with the requirement of Θ consisting of only Sahlqvist formulas. We see that there is at least one promising way of relaxing the finite diversity condition: requiring only finite diversity for each point-generated frame of finite depth.

Finally, we mention a broader question: can the theory of PQMLs inform the theory of modal μ-calculus [7] or vice versa, especially over completeness questions, noting that the μ operator is also a kind of propositional quantifier? For example, the recent work [21] utilized the fixpoint construction in NPQML, and we believe more needs to be done.

Acknowledgements

We thank the anonymous reviewers for their helpful suggestions that led to many improvements. We also thank Peter Fritz for commenting on a very early draft of this paper. The first author is supported by NSSF grant 22CZX066.

References

[1] Antonelli, G. A. and R. H. Thomason, *Representability in second-order propositional poly-modal logic*, The Journal of Symbolic Logic **67** (2002), pp. 1039–1054.

[2] Arló-Costa, H. and E. Pacuit, *First-order classical modal logic*, Studia Logica **84** (2006), pp. 171–210.

[3] Bednarczyk, B. and S. Demri, *Why does propositional quantification make modal and temporal logics on trees robustly hard?*, Logical Methods in Computer Science **18** (2022).

[4] Belardinelli, F., W. Van Der Hoek and L. B. Kuijer, *Second-order propositional modal logic: Expressiveness and completeness results*, Artificial Intelligence **263** (2018), pp. 3–45.

[5] Bílková, M., *Uniform interpolation and propositional quantifiers in modal logics*, Studia Logica **85** (2007), pp. 1–31.

[6] Blackburn, P., T. Braüner and J. L. Kofod, *An axiom system for basic hybrid logic with propositional quantifiers*, in: *International Workshop on Logic, Language, Information, and Computation*, Springer, 2023, pp. 118–134.

[7] Bradfield, J. and C. Stirling, *Modal mu-calculi*, in: *Handbook of Modal Logic*, Elsevier, 2007 pp. 721–756.

[8] ten Cate, B., *Expressivity of second order propositional modal logic*, Journal of Philosophical Logic **35** (2006), pp. 209–223.

[9] D'agostino, G. and M. Hollenberg, *Logical questions concerning the μ-calculus: Interpolation, Lyndon and Łoś-Tarski*, The Journal of Symbolic Logic **65** (2000), pp. 310–332.

[10] Dekker, P. M., *KD45 with propositional quantifiers*, Logic and Logical Philosophy **33** (2024), pp. 27–54.

[11] Ding, Y., *On the logics with propositional quantifiers extending S5Π*, in: G. Bezhanishvili, G. D'Agostino, G. Metcalfe and T. Studer, editors, *Advances in Modal Logic, Vol. 12*, College Publications, 2018 pp. 219–235.

[12] Ding, Y., *On the logic of belief and propositional quantification*, Journal of Philosophical Logic **50** (2021), pp. 1143–1198.

[13] Ding, Y. and W. H. Holliday, *Another problem in possible world semantics*, in: N. Olivetti and R. Verbrugge, editors, *Advances in Modal Logic, Vol. 13*, College Publications, 2020 pp. 149–168.

[14] D'Agostino, G. and G. Lenzi, *An axiomatization of bisimulation quantifiers via the μ-calculus*, Theoretical Computer Science **338** (2005), pp. 64–95.

[15] Fine, K., *Propositional quantifiers in modal logic*, Theoria **36** (1970), pp. 336–346.

[16] French, T., "Bisimulation quantifiers for modal logics," PhD thesis, The University of Western Australia (2006).

[17] French, T., *Idempotent transductions for modal logics*, in: *Frontiers of Combining Systems: 6th International Symposium, FroCoS 2007 Liverpool, UK, September 10-12, 2007 Proceedings 6*, Springer, 2007, pp. 178–192.

[18] French, T. and M. Reynolds, *A sound and complete proof system for QPTL*, in: P. Balbiani, N.-Y. Suzuki, F. Wolter and M. Zakharyaschev, editors, *Advances in Modal Logic, Vol. 4*, King's College Publications, 2002 pp. 127–148.

[19] Fritz, P., *Axiomatizability of propositionally quantified modal logics on relational frames*, The Journal of Symbolic Logic (2022), p. 1–36.

[20] Fritz, P., "The Foundations of Modality: From Propositions to Possible Worlds," Oxford University Press, 2023.
[21] Fritz, P., *Nonconservative extensions by propositional quantifiers and modal incompleteness*, manuscript (2023).
[22] Fritz, P., "Propositional Quantifiers," Elements in Philosophy and Logic, Cambridge University Press, 2024.
[23] Ghilardi, S. and M. Zawadowski, *Undefinability of propositional quantifiers in the modal system S4*, Studia Logica **55** (1995), pp. 259–271.
[24] Goldblatt, R. and E. D. Mares, *A general semantics for quantified modal logic.*, in: G. Governatori, I. Hodkinson and Y. Venema, editors, *Advances in modal logic, Vol. 6*, College Publications, 2006 pp. 227–246.
[25] Holliday, W. H., *A note on algebraic semantics for S5 with propositional quantifiers*, Notre Dame Journal of Formal Logic **60** (2019), pp. 311–332.
[26] Holliday, W. H., *Possibility semantics*, in: M. Fitting, editor, *Selected Topics from Contemporary Logics*, College Publications, 2021 pp. 363–476.
[27] Humberstone, I. L., *From worlds to possibilities*, Journal of Philosophical Logic (1981), pp. 313–339.
[28] Kaminski, M. and M. Tiomkin, *The expressive power of second-order propositional modal logic*, Notre Dame Journal of Formal Logic **37** (1996), pp. 35–43.
[29] Kremer, P., *On the complexity of propositional quantification in intuitionistic logic*, The Journal of Symbolic Logic **62** (1997), pp. 529–544.
[30] Kremer, P., *Completeness of second-order propositional S4 and H in topological semantics*, The Review of Symbolic Logic **11** (2018), pp. 507–518.
[31] Kuhn, S. et al., *A simple embedding of T into double S5*, Notre Dame Journal of Formal Logic **45** (2004), pp. 13–18.
[32] Kuusisto, A., *Second-order propositional modal logic and monadic alternation hierarchies*, Annals of Pure and Applied Logic **166** (2015), pp. 1–28.
[33] M. J. Cresswell, G. E. H., "A New Introduction to Modal Logic," Routledge, New York, 1996.
[34] Pitts, A. M., *On an interpretation of second order quantification in first order intuitionistic propositional logic*, The Journal of Symbolic Logic **57** (1992), pp. 33–52.
[35] Riba, C., *A model theoretic proof of completeness of an axiomatization of monadic second-order logic on infinite words*, in: *Theoretical Computer Science: 7th IFIP TC 1/WG 2.2 International Conference, TCS 2012, Amsterdam, The Netherlands, September 26-28, 2012. Proceedings 7*, Springer, 2012, pp. 310–324.
[36] Rumfitt, I., "The boundary stones of thought: An essay in the philosophy of logic," Oxford University Press, USA, 2015.
[37] Steinsvold, C., *Some formal semantics for epistemic modesty*, Logic and Logical Philosophy **29** (2020), pp. 381–413.
[38] van Benthem, J., *Two simple incomplete modal logics*, Theoria **44** (1978), pp. 25–37.
[39] ten Cate, B., "Model theory for extended modal languages," PhD thesis, University of Amsterdam (2004).
[40] Venema, Y., *Derivation rules as anti-axioms in modal logic*, The Journal of Symbolic Logic **58** (1993), pp. 1003–1034.
[41] Visser, A., *Bisimulations, model descriptions and propositinal quantifiers*, Logic Group Preprint Series **161** (1996).
[42] Zach, R., *Decidability of quantified propositional intuitionistic logic and S4 on trees of height and arity $\leqslant \omega$*, Journal of Philosophical Logic **33** (2004), pp. 155–164.

Toward the van Benthem Characterization Theorem for Non-Distributive Modal Logic

Yiwen Ding[a] Krishna Manoorkar[a] Mattia Panettiere[a]
Ruoding Wang[a, b]

[a] *Vrije Universiteit Amsterdam, The Netherlands*
[b] *Xiamen University, China*

Abstract

In this paper, we introduce the simulations and bisimulations on polarity-based semantics for non-distributive modal logic, which are natural generalizations of those notions on Kripke semantics for modal logic. We also generalize other important model-theoretic notions about Kripke semantics such as image-finite models, modally-saturated models, ultrafilter extension and ultrapower extension to the non-distributive setting. By using these generalizations, we prove the Hennessy-Milner theorem and the van Benthem characterization theorem for non-distributive modal logic based on polarity-based semantics.

Keywords: Polarity-based semantics, Non-distributive modal logics, Simulations, Bisimulations, Hennessy-Milner theorem, van Benthem characterization theorem.

1 Introduction

This paper focuses on advancing the model theory of polarity-based semantics for *normal non-distributive modal logics* [19,6,9]. These logics, also known as *LE-logics* (lattice expansion logics), are characterized by algebraic semantics provided by arbitrary lattices expanded with normal operators. The research on polarity-based semantics for normal non-distributive modal logics dates back to Gehrke's work on canonical extension of bounded lattices [22,21,17]. Leveraging the interpretation of polarities as formal contexts in Formal Concept Analysis [20], non-distributive modal logic is conceptualized as a logic of formal concepts or categories enriched with modal operators, such as approximation or knowledge operators [6,7]. Several model theoretic notions for polarity-based semantics like *disjoint unions*, *p-morphisms*, and *ultrafilter extensions* were introduced in prior work [10]. These concepts were used in [10] to prove

⋆ This project has received funding from the European Union's Horizon 2020 research and innovation programme under the Marie Skłodowska-Curie grant agreement No 101007627. The research of Krishna Manoorkar is supported by the NWO grant KIVI.2019.001. Yiwen Ding and Ruoding Wang are supported by the China Scholarship Council.

Goldblatt-Thomason theorem for LE-logics. Simulations on polarity-based models were defined and employed to demonstrate the preservation of lattice-based modal μ-calculus (an extension of basic LE-logic with the least and the greatest fixed point operators) under simulations [16, Section 6]. In this paper, we delve more into the problem of generalizing simulations and bisimulations from Kripke semantics to polarity-based semantics, and also prove fundamental results like Hennessy-Milner theorem and van Benthem characterization theorem for polarity-based semantics.

Bisimulations were initially defined by van Benthem [34] as p-relations on Kripke frames for classical modal logic. They aim to characterize points in Kripke models with equivalent behavior concerning modal logic formulas. Since their inception, bisimulations have played an important role in modal logic theory and its applications in computer science [34,32,25,23]. Hennessy and Milner [26] showed that on image-finite Kripke models, the semantic equivalence relation for modal logic forms a bisimulation relation, a result later extended to modally-saturated Kripke models [27]. Van Benthem [34] established the characterization theorem for modal logic, showing that classical modal logic is the bisimulation-invariant fragment of first-order logic.

Bisimulations have been extensively defined and studied across various semantic models for different logics. For instance, in topological semantics for (S4) modal logic, they are referred to as topo-bisimulations [1], where the Hennessy-Milner property for a class of topological models was established [11]. Similarly, in Kripke semantics for intuitionistic modal logic, they are known as asimulations, and results like the Hennessy-Milner theorem and the van Benthem theorem have been demonstrated [30]. Bezhanishvilli and Henke [3] proved the van Benthem theorem for models based on general descriptive frames using *Vietoris bisimulations* [2]. Similar studies have been carried out for different semantics of graded modal logics, dynamic modal logics, fuzzy modal logics, monotone modal logics, etc. [15,33,13,14,24]. De Groot [12] established a van Benthem theorem for the meet-semilattice semantics of non-distributive (positive) logic using the notion of meet simulations. In this work, we adopt a similar approach to prove the van Benthem theorem for the polarity-based semantics of normal non-distributive modal logic by defining appropriate notion of simulations. Along the way, we introduce essential model-theoretic notions like image-finite models, modally-saturated models, and ω-saturated models for this semantics, and prove significant results like the Hennessy-Milner theorem and the Hennessy-Milner property for the class of modally-saturated models.

Structure of the paper. In Section 2, we collect the required preliminaries. In Section 3, we define the simulations and bisimulations on the polarity-based semantics and prove the Hennessy-Milner theorem. In Section 4, we define the modally-saturated polarity-based models, filter-ideal extensions of polarity-based models and show that two pointed polarity-based models are modally equivalent if and only if their filter-ideal extensions are bisimilar. In Section 5, we define the standard translation and prove the van Benthem characterization theorem for polarity-based semantics of non-distributive modal logic. In Section 6, we summarize the main findings of the paper and suggest some possible future works.

2 Preliminaries

In this section, we gather useful preliminaries about normal non-distributive modal logic and its polarity-based semantics based on [9] and [7]. We assume readers are familiar with the basic concepts from the model theory of classical modal logic such as simulations, bisimulations, ultrafilter extensions, standard translations, Hennessy-Milner theorem, and van Benthem characterization theorem. For a detailed discussion on the model theory of modal logic, we refer to [4, Section 2].

2.1 Non-Distributive Modal Logic

Let $\mathcal{V}$ be a countable set of propositional variables. The language $\mathcal{L}$ (i.e. set of formulas) is defined as follows:

$$\varphi ::= p \mid \bot \mid \top \mid \varphi \wedge \varphi \mid \varphi \vee \varphi \mid \Box\varphi \mid \Diamond\varphi,$$

where $p \in \mathcal{V}$. Given any $\varphi, \psi \in \mathcal{L}$, $\varphi \vdash \psi$ is an *$\mathcal{L}$-sequent*. A *normal non-distributive modal logic* is a set of $\mathcal{L}$-sequents containing the following axioms:

$$p \vdash p, \quad p \vdash \top \quad \bot \vdash p, \quad p \vdash p \vee q, \quad q \vdash p \vee q, \quad p \wedge q \vdash p, \quad p \wedge q \vdash q,$$
$$\top \vdash \Box\top, \quad \Diamond\bot \vdash \bot, \quad \Box p \wedge \Box q \vdash \Box(p \wedge q), \quad \Diamond(p \vee q) \vdash \Diamond p \vee \Diamond q$$

and closed under the following inference rules:

$$\frac{\varphi\vdash\chi \quad \chi\vdash\psi}{\varphi\vdash\psi} \quad \frac{\varphi\vdash\psi}{\varphi(\chi/p)\vdash\psi(\chi/p)} \quad \frac{\chi\vdash\varphi \quad \chi\vdash\psi}{\chi\vdash\varphi\wedge\psi} \quad \frac{\varphi\vdash\chi \quad \psi\vdash\chi}{\varphi\vee\psi\vdash\chi} \quad \frac{\varphi\vdash\psi}{\Box\varphi\vdash\Box\psi} \quad \frac{\varphi\vdash\psi}{\Diamond\varphi\vdash\Diamond\psi}$$

The smallest such set $\mathbb{L}$ is called the *basic* normal non-distributive modal logic. For any sequent $\varphi \vdash \psi$ in $\mathbb{L}$ we denote it as $\varphi \vdash_{\mathbb{L}} \psi$.

2.2 Polarity-Based Semantics

Given any relation $R \subseteq A \times X$ and any $B \subseteq A$, $Y \subseteq X$, the maps $R^{(1)} : \mathcal{P}(A) \to \mathcal{P}(X)$ and $R^{(0)} : \mathcal{P}(X) \to \mathcal{P}(A)$ are defined as $R^{(1)}[B] := \{x \in X \mid \forall b(b \in B \Rightarrow bRx)\}$ and $R^{(0)}[Y] := \{a \in A \mid \forall y(y \in Y \Rightarrow aRy)\}$. For any relation $R \subseteq A \times X$, we $R^c \subseteq A \times X$ be the relation defined as follows: For any $a \in A$, $x \in X$, $(a, x) \in R^c$ iff $(a, x) \notin R$.

Definition 2.1 A *polarity* is a tuple $\mathfrak{P} = (A, X, I)$ such that A and X are sets and $I \subseteq A \times X$ is a binary relation. Given a polarity $\mathfrak{P} = (A, X, I)$, the maps $(\cdot)^\uparrow : \mathcal{P}(A) \to \mathcal{P}(X)$ and $(\cdot)^\downarrow : \mathcal{P}(X) \to \mathcal{P}(A)$ are defined as follows:

$$B^\uparrow := I^{(1)}[B] \text{ and } Y^\downarrow := I^{(0)}[Y], \text{ where } B \subseteq A \text{ and } Y \subseteq X.$$

The maps $(\cdot)^\uparrow$ and $(\cdot)^\downarrow$ form a *Galois connection* between posets $(\mathcal{P}(A), \subseteq)$ and $(\mathcal{P}(X), \subseteq)$, that is $Y \subseteq B^\uparrow$ iff $B \subseteq Y^\downarrow$ for all $B \in \mathcal{P}(A)$ and $Y \in \mathcal{P}(X)$.

Definition 2.2 A *formal concept* of the polarity $\mathfrak{P} = (A, X, I)$ is a pair $([\![c]\!], (\![c]\!))$ such that $[\![c]\!] \subseteq A$, $(\![c]\!) \subseteq X$, and $[\![c]\!]^\uparrow = (\![c]\!)$, $(\![c]\!)^\downarrow = [\![c]\!]$. It follows that $[\![c]\!]$ and $(\![c]\!)$ are *Galois-stable*, i.e. $[\![c]\!]^{\uparrow\downarrow} = [\![c]\!]$ and $(\![c]\!)^{\downarrow\uparrow} = (\![c]\!)$. The set $\mathcal{C}(\mathfrak{P})$ of all the formal concepts of $\mathfrak{P}$ can be partially ordered as follows: for any $c, d \in \mathcal{C}(\mathfrak{P})$, $c \leq d$ iff $[\![c]\!] \subseteq [\![d]\!]$ iff $(\![d]\!) \subseteq (\![c]\!)$. The poset $\mathfrak{P}^+ = (\mathcal{C}(\mathfrak{P}), \leq)$ is a complete lattice – called the *concept lattice* of $\mathfrak{P}$ – such that for any $\mathrm{K} \subseteq \mathcal{C}(\mathfrak{P})$,

$$\bigwedge \mathrm{K} = (\bigcap\{[\![c]\!] \mid c \in \mathrm{K}\}, (\bigcap\{[\![c]\!] \mid c \in \mathrm{K}\})^\uparrow),$$

$$\bigvee \mathrm{K} = ((\bigcap\{\llparenthesis c \rrparenthesis \mid c \in \mathrm{K}\})^{\downarrow}, \bigcap\{\llparenthesis c \rrparenthesis \mid c \in \mathrm{K}\}).$$

Proposition 2.3 *For any polarity* $\mathfrak{P} = (A, X, I)$*, the concept lattice* $\mathfrak{P}^+$ *is completely join-generated by the set* $\{\boldsymbol{a} := (a^{\uparrow\downarrow}, a^{\uparrow}) \mid a \in A\}$*, and is completely meet-generated by the set* $\{\boldsymbol{x} := (x^{\downarrow}, x^{\downarrow\uparrow}) \mid x \in X\}$*.*

Theorem 2.4 (Birkhoff's representation theorem) *Any complete lattice* **L** *is isomorphic to the concept lattice* $\mathfrak{P}^+$ *of some polarity* $\mathfrak{P}$*.*

Definition 2.5 An LE-*frame* is a tuple $\mathfrak{F} = (\mathfrak{P}, R_\Diamond, R_\Box)$, where $\mathfrak{P} = (A, X, I)$ is a polarity, and $R_\Diamond \subseteq X \times A$ and $R_\Box \subseteq A \times X$ are *I-compatible* relations, i.e., for all $a \in A$ and $x \in X$, $R_\Diamond^{(0)}[a]$ (resp. $R_\Box^{(0)}[x]$) and $R_\Box^{(1)}[a]$ (resp. $R_\Diamond^{(1)}[x]$) are Galois-stable.

Definition 2.6 For any LE-frame $\mathfrak{F} = (\mathfrak{P}, R_\Diamond, R_\Box)$, the *complex algebra* of $\mathfrak{F}$ is $\mathfrak{F}^+ := (\mathfrak{P}^+, \Diamond^{\mathfrak{P}^+}, \Box^{\mathfrak{P}^+})$ where $\mathfrak{P}^+$ is the concept lattice of $\mathfrak{P}$, $\Diamond^{\mathfrak{P}^+}$ and $\Box^{\mathfrak{P}^+}$ are unary operators on $\mathfrak{P}^+$ defined as follows: for every $c \in \mathfrak{P}^+$,

$$\Diamond^{\mathfrak{P}^+}(c) := (R_\Diamond^{(0)}[\llbracket c \rrbracket]^{\downarrow}, R_\Diamond^{(0)}[\llbracket c \rrbracket]) \quad \text{and} \quad \Box^{\mathfrak{P}^+}(c) := (R_\Box^{(0)}[\llparenthesis c \rrparenthesis], R_\Box^{(0)}[\llparenthesis c \rrparenthesis]^{\uparrow}).$$

Note that $\Diamond^{\mathfrak{P}^+}$ (resp. $\Box^{\mathfrak{P}^+}$) is completely join (resp. meet) preserving.

Definition 2.7 A *valuation* on an LE-frame $\mathfrak{F} = (\mathfrak{P}, R_\Diamond, R_\Box)$ is a map $V : \mathcal{V} \to \mathfrak{P}^+$. For each $p \in \mathcal{V}$, we let $\llbracket p \rrbracket := \llbracket V(p) \rrbracket$ (resp. $\llparenthesis p \rrparenthesis := \llparenthesis V(p) \rrparenthesis$) denote the extension (resp. intension) of the interpretation of p under V. A valuation can be homomorphically extended to an unique valuation $\overline{V} : \mathcal{L} \to \mathfrak{P}^+$ on all the $\mathcal{L}$-formulas. An LE-*model* is a pair $(\mathfrak{F}, V)$, where $\mathfrak{F}$ is an LE-frame, and V is a valuation on it.

Definition 2.8 For any LE-model $\mathfrak{M} = (\mathfrak{F}, V)$, the *modal satisfaction* relations $\Vdash$ and $\succ$ are defined inductively as follows:

$\mathfrak{M}, a \Vdash p$	iff $a \in \llbracket p \rrbracket_{\mathfrak{M}}$	$\mathfrak{M}, x \succ p$	iff $x \in \llparenthesis p \rrparenthesis_{\mathfrak{M}}$
$\mathfrak{M}, a \Vdash \top$	always	$\mathfrak{M}, x \succ \top$	iff $(\forall a \in A) aIx$
$\mathfrak{M}, x \succ \bot$	always	$\mathfrak{M}, a \Vdash \bot$	iff $(\forall x \in X) aIx$
$\mathfrak{M}, a \Vdash \varphi \wedge \psi$	iff $\mathfrak{M}, a \Vdash \varphi$ and $\mathfrak{M}, a \Vdash \psi$	$\mathfrak{M}, x \succ \varphi \wedge \psi$	iff $(\forall a \in A)$ $(\mathfrak{M}, a \Vdash \varphi \wedge \psi \Rightarrow aIx)$
$\mathfrak{M}, x \succ \varphi \vee \psi$	iff $\mathfrak{M}, x \succ \varphi$ and $\mathfrak{M}, x \succ \psi$	$\mathfrak{M}, a \Vdash \varphi \vee \psi$	iff $(\forall x \in X)$ $(\mathfrak{M}, x \succ \varphi \vee \psi \Rightarrow aIx)$
$\mathfrak{M}, a \Vdash \Box\varphi$	iff $(\forall x \in X)(\mathfrak{M}, x \succ \varphi \Rightarrow aR_\Box x)$	$\mathfrak{M}, x \succ \Box\varphi$	iff $(\forall a \in A)(\mathfrak{M}, a \Vdash \Box\varphi \Rightarrow aIx)$
$\mathfrak{M}, x \succ \Diamond\varphi$	iff $(\forall a \in A)(\mathfrak{M}, a \Vdash \varphi \Rightarrow xR_\Diamond a)$	$\mathfrak{M}, a \Vdash \Diamond\varphi$	iff $(\forall x \in X)(\mathfrak{M}, x \succ \Diamond\varphi \Rightarrow aIx)$

Note that unlike the classical modal logic, the modal operators $\Box$ and $\Diamond$ are not inter-definable in LE-logic. For any $\mathcal{L}$-sequent $\varphi \vdash \psi$, and an LE-model $\mathfrak{M} = (\mathfrak{F}, V)$, $\mathfrak{M} \models \varphi \vdash \psi$ iff $\llbracket V(\varphi) \rrbracket \subseteq \llbracket V(\psi) \rrbracket$ iff $\llparenthesis V(\psi) \rrparenthesis \subseteq \llparenthesis V(\varphi) \rrparenthesis$.

The following theorem states that the basic normal non-distributive modal logic defined in Section 2.1 is sound and complete w.r.t. the class of LE-models.

Theorem 2.9 (Proposition 3, [7]) *For any unprovable* $\mathcal{L}$*-sequent* $\varphi \vdash \psi$ *in the non-distributive modal logic, there is an* LE*-model* $\mathfrak{M}$ *such that* $\mathfrak{M} \not\models \varphi \vdash \psi$*.*

2.3 Two Sorted First-Order Logic

Given a countable set of propositional variables $\mathcal{V}$, we let $\mathcal{L}^1$ be the two sorted first-order language with equality built over two disjoint countable sets of variables G and M, three binary predicates I, $R_\Box$, $R_\Diamond$, and two unary predicates

P_A and P_X for every $p \in \mathcal{V}$. A *domain* for $\mathcal{L}^1$ is a pair of disjoint sets A and X. An *interpretation* I of $\mathcal{L}^1$ over domain (A, X) assigns binary predicates I, $R_\Box$, and $R_\Diamond$ to relations $I^{\mathrm{I}} \subseteq A \times X$, $R^{\mathrm{I}}{}_\Box \subseteq A \times X$, and $R^{\mathrm{I}}{}_\Diamond \subseteq X \times A$, such that $R^{\mathrm{I}}{}_\Box$ and $R^{\mathrm{I}}{}_\Diamond$ are I^{I}-compatible. For every propositional variable $p \in \mathcal{V}$, the unary predicates P_A and P_X are assigned to sets $P_A^{\mathrm{I}} \subseteq A$, $P_X^{\mathrm{I}} \subseteq X$, such that, $(I^{\mathrm{I}})^{(1)}[P_A^{\mathrm{I}}] = P_X^{\mathrm{I}}$, and $(I^{\mathrm{I}})^{(0)}[P_X^{\mathrm{I}}] = P_A^{\mathrm{I}}$, respectively. An $\mathcal{L}^1$-*structure* is a tuple $\mathfrak{M} = (A, X, \mathrm{I})$, where I is an interpretation of $\mathcal{L}^1$ over domain A and X. A *valuation* on an $\mathcal{L}^1$- structure $\mathfrak{M}$ is a map v which assigns every variable $g \in G$ (resp. $m \in M$) to an element $a \in A$ (resp. $x \in X$). The *satisfaction relation* for any $\mathcal{L}^1$-term w.r.t. any $\mathcal{L}^1$- structure $\mathfrak{M}$ and valuation v on it is defined as follows:

1. $\mathfrak{M}, v \models g_1 = g_2$ iff $v(g_1) = v(g_2)$
2. $\mathfrak{M}, v \models m_1 = m_2$ iff $v(m_1) = v(m_2)$
3. $\mathfrak{M}, v \models P_A(g_1)$ iff $v(g_1) \in P_A^{\mathrm{I}}$
4. $\mathfrak{M}, v \models P_X(m_1)$ iff $v(m_1) \in P_X^{\mathrm{I}}$
5. $\mathfrak{M}, v \models g_1 I m_1$ iff $v(g_1) I^{\mathrm{I}} v(m_1)$
6. $\mathfrak{M}, v \models g_1 R_\Box m_1$ iff $v(g_1) R_\Box^{\mathrm{I}} v(m_1)$
7. $\mathfrak{M}, v \models m_1 R_\Diamond g_1$ iff $v(m_1) R_\Diamond^{\mathrm{I}} v(g_1)$.

Any LE-model $\mathfrak{M} = (A, X, I^{\mathfrak{M}}, R_\Box^{\mathfrak{M}}, R_\Diamond^{\mathfrak{M}}, V)$ can be seen as an $\mathcal{L}^1$-structure, where the domain of interpretation is formed by sets A and X from the polarity. The binary predicate symbols I, $R_\Box$, $R_\Diamond$ are interpreted by the corresponding relations on the polarity, and unary predicate P_A (resp. P_X) is interpreted as the set $[\![V(p)]\!]$ (resp. $(\![V(p)]\!)$), for any $p \in \mathcal{V}$. For any LE-model $\mathfrak{M}$, we use the same name $\mathfrak{M}$ for the two sorted first-order structure when they are clearly distinguishable from the context.

3 Simulations and Bisimulations on LE-Models

In this section, we generalize simulations and bisimulations from Kripke models to LE-models and show invariance of $\mathcal{L}$-formulas under bisimulations. We give an example which shows that the Hennesy-Milner theorem does not hold with respect to this definition of bisimulation. Thus, this definition of bisimulation does not provide us notion of bisimilarity which has desired (or expected) properties. Hence, we give a more general definition of bisimilarity which is based on simulations, and prove Hennesy-Milner theorem for polarity-based semantics of non-distributive modal logic with respect to this generalized notion of bisimilarity.

Definition 3.1 Let $\mathfrak{M}_1 = (\mathfrak{F}_1, V_1)$ and $\mathfrak{M}_2 = (\mathfrak{F}_2, V_2)$ be any LE-models, a_1, x_1 in $\mathfrak{M}_1$ and a_2, x_2 in $\mathfrak{M}_1$, then

1. $\mathfrak{M}_1, a_1 \rightsquigarrow_A \mathfrak{M}_2, a_2$ if for any $\varphi \in \mathcal{L}$, $\mathfrak{M}_1, a_1 \Vdash \varphi$ implies $\mathfrak{M}_2, a_2 \Vdash \varphi$.
2. $\mathfrak{M}_1, x_1 \rightsquigarrow_X \mathfrak{M}_2, x_2$ if for any $\varphi \in \mathcal{L}$, $\mathfrak{M}_1, x_1 \succ \varphi$ implies $\mathfrak{M}_2, x_2 \succ \varphi$.
3. $\mathfrak{M}_1, a_1 \leftsquigarrow_A \mathfrak{M}_2, a_2$ if $\mathfrak{M}_2, a_2 \rightsquigarrow_A \mathfrak{M}_1, a_1$.
4. $\mathfrak{M}_1, x_1 \leftsquigarrow_X \mathfrak{M}_2, x_2$ if $\mathfrak{M}_2, x_2 \rightsquigarrow_X \mathfrak{M}_1, x_1$.
5. $\mathfrak{M}_1, a_1 \leftrightsquigarrow_A \mathfrak{M}_2, a_2$ if $\mathfrak{M}_1, a_1 \rightsquigarrow_A \mathfrak{M}_2, a_2$ and $\mathfrak{M}_1, a_1 \leftsquigarrow_A \mathfrak{M}_2, a_2$.
6. $\mathfrak{M}_1, x_1 \leftrightsquigarrow_X \mathfrak{M}_2, x_2$ if $\mathfrak{M}_1, x_1 \rightsquigarrow_X \mathfrak{M}_2, x_2$ and $\mathfrak{M}_1, x_1 \leftsquigarrow_X \mathfrak{M}_2, x_2$.

For ease of notation, we will use $a_1 \rightsquigarrow_A a_2$ instead of $\mathfrak{M}_1, a_1 \rightsquigarrow_A \mathfrak{M}_2, a_2$ when $\mathfrak{M}_1$ and $\mathfrak{M}_2$ are clear from the context. We use similar shorthands for the other relations defined in Definition 3.1. The following definition generalizes

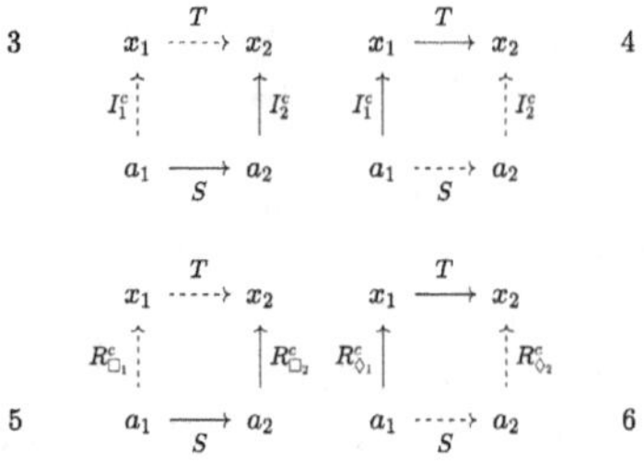

Fig. 1. The 'back and forth' conditions of a simulation (S, T)

simulations and bisimulations on Kripke models to LE-models.

Definition 3.2 Let $\mathfrak{M}_1 = (\mathfrak{F}_1, V_1)$ and $\mathfrak{M}_2 = (\mathfrak{F}_2, V_2)$ be any LE-models where $\mathfrak{F}_1 = (A_1, X_1, I_1, R_{\Box 1}, R_{\Diamond 1})$ and $\mathfrak{F}_2 = (A_2, X_2, I_2, R_{\Box 2}, R_{\Diamond 2})$. A *simulation* from $\mathfrak{M}_1$ to $\mathfrak{M}_2$ is a pair of relations (S, T) such that $S \subseteq A_1 \times A_2$ and $T \subseteq X_1 \times X_2$ satisfying the following conditions: for any $a_1 \in A_1$, $x_1 \in X_1$, $a_2 \in A_2$ and $x_2 \in X_2$,

1. If $a_1 S a_2$, then $\forall p \in \mathcal{V}$, if $a_1 \in [\![V_1(p)]\!]$, then $a_2 \in [\![V_2(p)]\!]$.
2. If $x_1 T x_2$, then $\forall p \in \mathcal{V}$, if $x_2 \in (\![V_2(p)]\!)$, then $x_1 \in (\![V_1(p)]\!)$.
3. If $a_1 S a_2$ and $a_2 I_2^c x_2$, then $\exists x_1 \in X_1$ such that $a_1 I_1^c x_1$ and $x_1 T x_2$.
4. If $x_1 T x_2$ and $a_1 I_1^c x_1$, then $\exists a_2 \in A_2$ such that $a_2 I_2^c x_2$ and $a_1 S a_2$.
5. If $a_1 S a_2$ and $a_2 R_{\Box_2}^c x_2$, then $\exists x_1 \in X_1$ such that $a_1 R_{\Box_1}^c x_1$ and $x_1 T x_2$.
6. If $x_1 T x_2$ and $x_1 R_{\Diamond_1}^c a_1$, then $\exists a_2 \in A_2$ such that $x_2 R_{\Diamond_2}^c a_2$ and $a_1 S a_2$.

The conditions 3-6 are called 'back and forth' conditions of the simulation, which can be represented by using diagrams in Figure 1. We write $\mathfrak{M}_1, a_1 \rightrightarrows \mathfrak{M}_2, a_2$ (resp. $\mathfrak{M}_1, x_1 \rightrightarrows \mathfrak{M}_2, x_2$) if there exists a simulation (S, T) from $\mathfrak{M}_1$ to $\mathfrak{M}_2$ such that $a_1 S a_2$ (resp. $x_1 T x_2$), and $\mathfrak{M}_1, a_1 \leftleftarrows \mathfrak{M}_2, a_2$ (resp. $\mathfrak{M}_1, x_1 \leftleftarrows \mathfrak{M}_2, x_2$) if $\mathfrak{M}_2, a_2 \rightrightarrows \mathfrak{M}_1, a_1$ (resp. $\mathfrak{M}_2, x_2 \rightrightarrows \mathfrak{M}_1, x_1$). A *bisimulation* between $\mathfrak{M}_1$ and $\mathfrak{M}_2$ is simulation (S, T) from $\mathfrak{M}_1$ to $\mathfrak{M}_2$, such that (S^{-1}, T^{-1}) is a simulation from $\mathfrak{M}_2$ to $\mathfrak{M}_1$.

Remark 3.3 The definition of bisimilarity implied by the above definition of bisimulation does not give us desired properties like the Hennessy-Milner theorem for polarity-based semantics. This will be illustrated in Example 3.8. Hence, a modified definition of bisimilarity is given in Definition 3.9 which allows us to prove Hennessy-Milner theorem.

This definition naturally generalizes sumulations and bisimulations on the Kripke models in the following sense. In [6, Section 3.3], it was showed that LE-models can be seen as a generalization of Kripke models by associating any Kripke model $\mathcal{M} = (W, R, V)$ with an LE-model $\mathrm{LE}(\mathcal{M}) = (W_A, W_X, \neq, R_\Box, R_\Diamond, V')$, where $W_A = W = W_B$ and $R_\Box = R_\Diamond \subseteq W \times W$ such that $w_1 R_\Box w_2$ iff $w_1 R^c w_2$ iff $w_1 R_\Diamond w_2$, and for any $p \in \mathcal{V}$, $V'(p) = (V(p), V(p)^c)$. This construction ensures that for any points in two Kripke models $\mathcal{M}$ and $\mathcal{M}'$ are modally equivalent, their liftings in the LE-models $\mathrm{LE}(\mathcal{M})$ and $\mathrm{LE}(\mathcal{M}')$ are modally equivalent w.r.t. the language $\mathcal{L}$. For more details, see [6]. The following lemma shows that any simulation (resp. bisimulation) on a Kripke model $\mathcal{M}$ naturally induces a simulation (resp. bisimulation) on LE($\mathcal{M}$).

Lemma 3.4 *Let $\mathcal{M}_1 = (W_1, R_1, V_1)$ and $\mathcal{M}_2 = (W_2, R_2, V_2)$ be any Kripke models,* LE($\mathcal{M}_1$) $= (W_{A_1}, W_{X_1}, \neq, R_{\Box_1}, R_{\Diamond_1}, V_1')$ *and* LE($\mathcal{M}_2$) $= (W_{A_2}, W_{X_2}, \neq, R_{\Box_2}, R_{\Diamond_2}, V_2')$ *be their corresponding* LE*-models. A relation $Z \subseteq W_1 \times W_2$ between the Kripke models $\mathcal{M}_1$ and $\mathcal{M}_2$ is a simulation (resp. bisimulation) iff the tuple (Z_A, Z_X), where $Z_A = Z = Z_X \subseteq W_{A_1} \times W_{A_2} = W_{X_1} \times W_{X_2}$ is a simulation (resp. bisimulation) between the* LE*-models* LE($\mathcal{M}_1$) *and* LE($\mathcal{M}_2$).

Proof. The proof follows straight-forwardly from the definitions of simulations and bisimulations on Kripke models and LE-models. □

The following theorem shows that the $\mathcal{L}$-formulas are preserved and reflected by simulations w.r.t. modal satisfaction relations $\Vdash$ and $\succ$, respectively.

Theorem 3.5 *Let $\mathfrak{M}_1 = (\mathfrak{F}_1, V_1)$ and $\mathfrak{M}_2 = (\mathfrak{F}_2, V_2)$ be any* LE*-models, a_1, x_1 in $\mathfrak{M}_1$ and a_2, x_2 in $\mathfrak{M}_2$, then*

1. *If $\mathfrak{M}_1, a_1 \rightrightarrows \mathfrak{M}_2, a_2$, then $\mathfrak{M}_1, a_1 \rightsquigarrow_A \mathfrak{M}_2, a_2$.*
2. *If $\mathfrak{M}_1, x_1 \rightrightarrows \mathfrak{M}_2, x_2$, then $\mathfrak{M}_1, x_1 \leftsquigarrow_X \mathfrak{M}_2, x_2$.*

Proof. We give a proof by induction on the complexity of the formula φ. Without loss of generality, let (S, T) be a simulation from $\mathfrak{M}_1$ to $\mathfrak{M}_2$. The initial step is when $\varphi \in \mathcal{V}$ is a propositional variable. For the first item, assume $a_1 S a_2$, then $a_1 \rightsquigarrow_A a_2$ follows directly from the item 1 of Definition 3.2. For the second item, assume $x_1 T a_2$, then $x_2 \rightsquigarrow_X x_1$ follows directly from the item 2 of Definition 3.2. Now, we consider induction step for all the connectives.

(1) Suppose φ is $\psi_1 \wedge \psi_2$.

To prove item 1, suppose $a_1 S a_2$, and $\mathfrak{M}_1, a_1 \Vdash \psi_1 \wedge \psi_2$. Then, $\mathfrak{M}_1, a_1 \Vdash \psi_1$ and $\mathfrak{M}_1, a_1 \Vdash \psi_2$, which by induction hypothesis implies $\mathfrak{M}_2, a_2 \Vdash \psi_1$ and $\mathfrak{M}_2, a_2 \Vdash \psi_2$, which implies $\mathfrak{M}, a_2 \Vdash \psi_1 \wedge \psi_2$.

To prove item 2, suppose $x_1 T x_2$ and $\mathfrak{M}_1, x_1 \not\succ \psi_1 \wedge \psi_2$. Then, there exists $a_1 \in A_1$ such that $a_1 I_1^c x_1$ and $\mathfrak{M}_1, a_1 \Vdash \psi_1 \wedge \psi_2$. Then, by item 4 of Definition 3.2, there exists $a_2 \in A_2$ such that $a_2 I_2^c x_2$ and $a_1 S a_2$. Therefore, $\mathfrak{M}_2, a_2 \Vdash \psi_1 \wedge \psi_2$, which implies $\mathfrak{M}_2, x_2 \not\succ \psi_1 \wedge \psi_2$.

(2) Suppose φ is $\psi_1 \vee \psi_2$. The proof is dual to the previous case.

(3) Suppose $\varphi = \Box\psi$.

To prove item 1, assume $a_1 S a_2$, and $\mathfrak{M}_2, a_2 \nVdash \Box\psi$. Then, there exists $x_2 \in X_2$ such that $\mathfrak{M}_2, x_2 \succ \psi$ and $a_2 R^c_{\Box_2} x_2$. By item 5 Definition 3.2, there exists $x_1 \in X_1$ such that $a_1 R^c_{\Box_1} x_1$, and $x_1 T x_2$. Therefore, by induction hypothesis $\mathfrak{M}_1, x_1 \succ \psi$. Hence, $\mathfrak{M}_1, a_1 \not\succ \Box\psi$.

To prove item 2, suppose $x_1 T x_2$, and $\mathfrak{M}_1, x_1 \not\succ \Box\psi$. Then, there exists $a_1 \in A_1$ such that $a_1 I_1^c x_1$ and $\mathfrak{M}_1, a_1 \Vdash \Box\psi$. By item 4 of Definition 3.2, there exists $a_2 \in A_2$ such that $a_2 I_2^c x_2$, and $a_1 S a_2$. Therefore, $\mathfrak{M}_2, a_2 \Vdash \Box\psi$, which implies $\mathfrak{M}_2, x_2 \not\succ \Box\psi$.

(4) Suppose φ is $\Diamond\psi$. In this case, the proof is dual to the previous case.

This concludes the proof.

□

As a corollary, we get that the $\mathcal{L}$-formulas are bisimulation-invariant.

Corollary 3.6 *Let* $\mathfrak{M}_1 = (\mathfrak{F}_1, V_1)$ *and* $\mathfrak{M}_2 = (\mathfrak{F}_2, V_2)$ *be any* LE*-models, where* $\mathfrak{F}_1 = (A_1, X_1, I_1, R_{\Box 1}, R_{\Diamond 1})$ *and* $\mathfrak{F}_2 = (A_2, X_2, I_2, R_{\Box 2}, R_{\Diamond 2})$*, and* (S,T) *be a bisimulation between them. Then*

1. *For any* $a_1 \in A_1$ *and* $a_2 \in A_2$*,* $a_1 S a_2$ *implies* $\mathfrak{M}_1, a_1 \leftrightsquigarrow_A \mathfrak{M}_2, a_2$.
2. *For any* $x_1 \in X_1$ *and* $x_2 \in X_2$*,* $x_1 T x_2$ *implies* $\mathfrak{M}_1, x_1 \leftrightsquigarrow_X \mathfrak{M}_2, x_2$.

Proof. Note that by definition for any bisimulation (S,T), both (S,T) and (S^{-1}, T^{-1}) are simulations. The proof follows immediately by Theorem 3.5. □

We now try to prove the Hennessy-Milner theorem for the polarity-based semantics of non-distributive modal logic. We begin by generalizing the notion of image-finite models from Kripke semantics to polarity-based semantics.

Definition 3.7 An LE-model $\mathfrak{M} = (A, X, I, R_\Box, R_\Diamond, V)$ is *image-finite* if for any $a \in A$ and $x \in X$, the sets $\{a' \mid a' I^c x\}, \{x' \mid a I^c x'\}, \{a' \mid a' R_\Box^c x\}$ and $\{x' \mid x' R_\Diamond^c a\}$ are all finite.

Note that a Kripke model $\mathcal{M}$ is image-finite iff LE($\mathcal{M}$) is image-finite. Thus, this definition can be seen as a generalization of image-finite models to polarity-based setting. However, as the language of non-distributive modal logic ($\mathcal{L}$) does not contain negation, the proof strategy for the classical Hennessy-Milner theorem cannot be adopted in the non-distributive setting. In fact, as the following example shows, Hennessy-Milner theorem does not hold for the polarity-based semantics of non-distributive modal logic for the notion of bisimulation given by Definition 3.2.

Example 3.8 Let $\mathfrak{M}_i = (A_i, X_i, I_i, R_{\Box i}, R_{\Diamond i}, V_i)$ for $i \in \{1,2\}$ be two image-finite models such that $A_1 = \{a_1, b_1\}$, $X_1 = \{x_1, y_1\}$, $I_1 = \{(b_1, x_1)\}$, $A_2 = \{a_2\}$, $X_2 = \{x_2\}$ and $I_2 = R_{\Box i} = R_{\Diamond i} = \emptyset$ for $i \in \{1,2\}$. Let p, q be any two fixed propositions. Suppose V_1 and V_2 are such that, $V_1(q) = (\{b_1\}, \{x_1\})$, $p \in \mathcal{V}$, $V_1(p) = (\{a_1, b_1\}, \emptyset)$, $V_2(q) = (\emptyset, \{x_2\})$ and $V_2(p) = (\{a_2\}, \emptyset)$. These models are depicted in Figure 2. It is easy to check that $a_1 \leftrightsquigarrow_A a_2$.

Suppose there exists a bisimulation (S,T) between $\mathfrak{M}_1$ and $\mathfrak{M}_2$, such that, $a_1 S a_2$. Then, by item 3 of Definition 3.2 for simulation (S^{-1}, T^{-1}) and $a_1 I_1^c y_1$, we must have $y_1 T x_2$. However, $\mathfrak{M}_1, y_1 \not\succ q$ and $\mathfrak{M}_2, x_2 \succ q$, which contradicts item 2 of Definition 3.2 for simulation (S,T). Hence, there can not be a bisimulation between a_1 and a_2.

Therefore, to obtain the Hennessy-Milner theorem for polarity-based semantics, we introduce the notion of *bisimilarity*, which is founded on the definition of simulations between LE-models. This notion naturally generalizes the definition of bisimulation outlined in Definition 3.2.

Definition 3.9 Let $\mathfrak{M}_1 = (\mathfrak{F}_1, V_1)$ and $\mathfrak{M}_2 = (\mathfrak{F}_2, V_2)$ be any LE-models, where $\mathfrak{F}_1 = (A_1, X_1, I_1, R_{\Box 1}, R_{\Diamond 1})$, $\mathfrak{F}_2 = (A_2, X_2, I_2, R_{\Box 2}, R_{\Diamond 2})$. For any a_1, x_1 in $\mathfrak{M}_1$ and a_2, x_2 in $\mathfrak{M}_2$, we say that a_1 and a_2 (resp. x_1 and x_2) are *bisimilar*,

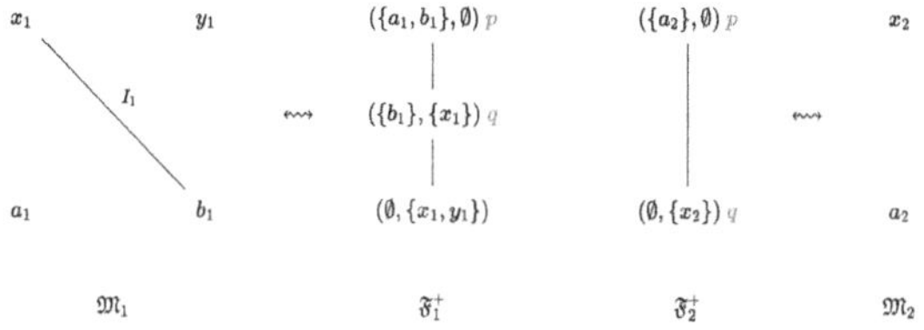

Fig. 2. Models $\mathfrak{M}_1$ and $\mathfrak{M}_2$ and their complex algebras

and denote it by $\mathfrak{M}_1, a_1 \rightleftarrows \mathfrak{M}_2, a_2$ (resp. $\mathfrak{M}_1, x_1 \rightleftarrows \mathfrak{M}_2, x_2$), if both $\mathfrak{M}_1, a_1 \rightrightarrows \mathfrak{M}_2, a_2$ and $\mathfrak{M}_1, a_1 \leftleftarrows \mathfrak{M}_2, a_2$ (resp. $\mathfrak{M}_1, x_1 \rightrightarrows \mathfrak{M}_2, x_2$ and $\mathfrak{M}_1, x_1 \leftleftarrows \mathfrak{M}_2, x_2$).

Henceforth, when we say that two elements are bisimilar, we refer to Definition 3.9, unless stated otherwise. We now show that even though Hennessy-Milner theorem for bisimulations does not generalize to the polarity-based setting, its counterpart for simulations does so.

Theorem 3.10 (Hennesy-Milner Theorem) *Let* $\mathfrak{M}_1 = (\mathfrak{F}_1, V_1)$ *and* $\mathfrak{M}_2 = (\mathfrak{F}_2, V_2)$ *be any image-finite* LE-*models,* a_1, x_1 *in* $\mathfrak{M}_1$ *and* a_2, x_2 *in* $\mathfrak{M}_2$*, then*

1. $\mathfrak{M}_1, a_1 \rightsquigarrow_A \mathfrak{M}_2, a_2$ *if and only if* $\mathfrak{M}_1, a_1 \rightrightarrows \mathfrak{M}_2, a_2$.
2. $\mathfrak{M}_1, x_1 \rightsquigarrow_X \mathfrak{M}_2, x_2$ *if and only if* $\mathfrak{M}_1, x_1 \leftleftarrows \mathfrak{M}_2, x_2$.

Proof. The right to left implication for both items follows from Theorem 3.5. We prove the left to right implication only for the first item, as the proof for the second is analogous.

For the left to right implication of the first item, we claim that $(\rightsquigarrow_A, \leftsquigarrow_X)$ is a simulation from $\mathfrak{M}_1$ to $\mathfrak{M}_2$, where $(a, a') \in \rightsquigarrow_A$ (denoted as $a \rightsquigarrow_A a'$) iff $\mathfrak{M}_1, a \rightsquigarrow_A \mathfrak{M}_2, a'$ and $(x, x') \in \leftsquigarrow_X$ (denoted as $x \leftsquigarrow_X x'$) iff $\mathfrak{M}_1, x \leftsquigarrow_X \mathfrak{M}_2, x'$. Assume $a_1 \rightsquigarrow_A a_2$ and $x_1 \leftsquigarrow_X x_2$, the items 1 and 2 of Definition 3.2 are satisfied immediately.

For item 3, assume $a_1 \rightsquigarrow_A a_2$ and $a_2 I_2^c x_2$. Therefore, $\mathfrak{M}_2, a_2 \not\Vdash \bot$. By definition of $\rightsquigarrow_A$, we have $\mathfrak{M}_1, a_1 \not\Vdash \bot$, which means that there exists x_1 such that $a_1 I_1^c x_1$. Therefore, the set $Y' := \{x \mid a_1 I_1^c x\}$ is a non-empty finite (by image-finiteness) set . Let Y' be the set $\{x'_1, \cdots, x'_n\}$. Assume for any $x \in Y'$, it is not the case that $x \leftsquigarrow_X x_2$. So, there exist finitely many formulas $\psi_1, \cdots, \psi_n$ such that for each i, $\mathfrak{M}_2, x_2 \succ \psi_i$ and $\mathfrak{M}_1, x'_i \not\succ \psi_i$. As $a_2 I_2^c x_2$, there is $\mathfrak{M}_2, a_2 \not\Vdash \psi_1 \vee \cdots \vee \psi_n$. On the other hand, let x' be any element in X_1. Thus, $\mathfrak{M}_1, x' \succ \psi_1 \vee \cdots \vee \psi_n$ implies $M_1, x' \succ \psi_i$ for each i, which means that $x' \notin Y'$, so $a_1 I_1 x'$. Hence, there is $\mathfrak{M}_1, a_1 \Vdash \psi_1 \vee ... \vee \psi_n$, which contradicts $a_1 \rightsquigarrow_A a_2$. So, there must exist x_1 such that $a_1 I_1^c x_1$ and $x_1 \leftsquigarrow_X x_2$.

Item 4 can be proved similarly to item 3.

For item 5, assume $a_1 \rightsquigarrow_A a_2$ and $a_2 R_{\Box_2}^c x_2$. Therefore, $\mathfrak{M}_2, a_2 \not\Vdash \Box\bot$, which implies $\mathfrak{M}_1, a_1 \not\Vdash \Box\bot$. Therefore, there exists $x_1 \in X_1$ such that $a_1 R_{\Box_1}^c x_1$. Thus, the set $Y' := \{x \mid a_1 R_{\Box_1}^c x\}$ is a non-empty finite (by image-finiteness) set. Let Y' be the set $\{x'_1, \cdots, x'_n\}$. Assume for any $x' \in Y'$, it is not the case that $x' \leftsquigarrow_X x_2$. Therefore, there exist finitely many formulas $\psi_1, \cdots, \psi_n$ such

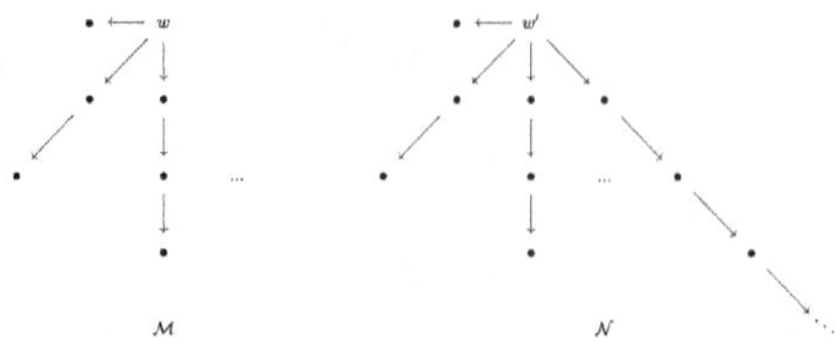

Fig. 3. Kripke models $\mathcal{M}$ and $\mathcal{N}$

that for each i, $\mathfrak{M}_2, x_2 \succ \psi_i$ and $\mathfrak{M}_1, x_i' \not\succ \psi_i$. Therefore, $\mathfrak{M}_2, x_2 \succ \psi_1 \vee \cdots \vee \psi_n$, which implies $\mathfrak{M}_2, a_2 \not\Vdash \Box(\psi_1 \vee \cdots \vee \psi_n)$. On the other hand, let y' be any element in X_1. $\mathfrak{M}_1, y' \succ \psi_1 \vee \cdots \vee \psi_n$ implies $\mathfrak{M}_1, y' \Vdash \psi_i$ for each i, which means $y' \notin Y'$, so $a_1 R_{\Box_1} y'$. Therefore, $\mathfrak{M}_1, a_1 \Vdash \Box(\psi_1 \vee \cdots \vee \psi_n)$, which contradicts $a_1 \rightsquigarrow_A a_2$. So, there must exist x_1 such that $a_1 R^c_{\Box_1} x_1$ and $x_1 \leftsquigarrow_X x_2$.

Item 6 can be proved similarly to item 5.

□

The above theorem can be seen as a generalization of Henessey-Milner theorem to non-distributive setting as it relates invariance under formulas to simulations. The following corollary is immediate from the above theorem.

Corollary 3.11 *Let* $\mathfrak{M}_1 = (\mathfrak{F}_1, V_1)$ *and* $\mathfrak{M}_2 = (\mathfrak{F}_2, V_2)$ *be any image-finite* LE-*models,* a_1, x_1 *in* $\mathfrak{M}_1$ *and* a_2, x_2 *in* $\mathfrak{M}_2$*, then*

1. $\mathfrak{M}_1, a_1 \leftrightsquigarrow_A \mathfrak{M}_2, a_2$ *if and only if* $\mathfrak{M}_1, a_1 \rightleftarrows \mathfrak{M}_2, a_2$.

2. $\mathfrak{M}_1, x_1 \leftrightsquigarrow_X \mathfrak{M}_2, x_2$ *if and only if* $\mathfrak{M}_1, x_1 \rightleftarrows \mathfrak{M}_2, x_2$.

As the simulations on LE-models can be seen as a generalizations of simulations on Kripke models, the converse of Theorem 3.5 does not always hold. That is, image-finiteness condition in Theorem 3.10 can not be dropped.

Example 3.12 Consider two Kripke models $\mathcal{M}$ and $\mathcal{N}$ in Figure 3, where any $p \in \mathcal{V}$ is false at all the points. It is easy to check that w and w' are modally equivalent in $\mathcal{M}$ and $\mathcal{N}$. Therefore ${w'}_A \rightsquigarrow_A w_A$ on their corresponding LE-models LE($\mathcal{N}$) and LE($\mathcal{M}$). Assume that there exists a simulation (S, T) from LE($\mathcal{N}$) to LE($\mathcal{M}$) such that ${w'}_A S w_A$. Then, S must be a simulation from $\mathcal{N}$ to $\mathcal{M}$. However, it is not possible to have simulations from $\mathcal{N}$ to $\mathcal{M}$, so there is a contradiction. Therefore, there is no simulation from LE($\mathcal{N}$) to LE($\mathcal{M}$) which can link ${w'}_A$ and w_A. This shows that converse of item 1 in Theorem 3.10 is not true in general. We can similarly show that converse of item 2 in Theorem 3.10 need not hold in general.

4 Modal Saturation via Filter-Ideal Extensions

In this section, we continue to investigate the relation between modal equivalence and bisimilarity. We first introduce the notion of modally-saturated (M-saturated) LE-models and the Hennessy-Milner property. Then, we show that the class of M-saturated LE-models has the Hennessy-Milner property. After that, we introduce the notion of filter-ideal extension of LE-models and show that any points in LE-models are modally equivalent if and only if they are bisimilar on their filter-ideal extensions.

4.1 Modally-saturated LE-models and Hennessy-Milner Property

Definition 4.1 Let K be a class of LE-models. K is a *Hennessy-Milner class* or has the *Hennessy-Milner property* if for any LE-models $\mathfrak{M}_1, \mathfrak{M}_2 \in$ K, and a_1, x_1 in $\mathfrak{M}_1$, and a_2, x_2 in $\mathfrak{M}_2$,

1. $\mathfrak{M}_1, a_1 \rightsquigarrow_X \mathfrak{M}_2, a_2$ implies $\mathfrak{M}_1, a_1 \rightrightarrows \mathfrak{M}_2, a_2$, and
2. $\mathfrak{M}_1, x_1 \rightsquigarrow_A \mathfrak{M}_2, x_2$ implies $\mathfrak{M}_1, x_1 \leftleftarrows \mathfrak{M}_2, x_2$.

Theorem 3.10 shows that image-finite LE-models have the Hennessy-Milner property; we now introduce a larger class of models which retains the property.

Definition 4.2 Let $\mathfrak{M} = (A, X, I, R_\Box, R_\Diamond, V)$ be an LE-model, $A' \subseteq A$ and $X' \subseteq X$. Let Σ be a set of $\mathcal{L}$-formulas. Σ is *satisfiable* in A' (resp. X') if for any $\varphi \in \Sigma$, there exists a point a (resp. x) such that, $\mathfrak{M}, a \Vdash \varphi$ (resp. $\mathfrak{M}, x \succ \varphi$). Σ is *finitely satisfiable* in A' (resp. X') if any finite subset of Σ is satisfiable in A' (resp. X'). $\mathfrak{M}$ is *modally-saturated* (M-saturated for short) if for any set of formulas Σ, $a \in A$ and $x \in X$ the following conditions hold:

1. If Σ is finitely satisfiable in $\{x' \mid aI^c x'\}$, then Σ is satisfiable in $\{x' \mid aI^c x'\}$.
2. If Σ is finitely satisfiable in $\{a' \mid a'I^c x\}$, then Σ is satisfiable in $\{a' \mid a'I^c x\}$.
3. If Σ is finitely satisfiable in $\{x' \mid aR^c_\Box x'\}$, then Σ is satisfiable in $\{x' \mid aR^c_\Box x'\}$.
4. If Σ is finitely satisfiable in $\{a' \mid xR^c_\Diamond a'\}$, then Σ is satisfiable in $\{a' \mid xR^c_\Diamond a'\}$.

Theorem 4.3 *The class of* M*-saturated models has Hennessy-Milner property.*

Proof. Let $\mathfrak{M}_1$ and $\mathfrak{M}_2$ be two M-saturated models, and let $a_1, x_1 \in \mathfrak{M}_1$ and $a_2, x_2 \in \mathfrak{M}_2$. We only prove the first item of Definition 4.1, and the proof of the second item is similar. Assume that $a_1 \rightsquigarrow_A a_2$, we claim that $(\rightsquigarrow_A, \leftsquigarrow_X)$ is a simulation from $\mathfrak{M}_1$ to $\mathfrak{M}_2$, where $(a, a') \in \rightsquigarrow_A$ (denoted as $a \rightsquigarrow_A a'$) iff $\mathfrak{M}_1, a \rightsquigarrow_A \mathfrak{M}_2, a'$ and $(x, x') \in \leftsquigarrow_X$ (denoted as $x \leftsquigarrow_X x'$) iff $\mathfrak{M}_1, x \leftsquigarrow_X \mathfrak{M}_2, x'$. Hence, it is sufficient to show that $(\rightsquigarrow_A, \leftsquigarrow_X)$ satisfies item 1-6 of Definition 3.2.

Items 1 and 2 are easy to check. To prove item 3, assume that $a'_1 \rightsquigarrow_A a'_2$ and $a'_2 I^c_2 x'_2$. Let Σ be the set of all the formulas satisfied at x'_2. Therefore, for any finite subset Δ of Σ, $\mathfrak{M}_2, x'_2 \succ \bigvee \Delta$. Therefore, $\mathfrak{M}_2, a'_2 \nVdash \bigvee \Delta$, which by $a'_1 \rightsquigarrow_A a'_2$ implies $\mathfrak{M}_1, a'_1 \nVdash \bigvee \Delta$. So, there exists x'_1, such that $\mathfrak{M}_1, x'_1 \succ \bigvee \Delta$ and $a'_1 I^c_1 x'_1$. Hence, Σ is finitely satisfiable in $\{x'_1 \mid a'_1 I^c_1 x'_1\}$. Therefore, by M-saturation of $\mathfrak{M}_1$, Σ is satisfiable in $\{x'_1 \mid a'_1 I^c_1 x'_1\}$. Therefore, there exists x_Σ such that $a'_1 I^c_1 x_\Sigma$ and all formulas in Σ are satisfied at x_Σ which means $x_\Sigma \leftsquigarrow_X x'_2$. The item 4 is proved similarly to item 3.

For item 5, assume that $a'_1 \rightsquigarrow_A a'_2$ and $a'_2 R^c_{\Box_2} x'_2$. Let Σ be the set of all the formulas satisfied at x'_2. Therefore, for any finite subset Δ of Σ, $\mathfrak{M}_2, x'_2 \succ \bigvee \Delta$. Therefore $\mathfrak{M}_2, a'_2 \nVdash \Box(\bigvee \Delta)$, which by $a'_1 \rightsquigarrow_A a'_2$ implies $\mathfrak{M}_1, a'_1 \nVdash \Box(\bigvee \Delta)$. So, there exists x'_1, such that $\mathfrak{M}_1, x'_1 \succ \bigvee \Delta$ and $a'_1 R^c_{\Box_1} x'_1$. Hence, Σ is finitely satisfiable in $\{x'_1 \mid a'_1 R^c_{\Box_1} x'_1\}$. Therefore, by M-saturation of $\mathfrak{M}_1$, Σ is satisfiable in $\{x'_1 \mid a'_1 R^c_{\Box_1} x'_1\}$. Therefore, there exists x_Σ such that $a'_1 R^c_{\Box_1} x_\Sigma$ and all formulas in Σ are satisfied at x_Σ, which means $x_\Sigma \leftsquigarrow_X x'_2$. The item 6 is proved similarly to item 5. This concludes the proof. □

4.2 Filter-Ideal Extension of LE-Models

Let $\mathbf{A} = (\mathbf{L}, \Box, \Diamond)$ be any LE-algebra (see [10, Definition 1]). $\mathbf{A}_+ = (\mathsf{Fi}_\mathbf{A}, \mathsf{Id}_\mathbf{A}, I, R_\Box, R_\Diamond)$ is the *filter-ideal frame* of $\mathbf{A}$ (c.f. [10, Definition 29]), where $\mathsf{Fi}_\mathbf{A}$ (resp. $\mathsf{Id}_\mathbf{A}$) is the set of all filters (resp. ideals) of $\mathbf{L}$ such that for any $F \in \mathsf{Fi}_\mathbf{A}$ and $J \in \mathsf{Id}_\mathbf{A}$,

1. FIJ if and only if $F \cap J \neq \emptyset$.
2. $FR_\Box J$ if and only if there exists $a \in J$ such that $\Box a \in F$.
3. $JR_\Diamond F$ if and only if there exists $a \in F$ such that $\Diamond a \in J$.

Definition 4.4 Let $\mathfrak{M} = (\mathfrak{F}, V)$ be any LE-model with $\mathfrak{F} = (A, X, I, R_\Box, R_\Diamond)$. The *filter-ideal extension* of $\mathfrak{M}$ is the tuple $\mathfrak{M}^{\mathsf{FI}} = (\mathfrak{F}^{\mathsf{FI}}, V^{\mathsf{FI}})$, where $\mathfrak{F}^{\mathsf{FI}} = (\mathfrak{F}^+)_+$ and for any $p \in \mathcal{V}$, $V^{\mathsf{FI}}(p) = ([\![V^{\mathsf{FI}}(p)]\!], (\![V^{\mathsf{FI}}(p)]\!)) = (\{F \mid ([\![p]\!], (\![p]\!)) \in F\}, \{J \mid ([\![p]\!], (\![p]\!)) \in J\})$.

Remark 4.5 In order to check that $\mathfrak{M}^{\mathsf{FI}}$ is well defined we need to check that for any $p \in \mathcal{V}$, $([\![V^{\mathsf{FI}}(p)]\!], (\![V^{\mathsf{FI}}(p)]\!))$ indeed forms a concept of $\mathfrak{F}$. We only prove $[\![V^{\mathsf{FI}}(p)]\!]^\uparrow = (\![V^{\mathsf{FI}}(p)]\!)$ here. The proof for $(\![V^{\mathsf{FI}}(p)]\!)^\downarrow = [\![V^{\mathsf{FI}}(p)]\!]$ is similar. The right to left inclusion is trivial. As for inclusion in other direction, suppose there exists $J \in [\![V^{\mathsf{FI}}(p)]\!]^\uparrow$ and $J \notin (\![V^{\mathsf{FI}}(p)]\!)$. Thus, $([\![p]\!], (\![p]\!)) \notin J$ and for any filter F which contains $([\![p]\!], (\![p]\!))$, $F \cap J \neq \emptyset$. As $([\![p]\!], (\![p]\!)) \notin J$, by the filter-ideal theorem, there exists a filter F' such that $([\![p]\!], (\![p]\!)) \in F'$ and $F' \cap J = \emptyset$. This contradicts our assumption. So, the left to right inclusion holds.

The following lemma shows that the filter-ideal extensions of LE-models preserve modal satisfaction.

Lemma 4.6 *Let* $\mathfrak{M} = (\mathfrak{F}, V)$ *be any* LE-*model where* $\mathfrak{F} = (A, X, I, R_\Box, R_\Diamond)$, *and* $\mathfrak{M}^{\mathsf{FI}} = (\mathfrak{F}^{\mathsf{FI}}, V^{\mathsf{FI}})$ *is the filter-ideal extension of* $\mathfrak{M}$. *Let* F *be a filter of* $\mathfrak{F}^+$ *and* J *be an ideal of* $\mathfrak{F}^+$. *Then*

$$\mathfrak{M}^{\mathsf{FI}}, F \Vdash \varphi \text{ iff } ([\![\varphi]\!], (\![\varphi]\!)) \in F \quad \text{and} \quad \mathfrak{M}^{\mathsf{FI}}, J \succ \varphi \text{ iff } ([\![\varphi]\!], (\![\varphi]\!)) \in J.$$

Proof. It is straight-forward by induction on the complexity of formulas. □

For any LE-frame $\mathfrak{F}$ and $\mathrm{K} \subseteq \mathfrak{F}^+$, let $\mathsf{Fi}(\mathrm{K})$ (resp. $\mathsf{Id}(\mathrm{K})$) denote the filter (resp. the ideal) generated by K. When $\mathrm{K} = \{k\} \subseteq \mathfrak{F}^+$, let $\mathsf{Fi}(k)$ (resp. $\mathsf{Id}(k)$) denote the principal filter (resp. principal ideal) $\mathsf{Fi}(\{k\})$ (resp. $\mathsf{Id}(\{k\})$).

Corollary 4.7 (truth lemma) *Let* $\mathfrak{M} = (\mathfrak{F}, V)$ *be any* LE-*model where* $\mathfrak{F} = (A, X, I, R_\Box, R_\Diamond)$, *and* $\mathfrak{M}^{\mathsf{FI}}$ *be the filter-ideal extension of* $\mathfrak{M}$. *For all* $\varphi \in \mathcal{L}$,

1. *For any* $a \in A$, $\mathfrak{M}, a \Vdash \varphi$ *if and only if* $\mathfrak{M}^{\mathsf{FI}}, \mathsf{Fi}(\mathbf{a}) \Vdash \varphi$, *where* $\mathbf{a} = (a^{\uparrow\downarrow}, a^\uparrow)$.
2. *For any* $x \in X$, $\mathfrak{M}, x \succ \varphi$ *if and only if* $\mathfrak{M}^{\mathsf{FI}}, \mathsf{Id}(\mathbf{x}) \succ \varphi$, *where* $\mathbf{x} = (x^\downarrow, x^{\downarrow\uparrow})$.

Proof.

$$\begin{array}{rclcrcl}
\mathfrak{M}, a \Vdash \varphi & \Leftrightarrow & \mathbf{a} \leq ([\![\varphi]\!], (\![\varphi]\!)) & \quad & \mathfrak{M}, x \succ \varphi & \Leftrightarrow & ([\![\varphi]\!], (\![\varphi]\!)) \leq \mathbf{x} \\
 & \Leftrightarrow & ([\![\varphi]\!], (\![\varphi]\!)) \in \mathsf{Fi}(\mathbf{a}) & & & \Leftrightarrow & ([\![\varphi]\!], (\![\varphi]\!)) \in \mathsf{Id}(\mathbf{x}) \\
 & \Leftrightarrow & \mathfrak{M}^{\mathsf{FI}}, \mathsf{Fi}(\mathbf{a}) \Vdash \varphi & & & \Leftrightarrow & \mathfrak{M}^{\mathsf{FI}}, \mathsf{Id}(\mathbf{x}) \succ \varphi
\end{array}$$

The first equivalences follow from the relationship between $\Vdash$ and $\succ$ on a polarity-based model and order $\leq$ on its complex algebra. The second equivalence follows from the definitions of $\mathsf{Fi}(\mathbf{a})$ and $\mathsf{Id}(\mathbf{x})$, and the third equivalence follows from Lemma 4.6. □

The above lemma implies that for any LE-model $\mathfrak{M}$ with filter-ideal extension $\mathfrak{M}^{\mathsf{FI}}$, for any $a \in A$, and $x \in X$, $a \leftrightsquigarrow_A \mathsf{Fi}(\mathbf{a})$ and $x \leftrightsquigarrow_X \mathsf{Id}(\mathbf{x})$.

Lemma 4.8 *The filter-ideal extension of an* LE*-model is* M*-saturated.*

Proof. Let $\mathfrak{M}$ be an LE-model and $\mathfrak{M}^{\mathsf{FI}}$ be its filter-ideal extension. We only prove items 1 and 3 of Definition 4.2 hold for $\mathfrak{M}^{\mathsf{FI}}$. The proofs for items 2 and 4 can be obtained dually.

1. Let F be any filter of $\mathfrak{F}^+$. Let Σ be a set of $\mathcal{L}$-formulas which is finitely satisfiable in $\{J \mid FI^cJ\}$. By definition of filter-ideal extension this set is same as $\{J \mid F \cap J = \emptyset\}$. By Lemma 4.6, for any formula φ, $\mathfrak{M}^{\mathsf{FI}}, J \succ \varphi$ iff $([\![\varphi]\!], (\![\varphi]\!)) \in J$. Thus, Σ is finitely satisfiable in $\{J \mid FI^cJ\}$ iff for any finite $\Delta \subseteq \Sigma$, the ideal generated by the set $\{([\![\varphi]\!], (\![\varphi]\!)) \mid \varphi \in \Delta\}$ (denoted by $Id(\Delta)$) is disjoint with F. Let

$$Y = \bigcup\{Id(\Delta) \mid \Delta \subseteq \Sigma,\ \Delta \text{ is finite}\}.$$

It is clear that $F \cap Y = \emptyset$ and $([\![\varphi]\!], (\![\varphi]\!)) \in Y$ for all $\varphi \in \Sigma$. We claim that Y is an ideal of $\mathfrak{F}^+$. Suppose $c \in Y$, and $d \leq c$. By definition of Y, c must be in $Id(\Delta)$ for some finite $\Delta \subseteq \Sigma$. Then, we must have $d \in Id(\Delta) \subseteq Y$. Finally, suppose $c_1, c_2 \in Y$. Then, we must have finite $\Delta_1, \Delta_2 \subseteq \Sigma$ such that, $c_1 \in Id(\Delta_1)$ and $c_2 \in Id(\Delta_2)$. Therefore, $c_1, c_2 \in Id(\Delta_1 \cup \Delta_2)$ implying $c_1 \vee c_2 \in Id(\Delta_1 \cup \Delta_2)$. As $\Delta_1 \cup \Delta_2 \subseteq \Sigma$ is finite, we must have $Id(\Delta_1 \cup \Delta_2) \subseteq Y$. Thus, we have $c_1 \vee c_2 \in Y$. So, Y is an ideal such that FI^cY and $\mathfrak{M}^{\mathsf{FI}}, Y \succ \Sigma$. Hence proved.

3. Let $\mathfrak{M} = (\mathfrak{F}, V)$ and F be any filter of $\mathfrak{F}^+$. Let Σ be a set of formulas which is finitely satisfiable in $\{J \mid FR^c_\Box J\}$. By definition of filter-ideal extension, this set is same as $\{J \mid F \cap \Box J = \emptyset\}$, where for any $C \subseteq \mathfrak{F}^+$, $\Box C = \{\Box c \mid c \in C\}$. By Lemma 4.6, for any formula φ, $\mathfrak{M}^{\mathsf{FI}}, J \succ \varphi$ iff $([\![\varphi]\!], (\![\varphi]\!)) \in J$. Thus, Σ is finitely satisfiable in $\{J \mid FR^c_\Box J\}$ iff for any finite $\Delta \subseteq \Sigma$, the ideal generated by the set $\{([\![\varphi]\!], (\![\varphi]\!)) \mid \varphi \in \Delta\}$ (denoted by $Id(\Delta)$) satisfies the condition $\Box Id(\Delta) \cap F = \emptyset$. Let

$$Y = \bigcup\{Id(\Delta) \mid \Delta \subseteq \Sigma,\ \Delta \text{ is finite}\}.$$

It is clear that $F \cap \Box Y = \emptyset$ and $([\![\varphi]\!], (\![\varphi]\!)) \in Y$ for all $\varphi \in \Sigma$. By same proof as in item 1, we have that Y is an ideal. Therefore, Y is an ideal such that $FR^c_\Box Y$ and $\mathfrak{M}^{\mathsf{FI}}, Y \succ \Sigma$. Hence proved. □

This lemma gives us the following theorem and corollary immediately, saying filter-ideal extensions LE-models have the Hennessy-Milner property.

Theorem 4.9 *For all* LE*-models* $\mathfrak{M}_1$ *and* $\mathfrak{M}_2$, a_1, x_1 *in* $\mathfrak{M}_1$, *and* a_2, x_2 *in* $\mathfrak{M}_2$,

1. $\mathfrak{M}_1, a_1 \rightsquigarrow_A \mathfrak{M}_2, a_2$ *if and only if* $\mathfrak{M}_1^{\mathsf{FI}}, \mathsf{Fi}(\mathbf{a_1}) \rightrightarrows \mathfrak{M}_2^{\mathsf{FI}}, \mathsf{Fi}(\mathbf{a_2})$.
2. $\mathfrak{M}_1, x_1 \rightsquigarrow_X \mathfrak{M}_2, x_2$ *if and only if* $\mathfrak{M}_1^{\mathsf{FI}}, \mathsf{Id}(\mathbf{x_1}) \leftleftarrows \mathfrak{M}_2^{\mathsf{FI}}, \mathsf{Id}(\mathbf{x_2})$.

Proof. Immediate by Corollary 4.7, Lemma 4.8, and Theorem 4.3. □

Corollary 4.10 *For all* LE-*models* $\mathfrak{M}_1$ *and* $\mathfrak{M}_2$, *any* a_1, x_1 *in* $\mathfrak{M}_1$ *and any* a_2, x_2 *in* $\mathfrak{M}_2$,

1. $\mathfrak{M}_1, a_1 \leftrightsquigarrow_A \mathfrak{M}_2, a_2$ *if and only if* $\mathfrak{M}_1^{\mathsf{FI}}, \mathsf{Fi}(\mathbf{a_1}) \rightleftarrows \mathfrak{M}_2^{\mathsf{FI}}, \mathsf{Fi}(\mathbf{a_2})$.
2. $\mathfrak{M}_1, x_1 \leftrightsquigarrow_X \mathfrak{M}_2, x_2$ *if and only if* $\mathfrak{M}_1^{\mathsf{FI}}, \mathsf{Id}(\mathbf{x_1}) \rightleftarrows \mathfrak{M}_2^{\mathsf{FI}}, \mathsf{Id}(\mathbf{x_2})$.

5 Toward van Benthem Characterization Theorem

In this section, we firstly introduce the standard translation for non-distributive modal logic based on polarity-based semantics, and then define the ultrapower extension for LE-models. Following that, we introduce ω-saturated LE-models and show that ω-saturated LE-models are M-saturated. Finally, we show that for any LE-model, there exists an ultrapower extension of it which is ω-saturated. We use this result to prove the van Benthem characterization theorem of non-distributive modal logic.

5.1 Standard Translation

The following definition of standard translation for non-distributive modal logic on polarity-based semantics is inspired from [8, Section 2.1].

Definition 5.1 Let $g \in G$ and $m \in M$ be two $\mathcal{L}_1$-variables and $\varphi \in \mathcal{L}$ (see Section 2.3). The *standard translations* $ST_g : \mathcal{L} \to \mathcal{L}^1$ and $ST_m : \mathcal{L} \to \mathcal{L}^1$ are defined as follows:

$ST_g(\bot) := \forall m(gIm)$ | $ST_m(\bot) := m = m$

$ST_g(\top) := g = g$ | $ST_m(\bot) := \forall g(gIm)$

$ST_g(p) := P_A(g)$ | $ST_m(p) := P_X(m)$

$ST_g(\varphi \vee \psi) := \forall m(ST_m(\varphi) \wedge ST_m(\psi) \to gIm)$ | $ST_m(\varphi \vee \psi) := ST_m(\varphi) \wedge ST_m(\psi)$

$ST_g(\varphi \wedge \psi) := ST_g(\varphi) \wedge ST_g(\psi)$ | $ST_m(\varphi \wedge \psi) := \forall g(ST_g(\varphi) \wedge ST_g(\psi) \to$

$ST_g(\Diamond\varphi) := \forall m(ST_m(\Diamond\varphi) \to gIm)$ | $ST_m(\Diamond\varphi) := \forall g(ST_g(\varphi) \to mR_\Diamond g)$

$ST_g(\Box\varphi) := \forall m(ST_m(\varphi) \to gR_\Box m)$ | $ST_m(\Box\varphi) := \forall g(ST_g(\Box\varphi) \to gIm)$

The following lemma is immediate from the definition above.

Lemma 5.2 *Let* $\mathfrak{M} = (A, X, I, R_\Box, R_\Diamond, V)$ *be an* LE-*model, then for any* $a \in A$ *and* $x \in X$, *and any* $\mathcal{L}_1$-*variables* $g \in G$ *and* $m \in M$:

1. $\mathfrak{M}, a \Vdash \varphi$ *iff* $\mathfrak{M} \models ST_g(\varphi)[a]$ *and* $\mathfrak{M}, x \succ \varphi$ *iff* $\mathfrak{M} \models ST_m(\varphi)[x]$.
2. $\mathfrak{M}, a \Vdash \varphi$ *iff* $\mathfrak{M} \models \forall g ST_g(\varphi)$ *and* $\mathfrak{M}, x \succ \varphi$ *iff* $\mathfrak{M} \models \forall m ST_m(\varphi)$.

Proof. It is straight-forward by induction on the complexity of formulas. □

5.2 Ultrapower Extension and ω-Saturated Model

Definition 5.3 Let K be a non-empty index set, and $\mathfrak{M} = (\mathfrak{F}, V)$ be an LE-model, where $\mathfrak{F} = (A, X, I, R_\Box, R_\Diamond)$ and for any $p \in \mathcal{V}$, $V(p) = ([\![p]\!], (\![p]\!))$. The K-*power* of $\mathfrak{M}$ is the LE-model $\mathfrak{M}^{\mathrm{K}} = (A^{\mathrm{K}}, X^{\mathrm{K}}, I^{\mathrm{K}}, R_\Box^{\mathrm{K}}, R_\Diamond^{\mathrm{K}}, V^{\mathrm{K}})$, where

1. A^{K} is the set of functions $s : \mathrm{K} \to A$.
2. X^{K} is the set of functions from $t : \mathrm{K} \to X$.
3. For any $s \in A^{\mathrm{K}}$ and $t \in X^{\mathrm{K}}$, $sI^{\mathrm{K}}t$ iff for any $k \in \mathrm{K}$, $s(k)It(k)$.
4. For any $s \in A^{\mathrm{K}}$ and $t \in X^{\mathrm{K}}$, $sR_\Box^{\mathrm{K}}t$ iff for any $k \in \mathrm{K}$, $s(k)R_\Box t(k)$.
5. For any $s \in A^{\mathrm{K}}$ and $t \in X^{\mathrm{K}}$, $tR_\Diamond^{\mathrm{K}}s$ iff for any $k \in \mathrm{K}$, $t(k)R_\Diamond s(k)$.

6. For any $p \in \mathcal{V}$, $V^{\mathrm{K}}(p) = (\llbracket p \rrbracket^{\mathrm{K}}, (\!|p|\!)^{\mathrm{K}})$ such that $s \in \llbracket p \rrbracket^{\mathrm{K}}$ iff for any $k \in \mathrm{K}$, $s(k) \in \llbracket p \rrbracket$, and $t \in (\!|p|\!)^{\mathrm{K}}$ iff for any $k \in \mathrm{K}$, $t(k) \in (\!|p|\!)$.

Let $\mathcal{U} \subseteq \mathcal{P}(\mathrm{K})$ be an ultrafilter over K. We define equivalence relations $\sim_A$ and $\sim_X$ on A^{K} and X^{K}, respectively, as follows:

1. $s_1 \sim_A s_2$ if and only if $\{k \in \mathrm{K} \mid s_1(k) = s_2(k)\} \in \mathcal{U}$.
2. $t_1 \sim_X t_2$ if and only if $\{k \in \mathrm{K} \mid t_1(k) = t_2(k)\} \in \mathcal{U}$.

The *ultrapower* of $\mathfrak{M}$ modulo K and $\mathcal{U}$ is a structure $\mathfrak{M}^{\mathrm{K}}_{\mathcal{U}} = (A^{\mathrm{K}}_{\mathcal{U}}, X^{\mathrm{K}}_{\mathcal{U}}, I^{\mathrm{K}}_{\mathcal{U}}, R^{\mathrm{K}}_{\Box\mathcal{U}}, R^{\mathrm{K}}_{\Diamond\mathcal{U}}, V^{\mathrm{K}}_{\mathcal{U}})$, where $A^{\mathrm{K}}_{\mathcal{U}}$ (resp. $X^{\mathrm{K}}_{\mathcal{U}}$) is quotient of A^{K} (resp. X^{K}) over $\sim_A$ (resp. $\sim_X$), and for any $[s] \in A^{\mathrm{K}}_{\mathcal{U}}$, $[t] \in X^{\mathrm{K}}_{\mathcal{U}}$,

1. $[s] I^{\mathrm{K}}_{\mathcal{U}} [t]$ if and only if $\{k \in \mathrm{K} \mid s(k) I t(k)\} \in \mathcal{U}$.
2. $[s] R^{\mathrm{K}}_{\Box\mathcal{U}} [t]$ if and only if $\{k \in \mathrm{K} \mid s(k) R_\Box t(k)\} \in \mathcal{U}$.
3. $[t] R^{\mathrm{K}}_{\Diamond\mathcal{U}} [s]$ if and only if $\{k \in \mathrm{K} \mid t(k) R_\Diamond s(k)\} \in \mathcal{U}$.
4. For any $p \in \mathcal{V}$, $V^{\mathrm{K}}_{\mathcal{U}}(p) = (\llbracket p \rrbracket^{\mathrm{K}}_{\mathcal{U}}, (\!|p|\!)^{\mathrm{K}}_{\mathcal{U}})$, where $[s] \in \llbracket p \rrbracket^{\mathrm{K}}_{\mathcal{U}}$ if and only if $\{k \in \mathrm{K} \mid s(k) \in \llbracket p \rrbracket\} \in \mathcal{U}$, and $[t] \in (\!|p|\!)^{\mathrm{K}}_{\mathcal{U}}$ if and only if $\{k \in \mathrm{K} \mid t(k) \in (\!|p|\!) \in \mathcal{U}\}$.

The following theorem is an immediate consequence of Łós's Theorem for first-order logic.

Theorem 5.4 (Łós's Theorem) *Let* K *be a non-empty index set and* $\mathcal{U}$ *be an ultrafilter over* K. *Let* $\mathfrak{M}$ *be an* LE*-model and* $\mathfrak{M}^{\mathrm{K}}_{\mathcal{U}}$ *be its ultrapower modulo* K *and* $\mathcal{U}$. *For any two-sorted first-order formula* $\varphi \in \mathcal{L}^1$ *(see Section 2.3),*

$$\mathfrak{M}^{\mathrm{K}}_{\mathcal{U}} \models \varphi[s,t] \quad \textit{if and only if} \quad \{k \in \mathrm{K} \mid \mathfrak{M} \models \varphi[s(k), t(k)]\} \in \mathcal{U}.$$

Remark 5.5 Any ultrapower $\mathfrak{M}^{\mathrm{K}}_{\mathcal{U}}$ of an LE-model $\mathfrak{M}$ is indeed an LE-model. In order to see this, we just need to check that relations $R^{\mathrm{K}}_{\Box\mathcal{U}}$ and $R^{\mathrm{K}}_{\Diamond\mathcal{U}}$ are I-compatible, and for any $p \in \mathcal{V}$, $\llbracket p \rrbracket^{\mathrm{K}}_{\mathcal{U}}$ and $(\!|p|\!)^{\mathrm{K}}_{\mathcal{U}}$ are Galois stable. This is true because all these conditions are first-order definable, and so they will be preserved in the ultrapower extension by Łós's Theorem.

We now define κ-saturated first-order models for any infinite cardinal κ.

Definition 5.6 Let κ be an infinite cardinal, and $\mathfrak{M}$ be an LE-model. For any set S of elements appearing in $\mathfrak{M}$, let $\mathcal{L}^1_S$ be the extension of $\mathcal{L}^1$ with constant symbols for every element of S. $\mathfrak{M}$ is *κ-saturated* if for any set of elements S appearing in $\mathfrak{M}$ with $|S| < \kappa$, and any set Σ of $\mathcal{L}^1_S$ formulas with finite many free variables, if Σ is finitely satisfiable in $\mathfrak{M}$, then Σ is satisfiable in $\mathfrak{M}$.

The following theorem follows from [5, Theorem 6.1.8].

Theorem 5.7 *For any* LE*-model* $\mathfrak{M}$, *there exist an non-empty index set* K *and an ultrafilter* $\mathcal{U}$ *over* K *such that* $\mathfrak{M}^{\mathrm{K}}_{\mathcal{U}}$ *is* ω *saturated.*

Proposition 5.8 *Any* ω*-saturated* LE*-model is* M*-saturated.*

Proof. Let $\mathfrak{M} = (\mathfrak{F}, V)$ be an ω-saturated LE-model, where $\mathfrak{F} = (A, X, I, R_\Box, R_\Diamond)$. We only prove the fist item of Definition 4.2, and other conditions can be proved similarly. Let Σ be a set of $\mathcal{L}$-formulas which are finitely satisfiable in set $\{x' \mid a I^c x'\}$. Define Σ' to be the set

$$\Sigma' = \{a I^c x\} \cup ST_x(\Sigma).$$

Clearly Σ' contains only finite many free variables (only x) and every finite subset of Σ' is satisfiable in $(\mathfrak{M}, x)_{x \in X}$. Hence, by ω-saturation Σ' is satisfiable

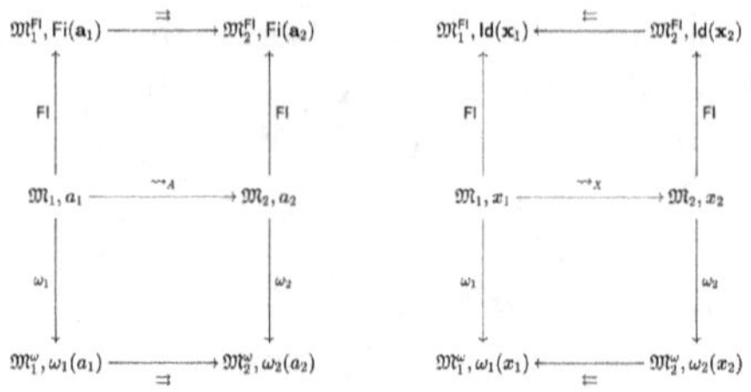

Fig. 4. Detour Lemma

in $(\mathfrak{M}, x)_{x \in X}$ at some x_0 with $aI^c x_0$. Then, by Lemma 5.2 we have $\mathfrak{M}, x_0 \succ \Sigma$. Therefore, Σ is satisfiable in $\{x' \mid aI^c x'\}$. □

It follows immediately from the above proposition and Theorem 4.3 that the class of ω-saturated LE-models satisfy Hennessy-Milner property.

5.3 The van Benthem Characterization Theorem

We are now ready to prove the van Benthem characterization theorem for non-distributive modal logic based on the polarity-based semantics, which follows from the detour lemma below.

Theorem 5.9 (Detour Lemma) *For all* LE-*models* $\mathfrak{M}_1$ *and* $\mathfrak{M}_2$, *any* a_1, x_1 *in* $\mathfrak{M}_1$ *and* a_2, x_2 *in* $\mathfrak{M}_2$,

1. $\mathfrak{M}_1, a_1 \rightsquigarrow_A \mathfrak{M}_2, a_2$ *if and only if there exist two ω-saturated* LE-*models* $\mathfrak{M}_1^{\omega}$ *and* $\mathfrak{M}_2^{\omega}$, *and two elementary embeddings* $\omega_1 : \mathfrak{M}_1 \to \mathfrak{M}_1^{\omega}$ *and* $\omega_2 : \mathfrak{M}_2 \to \mathfrak{M}_2^{\omega}$, *such that* $\mathfrak{M}_1^{\omega}, \omega_1(a_1) \rightrightarrows \mathfrak{M}_2^{\omega}, \omega_2(a_2)$.

2. $\mathfrak{M}_1, x_1 \rightsquigarrow_X \mathfrak{M}_2, x_2$ *if and only if there exist two ω-saturated* LE-*models* $\mathfrak{M}_1^{\omega}$ *and* $\mathfrak{M}_2^{\omega}$, *and two elementary embeddings* $\omega_1 : \mathfrak{M}_1 \to \mathfrak{M}_1^{\omega}$ *and* $\omega_2 : \mathfrak{M}_2 \to \mathfrak{M}_2^{\omega}$, *such that* $\mathfrak{M}_1^{\omega}, \omega_1(x_1) \leftleftarrows \mathfrak{M}_2^{\omega}, \omega_2(x_2)$.

Proof. We only prove the item 1 here, the proof of the item 2 is similar. The implication from right to left proved by contradiction. Suppose $\mathfrak{M}_1^{\omega}, \omega_1(a_1) \rightrightarrows \mathfrak{M}_2^{\omega}, \omega_2(a_2)$, and $\omega_1 : \mathfrak{M}_1 \to \mathfrak{M}_1^{\omega}$, $\omega_2 : \mathfrak{M}_2 \to \mathfrak{M}_2^{\omega}$ are two elementary embeddings. Suppose there exists $\varphi \in \mathcal{L}$, such that $\mathfrak{M}, a_1 \Vdash \varphi$, and $\mathfrak{M}, a_2 \nVdash \varphi$. Therefore, by Lemma 5.2, $\mathfrak{M}_1 \models ST_g(\varphi)[a_1]$ and $\mathfrak{M}_2 \nvDash ST_g(\varphi)[a_2]$, which implies that $\mathfrak{M}_1^{\omega} \models ST_g(\varphi)[\omega_1(a_1)]$ and $\mathfrak{M}_2^{\omega} \nvDash ST_g(\varphi)[\omega_2(a_2)]$. Therefore, by Lemma 5.2 again, $\mathfrak{M}_1^{\omega}, \omega_1(a_1) \Vdash \varphi$ and $\mathfrak{M}_2^{\omega}, \omega_2(a_2) \nVdash \varphi$. However, this contradicts $\mathfrak{M}_1^{\omega}, \omega_1(a_1) \rightrightarrows \mathfrak{M}_2^{\omega}, \omega_2(a_2)$ due to Theorem 3.10. Hence proved.

For the implication from left to right, assume that $\mathfrak{M}_1, a_1 \rightsquigarrow_A \mathfrak{M}_2, a_2$. We consider their respective ultrapower extensions $\mathfrak{M}^{\mathrm{K}}_{1\,\mathcal{U}}$, $\mathfrak{M}^{\mathrm{K}}_{2\,\mathcal{U}}$, and natural elementary embeddings $\omega_i : \mathfrak{M}_i \to \mathfrak{M}^{\mathrm{K}}_{i\,\mathcal{U}}$ where $i \in \{1, 2\}$ such that for any $a_i \in A_i$, $\omega_i(a_i) = [s^i_{a_i}]$, where $s^i_{a_i} : \mathrm{K} \to A_i$ are the constant functions which send all elements in K to a_i. Therefore, $\mathfrak{M}^{\mathrm{K}}_{1\,\mathcal{U}}, \omega_1(a_1) \rightsquigarrow_A \mathfrak{M}^{\mathrm{K}}_{2\,\mathcal{U}}, \omega_2(a_2)$. By Theorem 5.7, $\mathfrak{M}^{\mathrm{K}}_{1\,\mathcal{U}}$ and $\mathfrak{M}^{\mathrm{K}}_{2\,\mathcal{U}}$ are ω-saturated LE-models. Therefore, by Proposition 5.8 and Theorem 4.3, $\mathfrak{M}^{\mathrm{K}}_{1\,\mathcal{U}}, \omega_1(a_1) \rightrightarrows \mathfrak{M}^{\mathrm{K}}_{2\,\mathcal{U}}, \omega_2(a_2)$. □

Remark 5.10 Notice that in both items of the above lemma, "there exist" can be substituted by "for any". It is because the only if direction follows from Proposition 5.8 and the definition of elementary embeddings, and the if direction follows trivially from the existence of the ω-saturated elementary extensions.

Theorem 5.11 (van Benthem Characterization Theorem) *Let $\varphi(g)$ (resp. $\varphi(m)$) be any two sorted first-order formula in $\mathcal{L}^1$, then $\varphi(g)$ (resp. $\varphi(m)$) is preserved (resp. reflected) by simulation if and only if there exist finite many $\mathcal{L}$-formulas ψ_1, ψ_2, $\cdots$, ψ_n such that $\varphi(g)$ (resp. $\varphi(m)$) is equivalent to $\bigvee_{1\leq i\leq n} ST_g(\psi_i)$ (resp. $\bigvee_{1\leq i\leq n} ST_m(\psi_i)$).*

Proof. We only provide the proof for $\varphi(g)$, the the proof for $\varphi(m)$ is similar. The implication from right to left follows from Theorem 3.5. For the implication from left to right, assume that $\varphi(g)$ is preserved by simulation. We define a set of two sorted first-order formulas $\mathrm{JMOC}(\varphi) := \{\bigvee_{1\leq i\leq n} ST_g(\psi_i) \mid \psi_i \in \mathcal{L}, n \in \omega, \varphi(g) \models \bigvee_{1\leq i\leq n} ST_g(\psi_i)\}$. We first check that the following is true: if $\mathrm{JMOC}(\varphi) \models \varphi(g)$, then $\varphi(g)$ is equivalent to $\bigvee_j ST_g(\psi_j)$ for some finite set of $\mathcal{L}$-formulas $\{\psi_j \mid j \in \mathcal{J}\}$. Assume that $\mathrm{JMOC}(\varphi) \models \varphi(g)$. By compactness, there exists an finite set $\{\bigvee_{1\leq i\leq n} ST_g(\psi_{j,i}) \mid j \in \mathcal{J}, n \in \omega\} \subseteq \mathrm{JMOC}(\varphi)$ such that $\models \bigwedge_{j\in\mathcal{J}} \bigvee_{1\leq i\leq n} ST_g(\psi_{j,i}) \to \varphi(g)$. Therefore, we have $\models \varphi(g) \leftrightarrow \bigwedge_{j\in\mathcal{J}} \bigvee_{1\leq i\leq n} ST_g(\psi_{j,i})$. Note that $\bigwedge_{j\in\mathcal{J}} \bigvee_{1\leq i\leq n} ST_g(\psi_{j,i})$ is equivalent to $\bigvee_{1\leq i\leq n} \bigwedge_{j\in\mathcal{J}} ST_g(\psi_{j,i})$, and the latter is equivalent to $\bigvee_{1\leq i\leq n} ST_g(\bigwedge_{j\in\mathcal{J}} \psi_{j,i})$.

Therefore, it suffices to show that $\mathrm{JMOC}(\varphi) \models \varphi(g)$. Let $\mathfrak{M}$ be an LE-model such that $\mathfrak{M} \models \mathrm{JMOC}(\varphi)[a]$. Let $\mathrm{NT}(g) := \{\neg ST_g(\psi) \mid \psi \in \mathcal{L}, \mathfrak{M} \not\models ST_g(\psi)[a]\}$. We want to show that $\mathrm{NT}(g)\cup\{\varphi(g)\}$ is consistent. Suppose not. Then, by compactness there exists a finite subset $\mathrm{NT}_0(g) = \{\neg ST_g(\psi_k) \mid k \in \mathcal{K}\}$ of $\mathrm{NT}(g)$ such that $\mathrm{NT}_0(g)\cup\{\varphi(g)\}$ is inconsistent. Therefore, $\varphi(g) \to \neg \bigwedge_{k\in\mathcal{K}} \neg ST_g(\psi_i)$ is valid, which means that $\varphi(g) \to \bigvee_{k\in\mathcal{K}} ST_g(\psi_i)$ is valid. Therefore, $\bigvee_{k\in\mathcal{K}} ST_g(\psi_i) \in \mathrm{JMOC}(\varphi)$, which implies $\mathfrak{M} \models \bigvee_{k\in\mathcal{K}} ST_g(\psi_i)[a]$. However, $\mathfrak{M} \not\models ST_g(\psi_i)[a]$ for any $k \in \mathcal{K}$, which implies $\mathfrak{M} \not\models \bigvee_{k\in\mathcal{K}} ST_g(\psi_i)[a]$, so there is a contradiction. Hence, $\mathrm{NT}(g) \cup \{\varphi(g)\}$ must be consistent. Therefore, there exists an $\mathcal{L}^1$-structure $\mathfrak{N}$ and a' in $\mathfrak{N}$ such that $\mathfrak{N} \models \mathrm{NT}(g) \cup \{\varphi(g)\}[a']$.

For any formula ψ, $\mathfrak{M}, a \not\Vdash \psi$ iff $\neg ST_g(\psi) \in NT(g)$, which implies $\mathfrak{N}, a' \not\Vdash \psi$. Therefore, $\mathfrak{N}, a' \rightsquigarrow_A \mathfrak{M}, a$. By Theorem 5.9 (Detour Lemma), there exist two ω-saturated LE-models $\mathfrak{N}^\omega$ and $\mathfrak{M}^\omega$, and two elementary embeddings $\omega_n : \mathfrak{N} \to \mathfrak{N}^\omega$ and $\omega_m : \mathfrak{M} \to \mathfrak{M}^\omega$, such that $\mathfrak{N}^\omega, \omega_n(a') \rightrightarrows \mathfrak{M}^\omega, \omega_m(a)$. As ω_n is an elementary embedding and $\mathfrak{N} \models \varphi(g)[a']$, we have $\mathfrak{N}^\omega \models \varphi(g)[\omega_n(a')]$. Therefore, as $\varphi(g)$ is preserved by simulation, $\mathfrak{M}^\omega \models \varphi(g)[\omega_m(a)]$. As ω_m is an elementary embedding, we have $\mathfrak{M} \models \varphi(g)[a]$. This concludes the proof. □

6 Conclusion

In this paper, we introduced bisimulations on polarity-based semantics for non-distributive modal logic, demonstrating their application in proving the Hennessy-Milner and van Benthem theorems. These results show that the non-distributive modal logic is bisimulation invariant fragment (within a specific signature and axioms) of two-sorted first-order logic. This work suggests the following directions for future research.

Characterizing theorem for lattice-based modal μ-calculus: Lattice based modal μ-calculus, i.e. the extension of non-distributive modal logic with least and greatest fixed point operators, was defined and studied in [16]. In [16], it was showed that the formulas of modal μ-calculus are bisimulation-invariant. The classical modal μ-calculus is characterized as the bisimulation invariant fragment of monadic second-order logic. It would be interesting to see if a similar characterization holds in its lattice-based generalization.

Computational properties of bisimulations: In classical modal logic, computational complexity of different bisimulation related problems like checking if two models are bisimlar, computing the largest bisimulation on a model have been studied extensively [31]. It would be interesting to study similar complexity problems in a non-distributive setting and compare the results with the classical setting.

Bisimulations and network analysis: In Social Network Analysis (SNA), different notions of equivalences between nodes are defined to study the structural nodes in one or two social networks [28,18]. On of the most prominent structural equivalences used in SNA is the notion of *regular equivalence*, which is analogous to bisimulations on (multi-modal) Kripke frames [35,29]. Enriched formal contexts (polarities) can be seen as bi-partite networks; so, it would be interesting to study if simulations and bisimulations defined in this paper can be useful in bi-partite network analysis.

References

[1] Aiello, M., J. van Benthem and G. Bezhanishvili, *Reasoning about space: the modal way*, Journal of Logic and Computation **13** (2003), pp. 889–920.

[2] Bezhanishvili, N., G. Fontaine and Y. Venema, *Vietoris bisimulations*, Journal of Logic and Computation **20** (2010), pp. 1017–1040.

[3] Bezhanishvili, N. and T. Henke, *A model-theoretic approach to descriptive general frames: the van benthem characterization theorem*, Journal of Logic and Computation **30** (2020), pp. 1331–1355.

[4] Blackburn, P., M. de Rijke and Y. Venema, "Modal Logic," Cambridge Tracts in Theoretical Computer Science, Cambridge University Press, 2001.

[5] Chang, C. C. and H. J. Keisler, "Model theory," Elsevier, 1990.

[6] Conradie, W., S. Frittella, K. Manoorkar, S. Nazari, A. Palmigiano, A. Tzimoulis and N. M. Wijnberg, *Rough concepts*, Information Sciences **561** (2021), pp. 371–413.

[7] Conradie, W., S. Frittella, A. Palmigiano, M. Piazzai, A. Tzimoulis and N. M. Wijnberg, *Toward an epistemic-logical theory of categorization*, Electronic Proceedings in Theoretical Computer Science, EPTCS **251** (2017).

[8] Conradie, W. and A. Palmigiano, *Algorithmic correspondence and canonicity for non-distributive logics*, Annals of Pure and Applied Logic **170** (2019), pp. 923–974, the 23rd Workshop on Logic, Language, Information and Computation.

[9] Conradie, W., A. Palmigiano, C. Robinson and N. Wijnberg, *Non-distributive logics: from semantics to meaning*, in: A. Rezus, editor, *Contemporary Logic and Computing*, Landscapes in Logic **1**, College Publications, 2020 pp. 38–86.
URL `arXivpreprintarXiv:2002.04257`

[10] Conradie, W., A. Palmigiano and A. Tzimoulis, *Goldblatt-thomason for* LE-*logics*, arXiv preprint arXiv:1809.08225 (2018).

[11] Davoren, J., *On intuitionistic modal and tense logics and their classical companion logics: Topological semantics and bisimulations*, Annals of Pure and Applied Logic **161** (2009), pp. 349–367, papers presented at the Symposium on Logical Foundations of

Computer Science 2007.
URL `https://www.sciencedirect.com/science/article/pii/S0168007209001468`

[12] De Groot, J., *Non-distributive positive logic as a fragment of first-order logic over semilattices*, Journal of Logic and Computation **34** (2024), pp. 180–196.

[13] De Rijke, M., *A system of dynamic modal logic*, Journal of Philosophical Logic **27** (1998), pp. 109–142.
URL `http://www.jstor.org/stable/30227100`

[14] De Rijke, M., *A system of dynamic modal logic*, Journal of Philosophical Logic **27** (1998), pp. 109–142.
URL `http://www.jstor.org/stable/30227100`

[15] De Rijke, M., *A note on graded modal logic*, Studia Logica **64** (2000), pp. 271–283.

[16] Ding, Y., K. Manoorkar, M. Panettiere, A. Tzimoulis and R. Wang, *Game semantics for lattice-based modal {\mu}-calculus*, arXiv preprint arXiv:2310.13944 (2023).

[17] Dunn, J. M., M. Gehrke and A. Palmigiano, *Canonical extensions and relational completeness of some substructural logics*, The Journal of Symbolic Logic **70** (2005), pp. 713–740.

[18] Fan, T.-F. and C.-J. Liau, *Logical characterizations of regular equivalence in weighted social networks*, Artificial Intelligence **214** (2014), pp. 66–88.
URL `https://www.sciencedirect.com/science/article/pii/S000437021400068X`

[19] Frittella, S., K. Manoorkar, A. Palmigiano, A. Tzimoulis and N. Wijnberg, *Toward a dempster-shafer theory of concepts*, International Journal of Approximate Reasoning **125** (2020), pp. 14–25.

[20] Ganter, B. and R. Wille, *Applied lattice theory: Formal concept analysis*, in: *In General Lattice Theory, G. Grätzer editor, Birkhäuser*, Citeseer, 1997.

[21] Gehrke, M., *Generalized kripke frames*, Studia Logica **84** (2006), pp. 241–275.

[22] Gehrke, M. and J. Harding, *Bounded lattice expansions*, Journal of Algebra **238** (2001), pp. 345–371.

[23] Goranko, V. and M. Otto, *5 model theory of modal logic* (2007).

[24] Hansen, H. H., C. Kupke and E. Pacuit, *Neighbourhood structures: Bisimilarity and basic model theory*, Logical Methods in Computer Science **5** (2009).

[25] Hennessy, M. and H. Lin, *Symbolic bisimulations*, Theoretical Computer Science **138** (1995), pp. 353–389.

[26] Hennessy, M. and R. Milner, *Algebraic laws for nondeterminism and concurrency*, Journal of the ACM (JACM) **32** (1985), pp. 137–161.

[27] Hollenberg, M. J., "Hennessy-Milner classes and process algebra," Citeseer, 1994.

[28] Lorrain, F. and H. C. White, *Structural equivalence of individuals in social networks*, The Journal of mathematical sociology **1** (1971), pp. 49–80.

[29] Marx, M. and M. Masuch, *Regular equivalence and dynamic logic*, Social Networks **25** (2003), pp. 51–65.
URL `https://www.sciencedirect.com/science/article/pii/S0378873302000369`

[30] Olkhovikov, G. K., *Model-theoretic characterization of intuitionistic propositional formulas*, The Review of Symbolic Logic **6** (2013), pp. 348–365.

[31] Paige, R. and R. E. Tarjan, *Three partition refinement algorithms*, SIAM Journal on computing **16** (1987), pp. 973–989.

[32] Ponse, A., M. de Rijke and Y. Venema, "Modal Logic and Process Algebra: A Bisimulation Perspective," Center for the Study of Language and Information Publication Lecture Notes, Cambridge University Press, 1995.
URL `https://books.google.nl/books?id=PXPHQgAACAAJ`

[33] Stanković, M., M. Ćirić and J. Ignjatović, *Simulations and bisimulations for fuzzy multimodal logics over heyting algebras*, Filomat **37** (2023), pp. 711–743.

[34] Van Benthem, J., "Modal correspondence theory," Ph.D. thesis, University of Amsterdam (1977).

[35] White, D. R. and K. P. Reitz, *Graph and semigroup homomorphisms on networks of relations*, Social Networks **5** (1983), pp. 193–234.
URL `https://www.sciencedirect.com/science/article/pii/0378873383900254`

Frame Definability in Conditional Logic

Damiano Fornasiere

University of Barcelona

Johannes Marti

University of Zurich

Giovanni Varricchione

Utrecht University

Abstract

In this paper we investigate classes of finite partially ordered sets that are definable by non-nested formulas in conditional logic. We discuss examples of such definable classes and introduce the notion of a c-morphism between posets as a tool to show that a class of finite posets is not definable. Using an analogue of the Jankov-Fine formulas from modal logic, we show that a class of finite posets is definable by a set of formulas if and only if it is closed under c-morphic images. Lastly, we prove a Sahlqvist-like correspondence theorem stating that every class of finite posets that is definable by a formula without nested conditionals is also definable by a first-order formula.

Keywords: conditional logic, frame definability, non-monotonic logic, belief revision, semiorers

1 Introduction

Conditional logic is a non-normal modal logic that extends propositional logic with a binary modality $\rightsquigarrow$ that is called the (counterfactual) conditional. The guiding semantic intuition is that a conditional $\varphi \rightsquigarrow \psi$ is true if its consequent ψ is true at all worlds that are either most preferred, most plausible or maximally similar to the actual world among all the worlds that make the antecedent φ true. This intuition can be made precise considering an order over the set of worlds and then defining $\varphi \rightsquigarrow \psi$ to be true if ψ is true in all the minimal φ-worlds [13,17]. The same semantic clause is also used in other settings that are closely related to conditional logic, such as in default reasoning [28,15] and in belief revision theory [12,24].

The set of validities of conditional logic depends on what class of orders the semantics is based on. Lewis' conditional logic from [17] consists of all formulas that are valid with the above semantic clause over the class of all models based on weak orders. A generalization of this logic was obtained later by Burgess [5] and Veltman [31], who axiomatize the validities over the class of all partial orders. In the literature on default reasoning and belief revision theory the validities of further classes of posets have been investigated, such as the class of interval orders [16,19] or the class of semiorders [22,25].

In this paper we systematically investigate the relation between validity in conditional logic and classes of finite posets. To this aim we adapt ideas from frame correspondence theory for normal modal logics [3, ch. 3] to conditional logic. The central notion for this paper is that of a class of finite posets being definable by a set of formulas in conditional logic. A class of posets is definable by a set of formulas if the class contains precisely the posets over which all the formulas in the set are valid. As an example one has that Lewis' conditional logic defines the class of weak orders.

The main technical contributions of this paper as follows:

(i) We provide a formula in conditional logic that defines the class of semiorders. To our knowledge this is a novel result. Characterizations of semiorders have been given in the context of choice functions [14,8] and belief revision theory [22,25]. However, these characterizations are formulated in the metalanguage and it is not clear how to express them in conditional logic.

(ii) We characterize the definable classes of finite posets as those that are closed under c-morphic images. This results is similar to characterizations in modal logic, which state, roughly, that a class of frames is definable by modal formulas iff it is closed under generated subframes, coproducts and bounded morphic images [10]. As a consequence of our result we get that if a class of finite posets is not definable in conditional logic then there is a concrete counterexample of a c-morphism from a poset that is in the class to a poset that is not in the class.

(iii) We provide a procedure that, given a formula in conditional logic, computes a first-order formula that is true in exactly those posets where the conditional formula is valid. This result can be seen as a simple version of the Sahlqvist Theorem for conditional logic.

The statement of the second result makes use of the notion of a c-morphism. This notion is inspired by the notion of a bisimulation between preferential models that was studied in the context of default reasoning by Zhu in [32]. A c-morphism in the sense defined in this paper is a function whose graph is a bisimulation in the sense of [32]. The precise formulation of the conditions in the definition of a c-morphism is quite technical. One part of the definition is the familiar back-condition from the definition of a bounded morphism; however, in general c-morphisms are not order-preserving. The notion of a c-morphism plays a role that is comparable to the notion of a bounded morphism, also called p-morphism, in modal logic. In particular it holds that any two models that are connected by a c-morphism satisfy the same formulas.

To our knowledge this paper is the first study of frame definability in conditional logic. As such our approach still has the following limitations:

(i) We only consider formulas of conditional logic in which the conditional is not nested and all occurrences of propositional letters are in the scope of the conditional. This allows us to focus on the aspects of definability that are specific to the minimization semantics of conditionals.

(ii) We only consider finite posets. The main reason for this is that over non-wellfounded posets minimization does not yield a well-behaved conditional logic.

(iii) We only consider frame definability relative to posets, that are transitive, reflexive and anti-symmetric relations. If one gives up anti-symmetry one obtains a semantics of the condition over preorders, that is reflexive and transitive relations. However, it has been shown that the logic of the class of preorders is the same as the logic of the more restricted class of posets [5,31]. If one additionally also removes the assumptions of transitivity or reflexivity then one obtains conditional logics that are weaker than the logic on posets. Following Hansson [13], such logics were studied mainly in the context of deontic logic [23,20,9,21], but more recently they have attracted broader attention [11,6]. We did not manage to adapt our techniques, especially the central notion of a c-morphism, to such settings.

The structure of this paper is as follows: In Section 2 we discuss the syntax and semantics of conditional logic and the notions of validity and definable classes. In Section 3 we provide examples of definable classes of posets that have arisen in the literature. In Section 4 we introduce the notion of a c-morphism, which we then use in Section 5 to prove for some examples of classes of posets that they are not definable. Section 6 contains the proof of the characterization result that a class of finite posets is definable iff it is closed under c-morphic images. In Section 7 we show that there is a first-order correspondent for every non-nested formula of conditional logic.

2 Preliminaries

In this section we discuss the language of conditional logic and its semantics over posets. We also define the notion of a definable class of posets.

2.1 Syntax

Conditional logics are commonly formulated in a classical propositional modal language with one binary modality $\rightsquigarrow$. A formula of the form $\varphi \rightsquigarrow \psi$ is called a *conditional* with *antecedent* φ and *consequent* ψ. In conditional logic conditional can be nested within the scope of other conditionals, as for example in the formula $(((p \rightsquigarrow q) \rightsquigarrow r) \wedge q) \rightarrow r$. In this paper we, however, only consider formulas in which the conditional is not nested and all propositional letters are in the scope of a conditional. To make this precise fix an infinite set Prop of propositional letters and consider the grammar:

$$\varphi_0 ::= p \mid \top \mid \neg\varphi_0 \mid \varphi_0 \wedge \varphi_0, \qquad \text{where } p \in \mathsf{Prop},$$
$$\varphi_1 ::= \varphi_0 \rightsquigarrow \varphi_0 \mid \top \mid \neg\varphi_1 \mid \varphi_1 \wedge \varphi_1.$$

Let $\mathcal{L}_0$ be the set of formulas generated from φ_0 and $\mathcal{L}_1$ the set of formulas generated from φ_1. Note that $\mathcal{L}_0$ is just the language of classical propositional logic. In both $\mathcal{L}_0$ and $\mathcal{L}_1$ we use further Boolean connectives, such as $\bot$, $\vee$, $\to$, and $\leftrightarrow$, as abbreviations with their usual meaning in classical logic. To omit parenthesis we assume that $\neg$ binds stronger than $\wedge$ and $\vee$, which in turn bind stronger than $\rightsquigarrow$, $\to$ and $\leftrightarrow$.

We are going to focus on formulas from $\mathcal{L}_1$ that are of the shape

$$\bigwedge_{i=1}^{n}(\varphi_i \rightsquigarrow \psi_i) \to \bigvee_{j=1}^{m}(\gamma_j \rightsquigarrow \delta_j),$$

where $\varphi_i, \psi_i, \gamma_j, \delta_j \in \mathcal{L}_0$ for all i and j. We call such formulas *inference rules* or simply *inferences* or *rules* and suggestively write them as

$$\frac{\varphi_1 \rightsquigarrow \psi_1 \quad \dots \quad \varphi_n \rightsquigarrow \psi_n}{\gamma_1 \rightsquigarrow \delta_1 \quad \dots \quad \gamma_m \rightsquigarrow \delta_m},$$

or as Σ/Γ, where $\Sigma = \{\varphi_i \rightsquigarrow \psi_i \mid 1 \leq i \leq n\}$ and $\Gamma = \{\gamma_j \rightsquigarrow \delta_j \mid 1 \leq j \leq m\}$. The elements of Σ are called the *premises* of the inference Σ/Γ and the elements of Γ are its *conclusions*. We allow for the cases where Σ is empty, meaning that the inference corresponds to a formula of the form $\top \to \bigvee\Gamma$, and where Γ is empty, meaning that the inference corresponds to the formula $\bigwedge\Sigma \to \bot$.

It is a consequence of Corollary 2.6 below, that for the purpose of understanding classes of posets that are definable by a formula in $\mathcal{L}_1$ it suffices to only consider formulas that are in the shape of inference rules. Focusing on the presentation of formulas as inference rules also matches the presentation in the setting of non-monotonic consequence relations, where such rules between conditional, thought of as non-monotonic inference relations, are taken as basic [15]. In Section 3 we provide multiple natural examples of such inference rules that have been discussed in the literature.

2.2 Semantics

The semantics of the conditional in conditional logic can be given in terms of ternary similarity relations $\leq$ where $u \leq_w v$ holds if u is at least as similar to w as v [17, sec. 2.3]. A conditional then holds with respect to such a relation if the consequent is true at all the worlds that satisfy the antecedent and are maximally similar to the actual world among the wolds that satisfy the antecedent. In our setting, where we do not allow for the nesting of conditionals, one can however omit the relativization to the actual world and just minimize relative to a poset. This leads to the notion of a preferential model, which is the kind of semantic structure that is commonly used for non-monotonic consequence relations [15]:

Definition 2.1 Recall that a poset $P = (W, \leq)$ is a set W together with a partial order $\leq$ on W, where a partial order is just a reflexive, transitive and anti-symmetric relation $\leq \subseteq W \times W$. Here, anti-symmetry denotes the property that whenever $w \leq v$ and $v \leq w$ hold for some $w, v \in W$ then it follows that $w = v$. A *preferential model* $M = (W, \leq, V)$ is a poset $(W, \leq)$ together with a function $V : \mathsf{Prop} \to \mathcal{P}W$. The elements of W are called *worlds* and the function V is called the *valuation function.*

In this paper we assume that all models are finite, meaning that the set of worlds W is a finite set. We use the notation $w < v$ as a shorthand for the conjunction of statements that $w \leq v$ holds and that $v \leq w$ does not hold.

The semantics of formulas from $\mathcal{L}_0$ and $\mathcal{L}_1$ is defined in the standard way. The set of worlds $[\![\varphi]\!]_V \subseteq W$, where a formula $\varphi \in \mathcal{L}_0$ is true, is computed by the recursive clauses $[\![p]\!]_V = V(p)$, $[\![\top]\!]_V = W$, $[\![\neg\varphi]\!]_V = W \setminus [\![\varphi]\!]_V$ and $[\![\varphi \wedge \psi]\!]_V = [\![\varphi]\!]_V \cap [\![\psi]\!]_V$. If V is clear, we write $[\![\varphi]\!]$ instead of $[\![\varphi]\!]_V$.

The semantic clauses for the propositional connectives over $\mathcal{L}_1$ relative to the model $M = (W, \leq, V)$ are

$$M \models \neg\varphi \text{ iff not } M \models \varphi, \text{ and } \qquad M \models \varphi \wedge \psi \text{ iff } M \models \varphi \text{ and } M \models \psi.$$

For the semantics of the conditional we use the order $\leq$. A conditional is true if all of the minimal antecedent worlds satisfy the consequent:

$$M \models \varphi \rightsquigarrow \psi \quad \text{iff} \quad \min_{\leq}([\![\varphi]\!]) \subseteq [\![\psi]\!].$$

The minimal worlds of a set $A \subseteq W$ in a partial order $\leq$ over W can be defined as $\min_{\leq}(A) = \{x \in A \mid \forall y \in A(y \leq x \implies x \leq y)\}$.

Using a standard argument one can provide the following alternative formulation of the semantics for the conditional in finite orders:

Proposition 2.2 *Relative to all finite models* $(W, \leq, V)$ *it holds that*

$$M \models \varphi \rightsquigarrow \psi \quad \textit{iff} \quad \textit{for all } x \in [\![\varphi]\!] \textit{ there is a } y \leq x \textit{ with } y \in [\![\varphi]\!] \cap [\![\psi]\!].$$

The notions of validity and of a definable class are then defined analogously to how they are defined for normal modal logics:

Definition 2.3 A formula $\varphi \in \mathcal{L}_1$ is *valid* in a poset $P = (W, \leq)$ if for all valuations $V : \mathsf{Prop} \to \mathcal{P}W$ we have that $M \models \varphi$ for the model $M = (W, \leq, V)$. We can extend validity to sets of formulas and classes of posets: A set of formulas $\Sigma \subseteq \mathcal{L}_1$ is valid in a class of posets $\mathcal{C}$ if φ is valid in P for all formulas $\varphi \in \Sigma$ and P in the class $\mathcal{C}$. If a formula, or set of formulas, is not valid in a poset, or class of posets, we also say that the formula, or set of formulas, is *falsifiable* in the poset, or class of posets and that the poset, or class of posets, *falsifies* the formula, or set of formulas.

Definition 2.4 A formula $\varphi \in \mathcal{L}_1$ *defines* a class of finite posets $\mathcal{C}$ if and only if for all finite posets P it holds that

$$P \text{ is in } \mathcal{C} \quad \text{iff} \quad \varphi \text{ is valid in } P.$$

Similarly, a set of formulas $\Sigma \subseteq \mathcal{L}_1$ *defines* a class of finite posets $\mathcal{C}$ iff for all finite posets P it holds that P is in $\mathcal{C}$ if and only if Σ is valid in P. A class of finite posets $\mathcal{C}$ is *definable* if there is some set $\Sigma \subseteq \mathcal{L}_1$ such that Σ defines $\mathcal{C}$.

We conclude this section by showing that in order to study definable classes it suffices to consider formulas that are in the syntactic shape of inference rules.

Proposition 2.5 *Every formula $\varphi \in \mathcal{L}_1$ is equivalent to a conjunction of inference rules.*

Proof. This follows with propositional reasoning. To this aim consider $\varphi \in \mathcal{L}_1$ as a propositional formula where the conditionals are atoms. It is clear that we can rewrite φ into an equivalent conjunctive normal form, i.e, a conjunction of disjunctions of literals. It is also clear that every such disjunction of literals

$$\bigvee_{i=1}^{n} \neg(\varphi_i \rightsquigarrow \psi_i) \vee \bigvee_{j=1}^{m} (\gamma_j \rightsquigarrow \delta_j)$$

is propositionally equivalent to the inference

$$\bigwedge_{i=1}^{n} (\varphi_i \rightsquigarrow \psi_i) \rightarrow \bigvee_{j=1}^{m} (\gamma_j \rightsquigarrow \delta_j).$$

□

By observing that a conjunction is valid iff all if its conjuncts are valid we obtain the following Corollary:

Corollary 2.6 *For every formula $\varphi \in \mathcal{L}_1$ there is a finite set of inferences $\sigma = \{\Sigma_1/\Gamma_1, \ldots, \Sigma_n/\Gamma_1\}$ such that σ is valid in a poset $(W, \leq)$ if and only if φ is valid in $(W, \leq)$.*

3 Definable classes of posets

In this section we discuss examples of classes of finite posets that are definable by a formula in conditional logic. Figure 1 provides an overview of the examples from this section. The first-order formulas that describe the classes in the second column should be understood such that all free variables are universally quantified. The rules defining linear orders and orders with a minimum have an empty set of premises. Recall from Section 2.1 that we read the conclusion of these rules disjunctively.

Example 3.1 (Antichains) The class of antichains is defined by the rule

$$\frac{\top \rightsquigarrow p}{\neg p \rightsquigarrow \bot.}$$

In order to see this, suppose that $P = (W, \leq)$ is a poset which is not an antichain. Define the valuation V such that $V(p) = \min_{\leq}(W)$, thus p is true at all the minimal elements of P. Then $\top \rightsquigarrow p$ is true but we can show

Class	First-order formula	Defining rule
Antichains	$x \leq y \rightarrow y \leq x$	$\dfrac{\top \rightsquigarrow p}{\neg p \rightsquigarrow \bot}$
With minimum	$\exists x \forall y (x \leq y)$	$\overline{\top \rightsquigarrow p \quad \top \rightsquigarrow \neg p}$
Linear orders	$x \leq y \vee y \leq x$	$\overline{p \rightsquigarrow q \quad p \rightsquigarrow \neg q}$
Weak orders	$x < y \rightarrow z < y \vee x < z$	$\dfrac{p \rightsquigarrow q}{p \rightsquigarrow \neg r \quad p \wedge r \rightsquigarrow q}$
Interval orders	$x < y \wedge z < u \rightarrow x < u \vee z < y$	$\dfrac{p \vee r \rightsquigarrow q}{p \rightsquigarrow q \quad r \rightsquigarrow q}$
Semitransitive orders	$x < y \wedge y < z \rightarrow x < u \vee u < z$	$\dfrac{p \rightsquigarrow q \quad p \wedge \neg q \rightsquigarrow \neg r}{p \rightsquigarrow \neg r \quad p \wedge r \rightsquigarrow q}$

Fig. 1. Examples of definable classes of posets.

that $\neg p \rightsquigarrow \bot$ is false: because P is not an antichain, there is at least one element $y \in W$ that is not minimal, hence $[\![\neg p]\!] = W \setminus \min_{\leq}(W) \neq \emptyset$, and thus $\min_{\leq}([\![\neg p]\!]) \not\subseteq \emptyset = [\![\bot]\!]$.

Viceversa, if an antichain $P = (W, \leq)$ satisfies $\top \rightsquigarrow p$ it means that all the minimal elements of P satisfy p. But since P is an antichain, all its elements are minimal and because $[\![p]\!] \cap [\![\neg p]\!] = \emptyset$ it must be $[\![\neg p]\!] = \emptyset$, therefore $\emptyset \subseteq [\![\bot]\!]$.

Example 3.2 (Orders with a minimum) It is easy to check that

$$\overline{\top \rightsquigarrow p \quad \top \rightsquigarrow \neg p}$$

defines the class of all posets that have a unique minimal element.

Example 3.3 (Linear orders) We leave it to the reader to convince themself that the rule

$$\overline{p \rightsquigarrow q \quad p \rightsquigarrow \neg q}$$

defines the class of all linear orders. Note that this rule has no premises and thus expresses the formula $(p \rightsquigarrow q) \vee (p \rightsquigarrow \neg q)$. In [17, sec. 3.4] Lewis calls this formula "conditional excluded middle" and argues that its validity is characteristic of Stalnaker's account of conditionals from [29].

Example 3.4 (Weak orders) The class of weak orders is defined by the rule

$$\frac{p \rightsquigarrow q}{p \rightsquigarrow \neg r \quad p \wedge r \rightsquigarrow q}.$$

Note that weak orders are also called strict weak orders and they can also be represented as total preorders. They provide a semantics for conditional logic that is equivalent to Lewis' systems of spheres [17, sec. 2.3]. In belief revision theory they provide a semantics for the classic AGM postulates for belief revision [1,12]. In the guise of total preorders they are also behind the standard semantics of dynamic epistemic logic [2]. The inference rule used here to define weak orders is the rule of "rational monotonicity" from [16].

Example 3.5 (Interval orders) It is not hard to show that the following rule defines interval orders

$$\frac{p \vee r \rightsquigarrow q}{p \rightsquigarrow q \quad r \rightsquigarrow q}.$$

This rule has been called "disjunctive rationality" in the literature on non-monotonic reasoning [16,19,4]. Interval orders derive their name from the observation that they are precisely the orders that can be represented by a natural ordering of arbitrary length intervals on the line [7]. The interval order condition also plays a role in the semiorders that we discuss as the next example.

Example 3.6 (Semitransitive orders and semiorders) Semitransitive orders are defined by the inference rule

$$\frac{p \rightsquigarrow q \quad p \wedge \neg q \rightsquigarrow \neg r}{p \rightsquigarrow \neg r \quad p \wedge r \rightsquigarrow q}\,. \tag{1}$$

As semiorders are semitransitive interval orders, it follows that semiorders are defined by this rule together with the disjunctive rationality rule from Example 3.5. It has been argued that semiorders model human preference more adequately than weak orders [18,30]. It can be shown that they are representable, analogously to interval orders, by intervals of constant length [27]. Semiorders have been axiomatized in the context of choice functions [14,8] and in belief revision theory [22,25]. However, both settings use axioms that are not obviously expressible in the language of conditional logic.

To prove that the rule in (1) defines the class of semitransitive posets first assume that $P = (W, \leq)$ is a poset that fails to be semitransitive. This means that there are points x, y, z and u with $x < y < z$ such that u is incomparable to x and z. Thus, P has a subposet that looks as follows:

$$\begin{array}{ccc} z : p\bar{q}r & & \\ | & & \\ y : p\overline{qr} & \qquad & u : pqr \\ | & & \\ x : pq\bar{r} & & \end{array}$$

It is easy to see that in the model with a valuation V such that $V(p) = \{x, y, z, u\}$, $V(q) = \{x, u\}$ and $V(r) = \{z, u\}$ all the premises of the rule in (1) are true while all the conclusions are false.

For the other direction assume that the rule in (1) is valid in a poset $P = (W, \leq)$. To see that then P is semitransitive choose any points x, y, z and u in P such that $x < y < z$. To show that then either $x < u$ or $u < z$ consider a valuation V with $V(p) = \{x, y, z, u\}$, $V(q) = \{x, u\}$ and $V(r) = \{z, u\}$. It is clear that this makes the conditionals $p \rightsquigarrow q$ and $p \wedge \neg q \rightsquigarrow \neg r$ true because $\min_\leq([\![p]\!]) \subseteq \{x, u\} = [\![q]\!]$ and $\min_\leq([\![p \wedge \neg q]\!]) = \min_\leq(\{y, z\}) = \{y\} \subseteq [\![\neg r]\!]$. It follows that either $p \rightsquigarrow \neg r$ or $p \wedge r \rightsquigarrow q$ is true in $(W, \leq, V)$. The former means that $\min_\leq(\{x, y, z, u\}) \subseteq \{x, y\}$, which entails $x < u$, and the latter means that $\min_\leq(\{z, u\}) \subseteq \{x, u\}$, which means that $u < z$.

4 C-morphisms

In this section we introduce c-morphisms, which we will use to show that a class of posets is not definable by a formula in conditional logic. C-morphisms are precisely those functions whose graph is a bisimulation in the sense of [32].

Definition 4.1 A *c-morphism* f from a poset $P = (W, \leq)$ to a poset $P' = (W', \leq')$ is a function $f : W \to W'$ such that:

(i) For all $w \in W$ and $u' \leq' f(w)$ there is a $u \leq w$ such that $f(u) = u'$;

(ii) For all $w' \in W'$ there is a $w \in W$ such that $f(w) = w'$ and for all $u \leq w$ we have that $f(u) \leq' w'$.

A poset P' is a *c-morphic image* of a poset P if there is some c-morphism from P to P'. We can extend the notion of a c-morphism to models such that $f : W \to W'$ is a c-morphism from $M = (W, \leq, V)$ to $M' = (W', \leq', V')$ if f is a c-morphism from $(W, \leq)$ to $(W', \leq')$ and $V(p) = f^{-1}(V'(p))$ for all $p \in \mathsf{Prop}$.

Note that the first condition in the definition of c-morphisms is just the back condition for bounded morphisms in modal logic. Also note that it follows from the second condition that every c-morphism is surjective. However, the following examples shows that c-morphisms need not be order-preserving:

Example 4.2 The following is a picture of a c-morphism between two preferential models. The first models contains the worlds w_1, w_2 and w_3, and the second model contains the worlds v_1 and v_2. The mapping of the c-morphism is depicted with the dashed arrows.

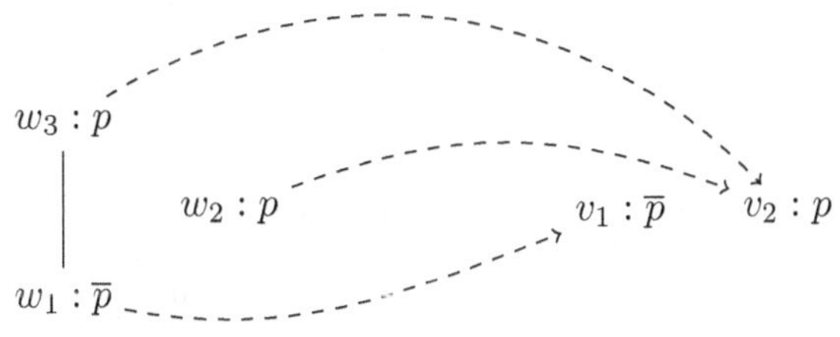

C-morphisms can be characterized via the preservation of minimal elements:

Proposition 4.3 *Let $P = (W, \leq)$ and $P' = (W', \leq')$ be finite posets, $f : W \to W'$ is a c-morphism if and only if for every subset $X' \subseteq W'$ we have that $f(\min_\leq(f^{-1}(X'))) = \min_{\leq'}(X')$.*

Proof. For the left-to-right direction, assume that $f : W \to W'$ is a c-morphism from P to P'. We prove that $f(\min_\leq(f^{-1}(X'))) = \min_{\leq'}(X')$ holds for every $X' \subseteq W'$:

The $\subseteq$-inclusion: Let $w' \in f(\min_{\leq}(f^{-1}(X')))$. Hence, there must be $w \in \min_{\leq}(f^{-1}(X'))$ such that $f(w) = w'$. To prove that $w' \in \min_{\leq'}(X')$ it suffices to show that, for any $u' \in X'$ with $u' \leq' w'$, $u' = w'$. Thus, let u' be any such world. By condition (i) from the definition of c-morphisms we have that there is $u \leq w$ with $f(u) = u'$. Hence, $u \in f^{-1}(X')$. As $w \in \min_{\leq}(f^{-1}(X'))$ we have that $u = w$, hence $u' = w'$, proving the claim.

The $\supseteq$-inclusion: Let $w' \in \min_{\leq'}(X')$ and assume for a contradiction that there is no $w \in \min_{\leq}(f^{-1}(X'))$ such that $f(w) = w'$. We use an induction to construct an infinite descending chain $w_0 > w_1 > \ldots$ in $f^{-1}(X')$, which contradicts the finiteness of W. By condition (ii) for c-morphisms there is some $w_0 \in W$ such that $f(w_0) = w'$ and for all $u \leq w_0$ we have that $f(u) \leq' w'$. Inductively, we then assume that we are given some $w_i \leq w_0$ such that $f(w_i) = w'$. Clearly, this inductive assumption is satisfied by w_0 in the base case. Since $f(w_i) = w'$ it follows by our assumption that $w_i \notin \min_{\leq}(f^{-1}(X'))$. This means that there is some $w_{i+1} < w_i$ such that $f(w_{i+1}) \in X'$. By instantiating the u from the condition on w_0 with w_{i+1} we get that $f(w_{i+1}) \leq w'$. Because $w' \in \min_{\leq'}(X')$ it follows from $f(w_{i+1}) \in X'$ and $f(w_{i+1}) \leq w'$ that $f(w_{i+1}) = w'$. Hence, w_{i+1} satisfies our inductive assumption as well.

For the right-to-left direction of the proposition, assume that f is such that for any $X' \subseteq W'$ we have that $f(\min_{\leq}(f^{-1}(X'))) = \min_{\leq'}(X')$. We show that f satisfies both conditions for c-morphisms:

Condition (i): Let $w \in W$ and $u' \leq' f(w) = w'$ be arbitrary. We show that for some $u \leq w$ we have that $f(u) = u'$. If $u' = w'$ then by reflexivity it trivially follows that $w \leq w$ and $f(w) = u'$. Hence, assume that $u' \neq w'$ and take $X' = \{u', w'\}$. By assumption we get that $f(\min_{\leq}(f^{-1}(X'))) = \min_{\leq'}(X') = \{u'\}$. Hence, as $w' \notin \min_{\leq'}(X')$, we have that $w \notin \min_{\leq}(f^{-1}(X'))$. Therefore, we deduce that there must be some $u \leq w$ with $u \in \min_{\leq}(f^{-1}(X'))$. As u is minimal in $f^{-1}(X')$, it follows that $f(u) = u'$, proving the claim.

Condition (ii): Let $w' \in W'$ be arbitrary and assume by way of contradiction that for all $w \in W$ with $f(w) = w'$ there is some $u \leq w$ such that $f(u) \nleq' w'$. We construct an infinite chain $X'_0 \subsetneq X'_1 \subsetneq \ldots$ of larger and larger subsets in W', leading to a contradiction with the assumption that W' is finite. This chain is constructed by an induction, where at every step i we guarantee that w' is a minimal element of X'_i. In the base case we set $X'_0 = \{w'\}$, for which we obviously have that w' is a minimal element of the set. In the inductive step assume that we are given X'_i such that w' is a minimal element of w'. Because $f(\min_{\leq}(f^{-1}(X'_i))) = \min_{\leq'}(X'_i)$ there is then some $w \in \min_{\leq}(f^{-1}(X'_i))$ such that $f(w) = w'$. It follows that there is some $u \leq w$ such that $f(u) \nleq' w'$. Define $X'_{i+1} = X'_i \cup \{f(u)\}$. Because $f(u) \nleq' w'$ it holds again that w' is a minimal element of X'_{i+1}. To show that $X'_i \subsetneq X'_{i+1}$ we argue that $f(u) \notin X'_i$. Assume for contradiction that $f(u) \in X'_i$. Then, $u \in f^{-1}(X'_i)$. Because $w \in \min_{\leq}(f^{-1}(X'_i))$ and $u \leq w$ it would follow that $u = w$, which is impossible because $f(u) \nleq' w' = f(w)$. □

The crucial property of c-morphisms is that they preserve the truth of formulas in conditional logic:

Proposition 4.4 *Let $M = (W, \leq, V)$ and $M' = (W', \leq', V')$ be finite preferential models and f a c-morphism from M to M'. Then, it holds for all formulas $\varphi \in \mathcal{L}_1$ that*

$$M \models \varphi \quad \textit{iff} \quad M' \models \varphi.$$

Proof. The proof is an induction on the complexity of formulas $\varphi \in \mathcal{L}_1$.

First, note that it follows with an easy induction on the complexity of formulas $\varphi \in \mathcal{L}_0$ that $[\![\varphi]\!]_V = f^{-1}([\![\varphi]\!]_{V'})$. The base case $V(p) = f^{-1}(V'(p))$ holds for all $p \in \mathsf{Prop}$ because f is a c-morphism of models.

The main inductive case is where $\varphi = \chi \rightsquigarrow \psi$ is a conditional, with $\chi, \psi \in \mathcal{L}_0$. First, observe that because the direct image map is left adjoint to the inverse image map we have that

$$f(\min_{\leq}([\![\varphi]\!]_V)) \subseteq [\![\psi]\!]_{V'} \quad \text{iff} \quad \min_{\leq}([\![\varphi]\!]_V) \subseteq f^{-1}([\![\psi]\!]_{V'}). \tag{2}$$

Because from the base case for formulas in $\mathcal{L}_0$ we have that $f^{-1}([\![\psi]\!]_{V'}) = [\![\psi]\!]_V$, it follows that the right side of (2) is equivalent to $\min_{\leq}([\![\varphi]\!]_V) \subseteq [\![\psi]\!]_V$, which means that $M \models \varphi \rightsquigarrow \psi$. On the other hand using that f is a c-morphism we get from Proposition 4.3 that $f(\min_{\leq}(f^{-1}([\![\varphi]\!]_{V'}))) = \min_{\leq'}([\![\varphi]\!]_{V'})$. Combining it with $f^{-1}([\![\psi]\!]_{V'}) = [\![\psi]\!]_V$, which follows from the base case, it implies that the left side of (2) is equivalent to $\min_{\leq'}([\![\varphi]\!]_{V'}) \subseteq [\![\psi]\!]_{V'}$. This is precisely the semantics of $M' \models \varphi \rightsquigarrow \psi$. The remaining cases for the propositional connectives in $\mathcal{L}_1$ follow by standard reasoning. □

By using the preservation result we have just shown, we can also prove that c-morphisms preserve validities in posets:

Theorem 4.5 *Let f be a c-morphism from a finite poset $P = (W, \leq)$ to a finite poset $P' = (W', \leq')$. Then, every formula $\varphi \in \mathcal{L}_1$ that is valid in P is also valid in P'.*

Proof. We reason by contraposition: assume that φ is not valid in P'. Then, there is some valuation $V' : \mathsf{Prop} \to \mathcal{P}W'$ such that for the derived model $M' = (W', \leq', V')$ we have that $M \not\models \varphi$. We lift V' to a valuation $V : \mathsf{Prop} \to \mathcal{P}W$ for P by exploiting f: for $p \in \mathsf{Prop}$, take $V(p) = f^{-1}(V'(p))$. Let $M = (W, \leq, V)$ be the derived model: by definition we now have that f is a c-morphism between the models M and M'. Hence, by Proposition 4.4, it follows that $M \not\models \varphi$, i.e. φ is not valid in P. □

One might rephrase Theorem 4.5 as stating that definable classes of posets are closed under c-morphic images. Thus, this result can be used show that some class of posets $\mathcal{C}$ is not definable by a formula, or set of formulas, in conditional logic: if there are two posets P and P' such that P is in $\mathcal{C}$ but P' is not in $\mathcal{C}$, and there is a c-morphism from P to P', then $\mathcal{C}$ is not definable.

5 Classes of posets that are not definable

In this section we will use Theorem 4.5 to give examples of classes of posets that are not definable by formulas in conditional logic.

We first make an observation about c-morphic images, which allows us to easily show for many classes of posets that they are not definable. The observation is that for every poset P with n minimal elements there is a c-morphism from P to the antichain with n elements A_n. Therefore, every class of posets that contains a poset with n minimal elements but does not contain A_n is not definable.

Lemma 5.1 *Let A_n be the antichain with n elements. Then, for every finite poset P with at least n minimal elements, we have that there is a c-morphism f from P to A_n.*

Proof. Let $P = (W, \leq)$ be a finite poset with an enumeration $m_1, \ldots, m_k$ of its k distinct minimal elements for $k \geq n$, and let $a_1, \ldots, a_n$ be an enumeration of the n elements in the antichain A_n. We define a surjective function g from $\{m_1, \ldots, m_k\}$ to $\{a_1, \ldots, a_n\}$ by setting $g(m_i) = a_i$ for $i \leq n$ and $g(m_i) = a_1$ for $i > n$. Then, observe that because P is finite we have for every $w \in W$ some minimal element m_w of P such that $m_w \leq w$. We then define the c-morphism f by mapping each $w \in W$ to the element $g(m_w)$ of A_n.

The definition of f satisfies condition (i) of Definition 4.1 because for every $w \in W$ and $u' \leq f(w)$ we have that $f(w) = u'$, as A_n is an antichain.

For condition (ii) note that for every $w' \in W$ we have $w' = f(m)$ for some minimal element m of P. If we then consider any $u \leq m$ we get that $u = m$, by the minimality of m, and hence it trivially holds that $f(u) = w'$. □

Together with Theorem 4.5 we obtain the following corollary:

Corollary 5.2 *If a class of finite posets contains a poset P with k minimal elements but does not contain the antichain A_n for some $n \leq k$ then the class is not definable.*

From this corollary we immediately obtain the following examples of classes that are not definable by formulas in conditional logic:

Example 5.3 The classes of posets with more than n elements for $n \geq 1$, with exactly n elements for $n > 1$, with more than n minimal/maximal elements for $n \geq 1$, and with chains longer than n for $n \geq 1$ are all not definable. In each of these classes we have some poset with at least 1 minimal element but all of these classes do not contain the antichain A_1 with exactly one element.

Example 5.4 The classes of connected posets, of posets with a maximum, and of join-semilattices, are not definable. All these classes contain the poset

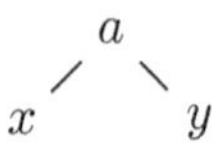

with exactly 2 minimal and 1 maximal element. But all of these classes do not contain the antichain A_2 with 2 elements.

Lastly, we provide an example of an undefinable class for which we could not use the above corollary to prove its undefinability:

Example 5.5 The class of finite meet-semilattices is not definable. Consider the following two posets:

Clearly, the left poset is a meet-semilattice, while the right one is not. One can verify that the function f with $f(x) = 4'$ and $f(n) = n'$ for $n \in \{1, 2, \ldots, 6\}$ defines a c-morphism from the left poset to the right one. To check condition (ii) of Definition 4.1 observe that for the element $4'$ on the right we have the element 4 on the left such that for each $n \leq 4$ it holds that $n' \leq 4'$.

6 The characterization theorem

We have already seen in Section 4 that every definable class of finite posets is closed under c-morphic images. In this Section we show that the converse is also true: every class of finite posets that is closed under c-morphic images is definable by a set of formulas in conditional logic. Thus, the definable classes of frames are precisely those that are closed under c-morphic images:

Theorem 6.1 *A class of finite posets $\mathcal{C}$ is definable by some set of formulas $\Sigma \subseteq \mathcal{L}_1$ if and only if $\mathcal{C}$ is closed under c-morphic images.*

To prove Theorem 6.1 we define an analogue of the Jankov-Fine formulas that can be used to show a similar characterization result for modal logic over finite transitive frames [3, sec. 3.4]. Thus, we define for every finite poset P a characteristic formula χ_P with the property that for every finite poset P' we get that P' falsifies χ_P iff P is a c-morphic image of P'.

Definition 6.2 Fix a finite poset $P = (W, \leq)$ such that $W = \{w_1, \ldots, w_n\}$. For every point $w_i \in W$ define the formula $\alpha_i \in \mathcal{L}_0$ such that

$$\alpha_i = \bigvee \{p_j \mid w_j \not\leq w_i\}.$$

We then define the *characteristic formula* $\chi_P \in \mathcal{L}_1$ of P as the rule

$$\frac{\neg \bigvee_{j=1}^{n} p_j \rightsquigarrow \bot \quad \{p_i \wedge p_j \rightsquigarrow \bot \mid i \neq j\} \quad \{p_i \vee p_j \rightsquigarrow p_i \mid w_i < w_j\}}{\{p_i \rightsquigarrow \bot \mid 1 \leq i \leq n\} \quad \{p_i \vee \alpha_i \rightsquigarrow \alpha_i \mid 1 \leq i \leq n\}} \, .$$

To gain some intuition about what it means that this formula is falsified in a poset, recall that if the formula is falsified in $P' = (W', \leq')$ then there is some valuation V' over W' such that in the resulting model $M' = (W', \leq', V')$ we have that all the conditional that are premises of χ_P are true in M', while all the conditionals that are conclusions of χ_P are false. First, observe that because of the truth of the premise $\neg \bigvee_{j=1}^{n} p_j \rightsquigarrow \bot$ and the truth of the premise $p_i \wedge p_j \rightsquigarrow \bot$, for $i \neq j$ the sets $[\![p_i]\!]_{V'}$, for $i \in \{1, \ldots, n\}$, partition W'. Because of the falsity of the conclusion $p_i \rightsquigarrow \bot$ there is at least one world that makes p_i true for every $i \in \{1, \ldots, n\}$. Moreover, we have that:

- The truth of the premise $p_i \vee p_j \rightsquigarrow p_i$ for $w_i \leq w_j$ means that every p_j-world in M' is above some p_i-world.
- The falsity of the conclusion $p_i \vee \alpha_i \rightsquigarrow \alpha_i$ entails that there is at least one p_i-world in M' that has no p_j-world for $w_j \not\leq w_i$ below it.

Lemma 6.3 *For every finite poset P, P falsifies χ_P.*

Proof. Let $P = (W, \leq)$ and $W = \{w_1, \ldots, w_n\}$. Define the valuation V such that $V(p_i) = \{w_i\}$ and consider $M = (W, \leq, V)$. It is clear that then the sets $[\![p_i]\!]_V$ partition W and thus the first two premises of χ_P are true in M and the first conclusion of χ_P is false. Also observe that if $w_i < w_j$ then $\min_{\leq}(\{w_i, w_j\}) = \{w_i\}$. Thus M makes the premise $p_i \vee p_j \rightsquigarrow p_i$ true. To see that M makes the conclusion $p_i \vee \alpha_i \rightsquigarrow \alpha_i$ false, for every fixed i, we use the alternative semantic clause from Proposition 2.2. Hence, we need to find some word $w \in [\![p_i \vee \alpha_i]\!]$ such that $v \not\leq w$ for all $v \in [\![\alpha_i]\!]$. Clearly we can take $w = w_i$ because every $v \in [\![\alpha_i]\!]$ is of the form $v = w_j$ for some $w_j \not\leq w_i$. □

Proposition 6.4 *For every finite poset P', P' falsifies χ_P if and only if there exists a c-morphism f from P' to P.*

Proof. The direction from right to left follows immediately from Lemma 6.3 together with Theorem 4.5.

For the direction from left to right, assume that $P' = (W', \leq')$ falsifies χ_P of some poset $P = (W, \leq)$ with $W = \{w_1, \ldots, w_n\}$. Let V' be the valuation such that $M' \not\models \chi_P$ for $M' = (P', \leq', V')$. As already described above, note that this means, by the truth of the first premises, that the $[\![p_i]\!]_{V'}$ partition W', and, by the falsity of the first conclusions, that none of the $[\![p_i]\!]_{V'}$ is empty. We then define a function $f : W' \to W$ such that $f(w') = w_i$ for the unique i such that $w' \in [\![p_i]\!]_{V'}$. We check that f is a c-morphism according to the conditions of Definition 4.1.

Condition (i): We need to show that for every $w' \in W'$ and $u \leq f(w')$ there is some $u' \leq' w'$ such that $f(u') = u$. Fix such w' and u and let $w_i = u$ and $w_j = f(w')$. If $u = f(w')$, then we can just let $u' = w'$ and thus we can assume that $u < f(w')$, implying that $w_i < w_j$. As the premise $p_i \vee p_j \rightsquigarrow p_i$ is true in M' and we have $w' \in [\![p_j]\!]_{V'}$, it follows that there must be some $u' \in [\![p_i]\!]_{V'}$ with $u' \leq' w'$. Then, from $u' \in [\![p_i]\!]_{V'}$, it follows that $f(u') = w_i = u$ and hence we have found a suitable u'.

Condition (ii): We need to show that for every $w_i \in W$ there is some $w' \in W'$ such that $f(w') = w_i$ and for every $u' \leq' w'$ we get $f(u') \leq w_i$. Fix w_i. Because the conclusion $p_i \vee \alpha_i \rightsquigarrow \alpha_i$ is false in M' it follows from Proposition 2.2 that there is some $w' \in [\![p_i]\!]_{V'}$ such that for all $u' \leq' w'$ it holds that $u' \notin [\![\alpha_i]\!]$. As $w' \in [\![p_i]\!]_{V'}$ we get that $f(w') = w_i$. Because α_i is defined as $\bigvee\{p_j \mid w_j \not\leq w_i\}$ the claim that $u' \notin [\![\alpha_i]\!]$ for all $u' \leq' w'$ is equivalent to the claim that $f(u') \leq w_i$ holds for all $u' \leq w'$. □

Proof. (Proof of Theorem 6.1) The direction from left to right follows from Theorem 4.5. For the direction from right to left, assume that $\mathcal{C}$ is closed under c-morphic images. Then, let $\Gamma_{\mathcal{C}}$ be the set of all formulas valid on the class $\mathcal{C}$, that is, $\Gamma_{\mathcal{C}} = \{\varphi \in \mathcal{L}_1 \mid \varphi$ is valid in P for all P in $\mathcal{C}\}$. It is obvious from the definition all the formulas in $\Gamma_{\mathcal{C}}$ are valid in all the posets from $\mathcal{C}$. To show that $\Gamma_{\mathcal{C}}$ defines $\mathcal{C}$ it suffices to show that if all of $\Gamma_{\mathcal{C}}$ is valid in some poset P, then $P \in \mathcal{C}$. Let P be such a poset. Consider its characteristic formula $\chi_P \in \mathcal{L}_1$. By Lemma 6.3 we know that P falsifies χ_P. Thus, it cannot be that $\chi_P \in \Gamma_{\mathcal{C}}$, which entails that there is some poset $P' \in \mathcal{C}$ that falsifies χ_P. From Proposition 6.4 it follows that P is a c-morphic image of P' and so P must also be in $\mathcal{C}$ as the latter is closed under c-morphic images. □

7 First-order correspondents

In this section we show that for every non-nested formula in conditional logic there is a formula in first-order logic that is true in precisely the finite posets where the formula of conditional logic is valid. Thus, every definable class of finite posets is elementary. For this result we are assuming a first-order language where the relations $\leq$ and $<$ of posets and the equality relation are expressible. We then obtain the following:

Theorem 7.1 *For all $\sigma \in \mathcal{L}_1$ one can compute a first-order formula φ_σ such that φ_σ is true in a finite poset P if and only if σ is valid in P.*

Remark 7.2 We do not expect that Theorem 7.1 still holds for formulas with nested conditionals. To consider an analogue of the theorem for the full language of conditional logic one would have to consider frames to be ternary relations $\leq$ such that for each world $w \in W$ the restricted relation $\leq_w$ is a poset. We conjecture that there are non-elementary classes of such frames that are definable in nested conditional logic. Applying the translation of the modal diamond $\Diamond\varphi$ as the conditional $\neg(\varphi \rightsquigarrow \bot)$ one can probably adapt examples of non-elementary classes that are definable in modal logic [3, sec. 3.2].

To prove Theorem 7.1, first observe that because of Corollary 2.6 it suffices to consider the case where σ is an inference of the form $\sigma = \Sigma/\Gamma$. We are going to define a first-order formula φ_σ that is true in some finite poset P iff σ is falsifiable in P. For the statement of the theorem we then need to consider the formula $\neg\varphi_\sigma$.

Let $\mathcal{A}$ the set of all propositional assignments to the propositional letters occurring in σ. It can be thought of as the set of all functions from these letters to the truth values 0 and 1. Note that $\mathcal{A}$ is a finite set. For every propositional formula $\alpha \in \mathcal{L}_0$ we define $[\![\alpha]\!] \subseteq \mathcal{A}$ to be the set of all assignments at which φ is true in the sense from classical propositional logic.

Define $\mathfrak{S} \subseteq \Gamma\Sigma^*$ to be the set of all sequences

$$(\gamma \rightsquigarrow \delta, \alpha_1 \rightsquigarrow \beta_1, \ldots, \alpha_n \rightsquigarrow \beta_n)$$

such that:

(i) $\gamma \rightsquigarrow \delta \in \Gamma$ is some conclusion from Γ;

(ii) $\alpha_j \rightsquigarrow \beta_j \in \Sigma$ is some premise from Σ for all $j \in \{1, \ldots, n\}$;

(iii) The Σ-part of the sequence does not contain repetitions. This means that $\alpha_j \rightsquigarrow \beta_j \neq \alpha_k \rightsquigarrow \beta_k$ for all $j, k \in \{1, \ldots, n\}$.

The *length* of a sequence $u \in \mathfrak{S}$ of the form $(\gamma \rightsquigarrow \delta, \alpha_1 \rightsquigarrow \beta_1, \ldots, \alpha_n \rightsquigarrow \beta_n)$ is the number n. We consider singleton sequences $(\gamma \rightsquigarrow \delta) \in \mathfrak{S}$, where $\gamma \rightsquigarrow \delta \in \Gamma$, to be of length 0. Note that because of the last clause the length of sequences in $\mathfrak{S}$ is bounded by the number of elements in Σ. Thus $\mathfrak{S}$ is a finite set.

Let $u = (\gamma \rightsquigarrow \delta, \alpha_1 \rightsquigarrow \beta_1, \ldots, \alpha_n \rightsquigarrow \beta_n)$ be some sequence from $\mathfrak{S}$ and let $\alpha \rightsquigarrow \beta \in \Sigma$ be such that $\alpha \rightsquigarrow \beta \neq \alpha_i \rightsquigarrow \beta_i$ for all $i \in \{1, \ldots, n\}$. We then write $u \cdot \alpha \rightsquigarrow \beta$ to denote the sequence $(\gamma \rightsquigarrow \delta, \alpha_1 \rightsquigarrow \beta_1, \ldots, \alpha_n \rightsquigarrow \beta_n, \alpha \rightsquigarrow \beta) \in \mathfrak{S}$.

We then consider partial functions S from the disjoint sum $\mathfrak{S}+\mathcal{P}\Gamma$ of $\mathfrak{S}$ and the powerset of Γ to the set $\mathcal{A}$. For any such partial function S let $D_S \subseteq \mathfrak{S}+\mathcal{P}\Gamma$ be the part of the domain on which S is defined. We call a partial function S from $\mathfrak{S} + \mathcal{P}\Gamma$ to $\mathcal{A}$ *coherent* if

(i) For all $\gamma \rightsquigarrow \delta \in \Gamma$ we have $(\gamma \rightsquigarrow \delta) \in D_S$ and $S((\gamma \rightsquigarrow \delta)) \in [\![\gamma \wedge \neg\delta]\!]$. Here, we consider $(\gamma \rightsquigarrow \delta)$ as a sequence in $\mathfrak{S}$ and explicitly distinguish it from the singleton set $\{\gamma \rightsquigarrow \delta\} \in \mathcal{P}\Gamma$;

(ii) If $u = (\gamma \rightsquigarrow \delta, \alpha_1 \rightsquigarrow \beta_1, \ldots, \alpha_n \rightsquigarrow \beta_n)$ and $u \in D_S$ then $S(u) \in [\![\alpha_n \wedge \beta_n]\!]$;

(iii) If $u = (\gamma \rightsquigarrow \delta, \alpha_1 \rightsquigarrow \beta_1, \ldots, \alpha_{n-1} \rightsquigarrow \beta_{n-1}, \alpha_n \rightsquigarrow \beta_n)$ and $u \in D_S$ then $S(u) \notin [\![\alpha_i]\!]$ for all $i \in \{1, \ldots, n-1\}$;

(iv) If $u \in \mathfrak{S} \cap D_S$ and $S(u) \in [\![\alpha \wedge \neg\beta]\!]$ for some $\alpha \rightsquigarrow \beta \in \Sigma$ then $u \cdot \alpha \rightsquigarrow \beta \in D_S$. Note that by the previous two conditions it is not possible that $\alpha \rightsquigarrow \beta$ is a premise that already occurs in u. Hence, $u \cdot \alpha \rightsquigarrow \beta$ is well-defined as an element of $\mathfrak{S}$;

(v) If $\Delta \in D_S$ for some $\Delta \subseteq \Gamma$ then $S(\Delta) \notin [\![\alpha \wedge \neg\beta]\!]$ for all $\alpha \rightsquigarrow \beta \in \Sigma$.

Let $\mathfrak{C}$ be the set of all coherent partial functions from $\mathfrak{S} + \mathcal{P}\Gamma$ to $\mathcal{A}$. Note that $\mathfrak{C}$ is finite.

We are working with a set of first-order variables $\{x_u \mid u \in \mathfrak{S} + \mathcal{P}\Gamma\}$.

The first-order formula φ_σ is defined such that it states the existence of some coherent partial function $S \in \mathfrak{C}$ and points in the poset that for each element of D_S:

$$\varphi_\sigma = \bigvee_{S \in \mathfrak{C}} \exists x_u \ldots u, v \in D_S \ldots \exists x_v (\kappa(S) \wedge \psi(S) \wedge \chi(S) \wedge \mu_1(S) \wedge \mu_2(S)).$$

We use the notation $\exists x_u \ldots u, v \in D_S \ldots \exists x_v$ to denote a chain of existential quantifiers that contains a quantifier for every variable x_u for $u \in D_S$. The points corresponding to the elements of D_S are further constrained by the formulas $\kappa(S)$, $\psi(S)$, $\chi(S)$, $\mu_1(S)$ and $\mu_2(S)$ that all depend on S and contain only free variables of the form x_u for $u \in D_S$.

The formula $\kappa(S)$ requires that any two variables that are interpreted as the same point must also map to the same assignment under S:

$$\kappa(S) = \bigwedge_{\substack{u,v \in D_S \\ S(u) \neq S(v)}} x_u \neq x_v.$$

The formula $\psi(S)$ requires that $\Delta \in D_S$ for every $\Delta \subseteq \Gamma$ for which there is some point in W that is below all the $x_{(\zeta)}$ for $\zeta \in \Delta$. Moreover, in this case x_Δ must be below all those $x_{(\zeta)}$:

$$\psi(S) = \bigwedge_{\Delta \subseteq \Gamma} (\exists y (\bigwedge_{\zeta \in \Delta} y < x_{(\zeta)}) \rightarrow \psi'(\Delta, S)),$$

where

$$\psi'(\Delta, S) = \begin{cases} \bot, & \text{if } \Delta \notin D_S, \\ \bigwedge_{\zeta \in \Delta} x_\Delta < x_{(\zeta)}, & \text{if } \Delta \in D_S. \end{cases}$$

The formula $\chi(S)$ states that if a sequence v from D_S extends another sequence u from D_S then x_v is below x_u in the poset:

$$\chi(S) = \bigwedge_{\substack{u \in D_S \cap \mathfrak{S} \\ v = u \cdot \xi \in D_S}} x_v < x_u.$$

The formula $\mu_1(S)$ requires that $x_{(\zeta)}$ for every $\zeta = \gamma \rightsquigarrow \delta \in \Gamma$ is minimal among all those points that interpret sequences that map under S to an assignment in $[\![\gamma]\!]$:

$$\mu_1(S) = \bigwedge_{\substack{\zeta \in \Gamma, u \in D_S \\ S(u) \in [\![\gamma]\!]}} \neg x_u < x_{(\zeta)}.$$

The formula $\mu_2(S)$ requires that x_u for every extended sequence $v = v' \cdot \alpha \rightsquigarrow \beta$, where $\alpha \rightsquigarrow \beta \in \Sigma$, is minimal among all those points that map to an assignment in $[\![\alpha]\!]$:

$$\mu_2(S) = \bigwedge_{\substack{v = v' \cdot \alpha \rightsquigarrow \beta \in D_S \\ u \in D_S, S(u) \in [\![\alpha]\!]}} \neg x_u < x_v.$$

It is clear that φ_σ can be computed from the inference σ. The remaining two lemmas of this section show that φ_σ does indeed express the falsifiability of σ.

Lemma 7.3 *If σ is falsifiable in a poset $(W, \leq)$ then $(W, \leq) \models \varphi_\sigma$.*

Proof. Let $V : \mathsf{Prop} \rightarrow \mathcal{P}W$ be a valuation such that $M = (W, \leq, V)$ makes all the premises in Σ true and all the conclusion in Γ false.

To show that φ_σ is true in the poset $(W, \leq)$ we define a coherent partial function $S \in \mathfrak{C}$ and for every $u \in D_S$ an interpretation $w_u \in W$ for the variable x_u. We use $a(w) \in \mathcal{A}$ for any $w \in W$ as a shorthand for the assignment

$$a(w)(p) = \begin{cases} 1, & \text{if } w \in V(p), \\ 0, & \text{if } w \notin V(p). \end{cases}$$

Note that with this definition $w \in [\![\varphi]\!]_V$ iff $a(w) \in [\![\varphi]\!]$ holds for all $\varphi \in \mathcal{L}_0$.

The definition of S and the selection of w_u, for $u \in D_S$, proceeds in three steps that extend the domain of S. In each of the steps our choice of the w_u is such that that $S(u) = a(w_u)$ for all $u \in D_S$.

(i) We first define S on sequences of the form $u = (\gamma \rightsquigarrow \delta)$ for all $\gamma \rightsquigarrow \delta \in \Gamma$. Fix a conclusion $\gamma \rightsquigarrow \delta \in \Gamma$. Because $\gamma \rightsquigarrow \delta$ is false in M, it follows that there is some world w that is minimal in $[\![\gamma]\!]_V$ such that $w \notin [\![\delta]\!]_V$. We let $w_u = w$ be this world and define $S(u) = a(w)$. Note that this ensures that S satisfies condition (i) from the definition of coherent functions.

(ii) In this step we inductively extend the definition of S to longer and longer sequences from $\mathfrak{S}$ such that condition (iv) becomes satisfied. This definition is by induction on the length of sequences $u \in \mathfrak{S}$. In every step, where we add a sequence $u = v \cdot \alpha \rightsquigarrow \beta$, we ensure that
(a) S and u satisfy conditions (ii) and (iii) for S being coherent,
(b) $w_u \leq w_v$, and
(c) $w_u \in \min_{\leq}([\![\alpha]\!]_V)$.
The base case consists simply of all the sequences of length 0 that were added in the previous step. In the inductive step assume that we have already added all required sequences of length n. Let $v \in D_S$ be a sequence of length n and assume that $S(v) \in [\![\alpha \wedge \neg\beta]\!]$ for some $\alpha \rightsquigarrow \beta \in \Sigma$. We are going to add the sequence $u = v \cdot \alpha \rightsquigarrow \beta$ to the definitional domain of S. Because $S(v) = a(w_v)$ it follows that $w_v \in [\![\alpha \wedge \neg\beta]\!]_V$. Since $M \models \alpha \rightsquigarrow \beta$ there must be some $w \in \min_{[\![\alpha]\!]_V}$ with $w \in [\![\beta]\!]_V$ and $w < w_v$. We set $w_u = w$ and $S(u) = a(w_u)$. This takes care of items (b) and (c) for u. This definition also satisfies condition (ii) for S being coherent because $S(u) \in [\![\alpha \wedge \beta]\!]$.
To check that u satisfies condition (iii) assume that $u' = v' \cdot \alpha' \rightsquigarrow \beta'$ is a proper initial segment of u. We need to show that $S(u) \notin [\![\alpha']\!]$, or equivalently that $w_u \notin [\![\alpha']\!]_V$. Inductively we can assume that all initial segments of v already satisfy item (b) from above. Thus, $w_v \leq w_{v'}$ and together with $w_u < w_v$ we obtain $w_u < w_{u'}$. Moreover, because u' satisfies item (c) we know that $w_{u'} \in \min_{\leq}([\![\alpha']\!]_V)$. Combining these facts we obtain that $w_u \notin [\![\alpha']\!]_V$.

(iii) Lastly, we consider any subset $\Delta \subseteq \Gamma$ such that there exists some $w'_\Delta \in W$ such that $w'_\Delta < w_{(\gamma \rightsquigarrow \delta)}$ for all $\gamma \rightsquigarrow \delta \in \Delta$. Fix such a Δ, define w_Δ to be any minimal element of $(W, \leq)$ that is below w'_Δ and set $S(\Delta) = a(w_\Delta)$. To see that this definition satisfies condition (v) on coherent partial functions we need to see that $w_\Delta \notin [\![\alpha \wedge \neg\beta]\!]_V$ for all $\alpha \rightsquigarrow \beta \in \Sigma$. If this was not the case then we would have that w_Δ is minimal in $[\![\alpha]\!]_V$, as it is minimal in W, but $w_\Delta \notin [\![\beta]\!]_V$. This would contradict the assumption that the model M makes the premise $\alpha \rightsquigarrow \beta \in \Sigma$ true.

It is clear that the partial function S that is defined in this way is coherent. It remains to be seen that the disjunct of φ_σ that corresponds to S is true in $(W, \leq)$. To this aim we interpret the existential variable x_u as the element

$w_u \in W$ for all $u \in D_S$. Because we have that $S(u) = a(w_u)$ for all $u \in D_S$ it is guaranteed that $\kappa(S)$ is true with this assignment. In the third step of the construction of S we make sure that $\psi(S)$ is true in $(W, \leq)$. The formula $\chi(S)$ holds because of item (b) from the second step. In the first step we chose $w_{(\gamma\rightsquigarrow\delta)} \in \min_{\leq}([\![\gamma]\!]_V)$ and hence $\mu_1(S)$ is true in $(W, \leq)$. Lastly, $\mu_2(S)$ holds because of item (c) from the second step. □

Lemma 7.4 *If* $(W, \leq) \models \varphi_\sigma$ *then* σ *is falsifiable in the poset* $(W, \leq)$.

Proof. Assume that the first-order formula φ_σ is true in the poset $(W, \leq)$. This means that there is some coherent $S \in \mathfrak{C}$ such that $\kappa(S)$, $\psi(S)$, $\chi(S)$, $\mu_1(S)$ and $\mu_2(S)$ hold for some interpretation of the existential variables from $\{x_u \mid u \in D_S\}$ in $(W, \leq)$. For all $u \in D_S$ let w_u be the value of the variable x_u for which this is the case. Define $X \subseteq W$ to be the set $X = \{w_u \in W \mid u \in D_S\}$.

Note that because $\kappa(S)$ holds for this interpretation of the existential variables it follows that $S(u) = S(v)$, whenever $w_u = w_v$ for some $u, v \in D_S$. For this reason the following function is well-defined $s : X \to \mathcal{A}, w_u \mapsto S(u)$.

Our next goal is to define a function $f : W \to X$ from which we then define the valuation $V : \mathsf{Prop} \to \mathcal{P}W$ by setting

$$V(p) = \{w \in W \mid s(f(w))(p) = 1\}.$$

To define the value of $f(w) \in X$ for some $w \in W$, we distinguish cases depending on how w is situated relative to the elements in X.

(i) If there is some $v \in X$ such that $v \leq w$ then we let $f(w) = y$ for some chosen $y \in X$ that is maximal among all $z \in X$ with $z \leq w$. Because X is finite such a maximal y always exists.

(ii) If there is no $v \in X$ such that $v \leq w$ then we consider the set $\Delta = \{\zeta \in \Gamma \mid w < w_{(\zeta)}\}$. Because $\psi(S)$ holds of our assignment of variables we have that $\Delta \in D_S$. Thus, we can set $f(w) = w_\Delta \in X$.

Note that because of the first clause f is the identity on all $w \in X \subseteq W$.

It remains to be proven that $M = (W, \leq, V)$ makes all conditionals in Σ true and all conditionals in Γ false.

Thus, consider any premise $\alpha \rightsquigarrow \beta \in \Sigma$. To show that $M \models \alpha \rightsquigarrow \beta$ we use the reformulation of the semantic clause from Proposition 2.2. To this aim take any $w \in [\![\alpha]\!]_V$. We need to find a $w' \leq w$ with $w' \in [\![\alpha \wedge \beta]\!]_V$. Distinguish cases depending on the definition of $f(w)$.

First consider the case where there is no $v \in X$ such that $v \leq w$. Then $f(w) = w_\Delta$ for some $\Delta \subseteq \Gamma$. Because of condition (v) of coherence it holds that $w_\Delta \notin [\![\alpha \wedge \neg\beta]\!]_V$. Note that the definition of V is such that w satisfies the same propositional letters as w_Δ because $f(w) = w_\Delta$. Thus $w \notin [\![\alpha \wedge \neg\beta]\!]_V$. Because $w \in [\![\alpha]\!]_V$ it follows that $w \in [\![\alpha \wedge \beta]\!]_V$ and we can take $w' = w$.

In the other case there is some $v \in X$ with $v \leq w$ then let $y \in X$ be such that $y = f(w)$ and $y \leq w$. Because $y \in X$ we have that $y = w_u$ for some $u \in \mathfrak{S} + \mathcal{P}\Gamma$. We distinguish further cases depending on whether $u \in \mathcal{P}\Gamma$ or $u \in \mathfrak{S}$. If $u \in \mathcal{P}\Gamma$ then $u = \Delta$ for some $\Delta \subseteq \Gamma$ and we can reason precisely as in the previous case. In the other case we have that $u \in \mathfrak{S}$. From $f(w) = w_u$ and $w \in [\![\alpha]\!]_V$ it follows that $w_u \in [\![\alpha]\!]_V$, because w and w_u satisfy the same propositional letters under V. Note that we can assume that $w_u \in [\![\alpha \wedge \neg\beta]\!]_V$ because if $w_u \in [\![\alpha \wedge \beta]\!]_V$ then also $w \in [\![\alpha \wedge \beta]\!]_V$ and we can set $w' = w$. But $w_u \in [\![\alpha \wedge \neg\beta]\!]_V$ means that $S(u) = s(w_u) \in [\![\alpha \wedge \neg\beta]\!]$. By condition (iv) from the definition of coherency this entails that $v = u \cdot \alpha \rightsquigarrow \beta \in D_S$. From condition (ii) we get that $S(v) \in [\![\alpha \wedge \beta]\!]$ and thus $w_v \in [\![\alpha \wedge \beta]\!]_V$. Using that $\chi(S)$ is true in $(W, \leq)$ we have that $w_v \leq w_u$. Using $w_u = y \leq w$ it follows that $w_v \leq w$. Thus we can take $w' = w_v$.

Lastly, we argue that the conclusions are false in M. Consider any conclusion $\gamma \rightsquigarrow \delta \in \Gamma$. By condition (i) on the coherent function S we have that $S(\gamma \rightsquigarrow \delta) \in [\![\gamma \wedge \neg\delta]\!]$. Thus $w_{(\gamma \rightsquigarrow \delta)} \in [\![\gamma \wedge \neg\delta]\!]_V$. We are going to show the claim that for all $w \in W$ with $w < w_{(\gamma \rightsquigarrow \delta)}$ we have $w \notin [\![\gamma]\!]_V$. From this it then follows by the alternative formulation of the semantics in Proposition 2.2 that $M \not\models \gamma \rightsquigarrow \delta$. To prove the claim, consider any $w < w_{(\gamma \rightsquigarrow \delta)}$. We distinguish cases depending on the clause defining $f(w)$.

If there is some $v \in X$ with $v \leq w$ then consider the $y \in X$ with $y \leq w$ such that $f(w) = y$. By transitivity it follows that $y \leq w_{(\gamma \rightsquigarrow \delta)}$ and hence we can use that $\mu_1(S)$ is true to derive that $y \notin [\![\gamma]\!]_V$. Because $f(w) = y$ we have that the valuation V is the same on w as on y and hence also $w \notin [\![\gamma]\!]_V$.

If there is no $v \in X$ with $v \leq w$ then $f(w) = w_\Delta$ for some $\Delta \subseteq \Gamma$ such that $\gamma' \rightsquigarrow \delta' \in \Delta$ whenever $w < w_{(\gamma' \rightsquigarrow \delta')}$. Because $w < w_{(\gamma \rightsquigarrow \delta)}$ this means that $\gamma \rightsquigarrow \delta \in \Gamma$. Because $\psi(S)$ is true in $(W, \leq)$ it follows that $w_\Delta < w_{(\gamma \rightsquigarrow \delta)}$ and because $\mu_1(S)$ is true it follows that $w_\Delta \notin [\![\gamma]\!]_V$. Because $f(w) = w_\Delta$ we can conclude that $w \notin [\![\gamma]\!]_V$. □

8 Conclusion

This paper provides results on frame definability in conditional logic. Definable classes of posets are characterized by being closed under c-morphic images and every definable class of posets is elementary.

An obvious direction for further research is to lift some of the limitations of the setting. First, one might be interested in studying definability of ternary relations by formulas in full conditional logic, where the conditional can occur nested. We expect that to obtain results in this direction one would have to combine ideas from this paper with ideas from the work on frame definability for normal modal logic. Second, one might try to generalize to the infinite case. We conjecture that most of our results generalize to wellfounded orders. For non-wellfounded orders, however, frame-definability seems to behave quite different than in the finite case. Third, one might try to adapt our approach to a setting that gives up some of the assumptions of anti-symmetry, transitivity or reflexivity that come from working with posets. We expect this to be quite challenging.

A further interesting open question is whether there are any general completeness results for conditional logics, similar to the Sahlqvist's completeness theorem for modal logic [26]. Many of the examples of formulas that we give in Section 3 were taken from literature that proves completeness results for the logic that is axiomatized by these formulas. One might hope that there is a general completeness result that gives a syntactic characterization a some class of formulas and then shows that if one adds formulas from this class as an additional axiom to the logic of Burgess [5] and Veltman [31], then one obtains a logic that is complete for the class of posets that the formula defines.

References

[1] Alchourrón, C. E., P. Gärdenfors and D. Makinson, *On the logic of theory change: Partial meet contraction and revision functions*, The Journal of Symbolic Logic **50** (1985), pp. 510–530.

[2] Baltag, A. and S. Smets, *Conditional doxastic models: A qualitative approach to dynamic belief revision*, Electronic Notes in Theoretical Computer Science **165** (2006), pp. 5–21.

[3] Blackburn, P., M. de Rijke and Y. Venema, "Modal Logic," Cambridge University Press, 2002.

[4] Booth, R. and I. Varzinczak, *Conditional inference under disjunctive rationality*, Proceedings of the AAAI Conference on Artificial Intelligence **35** (2021), pp. 6227–6234.

[5] Burgess, J., *Quick completeness proofs for some logics of conditionals*, Notre Dame Journal of Formal Logic **22** (1981), pp. 76–84.

[6] Ciabattoni, A., N. Olivetti and X. Parent, *Dyadic obligations: Proofs and countermodels via hypersequents*, in: *International Conference on Principles and Practice of Multi-Agent Systems*, 2022, pp. 54–71.

[7] Fishburn, P. C., *Intransitive indifference with unequal indifference intervals*, Journal of Mathematical Psychology **7** (1970), pp. 144–149.

[8] Fishburn, P. C., *Semiorders and choice functions*, Econometrica **43** (1975), pp. 975–977.

[9] Goble, L., *Axioms for hansson's dyadic deontic logics*, Filosofiska Notiser **6** (2019), pp. 13–61.

[10] Goldblatt, R. I. and S. K. Thomason, *Axiomatic classes in propositional modal logic*, in: *Algebra and Logic: Papers from the 1974 Summer Research Institute of the Australian Mathematical Society, Monash University, Australia*, 1975, pp. 163–173.

[11] Grossi, D., W. van der Hoek and L. B. Kuijer, *Reasoning about general preference relations*, Artificial Intelligence **313** (2022), p. 103793.

[12] Grove, A., *Two modellings for theory change*, Journal of Philosophical Logic **17** (1988), pp. 157–170.

[13] Hansson, B., *An analysis of some deontic logics*, Noûs **3** (1969), pp. 373–398.

[14] Jamison, D. T. and L. J. Lau, *Semiorders and the theory of choice*, Econometrica **41** (1973), pp. 901–912.

[15] Kraus, S., D. Lehmann and M. Magidor, *Nonmonotonic reasoning, preferential models and cumulative logics*, Artificial Intelligence **44** (1990), pp. 167–207.

[16] Lehmann, D. and M. Magidor, *What does a conditional knowledge base entail?*, Artificial Intelligence **55** (1992), pp. 1–60.

[17] Lewis, D., "Counterfactuals," Blackwell, Malden, MA, 1973.

[18] Luce, R. D., *Semiorders and a theory of utility discrimination*, Econometrica **24** (1956), pp. 178–191.

[19] Makinson, D., *General patterns in nonmonotonic reasoning*, in: D. M. Gabbay, C. J. Hogger and J. A. Robinson, editors, *Handbook of Logic in Artificial Intelligence and Logic Programming*, N/A **3**, Clarendon Press, 1994 pp. 35–110.

[20] Parent, X., *Completeness of Åqvist's systems e and f*, The Review of Symbolic Logic **8** (2015), pp. 164–177.

[21] Parent, X., *On some weakened forms of transitivity in the logic of conditional obligation*, Journal of Philosophical Logic **53** (2024), pp. 721–760.

[22] Peppas, P. and M.-A. Williams, *Belief change and semiorders*, in: *Fourteenth International Conference on the Principles of Knowledge Representation and Reasoning, KR 2014*, 2014.

[23] Åqvist, L., "Introduction to Deontic Logic and the Theory of Normative Systems," Humanities Press, Napoli, Italy, 1987.

[24] Rott, H., *Shifting priorities: Simple representations for twenty-seven iterated theory change operators*, in: D. Makinson, J. Malinowski and H. Wansing, editors, *Towards Mathematical Philosophy*, Springer, Dordrecht, 2009 pp. 269–296.

[25] Rott, H., *Four floors for the theory of theory change: The case of imperfect discrimination*, in: E. Fermé and J. Leite, editors, *Logics in Artificial Intelligence*, 2014, pp. 368–382.

[26] Sahlqvist, H., *Completeness and correspondence in the first and second order semantics for modal logic*, Studies in Logic and the Foundations of Mathematics **82** (1975), pp. 110–143.

[27] Scott, D. and P. Suppes, *Foundational aspects of theories of measurement*, The Journal of Symbolic Logic **23** (1958), pp. 113–128.

[28] Shoham, Y., "Reasoning about Change: Time and Causation from the Standpoint of Artificial Intelligence," MIT Press, 1988.

[29] Stalnaker, R. C., *A theory of conditionals*, in: N. Rescher, editor, *Studies in Logical Theory*, Blackwell, 1968 pp. 98–112.

[30] van Rooij, R., *Revealed preference and satisficing behavior*, Synthese **179** (2011), pp. 1–12.

[31] Veltman, F., "Logics for Conditionals," Ph.D. thesis, University of Amsterdam (1985).

[32] Zhu, Z., *Similarity between preferential models*, Theoretical Computer Science **353** (2006), pp. 26–52.

Birkhoff Style Proof Systems for Hybrid-Dynamic Quantum Logic

Daniel Găină [1]

Kyushu University

Abstract

We explore a simple approach to quantum logic based on hybrid and dynamic modal logic, where the set of states is given by some Hilbert space. In this setting, a notion of quantum clause is proposed in a similar way the notion of Horn clause is advanced in first-order logic, that is, to give logical properties for use in logic programming and formal specification. We propose proof rules for reasoning about quantum clauses and we investigate soundness and compactness properties that correspond to this proof calculus. Then we prove a Birkhoff completeness result for the fragment of hybrid-dynamic quantum logic determined by quantum clauses.

Keywords: quantum clause, Birkhoff completeness, hybrid-dynamic logic, algebraic specification, institution

1 Introduction

The logical framework for studying quantum theory was originally proposed by Birkhoff and von Neumann [8]. The truth values of this quantum logic are given by the elements of a Hilbert lattice, which is the lattice consisting of all closed subspaces of a Hilbert space. A more modern view on the semantics of quantum logic is based on Kripke structures and modal logics. One can express in a modal logic (based on a local Boolean satisfaction) quantum properties captured traditionally by (non-Boolean) Quantum Logic. For example, the *orthocomplement*, also called *quantum negation*, $\sim \varphi$, is defined as the set of all vectors orthogonal on the vectors where φ holds, while *quantum disjunction* $\varphi_1 \oplus \varphi_2$ is defined by $\sim (\sim \varphi_1 \wedge \sim \varphi_2)$. There are many approaches based on extensions of:

(i) dynamic logic such as Dynamic Quantum Logic [10], Logic of Quantum Actions [3] and Logic of Quantum Programs [4],

(ii) temporal logic such as Quantum Linear Temporal Logic [20] and Quantum Computation Tree Logic [7], and

[1] This work has been partially supported by Japan Society for the Promotion of Science, grant number 23K11048.

(iii) both dynamic and temporal logic such as Linear Temporal Quantum Logic proposed in [21].

In this paper, we study a variant of quantum logic with features from both hybrid and dynamic logics which was originally proposed in [17].

Dynamic propositional logic is suitable for reasoning about classical programs. Naturally, quantum versions of dynamic propositional logic were developed to reason about quantum programs. On the other hand, hybrid logics are known for their ability to name individual states of Kripke structures, which allow a more uniform proof theory [9,16] and model theory [19,15,18] than non-hybrid modal logics. Hybrid logics are equipped with features to distinguish states and reason about their properties, which in turn is important to applications in formal methods. In addition, one can express temporal properties using sentence operators inherent to hybrid logics such as store and retrieve [2]. Therefore, the variant of hybrid-dynamic quantum logic studied in this paper allows one to express both temporal and dynamical properties.

The users of a quantum programming language must describe the dynamics of quantum systems by relying on data types consisting of scalars and vectors of a concrete Hilbert space. Therefore, a necessary feature of a quantum programming language is the existence of some predefined data types for the scalars and vectors used in applications. From a model theoretic point of view, Hilbert spaces are vectorial spaces equipped with an inner product such that each Cauchy sequence of vectors has a limit. Since this is not a first-order property, the definition of Kripke structures whose states are given by some Hilbert space and the development of the logical results for them are significantly more difficult. Following the ideas advanced in [17], we employ the method of diagrams proposed by Robinson in model theory to define Hilbert spaces and Kripke structures over them. We use constant symbols to stand for the elements of the Hilbert space to be constructed, and we work within the theory which contains all the equations and relations satisfied by that Hilbert space. This means that the signature of nominals used to describe frames in hybrid logics is replaced by positive diagrams of concrete Hilbert spaces. The diagrams of Hilbert spaces can be regarded as the counterparts of libraries defining scalars and vectors from quantum programming, since individual elements of Hilbert spaces can be named in the hybrid-dynamic quantum logic studied in this paper. This is an important feature which brings the present work closer to applications in formal methods.

In this paper, we provide sound and complete proof calculi for a fragment of hybrid-dynamic quantum logic of [17]. The sentences are restricted to *quantum clauses*, which are obtained from propositional symbols by applying the following sentence operators: both quantum and classical implication, necessity over structural *actions* which, in turn, are constructed from *projective measurements* and *unitary transformations*, etc. In addition, we provide proof rules to reason formally about the properties of those Kripke structures that are specified using quantum clauses. To conclude, the main result of the paper is a completeness theorem for the fragment of hybrid-dynamic quantum logic obtained by restricting the sentences to quantum clauses.

A brief comparison with the work recently reported in [17] is also in order: both papers deal with properties of hybrid-dynamic quantum logic (however, [17] is the contribution in which we introduced the logic); and in both papers we examine quantum clauses; but the results that we develop are complementary: in [17], we focused on an initiality result and on Herbrand's theorem, whereas here we advance proof calculi for the logic. This latter endeavour is much more complex, because it deals with syntactic entailment instead of semantic entailment. In [13], the author proves a Birkhoff completeness result in the abstract framework given by notion of stratified institution [1,12], which is a category-based formalization of the notion of modal logic. Another completeness result for Horn clauses can be found in [14] for a variant of hybrid-dynamic first-order logic with user-defined sharing. It is worth noting that unlike [13] and [14], in this work, cut rule is not used to prove completeness, which means that lemma discovery is not needed in formal proofs.

According to [6], the correct semantics of quantum-logical connectives is in terms of dynamic modalities, rather than purely propositional operators. This philosophical interpretation is supported by some technical developments reported in [3,4,5]. The present work is based on the same ideas, but it departs from any of those studies due to the fact that the set of states is not the set of one-dimensional closed subspaces of some Hilbert space but the entire set of vectors of a Hilbert space. The present approach narrows the gap between theory and its applications (e.g., to formal methods), since it allows one to name concrete vectors and scalars and use them to build sentences.

2 Hilbert spaces

This section is dedicated to a brief presentation of Hilbert spaces from a first-order logic perspective. The signature of Hilbert spaces $\Sigma^{\mathrm{h}} = (S^{\mathrm{h}}, F^{\mathrm{h}}, P^{\mathrm{h}})$ is a first-order signature that consists of:

(i) a set of sorts $S^{\mathrm{h}} = \{\mathtt{c}, \mathtt{v}\}$, where $\mathtt{c}$ denotes the sort of complex numbers and $\mathtt{v}$ denotes the sorts of vectors.

(ii) a set of function symbols $F^{\mathrm{h}} = F^{\mathtt{c}} \cup F^{\mathtt{v}}$, where
   - (a) $F^{\mathtt{c}}$ is a set of the usual operations on complex numbers such as addition $\_+\_ : \mathtt{c}\ \mathtt{c} \to \mathtt{c}$, multiplication $\_*\_ : \mathtt{c}\ \mathtt{c} \to \mathtt{c}$, etc;
   - (b) $F^{\mathtt{v}} = \{\_+\_ : \mathtt{v}\ \mathtt{v} \to \mathtt{v}, 0 :\to \mathtt{v}, \_\_ : \mathtt{c}\ \mathtt{v} \to \mathtt{v}, \langle\_ \mid \_\rangle : \mathtt{v}\ \mathtt{v} \to \mathtt{c}\}$, where $\_+\_ : \mathtt{v}\ \mathtt{v} \to \mathtt{v}$ denotes the vector addition, 0 denotes the origin vector, the juxtaposition $\_\_ : \mathtt{c}\ \mathtt{v} \to \mathtt{v}$ denotes the scalar multiplication, and $\langle\_ \mid \_\rangle : \mathtt{v}\ \mathtt{v} \to \mathtt{c}$ denotes the inner product;

(iii) a singleton $P = \{\_<\_ : \mathtt{c}\ \mathtt{c}\}$ consisting of one relation symbol $\_<\_ : \mathtt{c}\ \mathtt{c}$ denoting the ordering among real numbers.

A Hilbert space is a first-order model $\mathcal{H}$ defined over Σ^{h} such that:

- $\mathcal{H}_{\mathtt{c}}$ is the set of complex numbers $\mathbb{C}$; the model $\mathcal{H}$ interprets all function symbols in $F^{\mathtt{c}}$ as the usual operations on complex numbers;

- $\mathcal{H}_v$ is a set of vectors, $+^{\mathcal{H}} : \mathcal{H}_v \times \mathcal{H}_v \to \mathcal{H}_v$ is the vector addition, $\_\_^{\mathcal{H}} : \mathcal{H}_c \times \mathcal{H}_v \to \mathcal{H}_v$ is the scalar multiplication, $\langle \_ \mid \_ \rangle^{\mathcal{H}} : \mathcal{H}_v \times \mathcal{H}_v \to \mathcal{H}_c$ is the inner product in which each Cauchy sequence of vectors has a limit. [2]
- $<^{\mathcal{H}}$ is the usual strict ordering on real numbers.

Two vectors $w_1, w_2 \in \mathcal{H}_v$ are orthogonal, in symbols, $w_1 \perp w_2$ if $\langle w_1 \mid w_2 \rangle^{\mathcal{H}} = 0^{\mathcal{H}}$. This notational convention can be extended to sets of vectors in the usual way. A closed subspace $\mathcal{X}$ of a Hilbert space $\mathcal{H}$ is a substructure $\mathcal{X} \subseteq \mathcal{H}$ such that (a) $\mathcal{X}_c = \mathbb{C}$ and (b) each Cauchy sequence of vectors from $\mathcal{X}$ has a limit in $\mathcal{X}$. In other words, a closed subspace is in particular a Hilbert space. In this paper, we identify a closed subspace by its set of vectors. The orthocomplement of $\mathcal{X}$ denoted $\mathcal{X}^{\perp}$ is defined by $\mathcal{X}^{\perp} = \{w \in \mathcal{H}_v \mid w \perp \mathcal{X}\}$. The orthocomplement of a set of vectors is a closed subspace. The direct sum of two closed subspaces $\mathcal{X}$ and $\mathcal{Y}$ is defined by $\mathcal{X} \oplus \mathcal{Y} := \{x +^{\mathcal{H}} y \mid x \in \mathcal{X} \text{ and } y \in \mathcal{Y}\}$. The following results are well-known.

Theorem 2.1 *Let $\mathcal{H}$ be a Hilbert space.*

(i) *A substructure $\mathcal{X} \subseteq \mathcal{H}$ is a closed subspace iff $\mathcal{X}^{\perp\perp} = \mathcal{X}$.*

(ii) $\mathcal{X} \oplus \mathcal{Y} = (\mathcal{X}^{\perp} \cap \mathcal{Y}^{\perp})^{\perp}$ *for all closed subspaces $\mathcal{X} \subseteq \mathcal{H}$ and $\mathcal{Y} \subseteq \mathcal{H}$.*

(iii) $\mathcal{X}^{\perp\perp} = \mathcal{Y} \cap (\mathcal{Y} \cap \mathcal{X}^{\perp})^{\perp}$ *for all closed subspaces $\mathcal{X} \subseteq \mathcal{Y} \subseteq \mathcal{H}$, which means that the (global) closure of $\mathcal{X}$ is equal to local closure of $\mathcal{X}$ relative to $\mathcal{Y}$.*

(iv) $\mathcal{H} = \mathcal{X} \oplus \mathcal{X}^{\perp}$ *for all closed subspaces $\mathcal{X} \subseteq \mathcal{H}$.*

By the fourth statement of the theorem above, for any closed subspace $\mathcal{X} \subseteq \mathcal{H}$ and any vector $w \in \mathcal{H}$, we have $w = w_1 +^{\mathcal{H}} w_2$, where $w_1 \in \mathcal{X}$ and $w_2 \in \mathcal{X}^{\perp}$. Since $\mathcal{X} \cap \mathcal{X}^{\perp} = \{0^{\mathcal{H}}\}$, the vectors $w_1 \in \mathcal{X}$ and $w_2 \in \mathcal{X}^{\perp}$ are uniquely determined. The projection $P_{\mathcal{X}} : \mathcal{H} \to \mathcal{X}$ is defined by $P_{\mathcal{X}}(w) = w_1$, where $w = w_1 + w_2$, $w_1 \in \mathcal{X}$ and $w_2 \in \mathcal{X}^{\perp}$.

3 Hybrid-Dynamic Quantum Logic

This section is dedicated to the presentation of Hybrid Dynamic Quantum Logic (HDQL) proposed in [17]. This logic is an extension of Hybrid-Dynamic Propositional Logic with some constraints on the possible world semantics.

3.1 Signatures

The signatures of HDQL are of the form $\Delta = (\Sigma, E, \mathtt{Prop})$, where:

(i) $\Sigma = (S^h, F^h \cup U \cup Q \cup D \cup C, P^h)$ is a first-order signature obtained from the signature of Hilbert spaces $\Sigma^h = (S^h, F^h, P^h)$ from Section 2 by adding:
   (a) a set U of unitary transformation symbols of the form $u : \mathrm{v} \to \mathrm{v}$;
   (b) a set Q of projective measurement symbols of the form $q : \mathrm{v} \to \mathrm{v}$;
   (c) a set D of constants of sort vector v; and
   (d) a set C of constants of sort scalar c;

[2] Recall that a sequence of vectors $\{w_n\}_{n\in\mathbb{N}}$ is a Cauchy sequence if for any $\varepsilon > 0$ there exists $n \in \mathbb{N}$ such that $||w_i - w_j|| < \varepsilon$ for all $i, j \geq n$, where $||w|| = \sqrt{\langle w \mid w \rangle}$ is the length/norm of a vector $w \in \mathcal{H}_v$.

(ii) E is a set of first-order sentences over Σ;

(iii) $\texttt{Prop}$ is a set of propositional symbols which contains a subset $\texttt{Prop}_c$ of *closed propositional symbols.*

An example of (Σ, E) is the positive Robinson diagram of some Hilbert space, that is, all equations and relations satisfied by some Hilbert space. See Section 4.1 for more details. We make the following notational conventions:

- We let Δ range over signatures of the form $(\Sigma, E, \texttt{Prop})$ as described above.
- Similarly, we let Δ_i range over signatures of the form $(\Sigma_i, E_i, \texttt{Prop}_i)$, where $\Sigma_i = (S^{\mathtt{h}}, F^{\mathtt{h}} \cup U_i \cup Q_i \cup D_i \cup C_i, P^{\mathtt{h}})$ and i is any index.

Signature morphisms $\chi : \Delta_1 \to \Delta_2$ consists of:

(i) a first-order theory morphism $\chi : (\Sigma_1, E_1) \to (\Sigma_2, E_2)$, which is the identity on $\Sigma^{\mathtt{h}}$, $\chi(U_1) \subseteq U_2$ and $\chi(Q_1) \subseteq Q_2$,[3] and

(ii) a mapping $\chi : \texttt{Prop}_1 \to \texttt{Prop}_2$ on propositional symbols, which preserves closed propositional symbols, that is, $\chi(\texttt{Prop}_1^{\mathtt{c}}) \subseteq \texttt{Prop}_2^{\mathtt{c}}$.

We overloaded the notation such that χ denotes not only the signature morphism $\chi : \Delta_1 \to \Delta_2$ but also its restrictions $\chi : \Sigma_1 \to \Sigma_2$ and $\chi : \texttt{Prop}_1 \to \texttt{Prop}_2$. We denote by $\texttt{Sig}$ the category of signatures in HDQL.

3.2 Models

A *quantum model* over a signature Δ is a Kripke structure (W, M) such that:

(i) W is a first-order model of the theory (Σ, E) such that
   (a) $W\restriction_{\Sigma^{\mathtt{h}}}$ is a Hilbert space, where $W\restriction_{\Sigma^{\mathtt{h}}}$ is the reduct of W to the first-order signature of Hilbert spaces $\Sigma^{\mathtt{h}}$,
   (b) for all symbols $\mathtt{u} : \mathtt{v} \to \mathtt{v} \in U$, the function $\mathtt{u}^W : W_{\mathtt{v}} \to W_{\mathtt{v}}$ is a unitary transformation, that is, $\mathtt{u}^W$ is a bounded linear operation which has an adjoint $(\mathtt{u}^W)^\dagger$ that is its inverse, $\mathtt{u}^W; (\mathtt{u}^W)^\dagger = (\mathtt{u}^W)^\dagger; \mathtt{u}^W = 1_{\mathcal{W}}$,
   (c) for all symbols $q : \mathtt{v} \to \mathtt{v} \in Q$, the function $q^W : W_{\mathtt{v}} \to W_{\mathtt{v}}$ is a projective measurement, that is, there exists a closed subspace $\mathcal{X} \subseteq W_{\mathtt{v}}$ such that $q^W(w) = P_{\mathcal{X}}(w)/\sqrt{\langle w \mid P_{\mathcal{X}}(w)\rangle}$ for all vectors $w \in W_{\mathtt{v}}$, where the function $P_{\mathcal{X}} : W_{\mathtt{v}} \to W_{\mathtt{v}}$ is the projection on $\mathcal{X}$.

(ii) $M : W_{\mathtt{v}} \to |\texttt{Mod}^{\mathsf{PL}}(\texttt{Prop})|$ is a mapping from the set of vectors $W_{\mathtt{v}}$ to the class of propositional models $|\texttt{Mod}^{\mathsf{PL}}(\texttt{Prop})|$ s.t. $r^{(W,M)} = \{w \in W_{\mathtt{v}} \mid r \in M_w\}$ is a closed subspace for all closed propositional symbols $r \in \texttt{Prop}_c$.[4]

We let (W, M) and (W^i, M^i) range over quantum models, where i is any index.

A homomorphism $h : (W, M) \to (W', M')$ in HDQL is a first-order homomorphism $h : W \to W'$ such that:

(i) $h_{\mathtt{c}} : W_{\mathtt{c}} \to W'_{\mathtt{c}}$ is the identity on complex numbers, and

[3] This means that $\chi : \Sigma_1 \to \Sigma_2$ is a first-order signature morphism such that E_2 satisfies $\chi(E_1)$, in symbols, $E_2 \models \chi(E_1)$.

[4] Notice that the propositional logic model M_w at the state w consists of a set of propositional symbols from $\texttt{Prop}$.

(ii) $M_w \subseteq M'_{h(w)}$ for all vectors $w \in W_{\mathtt{v}}$.

Lemma 3.1 *All homomorphisms of Hilbert spaces are injective. In particular, all homomorphisms of quantum models are injective.*

Proof. Let $h : (W, M) \to (V, N)$ be a homomorphism of quantum models. For all vectors $w \in W_{\mathtt{v}}$, the following are equivalent:

1	$h(w) = 0^V$	by inner product definition
2	$\langle h(w) \mid h(w)\rangle^V = 0$	since h is a first-order homomorphism
3	$h(\langle w \mid w\rangle^W) = 0$	h is the identity on complex numbers
4	$\langle w \mid w\rangle^W = 0$	by the definition of the inner product
5	$w = 0^W$	

Then for all $w_1, w_2 \in W_{\mathtt{v}}$, we have: $h(w_1) = h(w_2)$ iff $h(w_1) -^V h(w_2) = 0^V$ iff $h(w_1 -^W w_2) = 0^V$ iff $w_1 -^W w_2 = 0^W$ iff $w_1 = w_2$. Hence, h is injective. □

Remark 3.2 The class of quantum models over a signature Δ together with their homomorphisms forms a category denoted $\mathtt{Mod}(\Delta)$.

Given a signature morphism $\chi : \Delta \to \Delta'$, the reduct $(W', M')\upharpoonright_\chi$ of a Δ'-model (W', M') is a Δ-model defined by $(W', M')\upharpoonright_\chi = (W, M)$, where

(i) W is the reduct of W' across $\chi : \Sigma \to \Sigma'$ in first-order logic, in symbols, $W = W'\upharpoonright_\chi$, and

(ii) $M_w = \{p \in \mathtt{Prop} \mid \chi(p) \in M'_w\}$ is the reduct of M'_w across $\chi : \mathtt{Prop} \to \mathtt{Prop}'$ in propositional logic, in symbols, $M_w = M'_w \upharpoonright_\chi$, for all vectors $w \in W'_{\mathtt{v}}$.

The reduct $h'\upharpoonright_\chi$ of a homomorphism $h' \in \mathtt{Mod}(\Delta')$ is defined by $(h'\upharpoonright_\chi)_{\mathtt{v}} = h'_{\mathtt{v}}$ and $(h'\upharpoonright_\chi)_{\mathtt{c}} = h'_{\mathtt{c}}$.

Remark 3.3 For each signature morphism $\chi : \Delta \to \Delta'$ in $\mathtt{Sig}$, the model reduct $\_\upharpoonright_\chi : \mathtt{Mod}(\Delta') \to \mathtt{Mod}(\Delta)$ is a functor. Moreover, $\mathtt{Mod} : \mathtt{Sig} \to \mathbb{C}at^{op}$ defined by $\mathtt{Mod}(\chi)(h') = h'\upharpoonright_\chi$ for all signature morphisms $\chi : \Delta \to \Delta'$ and all homomorphisms $h' \in \mathtt{Mod}(\Delta')$, is a functor.

3.3 Sentences

The set of *actions* over a signature Δ is defined by the following grammar:

$$\mathfrak{a} ::= u \mid q \mid \mathfrak{a} \,\mathrm{;}\, \mathfrak{a} \mid \mathfrak{a} \cup \mathfrak{a} \mid \mathfrak{a}^*,$$

where u is a unitary transformation symbol and q is a quantum measurement symbol. We let $\mathfrak{a}$ and $\mathfrak{a}_i$ range over actions, where i is any index.

The set of *sentences*, $\mathtt{Sen}(\Delta)$, is defined by the following grammar:

$$\gamma ::= p \mid @_k\, \gamma \mid \gamma \wedge \gamma \mid \neg\gamma \mid\sim \gamma \mid [\mathfrak{a}]\gamma \mid {\downarrow} z \cdot \gamma,$$

where p is a propositional symbol, k is a term of sort vector, $\mathfrak{a}$ is an action, and z is a variable of sort vector. We refer to the sentence operators, in order, as *retrieve*, *conjunction*, *negation*, *quantum negation*, *necessity*, and *store*, respectively. Other quantum operators can be introduced as abbreviations. For example, quantum disjunction $\gamma_1 \oplus \gamma_2$ is defined by $\sim (\sim \gamma_1 \wedge \sim \gamma_2)$, for all sentences γ_1 and γ_2. We make the following notational conventions:

- We let k and k_i range over terms of sort vector, where i is any index.
- We let γ and γ_i range over hybrid-dynamic sentences, where i is any index.
- Similarly, we let Γ and Γ' range over sets of hybrid-dynamic sentences.

Each signature morphism $\chi : \Delta_1 \to \Delta_2$ induces a sentence translation $\chi : \mathtt{Sen}(\Delta_1) \to \mathtt{Sen}(\Delta_2)$ that replaces, in an inductive manner, in any sentence $\gamma \in \mathtt{Sen}(\Delta_1)$ the symbols from Δ_1 with symbols from Δ_2 according to $\chi : \Delta_1 \to \Delta_2$. As in ordinary multi-modal hybrid propositional logic, for each action $\mathfrak{a}$, one can define an operator *until* [2]:

$$Until_{\mathfrak{a}}(\gamma_1, \gamma_2) := {\downarrow} x \cdot \langle\mathfrak{a}\rangle {\downarrow} y \cdot \gamma_1 \wedge @_x ([\mathfrak{a}](\langle\mathfrak{a}\rangle y \Rightarrow \gamma_2)).$$

The current state is named x and then $\langle\mathfrak{a}\rangle$ is used to move to an accessible state, which is named y. The first argument of the conjunction says that the sentence γ_1 holds in the state y. The second argument of the conjunction sets the current state to x by applying $@_x$; then the sentence γ_2 holds in any state which succeeds x and precedes y.

Remark 3.4 The mapping $\mathtt{Sen} : \mathtt{Sig} \to \mathbb{S}et$ from the category of signatures $\mathtt{Sig}$ to the category of sets $\mathbb{S}et$ is a functor.

3.4 Local satisfaction relation

Let (W, M) be a quantum model over a signature Δ. The semantics of actions is defined in the standard way: (a) $(\mathfrak{a}_1 ⨾ \mathfrak{a}_2)^W = \mathfrak{a}_1^W ⨾ \mathfrak{a}_2^W$; (b) $(\mathfrak{a}_1 \cup \mathfrak{a}_2)^W = \mathfrak{a}_1^W \cup \mathfrak{a}_2^W$; (c) $(\mathfrak{a}^*)^W = \bigcup_{n\in\mathbb{N}} (\mathfrak{a}^n)^W$, where $\mathfrak{a}^0$ denotes the identity, and $\mathfrak{a}^{n+1} = \mathfrak{a} ⨾ \mathfrak{a}^n$ for all natural numbers $n \in \mathbb{N}$. The semantics of sentences is defined as follows:

- $p^{(W,M)} = \{w \in W_{\mathtt{v}} \mid p \in M_w\}$ for all propositional symbols p in Δ;
- $(@_k \gamma)^{(W,M)} = \begin{cases} W_{\mathtt{v}} & \text{if } k^W \in \gamma^{(W,M)}, \\ \emptyset & \text{if } k^W \notin \gamma^{(W,M)}; \end{cases}$
- $(\gamma_1 \wedge \gamma_2)^{(W,M)} = \gamma_1^{(W,M)} \cap \gamma_2^{(W,M)}$;
- $(\neg\gamma)^{(W,M)} = W_{\mathtt{v}} \setminus \gamma^{(W,M)}$;
- $(\sim \gamma)^{(W,M)} = (\gamma^{(W,M)})^{\perp}$;
- $([\mathfrak{a}]\gamma)^{(W,M)} = \{w \in W_{\mathtt{v}} \mid \mathfrak{a}^W(w) \subseteq \gamma^{(W,M)}\}$, where $\mathfrak{a}^W(w) = \{v \in W_{\mathtt{v}} \mid (w, v) \in \mathfrak{a}^W\}$;
- $({\downarrow} z \cdot \gamma)^{(W,M)} = \{w \in W_{\mathtt{v}} \mid w \in \gamma^{(W^{z\leftarrow w}, M)}\}$, where $(W^{z\leftarrow w}, M)$ is the unique expansion of (W, M) to $\Delta[z]$ which interprets z as w. Notice that $(W^{z\leftarrow w}, M)$ interprets all symbols in Δ as (W, M).

We say that (W, M) *satisfies* γ *in the state* w, in symbols, $(W, M) \models^w \gamma$, if $w \in \gamma^{(W,M)}$. The following result shows that the truth is invariant w.r.t. the change of notation, that is, HDQL is a stratified institution [1].

Theorem 3.5 (Local satisfaction condition) *For all signature morphisms $\chi : \Delta \to \Delta'$, all quantum models (W', M') defined over the signature Δ', all sentences γ defined over the signature Δ, and all vectors $w \in W'_{\mathtt{v}}$, we have:*

$(W', M') \models^w \chi(\gamma)$ *iff* $(W', M')\upharpoonright_\chi \models^w \gamma$.

Proof. Let (W, M) be the reduct of (W', M') across $\chi : \Delta \to \Delta'$. Since χ is the identity on the signature of Hilbert spaces $\Sigma^{\mathtt{h}}$, the Hilbert spaces $W\upharpoonright_{\Sigma^{\mathtt{h}}}$ and $W'\upharpoonright_{\Sigma^{\mathtt{h}}}$ coincide. It follows that $W_{\mathtt{v}} = W'_{\mathtt{v}}$ which means that the local satisfaction condition is well-defined. The proof of the local satisfaction condition is by well-founded (Noetherian) induction on the triple (n_1, n_2, n_3), where (i) n_1 is the number of occurrences of the operator $*$ in γ, (ii) n_2 is the number of occurrences of the operators $\fatsemi$ and $\cup$ in γ, and (iii) n_3 is the number of occurrences of the sentence operators in γ. We focus only on a few cases, since the rest of them can be discharged using the same arguments from the proof of the satisfaction condition in hybridized institutions [11, Theorem 3.2]. We denote by (W, M) the reduct of (W', M') across χ.

[$p \in \mathtt{Prop}$] The following are equivalent:

1	$(W', M') \models^w \chi(p)$	by semantics
2	$\chi(p) \in M'_w$	by the definition of M_w
3	$p \in \{p \in \mathtt{Prop} \mid \chi(p) \in M'_w\} = M_w$	by semantics
4	$(W, M) \models^w p$	

[$\sim\gamma$] By induction hypothesis, we have $(W', M') \models^w \chi(\gamma)$ iff $(W, M) \models^w \gamma$ for all vectors $w \in W'_{\mathtt{v}}$. Then:

1	$\chi(\gamma)^{(W',M')} = \gamma^{(W,M)}$	by semantics
2	$(\chi(\gamma)^{(W',M')})^\perp = (\gamma^{(W,M)})^\perp$	as $W'\upharpoonright_{\Sigma^{\mathtt{h}}} = W\upharpoonright_{\Sigma^{\mathtt{h}}}$
3	$(W', M') \models^w \sim \chi(\gamma)$ iff $(W, M) \models^w \sim \gamma$ for all $w \in W'_{\mathtt{v}}$	by semantics

[$[f]\gamma$] In this case, $f \in U \cup Q$. The following are equivalent:

1	$(W', M') \models^w [\chi(f)]\chi(\gamma)$	by semantics
2	$(W', M') \models^v \chi(\gamma)$, where $v = \chi(f)^{W'}(w)$	by induction hypothesis
3	$(W', M')\upharpoonright_\chi \models^v \gamma$	since $v = f^{(W'\upharpoonright_\chi)}(w)$
4	$(W', M')\upharpoonright_\chi \models^w [f]\gamma$	

[$[\mathfrak{a}_1 \fatsemi \mathfrak{a}_2]\gamma$]

The number of $*$ in $[\mathfrak{a}_1][\mathfrak{a}_2]\gamma$ is equal to the number of $*$ in $[\mathfrak{a}_1 \fatsemi \mathfrak{a}_2]\gamma$. The number of $\fatsemi$ in $[\mathfrak{a}_1][\mathfrak{a}_2]\gamma$ is strictly less than the number of $\fatsemi$ in $[\mathfrak{a}_1 \fatsemi \mathfrak{a}_2]\gamma$. The number of $\cup$ in $[\mathfrak{a}_1][\mathfrak{a}_2]\gamma$ is equal to the number of $\cup$ in $[\mathfrak{a}_1 \fatsemi \mathfrak{a}_2]\gamma$. By induction hypothesis, we have:

$$(W', M') \models^w [\chi(\mathfrak{a}_1)][\chi(\mathfrak{a}_2)]\chi(\gamma) \text{ iff } (W', M')\upharpoonright_\chi \models^w [\mathfrak{a}_1][\mathfrak{a}_2]\gamma.$$

The following are equivalent:

1	$(W', M') \models^w [\chi(\mathfrak{a}_1) \fatsemi \chi(\mathfrak{a}_2)]\chi(\gamma)$	by semantics
2	$(W', M') \models^w [\chi(\mathfrak{a}_1)][\chi(\mathfrak{a}_2)]\chi(\gamma)$	by induction hypothesis
3	$(W', M')\upharpoonright_\chi \models^w [\mathfrak{a}_1][\mathfrak{a}_2]\gamma$	by semantics
4	$(W', M')\upharpoonright_\chi \models^w [\mathfrak{a}_1 \fatsemi \mathfrak{a}_2]\gamma$	

[$[\mathfrak{a}_1 \cup \mathfrak{a}_2]\gamma$]

The number of $*$ in $[\mathfrak{a}_1]\gamma$ is less or equal to the number of $*$ in $[\mathfrak{a}_1 \cup \mathfrak{a}_2]\gamma$.

The number of $\mathbin{;}$ in $[\mathfrak{a}_1]\gamma$ is less or equal than the number of $\mathbin{;}$ in $[\mathfrak{a}_1 \cup \mathfrak{a}_2]\gamma$. The number of $\cup$ in $[\mathfrak{a}_1]\gamma$ is strictly less than the number of $\cup$ in $[\mathfrak{a}_1 \cup \mathfrak{a}_2]\gamma$. By induction hypothesis, we have:

$$(W', M') \models^w [\chi(\mathfrak{a}_1)]\chi(\gamma) \text{ iff } (W', M')\upharpoonright_\chi \models^w [\mathfrak{a}_1]\gamma.$$

Similarly, induction hypothesis holds for $[\mathfrak{a}_2]\gamma$ too.
The following are equivalent:

1	$(W', M') \models^w [\chi(\mathfrak{a}_1) \cup \chi(\mathfrak{a}_2)]\chi(\gamma)$	by semantics
2	$(W', M') \models^w [\chi(\mathfrak{a}_1)]\chi(\gamma)$ and $(W', M') \models^w [\chi(\mathfrak{a}_2)]\chi(\gamma)$	by induction hypothesis
3	$(W', M')\upharpoonright_\chi \models^w [\mathfrak{a}_1]\gamma$ and $(W', M')\upharpoonright_\chi \models^w [\mathfrak{a}_2]\gamma$	by semantics
4	$(W', M')\upharpoonright_\chi \models^w [\mathfrak{a}_1 \cup \mathfrak{a}_2]\gamma$	

[$[\mathfrak{a}^*]\gamma$] The number of $*$ in $[\mathfrak{a}^n]\gamma$ is strictly less than the number of $*$ in $[\mathfrak{a}^*]\gamma$, for all natural numbers $n \in \mathbb{N}$. By induction hypothesis, we have:

$$(W', M') \models^w [\chi(\mathfrak{a})^n]\chi(\gamma) \text{ iff } (W', M')\upharpoonright_\chi \models^w [\mathfrak{a}^n]\gamma \text{ for all } n \in \mathbb{N}.$$

The following are equivalent:

1	$(W', M') \models^w [\chi(\mathfrak{a})^*]\chi(\gamma)$	by semantics
2	$(W', M') \models^w [\chi(\mathfrak{a})^n]\chi(\gamma)$ for all $n \in \mathbb{N}$	by induction hypothesis
3	$(W', M')\upharpoonright_\chi \models^w [\mathfrak{a}^n]\gamma$ for all $n \in \mathbb{N}$	by semantics
4	$(W', M')\upharpoonright_\chi \models^w [\mathfrak{a}^*]\gamma$	

□

3.5 Global satisfaction relation

The global satisfaction relation between models and sentences is defined below:

- $(W, M) \models \gamma$, read (W, M) globally satisfies γ, when $\gamma^{(W,M)} = W_{\mathrm{v}}$.

In formal methods, the global satisfaction relation is at the core of formal verification, since the engineers need to model software and hardware systems with sets of sentences that need to be satisfied globally. The global satisfaction relation between models and sentences is naturally extended to a satisfaction relation between sets of sentences.

- $\Gamma \models \gamma$, read Γ globally satisfies γ, when $(W, M) \models \Gamma$ implies $(W, M) \models \gamma$, for all quantum models (W, M).

Notice that $\emptyset \models \bigwedge \Gamma \Rightarrow \gamma$ implies $\Gamma \models \gamma$ but the backward implication does not hold. This means that the semantics of the satisfaction relation $\models$ between sentences is different from the standard one used in modal logic literature.

- $\Gamma \models^k \gamma$, read Γ satisfies γ at k, when $(W, M) \models \Gamma$ implies $(W, M) \models^{(k^W)} \gamma$, for all models (W, M).

Notice that $\Gamma \models^k \gamma$ iff $\Gamma \models @_k\, \gamma$. Also, $\emptyset \models^k \bigwedge \Gamma \Rightarrow \gamma$ implies $\Gamma \models^k \gamma$ but the backward implication does not hold. The following result is a direct corollary of Theorem 3.5.

Corollary 3.6 $\Gamma \models^k \gamma$ *implies* $\chi(\Gamma) \models^{\chi(k)} \chi(\gamma)$, *for all signature morphisms*

$\chi : \Delta \to \Delta'$, *all sets of* Δ*-sentences* Γ *and all* Δ*-sentences* γ.

3.6 Closed sentences

In quantum logic literature, a distinguished class of sentences consist of all sentences that are interpreted as closed spaces. In this section, we will briefly look into some of their properties. The results presented in this subsection are from [17]. The set of *closed sentences* $\mathtt{Sen}_c(\Delta)$ over a signature Δ is defined by the following grammar:

$$\rho ::= r \mid\sim \rho \mid \rho \wedge \rho \mid [\mathfrak{b}]\rho,$$

where r is a closed propositional symbol, $\mathfrak{b}$ is a unitary action, that is, an action free of quantum measurement symbols. The following result shows that the interpretation of any closed sentence in a model is a closed subspace.

Theorem 3.7 *The semantics of any closed sentence is a closed subspace, i.e., for all models* (W, M) *and all closed sentences* ρ *defined over the same signature,* $\rho^{(W,M)}$ *is a closed subspace.*

Closed sentences have some unique features which distinguish them from the rest of the sentences. Some properties of closed sentences are stated in the next corollary.

Corollary 3.8 *For all models* (W, M) *and all closed sentences* ρ *defined over the same signature, we have:*

(i) $(W, M) \models^{(0^W)} \rho$.

(ii) *If* $(W, M) \models^{w} \rho$ *then* $(W, M) \models^{aw} \rho$, *for all vectors* $w \in W_{\mathtt{v}}$ *and all complex numbers* $a \in W_{\mathtt{c}}$.

(iii) *If* $(W, M) \models^{w_1} \rho$ *and* $(W, M) \models^{w_2} \rho$ *then* $(W, M) \models^{w_1+w_2} \rho$, *for all vectors* $w_1, w_2 \in W_{\mathtt{v}}$.

(iv) *Let* $\{w_n\}_{n\in\mathbb{N}}$ *be a Cauchy sequence of vectors, and let* w *be its limit.*
If $(W, M) \models^{w_n} \rho$ *for all* $n \in \mathbb{N}$ *then* $(W, M) \models^{w} \rho$.

Sasaki hook $\rho_1 \rightsquigarrow \rho_2$ is defined by $\sim (\rho_1 \wedge \sim (\rho_1 \wedge \rho_2))$, for all closed sentences ρ_1 and ρ_2. The following lemma shows that Sasaki hook can be viewed as an implication for closed sentences, that is, a quantum implication.

Lemma 3.9 (Quantum implication) *For all quantum models* (W, M) *and all closed sentences* ρ_1, ρ_2 *defined over the same signature, we have:*

(i) $\rho_1^{(W,M)} \cap (\rho_1 \rightsquigarrow \rho_2)^{(W,M)} \subseteq \rho_2^{(W,M)}$ *and*

(ii) $\rho_1^{(W,M)} \subseteq \rho_2^{(W,M)}$ *iff* $(W, M) \models \rho_1 \rightsquigarrow \rho_2$.

3.7 Entailment systems

The syntactic counterpart of the satisfaction relation for reasoning about the logical consequences of sentences is defined below.

Definition 3.10 [Entailment systems] An *entailment system* is a family of relations $\vdash = \{\vdash^k_\Delta\}_{\Delta\in|\mathtt{Sig}|, k\in T_\Sigma}$ between sets of sentences and sentences (that is,

$$(Monotonicity)\ \frac{\gamma \in \Gamma}{\Gamma \vdash^k \gamma} \qquad (Unions)\ \frac{\Gamma \vdash^k \gamma}{\Gamma \cup \Gamma' \vdash^k \gamma}$$

$$(Cut)\ \frac{\Gamma \vdash^k \gamma_2 \quad \Gamma \cup \{\gamma_2\} \vdash^k \gamma_1}{\Gamma \vdash^k \gamma_1} \qquad (Translation)\ \frac{\Gamma \vdash^k \gamma}{\chi(\Gamma) \vdash^{\chi(k)} \chi(\gamma)} \text{ where } \chi : \Delta \to \Delta'$$

Fig. 1. Entailment systems

$\vdash^k_\Delta \subseteq \mathcal{P}(\mathtt{Sen}(\Delta)) \times \mathtt{Sen}(\Delta)$ for all signatures Δ and all terms k of sort vector) satisfying (*Monotonicity*), (*Unions*) and (*Translation*) defined in Fig. 1.

We drop the subscript Δ from the notion $\vdash^k_\Delta$ when there is no danger of confusion. (*Cut*) is omitted from the definition of entailment system which shows that lemma discovery is not needed for the proof calculus developed in this paper.

Definition 3.11 [Entailment properties] Let $\vdash$ be an entailment system.

(i) $\vdash$ is *sound (complete)* if for all signatures Δ and all terms k, we have: $\vdash^k_\Delta \subseteq \models^k_\Delta$ ($\models^k_\Delta \subseteq \vdash^k_\Delta$).

(ii) $\vdash$ is *compact* if for all signatures Δ, all sets of sentences Γ, all sentences γ, and all terms k, we have: $\Gamma \vdash^k_\Delta \gamma$ implies $\Gamma_f \vdash^k_\Delta \gamma$ for some finite $\Gamma_f \subseteq \Gamma$.

4 Birkhoff completeness

We introduce two classes of sentences for which we study soundness, compactness and completeness.

Definition 4.1 [Clauses] Let Δ be any signature in HDQL.

(i) The set of *basic sentences* over Δ is defined by the following grammar:

$$\varphi ::= p \mid \varphi \wedge \varphi \mid @_k\, \varphi \mid [\mathfrak{a}]\varphi \mid {\downarrow} z \cdot \varphi,$$

where p is a propositional symbol, k is a term of sort vector, $\mathfrak{a}$ is an action, and z is a variable of sort vector.

(ii) The set of *quantum clauses* over Δ is defined by the following grammar:

$$\gamma ::= p \mid \rho_1 \rightsquigarrow \rho_2 \mid \varphi \Rightarrow \gamma \mid \gamma \wedge \gamma \mid @_k\, \gamma \mid [\mathfrak{a}]\gamma \mid {\downarrow} z \cdot \gamma,$$

where p is a propositional symbol, ρ_1 is a closed basic sentence, ρ_2 is a closed quantum clause, φ is a basic sentence, k is a term of sort vector, $\mathfrak{a}$ is an action, and z is a variable of sort vector.

4.1 Logical framework $\mathcal{L}$

We define the logical framework in which the results will be proved. This is a fragment of HDQL defined in Section 3.

Let $\Delta = (\Sigma, \emptyset, \mathtt{Prop})$ be a signature such that $\Sigma = (S^{\mathtt{h}}, F^{\mathtt{h}} \cup U \cup Q, P^{\mathtt{h}})$, where $\Sigma^{\mathtt{h}} = (S^{\mathtt{h}}, F^{\mathtt{h}}, P^{\mathtt{h}})$ is the signature of Hilbert spaces defined in Section 3, U is a set of unitary transformation symbols and Q is a set of measurement symbols. Let $\mathcal{W}$ be a first-order model defined over the signature Σ such that

(i) $\mathcal{W}\upharpoonright_{\Sigma^{\mathtt{h}}}$ is a Hilbert space,

(ii) $u^{\mathcal{W}}: \mathcal{W}_{\mathtt{v}} \to \mathcal{W}_{\mathtt{v}}$ is a unitary transformation for all $u: \mathtt{v} \to \mathtt{v} \in U$, and

(iii) $q^{\mathcal{W}}: \mathcal{W}_{\mathtt{v}} \to \mathcal{W}_{\mathtt{v}}$ is a quantum measurement for all $q: \mathtt{v} \to \mathtt{v} \in Q$.

We make the following notational conventions:

(i) Let $\Sigma_{\mathcal{W}}$ be the first-order signature obtained from Σ by adding all elements in $\mathcal{W}$ as constants, that is, $\Sigma_{\mathcal{W}} = (S^{\mathtt{h}}, F^{\mathtt{h}} \cup U \cup Q \cup D_{\mathcal{W}} \cup C_{\mathcal{W}}, P^{\mathtt{h}})$, where
   (a) $D_{\mathcal{W}} = \mathcal{W}_{\mathtt{v}}$, the set of vectors in $\mathcal{W}$, and
   (b) $C_{\mathcal{W}} = \mathbb{C}$, the set of complex numbers.

(ii) Let $\mathcal{W}_{\mathcal{W}}$ be the first-order model over $\Sigma_{\mathcal{W}}$ obtained from $\mathcal{W}$ by interpreting each constant $c \in C_{\mathcal{W}}$ as the complex number c and each constant $w \in D_{\mathcal{W}}$ as the vector w.

(iii) Let $E_{\mathcal{W}}$ be the set of (ground) equations and relations satisfied by $\mathcal{W}_{\mathcal{W}}$, which means that $(\Sigma_{\mathcal{W}}, E_{\mathcal{W}})$ is the positive diagram of $\mathcal{W}$.

In classical model theory, it is well-known that $\Sigma_{\mathcal{W}}$-models which satisfy $E_{\mathcal{W}}$ are in one-to-one correspondence with the Σ-homomorphisms with the domain $\mathcal{W}$, that is, there is an isomorphism of categories $\mathtt{Mod}(\Sigma_{\mathcal{W}}, E_{\mathcal{W}}) \cong \mathcal{W}/\mathtt{Mod}(\Sigma)$.

The underlying logic in which the subsequent results will be developed is an arbitrary fragment $\mathcal{L}$ of HDQL satisfying the following two properties:

(i) All signatures are of the form $(\Sigma_{\mathcal{W}}, E_{\mathcal{W}}, \mathtt{Prop})$ as described above.

(ii) Retrieve @ belongs to the vocabulary of $\mathcal{L}$ if classical implication $\Rightarrow$ or quantum implication $\leadsto$ belongs to the vocabulary of $\mathcal{L}$.

The fragment $\mathcal{L}$ is obtained from HDQL by restricting its syntax, that is, its signatures and sentences. Given a signature morphism $\chi: (\Sigma_{\mathcal{W}}, E_{\mathcal{W}}, \mathtt{Prop}) \to (\Sigma'_{\mathcal{W}'}, E_{\mathcal{W}'}, \mathtt{Prop}')$ in $\mathcal{L}$, since $(\Sigma_{\mathcal{W}}, E_{\mathcal{W}})$ and $(\Sigma'_{\mathcal{W}'}, E_{\mathcal{W}'})$ are the positive diagrams of the first-order models $\mathcal{W}$ and $\mathcal{W}'$, respectively, by Lemma 3.1, the first-order signature morphism $\chi: \Sigma_{\mathcal{W}} \to \Sigma'_{\mathcal{W}'}$ is injective. Moreover, $\chi: \Sigma_{\mathcal{W}} \to \Sigma'_{\mathcal{W}'}$ is the identity on complex numbers. In addition, one or more sentence operators from HDQL can be discarded. Of course, if retrieve is discarded from the grammar used to define sentences in HDQL then both classical and quantum implication must be dropped. This condition is necessary because the proof rules for both classical and quantum implication depend on the existence of retrieve. See Section 4.3. Other than that, the sentence operators are independent. For example, quantum implication is not required by any of the results in this paper. If quantum implication occurs, there are proof rules and arguments to deal with it. If quantum implication is not part of the vocabulary then the proofs still hold, since the cases corresponding to quantum implication can be, simply, discarded. $\mathcal{L}$ can be regarded as a parameter for the subsequent developments which can be adjusted depending on the applications. Fig. 2 contains the proof rules for reasoning about closed propositions. Notice that r ranges over closed propositional symbols, w ranges over vectors

$$(Origin)\ \frac{}{\Gamma \vdash^{0} r} \qquad (Mult)\ \frac{\Gamma \vdash^{w} r}{\Gamma \vdash^{aw} r}$$

$$(Add)\ \frac{\Gamma \vdash^{w_1} r \quad \Gamma \vdash^{w_2} r}{\Gamma \vdash^{w_1+w_2} r} \qquad (Cauchy)\ \frac{\Gamma \vdash^{w_n} r \text{ for all } n \in \mathbb{N}}{\Gamma \vdash^{w} r}$$

where $\{w_n\}_{n\in\mathbb{N}}$ is a Cauchy sequence such that $\lim_{n\to\infty} w_n = w$

Fig. 2. Closed propositional symbols

and a ranges over complex numbers. (*Cauchy*) is an infinitary rule, since it has a countably infinite number of premises, one for each natural number $n \in \mathbb{N}$. In applications, it is rarely used. Assume, for example, that each closed propositional symbol r is defined by an orthonormal basis $\{v_1, \ldots, v_n\}$ of a closed subspace, that is, $\{@_{v_1} r, \ldots, @_{v_n} r\} \subseteq \Gamma$ and r does not occur positively in any of the sentences from $\Gamma \setminus \{@_{v_1} r, \ldots, @_{v_n} r\}$. However, r can occur, for example, in the conditional part of the clauses from $\Gamma \setminus \{@_{v_1} r, \ldots, @_{v_n} r\}$. In this case, (*Cauchy*) can be discarded because the set of states where each r holds is the closed subspace generated by linear combinations of vectors from $\{v_1, \ldots, v_n\}$.

Definition 4.2 [Models defined by sentences] Let $\vdash$ be any sound entailment system of $\mathcal{L}$ closed under the proof rules defined in Fig. 2. Any set of sentences Γ over a signature $\Delta_{\mathcal{W}}$ defines a model (W^Γ, M^Γ) as follows:

(i) $W^\Gamma = \mathcal{W}_{\mathcal{W}}$, and

(ii) $M^\Gamma : W^\Gamma_{\mathsf{v}} \to |\mathtt{Mod}^{\mathsf{PL}}(\Sigma_{\mathcal{W}})|$ is defined by $M^\Gamma_w = \{p \in \mathtt{Prop} \mid \Gamma \vdash^w p\}$ for all vectors $w \in W^\Gamma_{\mathsf{v}}$.

The proof rules from Fig. 2 ensure that (W^Γ, M^Γ) is well-defined. We will show that (W^Γ, M^Γ) is the initial model of Γ if Γ is a set of quantum clauses.

4.2 Basic sentences

In this subsection, we define proof rules for reasoning about basic sentences and then we prove their completeness. (*EQ*) defined in Fig. 3 says that for each equation $k_1 = k_2$ in $E_{\mathcal{W}}$, if $\Gamma \vdash^{k_1} \gamma$ holds then one can deduce $\Gamma \vdash^{k_2} \gamma$. In practice, it is necessary to have an efficient way to compute projective measurements and unitary transformations for establishing the validity of the equation $k_1 = k_2$. See Section 5 for a concrete example. The sentence $\gamma[z \leftarrow k]$ used to define (*Store$_I$*) and (*Store$_E$*) in Fig. 3 is obtained from γ by substituting the term k for the variable z. Notice that (*Star$_I$*) is an infinitary proof rule, since it has a countably infinite number of premises, one for each natural number $n \in \mathbb{N}$. One needs inductive arguments to ensure that the premises of (*Star$_I$*) are satisfied. However, if $*$ is missing from the grammar which defines actions in $\mathcal{L}$ then completeness still holds.

$$(EQ)\ \frac{\Gamma \vdash^{k_1} \gamma}{\Gamma \vdash^{k_2} \gamma} \quad \text{for all } k_1 = k_2 \in E_{\mathcal{W}}$$

$$(Ret_I)\ \frac{\Gamma \vdash^{k_1} \gamma}{\Gamma \vdash^{k_2} @_{k_1} \gamma} \qquad (Ret_E)\ \frac{\Gamma \vdash^{k_2} @_{k_1} \gamma}{\Gamma \vdash^{k_1} \gamma}$$

$$(Store_I)\ \frac{\Gamma \vdash^{k} \gamma[z \leftarrow k]}{\Gamma \vdash^{k} \downarrow z \cdot \gamma} \qquad (Store_E)\ \frac{\Gamma \vdash^{k} \downarrow z \cdot \gamma}{\Gamma \vdash^{k} \gamma[z \leftarrow k]}$$

$$(Conj_I)\ \frac{\Gamma \vdash^{k} \gamma_1 \quad \Gamma \vdash^{k} \gamma_2}{\Gamma \vdash^{k} \gamma_1 \wedge \gamma_2} \qquad (Conj_E)\ \frac{\Gamma \vdash^{k} \gamma_1 \wedge \gamma_2}{\Gamma \vdash^{k} \gamma_i} \text{ for all } i \in \{1,2\}$$

$$(FT_I)\ \frac{\Gamma \vdash^{f(k)} \gamma}{\Gamma \vdash^{k} [f]\gamma} \qquad (FT_E)\ \frac{\Gamma \vdash^{k} [f]\gamma}{\Gamma \vdash^{f(k)} \gamma} \text{ for all } f \in U \cup Q$$

$$(Comp_I)\ \frac{\Gamma \vdash^{k} [\mathfrak{a}_1 \mathbin{;} \mathfrak{a}_2]\gamma}{\Gamma \vdash^{k} [\mathfrak{a}_1][\mathfrak{a}_2]\gamma} \qquad (Comp_E)\ \frac{\Gamma \vdash^{k} [\mathfrak{a}_1][\mathfrak{a}_2]\gamma}{\Gamma \vdash^{k} [\mathfrak{a}_1 \mathbin{;} \mathfrak{a}_2]\gamma}$$

$$(Union_I)\ \frac{\Gamma \vdash^{k} [\mathfrak{a}_1]\gamma \quad \Gamma \vdash^{k} [\mathfrak{a}_2]\gamma}{\Gamma \vdash^{k} [\mathfrak{a}_1 \cup \mathfrak{a}_2]\gamma} \qquad (Union_E)\ \frac{\Gamma \vdash^{k} [\mathfrak{a}_1 \cup \mathfrak{a}_2]\gamma}{\Gamma \vdash^{k} [\mathfrak{a}_i]\gamma} \text{ for all } i \in \{1,2\}$$

$$(Star_I)\ \frac{\Gamma \vdash^{k} [\mathfrak{a}^n]\gamma \text{ for any } n \in \mathbb{N}}{\Gamma \vdash^{k} [\mathfrak{a}^*]\gamma} \qquad (Star_E)\ \frac{\Gamma \vdash^{k} [\mathfrak{a}^*]\gamma}{\Gamma \vdash^{k} [\mathfrak{a}^n]\gamma} \text{ for all } n \in \mathbb{N}$$

Fig. 3. Basic proof rules

Lemma 4.3 (Basic soundness) *The least entailment system of $\mathcal{L}$ closed under the proof rules defined in Fig. 2 and 3 is sound.*

The proof of soundness relies on the closure of satisfaction relation under the proof rules defined in Fig. 2 and 3.

Proposition 4.4 (Basic compactness) *Assume that the signatures of $\mathcal{L}$ have no closed propositional symbols, and the actions of $\mathcal{L}$ are free of the operator $*$. Then the least entailment system of $\mathcal{L}$ closed under the proof rules defined in Fig. 3 is* compact.

Proof. Let $\vdash$ be the least entailment system closed under the proof rules defined in Fig. 3. Let $\Vdash$ be the compact entailment system defined by $\Gamma \Vdash^k \gamma$ if $\Gamma_f \vdash^k \gamma$ for some finite subset $\Gamma_f \subseteq \Gamma$. It suffices to show that $\Vdash$ is an entailment system (i.e., it is closed under (*Monotonicity*), (*Unions*) and (*Translation*)) closed under the proof rules defined in Fig. 3. We focus on two cases to illustrate the proof idea:

[(*Translation*)] Assume that $\Gamma \Vdash^k_{\Delta_{\mathcal{W}}} \gamma$. Let $\chi : \Delta_{\mathcal{W}} \to \Delta'_{\mathcal{W}'}$ be a signature morphism. By the definition of $\Vdash$, $\Gamma_f \vdash_{\Delta_{\mathcal{W}}} \gamma$ for some finite $\Gamma_f \subseteq \Gamma$. Since

$\vdash$ is closed under (*Translation*), $\chi(\Gamma_f) \vdash_{\Delta'_{\mathcal{W}'}} \chi(\gamma)$. Since $\chi(\Gamma_f)$ is finite, by the definition of $\Vdash$, we obtain $\chi(\Gamma) \Vdash_{\Delta'_{\mathcal{W}'}} \chi(\gamma)$.

[$(Union_I)$] Assume that $\Gamma \Vdash^k [\mathfrak{a}_1]\gamma$ and $\Gamma \Vdash^k [\mathfrak{a}_2]\gamma$. By the definition of $\Vdash$, we have $\Gamma' \vdash^k [\mathfrak{a}_1]\gamma$ and $\Gamma'' \vdash^k [\mathfrak{a}_2]\gamma$ for some finite subsets $\Gamma' \subseteq \Gamma$ and $\Gamma'' \subseteq \Gamma$. Let $\Gamma_f = \Gamma' \cup \Gamma''$. By $(Unions)$, $\Gamma_f \vdash^k [\mathfrak{a}_1]\gamma$ and $\Gamma_f \vdash^k [\mathfrak{a}_2]\gamma$. By $(Union_I)$, $\Gamma_f \vdash^k [\mathfrak{a}_1 \cup \mathfrak{a}_2]\gamma$. Since Γ_f is finite, by the definition of $\Vdash$, we get $\Gamma \Vdash^k [\mathfrak{a}_1 \cup \mathfrak{a}_2]\gamma$.

□

Proposition 4.4 establishes an important result, since it provides conditions for making deductions from a finite number of premises. The number of vectors is uncountable, but according to Proposition 4.4, for the formal proofs, it is enough to know the result of applying unitary transformations and projective measurements to the vectors occurring in the underlying specification.

Theorem 4.5 (Basic completeness) *Let $\vdash$ be any sound entailment system of $\mathcal{L}$ closed under the proof rules defined in Fig. 2 and 3. Let φ be a basic sentence defined over a signature $\Delta_{\mathcal{W}}$. Let k be a term of sort vector defined over $\Sigma_{\mathcal{W}}$. We denote by w the interpretation of k in W^Γ.*

(i) $\Gamma \vdash^k \varphi$ *iff* $(W^\Gamma, M^\Gamma) \models^w \varphi$*, for all sets of sentences* Γ.

(ii) $\Phi \vdash^k \varphi$ *iff* $\Phi \models^k \varphi$ *iff* $(W^\Phi, M^\Phi) \models^w \varphi$*, for all sets of basic sentences* Φ.

Proof. The first statement is proved by well-founded (Noetherian) induction on the triple (n_1, n_2, n_3), where (i) n_1 is the number of occurrences of the operator $*$ in φ, (ii) n_2 is the number of occurrences of the operators $⨟$ and $\cup$ in φ, and (iii) n_3 is the number of occurrences of the sentence operators in φ.

[$p \in$ `Prop`] The following are equivalent:

1	$\Gamma \vdash^k p$	by semantics
2	$p \in M^\Gamma_w$, where $w = k^{(W^\Gamma)}$	by semantics
3	$(W^\Gamma, M^\Gamma) \models^w p$	

[$@_j\,\varphi$] The following are equivalent:

1	$\Gamma \vdash^k @_j\,\varphi$	by (Ret_I) and (Ret_E)
2	$\Gamma \vdash^j \varphi$	by induction hypothesis
3	$(W^\Gamma, M^\Gamma) \models^v \varphi$, where $v = j^{(W^\Gamma)}$	by semantics
4	$(W^\Gamma, M^\Gamma) \models^w @_j\,\varphi$, where $w = k^{(W^\Gamma)}$	

[$\varphi_1 \wedge \varphi_2$] The following are equivalent:

1	$\Gamma \vdash^k \varphi_1 \wedge \varphi_2$	by $(Conj_I)$ and $(Conj_E)$
2	$\Gamma \vdash^k \varphi_1$ and $\Gamma \vdash^k \varphi_2$	by induction hypothesis
3	$(W^\Gamma, M^\Gamma) \models^w \varphi_1$ and $(W^\Gamma, M^\Gamma) \models^w \varphi_2$, where $w = k^{(W^\Gamma)}$	by semantics
4	$(W^\Gamma, M^\Gamma) \models^w \varphi_1 \wedge \varphi_2$	

[$[f]\varphi$] In this case $f \in U \cup Q$. The following are equivalent:

1	$\Gamma \vdash^{k} [f]\varphi$	by (FT_E) and (FT_I)
2	$\Gamma \vdash^{f(k)} \varphi$	by induction hypothesis
3	$(W^{\Gamma}, M^{\Gamma}) \models^{v} \varphi$, where $w = k^{(W^{\Gamma})}$ and $v = f^{(W^{\Gamma})}(w)$	by semantics
4	$(W^{\Gamma}, M^{\Gamma}) \models^{w} [f]\varphi$	

[$[\mathfrak{a}_1 ; \mathfrak{a}_2]\varphi$]
The number of $*$ in $[\mathfrak{a}_1][\mathfrak{a}_2]\varphi$ is equal to the number of $*$ in $[\mathfrak{a}_1 ; \mathfrak{a}_2]\varphi$. The number of $;$ in $[\mathfrak{a}_1][\mathfrak{a}_2]\varphi$ is strictly less than the number of $;$ in $[\mathfrak{a}_1;\mathfrak{a}_2]\varphi$. The number of $\cup$ in $[\mathfrak{a}_1][\mathfrak{a}_2]\varphi$ is is equal to the number of $\cup$ in $[\mathfrak{a}_1 ; \mathfrak{a}_2]\varphi$. By the induction hypothesis, for all vector terms k, we have:

$$\Gamma \vdash^{k} [\mathfrak{a}_1][\mathfrak{a}_2]\varphi \text{ iff } (W^{\Gamma}, M^{\Gamma}) \models^{w} [\mathfrak{a}_1][\mathfrak{a}_2]\varphi, \text{ where } w = k^{(W^{\Gamma})}.$$

The following are equivalent:

1	$\Gamma \vdash^{k} [\mathfrak{a}_1 ; \mathfrak{a}_2]\varphi$	by $(Comp_I)$ and $Comp_E)$
2	$\Gamma \vdash^{k} [\mathfrak{a}_1][\mathfrak{a}_2]\varphi$	by induction hypothesis
3	$(W^{\Gamma}, M^{\Gamma}) \models^{w} [\mathfrak{a}_1][\mathfrak{a}_2]\varphi$, where $w = k^{(W^{\Gamma})}$	by semantics
4	$(W^{\Gamma}, M^{\Gamma}) \models^{w} [\mathfrak{a}_1 ; \mathfrak{a}_2]\varphi$	

[$[\mathfrak{a}_1 \cup \mathfrak{a}_2]\varphi$]
The number of $*$ in $[\mathfrak{a}_1]\varphi$ is less or equal then the number of $*$ in $[\mathfrak{a}_1 \cup \mathfrak{a}_2]\varphi$. The number of $;$ in $[\mathfrak{a}_1]\varphi$ is less or equal than the number of $;$ in $[\mathfrak{a}_1 \cup \mathfrak{a}_2]\varphi$. The number of $\cup$ in $[\mathfrak{a}_1]\varphi$ is strictly less than the number of $\cup$ in $[\mathfrak{a}_1 \cup \mathfrak{a}_2]\varphi$. By the induction hypothesis, for all vector terms k, we have:

$$\Gamma \vdash^{k} [\mathfrak{a}_1]\varphi \text{ iff } (W^{\Gamma}, M^{\Gamma}) \models^{w} [\mathfrak{a}_1]\varphi, \text{ where } w = k^{(W^{\Gamma})}.$$

Similarly, the induction hypothesis holds for $[\mathfrak{a}_2]\varphi$ as well.
The following are equivalent:

1	$\Gamma \vdash^{k} [\mathfrak{a}_1 \cup \mathfrak{a}_2]\varphi$	by $(Union_I)$ and $(Union_E)$
2	$\Gamma \vdash^{k} [\mathfrak{a}_1]\varphi$ and $\Gamma \vdash^{k} [\mathfrak{a}_2]\varphi$	by induction hypothesis
3	$(W^{\Gamma}, M^{\Gamma}) \models^{w} [\mathfrak{a}_1]\varphi$ and $(W^{\Gamma}, M^{\Gamma}) \models^{w} [\mathfrak{a}_2]\varphi$, where $w = k^{(W^{\Gamma})}$	by semantics
4	$(W^{\Gamma}, M^{\Gamma}) \models^{w} [\mathfrak{a}_1 \cup \mathfrak{a}_2]\varphi$	

[$[\mathfrak{a}^*]\varphi$] The number of $*$ in $[\mathfrak{a}^n]\varphi$ is strictly less than the number of $*$ in $[\mathfrak{a}^*]\varphi$, for all natural numbers $n \in \mathbb{N}$. By the induction hypothesis, for all natural numbers $n \in \mathbb{N}$ and all vector terms k, we have:

$$\Gamma \vdash^{k} [\mathfrak{a}^n]\varphi \text{ iff } (W^{\Gamma}, M^{\Gamma}) \models^{w} [\mathfrak{a}^n]\varphi, \text{ where } w = k^{(W^{\Gamma})}.$$

The following are equivalent:

1	$\Gamma \vdash^{k} [\mathfrak{a}^*]\varphi$	by $(Star_I)$ and $(Star_E)$
2	$\Gamma \vdash^{k} [\mathfrak{a}^n]\varphi$ for all $n \in \mathbb{N}$	by induction hypothesis
3	$(W^{\Gamma}, M^{\Gamma}) \models^{w} [\mathfrak{a}^n]\varphi$ for all $n \in \mathbb{N}$, where $w = k^{(W^{\Gamma})}$	by semantics
4	$(W^{\Gamma}, M^{\Gamma}) \models^{w} [\mathfrak{a}^*]\varphi$	

[$\downarrow z \cdot \varphi$] The following are equivalent:

1	$\Gamma \vdash^k \downarrow z \cdot \varphi$	by $(Store_I)$ and $(Store_E)$
2	$\Gamma \vdash^k \varphi[z \leftarrow k]$	by induction hypothesis
3	$(W^\Gamma, M^\Gamma) \models^w \varphi[z \leftarrow k]$ where $w = k^{(W^\Gamma)}$	by semantics
4	$(W^\Gamma, M^\Gamma) \models^w \downarrow z \cdot \varphi$	

For the second statement, since all sentences in Φ are basic, by the first statement, we have $(W^\Phi, M^\Phi) \models \Phi$.

- Since $\vdash^k \subseteq \models^k$, we have that $\Phi \vdash^k \varphi$ implies $\Phi \models^k \varphi$.
- since $(W^\Phi, M^\Phi) \models \Phi$, we have that $\Phi \models^k \varphi$ implies $(W^\Phi, M^\Phi) \models^{k^{(W^\Phi)}} \varphi$.

By the first statement, $\Phi \vdash^k \varphi$ iff $(W^\Phi, M^\Phi) \models^{k^{(W^\Phi)}} \varphi$ iff $\Phi \models^k \varphi$. □

The sentences in Γ from the first statement of Theorem 4.5 are not necessarily basic. An application of the first statement is obtained in the following subsection where Γ is instantiated by a set of quantum clauses. The second statement of Theorem 4.5 is a direct consequence of the first due to the fact that (W^Φ, M^Φ) satisfies globally Φ. It says that in order to reason about basic sentences one needs to consider only its initial models.

4.3 Quantum clauses

In this section, we define proof rules for reasoning about quantum clauses and then we prove their completeness.

$$(MP)\ \frac{\Gamma \vdash^k \varphi \Rightarrow \gamma \quad \Gamma \vdash^k \varphi}{\Gamma \vdash^k \gamma} \qquad (MP_c)\ \frac{\Gamma \vdash^k \rho_1 \rightsquigarrow \rho_2 \quad \Gamma \vdash^k \rho_1}{\Gamma \vdash^k \rho_2}$$

$$(Imp)\ \frac{\Gamma \cup \{@_k \varphi\} \vdash^k \gamma}{\Gamma \vdash^k \varphi \Rightarrow \gamma} \qquad (Imp_c)\ \frac{\Gamma \cup \{@_k \rho_1\} \vdash^k \rho_2}{\Gamma \vdash^k \rho_1 \rightsquigarrow \rho_2}$$

Fig. 4. Implication

The proof rules for implication are defined in Fig. 4. Recall that retrieve @ belongs to the vocabulary of $\mathcal{L}$ if classical implication $\Rightarrow$ or quantum implication $\rightsquigarrow$ belongs to the vocabulary of $\mathcal{L}$. Therefore, (Imp) and (Imp_c) are well-defined.

Lemma 4.6 (Soundness) *The least entailment system of $\mathcal{L}$ closed under the proof rules defined in Fig. 2 – 4 is sound.*

As usual, the proof of soundness is based on the closure of satisfaction relation under the proof rules defined in Fig. 2 – 4.

Proposition 4.7 (Compactness) *Assume that the signatures of $\mathcal{L}$ have no closed propositional symbols, and the actions of $\mathcal{L}$ are free of the operator $*$.*

Then the least entailment system of $\mathcal{L}$ closed under the proof rules defined in Fig. 3 – 4 is compact.

Compactness for quantum clauses holds under the same conditions as for basic sentences. Initiality is proved in [17] using semantic arguments, while in the present contribution the proof of initiality relies on syntactic arguments based on proof rules.

Theorem 4.8 (Initiality) *Let $\vdash$ be any sound entailment system of $\mathcal{L}$ closed under the proof rules from Fig. 2 – 4. For all sets of quantum clauses Γ, all quantum clauses γ, all basic sentences φ and all terms k, we have:*

(i) $\Gamma \vdash^k \gamma$ *implies* $(W^\Gamma, M^\Gamma) \models^w \gamma$, *where* $w = k^{(W^\Gamma)}$.
In particular, (W^Γ, M^Γ) *is the initial model of* Γ.

(ii) $\Gamma \vdash^k \varphi$ *iff* $\Gamma \models^k \varphi$ *iff* $(W^\Gamma, M^\Gamma) \models^w \varphi$, *where* $w = k^{(W^\Gamma)}$.

Proof. Since the second statement is a direct consequence of the first, we prove only the first statement by well-founded (Noetherian) induction on (n_1, n_2, n_3), where (i) n_1 is the number of occurrences of the operator $*$ in γ, (ii) n_2 is the number of occurrences of the operators $\fatsemi$ and $\cup$ in γ, and (iii) n_3 is the number of occurrences of the sentence operators in γ. We focus only on implication since the remaining cases can be discharged using arguments from the proof of Theorem 4.5.

$[\,\varphi \Rightarrow \gamma\,]$ Assume that $\Gamma \vdash^k \varphi \Rightarrow \gamma$, where φ is basic and γ is a quantum clause.

1	assume that $(W^\Gamma, M^\Gamma) \models^{k^{(W^\Gamma)}} \varphi$	
2	$\Gamma \vdash^k \varphi$	by Theorem 4.5, since φ is basic
3	$\Gamma \vdash^k \gamma$	by (MP), as $\Gamma \vdash^k \varphi \Rightarrow \gamma$ and $\Gamma \vdash^k \varphi$
4	$(W^\Gamma, M^\Gamma) \models^{k^{(W^\Gamma)}} \gamma$	by induction hypothesis

It follows that $(W^\Gamma, M^\Gamma) \models^{k^{(W^\Gamma)}} \varphi \Rightarrow \gamma$.

$[\,\rho_1 \rightsquigarrow \rho_2\,]$ Assume that $\Gamma \vdash^k \rho_1 \rightsquigarrow \rho_2$, where ρ_1 is a closed basic sentence and ρ_2 is a closed quantum clause.

1	assume $(W^\Gamma, M^\Gamma) \models^{k^{(W^\Gamma)}} \rho_1$	
2	$\Gamma \vdash^k \rho_1$	by Theorem 4.5, since ρ_1 is basic
3	$\Gamma \vdash^k \rho_2$	by (MP_t), as $\Gamma \vdash^k \rho_1 \rightsquigarrow \rho_2$ and $\Gamma \vdash^k \rho_1$
4	$(W^\Gamma, M^\Gamma) \models^w \rho_2$	by induction hypothesis

It follows that $(W^\Gamma, M^\Gamma) \models^w \rho_1 \rightsquigarrow \rho_2$.

□

The first statement of Theorem 4.8 shows that any set of quantum clauses has an initial model which, in particular, means that any set of quantum clauses is satisfiable. The second statement of Theorem 4.8 shows that the satisfaction of basic sentences by a set of quantum clauses can be established by checking if the satisfaction holds in the initial model. It is worth mentioning that the proof of Theorem 4.8 does not rely on (Imp) and (Imp_c). The proof rules

(Imp) and (Imp_c) are needed for the proof of completeness.

Theorem 4.9 (Completeness) *Let $\vdash$ be any sound entailment system of $\mathcal{L}$ closed under the proof rules defined in Fig. 2 – 4. For all sets of quantum clauses Γ, all quantum clauses γ and all terms k, we have $\Gamma \models^k \gamma$ iff $\Gamma \vdash^k \gamma$.*

Proof. We prove the forward implication (completeness) by well-founded (Noetherian) induction on (n_1, n_2, n_3), where (i) n_1 is the number of occurrences of the operator $*$ in γ, (ii) n_2 is the number of occurrences of the operators $\S$ and $\cup$ in γ, and (iii) n_3 is the number of occurrences of the sentence operators in γ. We focus only on implication, since the remaining cases can be discharged using arguments from the proof of Theorem 4.5.

[$\varphi \Rightarrow \gamma$] $\Gamma \models^k \varphi \Rightarrow \gamma$, where φ is a basic sentence and γ is a quantum clause, iff $\Gamma \cup \{@_k \varphi\} \models^k \gamma$ iff (by induction hypothesis) $\Gamma \cup \{@_k \varphi\} \vdash^k \gamma$. By (Imp), $\Gamma \vdash^k \varphi \Rightarrow \gamma$.

[$\rho_1 \rightsquigarrow \rho_2$] $\Gamma \models^k \rho_1 \rightsquigarrow \rho_2$, where ρ_1 is closed basic sentence and ρ_2 is a closed quantum clause, iff $\Gamma \cup \{@_k \rho_1\} \models^k \rho_2$ iff (by induction hypothesis) $\Gamma \cup \{@_k \rho_1\} \vdash^k \rho_2$. By (Imp_c), $\Gamma \vdash^k \rho_1 \rightsquigarrow \rho_2$.

□

5 A case study

We show the practicality of the proof-theoretic infrastructure developed above. Working within a positive diagram of a Hilbert space means in practice that there is a library defining scalars, vectors and the operations on them which is available to the engineers to specify quantum programs. The projection on a closed space $\mathcal{X}$ with an orthonormal basis $\{v_1, \ldots, v_n\}$, $P_{\mathcal{X}} : \mathcal{W}_{\mathtt{v}} \to \mathcal{X}$, is defined by $P_{\mathcal{X}}(w) = a_1 v_1 + \cdots + a_n v_n$, where $a_i = \langle v_i \mid w \rangle$ for all $i \in \{1, \ldots, n\}$. Therefore, projective measurements are not difficult to define once an orthonormal basis for the corresponding closed space is provided. There are many examples of unitary transformations which are called quantum gates. In practice, all quantum gates used to define quantum circuits are obtained from the single qubit gates and the controlled-NOT gate in the 2-qubit system, $\mathsf{CNOT} = \begin{pmatrix} 1 & 0 & 0 & 0 \\ 0 & 1 & 0 & 0 \\ 0 & 0 & 0 & 1 \\ 0 & 0 & 1 & 0 \end{pmatrix}$. Examples of frequently used single qubit gates are the Hadamard gate $\mathsf{H} = \frac{1}{\sqrt{2}} \begin{pmatrix} 1 & 1 \\ 1 & -1 \end{pmatrix}$ and the Pauli gates $\mathsf{X} = \begin{pmatrix} 0 & 1 \\ 1 & 0 \end{pmatrix}$, $\mathsf{Y} = \begin{pmatrix} 0 & -i \\ i & 0 \end{pmatrix}$ and $\mathsf{Z} = \begin{pmatrix} 1 & 0 \\ 0 & -1 \end{pmatrix}$. Therefore, assuming that the quantum gates and the projective measurements are predefined, that is, working in a logical framework in which a positive diagram of a Hilbert space is given, is not a shortcoming.

5.1 Quantum teleportation

Quantum teleportation is a protocol for moving the state of a quantum system in the absence of a quantum communication channel linking a sender to a recipient. The sender and the recipient, traditionally called Alice and Bob, are separated in space. Each has one qubit of $|\beta_{00}\rangle = \dfrac{|00\rangle + |11\rangle}{\sqrt{2}}$. In addition to her part of β_{00}, Alice holds a qubit $\mathtt{w} = \alpha\,|0\rangle + \beta\,|1\rangle$, where α and β are unknown amplitudes. Alice "teleports" the qubit $|\mathtt{w}\rangle$ to Bob, that is, she performs a program that has the input $|\mathtt{w}_0\rangle = |\mathtt{w}\rangle \otimes |\beta_{00}\rangle$ and the output $|i\rangle \otimes |j\rangle \otimes |\mathtt{w}\rangle$, where $i, j \in \{0,1\}$. The quantum circuit is depicted in Fig. 5. The Hilbert space for this protocol is the 3-qubit system $\mathcal{H} \otimes \mathcal{H} \otimes \mathcal{H}$, where $\mathcal{H}$ denotes the 2-dimensional Hilbert space. The starting signature is $\Delta = (\Sigma, \emptyset, \mathtt{Prop})$, where

(i) Σ is obtained from the signature of Hilbert spaces Σ^{h} by adding the set of unitary transformation symbols $U = \{u_0, u_1, \sigma_0, \sigma_1, \delta_0, \delta_1\}$ and the set of measurement symbols $Q = \{q_{00}, q_{01}, q_{10}, q_{11}\}$, and

(ii) $\mathtt{Prop} = \{p\}$.

The first-order model $\mathcal{W}$ is obtained from the 3-qubit system $\mathcal{H} \otimes \mathcal{H} \otimes \mathcal{H}$ by interpreting

(i) the unitary transformation as follows: (a) $u_0^{\mathcal{W}} = CNOT \otimes I_2$, (b) $u_1^{\mathcal{W}} = H \otimes I_4$, (c) $\sigma_0^{\mathcal{W}} = I_4 \otimes X^0$, (d) $\sigma_1^{\mathcal{W}} = I_4 \otimes X^1$, (e) $\delta_0^{\mathcal{W}} = I_4 \otimes Z^0$, (f) $\delta_1^{\mathcal{W}} = I_4 \otimes Z^1$;

(ii) the quantum measurement symbols as follows: $q_{ij}^{\mathcal{W}}$ is the measurement corresponding to the projection on the closed subspace generated by the vectors $\{|ij\rangle \otimes |0\rangle, |ij\rangle \otimes |1\rangle\}$, where $i, j \in \{0,1\}$.

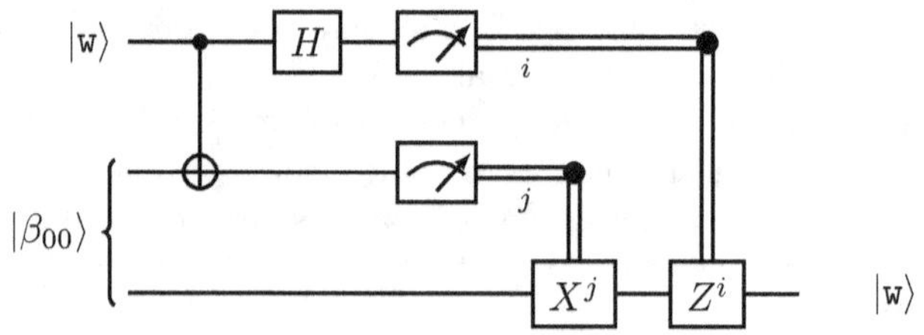

Fig. 5. Quantum teleportation

Let Φ be the set of basic sentences $\{@_{|ij\rangle\otimes|\mathtt{w}\rangle}\, p \mid i, j \leq 1\}$. We formally verify that the output is $|i\rangle \otimes |j\rangle \otimes |\mathtt{w}\rangle$, that is, $(W^{\Phi}, M^{\Phi}) \models^{|\mathtt{w}\rangle\otimes|\beta_{00}\rangle} [\bigcup_{i,j\in\{0,1\}} \mathfrak{a}_{ij}]p$, where $\mathfrak{a}_{ij} = u_0 \fatsemi u_1 \fatsemi q_{ij} \fatsemi \sigma_j \fatsemi \delta_i$ for all $i, j \in \{0,1\}$. By Theorem 4.8, we have:

(i) (W^{Φ}, M^{Φ}) is the initial model of Φ, and

(ii) $(W^{\Phi}, M^{\Phi}) \models^{|\mathtt{w}\rangle\otimes|\beta_{00}\rangle} [\bigcup_{i,j\in\{0,1\}} \mathfrak{a}_{ij}]p$ iff $\Phi \models^{|\mathtt{w}\rangle\otimes|\beta_{00}\rangle} [\bigcup_{i,j\in\{0,1\}} \mathfrak{a}_{ij}]p$.

Notice that since $@_{|ij\rangle\otimes|w\rangle}\, p \in \Phi$, by (*Monotonicity*), $\Phi \vdash^{k} @_{|ij\rangle\otimes|w\rangle}\, p$ for any term k of sort vector. By (Ret_E), $\Phi \vdash^{|ij\rangle\otimes|w\rangle} p$. The following are equivalent:

1	$\Phi \vdash^{|\mathtt{w}\rangle\otimes|\beta_{00}\rangle} [\bigcup_{i,j\in\{0,1\}} \mathfrak{a}_{ij}]p$	by ($Union_E$) and ($Union_I$)
2	$\Phi \vdash^{|\mathtt{w}\rangle\otimes|\beta_{00}\rangle} [\mathfrak{a}_{ij}]p$ for all $i, j \in \{0,1\}$	by the definition of $\mathfrak{a}_{ij}$

3	$\Phi \vdash^{\lvert \mathtt{w} \rangle \otimes \lvert \beta_{00} \rangle} [u_0 \fatsemi u_1 \fatsemi q_{ij} \fatsemi \sigma_j \fatsemi \delta_i]p$ for all $i, j \in \{0, 1\}$	by (FT_E) and (FT_I)
4	$\Phi \vdash^{k_{ij}} p$, where k_{ij} is the following term $\delta_i(\sigma_j(q_{ij}(u_1(u_0(\lvert \mathtt{w} \rangle \otimes \lvert \beta_{00} \rangle)))))$, for all $i, j \in \{0, 1\}$	by (EQ), since $k_{ij} = \lvert ij \rangle \otimes \lvert w \rangle \in E_{\mathcal{W}}$
5	$\Phi \vdash^{\lvert ij \rangle \otimes \lvert w \rangle} p$ for all $i, j \in \{0, 1\}$	which was proved above

Hence, $\Phi \vdash^{\lvert \mathtt{w} \rangle \otimes \lvert \beta_{00} \rangle} [\bigcup_{i,j \in \{0,1\}} \mathfrak{a}_{ij}]p$ holds. We reiterate the comment made for (EQ) in Section 4.2: in order to establish the validity of $\delta_i(\sigma_j(q_{ij}(u_1(u_0(\lvert \mathtt{w} \rangle \otimes \lvert \beta_{00} \rangle))))) = \lvert ij \rangle \otimes \lvert w \rangle$ one needs an efficient tool to compute the application of unitary transformations and projective measurements.

6 Conclusions

The hybrid-dynamic quantum logic studied in this contribution is obtained by enriching hybrid-dynamic propositional logic using a two-layered approach to the design and analysis of quantum systems: (a) *a local view* concerning the structural properties of the states which are vectors of a Hilbert space, and (b) *a global view*, which corresponds to a specialized language for capturing the dynamics of quantum systems. In this way, quantum systems can be modeled as Kripke structures whose frames are Hilbert spaces together with a set of unitary transformations and projective measurements. We use Robinson diagrams to define the frames which allows formal methods practitioners to focus on the dynamics of quantum systems assuming that the vectors and the operations on them are already defined. Similarly, quantum programs are built by relying on libraries that provide data types such as scalars and vectors.

We have developed a layered approach towards a Birkhoff completeness result: first, the basic layer, which deals with entailments where both the premises and the conclusion are basic sentences; second, a mixed layer, which deals with entailments where the premises are quantum clauses, but the conclusion is a basic sentence; and third, a quantum clause layer, which deals with entailments where both the premises and the conclusion are quantum clauses.

The set of states is not the set of one-dimensional closed subspaces of a Hilbert space. Also, the set of states is not constrained to the set of pure states. The responsibility of correct modeling is passed to the specifier. These ideas led to a proof of initiality which was reported in [17], the contribution where hybrid-dynamic quantum logic was originally defined and the foundation of the present work. We are not aware of any other Birkhoff completeness result for quantum logics. Recent studies show that classical computing can aid unreliable quantum processors to solve large problems reliably. In the future, we are planning to equip our hybrid-dynamic quantum logic with features that support the description of classical programs with quantum subroutines. With the development of quantum hardware devices, it becomes increasingly important to develop high-quality and trustworthy quantum software. This is possible only by applying formal methods based on solid logical foundations.

References

[1] Aiguier, M. and R. Diaconescu, *Stratified institutions and elementary homomorphisms*, Information Processing Letters **103** (2007), pp. 5–13.

[2] Areces, C., P. Blackburn and M. Marx, *Hybrid logics: Characterization, interpolation and complexity*, J. Symb. Log. **66** (2001), pp. 977–1010.

[3] Baltag, A. and S. Smets, *Complete axiomatizations for quantum actions*, International Journal of Theoretical Physics **44** (2005), pp. 2267–2282.

[4] Baltag, A. and S. Smets, *Lqp: The dynamic logic of quantum information*, Math. Struct. Comput. Sci. **16** (2006), pp. 491–525.

[5] Baltag, A. and S. Smets, *A dynamic-logical perspective on quantum behavior*, Stud Logica **89** (2008), pp. 187–211.

[6] Baltag, A. and S. Smets, *Quantum logic as a dynamic logic*, Synthese **179** (2011), pp. 285–306.

[7] Baltazar, P., R. Chadha and P. Mateus, *Quantum computation tree logic: Model checking and complete calculus*, International Journal of Quantum Information **6** (2008), pp. 219–236.

[8] Birkhoff, G. and J. von Neumann, *The logic of quantum mechanics*, The Annals of Mathematics **37** (1936), pp. 823–843.

[9] Braüner, T., "Hybrid logic and its Proof-Theory," Applied Logic Series **37**, Springer, 2011.

[10] Brunet, O. and P. Jorrand, *Dynamic quantum logic for quantum programs*, International Journal of Quantum Information **2** (2004), pp. 45–54.

[11] Diaconescu, R., *Quasi-varieties and initial semantics for hybridized institutions*, J. Log. Comput. **26** (2016), pp. 855–891.

[12] Diaconescu, R., *Implicit kripke semantics and ultraproducts in stratified institutions*, J. Log. Comput. **27** (2017), pp. 1577–1606.

[13] Găină, D., *Birkhoff style calculi for hybrid logics*, Formal Aspects Comput. **29** (2017), pp. 805–832.

[14] Găină, D. and I. Tutu, *Birkhoff completeness for hybrid-dynamic first-order logic*, in: S. Cerrito and A. Popescu, editors, *Automated Reasoning with Analytic Tableaux and Related Methods - 28th International Conference, TABLEAUX 2019, London, UK, September 3-5, 2019, Proceedings*, Lecture Notes in Computer Science **11714** (2019), pp. 277–293.

[15] Găină, D., *Foundations of logic programming in hybrid logics with user-defined sharing*, Theoretical Computer Science **686** (2017), pp. 1–24.

[16] Găină, D., *Forcing and calculi for hybrid logics*, J. ACM **67** (2020), pp. 25:1–25:55.

[17] Găină, D., *Foundations of logic programming in hybrid-dynamic quantum logic* (2024). URL `https://doi.org/10.48550/arXiv.2406.02085`

[18] Găină, D., G. Badia and T. Kowalski, *Robinson consistency in many-sorted hybrid first-order logics*, in: D. Fernández-Duque, A. Palmigiano and S. Pinchinat, editors, *Advances in Modal Logic, AiML 2022, Rennes, France, August 22-25, 2022* (2022), pp. 407–428.

[19] Găină, D., G. Badia and T. Kowalski, *Omitting types theorem in hybrid dynamic first-order logic with rigid symbols*, Ann. Pure Appl. Log. **174** (2023), p. 103212.

[20] Mateus, P., J. Ramos, A. Sernadas and C. Sernadas, "Temporal Logics for Reasoning about Quantum Systems," Cambridge University Press, 2009 p. 389–413.

[21] Takagi, T., *Semantic analysis of a linear temporal extension of quantum logic and its dynamic aspect*, ACM Trans. Comput. Logic **24** (2023).

Varieties of Distributed Knowledge

Rustam Galimullin

University of Bergen
Bergen, Norway

Louwe B. Kuijer

University of Liverpool
Liverpool, U.K.

Abstract

Distributed knowledge is one of the better known group knowledge modalities. While its intuitive idea is relatively clear, there is ample room for interpretation of details. We investigate 12 definitions of distributed knowledge that differ from each other in the kinds of information sharing the agents can perform in order to achieve mutual knowledge of a proposition. We then show which kinds of distributed knowledge are equivalent, and which kinds imply each other, i.e., for any two variants τ_1 and τ_2 of distributed knowledge we show whether a proposition φ being distributed knowledge under definition τ_1 implies that φ is distributed knowledge under definition τ_2.

Keywords: Epistemic Logic, Distributed Knowledge

1 Introduction

Epistemic logic (see, e.g., [7,13]) can be used to describe the knowledge of one or more agents. If multiple agents are involved, one can then study various kinds of group knowledge. On the one hand, we may consider types of group knowledge that are stronger than individual knowledge; for example, one may wonder whether a particular proposition φ is known by all members of the group, or even whether φ is so obvious (to the group members) as to be common knowledge. On the other hand, we can also consider a type of group knowledge that is weaker than individual knowledge; even if φ is not currently known by any individual group member, the group might be able to learn φ if they combine their information. For example, perhaps agent a knows that $\varphi \to \psi$ and b knows φ. Neither of them knows ψ, yet if they pool their knowledge they would be able to get to know it.

This latter kind of group knowledge is typically known as *distributed knowledge* (see, e.g., [11,13,7,16,14,4]). Distributed knowledge is a kind of hypothetical knowledge: φ is distributed knowledge among a group G if the members

E-mails: Rustam.Galimullin@uib.no and Louwe.Kuijer@liverpool.ac.uk of G could, if they combined their knowledge, learn that φ is true. Note that we do not require the agents to actually combine their knowledge in this way, it suffices that they *could* do so and learn φ.

While the general idea of distributed knowledge is reasonably clear, there is no consensus about how to formally define it. Broadly speaking, there have historically been two main approaches.We will refer to these approaches as the *intersection* approach and the *full communication* approach, with the latter term being derived from [16]. We should note, however, that these are not standardised terms. In fact, even "distributed knowledge" is not fully standard, with other terms such as "group knowledge", "collective knowledge" and "implicit knowledge" also being used.

In both approaches, the distributed knowledge of a group G of agents depends on what information the group members possess. In epistemic logic, the information state of an agent a is generally represented by an accessibility relation $\sim_a$, and a knows a formula φ, denoted $\Box_a\varphi$, in world s if and only if φ is true in every world s' such that $s \sim_a s'$.

The intersection approach is the most common one, and is used in [11,13,7,16,10,14,4], among many others. Here, φ is distributed knowledge in world s if φ is true in every world s' such that $s \sim_G s'$, where $\sim_G = \bigcap_{a\in G} \sim_a$. The intuition behind this approach is that the group G is collectively capable of distinguishing between s and s' if any of its members can.

The full communication approach is less common, but still used in many places, including [12,16,10,14]. In this approach, φ is distributed knowledge among G if and only if the set of formulas known by any of the agents entails φ, i.e., if $\{\psi \in \mathcal{L}_0 \mid \exists a \in G : s \models \Box_a\psi\} \models \varphi$. In order to avoid circularity we do have to be careful to specify that the known formulas ψ must not reference distributed knowledge, i.e. they must be from the basic epistemic logic $\mathcal{L}_0$.

Observant readers may notice that there is significant overlap between the list of papers using the intersection approach and those using the full communication approach. This is because one of the topics studied has been the relation between the intersection and full communication versions of distributed knowledge. The outcomes of this comparison are that (i) if φ is full communication distributed knowledge then it is also intersection distributed knowledge, (ii) φ can be intersection intersection distributed knowledge without being full communication distributed knowledge, and (iii) on certain types of models, the two kinds of distributed knowledge are equivalent.

The reason the two approaches to distributed knowledge persist side by side, albeit with the intersection approach being more popular, is that they appeal to different intuitions. Specifically, the issue is whether agents share information in a way that can be expressed in epistemic logic. Suppose that a considers a world s_1 possible but s_2 impossible, while b considers s_1 to be impossible and s_2 to be possible, but that s_1 and s_2 are not distinguishable by any formula of epistemic logic. Can a and b, when working together, discover that neither s_1 nor s_2 is possible?

The full communication approach says "no, they cannot exclude s_1 and

s_2". After all, while a does not consider s_2 to be possible, there is nothing they can say to b that would communicate this impossibility, and b is likewise incapable of communicating the impossibility of s_1. The intersection approach, on the other hand, says "yes, they can exclude s_1 and s_2". Even if neither of them can express the difference (in epistemic logic, at least), a knows that s_2 is impossible and b knows that s_1 is impossible, so together they know both are impossible. Perhaps they communicate this impossibility to each other in a language other than epistemic logic, such as first-order logic. Perhaps they simply point at the worlds they consider impossible. Perhaps they perform a Vulcan mind-meld, or somehow merge their databases or neural networks. What matters, to the intersection approach, is not how the agents share their information, but only that the agents possess the required information.

In this paper, we will not try to settle the debate in favour of one variant. On the contrary, we will introduce several further variants of distributed knowledge. This is because, in addition to the form of information sharing (formulas or mind-meld) which makes the difference between the intersection and full communication variants, there are several more questions one can ask about how distributed knowledge is established.

How much information do the agents share? Do they share information simultaneously, or is there an order? Are all agents required to know that φ is true after the knowledge sharing, or does it suffice if one agent knows φ?

Different answers to these questions may lead to different notions of distributed knowledge. That is not to say that all possible combinations lead to different kinds of distributed knowledge. For example, suppose that each agent shares a single proposition known to that agent. Then it does not matter whether the agents share simultaneously or in order (Proposition 4.4), and if one of the agents can learn φ then all of them can learn it (Proposition 4.3). But in other cases, the difference does matter.

We will consider 12 possible definitions of distributed knowledge. One of these, which we label $(\cap, \epsilon, \epsilon, \forall)$, is the intersection definition of distributed knowledge. None of our definitions is exactly the same as the full communication definition, but the variant that we label $(\mathcal{L}_0, \odot, \Uparrow, \forall)$ is equivalent to full communication (Proposition 3.2).[1] To the best of our knowledge, other definitions of distributed knowledge in our taxonomy have not been considered in the literature before.

After introducing the basic technical definitions in Section 2, we will define all variants of distributed knowledge in Section 3 and compare them to the existing approaches. Then we compare the introduced definitions of distributed knowledge to each other in Section 4. Finally, we conclude and outline the directions for further research in Section 5.

[1] Furthermore, several other variants are equivalent to $(\mathcal{L}_0, \odot, \Uparrow, \forall)$, and therefore, by transitivity, also equivalent to full communication.

2 Basic Definitions

Each of the variants of distributed knowledge that we consider will use the same language, which is basic epistemic logic with an additional operator D_G, indicating distributed knowledge among group G.

Definition 2.1 Let $\mathcal{P}$ be a countable set of propositional atoms and $\mathcal{A}$ a finite set of agents. The language $\mathcal{L}$ is given by the following normal form

$$\mathcal{L} \ni \varphi ::= p \mid \neg\varphi \mid (\varphi \vee \varphi) \mid \Box_a\varphi \mid D_G\varphi$$

where $p \in \mathcal{P}$, $a \in \mathcal{A}$ and $\emptyset \neq G \subseteq \mathcal{A}$. We denote the fragment of $\mathcal{L}$ that does not contain D_G by $\mathcal{L}_0$. We omit parentheses where this should not cause confusion.

We use $\wedge, \rightarrow, \leftrightarrow$ and $\Diamond_a$ in the usual way as abbreviations. Similarly, we use $\bigwedge$ and $\bigvee$ for n-ary conjunction and disjunction, respectively.

Because we are describing (distributed) *knowledge*, we will use S5 models. We should stress, however, that our results also hold for K models.

Definition 2.2 A *model* $\mathcal{M}$ is a tuple $(S, \sim, V)$, where S is a non-empty set of *worlds*, $\sim: \mathcal{A} \rightarrow 2^{S\times S}$ assigns to each agent $a \in \mathcal{A}$ an equivalence relation $\sim_a \subseteq S \times S$ and $V : \mathcal{P} \rightarrow 2^S$ is the valuation function. If necessary, we will refer to the elements of the tuple as $S^{\mathcal{M}}$, $\sim^{\mathcal{M}}$, and $V^{\mathcal{M}}$. A *pointed model* is a pair $\mathcal{M}, s$ where s is a world of $\mathcal{M}$.

All operators other than D_G are given their normal semantics. The semantics of $\mathcal{L}_0$ is therefore as follows.

Definition 2.3 Let $\mathcal{M} = (S, \sim, V)$ be a model and $s \in S$. Then

$\mathcal{M}, s \models p$	iff	$s \in V(p)$
$\mathcal{M}, s \models \neg\varphi$	iff	$\mathcal{M}, s \not\models \varphi$
$\mathcal{M}, s \models \varphi \vee \psi$	iff	$\mathcal{M}, s \models \varphi$ or $\mathcal{M}, s \models \psi$
$\mathcal{M}, s \models \Box_a\varphi$	iff	$\mathcal{M}, t \models \varphi$ for all t such that $s \sim_a t$.

For D_G, the semantics will depend on the type of distributed knowledge under consideration, which we discuss in the next section.

In several of the proofs throughout this paper, we will make use of the concept of Q-bisimilarity, where $Q \subseteq \mathcal{P}$.

Definition 2.4 Let $Q \subseteq \mathcal{P}$, and $\mathcal{M} = (S^{\mathcal{M}}, \sim^{\mathcal{M}}, V^{\mathcal{M}})$ and $\mathcal{N} = (S^{\mathcal{N}}, \sim^{\mathcal{N}}, V^{\mathcal{N}})$ be models. We say that $\mathcal{M}$ and $\mathcal{N}$ are *Q-bisimilar* (denoted $\mathcal{M} \approx_Q \mathcal{N}$) if there is a non-empty relation $B \subseteq S^{\mathcal{M}} \times S^{\mathcal{N}}$, called *Q-bisimulation*, such that for all $B(s,t)$, the following conditions are satisfied:

Atoms for all $p \in Q$: $s \in V^{\mathcal{M}}(p)$ if and only if $t \in V^{\mathcal{N}}(p)$,

Forth for all $a \in \mathcal{A}$ and $u \in S^{\mathcal{M}}$ such that $s \sim_a^{\mathcal{M}} u$, there is a $v \in S^{\mathcal{N}}$ such that $t \sim_a^{\mathcal{N}} v$ and $B(u,v)$,

Back for all $a \in \mathcal{A}$ and $v \in S^{\mathcal{N}}$ such that $t \sim_a^{\mathcal{N}} v$, there is a $u \in S^{\mathcal{M}}$ such that $s \sim_a^{\mathcal{M}} u$ and $B(u,v)$.

We say that $\mathcal{M}, s$ and $\mathcal{N}, t$ are Q-bisimilar and denote this by $\mathcal{M}, s \approx_Q \mathcal{N}, t$ if there is a Q-bisimulation linking worlds s and t.

In the paper, we will make use of the classic result that bisimilar models satisfy the same formulas of epistemic logic.

Theorem 2.5 *Given $\mathcal{M}, s$ and $\mathcal{N}, t$, if $\mathcal{M}, s \approx_Q \mathcal{N}, t$, then for all $\varphi \in \mathcal{L}_0$ that include atoms only from Q, we have that $\mathcal{M}, s \models \varphi$ if and only if $\mathcal{N}, t \models \varphi$.*

3 Varieties of distributed knowledge

In Section 1 we mentioned a number of questions regarding the exact workings of distributed knowledge. Here we discuss these questions in more detail, and use the potential answers to define types of distributed knowledge. Before we get into these details, however, we should remark on one aspect of distributed knowledge that will hold for every variant, namely that distributed knowledge is backward-looking.

In both the intersection and full communication definitions of distributed knowledge, a proposition φ is distributed knowledge among G if, by combining their knowledge, G can discover that φ *was* true *before* they combined their knowledge.[2] In this paper we also follow this tradition. The past tense is important because φ may contain claims that certain group members are ignorant of some fact, and this ignorance can be broken when agents in G share their knowledge.

For example, let φ be the famous Moore-sentence $p \wedge \neg\Box_a p$, i.e., "p is true but a does not know that p is true." This sentence cannot be known by agent a, yet it can be distributed knowledge between a and b. Perhaps a knows that $\neg\Box_a p$ while b knows that p. When a and b combine their knowledge, they will learn that $p \wedge \neg\Box_a p$ used to be true, but that very same communication will render the formula false, since a will learn that p is true.

Because of this backward looking nature, distributed knowledge is a *static* operator, as opposed to the *dynamic* operators from dynamic epistemic logic (DEL) [17]. A dynamic take on distributed knowledge is also possible, and would likely correspond to what agents may learn through communication with each other[3]. Some of the known approaches include a single agent sharing all her information with everyone [5], a group of agents sharing everything they know among themselves [4,6], topic-based communication within a group of agents [9], and various forms of public communication by agents and their effects [2,1,3]. While such dynamic treatment of distributed knowledge is interesting, it is outside the scope of this contribution.

We now continue with a detailed discussion of each of our questions regarding the meaning of distributed knowledge.

[2] See [4] for a more thorough discussion of this aspect of distributed knowledge.

[3] As with static distributed knowledge, we need to account for many small but important implementation decisions, for example the extent to which agents that are not part of the group will be aware of the discussion among the group members.

Forms of information The first important question is the form of information shared by the group members in their attempt to establish knowledge of a proposition. We consider two answers to this question. Firstly, agents may be able to share *formulas of $\mathcal{L}_0$ that they know.* So if $\Box_a\psi$ holds, for some $\psi \in \mathcal{L}_0$ and $a \in G$, then in a group discussion among G, agent a can contribute ψ. The restriction to formulas that agents actually know arises from the intuition that distributed knowledge is about combining individual *knowledge* rather than arbitrary formulas.

The alternative is that the agents may be sharing information in a way that is either entirely non-lingual, or at least phrased in a language stronger than $\mathcal{L}_0$. Importantly, such information sharing is not bound to respect bisimilarity in the sense of Theorem 2.5 (see, e.g., [14]).

Note that we restrict ψ to $\mathcal{L}_0$, so the formulas that the agents can share in their deliberation cannot include the D_G operator. This is required in order to avoid vicious circularity; if we allow the D_G operator to be used during the deliberations, there are situations where agents can learn φ if they combine their knowledge, but only if that knowledge includes the fact that φ is distributed knowledge. Hence φ would be distributed knowledge if and only if... φ is distributed knowledge. As a result of this circularity, the semantics would become underdetermined, i.e., there would be pointed models where both $D_G\varphi$ and $\neg D_G\varphi$ are consistent with the semantics (see Appendix A).

Recall from the introduction that the intersection definition of distributed knowledge assumes the non-lingual answer to this question, whereas the full communication definition assumes the information being communicated is in the form of $\mathcal{L}_0$ formulas.

Amount of information The next question is how much information the agents share. In principle, any measure could be used here. For example, one could imagine a situation where each agent has, say, 5 seconds to contribute their share. Or perhaps agents are limited to statements of a given maximum complexity.

Here, however, we will restrict ourselves to a coarser distinction: agents will be able to share either a single formula, or an infinite set of formulas. Note that since we are limiting only the amount of formulas, not their complexity, it would not make sense to restrict to a given finite number of formulas, since any finite number of formulas can be combined into one using conjunctions.

Note that this distinction only makes sense if information is shared as $\mathcal{L}_0$ formulas; if information is shared non-linguistically we do not have a sensible measure of the amount of information shared.

Order and turn-taking Another consideration is whether the agents share all information simultaneously, or in some order (with agents taking a single turn if they share one formula, or multiple turns if they share a set of formulas). This distinction is relevant because agents can only share formulas that they know; if the agents share their information in some order, then the later agents may be able to contribute some formulas that they did not know initially but that they have come to know based on the information provided by the agents before them. Again, this distinction only works if agents share their information in the form of formulas.

Collective or individual success Finally, we can distinguish between a type of distributed knowledge where all group members need to learn the truth of a formula and a type where only one group member needs to learn it. In other words, if there is a possible communication between a and b that would result in a knowing φ while b remains ignorant of it, would φ be distributed knowledge?

Equivalence among variants Not all of the combinations of answers to questions from the previous section make sense. Still, the answers allow us to define 12 variations of distributed knowledge.

We should stress, however, that not all these variations are truly different. For example, suppose that, in their communication, every agent shares a single formula, and they do so simultaneously. Then it is possible for the agents to communicate in such a way that a single agent learns φ if and only if it is possible for them to communicate in a way where all agents learn φ (see Proposition 4.2).

In fact, our main contributions in this paper are (1) formal definitions of the various kinds of distributed knowledge, and (2) the results on which of the variants are equivalent to each other.

3.1 Semantics for distributed knowledge

In our taxonomy, a type of distributed knowledge can be identified by the form of information shared, the amount of information, whether there is an order, and how many agents need to learn the target formula. A type τ is therefore a tuple $\tau = (f, a, o, q)$, where $f \in \{\cap, \mathcal{L}_0\}$ indicates whether the information is presented in the form of formulas ($\mathcal{L}_0$) or not ($\cap$), $a \in \{\odot, \bullet, \epsilon\}$ indicates whether a single formula is shared ($\odot$) or a set of formulas ($\bullet$), while $a = \epsilon$ is used for the case where $f = \cap$ and therefore no formulas are shared at all. The parameter $o \in \{\Uparrow, \omega, \Omega, \epsilon\}$ indicates whether the agents share their knowledge simultaneously ($\Uparrow$), in a sequence with a length α bounded by the first infinite ordinal (ω), in a sequence that can have any ordinal α as its length (Ω), or whether $f = \cap$ and therefore the question of an order doesn't make sense (ϵ). Finally, $q \in \{\exists, \forall\}$ indicates whether at least one agent must learn φ ($\exists$) or all of them must learn it ($\forall$).

We use $f(\tau)$, $a(\tau)$, $o(\tau)$ and $q(\tau)$ to denote the values of f, a, o and q, respectively, in τ. Each type τ of distributed knowledge induces semantics for the language $\mathcal{L}$, which we denote by $\models_\tau$.

The main idea of each of the semantics is that φ is distributed knowledge among G if there is some way for G to share information among themselves that would result in them learning the truth of φ. Communication by G will change the current information state, which is encoded by the set $\sim$ of relations, into a new information state $\sim'$. Often there are different things that G could communicate, and each such possible communication will lead to a new information state. Hence we will, in general, need to consider not one new information state $\sim'$ but a set of such information states.

We will denote the set of information states that can be reached by group G, when discussing in world s, using the communication type τ, as $\mathcal{R}_{G,s,\tau}$.

The semantics for the distributed knowledge operator are then given by $\mathcal{M}, s \models_\tau D_G\varphi$ iff

$$\exists \sim' \in \mathcal{R}_{G,s,\tau} \exists a \in G : \mathcal{M}, t \models \varphi \text{ for all } t \text{ such that } s \sim'_a t$$

if $q(\tau) = \exists$, or $\mathcal{M}, s \models_\tau D_G\varphi$ iff

$$\exists \sim' \in \mathcal{R}_{G,s,\tau} \forall a \in G : \mathcal{M}, t \models \varphi \text{ for all } t \text{ such that } s \sim'_a t$$

if $q(\tau) = \forall$.

We should note that, while the agents are generally not *required* to share as much information as they can, distributed knowledge is about whether the agents are able to achieve knowledge of φ. The more information is shared, the more likely it is that the agents will learn φ.[4] In particular, if there is a unique "maximal communication", it suffices to consider only that communication.

For example, in the intersection definition of distributed knowledge, agents are capable of explaining, in a non-linguistic way, exactly which worlds they consider possible. Conceptually, it seems reasonable that agents could, instead of communicating their exact set of possible worlds, communicate a superset of it. We need not consider this possibility, however, since sharing the exact set of worlds they consider possible is the optimal strategy. In this case, it therefore suffices to consider the singleton set $\mathcal{R}_{G,s,\tau} = \{\sim'\}$ where, for every $a \in G$, $\sim'_a = \bigcap_{b \in G} \sim_b$.

In other cases there may be no single most informative communication, so we cannot restrict ourselves to a single information state in this way. For example, if every agent can communicate a single formula that is known to them, there is not, in general, a single most informative formula for them to state. For every formula $\psi_G = \bigwedge_{a \in G} \psi_a$ with the property that $\mathcal{M}, s \models \Box_a \psi_a$ for every a, we therefore need to consider the information state $\sim^{\psi_G}$.

Based on the considerations from the previous section, we can define the following 12 variants of distributed knowledge (see Figure 1 for the full list of variants and their relative strength). Recall, however, that some of these variants are equivalent to one another.

Non-linguistic sharing Suppose information is shared non-linguistically. Then our method of restricting the amount of information shared is inapplicable. Furthermore, because we do not know how information is shared we also cannot speak of an ordering in which information is presented.

The only further distinction that is available is whether one agent needs to learn the formula or all of them do. We therefore need to consider the variants $\tau = (\cap, \epsilon, \epsilon, \exists)$ and $\tau = (\cap, \epsilon, \epsilon, \forall)$, respectively.

---

[4] This does rely on the fact that we are looking at *static* communication, i.e. φ is distributed knowledge if the agents can learn that φ *used to be* true. In *dynamic* communication, where agents are trying to learn that φ *is* true, sharing as much information as possible may not be an optimal strategy since φ can contain ignorance conditions that become false when more information is shared (e.g. the Moore formula).

As stated above, for either of these cases we have $\mathcal{R}_{G,s,\alpha} = \{\sim'\}$ where $\sim'_a = \bigcap_{b \in G} \sim_b$. Note, also, that at this point it is already easy to see that $(\cap, \epsilon, \epsilon, \exists)$ and $(\cap, \epsilon, \epsilon, \forall)$ are equivalent. This is because all agents end up with the same accessibility relation, so if one of them learns φ then they all do.

Proposition 3.1 *$\mathcal{M}, s \models_{(\cap,\epsilon,\epsilon,\exists)} D_G\varphi$ if and only if $\mathcal{M}, s \models_{(\cap,\epsilon,\epsilon,\forall)} D_G\varphi$.*

Simultaneous sharing of formulas Suppose the information sharing happens by the agents stating one or more formulas each, and that this happens simultaneously. So we are considering the variants $\tau \in \{(\mathcal{L}_0, \odot, \Uparrow, \exists), (\mathcal{L}_0, \odot, \Uparrow, \forall), (\mathcal{L}_0, \bullet, \Uparrow, \exists), (\mathcal{L}_0, \bullet, \Uparrow, \forall)\}$, where the first two assume that each agent contributes a single formula while the last two let each agent contribute a (potentially infinite) set of formulas.

We specify which information states the agents can achieve by sharing information in two steps. First, we specify all of the ways the agents could share information. In effect, this acts as a set of indices used to identify the various outcome information states. Then, for each index we specify what that outcome information state is.

The important condition on information sharing is that each agent must contribute one or more formulas that they know.[5] Hence if $\tau = (\mathcal{L}_0, \odot, \Uparrow, \exists)$ or $\tau = (\mathcal{L}_0, \odot, \Uparrow, \forall)$ then

$$\mathcal{R}_{G,s,\tau} = \{\sim^{\{\psi_a | a \in G\}} \mid \forall a \in G : \psi_a \in \mathcal{L}_0 \text{ and } \mathcal{M}, s \models \Box_a \psi_a\}$$

and

$$\sim_b^{\{\psi_a | a \in G\}} = \{(x, y) \in S \times S \mid (x, y) \in \sim_b \text{ and } \mathcal{M}, y \models \bigwedge_{a \in G} \psi_a\}.$$

If each agent shares a set of formulas, i.e. if $\tau = (\mathcal{L}_0, \bullet, \Uparrow, \exists)$ or $\tau = (\mathcal{L}_0, \bullet, \Uparrow, \forall)$, then they need to know each of the formulas they provide, so

$$\mathcal{R}_{G,s,\tau} = \{\sim^{\{\Psi_a | a \in G\}} \mid \forall a \in G : \Psi_a \subseteq \mathcal{L}_0 \text{ and } \forall \psi \in \Psi_a : \mathcal{M}, s \models \Box_a \psi\}$$

and

$$\sim_b^{\{\Psi_a | a \in G\}} = \{(x, y) \in S \times S \mid (x, y) \in \sim_b \text{ and } \forall a \forall \psi \in \Psi_a : \mathcal{M}, y \models \psi\}.$$

Taking turns Suppose that, as in the previous case, agents share one or more formulas, but now they do so sequentially. The crucial difference with simultaneous communication is that in the sequential case agents can state a formula that they only know because of the information provided to them by the previous speakers.

Assume, for example, that a knows p and b knows $p \to q$. If a speaks first and tells b that p is true, b can then, when it is their turn to speak, say that q is true, which they only know because a told them p.

[5] Note that we can assume without loss of generality that every agent shares at least one formula because agents can always use the uninformative formula $\top$ that is known by everyone.

If every agent shares exactly one formula, then the sequential sharing of information means every agent takes a single turn, where the later agents can use the information provided by the earlier ones.

If each agent provides a set of formulas, we still need to specify the order among the agents, but this will generally have to be an infinite order. Note that we do not have to assume that this order is "fair", since agents can skip their turn by providing the trivial formula $\top$. What is potentially important, however, is whether the turn-taking is limited to ω rounds (where ω is the first infinite ordinal), or whether any ordinal can be used. This gives us six variants: $\tau \in \{(\mathcal{L}_0, \odot, \omega, \exists), (\mathcal{L}_0, \odot, \omega, \forall), (\mathcal{L}_0, \bullet, \omega, \exists), (\mathcal{L}_0, \bullet, \omega, \forall), (\mathcal{L}_0, \bullet, \Omega, \exists), (\mathcal{L}_0, \bullet, \Omega, \forall)\}$. Observe that we do not consider types $(\mathcal{L}_0, \odot, \Omega, \forall)$ and $(\mathcal{L}_0, \odot, \Omega, \exists)$ since a finite number of agents communicating one formula each will never step on the trans-finite territory.

Before we can define the possible effects of sequential communication, we first need a little bit more notation. We want to consider finite sequences, infinite sequences, and even trans-finite sequences of statements. Therefore, let α be any ordinal. At each ordinal $\delta < \alpha$, one of the agents will state the truth of one formula; let us write $f(\delta)$ for the agent and $g(\delta)$ for the formula. As α can be considered to be identical to the set of all ordinals less than it, this means f and g are functions of type $f : \alpha \to G$ and $g : \alpha \to \mathcal{L}_0$.

Agent $f(\delta)$ needs to know formula $g(\delta)$ at time δ, since otherwise they would not be able to state the truth of the formula. As such, we need to keep track of the information state at each point in the process. Formally, this means that we are interested in the final information state $\sim^{\alpha,f,g}$, but we also need to define $\sim^{\alpha,f,g,\delta}$ for $\delta < \alpha$, which represents the information state immediately after the announcement that takes place at time δ. We do this by defining

$$\sim_a^{\alpha,f,g,0} = \{(x,y) \in S \times S \mid (x,y) \in \sim_a \text{ and } \mathcal{M}, y \models g(0)\}$$

and

$$\sim_a^{\alpha,f,g,\delta} = \{(x,y) \in S \times S \mid (x,y) \in \bigcap_{\epsilon<\delta} \sim_a^{\alpha,f,g,\epsilon} \text{ and } \mathcal{M}, y \models g(\delta)\}$$

for $0 < \delta < \alpha$. Finally, we define $\sim_a^{\alpha,f,g} = \bigcap_{\delta<\alpha} \sim_a^{\alpha,f,g,\delta}$.

Note that $\sim^{\alpha,f,g,\delta}$ is the information state *after* the communication at time δ, and that there is such a communication at every time $\delta < \alpha$. Hence $\sim_a^{\alpha,f,g,0}$ is generally not identical to $\sim_a$, since the latter represents the information state before any communication takes place.

Furthermore, communication only happens at $\delta < \alpha$. Hence, in particular, if $\alpha = \omega$ then communication takes place at every finite time step, but there is no "infinity-th" communication at ω.

Importantly, the definition of $\sim^{\alpha,f,g}$ does not check whether agent $f(\delta)$ actually knows $g(\delta)$ at time δ. So while the definition determines the effect that a given communication sequence would have, it does not determine whether the agents are actually capable of saying the formulas included in the sequence. For this, we need to look at another property that we will refer to as *correctness*. We say that α, f and g are *correct* for group G and world s if

$$\forall s' : \text{ if } (s, s') \in \sim_{f(\delta)} \cap \bigcap_{\epsilon<\delta} \sim_{f(\delta)}^{\alpha,f,g,\epsilon}, \text{ then } \mathcal{M}, s' \models g(\delta).$$

In other words, if s' was accessible originally (for agent $f(\delta)$) and has not been excluded by any of the preceding statements at times $\epsilon < \delta$, then $g(\delta)$ must be true in s'.

Now, we can formally define $\mathcal{R}_{G,s,\tau}$ for the sequential types of communication. If $\tau = (\mathcal{L}_0, \odot, \omega, \exists)$ or $\tau = (\mathcal{L}_0, \odot, \omega, \forall)$, then we require every agent to have exactly one turn, and hence $\alpha = |G|$. Moreover, f is a bijection, so

$$\mathcal{R}_{G,s,\tau} = \{\sim^{|G|,f,g} | \ f \text{ is a bijection and } |G|, f \text{ and } g \text{ are correct for } G, s\}.$$

If $\tau = (\mathcal{L}_0, \bullet, \omega, \exists)$ or $\tau = (\mathcal{L}_0, \bullet, \omega, \forall)$, then we take $\alpha = \omega$, i.e., we allow infinite statements, but there is no "infinity-th statement". We could demand that the turn-taking by the agents is "fair" in some way, but that is pointless; if there is an unfair turn-taking the agents could use for their communication, this can be transformed into a fair one where some agents give the trivial statement $\top$. As such, we get

$$\mathcal{R}_{G,s,\tau} = \{\sim^{\omega,f,g} | \ \omega, f \text{ and } g \text{ are correct for } G, s\}.$$

Finally, we can allow any ordinal number α of statements. If $\tau = (\mathcal{L}_0, \bullet, \Omega, \exists)$ or $\tau = (\mathcal{L}_0, \bullet, \Omega, \forall)$, then

$$\mathcal{R}_{G,s,\tau} = \{\sim^{\alpha,f,g} | \ \alpha, f \text{ and } g \text{ are correct for } G, s\}.$$

Note that all $\mathcal{R}_{G,s,\tau}$'s are sets, and we essentially quantify over all possible sequences of announcements, and, according to the definition of distributed knowledge, it is enough that at least one knowledge state induced by any sequence satisfies φ.

3.2 Connections to the traditional definitions

Our variant $(\cap, \epsilon, \epsilon, \forall)$ of distributed knowledge is, modulo some notation, identical to the traditional definition of distributed knowledge based on intersection. Our variant that most closely matches the full communication definition of distributed knowledge is $(\mathcal{L}_0, \odot, \Uparrow, \forall)$.

In fact, as mentioned in the introduction, $(\mathcal{L}_0, \odot, \Uparrow, \forall)$ is equivalent to the full communication definition, but this is not entirely obvious and therefore requires a short proof.

Proposition 3.2 *We have*

$$\mathcal{M}, s \models_{(\mathcal{L}_0,\odot,\Uparrow,\forall)} D_G\varphi$$

if and only if

$$\{\psi \in \mathcal{L}_0 \mid \exists a \in G : \mathcal{M}, s \models \Box_a\psi\} \models \varphi.$$

Proof. Suppose $\mathcal{M}, s \models_{(\mathcal{L}_0,\odot,\Uparrow,\forall)} D_G\varphi$. Then there are $\{\psi_b \mid b \in G\}$ such that (1) for all $b \in G$, $\mathcal{M}, s \models \Box_b\psi_b$ and (2) for all $a \in G$ and every s', if $s \sim_a^{\{\psi_b \mid b \in G\}} s'$ then $\mathcal{M}, s' \models \varphi$. Furthermore, $s \sim_a^{\{\psi_b \mid b \in G\}} s'$ holds if and only if $s \sim_a s'$ and $\mathcal{M}, s' \models \bigwedge\{\psi_b \mid b \in G\}$.

This implies that for every s', if $s \sim_a s'$, then $\mathcal{M}, s' \models \bigwedge\{\psi_b \mid b \in G\} \to \varphi$. Furthermore, since $\mathcal{M}, s \models \Box_a\psi_a$, we also have $\mathcal{M}, s' \models \psi_a$. It follows that $\mathcal{M}, s \models \Box_a(\psi_a \wedge (\bigwedge\{\psi_b \mid b \in G\} \to \varphi))$.

Now, note that $\{\psi_a \wedge (\bigwedge\{\psi_b \mid b \in G\} \to \varphi) \mid a \in G\} \models \varphi$. As such, $\{\psi \mid \exists a \in G : \mathcal{M}, s \models \Box_a\psi\} \models \varphi$.

For the other direction, suppose that $\{\psi \mid \exists a \in G : \mathcal{M}, s \models \Box_a\psi\} \models \varphi$. Since epistemic logic is compact, there is a finite $\Psi \subseteq \{\psi \mid \exists a \in G : \mathcal{M}, s \models \Box_a\psi\}$ such that $\Psi \models \varphi$. For every $a \in G$, let $\psi_a = \bigwedge\{\psi \in \Psi \mid \mathcal{M}, s \models \Box_a\psi\}$.

We now have $\mathcal{M}, s \models \Box_a\psi_a$ for every $a \in G$. Furthermore, since $\Psi \models \varphi$, we also have $\mathcal{M}, s' \models \bigwedge\{\psi_a \mid a \in G\} \to \varphi$ for every $s' \in S$ such that $s \sim_a s'$. It follows that $\mathcal{M}, s \models_{(\mathcal{L}_0,\odot,\Uparrow,\forall)} D_G\varphi$. □

Note that the proof critically depends on the compactness of epistemic logic. If we used a non-compact base logic, such as epistemic logic with common knowledge, the equivalence would not hold.

4 Relative Strength

Now that we have formally defined the various types of distributed knowledge that we are interested in, we can investigate their properties. In particular, we are interested in which variants imply each other.

In most cases, it is clear that one variant τ_1 is at least as strong as another variant τ_2, in the sense that $\mathcal{M}, s \models_{\tau_1} D_G\varphi$ implies $\mathcal{M}, s \models_{\tau_2} D_G\varphi$. In particular, it is easy to see that the following hold.

Proposition 4.1

- *For every τ, $\mathcal{M}, s \models_\tau D_G\varphi$ implies $\mathcal{M}, s \models_{(\cap,\epsilon,\epsilon,\forall)} D_G\varphi$.*
- *For every f, a and o, $\mathcal{M}, s \models_{(f,a,o,\forall)} D_G\varphi$ implies $\mathcal{M}, s \models_{(f,a,o,\exists)} D_G\varphi$.*
- *For all o and q, $\mathcal{M}, s \models_{(\mathcal{L}_0,\odot,o,q)} D_G\varphi$ implies $\mathcal{M}, s \models_{(\mathcal{L}_0,\bullet,o,q)} D_G\varphi$.*
- *For all a and q, $\mathcal{M}, s \models_{(\mathcal{L}_0,a,\Uparrow,q)} D_G\varphi$ implies $\mathcal{M}, s \models_{(\mathcal{L}_0,a,\omega,q)} D_G\varphi$.*
  Furthermore, $\mathcal{M}, s \models_{(\mathcal{L}_0,\bullet,\omega,q)} D_G\varphi$ implies $\mathcal{M}, s \models_{(\mathcal{L}_0,\bullet,\Omega,q)} D_G\varphi$.

Furthermore, it follows from [16] that $(\cap,\epsilon,\epsilon,\forall)$ and $(\cap,\epsilon,\epsilon,\exists)$ do not imply $(\mathcal{L}_0,\odot,\Uparrow,\forall)$.This, however, leaves many comparisons open, which we solve here.

4.1 Single formula

First, let us compare variants that use a single formula. In particular, we will show that all the four variants, i.e. $(\mathcal{L}_0,\odot,\Uparrow,\exists)$, $(\mathcal{L}_0,\odot,\Uparrow,\forall)$, $(\mathcal{L}_0,\odot,\omega,\exists)$, and $(\mathcal{L}_0,\odot,\omega,\forall)$, are equivalent

Proposition 4.2 *$\mathcal{M}, s \models_{(\mathcal{L}_0,\odot,\Uparrow,\exists)} D_G\varphi$ if and only if $\mathcal{M}, s \models_{(\mathcal{L}_0,\odot,\Uparrow,\forall)} D_G\varphi$.*

Proof. We know from Proposition 4.1 that "all" implies "single", so it suffices to show that the reverse also holds. Suppose, therefore, that $\mathcal{M}, s \models_{(\mathcal{L}_0,\odot,\Uparrow,\exists)}$

$D_G\varphi$. Let $\{\psi_a \mid a \in G\}$ be the witnessing formulas that, if communicated among the group, would make one agent, let's call them x, learn that φ is true.

This means that for every s', if $(s, s') \in\sim_x$ and $\mathcal{M}, s' \models \bigwedge_{a\in G}\psi_a$, then $\mathcal{M}, s' \models \varphi$. This implies that for every $(s, s') \in\sim_x$, we have $\mathcal{M}, s' \models \bigwedge_{a\in G}\psi_a \to \varphi$. Furthermore, since x was able to provide the formula ψ_x, we also have $\mathcal{M}, s' \models \psi_x$ for each such s'. Hence $\mathcal{M}, s \models \Box_x(\psi_x \wedge (\bigwedge_{a\in G}\psi_a \to \varphi))$.

Consider then the alternative communication $\{\psi'_a \mid a \in G\}$ where $\psi'_a = \psi_a$ for $a \neq x$ and $\psi'_x = \psi_x \wedge (\bigwedge_{a\in G}\psi_a \to \varphi)$. For every agent $a \in G$ and every s', if $(s, s') \in\sim_a$ and $\mathcal{M}, s' \models \bigwedge_{a\in G}\psi'_a$, we then have, in particular, $\mathcal{M}, s' \models \bigwedge_{a\in G}\psi_a$ and $\mathcal{M}, s' \models \bigwedge_{a\in G}\psi_a \to \varphi$, and hence $\mathcal{M}, s' \models \varphi$. This implies that all agents learn φ, and therefore $\mathcal{M}, s \models_{(\mathcal{L}_0,\odot,\Uparrow,\forall)} D_G\varphi$. □

In effect, the single agent x that learns φ can include hypothetical reasoning in the formula that they provide to the group. Instead of saying "ψ_x is true", they can say "ψ_x is true, and if you were to tell me $\{\psi_a \mid a \in G\}$, then I would learn that φ is true". This suffices for all the other agents to learn the truth of φ, if all $a \neq x$ do indeed provide formulas ψ_a. The same trick can be used in the sequential version.

Proposition 4.3 $\mathcal{M}, s \models_{(\mathcal{L}_0,\odot,\omega,\exists)} D_G\varphi$ *if and only if* $\mathcal{M}, s \models_{(\mathcal{L}_0,\odot,\omega,\forall)} D_G\varphi$.

Proof. Let $(a_1, \cdots, a_n)$ be an ordering of G such that a_i takes their turn before a_j iff $i < j$, and let $(\psi_1, \cdots, \psi_n)$ be the corresponding formulas that witness $\mathcal{M}, s \models_{(\mathcal{L}_0,\odot,\omega,\exists)} D_G\varphi$, where a_x is the agent that learns φ.

Then a_x already knows, before the communication starts, that $\bigwedge\{\psi_i \mid 1 \leq i \leq n\} \to \varphi$. Furthermore, once it is their turn, they have also learned that ψ_{a_x}. Hence they could instead say $\psi'_{a_x} = \psi_{a_x} \wedge (\bigwedge\{\psi_i \mid 1 \leq i \leq n\} \to \varphi)$, which would result in all agents learning φ. Note that since we work with the static notion of distributed knowledge, rather than a dynamic one, knowledge of agents is monotonic under announcements and they can always announce their respective formulas. □

Additionally, a similar kind of hypothetical reasoning can be used to remove the reliance on sequential communication.

Proposition 4.4 $\mathcal{M}, s \models_{(\mathcal{L}_0,\odot,\omega,\forall)} D_G\varphi$ *if and only if* $\mathcal{M}, s \models_{(\mathcal{L}_0,\odot,\Uparrow,\forall)} D_G\varphi$.

Proof. By Proposition 4.1, $\mathcal{M}, s \models_{(\mathcal{L}_0,\odot,\Uparrow,\forall)} D_G\varphi$ implies $\mathcal{M}, s \models_{(\mathcal{L}_0,\odot,\omega,\forall)} D_G\varphi$. Left to show is the other direction.

Suppose therefore that $\mathcal{M}, s \models_{(\mathcal{L}_0,\odot,\omega,\forall)} D_G\varphi$, as witnessed by order $(a_1, \cdots, a_n)$ and formulas $(\psi_1, \cdots, \psi_n)$. Then, at stage i, the preceding communications $\{\psi_j \mid j < i\}$ suffice for agent i to learn that ψ_i holds, in the sense that for all s', if $(s, s') \in\sim_{a_i}$ and $\mathcal{M}, s' \models \bigwedge_{j<i}\psi_j$ then $\mathcal{M}, s' \models \psi_i$.

It follows that, before the communication started, a_i already knew $\bigwedge_{j<i}\psi_j \to \psi_i$. So, in the simultaneous version, a_i could provide that formula.

Furthermore, collectively, communicating $\{\psi_i \mid 1 \leq i \leq n\}$ has the same effect as communicating $\{\bigwedge_{j<i}\psi_j \to \psi_i \mid 1 \leq i \leq n\}$. So we also have $\mathcal{M}, s \models_{(\mathcal{L}_0,\odot,\Uparrow,\forall)} D_G\varphi$. □

The equivalence of other pairs of the single formula variants follows immediately from the transitivity of the equivalence relation.

4.2 Sets of formulas

Let us now consider the variants where each agent may provide a set of formulas. As a first step, we will show that none of them imply the single formula variants. For this, it suffices to show that any one of the single formula variants is not implied by the strongest set variant (i.e., the set variant that is the hardest to satisfy, which is $(\mathcal{L}_0, \bullet, \Uparrow, \forall)$).

Proposition 4.5 $\mathcal{M}, s \models_{(\mathcal{L}_0,\bullet,\Uparrow,\forall)} D_G\varphi$ *does not necessarily imply that* $\mathcal{M}, s \models_{(\mathcal{L}_0,\odot,\Uparrow,\forall)} D_G\varphi$.

Proof. (*Sketch; full proof in Appendix B*). Suppose that, for every $i \in \mathbb{N}$, agent a knows whether p_i holds while b knows whether q_i holds. Furthermore, suppose both agents know that r is true if and only if, for every i, $p_i \leftrightarrow q_i$ holds. Consider the case where all p_i and q_i are, in fact, false, so r holds.

Under the $(\mathcal{L}_0, \bullet, \Uparrow, \forall)$ definition, we then have $D_{\{a,b\}}r$. After all, a can tell b that all p_i are false, while b can tell a that all q_i are false. This suffices for both of them to discover that r is true.

Under the $(\mathcal{L}_0, \odot, \Uparrow, \forall)$ definition, however, we have $\neg D_{\{a,b\}}r$. This is because the agents can only learn that r is true if $p_i \leftrightarrow q_i$ for all i, which cannot be expressed in a finite set of formulas. □

Next, let us note that $(\mathcal{L}_0, \bullet, \Uparrow, \exists)$ does not imply $(\mathcal{L}_0, \bullet, \Uparrow, \forall)$.

Proposition 4.6 $\mathcal{M}, s \models_{(\mathcal{L}_0,\bullet,\Uparrow,\exists)} D_G\varphi$ *does not necessarily imply* $\mathcal{M}, s \models_{(\mathcal{L}_0,\bullet,\Uparrow,\forall)} D_G\varphi$.

Proof. (*Sketch; full proof in Appendix B*). As in Proposition 4.5, suppose a knows whether p_i is true and b knows whether q_i is true. Now, however, suppose that b knows that whether r is true depends on the parity of the number of indices i such that p_i and q_i differ in value. Specifically, b knows that r is true if that number is even, while a is uncertain whether r holds if the number is even, or if it is odd. Both agents know r is false if p_i and q_i differ infinitely often. As before, suppose that all p_i and q_i happen to be false.

With the $(\mathcal{L}_0, \bullet, \Uparrow, \exists)$ definition of distributed knowledge, we then have $D_{\{a,b\}}r$. This is because, when a tells b that all p_i are false, agent b will learn that there are 0 indices where p_i and q_i differ, so r is true.

Yet r is not distributed knowledge under the $(\mathcal{L}_0, \bullet, \Uparrow, \forall)$ definition of distributed knowledge, since only b can learn that r is true. The reason a can't learn this is that "r is true iff there is an even number of i such that p_i and q_i disagree" cannot be expressed in epistemic logic. Furthermore, while b learns that r is true once the communication is complete, $(\mathcal{L}_0, \bullet, \Uparrow, \forall)$ requires simultaneous communication, so b cannot simply say that r is true. □

What does not make a difference, however, is simultaneous statements or ω-sequential ones.

Proposition 4.7 *$\mathcal{M}, s \models_{(\mathcal{L}_0,\bullet,\omega,\exists)} D_G\varphi$ iff $\mathcal{M}, s \models_{(\mathcal{L}_0,\bullet,\Uparrow,\exists)} D_G\varphi$, and $\mathcal{M}, s \models_{(\mathcal{L}_0,\bullet,\omega,\forall)} D_G\varphi$ iff $\mathcal{M}, s \models_{(\mathcal{L}_0,\bullet,\Uparrow,\forall)} D_G\varphi$*

Proof. At every step in the ω-sequential communication, when ψ_i is stated, a finite set $\{\psi_j \mid j < i\}$ preceded it. We can replace sequential announcement of $\{\psi_i \mid i \in \mathbb{N}\}$ by simultaneous announcement of $\{\bigwedge_{j<i} \psi_j \to \psi_i \mid i \in \mathbb{N}\}$. □

The Ω-sequential variant, on the other hand, is strictly weaker than ω-sequential or simultaneous ones.

Proposition 4.8 *$\mathcal{M}, s \models_{(\mathcal{L}_0,\bullet,\Omega,\exists)} D_G\varphi$ does not necessarily imply $\mathcal{M}, s \models_{(\mathcal{L}_0,\bullet,\Uparrow,\exists)} D_G\varphi$.*

Proof. (*Sketch; full proof in Appendix B*). Suppose that a knows, for all i and j, whether $p_{i,j}$ and $q_{i,j}$ hold. Furthermore, suppose that the value of x_i depends on the number of j such that $p_{i,j}$ differs from $q_{i,j}$, in a way known to b but not to a and c, and that this dependence cannot be expressed in epistemic logic. (See the full proof in the appendix for one way to create such an inexpressible dependence.) Similarly, y_i depends on the number of j such that $p_{i,j}$ and $q_{i,j}$ differ, in an inexpressible way that is known to c but not to a and b. Finally, all three agents know that z is true iff there is an even number of i such that x_i and y_i differ.

Then if z is true, that is distributed knowledge using the $(\mathcal{L}_0, \bullet, \Omega, \exists)$ definition, since a can tell b and c which $p_{i,j}$ and $q_{i,j}$ hold, at which point they can say which x_i and y_i are true, allowing all of them to determine that z holds.

With the $(\mathcal{L}_0, \bullet, \Uparrow, \exists)$ definition z is not distributed knowledge, however. This is because, in order for any of the three agents to learn that z is true, all three agents need to contribute their information. But b and c cannot initially contribute any non-trivial formulas, since their only private information is the way in which x_i or y_j depends on the values of $p_{i,j}$ and $q_{i,j}$, and this dependence is not expressible in epistemic logic.

It is only after a has informed b and c about the values of every $p_{i,j}$ and $q_{i,j}$ that b and c can apply their knowledge in order to determine the truth of x_i and y_j, respectively, which they can then communicate to the other agents. So a first needs to contribute at least ω formulas before b and c can get involved, which is not possible if $\tau = (\mathcal{L}_0, \bullet, \Uparrow, \exists)$. □

However, if we allow any ordinal number of communication steps, the difference between a single agent learning the formula or all of them doing so disappears.

Proposition 4.9 *$\mathcal{M}, s \models_{(\mathcal{L}_0,\bullet,\Omega,\exists)} D_G\varphi$ if and only if $\mathcal{M}, s \models_{(\mathcal{L}_0,\bullet,\Omega,\forall)} D_G\varphi$.*

Proof. Suppose that $\mathcal{M}, s \models_{(\mathcal{L}_0,\bullet,\Omega,\exists)} D_G\varphi$. So there is a sequence of communications of length α after which one agent a knows φ. Now, consider the sequence of length $\alpha + 1$ where, in the last step, agent a states that φ is true. This suffices for all agents to learn that φ is true. □

The above suffices to determine the comparative strength of each of the variants we discussed.

5 Discussion

We have considered 12 natural interpretations of the idea behind distributed knowledge. For these interpretations, we have analysed which ones are equivalent to each other, and which ones are different. The complete landscape of distributed knowledge is shown in Figure 1.

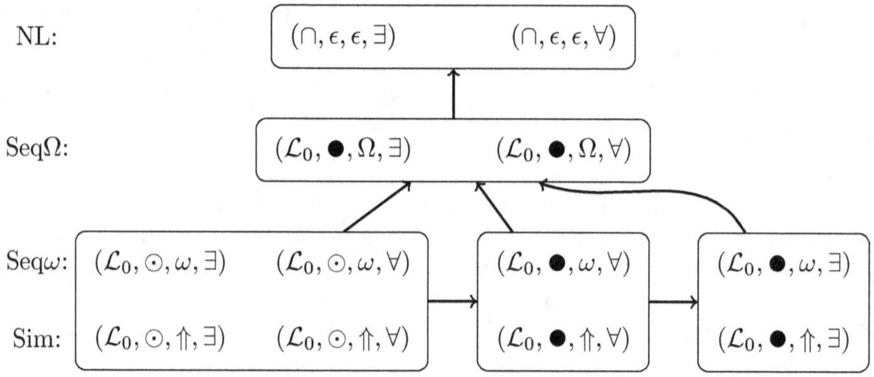

Fig. 1. The expressivity landscape of distributed knowledge, where 'NL' stands for 'non-linguistic sharing', 'SeqΩ' denotes unlimited sequential sharing, 'Seqω' stands for sequential sharing limited to ordinal ω, and 'Sim' denotes simultaneous sharing. Equivalent variations of distributed knowledge are enclosed in a box. Arrows point from stronger variants to weaker ones. Some arrows that follow from transitivity have been omitted for the sake of clarity.

Out of all variants of distributed knowledge, only the classic one (i.e. $(\cap, \epsilon, \epsilon, \forall)$ and $(\cap, \epsilon, \epsilon, \exists)$ in our taxonomy) was axiomatised [8,15]. The task of providing axiomatisations for the remaining variants seems to be both colossal and irresistibly tempting, and we thus leave it for future work.

Acknowledgements

A preliminary version of this work has benefited from discussion with Fernando R. Velázquez-Quesada. We would also like to thank the anonymous reviewers of AiML for their attention to details and constructive criticism that helped us to improve the presentation of this work.

Appendix

A Vicious Circularity

Suppose that we define the type $(\mathcal{L}, \odot, \Uparrow, \forall)$ in the same way as the type $(\mathcal{L}_0, \odot, \Uparrow, \forall)$, except that we now allow $a \in G$ to contribute any formula $\psi_a \in \mathcal{L}$ that they know, as opposed to any known formula from $\mathcal{L}_0$.

Clearly, such a definition would be circular. Here, we show that this circularity is vicious. More specifically, it is under-determined, in the sense that there are models where both $D_G\varphi$ and $\neg D_G\varphi$ are consistent with the semantics.

Proposition A.1 *There are $\mathcal{M}, s$ and φ such that both $\mathcal{M}, s \models_{(\mathcal{L}, \odot, \Uparrow, \forall)} D_G\varphi$*

and $\mathcal{M}, s \models_{(\mathcal{L},\odot,\Uparrow,\forall)} \neg D_G\varphi$ *are consistent with the* $(\mathcal{L},\odot,\Uparrow,\forall)$*-semantics.*

Proof. Consider the following model:

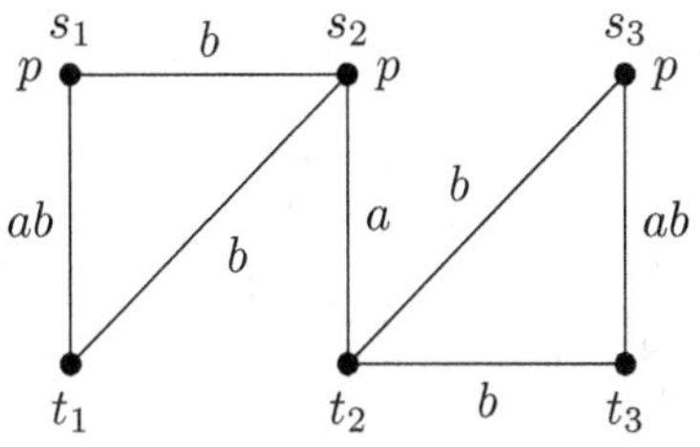

We will show that both $\mathcal{M}, s_2 \models D_{\{a,b\}}p$ and $\mathcal{M}, s_2 \not\models D_{\{a,b\}}p$ are consistent with the circular semantics. To see why this is the case, first note that the semantics are extensional, so while there are infinitely many formulas it suffices to consider only those that have different extensions. Let use denote the extension of φ by $[\![\varphi]\!]$.

Let E be the set of all extension on this model. In order for E to be consistent with the $(\mathcal{L},\odot,\Uparrow,\forall)$-semantics, it has to be "self-fulfilling", in the sense that, if we assume that E is the set of all extensions, then we should have $[\![\varphi]\!] \in E$ for all $\varphi \in \mathcal{L}$, and for every $e \in E$ there should be some $\varphi_e \in \mathcal{L}$ such that $[\![\varphi_e]\!] = e$.

We will show that there are two different sets of extensions that satisfy this criterion: we can take $E = 2^S$, in which case we have $\mathcal{M}, s_2 \models D_{\{a,b\}}p$, and we can take $E = \{\emptyset, \{s_1, s_2, s_3\}, \{t_1, t_2, t_3\}, S\}$, in which case $\mathcal{M}, s_2 \not\models D_{\{a,b\}}p$.

Suppose therefore that $E = 2^S$. It is immediate that $[\![\varphi]\!] \in E$ for all $\varphi \in \mathcal{L}$. Left to show is that every extension $e \in E$ is witnessed by some formula φ_e. To this purpose, we first note that we have $\mathcal{M}, s_2 \models D_{\{a,b\}}p$. This is because, by assumption, $E = 2^S$ is the set of extensions, so there are formulas φ_1 and φ_2 such that $[\![\varphi_1]\!] = \{s_2, t_2\}$ and $[\![\varphi_2]\!] = \{s_1, s_2, t_1\}$. Then $\mathcal{M}, s_2 \models \Box_a\varphi_1$ and $\mathcal{M}, s_2 \models \Box_b\varphi_2$, so the agents can share φ_1 and φ_2. Furthermore, by putting φ_1 and φ_2 together, the agents discover that s_2 is the only possible world. As p is true there, we have $\mathcal{M}, s_2 \models D_{\{a,b\}}p$.

In every other world, $D_{\{a,b\}}p$ is false. For t_1, t_2 and t_3 this follows from the fact that distributed knowledge is truthful and p is false in t_1, t_2 and t_3. For s_1 and s_3 it follows from the fact that there is an ab-successor (t_1 or t_3, respectively) where p is false. Since this world is an ab-successor, it can never be excluded by any formula known to a or b, so the agents cannot exclude the possibility of $\neg p$ by combining their information.

We have now shown that the formula $D_{\{a,b\}}p$ uniquely identifies the world s_2. It follows that there are also formulas uniquely identifying every other world:

s_1**:** $p \land \neg D_{\{a,b\}}p \land \Diamond_b D_{\{a,b\}}p$

s_3**:** $p \land \neg D_{\{a,b\}}p \land \neg\Diamond_b D_{\{a,b\}}p$

t_1**:** $\neg p \land \Diamond_b D_{\{a,b\}}p$

t_2: $\neg p \wedge \Diamond_a D_{\{a,b\}} p$

t_3: $\neg p \wedge \neg \Diamond_b D_{\{a,b\}} p \wedge \neg \Diamond_a D_{\{a,b\}} p$.

Let us denote the formula for any world by φ_{s_i} or φ_{t_i}. Any $e \in E$ is the extension of some disjunction of the relevant φ_{s_i} and/or φ_{t_i}.

We have now shown that $E = 2^S$ being the set of extensions is consistent with the semantics, and that we then have $\mathcal{M}, s_2 \models D_{\{a,b\}} p$. The witnessing formulas for this distributed knowledge had to have extensions $\{s_2, t_2\}$ and $\{s_1, t_1, s_2\}$, so we can take $\varphi_a = \{\varphi_{s_2} \vee \varphi_{t_2}\}$ and $\varphi_b = \{\varphi_{s_1} \vee \varphi_{s_2} \vee \varphi_{t_1}\}$.

Next, we will show that it is consistent with the semantics to have $E = \{\emptyset, \{s_1, s_2, s_3\}, \{t_1, t_2, t_3\}, S\}$, in which case $\mathcal{M}, s_2 \not\models D_{\{a,b\}} p$. In this case, it is easy to see that for every $e \in E$ there is some φ such that $e = [\![\varphi]\!]$; we have $\emptyset = [\![\bot]\!]$, $\{s_1, s_2, s_3\} = [\![p]\!]$, $\{t_1, t_2, t_3\} = [\![\neg p]\!]$ and $S = [\![\top]\!]$.

Left to show, therefore, is that for every φ, we have $[\![\varphi]\!] \in E$. Because the s_i are bisimilar to each other, as are the t_j, the bisimulation-invariance of modal logic implies that every $\psi \in \mathcal{L}_0$ will have one of the four extentions in E.

Furthermore, the D_G operator cannot break this symmetry. This is because, in every world, both a and b consider at least one t_i world and at least one s_i world possible. It follows that, for every world x, if $\mathcal{M}, x \models \Box_a \varphi$ or $\mathcal{M}, x \models \Box_b \varphi$, then $[\![\varphi]\!]$ contains at least one t_i and at least one s_i. Among the four extensions in E, the only one with this property is S.

In their communication, a nor b can therefore only contribute formulas with extension S, so neither of them provides non-trivial information. It follows that φ is distributed knowledge if and only if one of the agents already knew φ before the agents started sharing information. Hence $\mathcal{M}, x \models D_{\{a,b\}} \varphi$ if and only if $\mathcal{M}, x \models \Box_a \varphi \vee \Box_b \varphi$. It follows that, for every $\varphi \in \mathcal{L}$ there is a $\varphi_0 \in \mathcal{L}_0$ such that $[\![\varphi]\!] = [\![\varphi_0]\!]$, which implies that $[\![\varphi]\!] \in E$.

Furthermore, because $\mathcal{M}, s_2 \not\models \Box_a p \vee \Box_b p$, we have $\mathcal{M}, s_2 \not\models D_{\{a,b\}} p$. □

B Full versions of proofs

Proposition B.1 *$\mathcal{M}, s \models_{(\mathcal{L}_0, \bullet, \Uparrow, \forall)} D_G \varphi$ does not necessarily imply that $\mathcal{M}, s \models_{(\mathcal{L}_0, \odot, \Uparrow, \forall)} D_G \varphi$.*

Proof. Let $\mathcal{M} = (S, \sim, V)$, where S and $\sim$ are given as follows.

- $S = \{s_{x,y} \mid (x, y) \in [0,1) \times [0,1)\}$,
- $\sim_a = \{(s_{x,y}, s_{x,y'})\}$
- $\sim_b = \{(s_{x,y}, s_{x',y})\}$

Each x or y can be interpreted, written in binary notation, as a set of natural numbers. Let us write $\ulcorner x \urcorner$ for the set of natural numbers represented by x. Then we take $V(p_i) = \{s_{x,y} \mid i \in \ulcorner x \urcorner\}$, $V(q_i) = \{s_{x,y} \mid i \in \ulcorner y \urcorner\}$ and $V(r) = \{s_{x,y} \mid x = y\}$.

In every world, a knows the x-coordinate while being uncertain about the y-coordinate, while b is uncertain about the x-coordinate while knowing the y-coordinate. Since the value of p_i depends only on the x-coordinate, this means that a knows, for every i, whether p_i is true. Similarly, b knows whether q_i is true.

Furthermore, it is true throughout the model, and therefore known to both agents, that r holds if and only if $x = y$, and therefore if and only if $p_i \leftrightarrow q_i$ for all i.

We have $\mathcal{M}, s_{0,0} \models_{(\mathcal{L}_{\mathrm{O}},\bullet,\Uparrow,\forall)} D_{\{a,b\}}r$, which is witnessed by the sets $\Psi_a = \{\neg p_i \mid i \in \mathbb{N}\}$ and $\Psi_b = \{\neg q_i \mid i \in \mathbb{N}\}$. After all, the only world where all p_i and q_i are false is $s_{0,0}$, where r is true.

Suppose now, towards a contradiction, that there are formulas ψ_a and ψ_b that are known by their respective agents in $s_{0,0}$, and that would allow the agents to learn that r is true. Let Q be the set of atoms that occur in ψ_a and ψ_b. Note that this is a finite set.

Because ψ_a is known by a, it must hold in every $s_{0,y}$. Similarly, ψ_b holds on every $s_{x,0}$. Let x and y be such that $x \neq 0$, $y \neq 0$, $\ulcorner x \urcorner \cap Q = \ulcorner y \urcorner \cap Q = \emptyset$ and $x \neq y$. Now, consider the world $s_{x,y}$. We will show that it is Q-bisimilar to both $s_{x,0}$ and $s_{0,y}$.

To this end, consider the relation $\approx \subseteq S \times S$, such that $s_{x,y} \approx s_{x',y'}$ iff $s_{x,y}$ and $s_{x',y'}$ agree on $Q \cup \{r\}$. We claim that this relation is a bisimulation. Atomic agreement (when restricted to Q) between any two $\approx$-related worlds is immediate from the construction. We show forth for b, the other cases can be shown similarly.

So suppose that $s_{x,y} \approx s_{x',y'}$ and that $s_{x,y} \sim_b s_{u,v}$. Then $y = v$. Now, let $v' = y'$ and

- if $u = v$ then $u' = v'$,
- if $u \neq v$ then u' is any number such that $u' \neq v'$ and $\ulcorner u' \urcorner \cap Q = \ulcorner u \urcorner \cap Q$.

Then $s_{u,v}$ and $s_{u',v'}$ agree on all atoms in $Q \cup \{r\}$. Hence $s_{u,v} \approx s_{u',v'}$ by our assumption. Furthermore, because $y' = v'$, we also have $s_{x',y'} \sim_b s_{u',v'}$ by the construction of the model. So the forth condition is satisfied.

From this bisimilarity and the fact that $\mathcal{M}, s_{x,0} \models \psi_b$, it follows that $\mathcal{M}, s_{x,y} \models \psi_b$, and therefore also that $\mathcal{M}, s_{0,y} \models \psi_b$. For the same reason, $\mathcal{M}, s_{x,0} \models \psi_a$.

We can thus conclude that ψ_a and ψ_b cannot exclude the worlds $s_{x,0}$ and $s_{0,y}$. In both these worlds r is false, so neither a nor b learns that r is true. This contradicts our assumptions, so we have that $\mathcal{M}, s \not\models_{(\mathcal{L}_{\mathrm{O}},\odot,\Uparrow,\forall)} D_{\{a,b\}}r$. □

Proposition B.2 *$\mathcal{M}, s \models_{(\mathcal{L}_{\mathrm{O}},\bullet,\Uparrow,\exists)} D_G\varphi$ does not necessarily imply $\mathcal{M}, s \models_{(\mathcal{L}_{\mathrm{O}},\bullet,\Uparrow,\forall)} D_G\varphi$.*

Proof. Let

- $S = \{s_{x,y} \mid (x,y) \in [0,1) \times [0,1)\} \cup \{t_{x,y} \mid (x,y) \in [0,1) \times [0,1)\}$,
- $\sim_a = \{(u_{x,y}, v_{x,y'}) \mid u, v \in \{s,t\}, x, y, y' \in [0,1)\}$
- $\sim_b = \{(s_{x,y}, s_{x',y}) \mid x, x', y \in [0,1)\} \cup \{(t_{x,y}, t_{x',y}) \mid x, x', y \in [0,1)\}$
- $V(p_i) = \{s_{x,y} \mid i \in \ulcorner x \urcorner\} \cup \{t_{x,y} \mid i \in \ulcorner x \urcorner\}$
- $V(q_i) = \{s_{x,y} \mid i \in \ulcorner y \urcorner\} \cup \{t_{x,y} \mid i \in \ulcorner y \urcorner\}$

- $V(r) = \{s_{x,y} \mid |\ulcorner x \urcorner \triangle \ulcorner y \urcorner| \text{ is even}\} \cup \{t_{x,y} \mid |\ulcorner x \urcorner \triangle \ulcorner y \urcorner| \text{ is odd}\}$ [6]

Essentially, the model consists of two grids, one with $s_{x,y}$ worlds and another one with $t_{x,y}$ worlds. Agent a does not know which grid the current world belongs to, while agent b does. On $s_{x,y}$ worlds r is true if the number atoms on which x and y differs is even, whereas on $t_{x,y}$ worlds r is true if that number is odd. As before, a knows the value of the p_i atoms, while b knows the value of q_i. Finally, b can tell the difference between the s and t worlds, and therefore the required parity, while a does not.

We have $\mathcal{M}, s_{0,0} \models_{(\mathcal{L}_0,\bullet,\Uparrow,\exists)} D_{\{a,b\}}r$ because, as in the previous proof, a can announce which p_i hold and b can announce which q_i hold, which suffices for b to determine that r holds.

There is no way for a to learn that r is true by simultaneous statements, however. This follows from another Q-bisimilarity argument. □

Proposition B.3 *$\mathcal{M}, s \models_{(\mathcal{L}_0,\bullet,\Omega,\exists)} D_G\varphi$ does not necessarily imply $\mathcal{M}, s \models_{(\mathcal{L}_0,\bullet,\Uparrow,\exists)} D_G\varphi$.*

Proof. Let

$$\begin{aligned} S = \{s_{e,f,g} \mid\ & e : \mathbb{N} \times \mathbb{N} \times \{0,1\} \to \{0,1\} \\ & f : \mathbb{N} \to \mathbb{N} \\ & g : \mathbb{N} \to \mathbb{N}\} \end{aligned}$$

and

- $s_{e,f,g} \sim_a s_{e',f',g'}$ iff $e = e'$,
- $s_{e,f,g} \sim_b s_{e',f',g'}$ iff $f = f'$,
- $s_{e,f,g} \sim_c s_{e',f',g'}$ iff $g = g'$.

In other words, in $s_{e,f,g}$ agent a knows e, agent b knows f and agent c knows g.

Furthermore, for $i, j \in \mathbb{N}$ let

- $V(p_{i,j}) = \{s_{e,f,g} \mid e(i,j,0) = 1\}$,
- $V(q_{i,j}) = \{s_{e,f,g} \mid e(i,j,1) = 1\}$,
- $V(x_j) = \{s_{e,f,g} \mid$ the number of indices $i \in \mathbb{N}$ s.t. $e(i,j,0) \neq e(i,j,1)$ is divisible by $f(j)\}$
- $V(y_j) = \{s_{e,f,g} \mid$ the number of indices $i \in \mathbb{N}$ s.t. $e(i,j,0) \neq e(i,j,1)$ is divisible by $g(j)\}$
- $V(z) = \{s_{e,f,g} \mid$ there is an even number of indices $j \in \mathbb{N}$ s.t. exactly one of x_j, y_j holds on $s_{e,f,g}\}$

So e simply determines which $p_{i,j}$ and $q_{i,j}$ hold. The function f, meanwhile, determines for each j the number, $f(j)$, that must divide the amount of indices i on which $p_{i,j}$ and $q_{i,j}$ are different, in order for x_j to be true. Similarly,

[6] For sets X and Y, the symmetric difference $X \triangle Y$ is defined as $(X \setminus Y) \cup (Y \setminus X)$.

g determines the number of indices that must be different for y_j to be true. Finally, z holds if and only if x_j and y_j differ in value for an even number of indices j.

It is now easy to see that for any $s_{e,f,g}$ such that $\mathcal{M}, s_{e,f,g} \models z$ we have $\mathcal{M}, s_{e,f,g} \models_{(\mathcal{L}_0,\bullet,\Omega,\exists)} D_{\{a,b,c\}}z$. The communication the agents can perform is as follows:

- In the time steps before ω, agent a tells the other two exactly which $p_{i,j}$ and $q_{i,j}$ hold.
- At time ω, agents b and c know which $p_{i,j}$ and $q_{i,j}$ hold. As a consequence, b knows which x_j hold while c knows which y_j hold.
- In the time steps between ω and $2 \times \omega$, agents b and c tell the other two which x_j and y_j hold.
- At time $2 \times \omega$, all agents know exactly which x_j and y_j hold, and therefore whether z holds.
- We assumed we were in a world where z is true, so the agents learn that z is true.

Now, to show that $\mathcal{M}, s_{e,f,g} \not\models_{(\mathcal{L}_0,\bullet,\Uparrow,\exists)} D_{\{a,b,c\}}z$.

The key observation here is that b and c cannot provide any non-trivial announcements. Suppose towards a contradiction that $\mathcal{M}, s_{e,f,g} \models \Box_b\psi_b$ and $\mathcal{M}, s_{e',f',g'} \not\models \psi_b$. Let Q be the set of atoms in ψ_b. Then there is some $s_{e'',f,g''}$ that is Q-bisimilar to $s_{e',f',g'}$. But, by $\mathcal{M}, s_{e,f,g} \models \Box_b\psi_b$, we have $\mathcal{M}, s_{e'',f,g''} \models \psi_b$, which by bisimilarity implies $\mathcal{M}, s_{e',f',g'} \models \psi_b$, contradicting our assumption.

Hence the only announcements b can provide hold in every world of the model, and are therefore uninformative. That c cannot make a non-trivial announcement is shown similarly.

This means that only a can provide information, so after the communication all worlds of the form $s_{e,f',g'}$ are still accessible. Each agent then still considers both z worlds and $\neg z$ worlds to be possible, so z is not distributed knowledge.□

References

[1] Ågotnes, T., N. Alechina and R. Galimullin, *Logics with group announcements and distributed knowledge: Completeness and expressive power*, Journal of Logic, Language and Information **31** (2022), pp. 141–166.

[2] Ågotnes, T., P. Balbiani, H. van Ditmarsch and P. Seban, *Group announcement logic*, Journal of Applied Logic **8** (2010), pp. 62–81.

[3] Ågotnes, T. and R. Galimullin, *Quantifying over information change with common knowledge*, Autonomous Agents Multi-Agent Systems **37** (2023), p. 40.

[4] Ågotnes, T. and Y. N. Wáng, *Resolving distributed knowledge*, Artificial Intelligence **252** (2017), pp. 1–21.

[5] Baltag, A., *What is DEL good for?* (2010), Workshop on Logic, Rationality and Intelligent Interaction.
URL `http://ai.stanford.edu/~epacuit/lograt/esslli2010-slides/copenhagenesslli.pdf`

[6] Baltag, A. and S. Smets, *Learning what others know*, in: E. Albert and L. Kovács, editors, *Proceedings of the 23rd LPAR*, EPiC Series in Computing **73** (2020), pp. 90–119.

[7] Fagin, R., J. Y. Halpern, Y. Moses and M. Vardi, "Reasoning About Knowledge," The MIT Press, 1995.

[8] Fagin, R., J. Y. Halpern and M. Y. Vardi, *What can machines know? On the properties of knowledge in distributed systems*, Journal of the ACM **39** (1992), pp. 328–376.

[9] Galimullin, R. and F. R. Velázquez-Quesada, *(Arbitrary) partial communication*, in: N. Agmon, B. An, A. Ricci and W. Yeoh, editors, *Proceedings of the 22nd AAMAS* (2023), pp. 400–408.

[10] Gerbrandy, J., "Bisimulations on Planet Kripke," Ph.D. thesis, University of Amsterdam (1999).

[11] Halpern, J. Y. and Y. Moses, *Knolwedge and common knolwedge in a distributed environment*, Journal of the Association for Computing Machinery **37** (1990), pp. 549–587.

[12] Humberstone, I. L., *The formalities of collective omniscience*, Philosophical Studies **48** (1985), pp. 401–423.

[13] Meyer, J. C. and W. van der Hoek, "Epistemic logic for AI and computer science," Cambridge tracts in theoretical computer science **41**, CUP, 1995.

[14] Roelofsen, F., *Distributed knowledge*, Journal of Applied Non-Classical Logics **17** (2007), pp. 255–273.

[15] van der Hoek, W. and J. C. Meyer, *Making some issues of implicit knowledge explicit*, International Journal of Foundations of Computer Science **3** (1992), pp. 193–223.

[16] van der Hoek, W., B. van Linder and J.-J. Meyer, *Group knowledge is not always distributed (neither is it always implicit)*, Mathematical Social Sciences **38** (1999), pp. 215–240.

[17] van Ditmarsch, H., W. van der Hoek and B. Kooi, "Dynamic Epistemic Logic," Synthese Library **337**, Springer, 2008.

Deducibility in the Full Lambek Calculus With Weakening Is HAck-Complete

Vitor Greati

Bernoulli Institute, University of Groningen, The Netherlands

Revantha Ramanayake [1]

Bernoulli Institute and CogniGron, University of Groningen, The Netherlands

Abstract

We show that the problem of deciding the consequence relation (deducibility) for the full Lambek calculus with weakening ($\mathbf{FL_w}$) is complete for the class HACK of hyper-Ackermannian problems (i.e., $\mathbf{F}_{\omega^\omega}$ in the ordinal-indexed hierarchy of fast-growing complexity classes). Provability was already known to be PSPACE-complete. We prove that deducibility is HACK-complete even for the multiplicative fragment. Lower bounds are obtained via a novel reduction from reachability in lossy channel systems. Upper bounds are obtained via structural proof theory (forward proof search over sequent calculi) and well-quasi-order theory (length theorems for Higman's ordering on words over a finite alphabet).

Keywords: substructural logics, hyper-Ackermannian problems, Lambek calculus, Higman's Lemma, well-quasi-order theory, sequent calculi

1 Introduction

Substructural logics are obtained from intuitionistic and classical logic by removing familiar structural properties like exchange (e), contraction (c) and weakening (w), and extending with further axioms and language connectives. A prominent example is the family of *linear logics* where a propositional substructural base is extended with modalities ("exponentials") for weakening and contraction. The absence of structural properties leads to highly expressive and complex logics, and this is evident even in the most fundamental of substructural logics (see Table 1), namely the extensions of the full Lambek calculus **FL** by subsets of $\{(e), (c), (w)\}$. A significant gap in this table persisted at $\mathbf{FL_w}$ since the deducibility problem (provability of a formula from a finite set of formulas) was, thus far, only known to be decidable, a result obtained by Blok and van Alten [2] via the finite embeddability property for $\mathsf{FL_w}$-algebras,

[1] The financial support of the CogniGron research center, the Ubbo Emmius Funds (University of Groningen), and FWF project P33548 is acknowledged.

the algebraic semantics of $\mathbf{FL_w}$. We fill this gap by establishing completeness with respect to the fast-growing hyper-Ackermannian complexity class HACK, placing $\mathbf{FL_w}$ as the hardest for deducibility among the decidable basic substructural logics (taking the crown from $\mathbf{FL_{ec}}$, where deducibility has Ackermannian complexity). Let $\mathbf{FL}_\mathbf{w}^\Sigma$ denote the restriction of $\mathbf{FL_w}$ to connectives in Σ.

Theorem 1.1 (Main theorem). *For $\{\otimes, \backslash, 0, 1\} \subseteq \Sigma \subseteq \Sigma_{\mathbf{FL}}$, deducibility in $\mathbf{FL}_\mathbf{w}^\Sigma$ is HACK-complete. In particular, $\mathbf{FL_w}$ and its multiplicative fragment $\mathbf{L_w}$ are HACK-complete.*

Logic	Deducibility	Provability
$\mathbf{FL}$	Undecidable [13]	PSPACE-complete [12]
$\mathbf{FL_e}$	Undecidable [14]	PSPACE-complete [12]
$\mathbf{FL_{ec}}$	ACK-complete [21]	ACK-complete [21]
$\mathbf{L_{ec}}$	2-EXPTIME-complete [17]	2-EXPTIME-complete [17]
$\mathbf{FL_{ew}}$	TOWER-complete [20]	PSPACE-complete [12]
$\mathbf{FL_c}$	Undecidable [5]	Undecidable [5]
$\mathbf{FL_w}$	Decidable [2]. Complexity: **this paper**	PSPACE-complete [12]
$\mathbf{L_w}$	Decidable [10]. Complexity: **this paper**	PSPACE [12]

Table 1
The computational status of the basic substructural logics.

The above statement references Schmitz's ordinal-indexed *fast-growing complexity classes* $\{\mathbf{F}_\alpha\}_\alpha$ [16]. Important members of this family are $\mathbf{F}_2$, problems solvable in time expressed by an elementary function (such as polynomial and exponential); $\mathbf{F}_3$, the first class of non-elementary problems, known as TOWER; $\mathbf{F}_\omega$, the first class of non-primitive recursive problems, known as ACK (short for 'Ackermann'); and $\mathbf{F}_{\omega^\omega}$, the first class of non-multiply recursive problems. The complexity of the latter cannot be expressed by a function defined by multiple nested recursion [15, §10], and it is known as HACK ('hyper-Ackermann').

An overview of the lower bound argument Inspired by Urquhart's [21] lower bound for $\mathbf{FL_{ec}}$ where *acceptance* for expansive additive counter machines is reduced to deducibility, we reduce *reachability* for *lossy channel systems* (LCS) to deducibility in $\mathbf{FL_w}$.

An LCS is a computational model whose configurations are tuples $(q, u_1, \ldots, u_n)$ where q is the state and u_k is the kth channel's contents i.e., some word over a finite alphabet $\mathtt{M}$. Its instructions take the following form: *read* letter $\mathtt{a}$ from the front of channel k in state q_i, remove it and transit to state q_j; *write* letter $\mathtt{a}$ to the back of channel k in state q_i and transit to state q_j; perform a *lossy* step by deleting some letters from any channel (the state remains unchanged). Chambart and Schnoebelen [4] show that deciding reachability of a configuration v from a configuration u is HACK-complete.

We code the set $\mathtt{I}$ of instructions of an LCS $\mathcal{L}$ as a finite set $\mathcal{T}(\mathcal{L})$ of sequents, and simulate read and write instructions by cuts with these sequents, and lossy steps by left weakening. For instance, reachability of $(q_2, \mathtt{a}, \mathtt{b})$ from

$(q_1, \mathtt{aa}, \mathtt{b})$ in $\mathcal{L}$ is coded as deducibility of the following sequent from $\mathcal{T}(\mathcal{L})$:

$$Q_1, \#_1^s, A, A, \#_1^e, \#_2^s, B, \#_2^e \Rightarrow Q_2 \otimes (\#_1^s \otimes (A \otimes (\#_1^e \otimes (\#_2^s \otimes (B \otimes \#_2^e)))))$$

Reading upwards from the endsequent, the following deduction in $\mathbf{FL_w}$ simulates an instruction $(q_i, c_1, \mathtt{a}, ?, q_j) \in \mathtt{I}$; i.e., read $\mathtt{a}$ from the front of channel c_1 in state q_i, remove it, then transit to q_j.

$$\dfrac{\begin{array}{c}\text{sequent in } \mathcal{T}(\mathcal{L}) \text{ coding instruction}\\ Q_i, \#_1^s, A \Rightarrow Q_j \otimes \#_1^s\end{array} \qquad \dfrac{\dfrac{\cdots}{Q_j, \#_1^s, B, C, \#_1^e \Rightarrow v}}{Q_j \otimes \#_1^s, B, C, \#_1^e \Rightarrow v}\,(\mathrm{L}\otimes)}{Q_i, \#_1^s, A, B, C, \#_1^e \Rightarrow v}\,(\mathrm{cut})$$

Every computation in the LCS is shown to correspond to a deduction of its reachability encoding. The converse direction is far more challenging, requiring a careful analysis of deductions in $\mathbf{FL_w}$. The starting point is a cut-elimination style argument that establishes a normal form (of "standard deductions") where the left premise of every cut is an element of the theory. The computation is read off the standard deduction, starting at the endsequent.

The upper bound argument We define a forward proof procedure extending the construction in Balasubramanian et al. [1]. The idea is to start from the set D_0 of initial sequent instances and form an increasing chain $D_0 \subset D_1 \subset \ldots$ of finite sets of deducible sequents (each step corresponds to limited weakening and then a single rule application), taking care that each D_i contains only minimal elements under the Higman ordering on words (omit words that can be obtained from another by inserting letters); the ordering is a well-quasi-order (wqo) [11] so the chain must be finite. By construction, the bad sequence underlying the chain is controlled (no element in D_{i+1} is much larger than some element in D_i), and the complexity is dominated by the maximum length of a controlled bad sequence. This is a function in HACK as shown by Schmitz and Schnoebelen [17]. The argument applies to infinitely-many structural rule extensions of $\mathbf{FL_w}$.

Contributions We show that (1) sequent-deducibility in fragments of $\mathbf{FL_w}$ containing $\otimes$ is HACK-hard and that (2) deducibility for any $\mathcal{N}_2$-analytic structural rule extension [6] of any fragment of $\mathbf{FL_w}$ is in HACK. Our main theorem follows from (1) and (2). Algebraizability of $\mathbf{FL_w}$ implies that (3) the word problem and the quasiequational theory of the variety of integral zero-bounded full Lambek algebras are HACK-complete.

2 Noncommutative substructural logics with weakening

Let Σ be a propositional signature and fix a countable set $\mathcal{P}$ of *propositional variables*. Let $\mathrm{Fm}_\Sigma(\mathcal{P})$ denote the set of *formulas* over Σ generated by $\mathcal{P}$. We consider subsignatures of $\Sigma_{\mathbf{FL}} := \{0, 1, \top, \bot, \otimes, \wedge, \vee, /, \backslash\}$, where $0, 1, \top$ and $\bot$ are constants and the other connectives are binary. A subsignature is specified as $\Sigma_{\mathbf{FL}}^{\star_1 \cdots \star_k} := \{\star_1, \ldots, \star_k\} \subseteq \Sigma_{\mathbf{FL}}$. For example, $\Sigma_{\mathbf{FL}}^{\otimes\backslash} = \{\otimes, \backslash\}$. Sequences (or lists) of formulas will be written without delimiters, with commas separating the formulas. We write $\bigotimes \varphi_1, \ldots, \varphi_m$ as abbreviation for $\varphi_1 \otimes (\varphi_2 \otimes (\ldots (\varphi_{m-1} \otimes$

$\varphi_m)\ldots))$, where $m \geq 1$ and $\varphi_1, \ldots, \varphi_m$ is a list of formulas. If the constant 1 is present in the language, we allow $m = 0$ (empty fusion), the result being 1.

A *sequent over* $\mathrm{Fm}_\Sigma(\mathcal{P})$ is an expression $\Gamma \Rightarrow \Pi$, where Γ is a finite sequence of formulas and Π is an empty sequence or a sequence with a single formula. Given a set $\mathcal{T}$ of sequents, let $\mathsf{subf}(\mathcal{T})$ denote the set of subformulas of the formulas appearing in it. We assume familiarity with the sequent calculus (see e.g., [9, Sec. 2.1.3]): a *sequent calculus* $\mathbf{S}$ *over* $\mathrm{Fm}_\Sigma(\mathcal{P})$ is a collection of schematic sequent rules. *Instantiations* of their schematic variables lead to *rule instances. Derivations* in $\mathbf{S}$ are finite rooted trees labelled with sequents over $\mathrm{Fm}_\Sigma(\mathcal{P})$ such that a node and its children are the conclusion and premises of a rule instance. For a finite set $\mathcal{T}$ of sequents, a $\mathcal{T}$*-deduction* in $\mathbf{S}$ is a derivation in which every leaf is a rule with no premises (*axiom* or *initial sequent*) or a sequent in $\mathcal{T}$. A *proof* in $\mathbf{S}$ is a $\varnothing$-deduction in $\mathbf{S}$. Any $\mathbf{S}' \subseteq \mathbf{S}$ is called a *subcalculus* of $\mathbf{S}$, and any $\mathbf{S}'' \supseteq \mathbf{S}$ is a *rule extension* of $\mathbf{S}$. *Structural rule extensions* are extensions by *structural rules* (rules without logical connectives).

Subcalculi of the sequent calculus $\mathbf{FL_w}$ displayed in Fig. 1 (*full Lambek calculus with weakening*) are induced by a subsignature of $\Sigma_{\mathbf{FL}}$ as follows. Let $\mathbf{FL}_\mathbf{w}^{\star_1 \cdots \star_k}$ be the sequent calculus consisting of rules that mention no connective (i.e., rule (id) and structural rules), or that mentions a connective in $\{\star_1, \ldots, \star_k\}$. For instance, $\mathbf{FL}_\mathbf{w}^{\otimes\backslash}$ is the calculus over $\mathrm{Fm}_{\Sigma_{\mathbf{FL}}^{\otimes\backslash}}(\mathcal{P})$ consisting of the axioms, all structural rules and the rules mentioning $\otimes$ and $\backslash$.

A sequent calculus $\mathbf{S}$ over $\mathrm{Fm}_\Sigma(\mathcal{P})$ canonically determines two *consequence relations*, one over sequents and another over formulas. The *sequent-deduction relation* $\vdash_{\mathbf{S}\Rightarrow}$ is such that, for all finite sets $\mathcal{T} \cup \{s\}$ of sequents over $\mathrm{Fm}_\Sigma(\mathcal{P})$, $\mathcal{T} \vdash_{\mathbf{S}\Rightarrow} s$ iff there is a $\mathcal{T}$-deduction of s in $\mathbf{S}$. The *formula-deduction relation* $\vdash_\mathbf{S}$ is such that, for all finite sets $G \cup \{\varphi\}$ of formulas over $\mathrm{Fm}_\Sigma(\mathcal{P})$, $G \vdash_\mathbf{S} \varphi$ iff there is a $\{\ \Rightarrow \psi \mid \psi \in G\}$-deduction of $\ \Rightarrow \varphi$ in $\mathbf{S}$. Given a sequent calculus $\mathbf{S}$, the problem of deciding $\vdash_{\mathbf{S}\Rightarrow}$ is called *sequent-deducibility* and the problem of deciding $\vdash_\mathbf{S}$ is called *formula-deducibility* or simply *deducibility.* We know that $\vdash_{\mathbf{FL_w}}$ is the consequence relation of the full Lambek logic with weakening (see [9, 2.1.4] for the Hilbert-calculus presentation and residuated lattices semantics of these logics). The reader is also referred to [9, 2.1.4] for the notion of *axiomatic extensions* of $\vdash_{\mathbf{FL_w}}$.

The following lemma says that $\vdash_{\mathbf{S}\Rightarrow}$ reduces to $\vdash_\mathbf{S}$ when $\mathbf{S}$ is a subcalculus of $\mathbf{FL_w}$ containing the rules for $\otimes, \backslash, 0$ and 1. For that, consider the translation $(\cdot)^\rightarrow$ of sequents over $\mathrm{Fm}_\Sigma(\mathcal{P})$ into formulas over $\mathrm{Fm}_{\Sigma\cup\{0,1,\otimes,\backslash\}}(\mathcal{P})$ defined $(\Gamma \Rightarrow \Pi)^\rightarrow := \bigotimes\Gamma\backslash\Pi^\star$, where $\Pi^\star$ is φ if $\Pi = \varphi$ and 0 if Π is empty.

Lemma 2.1. *Let* $\Sigma \supseteq \{\otimes, \backslash, 0, 1\}$ *and* $\mathbf{S} := \mathbf{FL}_\mathbf{w}^\Sigma$. *For all finite sets of sequents* $\mathcal{T}$ *and sequents* s *over the language of* $\mathbf{S}$, $\mathcal{T} \vdash_{\mathbf{S}\Rightarrow} s$ *iff* $\{t^\rightarrow \mid t \in \mathcal{T}\} \vdash_\mathbf{S} s^\rightarrow$.

Proof. From left to right, let δ be a deduction witnessing $\mathcal{T} \vdash_{\mathbf{S}\Rightarrow} s$. We show how to convert δ into a witness of $\{t^\rightarrow \mid t \in \mathcal{T}\} \vdash_\mathbf{S} s^\rightarrow$ by structural induction. In the base case, s is either the result of an axiomatic rule or an element of $\mathcal{T}$. To translate $t \in \mathcal{T}$ into $t^\rightarrow$, repeatedly use the rules (L$\otimes$) and (R$\backslash$), and (L1) and (R0) if required. In the inductive step, assume $\Gamma \Rightarrow$

Π was derived by an application of a k-ary rule r, whose premises are the sequents $\Gamma_1 \Rightarrow \Pi_1, \ldots, \Gamma_k \Rightarrow \Pi_k$. Note first that the rule $\dfrac{\Rightarrow \bigotimes \Gamma \backslash \Pi}{\Gamma \Rightarrow \Pi}$ (FS) is derivable in presence of cut. By IH, the sequents $\Rightarrow \bigotimes \Gamma_1 \backslash \Pi_1^\star, \ldots, \Rightarrow \bigotimes \Gamma_k \backslash \Pi_k^\star$ are provable from $\mathcal{T}^{\rightarrow}$. Apply to these premises the derived rule (FS) thus obtaining the premises $\Gamma_1 \Rightarrow \Pi_1^\star, \ldots, \Gamma_k \Rightarrow \Pi_k^\star$ (maybe cuts with $\Rightarrow 1$ will be necessary to recover empty antecedents). Use cuts with $0 \Rightarrow$ whenever necessary to obtain from those the premises $\Gamma_1 \Rightarrow \Pi_1, \ldots, \Gamma_k \Rightarrow \Pi_k$. Apply the same rule r to the latter, obtaining $\Gamma \Rightarrow \Pi$, from which $\Rightarrow \bigotimes \Gamma \backslash \Pi^\star$ is easily derivable (if Γ is empty, use (L1)), and we are done.

From right to left, first transform all the leaves with translated sequents from $\mathcal{T}$ into trees with leaves being either instances of axioms or the sequents themselves. This is easy to do using rules (R$\backslash$), (L$\otimes$), (L1) and (R0). Then it is enough to transform the root $s^{\rightarrow}$ into s using (FS) as we did in the converse direction. □

$$\frac{}{p \Rightarrow p}\,(\mathrm{id}) \qquad \frac{}{\Gamma, \bot, \Delta \Rightarrow \Pi}\,(\mathrm{L}\bot) \qquad \frac{}{\Gamma \Rightarrow \top}\,(\mathrm{R}\top) \qquad \frac{}{0 \Rightarrow}\,(\mathrm{L}0) \qquad \frac{}{\Rightarrow 1}\,(\mathrm{R}1)$$

$$\frac{\Gamma, \Delta \Rightarrow \Pi}{\Gamma, \varphi, \Delta \Rightarrow \Pi}\,(\mathrm{lw}) \qquad \frac{\Gamma \Rightarrow}{\Gamma \Rightarrow \varphi}\,(\mathrm{rw}) \qquad \frac{\Theta \Rightarrow \varphi \quad \Gamma, \varphi, \Delta \Rightarrow \Pi}{\Gamma, \Theta, \Delta \Rightarrow \Pi}\,(\mathrm{cut})$$

$$\frac{\Gamma \Rightarrow}{\Gamma \Rightarrow 0}\,(\mathrm{R}0) \qquad \frac{\Gamma, \Delta \Rightarrow \Pi}{\Gamma, 1, \Delta \Rightarrow \Pi}\,(\mathrm{L}1) \qquad \frac{\Gamma, \varphi, \psi, \Delta \Rightarrow \Pi}{\Gamma, \varphi \otimes \psi, \Delta \Rightarrow \Pi}\,(\mathrm{L}\otimes)$$

$$\frac{\Gamma \Rightarrow \varphi \quad \Delta \Rightarrow \psi}{\Gamma, \Delta \Rightarrow \varphi \otimes \psi}\,(\mathrm{R}\otimes) \qquad \frac{\Gamma, \varphi, \Delta \Rightarrow \Pi \quad \Gamma, \psi, \Delta \Rightarrow \Pi}{\Gamma, \varphi \vee \psi, \Delta \Rightarrow \Pi}\,(\mathrm{L}\vee) \qquad \frac{\Gamma \Rightarrow \varphi_i}{\Gamma \Rightarrow \varphi_1 \vee \varphi_2}\,(\mathrm{R}\vee)$$

$$\frac{\Gamma, \varphi_i, \Delta \Rightarrow \Pi}{\Gamma, \varphi_1 \wedge \varphi_2, \Delta \Rightarrow \Pi}\,(\mathrm{L}\wedge) \qquad \frac{\Gamma \Rightarrow \varphi \quad \Gamma \Rightarrow \psi}{\Gamma \Rightarrow \varphi \wedge \psi}\,(\mathrm{R}\wedge) \qquad \frac{\Gamma \Rightarrow \varphi \quad \Delta, \psi, \Theta \Rightarrow \Pi}{\Delta, \psi/\varphi, \Gamma, \Theta \Rightarrow \Pi}\,(\mathrm{L}/)$$

$$\frac{\Gamma, \varphi \Rightarrow \psi}{\Gamma \Rightarrow \psi/\varphi}\,(\mathrm{R}/) \qquad \frac{\Gamma \Rightarrow \varphi \quad \Delta, \psi, \Theta \Rightarrow \Pi}{\Delta, \Gamma, \varphi\backslash\psi, \Theta \Rightarrow \Pi}\,(\mathrm{L}\backslash) \qquad \frac{\varphi, \Gamma \Rightarrow \psi}{\Gamma \Rightarrow \varphi\backslash\psi}\,(\mathrm{R}\backslash)$$

Fig. 1. The sequent calculus $\mathbf{FL_w}$.

Every $\varnothing$-deduction (i.e., every proof) in $\mathbf{FL}_\mathbf{w}^\Sigma$ can be converted to a proof without cuts (by cut elimination [9, 4.1.1]), so these calculi satisfy the *subformula property* (every formula that occurs in the proof is a subformula of the endsequent). However, not all cuts can be eliminated from a $\mathcal{T}$-deduction. Still, for suitable $\mathcal{T}$ identified below, the applications of cut can be restricted to obtain a generalized subformula property.

Definition 2.2. *A sequent is said to be* regular *if its antecedent contains only propositional variables and the succedent is not empty. A* regular theory *is a finite collection of regular sequents. For a regular theory $\mathcal{T}$, a* standard cut *(over $\mathcal{T}$) has as left premise a leaf that is a sequent from $\mathcal{T}$. A* standard deduction *(from $\mathcal{T}$) is one in which all cuts are standard. We write $\mathcal{T} \vdash^{\mathsf{std}}_{\mathbf{FL}_\mathbf{w}^\Sigma \Rightarrow} \Gamma \Rightarrow \varphi$ to denote that $\Gamma \Rightarrow \varphi$ has a standard deduction from $\mathcal{T}$ in $\mathbf{FL}_\mathbf{w}^\Sigma$.*

Lemma 2.3. *Let $\Sigma \subseteq \Sigma_{\mathbf{FL}}$. For a regular theory $\mathcal{T}$ and sequent s, $\mathcal{T} \vdash_{\mathbf{FL}_\mathbf{w}^\Sigma \Rightarrow} s$ iff $\mathcal{T} \vdash^{\mathsf{std}}_{\mathbf{FL}_\mathbf{w}^\Sigma \Rightarrow} s$. Moreover, every formula in a standard deduction in $\mathbf{FL}_\mathbf{w}^\Sigma$ is in* $\mathsf{subf}(\mathcal{T} \cup \{s\})$.

Proof.

We establish the claim for $\Sigma = \Sigma_{\mathbf{FL}}$. The reader will observe that removing rules means less cases to be checked, and thus the argument here applies to any $\Sigma \subseteq \Sigma_{\mathbf{FL}}$. Let us call the instantiation of the schematic-variable φ in an instance of the (cut) rule (cf. Fig. 1) as the *cut formula* of that instantiation.

For first assertion, the argument is based on the usual cut-elimination procedure for $\mathbf{FL_w}$ [9, 4.1.1]; the difference is that we now have to deal with cuts with a premise in $\mathcal{T}$. We repeatedly eliminate topmost non-standard cuts, by primary induction on the grade of the cut (the length of the cut formula), and secondary induction on the cut-height (the sum of the number of sequents appearing in the proofs of the premises of the cut).

Consider a topmost non-standard cut with cut-height 2 (note that < 2 is not possible). The premises of the cut are instances of axiomatic rules or elements from $\mathcal{T}$. If no element from $\mathcal{T}$ appears as a premise, the cut is eliminated in the usual way as for $\mathbf{FL_w}$. If the left premise is from $\mathcal{T}$, nothing needs to be done since we admit standard cuts. If the right premise is from $\mathcal{T}$, the left premise can only be an instance either of (id) or (L$\bot$); i.e., the cut has one of these forms:

$$\dfrac{\dfrac{}{p \Rightarrow p}\,(\text{id}) \qquad \dfrac{}{\Gamma_1, p, \Gamma_2 \Rightarrow \varphi}\,\mathcal{T}}{\Gamma_1, p, \Gamma_2 \Rightarrow \varphi}\,(\text{cut}) \qquad \dfrac{\dfrac{}{\Delta_1, \bot, \Delta_2 \Rightarrow p}\,(\bot\text{L}) \qquad \dfrac{}{\Gamma_1, p, \Gamma_2 \Rightarrow \varphi}\,\mathcal{T}}{\Gamma_1, \Delta_1, \bot, \Delta_2, \Gamma_2 \Rightarrow \varphi}\,(\text{cut})$$

The cut above left is clearly eliminable since the second premise matches the conclusion; the cut above right is replaced by an instance of (L$\bot$).

Inductive step. Assume the cut has cut-height ≥ 3. If neither of the premises of cut are from $\mathcal{T}$, the usual cut-reduction steps apply [9, 4.1.1]. Else, if the topmost cut is standard, nothing needs to be done since we admit such cuts. The remaining case is that the right premise is from $\mathcal{T}$ and the left premise is from some other rule. It follows that the topmost non-standard cut occurs as follows.

$$\dfrac{\overset{\delta_1}{\Gamma \Rightarrow p} \qquad \dfrac{}{\Delta_1, p, \Delta_2 \Rightarrow \Pi}\,\mathcal{T}}{\Delta_1, \Gamma, \Delta_2 \Rightarrow \Pi}\,(\text{cut})$$

Since the cut-height is ≥ 3, we have that the left premise is not an instance of an axiomatic rule. Consider the last rule in δ_1. We transform each case so the non-standard cut is replaced with a cut that is smaller under the induction measure. We indicate the transformations with $\rightsquigarrow$. Since the cut formula is a propositional variable (this is forced by the right premise being from the regular theory $\mathcal{T}$), the last rule in δ_1 cannot be a right-introduction rule. We provide some representative cases.

- The left premise of the cut under consideration is the conclusion of a standard cut:

$$\dfrac{\dfrac{\dfrac{}{\Theta \Rightarrow \varphi}\,\mathcal{T} \qquad \overset{\delta_1}{\Gamma_1, \varphi, \Gamma_2 \Rightarrow p}}{\Gamma_1, \Theta, \Gamma_2 \Rightarrow p}\,(\text{cut}) \qquad \dfrac{}{\Delta_1, p, \Delta_2 \Rightarrow \Pi}\,\mathcal{T}}{\Delta_1, \Gamma_1, \Theta, \Gamma_2, \Delta_2 \Rightarrow \Pi}\,(\text{cut})$$

$$\rightsquigarrow$$

$$\dfrac{\dfrac{}{\Theta \Rightarrow \varphi}\,\mathcal{T} \qquad \dfrac{\overset{\delta_1}{\Gamma_1, \varphi, \Gamma_2 \Rightarrow p} \qquad \dfrac{}{\Delta_1, p, \Delta_2 \Rightarrow \Pi}\,\mathcal{T}}{\Delta_1, \Gamma_1, \varphi, \Gamma_2, \Delta_2 \Rightarrow p}\,(\text{cut})}{\Delta_1, \Gamma_1, \Theta, \Gamma_2, \Delta_2 \Rightarrow \Pi}\,(\text{cut})$$

- (L$\otimes$)

$$\dfrac{\dfrac{\overset{\delta_2}{\Gamma_1, \varphi_1, \varphi_2, \Gamma_2 \Rightarrow p}}{\Gamma_1, \varphi_1 \otimes \varphi_2, \Gamma_2 \Rightarrow p}\,(\text{L}\otimes) \qquad \dfrac{}{\Delta_1, p, \Delta_2 \Rightarrow \Pi}\,\mathcal{T}}{\Delta_1, \Gamma_1, \varphi_1 \otimes \varphi_2, \Gamma_2, \Delta_2 \Rightarrow \Pi}\,(\text{cut})$$

$$\rightsquigarrow$$

$$\dfrac{\dfrac{\overset{\delta_2}{\Gamma_1, \varphi_1, \varphi_2, \Gamma_2 \Rightarrow p} \qquad \dfrac{}{\Delta_1, p, \Delta_2 \Rightarrow \Pi}\,\mathcal{T}}{\Delta_1, \Gamma_1, \varphi_1, \varphi_2, \Gamma_2, \Delta_2 \Rightarrow \Pi}\,(\text{cut})}{\Delta_1, \Gamma_1, \varphi_1 \otimes \varphi_2, \Gamma_2, \Delta_2 \Rightarrow \Pi}\,(\text{L}\otimes)$$

The second assertion follows by structural induction on standard deductions. The base case is obvious. Inductive step. The claim is immediate for a logical rule, (lw), and (rw), by the induction hypothesis applied to the deduction concluding the premise, and since every formula in the premise must be a subformula of a formula in the conclusion. It remains to check the case of a standard cut.

$$\dfrac{\dfrac{}{\Delta \Rightarrow \psi}\,\mathcal{T} \qquad \overset{\delta_1}{\Gamma_1, \psi, \Gamma_2 \Rightarrow \varphi}}{\Gamma_1, \Delta, \Gamma_2 \Rightarrow \varphi}\,\text{cut}$$

The claim follows since ψ is a formula present in one of the sequents in $\mathcal{T}$. □

The next lemma is a direct consequence of Lem. 5 and tells us that sequent-deducibility in $\mathbf{FL}_{\mathbf{w}}^{\Sigma'}$ reduces to the same problem over an expanded signature $\Sigma_{\mathbf{FL}} \supseteq \Sigma \supseteq \Sigma'$. In other words, $\vdash_{\mathbf{FL}_{\mathbf{w}}^{\Sigma}\Rightarrow}$ is a *conservative expansion* of $\vdash_{\mathbf{FL}_{\mathbf{w}}^{\Sigma'}\Rightarrow}$.

Lemma 2.4. *Let* $\Sigma' \subseteq \Sigma \subseteq \Sigma_{\mathbf{FL}}$. *If* $\mathcal{T} \cup \{s\}$ *are sequents over* $\mathrm{Fm}_{\Sigma'}(\mathcal{P})$, *then* $\mathcal{T} \vdash_{\mathbf{FL}_{\mathbf{w}}^{\Sigma'}\Rightarrow} s$ *iff* $\mathcal{T} \vdash_{\mathbf{FL}_{\mathbf{w}}^{\Sigma}\Rightarrow} s$.

Proof. The nontrivial direction is right-to-left, so suppose that $\mathcal{T} \vdash_{\mathbf{FL}_{\mathbf{w}}^{\Sigma}\Rightarrow} s$ is witnessed by a deduction δ. Obtain a standard deduction δ' from δ by Lem. 5. The new proof will have sequents over $\mathrm{Fm}_{\Sigma'}(\mathcal{P})$ only (by the subformula property proved also in Lem. 5), and thus must be a deduction in $\mathbf{FL}_{\mathbf{w}}^{\Sigma'}$. □

3 Hyper-Ackermannian lower bounds

We obtain hyper-Ackermannian lower bounds for the sequent-deducibility problem in Σ-fragments of $\mathbf{FL_w}$ where $\otimes \in \Sigma$, via a reduction from the reachability problem in lossy channel systems. The result extends to the problem of deducibility if Σ also contains $\backslash$ (or $/$), 0 and 1, in view of Lem. 4. We begin by introducing lossy channel systems and their reachability problem, and then show how to encode it into sequent-deducibility in $\mathbf{FL}_\mathbf{w}^\otimes$.

3.1 Lossy channel systems

A channel system is a computational model with unbounded FIFO channels (queues), defined as follows.

Definition 3.1. *A* channel system *(CS) is a structure* $\mathcal{L} := \langle \mathtt{Q}, \mathtt{C}, \mathtt{M}, \mathtt{I} \rangle$, *where*

(i) $\mathtt{Q} := \{q_1, \ldots, q_m\}$ *is a finite set of* states*;*

(ii) $\mathtt{C} := \{c_1, \ldots, c_n\}$ *is a finite set of* channels*;*

(iii) $\mathtt{M} := \{\mathtt{a}_1, \ldots, \mathtt{a}_k\}$ *is a finite* message alphabet*; and*

(iv) $\mathtt{I} \subseteq \mathtt{Q} \times \mathtt{C} \times \mathtt{M} \times \{!, ?\} \times \mathtt{Q}$ *is a finite set of* instructions.

We denote by $\mathsf{Conf}(\mathcal{L})$ *the set* $\mathtt{Q} \times (\mathtt{M}^*)^n$ *of* configurations *of* $\mathcal{L}$, *where* $\mathtt{M}^*$ *is the set of all finite sequences of elements in* $\mathtt{M}$.

Definition 3.2. *The* perfect steps *of a CS* $\mathcal{L}$ *are given by a binary relation* $\rightarrow_{\mathcal{L}\ \mathsf{perf}} \subseteq \mathsf{Conf}(\mathcal{L}) \times \mathsf{Conf}(\mathcal{L})$ *such that* $(q_i, u_1, \ldots, u_r, \ldots, u_n) \rightarrow_{\mathcal{L}\ \mathsf{perf}} (q_j, u_1, \ldots, v_r, \ldots, u_n)$ *if either*

(i) $u_r = \mathtt{a}_p v_r$ *and* $(q_i, c_r, \mathtt{a}_p, ?, q_j) \in \mathtt{I}$ *(read and remove* $\mathtt{a}_p$ *from the* front *of channel* c_r *and change state from* q_i *to* q_j*)*

(ii) $v_r = u_r \mathtt{a}_p$ *and* $(q_i, c_r, \mathtt{a}_p, !, q_j) \in \mathtt{I}$ *(write* $\mathtt{a}_p$ *at the* back *of channel* c_r *and change state from* q_i *to* q_j*).*

Let $\rightarrow^?_{\mathcal{L}\ \mathsf{perf}}$ *and* $\rightarrow^!_{\mathcal{L}\ \mathsf{perf}}$ *denote the perfect steps given by items 1 and 2 respectively. Also let* $\rightarrow^*_{\mathcal{L}\ \mathsf{perf}}$ *denote the reflexive and transitive closure of* $\rightarrow_{\mathcal{L}\ \mathsf{perf}}$.

Clearly, $u \rightarrow^*_{\mathcal{L}\ \mathsf{perf}} v$ iff there is a finite sequence $w_1, \ldots, w_\eta$ ($\eta \geq 1$) of configurations such that $u = w_1$, $v = w_\eta$ and $w_i \rightarrow_{\mathcal{L}\ \mathsf{perf}} w_{i+1}$ for each i ($1 \leq i < \eta$). This sequence is called a *perfect computation in* $\mathcal{L}$ of length η and it witnesses that u *reaches* v in $\mathcal{L}$.

Reachability (as well as many other verification problems for CSs) is undecidable [3], but adding lossy behaviour makes it decidable. This behaviour is introduced in the operational semantics of a CS as follows.

Definition 3.3. *The* lossy semantics *of a CS* $\mathcal{L}$ *is given by a relation* $\rightarrow_{\mathcal{L}\ \mathsf{lossy}} \subseteq \mathsf{Conf}(\mathcal{L}) \times \mathsf{Conf}(\mathcal{L})$ *that extends* $\rightarrow_{\mathcal{L}\ \mathsf{perf}}$ *with all transitions (called* lossy steps*) of the form* $(q_i, u_1, \ldots, u_r, \ldots, u_n) \rightarrow^l_{\mathcal{L}\ \mathsf{lossy}} (q_i, u_1, \ldots, v_r, \ldots, u_n)$ *such that* $u_r = w_1 \mathtt{a}_p w_2$ *and* $v_r = w_1 w_2$. *Let* $\rightarrow^*_{\mathcal{L}\ \mathsf{lossy}}$ *be the reflexive and transitive closure of* $\rightarrow_{\mathcal{L}\ \mathsf{lossy}}$. *A channel system with a lossy semantics is called a* lossy channel system (LCS).

Intuitively, in an LCS the lossy steps permit any channel to lose a message (a symbol) from any position at any moment without changing the state. No instruction in the machine is required for this. The relation $\rightarrow^*_{\mathcal{L}\ \mathsf{lossy}}$ induces a notion of *lossy computation in* $\mathcal{L}$ and reachability is defined analogously to perfect computations with $\rightarrow^*_{\mathcal{L}\ \mathsf{perf}}$. Since a step can be either read, write, or lossy, we use $\rightarrow^?_{\mathcal{L}\ \mathsf{lossy}}$, $\rightarrow^!_{\mathcal{L}\ \mathsf{lossy}}$ or $\rightarrow^l_{\mathcal{L}\ \mathsf{lossy}}$ to indicate precisely which step was performed in a lossy computation.

Definition 3.4 (Reachability problem in LCSs). *Given a LCS* $\mathcal{L}$ *and configurations* $u, v \in \mathsf{Conf}(\mathcal{L})$, *does* $u \rightarrow^*_{\mathcal{L}\ \mathsf{lossy}} v$*?*

Remark 3.5. *The relation* $\rightarrow^*_{\mathcal{L}\ \mathsf{lossy}}$ *has an equivalent definition in the literature [19, Sec. 2] as* $u \rightsquigarrow^*_{\mathcal{L}} v$ *if, and only if, there are* $u', v' \in \mathsf{Conf}(\mathcal{L})$ *such that* $u \rightarrow^{l*}_{\mathcal{L}\ \mathsf{lossy}} u' \rightarrow_{\mathcal{L}\ \mathsf{perf}} v' \rightarrow^{l*}_{\mathcal{L}\ \mathsf{lossy}} v$ *(the* $*$ *denotes the reflexive transitive closure). That is, messages may be lost before and after performing a perfect step. It can be shown that* $\rightarrow^*_{\mathcal{L}\ \mathsf{lossy}} = \rightarrow^{l*}_{\mathcal{L}\ \mathsf{lossy}} \circ \rightarrow_{\mathcal{L}\ \mathsf{perf}} \circ \rightarrow^{l*}_{\mathcal{L}\ \mathsf{lossy}} = \rightsquigarrow^*_{\mathcal{L}}$. *The latter relation is shown [4, App. A] to have no impact on the complexity of reachability compared to the "write-lossy" mechanism employed in the proof of Thm. 9 [4] stated below.*

Theorem 3.6 ([4, Thm. 5.5, Obs. 6.1, App. A]). *Reachability in LCSs is* HACK*-complete.*

3.2 Encoding reachability in LCSs into sequent-deducibility

We code each instruction of a lossy channel system into a sequent. The collection of these sequents is the *theory* associated to the system.

Definition 3.7. *For an LCS* $\mathcal{L} = \langle \mathtt{Q}, \mathtt{C}, \mathtt{M}, \mathtt{I} \rangle$, *define these sets of propositional variables:*

(i) $\mathsf{Q} := \{Q_i \mid q_i \in \mathtt{Q}\}$ *(state variables)*

(ii) $\# := \{\#^s_i \mid c_i \in \mathtt{C}\} \cup \{\#^e_i \mid c_i \in \mathtt{C}\}$ *(channel markers)*

(iii) $\mathsf{M} := \{P_a \mid a \in \mathtt{M}\}$ *(alphabet variables)*

Moreover, define $\mathsf{Props}(\mathcal{L}) := \mathsf{Q} \cup \# \cup \mathsf{M}$.

Definition 3.8 (Theory of an LCS). *Given an LCS* $\mathcal{L} = \langle \mathtt{Q}, \mathtt{C}, \mathtt{M}, \mathtt{I} \rangle$, *the* theory *of* $\mathcal{L}$, *denoted by* $\mathcal{T}(\mathcal{L})$, *is defined as the union of the following finite sets of sequents:*

(i) $\mathcal{T}(\mathcal{L})^! := \{\#^e_l, Q_i \Rightarrow P_a \otimes (\#^e_l \otimes Q_j) \mid (q_i, c_l, a, !, q_j) \in \mathtt{I}\}$;

(ii) $\mathcal{T}(\mathcal{L})^? := \{\#^s_l, P_a, Q_i \Rightarrow \#^s_l \otimes Q_j \mid (q_i, c_l, a, ?, q_j) \in \mathtt{I}\}$;

(iii) $\mathcal{T}(\mathcal{L})^{\mathtt{Q}} := \bigcup_{\substack{Q \in \mathsf{Q} \\ R \in \mathsf{Props}(\mathcal{L}) \setminus \mathsf{Q}}} \{R, Q \Rightarrow Q \otimes R\} \cup \{Q, R \Rightarrow R \otimes Q\}$.

Note that $\mathcal{T}(\mathcal{L})$ is a regular theory in the sense of Def. 4.1.

Recall that $\bigotimes \varphi_1, \ldots, \varphi_m := \varphi_1 \otimes (\varphi_2 \otimes (\ldots (\varphi_{m-1} \otimes \varphi_m) \ldots))$, where $m \geq 1$ and $\varphi_1, \ldots, \varphi_m$ is a sequence of formulas. We now translate a reachability instance in a fixed LCS $\mathcal{L} = \langle \mathtt{Q}, \mathtt{C}, \mathtt{M}, \mathtt{I} \rangle$ to a sequent-deducibility instance in $\mathbf{FL}^{\otimes}_{\mathbf{w}}$. We begin by coding a configuration as a sequence.

Definition 3.9. *Given* $u := (q_i, u_1, \ldots, u_n) \in \mathsf{Conf}(\mathcal{L})$, *define the sequence* $E(u) := Q_i, \#_1^s, U_1, \#_1^e, \ldots, \#_n^s, U_n, \#_n^e$, *where, for* $u_i = a_1 \cdots a_{k_i} \in \mathtt{M}^*$, U_i *is the sequence* $P_{a_1}, \ldots, P_{a_{k_i}}$ *of propositional variables.*

Definition 3.10. *For* $u := (q_i, u_1, \ldots, u_n)$, $v := (q_j, v_1, \ldots, v_n) \in \mathsf{Conf}(\mathcal{L})$, *define the sequent*

$$S_{u \to v} := E(u) \Rightarrow \bigotimes E(v).$$

Define also $S^{\bowtie}_{u \to v}$ *as the set of sequents of the above form where the occurrence of* Q_i *in the antecedent may appear in any position (i.e., not necessarily at the front as in* $E(u)$*).*

We note that sequents in standard deductions in $\mathbf{FL}_\mathbf{w}^{\otimes}$ from a regular theory have a nonempty succedent coming from the observation that every leaf has this property:

Lemma 3.11. *Let* $\mathcal{T}$ *be a regular theory. It is never the case that* $\mathcal{T} \vdash^{\mathsf{std}}_{\mathbf{FL}_\mathbf{w}^{\otimes}\Rightarrow} \Gamma \Rightarrow$.

Proof. It follows because no axiomatic rule has conclusion with empty succedent and no other rule allows to derive a sequent with empty succedent from one with nonempty succedent. □

The *flattening* $[\varphi]$ is the sequence of propositional variables obtained from a formula φ by replacing '$\otimes$' with ','. E.g., $[p \otimes ((q \otimes s) \otimes r)] = p, q, s, r$. The flattening $[\Gamma]$ of a sequence Γ of formulas is the concatenation of the flattening of the formulas in Γ in the order they appear in the sequence. Let $\#\Delta$ denote the subsequence of Δ having only variables from the set $\#$ (that is, only channel markers). The Q-*free subsequence* of a sequence of propositional variables is obtained by deleting all occurrences of state variables.

The following lemmas express that in standard deductions from a regular theory, in various situations, propositional variables occurring positively must also occur negatively. They abstract what is needed for the right-to-left direction of Lem. 14 below. The proofs are by structural induction on standard deductions.

Lemma 3.12. *Let* Γ *be a sequence of formulas, none having a state variable. If* $\mathcal{T}(\mathcal{L}) \vdash^{\mathsf{std}}_{\mathbf{FL}_\mathbf{w}^{\otimes}\Rightarrow} \Gamma \Rightarrow \varphi$, *then the* Q*-free subsequence of* $[\varphi]$ *is a subsequence of* $[\Gamma]$.

Proof. Structural induction on a standard deduction witnessing $\mathcal{T}(\mathcal{L}) \vdash_{\mathbf{FL}_\mathbf{w}^{\otimes}\Rightarrow} \Gamma \Rightarrow \varphi$. *Base case.* For elements from $\mathcal{T}(\mathcal{L})$ it holds vacuously since each sequent contains a state variable in the antecedent; for an identity sequent the claim is immediate. *Inductive step.* Suppose that the last rule instance applied is (lw), (R$\otimes$) or (L$\otimes$). Applying the IH to its premise(s), we have that the Q-free subsequence of $[\varphi]$ is a subsequence of the flattening of the premise antecedent(s), and hence also of $[\Gamma]$. The last rule instance cannot be a standard cut as the antecedent of each element in $\mathcal{T}(\mathcal{L})$ contain a state variable hence Γ would contains a state variable, contradicting the hypotheses. The rule (rw) cannot occur due to Lem. 10. □

Lemma 3.13. *Let Γ be a sequence of formulas, $u \in \mathsf{Conf}(\mathcal{L})$ and L a nonempty subsequence of $E(u)$. Then $\mathcal{T}(\mathcal{L}) \vdash^{\mathsf{std}}_{\mathbf{FL}^{\otimes}_{\mathbf{w}}\Rightarrow} \Gamma \Rightarrow \bigotimes L$ implies that the variables in $\#L$ appear at least once in $[\Gamma]$.*

Proof. Induction on the structure of a standard deduction witnessing $\mathcal{T}(\mathcal{L}) \vdash_{\mathbf{FL}^{\otimes}_{\mathbf{w}}\Rightarrow} \Gamma \Rightarrow \bigotimes L$. *Base case.* Straightforward: identity sequent $p \Rightarrow p$ (trivial) and an element from $\mathcal{T}(\mathcal{L})$ (by inspection, it contains no channel variables or the same channel variable occurs in the antecedent and succedent). *Inductive step.* Consider the last rule instance in the standard deduction. For (R$\otimes$), (L$\otimes$) and (lw), we have by IH that each variable in $\#L$ appears in $[\Gamma']$ for the antecedent Γ' of some premise of the rule instance. Those variables will be carried to the conclusion antecedent. If the last rule applied is a standard cut, its right premise has the form $\Gamma_1, \varphi, \Gamma_2 \Rightarrow \bigotimes L$ where φ is the cut formula. By IH, every variable in $\#L$ appears in $[\Gamma_1, \varphi, \Gamma_2]$. Moreover, since the left premise is from $\mathcal{T}(\mathcal{L})$, any channel variable occurring in the cut formula will occur in its antecedent, and hence also in $[\Gamma]$. The case of (rw) does not need to be considered in view of Lem. 10. □

Lemma 3.14. *Let Γ be a sequence of formulas and φ be a formula. If $\mathcal{T}(\mathcal{L}) \vdash^{\mathsf{std}}_{\mathbf{FL}^{\otimes}_{\mathbf{w}}\Rightarrow} \Gamma \Rightarrow Q_i \otimes \varphi$ or $\mathcal{T}(\mathcal{L}) \vdash^{\mathsf{std}}_{\mathbf{FL}^{\otimes}_{\mathbf{w}}\Rightarrow} \Gamma \Rightarrow Q_i$, then $[\Gamma]$ must contain a state variable.*

Proof. Structural induction on a derivation witnessing $\mathcal{T}(\mathcal{L}) \vdash_{\mathbf{SL}^{\cdot}_{\mathbf{w}}\Rightarrow} \Gamma \Rightarrow Q_i \otimes \varphi$. *Base case.* The case of an initial sequent is trivial, and the antecedent of any sequent from $\mathcal{T}(\mathcal{L})$ contains a state variable. *Inductive step.* Suppose that the last rule is (R$\otimes$). The left premise has Q_i as succedent, and hence by IH, its antecedent will contain a state variable that is carried down to the conclusion. The argument is analogous for (lw) and (L$\otimes$). If the last rule instance is a standard cut, its left premise is a sequent in $\mathcal{T}(\mathcal{L})$. As already observed, the antecedent of such a sequent contains a state variable and that will be carried down to the conclusion. The case of (rw) does not need to be considered in view of Lem. 10.

The case of $\mathcal{T}(\mathcal{L}) \vdash_{\mathbf{FL}^{\otimes}_{\mathbf{w}}\Rightarrow} \Gamma \Rightarrow Q_i$ is similar. □

We are ready to reduce reachability in lossy channel systems to sequent-deducibility.

Lemma 3.15. *For all LCS $\mathcal{L}$, given $u, v \in \mathsf{Conf}(\mathcal{L})$, $u \to^{*}_{\mathcal{L}\ \mathsf{lossy}} v$ iff $\mathcal{T}(\mathcal{L}) \vdash^{\mathsf{std}}_{\mathbf{FL}^{\otimes}_{\mathbf{w}}\Rightarrow} s$ for some $s \in S^{\bowtie}_{u \to v}$.*

Proof. In the left-to-right direction, we work by induction on the length $k \geq 1$ of a computation witnessing $u \to^{*}_{\mathcal{L}\ \mathsf{lossy}} v$. *Base case.* If $k = 1$, $u = v$, and $S_{u \to v}$ is provable by (id) and (R$\otimes$). *Inductive step.* The computation decomposes as $u \to^{\sigma}_{\mathcal{L}\ \mathsf{lossy}} v' \to^{*}_{\mathcal{L}\ \mathsf{lossy}} v$ with length $k+1$. As $v' \to^{*}_{\mathcal{L}\ \mathsf{lossy}} v$ is witnessed by a computation of length k, IH yields $\mathcal{T}(\mathcal{L}) \vdash_{\mathbf{FL}^{\otimes}_{\mathbf{w}}\Rightarrow} s'$ for some $s' \in S^{\bowtie}_{v' \to v}$. We extend the latter deduction downwards to s, using a case analysis on σ to determine the form of s' and s:

(i) if $\sigma = ?$, by effect of $(q_i, c_l, a, ?, q_j) \in \mathtt{I}$: then $u_l = av'_l$, all the other channels have the same content, and the state transits from q_i to q_j. Thus s' has the following form, possibly following repeated standard cuts with sequents in $\mathcal{T}(\mathcal{L})^{\mathtt{Q}}$.

$$\#^s_1, U_1, \#^e_1, \ldots, \#^s_{l-1}, U_{l-1}, \#^e_{l-1}, \#^s_l, Q_j, V'_l, \#^e_l, \ldots, \#^s_n, U_n, \#^e_n \Rightarrow \bigotimes E(v)$$

Apply (L$\otimes$) and a standard cut with $\#^s_l, P_a, Q_i \Rightarrow \#^s_l \otimes Q_j$ to obtain s.

(ii) if $\sigma = l$: the state is unchanged, and $v'_l = z_1 z_2$, for $u = z_1 a z_2$. Thus s' has the following form, possibly following repeated standard cuts with sequents in $\mathcal{T}(\mathcal{L})^{\mathtt{Q}}$.

$$Q_j, \#^s_1, U_1, \#^e_1, \ldots, \#^s_l, Z_1, Z_2, \#^e_l, \ldots, \#^s_n, U_n, \#^e_n \Rightarrow \bigotimes E(v)$$

By (lw) $Q_j, \#^s_1, U_1, \#^e_1, \ldots, \#^s_l, Z_1, P_a, Z_2, \#^e_l, \ldots, \#^s_n, U_n, \#^e_n \Rightarrow \bigotimes E(v)$

(iii) if $\sigma = !$, by effect of $(q_i, c_l, a, !, q_j) \in \mathtt{I}$: by definition of this step, $v'_l = u_l a$, the content in other channels of u and v' is unchanged, and the state transits from q_i to q_j. Thus s' has the following form modulo the position of Q_j. In any case, the s' can be moved to the displayed position through repeated standard cuts with sequents in $\mathcal{T}(\mathcal{L})^{\mathtt{Q}}$.

$$\#^s_1, U_1, \#^e_1, \ldots, \#^s_l, U_l, P_a, \#^e_l, Q_j, \#^s_{l+1}, \ldots, \#^s_n, U_n, \#^e_n \Rightarrow \bigotimes E(v)$$

It suffices now to apply (L$\otimes$) twice and a standard cut with $\#^e_l, Q_i \Rightarrow P_a \otimes \#^e_l \otimes Q_j$ to obtain the sequent s: $\#^s_1, U_1, \#^e_1, \ldots, \#^s_l, U_l, \#^e_l, Q_j, \#^s_{l+1}, \ldots, \#^s_n, U_n, \#^e_n \Rightarrow \bigotimes E(v)$.

For the right-to-left direction, we establish property $\mathcal{P}(\delta)$:=: "for all $u, v \in \mathsf{Conf}(\mathcal{L})$, if δ is a standard deduction witnessing $\mathcal{T}(\mathcal{L}) \vdash_{\mathbf{FL}^{\otimes}_{\mathbf{w}\Rightarrow}} s$ for some $s \in S^{\bowtie}_{u\to v}$, then $u \to^*_{\mathcal{L}\ \mathsf{lossy}} v$". Induction on the structure of δ.

Base case. No initial sequent nor any sequent from $\mathcal{T}(\mathcal{L})$ is an image of an encoding of a configuration since it does not contain any pair $\{\#^s_i, \#^e_i\}$ of channel markers.

Inductive step. Let $u = (q_i, u_1, \ldots, u_n)$ and $v = (q_j, v_1, \ldots, v_n)$. By cases on the last rule applied in δ. It cannot be (L$\otimes$) since the antecedent of s is a sequence of propositional variables. Then it could be:

(i) (rw): Not possible by Lem. 10 since δ is a standard deduction.

(ii) (R$\otimes$): Write s as $\Gamma_1, \Gamma_2 \Rightarrow Q_j \otimes \varphi_2$. The premises then are $\Gamma_1 \Rightarrow Q_j$ and $\Gamma_2 \Rightarrow \varphi_2$. By Lem. 13, Γ_1 must have a formula with a state variable, and thus Γ_2 lacks a state variable. If Γ_1 contained anything more, it must contain a channel marker. By Lem. 11, Γ_2 must contain all the channel markers to fulfil the requirement that $[\varphi_2]$ must be a subsequence of $[\Gamma_2]$. Hence Γ_1 is the singleton sequence Q_l. Now, for $Q_l \Rightarrow Q_j$ to be provable, we must have $l = j$. Also, $[\varphi_2]$ a subsequence of $[\Gamma_2]$ implies that the channel portions in the succedent are subsequences of the corresponding ones in the antecedent. Thus, u can reach v by losing messages. Observe: the IH is not employed.

(iii) (lw): We claim that the weakening introduces a variable into the channel portion. This suffices since the premise would then be the image of a configuration that, by IH, reaches v, and is itself reachable from u by a lossy step. To establish the claim, first note that the sequents involved in an application of (lw) have the same succedent, in this case $\bigotimes E(v)$. Let Γ' be the premise antecedent. By Lem. 12, we have that all channel variables appear in Γ'. Since these symbols appear only once in conclusion antecedent, weakening could not have introduced them. If the weakening introduced the state variable, then Γ' would have no state variables; that is impossible due to Lem. 13 since the succedent does contain a state variable. The remaining possibility is that weakening introduced a variable in the channel portion, so the claim is proved.

(iv) Standard cut with a sequent in $\mathcal{T}(\mathcal{L})^?$: The deduction has the form

$$\dfrac{\dfrac{}{\#_l^s, P_a, Q_i \Rightarrow \#_l^s \otimes Q_p}\,\mathcal{T}(\mathcal{L})^? \qquad \dfrac{\delta_1}{\#_1^s, U_1, \#_1^e, \ldots, \#_l^s \otimes Q_p, U_l, \#_l^e, \ldots, \#_n^s, U_n, \#_n^e \Rightarrow \bigotimes E(v)}}{\#_1^s, U_1, \#_1^e, \ldots, \#_l^s, P_a, Q_i, U_l, \#_l^e, \ldots, \#_n^s, U_n, \#_n^e \Rightarrow \bigotimes E(v)}$$

Consider the last rule in δ_1. It cannot be a (standard) cut as that would have introduced a state variable on its own (i.e., unfused) into the conclusion. If it was (R$\otimes$) then the left premise has the form $\Gamma_1 \Rightarrow Q_j$ and the second premise has the form $\Gamma_2 \Rightarrow \varphi_2$. By Lem. 13, $\#_l^s \otimes Q_p \in \Gamma_1$, and thus Γ_2 consists only of propositional variables. Since $\#_l^s$ is not in Γ_2 but φ_2 contains all channel markers, Lem. 12 gives a contradiction. So (R$\otimes$) was not the last rule applied. Neither can it be (rw) due to Lem. 10.

We are left with two alternatives for the last rule in δ_1: (lw) and (L$\otimes$). If (lw), arguing as in the previous item, the introduced variable must be in the channel portion by Lem. 12 and Lem. 13. Iterating the entire argument up to this point (implicit induction), after multiple applications of (lw) we must ultimately encounter (L$\otimes$) (we have ruled out all the other possibilities). The situation is therefore the following.

$$\dfrac{\dfrac{}{\#_l^s, P_a, Q_i \Rightarrow \#_l^s \otimes Q_p}\,\mathcal{T}(\mathcal{L})^? \qquad \dfrac{\dfrac{\dfrac{\delta_1'}{\#_1^s, U_1', \#_1^e, \ldots, \#_l^s, Q_p, U_l', \#_l^e, \ldots, \#_n^s, U_n', \#_n^e \Rightarrow \bigotimes E(v)}}{\#_1^s, U_1', \#_1^e, \ldots, \#_l^s \otimes Q_p, U_l', \#_l^e, \ldots, \#_n^s, U_n', \#_n^e \Rightarrow \bigotimes E(v)}}{\#_1^s, U_1, \#_1^e, \ldots, \#_l^s \otimes Q_p, U_l, \#_l^e, \ldots, \#_n^s, U_n, \#_n^e \Rightarrow \bigotimes E(v)}}{\#_1^s, U_1, \#_1^e, \ldots, \#_l^s, P_a, Q_i, U_l, \#_l^e, \ldots, \#_n^s, U_n, \#_n^e \Rightarrow \bigotimes E(v)}$$

From the IH, we obtain that $(q_p, u_1', \ldots, u_l', \ldots, u_n') \rightarrow^*_{\mathcal{L}\ \mathsf{lossy}} v$. Since $(q_i, u_1, \ldots, au_l, \ldots, u_n) \rightarrow^?_{\mathcal{L}\ \mathsf{lossy}} (q_p, u_1, \ldots, u_l, \ldots, u_n) \rightarrow^*_{\mathcal{L}\ \mathsf{lossy}} (q_p, u_1', \ldots, u_l', \ldots, u_n')$, we are done.

(v) Standard cut with a sequent in $\mathcal{T}(\mathcal{L})^!$: In this case the deduction has the form

$$\dfrac{\dfrac{}{\#^e_l, Q_i \Rightarrow P_a \otimes \#^e_l \otimes Q_p}\mathcal{T}(\mathcal{L})^! \qquad \dfrac{\delta_1}{\#^s_1, U_1, \#^e_1, \ldots, \#^s_l, U_l, P_a \otimes \#^e_l \otimes Q_p, \ldots, \#^s_n, U_n, \#^e_n \Rightarrow}}{\#^s_1, U_1, \#^e_1, \ldots, \#^s_l, U_l, \#^e_l, Q_i, \ldots, \#^s_n, U_n, \#^e_n \Rightarrow \bigotimes E(v)}$$

Consider the last rules in δ_1. Arguing as in the $\mathcal{T}(\mathcal{L})^?$ case, it is a sequence of (lw) rules introducing variables in channel portions, then (L$\otimes$). The latter premise is

$$\#^s_1, U'_1, \#^e_1, \ldots, \#^s_l, U'_l, P_a, \#^e_l \otimes Q_p, \ldots, \#^s_n, U'_n, \#^e_n \Rightarrow \bigotimes E(v)$$

Since no state variable occurs by itself in the antecedent, the above sequent was not derived by a (standard) cut. Nor was it by (R$\otimes$), since its left premise would contain $\#^e_l \otimes Q_p$ hence depriving the right premise antecedent of $\#^e_l$ leading to a contradiction with Lem. 12. Nor was it (rw) due to Lem. 10. Hence we encounter once again a sequence of (lw) rules followed by an application of (L$\otimes$), resulting in the sequent

$$\#^s_1, U''_1, \#^e_1, \ldots, \#^s_l, U''_l, P_a, \#^e_l, Q_p, \ldots, \#^s_n, U''_n, \#^e_n \Rightarrow \bigotimes E(v)$$

By IH, we obtain that $(q_p, u''_1, \ldots, u''_l a, \ldots, u''_n) \rightarrow^*_{\mathcal{L}\text{ lossy}} v$. Since $(q_i, u_1, \ldots, u_l, \ldots, u_n) \rightarrow^!_{\mathcal{L}\text{ lossy}} (q_p, u_1, \ldots, u_l a, \ldots, u_n) \rightarrow^*_{\mathcal{L}\text{ lossy}} (q_p, u''_1, \ldots, u''_l a, \ldots, u''_n)$, we conclude $u \rightarrow^*_{\mathcal{L}\text{ lossy}} v$.

(vi) Standard cut with $Q_i, R \Rightarrow R \otimes Q_i \in \mathcal{T}(\mathcal{L})^{\mathtt{Q}}$ (sequent $R, Q_i \Rightarrow Q_i \otimes R$ is similar):

$$\dfrac{\dfrac{}{Q_i, R \Rightarrow R \otimes Q_i}\mathcal{T}(\mathcal{L})^{\mathtt{Q}} \qquad \dfrac{\delta_1}{\Gamma_1, R \otimes Q_i, \Gamma_2 \Rightarrow \bigotimes E(v)}}{\Gamma_1, Q_i, R, \Gamma_2 \Rightarrow \bigotimes E(v)}\text{(cut)}$$

As in the previous cases, the last rule applied in δ_1 could not have been a standard cut nor (rw). Suppose it is (R$\otimes$). Writing the endsequent of δ_1 as $\Gamma_3, \Gamma_4 \Rightarrow Q_j \otimes \varphi_2$, the left premise would have the form $\Gamma_3 \Rightarrow Q_i$ where Γ_3 must contain $R \otimes Q_i$ (Lem. 13) and hence also Γ_1. If Γ_1 is non-empty then it must contain a channel marker and hence in the right premise $\Gamma_4 \Rightarrow \varphi_2$, the Γ_4 would lack a channel marker contradicting Lem. 12. If Γ_1 is empty then R must be a channel marker and once again Γ_4 would lack a channel marker leading to a contradiction. Therefore the rule is not (R$\otimes$).

The remaining possibility is a sequence of (lw) followed by (L$\otimes$). The premise of the latter is in the image of an encoding, thus the IH applies and we are done.

□

Theorem 3.16. *Let $\Sigma \subseteq \Sigma_{\mathbf{FL}}$. If $\otimes \in \Sigma$, then sequent-deducibility in $\mathbf{FL}^\Sigma_\mathbf{w}$ is* HACK*-hard. Moreover, if $\{\otimes, \backslash, 0, 1\} \subseteq \Sigma$, then deducibility is* HACK*-hard.*

Proof. The first assertion follows from Lem. 14, plus the fact that the reduction established there is polynomial-time, and Lem. 6. The second one follows from the first and Lem. 4. □

4 Hyper-Ackermannian upper bounds

We give an algorithm to decide the relation $\vdash_{\mathbf{S}\Rightarrow}$ for regular theories, for a sequent calculus $\mathbf{S}$ containing the rule (lw) and satisfying a generalized subformula property as defined below. In what follows, given a finite $\Omega \subseteq \mathrm{Fm}_{\Sigma}(\mathcal{P})$, an *$\Omega$-sequent* is a sequent in which only formulas in Ω appear.

Definition 4.1. *A sequent calculus that is a structural rule extension of* $\mathbf{FL}_{\mathbf{w}}^{\Sigma}$ *($\Sigma \subseteq \Sigma_{\mathbf{FL}}$) is called* amenable *provided it satisfies the* generalized subformula property*: for every regular theory $\mathcal{T}$, it is the case that $\mathcal{T} \vdash_{\mathbf{S}\Rightarrow} s$ iff there is a $\mathcal{T}$-deduction of s where only $\mathsf{subf}(\mathcal{T} \cup \{s\})$-sequents appear (such a deduction is called* analytic*).*

$\mathbf{FL}_{\mathbf{w}}^{\Sigma}$ is amenable for any $\Sigma \subseteq \Sigma_{\mathbf{FL}}$; Lem. 5 gives the generalized subformula property. Indeed, any *analytic structural rule* extension of an amenable calculus is amenable (see results for the substructural hierarchy, Ciabattoni et al. [6]).

4.1 Just enough on well-quasi-order theory

A *quasi-ordered set (qo-set)* is a structure $\mathbf{W} := \langle W, \preccurlyeq_{\mathbf{W}} \rangle$, where W is a set and $\preccurlyeq_{\mathbf{W}} \subseteq W \times W$ is reflexive and transitive (a *quasi order*, for short). Abusing notation, we write $a \in \mathbf{W}$ for $a \in W$. We denote by $\mathbb{N}$ the set of natural numbers; by $\mathbb{N}^{+}$ the set $\mathbb{N}\backslash\{0\}$; and, given $\ell \in \mathbb{N} \cup \{\omega\}$, we write $\mathbb{N}_{<\ell}$ for the set $\{a \in \mathbb{N} \mid a < \ell\}$. Given a qo-set $\mathbf{W}$, a *bad sequence over* $\mathbf{W}$ (of length ℓ) is a sequence $(a_i)_{i \in \mathbb{N}_{<\ell}}$ of elements of $\mathbf{W}$ such that, for all $i < j$, $a_i \not\preccurlyeq_{\mathbf{W}} a_j$. A qo-set $\mathbf{W}$ is a *well-quasi-ordered set (wqo-set)* if every bad sequence over it is finite (see [11, Sec. 2] for equivalent definitions). Examples include $\mathbb{N}^k$ under component-wise ordering (Dickson's Lemma [7]), and sequences under the subword embedding ordering:

Theorem 4.2 (Higman's Lemma [11])**.** *For any finite set Ω, $\langle \Omega^*, \preccurlyeq_* \rangle$ is a wqo, where, recall, Ω^* is the set of all finite sequences of elements from Ω and $w_1 \preccurlyeq_* w_2$ iff w_1 is obtainable from w_2 by deleting some elements.*

Bad sequences over a wqo, though finite, do not in general have a maximum length. For example, $(1,0),(0,k),(0,k-1),\ldots,(0,0)$ is a bad sequence on $\mathbb{N}^2$ under the component-wise ordering with length $k+2$ for every $k \in \mathbb{N}$. Nevertheless, a maximum length can be ensured by limiting the size of each element in the sequence to some (fixed) function of the size of the preceding element. To achieve this, the wqo is enriched with more structure. A *normed (well-)quasi-ordered set (n(w)qo-set)* [18] is a structure $\mathbf{W} := \langle W, \preccurlyeq_{\mathbf{W}}, [\![\cdot]\!]_{\mathbf{W}} \rangle$, where $\langle W, \preccurlyeq_{\mathbf{W}} \rangle$ is a (w)qo-set and $[\![\cdot]\!]_{\mathbf{W}} : W \to \mathbb{N}$ is a *proper norm*, meaning that, for all $n \in \mathbb{N}$, $\{a \in W : [\![a]\!]_{\mathbf{W}} < n\}$ is finite. It is easy to check that the following is a nwqo.

Definition 4.3. *Given a finite set Ω, let $\mathbf{\Omega}^* := \langle \Omega^*, \preccurlyeq_*, [\![\cdot]\!]_* \rangle$, where $[\![w]\!]_*$ is the length of w (and $\langle \Omega^*, \preccurlyeq_* \rangle$ is as in Thm. 17).*

Indeed, the *disjoint sum* of these nwqos, defined below, is also an nwqo [18].

Definition 4.4. *Given a finite set Ω and $m \in \mathbb{N}$, let $m \cdot \mathbf{\Omega}^* :=$*

$\langle m \cdot \Omega^*, \preccurlyeq_{m \cdot \Omega^*}, [\![\cdot]\!]_{m \cdot \Omega^*} \rangle$. *Here,* $m \cdot \Omega^* := \{1, \ldots, m\} \times \Omega^*$ *is the disjoint sum of* m *copies of* Ω^*, *and* $(i, w_1) \preccurlyeq_{m \cdot \Omega^*} (j, w_2)$ *iff* $i = j$ *and* $w_1 \preccurlyeq_* w_2$. *Also, define* $[\![(i, w)]\!]_{m \cdot \Omega^*} := [\![w]\!]_*$.

A *control function* is a mapping $g : \mathbb{N} \to \mathbb{N}$ that is strictly increasing (for all $a, b \in \mathbb{N}$, if $a < b$, then $g(a) < g(b)$) and strictly inflationary ($a < g(a)$ for all $a \in \mathbb{N}$).

For an nwqo-set $\mathbf{W}$, control function g, and *initial parameter* $t \in \mathbb{N}$, a *(g, t)-controlled bad sequence over* $\mathbf{W}$ is a bad sequence $a_0, \ldots, a_{\ell-1}$ over $\mathbf{W}$ where $[\![a_i]\!]_{\mathbf{W}} < g^i(t)$ for all $0 \leq i < \ell$. The (g, t)-controlled bad sequences have a maximum length [8] using König's Lemma. Denote by $L^g_{\mathbf{W}}(t)$ the maximum length of a (g, t)-controlled bad sequence over $\mathbf{W}$. We call $L^g_{\mathbf{W}} : \mathbb{N} \to \mathbb{N}$ the *length function* of $\mathbf{W}$ (for g). Bounds for (g, t)-controlled bad sequences over $\mathbf{W}$ reduce to finding bounds for $L^g_{\mathbf{W}}$ ('*length theorems*'), expressed in terms of the ordinal-indexed *extended Grzegorczyk hierarchy* $\{\mathfrak{F}_\alpha\}_\alpha$, from which the hierarchy $\{\mathbf{F}_\alpha\}_\alpha$ of *fast-growing complexity classes* are defined (the reader is referred to [16, Sec. 2] for detailed definitions). What matters here is that to classify a problem in $\mathbf{F}_\alpha$, it suffices to show that the complexity of every instance is upper bounded by a function in $\mathfrak{F}_\beta$ for some $\beta < \alpha$. Schmitz and Schnoebelen obtained the following length theorem for Higman's lemma.

Theorem 4.5 ([17, Thm. 5.3]). *For any finite set* Ω, $m \in \mathbb{N}$, *and primitive recursive control function* g, *the length function* $L^g_{m \cdot \Omega^*}$ *is upper bounded by a function in* $\mathfrak{F}_{\omega^{|\Omega|-1}}$.

A *reflection* is a map between nqo-sets that preserves bad sequences. Hence, its existence implies the reverse transfer of the wqo-property and upper bounding of length theorem. In the next subsection, we obtain a length theorem for a nwqo over sequents in this way.

Definition 4.6 (Reflection). *Let* $\mathbf{W}_1 := \langle W_1, \preccurlyeq_1, [\![\cdot]\!]_1 \rangle$ *and* $\mathbf{W}_2 := \langle W_2, \preccurlyeq_2, [\![\cdot]\!]_2 \rangle$ *be nqo-sets. A* reflection *between* $\mathbf{W}_1$ *and* $\mathbf{W}_2$ *is a mapping* $f : W_1 \to W_2$ *such that, for all* $a, b \in W_1$, *if* $f(a) \preccurlyeq_2 f(b)$, *then* $a \preccurlyeq_1 b$ *and, for all* $a \in W_1$, $[\![f(a)]\!]_2 \leq [\![a]\!]_1$. *We write* $\mathbf{W}_1 \hookrightarrow^f \mathbf{W}_2$ *(sometimes omitting the superscript* f*).*

Lemma 4.7 ([17, p. 446]). *Whenever* $\mathbf{W}_1 \hookrightarrow \mathbf{W}_2$, *if* $\mathbf{W}_2$ *is a nwqo, then so is* $\mathbf{W}_1$*; and if* g *is a control function, then* $L^g_{\mathbf{W}_1}(t) \leq L^g_{\mathbf{W}_2}(t)$ *for all* $t \in \mathbb{N}$.

4.2 A well-quasi-order over sequents based on weakening

Observe that the antecedent of the premise of the left-weakening rule (lw) is obtainable from the antecedent of the conclusion by deleting letters, and the succedents are identical. This motivates an nwqo over sequents, for which we obtain a length theorem via a reflection into a disjoint sum of Higman's orderings. In what follows, let Σ be an arbitrary signature.

Definition 4.8. *For finite* $\Omega \subseteq \mathrm{Fm}_\Sigma(\mathcal{P})$, *let* $\mathbf{W}_\Sigma(\Omega) := \langle \mathsf{Seq}_\Sigma(\Omega), \rightsquigarrow^\Omega_\Sigma, [\![\cdot]\!] \rangle$, *where*

- $\mathsf{Seq}_\Sigma(\Omega)$ *is the set of all* Ω*-sequents over* $\mathrm{Fm}_\Sigma(\mathcal{P})$;

- $\rightsquigarrow^{\Omega}_{\Sigma} \subseteq \mathsf{Seq}_{\Sigma}(\Omega) \times \mathsf{Seq}_{\Sigma}(\Omega)$ *is such that* $s_1 \rightsquigarrow^{\Omega}_{\Sigma} s_2$ *iff* s_2 *is obtained by successive applications of* (lw) *starting from* s_1*;*
- $[\![\cdot]\!] : \mathsf{Seq}_{\Sigma}(\Omega) \to \mathbb{N}$ *is such that* $[\![\Gamma \Rightarrow \Pi]\!] := |\Gamma|$ *(i.e., the length of the antecedent).*

Theorem 4.9. *For all finite* $\Omega \subseteq \mathrm{Fm}_{\Sigma}(\mathcal{P})$*,* $\mathbf{W}_{\Sigma}(\Omega)$ *is a nwqo; and if* g *is primitive recursive, then* $L^{g}_{\mathbf{W}_{\Sigma}(\Omega)}$ *is upper bounded by a function in* $\mathfrak{F}_{\omega^{|\Omega|-1}}$*.*

Proof. Fix an enumeration $\varphi_1, \ldots, \varphi_m$ of the formulas in Ω. Then the mapping $f : \mathsf{Seq}_{\Sigma}(\Omega) \to (|\Omega|{+}1)\cdot\Omega^*$ such that $f(\Gamma \Rightarrow\ ) := (0, \Gamma)$ and $f(\Gamma \Rightarrow \varphi_j) := (j, \Gamma)$ is a reflection; hence the result follows from Thm. 18 and Lem. 19. □

4.3 Decision procedure for amenable sequent calculi

Fix an amenable sequent calculus $\mathbf{S}$ (see Def. 16.1) over a signature Σ.

Definition 4.10. *For* $\mathcal{X}$ *a formula, a sequent, or a sequent rule, let* $\langle\mathcal{X}\rangle$ *be the length of the written representation of* $\mathcal{X}$*. If* X *is a finite set of formulas, of sequents, or of sequent rules, let* $\langle X\rangle := \sum_{\mathcal{X}\in X}\langle\mathcal{X}\rangle$ *and* $\langle_{\mathsf{max}} X\rangle := \max_{\mathcal{X}\in X}\langle\mathcal{X}\rangle$*. In particular,* $\langle_{\mathsf{max}} \mathbf{S}\rangle := \max_{\mathsf{r}\in\mathbf{S}}\langle\mathsf{r}\rangle$ *i.e., the largest among the representation lengths of the rules in the calculus.*

The construction in this subsection extends the commutative setting that appears in Balasubramanian et al. [1]. For a finite set D of sequents, define $[\![_{\mathsf{max}}D]\!] := \max_{s\in D} [\![s]\!]$ ($[\![s]\!]$ is the length of the antecedent of s, see Def. 19.1).

Definition 4.11. *Let* Ω *be a finite set of formulas closed under subformulas and* $\mathcal{T}$ *be a regular theory containing only* Ω*-sequents. Define* D_0 *as the set of minimal elements with respect to* $\rightsquigarrow^{\Omega}_{\Sigma}$ *(i.e., not obtainable from another element by repeated* (lw)*) from the following finite set: the union of* $\mathcal{T}$ *with the set of all instances of initial sequents in* $\mathbf{S}$ *that satisfy the following:*

a) formula-variables are instantiated to elements of Ω*;*

b) succedent-variables are instantiated to an element in Ω *or as empty;*

c) sequence-variables are instantiated to the empty sequence.

Define $D_{i+1} := D_i \cup \partial D_i$ $(i \geq 0)$*, where* ∂D_i *is the set of* Ω*-sequents* s *satisfying:*

(i) $s_1 \cdots s_k / s$ *is a rule instance of* $\mathbf{S}$ *such that, for all* $1 \leq j \leq k$*, there is* $s_j{}' \in D_i$ *with* $s_j{}' \rightsquigarrow^{\Omega}_{\Sigma} s_j$*;*

(ii) *the antecedent of* s *has length* $\leq (\langle\mathbf{S}\rangle[\![_{\mathsf{max}}D_i]\!]) \cdot \langle\mathbf{S}\rangle$*;*

(iii) *there is no* $s' \in D_i$ *such that* $s' \rightsquigarrow^{\Omega}_{\Sigma} s$*.*

The above construction yields a chain $D_0 \subseteq D_1 \subseteq \ldots$ of sets of sequents. The following shows that the chain stabilizes at a finite index. Indeed, if that were not the case, Def. 20.2 (iii) forces the existence of an infinite bad sequence $(s_i)_{i<\omega}$ from $s_i \in D_i \setminus D_{i-1}$ $(i < \omega)$, and that is impossible.

Theorem 4.12. D_{i+1} *is computable from* D_i *and there is* $N \in \mathbb{N}$ *such that* $D_N = D_{N+i}$ *for all* $i \geq 0$*.*

Proof. Each D_i is computable because (a) there are only finitely-many rules in **S**, (b) finitely-many formulas in Ω, (c) Def. 20.2 (ii) restricts the length of antecedents (thus there are only finitely-many rule instances to consider), and (d) the relation $\rightsquigarrow^{\Omega}_{\Sigma}$ is computable.

If there is some N such that $D_N = D_{N+1}$ then, by the construction of these sets, $D_N = D_{N+i}$ for every $i \geq 0$. Suppose that no such N exists. It follows that $D_0 \subset D_1 \subset D_2 \subset \ldots$. Choose any $s_i \in D_i \setminus D_{i-1}$ for each $i \in \mathbb{N}$. By Def. 20.2 (iii), $(s_i)_{i<\omega}$ is an infinite bad sequence over $\mathbf{W}_{\Sigma}(\Omega)$, contradicting that the latter is a wqo (Thm. 20). □

We now establish (Thm. 23) that every deducible sequent is obtainable by weakening from a sequent at the stabilization point. First, a technical lemma.

Lemma 4.13. *If $s' \rightsquigarrow^{\Omega}_{\Sigma} s$ with $s = L_1 \cdots L_m \Rightarrow \Pi$, then there are sequences $L'_1, \ldots, L'_m$ such that $s' = L'_1 \cdots L'_m \Rightarrow \Pi$ and $L'_i \preccurlyeq_* L_i$ for all $1 \leq i \leq m$.*

Proof. By induction on a deduction δ witnessing $s' \rightsquigarrow^{\Omega}_{\Sigma} L_1 \cdots L_m \Rightarrow \Pi$. If it has a single node, the involved sequents are the same, so take $L'_i := L_i$. In the inductive step, assume $s = L_1 \cdots L^1_i \varphi L^2_i \cdots L_m \Rightarrow \Pi$ was obtained by weakening from $s'' = L_1 \cdots L^1_i L^2_i \cdots L_m \Rightarrow \Pi$, where $L_i = L^1_i L^2_i$. Then, by the (IH), $s' = L'_1 \cdots L'_i \cdots L'_m \Rightarrow \Pi$ with $L'_j \preccurlyeq_* L_j$ for all $1 \leq j \leq m$. In particular, $L'_i \preccurlyeq_* L^1_i L^2_i \preccurlyeq_* L^1_i \varphi L^2_i$, and we are done. □

Theorem 4.14. *Let $\mathcal{T} \cup \{s\}$ be a finite set of Ω-sequents such that $\mathcal{T}$ is a regular theory (as in Def. 20.2). Then $\mathcal{T} \vdash_{\mathbf{S}_{\Rightarrow}} s$ iff there is $M \in \mathbb{N}$ such that $s' \rightsquigarrow^{\Omega}_{\Sigma} s$ for some $s' \in D_M$.*

Proof. The left-to-right direction is the non-trivial one. Since **S** is amenable, $\mathcal{T} \vdash_{\mathbf{S}_{\Rightarrow}} s$ is witnessed by an analytic deduction δ. We argue by induction on the structure of δ.

Base case. Then δ is a single node, so the node is a sequent in $\mathcal{T}$ or an instance of an initial sequent. So s is in D_0, or s is obtainable from the node by applications of (lw).

Inductive step. The last rule instance in δ is an instance $s_1 \cdots s_k / s$ of r. Since each s_i is an Ω-sequent by virtue of the deduction being analytic, by applying IH for each $1 \leq i \leq k$, there is $N_i \in \mathbb{N}$ with $s'_i \rightsquigarrow^{\Omega}_{\Sigma} s_i$ for some $s'_i \in D_{N_i}$. Let $M := \max_i N_i$, which gives us $s'_i \in D_M$ for all $1 \leq i \leq k$. Hence Def. 20.2 (i) is satisfied for s. The situation is the following.

$$\dfrac{\begin{matrix} s'_1 \in D_M & \cdots & s'_k \in D_M \\ \wr\!\downarrow^{\Omega}_{\Sigma} & & \wr\!\downarrow^{\Omega}_{\Sigma} \\ s_1 & \cdots & s_k \end{matrix}}{s}\ \mathsf{r}$$

We now show by cases that there is $s' \in D_{M+1}$ such that $s' \rightsquigarrow^{\Omega}_{\Sigma} s$.

If Def. 20.2 (ii) is satisfied, then $s \in D_{M+1}$, or (due to Def. 20.2 (iii)) there is some $s' \in D_{M+1}$ such that $s' \rightsquigarrow^{\Omega}_{\Sigma} s$. In either case we are done, so assume that Def. 20.2 (ii) fails. Thus s has antecedent of length $> (\langle \mathbf{S} \rangle [\![{}_{\mathsf{max}} D_M]\!]) \cdot \langle \mathbf{S} \rangle$. Thus there are sequence-variables $\Gamma_1, \ldots, \Gamma_m$ ($m > 0$) in the conclusion of r instantiated with sequences of length $> \langle \mathbf{S} \rangle [\![{}_{\mathsf{max}} D_M]\!]$ (if $m = 0$ then the antecedent would be instantiations of formula-variables and hence its length would be $\leq \langle \mathbf{S} \rangle [\![_{\mathsf{max}} D_0]\!]$). Assume wlog that $\Gamma_1, \ldots, \Gamma_{m'}$ with $m' \leq m$ appear in the premises of r, while the other variables appear exclusively in the conclusion.

Write $s_i = L_{i1} \cdots L_{iu_i} \Rightarrow \Pi_i$ and $s'_i = L'_{i1} \cdots L'_{iu_i} \Rightarrow \Pi_i$ where each L_i is an instantiation of a sequence-variable or a singleton sequence of a formula, such that $L'_{ij} \preccurlyeq_* L_{ij}$ for each $1 \leq j \leq u_i$ (the existence of such sequences is guaranteed by Lem. 22). The length of $L'_{i1} \cdots L'_{iu_i}$ is $\leq [\![_{\mathsf{max}} D_M]\!]$ by the fact that $s'_i \in D_M$, hence each L'_{ij} has length $\leq [\![_{\mathsf{max}} D_M]\!]$.

We construct a new instance I' of r, with premises s''_i instead of s_i, and a conclusion s' whose antecedent has size satisfying Def. 20.2(ii), and $s' \rightsquigarrow^{\Omega}_{\Sigma} s$:

$$\dfrac{\begin{array}{ccc} s'_1 \in D_M & \cdots & s'_k \in D_M \\ \wr\, {\scriptstyle \Omega\Sigma} & & \wr\, {\scriptstyle \Omega\Sigma} \\ s''_1 & \cdots & s''_k \end{array}}{s' \text{ (smaller than } s)}\ \mathsf{r}$$

Construct I' by consideration of where each Γ_j $(1 \leq j \leq m')$ occurs. For Γ_j occurring in a single premise, instantiate it with the sequence L'_{jl} instead of the L_{jl} that was used before.

For each Γ_j $(1 \leq j \leq m')$ occurring in multiple premises, we need a single instantiation that can be used for each occurrence (reflecting the fact that we are dealing with additive rules here). Without loss of generality, assume that $s_1, \ldots, s_l$ are all the premises where Γ_j occurs. Assume that Γ_j was instantiated with the sequence L. Let $L_{1j_1}, \ldots, L_{lj_l}$ be the sequences in these premises that instantiate Γ_j (each is equal to L). We want a subsequence of L reachable by weakening from each $L'_{1j_1}, \ldots, L'_{lj_l}$ (these need not be identical): start with L and cross off each of the sequences $L'_{1j_1}, \ldots, L'_{lj_l}$ (letter by letter, respecting the order). Use the sequence crossed-off to instantiate Γ_j. It has size $\leq k \cdot [\![_{\mathsf{max}} D_M]\!] < \langle \mathbf{S} \rangle \cdot [\![_{\mathsf{max}} D_M]\!]$.

Finally, for sequence-variables that appear only in the conclusion of the rule, i.e., $\Gamma_{m'+1}, \ldots, \Gamma_m$, instantiate each of these with the empty sequence.

The new instantiation I' has premises s''_i where $s'_i \rightsquigarrow^{\Omega}_{\Sigma} s''_i$ for each $1 \leq i \leq k$, and its conclusion is $s' \rightsquigarrow^{\Omega}_{\Sigma} s$ since the instantiation of each Γ_j is a subsequence of the original instantiation. Moreover, the number of variables in the conclusion of r is $< \langle \mathbf{S} \rangle$ and each variable instantiation under I' has size $< \langle \mathbf{S} \rangle [\![_{\mathsf{max}} D_M]\!]$. Hence it follows that $s' \in D_{M+1}$ or there is some $s'_0 \in D_M$ with $s'_0 \rightsquigarrow^{\Omega}_{\Sigma} s'$. In either case, the claim is proved. □

4.4 Complexity analysis

The fast-growing complexity classes are closed under exponentiation so the distinction between space and time (also determinism and non-determinism) is unimportant [16]. We undertake a space analysis of the algorithm based on Thm. 21 and Thm. 23. It suffices to write down each D_i in turn at the same location, computing each rule instance in turn; when the stabilization index D_N is reached, check whether $s' \rightsquigarrow^{\Omega}_{\Sigma} s$ for some $s' \in D_N$. Each $D_i \subseteq D_N$, and writing down a rule instance takes space bounded by the size of an element from $D_N^{\langle \mathbf{S} \rangle} \times D_N$. So the space requirements are an elementary function in $|D_N|$ and the size of an element from D_N. We upper bound each of these.

Lemma 4.15. *Let $\mathcal{T}$ be a regular theory, s be a sequent and $\Omega := \mathsf{subf}(\mathcal{T} \cup \{s\})$.*

1. $|\Omega| \leq \langle s \rangle + \langle \mathcal{T} \rangle$;
2. *if $s' \in D_i$ then $[\![s']\!] < (\langle \mathcal{T} \rangle + 1)\langle \mathbf{S} \rangle^{2i+1}$, and $[\![_{\mathsf{max}} D_i]\!] < (\langle \mathcal{T} \rangle + 1)\langle \mathbf{S} \rangle^{2i+1}$;*
3. *if $s' \in D_i$ then $\langle s' \rangle < 4(\langle \mathcal{T} \rangle + \langle s \rangle)^2 \langle \mathbf{S} \rangle^{2N+1}$;*
4. *for all $i \geq 0$, $|D_i| < E(\langle \mathcal{T} \rangle + \langle s \rangle, N)$ for E an elementary function (i.e., in $\mathfrak{F}_2$).*

Proof.

1. It is easy to see by structural induction that $|\mathsf{subf}(\varphi)| \leq \langle \varphi \rangle$ for all formulas φ. Then, for any sequent $t := \psi_1, \ldots, \psi_m \Rightarrow \Pi$, we have $|\mathsf{subf}(t)| \leq |\mathsf{subf}(\Pi)| + \sum_{i=1}^m |\mathsf{subf}(\psi_i)| \leq \langle \Pi \rangle + \sum_{i=1}^m \langle \psi_i \rangle \leq \langle t \rangle$. The result then follows because $|\Omega| \leq |\mathsf{subf}(s)| + \sum_{t \in \mathcal{T}} |\mathsf{subf}(t)| \leq \langle s \rangle + \langle \mathcal{T} \rangle$.
2. By induction on $i \geq 0$. Note that if $s' \in D_0$, since D_0 is a subset of the union of $\mathcal{T}$ and instantiations from Ω of initial sequents in $\mathbf{S}$, $[\![s']\!] < [\![_{\mathsf{max}} \mathcal{T}]\!] + \langle \mathbf{S} \rangle \leq ([\![_{\mathsf{max}} \mathcal{T}]\!] + 1)\langle \mathbf{S} \rangle \leq (\langle \mathcal{T} \rangle + 1)\langle \mathbf{S} \rangle$. Inductive step: if $s' \in D_{i+1}$, then $[\![s']\!] < \langle \mathbf{S} \rangle^2 [\![_{\mathsf{max}} D_i]\!] < \langle \mathbf{S} \rangle^2 (\langle \mathcal{T} \rangle + 1)\langle \mathbf{S} \rangle^{2i+1} = (\langle \mathcal{T} \rangle + 1)\langle \mathbf{S} \rangle^{2(i+1)+1}$. Since $[\![_{\mathsf{max}} D]\!] = \max_{s \in D} [\![s]\!]$, it follows that $[\![_{\mathsf{max}} D_i]\!] < (\langle \mathcal{T} \rangle + 1)\langle \mathbf{S} \rangle^{2i+1}$.
3. It is enough to observe that $\langle s' \rangle$ is upper bounded by the sum of the sizes of each formula in the antecedent, plus 1 for each comma, plus 1 to account for the sequent symbol, plus the size of the succedent. This gives us $\langle s' \rangle \leq [\![s']\!]\langle \Omega \rangle + [\![s']\!] + 1 + \langle \Omega \rangle \leq 4([\![s']\!] + 1)\langle \Omega \rangle$. From the previous items, the latter is $\leq 4(((\langle \mathcal{T} \rangle + 1)\langle \mathbf{S} \rangle^{2N+1} + 1)(\langle s \rangle + \langle \mathcal{T} \rangle) \leq 4(((\langle \mathcal{T} \rangle + \langle s \rangle)\langle \mathbf{S} \rangle^{2N+1})(\langle s \rangle + \langle \mathcal{T} \rangle)$ where we have used that $\langle s \rangle \geq 2$ and $\langle \mathbf{S} \rangle \geq 1$.
4. Let $\eta, \rho \in \mathbb{N}$. Then define $\left[{\eta \atop \rho}\right]_{\mathsf{seq}} := (\sum_{l=0}^{\rho} \eta^l) \cdot (\eta + 1) \leq (\eta + 1)^{\rho+1} \cdot (\eta + 1)$, the number of distinct sequents over η formulas and having norm $\leq \rho$. Note that $|D_0| \leq \left[{\langle s \rangle + \langle \mathcal{T} \rangle \atop \langle s \rangle + \langle \mathcal{T} \rangle}\right]_{\mathsf{seq}}$ and $|D_{i+1}| \leq \left[{|\Omega| \atop \langle \mathbf{S} \rangle^2 [\![_{\mathsf{max}} D_i]\!]}\right]_{\mathsf{seq}} < \left[{\langle s \rangle + \langle \mathcal{T} \rangle \atop \langle \mathbf{S} \rangle^2 (\langle \mathcal{T} + 1 \rangle) \langle \mathbf{S} \rangle^{2N+1}}\right]_{\mathsf{seq}} \leq \left[{\langle s \rangle + \langle \mathcal{T} \rangle \atop \langle \mathbf{S} \rangle^2 (\langle s \rangle + \langle \mathcal{T} \rangle) \langle \mathbf{S} \rangle^{2N+1}}\right]_{\mathsf{seq}}$. Thus, for all $i \geq 0$, $|D_i| < E(\langle s \rangle + \langle \mathcal{T} \rangle, N)$ for E an elementary function (i.e. in $\mathfrak{F}_2$).

□

Theorem 4.16. *Deciding whether $\mathcal{T} \vdash_{\mathbf{S}_\Rightarrow} s$ for a regular theory $\mathcal{T}$ is in* HACK.

Proof. Lem. 24 (4) shows that $|D_N| < E_1(\langle \mathcal{T} \rangle + \langle s \rangle, N)$, for an elementary function E_1. Also, Lem. 24 (3) shows that that $s' \in D_N$ implies that its size $\langle s' \rangle$ is bounded by $E_2(\langle \mathcal{T} \rangle + \langle s \rangle, N)$, for an elementary function E_2. It remains to upper bound N, and it is this that forces the fast-growing complexity. Let $\Omega := \mathsf{subf}(\mathcal{T} \cup \{s\})$. In the proof of Thm. 21, we extracted a bad sequence $(s_i)_{i \leq N}$ over $\mathbf{W}_\Sigma(\Omega)$ from the chain of sets $D_0 \subset \ldots \subset D_N$ by choosing any $s_i \in D_i \setminus D_{i-1}$. We claim that this bad sequence is (g, n)-controlled for the control function $g(x) := \langle \mathbf{S} \rangle^2 x$ and $n := (\langle \mathcal{T} \rangle + 1)\langle \mathbf{S} \rangle$, allowing us to conclude that $N \leq L^g_{\mathbf{W}_\Sigma(\Omega)}(n) \leq L^g_{\mathbf{W}_\Sigma(\Omega)}(((\langle \mathcal{T} \rangle + \langle s \rangle)\langle \mathbf{S} \rangle)$. By induction on i. *Base case*: $s_0 \in D_0$, so Lem. 24(2) yields $[\![s_0]\!] < ((\langle \mathcal{T} \rangle + 1)\langle \mathbf{S} \rangle$. *Inductive step*: $[\![s_{i+1}]\!] < ((\langle \mathcal{T} \rangle + 1)\langle \mathbf{S} \rangle^{2(i+1)+1} = \langle \mathbf{S} \rangle^2((\langle \mathcal{T} \rangle + 1)\langle \mathbf{S} \rangle^{2i+1} = \langle \mathbf{S} \rangle^2 g^i((\langle \mathcal{T} \rangle + 1)\langle \mathbf{S} \rangle) = g^{i+1}(n)$, so we are done. So the space to run the algorithm is a composition of elementary functions with a function in $\mathfrak{F}_{\omega^{|\Omega|-1}}$ by Thm. 20, thus it is bounded by a function in $\mathfrak{F}_{\omega^{|\Omega|-1}}$. So, for this particular $|\Omega|$, the problem is in $\mathbf{F}_{\omega^{|\Omega|}}$. As Ω varies with the input, we wish to eliminate its dependence. Upper bounding over all Ω, we have that the problem is in $\mathbf{F}_{\omega^\omega} = $ HACK. □

Corollary 4.17. *Any subcalculus of* $\mathbf{FL_w}$ *containing* (lw) *has deducibility in* HACK. *This holds in particular for* $\mathbf{FL}^\Sigma_\mathbf{w}$ *for any* $\Sigma \subseteq \Sigma_{\mathbf{FL}}$.

5 Final considerations

Combining Sec. 7 and Sec. 16, we finally obtain the promised result:

Theorem 1 (Main theorem). *For* $\{\otimes, \backslash, 0, 1\} \subseteq \Sigma \subseteq \Sigma_{\mathbf{FL}}$, *deducibility in* $\mathbf{FL}^\Sigma_\mathbf{w}$ *is* HACK-*complete. In particular,* $\mathbf{FL_w}$ *and its multiplicative fragment* $\mathbf{L_w}$ *are* HACK-*complete.*

Complexity of the word problem in integral FL-algebras Let V be an equational class of algebras (a *variety*) over a signature Σ. The *word problem of* V [9, Sec. 4.4.2] asks whether, given fixed finite sets of variables X (generators) and equations E over $\mathrm{Fm}_\Sigma(X)$, the quasiequation $\&E \Longrightarrow e$ is valid in V, where e is an equation over $\mathrm{Fm}_\Sigma(X)$. Note that deciding this problem allows for one algorithm per pair (X, E), as opposed to deciding the *quasiequational theory of* V, which asks for a single algorithm that applies to every (X, E). In view of the algebraizability of $\vdash_{\mathbf{FL_w}}$ [9, Sec. 2.6] w.r.t. the variety of integral zero-bounded FL-algebras $\mathsf{FL_w}$, our results imply that the word problem and the quasiequational theory of $\mathsf{FL_w}$ are HACK-complete. In fact, because $\mathsf{Props}(\mathcal{L})$ (cf. Def. 9.1) is a finite set of generators and $\mathcal{T}(\mathcal{L})^{\rightarrow}$ (cf. Def. 9.2) is a finite set of formulas over $\mathrm{Fm}_{\Sigma_{\mathbf{FL}}}(\mathsf{Props}(\mathcal{L}))$, we have that the word problem in this variety is HACK-hard. The quasiequational theory is in HACK in view of the proof search procedure in the previous section. Since the word problem reduces to the quasiequational theory, we obtain that both problems are HACK-complete. This also applies to integral FL-algebras and integral residuated lattices since the constant 0 does not play any essential role in the arguments.

Non-existence of deduction theorem Our lower bounds imply that fragments of $\mathbf{FL_w}$ covered by Thm. 15 have no deduction theorem (DT); else deducibility would reduce to provability, yet provability in these logics is PSPACE.

Upper bound for axiomatic extensions The obtained upper-bounds apply to deducibility for $\mathcal{N}_2$-analytic structural rule extensions of $\mathbf{FL_w}$, and to the corresponding axiomatic extensions (refer to the substructural hierarchy [6]).

Undecidability of deducibility in FL The encoding in Sec. 9.1 offers a new proof of the undecidability of the deducibility problem in **FL** (remove lossiness; reachability in channel systems is undecidable). This might be useful to prove undecidability of deducibility in some axiomatic extensions of **FL**.

References

[1] Balasubramanian, A. R., T. Lang and R. Ramanayake, *Decidability and complexity in weakening and contraction hypersequent substructural logics*, in: *36th Annual ACM/IEEE Symposium on Logic in Computer Science, LICS 2021, Rome, Italy, June 29 - July 2, 2021* (2021), pp. 1–13.

[2] Blok, W. J. and C. J. V. Alten, *On the finite embeddability property for residuated ordered groupoids*, Transactions of the American Mathematical Society **357** (2005), pp. 4141–4157.

[3] Brand, D. and P. Zafiropulo, *On communicating finite-state machines*, J. ACM **30** (1983), p. 323–342.

[4] Chambart, P. and P. Schnoebelen, *The ordinal recursive complexity of lossy channel systems*, in: *2008 23rd Annual IEEE Symposium on Logic in Computer Science*, 2008, pp. 205–216.

[5] Chvalovský, K. and R. Horčík, *Full Lambek calculus with contraction is undecidable*, The Journal of Symbolic Logic **81** (2016), pp. 524–540.

[6] Ciabattoni, A., N. Galatos and K. Terui, *Algebraic proof theory: Hypersequents and hypercompletions*, Ann. Pure Appl. Logic **168** (2017), pp. 693–737.

[7] Dickson, L. E., *Finiteness of the odd perfect and primitive abundant numbers with n distinct prime factors*, American Journal of Mathematics **35** (1913), pp. 413–422.

[8] Figueira, D., S. Figueira, S. Schmitz and P. Schnoebelen, *Ackermannian and primitive-recursive bounds with Dickson's Lemma*, in: *Proceedings - Symposium on Logic in Computer Science*, 2011, pp. 269–278.

[9] Galatos, N., P. Jipsen, T. Kowalski and H. Ono, "Residuated lattices: an algebraic glimpse at substructural logics," Studies in Logic and the Foundations of Mathematics **151**, Elsevier, 2007, xxii+509 pp.

[10] Galatos, N. and H. Ono, *Cut elimination and strong separation for substructural logics: An algebraic approach*, Annals of Pure and Applied Logic **161** (2010), pp. 1097–1133.

[11] Higman, G., *Ordering by divisibility in abstract algebras*, Proceedings of the London Mathematical Society **s3-2** (1952), pp. 326–336.

[12] Horčík, R. and K. Terui, *Disjunction property and complexity of substructural logics*, Theoretical Computer Science **412** (2011), pp. 3992–4006.

[13] Jipsen, P. and C. Tsinakis, "A Survey of Residuated Lattices," Springer US, Boston, MA, 2002 pp. 19–56.

[14] Lincoln, P., J. Mitchell, A. Scedrov and N. Shankar, *Decision problems for propositional linear logic*, Annals of Pure and Applied Logic **56** (1992), pp. 239–311.

[15] Péter, R., "Recursive Functions," Academic Press, 1967.

[16] Schmitz, S., *Complexity hierarchies beyond elementary*, ACM Trans. Comput. Theory **8** (2016).

[17] Schmitz, S. and P. Schnoebelen, *Multiply-recursive upper bounds with Higman's Lemma*, in: L. Aceto, M. Henzinger and J. Sgall, editors, *Automata, Languages and Programming* (2011), pp. 441–452.

[18] Schmitz, S. and P. Schnoebelen, *Algorithmic Aspects of WQO Theory* (2012), lecture.

[19] Schnoebelen, P., *Verifying lossy channel systems has nonprimitive recursive complexity*, Information Processing Letters **83** (2002), pp. 251–261.

[20] Tanaka, H., *Tower-complete problems in contraction-free substructural logics*, in: B. Klin and E. Pimentel, editors, *31st EACSL Annual Conference on Computer Science Logic, CSL 2023, February 13-16, 2023, Warsaw, Poland*, LIPIcs **252** (2023), pp. 34:1–34:19.

[21] Urquhart, A., *The complexity of decision procedures in relevance logic II*, The Journal of Symbolic Logic **64** (1999), pp. 1774–1802.

Modal Logic, Fundamentally

Wesley H. Holliday

University of California, Berkeley

Abstract

Non-classical generalizations of classical modal logic have been developed in the contexts of constructive mathematics and natural language semantics. In this paper, we discuss a general approach to the semantics of non-classical modal logics via algebraic representation theorems. We begin with complete lattices L equipped with an antitone operation $\neg$ sending 1 to 0, a completely multiplicative operation $\Box$, and a completely additive operation $\Diamond$. Such lattice expansions can be represented by means of a set X together with binary relations $\lhd$, R, and Q, satisfying some first-order conditions, used to represent $(L, \neg)$, $\Box$, and $\Diamond$, respectively. Indeed, any lattice L equipped with such a $\neg$, a multiplicative $\Box$, and an additive $\Diamond$ embeds into the lattice of propositions of a frame $(X, \lhd, R, Q)$. Building on our recent study of *fundamental logic*, we focus on the case where $\neg$ is dually self-adjoint ($a \leq \neg b$ implies $b \leq \neg a$) and $\Diamond\neg a \leq \neg\Box a$. In this case, the representations can be constrained so that $R = Q$, i.e., we need only add a single relation to $(X, \lhd)$ to represent both $\Box$ and $\Diamond$. Using these results, we prove that a system of fundamental modal logic is sound and complete with respect to an elementary class of bi-relational structures $(X, \lhd, R)$.

Keywords: non-classical modal logic, orthologic, intuitionistic logic, fundamental logic, lattices, weak pseudocomplementation, necessity, possibility, representation

1 Introduction

In classical modal logic, necessity and possibility are duals in the sense that $\Box a = \neg\Diamond\neg a$ and $\Diamond a = \neg\Box\neg a$, putting the point algebraically, so typically just one is taken as primitive and the other is treated as defined. The same is true in certain non-classical modal logics, such as the *epistemic orthologic* of [44]. However, in standard treatments of *intuitionistic* modal logic [24,9,56], $\neg\Box a$ does not entail $\Diamond\neg a$, just as in intuitionistic predicate logic, $\neg\forall x P(x)$ does not entail $\exists x\neg P(x)$. In this setting, both $\Box$ and $\Diamond$ must be taken as primitive. Thus, a general approach to non-classical modal logic should do the same. In this paper, building on [41,42], we study an approach to the semantics of non-classical modal logics incorporating Ploščica's [51] approach to the representation of lattices, Birkhoff's [7] approach to the representation of negation, and the Jónsson-Tarski [45] approach to the representation of modal operations; a similar approach without negation was earlier investigated in [10]. Here we add to our treatment of $\neg$ and $\Box$ in [41,42] a new representation of $\Diamond$.

Our motivation for doing so comes from our recent study of *fundamental logic*, a sublogic of both intuitionistic logic [37] and orthologic [34]. Fundamental propositional logic is defined in [42] in terms of a Fitch-style natural deduction system containing only introduction and elimination rules for the logical connectives $\wedge$, $\vee$, and $\neg$. Thus, unlike Fitch's [25,26] proof system for classical logic, the Fitch-style proof system for fundamental logic does not contain the rule of Reductio Ad Absurdum (if assuming $\neg\varphi$ leads to a contradiction, conclude φ) or the rule of Reiteration (which allows pulling previously derived formulas into a subproof). Motivations for dropping Reductio Ad Absurdum include the usual constructive ones, while motivations for dropping Reiteration come from applications to natural language [44], as well as quantum logic [13]. Adding Reductio Ad Absurdum to fundamental logic yields orthologic, while adding Reiteration yields intuitionistic logic in the $\{\wedge, \vee, \neg\}$-fragment. Adding both Reductio and Reiteration gives us back classical logic.

In light of arguments that reasoning with epistemic modals motivates moving from classical logic to orthologic [44] and arguments that reasoning with vague predicates motivates moving from classical to intuitionistic modal logic [8], it is natural to inquire into extending fundamental logic with modalities. Doing so calls for taking both $\Box$ and $\Diamond$ as primitive, as in intuitionistic modal logic. To accomplish this, we can use *two* accessibility relations, say R for $\Box$ and Q for $\Diamond$. However, we shall see that in the setting of fundamental logic (in which $\neg$ is dually self-adjoint, i.e., $a \leq \neg b$ implies $b \leq \neg a$), just one natural assumption about the interaction of possibility, necessity, and negation, namely that $\Diamond\neg a \leq \neg\Box a$, enables us to use a *single* accessibility relation for both $\Box$ and $\Diamond$. Thus, we will give a simple semantics for fundamental modal logic using bi-relational structures $(X, \lhd, R)$ in which $(X, \lhd)$ determines a lattice of propositions with negation, and R determines both $\Box$ and $\Diamond$ on the lattice.

In § 2, we review the background of this project: the system of fundamental logic (§ 2.1), its algebraic semantics (§ 2.2), and its relational semantics (§ 2.3). In § 3, we add modalities to the picture and present two representation theorems for lattices with weak negations and independent $\Box$ and $\Diamond$ operations. At this stage, no interaction axioms between $\neg$, $\Box$, and $\Diamond$ are assumed. We study such interactions in § 4, which leads in § 5 to the appealing simplification mentioned above: in the setting of fundamental logic, assuming $\Diamond\neg a \leq \neg\Box a$ allows us to unify the two accessibility relations for $\Box$ and $\Diamond$. Then from a representation theorem in § 5, we obtain the completeness of fundamental modal logic with respect to our bi-relational semantics in § 6. We conclude in § 7.

2 Background

2.1 Fundamental logic

As noted in § 1, the primary definition of fundamental logic in [42] is in terms of a Fitch-style proof system with introduction and elimination rules for $\wedge, \vee, \neg$. For the sake of space, here we will use a secondary but equivalent definition of fundamental logic from [42] as a certain *binary logic* in the sense of [34].

Let $\mathcal{L}$ be the language of propositional logic generated from a countably infinite set Prop of propositional variables by $\wedge$, $\vee$, and $\neg$.

Definition 2.1 An *intro-elim logic* is a binary relation $\vdash \subseteq \mathcal{L} \times \mathcal{L}$ such that for all $\varphi, \psi, \chi \in \mathcal{L}$:

1. $\varphi \vdash \varphi$	8. if $\varphi \vdash \psi$ and $\psi \vdash \chi$, then $\varphi \vdash \chi$
2. $\varphi \wedge \psi \vdash \varphi$	
3. $\varphi \wedge \psi \vdash \psi$	9. if $\varphi \vdash \psi$ and $\varphi \vdash \chi$, then $\varphi \vdash \psi \wedge \chi$
4. $\varphi \vdash \varphi \vee \psi$	
5. $\varphi \vdash \psi \vee \varphi$	10. if $\varphi \vdash \chi$ and $\psi \vdash \chi$, then $\varphi \vee \psi \vdash \chi$
6. $\varphi \vdash \neg\neg\varphi$	
7. $\varphi \wedge \neg\varphi \vdash \psi$	11. if $\varphi \vdash \psi$, then $\neg\psi \vdash \neg\varphi$.

We call the smallest intro-elim logic *fundamental logic*, denoted $\vdash_{\mathsf{F}}$.

Orthologic [34], denoted $\vdash_{\mathsf{O}}$, is obtained from fundamental logic by adding *double negation elimination*: $\neg\neg\varphi \vdash \varphi$. *Intuitionistic logic* in the $\{\wedge, \vee, \neg\}$-fragment [53] is obtained from fundamental logic by strengthening Definition 2.1.6/11 to the *psuedocomplementation* rule that if $\varphi \wedge \psi \vdash \varphi \wedge \neg\varphi$, then $\varphi \vdash \neg\psi$, and strengthening proof-by-cases in Definition 2.1.10 to proof-by-cases *with side assumptions*: if $\alpha \wedge \varphi \vdash \chi$ and $\alpha \wedge \psi \vdash \chi$, then $\alpha \wedge (\varphi \vee \psi) \vdash \psi$. *Classical logic*, denoted $\vdash_{\mathsf{C}}$, is obtained by strengthening fundamental logic with double negation elimination and either of the intuitionistic rules just mentioned [44, Prop. 3.7]. Of course, there are also weaker logics (in their common signature) than fundamental logic (see [4] and Remark 1.2 of [42]).

Aguilera and Bydžovský [1] show that fundamental logic can also be presented in terms of a Gentzen-style sequent calculus where sequents can have at most one formula on the right, as for intuitionistic logic [30], and at most two formulas altogether, as for orthologic [49]. By analyzing this sequent calculus, they show that unlike classical and intuitionistic logic, but like orthologic, fundamental logic is decidable in polynomial time.

Theorem 2.2 **([1])** *It is decidable in polynomial time whether* $\varphi \vdash_{\mathsf{F}} \psi$.

Recall the negative translation of classical into intuitionistic logic [32,31]:

$$g(p) = \neg\neg p \qquad g(\varphi \wedge \psi) = (g(\varphi) \wedge g(\psi))$$
$$g(\neg\varphi) = \neg g(\varphi) \quad g(\varphi \vee \psi) = g(\neg(\neg\varphi \wedge \neg\psi)).$$

As shown in [42], this translation is also a full and faithful embedding of orthologic into fundamental logic.

Proposition 2.3 **([42])** *For all* $\varphi, \psi \in \mathcal{L}$, *we have* $\varphi \vdash_{\mathsf{O}} \psi$ *iff* $g(\varphi) \vdash_{\mathsf{F}} g(\psi)$.

We can also carry out classical reasoning inside fundamental logic, but given that the problem of checking $\varphi \vdash_{\mathsf{C}} \psi$ is co-NP-complete and that of checking $\varphi \vdash_{\mathsf{F}} \psi$ is in P, we cannot hope for a polynomial-time reduction. Yet we can carry out a reduction at the expense of an exponential blowup in formula length. Given a propositional formula φ, let $\mathsf{Prop}(\varphi)$ be the set of variables occurring in φ. Given a set $P = \{p_1, \ldots, p_n\}$ of propositional variables, we define the set of *state descriptions over* P, $sd(P)$, as the set of all conjunctions of the form $\pm_1 p_1 \wedge \cdots \wedge \pm_n p_n$ where $\pm_i$ is $\neg$ or empty. The following result shows that ψ is classically derivable from φ iff ψ is fundamentally derivable from the assumption that "there is some determinate way that reality is (in the relevant respects) together with φ." Hence classical propositional logic can be seen as obtained from a logical core of fundamental propositional logic by strengthening the premises of arguments with certain metaphysical assumptions.

Proposition 2.4 *For any $\varphi, \psi \in \mathcal{L}$, the following are equivalent:*

(i) $\varphi \vdash_{\mathsf{C}} \psi$;

(ii) $\bigvee_{\delta \in sd(\mathsf{Prop}(\varphi) \cup \mathsf{Prop}(\psi))} (\delta \wedge \varphi) \vdash_{\mathsf{F}} \psi$.

We will prove Proposition 2.4 using the relational semantics in § 2.3.

2.2 Algebraic semantics

Algebraic semantics for fundamental logic can be given using bounded lattices—crucially not assumed to be distributive—equipped with what [22,23,3] call a *weak pseudocomplementation.*

Definition 2.5 A unary operation $\neg$ on a bounded lattice is a *weak pseudocomplementation* if it satisfies:

(i) semicomplementation: $a \wedge \neg a = 0$;

(ii) dual self-adjointness: $a \leq \neg b$ implies $b \leq \neg a$.

The following easy folklore lemma relates Definitions 2.5 and 2.1.

Lemma 2.6 *$\neg$ is dually self-adjoint iff it is antitone ($a \leq b$ implies $\neg b \leq \neg a$) and double inflationary ($a \leq \neg\neg a$).*

We assume familiarity with how a class $\mathbb{C}$ of lattices with a unary operation $\neg$ provides algebraic semantics for $\mathcal{L}$ and a consequence relation $\vDash_{\mathbb{C}} \subseteq \mathcal{L} \times \mathcal{L}$. Standard techniques of algebraic logic then yield the following (see [42]).

Proposition 2.7 *Fundamental logic is sound and complete with respect to the class $\mathbb{WPL}$ of bounded lattices equipped with a weak pseudocomplementation: for all $\varphi, \psi \in \mathcal{L}$, $\varphi \vdash_{\mathsf{F}} \psi$ iff $\varphi \vDash_{\mathbb{WPL}} \psi$.*

2.3 Relational semantics

A relational semantics for fundamental logic [42] can be given using Ploščica's [51] approach to the representation of lattices, Birkhoff's [7] approach to the representation of negation, and appropriate additional first-order conditions on the relations. For comparisons with related works [3,13,15,16,17,18,19,20,21,22,23,47,55,40,57], as well as examples, see [42, § 4]. Here we only quickly summarize the key facts concerning this semantics.

Definition 2.8 A *relational frame* is a pair $(X, \lhd)$ of a nonempty set X and binary relation $\lhd$ on X.

If $x \lhd y$, we say that x is *open to* y. As explained in Remark 4.2 of [42], this reading of $\lhd$ is associated with a four-way distinction between acceptance, non-acceptance, rejection, and acceptance of the negation of a proposition A, where propositions are fixpoints of the operation $c_{\lhd}$ in Proposition 2.9 below:

- x *accepts* A if $x \in A$;
- x does not accept A if $x \notin A$;
- x *rejects* A if for all $y \rhd x$, $y \notin A$;
- x accepts *the negation of* A if for all $y \lhd x$, $y \notin A$.

Non-acceptance should not entail rejection, since a state may be completely noncommittal about A; and rejection should not entail acceptance of the negation, since we would like to accommodate, e.g., intuitionists who reject instances of excluded middle but of course do not accept their negations. Given these distinctions, we can provide more intuition to the notion of "openness" intended for $\lhd$: x is open to y iff x *does not reject any proposition that* y *accepts.*

Proposition 2.9 *Let* $(X, \lhd)$ *be a relational frame.*

(i) *The following operation* $c_\lhd : \wp(X) \to \wp(X)$ *is a closure operator:*

$$c_\lhd(A) = \{x \in X \mid \forall y \lhd x \, \exists z \rhd y : z \in A\}.$$

(ii) *The fixpoints of* $c_\lhd$*, i.e., those* $A \subseteq X$ *with* $c_\lhd(A) = A$*, form a complete lattice* $\mathfrak{L}(X, \lhd)$ *with meet as intersection and join as* closure of *union:*

$$\bigvee_{i \in I} A_i = \{x \in X \mid \forall y \lhd x \, \exists z \rhd y : z \in \bigcup_{i \in I} A_i\}.$$

(iii) *The operation* $\neg_\lhd : \wp(X) \to \wp(X)$ *defined by*

$$\neg_\lhd(A) = \{x \in X \mid \forall y \lhd x, y \notin A\}$$

sends $c_\lhd$*-fixpoints to* $c_\lhd$*-fixpoints, is antitone with respect to* $\subseteq$*, and sends the* 1 *of* $\mathfrak{L}(X, \lhd)$*, namely* X*, to the* 0 *of* $\mathfrak{L}(X, \lhd)$*, namely* $c_\lhd(\varnothing)$*.*

For fundamental logic, we want $\neg_\lhd$ to have additional properties. For the following result, say that an $x \in X$ is *non-absurd* if there is some $y \lhd x$. Given $x, z \in X$, say that z *pre-refines* x if for all $w \lhd z$, we have $w \lhd x$. It follows that for all $c_\lhd$-fixpoints A, if $x \in A$, then $z \in A$ [42, Lemma 4.12].

Proposition 2.10 ([42], Proposition 4.14.1-2) *For any relational frame* $(X, \lhd)$*, in each of the following pairs,* (a) *and* (b) *are equivalent:*

(i) (a) *for all* $c_\lhd$*-fixpoints* A*, we have* $A \cap \neg_\lhd A = 0$*;*
   (b) pseudo-reflexivity*: for all non-absurd* $x \in X$*, there is a* $z \lhd x$ *that pre-refines* x*.*

(ii) (a) *for all* $c_\lhd$*-fixpoints* A*, we have* $A \subseteq \neg_\lhd \neg_\lhd A$*;*
   (b) pseudo-symmetry*: for all* $x \in X$ *and* $y \lhd x$*, there is a* $z \lhd y$ *that pre-refines* x*.*

The facts above yield the soundness of fundamental logic with respect to relational frames that are pseudo-reflexive and pseudo-symmetric, interpreting $\mathcal{L}$ in the algebras $(\mathfrak{L}(X, \lhd), \neg_\lhd)$. That is, a *relational model* adds to a relational frame $(X, \lhd)$ a valuation V interpreting propositional variables as fixpoints of

$c_{\lhd}$. The forcing relation $\Vdash$ between states $x \in X$ and formulas is defined in the obvious way in light of Proposition 2.9(ii)-(iii) [42, Definition 4.19].

For completeness, we use the following representation theorem.

Theorem 2.11 ([42], Theorems 4.24 and 4.30) *Any bounded (resp. complete) lattice L equipped with an antitone operation $\neg$ sending 1 to 0 embeds into (resp. is isomorphic to) $(\mathfrak{L}(X, \lhd), \neg_{\lhd})$ for some relational frame $(X, \lhd)$.*

Moreover, if $\neg$ satisfies $a \leq \neg\neg a$ (resp. $a \wedge \neg a = 0$) for all $a \in L$, then $\lhd$ may be taken to be pseudo-symmetric (resp. pseudo-reflexive—in fact, reflexive).

Theorem 2.12 ([42]) *Fundamental logic is sound and complete with respect to the class of relational frames that are pseudo-reflexive and pseudo-symmetric.*

As an example application of Theorem 2.12, let us prove Proposition 2.4.

Proof. From 2 to 1, $\varphi \vdash_{\mathsf{C}} \bigvee_{\delta \in sd(\mathsf{Prop}(\varphi) \cup \mathsf{Prop}(\psi))} (\delta \wedge \varphi) \vdash_{\mathsf{F}} \psi$ and hence $\varphi \vdash_{\mathsf{C}} \psi$.

Now suppose 2 does not hold. Then there is $\delta \in sd(\mathsf{Prop}(\varphi) \cup \mathsf{Prop}(\psi))$ such that $\delta \wedge \varphi \nvdash_{\mathsf{F}} \psi$, for otherwise 2 holds using Definition 2.1.10. Then by Theorem 2.12, there is a pseudo-reflexive and pseudo-symmetric model $\mathcal{M} = (X, \lhd, V)$ and $x \in X$ such that $\mathcal{M}, x \Vdash \delta \wedge \varphi$ but $\mathcal{M}, x \nVdash \psi$. Define a valuation $\pi : \mathsf{Prop} \to \{0, 1\}$ by $\pi(p) = 1$ if $\mathcal{M}, x \Vdash p$ and 0 otherwise. Let $\tilde{\pi} : \mathcal{L} \to \{0, 1\}$ be the usual recursively defined extension of π as in classical semantics.

We prove by induction that for any propositional formula χ with $\mathsf{Prop}(\chi) \subseteq \mathsf{Prop}(\varphi) \cup \mathsf{Prop}(\psi)$, we have:

(a) $\tilde{\pi}(\varphi) = 1$ iff $\mathcal{M}, x \Vdash \chi$; (b) $\tilde{\pi}(\varphi) = 0$ iff $\mathcal{M}, x \Vdash \neg\chi$.

Suppose χ is a propositional variable p. Then since $\chi \in \mathsf{Prop}(\varphi) \cup \mathsf{Prop}(\psi)$, from $\mathcal{M}, x \Vdash \delta$ it follows that either $\mathcal{M}, x \Vdash p$, in which case $\tilde{\pi}(p) = 1$ by definition of π, or $\mathcal{M}, x \Vdash \neg p$, which implies $\mathcal{M}, x \nVdash p$ by the pseudo-reflexivity of $\lhd$, so $\tilde{\pi}(p) = 0$ by definition of π. This establishes (a) and (b) for p.

Suppose χ is $\neg\alpha$. If $\tilde{\pi}(\alpha) = 0$, then by the inductive hypothesis, $\mathcal{M}, x \Vdash \neg\alpha$, in line with $\tilde{\pi}(\neg\alpha) = 1$. On the other hand, if $\tilde{\pi}(\alpha) = 1$, then by the inductive hypothesis, $\mathcal{M}, x \Vdash \alpha$, which by the pseudo-symmetry of $\lhd$ implies $\mathcal{M}, x \Vdash \neg\neg\alpha$, in line with $\tilde{\pi}(\neg\alpha) = 0$. Thus, (a) and (b) hold for $\neg\alpha$.

Suppose χ is $\alpha \wedge \beta$. Simply consider the four possible truth assignments to α, β by $\tilde{\pi}$ and use the fact that $\mathcal{M}, x \Vdash \neg\gamma_i$ implies $\mathcal{M}, x \Vdash \neg(\gamma_1 \wedge \gamma_2)$.

Finally, suppose χ is $\alpha \vee \beta$. Again consider the four possible truth assignments to α, β by $\tilde{\pi}$ and use the fact that $\mathcal{M}, x \Vdash \neg\alpha \wedge \neg\beta$ implies $\mathcal{M}, x \Vdash \neg(\alpha \vee \beta)$. To see this, suppose $\mathcal{M}, x \Vdash \neg\alpha \wedge \neg\beta$ and $y \lhd x$. For contradiction, suppose $\mathcal{M}, y \Vdash \alpha \vee \beta$. Given $y \lhd x$ and the pseudo-symmetry of $\lhd$, there is a $z \lhd y$ that pre-refines x, so $\mathcal{M}, z \Vdash \neg\alpha \wedge \neg\beta$. Since $\mathcal{M}, y \Vdash \alpha \vee \beta$ and $z \lhd y$, there is a $w \rhd z$ with $\mathcal{M}, w \Vdash \alpha$ or $\mathcal{M}, w \Vdash \beta$. Given $w \rhd z$ and the pseudo-symmetry of $\lhd$, there is a $u \lhd z$ that pre-refines w, so $\mathcal{M}, u \Vdash \alpha$ or $\mathcal{M}, u \Vdash \beta$. But this contradicts $u \lhd z$ together with $\mathcal{M}, z \Vdash \neg\alpha \wedge \neg\beta$. Thus, we conclude $\mathcal{M}, y \nVdash \alpha \vee \beta$, which shows that $\mathcal{M}, x \Vdash \neg(\alpha \vee \beta)$.

Now given (a), $\mathcal{M}, x \Vdash \varphi$ implies $\tilde{\pi}(\varphi) = 1$, and $\mathcal{M}, x \nVdash \psi$ implies $\tilde{\pi}(\psi) \neq 1$. Then by the soundness of classical logic with respect to its standard valuation semantics, $\varphi \nVdash_{\mathsf{C}} \psi$, so we are done. □

3 Modalities

3.1 Algebras and frames

To add modalities to our story, let us recall the following standard definitions.

Definition 3.1 A unary operation f on a lattice L is *monotone* if $a \leq b$ implies $f(a) \leq f(b)$. We say that f is *multiplicative* (resp. *completely multiplicative*) if for any finite (resp. arbitrary) subset S of elements of L (such that $\bigwedge S$ exists),

$$f(\bigwedge S) = \bigwedge\{f(a) \mid a \in S\}.$$

Dually, f is *additive* (resp. *completely additive*) if for any finite (resp. arbitrary) subset S of elements of L (such that $\bigvee S$ exists),

$$f(\bigvee S) = \bigvee\{f(a) \mid a \in S\}.$$

As suggested in § 1, the first idea for extending the relational semantics of § 2.3 to handle necessity and possibility modals is to add two accessibility relations to $(X, \lhd)$; see [10] for a similar approach but without negation in the signature. Other related approaches to representing lattices with modalities can be found in, e.g., [5,11,14,27,33,35,36,38,39,50].

Definition 3.2 A *modal frame* is a triple $(X, \lhd, R, Q)$ such that $\lhd$, R, and Q are binary relations on X, and for all $x, y, z \in X$,

$$\text{if } xRy \rhd z, \text{ then } \exists x' \lhd x\ \forall x'' \rhd x'\ \exists y'' \colon x''Ry'' \rhd z.$$

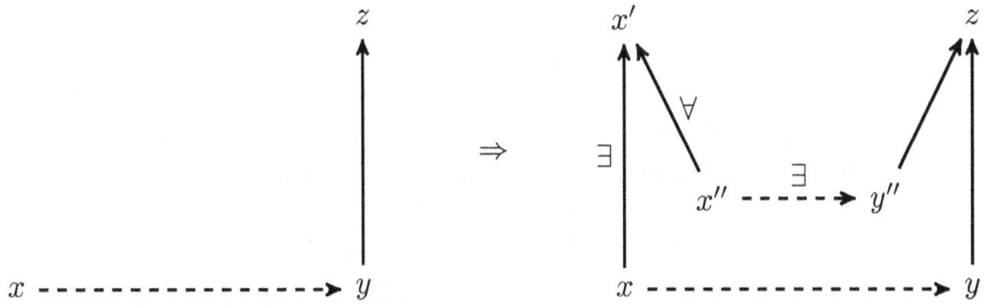

Fig. 1. Illustration of the modal frame condition in Definition 3.2. A solid line from w to v indicates $w \rhd v$, and a dashed line from w to v indicates wRv.

From R we define a necessity modality $\Box_R$ as usual. However, from Q we will define our possibility modality $\Diamond_Q$ using a more intricate quantificational pattern, as shown in Fig. 2 below. For any $S \subseteq X^2$, let $S(x) = \{y \in X \mid xSy\}$.

Proposition 3.3 *Given a modal frame $\mathcal{F} = (X, \lhd, R, Q)$, define operations $\Box_R$ and $\Diamond_Q$ on the lattice of $c_\lhd$-fixpoints of $\mathcal{F}$ as follows:*

$$\Box_R A = \{x \in X \mid R(x) \subseteq A\};$$
$$\Diamond_Q A = \{x \in X \mid \forall x' \lhd x\, \exists y' \in Q(x')\, \exists y \rhd y' : y \in A\}.$$

Then $\Box_R$ and $\Diamond_Q$ send $c_\lhd$-fixpoints to $c_\lhd$-fixpoints, $\Box_R$ is completely multiplicative, and $\Diamond_Q$ is monotone.

Proof. First, we show that $\Box_R A$ is a $c_\lhd$-fixpoint for any $c_\lhd$-fixpoint A. Equivalently, we show that if $x \in X \setminus \Box_R A$, then $\exists x' \lhd x\, \forall x'' \rhd x'\; x'' \notin \Box_R A$. Suppose $x \notin \Box_R A$, so there is some y such that $xRy \notin A$. Then since A is a $c_\lhd$-fixpoint, there is a $z \lhd y$ such that ($\star$) for all $z' \rhd z$, we have $z' \notin A$. Since $xRy \rhd z$, by the modal frame condition we have $\exists x' \lhd x\, \forall x'' \rhd x'\, \exists y'' : x''Ry'' \rhd z$. Now $z \lhd y''$ implies $y'' \notin A$ by ($\star$), which with $x''Ry''$ implies $x'' \notin \Box_R A$.

Next, we show that $\Diamond_Q A$ is a $c_\lhd$-fixpoint for any $c_\lhd$-fixpoint A. Suppose $x \notin \Diamond_Q A$, so $\exists x' \lhd x\; \forall y' \in Q(x')\; \forall y \rhd y' : y \notin A$. Then clearly there is no $x'' \rhd x'$ with $x'' \in \Diamond_Q A$. Hence $\exists x' \lhd x\; \forall x'' \rhd x'\; x'' \notin \Diamond_Q A$, as desired.

That $\Box_R$ is completely additive and $\Diamond_Q$ monotone is obvious from the definitions. $\Box$

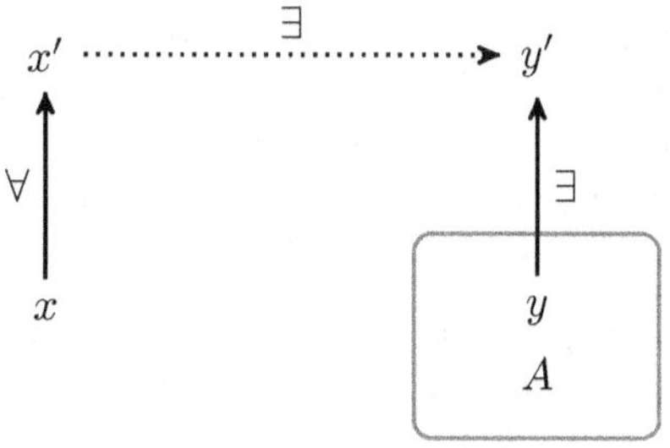

Fig. 2. Illustration of the condition for $x \in \Diamond_Q A$ from Proposition 3.3. The dotted line from x' to y' indicates $x'Qy'$.

The definition of $\Diamond_Q$ can be understood intuitively as follows, using notions from § 2.3. First, let $\Box_Q$ be the necessity modality defined from Q in the usual way, so $\Box_Q A = \{x \in X \mid Q(x) \subseteq A\}$. (So when we take $Q = R$ in § 5, this is just $\Box_R$.) Then $x \in \Diamond_Q A$ in effect means that according to x,

it's not the case that A is necessarily (relative to Q) rejected.

For given our interpretation of 'not', the displayed condition means there is some $x' \lhd x$ that does not accept that A *is necessarily (relative to Q) rejected*, which in turn means there is some y' that is Q-accessible from x' and does not reject A, which in turns means there is a some $y \rhd y'$ with $y \in A$.

Remark 3.4 Without further conditions, $\Diamond_Q$ is not guaranteed to be additive. But this is a feature, rather than a bug, of the above approach to possibility, since there are contexts in which additivity is not desired for $\Diamond$. For example, Wijesekera [56] intentionally designed his system of intuitionistic modal logic so that $\Diamond$ does not distributive over $\vee$, since this is not wanted for some applications of intuitionistic modal logic in computer science. For another example, we recall Kenny's [46] argument that the *ability* modality does not distribute over $\vee$: you may be able to ensure that your dart hits the top half of the dart board or your dart hits the bottom half of the board; it does not follow that you are able to ensure that your dart hits the top half or that you are able to ensure that your dart hits the bottom half, since that may be beyond your skill.

When we want $\Diamond_Q$ to be completely additive, as we now do, we simply impose a condition analogous to that of Definition 3.2 for Q but with $\lhd$ flipped.

Definition 3.5 A modal frame $(X, \lhd, R, Q)$ is *additive* if for all $x, y, z \in X$,

$$\text{if } xQy \lhd z, \text{ then } \exists x' \rhd x\ \forall x'' \lhd x'\ \exists y''\colon x''Qy'' \lhd z.$$

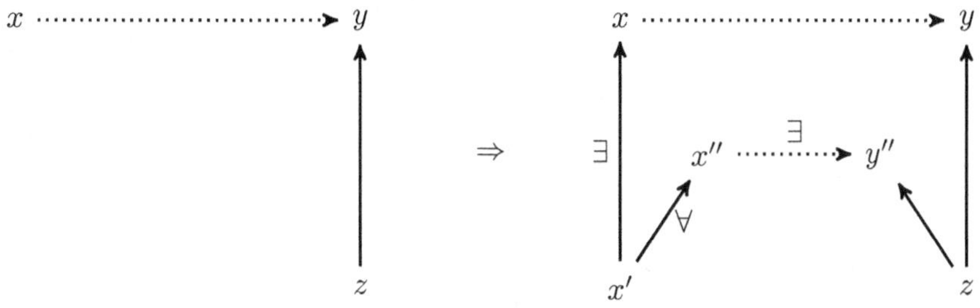

Fig. 3. Illustration of the additivity condition in Definition 3.5.

Proposition 3.6 *If $(X, \lhd, R, Q)$ is an additive modal frame, then the operation $\Diamond_Q$ is completely additive.*

Proof. Suppose $w \in \Diamond \bigvee\{A_i \mid i \in I\}$. Toward showing $w \in \bigvee\{\Diamond A_i \mid i \in I\}$, consider some $x \lhd w$. Then since $w \in \Diamond \bigvee\{A_i \mid i \in I\}$, there are y, u such that $xQy \lhd u \in \bigvee\{A_i \mid i \in I\}$. Since $u \in \bigvee\{A_i \mid i \in I\}$ and $y \lhd u$, it follows that there is some $z \rhd y$ such that $z \in A_i$ for some $i \in I$. Then picking x' as in Definition 3.5, we have $x' \in \Diamond A_i$. Thus, for every $x \lhd w$, there is an $x' \rhd x$ and $i \in I$ such that $x' \in \Diamond A_i$, which shows that $w \in \bigvee\{\Diamond A_i \mid i \in I\}$. $\Box$

In summary, for $\Box_R$ we impose an interaction condition on R and $\lhd$ to ensure that $\Box_R$ send $c_\lhd$-fixpoints to $c_\lhd$-fixpoints, but no interaction condition is required for $\Box_R$ to be completely multiplicative. By contrast, for $\Diamond_Q$ no interaction condition on Q and $\lhd$ is required for $\Diamond_Q$ to send $c_\lhd$-fixpoints to $c_\lhd$-fixpoints, but we impose an interaction condition when we want $\Diamond_Q$ to be completely additive.

3.2 Representation

The representational power of the relational frames from the previous subsection is shown by the following result.

Theorem 3.7 *Let L be a complete lattice equipped with $\neg$, $\Box$, and $\Diamond$ where*

- *$\neg$ is an antitone unary operation on L with $\neg 1 = 0$,*

- *$\Box$ is a completely multiplicative unary operation on L, and*
- *$\Diamond$ is a completely additive unary operation on L.*

Then define:

- $X = \{(a,b) \mid a,b \in L, \neg a \leq b\}$, *and for* $x = (a,b) \in X$, $x_0 = a$, *and* $x_1 = b$;
- $x \lhd y$ *iff* $x_1 \not\geq y_0$;
- xRy *iff for all* $a \in L$, $x_0 \leq \Box a \Rightarrow y_0 \leq a$;
- xQy *iff for all* $a \in L$, $\Diamond a \leq x_1 \Rightarrow a \leq y_1$.

Then $(X, \lhd, R, Q)$ *is an additive modal frame, and* $(L, \neg, \Box, \Diamond)$ *is isomorphic to* $(\mathfrak{L}(X, \lhd), \neg_\lhd, \Box_R, \Diamond_Q)$.

To prove Theorem 3.7, we make use of the following definition and proposition from [41,42].

Definition 3.8 Let L be a lattice and P a set of pairs of elements of L. Define a binary relation $\lhd$ on P by $(a,b) \lhd (c,d)$ if $c \not\leq b$. Then we say P is *separating* if for all $a, b \in L$:

(i) if $a \not\leq b$, then there is a $(c,d) \in P$ with $c \leq a$ and $c \not\leq b$;

(ii) for all $(c,d) \in P$, if $c \not\leq b$, then there is a $(c',d') \lhd (c,d)$ such that for all $(c'',d'') \rhd (c',d')$, we have $c'' \not\leq b$.

Proposition 3.9 *Let L be a lattice and P a separating set of pairs of elements of L. For $a \in L$, define $f(a) = \{(x,y) \in P \mid x \leq a\}$. Then:*

(i) *f is a complete embedding of L into $\mathfrak{L}(P, \lhd)$;*

(ii) *if L is complete, then f is an isomorphism from L to $\mathfrak{L}(P, \lhd)$.*

In (i), $\mathfrak{L}(P, \lhd)$ is the MacNeille completion of L (see [29, Thm. 2.2]). For a proof of Proposition 3.9, see [42, Proposition 4.23]. We now prove Theorem 3.7.

Proof. We first prove a preliminary lemma that we will use repeatedly. For any $x \in X$, let

$$\rho(x) = (\bigwedge\{b \mid x_0 \leq \Box b\}, \neg \bigwedge\{b \mid x_0 \leq \Box b\})$$
$$\sigma(x) = (1, \bigvee\{b \mid \Diamond b \leq x_1\}),$$

so $\rho(x), \sigma(x) \in X$. Then obviously (a) $xR\rho(x)$ and (b) $xQ\sigma(x)$. In addition:

(c) if $x_0 \not\leq \Box a$, then $\rho(x)_0 \not\leq a$. Contrapositively,

$$\begin{aligned} & \bigwedge\{b \mid x_0 \leq \Box b\} \leq a \\ \Rightarrow\ & \Box \bigwedge\{b \mid x_0 \leq \Box b\} \leq \Box a \quad \text{by monotonicity of } \Box \\ \Rightarrow\ & \bigwedge\{\Box b \mid x_0 \leq \Box b\} \leq \Box a \quad \text{by complete multiplicativity of } \Box \\ \Rightarrow\ & x_0 \leq \Box a. \end{aligned}$$

(d) if $\Diamond a \not\leq x_1$, then $a \not\leq \sigma(x)_1$. Contrapositively,

$$\begin{aligned} & a \leq \bigvee\{b \mid \Diamond b \leq x_1\} \\ \Rightarrow\ & \Diamond a \leq \Diamond \bigvee\{b \mid \Diamond b \leq x_1\} \quad \text{by the monotonicity of } \Diamond \\ \Rightarrow\ & \Diamond a \leq \bigvee\{\Diamond b \mid \Diamond b \leq x_1\} \quad \text{by the complete additivity of } \Diamond \\ \Rightarrow\ & \Diamond a \leq x_1. \end{aligned}$$

Now we show that $(X, \lhd, R, Q)$ is a modal frame as in Definition 3.2:

$$\text{if } xRy \rhd z, \text{ then } \exists x' \lhd x\ \forall x'' \rhd x'\ \exists y''\colon x''Ry'' \rhd z.$$

Suppose $xRy \rhd z$. Then $x_0 \not\leq \Box z_1$, for otherwise xRy implies $y_0 \leq z_1$, contradicting $y \rhd z$. Now let $x' = (1, \Box z_1)$, so $x' \lhd x$. Consider any $x'' \rhd x'$, so $x''_0 \not\leq x'_1 = \Box z_1$. Let $y'' = \rho(x'')$, so $x''Ry''$ by (a). Then $x''_0 \not\leq \Box z_1$ implies $y''_0 \not\leq z_1$ by (c), so $y'' \rhd z$, which establishes the modal frame condition.

Next we show that $(X, \lhd, R, Q)$ is additive as in Definition 3.5:

$$\text{if } xQy \lhd z, \text{ then } \exists x' \rhd x\ \forall x'' \lhd x'\ \exists y''\colon x''Qy'' \lhd z.$$

Suppose $xQy \lhd z$. Since $y \lhd z$, we have $z_0 \not\leq y_1$, which with xQy implies $\Diamond z_0 \not\leq x_1$. Then where $x' = (\Diamond z_0, \neg\Diamond z_0)$, we have $x' \rhd x$. Now consider any $x'' \lhd x'$, so $\Diamond z_0 \not\leq x''_1$. Let $y'' = \sigma(x'')$, so $x''Qy''$ by (b). Then $\Diamond z_0 \not\leq x''_1$ implies $z_0 \not\leq y''_1$ by (d), so $y'' \lhd z$, which shows that the frame is additive.

Now we prove that $(L, \neg, \Box, \Diamond)$ is isomorphic to $(\mathfrak{L}(X, \lhd), \neg_{\lhd}, \Box_R, \Diamond_Q)$. First, we claim that X is separating as in Definition 3.8. For part (i) of Definition 3.8, take $(c, d) = (a, \neg a)$. For (ii), suppose $(c, d) \in X$ and $c \not\leq b$. Let $(c', d') = (1, b)$. Since $b \neq 1$ and $\neg 1 = 0 \leq b$, $(1, b) \in X$, and since $c \not\leq b$, $(c', d') \lhd (c, d)$. Now consider any $(c'', d'') \in X$ with $(c', d') \lhd (c'', d'')$. Then $c'' \not\leq d' = b$, so (ii) holds. Thus, by Proposition 3.9, the f defined there is an isomorphism from L to $\mathfrak{L}(P, \lhd)$. Next, we show that f preserves $\neg$, $\Box$, and $\Diamond$.

To show $f(\neg a) = \neg_{\lhd} f(a)$, first suppose $(x, y) \in f(\neg a)$, so $x \leq \neg a$, and $(x', y') \lhd (x, y)$. If $x' \leq a$, then $\neg a \leq \neg x'$, which with $x \leq \neg a$ implies $x \leq \neg x'$, which with $\neg x' \leq y'$ implies $x \leq y'$, contradicting $(x', y') \lhd (x, y)$. Thus, we have $x' \not\leq a$, so $(x', y') \notin f(a)$. Hence $(x, y) \in \neg_{\lhd} f(a)$. Conversely, let $(x, y) \in X \setminus f(\neg a)$, so $x \not\leq \neg a$. Then $(a, \neg a) \lhd (x, y)$, so $(x, y) \notin \neg_{\lhd} f(a)$.

To show $f(\Box b) = \Box_R f(b)$, first suppose $x \in f(\Box a)$, so $x_0 \leq \Box a$. Then xRy implies $y_0 \leq a$ and hence $y_0 \in f(a)$. Thus, $x \in \Box_R f(a)$. Conversely, suppose $x \notin f(\Box a)$, so $x_0 \not\leq \Box a$. Let $y = \rho(x)$, so xRy by (a). Then $x_0 \not\leq \Box a$ implies $y_0 \not\leq a$ by (c), so $y \notin f(a)$ and hence $x \notin \Box_R f(a)$.

To show $f(\Diamond a) = \Diamond f(a)$, first suppose $x \in f(\Diamond a)$, so $x_0 \leq \Diamond a$. Further suppose $x' \lhd x$, so $x_0 \not\leq x'_1$ and hence $\Diamond a \not\leq x'_1$. Let $y' = \sigma(x')$, so $x'Qy'$ by (b). The $\Diamond a \not\leq x'_1$ implies $a \not\leq y'_1$ by (d), so $y' \lhd (a, \neg a)$. Since $(a, \neg a) \in f(a)$, this shows that $x \in \Diamond f(a)$.

Conversely, suppose $x \notin f(\Diamond a)$, so $x_0 \not\leq \Diamond a$. Let $x' = (1, \Diamond a)$, so $x' \lhd x$. Now consider any y' such that $x'Qy'$, which with $\Diamond a \leq x'_1$ implies $a \leq y'_1$. Then for any $y \rhd y'$, we have $y_0 \not\leq y'_1$ and hence $y_0 \not\leq a$, so $y \notin f(a)$. Thus, $\exists x' \lhd x\ \forall y' \in Q(x')\ \forall y \rhd y'\ y \notin f(a)$, which shows $x \notin \Diamond f(a)$. □

Dropping the completeness of L, we can prove the following result (note the proof can be carried out in ZF without the Axiom of Choice, in the spirit of [6]), which embeds L into its canonical extension (see [28]). This is closely related to the topological representations of bounded lattices in [51] and [12], building on [54] and [2], and the topological representation of Boolean algebras in [6]. The treatment of negation was added in [41].

Theorem 3.10 *Let L be a bounded lattice with $\neg$, $\Box$, and $\Diamond$ where*

- *$\neg$ is an antitone unary operation on L with $\neg 1 = 0$,*
- *$\Box$ is a multiplicative unary operation on L, and*
- *$\Diamond$ is an additive unary operation on L.*

Then define:

- *$X = \{(F, I) \mid F$ is a filter in L, I is an ideal in L, and $\{\neg a \mid a \in F\} \subseteq I\}$;*
- *$(F, I) \lhd (F', I')$ iff $I \cap F' = \varnothing$;*
- *$(F, I)R(F', I')$ iff for all $a \in L$, $\Box a \in F \Rightarrow a \in F'$;*
- *$(F, I)Q(F', I')$ iff for all $a \in L$, $\Diamond a \in I \Rightarrow a \in I'$.*

Then $(X, \lhd, R, Q)$ is an additive modal frame; $(L, \neg, \Box, \Diamond)$ embeds into $(\mathfrak{L}(X, \lhd), \neg_{\lhd}, \Box_R, \Diamond_Q)$; and $(L, \neg, \Box, \Diamond)$ is isomorphic to the subalgebra of $(\mathfrak{L}(X, \lhd), \neg_{\lhd}, \Box_R, \Diamond_Q)$ consisting of $c_{\lhd}$-fixpoints that are compact open in the topology on X generated by $\{\widehat{a} \mid a \in L\}$, where $\widehat{a} = \{(F, I) \mid a \in F\}$.

Proof. For the following, given an element a of a lattice, let $\uparrow a$ (resp. $\downarrow a$) be the principal filter (resp. ideal) generated by a.

We first show that $(X, \lhd, R, Q)$ is a modal frame as in Definition 3.2:

$$\text{if } xRy \rhd z, \text{ then } \exists x' \lhd x \; \forall x'' \rhd x' \; \exists y'' \colon x''Ry'' \rhd z.$$

The proof slightly adapts that of Proposition 4.10 of [41] to account for the role of $\neg$ in the definition of X. Suppose $(F, I)R(G, H) \rhd (J, K)$, which implies $K \cap G = \varnothing$ and hence $K \cap \{a \mid \Box a \in F\} = \varnothing$. Then where $F' = \uparrow 1$ and I' is the ideal generated by $\{\Box a \mid a \in K\}$, we claim that $I' \cap F = \varnothing$, so $(F', I') \lhd (F, I)$. For if $b \in I' \cap F$, then for some $a_1, \dots, a_n \in K$, we have $b \leq \Box a_1 \vee \dots \vee \Box a_n$, which implies $b \leq \Box(a_1 \vee \dots \vee a_n)$, so $\Box(a_1 \vee \dots \vee a_n) \in F$, whence $a_1 \vee \dots \vee a_n \notin K$, contradicting $a_1, \dots, a_n \in K$. Now suppose $(F', I') \lhd (F'', I'')$, so $I' \cap F'' = \varnothing$. Let $G'' = \{b \mid \Box b \in F''\}$, which is a filter, and let H'' be the ideal generated by $\{\neg a \mid a \in G''\}$. We claim $K \cap G'' = \varnothing$, so $(J, K) \lhd (G'', H'')$. For if $a \in G''$, then $\Box a \in F''$, so $\Box a \notin I'$, whence $a \notin K$. Thus, $(F'', I'')R(G'', H'') \rhd (J, K)$, which establishes the condition.

Now we show that $(X, \lhd, R, Q)$ is additive as in Definition 3.5:

$$\text{if } xQy \lhd z, \text{ then } \exists x' \rhd x \; \forall x'' \lhd x' \; \exists y'' \colon x''Qy'' \lhd z.$$

Suppose $(F, I)Q(G, H) \lhd (J, K)$, which implies $H \cap J = \varnothing$ and hence $\{a \mid \Diamond a \in I\} \cap J = \varnothing$. Let F' be the filter generated by $\{\Diamond a \mid a \in J\}$ and I' the ideal generated by $\{\neg a \mid a \in F'\}$, so $(F', I') \in X$. We claim that

$I \cap F' = \varnothing$, so $(F', I') \rhd (F, I)$. If $b \in F'$, then there are $a_1, \dots, a_n \in J$ such that $\Diamond a_1 \wedge \dots \wedge \Diamond a_n \leq b$, which implies $\Diamond(a_1 \wedge \dots \wedge a_n) \leq b$. If in addition $b \in I$, then $\Diamond(a_1 \wedge \dots \wedge a_n) \in I$, so $a_1 \wedge \dots \wedge a_n \notin J$, contradicting $a_1, \dots, a_n \in J$. Now consider any $(F'', I'') \lhd (F', I')$, so $I'' \cap F' = \varnothing$. Let $G'' = {\uparrow}1$ and $H'' = \{a \mid \Diamond a \in I''\}$, which is an ideal, so $(F'', I'')Q(G'', H'')$. We claim that $H'' \cap J = \varnothing$, so $(G'', H'') \lhd (J, K)$. For if $b \in H''$, then $\Diamond b \in I''$, which implies $\Diamond b \notin F'$ and hence $b \notin J$. This completes the proof of the condition.

The claimed embedding sends a to $\widehat{a}$. We verify that it preserves $\Box$ and $\Diamond$. For its other claimed properties, see the proof of Theorem 4.30 in [42].[1]

Let us show $\widehat{\Box a} = \Box_R \widehat{a}$. Suppose $(F, I) \in \widehat{\Box a}$, so $\Box a \in F$. Then if $(F, I)R(F', I')$, we have $a \in F'$ and hence $(F', I') \in \widehat{a}$. Thus, $(F, I) \in \Box_R \widehat{a}$. Conversely, suppose $(F, I) \notin \widehat{\Box a}$, so $\Box a \notin F$. Let $F' = \{b \mid \Box b \in F\}$, which is a filter, and let I' be the ideal generated by $\{\neg b \mid b \in F'\}$, so $(F', I') \in X$. Then $\Box a \notin F$ implies $a \notin F'$ and hence $(F', I') \notin \widehat{a}$, and by construction of F', we have $(F, I)R(F', I')$. Thus, $(F, I) \notin \Box_R \widehat{a}$.

Finally, we show $\widehat{\Diamond a} = \Diamond_Q \widehat{a}$. Suppose $(F, I) \in \widehat{\Diamond a}$, so $\Diamond a \in F$. Consider any $(F', I') \lhd (F, I)$, so $\Diamond a \notin I'$. Let $G' = {\uparrow}1$ and $H' = \{b \in L \mid \Diamond b \in I'\}$, which is an ideal, so $(F', I')Q(G', H')$. Let $G = {\uparrow}a$ and $H = {\downarrow}\neg a$, so $(G, H) \in \widehat{a}$. We claim that $H' \cap G = \varnothing$, so $(G', H') \lhd (G, H)$. For otherwise $a \in H'$, so $\Diamond a \in I'$, contradicting what we derived above. This proves that $(F, I) \in \Diamond_Q \widehat{a}$. Conversely, suppose $(F, I) \notin \widehat{\Diamond a}$, so $\Diamond a \notin F$. Let $F' = {\uparrow}1$ and $I' = {\downarrow}\Diamond a$, so $(F', I') \lhd (F, I)$. Consider any (G', H') and (G, H) such that $(F', I')Q(G', H') \lhd (G, H)$. Then since $\Diamond a \in I'$, we have $a \in H'$ and hence $a \notin G$, so $(G, H) \notin \widehat{a}$. This proves that $(F, I) \notin \Diamond_Q \widehat{a}$. $\Box$

4 Interactions

We now consider the interaction of $\Box$ and $\Diamond$ via $\neg$. Of course, if $\Box$ and $\Diamond$ come from different flavors of modality, e.g., $\Box a$ means that *the agent believes a* and $\Diamond a$ means that *a will hold sometime in the future*, there need be no interaction between them via $\neg$. But even if $\Box$ and $\Diamond$ are of the same flavor of modality, the interactions between them via $\neg$ may be subtle (see Remark 4.3).

One may also consider interactions between $\Box$ and $\Diamond$ via other operations, such as $\wedge$. In classical modal logic, we have $\Box a \wedge \Diamond b \leq \Diamond(a \wedge b)$. However, this is not desirable in epistemic orthologic [44, Example 3.39], so we do not wish to impose this constraint. Of course, the simplest interaction to consider is $\Box a \leq \Diamond a$, but this cannot be imposed for doxastic logic with possibly inconsistent agents. By contrast, some interactions between $\Box$ and $\Diamond$ via $\neg$ seem generally acceptable—when $\Box$ and $\Diamond$ come from the same flavor of modality—and will allow us to simplify our semantics by setting $R = Q$ in § 5 (cf. [52] on when a single relation suffices for distributive modal logics). Thus, here we focus only on interactions via $\neg$ and leave the study of further interactions for future work.

[1] That proof assumes $F \cap I = \varnothing$ for each filter-ideal pair, in which case $\lhd$ is reflexive, but the proof easily adapts to drop that assumption. Cf. the proof of Theorem B.7 of [42].

4.1 Lattice inequalities

Consider the following axioms, implicitly universally quantified:

$$\Diamond\neg a \leq \neg\Box a \tag{$\Diamond\neg$}$$
$$\Box\neg a \leq \neg\Diamond a \tag{$\Box\neg$}$$
$$\neg\Diamond a \leq \Box\neg a \tag{$\neg\Diamond$}$$
$$\neg\Box a \leq \Diamond\neg a. \tag{$\neg\Box$}$$

First we observe that over the most general algebras considered in § 3.1, the above axioms are all independent.

Proposition 4.1 *Each of* $(\Diamond\neg)$, $(\Box\neg)$, $(\neg\Diamond)$, *and* $(\neg\Box)$ *is independent of all the others over finite lattices equipped with an antitone* $\neg$ *sending* 1 *to* 0, *multiplicative* $\Box$, *and additive* $\Diamond$.

Proof. The independence of $(\neg\Diamond)$ and $(\neg\Box)$ will be shown in Propositions 4.4.(ii) and 4.5, respectively. For $(\Diamond\neg)$, consider the following four-element lattice equipped with the following $\neg$, $\Box$, and $\Diamond$:

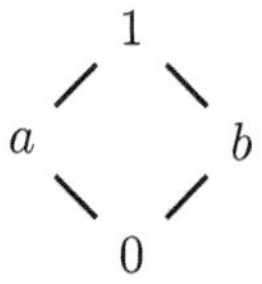

x	$\neg x$	$\Box x$	$\Diamond x$
1	0	1	1
a	0	a	1
b	a	b	b
0	1	0	0

Then $\neg$ is antitone and sends 1 to 0, $\Box$ is multiplicative, $\Diamond$ is additive, and $(\Box\neg)$, $(\neg\Diamond)$, and $(\neg\Box)$ hold. However, $\Diamond\neg b = \Diamond a = 1 \not\leq a = \neg b = \neg\Box b$, so $(\Diamond\neg)$ does not hold. For $(\Box\neg)$, consider the same lattice as above with the same $\neg$ but with the following $\Box$ and $\Diamond$:

x	$\neg x$	$\Box x$	$\Diamond x$
1	0	1	1
a	0	1	1
b	a	0	1
0	1	0	0

Then $\neg$ is antitone and sends 1 to 0, $\Box$ is multiplicative, $\Diamond$ is additive, and $(\Diamond\neg)$, $(\neg\Diamond)$, and $(\neg\Box)$ hold. However, $\Box\neg b = \Box a = 1 \not\leq 0 = \neg 1 = \neg\Diamond b$, so $(\Box\neg)$ does not hold. □

4.1.1 Interactions in fundamental logic

In the context of fundamental logic, we collapse one distinction from § 4.1.

Proposition 4.2 *If* $\neg$ *is dually self-adjoint and* $\Box$ *and* $\Diamond$ *are monotone, then* $(\Diamond\neg)$ *is equivalent to* $(\Box\neg)$.

Proof. Assume $\neg$ is dually self-adjoint and hence antitone and double inflationary by Lemma 2.6. Assuming f is a monotone unary operation and g an arbitrary unary operation, we prove that if for all $a \in L$, $f(\neg a) \leq \neg g(a)$, then for all $a \in L$, $g(\neg a) \leq \neg f(a)$, from which the statement in the lemma follows.

Assume $f(\neg a) \leq \neg g(a)$ for all $a \in L$. By dual self-adjointness, $f(\neg a) \leq \neg g(a)$ implies $g(a) \leq \neg f(\neg a)$, so for all $a \in L$, we have $g(\neg a) \leq \neg f(\neg\neg a)$. Then since $a \leq \neg\neg a$, we have $f(a) \leq f(\neg\neg a)$ by the monotonicity of f, so $\neg f(\neg\neg a) \leq \neg f(a)$ by the antitonicity of $\neg$. Hence $g(\neg a) \leq \neg f(a)$. □

Remark 4.3 Consider the interpretation of the modalities where $\Box a$ means *the agent is certain that* a and $\Diamond a$ means *the agent considers it possible that* a. Then $(\Diamond\neg)$ and $(\Box\neg)$ are plausible. Yet $(\neg\Diamond)$ is questionable: from the assumption that an agent does not consider it possible that a, it does not follow that the agent is certain that $\neg a$; for the agent may be totally *unaware* of a, neither entertaining the possibility of a nor having any attitude with the content $\neg a$. Similarly, $(\neg\Box)$ is questionable: from the fact that the agent is not certain that a, it does not follow that the agent considers it possible that $\neg a$, again because the agent may have unawareness. Thus, neither $(\neg\Diamond)$ nor $(\neg\Box)$ belongs in a base system of fundamental modal logic.

4.1.2 Interactions in intuitionistic logic

In intuitionistic modal logic, where $\neg$ is *pseudocomplementation* ($a \wedge b = 0$ implies $b \leq \neg a$, and $a \wedge \neg a = 0$) and hence dually self-adjoint, it is standard to have not only $(\Diamond\neg)$ and $(\Box\neg)$ but also $(\neg\Diamond)$, despite the concern about $(\neg\Diamond)$ in Remark 4.3. Note that $(\neg\Diamond)$ is an additional condition, even classically.

Proposition 4.4

(i) *$(\neg\Diamond)$ is independent of $(\Diamond\neg)$ and $(\Box\neg)$ over Boolean algebras equipped with a multiplicative $\Box$ and additive $\Diamond$.*

(ii) *$(\neg\Diamond)$ is independent of $(\Diamond\neg)$, $(\Box\neg)$, and $(\neg\Box)$ over Heyting algebras equipped with a multiplicative $\Box$ and additive $\Diamond$.*

Proof. For part (i), consider the four-element Boolean algebra equipped with the following $\Box$ and $\Diamond$:

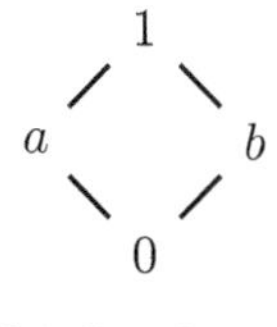

x	$\Box x$	$\Diamond x$
1	1	1
a	0	1
b	0	0
0	0	0

Then $\Box$ is multiplicative, $\Diamond$ is additive, $\Box x \leq \Diamond x$, and $(\Diamond\neg)$ and $(\Box\neg)$ hold. However, we have $\neg\Diamond b = \neg 0 = 1 \not\leq 0 = \Box a = \Box\neg b$, so $(\neg\Diamond)$ does not hold.

For part (ii), consider the following five-element Heyting algebra with $\Box$ and $\Diamond$ operations:

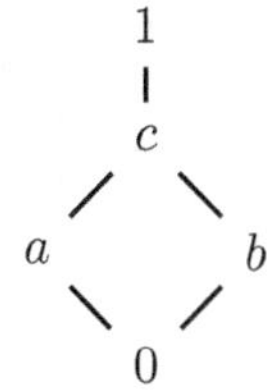

x	$\Box x$	$\Diamond x$
1	1	1
c	1	1
a	c	1
b	0	0
0	0	0

Then $\Box$ is multiplicative, $\Diamond$ is additive, $\Box x \leq \Diamond x$, and not only $(\Diamond\neg)$ and $(\Box\neg)$ but also $(\neg\Box)$ holds:

- $\neg\Box 1 = \neg 1 = 0 \leq \Diamond\neg 1$; $\neg\Box c = \neg 1 = 0 \leq \Diamond\neg c$; $\neg\Box a = \neg c = 0 \leq \Diamond\neg a$;
- $\neg\Box b = \neg 0 = 1 = \Diamond a = \Diamond\neg b$; $\neg\Box 0 = \neg 0 = 1 = \Diamond 1 = \Diamond\neg 0$.

However, we have $\neg\Diamond b = \neg 0 = 1 \not\leq c = \Box a = \Box\neg b$, so $(\neg\Diamond)$ does not hold. $\Box$

As noted in § 1, $(\neg\Box)$ is generally not assumed in intuitionistic modal logic.

Proposition 4.5 *$(\neg\Box)$ is independent of $(\Diamond\neg)$, $(\Box\neg)$, and $(\neg\Diamond)$ over Heyting algebras H equipped with a multiplicative $\Box$ and additive $\Diamond$, even assuming that $\Box a \leq \Diamond a$ for all $a \in H$.*

Proof. Consider the three-element Heyting algebra equipped with the following $\Box$ and $\Diamond$:

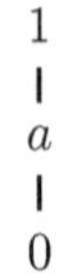

x	$\Box x$	$\Diamond x$
1	1	1
a	0	1
0	0	0

Then $\Box$ is multiplicative, $\Diamond$ is additive, and $(\Diamond\neg)$, $(\Box\neg)$, and $(\neg\Diamond)$ hold. But $\neg\Box a = \neg 0 = 1 \not\leq 0 = \Diamond 0 = \Diamond\neg a$, so $(\neg\Box)$ does not hold. $\Box$

4.1.3 Interactions in orthologic

Finally, in the context of modal orthologic, where $\neg$ is involutive, it is natural to take $\Diamond$ and $\Box$ to be duals in the following sense:

$$\Diamond a = \neg\Box\neg a \qquad (\Diamond \text{ def})$$
$$\Box a = \neg\Diamond\neg a. \qquad (\Box \text{ def})$$

Proposition 4.6 *If $\neg$ is antitone and involutive and $\Box$, $\Diamond$ are monotone, then:*

(i) *$(\Diamond\neg)$ and $(\Box\neg)$ are equivalent;*

(ii) *$(\neg\Diamond)$ and $(\neg\Box)$ are equivalent;*

(iii) *$(\Diamond$ def$)$ and $(\Box$ def$)$ are equivalent to each other and to the conjunction of $(\Diamond\neg)$, $(\Box\neg)$, $(\neg\Diamond)$, and $(\neg\Box)$.*

Proof. For part (i), since $\neg$ is antitone and involutive, it is dually self-adjoint by Lemma 2.6, so $(\Diamond\neg)$ and $(\Box\neg)$ are equivalent by Lemma 4.2. For part (ii), assume $(\neg\Diamond)$. As an instance, we have $\neg\Diamond\neg a \leq \Box\neg\neg a$, which implies $\neg\Box a \leq \Diamond\neg a$ by antitonicity and involution, so $(\neg\Box)$ holds. Now assume $(\neg\Box)$. As an instance, we have $\neg\Box\neg a \leq \Diamond\neg\neg a$, which implies $\neg\Diamond a \leq \Box\neg a$ by antitonicity and involution, so $(\neg\Diamond)$ holds. For part (iii), clearly $(\Diamond$ def$)$ and $(\Box$ def$)$ are equivalent given involution and imply $(\Diamond\neg)$, $(\Box\neg)$, $(\neg\Diamond)$, and $(\neg\Box)$ given involution. Conversely, assume $(\Diamond\neg)$, $(\Box\neg)$, $(\neg\Diamond)$, and $(\neg\Box)$. By $(\neg\Diamond)$, we have $\neg\Diamond a \leq \Box\neg a$, which implies $\neg\Box\neg a \leq \Diamond a$ by antitonicity and involution. By $(\Box\neg)$, we have $\Box\neg a \leq \neg\Diamond a$, which implies $\Diamond a \leq \neg\Box\neg a$ by antitonicity and involution. Hence $\Diamond a = \neg\Box\neg a$, so $(\Diamond$ def$)$ holds. $\Box$

4.2 Frame conditions

Let us now identify frame conditions sufficient for the principle ($\Diamond\neg$), which seems unobjectionable (when the same flavor of modality is involved on both sides), and ($\neg\Diamond$), which is typically assumed in intuitionistic modal logic.

Proposition 4.7 *Let $(X, \lhd, R, Q)$ be a modal frame in which $Q \subseteq R$. Then for every $c_\lhd$-fixpoint A,*

$$\Diamond\neg A \subseteq \neg\Box A.$$

Proof. Suppose $x \in \Diamond\neg A$. Toward a contradiction, suppose $x \notin \neg\Box A$, so there is a $y \lhd x$ such that $y \in \Box A$. Since $y \lhd x$ and $x \in \Diamond\neg A$, there is a $z \in Q(y)$ and $w \rhd z$ with $w \in \neg A$, which implies $z \notin A$, which contradicts the facts that $y \in \Box A$ and yQz, since by our assumption yQz implies yRz. □

Definition 4.8 A modal frame $(X, \lhd, R, Q)$ is *negative* if for all $x, y, z \in X$,

$$\text{if } xRy \rhd z, \text{ then } \exists x' \lhd x\ \forall x'' \lhd x'\ \exists y'':\ x''Qy'' \lhd z.$$

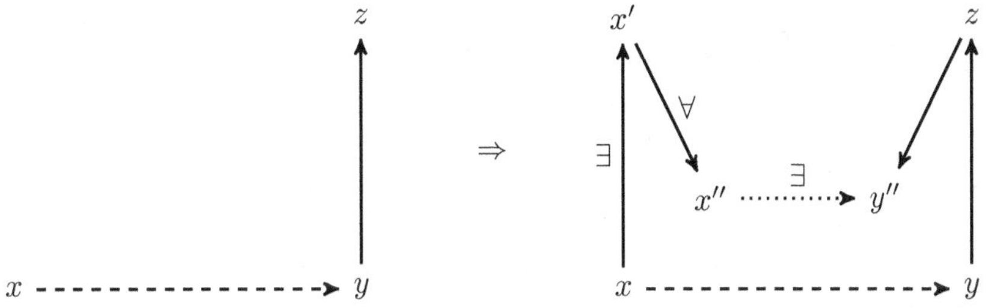

Fig. 4. Illustration of the negativity condition in Definition 4.8.

Proposition 4.9 *If $(X, \lhd, R, Q)$ is a negative modal frame, then for every $c_\lhd$-fixpoint A,*

$$\neg\Diamond A \subseteq \Box\neg A.$$

Proof. Suppose $x \notin \Box\neg A$, so there is a $y \in X$ with xRy and $z \lhd y$ with $z \in A$. Then where x' is as in Definition 4.8, we have $x' \in \Diamond A$, so $x \notin \neg\Diamond A$. □

5 Unification

By assuming ($\Diamond\neg$) and the dual self-adjointness of $\neg$, we can simplify the representation from Theorem 3.7.

Definition 5.1 A modal frame $(X, \lhd, R, Q)$ is *unified* if $R = Q$.

Theorem 5.2 *Let L be a complete lattice with $\neg$, $\Box$, and $\Diamond$ where*

- *$\neg$ is a dually self-adjoint unary operation on L with $\neg 1 = 0$,*
- *$\Box$ is a completely multiplicative unary operation on L,*
- *$\Diamond$ is a completely additive unary operation on L, and*

- *$\Diamond\neg a \leq \neg\Box a$ for all $a \in L$.*

Then define:

- *$X = \{(a, b) \mid a, b \in L, \neg a \leq b\}$;*
- *$x \lhd y$ iff $x_1 \not\geq y_0$;*
- *xRy iff for all $a \in L$, $x_0 \leq \Box a \Rightarrow y_0 \leq a$ and $\Diamond a \leq x_1 \Rightarrow a \leq y_1$;*
- *$Q = R$.*

Then:

(i) *$\mathcal{F} = (X, \lhd, R, Q)$ is a unified, additive modal frame with $\lhd$ pseudo-symmetric;*

(ii) *$(L, \neg, \Box, \Diamond)$ is isomorphic to $(\mathfrak{L}(X, \lhd), \neg_{\lhd}, \Box_R, \Diamond_Q)$;*

(iii) *if $a \wedge \neg a = 0$ for all $a \in L$, then $\lhd$ is pseudo-reflexive;*

(iv) *if $\neg\Diamond a \leq \Box\neg a$ for all $a \in L$, then $\mathcal{F}$ is negative.*

Proof. Given any $x \in X$, let

$$\tau(x) = (\bigwedge\{b \mid x_0 \leq \Box b\}, \bigvee\{b \mid \Diamond b \leq x_1\} \vee \neg \bigwedge\{b \mid x_0 \leq \Box b\}).$$

Then observe the following:

(a) We have $xR\tau(x)$. For $x_0 \leq \Box c$ implies $\tau(x)_0 \leq c$; and if $c \not\leq \tau(x)_1$, then $c \not\leq \bigvee\{b \mid \Diamond b \leq x_1\}$ and hence $\Diamond c \not\leq x_1$.

(b) If $x_0 \not\leq \Box a$, then $\tau(x)_0 \not\leq a$, by the same reasoning as in the proof of Theorem 3.7.

(c) If $\Diamond a \not\leq x_1$, then $a \not\leq \tau(x)_1$. Contrapositively,

$$\begin{aligned}
& a \leq \bigvee\{b \mid \Diamond b \leq x_1\} \vee \neg \bigwedge\{b \mid x_0 \leq \Box b\} \\
\Rightarrow\ & \Diamond a \leq \Diamond\Big(\bigvee\{b \mid \Diamond b \leq x_1\} \vee \neg \bigwedge\{b \mid x_0 \leq \Box b\}\Big) \\
& \quad \text{by monotonicity of } \Diamond \\
\Rightarrow\ & \Diamond a \leq \bigvee\{\Diamond b \mid \Diamond b \leq x_1\} \vee \Diamond\neg \bigwedge\{b \mid x_0 \leq \Box b\} \\
& \quad \text{by complete additivity of } \Diamond \\
\Rightarrow\ & \Diamond a \leq x_1 \vee \Diamond\neg \bigwedge\{b \mid x_0 \leq \Box b\} \\
\Rightarrow\ & \Diamond a \leq x_1 \vee \neg\Box \bigwedge\{b \mid x_0 \leq \Box b\} \quad \text{since } \Diamond\neg d \leq \neg\Box d \\
\Rightarrow\ & \Diamond a \leq x_1 \vee \neg \bigwedge\{\Box b \mid x_0 \leq \Box b\} \quad \text{by complete multiplicativity of } \Box \\
\Rightarrow\ & \Diamond a \leq x_1 \vee \neg x_0 \quad \text{by antitonicity of } \neg \\
\Rightarrow\ & \Diamond a \leq x_1 \quad \text{since } \neg v_0 \leq v_1 \text{ for all } v \in X.
\end{aligned}$$

Now for part (i), that $\mathcal{F}$ is unified is immediate from the definition. The proof that $\mathcal{F}$ is an additive modal frame is almost exactly as in the proof of Theorem 3.7, only using τ instead of ρ and σ:

First we show that $\mathcal{F}$ is a modal frame: if $xRy \rhd z$, then $\exists x' \lhd x\ \forall x'' \rhd x'\ \exists y''$: $x''Ry'' \rhd z$. Suppose $xRy \rhd z$. Then $x_0 \not\leq \Box z_1$, for otherwise xRy implies $y_0 \leq z_1$, contradicting $y \rhd z$. Now let $x' = (1, \Box z_1)$, so $x' \lhd x$. Consider any $x'' \rhd x'$, so $x''_0 \not\leq x'_1 = \Box z_1$. Let $y'' = \tau(x'')$, so $x''Ry''$ by (a) above. Then $x''_0 \not\leq \Box z_1$ implies $y''_0 \not\leq z_1$ by (b) above, so $y'' \rhd z$.

Next we show that $\mathcal{F}$ is additive: if $xQy \lhd z$, then $\exists x' \rhd x\ \forall x'' \lhd x'\ \exists y''$: $x''Qy'' \lhd z$. Suppose $xQy \lhd z$. Since $y \lhd z$, we have $z_0 \not\leq y_1$, which with xQy implies $\Diamond z_0 \not\leq x_1$. Then where $x' = (\Diamond z_0, \neg\Diamond z_0)$, we have $x' \rhd x$. Now consider any $x'' \lhd x'$, so $\Diamond z_0 \not\leq x''_1$. Let $y'' = \tau(x'')$, so $x''Ry''$ by (a) above. Then since $\Diamond z_0 \not\leq x''_1$, we have $z_0 \not\leq y''_1$ by (c) above. Hence $y'' \lhd z$.

That $\lhd$ is pseudo-symmetric follows from the isomorphism in part (ii) and Proposition 2.10(ii).

The proof of part (ii) is almost exactly as in the proof of Theorem 3.7, only using τ in place of ρ and σ:

Exactly as in the proof of Theorem 3.7, the function f defined by $f(a) = \{x \in P \mid x_0 \leq a\}$ is an isomorphism from L to $\mathfrak{L}(X, \lhd)$ that also preserves $\neg$. It only remains to check $\Box$ and $\Diamond$.

To show that $f(\Box a) = \Box_R f(a)$, first suppose $x \in f(\Box a)$, so $x_0 \leq \Box a$. Then xRy implies $y_0 \leq a$ and hence $y_0 \in f(a)$. Thus, $x \in \Box_R f(a)$. Conversely, suppose $x \notin f(\Box a)$, so $x_0 \not\leq \Box a$. Let $y = \tau(x)$, so xRy by (a) above. Then $x_0 \not\leq \Box a$ implies $y_0 \not\leq a$ by (b) above, so $y \notin f(a)$ and hence $x \notin \Box_R f(a)$.

To show that $f(\Diamond a) = \Diamond_Q f(a)$, first suppose $x \in f(\Diamond a)$, so $x_0 \leq \Diamond a$. Further suppose $x' \lhd x$, so $x_0 \not\leq x'_1$ and hence $\Diamond a \not\leq x'_1$. Let $y' = \tau(x')$, so $x'Qy'$ by (a) above. Then since $\Diamond a \not\leq x'_1$, we have $a \not\leq y'_1$ by (c) above, so $y' \lhd (a, \neg a)$. Thus, for all $x' \lhd x$, there are y', z' such that $x'Qy' \lhd z' \in f(a)$. Hence $x \in \Diamond_Q f(a)$. Conversely, suppose $x \notin f(\Diamond a)$, so $x_0 \not\leq \Diamond a$. Let $x' = (1, \Diamond a)$, so $x' \lhd x$. Now consider any y' such that $x'Qy'$, which with $\Diamond a \leq x'_1$ implies $a \leq y'_1$. Then for any $y \rhd y'$, we have $y_0 \not\leq y'_1$ and hence $y_0 \not\leq a$, so $y \notin f(a)$. Thus, $\exists x' \lhd x\ \forall y' \in Q(x')\ \forall y \rhd y'$: $y \notin f(a)$, which shows $x \notin \Diamond f(a)$.

For part (iii), assuming $a \wedge \neg a = 0$ for all $a \in L$, that $\lhd$ is pseudo-reflexive follows from the isomorphism in part (ii) and Proposition 2.10(i).

For part (iv), assuming $\neg\Diamond a \leq \Box\neg a$ for all $a \in L$, we must show negativity: if $xRy \rhd z$, then $\exists x' \lhd x\ \forall x'' \lhd x'\ \exists y''$: $x''Qy'' \lhd z$. Suppose $xRy \rhd z$, which implies $x_0 \not\leq \Box z_1$. Since $\neg z_0 \leq z_1$, we have $\Box\neg z_0 \leq \Box z_1$ by the monotonicity of $\Box$, so $x_0 \not\leq \Box z_1$ implies $x_0 \not\leq \Box\neg z_0$. Hence by our initial assumption, $x_0 \not\leq \neg\Diamond z_0$. Then where $x' = (\Diamond z_0, \neg\Diamond z_0)$, we have $x' \lhd x$. Then since $x' \in f(\Diamond z_0)$ and f preserves $\Diamond$, we have $x' \in \Diamond_Q f(z_0)$, which implies that $\forall x'' \lhd x'\ \exists y'' \in Q(x'')\ \exists y' \rhd y''$: $y' \in f(z_0)$, so $y'_0 \leq z_0$. From $y' \rhd y''$, we have $y'_0 \not\leq y''_1$, which with $y'_0 \leq z_0$ implies $z_0 \not\leq y''_1$, so $y'' \lhd z$. Thus, we have shown that $\forall x'' \lhd x'\ \exists y''$: $x''Qy'' \lhd z$, as desired. $\Box$

Similarly, we have the following unified analogue of Theorem 3.10.

Theorem 5.3 *Let L be a bounded lattice with $\neg$, $\Box$, and $\Diamond$ where*

- *$\neg$ is a dually self-adjoint unary operation on L with $\neg 1 = 0$,*
- *$\Box$ is a multiplicative unary operation on L,*

- *$\Diamond$ is an additive unary operation on L, and*
- *$\Diamond\neg a \leq \neg\Box a$ for all $a \in L$.*

Then define:

- *$X = \{(F, I) \mid F$ is a filter in L, I is an ideal in L, and $\{\neg a \mid a \in F\} \subseteq I\}$;*
- *$(F, I) \lhd (F', I')$ iff $I \cap F' = \varnothing$;*
- *$(F, I)R(F', I')$ iff for all $a \in L$, $\Box a \in F \Rightarrow a \in F'$ and $\Diamond a \in I \Rightarrow a \in I'$;*
- *$Q = R$.*

Then:

(i) *$\mathcal{F} = (X, \lhd, R, Q)$ is a unified, additive modal frame with $\lhd$ pseudo-symmetric;*

(ii) *there is an embedding of $(L, \neg, \Box, \Diamond)$ into $(\mathfrak{L}(X, \lhd), \neg_{\lhd}, \Box_R, \Diamond_Q)$ and an isomorphism between $(L, \neg, \Box, \Diamond)$ and the subalgebra of $(\mathfrak{L}(X, \lhd), \neg_{\lhd}, \Box_R, \Diamond_Q)$ consisting of $c_{\lhd}$-fixpoints that are compact open in the topology on X generated by $\{\widehat{a} \mid a \in L\}$, where $\widehat{a} = \{(F, I) \mid a \in F\}$;*

(iii) *if $a \wedge \neg a = 0$ for all $a \in L$, then $\mathcal{F}$ is pseudo-reflexive;*

(iv) *if $\neg\Diamond a \leq \Box\neg a$ for all $a \in L$, then $\mathcal{F}$ is negative.*

Proof. For part (i), that $\mathcal{F}$ is unified is immediate from the definition. For the pseudo-symmetry of $\lhd$, the proof is the same as in the proof of Proposition 4.32 in [42] (only without claiming $F \cap I'' = \varnothing$, which we do not need here).

For the other properties, we explain how to modify the proof of Theorem 3.10 in light of the modified definitions of R and Q in Theorem 5.3. To show that $\mathcal{F}$ is a modal frame, modify the proof of the modal frame condition for Theorem 3.10 as follows: let $H'' = \{a \mid \Diamond a \in I''\}$, which is an ideal. Then if $a \in G''$, we have $\Box b \in F''$, so $\neg\Box b \in I''$ and hence $\Diamond\neg b \in I''$ by the fourth bullet point of Theorem 5.3, so $\neg b \in H''$. Thus, $(G'', H'') \in X$.

To show $\mathcal{F}$ is additive, modify the proof of additivity for Theorem 3.10 as follows: let $G'' = \{a \mid \Box a \in F''\}$, which is a filter. Where $H'' = \{a \mid \Diamond a \in I''\}$, which is an ideal, we have $(G'', H'') \in X$ by the same reasoning as in the previous paragraph. Then we have $(F'', I'')Q(G'', H'')$ by construction.

For the proof that $\widehat{\Box a} = \Box_R \widehat{a}$, modify the proof in Theorem 3.10 by setting $I' = \{b \mid \Diamond b \in I\}$. For the proof that $\widehat{\Diamond a} = \Diamond_Q \widehat{a}$, modify the proof in Theorem 3.10 by setting $G' = \{b \mid \Box b \in F'\}$.

For part (iii), assuming $a \wedge \neg a = 0$ for all $a \in L$, consider a non-absurd (F, I), so there is some $(G, H) \lhd (F, I)$. Hence $H \cap F = \varnothing$, so $0 \notin F$. Let $I' = \{a \mid \neg a \in F\}$. Then $F \cap I' = \varnothing$, for otherwise we have $a, \neg a \in F$ and hence $0 \in F$, contradicting what we previously derived. Thus, $(F, I') \lhd (F, I)$, and (F, I') pre-refines (F, I). This shows that $\lhd$ is pseudo-reflexive.

For part (iv), assuming $\neg\Diamond a \leq \Box\neg a$, we show that $\mathcal{F}$ is negative:

$$\text{if } xRy \rhd z, \text{ then } \exists x' \lhd x\ \forall x'' \lhd x'\ \exists y''\colon x''Qy'' \lhd z.$$

Suppose $(F, I)R(G, H) \rhd (J, K)$. Let F' be the filter generated by $\{\Diamond a \mid a \in J\}$

and I' the ideal generated by $\{\neg\Diamond a \mid a \in J\}$, which is equal to $\{\Box\neg a \mid a \in J\}$ given $\neg\Diamond a \leq \Box\neg a$ and the converse from Proposition 4.2. We claim that $I' \cap F = \varnothing$. For if $b \in I'$, then $b \leq \Box\neg a_1 \vee \cdots \vee \Box\neg a_n \leq \Box\neg(a_1 \wedge \cdots \wedge a_n)$ for some $a_1, \ldots, a_n \in J$; and then if $b \in F$, we have $\Box\neg(a_1 \wedge \cdots \wedge a_n) \in F$, so $\neg(a_1 \wedge \cdots \wedge a_n) \in G$, which implies $\neg(a_1 \wedge \cdots \wedge a_n) \notin K$ and hence $a_1 \wedge \cdots \wedge a_n \notin J$, contradicting $a_1, \ldots, a_n \in J$. Thus, $I' \cap F = \varnothing$ and hence $(F', I') \lhd (F, I)$. Now consider any $(F'', I'') \lhd (F', I')$, so $I'' \cap F' = \varnothing$. Let $G'' = \{a \mid \Box a \in F''\}$ and $H'' = \{a \mid \Diamond a \in I''\}$, so $(G'', H'') \in X$ as in the second paragraph of the proof above, and $(F'', I'')Q(G'', H'')$. We claim that $H'' \cap J = \varnothing$. For if $a \in J$, then $\Diamond a \in F'$, which implies $\Diamond a \notin I''$, which in turn implies $a \notin H''$. Thus, $(G'', H'') \lhd (J, K)$, which completes the proof. $\Box$

6 Fundamental modal logic

Let $\mathcal{ML}$ be the propositional modal language with $\wedge, \vee, \neg, \Box, \Diamond$ and now also $\bot, \top$. At last, we define our proposed system of fundamental modal logic.

Definition 6.1 *Fundamental modal logic* is the smallest binary relation $\vdash \subseteq \mathcal{ML} \times \mathcal{ML}$ such that for all $\varphi, \psi, \chi \in \mathcal{ML}$, not only conditions 1-11 of Definition 2.1 but also the following hold:

12. $\bot \vdash \varphi \vdash \top$
13. $\neg\top \vdash \bot$
14. $\Box\varphi \wedge \Box\psi \vdash \Box(\varphi \wedge \psi)$
15. $\Diamond(\varphi \vee \psi) \vdash \Diamond\varphi \vee \Diamond\psi$
16. $\Diamond\neg\varphi \vdash \neg\Box\varphi$
17. $\top \vdash \Box\top$
18. $\Diamond\bot \vdash \bot$
19. if $\varphi \vdash \psi$, then $\Box\varphi \vdash \Box\psi$
20. if $\varphi \vdash \psi$, then $\Diamond\varphi \vdash \Diamond\psi$.

The forcing clauses for $\Box\varphi$ and $\Diamond\varphi$ in relational models are as in Proposition 3.3; $\bot$ is forced only at absurd states (recall § 2.3), while $\top$ is forced at all states.

Theorem 6.2 *Fundamental modal logic is sound and complete with respect to the class of unified, additive modal frames $(X, \lhd, R)$ in which $\lhd$ is pseudo-reflexive and pseudo-symmetric.*

Proof. Soundness is by Propositions 2.9, 3.3, 3.6, and 4.7. For completeness, apply Theorem 5.3 to the Lindenbaum-Tarski algebra of the logic. $\Box$

7 Conclusion

We have proposed a way of adding modalities to fundamental logic, both axiomatically and semantically. Our representation theorems raise obvious questions about associated categorical dualities (see [48] for morphisms), and the interactions between $\Box$ and $\Diamond$ via $\neg$ cry out for systematic correspondence theory. Also conspicuously absent has been "the" conditional $\to$. Weak conditionals possibly appropriate for fundamental logic are discussed in [42, § 6] and [43]. Treating a language with both modalities and conditionals is a natural next step, especially in connection with applications to natural language as in [44]. Finally, our focus here has been entirely semantical. Yet we hope that in light of recent proof-theoretic successes with fundamental logic [1], proof theorists might also find fundamental modal logic to be a worthy object of study.

Acknowledgements

I thank Yifeng Ding, Daniel Gonzalez, Guillaume Massas, and Yanjing Wang for helpful discussion and the three anonymous referees for helpful comments.

References

[1] Aguilera, J. P. and J. Bydžovský, *Fundamental logic is decidable*, ACM Transactions on Computational Logic (Forthcoming), https://doi.org/10.1145/3665328.

[2] Allwein, G. and C. Hartonas, *Duality for bounded lattices* (1993), Indiana University Logic Group, Preprint Series, IULG-93-25 (1993).

[3] Almeida, A., *Canonical extensions and relational representations of lattices with negation*, Studia Logica **91** (2009), pp. 171–199.

[4] Battilotti, G. and G. Sambin, *Basic logic and the cube of its extensions*, in: A. Cantini, E. Casari and P. Minari, editors, *Logic and Foundations of Mathematics*, Synthese Library **280**, Kluwer Academic Publishers, 1999 pp. 165–186.

[5] Bezhanishvili, N., A. Dmitrieva, J. de Groot and T. Moraschini, *Positive modal logic beyond distributivity*, Annals of Pure and Applied Logic **175** (2024), p. 103374.

[6] Bezhanishvili, N. and W. H. Holliday, *Choice-free Stone duality*, The Journal of Symbolic Logic **85** (2020), pp. 109–148.

[7] Birkhoff, G., "Lattice Theory," American Mathematical Society, New York, 1940.

[8] Bobzien, S. and I. Rumfitt, *Intuitionism and the modal logic of vagueness*, Journal of Philosophical Logic **49** (2020), pp. 221–248.

[9] Božić, M. and K. Došen, *Models for normal intuitionistic modal logics*, Studia Logica **43** (1984), pp. 217–245.

[10] Conradie, W., A. Craig, A. Palmigiano and N. M. Wijnberg, *Modelling informational entropy*, in: R. Iemhoff, M. Moortgat and R. Queiroz, editors, *Logic, Language, Information, and Computation. WoLLIC 2019*, Lectures Notes in Computer Science **11541**, 2019, pp. 140–160.

[11] Conradie, W., S. Frittella, A. Palmigiano, M. Piazzai, A. Tzimoulis and N. M. Wijnberg, *Categories: How I learned to stop worrying and love two sorts*, in: J. Väänänen, A. Hirvonen and R. de Queiroz, editors, *Logic, Language, Information, and Computation. WoLLIC 2016*, Lectures Notes in Computer Science **9803**, 2016, pp. 145–164.

[12] Craig, A. P. K., M. Haviar and H. A. Priestley, *A fresh perspective on canonical extensions for bounded lattices*, Applied Categorical Structures **21** (2013), pp. 725–749.

[13] Dalla Chiara, M. L. and R. Giuntini, *Quantum logics*, in: D. Gabbay and F. Guenthner, editors, *Handbook of Philosophical Logic*, Springer, 2002 pp. 129–228.

[14] Dmitrieva, A., "Positive modal logic beyond distributivity: duality, preservation and completeness," Master's thesis, University of Amsterdam (2021).

[15] Došen, K., *Negative modal operators in intuitionistic logic*, Publications de l'Institut Mathématique. Nouvelle Série **35** (1984), pp. 3–14.

[16] Došen, K., *Negation as a modal operator*, Reports on Mathematical Logic **20** (1986), pp. 15–27.

[17] Došen, K., *Negation in the light of modal logic*, in: D. M. Gabbay and H. Wansing, editors, *What is Negation?*, Kluwer, Dordrecht, 1999 pp. 77–86.

[18] Dunn, J. M., *Star and perp: Two treatments of negation*, Philosophical Perspectives **7** (1993), pp. 331–357.

[19] Dunn, J. M., *Generalized ortho negation*, in: H. Wansing, editor, *Negation. A Notion in Focus*, de Gruyter, Berlin, 1996 pp. 3–26.

[20] Dunn, J. M., *A comparative study of various model-theoretic treatments of negation: a history of formal negation*, in: D. M. Gabbay and H. Wansing, editors, *What is Negation?*, Kluwer, Dordrecht, 1999 pp. 23–51.

[21] Dunn, J. M. and C. Zhou, *Negation in the context of gaggle theory*, Studia Logica **80** (2005), pp. 235–264.

[22] Dzik, W., E. Orlowska and C. van Alten, *Relational representation theorems for general lattices with negations*, in: *Relations and Kleene Algebra in Computer Science. RelMiCS 2006*, Lecture Notes in Computer Science **4136** (2006), pp. 162–176.

[23] Dzik, W., E. Orlowska and C. van Alten, *Relational representation theorems for lattices with negations: A survey*, Lecture Notes in Artificial Intelligence **4342** (2006), pp. 245–266.

[24] Fischer Servi, G., *On modal logic with an intuitionistic base*, Studia Logica **36** (1977), pp. 141–149.

[25] Fitch, F. B., "Symbolic Logic: An Introduction," The Ronald Press Company, New York, 1952.

[26] Fitch, F. B., *Natural deduction rules for obligation*, American Philosophical Quarterly **3** (1966), pp. 27–38.

[27] Gehrke, M., *Generalized Kripke frames*, Studia Logica **84** (2006), pp. 241–275.

[28] Gehrke, M. and J. Harding, *Bounded lattice expansions*, Journal of Algebra **238** (2001), pp. 345–371.

[29] Gehrke, M., J. Harding and Y. Venema, *MacNeille completions and canonical extensions*, Transactions of the American Mathematical Society **358** (2005), pp. 573–590.

[30] Gentzen, G., *Untersuchungen über das logische Schließen*, Mathematische Zeitschrift **39** (1935), pp. 176–210, 405–431.

[31] Gentzen, G., *Die Widerspruchsfreiheit der reinen Zahlentheorie*, Mathematische Annalen **112** (1936), pp. 493–565.

[32] Gödel, K., *Zur intuitionistischen Arithmetik und Zahlentheorie*, Ergebnisse eines Mathematischen Kolloquiums **4** (1933), pp. 34–38.

[33] Goldblatt, R., *Morphisms and duality for polarities and lattices with operators* (2019), arXiv:1902.09783 [math.LO].

[34] Goldblatt, R. I., *Semantic analysis of orthologic*, Journal of Philosophical Logic **3** (1974), pp. 19–35.

[35] Hartonas, C., *Discrete duality for lattices with modal operators*, Journal of Logic and Computation **29** (2018), pp. 71–89.

[36] Hartonas, C. and E. Orłowska, *Representation of lattices with modal operators in two-sorted frames*, Fundamenta Informaticae **166** (2019), pp. 29–56.

[37] Heyting, A., *Die formalen Regeln der intuitionistischen Logik I*, Sitzungsberichte der Preussischen Akademie der Wissenschaften **49** (1930), pp. 42–65.

[38] Holliday, W. H., *Possibility frames and forcing for modal logic* (2015), forthcoming in *The Australasian Journal of Logic*, UC Berkeley Working Paper in Logic and the Methodology of Science, https://escholarship.org/uc/item/0tm6b30q.

[39] Holliday, W. H., *Possibility semantics*, in: M. Fitting, editor, *Selected Topics from Contemporary Logics*, Landscapes in Logic, College Publications, 2021 pp. 363–476, arXiv:2405.06852 [math.LO].

[40] Holliday, W. H., *Three roads to complete lattices: Orders, compatibility, polarity*, Algebra Universalis **82** (2021), article number 26.

[41] Holliday, W. H., *Compatibility and accessibility: lattice representations for semantics of non-classical and modal logics*, in: D. F. Duque and A. Palmigiano, editors, *Advances in Modal Logic, Vol. 14*, College Publications, London, 2022 pp. 507–529, arXiv:2201.07098 [math.LO].

[42] Holliday, W. H., *A fundamental non-classical logic*, Logics **1** (2023), pp. 36–79.

[43] Holliday, W. H., *Preconditionals*, in: I. Sedlár, editor, *The Logica Yearbook 2023*, College Publications, Forthcoming arXiv:2402.02296 [math.LO].

[44] Holliday, W. H. and M. Mandelkern, *The orthologic of epistemic modals*, Journal of Philosophical Logic (Forthcoming), arXiv:2203.02872 [cs.LO].

[45] Jónsson, B. and A. Tarski, *Boolean Algebras with Operators. Part I.*, American Journal of Mathematics **73** (1951), pp. 891–939.

[46] Kenny, A., *Human abilities and dynamic modalities*, in: J. Manninen and R. Tuomela, editors, *Essays on Explanation and Understanding*, Synthese Library **72**, Springer, Dordrecht, 1976 pp. 209–232.

[47] Massas, G., *B-frame duality*, Annals of Pure and Applied Logic **174** (2023), p. 103245.

[48] Massas, G., *Goldblatt-thomason theorems for fundamental (modal) logic*, in: A. Ciabattoni and D. Gabelaia, editors, *Advances in Modal Logic, Vol. 14*, College Publications, London, Forthcoming ArXiv:2406.10182 [math.LO].

[49] Mönting, J. S., *Cut elimination and word problems for varieties of lattices*, Algebra Universalis **12** (1981), pp. 290–321.

[50] Orłowska, E. and D. Vakarelov, *Lattice-based modal algebras and modal logics*, in: P. Hájek, L. Valdés-Villanueva and D. Westerståhl, editors, *Logic, methodology and philosophy of science. Proceedings of the 12th international congress*, College Publications, London, 2005 pp. 147–170.

[51] Ploščica, M., *A natural representation of bounded lattices*, Tatra Mountains Mathematical Publication **5** (1995), pp. 75–88.

[52] Přenosil, A., *Compatibility between modal operators in distributive modal logic* (2023), arXiv:2311.10017 [math.LO].

[53] Rebagliato, J. and V. Verdú, *On the algebraization of some Gentzen systems*, Fundamenta Informaticae **17** (1993), pp. 319–338.

[54] Urquhart, A., *A topological representation theory for lattices*, Algebra Universalis **8** (1978), pp. 45–58.

[55] Vakarelov, D., *Consistency, completeness and negation*, in: G. Priest, R. Routley and J. Norman, editors, *Paraconsistent Logic: Essays on the Inconsistent*, Philosophia Verlag, Munich, 1989 pp. 328–368.

[56] Wijesekera, D., *Constructive modal logics I*, Annals of Pure and Applied Logic **50** (1990), pp. 271–301.

[57] Zhong, S., *A general relational semantics of propositional logic: Axiomatization*, in: A. Silva, R. Wassermann and R. Queiroz, editors, *Logic, Language, Information, and Computation. WoLLIC 2021*, Lecture Notes in Computer Science **13038** (2021), pp. 82–99.

Logics of Knowability

Ahmee Christensen [1]

University of California, Berkeley

Abstract

Knowability has a significant history in the field of epistemic logic dating back to the introduction of Fitch's paradox in 1963. More recently it has received attention from the perspectives of dynamic epistemic logic and epistemic temporal logic. However, existing treatments of knowability either address a knowability-like concept that deviates from an intuitive understanding of knowability in some important way or take knowability to be a complex notion merely definable in the language. In this paper, we take 'it is knowable that' as a primitive modality and offer axiomatizations of two knowability logics—one of knowledge and knowability and one of pure knowability. We prove that these logics are sound and complete with respect to a distinguished subclass of birelational frames. These proofs are of technical interest, as well; because knowability is, we argue, best understood as a non-normal modality, the completeness proofs require novel constructions.

Keywords: epistemic logic, knowability, non-normal modal logic, completeness

1 Introduction

In the setting of dynamic epistemic logic, knowability has recently received treatment in a number of papers (see, e.g., [1,3,4,10,14,15]). From this perspective, a proposition φ is knowable for an agent if the agent would come to know φ upon receiving some information or after the successful execution of some action, like a true public announcement. One might be interested in an approach that is, on the face of it, simpler, however. Namely, much in the way that one project of epistemic logic has been to work out the logic of the knowledge operator K (see, e.g., [9,11,12]), it seems natural to introduce a knowability operator $\Box$ and propose some candidate logic for it.

Something like this is done in [5] where a logic of knowability, knowledge, and belief is introduced in order to refine the picture of knowledge and belief that Stalnaker presents in [12]. The notion captured in [5] is not quite knowability as one might naïvely conceive of it, however. Bjorndahl and Özgün propose the following gloss for $\Box\varphi$: "one could come to know what φ used to express (before you came to know it)" [5, p. 753]. Their knowability operator in a sense deindexicalizes the proposition on which it acts. Though this notion is

[1] achris@berkeley.edu

interesting in its own right, it seems to exhibit significant logical deviation from knowability simpliciter.

For example, after accepting positive introspection of knowledge, it seems as though one ought to be committed to the inference that if φ is knowable, then φ is knowably known. But $\Box\varphi \to \Box K\varphi$ is not a theorem of logic found in [5]. (This fact is intuitive given their gloss and can be easily shown by finding a countermodel and appealing to their soundness result.) Perhaps even more glaring is that this notion of knowability seems to commit us to the principle $\Box\varphi \wedge \Box\psi \to \Box(\varphi \wedge \psi)$. On the simpler reading of *knowable* this sounds wrong: presumably we commonly find ourselves in situations where we must choose to direct our attention in one direction or the other. Upon attending to one matter we may come to know φ, but in doing so preclude ourselves from knowing ψ, which we would have known had we otherwise directed our attention.

The approach that we take in this paper to capture the simpler notion of knowability is heavily inspired by the considerable literature on Fitch's paradox of knowability, introduced in [8]. The paradox is typically rendered in a bimodal logic where the *knowable* is decomposed as *possibly known*. Without significant logical assumptions imposed, such a treatment of knowability will avoid validating principles like $\Box\varphi \wedge \Box\psi \to \Box(\varphi \wedge \psi)$, but will, with rather minor logical assumptions, validate principles like $\Box\varphi \to \Box K\varphi$. We also take inspiration from Burgess's approach to knowability in [7] where the possibility part of the decomposition has a temporal flavor, though unlike Burgess, we do not assume linearity of time.

The goal of this paper is to propose two logics—one of knowledge and knowability and one of pure knowability—and establish their completeness with respect to a class of models that have an intuitive appeal. In order to achieve this, a third auxiliary logic in a more expressive language will be introduced. Completeness of that logic with respect to the restricted model class will be established and the result transferred over to the two knowability logics by way of establishing full and faithful embeddings of them into the auxiliary one. While the paper is motivated by epistemic considerations, the results and proofs are also of technical interest, as the completeness theorems are proved with novel canonical model constructions; more interesting techniques are required given that we are producing relational models for logics with a non-normal modal operator.

The outline of the paper is as follows. In Section 2 we introduce the three languages that we are going to work with, as well as a few syntactic notions, including the translations. Birelational semantics are then presented in Section 3. In Section 4 we establish the auxiliary logic and prove that it is complete with respect to the smaller model class defined in Section 3. Finally, Sections 5 and 6 propose the logics of knowledge and knowability and of pure knowability, proving the expected soundness and completeness results.

2 Languages

2.1 The Languages of Knowability

Fix a countably infinite set $\mathbb{P} = \{p, q, \ldots\}$ of propositional variables. The language $\mathcal{L}_{K\Box}$ is given by the formal grammar

$$\varphi ::= p \mid \neg\varphi \mid (\varphi \wedge \varphi) \mid K\varphi \mid \Box\varphi$$

where $p \in \mathbb{P}$. We will use the following standard bits of shorthand: $\bot$ for $p \wedge \neg p$, $\top$ for $\neg(p \wedge \neg p)$, $\varphi \vee \psi$ for $\neg(\neg\varphi \wedge \neg\psi)$, $\varphi \to \psi$ for $\neg(\varphi \wedge \neg\psi)$, $\hat{K}\varphi$ for $\neg K \neg\varphi$, and $\Diamond\varphi$ for $\neg\Box\neg\varphi$. Additionally, we write $\mathcal{L}_{\Box}$ for the K-free fragment of $\mathcal{L}_{K\Box}$. The proposed readings of the modalities are *the agent knows that* φ for $K\varphi$ and *it is knowable for the agent that* φ for $\Box\varphi$.

Definition 2.1 We define *complexity* $c : \mathcal{L}_{K\Box} \to \mathbb{N}$ recursively as follows:

- $c(p) = 0$
- $c(\neg\varphi) = c(\varphi) + 1$
- $c(\varphi \wedge \psi) = \max(c(\varphi), c(\psi)) + 1$
- $c(K\varphi) = c(\varphi) + 1$
- $c(\Box\varphi) = c(\varphi) + 2$

Note the +2 in the last clause. This definition allows us to apply the inductive hypothesis to the formula $K\varphi$ when checking the $\Box\varphi$ case. We also distinguish a special class formulas that play a central roll in our axiomatizations.

Definition 2.2 We define the set of *sharp formulas* inductively as follows:

- $\top$ and $\bot$ are sharp;
- For any φ, $\Diamond\varphi$ is sharp;
- If φ and ψ are sharp, then $\varphi \wedge \psi$ and $\varphi \vee \psi$ are sharp.
- If φ is sharp, then $K\varphi$ is sharp;

We have the dual notion of a flat formula, which is not strictly speaking necessary to introduce but will make some definitions and arguments later on a bit cleaner.

Definition 2.3 We define the set of *flat formulas* inductively as follows:

- $\top$ and $\bot$ are flat;
- For any φ, $\Box\varphi$ is flat;
- If φ and ψ are flat, then $\varphi \wedge \psi$ and $\varphi \vee \psi$ are flat.
- If φ is flat, then $\hat{K}\varphi$ is flat;

With just classical reasoning, one can see that every sharp formula is equivalent to the negation of a flat one and vice versa. (This holds even in the restricted language $\mathcal{L}_{\Box}$.) For a set of formulas Γ, we write $\Gamma^{\sharp}$ for $\{\varphi \in \Gamma : \varphi \text{ is sharp}\}$ and $\Gamma^{\flat}$ for $\{\varphi \in \Gamma : \varphi \text{ is flat}\}$.

2.2 The Language of Knowledge and the Future

Our third language of interest $\mathcal{L}_{KG}$ is given by the formal grammar

$$\varphi ::= p \,|\, \neg\varphi \,|\, (\varphi \wedge \varphi) \,|\, K\varphi \,|\, G\varphi$$

where $p \in \mathbb{P}$. We will use the same bits of shorthand as before in addition to writing $F\varphi$ for $\neg G \neg \varphi$. $G\varphi$ is to be read as *from now on,* φ. As a warning, this reading differs slightly from the usual one found in tense logic. We want to hear our reading of G as naturally satisfying the **T** axiom.

For the rest of this paper, we fix the translation $t : \mathcal{L}_{K\Box} \to \mathcal{L}_{KG}$, which is defined as follows:

- $t(p) = p$;
- $t(\neg\varphi) = \neg t(\varphi)$;
- $t(\varphi \wedge \psi) = t(\varphi) \wedge t(\psi)$;
- $t(K\varphi) = Kt(\varphi)$;
- $t(\Box\varphi) = FKt(\varphi)$.

We will write $t_\Box$ for $t|_{\mathcal{L}_\Box}$.

2.3 Logics

First we remark that all of the logics considered this paper will be classical in the sense that this will contain every substitution instance of every theorem of the classical propositional calculus (CPC). Given some logic L in one of the above languages, we will write $\vdash_{\mathsf{L}} \varphi$ to mean that φ is a theorem of L. Further, we write $\Gamma \vdash_{\mathsf{L}} \varphi$ to mean that for some finite set $\{\gamma_i\}_i \subseteq \Gamma$ we have $\vdash_{\mathsf{L}} \bigwedge_i \gamma_i \to \varphi$. In either case, we might suppress the subscript when it is clear from context which logic is being discussed.

Definition 2.4 A set of formulas Γ is an L-*theory* just in case whenever $\Gamma \vdash_{\mathsf{L}} \varphi$, we have $\varphi \in \Gamma$.

Again, the L- prefix might be omitted when there is only one logic in play. As one final piece of notation, we write $\mathrm{cl}(\Gamma)$ for the smallest theory containing Γ. This is sure to exist as the property of being a theory is evidently closed under intersection. We also have the notion of maximal consistency.

Definition 2.5 An L-theory Γ is called L-*maximally consistent* if $\Gamma \nvdash_{\mathsf{L}} \bot$, but for any $\varphi \notin \Gamma$, $\Gamma \cup \{\varphi\} \vdash_{\mathsf{L}} \bot$.

Accordingly, we will be able to make free use of Lindenbaum's lemma.

Lemma 2.6 (Lindenbaum's Lemma) *Every consistent* L*-theory can be extended to maximally consistent* L*-theory.*

Proof. See, e.g., [6]. □

3 Semantics

3.1 Binary Relations

As the languages introduced in the previous section naturally have both an epistemic and a temporal flavor, this paper will be primarily concerned with birelational structures.

Definition 3.1 A *frame* is a triple $(W, R, \leqslant)$ such that W is a set and R and $\leqslant$ are binary relations on W.

We will typically refer to the first relation of a frame (R in the above definition) as the *epistemic relation* and the second relation of a frame ($\leqslant$ in the above definition) as the *temporal relation.* Of course, we will be interested in frames where the relations satisfy some additional properties. Recall the following standard definition.

Definition 3.2 A binary relation $\lhd$ on a set A is a *quasi-order* if it satisfies both of the following conditions.

- (Reflexivity) For all $x \in A$, $x \lhd x$.
- (Transitivity) For all $x, y, z \in A$, if $x \lhd y$ and $y \lhd z$, then $x \lhd z$.

A set equipped with a quasi-order is called a *quasi-ordered set*, which we might shorten to *qoset.* We also have distinguished maps of qosets.

Definition 3.3 Where $Q = (A, \lhd_A)$ and $P = (B, \lhd_B)$ an *order-preserving map* from Q to P is a function $f : A \to B$ such that for all $a, a' \in A$, $a \lhd_A a'$ implies $f(a) \lhd_B f(a')$.

We will also make use of linear orders as index sets for times.

Definition 3.4 A relation $\lhd$ is a *linear order* on a set A if it is quasi-order that moreover satisfies the following:

- (Antisymmtery) For all $x, y \in A$, if $x \lhd y$ and $y \lhd x$, then $x = y$.
- (Totality) For all $x, y \in A$, $x \lhd y$ or $y \lhd x$.

3.2 ET Frames and ET Diagrams

In our pursuit of frames suitable for understanding knowability, we begin with a frame-like notion that hopefully has some intuitive pull.

Definition 3.5 An *ET diagram* is a triple $\mathscr{D} = (\mathcal{I}, Q_{-}, P_{-,-})$ such that the following hold:

- $\mathcal{I} = (I, \preccurlyeq)$ is a linearly ordered set;
- Q_{-} assigns to each $\tau \in I$ a quasi-order Q_τ;
- $P_{-,-}$ assigns to each pair (τ, σ) with $\tau \preccurlyeq \sigma$ an order-preserving map $P_{\tau,\sigma} : Q_\sigma \to Q_\tau$ where $P_{\tau,\tau} = \mathrm{id}_{Q_\tau}$ for all $\tau \in I$ and $P_{\tau,\sigma} \circ P_{\sigma,\rho} = P_{\tau,\rho}$.

In such a diagram, I is a collection of times with $\preccurlyeq$ serving as the natural *at least as late as* order, Q assigns to each time τ a qoset meant to capture the agent's epistemic state at the different possible worlds at τ, and $P_{\tau,\sigma}$ maps each possible world at time σ to the possible world in its past at time τ. The order-preserving condition is a perfect-recall-like condition, enforcing that as the agent's epistemic state evolves in time, it only rules out possibilities, never ruling in a possibility already ruled out. To shed light on the definition of an ET diagram, we will present an example.

Example 3.6 Consider an urn containing two marbles, and assume that Betty knows that each marble is exactly black or white. She then retrieves the marbles from the urn, one at a time. The relevant times for us will be before she has retrieved any marbles (t_0), after she has retrieved one (t_1), and after she has retrieved both (t_2). Accordingly, the ET diagram will contain three qosets[2]. At t_0, Betty knows nothing about the colors of the marbles in the urn, so the the qoset will be one with three states—there are either zero, one, or two white marbles—that are all in the same equivalence class. At t_1, Betty has drawn either a white marble or a black marble, so there will be two equivalence classes with two states each; if she drew a black marble, then she is either in the case where there are no white marbles or the case where there is one (and similarly if she drew white). Finally at t_2, Betty has reached the end of inquiring about the urn, so the equivalence relation will just be identity. Note that we will still have four states, since in the case where there was a marble of each color in the urn, there were two possible histories.

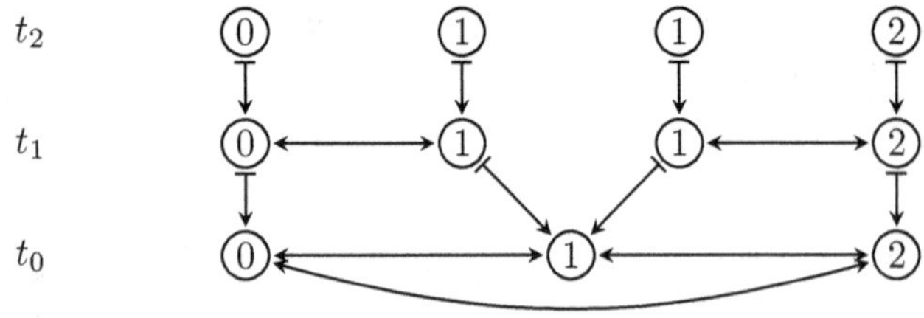

Fig. 1. The ET diagram for Example 3.6, where each state is marked with the number of white marbles in the urn. The identity maps have been omitted, as has the map from t_2 to t_0, which can be computed by composition. The reflexive arrows have also been omitted for each quasi-order.

Though it would be possible to work with ET diagrams directly, it will be easier to extract birelational structures so that we can use standard relation semantics.

Definition 3.7 Given an ET diagram $\mathscr{D} = (I, Q_-, P_{-,-})$, its *induced frame* is $F(\mathscr{D}) = (W_\mathscr{D}, R_\mathscr{D}, \leqslant_\mathscr{D})$, where $(W_\mathscr{D}, R_\mathscr{D})$ is the qoset obtained by taking the disjoint union of the Q_τs and $w \leqslant_\mathscr{D} v$ just in case $P_{\tau,\sigma}(v) = w$ where $w \in Q_\tau$ and $v \in Q_\sigma$.

[2] In this case, Betty's information is completely transparent to her, so the quasi-orders will in fact be equivalence relations.

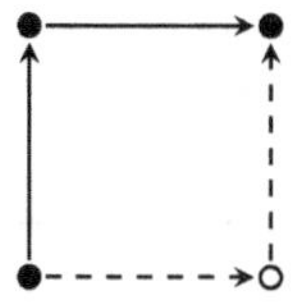

Fig. 2. A diagram representing the progressivity condition from Definition 3.9

Definition 3.8 A birelational frame $\mathscr{F}$ is an *ET frame* if $\mathscr{F} = F(\mathscr{D})$ for some ET diagram $\mathscr{D}$.

We call the class of ET frames $\mathfrak{E}$. If one is convinced by the naturality of ET diagrams, then $\mathfrak{E}$ seems to be the appropriate frame class for modeling knowability. One of the key moves that we want to make is recognizing that the class of ET frames sits inside a larger frame class that is more readily defined, easier to work with, and as it turns out, has the same logic. We now define these frames and verify that they constitute a superclass of $\mathfrak{E}$.

Definition 3.9 A *weak ET frame* is a triple $(W, R, \leqslant)$ where W is a set, R and $\leqslant$ are quasi-orders, and the following condition holds:

- (Progressivity) If $w \leqslant x \; R \; y$, then there exists $z \in W$ such that $w \; R \; z \leqslant y$.

We call the class of weak ET frames $\mathfrak{W}$.

Proposition 3.10 *If $\mathscr{F}$ is an ET frame, then it is a weak ET frame.*

Proof. By assumption, we have $\mathscr{F} = F(\mathscr{D}) = (W_{\mathscr{D}}, R_{\mathscr{D}}, \leqslant_{\mathscr{D}})$ for some ET diagram $\mathscr{D}$. $R_{\mathscr{D}}$ is a quasi-order since disjoint unions of quasi-orders are quasi-orders. $\leqslant_{\mathscr{D}}$ is also a quasi-order by the assumptions on $P_{-,-}$. Namely, $w \leqslant w$ since $P_{\tau,\tau}(w) = \mathrm{id}_{Q_\tau}(w) = w$, and if $w \leqslant_{\mathscr{D}} v \leqslant_{\mathscr{D}} u$, we have $P_{\tau,\rho}(u) = P_{\tau,\sigma}(P_{\sigma,\rho}(u)) = P_{\tau,\sigma}(v) = w$, so $w \leqslant_{\mathscr{D}} u$. Progression holds due to each $P_{\tau,\sigma}$ being order-preserving. If $w \leqslant_{\mathscr{D}} x \; R_{\mathscr{D}} \; y$, then the desired z is just $P_{\tau,\sigma}(y)$ where $w \in Q_\tau$ and $x, y \in Q_\sigma$. □

Definition 3.11 A *weak ET model* is a 4-tuple $(W, R, \leqslant, V)$ where $(W, R, \leqslant)$ is a weak ET frame and $V : \mathbb{P} \to \mathcal{P}(W)$ is a valuation.

Definition 3.12 Truth of a formula (in any of the aforementioned languages) at a world w in a weak ET model $\mathscr{M} = (W, R, \leqslant, V)$ is defined as recursively as follows:

$$
\begin{array}{lll}
\mathscr{M}, w \models p & \text{iff} & w \in V(p) \\
\mathscr{M}, w \models \neg\varphi & \text{iff} & \mathscr{M}, w \not\models \varphi \\
\mathscr{M}, w \models \varphi \wedge \psi & \text{iff} & \mathscr{M}, w \models \varphi \text{ and } \mathscr{M}, w \models \psi \\
\mathscr{M}, w \models K\varphi & \text{iff} & \text{for all } v \text{ such that } w \; R \; v,\ \mathscr{M}, v \models \varphi \\
\mathscr{M}, w \models G\varphi & \text{iff} & \text{for all } v \text{ such that } w \leqslant v,\ \mathscr{M}, v \models \varphi \\
\mathscr{M}, w \models \Box\varphi & \text{iff} & \text{there exists } v \text{ such that } w \leqslant v \\
 & & \text{and for all } u \text{ such that } v \; R \; u,\ \mathscr{M}, u \models \varphi
\end{array}
$$

Just by inspecting the truth clauses, we can see that our translation t preserves truth.

Lemma 3.13 *Let $\mathscr{M} = (W, R, \leqslant, V)$ be a weak ET model, $w \in W$, and $\varphi \in \mathcal{L}_{K\Box}$. Then $\mathscr{M}, w \vDash \varphi$ if and only if $\mathscr{M}, w \vDash t(\varphi)$.*

Proof. Induction on formula complexity. The only step that is not completely trivial simply requires recognizing that $FK\varphi$ and $\Box\varphi$ have the same truth conditions. □

We conclude the section by proving that sharp (respectively, flat) formulas are persistent (respectively, conversely persistent) along the temporal relation. This should make the axiomatizations of the knowability logics in the later sections more perspicuous.

Proposition 3.14 *Let $\mathscr{M}$ be a weak ET model and suppose that $w \leqslant x$. For a sharp formula χ, if $w \vDash \chi$, then $x \vDash \chi$. Dually, for a flat formula ε, if $x \vDash \varepsilon$ then $w \vDash \varepsilon$.*

Proof. We proceed by induction on rank. If $\chi = \Diamond\varphi$ for some φ, then $w \vDash \chi$ means that whenever $y \geq w$, $y \mathrel{R} z$ for some z such that $z \vDash \varphi$. Clearly x will satisfy this condition, as well, since $y \geq x$ implies $y \geq w$ by transitivity. The case where χ is the conjunction or disjunction of two sharp formulas follows from the fact that the intersection and union of two upwardly closed sets is again upwardly closed. Finally suppose that $\chi = K\chi'$ where χ' is sharp. We want to show that if $x \mathrel{R} y$, then $y \vDash \chi'$. For any such y, however, by excision we have a z such that $w \mathrel{R} z \leqslant y$. Since $w \vDash K\chi'$, $z \vDash \chi'$. By induction, $y \vDash \chi'$, so indeed $x \vDash K\chi'$.

The converse persistence of flat formulas can be easily recovered from the persistence of sharp formulas together with the observation that the negation of a flat formula is equivalent to a sharp formula. □

4 Knowledge and the Future

The first logic that we will consider is LKF (presented in Figure 3), a logic of knowledge and the future, which we will show is sound and complete with respect to both of the model classes introduced in the previous section. This logic is fairly straightforward, only consisting in the fusion of S4 with itself plus one additional axiom to capture the progressivity condition. We begin with the following easy result, which is almost certainly known.

Theorem 4.1 LKF *is sound with respect to $\mathfrak{W}$.*

Proof. A bulk of the work is just checking the soundness of S4 with respect to quasi-orders, which is known (see, e.g., [6]). To verify the soundness of **P**, let $\mathscr{M} = (W, R, \leqslant, V)$ be a weak ET model, and suppose that some $w \in W$ satisfies $w \vDash KG\varphi$. Let x be arbitrary such that $w \leqslant x$, and let y be arbitrary such that $x \mathrel{R} y$. By progressivity, there is some z such that $w \mathrel{R} z \leqslant y$. Since $w \vDash KG\varphi$, $z \vDash G\varphi$, so $y \vDash \varphi$. As y was an arbitrary R-successor of x, $x \vDash K\varphi$, and since x was an arbitrary $\leqslant$-successor of w, $w \vDash GK\varphi$. □

Corollary 4.2 LKF *is sound with respect to $\mathfrak{E}$*

The Logic LKF	
C	any substitution instance of a theorem of CPC
$\mathbf{K}_K$	$K(\varphi \to \psi) \to (K\varphi \to K\psi)$
$\mathbf{T}_K$	$K\varphi \to \varphi$
$\mathbf{4}_K$	$K\varphi \to KK\varphi$
$\mathbf{K}_G$	$G(\varphi \to \psi) \to (G\varphi \to G\psi)$
$\mathbf{T}_G$	$G\varphi \to \varphi$
$\mathbf{4}_G$	$G\varphi \to GG\varphi$
P	$KG\varphi \to GK\varphi$
MP	from $\varphi \to \psi$ and φ infer ψ
$\mathtt{Nec}_K$	from φ infer $K\varphi$
$\mathtt{Nec}_G$	from φ infer $G\varphi$

Fig. 3.

Now we verify completeness with respect to $\mathfrak{E}$. We will in fact be able to get a slightly stronger result, which is that LKF is complete with respect to $\mathfrak{E}_\omega$, the class of frames induced by an ET diagram where the linear order is ω. Our approach will be to define a canonical ET diagram, from which a model can be easily recovered.

Definition 4.3 Let C and D be sets of $\mathcal{L}_{KG}$-formulas. We say that the pair (C, D) is *coherent* just in case D is LKF-consistent and for all formulas φ, if $C \vdash_{\mathsf{LKF}} G\varphi$, then $D \vdash_{\mathsf{LKF}} \varphi$.

Lemma 4.4 *Suppose that (C, Δ) is coherent, where Δ is a LKF-maximally consistent theory. Then there exists a LKF-maximally consistent theory $\Gamma \supseteq C$ such that (Γ, Δ) is coherent.*

Proof. Set $\Gamma_0 = C \cup \{\neg G\varphi \in \mathsf{LKF} : \neg\varphi \in \Delta\}$. We now check the consistency of Γ_0. If it were not consistent, then $C \vdash \neg(\bigwedge_{i=1}^n \neg G\varphi_i)$ where each $\neg\varphi_i \in \Delta$ for each i. By propositional logic, $C \vdash \bigvee_{i=1}^n G\varphi_i$. Since $\bigvee_{i=1}^n G\varphi_i \to G(\bigvee_{i=1}^n G\varphi_i)$ is a theorem of S4, we in fact have $C \vdash G(\bigvee_{i=1}^n G\varphi_i)$. By the coherence of (C, Δ), $\Delta \vdash \bigvee_{i=1}^n G\varphi_i$. The maximality of Δ ensure that $\Delta \vdash G\varphi_k$ for some k. By $\mathbf{T}_G$, $\Delta \vdash \varphi_k$, which contradicts the consistency of Δ. Let Γ be some maximally consistent theory extending Γ_0. It is clear that (Γ, Δ) is coherent, since whenever $\Delta \nvdash \varphi$, $\Gamma \nvdash G\varphi$ by construction. □

Definition 4.5 A *history* $h = \langle \Gamma_1, \ldots, \Gamma_n \rangle$ is a finite sequence of complete theories such that for any $1 \leq i < n$ the pair (Γ_i, Γ_{i+1}) is coherent.

We introduce some pieces of notation to facilitate talking about histories.

- We write $\ell(h)$ for the length of a history.
- For $n \leq \ell(h)$, we write $h|_n$ for the history obtained by keeping only the first n entries h.
- For $i \leq \ell(h)$ we write $h(i)$ for the ith entry of h.
- We write $h_\bullet$ for $h(\ell(h))$.

- The set of all history of length n is denoted H_n.
- For two histories h and h' with $\ell(h) = \ell(h') = n$, we write $h \; R_n \; h'$ if for all $1 \leq i \leq n$, whenever $K\varphi \in h(i)$, $\varphi \in h'(i)$.

Lemma 4.6 *For each n, (H_n, R_n) is a qoset.*

Proof. Where Γ and Δ are maximally consistent theories, define a relation r by $\Gamma \; r \; \Delta$ just in case whenever $K\varphi \in \Gamma$, $\varphi \in \Delta$. It follows from the logic of the modality K being S4 that r is a quasi-order. (This is checked when one verifies that the canonical frame for S4 is in the desired frame class, for example. For a reference, see [6].) It is easy to see that R_n is just the product r^n restricted to the set of histories of length n. Quasi-orders are closed under products and restriction, so we are done. □

Definition 4.7 The *canonical diagram for* LKF is $D^{\mathsf{LKF}} = (\omega, Q^{\mathsf{LKF}}_{-}, P^{\mathsf{LKF}}_{-,-})$ where

- Q^{LKF}_n is the set of all histories of length n ordered by R_n, and
- $P^{\mathsf{LKF}}_{n,m} : H_m \to H_n$ is given by $h \mapsto h|_n$.

Proposition 4.8 *The canonical diagram for* LKF *is an ET diagram.*

Proof. This follows from the previous lemma and the facts that $h|_{\ell(h)} = h$ and that $(h|_n)_m = h_m$ where $m \leq n \leq \ell(h)$. □

Lemma 4.9 *Suppose $h \in H_n$ and $K\varphi \notin h(n)$. Then there exists $h' \in H_n$ such that $\varphi \notin h'(n)$ and $h \; R_n \; h'$.*

Proof. Write $h = \langle \Gamma_1, \ldots, \Gamma_n \rangle$. We first construct a sequence

$$\langle \Gamma_1^K, \ldots, \Gamma_{n-1}^K, \Gamma_n^K \cup \{\neg\varphi\} \rangle$$

where Γ_i^K is just notation for $\{\psi \in \mathcal{L}_{KG} : K\psi \in \Gamma_i\}$. Now we check that each consecutive pair is coherent. By $\mathbf{T}_K$, for each $1 \leq i \leq n-1$, we have $\Gamma_i^K \subseteq \Gamma_i$, so Γ_i^K is consistent. To see that the last entry is consistent, we assume for the sake of contradiction that $\Gamma_n^K \vdash \varphi$. Then for some $\psi \in \Gamma_n^K$, we have $\vdash \psi \to \varphi$. Applying $\mathtt{Nec}_K$, $\vdash K(\psi \to \varphi)$. Then by $\mathbf{K}_K$, $\vdash K\psi \to K\varphi$. Since $K\psi \in \Gamma_n$, we have $K\varphi \in \Gamma_n$. Contradiction. Now let $1 \leq i \leq n-1$ and suppose $\Gamma_i^K \vdash G\psi$. Then $\Gamma_i \vdash KG\psi$. By $\mathbf{P}$, $\Gamma_i \vdash GK\psi$. Because (Γ_i, Γ_{i+1}) is coherent, $\Gamma_{i+1} \vdash K\psi$ so $\Gamma_{i+1}^K \vdash \psi$.

Let Δ_n be some maximally consistent theory extending $\Gamma_n^K \cup \{\neg\varphi\}$. We apply Lemma 4.4 to $(\Gamma_{n-1}^K, \Delta_n)$ to obtain Δ_{n-1} extending Γ_{n-1}^K such that (Δ_{n-1}, Δ_n) is coherent. Next, apply the lemma to $(\Gamma_{n-2}^K, \Delta_{n-1})$ to obtain Δ_{n-2}. Continue in this way until Δ_1 has been defined, and then set $h' = \langle \Delta_1, \ldots, \Delta_n \rangle$. By construction $h \; R_n \; h'$. □

Definition 4.10 The *canonical ET model for* LKF is $\mathscr{M}^{\mathsf{LKF}} = (W^{\mathsf{LKF}}, R^{\mathsf{LKF}}, \leqslant^{\mathsf{LKF}}, V^{\mathsf{LKF}})$ where $W^{\mathsf{LKF}} = W_{D^{\mathsf{LKF}}}$, $R^{\mathsf{LKF}} = R_{D^{\mathsf{LKF}}}$, $\leqslant^{\mathsf{LKF}} = \leqslant_{D^{\mathsf{LKF}}}$, and $V(p) = \{h \in W^{\mathsf{LKF}} : p \in h_\bullet\}$.

Lemma 4.11 *For any $\varphi \in \mathcal{L}_{KG}$ and $h \in W^{\mathsf{LKF}}$, $\mathscr{M}^{\mathsf{LKF}}, h \vDash \varphi$ if and only if $\varphi \in h_\bullet$.*

Proof. The proof is by induction on formula complexity. The atomic and boolean connective cases are trivial, so we restrict our attention to the two modal cases.

First we consider when $\varphi = K\psi$ for some ψ. Suppose that $K\psi \in h_\bullet$. Let h' be arbitrary such that $h\ R^{\mathsf{LKF}}\ h'$. This means that $h\ R_{\ell(h)}\ h'$, so by the definition of this relation, $\psi \in h'(\ell(h))$. By induction, $h' \vDash \psi$, so $h \vDash K\psi$. In the other direction, suppose that $K\psi \notin h_\bullet$. By Lemma 4.9, there is another history $h' \in H_{\ell(h)}$ such that $h\ R_{\ell(h)}\ h'$ and $\psi \notin h'(\ell(h))$. By induction $h' \nvDash \psi$, so $h \nvDash K\psi$.

Now we consider the case where $\varphi = G\psi$. If $h \leqslant^{\mathsf{LKF}} h'$, then $G\psi \in h_\bullet = h'(\ell(h))$. By $\mathbf{4}_G$ and $\mathbf{T}_G$[3], $G^k\psi \in h'(\ell(h))$ where $k = \ell(h') - \ell(h)$. Since h' is a history, we will have $G^{k-l}\psi \in h'(\ell(h)+l)$ for all $0 \leq l \leq k$. In particular, $\psi \in h'(\ell(h)+k) = h'_\bullet$. By induction, $h' \vDash \psi$, so $h \vDash G\psi$. If $G\psi \notin h_\bullet$, a standard argument shows that $h_\bullet^G \cup \{\neg\psi\}$ is consistent. Let Γ be some maximally consistent theory extending this set. $(h_\bullet, \Gamma)$ is coherent by construction, so we set $h' = h^\frown\Gamma$. Since $\psi \notin h'_\bullet$, by induction $h' \nvDash \psi$. But $h \leqslant^{\mathsf{LKF}} h'$, so $h \nvDash G\psi$. □

Theorem 4.12 LKF *is strongly complete with respect to* $\mathfrak{C}$.

Proof. Suppose that $\Gamma \nvdash_{\mathsf{LKF}} \varphi$. Then $\Gamma \cup \{\neg\varphi\}$ is consistent and can be extended to a maximally consistent theory Γ'. Rather trivially $\langle\Gamma'\rangle$ is a history, so we have $\mathscr{M}^{\mathsf{LKF}}, \langle\Gamma'\rangle \nvDash \varphi$ by the truth lemma. □

Corollary 4.13 LKF *is complete with respect to* $\mathfrak{W}$.

5 The Logic of Knowledge and Knowability

We now present our first knowability logic LKKa, which is more specifically a logic of knowledge and knowability. The axioms are presented in Figure 4.

The Logic LKKa		
$\mathbf{C}$	any substitution instance of a theorem of CPC	
$\mathbf{K}_K$	$K(\varphi \to \psi) \to (K\varphi \to K\psi)$	
$\mathbf{T}_K$	$K\varphi \to \varphi$	
$\mathbf{4}_K$	$K\varphi \to KK\varphi$	
$\mathbf{A}$	$K\varphi \to \Box\varphi$	
MP	from $\varphi \to \psi$ and φ infer ψ	
$\mathtt{Nec}_K$	from φ infer $K\varphi$	
Sh	from $\chi \to \hat{K}\psi$ infer $\chi \to \Diamond\psi$	χ is sharp

Fig. 4.

Theorem 5.1 LKKa *is sound with respect to* $\mathfrak{W}$.

Proof. The only interesting case to check is the validity of Sh. Suppose that $\chi \to \hat{K}\psi$ is globally true in some model $\mathscr{M} = (W, R, \leqslant, V)$. Then if $w \vDash \chi$,

[3] You need to use $\mathbf{T}_G$ only in the case where $h = h'$.

we have $v \vDash \chi$ for all $v \geqslant w$ by Proposition 3.14. This means that every $\leqslant$-successor of w must make true $\hat{K}\psi$, implying $w \vDash \Diamond\psi$, as desired. □

Definition 5.2 The *canonical model for* LKKa is defined as $(W^{\mathsf{LKKa}}, R^{\mathsf{LKKa}}, \leqslant^{\mathsf{LKKa}}, V^{\mathsf{LKKa}})$ where

(i) W^{LKKa} is the set of all LKKa-maximally consistent theories;

(ii) $\Gamma\ R^{\mathsf{LKKa}}\ \Delta$ just in case for all φ, if $K\varphi \in \Gamma$ then $\varphi \in \Delta$;

(iii) $\Gamma \leqslant^{\mathsf{LKKa}} \Delta$ just in case for all flat formulas φ, if $\varphi \in \Delta$, then $\varphi \in \Gamma$;

(iv) $V^{\mathsf{LKKa}}(p) = \{\Gamma \in W^{\mathsf{LKKa}} : p \in \Gamma\}$.

Note that item (iii) in the above definition could be replaced by declaring $\Gamma \leqslant^{\mathsf{LKKa}} \Delta$ just in case for all sharp formulas φ, if $\varphi \in \Gamma$, then $\varphi \in \Delta$.

Lemma 5.3 *Suppose that* $\Box\varphi \in \Gamma$. *Then there exists a* Δ *such that* $\Gamma \leqslant^{\mathsf{LKKa}} \Delta$ *and* $K\varphi \in \Delta$.

Proof. We want to show that $\Gamma^{\sharp} \cup \{K\varphi\}$ is consistent. If not, we have some formula $\psi \in \Gamma^{\sharp}$ such that $\vdash \psi \to \hat{K}\neg\varphi$. By Sh, $\vdash \psi \to \Diamond\neg\varphi$. Therefore $\Gamma \vdash \Diamond\neg\varphi$, which is a contradiction. □

Lemma 5.4 *The frame* $(W^{\mathsf{LKKa}}, R^{\mathsf{LKKa}}, \leqslant^{\mathsf{LKKa}})$ *is progressive.*

Proof. Suppose that $\Gamma \leqslant^{\mathsf{LKKa}} \Delta$ and $\Delta\ R^{\mathsf{LKKa}}\ \Theta$. We wish to construct a Λ such that $\Gamma\ R^{\mathsf{LKKa}}\ \Lambda$ and $\Lambda \leqslant^{\mathsf{LKKa}} \Theta$. To this end, we will show that $\Gamma^K \cup \Theta^\flat$ is consistent. If not, then for some $\varphi \in \Gamma^K$ and some $\psi \in \Theta^\flat$, we have $\vdash \varphi \to \neg\psi$. By $\mathtt{Nec}_K$ and $\mathbf{K}_K$, $\vdash K\varphi \to K\neg\psi$, so $\Gamma \vdash K\neg\psi$. On the other hand, since $\Delta \leqslant^{\mathsf{LKKa}} \Theta$, we have $\Delta \vdash \hat{K}\psi$. Since ψ is flat, $\hat{K}\psi$ is also flat, so $\Gamma \vdash \hat{K}\psi$. Contradiction. We take Λ to be a maximally consistent extension of $\Gamma^K \cup \Theta^\flat$. By construction, $\Gamma\ R^{\mathsf{LKKa}}\ \Lambda$ and $\Lambda \leqslant^{\mathsf{LKKa}} \Theta$. □

Lemma 5.5 $\varphi \in \Gamma$ *if and only if* $\Gamma \vDash \varphi$.

Proof. We proceed by induction on complexity. The atomic and boolean connective cases are trivial. The K case follows from a standard argument. Now we suppose that $\varphi = \Box\psi$. If $\Box\psi \in \Gamma$, then by Lemma 5.3 there is some Δ such that $K\psi \in \Delta$ and $\Gamma \leqslant^{\mathsf{LKKa}} \Delta$. By induction, $\Delta \vDash K\psi$, so $\Gamma \vDash \Box\psi$. On the other hand, if $\Box\psi \notin \Gamma$, which means that $\Diamond\neg\psi \in \Gamma$. Let Δ be arbitrary such that $\Gamma \leqslant^{\mathsf{LKKa}} \Delta$. Since $\Diamond\neg\psi$ is sharp, we have $\Diamond\neg\psi \in \Delta$. This implies that $\hat{K}\neg\psi \in \Delta$. □

Theorem 5.6 LKKa *is strongly complete with respect to* $\mathfrak{W}$.

Corollary 5.7 *The translation* t *fully and faithfully embeds* LKKa *into* LKF.

Proof. Suppose that φ is a theorem of LKKa. By soundness of LKKa, φ is true in all weak ET models. Applying Lemma 3.13, $t(\varphi)$ is also true in all weak ET models. By completeness of LKF, $\vdash_{\mathsf{LKF}} t(\varphi)$. In the other direction, if φ is not a theorem of LKKa, because LKKa is complete with respect to $\mathfrak{W}$, φ is refuted in some model in the class. Lemma 3.13 then tells us that $t(\varphi)$ is refuted in the same model. By the soundness of LKF, $\nvdash_{\mathsf{LKF}} t(\varphi)$. □

Corollary 5.8 LKKa *is complete with respect to* $\mathfrak{E}$.

6 Pure Knowability

The final logic we introduce is the logic of pure knowability LPKa. An axiomatization is presented in Figure 5

The Logic of Knowability		
C	any substitution instance of a theorem of CPC	
$\mathbf{Ka}_\Box$	$\Box\top$	
$\mathbf{4}^*_\Box$	$\Box\varphi \to \Box(\varphi \wedge \Box\varphi)$	
$\mathbf{T}^\flat_\Box$	$\Box\varepsilon \to \varepsilon$	ε is flat
MP	from $\varphi \to \psi$ and φ infer ψ	
Reg	from $\varphi \to \psi$ infer $\Box\varphi \to \Box\psi$	
PR	from $\chi \to (\varphi \to \psi)$ infer $\Box(\theta \wedge \chi) \to \Box(\theta \wedge (\Box\varphi \to \Box\psi))$	χ is sharp

Fig. 5.

The only deductive fact beyond the axioms and rules that we will need is that $\vdash_{\mathsf{LPKa}} \neg\Box\bot$, which follows from taking $\varepsilon = \bot$ in $\mathbf{T}^\flat_\Box$.

Theorem 6.1 LPKa *is sound with respect to* $\mathfrak{E}$.

Proof. **C**, $\mathbf{Ka}_\Box$, and $\mathbf{4}^*_\Box$ are all straightforward to verify. $\mathbf{T}^\flat_\Box$ follows quickly from Lemma 3.14. MP and Reg are also easy. This leaves PR. Suppose that $\vDash \chi \to (\varphi \to \psi)$, and let x be an arbitrary world in a model $\mathscr{M} = (W, R, \leqslant)$. Suppose that $x \vDash \Box(\theta \wedge \chi)$. Then there exists a world $x' \geq x$ such that each R-successor y' of x' satisfies $y' \vDash \theta \wedge \chi$. Suppose that such a y' also has $y' \vDash \Box\varphi$. Then there is some $y'' \geq y'$ such that each R-successor z'' of y'' has $z'' \vDash \varphi$. By progressivity, there is a $z' \leqslant z''$ such that $y' \, R \, z'$. By the transitivity of R, $x' \, R \, z'$, so $z' \vDash \chi$. Applying Lemma 3.14, $z'' \vDash \chi$. Therefore $z'' \vDash \psi$. Since z'' was an arbitrary R-successor of y'' and $y' \leqslant y''$, we must have $y' \vDash \Box\psi$. As $y' \vDash \theta$ and y' is an arbitrary R-successor of x' satisfying $\Box\varphi$, so $x \leqslant x'$ implies that $x \vDash \Box(\theta \wedge \chi) \to \Box(\theta \wedge (\Box\varphi \to \Box\psi))$. □

In establishing completeness, one might seek a model construction similar to the one found for LKKa in the previous section. The obvious obstacle to doing this, however, is that we no longer have a K modality with which to define the epistemic relation. As a workaround, we take inspiration from a standard canonical model argument. In a canonical model, we settle on using maximally consistent theories as points because they are exactly the kinds of theories that can appear as the set of all true formulas at a world in a model. We will attempt to do the same thing for *known formulas*. That is, we will offer a characterization of theories which can appear as the set of all formulas that are true at all epistemically accessible worlds from some world in some model.

Definition 6.2 A theory $\mathcal{E} \subseteq \mathcal{L}_\Box$ is called *epistemic* if it satisfies the following conditions

- (Consistency) $\bot \notin \mathcal{E}$;

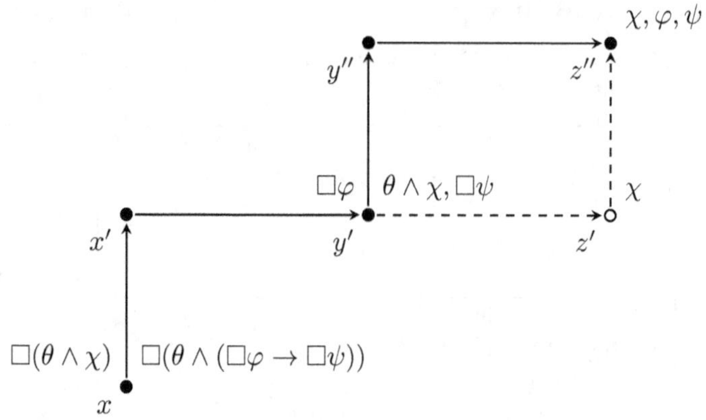

Fig. 6. A diagram illustrating the argument in the proof of Theorem 6.1.

- (Box closure) If $\alpha \in \mathcal{E}$, then $\Box\alpha \in \mathcal{E}$;
- (Sharp regularity) If $\chi \in \mathcal{E}^\sharp$ and $\vdash \chi \to (\beta \to \gamma)$, then $\Box\beta \to \Box\gamma \in \mathcal{E}$.

The points in the model we construct will then be pairs of maximally consistent theories (the truths) and epistemic theories (the known truths). But which such pairs will we use? As it turns out, the only compatibility condition we need is that the epistemic theory is contained in the maximally consistent theory, guaranteeing factivity of knowledge.

Definition 6.3 A pair of theories $(\Gamma, \mathcal{E})$ is called *concordant* if Γ is a maximally consistent theory, $\mathcal{E}$ is an epistemic theory, and $\mathcal{E} \subseteq \Gamma$.

We are now positioned to define the model. We use upward preservation of sharp formulas and of known sharp formulas to define the temporal relation, while the epistemic relation now has a simple definition that we extract from the epistemic theories.

Definition 6.4 We define the *canonical model for* LPKa $\mathscr{M}^{\mathsf{LPKa}} = (W^{\mathsf{LPKa}}, R^{\mathsf{LPKa}}, \leqslant^{\mathsf{LPKa}}, V^{\mathsf{LPKa}})$ as follows:

(i) W^{LPKa} is the set of concordant pairs.

(ii) $(\Gamma, \mathcal{E})\ R^{\mathsf{LPKa}}\ (\Delta, \mathcal{F})$ just in case $\mathcal{E} \subseteq \mathcal{F}$.

(iii) $(\Gamma, \mathcal{E}) \leqslant^{\mathsf{LPKa}} (\Delta, \mathcal{F})$ just in case $\Gamma^\sharp \subseteq \Delta^\sharp$ and $\mathcal{E}^\sharp \subseteq \mathcal{F}^\sharp$.

(iv) $V^{\mathsf{LPKa}}(p) = \{(\Gamma, \mathcal{E}) \in W : p \in \Gamma\}$.

The way that we have defined $\mathscr{M}^{\mathsf{LPKa}}$ makes it fairly easy to check that it is in the correct frame class.

Proposition 6.5 $(W^{\mathsf{LPKa}}, R^{\mathsf{LPKa}}, \leqslant^{\mathsf{LPKa}})$ *is in* $\mathfrak{C}$.

Proof. It is immediate that R^{LPKa} and $\leqslant^{\mathsf{LPKa}}$ are quasi-orders, so what remains is to check progressivity. Suppose that $(\Gamma, \mathcal{E}) \leqslant^{\mathsf{LPKa}} (\Delta, \mathcal{F})\ R^{\mathsf{LPKa}}\ (\Theta, \mathcal{G})$. We first show that $\mathcal{E} \cup \Theta^\flat$ is consistent. If not, then $\vdash \alpha \to \neg\varepsilon$ for some $\alpha \in \mathcal{E}$ and

$\varepsilon \in \Theta^\flat$. This implies that $\neg\varepsilon \in \mathcal{E}$, and because $\neg\varepsilon$ is equivalent to some sharp formula, we have $\neg\varepsilon \in \mathcal{F}$, which implies $\neg\varepsilon \in \mathcal{G}$, contradicting the consistency of Θ. Let Λ be some maximally consistent theory extending $\mathcal{E} \cup \Theta^\flat$. We now claim that $(\Gamma, \mathcal{E})\ R^{\mathsf{LPKa}}\ (\Lambda, \mathcal{E}) \leqslant^{\mathsf{LPKa}} (\Theta, \mathcal{G})$. It is trivial that $(\Gamma, \mathcal{E})\ R^{\mathsf{LPKa}}\ (\Lambda, \mathcal{E})$ since $\mathcal{E} = \mathcal{E}$. We also have $\Lambda^\sharp \subseteq \Theta^\sharp$ by construction, as this is equivalent to $\Theta^\flat \subseteq \Lambda^\flat$. What remains to show is that $\mathcal{E}^\sharp \subseteq \mathcal{G}^\sharp$. But this is immediate, since the assumptions imply $\mathcal{E}^\sharp \subseteq \mathcal{F}^\sharp \subseteq \mathcal{G}^\sharp$. □

We now establish some lemmas that will facilitate the proof of the truth lemma. The first is rather quick, and intuitively just says that formulas omitted from the epistemic theory of a concordant are in fact not know there.

Lemma 6.6 *Suppose that $(\Gamma, \mathcal{E}) \in W$ has $\varphi \notin \mathcal{E}$. Then there exists a concordant pair $(\Delta, \mathcal{F})$ such that $\varphi \notin \Delta$ and $(\Gamma, \mathcal{E})\ R^{\mathsf{LPKa}}\ (\Delta, \mathcal{F})$.*

Proof. It is immediate that $\mathcal{E} \cup \{\neg\varphi\}$ is consistent. Let Δ be a maximally consistent theory extending $\mathcal{E} \cup \{\neg\varphi\}$, and let $\mathcal{F} = \mathcal{E}$. □

It will be much more difficult to verify that if the maximally consistent theory of a concordant pair contains a knowability claim, then the purportedly knowable proposition actually appears in the epistemic theory of some temporal successor of the pair. This will require a recursive construction whose core machinery is established in the following three claims.

Claim 6.7 *Suppose that $(\Gamma, \mathcal{E})$ is concordant and that $\Box\alpha \in \Gamma$. Then $\mathcal{E}^\sharp \cup \{\alpha \wedge \Box\alpha\}$ is consistent.*

Proof. Suppose not. Then $\vdash \chi \to \neg(\alpha \wedge \Box\alpha)$ for some $\chi \in \mathcal{E}^\sharp$. This can be equivalently written as $\vdash \chi \to ((\alpha \wedge \Box\alpha) \to \bot)$. By sharp regularity, $\Box(\alpha \wedge \Box\alpha) \to \Box\bot \in \mathcal{E}$. Contraposing and using $\vdash \neg\Box\bot$, we see that $\neg\Box(\alpha \wedge \Box\alpha) \in \mathcal{E} \subseteq \Gamma$. But since $\Box\alpha \in \Gamma$, by $\mathbf{4}^*_\Box$, $\Box(\alpha \wedge \Box\alpha) \in \Gamma$, which is a contradiction. □

Claim 6.8 *Suppose that $(\Gamma, \mathcal{E})$ is concordant and that $\Box\alpha \in \Gamma$. If $\mathcal{E}^\sharp \vdash (\alpha \wedge \Box\alpha) \to \beta$, then $\Box(\alpha \wedge \beta) \in \Gamma$.*

Proof. As $\mathcal{E}^\sharp$ is closed under conjunction, the hypothesis gives us $\vdash \chi \to ((\alpha \wedge \Box\alpha) \to \beta)$ for some $\chi \in \mathcal{E}^\sharp$. This is classically equivalent to $\vdash \chi \to ((\alpha \wedge \Box\alpha) \to (\alpha \wedge \beta))$. By sharp regularity, $\Box(\alpha \wedge \Box\alpha) \to \Box(\alpha \wedge \beta) \in \mathcal{E}$. Since $\vdash \Box\alpha \to \Box(\alpha \wedge \Box\alpha)$, we have $\Box\alpha \to \Box(\alpha \wedge \beta) \in \mathcal{E}$. Finally, since $\mathcal{E} \subseteq \Gamma$ and $\Box\alpha \in \Gamma$, we have $\Box(\alpha \wedge \beta) \in \Gamma$. □

Claim 6.9 *Suppose that $(\Gamma, \mathcal{E})$ is concordant and that $\Box\alpha \in \Gamma$. If $\mathcal{E}^\sharp \vdash (\alpha \wedge \Box\alpha) \to \chi$ and $\vdash \chi \to (\beta \to \gamma)$ for some sharp formula χ, then $\Box(\alpha \wedge (\Box\beta \to \Box\gamma)) \in \Gamma$.*

Proof. First note that by Claim 6.8, $\Box(\alpha \wedge \chi) \in \Gamma$. Applying PR to the assumption $\vdash \chi \to (\beta \to \gamma)$, we obtain $\vdash \Box(\alpha \wedge \chi) \to \Box(\alpha \wedge (\Box\beta \to \Box\gamma))$. These facts entail $\Box(\alpha \wedge (\Box\beta \to \Box\gamma)) \in \Gamma$, as desired. □

We can now prove the key lemma for the completeness argument.

Lemma 6.10 *Suppose that $(\Gamma, \mathcal{E}) \in W$ and $\Box\varphi \in \Gamma$. Then there exists a concordant pair $(\Delta, \mathcal{F})$ such that $\varphi \in \mathcal{F}$ and $(\Gamma, \mathcal{E}) \leqslant^{\mathsf{LPKa}} (\Delta, \mathcal{F})$.*

Proof. First we construct $\mathcal{F}$. Fix an enumeration $(\psi_i)_{i<\omega}$ of all formulas and an enumeration $(\theta_i)_{i<\omega}$ of all formulas whose outermost connective is a conditional. For convenience, write $\theta_i = \beta_i \to \gamma_i$. We will recursively define a sequence of formulas $(\alpha_i)_{i<\omega}$ as follows:

- $\alpha_0 = \varphi$.
- If $n > 0$ is odd, let k_n be the least natural such that $\mathcal{E}^\sharp \cup \{\alpha_{n-1} \wedge \Box\alpha_{n-1}\} \vdash \psi_{k_n}$ but $\mathcal{E}^\sharp \cup \{\alpha_{n-1} \wedge \Box\alpha_{n-1}\} \nvdash \Box\psi_{k_n}$. Set $\alpha_n = \alpha_{n-1} \wedge \psi_{k_n}$. If there is no such k_n, take $\alpha_n = \alpha_{n-1}$.
- If $n > 0$ is even, let k_n be the least natural such that $\mathcal{E}^\sharp \cup \{\alpha_{n-1} \wedge \Box\alpha_{n-1}\} \vdash \chi$ and $\vdash \chi \to \theta_{k_n}$ but $\mathcal{E}^\sharp \cup \{\alpha_{n-1} \wedge \Box\alpha_{n-1}\} \nvdash \Box\beta_i \to \Box\gamma_i$. Set $\alpha_n = \alpha_{n-1} \wedge (\Box\beta_i \to \Box\gamma_i)$. If there is no such k_n, take $\alpha_n = \alpha_{n-1}$.

First we claim that for each n, $\Box\alpha_n \in \Gamma$. The proof is by induction. The base case is by assumption. Now consider the case where $n > 0$ is odd. If $\alpha_n = \alpha_{n-1}$ we are done by induction. Otherwise, by Claim 6.8, $\Box(\alpha_{n-1} \wedge \psi_{k_n}) \in \Gamma$. Since $\alpha_n = \alpha_{n-1} \wedge \psi_{k_n}$, we are done. Last is the case where $n > 0$ is even. Again, if $\alpha_n = \alpha_{n-1}$, we are done by induction. Otherwise, Claim 6.8 tells us that $\Box(\alpha_{n-1} \wedge (\Box\beta_{k_n} \to \Box\gamma_{k_n})) \in \Gamma$.

For each n, let $\mathcal{F}_n = \mathrm{cl}(\mathcal{E}^\sharp \cup \{\alpha_n \wedge \Box\alpha_n\})$. Note that by Claim 6.7, each $\mathcal{F}_n$ is consistent. Now we claim that for any n, $\mathcal{F}_n \subseteq \mathcal{F}_{n+1}$. It suffices to show that $\alpha_{n+1} \wedge \Box\alpha_{n+1} \vdash \alpha_n \wedge \Box\alpha_n$. It is clear that $\alpha_{n+1} \vdash \alpha_n$, since α_{n+1} is either α_n or the conjunction of α_n with another formula. By Reg, we then also have $\Box\alpha_{n+1} \vdash \Box\alpha_n$.

Now we set $\mathcal{F} = \bigcup \mathcal{F}_n$. We want to check that $\mathcal{F}$ is an epistemic theory.

- $\mathcal{F}$ is an increasing union of consistent theories, so it is a consistent theory.
- Suppose that $\alpha \in \mathcal{F}$. If $\Box\alpha \notin \mathcal{F}$, then at some odd stage $n > 0$, we would have set $\alpha_n = \alpha_{n-1} \wedge \alpha$. But then $\Box\alpha \in \mathrm{cl}(\mathcal{E}^\sharp \cup \{\alpha_n \wedge \Box\alpha_n\}) = \mathcal{F}_n$, which is a contradiction. Therefore $\Box\alpha \in \mathcal{F}$, so $\mathcal{F}$ is box closed.
- Suppose that $\chi \in \mathcal{F}^\sharp$ and $\vdash \chi \to (\beta \to \gamma)$. If $\Box\beta \to \Box\gamma \notin \mathcal{F}$, then at some even stage $n > 0$, we would have set $\alpha_n = \alpha_{n_1} \wedge (\Box\beta \to \Box\gamma)$. Again, this implies that $\Box\beta \to \Box\gamma \in \mathcal{F}_n$, which is a contradiction, so indeed $\Box\beta \to \Box\gamma \in \mathcal{F}$, and we can conclude that $\mathcal{F}$ is sharply regular.

To finish the argument, we need to find a maximally consistent theory Δ such that $(\Delta, \mathcal{F})$ is concordant and $(\Gamma, \mathcal{E}) \leqslant^{\mathsf{LPKa}} (\Delta, \mathcal{F})$. To this end, we verify the consistency of $\Gamma^\sharp \cup \mathcal{F}$. First, we claim that if $\alpha \in \mathcal{F}$ then $\Box\alpha \in \Gamma$. If $\alpha \in \mathcal{F}$, then $\alpha \in \mathcal{F}_n$ for some n. For such an n, $\vdash \chi \to ((\alpha_n \wedge \Box\alpha_n) \to \alpha)$ for some $\chi \in \mathcal{E}^\sharp$. By sharp regularity, $\Box(\alpha_n \wedge \Box\alpha_n) \to \Box\alpha \in \mathcal{E}$, so by $\mathbf{4}^*_\Box$, $\Box\alpha_n \to \Box\alpha \in \mathcal{E}$. We already established that $\Box\alpha_n \in \Gamma$ for each n, so it must be that $\Box\alpha \in \mathcal{E}$. Now suppose that $\Gamma^\sharp \cup \mathcal{F}$ is inconsistent. This implies that $\vdash \alpha \to \neg\chi$ for some $\alpha \in \mathcal{F}$ and $\chi \in \Gamma^\sharp$. By Reg, $\vdash \Box\alpha \to \Box\neg\chi$, and since the claim tells us that $\Box\alpha \in \Gamma$, we have $\Box\neg\chi \in \Gamma$. But $\chi \in \Gamma^\sharp$, so $\Diamond\chi \in \Gamma$ by $\mathbf{T}^\sharp_\Diamond$, contradicting the consistency of Γ. Let Δ be some maximally consistent theory extending Γ. We have $\Gamma^\sharp \subseteq \Delta^\sharp$ and $\mathcal{E}^\sharp \subseteq \mathcal{F}^\sharp$, so $(\Gamma, \mathcal{E}) \leqslant^{\mathsf{LPKa}} (\Delta, \mathcal{F})$, as desired. □

Lemma 6.11 *For any concordant pair* $(\Gamma, \mathcal{E})$, $\varphi \in \Gamma$ *if and only if* $\mathscr{M}, (\Gamma, \mathcal{E}) \models \varphi$.

Proof. The proof is by induction on formula complexity. The only non-trivial case is when $\varphi = \Box\psi$ for some ψ.

Suppose $\Box\psi \in \Gamma$. By Lemma 6.10, there exists a concordant pair $(\Delta, \mathcal{F})$ such that $\psi \in \mathcal{F}$. Now let $(\Theta, \mathcal{G})$ be an arbitrary concordant pair such that $(\Delta, \mathcal{F})\ R^{\mathsf{LPKa}}\ (\Theta, \mathcal{G})$. Since $\mathcal{F} \subseteq \mathcal{G}$, $\psi \in \mathcal{G}$. But then $\psi \in \Theta$, so by the induction hypothesis, $\mathscr{M}, (\Theta, \mathcal{G}) \models \psi$. We conclude that $\mathscr{M}, (\Gamma, \mathcal{E}) \models \Box\psi$.

In the other direction, suppose that $\Box\psi \notin \Gamma$. Then $\Diamond\neg\psi \in \Gamma$. Let $(\Delta, \mathcal{F})$ be arbitrary such that $(\Gamma, \mathcal{E}) \leqslant^{\mathsf{LPKa}} (\Delta, \mathcal{F})$. Since $\Diamond\neg\psi$ is sharp, $\Diamond\neg\psi \in \Delta$. This implies $\psi \notin \mathcal{F}$, as otherwise we would have $\Box\psi \in \mathcal{F} \subseteq \Delta$, contradicting the consistency of Δ. By Lemma 6.6, there is a concordant pair $(\Theta, \mathcal{G})$ such that $\psi \notin \Theta$. By induction, $(\Theta, \mathcal{G}) \not\models \psi$. Therefore, $(\Gamma, \mathcal{E}) \not\models \Box\psi$. □

Theorem 6.12 LPKa *is strongly complete with respect to* $\mathfrak{W}$.

Proof. Suppose that $\Gamma \not\vdash \varphi$. Then $\Gamma \cup \{\neg\varphi\}$ is consistent and can be extended to some maximally consistent Γ'. We first want to find some theory $\mathcal{E}$ such that $(\Gamma', \mathcal{E})$ is concordant. We propose $\mathcal{E} = \mathrm{cl}(\varnothing)$. Clearly, we have $\mathcal{E} \subseteq \Gamma'$, so we just need to check that $\mathcal{E}$ is epistemic.

- Consistency of $\mathcal{E}$ follows from soundness and the nonemptiness of $\mathfrak{W}$.
- Box closure follows from $\mathbf{Ka}_\Box$.
- Sharp regularity follows from Reg.

Since $(\Gamma', \mathcal{E})$ is concordant, $\mathscr{M}, (\Gamma', \mathcal{E}) \models \Gamma$ but $\mathscr{M}, (\Gamma', \mathcal{E}) \not\models \varphi$ by the truth lemma, so $\Gamma \not\models \varphi$. □

Corollary 6.13 *The translation* $t_\Box$ *fully and faithfully embeds* LPKa *into* LKF.

Proof. The proof is essentially identical to that of Corollary 5.7. □

Corollary 6.14 LPKa *is complete with respect to* $\mathfrak{E}$.

7 Future Work

The ideas in this paper suggests a systematic approach to thinking about knowability after one settles on a particular logic for knowledge. Namely, provided the logic of knowledge is Kripke complete, one could redefine an ET diagram so that each Q_τ is of the appropriate frame class. Then the logics of knowledge and knowability and of pure knowability coming from that notion of ET diagram could be axiomatized, providing an understanding how knowability behaves relative to that particular conception of knowledge. Interesting cases to treat include S5, which is often used in applications of epistemic logic to game theory (see, e.g., [2] and [13]) and S4.2, which Stalnaker claims to be the correct logic of knowledge in [12]. It might also be of interest to investigate the analogs of these knowability logics that come from removing or relaxing the progressivity condition on frames.

References

[1] Balbiani, P., A. Baltag, H. van Ditmarsch, A. Herzig, T. Hoshi and T. de Lima, *'knowable' as 'known after an announcement'*, Review of Symbolic Logic **1** (2008), pp. 305–334.

[2] Battigalli, P. and G. Bonanno, *Recent results on belief, knowledge and the epistemic foundations of game theory*, Research in Economics **53** (1999), pp. 149–225.
URL `https://www.sciencedirect.com/science/article/pii/S1090944399901874`

[3] Berto, F. and P. Hawke, *Knowability relative to information*, Mind **130** (2021), pp. 1–33.

[4] Bjorndahl, A., *Topological subset space models for public announcements*, in: H. van Ditmarsch and G. Sandu, editors, *Jaakko Hintikka on Knowledge and Game Theoretical Semantics*, Springer, 2018 pp. 165–186.

[5] Bjorndahl, A. and A. Özgün, *Logic and topology for knowledge, knowability, and belief*, The Review of Symbolic Logic **13** (2020), p. 748–775.

[6] Blackburn, P., J. van Benthem and F. Wolter, editors, "Handbook of Modal Logic," Elsevier, 2006.

[7] Burgess, J., *Can truth out?*, in: J. Salerno, editor, *New Essays on the Knowability Paradox*, Oxford University Press, 2009 .

[8] Fitch, F. B., *A logical analysis of some value concepts*, The Journal of Symbolic Logic **28** (1963), pp. 135–142.
URL `http://www.jstor.org/stable/2271594`

[9] Hintikka, J., "Knowledge and Belief," Cornell University Press, Ithaca, N.Y.,, 1962.

[10] Liu, M., J. Fan, H. Ditmarsch and L. Kuijer, *Logics for knowability*, Logic and Logical Philosophy (2021), pp. 1–42.

[11] Schwarz, G., *In search of a "true" logic of knowledge: the nonmonotonic perspective*, Artificial Intelligence **79** (1995), pp. 39–63.
URL `https://www.sciencedirect.com/science/article/pii/0004370294000670`

[12] Stalnaker, R., *On logics of knowledge and belief*, Philosophical Studies: An International Journal for Philosophy in the Analytic Tradition **128** (2006), pp. 169–199.
URL `http://www.jstor.org/stable/4321718`

[13] van Benthem, J., *Games in dynamic-epistemic logic*, Bulletin of Economic Research **53** (2001), pp. 219–248.
URL `https://onlinelibrary.wiley.com/doi/abs/10.1111/1467-8586.00133`

[14] van Benthem, J., *What one may come to know*, Analysis **64** (2004), pp. 95–105.
URL `http://www.jstor.org/stable/3329111`

[15] van Ditmarsch, H., W. van der Hoek and P. Iliev, *Everything is knowable ? how to get to know whether a proposition is true*, Theoria **78** (2012), pp. 93–114.

The Interpolant Existence Problem for Weak K4 and Difference Logic

Agi Kurucz

Department of Informatics
King's College London, U.K.

Frank Wolter

Department of Computer Science
University of Liverpool, U.K.

Michael Zakharyaschev

School of Computing and Mathematical Sciences
Birkbeck, University of London, U.K.

Abstract

As well known, weak K4 and the difference logic DL do not enjoy the Craig interpolation property. Our concern here is the problem of deciding whether any given implication does have an interpolant in these logics. We show that the nonexistence of an interpolant can always be witnessed by a pair of bisimilar models of polynomial size for DL and of triple-exponential size for weak K4, and so the interpolant existence problems for these logics are decidable in coNP and coN3ExpTime, respectively. We also establish coNExpTime-hardness of this problem for weak K4, which is higher than the PSpace-completeness of its decision problem.

Keywords: Craig interpolant, propositional modal logic, computational complexity.

1 Introduction

Weak K4 is the modal logic one obtains when the $\Diamond$-operator of the propositional classical (uni)modal language is interpreted by the derivative operation [1] in topological spaces [7]—rather than the more conventional topological closure, which results in classical S4 [17]. In terms of Kripke semantics, weak K4 is characterised [7] by the class of weakly transitive frames, i.e., those $\mathfrak{F} = (W, R)$ that satisfy the condition

$$\forall x, y, z \in W \big(xRyRz \to (x = z) \vee xRz\big), \qquad (1)$$

[1] The *derived set* of a subset X of a topological space comprises all limit points of X, i.e., those x all of whose neighbourhoods contain a point in X different from x.

which explains the moniker 'weak K4' or wK4 for this logic. Syntactically, wK4 is obtained by adding the axiom $\Diamond\Diamond p \to (p \vee \Diamond p)$ to the basic normal modal logic K. A notable extension of wK4 is the difference logic DL, which goes back to the 'logic of elsewhere' [23,19] and can be axiomatised by adding the Brouwersche axiom $p \to \Box\Diamond p$ to wK4. While an arbitrary frame for DL is a symmetric and weakly transitive relation, DL is also known to be characterised by the class of *difference frames* (that is, Kripke frames of the form $(W, \neq)$) [19].

Despite their apparent similarity to K4, S4 and S5, the logics wK4 and DL have—or rather lack—one important feature: they do not enjoy the Craig interpolation property (CIP) [12], according to which each valid implication $\varphi \to \psi$ in a logic L has an interpolant in L, viz. a formula ι built from common variables of φ and ψ such that $(\varphi \to \iota) \in L$ and $(\iota \to \psi) \in L$; if φ and ψ have no variables in common, variable-free interpolant ι is built from the logical constants $\top$ and $\bot$.

Example 1.1 In the pictures below, $\bullet$ always denotes an irreflexive point, $\circ$ a reflexive one, and an ellipse represents a cluster (a set of points, in which any two distinct ones 'see' each other).

(*i*) Consider the following formulas without common variables:

$$\varphi = \Diamond\Diamond p \wedge \neg\Diamond p, \qquad \psi = \Diamond\Diamond\neg q \vee q \; (\equiv \Box\Box q \to q).$$

It is easy to see that $\varphi \to \psi$ is true in all models based on weakly transitive frames, and so $(\varphi \to \psi) \in$ wK4. On the other hand, the picture below shows models $\mathfrak{M}_\varphi$ and $\mathfrak{M}_\psi$ based on weakly transitive frames such that $\mathfrak{M}_\varphi, r_\varphi \models \varphi$, $\mathfrak{M}_\psi, r_\psi \models \neg\psi$, and the universal relation $\boldsymbol{\beta}$ between the points of $\mathfrak{M}_\varphi$ and $\mathfrak{M}_\psi$ is a ϱ-bisimulation for the shared signature $\varrho = \emptyset$ of φ and ψ, with $r_\varphi \boldsymbol{\beta} r_\psi$ (see Sec. 2 for definitions).

It follows that there is no variable-free formula ι with $(\varphi \to \iota) \in$ wK4 and $(\iota \to \psi) \in$ wK4 because ϱ-bisimulations preserve the truth-values of ϱ-formulas.

(*ii*) Consider the formulas

$$\varphi = \Diamond\Diamond p \wedge \neg\Diamond p \wedge \neg\Diamond\Diamond\Diamond p, \qquad \psi = \Diamond\Diamond s \wedge \neg\Diamond s \to \neg\big(\Diamond(q \wedge \Diamond s) \wedge \Diamond(\neg q \wedge \Diamond s)\big).$$

It is easy to see that $\varphi \to \psi$ is true in all models based on a weakly transitive frame, and so $(\varphi \to \psi) \in$ wK4.

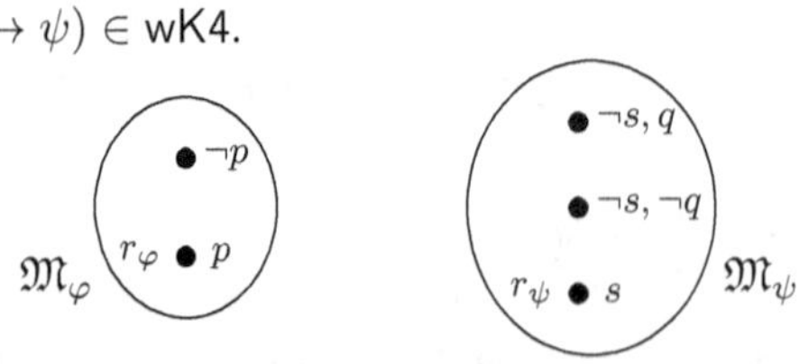

On the other hand, the models $\mathfrak{M}_\varphi$ and $\mathfrak{M}_\psi$ above are based on difference frames with $\mathfrak{M}_\varphi, r_\varphi \models \varphi$, $\mathfrak{M}_\psi, r_\psi \models \neg\psi$, and the universal relation $\boldsymbol{\beta}$ between the points of $\mathfrak{M}_\varphi$ and $\mathfrak{M}_\psi$ is a ϱ-bisimulation with $r_\varphi \boldsymbol{\beta} r_\psi$, for the shared signature $\varrho = \emptyset$ of φ, ψ. Therefore, φ and ψ do not have an interpolant in any logic between wK4 and DL. ⊣

Our concern in this paper is the following

***interpolant existence problem (IEP)* for $L \in \{\mathsf{wK4}, \mathsf{DL}\}$:** given formulas φ and ψ, decide whether $\varphi \to \psi$ has an interpolant in L.

We show that the IEP for wK4 is decidable in coN3ExpTime, being coNExpTime-hard (harder than its decision problem), while the IEP for DL is coNP-complete (as is its decision problem).

In Section 2, we introduce the necessary technical tools and demonstrate them in the case of DL. Then we focus on the much more involved IEP for wK4, establishing the upper bound in Section 3 and the lower one in Section 4. Finally, we discuss related and open problems in Section 5.

2 Preliminaries

All (Kripke) frames $\mathfrak{F} = (W, R)$ we deal with in this paper are assumed to be *weakly transitive* (1). By a *cluster* in $\mathfrak{F}$ we mean any set of the form

$$C(x) = \{x\} \cup \{y \in W \mid xRy \wedge yRx\}, \quad x \in W.$$

A cluster $\mathfrak{F}$ can contain *reflexive* points y, for which yRy, as well as *irreflexive* points z, for which $\neg(zRz)$. A cluster with a single point, which is irreflexive, is said to be *degenerate*. Given $x, y \in W$, we write $xR^s y$ iff xRy and $C(x) \neq C(y)$.

Suppose C and C' are clusters in $\mathfrak{F}$ and $x \in W$. We write CRx if there exists $y \in C$ such that yRx, and CRC' if there are $x \in C$ and $y \in C'$ with xRy. Observe that CRC iff C is non-degenerate. We write $CR^s x$ if CRx and $x \notin C$, and $CR^s C'$ if CRC' and $C \neq C'$. Thus, R^s is a strict partial order on the set W_c of clusters in $\mathfrak{F}$. The reflexive closure of R^s is denoted by R^r. The frame $\mathfrak{F}$ is called *rooted* if there is $r \in W$, a *root* of $\mathfrak{F}$, with $W_c = \{C(x) \mid C(r) R^r C(x)\}$.

By a *signature* we mean any *finite* set of propositional variables, p_i. Given a signature σ, a *σ-formula* is built from variables in σ and logical constants $\bot$, $\top$ using the Boolean connectives $\wedge$, $\neg$ and the modal possibility operator $\Diamond$. The other Boolean connectives and the necessity operator $\Box$ are regarded as standard abbreviations. We denote by $sig(\varphi)$ the set of variables in a formula φ and by $sub(\varphi)$ the set of subformulas of φ together with their negations, setting $|\varphi| = |sub(\varphi)|$. We also use abbreviations $\Diamond^+\varphi = \varphi \vee \Diamond\varphi$, $\Box^+\varphi = \varphi \wedge \Box\varphi$ and $\Diamond\Gamma = \{\Diamond\varphi \mid \varphi \in \Gamma\}$, for a set Γ of formulas.

A *σ-model* based on a frame $\mathfrak{F} = (W, R)$ is a pair $\mathfrak{M} = (\mathfrak{F}, \mathfrak{v})$ with a *valuation* $\mathfrak{v} \colon \sigma \to 2^W$. The *truth-relation* $\mathfrak{M}, x \models \varphi$, for $x \in W$ and a σ-formula φ, is defined by induction as usual in Kripke semantics (e.g., $\mathfrak{M}, x \models \Diamond\varphi$ iff $\mathfrak{M}, y \models \varphi$, for some $y \in W$ with xRy). For any $\varrho \subseteq \sigma$, the *ϱ-type of $x \in W$ in $\mathfrak{M}$* is the set $t^\varrho_\mathfrak{M}(x)$ of all ϱ-formulas that are true at x in $\mathfrak{M}$, and the *atomic ϱ-type of $x \in W$ in $\mathfrak{M}$* is $at^\varrho_\mathfrak{M}(x) = t^\varrho_\mathfrak{M}(x) \cap \varrho$. For a set X of points in $\mathfrak{M}$, we let $t^\varrho_\mathfrak{M}(X) = \{t^\varrho_\mathfrak{M}(x) \mid x \in X\}$ and $at^\varrho_\mathfrak{M}(X) = \{at^\varrho_\mathfrak{M}(x) \mid x \in X\}$. A set Γ of

σ-formulas is *finitely satisfiable* in $\mathfrak{M}$ if, for every finite $\Gamma' \subseteq \Gamma$, there is $x' \in W$ such that $\Gamma' \subseteq t^{\sigma}_{\mathfrak{M}}(x')$; Γ is *satisfiable in* $\mathfrak{M}$ if $\Gamma \subseteq t^{\sigma}_{\mathfrak{M}}(x)$, for some $x \in W$.

A σ-model $\mathfrak{M}$ is *descriptive* if, for any $x, y \in W$ and any set Γ of σ-formulas,

(dif) $x = y$ iff $t^{\sigma}_{\mathfrak{M}}(x) = t^{\sigma}_{\mathfrak{M}}(y)$,

(ref) xRy iff $\Diamond t^{\sigma}_{\mathfrak{M}}(y) \subseteq t^{\sigma}_{\mathfrak{M}}(x)$ iff $\{\varphi \mid \Box\varphi \in t^{\sigma}_{\mathfrak{M}}(x)\} \subseteq t^{\sigma}_{\mathfrak{M}}(y)$,

(com) if Γ is finitely satisfiable in $\mathfrak{M}$, then Γ is satisfiable in $\mathfrak{M}$.

(In other words, descriptive models are based on finitely generated descriptive frames for wK4 [5].) We remind the reader that, for any σ-formula φ, we have $\varphi \in$ wK4 iff $\neg\varphi$ is not satisfiable in a (finite) σ-model iff $\neg\varphi$ is not satisfiable in a (finite) descriptive σ-model. The finite model property of wK4 was established in [3,4,13]. The decision problem for wK4 is PSPACE-complete [20]. For the *difference logic* DL, we have $\varphi \in$ DL iff $\neg\varphi$ is not satisfiable in a polynomial-size σ-model based on a frame for DL; the decision problem for DL is coNP-complete [6].

In the remainder of this section, we present the technical tools and results we need for deciding the interpolant existence problem for DL and wK4. Given $\varrho \subseteq \sigma$, we call a cluster C *ϱ-maximal* in $\mathfrak{M}$ if, for any $x \in C$ and $y \in W$, whenever CRy and $t^{\varrho}_{\mathfrak{M}}(x) = t^{\varrho}_{\mathfrak{M}}(y)$, then $y \in C$. The following fundamental properties of descriptive σ-models for wK4 are similar to the corresponding well-known properties of finitely generated descriptive frames for K4 [8,5]:

Lemma 2.1 *Suppose $\mathfrak{M}$ is a descriptive σ-model based on some* wK4*-frame (W,R), $\varrho \subseteq \sigma$, C is a cluster in $\mathfrak{M}$, and Γ a set of σ-formulas. Then the following hold:*

(a) $|C| \leq 2^{|\sigma|}$;

(b) if $\mathfrak{M}, x \models \Diamond \bigwedge \Gamma'$ for every finite $\Gamma' \subseteq \Gamma$, then there is y such that xRy and $\mathfrak{M}, y \models \Gamma$;

(c) if $\Diamond t^{\varrho}_{\mathfrak{M}}(z) \subseteq t^{\varrho}_{\mathfrak{M}}(x)$, then there is y such that xRy and $t^{\varrho}_{\mathfrak{M}}(y) = t^{\varrho}_{\mathfrak{M}}(z)$;

(d) there exists a ϱ-maximal cluster C' such that $CR^{r}C'$ and $t^{\varrho}_{\mathfrak{M}}(C) \subseteq t^{\varrho}_{\mathfrak{M}}(C')$.

Proof. *(a)* It is easy to see that if $x, y \in C$ and $at^{\sigma}_{\mathfrak{M}}(x) = at^{\sigma}_{\mathfrak{M}}(y)$, then $t^{\sigma}_{\mathfrak{M}}(x) = t^{\sigma}_{\mathfrak{M}}(y)$. It follows by **(dif)** that $|C| \leq 2^{|\sigma|}$.

(b) By the assumption, the set $\Xi = \Gamma \cup \{\varphi \mid \Box\varphi \in t^{\sigma}_{\mathfrak{M}}(x)\}$ is finitely satisfiable, and so, by **(com)**, $\mathfrak{M}, y \models \Xi$, for some y, with xRy by **(ref)**.

(c) is a straightforward consequence of *(b)* with $\Gamma = t^{\varrho}_{\mathfrak{M}}(z)$.

(d) Take any $x \in C$ and consider the set

$$\mathcal{X} = \{X' \subseteq W \mid x \in X',\ t^{\varrho}_{\mathfrak{M}}(x') = t^{\varrho}_{\mathfrak{M}}(x) \text{ for all } x' \in X', \text{ and } R^{s} \cap (X' \times X') \text{ is a strict linear order with smallest element } x\}.$$

By Zorn's lemma, there is a $\subseteq$-maximal set $X^{\dagger}$ in $\mathcal{X}$. We claim that there is an R^{s}-maximal point in $X^{\dagger}$. Indeed, suppose otherwise. Then the set

$$\Xi = t^{\varrho}_{\mathfrak{M}}(x) \cup \{\varphi \mid \Box\varphi \in t^{\sigma}_{\mathfrak{M}}(y) \text{ for some } y \in X^{\dagger}\}$$

is finitely satisfiable. By **(com)**, there is z satisfying Ξ, so $t^{\varrho}_{\mathfrak{M}}(z) = t^{\varrho}_{\mathfrak{M}}(x)$ and, by **(ref)**, yRz for all $y \in X^{\dagger}$. Two cases are possible now. (i) If $yR^{s}z$ for all $y \in X^{\dagger}$, then $X^{\dagger} \cup \{z\} \in \mathcal{X}$, and so $z \in X^{\dagger}$ by the $\subseteq$-maximality of $X^{\dagger}$ in $\mathcal{X}$. Thus, z is an R^{s}-maximal point in $X^{\dagger}$, contrary to our assumption. (ii) If zRy, for some $y \in X^{\dagger}$, then by our assumption there is $y' \in X^{\dagger}$ with $yR^{s}y'$, and so $y'Rz$. By (1), either $y' = y$ or $y'Ry$, contrary to $yR^{s}y'$.

Now, let $x^{\dagger}$ be an R^{s}-maximal point in $X^{\dagger}$. We claim that $C(x^{\dagger})$ is ϱ-maximal. Suppose otherwise and there exists $z \in C(x^{\dagger})$ with an R-successor $z' \notin C(x^{\dagger})$ such that $t^{\varrho}_{\mathfrak{M}}(z) = t^{\varrho}_{\mathfrak{M}}(z')$. Then $x^{\dagger}R^{s}z'$. Since $z \in C(x^{\dagger})$, either (i) $z = x^{\dagger}$, and so $t^{\varrho}_{\mathfrak{M}}(z') = t^{\varrho}_{\mathfrak{M}}(x^{\dagger}) = t^{\varrho}_{\mathfrak{M}}(x)$, or (ii) $zRx^{\dagger}$, and so $\Diamond t^{\varrho}_{\mathfrak{M}}(x) = \Diamond t^{\varrho}_{\mathfrak{M}}(x^{\dagger}) \subseteq t^{\varrho}_{\mathfrak{M}}(z) = t^{\varrho}_{\mathfrak{M}}(z')$. By (c), this implies that there is an R-successor z'' of z' with $t^{\varrho}_{\mathfrak{M}}(z'') = t^{\varrho}_{\mathfrak{M}}(x)$. So in both (i) and (ii), there is an R^{s}-successor z'' of $x^{\dagger}$ with $t^{\varrho}_{\mathfrak{M}}(z'') = t^{\varrho}_{\mathfrak{M}}(x)$. By the R^{s}-maximality of $x^{\dagger}$ in $X^{\dagger}$, it follows that $X^{\dagger} \cup \{z''\} \in \mathcal{X}$ and $z'' \notin X^{\dagger}$, contrary to the $\subseteq$-maximality of $X^{\dagger}$ in $\mathcal{X}$.

A similar argument shows that $t^{\varrho}_{\mathfrak{M}}(C(x)) \subseteq t^{\varrho}_{\mathfrak{M}}(C(x^{\dagger}))$. Let $z \in C(x)$. If $z = x$, then clearly $t^{\varrho}_{\mathfrak{M}}(z) \in t^{\varrho}_{\mathfrak{M}}(C(x^{\dagger}))$. So suppose $z \neq x$ and hence $\Diamond t^{\varrho}_{\mathfrak{M}}(z) \subseteq t^{\varrho}_{\mathfrak{M}}(x) = t^{\varrho}_{\mathfrak{M}}(x^{\dagger})$. By (c), there is an R-successor z' of $x^{\dagger}$ with $t^{\varrho}_{\mathfrak{M}}(z') = t^{\varrho}_{\mathfrak{M}}(z)$. We claim that $z' \in C(x^{\dagger})$. Indeed, otherwise we have $x^{\dagger}R^{s}z'$. As $\Diamond t^{\varrho}_{\mathfrak{M}}(x) \subseteq t^{\varrho}_{\mathfrak{M}}(z) = t^{\varrho}_{\mathfrak{M}}(z')$, by (c) there is an R-successor z'' of z' with $t^{\varrho}_{\mathfrak{M}}(z'') = t^{\varrho}_{\mathfrak{M}}(x)$. By (1), it follows that $x^{\dagger}R^{s}z''$. Thus, by the R^{s}-maximality of $x^{\dagger}$ in $X^{\dagger}$, we have $X^{\dagger} \cup \{z''\} \in \mathcal{X}$ and $z'' \notin X^{\dagger}$, contrary to the $\subseteq$-maximality of $X^{\dagger}$ in $\mathcal{X}$.

As $C(x)R^{r}C(x^{\dagger})$, cluster $C' = C(x^{\dagger})$ is as required. □

Let $\mathfrak{M}_i$, $i = 1, 2$, be σ-models based on frames $\mathfrak{F}_i = (W_i, R_i)$ for wK4 and let $\varrho \subseteq \sigma$. A relation $\boldsymbol{\beta} \subseteq W_1 \times W_2$ is called a *ϱ-bisimulation* between $\mathfrak{M}_1$ and $\mathfrak{M}_2$ in case the following conditions hold: whenever $x_1 \boldsymbol{\beta} x_2$,

(atom) $at^{\varrho}_{\mathfrak{M}_1}(x_1) = at^{\varrho}_{\mathfrak{M}_2}(x_2)$;

(move) if $x_1 R_1 y_1$, then there is y_2 such that $x_2 R_2 y_2$ and $y_1 \boldsymbol{\beta} y_2$; and, conversely, if $x_2 R_2 y_2$, then there is y_1 with $x_1 R_1 y_1$ and $y_1 \boldsymbol{\beta} y_2$.

If there is such $\boldsymbol{\beta}$ with $z_1 \boldsymbol{\beta} z_2$, we write $\mathfrak{M}_1, z_1 \sim^{\varrho} \mathfrak{M}_2, z_2$. The following characterisation of bisimulations between descriptive models in terms of types is well-known; see [10] and references therein:

Lemma 2.2 *For any $\varrho \subseteq \sigma$, descriptive σ-models $\mathfrak{M}_i$, $i = 1, 2$, and $x_i \in W_i$,*

$$t^{\varrho}_{\mathfrak{M}_1}(x_1) = t^{\varrho}_{\mathfrak{M}_2}(x_2) \quad \textit{iff} \quad \mathfrak{M}_1, x_1 \sim^{\varrho} \mathfrak{M}_2, x_2.$$

The implication ($\Leftarrow$) holds for arbitrary (not necessarily descriptive) models.

Variations of the next criterion of interpolant (non-)existence are implicit in various (dis-)proofs of the CIP in modal logics [16,10]:

Lemma 2.3 *Let $\sigma = sig(\varphi) \cup sig(\psi)$ and $L \in \{\mathsf{wK4}, \mathsf{DL}\}$. Then $\varphi \to \psi$ has no interpolant in L iff there are descriptive σ-models $\mathfrak{M}_\varphi$ and $\mathfrak{M}_\psi$ based on frames for L with roots r_φ and r_ψ, respectively, such that*

(a) $\mathfrak{M}_\varphi, r_\varphi \models \varphi$ *and* $\mathfrak{M}_\psi, r_\psi \models \neg\psi$;

(b) $\mathfrak{M}_\varphi, r_\varphi \sim^\varrho \mathfrak{M}_\psi, r_\psi$, *where* $\varrho = sig(\varphi) \cap sig(\psi)$.

As both wK4 *and* DL *are canonical, the requirement that models* $\mathfrak{M}_\varphi$ *and* $\mathfrak{M}_\psi$ *be descriptive can be omitted.*

We first apply this criterion to decide the IEP for the difference logic DL. A key observation is that, from any two σ-models $\mathfrak{M}_\varphi$ and $\mathfrak{M}_\psi$ witnessing the nonexistence of an interpolant for a given $\varphi \to \psi$ in DL in the sense of Lemma 2.3, we can extract sub-models of polynomial size in $|\varphi|$ and $|\psi|$ that also satisfy the above criterion. We call this phenomenon the *polysize bisimilar model property* of DL, which clearly implies that the IEP for DL is decidable in coNP. Indeed, to check that $\varphi \to \psi$ has no interpolant in DL, we can guess polynomial-size $\mathfrak{M}_\varphi$ and $\mathfrak{M}_\psi$ together with a relation $\boldsymbol{\beta}$ between them and then check whether they satisfy the criterion of Lemma 2.3.

We remind the reader that rooted frames for DL (and so the frames $\mathfrak{M}_\varphi$ and $\mathfrak{M}_\psi$ are based on) are clusters, containing possibly both reflexive and irreflexive points. To show the polysize bisimilar model property of DL, we proceed in two steps. First, for every $\alpha \in sub(\varphi)$ ($\alpha \in sub(\psi)$) satisfiable in $\mathfrak{M}_\varphi$ (respectively, $\mathfrak{M}_\psi$), we pick two points x_α, x'_α satisfying α in $\mathfrak{M}_\varphi$ (in $\mathfrak{M}_\psi$) if they exist, otherwise a single such point x_α. Denote the set of the points selected this way by M_φ (M_ψ), assuming that $r_\varphi \in M_\varphi$ and $r_\psi \in M_\psi$. Let

$$T = \{t^\varrho_{\mathfrak{M}_\varphi}(x) \mid x \in M_\varphi\} \cup \{t^\varrho_{\mathfrak{M}_\psi}(x) \mid x \in M_\psi\}.$$

As $\mathfrak{M}_\varphi, r_\varphi \sim^\varrho \mathfrak{M}_\psi, r_\psi$, every ϱ-type $t \in T$ is satisfied in both $\mathfrak{M}_\varphi$ and $\mathfrak{M}_\psi$. Now, for each $t \in T$, we pick two distinct points satisfying t in $\mathfrak{M}_\varphi$, if they exist, and otherwise a single such point, and add them to M_φ if they were not already there. We do the same for $\mathfrak{M}_\psi$ and M_ψ. Let $\mathfrak{M}^\dagger_\varphi$ and $\mathfrak{M}^\dagger_\psi$ be the restrictions of $\mathfrak{M}_\varphi$ and $\mathfrak{M}_\psi$ to the resulting M_φ and M_ψ, and let

$$\boldsymbol{\beta}^\dagger = \{(x, x') \in M_\varphi \times M_\psi \mid t^\varrho_{\mathfrak{M}_\varphi}(x) = t^\varrho_{\mathfrak{M}_\psi}(x')\}.$$

Lemma 2.4 *(a)* $\mathfrak{M}^\dagger_\varphi, r_\varphi \models \varphi$, $\mathfrak{M}^\dagger_\psi, r_\psi \models \neg\psi$, *and (b)* $\boldsymbol{\beta}^\dagger$ *is a* ϱ*-bisimulation between* $\mathfrak{M}^\dagger_\varphi$ *and* $\mathfrak{M}^\dagger_\psi$ *with* $r_\varphi \boldsymbol{\beta}^\dagger r_\psi$.

Proof. Suppose that $\mathfrak{M}_\varphi$ and $\mathfrak{M}_\psi$ are based on the respective clusters (W_φ, R_φ) and (W_ψ, R_ψ).

(a) $\mathfrak{M}^\dagger_\varphi, r_\varphi \models \varphi$ follows from the fact that, for any $\chi \in sub(\varphi)$ and $x \in M_\varphi$, $\mathfrak{M}_\varphi, x \models \chi$ iff $\mathfrak{M}^\dagger_\varphi, x \models \chi$, which can be established by a straightforward induction on the construction of φ. We only show ($\Rightarrow$) for $\chi = \Diamond\alpha$. Suppose that $\Diamond\alpha \in sub(\varphi)$, $x \in M_\varphi$, and $\mathfrak{M}_\varphi, x \models \Diamond\alpha$. If there exist two points in $\mathfrak{M}_\varphi$ satisfying α, then either x_α or x'_α is distinct from x, and so $\mathfrak{M}^\dagger_\varphi, x \models \Diamond\alpha$. Otherwise, $x R_\varphi x_\alpha$ must hold, and so we also have $\mathfrak{M}^\dagger_\varphi, x \models \Diamond\alpha$. (A similar argument shows that $\mathfrak{M}^\dagger_\psi, r_\psi \models \neg\psi$.)

(b) Suppose $x, y \in M_\varphi$, $x' \in M_\psi$, $x \boldsymbol{\beta}^\dagger x'$, and $x R_\varphi y$. Then $t^\varrho_{\mathfrak{M}_\varphi}(x) = t^\varrho_{\mathfrak{M}_\psi}(x')$ and $t^\varrho_{\mathfrak{M}_\varphi}(y) \in T$. There are two cases: *(i)* If $t^\varrho_{\mathfrak{M}_\varphi}(x) \neq t^\varrho_{\mathfrak{M}_\varphi}(y)$, then there is $y' \in M_\psi$ such that $t^\varrho_{\mathfrak{M}_\varphi}(y) = t^\varrho_{\mathfrak{M}_\varphi}(y')$ and $y' \neq x'$. *(ii)* If $t^\varrho_{\mathfrak{M}_\varphi}(x) = t^\varrho_{\mathfrak{M}_\varphi}(y) = t$,

then $\Diamond t \subseteq t$, and so either (*ii*.1) there are two distinct points in M_ψ satisfying t in $\mathfrak{M}_\psi$ or (*ii*.2) there is a single reflexive point in M_ψ satisfying t in $\mathfrak{M}_\psi$. In case (*ii*.1) one of these two points must be different from x', and in case (*ii*.2) this single reflexive point must be x'. Thus, in all cases, we have a point $y' \in M_\psi$ with $x' R_\psi y'$ and $t^\varrho_{\mathfrak{M}_\psi}(y) = t^\varrho_{\mathfrak{M}_\psi}(y')$, and so $y \boldsymbol{\beta}^\dagger y'$. □

By the construction, $|M_\varphi|$ and $|M_\psi|$ are polynomial in $|\varphi|$ and $|\psi|$. Thus, we obtain:

Theorem 2.5 (*a*) DL *enjoys the polysize bisimilar model property.*
(*b*) *The interpolant existence property for* DL *is* coNP*-complete.*

3 Deciding Interpolant Existence for wK4

In this section we show that wK4 has the 3-exponential-size bisimilar model property, which means that the IEP is decidable in coN3ExpTime.

Given formulas φ and ψ, let $sub(\varphi,\psi) = sub(\varphi) \cup sub(\psi)$, $\sigma = sig(\varphi) \cup sig(\psi)$ and $\varrho = sig(\varphi) \cap sig(\psi)$. If $\varphi \to \psi$ does not have an interpolant in wK4, then Lemma 2.3 provides two pointed descriptive σ-models $\mathfrak{M}_\varphi, r_\varphi$ and $\mathfrak{M}_\psi, r_\psi$ based on weakly transitive frames. To simplify notation, we will operate with a single descriptive σ-model $\mathfrak{M}$ based on a weakly transitive frame $\mathfrak{F} = (W, R)$—the disjoint union of $\mathfrak{M}_\varphi$ and $\mathfrak{M}_\psi$—containing two points r_φ, r_ψ such that $\mathfrak{M}, r_\varphi \models \varphi$, $\mathfrak{M}, r_\psi \models \neg\psi$ and $\mathfrak{M}, r_\varphi \sim^\varrho \mathfrak{M}, r_\psi$. Our aim is to convert $\mathfrak{M}$ into a model $\mathfrak{M}^\dagger$ based on a weakly transitive frame $\mathfrak{F}^\dagger = (W^\dagger, R^\dagger)$ that still witnesses the lack of an interpolant for $\varphi \to \psi$ in the above sense, and has $W^\dagger$ of triple-exponential size in $|sub(\varphi,\psi)|$.

Given a point x in $\mathfrak{M}$, we define the φ,ψ-*type* $\boldsymbol{t}^{\varphi,\psi}_{\mathfrak{M}}(x) = t^\sigma_{\mathfrak{M}}(x) \cap sub(\varphi,\psi)$. For a set X of points in $\mathfrak{M}$, we let $\boldsymbol{t}^{\varphi,\psi}_{\mathfrak{M}}(X) = \{\boldsymbol{t}^{\varphi,\psi}_{\mathfrak{M}}(x) \mid x \in X\}$. Our construction is an elaborate $sub(\varphi,\psi)$-filtration. (For the well-known *filtration* techniques in modal logic see, e.g., [5].) As usual, it keeps track of the φ,ψ-types of points in $\mathfrak{M}$ to satisfy condition (*a*) of Lemma 2.3. In addition, some connections of these φ,ψ-types with the ϱ-types of points in $\mathfrak{M}$ are also noted in order to satisfy condition (*b*) of Lemma 2.3. When applied to satisfiability checking, our construction reduces to the filtration of [13]; see Remark 3.12.

To begin with, we define an equivalence relation $\approx$ between clusters in $\mathfrak{M}$ by taking $C \approx C'$ iff there exists a sequence $C = C_0, \dots, C_n = C'$ of clusters in $\mathfrak{M}$ such that, for each $i < n$ there are $x_i \in C_i$ and $y_{i+1} \in C_{i+1}$ with $t^\varrho_{\mathfrak{M}}(x_i) = t^\varrho_{\mathfrak{M}}(y_{i+1})$. Let $[C] = \{C' \mid C' \approx C\}$ and $T^\varrho[C] = \bigcup_{C' \in [C]} t^\varrho_{\mathfrak{M}}(C')$. It follows from the definition that $[C] = \{C' \mid t^\varrho_{\mathfrak{M}}(C') \cap T^\varrho[C] \neq \emptyset\}$. By Lemma 2.1 (*d*), for every $C' \in [C]$, there is a ϱ-maximal $D \in [C]$ with $C' R^r D$ and $t^\varrho_{\mathfrak{M}}(C') \subseteq t^\varrho_{\mathfrak{M}}(D)$.

Lemma 3.1 *If $D \in [C]$ is a ϱ-maximal cluster, then $T^\varrho[C] = t^\varrho_{\mathfrak{M}}(D)$, and so $|T^\varrho[C]| \leq 2^{|\varrho|}$.*

Proof. Let $D \in [C]$ be a ϱ-maximal cluster. Clearly, it is enough to prove that $T^\varrho[C] \subseteq t^\varrho_{\mathfrak{M}}(D)$. Let $x \in C'$, for some $C' \in [C]$. We need to show that there exists $y \in D$ with $t^\varrho_{\mathfrak{M}}(y) = t^\varrho_{\mathfrak{M}}(x)$. As $D \approx C'$, by the definition of $\approx$, we have $z \in D$ such that $\Diamond^n t^{\varrho}_{\mathfrak{M}}(z) \subseteq t^{\varrho}_{\mathfrak{M}}(x)$ and $\Diamond^n t^{\varrho}_{\mathfrak{M}}(x) \subseteq t^{\varrho}_{\mathfrak{M}}(z)$, for some $n \geq 0$. If $n = 0$, then we take $y = z$. If $n > 0$, then by repeated applications of Lemma 2.1 (c), we obtain $u, v \in W$ with $zR^n uR^n v$, $t^{\varrho}_{\mathfrak{M}}(u) = t^{\varrho}_{\mathfrak{M}}(x)$ and $t^{\varrho}_{\mathfrak{M}}(v) = t^{\varrho}_{\mathfrak{M}}(z)$. By weak transitivity, $z = v$ or zRv. In either case, by the ϱ-maximality of D, we must have $v \in D$, and so $u \in D$, yielding $y = u$. □

For every equivalence class $[C]$ in $\mathfrak{M}$, we let $AT^{\varrho}[C] = \{t \cap \varrho \mid t \in T^{\varrho}[C]\}$. Given a cluster C, we define the *cluster-type* of C in $\mathfrak{M}$ as the function $\tau_C \colon AT^{\varrho}[C] \to 2^{\boldsymbol{t}^{\varphi,\psi}_{\mathfrak{M}}(C)}$ where, for any $a \in AT^{\varrho}[C]$,

$$\tau_C(a) = \{\boldsymbol{t}^{\varphi,\psi}_{\mathfrak{M}}(x) \mid x \in C,\ \boldsymbol{t}^{\varphi,\psi}_{\mathfrak{M}}(x) \cap \varrho = at^{\varrho}_{\mathfrak{M}}(x) = a\}.$$

Observe that $AT^{\varrho}[C'] = AT^{\varrho}[C]$ for every $C' \in [C]$, and so τ_C and $\tau_{C'}$ have the same domain $\mathsf{dom}\,\tau_C = \mathsf{dom}\,\tau_{C'} = AT^{\varrho}[C]$. As $\bigcup_{a \in \mathsf{dom}\,\tau_C} \tau_C(a) = \boldsymbol{t}^{\varphi,\psi}_{\mathfrak{M}}(C)$, τ_C keeps a record of both φ,ψ-types and atomic ϱ-types of points in C in the context of the whole equivalence class $[C]$. Also, $\tau_C(a)$ might be empty for some $a \in \mathsf{dom}\,\tau_C$, but if C is ϱ-maximal then $\tau_C(a) \neq \emptyset$ for all $a \in \mathsf{dom}\,\tau_C$. Note that the number of pairwise distinct cluster-types in $\mathfrak{M}$ does not exceed $2^{|\varrho|} \cdot \left(2^{2^{|sub(\varphi,\psi)|}}\right)^{2^{|\varrho|}} = 2^{|\varrho| + 2^{|sub(\varphi,\psi)| + |\varrho|}}$.

By the *mosaic of* $[C]$ *in* $\mathfrak{M}$ we mean the set $M_{[C]} = \{\tau_{C'} \mid C' \in [C]\}$ of cluster-types of the same domain (also called as the *domain of* $M_{[C]}$ and denoted by $\mathsf{dom}\,M_{[C]}$). M is a *mosaic in* $\mathfrak{M}$ if $M = M_{[C]}$ for some C. Clearly, the number of pairwise distinct mosaics in $\mathfrak{M}$ is $\mathcal{O}\big(2^{2^{2^{|sub(\varphi,\psi)|}}}\big)$.

We are now in a position to define the model $\mathfrak{M}^{\dagger} = (\mathfrak{F}^{\dagger}, \mathfrak{v}^{\dagger})$ and its underlying frame $\mathfrak{F}^{\dagger} = (W^{\dagger}, R^{\dagger})$. Suppose $x \in W$. Then we set

$$\boldsymbol{w}(x) = \big(\boldsymbol{t}^{\varphi,\psi}_{\mathfrak{M}}(x), at^{\varrho}_{\mathfrak{M}}(x), \tau_{C(x)}, M_{[C(x)]}\big) \quad \text{and} \quad W^{\dagger} = \{\boldsymbol{w}(x) \mid x \in W\}.$$

Observe that if $\boldsymbol{w} \in W^{\dagger}$ and $\boldsymbol{w} = (\boldsymbol{t}, a, \tau, M)$, then $a = \boldsymbol{t} \cap \varrho$, $\tau \in M$ and $\boldsymbol{t} \in \tau(a)$ always hold. Moreover,

$$(\boldsymbol{t}, a, \tau, M) \in W^{\dagger}, \text{ for all mosaics } M,\ \tau \in M,\ a \in \mathsf{dom}\,M, \text{ and } \boldsymbol{t} \in \tau(a). \quad (2)$$

We call M and a the *mosaic* and the *ϱ-index* of $\boldsymbol{w}$, respectively. Later on, we shall see that $\boldsymbol{w}$ and $\boldsymbol{w}'$ are ϱ-bisimilar if they share the same mosaic and ϱ-index. By the above calculations, $|W^{\dagger}| = \mathcal{O}\big(2^{2^{2^{|sub(\varphi,\psi)|}}}\big)$. The valuation $\mathfrak{v}^{\dagger}$ on $W^{\dagger}$ is inherited from $\mathfrak{v}$ in $\mathfrak{M}$: for every $p \in \sigma$,

$$\mathfrak{v}^{\dagger}(p) = \{\boldsymbol{w}(x) \mid x \in \mathfrak{v}(p)\} = \{\boldsymbol{w}(x) \mid p \in \boldsymbol{t}^{\varphi,\psi}_{\mathfrak{M}}(x),\ x \in W\}.$$

To define the accessibility relation $R^{\dagger}$ in $\mathfrak{F}^{\dagger}$, we require some new notions. Let $\boldsymbol{t}$ and $\boldsymbol{t}'$ be φ,ψ-types, $\tau_C \colon AT^{\varrho}[C] \to 2^{\boldsymbol{t}^{\varphi,\psi}_{\mathfrak{M}}(C)}$ and $\tau_{C'} \colon AT^{\varrho}[C'] \to 2^{\boldsymbol{t}^{\varphi,\psi}_{\mathfrak{M}}(C')}$ cluster-types, M and M' mosaics. Define a relation $\twoheadrightarrow$ between such pairs by taking:

- $\boldsymbol{t} \twoheadrightarrow \boldsymbol{t}'$ iff, for every $\Diamond\psi \in sub(\varphi,\psi)$, whenever χ or $\Diamond\chi$ is in $\boldsymbol{t}'$, then $\Diamond\chi \in \boldsymbol{t}$;

- $\tau_C \twoheadrightarrow \tau_{C'}$ iff $\boldsymbol{t} \twoheadrightarrow \boldsymbol{t}'$ for all $\boldsymbol{t} \in \boldsymbol{t}_{\mathfrak{M}}^{\varphi,\psi}(C)$ and $\boldsymbol{t}' \in \boldsymbol{t}_{\mathfrak{M}}^{\varphi,\psi}(C')$;
- $M \twoheadrightarrow M'$ iff for every $\tau \in M$ there is $\tau' \in M'$ with $\tau \twoheadrightarrow \tau'$.

It is readily checked that the defined relation $\twoheadrightarrow$ has the following properties:

$$\twoheadrightarrow \text{ is transitive in all three settings;} \tag{3}$$

$$\text{if } xRy, \text{ then } \boldsymbol{t}_{\mathfrak{M}}^{\varphi,\psi}(x) \twoheadrightarrow \boldsymbol{t}_{\mathfrak{M}}^{\varphi,\psi}(y); \tag{4}$$

$$\text{if } CRC' \text{ and } C \neq C', \text{ then } \tau_C \twoheadrightarrow \tau_{C'}. \tag{5}$$

Note that CRC does not necessarily imply $\tau_C \twoheadrightarrow \tau_C$; see Example 3.2. By (5), Lemmas 2.1 (d) and 3.1, we also have that

$$\text{for all } C \text{ there is a } \varrho\text{-maximal } D \in [C] \text{ such that } C = D \text{ or } \tau_C \twoheadrightarrow \tau_D. \tag{6}$$

Next, for any mosaic M, we define a subset I_M of $\mathsf{dom}\, M$ by taking

$$I_M = \{a \in \mathsf{dom}\, M \mid \Diamond t \not\subseteq t, \text{ for all clusters } C \text{ with } M = M_{[C]} \text{ and all } t \in T^{\varrho}[C] \text{ with } t \cap \varrho = a\}.$$

As $a \in I_M$ implies that for every cluster C with $M = M_{[C]}$ there is at most one $x \in C$ such that $\boldsymbol{t}_{\mathfrak{M}}^{\varphi,\psi}(x) \in \tau_C(a)$ and such an x is irreflexive, it follows that

$$\text{if } \tau_C \in M, \text{ then } |\tau_C(a)| \leq 1 \text{ for every } a \in I_M. \tag{7}$$

We also claim that

$$\text{if } C \text{ is not } \varrho\text{-maximal and } \tau_C \in M, \text{ then } \tau_C(a) = \emptyset \text{ for every } a \in I_M. \tag{8}$$

Indeed, by Lemmas 2.1 (d) and 3.1, there is a ϱ-maximal $D \in [C(x)]$ with $C \neq D$ and CRD. Let $a \in \mathsf{dom}\, M_{[C]}$ be such that $\tau_C(a) \neq \emptyset$. There there is $x \in C$ with $t_{\mathfrak{M}}^{\varrho}(x) \cap \varrho = a$. By Lemma 3.1, there is $y \in D$ with $t_{\mathfrak{M}}^{\varrho}(y) = t_{\mathfrak{M}}^{\varrho}(x)$. Thus, $\Diamond t_{\mathfrak{M}}^{\varrho}(x) \subseteq t_{\mathfrak{M}}^{\varrho}(x)$, and so $a \notin I_{M_{[C]}}$.

Now, to define $R^\dagger$ on $W^\dagger$, suppose $\boldsymbol{w}, \boldsymbol{w}' \in W^\dagger$, where

$$\boldsymbol{w} = (\boldsymbol{t}, a, \tau, M), \qquad \boldsymbol{w}' = (\boldsymbol{t}', a', \tau', M').$$

The definition of $R^\dagger$ for $\boldsymbol{w}$, $\boldsymbol{w}'$ depends on whether $M = M'$ and $M \twoheadrightarrow M$:

Case $M \neq M'$: then $\boldsymbol{w} R^\dagger \boldsymbol{w}'$ iff $M \twoheadrightarrow M'$ and $\tau \twoheadrightarrow \tau'$.

Case $M = M'$, $M \twoheadrightarrow M$: then $\boldsymbol{w} R^\dagger \boldsymbol{w}'$ iff
- either $\boldsymbol{w} \neq \boldsymbol{w}'$ and ($\tau = \tau'$ or $\tau \twoheadrightarrow \tau'$),
- or $\boldsymbol{w} = \boldsymbol{w}'$ and $\boldsymbol{t} \twoheadrightarrow \boldsymbol{t}$.

Case $M = M'$, $M \not\twoheadrightarrow M$: then $\boldsymbol{w} R^\dagger \boldsymbol{w}'$ iff
- either $\boldsymbol{w} \neq \boldsymbol{w}'$ and $\big(\tau = \tau'$ or $(\tau \twoheadrightarrow \tau'$ and $\tau(b) = \emptyset$ for all $b \in I_M)\big)$,
- or $\boldsymbol{w} = \boldsymbol{w}'$, $\boldsymbol{t} \twoheadrightarrow \boldsymbol{t}$, and $a \notin I_M$.

Example 3.2 Consider φ, ψ, $\mathfrak{M}_\varphi$ and $\mathfrak{M}_\psi$ with $\sigma = \{p, q\}$ and $\varrho = \emptyset$ from Example 1.1 (i). Then $sub(\varphi, \psi)$ consists of the formulas p, q, $\Diamond p$, $\Diamond\neg q$, $\Diamond\Diamond p$,

$\Diamond\Diamond\neg q$, φ, ψ and their negations. Let $\mathfrak{M}$ be the disjoint union of $\mathfrak{M}_\varphi$ and $\mathfrak{M}_\psi$. Then $C(r_\varphi) = \{r_\varphi, x_1\} = C(x_1)$, $C(r_\psi) = \{r_\psi\}$, $C(x_2) = \{x_2\}$, with all of these clusters being $\approx$-equivalent, $T^\varrho[C(r_\varphi)] = \{t\}$ for $t = t^\varrho_{\mathfrak{M}}(r_\varphi) = \{\Diamond^n\top \mid n < \omega\}$, and $\mathsf{dom}\,\tau_{C(r_\varphi)} = \mathsf{dom}\,\tau_{C(r_\psi)} = \mathsf{dom}\,\tau_{C(x_2)} = AT^\varrho[C(r_\varphi)] = \{\emptyset\}$. This gives $\tau_{C(r_\varphi)}(\emptyset) = \{\boldsymbol{t}^{\varphi,\psi}_{\mathfrak{M}}(r_\varphi), \boldsymbol{t}^{\varphi,\psi}_{\mathfrak{M}}(x_1)\}$, where

$$\boldsymbol{t}^{\varphi,\psi}_{\mathfrak{M}}(r_\varphi) = \{p, \neg q, \neg\Diamond p, \Diamond\neg q, \Diamond\Diamond p, \Diamond\Diamond\neg q, \varphi, \psi\},$$
$$\boldsymbol{t}^{\varphi,\psi}_{\mathfrak{M}}(x_1) = \{\neg p, \neg q, \Diamond p, \Diamond\neg q, \Diamond\Diamond p, \Diamond\Diamond\neg q, \neg\varphi, \psi\}.$$

On the other hand, $\tau_{C(r_\psi)}(\emptyset) = \{\boldsymbol{t}^{\varphi,\psi}_{\mathfrak{M}}(r_\psi)\}$ and $\tau_{C(x_2)}(\emptyset) = \{\boldsymbol{t}^{\varphi,\psi}_{\mathfrak{M}}(x_2)\}$, where

$$\boldsymbol{t}^{\varphi,\psi}_{\mathfrak{M}}(r_\psi) = \{\neg p, \neg q, \neg\Diamond p, \neg\Diamond\neg q, \neg\Diamond\Diamond p, \neg\Diamond\Diamond\neg q, \neg\varphi, \neg\psi\},$$
$$\boldsymbol{t}^{\varphi,\psi}_{\mathfrak{M}}(x_2) = \{\neg p, q, \neg\Diamond p, \neg\Diamond\neg q, \neg\Diamond\Diamond p, \neg\Diamond\Diamond\neg q, \neg\varphi, \psi\}.$$

Then $\boldsymbol{w}(r_\varphi)$, $\boldsymbol{w}(x_1)$, $\boldsymbol{w}(r_\psi)$, and $\boldsymbol{w}(x_2)$ are all different, but they share the same ϱ-index $\emptyset$ and the same mosaic $M_{[C(r_\varphi)]} = \{\tau_{C(r_\varphi)}, \tau_{C(r_\psi)}, \tau_{C(x_2)}\}$ with $\mathsf{dom}\,M_{[C(r_\varphi)]} = \{\emptyset\}$. As $\Diamond t \subseteq t$, we have $I_{M_{[C(r_\varphi)]}} = \emptyset$.

We have $C(r_\varphi)RC(r_\varphi)$ but $\tau_{C(r_\varphi)} \not\twoheadrightarrow \tau_{C(r_\varphi)}$ as $\boldsymbol{t}^{\varphi,\psi}_{\mathfrak{M}}(r_\varphi) \not\twoheadrightarrow \boldsymbol{t}^{\varphi,\psi}_{\mathfrak{M}}(r_\varphi)$ because $p \in \boldsymbol{t}^{\varphi,\psi}_{\mathfrak{M}}(r_\varphi)$ but $\Diamond p \notin \boldsymbol{t}^{\varphi,\psi}_{\mathfrak{M}}(r_\varphi)$. So $\boldsymbol{w}(r_\varphi)R^\dagger\boldsymbol{w}(r_\varphi)$ does not hold. In fact, it is not hard to check that $\mathfrak{M}^\dagger$ is isomorphic to $\mathfrak{M}$. For example, we do not have $\boldsymbol{w}(r_\psi)R^\dagger\boldsymbol{w}(r_\psi)$ as $\boldsymbol{t}^{\varphi,\psi}_{\mathfrak{M}}(r_\psi) \not\twoheadrightarrow \boldsymbol{t}^{\varphi,\psi}_{\mathfrak{M}}(r_\psi)$ because $\neg q \in \boldsymbol{t}^{\varphi,\psi}_{\mathfrak{M}}(r_\psi)$ but $\Diamond\neg q \notin \boldsymbol{t}^{\varphi,\psi}_{\mathfrak{M}}(r_\psi)$. But we obtain $\boldsymbol{w}(r_\varphi)R^\dagger\boldsymbol{w}(x_1)$ because they share the same cluster-type $\tau_{C(r_\varphi)}$, and $\boldsymbol{w}(x_1)R^\dagger\boldsymbol{w}(x_1)$ because $\boldsymbol{t}^{\varphi,\psi}_{\mathfrak{M}}(x_1) \twoheadrightarrow \boldsymbol{t}^{\varphi,\psi}_{\mathfrak{M}}(x_1)$. ⊣

Example 3.3 We now illustrate the role of the set I_M in the definition of $R^\dagger$. Consider the model $\mathfrak{M}$ below and $\varphi = p \wedge \neg\Diamond q$ and $\psi = \neg p \wedge \Diamond r$ with $\varrho = \{p\}$ and $\sigma = \{p, q, r\}$. (The form of φ, ψ is not important here, but $sub(\varphi, \psi)$ is.)

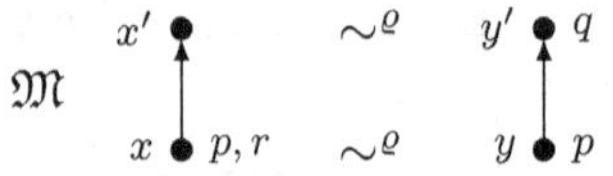

Then $t^\varrho_{\mathfrak{M}}(x) = t^\varrho_{\mathfrak{M}}(y) = t$ with $\Diamond t \not\subseteq t$. Let $\boldsymbol{t}^{\varphi,\psi}_{\mathfrak{M}}(x) = \boldsymbol{t}_x$ and $\boldsymbol{t}^{\varphi,\psi}_{\mathfrak{M}}(y) = \boldsymbol{t}_y$. Then $\boldsymbol{t}_x \not\twoheadrightarrow \boldsymbol{t}_x$ because $r \in \boldsymbol{t}^{\varphi,\psi}_{\mathfrak{M}}(x)$ and $\Diamond r \notin \boldsymbol{t}^{\varphi,\psi}_{\mathfrak{M}}(x)$, while $\boldsymbol{t}_y \twoheadrightarrow \boldsymbol{t}_y$ because $\Diamond p \notin sub(\varphi, \psi)$; we also have $\boldsymbol{t}_x \not\twoheadrightarrow \boldsymbol{t}_y$ and $\boldsymbol{t}_y \not\twoheadrightarrow \boldsymbol{t}_x$. Now, consider

$$\boldsymbol{w}_x = (\boldsymbol{t}_x, \{p\}, \tau_{C(x)}, M), \quad \boldsymbol{w}_y = (\boldsymbol{t}_y, \{p\}, \tau_{C(y)}, M),$$

where $T^\varrho[C(x)] = \{t\}$, $[C(x)] = \{C(x), C(y)\}$, and $M = \{\tau_{C(x)}, \tau_{C(y)}\}$ with $\mathsf{dom}\,M = AT^\varrho[C(x)] = \{\{p\}\}$, $\tau_{C(x)}(\{p\}) = \{\boldsymbol{t}_x\}$, and $\tau_{C(y)}(\{p\}) = \{\boldsymbol{t}_y\}$. Then $M \not\twoheadrightarrow M$ and $I_M = \{\{p\}\}$. By the last item in the definition of $R^\dagger$, neither $\boldsymbol{w}_xR^\dagger\boldsymbol{w}_x$ nor $\boldsymbol{w}_yR^\dagger\boldsymbol{w}_y$ holds because $\{p\} \in I_M$. However, without the condition $a \notin I_M$ in the definition, we would have $\boldsymbol{w}_yR^\dagger\boldsymbol{w}_y$ but still not $\boldsymbol{w}_xR^\dagger\boldsymbol{w}_x$, which would destroy the ϱ-bisimilarity of $\boldsymbol{w}_x$ and $\boldsymbol{w}_y$. ⊣

Lemma 3.4 *The relation $R^\dagger$ on $W^\dagger$ is weakly transitive.*

Proof. Suppose $\boldsymbol{w}R^{\dagger}\boldsymbol{w}'R^{\dagger}\boldsymbol{w}''$ and $\boldsymbol{w} \neq \boldsymbol{w}''$, where

$$\boldsymbol{w} = (\boldsymbol{t}, a, \tau, M), \quad \boldsymbol{w}' = (\boldsymbol{t}', a', \tau', M'), \quad \boldsymbol{w}'' = (\boldsymbol{t}'', a'', \tau'', M'').$$

We need to show, by a straightforward checking of all the cases in the definition of $R^{\dagger}$, that $\boldsymbol{w}R^{\dagger}\boldsymbol{w}''$. This is trivial if $\boldsymbol{w} = \boldsymbol{w}'$ or $\boldsymbol{w}' = \boldsymbol{w}''$, so we assume that $\boldsymbol{w} \neq \boldsymbol{w}'$ and $\boldsymbol{w}' \neq \boldsymbol{w}''$.

Suppose first that $M \neq M''$. We need to show that $M \twoheadrightarrow M''$ and $\tau \twoheadrightarrow \tau''$.

Case $M \neq M'$, $M' \neq M''$: By the definition of $R^{\dagger}$, we have $M \twoheadrightarrow M' \twoheadrightarrow M''$ and $\tau \twoheadrightarrow \tau' \twoheadrightarrow \tau''$, which gives $M \twoheadrightarrow M''$ and $\tau \twoheadrightarrow \tau''$ by (3).

Case $M \neq M'$, $M' = M''$: Then $M \twoheadrightarrow M'$ and $\tau \twoheadrightarrow \tau'$ because $\boldsymbol{w}R^{\dagger}\boldsymbol{w}'$, and so $M \twoheadrightarrow M''$. Then no matter whether $M' \twoheadrightarrow M''$ or $M' \not\twoheadrightarrow M''$, we have either $\tau' = \tau''$ or $\tau' \twoheadrightarrow \tau''$ as $\boldsymbol{w}'R^{\dagger}\boldsymbol{w}''$ and $\boldsymbol{w}' \neq \boldsymbol{w}''$. Thus, $\tau \twoheadrightarrow \tau''$ by (3).

Case $M = M'$, $M' \neq M''$: Then $M' \twoheadrightarrow M''$ and $\tau' \twoheadrightarrow \tau''$ because $\boldsymbol{w}'R^{\dagger}\boldsymbol{w}''$, and so $M \twoheadrightarrow M''$. Then no matter whether $M \twoheadrightarrow M'$ or $M \not\twoheadrightarrow M'$, either $\tau = \tau'$ or $\tau \twoheadrightarrow \tau'$, because $\boldsymbol{w}R^{\dagger}\boldsymbol{w}'$ and $\boldsymbol{w} \neq \boldsymbol{w}'$. Thus, $\tau \twoheadrightarrow \tau''$ by (3).

Now suppose $M = M''$. If $M \neq M'$, then $M' \neq M''$. By the definition of $R^{\dagger}$, we have $M \twoheadrightarrow M' \twoheadrightarrow M$ and $\tau \twoheadrightarrow \tau' \twoheadrightarrow \tau''$, which gives $M \twoheadrightarrow M$ and $\tau \twoheadrightarrow \tau''$ by (3). As $\boldsymbol{w} \neq \boldsymbol{w}''$, these imply $\boldsymbol{w}R^{\dagger}\boldsymbol{w}''$.

Finally, suppose $M = M' = M''$. Then two cases are possible.

Case $M \twoheadrightarrow M$: As $\boldsymbol{w} \neq \boldsymbol{w}''$, we need to show that $\tau = \tau''$ or $\tau \twoheadrightarrow \tau''$. As $\boldsymbol{w}R^{\dagger}\boldsymbol{w}'$ and $\boldsymbol{w} \neq \boldsymbol{w}'$, we have $\tau = \tau'$ or $\tau \twoheadrightarrow \tau'$. Similarly, as $\boldsymbol{w}'R^{\dagger}\boldsymbol{w}''$ and $\boldsymbol{w}' \neq \boldsymbol{w}''$, we have $\tau' = \tau''$ or $\tau' \twoheadrightarrow \tau''$, which yields the required.

Case $M \not\twoheadrightarrow M$: We need to show that $\tau = \tau''$ or ($\tau \twoheadrightarrow \tau''$ and $\tau(b) = \emptyset$ for all $b \in I_M$). As $\boldsymbol{w}R^{\dagger}\boldsymbol{w}'$ and $\boldsymbol{w} \neq \boldsymbol{w}'$, we have $\tau = \tau'$ or ($\tau \twoheadrightarrow \tau'$ and $\tau(b) = \emptyset$ for all $b \in I_M$). Similarly, as $\boldsymbol{w}'R^{\dagger}\boldsymbol{w}''$ and $\boldsymbol{w}' \neq \boldsymbol{w}''$, we have $\tau' = \tau''$ or ($\tau' \twoheadrightarrow \tau''$ and $\tau'(b) = \emptyset$ for all $b \in I_M$). By (3), it follows that $\tau = \tau''$ or ($\tau \twoheadrightarrow \tau''$ and $\tau(b) = \emptyset$ for all $b \in I_M$). As $\boldsymbol{w} \neq \boldsymbol{w}''$, these imply $\boldsymbol{w}R^{\dagger}\boldsymbol{w}''$.

This completes the proof of the lemma. □

The next lemma says that $R^{\dagger}$ contains the smallest $sub(\varphi, \psi)$-filtration:

Lemma 3.5 *For all $x, y \in W$, if xRy, then $\boldsymbol{w}(x)R^{\dagger}\boldsymbol{w}(y)$.*

Proof. Suppose we have xRy. To begin with, we claim that

$$\text{if } [C(x)] \neq [C(y)] \text{ then } M_{[C(x)]} \twoheadrightarrow M_{[C(y)]}. \tag{9}$$

Indeed, take any $\tau \in M_{[C(x)]}$ and let $C \in [C(x)]$ be such that $\tau_C = \tau$. By (6), there is a ϱ-maximal $D \in [C(x)]$ with $C = D$ or $\tau_C \twoheadrightarrow \tau_D$. So, by Lemma 3.1, there is $x' \in D$ with $t^{\varrho}_{\mathfrak{M}}(x) = t^{\varrho}_{\mathfrak{M}}(x')$. As xRy and $\mathfrak{M}$ is descriptive, it follows from Lemma 2.2 that there is y' with $x'Ry'$ and $t^{\varrho}_{\mathfrak{M}}(y') = t^{\varrho}_{\mathfrak{M}}(y)$. Then $C(y') \in [C(y)] \neq [C(x)] = [C(x')] = [D]$. Therefore, $\tau_D \twoheadrightarrow \tau_{C(y')}$ by (5), and so $\tau_C \twoheadrightarrow \tau_{C(y')}$ by (3), as required.

Now we show that $\boldsymbol{w}(x)R^{\dagger}\boldsymbol{w}(y)$ follows from (9) in all cases of the definition of $R^{\dagger}$. Assume first that $M_{[C(x)]} \neq M_{[C(y)]}$. Then $[C(x)] \neq [C(y)]$, and so $M_{[C(x)]} \twoheadrightarrow M_{[C(y)]}$ by (9), and $\tau_{C(x)} \twoheadrightarrow \tau_{C(y)}$ by (5), as required.

Assume next that $M_{[C(x)]} = M_{[C(y)]}$ and $M_{[C(x)]} \twoheadrightarrow M_{[C(y)]}$. By (5), $\tau_{C(x)} \not\twoheadrightarrow \tau_{C(y)}$ implies $C(x) = C(y)$, and so $\tau_{C(x)} = \tau_{C(y)}$. So we have $\boldsymbol{w}(x) R^{\dagger} \boldsymbol{w}(y)$ when $\boldsymbol{w}(x) \neq \boldsymbol{w}(y)$. If $\boldsymbol{w}(x) = \boldsymbol{w}(y)$ then $t^{\varphi,\psi}_{\mathfrak{M}}(x) \twoheadrightarrow t^{\varphi,\psi}_{\mathfrak{M}}(y)$ follows by (4), and so we also have $\boldsymbol{w}(x) R^{\dagger} \boldsymbol{w}(y)$.

Finally, assume that $M_{[C(x)]} = M_{[C(y)]}$, $M_{[C(x)]} \not\twoheadrightarrow M_{[C(y)]}$. Then (9) implies that $[C(x)] = [C(y)]$. There are two cases:

Case $C(x)$ is ϱ-maximal: Then $C(x) = C(y)$, and so $\tau_{C(x)} = \tau_{C(y)}$. Thus, we have $\boldsymbol{w}(x) R^{\dagger} \boldsymbol{w}(y)$ if $\boldsymbol{w}(x) \neq \boldsymbol{w}(y)$. If $\boldsymbol{w}(x) = \boldsymbol{w}(y)$, then $\boldsymbol{t}^{\varphi,\psi}_{\mathfrak{M}}(x) \twoheadrightarrow \boldsymbol{t}^{\varphi,\psi}_{\mathfrak{M}}(x)$ by (4). As $at^{\varrho}_{\mathfrak{M}}(x) = at^{\varrho}_{\mathfrak{M}}(y)$ and $C(x) = C(y)$, we have $t^{\varrho}_{\mathfrak{M}}(x) = t^{\varrho}_{\mathfrak{M}}(y)$, and so $\Diamond t^{\varrho}_{\mathfrak{M}}(x) \subseteq t^{\varrho}_{\mathfrak{M}}(x)$ by xRy. Therefore, $at^{\varrho}_{\mathfrak{M}}(x) \notin I_{M_{[C(x)]}}$, and so $\boldsymbol{w}(x) R^{\dagger} \boldsymbol{w}(y)$.

Case $C(x)$ is not ϱ-maximal: Then, by (8), $\tau_{C(x)}(a) = \emptyset$ for every $a \in I_{M_{[C(x)]}}$. In particular, as $\boldsymbol{t}^{\varphi,\psi}_{\mathfrak{M}}(x) \in \tau_{C(x)}(at^{\varrho}_{\mathfrak{M}}(x))$, it follows that $at^{\varrho}_{\mathfrak{M}}(x) \notin I_{M_{[C(x)]}}$. If $\boldsymbol{w}(x) = \boldsymbol{w}(y)$ then we have $\boldsymbol{t}^{\varphi,\psi}_{\mathfrak{M}}(x) \twoheadrightarrow \boldsymbol{t}^{\varphi,\psi}_{\mathfrak{M}}(x)$ by (4), and so $\boldsymbol{w}(x) R^{\dagger} \boldsymbol{w}(y)$. If $\boldsymbol{w}(x) \neq \boldsymbol{w}(y)$ and $C(x) = C(y)$, then $\tau_{C(x)} = \tau_{C(y)}$, and so $\boldsymbol{w}(x) R^{\dagger} \boldsymbol{w}(y)$. And if $\boldsymbol{w}(x) \neq \boldsymbol{w}(y)$ and $C(x) \neq C(y)$, then $\tau_{C(x)} \twoheadrightarrow \tau_{C(y)}$ by (5), and so $\boldsymbol{w}(x) R^{\dagger} \boldsymbol{w}(y)$ again.

This completes the proof of the lemma. □

The next lemma says that $R^{\dagger}$ is contained in the largest $sub(\varphi, \psi)$-filtration:

Lemma 3.6 *If* $\boldsymbol{w}(x) R^{\dagger} \boldsymbol{w}(y)$, *then* $\chi \in \boldsymbol{t}^{\varphi,\psi}_{\mathfrak{M}}(y)$ *implies* $\Diamond\chi \in \boldsymbol{t}^{\varphi,\psi}_{\mathfrak{M}}(x)$, *for every* $\Diamond\chi \in sub(\varphi, \psi)$.

Proof. As $\boldsymbol{t}^{\varphi,\psi}_{\mathfrak{M}}(x) \twoheadrightarrow \boldsymbol{t}^{\varphi,\psi}_{\mathfrak{M}}(y)$ implies that, for every $\Diamond\chi \in sub(\varphi, \psi)$, whenever $\chi \in \boldsymbol{t}^{\varphi,\psi}_{\mathfrak{M}}(y)$ then $\Diamond\chi \in \boldsymbol{t}^{\varphi,\psi}_{\mathfrak{M}}(x)$, and $\tau_{C(x)} \twoheadrightarrow \tau_{C(y)}$ implies $\boldsymbol{t}^{\varphi,\psi}_{\mathfrak{M}}(x) \twoheadrightarrow \boldsymbol{t}^{\varphi,\psi}_{\mathfrak{M}}(y)$, we only need to check those cases where we have $\boldsymbol{w}(x) R^{\dagger} \boldsymbol{w}(y)$ but neither $\boldsymbol{t}^{\varphi,\psi}_{\mathfrak{M}}(x) \twoheadrightarrow \boldsymbol{t}^{\varphi,\psi}_{\mathfrak{M}}(y)$ nor $\tau_{C(x)} \twoheadrightarrow \tau_{C(y)}$ holds.

An inspection of the definition of $R^{\dagger}$ shows that this can only happen when $\boldsymbol{w}(x) \neq \boldsymbol{w}(y)$, $M_{[C(x)]} = M_{[C(y)]}$ and $\tau_{C(x)} = \tau_{C(y)}$. In this case, we have $\boldsymbol{t}^{\varphi,\psi}_{\mathfrak{M}}(x) \in \tau_{C(x)}\big(at^{\varrho}_{\mathfrak{M}}(x)\big) = \tau_{C(y)}\big(at^{\varrho}_{\mathfrak{M}}(x)\big)$, and so there is $y' \in C(y)$ with $\boldsymbol{t}^{\varphi,\psi}_{\mathfrak{M}}(y') = \boldsymbol{t}^{\varphi,\psi}_{\mathfrak{M}}(x)$. Then $y' \neq y$, as otherwise we would have $\boldsymbol{t}^{\varphi,\psi}_{\mathfrak{M}}(x) = \boldsymbol{t}^{\varphi,\psi}_{\mathfrak{M}}(y)$, and so $at^{\varrho}_{\mathfrak{M}}(x) = at^{\varrho}_{\mathfrak{M}}(y)$ as well, contradicting $\boldsymbol{w}(x) \neq \boldsymbol{w}(y)$. Now it follows that, for every $\Diamond\chi \in sub(\varphi, \psi)$, if $\chi \in \boldsymbol{t}^{\varphi,\psi}_{\mathfrak{M}}(y)$, then $\Diamond\chi \in \boldsymbol{t}^{\varphi,\psi}_{\mathfrak{M}}(y') = \boldsymbol{t}^{\varphi,\psi}_{\mathfrak{M}}(x)$, as required. □

As a consequence of Lemmas 3.5 and 3.6, we obtain the usual 'filtration lemma' for $\mathfrak{M}^{\dagger}$ that can be proved by induction on χ:

Lemma 3.7 *For any* $\chi \in sub(\varphi, \psi)$ *and any* $\boldsymbol{w} = (\boldsymbol{t}, a, \tau, M) \in W^{\dagger}$, *we have* $\mathfrak{M}^{\dagger}, \boldsymbol{w} \models \chi$ *iff* $\chi \in \boldsymbol{t}$.

As a consequence of Lemma 3.7 we obtain:

Corollary 3.8 $\mathfrak{M}^{\dagger}, \boldsymbol{w}(r_{\varphi}) \models \varphi$ *and* $\mathfrak{M}^{\dagger}, \boldsymbol{w}(r_{\psi}) \models \neg\psi$.

Define a binary relation $\boldsymbol{\beta}^\dagger$ on $W^\dagger$ by taking

$$(\boldsymbol{t}, a, \tau, M)\, \boldsymbol{\beta}^\dagger\, (\boldsymbol{t}', a', \tau', M') \quad \text{iff} \quad a = a',\ M = M'.$$

Lemma 3.9 *The relation $\boldsymbol{\beta}^\dagger$ is a ϱ-bisimulation on $\mathfrak{M}^\dagger$ with $\boldsymbol{w}(r_\varphi)\boldsymbol{\beta}^\dagger\boldsymbol{w}(r_\psi)$.*

Proof. As $\mathfrak{M}, r_\varphi \sim^\varrho \mathfrak{M}, r_\psi$, we have $t^\varrho_{\mathfrak{M}}(r_\varphi) = t^\varrho_{\mathfrak{M}}(r_\psi)$. So $at^\varrho_{\mathfrak{M}}(r_\varphi) = at^\varrho_{\mathfrak{M}}(r_\psi)$ and $[C(r_\varphi)] = [C(r_\psi)]$, and hence $\boldsymbol{w}(r_\varphi)\boldsymbol{\beta}^\dagger\boldsymbol{w}(r_\psi)$.

Condition **(atom)** follows from Lemma 3.7. To prove **(move)**, suppose $\boldsymbol{w}_1\boldsymbol{\beta}^\dagger\boldsymbol{w}_1'$ and $\boldsymbol{w}_1 R^\dagger \boldsymbol{w}_2$, for $\boldsymbol{w}_1 = (\boldsymbol{t}_1, a_1, \tau_1, M_1)$, $\boldsymbol{w}_1' = (\boldsymbol{t}_1', a_1, \tau_1', M_1)$, and $\boldsymbol{w}_2 = (\boldsymbol{t}_2, a_2, \tau_2, M_2)$. We show that there is $\boldsymbol{w}_2'$ with $\boldsymbol{w}_2\boldsymbol{\beta}^\dagger\boldsymbol{w}_2'$ and $\boldsymbol{w}_1' R^\dagger \boldsymbol{w}_2'$, that is, there exist $\boldsymbol{t}_2'$ and τ_2' such that $\boldsymbol{w}_2' = (\boldsymbol{t}_2', a_2, \tau_2', M_2) \in W^\dagger$ and $\boldsymbol{w}_1' R^\dagger \boldsymbol{w}_2'$. We proceed by case distinction.

Case $M_1 \neq M_2$: As $\boldsymbol{w}_1 R^\dagger \boldsymbol{w}_2$, we have $M_1 \twoheadrightarrow M_2$. As $\tau_1' \in M_1$, there is some $\tau_2' \in M_2$ with $\tau_1' \twoheadrightarrow \tau_2'$. By (3) and (6), we may assume that $\tau_2' = \tau_D$ for some ϱ-maximal D, and so $\tau_2'(a_2) \neq \emptyset$. Take any $\boldsymbol{t}_2' \in \tau_2'(a_2)$. Then τ_2' and $\boldsymbol{t}_2'$ are as required, by (2).

Case $M_1 = M_2$, $M_1 \twoheadrightarrow M_1$: As $\tau_1' \in M_1$, there is $\tau_2' \in M_1$ with $\tau_1' \twoheadrightarrow \tau_2'$. By (3) and (6), we may assume that $\tau_2' = \tau_D$ for some ϱ-maximal D, and so $\tau_2'(a_2) \neq \emptyset$. Take any $\boldsymbol{t}_2' \in \tau_2'(a_2)$. Then $\boldsymbol{w}_2' = (\boldsymbol{t}_2', a_2, \tau_2', M_1) \in W^\dagger$, by (2). As $\tau_1' \twoheadrightarrow \tau_2'$ implies $\boldsymbol{t}_1' \twoheadrightarrow \boldsymbol{t}_2'$, we have $\boldsymbol{w}_1' R^\dagger \boldsymbol{w}_2'$ if $\boldsymbol{w}_1' = \boldsymbol{w}_2'$ or $\boldsymbol{w}_1' \neq \boldsymbol{w}_2'$.

Case $M_1 = M_2$, $M_1 \not\twoheadrightarrow M_2$, $a_1 \neq a_2$: If $\tau_1'(a_2) \neq \emptyset$, then take any $\boldsymbol{t}_2' \in \tau_2'(a_2)$. Then $\boldsymbol{w}_2' = (\boldsymbol{t}_2', a_2, \tau_1', M_1)$ is as required. If $\tau_1'(a_2) = \emptyset$, then $\tau_1' \neq \tau_D$ for any ϱ-maximal D. Thus, by (8), $\tau_1'(a) = \emptyset$ for every $a \in I_{M_1}$. Also, by (6), there is some ϱ-maximal D with $\tau_D \in M_1$ and $\tau_1' \twoheadrightarrow \tau_D$. Take any $\boldsymbol{t}_2' \in \tau_D(a_2)$. By (2), $\boldsymbol{w}_2' = (\boldsymbol{t}_2', a_2, \tau_D, M_1)$ is as required.

Case $M_1 = M_2$, $M_1 \not\twoheadrightarrow M_2$, $a_1 = a_2$: We claim that $a_1 \notin I_{M_1}$. Indeed, suppose $a_1 \in I_{M_1}$. As $\boldsymbol{w}_1 R^\dagger \boldsymbol{w}_2$, $\boldsymbol{w}_1 \neq \boldsymbol{w}_2$ follows. If $\tau_1 = \tau_2$ held, then $\{\boldsymbol{t}_1\} = \tau_1(a_1) = \tau_2(a_1) = \{\boldsymbol{t}_2\}$ by (7), and so $\boldsymbol{t}_1 = \boldsymbol{t}_2$ would follow, contradicting $\boldsymbol{w}_1 \neq \boldsymbol{w}_2$. So $\tau_1 \neq \tau_2$, and thus $\boldsymbol{w}_1 R^\dagger \boldsymbol{w}_2$ implies that $\tau_1(a) = \emptyset$ for all $a \in I_{M_1}$. As $\boldsymbol{t}_1 \in \tau_1(a_1)$, $a_1 \notin I_{M_1}$ follows, as required.

As $a_1 \notin I_{M_1}$, there exist C with $M_1 = M_{[C]}$ and $t \in T^\varrho[C]$ such that $t \cap \varrho = a_1$ and $\Diamond t \not\subseteq t$. By Lemma 3.1, it follows that, for every ϱ-maximal $D \in [C]$, either (i) there are at least two $x \in D$ with $t^\varrho_{\mathfrak{M}}(x) = t$ or (ii) xRx for the single $x \in D$ with $t^\varrho_{\mathfrak{M}}(x) = t$.

If $\tau_1' = \tau_D$ for a ϱ-maximal $D \in [C]$, then in case (ii) we have $\tau_1'(a_1) = \{\boldsymbol{t}^{\varphi,\psi}_{\mathfrak{M}}(x)\} = \{\boldsymbol{t}_1'\}$ and $\boldsymbol{t}_1' \twoheadrightarrow \boldsymbol{t}_1'$. Therefore, $\boldsymbol{w}_2' = (\boldsymbol{t}_1', a_1, \tau_1', M_1)$ is as required. In case (i), if $\boldsymbol{t}_1' \not\twoheadrightarrow \boldsymbol{t}_1'$, then there exists $\boldsymbol{t} \in \tau_1'(a_1)$ with $\boldsymbol{t} \neq \boldsymbol{t}_1'$, and so $\boldsymbol{w}_2' = (\boldsymbol{t}, a_1, \tau_1', M_1)$ is as required.

If $\tau_1' \neq \tau_D$ for any ϱ-maximal $D \in [C]$, then $\tau_1'(a) = \emptyset$ for every $a \in I_{M_1}$, by (8). By (6), there is a ϱ-maximal D such that $\tau_D \in M_1$ and $\tau_1' \twoheadrightarrow \tau_D$. Take any $\boldsymbol{t}_2' \in \tau_D(a_1)$. By (2), $\boldsymbol{w}_2' = (\boldsymbol{t}_2', a_1, \tau_D, M_1)$ is as required,

completing the proof of the lemma. □

The results obtained above yield the following:

Theorem 3.10 *Any given implication $\varphi \to \psi$ does not have an interpolant in* wK4 *iff there are models $\mathfrak{M}_\varphi$ and $\mathfrak{M}_\psi$ satisfying the criterion of Lemma 2.3 and having size triple-exponential in $|\varphi|$ and $|\psi|$.*

Thus, to decide whether $\varphi \to \psi$ does not have an interpolant in wK4, we can guess models $\mathfrak{M}_\varphi$ and $\mathfrak{M}_\psi$ of triple-exponential size in $|\varphi|$ and $|\psi|$ together with a binary relation $\sim^\varrho$ between their points and check in polynomial time in the size of $\mathfrak{M}_\varphi$ and $\mathfrak{M}_\psi$ whether the conditions of Lemma 2.3 are met.

Theorem 3.11 *The IEP for* wK4 *is decidable in* coN3ExpTime.

Remark 3.12 We can use the above construction to check whether a formula ψ is in wK4 as follows. We clearly have $\psi \notin$ wK4 iff $\varphi \to \psi$ has no interpolant in wK4, for $\varphi = \psi \vee \neg\psi$. In this case, $\sigma = \varrho = sig(\psi)$, and if $\mathfrak{M}$ is a descriptive σ-model, then $\boldsymbol{w}(x) = \boldsymbol{w}(y)$ iff $\boldsymbol{t}^{\varphi,\psi}_{\mathfrak{M}}(x) = \boldsymbol{t}^{\varphi,\psi}_{\mathfrak{M}}(y)$ and $\boldsymbol{t}^{\varphi,\psi}_{\mathfrak{M}}(C(x)) = \boldsymbol{t}^{\varphi,\psi}_{\mathfrak{M}}(C(y))$. Also, our filtration becomes the $sub(\psi)$-filtration given in [13]. Note that this filtration gives a double-exponential bound on the size of the model satisfying $\neg\psi$, which is not optimal as the decision problem for wK4 is PSpace-complete [20].

4 Lower Bound

Theorem 4.1 *The IEP for* wK4 *is* coNExpTime*-hard.*

Proof. We show NExpTime-hardness of interpolant non-existence by a reducion of the *exponential torus tiling problem*. A *tiling system* is a triple $P = (T, H, V)$, where T is a finite set of *tile types* and $H, V \subseteq T \times T$ are the *horizontal* and *vertical* matching conditions, respectively. An *initial condition* for P and $n > 0$ takes the form $\bar{t} = (t_0, \dots, t_{n-1}) \in T^n$. A map $\tau \colon \{0, \dots, 2^n - 1\} \times \{0, \dots, 2^n - 1\} \to T$ is a *solution to P and $\bar{t}$* if $\tau(i, 0) = t_i$ for all $i < n$, and for all $i, j < 2^n$, the following conditions hold (where $\oplus$ denotes addition modulo 2^n):

- if $\tau(i,j) = t$ and $\tau(i \oplus 1, j) = t'$, then $(t, t') \in H$;
- if $\tau(i,j) = t$ and $\tau(i, j \oplus 1) = t'$, then $(t, t') \in V$.

It is well-known that the problem of deciding whether there is a solution to given P and $\bar{t}$ is NExpTime-hard [2, Section 5.2.2].

Given a tiling system P and an initial condition $\bar{t}$ of length $n > 0$, we define formulas φ, ψ of size polynomial in $|P|$ and n, such that, for $\sigma = sig(\varphi) \cup sig(\psi)$ and $\varrho = sig(\varphi) \cap sig(\psi)$,

there is a solution for P and $\bar{t}$ iff

there exist σ-models $\mathfrak{M}_\varphi$ and $\mathfrak{M}_\psi$ based on frames for wK4 with $\mathfrak{M}_\varphi, r_\varphi \sim^\varrho \mathfrak{M}_\psi, r_\psi$, $\mathfrak{M}_\varphi, r_\varphi \models \varphi$ and $\mathfrak{M}_\psi, r_\psi \models \neg\psi$. (10)

The shared signature ϱ consists of

- a variable t, for each tile type $t \in T$;
- variables $\mathsf{b}_0, \dots, \mathsf{b}_{2n-1}$, that serve as bits in the binary representation of *grid positions* (i, j) with $i, j < 2^n$. We will use $[\mathsf{b} = (i, j)]$ as a short-

hand for the formula where $\mathsf{b}_0, \dots, \mathsf{b}_{n-1}$ represent the horizontal coordinate i and $\mathsf{b}_n, \dots, \mathsf{b}_{2n-1}$ the vertical coordinate j (with b_0 and b_n being the respective least significant bits); for instance, $[\mathsf{b} = (2,3)]$ stands for $\neg\mathsf{b}_0 \wedge \mathsf{b}_1 \wedge \big(\bigwedge_{1<k<n} \neg\mathsf{b}_k\big) \wedge \mathsf{b}_n \wedge \mathsf{b}_{n+1} \wedge \bigwedge_{n+1<\ell<2n} \neg\mathsf{b}_\ell$;

- a variable e that will be used to establish connections between the two ϱ-bisimilar models that force the tiling matching conditions.

The formula φ is defined as

$$\varphi = \mathsf{e} \wedge (\Diamond\Diamond p \wedge \neg\Diamond p) \wedge \Box(\mathsf{e} \to \Diamond p),$$

to which we add the other symbols in ϱ using tautologies. Observe that φ has the first formula of Example 1.1 (i) as a conjunct. Thus, if $\mathfrak{M}$ is based on a frame for $\mathsf{wK4}$ and $\mathfrak{M}, x \models \varphi$, then x is irreflexive, $C(x)$ contains a point different from x, and e is true everywhere in $C(x)$ and nowhere else R-accessible from x.

Our formula ψ takes form $\chi \to (\Box\Box q \to q)$ (cf. the second formula in Example 1.1 (i)). We next define χ. To begin with, χ has conjuncts that use variables $\mathsf{a}_0, \dots, \mathsf{a}_{2n-1}$ and variables $\mathsf{level}_0, \dots, \mathsf{level}_{2n}$ to generate a binary tree of depth $2n$ on nodes satisfying e such that a counter implemented using $\mathsf{a}_0, \dots, \mathsf{a}_{2n-1}$ is realised at its leaves:

$$\mathsf{level}_0 \wedge \Box^+ \bigwedge_{i<j\leq 2n} \neg(\mathsf{level}_i \wedge \mathsf{level}_j), \tag{11}$$

$$\Box^+\big(\mathsf{level}_i \to \Diamond(\mathsf{level}_{i+1} \wedge \mathsf{a}_i) \wedge \Diamond(\mathsf{level}_{i+1} \wedge \neg\mathsf{a}_i)\big), \quad \text{for } i < 2n, \tag{12}$$

$$\Box\big(\mathsf{level}_{i+1} \wedge \mathsf{a}_i \to \Box(\mathsf{level}_j \to \mathsf{a}_i)\big) \wedge \Box\big(\mathsf{level}_{i+1} \wedge \neg\mathsf{a}_i \to \Box(\mathsf{level}_j \to \neg\mathsf{a}_i)\big), \quad \text{for } i < j \leq 2n, \tag{13}$$

$$\Box^+(\mathsf{level}_i \to \mathsf{e}), \quad \text{for } i \leq 2n. \tag{14}$$

Next, we express that any leaf making $[\mathsf{a} = (i,j)]$ true has an R-successor making $\neg\mathsf{e} \wedge [\mathsf{b} = (i,j)]$ and a unique tile-variable t true, by defining

$$\mathsf{grid}: \quad \bigwedge_{k<2n} (\mathsf{a}_k \leftrightarrow \mathsf{b}_k),$$

and then adding the following conjuncts to χ:

$$\Box\Big(\mathsf{level}_{2n} \to \Diamond\big(\neg\mathsf{e} \wedge \mathsf{grid} \wedge \bigvee_{t\in T} \mathsf{t}\big)\Big), \tag{15}$$

$$\Box\big(\mathsf{level}_{2n} \wedge \mathsf{a}_i \to \Box(\neg\mathsf{e} \to \mathsf{a}_i)\big) \wedge \Box\big(\mathsf{level}_{2n} \wedge \neg\mathsf{a}_i \to \Box(\neg\mathsf{e} \to \neg\mathsf{a}_i)\big), \quad \text{for } i < 2n, \tag{16}$$

$$\Box \bigwedge_{t\neq t'\in T} \neg(\mathsf{t} \wedge \mathsf{t}'), \tag{17}$$

$$\Box\Big(\mathsf{level}_{2n} \to \bigwedge_{t\in T} \big(\Diamond(\neg\mathsf{e} \wedge \mathsf{grid} \wedge \mathsf{t}) \to \Box(\neg\mathsf{e} \wedge \mathsf{grid} \to \mathsf{t})\big)\Big). \tag{18}$$

Recall that computing $\oplus$ for numbers $i < 2^n$ in binary on n bits is as follows: If $i = 2^n - 1$ then flip all 1-bits to 0; otherwise, flip the first (when starting from the least significant bit) 0-bit of i to 1, flip all 1-bits of i before the first 0-bit to 0, and leave all other bits of i the same. So the following formulas, respectively, express that '$[\mathsf{a} = (i, j)]$ and $[\mathsf{b} = (i \oplus 1, j)]$' and '$[\mathsf{a} = (i, j)]$ and $[\mathsf{b} = (i, j \oplus 1)]$':

$$\mathsf{succ}_x : \Big(\bigvee_{m<n} (\mathsf{b}_m \wedge \neg \mathsf{a}_m \wedge \bigwedge_{k<m} (\neg \mathsf{b}_k \wedge \mathsf{a}_k) \wedge \bigwedge_{m<k<n} (\mathsf{b}_k \leftrightarrow \mathsf{a}_k)) \vee \bigwedge_{m<n} (\neg \mathsf{b}_m \wedge \mathsf{a}_m) \Big) \wedge \bigwedge_{n \le k < 2n} (\mathsf{b}_k \leftrightarrow \mathsf{a}_k),$$

$$\mathsf{succ}_y : \Big(\bigvee_{n \le m < 2n} (\mathsf{b}_m \wedge \neg \mathsf{a}_m \wedge \bigwedge_{n \le k < m} (\neg \mathsf{b}_k \wedge \mathsf{a}_k) \wedge \bigwedge_{m<k<2n} (\mathsf{b}_k \leftrightarrow \mathsf{a}_k)) \vee \bigwedge_{n \le m < 2n} (\neg \mathsf{b}_m \wedge \mathsf{a}_m) \Big) \wedge \bigwedge_{k<n} (\mathsf{b}_k \leftrightarrow \mathsf{a}_k).$$

We add the following conjuncts to χ to ensure the tiling matching conditions:

$$\Box \Big(\mathsf{level}_{2n} \to \bigwedge_{t \in T} (\Diamond(\neg \mathsf{e} \wedge \mathsf{grid} \wedge \mathsf{t}) \to \Box(\neg \mathsf{e} \wedge \mathsf{succ}_x \to \bigvee_{(t,t)' \in H} \mathsf{t}')) \Big), \qquad (19)$$

$$\Box \Big(\mathsf{level}_{2n} \to \bigwedge_{t \in T} (\Diamond(\neg \mathsf{e} \wedge \mathsf{grid} \wedge \mathsf{t}) \to \Box(\neg \mathsf{e} \wedge \mathsf{succ}_y \to \bigvee_{(t,t)' \in V} \mathsf{t}')) \Big). \qquad (20)$$

Finally, we ensure that the initial condition $\bar{t}$ holds, that is $\tau(i, 0) = t_i$ for $i < n$. To this end, we add to χ the conjuncts

$$\Box \big(\mathsf{level}_{2n} \wedge [\mathsf{a} = (i, 0)] \to \Box(\neg \mathsf{e} \wedge \mathsf{grid} \to \mathsf{t}_i) \big), \quad \text{for } i < n. \qquad (21)$$

It follows from the argument in Example 1.1 (i) that $(\varphi \to \psi) \in \mathsf{wK4}$. Below we show that (10) holds.

($\Rightarrow$) Suppose τ is a solution to P and $\bar{t}$. We define σ-models $\mathfrak{M}_\varphi$ and $\mathfrak{M}_\psi$ as follows. The underlying wK4-frame of $\mathfrak{M}_\varphi$ consist of a two-element cluster C having an irreflexive point r_φ and a reflexive point x, and C has 2^{2n} irreflexive and pairwise R-incomparable R-successors $w_{k,\ell}$, $k, \ell < 2^n$. The valuation in $\mathfrak{M}_\varphi$ is such that p holds at r_φ, e holds everywhere in C, and for $k, \ell < 2^n$, $\mathfrak{M}_\varphi, w_{k,\ell} \models [\mathsf{b} = (k, \ell)] \wedge \mathsf{t}$, where $t = \tau(k, \ell)$. The underlying wK4-frame of $\mathfrak{M}_\psi$ is the transitive closure of the following frame: First, take a full binary tree (T, R_T) of depth $2n$, with an irreflexive root r_ψ, all other nodes being reflexive, and having 2^{2n} leaves $e_{i,j}$, $i, j < 2^n$. Then, for each leaf $e_{i,j} \in T$, add 2^{2n} irreflexive R-successors $u_{i,j}^{k,\ell}$, $k, \ell < 2^n$, such that $u_{i,j}^{k,\ell}$ and $u_{i',j'}^{k',\ell'}$ are R-incomparable whenever $(i, j, k, \ell) \neq (i', j', k', \ell')$. The valuation in $\mathfrak{M}_\psi$ is such that q holds everywhere apart from r_ψ, e holds everywhere in T, $\mathsf{level}_0, \dots, \mathsf{level}_{2n}$ and $\mathsf{a}_0, \dots, \mathsf{a}_{2n-1}$ 'mark' the nodes of the tree (T, R_T) in such a way that, for $i, j < 2^n$, $\mathfrak{M}_\psi, e_{i,j} \models \mathsf{level}_{2n} \wedge [\mathsf{a} = (i, j)]$, and for $i, j, k, \ell < 2^n$, $\mathfrak{M}_\psi, u_{i,j}^{k,\ell} \models [\mathsf{a} = (i, j)] \wedge [\mathsf{b} = (k, \ell)] \wedge \mathsf{t}$, where $t = \tau(k, \ell)$.

It is straightforward to check that $\mathfrak{M}_\varphi, r_\varphi \models \varphi$ and $\mathfrak{M}_\psi, r_\psi \models \neg\psi$, and the relation

$$\boldsymbol{\beta} = (C \times T) \cup \{(w_{k,\ell}, u_{i,j}^{k,\ell}) \mid i, j, k, \ell < 2^n\}$$

is a ϱ-bisimulation between $\mathfrak{M}_\varphi$ and $\mathfrak{M}_\psi$ with $r_\varphi \boldsymbol{\beta} r_\psi$. ⊣

($\Leftarrow$) Suppose $\mathfrak{M}_\varphi, r_\varphi \sim^\varrho \mathfrak{M}_\psi, r_\psi$ with $\mathfrak{M}_\varphi, r_\varphi \models \varphi$ and $\mathfrak{M}_\psi, r_\psi \models \neg\psi$, for some σ-models $\mathfrak{M}_\varphi$ and $\mathfrak{M}_\psi$ based on frames for wK4. Thus, r_φ is irreflexive, the cluster C of r_φ contains a point different from r_φ, and e is true everywhere in C and nowhere else in $\mathfrak{M}_\varphi$. As $\mathfrak{M}_\psi, r_\psi \models \neg\psi$ and by (11)–(13), r_ψ is the root of a full binary tree of depth $2n$ having its leaves $e_{i,j}$, $i, j < 2^n$, marked by level_{2n} and the corresponding formula $[\mathsf{a} = (i,j)]$. (Note that, as $\mathfrak{M}_\psi, r_\psi \models \Box\Box q \wedge \neg q$, none of $e_{i,j}$ is in $C(r_\psi)$.) By (14), for all $i, j < 2^n$, $\mathfrak{M}_\psi, e_{i,j} \models \mathsf{e}$, and so we must have $\mathfrak{M}_\varphi, x_{i,j} \sim_\varrho \mathfrak{M}_\psi, e_{i,j}$ for some $x_{i,j} \in C$. Thus, by (13) and (15),

(t1) for all $i, j < 2^n$, C has an R-successor w such that $\mathfrak{M}_\varphi, w \models [\mathsf{b} = (i,j)] \wedge \mathsf{t}$ for some $t \in T$.

We claim that

(t2) for all $i, j < 2^n$, if w, w' are R-successors of C with $\mathfrak{M}_\varphi, w \models [\mathsf{b} = (i,j)] \wedge \mathsf{t}$ and $\mathfrak{M}_\varphi, w' \models [\mathsf{b} = (i,j)] \wedge \mathsf{t}'$, then $t = t'$.

Indeed, by ϱ-bisimilarity, there exist R-successors u, u' of $e_{i,j}$ such that $\mathfrak{M}_\psi, u \models \neg\mathsf{e} \wedge [\mathsf{b} = (i,j)] \wedge \mathsf{t}$ and $\mathfrak{M}_\psi, u' \models \neg\mathsf{e} \wedge [\mathsf{b} = (i,j)] \wedge \mathsf{t}'$. By (16), $[\mathsf{a} = (i,j)]$ is true at both u and u', and so grid is true at both u and u' as well. Thus, $t = t'$ follows from (17) and (18).

Now we define a map τ by taking, for all $i, j < 2^n$, $\tau(i,j) = t$ iff C has an R-successor w with $\mathfrak{M}_\varphi, w \models [\mathsf{b} = (i,j)] \wedge \mathsf{t}$. By **(t1)** and **(t2)**, τ is well-defined. We claim that

(t3) for all $i, j < 2^n$, if $\tau(i,j) = t$ and $\tau(i \oplus 1, j) = t'$ then $(t, t') \in H$, and

(t4) for all $i, j < 2^n$, if $\tau(i,j) = t$ and $\tau(i, j \oplus 1) = t'$ then $(t, t') \in V$.

Indeed, for **(t3)**, by the definition of τ and ϱ-bisimilarity, there are R-successors u, u' of $e_{i,j}$ such that

$$\mathfrak{M}_\psi, u \models \neg\mathsf{e} \wedge [\mathsf{b} = (i,j)] \wedge \mathsf{t}, \quad \mathfrak{M}_\psi, u' \models \neg\mathsf{e} \wedge [\mathsf{b} = (i \oplus 1, j)] \wedge \mathsf{t}'.$$

By (16), $[\mathsf{a} = (i,j)]$ is true at both u and u'. So grid is true at u and succ_x is true at u', and so $(t, t') \in H$ follows from (19). Condition **(t4)** can be shown similarly, using (20) in place of (19).

It follows from **(t3)**, **(t4)** and (21) that τ is a solution to P and $\bar{t}$, completing the proof of (10). Now the theorem follows by Lemma 2.3. □

5 Discussion

Our investigation of the interpolant existence problem for weak K4 and the difference logic DL is part of a research programme that aims to understand Craig interpolants for logics not enjoying the CIP. It turns out that wK4 shares with standard modal logics with nominals [1], decidable fragments of first-order modal logics [14], and the guarded and two-variable fragment of first-order logic [11] that interpolant existence is still decidable but computationally harder than validity. In contrast, the difference logic DL shares with normal extensions of K4.3 [15] that interpolant existence has the same complexity as validity. Linear temporal logic LTL is another example of a logic without the CIP, for which interpolant existence is decidable [18]. For both wK4 and LTL, establishing tight complexity bounds remains an interesting open problem. Further open problems include the following. As our decision procedures are non-constructive, it would be of interest to develop algorithms that compute interpolants whenever they exist. Also, is it possible to establish general decidability results for interpolant existence for families of extensions of wK4? Another related question is to find out which additional logical connectives would repair the CIP for wK4; see [21,22] for elegant answers to such questions for modal logics with nominals and other fragments of first-order logic.

The results of this paper are also relevant to the *explicit definition existence problem* (EDEP) for $L \in \{\mathsf{DL}, \mathsf{wK4}\}$: given formulas φ, ψ and a signature ϱ, decide whether there is a ϱ-formula χ with $\varphi \to (\psi \leftrightarrow \chi) \in L$, called an *explicit* ϱ*-definition of* ψ *modulo* φ *in* L. The EDEP reduces trivially to validity for logics enjoying the projective Beth definability property [9], which is not the case for DL and wK4. In fact, one can prove in exactly the same way as for fragments of first-order modal logics [14] that, for $L \in \{\mathsf{DL}, \mathsf{wK4}\}$, the IEP and EDEP are polynomial-time reducible to each other. Thus, our results above also provide complexity bounds for the EDEP in DL and wK4.

Acknowledgements. Thanks are due to the anonymous referees for their comments on the draft version of this paper.

References

[1] Artale, A., J. C. Jung, A. Mazzullo, A. Ozaki and F. Wolter, *Living without Beth and Craig: Definitions and interpolants in description and modal logics with nominals and role inclusions*, ACM Trans. Comput. Log. **24** (2023), pp. 34:1–34:51.

[2] Baader, F., I. Horrocks, C. Lutz and U. Sattler, "An Introduction to Description Logic," Cambridge University Press, 2017.

[3] Bezhanishvili, G., L. Esakia and D. Gabelaia, *Spectral and* T_0*-spaces in d-semantics*, in: N. Bezhanishvili, S. Löbner, K. Schwabe and L. Spada, editors, *Logic, Language, and Computation - 8th International Tbilisi Symposium on Logic, Language, and Computation, TbiLLC 2009. Revised Selected Papers*, Lecture Notes in Computer Science **6618** (2009), pp. 16–29.

[4] Bezhanishvili, G., S. Ghilardi and M. Jibladze, *An algebraic approach to subframe logics. Modal case*, Notre Dame J. Formal Log. **52** (2011), pp. 187–202.

[5] Chagrov, A. V. and M. Zakharyaschev, "Modal Logic," Oxford logic guides **35**, Oxford University Press, 1997.

[6] de Rijke, M., *The modal logic of inequality*, J. Symb. Log. **57** (1992), pp. 566–584.

[7] Esakia, L., *Weak transitivity—a restitution*, Logical Investigations **8** (2001), pp. 244–255, (In Russian).

[8] Fine, K., *Logics containing* K4*. part I*, J. Symb. Log. **39** (1974), pp. 31–42.

[9] Gabbay, D. M. and L. Maksimova, "Interpolation and Definability: Modal and Intuitionistic Logics," Oxford University Press, 2005.

[10] Goranko, V. and M. Otto, *Model theory of modal logic*, in: P. Blackburn, J. van Benthem and F. Wolter, editors, *Handbook of Modal Logic*, Studies in logic and practical reasoning **3**, Elsevier, 2007 pp. 249–329.

[11] Jung, J. C. and F. Wolter, *Living without Beth and Craig: Definitions and interpolants in the guarded and two-variable fragments*, in: *Proceedings of the 36th Annual ACM/IEEE Symposium on Logic in Computer Science, LICS 2021* (2021), pp. 1–14.

[12] Karpenko, A. and L. Maksimova, *Simple weakly transitive modal algebras*, Algebra and Logic **49** (2010), pp. 233–245.

[13] Kudinov, A. and I. Shapirovsky, *Filtrations for* wK4 *and its relatives*, CoRR **abs/2401.00457** (2024).

[14] Kurucz, A., F. Wolter and M. Zakharyaschev, *Definitions and (uniform) interpolants in first-order modal logic*, in: P. Marquis, T. C. Son and G. Kern-Isberner, editors, *Proceedings of the 20th International Conference on Principles of Knowledge Representation and Reasoning, KR 2023*, 2023, pp. 417–428.

[15] Kurucz, A., F. Wolter and M. Zakharyaschev, *A non-uniform view of Craig interpolation in modal logics with linear frames*, CoRR **abs/2312.05929** (2023).

[16] Marx, M., *Interpolation in modal logic*, in: *Proceedings of the 7th International Conference on Algebraic Methodology and Software Technology, AMAST 1998* (1998), pp. 154–163.

[17] McKinsey, J. and A. Tarski, *The algebra of topology*, Annals of Mathematics **45** (1944), pp. 141–191.

[18] Place, T. and M. Zeitoun, *Separating regular languages with first-order logic*, Log. Methods Comput. Sci. **12** (2016).

[19] Segerberg, K., *A note on the logic of elsewhere*, Theoria **46** (1980), pp. 183–187.

[20] Shapirovsky, I., *Satisfiability problems on sums of Kripke frames*, ACM Trans. Comput. Log. **23** (2022), pp. 15:1–15:25.

[21] ten Cate, B., "Model theory for extended modal languages," Phd, Universiteit van Amsterdam (2005).

[22] ten Cate, B. and J. Comer, *Craig interpolation for decidable first-order fragments*, CoRR **abs/2310.08689** (2024).

[23] von Wright, G., *A modal logic of place*, in: E. Sosa, editor, *The philosophy of Nicolas Rescher*, Dordrecht, 1979 pp. 65–73.

The Complexity of Reasoning about Classifiers

Xinghan Liu [1]

Vienna University of Technology
Vienna, Austria

Emiliano Lorini [2]

IRIT, CNRS, University of Toulouse
Toulouse, France

Abstract

The field of product modal logics was extensively developed in the last two decades. One of the simplest logic in this family is the product S5 × S5 ($S5^2$). In this paper, we explore the complexity of a specific instance of $S5^2$, called PLC (Product modal Logic of Classifiers), which was recently proposed in the field of logic-based explainable AI to model "black box" classifiers and their explanations. The PLC semantics gives concrete meanings to the two dimensions: the first ranges over the possible input instances to be classified and the other over the possible classifiers. We will show that, unlike $S5^2$ whose satisfiability problem is known to be NEXPTIME-complete, the satisfiability problem of PLC is simply EXPTIME-complete. This is due to the computationally grounded nature of PLC in comparison with $S5^2$ and, in particular, to the fact that along one of its two dimensions only classifications can vary while values of atomic propositions representing input instances stay the same.

1 Introduction

Product of modal logics [8] is one of the most fascinating and mathematically interesting research topics in the field of non-classical logics, given its tight connection with geometry and its level of generality. In a product modal logic the truth of a formula is evaluated with respect to a point in a multi-dimensional space, where each dimension has its own precise interpretation. One of the conceptually simplest logics in this area is S5×S5 ($S5^2$ for short) in which each of the two dimensions ranges over an S5 model, namely, a set of points (or possible worlds) which includes the actual point and which remains constant as the evaluation point of a formula varies. But the mathematical elegance of product modal logics is at odds with their computational complexity. In fact, they easily become undecidable as the number of dimensions increases.

[1] xinghan.liiu@tuwien.ac.at

[2] Emiliano.Lorini@irit.fr

	f_F	f_D	f_A
$s_1 = \emptyset$	1	1	0
$s_2 = \{sp\}$	1	1	1
$s_3 = \{al\}$	2	2	2
$s_4 = \{gr\}$	0	0	0
$s_5 = \{sp, al\}$	2	2	2
$s_6 = \{sp, gr\}$	1	0	1
$s_7 = \{al, gr\}$	2	2	2
$s_8 = \{sp, al, gr\}$	2	2	2

Table 1
Three classifiers of traffic regulations in different EU countries.

For example, the satisfiability of products of S5 becomes undecidable when the number of dimensions is strictly greater than 2 [8], while it is NEXPTIME-complete for $S5^2$ [17, Corollary 3.1].

In a recent paper [13] we have proposed a variant of $S5^2$ logic, called PLC (Product modal Logic of binary-input Classifiers), that can be used to reason about classifiers with binary inputs as well as to formally represent different kinds of explanation of their decisions. A classifier is simply an algorithm, or more generally a function, that categorises data into distinct classes. It is one of the most general and widely used concepts in AI and machine learning. Explaining classifiers became increasingly vital in the recent years given the emphasis put on the ethical and safety aspects of AI [18,11]. Nowadays explaining classifiers is one of the main topics in the field of explainable AI (XAI) [6,16,1,14,7,20,5,15,2,10]. Given their central role in AI and in the current debate in the XAI field, reasoning about classifiers with the help of modal logic has a similar status as reasoning about beliefs, knowledge or actions of agents.

The idea of our semantics is illustrated in Table 1 for Example 1.1. The first S5-dimension ranges over a set of instances that a classifier needs to classify (the vertical dimension in Table 1, i.e. the possible states of a passenger car driver). Each instance is nothing but a valuation of the atomic propositions from a countable set, each of them representing an input feature. The second S5-dimension ranges over a set of possible classifiers (the horizontal dimension in Table 1) which is open to a number of different interpretations. For instance, it can be conceived as the set of functions that are compatible with a learner's (or inquirer's) partial information of the actual function (epistemic interpretation), or as the set of classifiers complying with a number of hard requirements (nomological interpretation), or finally as the set of classifiers that are evaluated and compared, e.g., with respect to their ideality (evaluative interpretation).

Example 1.1 [Traffic regulations] EU countries have varying traffic regulations. Let *sp* stand for "*sp*eeding over 130 km/h on the freeway", *al* for "blood *al*cohol over 0.5g/L" and *gr* for "displaying a *gr*een environmental sticker while driving in the city". The classification $0, 1, 2$ stand resp. for *no sanction*, *weak sanction* (paying a fine) and *strong sanction* (suspension/revocation of the driving licence besides a fine). Table 1 shows a simplified version of three possible regulations in EU countries on passenger cars, whose representatives are France, Germany and Austria. We omit some details, e.g. actually only some freeways in Germany have no speed limit. In all three of them, high blood alcohol is classified as a strong sanction. However, in f_D there is no speed limit therefore $f_D(s_6) = 0 < 1 = f_F(s_6) = f_A(s_6)$. In f_A passenger cars are not required to display an environmental sticker, unlike the other two, and therefore $f_F(s_1) = f_G(s_1) = 1 > 0 = f_A(s_1)$. Last but not least, even for the same classification there are different reasons: $f_D(s_2) = 1$ because of no green environmental sticker displayed, while $f_A(s_2) = 1$ because of speeding.

At the language level, PLC uses two modal operators $\Box$ and $\blacksquare$ to represent the two dimensions. The modal operator $\Box$ moves along the first dimension by keeping the second one fixed: $\Box\varphi$ means that property φ holds for all input instances in the actual classifier. The modal operator $\blacksquare$ moves along the second dimension by keeping the first one fixed: $\blacksquare\varphi$ means that property φ holds at the actual instance for all possible classifiers.

In [13] we provided an axiomatics for PLC as well as an NEXPTIME complexity upper bound for its satisfiability problem using a filtration-based argument. But the tight complexity bounds for this logic remained an open problem which we tackle in the present article. We will show that, unlike $S5^2$ whose satisfiability problem is known to be NEXPTIME-complete, the satisfiability problem of PLC is simply EXPTIME-complete. This is due to the computationally grounded nature of PLC in comparison with $S5^2$ and, in particular, to the fact that along the $\blacksquare$-dimension only classifications can vary while values of atomic propositions representing input features stay the same. The proof of the upper bound relies on a non-trivial decision procedure based on a normal form, while the proof of the lower bound relies on a reduction of the tiling problem into the PLC-satisfiability problem.

The paper is organized as follows. In Section 2 we introduce the semantics of PLC based on the model class **MCM** (multi-classifier models). Section 3 discusses some special cases of the satisfiability problem in **MCM**. Sections 4 and 5 deal with the upper and lower bound of the problem in general respectively. In Section 6 we conclude.

2 Language and Semantics

Let $Atm_0 = \{p, q, \dots\}$ be a countable set of atomic propositions which intends to denote input variables (features) of a classifier. We introduce a finite set Val to denote the possible classifications (decisions, output values) of the classifier. Elements of Val are denoted by $c, c', \dots$ For any $c \in Val$, we call $\mathsf{t}(c)$ a decision atom, read it as "the current classification takes value c", and have $Dec = \{\mathsf{t}(c) : c \in Val\}$.[3] Finally let $Atm = Atm_0 \cup Dec$.

The modal language $\mathcal{L}$ is defined by the following grammar:

[3] Notice that p denotes an input *variable*, while c is an output *value* rather than the output *variable*, which justifies the symbolic difference between p and $\mathsf{t}(c)$.

$$\varphi \quad ::= \quad p \mid \mathsf{t}(c) \mid \neg\varphi \mid \varphi \wedge \varphi \mid \Box\varphi \mid \blacksquare\varphi,$$

where p ranges over Atm_0 and c over Val. Other Boolean connectives are defined as abbreviations, as usual. We call φ an $\mathcal{L}$-formula, if $\varphi \in \mathcal{L}$.

Definition 2.1 A multi-classifier model (MCM) is a pair $\Gamma = (S, F_S)$ where $S \subseteq 2^{Atm_0}$ and $F_S \subseteq Val^S$, namely $F_S \subseteq \{f : dom(f) = S, cod(S) = Val\}$. A pointed MCM is a triple (Γ, s, f) with $\Gamma = (S, F_S)$ an MCM, $s \in S$ and $f \in F_S$. The class of all MCMs is denoted by **MCM**.

Each $s \in S$ is called a state which intends to denote an input instance; each $f \in F_S$ is called a (classification) function which intends to denote a classifier.[4]

Example 2.2 Recall Example 1.1 in the introduction. It can be represented as an MCM $\Gamma = (S, F_S)$ s.t.

(i) $S = 2^{\{sp,al,gr\}}$;

(ii) $\forall f \in \{0,1,2\}^{\{sp,al,gr\}}$, $f \in F_S$ if and only if
  (a) $f(s) = 2 \iff al \in s$;
  (b) if $f(s) = 1$, then either $sp \in s$ or $gr \notin s$;
  (c) $f(s) = 0$, if $s = \{gr\}$.

Indeed, all three classifiers f_F, f_D, f_A have the common domain $2^{\{sp,al,gr\}}$. And it is not hard to check that f_F, f_D, f_A are all and only classifiers in F_S.

Formulas in $\mathcal{L}$ are interpreted relative to a pointed MCM as follows.

Definition 2.3 Let $\Gamma = (S, F_S)$ be an MCM, $s \in S$ and $f \in F_S$. Then,

$$\begin{aligned}
(\Gamma, s, f) \models p &\iff p \in s,\\
(\Gamma, s, f) \models \mathsf{t}(c) &\iff f(s) = c,\\
(\Gamma, s, f) \models \neg\varphi &\iff (\Gamma, s, f) \not\models \varphi,\\
(\Gamma, s, f) \models \varphi \wedge \psi &\iff (\Gamma, s, f) \models \varphi \text{ and } (\Gamma, s, f) \models \psi,\\
(\Gamma, s, f) \models \Box\varphi &\iff \forall s' \in S : (\Gamma, s', f) \models \varphi,\\
(\Gamma, s, f) \models \blacksquare\varphi &\iff \forall f' \in F_S : (\Gamma, s, f') \models \varphi.
\end{aligned}$$

A formula φ of $\mathcal{L}$ is said to be satisfiable relative to the class **MCM** if there exists a pointed multi-classifier model (Γ, s, f) with $\Gamma \in \mathbf{MCM}$ such that $(\Gamma, s, f) \models \varphi$. It is said to be valid relative to **MCM**, denoted $\models_{\mathbf{MCM}} \varphi$, if $\neg\varphi$ is not satisfiable relative to **MCM**.

Let us elaborate on the satisfaction relation. The $\Box$ operator ranges over *all* states while keeping the actual classifier fixed, namely in valuating $\Box\varphi$ we consider (Γ, s', f) for all $s' \in S$. Similarly, the $\blacksquare$ operator ranges over *all* classifiers while keeping the actual state fixed, namely in valuating $\blacksquare\varphi$ we consider (Γ, s, f') for all $f' \in F_S$. This clearly shows that every MCM is a model of $S5^2$. The interested reader can verify it by referring to e.g. [17] where $S5^2$ is introduced in detail.

---

[4] Rigorously, an input instance is a binary sequence rather than a state, i.e. a set of variables. But fixing the vocabulary it is common to identify them by viewing $p \notin s$ as $p = 0$ in s.

Moreover, it shall be clear that **MCM** is a *proper* subclass of the model class of $S5^2$. Firstly, the satisfaction relation for $\mathsf{t}(c)$ implies that for any pointed MCM exactly one decision atom is satisfied. Actually, since we require *Dec* to be a finite set, we have the following validities:

$$\models_{\mathbf{MCM}} \bigvee_{c \in Val} \mathsf{t}(c),$$
$$\models_{\mathbf{MCM}} \bigwedge_{c,c' \in Val, c \neq c'} (\mathsf{t}(c) \to \neg \mathsf{t}(c')).$$

Secondly and more subtly, for any pointed MCM (Γ, s, f) the value of p is locally determined by s, namely whether $p \in s$. That means the truth value of p is *independent* of the classifier dimension. (By contrast obviously the value of $\mathsf{t}(c)$ depends on both the classifier and the state dimension.) Formally this is captured by the following two validities:

$$\models_{\mathbf{MCM}} p \to \blacksquare p,$$
$$\models_{\mathbf{MCM}} \neg p \to \blacksquare \neg p.$$

As a consequence, intuitively speaking the $\blacksquare$-dimension is "fake" for any formula which is *Dec*-free.

These four validities were originally presented as axioms in [13]. As we will see, they are key to establish the complexity results here.

We conclude this section by illustrating the language $\mathcal{L}$ with the help of our running example.

Example 2.4 Recall the MCM in Example 2.2. For all $s \in S, f \in F_S$, we have

$$(\Gamma, s, f) \models \bigwedge_{X \subseteq Atm_0} \Diamond \big(\bigwedge_{p \in X} p \wedge \bigwedge_{q \in Atm_0 \setminus X} \neg q \big),$$

since $S = 2^{\{sp,al,gr\}}$. Furthermore, we have

$$(\Gamma, s, f) \models \blacksquare \Box \Big((\mathsf{t}(2) \leftrightarrow al) \wedge \big((\neg sp \wedge \neg al \wedge gr) \to \mathsf{t}(0) \big) \Big) \wedge$$
$$\blacksquare \Diamond \mathsf{t}(1) \wedge \bigwedge_{\psi \in \{sp, \neg gr\}} \blacklozenge \Box (\mathsf{t}(1) \to \psi),$$

which we know from constraints (a), (b), (c) in Example 2.2.

3 Special Cases: $|Atm_0|$ is finite or $|Dec| = 1$

As we have seen, **MCM** is a proper subclass of the model class of $S5^2$, as PLC is a proper normal extension of $S5^2$. In [3], Corollary 4.3 has shown that the satifiability problem for any proper normal extension of $S5^2$ is NP-complete. However, our logic is a *non-normal* extension of $S5^2$, since the additional axioms are not schemata but rather restricted to Atm_0 and *Dec* respectively.

It turns out to be that the complexity of satisfiability in the class **MCM** is sensitive to the cardinalities of both Atm_0 and Dec. Before proving the general case, let us first deal with two special and easy cases as a starter.

When $|Atm_0|$ is finite, the class **MCM** is bounded by some integer (since there are $|\wp(\wp(Atm_0))|$ many sets of states S and $|\wp(Val^S)|$ many sets of classifiers F_S for each S). Hence following [9], to decide whether some formula φ is satisfiable, one simply needs to do model checking for all pointed MCMs. This can be done in polynomial time in the size of φ.

Fact 3.1 *Let $|Atm_0|$ be finite. Then, deciding satisfiability of $\mathcal{L}$-formulas relative to the class* **MCM** *can be done in polynomial time.*

When $|Dec| = 1$, let us simply write $Dec = \{\mathtt{t}(1)\}$ (after renaming). One easily sees that $\mathtt{t}(1) \leftrightarrow \top$ becomes a validity. As a result, the whole $\blacksquare$-dimension collapses, such that for any $S \subseteq 2^{Atm_0}$ there is exactly one MCM with one function $\Gamma = \{S, \{f\}\}$ s.t. $f(s) = 1 \iff s \in S$. That means the logic collapses to S5, whose satisfiability problem is NP-complete (see, e.g. [12]).

Fact 3.2 *Let $|Atm_0|$ be infinite and $|Dec| = 1$. Then deciding satisfiability of $\mathcal{L}$-formulas relative to the class* **MCM** *is NP-complete.*

4 Upper Bound via Hintikka Set Construction

We present a decision procedure SATPLC as the pseudo-code in Algorithm 1. Its main subprocedure WITRECT can be considered as a variant of a well-known Hintikka set construction in the complexity theory for modal logics called WITNESS (see e.g. [4, Chap. 6.7]). The original witness procedure establishes the PSPACE upper bound of modal logic K by generating a tree structure. Here on one hand, a tree structure is by no means enough, since we need a (quasi-)rectangular structure in order to build the intended MCM. Therefore the complexity cannot be lower than EXPTIME. On the other hand, due to the special properties of PLC the idea of witness nevertheless works, which is generally not the case in product modal logics. A key point is a special negation normal form of PLC which we are going to define. Below we simply refer to it as negation normal form.

Algorithm 1: SATPLC(φ) returns YES/NO

Let Conflict(φ) $= \emptyset$. **forall** $\Box(\neg\omega \vee \delta), \Diamond(\omega' \wedge \delta') \in sub^+(\varphi)$ **do**
    if SAT($\omega' \wedge \neg\omega$) $=$ *NO*, SATDEC($\delta \wedge \delta'$) $=$ *NO* **then**
        add $(\Box(\neg\omega \vee \delta), \Diamond(\omega' \wedge \delta'))$ into Conflict(φ).
    end
end
if $\exists x_0$ *a Hintikka set over* $sub^+(\varphi)$, WITRECT(x_0, φ) $=$ *YES* **then**
    Return YES **else** Return NO.
end

Negation normal form To express the negation normal form in a readable way, let us start with the following two basic types of formulas.

$$\omega \quad ::= \quad p \mid \neg\omega \mid \omega \wedge \omega;$$
$$\delta \quad ::= \quad \mathsf{t}(c) \mid \neg\delta \mid \delta \wedge \delta \mid \Box\delta \mid \blacksquare\delta.$$

They are nothing but the *propositional* fragment of $\mathcal{L}$ on Atm_0, and the fragment of all Atm_0-free formulas, noted $\mathcal{L}_{Dec}$.

Based on these basic formulas, we define the following types of formulas.

$$\varphi^0 \quad ::= \quad \omega \mid \delta \mid \varphi^0 \wedge \varphi^0 \mid \varphi^0 \vee \varphi^0;$$
$$\varphi^1 \quad ::= \quad \varphi^0 \mid \Box\varphi^1 \mid \Diamond\varphi^1 \mid \blacksquare\varphi^1 \mid \blacklozenge\varphi^1 \mid \varphi^1 \wedge \varphi^1 \mid \varphi^1 \vee \varphi^1.$$

The set of all φ^i formulas denoted by $\mathcal{L}^i$ for $i \in \{0,1\}$. Clearly $\mathcal{L}^0 \subset \mathcal{L}^1$. We call formulas in $\mathcal{L}^1$ lower-level formulas, and others higher-level. Any formula in $\mathcal{L}$ can be transformed to some lower-level formula by recursively using the function $nnf(\cdot)$ as below (the propositional cases are omitted). Use the following "variables" for connectives: $\circ, \circ' \in \{\Box, \Diamond\}, \bullet, \bullet' \in \{\blacksquare, \blacklozenge\}$, $\mathbin{N} \in \{\wedge, \vee\}$.

$$nnf(\{\Box, \Diamond, \blacksquare, \blacklozenge\}^n \circ \bullet \psi) = nnf(\circ \bullet \psi) \tag{1}$$
$$nnf(\{\Box, \Diamond, \blacksquare, \blacklozenge\}^n \bullet \circ \psi) = nnf(\bullet \circ \psi) \tag{2}$$
$$nnf(\circ(\bullet\circ'\psi \mathbin{N} \xi)) = nnf(\bullet\circ'\psi \mathbin{N} \circ\xi) \tag{3}$$
$$nnf(\bullet(\circ\bullet'\psi \mathbin{N} \xi)) = nnf(\circ\bullet'\psi \mathbin{N} \bullet\xi) \tag{4}$$
$$nnf(\bullet(\psi \mathbin{N} \xi)) = nnf(\psi \mathbin{N} \bullet\xi), \ \textit{if } \psi \textit{ is Dec-free} \tag{5}$$
$$nnf(\bullet\circ(\psi \mathbin{N} \xi)) = \begin{cases} nnf(\circ\psi \mathbin{N} \bullet\circ\xi) & \textit{if } (*) \textit{ and } \psi \text{ is } \textit{Dec}\text{-free} \\ nnf(\bullet\circ(\psi \mathbin{N} \xi)) & \textit{otherwise,} \end{cases} \tag{6}$$

where $(*)$ is *either* $\bullet\circ \in \{\blacksquare\Box, \blacklozenge\Diamond\}$, *or* $(\bullet\circ = \blacksquare\Diamond, \mathbin{N} = \vee)$, *or* $(\bullet\circ = \blacklozenge\Box, \mathbin{N} = \wedge)$. Notice the special case of ψ or ξ being $\top$ or $\bot$. Clearly, (1-4) is by virtue of the S5 operators so that the unnecessary modal depths are reduced. (5-6) are more crucial where we use the independency properties of p regarding $\blacksquare$. Therefore, for any φ, not only $nnf(\varphi) \in \mathcal{L}^1$, but also $nnf(\varphi)$ can not have any subformula of the form $\bullet\psi$ with $\psi \in \mathcal{L}^0$.

Lemma 4.1 *For any* $\varphi \in \mathcal{L}, \models_{\mathbf{MCM}} \varphi \leftrightarrow nnf(\varphi)$.

The key virtue of $nnf(\cdot)$ is that $\mathcal{L}^0$ formulas are "basic", since for both ω and δ the satisfiability checking can be done "locally" (recall the finiteness of $|Dec|$). Consequently, as we will see to ensure the non-contradiction of Hintikka set construction we only need to consider formulas of form $\circ\psi$ with $\psi \in \mathcal{L}^0$.

Hintikka set for PLC Let Φ be a set of $\mathcal{L}$-formulas. We define the generalized subformula closure of Φ, denoted $sub^+(\Phi)$, as the closure under subformulas, single negation, and additionally (recall our usage of ω and δ):

i) if $\omega \in sub^+(\Phi)$, then $\blacksquare\omega \in sub^+(\Phi)$;

ii) if $\Phi \cap Dec = \emptyset$, then designate a $\mathsf{t}(c)$, s.t. $\mathsf{t}(c) \in sub^+(\Phi)$;

iii) $\circ\bullet\varphi \in sub^+(\Phi) \iff \bullet\circ\varphi \in sub^+(\Phi), \circ\bullet \in \{\Box\blacksquare, \Diamond\blacklozenge\}$.

As usual when Φ is a singleton $\{\varphi\}$, we write $sub^+(\varphi)$ instead of $sub^+(\{\varphi\})$.

Definition 4.2 [Hintikka sets (for PLC)] Let Φ be a set of formulas. A Hintikka set x over Φ is a subset of $sub^+(\Phi)$ s.t. for all $\varphi \in sub^+(\Phi)$,

(i) $\varphi \notin x \iff \neg\varphi \in x$;

(ii) if $\varphi \wedge \psi \in x$, then $\varphi, \psi \in x$;

(iii) if $\boxplus\varphi \in x$ then $\varphi \in x$, with $\boxplus \in \{\Box, \blacksquare\}$;

(iv) $\blacksquare\Box\varphi \in x \iff \Box\blacksquare\varphi \in x$;

(v) if φ is some p, then $p \in x \iff \blacksquare p \in x \iff \blacklozenge p \in x$;

(vi) $\exists! \mathsf{t}(c) \in Dec, \mathsf{t}(c) \in x$;

(vii) $\{\delta : \delta \in x\}$ is satisfiable.

We write $x \equiv_{\boxplus} y$, if $\forall \boxplus\psi \in sub^+(\Phi), \boxplus\psi \in x \iff \boxplus\psi \in y$ for $\boxplus \in \{\blacksquare, \Box\}$.

SATplc and WitRect Now we are able to introduce the decision procedure. Let us start with the intuitive idea. Fix a $\varphi \in \mathcal{L}$ which we assume is already in our negation normal form. First we obtain from φ the set of conflicting rule-instance pairs, whose meaning is introduced below. Then for any Hintikka set x_0 with $\varphi \in x_0$, the subprocedure WitRect(x_0, φ) constructs Hintikka sets step by step which 1) witnesses all the diamond formulas in $sub^+(\varphi)$ for each already constructed Hintikka set; 2) aligns the depth and width in order to maintain the quasi-rectangular shape. The resulting set induces the intended MCM where φ is satisfied.

Then we explain Algorithms 1 and 2 in detail. Let us start with what we call a conflicting rule-instance pair. It has to do with two types of formulas: $\Box(\neg\omega \vee \delta)$ and $\Diamond(\omega \wedge \delta)$. Intuitively, $\Box(\neg\omega \vee \delta)$ can be seen as a (quasi-)rule $\delta \leftarrow \omega$, saying that for all input instances, if they verify ω, then their classifications necessarily verify δ. And $\Diamond(\omega \wedge \delta)$ can be seen as a (quasi-)instance (ω, δ), whose input part is ω and output part is δ.[5] A rule $\delta \leftarrow \omega$ and an instance (ω', δ') conflict, if the rule is applicable to the instance, namely $\omega' \to \omega$ is valid, but their output parts conflict, namely $\delta \wedge \delta'$ is unsatisfiable. We thus call $(\Box(\neg\omega \vee \delta), \Diamond(\omega' \wedge \delta'))$ a conflicting rule-instance pair. Any function in the intended MCM shall contain no such pair. Therefore, in Lines 1-6 of Algorithm 1, we find all these conflicting pairs in $sub^+(\varphi)$ and put them in $\texttt{Conflict}(\varphi)$.

Now we turn to Algorithm 2. For any Hintikka set x_0 over $sub^+(\varphi)$, the procedure WitRect(x_0, φ) works as follows. Initially we do two checks. Obviously, if $\varphi \notin x_0$ then the procedure shall terminate with a NO. In addition, we want x_0 to not contain any conflicting pairs, therefore define:

$$x_0 \uparrow \texttt{Conflict}(\varphi) \iff \forall(\psi, \psi') \in \texttt{Conflict}(\varphi), \{\psi, \psi'\} \nsubseteq x_0.$$

Clearly, if $x \uparrow \texttt{Conflict}(\varphi)$ does not hold, then φ cannot be satisfied, therefore the procedure also terminates and outputs NO.

---

[5] "Quasi-", because δ is *any* $\mathcal{L}_{Dec}$ formula, and ω not necessarily a conjunction of literals. But this is not a vital point and we omit the "quasi-" below.

Algorithm 2: WITRECT(x_0, φ) returns YES/NO

1 Check $\varphi \in x_0$ and $x_0 \uparrow \texttt{Conflict}(\varphi)$. Return NO if it is not.
2 Let $\mathcal{H}^0 = \{x_0\}$, **forall** $\Diamond\psi \in x_0$ *with* $\psi \notin x_0$ **do**
3   construct a Hintikka set x, s.t. $x \equiv_{\Box} x_0 \equiv_{\blacklozenge\Box} x$, $\psi \in x$, and add it into $\mathcal{H}^0$. Return NO if fails.
4 **end**
5 **while** $\mathcal{H}^n \neq \mathcal{H}^{n-1}$ $(n \geq 1)$ **do**
6   **forall** $x \in \mathcal{H}^n, \blacklozenge\psi \in x$ *if not* $\exists x' \in \mathcal{H}^n, x \equiv_{\blacksquare} x' \equiv_{\Diamond\blacksquare} x, \psi \in x'$ *and* $x' \uparrow \texttt{Conflict}(\varphi)$ **do**
7   construct such an x', add it into $\mathcal{H}^n$. Return NO if fails.
8   **end**
9   **forall** $x \in \mathcal{H}^n, \Diamond\psi \in x$ *if not* $\exists x' \in \mathcal{H}^n, x \equiv_{\Box} x' \equiv_{\blacklozenge\Box} x, \psi \in x'$ **do**
10   construct such an x', add it into $\mathcal{H}^n$. Return NO if fails.
11   **end**
12   **forall** $x \in \mathcal{H}^n$, *if not* $\exists y \in \mathcal{H}^0, x \equiv_{\blacksquare} y$ **do**
13   construct such a y s.t. $x \equiv_{\blacksquare} y \equiv_{\Box} x_0$, add it into $\mathcal{H}^0$ and $\mathcal{H}^n$. Return NO if fails.
14   **end**
15   **forall** $x \in \mathcal{H}^n, y \in \mathcal{H}^0$, *if not* $\exists z \in \mathcal{H}^n, x \equiv_{\Box} z \equiv_{\blacksquare} y$ **do**
16   construct such a z, add it into $\mathcal{H}^n$ and rename $\mathcal{H}^n$ as $\mathcal{H}^{n+1}$. Return NO if fails.
17   **end**
18 **end**
19 **return** YES.

If both checks are positive, we shall move on. The goal is to generate a structure with a quasi-rectangular shape, such that all the diamond sentences in every constructed Hintikka set is witnessed.

Lines 2-4 of Algorithm 2 describe the first stage construction, which only involves formulas of form $\Diamond\psi$ in x_0. Originally, to *witness* a $\Diamond\psi$ it is enough to construct some x s.t. $\psi \in x$ and $x \equiv_{\Box} x_0$. But here we need in addition $x \equiv_{\blacklozenge\Box} x_0 \equiv_{\blacksquare\Box} x$, required by the product modal logic. The resulting set of Hintikka sets is denoted by $\mathcal{H}^0$. It is the initial "row" of the structure.

Lines 5-17 recursively enlarge the rectangular such that, not only the two sorts of diamond formulas are witnessed, but also the voids between rows and columns are filled. At each stage $\mathcal{H}^n$, the following four types of constructions are needed (rather than a single construction in the standard case of K):

6-8. (**$\blacklozenge$-witness**) ensure every $\blacklozenge\psi$ in each $x \in \mathcal{H}^n$ is witnessed by a conflict-free x' in the same "column", [6] and expand the "depth" of the structure;

9-11. (**$\Diamond$-witness**) ensure every $\Diamond\psi$ in each $x \in \mathcal{H}^n$ is witnessed by a conflict-free x' in the same "row", and extend the "width" of the structure;

12-14. (**$\mathcal{H}^0$-fill**) align the widths of all the rows by updating $\mathcal{H}^0$ as baseline;

[6] Only the $\blacklozenge$-witnesses must check $\texttt{Conflict}(\varphi)$. Others do not need because of $\equiv_{\Box}$.

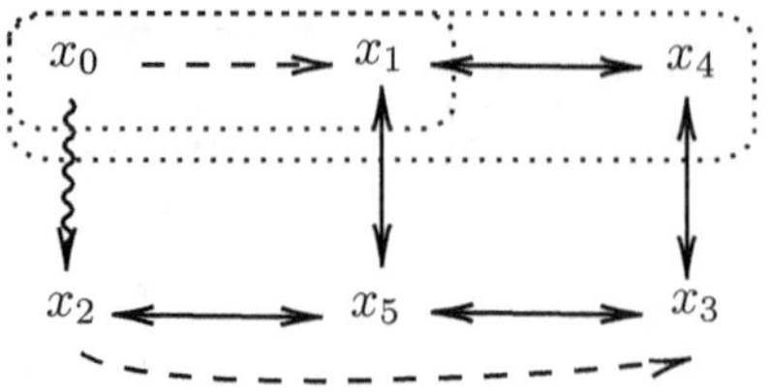

Fig. 1. A construction of $\text{WitRect}(x_0, \blacklozenge\mathtt{t}(0) \wedge \Box(\neg p \vee \mathtt{t}(1)) \wedge \blacksquare\Diamond(\neg q \wedge \mathtt{t}(0))$, where $x_0 \cap Atm = \{p, q, \mathtt{t}(1)\}$. Dash arrow for $\Diamond$-witness, wave arrow for $\blacklozenge$-witness, bidirectional arrow for $\mathcal{H}^i$-fill with $i \in \{0, n\}$, and dotted circle for $\mathcal{H}^0$ in Stage 1 and 2.

15-17. (**$\mathcal{H}^n$-fill**) align the depths of all the columns to finish the construction at current stage n.

It terminates if the construction fails somewhere and outputs NO, otherwise in the first n s.t. $\mathcal{H}^n = \mathcal{H}^{n-1}$ and outputs YES. An example is given in Figure 1.

Fact 4.3 *Let $\mathcal{H}$ result from the construction procedure for some φ. Then $\forall x, y \in \mathcal{H}, \exists z \in \mathcal{H}, x \equiv_{\Box} z \equiv_{\blacksquare} y$. In addition, $x \uparrow \mathtt{Conflict}(\varphi)$.*

Construct the intended model Let $\mathcal{H}$ result from some $\text{WitRect}(x_0, \varphi)$. We need to construct an MCM $\Gamma = (S, F_S)$ out of the $\mathcal{H}$. Since $\mathcal{H}$ already has a quasi-rectangular shape, the simple thought is to define F_S such that each $x \in \mathcal{H}$ stands for a $(s, f), s \in S, f \in F_S$. But it has flaws. We may have $x, y \in \mathcal{H}$ s.t. $x \equiv_{\Box} y$, $x \cap Atm_0 = y \cap Atm_0$, yet $x \cap dec \neq y \cap Dec$, which makes $f(x)$ not well-defined. More generally we may have $x \equiv_{\Box} y \equiv_{\blacksquare} x$, yet $x \neq y$.

Nevertheless, it can be fixed. A Hintikka set may represent more than one pair (s, f). We will introduce the notion of *path*, which helps identify the same state in different functions. Every path will be marked by a fresh Atm_0 variable, so that all the intended f are eventually well-defined.

Definition 4.4 [MCM upon $\mathcal{H}$] Let $\mathcal{H}$ be the set of Hintikka sets resulting from some $\text{WitRect}(x_0, \varphi)$. A path π in $\mathcal{H}$ is a maximal subset of $\mathcal{H}$ s.t.

$$\forall x, y \in \pi, x \equiv_{\blacksquare} y \ \& \ \big(x \equiv_{\Box} y \Longrightarrow x = y\big).$$

Let $\mathcal{P}_{\mathcal{H}}$ denote all the paths in $\mathcal{H}$ and enumerate them as $\pi_1, \pi_2, \ldots \pi_m$. Since $\mathcal{H}$ is finite, the number of paths is finite. Now for any path π_i, we name a $q_i \in Atm_0 \setminus Atm(\varphi)$. The $\Gamma = (S, F_S)$ constructed upon $\mathcal{H}$ is thus:

$$\begin{aligned} S =& \{s_i^x : \pi_i \in \mathcal{P}_{\mathcal{H}}, x \in \pi_i\}, \text{ where } s_i^x = (x \cap Atm_0) \cup \{q_i\}; \\ F_S =& \{f^x : x \in \mathcal{H}\}, \text{ where } \forall s_j^z \in S, \\ & f^x(s_j^z) = c \text{ iff } \exists y \in \pi_j, y \equiv_{\Box} x \ \& \ \mathtt{t}(c) \in y. \end{aligned}$$

The key point of path is that $\forall \pi_i \in \mathcal{P}_{\mathcal{H}}, \forall f^x \in F_S, \exists! y \in \pi_i$ s.t. $x \equiv_{\Box} y$. Therefore $f^x(s_i^z)$ is uniquely determined by such y, which makes the MCM above well-defined, as the following lemma highlights.

Lemma 4.5 *Let Γ be the model defined above. For all $x, y \in \mathcal{H}$ we have*

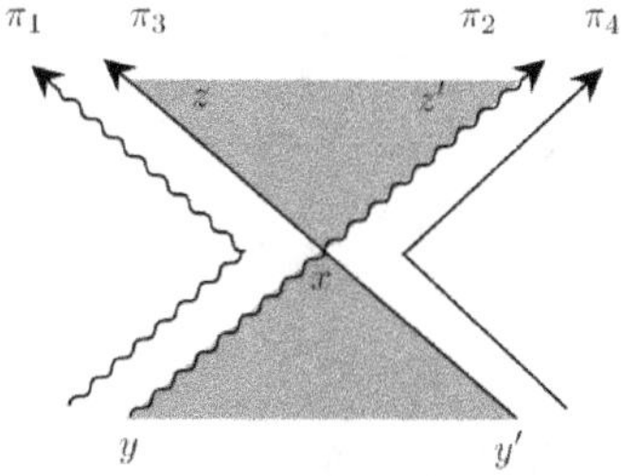

Fig. 2. Let $\mathcal{H} = \{x, y, y', z, z'\}$, where $y \equiv_{\Box} y' \not\equiv_{\Box} x \not\equiv_{\Box} z \equiv_{\Box} z'$, and $w \equiv_{\blacksquare} v$ for all $w, v \in \mathcal{H}$. Then we have overall four paths $\mathcal{P}_{\mathcal{H}} = \{\pi_1, \ldots \pi_4\}$.

(i) $\forall \pi \in \mathcal{P}_{\mathcal{H}}, \exists! z \in \pi, z \equiv_{\Box} x$;

(ii) $\forall \pi \in \mathcal{P}_{\mathcal{H}}$ *with* $x \in \pi, \exists! z \in \pi, z \equiv_{\Box} y$;

(iii) *if* $s_i^x, s_j^y \in S$, *then* $s_i^x = s_j^y \iff x \equiv_{\blacksquare} y$ *and* $i = j$;

(iv) $f^x = f^y \Longleftarrow x \equiv_{\Box} y$.

Lemma 4.6 *The* $\Gamma = (S, F_S)$ *upon* $\mathcal{H}$ *is indeed an MCM.*

The last one is a truth lemma, which is stronger than the usual one.

Lemma 4.7 (Truth) *Let* $\Gamma = (S, F_S)$ *be the MCM upon* $\mathcal{H}$*, where* $\mathcal{H}$ *results from some* WITRECT(x_0, φ). *Then* $\forall \psi \in sub^+(\varphi)$,

$$\begin{aligned} \psi \in x &\iff \forall \pi_i \in \mathcal{P}_{\mathcal{H}} \text{ with } x \in \pi_i, (\Gamma, s_i^x, f^x) \models \psi \\ &\iff \exists \pi_j \in \mathcal{P}_{\mathcal{H}} \text{ with } x \in \pi_j, (\Gamma, s_j^x, f^x) \models \psi. \end{aligned}$$

Theorem 4.8 *Deciding satisfiability of* $\mathcal{L}$*-formulas relative to the class* **MCM** *is in EXPTIME.*

Proof. For any formula $\varphi' \in \mathcal{L}$, we first let $nnf(\varphi') = \varphi$. Hence by Lemma 4.1 $\models_{\mathbf{MCM}} \varphi' \leftrightarrow \varphi$. We shall prove φ is satisfiable iff SATPLC(φ) returns YES, which further means there exists a Hintikka set x_0 over $sub^+(\varphi)$ s.t. WITRECT(x_0, φ) terminates with a set of Hintikka sets $\mathcal{H}$ and returns YES.

If there is the case, then easy to see that the whole procedure runs in EXPTIME: transforming φ' to φ is in polynomial time; generating `Conflict`(φ) in co-NP; both constructing all Hintikka sets over $sub^+(\varphi)$ and running a single time WITRECT are in EXPTIME. So in the worst case we have $\mathcal{O}(2^n) \times \mathcal{O}(2^n)$, which is still $\mathcal{O}(2^n)$ where n is the size of φ'.

The right to left direction is now obvious in light of the lemmas above. For the other direction, we show that if x_0 is satisfiable, then WITRECT(x, φ) outputs YES. This is to show if x_0 is satisfiable, then any construction required by WITRECT(x_0, φ) will not fail. A construction fails when it requires a *clash*, by which we mean an explicit contradiction. Since every Hintikka set is locally consistent, only modal formulas may give rise to contradiction (in the next step construction). To show that this is impossible given the satisfiable x_0, for any stage i of construction, $\forall x, y \in \mathcal{H}^i$, do induction on modal formulas. 1) $\delta \in x$ where δ is unsatisfiable, which is impossible because of our definition of Hintiika set for PLC. 2) $\Box(\neg\omega \vee \delta), \Diamond(\omega' \wedge \delta') \in x$ but their conjunction is unsatisfiable, which is exactly prohibited by the check of $x \uparrow \texttt{Conflict}(\varphi)$. 3) $\blacksquare(\neg\omega \vee \delta), \blacklozenge(\omega' \wedge \delta') \in x$, which is impossible since it would be transformed by nnf as $\neg\omega \vee \blacksquare\delta$ and $\omega' \wedge \blacksquare\delta'$. 4) $\dagger\ddagger\psi \in x, \dagger\ddagger\neg\psi \in y$ and we fail to construct some z with $x \equiv_\dagger z \equiv_\ddagger y$, with $\dagger\ddagger \in \{\blacksquare\Box, \blacklozenge\Box, \Diamond\blacksquare\}$, which is impossible since we have $x \equiv_{\dagger\ddagger} x_0 \equiv_{\dagger\ddagger} y$. □

5 Lower Bound: A Tiling Approach

Now we turn to the lower bound, and assume $|Atm_0|$ is infinite and $|Dec| > 1$.

5.1 Tiling problems

Tiling problem, or domino problem, is a mathematical problem widely used in modal logics, e.g. [22]. Recall that a tile (domino) type has four colors, one for each of its four sides. Tiling problem aims to tile a plane with a set of tile types s.t. the adjacent sides of any two tiles share the same color. Different tiling problems have different computability and complexity. The particular tiling problem that we will focus on is a two-player game between Eloise ($\exists$) and Abelard ($\forall$) to tile a $2^n \times n$ corridor, namely 2^n rows and n columns, n given by the input. Moreover, players take turn to tile a role at each round, starting with Eloise from Row 1. For technical reasons it is standard (see e.g. [4]) to assume Row 0, Column 0 and Column n+1, which are previously tiled by a white domino type. The complexity of deciding whether Eloise has a winning strategy is EXPTIME-complete, see e.g. [19].

Recall that Eloise has a winning strategy, if there is a game tree such that:

(i) each node denotes a tiling of a row, s.t. all nodes in the odd levels belong to Eloise, and the rest to Abelard;

(ii) each non-root node encodes a tiling of a row s.t. the tiling is compatible with the tiling of its predecessor;

(iii) at each level, every possible tiling of Abelard is encoded by some node;

(iv) the tree has 2^n depth.

Every branch from the root to a leaf is called a game. The game tree guarantees that Eloise always finds a branch whatever Abelard plays. We shall prove that deciding whether Eloise has a winning strategy in this tiling game can be reduced to the satisfiability of an $\mathcal{L}$-formula φ.

5.2 Preparation and Intuition

Fix a $2^n \times n$ corridor and a set T a set of domino tile types. We need the following sets of Atm_0-variables: $\{eloise, aberlard\}$ for the players; $\{p_\#, p_{til}\}$ for marking whether it is a counting or tiling state; $\{p_i : 1 \leq i \leq n\}$ for counting; $\{pos_i : 0 \leq i \leq n+1\}$ for positions/cells in a row; $\{p_d : d \in T\} \cup \{p_{white}\}$ for domino tile types where p_{white} is the white type; $\{col_i(d) : 0 \leq i \leq n+1, d \in T\}$ read as "the *previous* tiling of position i is domino type

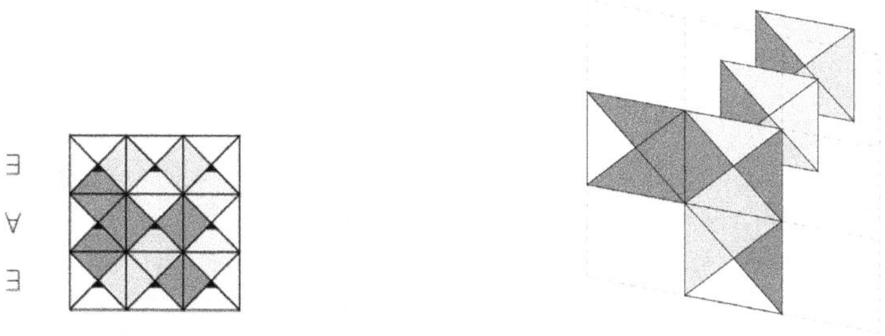

Fig. 3. Left: a two-player tiling game example in [19]; Right: different options of tiling the middle cell.

d".[7] We merely need two *Dec*-variables, $\mathsf{t}(0), \mathsf{t}(1)$ and they play double roles regarding counting- and tiling-states.

Before giving the reduction formula, let us give the intuition for the desired model $\Gamma = (S, F_S)$ where the formula is supposed to be satisfied. S consists of two disjoint sets of states, *tiling states* and *counting states*, telling which tile in a position/cell of a row and which step the current row is respectively. Each member of F_S must decide which tiling states and counting states are selected.

For example, the right figure of Figure 5.1 shows that, according to the tiles on its left and down side, Abelard has three options to tile Position 2 of Row 2. The three options will be represented as follows:

$$s = \{p_{til}, abelard, pos_2, col_1(p_{rrrw}), col_2(p_{grwg}), p_{yrgr}\},$$
$$s' = \{p_{til}, abelard, pos_2, col_1(p_{rrrw}), col_2(p_{grwg}), p_{yygr}\},$$
$$s'' = \{p_{til}, abelard, pos_2, col_1(p_{rrrw}), col_2(p_{grwg}), p_{yggr}\},$$

where domino types here are named after their colors clockwise as up-right-down-left. Thus p_{yrgr} means the up-color is yellow, right-color red, and so on. So s means that it is a tiling state for Aberlard's round, where in Position 2, p_{yrgr} is tiled regarding $col_1(p_{rrrw})$ and $col_2(_{grwg})$. All three states are legal, since the colors match. However, in any classifier, which encodes a row in a game, at most one of them can be "activated" to actually tile. In our case it is $f(s) = 1$ while $f(s') = f(s'') = 0$.

Moreover, f must also inform which round it is at. Hence we shall have two counting states $\{p_\#, p_1\}, \{p_\#, p_2\}$ s.t. $f(\{p_\#, p_1\}) = 1$ and $f(\{p_\#, p_2\}) = 0$ encoding binary 01. It is the decimal 1 which represents Round 2. Therefore, $\mathsf{t}(1)$ and $\mathsf{t}(0)$ means *active* and *inactive* in classifying tiling states, while meaning binary values 0 and 1 in counting states.

Hence, the desired MCM is a set of such f, each f stands for a node in the game tree, encoding the information of the actual tiling and the actual round.

[7] The difference of i ranging from 1 through n and from 0 through $n+1$ is only for the technical trick of letting the 0th row, the 0th and $n+1$st columns be tiled by p_{white}.

5.3 Reduction

We will reduce the aforementioned tiling game to an $\mathcal{L}$-formula. Following [17], we define a modal logic K operator $[K]\varphi := \blacksquare(\mathtt{t}(0) \to \Box(\mathtt{t}(1) \to \varphi))$ to denote "in all the possible tilings for the next round, φ holds". For the sake of space we use $\mathtt{T}(i, d_1, d_2, d_3)$ to denote $\mathtt{t}(1) \wedge pos_i \wedge col_{i-1}(d_1) \wedge col_i(d_2) \wedge p_{d_3}$, meaning that one tiles d_3 at Position i (of some row), where its left neighbor is tiled d_1 and its down neighbor tiled d_2. Similarly, let $\mathtt{T}(i, d_1, d_2, ?)$ denote $\mathtt{t}(1) \wedge pos_i \wedge col_{i-1}(d_1) \wedge col_i(d_2)$ meaning that what is tiled is not specified. The eventual reduction formula φ is the conjunction of the following formulas.

(i) Each state is either a tiling state or a counting state

$$\blacksquare\Box\big((p_{\#} \vee p_{til}) \wedge \neg(p_{\#} \wedge p_{til})\big). \tag{7}$$

(ii) Tiling

(a) each tiling state encodes exactly one tile at position i, one previous tile at i and one previous tile at $i-1$ [8]

$$\blacksquare \bigwedge_i \Big(\bigvee_{d_1, d_2, d_3} \Diamond(p_{til} \wedge \mathtt{T}(i, d_1, d_2, d_3)) \wedge \bigwedge_{(d_1', d_2', d_3') \neq (d_1, d_2, d_3)} \Box\neg\mathtt{T}(i, d_1', d_2', d_3')\Big) \tag{8}$$

(b) start configuration: all previous tiles are p_{white}

$$\bigwedge_i \Diamond(p_{til} \wedge \mathtt{t}(1) \wedge col_i(p_{white})) \tag{9}$$

(c) colors match, where $right(d')$ denotes the set of dominos whose left-color is the right-color of d', and similar to $top(d'')$

$$\blacksquare\Box \bigwedge_{i, d', d''} \big(\mathtt{T}(i, d', d'', ?) \to \bigvee_{d \in right(d') \cap top(d'')} p_d\big) \tag{10}$$

(d) The *current* tiling of the current position in the current row is the *previous* tiling of the current position in whatever *next* row

$$\blacksquare\Box \bigwedge_{i,d} \big((p_{til} \wedge pos_i \wedge \mathtt{t}(1) \wedge p_d) \to [K]\Box((p_{til} \wedge pos_i \wedge \mathtt{t}(1)) \to col_i(d))\big). \tag{11}$$

(iii) Counting: we use $p_{i,\#,c}$ to denote $(p_i \wedge p_{\#} \wedge \mathtt{t}(c))$ with $1 \leq i \leq n, c \in \{1, 0\}$. The following mean that every row has a number; counting starts with 0, increases and not terminates before $2^n - 1$. [9]

$$\blacksquare \bigwedge_{1 \leq i \leq n} \big(\Diamond(p_{i,\#,1} \vee p_{i,\#,0}) \wedge \Box\neg(p_{i,\#,1} \wedge p_{i,\#,0})\big) \tag{12}$$

---

[8] To save space we write $\bigvee_i$ instead of $\bigvee_{0 \leq i \leq n+1}$, $\bigvee_d$ instead of $\bigvee_{d \in T}$ for all the formulas below, and similar for the conjunction cases. We will write explicitly $\bigwedge_{1 \leq i \leq n}$ when i ranges from 1 through n instead of from 0 through $n+1$.

[9] A detail is that counting starts from 0 but denotes Row 1 since Row 0 is for white tiles.

$$\Box \bigwedge_{1\leq i\leq n} ((p_i \wedge p_{\#}) \rightarrow \mathtt{t}(0)) \tag{13}$$

$$\blacksquare \bigwedge_{1\leq k\leq n} \Big(\Diamond p_{k,\#,0} \wedge \bigwedge_{1\leq i<k} \Diamond p_{k,\#,1}) \rightarrow \Big([K]\Box\big((p_k \wedge p_{\#}) \rightarrow \mathtt{t}(1)\big) \\ \wedge \bigwedge_{1\leq i<k} \Box\big((p_k \wedge p_{\#}) \rightarrow \mathtt{t}(0)\big) \wedge \bigwedge_{n\geq i>k} store(p_i)\Big)\Big) \tag{14}$$

where $store(p_i) := \bigwedge_{c\in\{0,1\}} \big(\Diamond p_{i,\#,c} \rightarrow [K]\Box\big((p_k \wedge p_{\#}) \rightarrow \mathtt{t}(c)\big)\big)$

$$\blacksquare([K]\bot \rightarrow \bigwedge_{1\leq i\leq n} \Diamond p_{i,\#,1}). \tag{15}$$

(iv) Players: the following mean that each row belongs to a player; Eloise starts; players alternate; Abelard can play whatever possible move until counting to $2^n - 1$ (viz. Eloise has a winning strategy).

$$\blacksquare\big(\Diamond(eloise \wedge \mathtt{t}(1)) \rightarrow \Box(abelard \rightarrow \mathtt{t}(0))\big) \tag{16}$$

$$\Diamond(eloise \wedge \mathtt{t}(1)) \tag{17}$$

$$\blacksquare \bigwedge_{\pi,\pi'\in\{eloise,aberlard\},\pi\neq\pi'} (\Diamond(\pi \wedge \mathtt{t}(1)) \rightarrow [K]\Box(\pi' \rightarrow \mathtt{t}(1))) \tag{18}$$

$$\blacksquare\Box \bigwedge_{i,d',d''} \Big(\big(abelard \wedge p_{til} \wedge \mathtt{T}(i,d',d'',?)\big) \rightarrow \\ \big(\bigwedge_{d\in right(d')\cap top(d'')} \neg[K]\neg col_i(d) \vee \bigwedge_{1\leq i\leq n} \Diamond p_{i,\#,1}\big)\Big). \tag{19}$$

The proof of the following lemma is in the Appendices.

Lemma 5.1 *Eloise has a winning strategy in the two-player $2^n \times n$ corridor tiling game if and only if the reduction formula is satisfiable in* **MCM**.

With the lemma and the EXPTIME-completeness of the tiling game, we have the main result of this section.

Theorem 5.2 *Let Atm_0 be infinite and $|Dec| > 1$. Then deciding satisfiability of $\mathcal{L}$-formulas relative to the class* **MCM** *is EXPTIME-hard.*

The main result of the paper now turns to be a corollary.

Corollary 5.3 *Let Atm_0 be infinite and $|Dec| > 1$. Then deciding satisfiablity of $\mathcal{L}$-formulas relative to the class* **MCM** *is EXPTIME-complete.*

6 Conclusion

In this paper, we explored the complexity of satisfiability checking for the two-dimensional modal language $\mathcal{L}$, the language of the logic PLC, relative to the class **MCM**. We showed that when the set of input features is countably infinite and there is more than one decision variable, this problem is EXPTIME-complete. This result is quite surprising since the product modal logic $S5^2$, of which PLC is a special instance, is known to be NEXPTIME-complete. But, we believe, there is an intuitive explanation of this result. In the language $\mathcal{L}$ the $\blacksquare$-dimension only matters when some decision variable from *Dec* is involved. Indeed, any *Dec*-free formula φ in $\mathcal{L}$ is equivalent to some formula ψ, which results from the removal of all the occurrences of $\blacksquare$ and $\blacklozenge$ from φ.

A crucial assumption of our analysis is that the set of decision variables *Dec* is finite, namely, we only consider classifiers with finitely many possible outputs. Our conjecture is that by dropping this assumption we would make the satisfiability problem back to NEXPTIME-complete.

Directions of future work are manifold. On the automated reasoning side, we plan to explore methods for automating satisfiability checking of PLC including tableaux or sequent calculi. Such methods could also be useful to deepen our understanding of this special type of product modal logic.

On the conceptual side, as mentioned in the introduction, the semantics of PLC based on multi-classifier models (MCMs) (Definition 2.1) can serve as a stage to evaluate and compare classifiers according to their ideality. This goes into the direction of deontic logic. But to be able to do this, an MCM $\Gamma = (S, F_S)$ should be equipped with a partial preorder $\preceq \subseteq F_S \times F_S$, with $f \preceq f'$ meaning that the classifier f' is at least as ideal (in the deontic sense) as the classifier f. The resulting structure $\Gamma = (S, F_S, \preceq)$ could be reasonably called MCM with ideality (MCMI). Different deontic extensions of the language $\mathcal{L}$, exploiting this new class of structures, are worth to be explored. Probably the conceptually simplest one is the extension with the so-called S4 "betterness" modality $[\preceq]$ [21], whose semantic interpretation with respect to an MCMI $\Gamma = (S, F_S, \preceq)$, $s \in S$ and $f \in F_S$ would go as follows:

$$(\Gamma, s, f) \models [\preceq]\varphi \iff \forall f' \in F_S : \text{ if } f \preceq f' \text{ then } (\Gamma, s, f') \models \varphi.$$

The formula $[\preceq]\varphi$ has to read "φ holds for all classifiers that are at least as ideal as the actual one". In future work, we plan to study the proof-theoretic and complexity aspects of the system combining the two modalities $\Box$ and $[\preceq]$ as well as of the more general system combining the three modalities $\Box$, $\blacksquare$ and $[\preceq]$. Note that the former is a so-called semi-product.

The interest of studying deontic extensions of PLC lies in the possibility of using them to formalize a variety of normative concepts that are discussed in the area of ethical machine learning [11], including the prohibition for a classifier to discrimate or to make biased decisions, the obligation to decide in a fair way, and the more subtle notion of positive discrimination which can be seen as a kind of conditional permission (i.e., the permission to discriminate only on the condition that it favors a disadvantaged or protected individual).

References

[1] Amgoud, L. and J. Ben-Naim, *Axiomatic foundations of explainability*, in: *Proceedings of the 31st International Joint Conference on Artificial Intelligence (IJCAI 2022)*, 2022, pp. 636–642.

[2] Audemard, G., S. Bellart, L. Bounia, F. Koriche, J.-M. Lagniez and P. Marquis, *On the computational intelligibility of boolean classifiers*, in: *Proceedings of the International Conference on Principles of Knowledge Representation and Reasoning*, 2021, pp. 74–86.

[3] Bezhanishvili, N. and I. M. Hodkinson, *All normal extensions of s5-squared are finitely axiomatizable*, Studia Logica **78** (2004), pp. 443–457.

[4] Blackburn, P., M. de Rijke and Y. Venema, "Modal Logic," Cambridge University Press, Cambridge, Massachusetts, 2001.

[5] Darwiche, A. and A. Hirth, *On the reasons behind decisions*, in: *24th European Conference on Artificial Intelligence (ECAI 2020)*, Frontiers in Artificial Intelligence and Applications **325** (2020), pp. 712–720.

[6] Darwiche, A. and P. Marquis, *On quantifying literals in boolean logic and its applications to explainable AI*, Journal of Artificial Intelligence Research **72** (2021), pp. 285–328.

[7] Eiter, T., T. Geibinger and J. Oetsch, *Contrastive explanations for answer-set programs*, in: *European Conference on Logics in Artificial Intelligence*, Springer, 2023, pp. 73–89.

[8] Gabbay, D. M., A. Kurucz, F. Wolter and M. Zakharyaschev, "Many-dimensional modal logics: theory and applications," Elsevier, 2003.

[9] Halpern, J. Y., *The effect of bounding the number of primitive propositions and the depth of nesting on the complexity of modal logic*, Artificial Intelligence **75** (1995), pp. 361–372.

[10] Ignatiev, A., N. Narodytsk and J. Marques-Silva, *Abduction-based explanations for machine learning models*, in: *Proceedings of the Thirty-third AAAI Conference on Artificial Intelligence (AAAI-19)*, 2019, pp. 1511–1519.

[11] Kearns, M. and A. Roth, "The ethical algorithm: The science of socially aware algorithm design," Oxford University Press, 2019.

[12] Ladner, R. E., *The computational complexity of provability in systems of modal propositional logic*, SIAM journal on computing **6** (1977), pp. 467–480.

[13] Liu, X. and E. Lorini, *A logic of "black box" classifier systems*, in: *Logic, Language, Information, and Computation: 28th International Workshop, WoLLIC 2022, Iași, Romania, 2022, Proceedings*, Springer, 2022, pp. 158–174.

[14] Liu, X. and E. Lorini, *A unified logical framework for explanations in classifier systems*, Journal of Logic and Computation **33** (2023), pp. 485–515.

[15] Marques-Silva, J., T. Gerspacher, M. C. Cooper, A. Ignatiev and N. Narodytska, *Explaining naive bayes and other linear classifiers with polynomial time and delay*, in: *Advances in Neural Information Processing Systems 33: Annual Conference on Neural Information Processing Systems 2020, NeurIPS 2020*, 2020.

[16] Marques-Silva, J. and A. Ignatiev, *Delivering trustworthy AI through formal XAI*, in: *Proceedings of the AAAI conference on Artificial Intelligence*, 2022, pp. 12342–12350.

[17] Marx, M., *Complexity of products of modal logics*, Journal of Logic and Computation **9** (1999), pp. 197–214.

[18] Russell, S. J. and P. Norvig, "Artificial intelligence: a modern approach," Pearson, 2021, 4th edition.

[19] Schwarzentruber, F., *The complexity of tiling problems*, arXiv preprint arXiv:1907.00102 (2019).

[20] Shih, A., A. Choi and A. Darwiche, *A symbolic approach to explaining bayesian network classifiers*, in: *Proceedings of the Twenty-Seventh International Joint Conference on Artificial Intelligence (IJCAI 2018)* (2018), pp. 5103–5111.
URL `https://doi.org/10.24963/ijcai.2018/708`

[21] van Benthem, J. and F. Liu, *Dynamic logic of preference upgrade*, Journal of Applied Non Classical Logics **17** (2007), pp. 157–182.

[22] van Emde Boas, P., *The convenience of tilings*, in: *Complexity, Logic, and Recursion Theory*, CRC Press, 1997 pp. 331–363.

Appendix

A Proof of Lemma 4.5

Proof. 1. The existence is guaranteed by the maximality of path, with a special case that $x = z$ when $x \in \pi$. For the uniqueness, if $\exists z' \in \pi$ s.t. $z \neq z' \equiv_{\square} x$, then $z \equiv_{\square} z' \equiv_{\blacksquare} z$, contradicting the definition of path.

2. Special case of 1, s.t. if $x \equiv_{\square} y$ or $x \equiv_{\blacksquare} y$ then $z = x$.

3. Since $q_i \in s_i^x, q_j \in s_j^y$ and q_i and q_j are distinct, it must be $i = j$. But then it means $x, y \in \pi_i = \pi_j$. Then by definition $x \equiv_{\blacksquare} y$.

4. Straightforward from definition. □

B Proof of Lemma 4.6

Proof. Clearly S is the common domain. For any f^x, Lemma 4.5 shows that it is indeed a function. More precisely, suppose towards a contradiction $(s_i^y, c), (s_j^z, c') \in f^x$ for some f^x with $s_i^y = s_j^z, c \neq c'$. From $s_i^y = s_j^z$ we know $i = j$ and $y \equiv_{\blacksquare} z$.

Moreover, by the lemma above, $\exists! z' \in \pi_i$ s.t. $z' \equiv_{\square} x$, and therefore $f^x(s_i^y) = f^x(s_j^z) = f^x(s_i^{z'})$, the contradiction. Hence the $\Gamma = (S, F_S)$ is well-defined. □

C Proof of Lemma 4.7

Proof. Prove by induction. The base cases of ψ being some p or $\mathsf{t}(c)$ are straightforward. Notice crucially that for any π_i with $x \in \pi_i, x$ is the unique member in π_i s.t. $x \equiv_{\square} y$, hence $f^x(s_i^x) = c$ iff $\mathsf{t}(c) \in x$. The Boolean cases are obvious by the definition of the Hintikka sets. The modal cases are more complex.

For the right direction, If ψ is some $\blacksquare\psi'$, by $\blacksquare\psi' \in x$ we have $\psi' \in x$. Suppose towards a contradiction $(\Gamma, s_i^x, f^x) \models \blacklozenge\neg\psi'$. That means $\exists f^y \in F_S, (\Gamma, s_i^x, f^y) \models \neg\psi'$. Lemma 4.5 gives us $\exists! z \in \pi_i, z \equiv_{\square} y$, viz. $s_i^z = s_i^x, f^z = f^y$. Therefore we have $(\Gamma, s_i^z, f^z) \models \neg\psi'$. However from $z, x \in \pi_i$ we have $z \equiv_{\blacksquare} x$, which means $\blacksquare\psi' \in z$, and consequently $\psi' \in z$. By induction hypothesis $(\Gamma, s_i^z, f^z) \models \psi'$, the contradiction we want. If ψ is some $\blacklozenge\psi'$, since $\blacklozenge\psi' \in x$, by the elimination procedure either $\psi' \in x$, or $\exists z \in \mathcal{H}, z \equiv_{\blacksquare} x \not\equiv_{\square} z, \psi' \in z$. By induction hypothesis the former gives $(\Gamma, s_i^x, f^x) \models \blacklozenge\psi'$, which is what we want; the latter gives $(\Gamma, s_j^z, f^z) \models \psi'$ for all $\pi_j \ni z$. Again by Lemma 4.5, we have $\exists! y \in \pi_i, y \equiv_{\square} z$. Hence $(\Gamma, s_i^z, f^y) \models \psi$. Moreover, since $s_i^z = s_i^x, f^y = f^z$, we have $(\Gamma, s_i^x, f^y) \models \psi$, which is sufficient to have $(\Gamma, s_i^x, f^x) \models \blacklozenge\psi$. The case of ψ being some $\circ$ formula is similar and easier, and therefore omitted.

For the left direction, again we only prove the $\bullet$ cases. Let ψ be some $\blacksquare\psi'$. Suppose towards a contradiction that the antecedent holds but $\blacksquare\psi' \notin x$, then $\blacklozenge\neg\psi' \in x$. If $\neg\psi' \in x$ then we have an obvious contradiction. Otherwise, the elimination procedure ensures some $y \equiv_{\blacksquare} x \not\equiv_{\square} y$ s.t. $\neg\psi' \in y$. By the right direction we have $(\Gamma, s_i^y, f^y) \models \neg\psi'$ for all $\pi_i \ni y$. Now since $x \equiv_{\blacksquare} y \not\equiv_{\square} y$, there is some π_j s.t. $x, y \in \pi_j$ and by Lemma 4.5 we have $s_j^y = s_j^x$. It then means $(\Gamma, s_j^x, f^y) \models \neg\psi'$, contradicting the antecedent that $(\Gamma, s_j^x, f^x) \models \blacksquare\psi'$.

Let ψ be some $\blacklozenge\psi'$. Again suppose $\forall\pi_i \ni x, (\Gamma, s_i^x, f^x) \models \blacklozenge\psi'$ but $\blacklozenge\psi' \notin x$. Clearly $\psi' \notin x$, which means $\exists\pi_j \ni x, (\Gamma, s_j^x, f^x) \not\models \psi'$. Hence $\exists f^y \neq f^x$ s.t. $(\Gamma, s_j^x, f^y) \models \neg\psi$. By Lemma 4.5 this means $x \not\equiv_\Box y$. Then also from Lemma 4.5 we know $\exists! z \in \pi_i$, s.t $z \equiv_\Box y$. Therefore $s_i^x = s_i^z, f^y = f^z$, which gives $(\Gamma, s_i^z, f^z) \models \psi$. By induction hypothesis $\psi \in z$. But since $z, x \in \pi_i$, we have $z \equiv_\blacksquare x$, hence $\neg\psi \in z$ since $\blacksquare\neg\psi \in x$, a contradiction.

Now, by definition $\exists\pi_i \in \mathcal{P}_\mathcal{H}$ s.t. $x, y \in \pi_i$ and therefore $s_i^x = s_i^y$. That means $(\Gamma, s_i^x, f^y) \models \neg\psi'$, which is sufficient for $(\Gamma, s_i^x, f^y) \models \neg\blacksquare\psi'$. Finally let ψ be some $\blacklozenge\psi'$ and $(\Gamma, s_i^x, f^x) \models \blacklozenge\psi'$. Suppose towards a contradiction that $\blacklozenge\psi' \notin x$, which means $\blacksquare\neg\psi' \in x$. Clearly $\psi \notin x$. Now from $(\Gamma, s_i^x, f^x) \models \blacklozenge\psi'$ we have $(\Gamma, s_i^x, f^y) \models \psi'$ for some y. Notice $y \not\equiv_\Box x$, otherwise $(\Gamma, s_i^x, f^x) \models \psi'$ and by induction hypothesis $\psi' \in x$. Therefore by the lemma $\exists! z \in \pi_i, z \equiv_\Box y$, which makes $(\Gamma, s_i^z, f^y) \models \psi$. Again notice $s_i^z = s_i^x, f^y = f^z$, and by induction hypothesis $\psi \in z$, contradicting the fact that $\blacksquare\neg\psi \in z$ since $z \equiv_\blacksquare x$. □

D Proof of Lemma 5.1

Proof. Suppose Eloise has a winning strategy. Then there shall be a game tree for Eloise, which is satisfied by a desired model for it. We construct such a model $\Gamma = (S, F_S)$ as follows

(i) $S = S_{til} \cup S_\#$ is partitioned as sets of tiling and counting states where
  (a) $S_{til} = \{\{p_{til}, \pi, pos_i, p_d, col_{i-1}(d'), col_i(d'')\} : \pi \in \{eloise, abelard\}, 0 \leq i \leq n+1, d, d', d'' \in T\}$
  (b) $S_\# = \{\{p_\#, p_i\} : 1 \leq i \leq n\}$

(ii) $f \in F_S$, iff there is a node in the game tree, which encodes a row s.t.
  (a) it is played by player π iff $(\exists s \in S, \pi \in s, f(s) = 1)$ implies that $(\forall s \in S$, if $\pi' \in s$ then $f(s) = 0)$, for $\pi, \pi' \in \{eloise, abelard\}$ and $\pi \neq \pi'$.
  (b) for every row f which is tiled by some π, $\mathtt{T}(i, d, d', d'')$ holds iff $f(\{p_{til}, \pi, pos_i, p_d, col_{i-1}(d'), col_i(d'')\}) = 1$
  (c) it is at the k-th level of the tree, iff the sequence $f(p_n)f(p_{n-1})\ldots f(p_1)$ binary encodes $k-1$. [10]

We need show that the model satisfies the reduction formula φ. Obviously, the tiling and counting conjuncts are satisfied, otherwise either some row is not legally tiled or some player cannot tile after some k-round for $1 \leq k \leq 2^n$. In both cases the game tree would not exist. For the player conjuncts, that each player holds a whole row and players alternate are clearly satisfied. To see the winning strategy formula is satisfied, notice that if it is not satisfied in the desired model, then either Abelard has some move in some k-th round $(1 < k \leq 2^n)$ that is "unexpected" by the game tree, or the game does not terminate after 2^n rounds. In both case Eloise would have no winning strategy.

Let the formula be satisfiable. We first reduce any model satisfying the formula to a tree model in terms of the K-accessibility relation. Then Eloise

[10] Notice again we start counting rounds from 1 while the binary encoding starts from 0.

starts the game from the pointed model that satisfies the formula. Whatever Abelard plays in his round, say k with $1 < k < 2^n$, what Eloise needs to do is to find a classifier f' in the model whose number is k and whose tiling is exactly what Abelard tiles. Then, she chooses any of its K-successor f'', and tiles as f'''s instruction. The formula guarantees that whatever Abelard tiles, Eloise does not stop tiling before the 2^n-th round. □

Goldblatt-Thomason Theorems for Fundamental (Modal) Logic

Guillaume Massas [1]

Scuola Normale Superiore, Pisa

Abstract

Holliday recently introduced a non-classical logic called Fundamental Logic, which intends to capture exactly those properties of the connectives "and", "or" and "not" that hold in virtue of their introduction and elimination rules in Fitch's natural deduction system for propositional logic. Holliday provides an intuitive relational semantics for fundamental logic which generalizes both Goldblatt's semantics for orthologic and Kripke semantics for intuitionistic logic. In this paper, we further the analysis of this semantics by providing a Goldblatt-Thomason theorem for Fundamental Logic. We identify necessary and sufficient conditions on a class K of fundamental frames for it to be axiomatic, i.e., to be the class of frames satisfying some logic extending Fundamental Logic. As a straightforward application of our main result, we also obtain a Goldblatt-Thomason theorem for Fundamental Modal Logic, which extends Fundamental Logic with standard $\Box$ and $\Diamond$ operators.

Keywords: Non-classical Logic, Relational Semantics, Goldblatt-Thomason Theorem

1 Introduction

The celebrated Goldblatt-Thomason theorem [8] for modal logic characterizes when an elementary class $\mathfrak{K}$ of Kripke frames is *axiomatic* (i.e., *modally definable*) in terms of natural closure properties imposed on $\mathfrak{K}$. From a categorical perspective, Goldblatt and Thomason's result relies on transferring results and techniques from universal algebra (particularly Birkhoff's HSP theorem [1]) into the setting of relational semantics. This can typically be achieved by *canonical extensions* [4,11,12] which, in the context of Kripke semantics for modal logic, are closely related to ultrafilter extensions. Because the mathematical core of the Goldblatt-Thomason theorem is algebraic in nature, the result straightforwardly generalizes to many relational semantics for non-classical logics, including Kripke semantics for (modal) intuitionistic logic [3,6,17], polarity-based semantics for lattice-expansion logics [2], or the more general setting of coalgebras [15].

[1] guillaume.massas@sns.it

Our goal here is to present such a generalization of the Goldblatt-Thomason theorem to Holliday's relational semantics for Fundamental Logic [9]. Holliday's semantics generalizes both Goldblatt's semantics for orthologic [7] and Kripke semantics for intuitionistic logic [13], while arguably preserving the intuitiveness and simplicity of both. Accordingly, this makes it a promising framework for the study of a large class of non-classical logics, towards which our result can be seen as a first step. We proceed as follows. In Section 2, we review some background on Fundamental Logic, its algebraic semantics in terms of fundamental lattices and its relational semantics in terms of fundamental frames. We also introduce the relevant notion of morphism between fundamental frames. In Section 3 we identify the relational duals of subalgebras, homomorphic images and products of fundamental lattices. Section 4 introduces a particular kind of construction on fundamental frames called the filter extension, which is used to prove our main result, a Goldblatt-Thomason theorem for Fundamental Logic. As a corollary, we also obtain a characterization of those classes of fundamental frames that are axiomatized by a canonical logic extending Fundamental Logic. Finally, Section 5 generalizes our results to the setting of Fundamental Modal Logic [10].

2 Background

In this section, we first provide some background on Fundamental Logic and on two semantics for it presented in [9], an algebraic semantics in terms of fundamental lattices and a relational semantics in terms of fundamental frames. We then define a notion of morphism between fundamental frames that is the relational analogue of morphisms between fundamental lattices. We assume some familiarity with basic notions of lattice theory and algebraic logic.

2.1 Fundamental Logic

Introduced by Holliday in [9], Fundamental Logic is the logic of a propositional language containing connectives $\wedge$, $\vee$, $\neg$ and constants $\bot$ and $\top$ determined uniquely by the introduction and elimination rules for each connective in Fitch's natural deduction system [5]. Equivalently, letting $\mathcal{L}$ be such a propositional language, $\vdash_{FL}$ is the smallest reflexive and transitive relation on $\mathcal{L}\times\mathcal{L}$ satisfying the following closure conditions:

$$
\begin{array}{lll}
\bot \vdash \varphi; & \varphi \vdash \top; & \top \vdash \neg\bot; \\
\varphi\wedge\psi \vdash \varphi; & \varphi\wedge\psi \vdash \psi; & \chi \vdash \varphi \,\&\, \chi \vdash \psi \Rightarrow \chi \vdash \varphi\wedge\psi; \\
\varphi \vdash \varphi\vee\psi; & \psi \vdash \varphi\vee\psi; & \varphi \vdash \chi \,\&\, \psi \vdash \chi \Rightarrow \varphi\vee\psi \vdash \chi; \\
\varphi\wedge\neg\varphi \vdash \bot; & \varphi \vdash \neg\neg\varphi; & \varphi \vdash \psi \Rightarrow \neg\psi \vdash \neg\varphi.
\end{array}
$$

Fundamental Logic generalizes both the $\rightarrow$-free fragment of intuitionistic logic and orthologic. In Fitch's natural deduction system, the former can be recovered from Fundamental Logic by adding the Reiteration rule, and the latter by adding the double negation elimination rule.

Fundamental Logic has a natural algebraic semantics in terms of fundamental lattices. Recall first the following definition.

Definition 2.1 An antitone map $f : L \to L$ on a lattice L is *dually self-adjoint* if for any $a, b \in L$, $a \leq_L f(b)$ iff $b \leq_L f(a)$.

Definition 2.2 A *fundamental lattice* is a pair $(L, \neg)$ such that L is a bounded lattice and $\neg : L \to L$ is a dually self-adjoint map such that $a \wedge_L \neg a = 0_L$ for any $a \in L$.

Given a fundamental lattice $(L, \neg)$, a valuation V maps any propositional letter $p \in \mathcal{L}$ to some $a \in L$, and is then recursively extended to any formula in $\mathcal{L}$ by using the operations on $(L, \neg)$ in the obvious way. For any two formulas $\varphi, \psi \in \mathcal{L}$, ψ is an algebraic consequence of φ iff $V(\varphi) \leq_L V(\psi)$ for any fundamental lattice $(L, \neg)$ and any valuation V on L. As shown in Holliday [9], Fundamental Logic is sound and complete with respect to the algebraic semantics thus obtained.

2.2 Fundamental Frames

By a relational frame, we simply mean a pair (X, R) such that R is a co-serial relation on X, meaning that for any $x \in X$ there is $x' \in X$ such that $x'\mathsf{R}x$. Points in a relational frame can be viewed as partial states of information, situations, or positions in discourse, at which propositions may be either *accepted* or *rejected.* The relation R can be interpreted as a relation of *openness* between such states, where $x\mathsf{R}y$ is interpreted as "x is open to y". In what follows, we write $\reflectbox{$\mathsf{R}$}$ for the converse of the relation R.

Given a relation frame (X, R), the relation R induces two antitone operations $\neg_{\mathsf{R}}, \neg_{\reflectbox{R}} : \mathcal{P}(X) \to \mathcal{P}(X)$ given by $\neg_{\mathsf{R}} A = \{x \in X \mid \forall x'\mathsf{R}x : x' \notin A\}$ and $\neg_{\reflectbox{$\mathsf{R}$}} A = \{x \in X \mid \forall x'\reflectbox{R}x : x' \notin A\}$ for any $A \subseteq X$. Clearly, these two maps form a contravariant adjunction, so they induce two anti-isomorphic complete lattices $\chi_{\mathsf{R}}(X)$ and $\chi_{\reflectbox{$\mathsf{R}$}}(X)$ with domains $\{\neg_{\mathsf{R}} A \mid A \subseteq X\}$ and $\{\neg_{\reflectbox{$\mathsf{R}$}} A \mid A \subseteq X\}$ respectively, which are the fixpoints of the operations $\neg_{\mathsf{R}}\neg_{\reflectbox{R}}$ and $\neg_{\reflectbox{$\mathsf{R}$}}\neg_{\mathsf{R}}$ respectively. The composition of the two maps $\neg_{\mathsf{R}}$ and $\neg_{\reflectbox{$\mathsf{R}$}}$ yields a closure operator C_{R}, which can be explicitly described as

$$C_{\mathsf{R}}(A) = \{x \in X \mid \forall y\mathsf{R}x \exists z \reflectbox{R} y : z \in A\}$$

for any $A \subseteq X$. Finally, we call $\chi_{\mathsf{R}}(X)$ the *positive algebra* of the frame (X, R), and $\chi_{\reflectbox{$\mathsf{R}$}}$ its *negative algebra.*

The following relations can always be defined from an openness relation R, and will be repeatedly used throughout.

Definition 2.3 Let (X, R) be a relational frame. For any two elements $x, x' \in X$, x *positively refines* x' (noted $x \leq_{\mathsf{R}} x'$) if for any $z \in X$: $z\mathsf{R}x \Rightarrow z\mathsf{R}x'$, and x *negatively refines* x' (noted $x \leq_{\reflectbox{$\mathsf{R}$}} x'$) if for any $z \in X$: $x\mathsf{R}z \Rightarrow x'\mathsf{R}z$.

We will also often appeal to some straightforward facts about relational frames whose proofs we omit.

Fact 2.4 *The following hold for any relational frame* (X, R)*:*

(i) *For any* $A \subseteq X$, $A \in \chi_{\mathsf{R}}(X)$ *iff* $\forall x \in X : x \in A \Leftrightarrow \forall x'\mathsf{R}x \exists y \reflectbox{R} x' : y \in A$*;*

(ii) *For any* $A \subseteq X$, $A \in \chi_{\reflectbox{$\mathsf{R}$}}(X)$ *iff* $\forall x \in X$*:* $x \in A \Leftrightarrow \forall x'\reflectbox{R}x \exists y \mathsf{R} x' : y \in A$*;*

(iii) *For any* $x \in X$, $\downarrow_{\mathsf{R}}(x) = \{y \in X \mid y \leq_{\mathsf{R}} x\}$ *and* $\overline{\mathsf{R}}(x) = \{y \in X \mid \neg x\mathsf{R}y\}$ *are elements in* $\chi_{\mathsf{R}}(X)$*;*

(iv) *For any* $x \in X$, $\downarrow_{\mathsf{Я}}(x) = \{y \in X \mid y \leq_{\mathsf{Я}} x\}$ *and* $\overline{\mathsf{Я}}(x) = \{y \in X \mid \neg x\mathsf{Я}y\}$ *are elements in* $\chi_{\mathsf{Я}}(X)$*;*

(v) $\leq_{\mathsf{R}}$ *and* $\leq_{\mathsf{Я}}$ *are preorders on* X*;*

(vi) *For any* $x, x' \in X$, $x \leq_{\mathsf{R}} x'$ *iff for any* $A \in \chi_{\mathsf{R}}(X)$, $x' \in A$ *implies* $x \in A$*;*

(vii) *For any* $x, x' \in X$, $x \leq_{\mathsf{Я}} x'$ *iff for any* $B \in \chi_{\mathsf{Я}}(X)$, $x' \in B$ *implies* $x \in B$.

For any relational frame (X, R), its positive algebra $\chi_{\mathsf{R}}(X)$ can be equipped with the unary antitone map $\neg_{\mathsf{R}}$. In order to ensure that the resulting pair $(\chi_{\mathsf{R}}(X), \neg_{\mathsf{R}})$ is a fundamental lattice, we need to impose two conditions on R:

Definition 2.5 Let (X, R) be a relational frame. Let $\mathsf{R}(x)$ and $\mathsf{Я}(x)$ be the sets $\{y \in X \mid x\mathsf{R}y\}$ and $\{y \in X \mid y\mathsf{R}x\}$ respectively.

- R is *pseudo-reflexive* if $\mathsf{Я}(x) \cap \downarrow_{\mathsf{R}}(x) \neq \emptyset$ for any $x \in X$;
- R is *pseudo-symmetric* if for any $x, x' \in X$, $x' \in \mathsf{R}(x)$ implies $\mathsf{Я}(x) \cap \downarrow_{R}(x') \neq \emptyset$.

A *fundamental frame* is a relational frame (X, R) such that R is pseudo-reflexive and pseudo-symmetric.

Fundamental frames provide a relational semantics for fundamental logic in a straightforward way. Given a fundamental frame (X, R), a valuation V maps any propositional letter $p \in \mathcal{L}$ to some $A \in \chi_{\mathsf{R}}(X)$, and is recursively extended to all formulas $\varphi \in \mathcal{L}$ as follows:

- $V(\neg\varphi) = \neg_{\mathsf{R}} V(\varphi)$;
- $V(\varphi \wedge \psi) = V(\varphi) \cap V(\psi)$;
- $V(\varphi \vee \chi) = C_{\mathsf{R}}(V(\varphi) \cup V(\psi))$.

This ensures that formulas are always evaluated as elements in the positive algebra of a fundamental frame. As usual, given a fundamental frame (X, R), we write $\varphi \models_{(X,\mathsf{R})} \psi$ if $V(\varphi) \subseteq V(\psi)$ for any valuation V on (X, R). The following establishes the soundness of this semantics for Fundamental Logic.

Theorem 2.6 ([9], Prop. 4.14) *For any relational frame* (X, R), $\chi_{\mathsf{R}}(X)$ *is a fundamental lattice iff* (X, R) *is a fundamental frame.*

As is standard in relational semantics, completeness is established via a canonical frame construction. Since this construction plays a central role in this paper, we briefly review it now.

Definition 2.7 Let $(L, \neg)$ be a fundamental lattice. The *canonical frame* of $(L, \neg)$ is the relational frame $\mathcal{F}(L) = (X, \mathsf{R})$ given by the following data:

- X is the set of all pairs (F, I) such that F and I are a proper filter and a proper ideal on L respectively, and $F \subseteq \neg^{-1}[I]$.
- For any $(F, I), (G, J) \in X$, $(F, I)\mathsf{R}(G, J)$ iff $G \cap I = \emptyset$.

It is straightforward to verify that, if $(X, \mathsf{R}) = \mathcal{F}(L)$ for some fundamental lattice $(L, \neg)$, then the positive refinement relation $\leq_{\mathsf{R}}$ is given by converse inclusion on filters and the negative refinement relation $\leq_{\text{Я}}$ is given by converse inclusion on ideals. In other words, for any two points $(F, I), (G, J) \in X$, $(F, I) \leq_{\mathsf{R}} (G, J)$ iff $F \supseteq G$, and $(F, I) \leq_{\text{Я}} (G, J)$ iff $I \supseteq J$. Moreover, any fundamental lattice embeds into the positive algebra of its canonical frame via a standard Stone-like map.

Theorem 2.8 ([9], Thm. B.7) *For any fundamental lattice* $(L, \neg)$, $\mathcal{F}(L) = (X, R)$ *is a fundamental frame, and the map* $\widehat{\cdot} : (L, \neg) \to (\chi_{\mathsf{R}}(X), \neg_{\mathsf{R}})$ *given by* $a \mapsto \{(F, I) \in X \mid a \in F\}$ *is a lattice embedding such that* $\widehat{\neg a} = \neg_{\mathsf{R}}\widehat{a}$ *for all* $a \in L$.

Given a class $\mathfrak{K}$ of fundamental frames, we let $Log(\mathfrak{K})$ be the set $\{\varphi \in \mathcal{L} \mid \forall (X, \mathsf{R}) \in \mathfrak{K} : \top \models_{(X,\mathsf{R})} \varphi\}$. Similarly, given some $\Gamma \subseteq \mathcal{L}$, we let $Mod(\Gamma)$ be the class of fundamental frames $\{(X, \mathsf{R}) \mid \forall \varphi \in \Gamma : \top \models_{(X,\mathsf{R})} \varphi\}$. As usual, Log and Mod form a Galois connection between the set of subsets of $\mathcal{L}$ and the class of all classes of fundamental frames, both ordered by inclusion.

2.3 F-Morphisms

In this section, we introduce the notion of a fundamental morphism between fundamental frames. We show first that such morphisms induce fundamental lattice homomorphisms between positive algebras in a natural way, before establishing that fundamental lattice homomorphisms induce fundamental morphisms between canonical frames.

Definition 2.9 Let (X, R) and (Y, S) be two fundamental frames. A *fundamental morphism* (f-morphism for short) is a map $h : X \to Y$ satisfying the following four properties for any $x, x' \in X$ and $y, y' \in Y$:

(i) $x\mathsf{R}x'$ implies $h(x)\mathsf{S}h(x')$;

(ii) $h(x)\mathsf{S}y$ implies $\exists x' \in X : x\mathsf{R}x'$ and $h(x') \leq_{\mathsf{S}} y$;

(iii) $y\mathsf{S}h(x)$ implies $\exists x' \in X : x'\mathsf{R}x$ and $h(x') \leq_{\text{Ƨ}} y$;

(iv) $y\mathsf{S}h(x)$ implies $\exists x'' \in X : x''\mathsf{R}x$ and $h(x'') \leq_{\mathsf{S}} y$.

The following is a diagrammatic representation of the conditions required on f-morphisms. Single arrows are labelled according to which relation they represent, and double arrows are implications. From left to right, each diagram corresponds to conditions 1, 2 and 3 – 4 respectively.

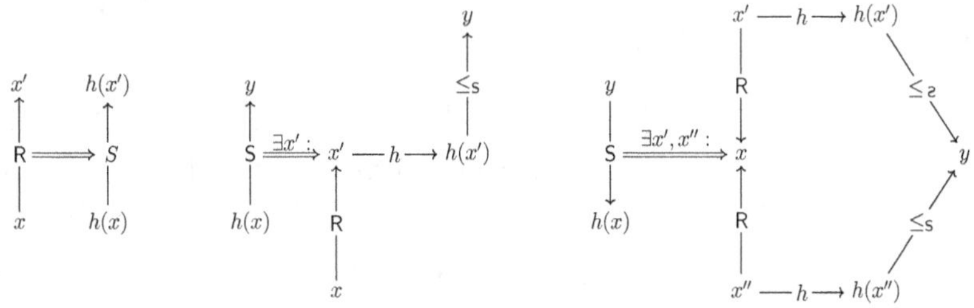

Definition 2.10 Let $(L, \neg), (M, \sim)$ be two fundamental lattices. A *fundamental lattice homomorphism* is a lattice homomorphism $f : (L, \neg) \to (M, \sim)$ such that $f(\neg a) = \sim f(a)$ for any $a \in L$.

Lemma 2.11 *For any f-morphism* $h : (X, \mathsf{R}) \to (Y, \mathsf{S})$, *the map* $\chi(h) : \chi_{\mathsf{S}}(Y) \to \chi_{\mathsf{R}}(X)$ *given by* $A \mapsto h^{-1}[A]$ *is a fundamental lattice homomorphism.*

Proof. We claim that for any $A \subseteq Y$, $C_{\mathsf{R}}(h^{-1}[A]) = h^{-1}[C_{\mathsf{S}}(A)]$. It is routine to check that this implies that $\chi(h) : \chi_{\mathsf{S}}(Y) \to \chi_{\mathsf{R}}(X)$ is well-defined and a lattice homomorphism. Fix some $A \subseteq Y$. For the right-to-left inclusion, assume that $h(x) \in C_{\mathsf{S}}(A)$, and let $x'\mathsf{R}x$. Then by property 1 of f-morphisms, $h(x')\mathsf{S}h(x)$. Since $h(x) \in C_{\mathsf{S}}(A)$, this means that there is $y \in A$ such that $h(x')\mathsf{S}y$. By property 2 of f-morphisms, there is $x'' \in X$ such that $x'\mathsf{R}x''$ and $h(x'') \leq_{\mathsf{S}} y$. But the latter implies that $h(x'') \in A$ and thus that $x'' \in h^{-1}[A]$. This shows that $x \in C_{\mathsf{R}}(h^{-1}[A])$.

For the converse inclusion, assume that $x \in C_{\mathsf{R}}(h^{-1}[A])$, and let $y\mathsf{S}h(x)$. By property 3 of f-morphisms, there is $x' \in X$ such that $x'\mathsf{R}x$ and $h(x') \leq_{\mathsf{S}} y$. Since $x \in C_{\mathsf{R}}(h^{-1}[A])$, there is $x'' \in h^{-1}[A]$ such that $x'\mathsf{R}x''$. By property 1 of f-morphisms, we have that $h(x')\mathsf{S}h(x'')$. As $h(x') \leq_{\mathsf{S}} y$, we also have $y\mathsf{S}h(x'')$. But this shows that $x \in C_{\mathsf{S}}(h^{-1}[A])$, as desired.

Finally, let us show that $\chi(h)$ preserves fundamental complements. Fix some $A \in \chi_{\mathsf{S}}(Y)$. We need to show that for any $x \in X$,

$$x \in h^{-1}[\neg_{\mathsf{S}} A] \Leftrightarrow x \in \neg_{\mathsf{R}} h^{-1}[A].$$

Suppose first that $h(x) \in \neg_{\mathsf{S}} A$, and let $x'\mathsf{R}x$. Then $h(x')\mathsf{S}h(x)$, hence $h(x') \notin A$. This shows the left-to-right direction of the biconditional. For the converse, assume that $x \in \neg_{\mathsf{R}} h^{-1}[A]$, and let $y\mathsf{S}h(x)$. By property 4 of f-morphisms, there is $x'' \in X$ such that $x''\mathsf{R}x$ and $h(x'') \leq_{\mathsf{S}} y$. Since $x \in \neg_{\mathsf{R}} h^{-1}[A]$, we have that $h(x'') \notin A$, from which it also follows that $y \notin A$. This completes the proof. □

One can verify that the definition of an f-morphism is slightly stronger than what would need to be required of a map h to ensure that its inverse image is a fundamental lattice homomorphism.[2] Nonetheless, our definition is easy to state and general enough, as established by the following lemma.

Lemma 2.12 *Let* $f : (L, \neg) \to (M, \sim)$ *be a fundamental lattice homomorphism. Then the map* $(F, I) \mapsto (f^{-1}[F], f^{-1}[I])$ *yields a f-morphism* $\mathcal{F}(f) : \mathcal{F}(M) \to \mathcal{F}(L)$ *such that* $\chi(\mathcal{F}(f))(\widehat{a}) = \widehat{f(a)}$ *for any* $a \in L$.

Proof. Given a pair (F, I) in $\mathcal{F}(M)$, let $\mathcal{F}(f)(F, I) = (f^{-1}[F], f^{-1}[I])$. To see that $\mathcal{F}(f)(F, I) \in \mathcal{F}(L)$, it is enough to notice that

$$a \in f^{-1}[F] \Rightarrow f(a) \in F \Rightarrow \sim f(a) \in I \Rightarrow f(\neg a) \in I \Rightarrow \neg a \in f^{-1}[I].$$

Let us now check that $\mathcal{F}(f)$ is a f-morphism. In what follows, we will write points in $\mathcal{F}(M)$ and $\mathcal{F}(L)$ as pairs of the form $x : (x_F, x_I)$. Let R be the

[2] See Holliday [9], footnote 15 for more on this.

canonical relation on $\mathcal{F}(M)$ given by $x\mathsf{R}x'$ iff $x_I \cap x'_F = \emptyset$, and S the canonical relation on $\mathcal{F}(L)$ defined similarly.

(i) For condition 1, we claim that for any $x, x' \in \mathcal{F}(M)$, $x_I \cap x'_F = \emptyset$ implies that $\mathcal{F}(f)(x)_I \cap \mathcal{F}(f)(x')_F = \emptyset$. This clearly follows from the fact that $\mathcal{F}(f)(x)_I = f^{-1}[x_I]$ and that $\mathcal{F}(f)(x')_F = f^{-1}[x'_F]$. As a consequence, we have that $x\mathsf{R}x'$ implies $\mathcal{F}(f)(x)\mathsf{S}\mathcal{F}(f)(x')$.

(ii) Now suppose that $\mathcal{F}(f)(x)\mathsf{S}y$, i.e., $f^{-1}[x_I] \cap y_F = \emptyset$, and let $x' = (\uparrow f[y_F], I')$, where $I' = \{\neg c \mid c \in \uparrow f[y_F]\}$. To verify that $x' \in \mathcal{F}(M)$, it is enough to show that $0 \notin \uparrow f[y_F]$. But if there is $c \in y_F$ such that $f(c) \leq 0$, then $c \in f^{-1}[x_I]$, contradicting our assumption. Note that $y_F \subseteq f^{-1}[\uparrow f[y_F]]$, from which it follows that $\mathcal{F}(f)(x') \leq_{\mathsf{S}} y$. Hence we only need to verify that $x\mathsf{R}x'$, i.e., that $x_I \cap \uparrow f[y_F] = \emptyset$. But if there are $a \in x_I, b \in y_F$ with $f(b) \leq a$, it follows that $b \in f^{-1}[x_I] \cap y_F$, contradicting our assumption. Hence x' is the required point in $\mathcal{F}(M)$.

(iii) For condition 3, suppose that $y\mathsf{S}\mathcal{F}(f)(x)$, i.e., $y_I \cap f^{-1}[x_F] = \emptyset$. Let $x' = (\{1\}, \downarrow f[y_I])$. Let us first verify that $x' \in \mathcal{F}(M)$. Clearly, x'_F is a filter and x'_I is an ideal, and if there is $b \in y_I$ such that $1 \leq f(b)$, then $b \in f^{-1}[x_F]$, contradicting our assumption. Moreover, $y_I \subseteq f^{-1}[\downarrow f[y_I]]$, which means that $\mathcal{F}(f)(x') \leq_{\mathsf{2}} y$. Hence it only remains to check that $x'\mathsf{R}x$, i.e., that $\downarrow f[y_I]] \cap x_F = \emptyset$. To see this, suppose that there is $a \in x_F$ and $b \in y_I$ such that $a \leq f(b)$. Then $b \in f^{-1}[x_F]$, contradicting our assumption.

(iv) Finally, we check condition 4. Once again, suppose that $y\mathsf{S}\mathcal{F}(f)(x)$, i.e., $y_I \cap f^{-1}[x_F] = \emptyset$. Let $x'' = (\uparrow f[y_F], I')$, where $I' = \{\neg c \mid c \in \uparrow f[y_F]\}$. To see that $x'' \in \mathcal{F}(M)$, it is enough to verify that $0 \notin \uparrow f[y_F]$. But if there is $c \in y_F$ such that $0 = f(c)$, then $f(\neg c) = \neg f(c) = 1$, so $\neg c \in y_I \cap f^{-1}[x_F]$, contradicting our assumption. Once again, it is easy to verify that $\mathcal{F}(x'') \leq_{\mathsf{S}} y$, so we only check that $x''\mathsf{R}x$. This amounts to verifying that $\{\neg c \mid c \in \uparrow f[y_F]\} \cap x_F = \emptyset$. Suppose, towards a contradiction, that there is $a \in x_F$ and $c \in \uparrow f[y_F]$ such that $a \leq \neg c$. Then there is $b \in y_F$ such that $f(b) \leq c$, which implies that $a \leq \neg c \leq \neg f(b) \leq f(\neg b)$, and thus $\neg b \in f^{-1}[x_F]$. At the same time, $b \in y_F$ implies $\neg b \in y_I$, which means that $y_I \cap f^{-1}[x_F] \neq \emptyset$, contradicting our assumption.

We conclude by showing that $\chi(\mathcal{F}(f))(\widehat{a}) = \widehat{f(a)}$ for any $a \in L$. It is enough to show that for any $x \in \mathcal{F}(M)$, $f(a) \in x_F$ iff $x \in \mathcal{F}(f)^{-1}[\widehat{a}]$. But the latter is equivalent to $a \in f^{-1}[x_F]$, which clearly holds iff $f(a) \in x_F$. This concludes the proof. □

The results gathered so far can be conveniently summed up with the following categorical perspective. Let **FL** be the category of fundamental lattices and fundamental lattice homomorphisms between them, and let **FFrm** be the category of fundamental frames and f-morphisms between them. Then we have two contravariant functors $\mathcal{F} : \mathbf{FL} \to \mathbf{FFrm}$ and $\chi : \mathbf{FFrm} \to \mathbf{FL}$ defined as follows:

- For any $L \in \mathbf{FL}$, $\mathcal{F}(L)$ is the canonical frame of L as defined in Definition 2.7;
- For any fundamental lattice homomorphism $f : L \to M$, $\mathcal{F}(f) : \mathcal{F}(M) \to \mathcal{F}(L)$ is given by $\mathcal{F}(h)(x_F, x_I) = (f^{-1}[x_F], f^{-1}[x_I])$ for any $(x_F, x_I) \in \mathcal{F}(M)$, and is a f-morphism by Lemma 2.12;
- For any $(X, \mathsf{R}) \in \mathbf{FFrm}$, $\chi(X, \mathsf{R}) = (\chi_{\mathsf{R}}(X), \neg_{\mathsf{R}})$, which is a fundamental lattice by Theorem 2.6;
- For any f-morphism $h : (X, \mathsf{R}) \to (Y, \mathsf{S})$, $\chi(h) : \chi(Y, \mathsf{S}) \to \chi(X, \mathsf{R})$ is given by $\chi(h)(A) = h^{-1}[A]$ for any $A \in \chi_{\mathsf{S}}(Y)$ and is a fundamental lattice homomorphism by Lemma 2.11.

Finally, for any fundamental lattice $(L, \neg)$, the map $\widehat{\cdot} : L \to \chi\mathcal{F}(L)$ is an embedding, a fact that we will use several times below.

3 Subframes, Dense Images and Coproducts

In this section, we identify the relational duals of subalgebras, homomorphic images and products of fundamental lattices. For the first two, we will give necessary and sufficient conditions on a f-morphism h for its dual $\chi(h)$ to be injective (resp. surjective), as well as necessary and sufficient conditions on $\mathcal{F}(f)$ when a fundamental homomorphism f is injective (resp. surjective). Because the class of relational frames we consider is larger than the class of dual frames of fundamental lattices, the two conditions do not coincide. However, as we shall see in the next section, the characterization given here will be enough to yield a version of the Goldblatt-Thomason theorem.

We start by identifying when the dual of a f-morphism is injective or surjective.

Definition 3.1 Let $h : (X, \mathsf{R}) \to (Y, \mathsf{S})$ be a f-morphism.

- h is *dense* if for any $y, y' \in Y$:

$$y' \mathsf{S} y \Rightarrow \exists x \in X : h(x) \leq_{\mathsf{S}} y \text{ and } y' \mathsf{S} h(x).$$

- h is an *embedding* if for any $x, x' \in X$:

$$h(x) \mathsf{S} h(x') \Rightarrow \exists z \in X : x \mathsf{R} z \text{ and } z \leq_{\mathsf{R}} x'.$$

The definitions of dense f-morphisms and embeddings can be given mirroring diagrammatic representations, with dense f-morphisms represented on the left and embeddings on the right:

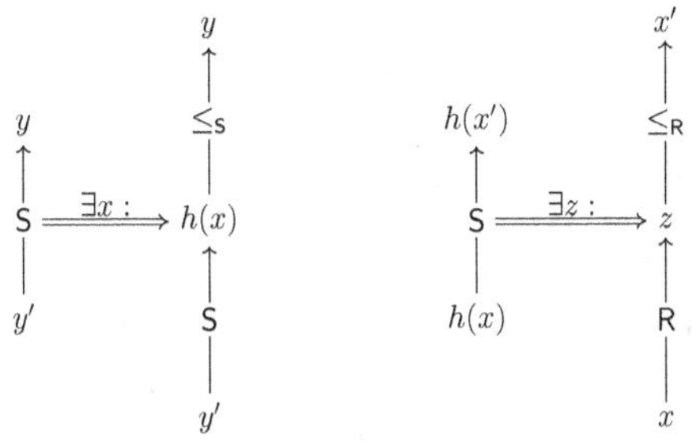

Let us show first that dense f-morphisms induce injective homomorphisms.

Lemma 3.2 *For any f-morphism $h : (X, \mathsf{R}) \to (Y, \mathsf{S})$, $\chi(h)$ is injective if and only if h is dense.*

Proof. Assume first that h is dense, and fix $A, B \in \chi_{\mathsf{S}}(Y)$. We claim that $h^{-1}[A] \subseteq h^{-1}[B]$ implies $A \subseteq B$. To see this, fix $y \in A$ and $y' \in Y$ such that $y'\mathsf{S}y$. Since h is dense, there is $x \in X$ such that $y'\mathsf{S}h(x) \leq_{\mathsf{S}} y$. Since $y \in A$, we also have that $h(x) \in A$, hence $x \in h^{-1}[A] \subseteq h^{-1}[B]$, from which it follows that $h(x) \in B$. But this implies that $y \in C_{\mathsf{R}}(B) = B$.

Conversely, let us now assume that h is not dense. This means that we have $y, y' \in Y$ such that $y'\mathsf{S}y$ and $h(x) \leq_{\mathsf{S}} y$ implies $\neg y'\mathsf{S}h(x)$ for any $x \in X$. Now consider $A = \downarrow_S(y)$ and $B = \overline{\mathsf{S}}(y')$. By choice of y and y', we have that $h^{-1}[A] \subseteq h^{-1}[B]$, but also note that $y \in A \setminus B$. Hence h is not injective. □

Let us now show that embeddings induce surjective homomorphisms.

Lemma 3.3 *For any f-morphism $h : (X, \mathsf{R}) \to (Y, \mathsf{S})$, $\chi(h)$ is surjective if and only if h is an embedding.*

Proof. Suppose first that h is an embedding, and fix $A \in \chi_{\mathsf{R}}(X)$. We claim that $A = h^{-1}h[A]$. Note that this implies that

$$\chi(h)(C_S(h[A])) = h^{-1}[C_S(h[A])] = C_R(h^{-1}h[A]) = C_R(A) = A,$$

and thus that $\chi(h)$ is surjective. For the proof of the claim, note first that the inclusion $A \subseteq h^{-1}h[A]$ is clear. For the converse, assume $h(x) \in h[A]$ for some $x \in X$. This means that $h(x) = h(x')$ for some $x' \in A$. Now let $z\mathsf{R}x$. This implies that $h(z)\mathsf{S}h(x')$, so, since h is an embedding, there is $w \in X$ with $z\mathsf{R}w \leq_{\mathsf{R}} x'$. But this means that $w \in A$, and therefore that $x \in C_R(A) = A$.

Conversely, assume that $\chi(h)$ is surjective and fix $x, x' \in X$ such that $h(x)\mathsf{S}h(x')$. Since h is an f-morphism, there is $z \in X$ such that $x\mathsf{R}z$ and $h(z) \leq_{\mathsf{S}} h(x')$. Now since $\chi(h)$ is surjective, $\downarrow_{\mathsf{R}}(x') = h^{-1}[A]$ for some $A \in \chi_{\mathsf{S}}(Y)$. Since $h(z) \leq_{\mathsf{S}} h(x')$ and $h(x') \in A$, it follows from Fact 2.4(vi) that $h(z) \in A$, and therefore $z \in \downarrow_{\mathsf{R}}(x')$. But this means that $z \leq_{\mathsf{R}} x'$, and therefore that h is an embedding. □

Dense f-morphisms therefore induce injective fundamental homomorphisms, and embeddings induce surjective fundamental homomorphisms. Interestingly, this means that the dual of an f-morphism may be an isomorphism without f itself being an isomorphism of fundamental frames. This situation is not uncommon. A similar phenomenon occurs for dense embeddings in the so-called "forcing duality" between complete Boolean algebras and separative posets [14], and in its generalization to complete lattices given by the b-frame duality presented in [16].

Starting from fundamental homomorphisms instead of f-morphisms, let us now identify conditions on $\mathcal{F}(h)$ that are equivalent to h being injective or surjective for a fundamental homomorphism h.

Definition 3.4 Let $h : (X, \mathsf{R}) \to (Y, \mathsf{S})$ be a f-morphism. Then h is *strongly dense* if for any $y \in Y$, there is $x \in X$ such that $h(x) \leq_{\mathsf{S}} y$ and $y \leq_{\mathsf{S}^{\smile}} h(x)$, and it is a *strong embedding* if for any $x, x' \in X$, $h(x)\mathsf{S}h(x')$ implies $x\mathsf{R}x'$.

Clearly, these two conditions are strengthenings of those in Definition 3.1. As the result below establishes, they are the frame correspondents of injectivity and surjectivity in the case of canonical frames.

Lemma 3.5 *Let $f : (L, \neg) \to (M, \sim)$ be a fundamental lattice homomorphism. Then:*

(i) *f is injective iff $\mathcal{F}(f)$ is strongly dense.*

(ii) *f is surjective iff $\mathcal{F}(f)$ is a strong embedding.*

Proof. We prove both items in turn. For convenience, we write R for the relation on $\mathcal{F}(M)$ and S for the relation on $\mathcal{F}(L)$.

(i) Assume first that f is injective, and let $x \in \mathcal{F}(L)$. Let $y = (\uparrow f[x_F], \downarrow f[x_I])$. Note that $f(a) \leq b$ for some $a \in F$ implies that ${\sim}b \leq {\sim}f(a) = f(\neg a)$, which shows that $b \in \uparrow f[x_F]$ implies ${\sim}b \in \downarrow f[x_I]$. Hence $y \in \mathcal{F}(M)$. Moreover, since f is injective, we have that $x_F = f^{-1}[\uparrow f[x_F]]$ and $x_I = f^{-1}[\uparrow f[x_I]]$, which shows that $y \leq_{\mathsf{S}} x$ and $x \leq_{\mathsf{S}^{\smile}} y$. Conversely, suppose that $\mathcal{F}(f)$ is strongly dense, and let $a, b \in L$ such that $a \nleq b$. Since strong density implies density, by Lemma 3.2, we have that $\chi\mathcal{F}(f)$ is injective. Moreover, $a \nleq b$ implies that $\widehat{a} \nsubseteq \widehat{b}$. Hence $\chi(\mathcal{F}(f))(\widehat{a}) \nsubseteq \chi(\mathcal{F}(f))(\widehat{b})$. But by Lemma 2.12, we have that $\chi(\mathcal{F}(f))(\widehat{a}) = \widehat{f(a)}$ and $\chi(\mathcal{F}(f))(\widehat{b}) = \widehat{f(b)}$, which means that $\widehat{f(a)} \nsubseteq \widehat{f(b)}$. But this implies that $f(a) \nleq f(b)$, establishing that f is injective.

(ii) Assume first that f is surjective, and let $x, x' \in \mathcal{F}(M)$ be such that $\mathcal{F}(f)(x)\mathsf{S}\mathcal{F}(f)(x')$, i.e., $f^{-1}[x_I] \cap f^{-1}[x'_F] = \emptyset$. Suppose towards a contradiction that there is $a \in x_I \cap x'_F$. Then $a = f(b)$ for some $b \in L$, so $b \in f^{-1}[x_I] \cap f^{-1}[x'_F]$, contradicting our assumption. Hence $x\mathsf{R}x'$. Conversely, suppose now that f is not surjective, and let $a \in M$ be such that $f(b) \neq a$ for all $b \in L$. Note in particular that $a \neq 1$, so $x := (1, \downarrow a)$ and $x' := (\uparrow a, \downarrow \neg a)$ are points in $\mathcal{F}(M)$. Clearly we have $\neg x\mathsf{R}x'$. However, by choice of a, $f^{-1}[\downarrow a] \cap f^{-1}[\uparrow a] = \emptyset$, from which it follows that $\mathcal{F}(f)(x)\mathsf{S}\mathcal{F}(f)(x')$. Hence $\mathcal{F}(f)$ is not strongly dense.

□

The previous results motivate the following definitions.

Definition 3.6 Let (X, R) and (Y, S) be fundamental frames. Then (X, R) is a *subframe* of (Y, S) if there is an embedding $h : (X, \mathsf{R}) \to (Y, \mathsf{S})$, and it is a *dense image* of (Y, S) if there is a dense f-morphism $g : (Y, \mathsf{S}) \to (X, \mathsf{R})$. Moreover, (X, R) is a *strong subframe* (resp. a *strongly dense image*) of (Y, S) if h is a strong embedding (resp. g is strongly dense).

Lemma 3.7 *Let (X, R) and (Y, S) be two fundamental frames. Then:*

- *If (X, R) is a subframe of (Y, S), then $\chi_{\mathsf{R}}(X)$ is a homomorphic image of $\chi_{\mathsf{S}}(Y)$;*
- *If (X, R) is a dense image of (Y, S), then $\chi_{\mathsf{R}}(X)$ is a subalgebra of $\chi_{\mathsf{S}}(Y)$.*

Moreover, for any two fundamental lattices $(L, \neg)$ and $(M, \sim)$:

- *If $(L, \neg)$ is a subalgebra of $(M, \sim)$, then $\mathcal{F}(L)$ is a strongly dense image of $\mathcal{F}(M)$;*
- *If $(L, \neg)$ is a homomorphic image of $(M, \sim)$, then $\mathcal{F}(L)$ is a strong subframe of $\mathcal{F}(M)$.*

Proof. The first part of the lemma follows directly from Lemmas 3.2 and 3.3, and the second part is an immediate consequence of Lemma 3.5. □

We conclude this section by identifying the frame-theoretic notion that corresponds to products of fundamental lattices. Unsurprisingly, this is given by a disjoint union construction.

Definition 3.8 Let $\{(X_i, \mathsf{R}_i)\}_{i\in I}$ be a family of fundamental frames. The *coproduct* of the family $\{(X_i, \mathsf{R}_i)\}_{i\in I}$ is the relation frame (X_I, R_I), where X_I is the disjoint union of the sets X_i, and the relation R_I is given by $(i, x)\mathsf{R}_I(i', x')$ iff $i = i'$ and $x\mathsf{R}_i x'$ for any $x \in X_i$, $x' \in X_{i'}$.

It is straightforward to verify that the coproduct of any family of fundamental frames is also a fundamental frame. In fact, the dual fundamental lattice of the coproduct of a family of frames is easily seen to be isomorphic to the product of the corresponding family of fundamental lattices.

Lemma 3.9 *For any family $\{(X_i, \mathsf{R}_i)\}_{i\in I}$ of fundamental frames with coproduct (X_I, R_I), $\chi_{\mathsf{R}_I}(X_I)$ is isomorphic to $\prod_{i\in I} \chi_{\mathsf{R}_i}(X_i)$.*

Proof. Define the map $f : \prod_{i\in I} X_i \to X_I$ by $f(\{A_i\}_{i\in I}) = \bigcup_{i\in I}(i, A_i)$, where $(i, A_i) = \{(i, x) \in X_I \mid x \in A_i\}$. Since all relations R_i are disjoint from one another, it is easy to check that for any $A \subseteq X_I$, $A \in \chi_{\mathsf{R}_I}(X_I)$ iff $\{x \in X_i \mid (i, x) \in A\} \in \chi_{\mathsf{R}_i}(X_i)$ for every $i \in I$. This shows that the image of the restriction of f to $\prod_{i\in I} \chi_{\mathsf{R}_i}(X_i)$ is exactly $\chi_{\mathsf{R}_I}(X_I)$. Moreover, f clearly preserves and reflects the inclusion order, and hence its restriction to $\prod_{i\in I} \chi_{\mathsf{R}_i}(X_i)$ is an isomorphism. This completes the proof. □

4 The Goldblatt-Thomason Theorem

In this section, we prove our main result, namely, a version of the Goldblatt-Thomason Theorem for Fundamental Logic. In the original setting of modal logic, ultrafilter extensions play a central role in bridging the gap between algebraic and relational structures. In our setting, a similar role is played by filter extensions, to which we now turn.

Definition 4.1 Let (X, R) be a fundamental frame. The *filter extension* of (X, R) is the fundamental frame $\mathcal{F}(\chi_R(X))$.

The filter extension of a fundamental frame (X, R) can be characterized more concretely as follows. Points are pairs (F, I) such that F is a proper filter on the positive algebra $\chi_{\mathsf{R}}(X)$ and I is a proper filter on the negative algebra $\chi_{\reflectbox{$\mathsf{R}$}}(X)$ such that $A \in F$ implies $\neg_{\reflectbox{$\mathsf{R}$}}\neg_{\mathsf{R}}A \in I$. Moreover, for any two such points $(F, I), (G, J)$, we let $(F, I)\mathsf{S}(G, J)$ iff there is no $A \in G$ such that $\neg_{\reflectbox{$\mathsf{R}$}}A \in I$. Note that, since $\neg_{\mathsf{R}}$ and $\neg_{\reflectbox{$\mathsf{R}$}}$ are inverse anti-isomorphisms between $\chi_{\reflectbox{$\mathsf{R}$}}(X)$ and $\chi_{\mathsf{R}}(X)$, this is equivalent to taking exactly the points in the canonical frame of $\chi_{\mathsf{R}}(X)$.

It is worth commenting on the relationship between the positive algebra of a fundamental frame and the positive algebra of its filter extension. As it turns out, the latter is the π-canonical extension of the former (see [4] for the general definition of π-canonical extensions of lattice expansions). This follows from the following general fact about fundamental lattices.

Lemma 4.2 *Let $(L, \neg)$ be a fundamental lattice. Then $\chi(\mathcal{F}(L))$ is the π-canonical extension of L, as witnessed by the embedding* $\widehat{\cdot} : L \to \chi(\mathcal{F}(L))$.

Proof. Recall first that the canonical extension of a lattice L is characterized up to isomorphism as a complete lattice L and an embedding $\alpha : L \to C$ with the following properties:

- L is doubly-dense in C: for any $A \in C$, there are families $\{I_k\}_{k \in K}$ and $\{J_h\}_{h \in H}$ of subsets of L such that $\bigwedge_C\{\bigvee_C\{\alpha(a) \mid a \in I_k \mid k \in K\} = A = \bigvee_C\{\bigwedge_C\{\alpha(b) \mid b \in J_h \mid h \in H\}$;
- L sits compactly inside of C: for any $A, B \subseteq L$ such that $\bigwedge_C\{\alpha(a) \mid a \in A\} \leq_C \bigvee_C\{\alpha(b) \mid b \in B\}$, there are finite sets $A' \subseteq A$ and $B' \subseteq B$ such that $\bigwedge_L A' \leq_L \bigvee_L B'$.

Let $\mathcal{F}(L) = (X, \mathsf{R})$. We prove that the embedding $\widehat{\cdot} : L \to \chi_{\mathsf{R}}(L)$ satisfies these two properties. Let $A \in \chi_{\mathsf{R}}(L)$. First, we claim that $A = \bigcup_{x \in A} \bigcap_{a \in x_F} \widehat{a}$. Since $\neg_{\mathsf{R}}\neg_{\reflectbox{R}}A = A$, this will establish that $A = \bigvee_{x \in A} \bigwedge_{a \in x_F} \widehat{a}$. For the proof of the claim, note that the left-to-right inclusion is immediate. For the converse, let $x \in X$ and suppose that there is $x' \in A$ such that $x' \in \bigcap_{a \in x_F} \widehat{a}$. Then $x' \leq_{\mathsf{R}} x$, so, since $A \in \chi_{\mathsf{R}}$, it follows from Fact 2.4.(vi) that $x' \in A$. This shows that any element in $\chi_{\mathsf{R}}(X)$ is a join of meets of images of elements of L. To show that it is also a meet of joins of images of elements of L, note first that a completely similar argument shows that, for any $B \in \chi_{\reflectbox{$\mathsf{R}$}}(X)$, $B = \bigvee_{\chi_{\reflectbox{R}}(X)}\{\bigwedge_{\chi_{\reflectbox{R}}(X)}\{\check{b} \mid b \in x_I\} \mid x \in B\}$, where $\check{b} = \{x \in X \mid b \in x_I\}$. Now for any $A \in \chi_{\mathsf{R}}(X)$, there is $B \in \chi_{\reflectbox{$\mathsf{R}$}}(X)$ such that $A = \neg_{\mathsf{R}}B$. Fix $A \in \chi_{\mathsf{R}}(X)$ and such a B. Since $\neg_{\mathsf{R}}$ is an anti-isomorphism, it follows that $A = \bigwedge_{\chi_{\mathsf{R}}(X)}\{\bigvee_{\chi_{\mathsf{R}}(X)}\{\neg_{\mathsf{R}}\check{b} \mid b \in x_I\} \mid x \in B\}$. Since $\neg_{\mathsf{R}}\check{b} = \widehat{b}$ for any $b \in L$, this completes the proof that L is doubly-dense in $\chi_{\mathsf{R}}(X)$.

Let us now prove that L sits compactly inside of $\chi_{\mathsf{R}}(X)$. Let $A, B \subseteq L$ such that for any finite $A' \subseteq A$, $B' \subseteq B$, $\bigwedge_L A' \nleq_L \bigvee_L B'$. This means that there exist F and I, respectively a filter and an ideal on L, such that $A \subseteq F$, $B \subseteq I$ and $F \cap I = \emptyset$. Let $x \in X$ be the point (F, I). Note that $x \in \bigcap_{a \in A} \widehat{a}$. However, $x\mathsf{R}x$, and for any $y \in X$ such that $x\mathsf{R}y$, $y \notin \bigcup_{b \in B} \widehat{b}$, since otherwise $I \cap y_F \neq \emptyset$, contradicting $x\mathsf{R}y$. Hence $x \notin \neg_{\mathsf{R}}\neg_{\mathsf{Я}}(\bigcup_{b \in B} \widehat{b})$, which shows the compactness property.

To complete the proof that $(\chi_{\mathsf{R}}(\mathcal{F}(L), \neg_{\mathsf{R}})$ is the π-canonical extension of $(L, \neg)$, it remains to show that $\neg_{\mathsf{R}}$ is the π-extension of $\neg$. Recall that, for $f : L \to L$ an antitone map, its π-extension $f^\pi : L^\sigma \to L^\sigma$ is given by:

$$f^\pi(a) = \bigwedge\{\bigvee_{b \in F} f(b) \mid F \subseteq L : \text{ is a filter and } \bigwedge F \leq a\}$$

for any $a \in L^\sigma$.

In our setting this means that we must show that for any $A \in \chi_{\mathsf{R}}(\mathcal{F}(L))$, $\neg_{\mathsf{R}} A = \bigcap_{x \in A} \bigvee_{a \in x_F} \widehat{\neg a}$. For the left-to-right direction, fix some $y \in \neg_{\mathsf{R}} A$ and some $x \in A$. It is enough to show that $\neg[x_F] \cap y_I \neq \emptyset$, where $\neg[x_F] = \{\neg a \mid a \in x_F\}$. Suppose towards a contradiction that $\neg[x_F] \cap y_I = \emptyset$. Then consider the pair $z = (x_F, \neg[x_F]) \in \mathcal{F}(L)$, and notice that $z\mathsf{R}y$. But $z \leq_{\mathsf{R}} x$, so $z \in A$. This contradicts $y \in \neg_{\mathsf{R}} A$. For the converse direction, suppose that $y \notin \neg_{\mathsf{R}} A$. Then there is $x \in A$ such that $x\mathsf{R}y$. We claim that for all $z\mathsf{Я}x$ and for all $a \in x_F$, $z \notin \widehat{\neg a}$. Indeed, if there is $a \in x_F$ such that $z \in \widehat{\neg a}$, then $z_F \cap x_I \neq \emptyset$, so $\neg x\mathsf{R}z$. But this means that $x \in \neg_{\mathsf{Я}} \bigcup_{a \in x_F} \widehat{\neg a}$, hence $y \notin \neg_{\mathsf{R}}\neg_{\mathsf{Я}} \bigcup_{a \in x_F} \widehat{\neg a} = \bigvee_{a \in x_F} \widehat{\neg a}$. This completes the proof.

□

The following is a standard definition in the literature on Goldblatt-Thomason theorems.

Definition 4.3 A class $\mathfrak{K}$ of fundamental frames is *axiomatic* if there is a set Γ of formulas of FL such that $\mathfrak{K} = Mod(\Gamma)$.

Note that a class $\mathfrak{K}$ is axiomatic iff $\mathfrak{K} \supseteq Mod(Log(\mathfrak{K}))$, as an easy argument shows. The following is a routine fact.

Lemma 4.4 *Let $\mathfrak{K}$ be an axiomatic class of fundamental frames. Then $\mathfrak{K}$ is closed under subframes, dense images and coproducts, and it reflects filter extensions.*

Proof. This is a routine argument. Suppose that $\mathfrak{K} = Mod(\Gamma)$ for some set Γ of formulas of $\mathcal{L}$, and let (X, R) and (Y, S) be fundamental frames. If (X, R) is a subframe (resp. dense image) of (Y, S), then $\chi_{\mathsf{R}}(X)$ is a homomorphic image of (resp. embeds into) $\chi_{\mathsf{S}}(Y)$ by Lemma 3.7. Now if (Y, S) is in $\mathfrak{K}$, then Γ is valid on $\chi_{\mathsf{S}}(Y)$. But then Γ is also valid on any subalgebra or homomorphic image of $\chi_{\mathsf{S}}(Y)$. This shows that $\mathfrak{K}$ is closed under subframes and dense images.

Moreover, if $\{(X_i, \mathsf{R}_i)\}_{i \in I}$ is a family of fundamental frames in $\mathfrak{K}$, then Γ is valid on $\chi_{\mathsf{R}_i}(X_i)$ for any $i \in I$, hence also on $\prod_{i \in I} \chi_{\mathsf{R}_i}(X_i)$. By Lemma 3.9, the latter is isomorphic to $\chi_{\mathsf{R}_I}(X_I)$, the positive algebra of the coproduct (X_I, R_I) of the family $\{(X_i, \mathsf{R}_i)\}_{i\in I}$. Hence $(X_I, \mathsf{R}_I) \in \mathfrak{K}$, which shows that $\mathfrak{K}$ is closed under coproducts.

Finally, assume that (X, R) is a fundamental frame such that its filter extension $\mathcal{F}(\chi_{\mathsf{R}}(X))$ is in $\mathfrak{K}$. This means that Γ is valid on the positive algebra of $\mathcal{F}(\chi_{\mathsf{R}}(X))$. But by Lemma 4.2, the latter is the canonical extension of $\chi_{\mathsf{R}}(X)$, hence $\chi_{\mathsf{R}}(X)$ embeds into it. It follows that Γ is valid on $\chi_{\mathsf{R}}(X)$, and therefore $(X, \mathsf{R}) \in \mathfrak{K}$. Hence $\mathfrak{K}$ reflects filter extensions. □

Our key lemma towards a Goldblatt-Thomason theorem is the following.

Lemma 4.5 *Let $\mathfrak{K}$ be a class of fundamental frames closed under filter extensions. If $\mathfrak{K}$ is closed under strong subframes, strongly dense images and coproducts, and it reflects filter extensions, then $\mathfrak{K}$ is axiomatic.*

Proof. Suppose $\mathfrak{K}$ satisfies the conditions of the lemma and let $\Gamma = Log(\mathfrak{K})$. Clearly $\mathfrak{K} \subseteq Mod(\Gamma)$, so we only need to show the converse. So assume that (X, R) is a fundamental frame such that Γ is valid on (X, R). By Birkhoff's HSP theorem, it follows that $\chi_{\mathsf{R}}(X) \in \mathbb{HSP}(\{\chi_{\mathsf{S}}(Y) \mid (Y, \mathsf{S}) \in \mathfrak{K}\})$. This means that there is a fundamental lattice $(L, \neg)$ and a family $\{(X_i, \mathsf{R}_i)\}_{i\in I}$ of frames in $\mathfrak{K}$ such that $\chi_{\mathsf{R}}(X)$ is a homomorphic image of $(L, \neg)$, and $(L, \neg)$ is a subalgebra of $\prod_{i\in I} \chi_{\mathsf{R}_i}(X_i)$. By Lemma 3.9, $(L, \neg)$ embeds into $\chi_{\mathsf{R}_I}(X_I)$, where (X_I, R_I) is the coproduct of the family $\{(X_i, \mathsf{R}_i)\}_{i\in I}$. By Lemma 3.7, it follows that $\mathcal{F}(L)$ is a strongly dense image of $\mathcal{F}(\chi_{\mathsf{R}_I}(X_I))$ and that $\mathcal{F}(\chi_{\mathsf{R}}(X))$ is a strong subframe of $\mathcal{F}(L)$, as shown in the diagram below.

$$\begin{array}{ccc} L & \hookrightarrow & \chi_{\mathsf{R}_I}(X_I) \\ \downarrow\!\!\downarrow & & \\ \chi_{\mathsf{R}}(X) & & \end{array} \qquad\qquad \begin{array}{ccc} \mathcal{F}(L) & \twoheadleftarrow & \mathcal{F}(\chi_{\mathsf{R}_I}(X_I)) \\ \uparrow & & \\ \mathcal{F}(\chi_{\mathsf{R}}(X)) & & \end{array}$$

Now since $\mathfrak{K}$ is closed under coproducts and filter extensions, it follows that $\mathcal{F}(\chi_{\mathsf{R}_I}(X_I)) \in \mathfrak{K}$. Since $\mathfrak{K}$ is closed under strongly dense images, $\mathcal{F}(L) \in \mathfrak{K}$, and since $\mathfrak{K}$ is closed under strong subframes, $\mathcal{F}(\chi_{\mathsf{R}}(X)) \in \mathfrak{K}$. Finally, since $\mathfrak{K}$ reflects filter extensions, we can conclude that $(X, \mathsf{R}) \in \mathfrak{K}$. Hence $\mathfrak{K}$ is axiomatic. □

Recall that we identified in the previous section two distinct notions of subframes and two distinct notions of dense images. In the case of axiomatic classes, we can see that the two notions coincide in a sense spelled out by our main result, which we are now in a position to prove.

Theorem 4.6 *Let $\mathfrak{K}$ be a class of fundamental frames closed under filter extensions. Then the following are equivalent:*

(i) *$\mathfrak{K}$ is axiomatic;*

(ii) *$\mathfrak{K}$ is closed under subframes, dense images and coproducts, and it reflects filter extensions;*

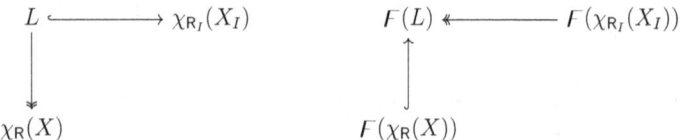

(iii) *$\mathfrak{K}$ is closed under strong subframes, strongly dense images and coproducts, and it reflects filter extensions.*

Proof. We have the following chain of implications:

$$(i) \Rightarrow (ii) \Rightarrow (iii) \Rightarrow (i),$$

where the first implication follows from Lemma 4.4, the second implication is immediate, and the third one follows from Lemma 4.5. □

Finally, we can also characterize classes of fundamental frames of the form $Mod(L)$ for L a canonical superfundamental logic, i.e., a logic extending Fundamental Logic whose corresponding variety of fundamental lattices is closed under π-canonical extensions.

Corollary 4.7 *Let $\mathfrak{K}$ be a class of fundamental frames. The following are equivalent:*

(i) *$\mathfrak{K} = Mod(L)$, for L a canonical superfundamental logic;*

(ii) *$\mathfrak{K}$ is closed under filter extensions, subframes, dense images and coproducts, and it reflects filter extensions;*

(iii) *$\mathfrak{K}$ is closed under filter extensions, strong subframes, strongly dense images and coproducts, and it reflects filter extensions.*

Proof. The equivalence between (ii) and (iii) clearly follows from Theorem 4.6, so we only need to check that $(i) \Leftrightarrow (ii)$. Suppose first that $\mathfrak{K}$ is axiomatic and that $L = Log(\mathfrak{K})$ is canonical. By Theorem 4.6, it is enough to show that $\mathfrak{K}$ is closed under filter extensions. So suppose $(X, \mathsf{R}) \in \mathfrak{K}$. Then $\chi_{\mathsf{R}}(X)$ is in the variety corresponding to L. Since L is canonical, this implies that $(\chi_{\mathsf{R}}(X)^\sigma, \neg_{\mathsf{R}}^{\pi})$, the π-canonical extension of $(\chi_{\mathsf{R}}(X), \neg_{\mathsf{R}})$, is also in that variety. But since the positive algebra of the filter extension of (X, R) is isomorphic to $(\chi_{\mathsf{R}}(X)^\sigma, \neg_{\mathsf{R}}^{\pi})$ by Lemma 4.2, it follows that L is valid on $\mathcal{F}(\chi_{\mathsf{R}}(X))$, hence $\mathcal{F}(\chi_{\mathsf{R}}(X)) \in Mod(L) = \mathfrak{K}$ since $\mathfrak{K}$ is axiomatic. Hence $\mathfrak{K}$ is closed under filter extensions.

Conversely, let us now assume that $\mathfrak{K}$ is closed under filter extensions, subframes, dense images and coproducts, and that it reflects filter extensions. By Theorem 4.6, it follows that $\mathfrak{K}$ is axiomatic, i.e., $\mathfrak{K} = Mod(Log(\mathfrak{K}))$. It remains to show that $L = Log(\mathfrak{K})$ is canonical. Let A be a fundamental lattice in the variety corresponding to L. Then there is a family $\{(X_i, \mathsf{R}_i)\}_{i \in I}$ of frames in $\mathfrak{K}$ and a fundamental lattice B such that A is a homomorphic image of B and B embeds into $\prod_{i \in I} \chi_{\mathsf{R}_i}(X_i)$, which is isomorphic to $\chi_{\mathsf{R}_I}(X_I)$, the positive algebra of the coproduct of the family $\{(X_i, \mathsf{R}_i)\}_{i \in I}$. This yields the following two diagrams, where the right one is obtained from the left one by applying successively the functors $\mathcal{F}$ and χ:

$$\begin{array}{ccc} B & \hookrightarrow & \chi_{\mathsf{R}_I}(X_I) \\ \downarrow & & \\ A & & \end{array} \qquad \begin{array}{ccc} \chi(\mathcal{F}(B)) & \hookrightarrow & \chi(\mathcal{F}(\chi_{\mathsf{R}_I}(X_I))) \\ \downarrow & & \\ \chi(\mathcal{F}(A)) & & \end{array}$$

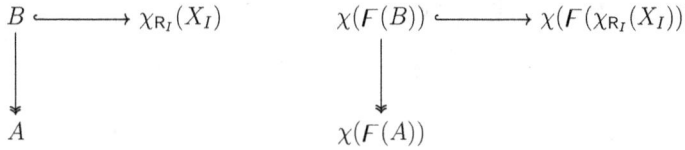

Now $F(\chi_{R_I}(X_I))$ is the filter extension of the coproduct of the family $\{(X_i, \mathsf{R}_i)\}_{i\in I}$, so since $\mathfrak{K}$ is closed under coproducts and filter extensions, it follows that $F(\chi_{\mathsf{R}_I}(X_I)) \in \mathfrak{K}$, hence L is valid on $\chi(F(\chi_{\mathsf{R}_I}(X_I)))$. But since $\chi(F(A))$ is a homomorphic image of a subalgebra of $\chi(F(\chi_{\mathsf{R}_I}(X_I)))$, L is also valid on $\chi(F(A))$. Finally, since $\chi(F(A))$ is isomorphic to the π-canonical extension of A by Lemma 4.2, it follows that L is a canonical. This completes the proof. □

5 Adding Modal Operators

In this final section, we generalize the results obtained in the previous section to Fundamental Modal Logic, which was recently developed by Holliday in [10]. We will focus on Holliday's *additive unified* setting, in which the two modalities $\Box$ and $\Diamond$ are dual notions (even though, just like in modal intuitionistic logic, the two modalities may not be interdefinable), and $\Diamond$ is an additive operator. Moreover, we will only consider the case in which a single pair $(\Box, \Diamond)$ is added to the language of fundamental logic, but generalizations to a polymodal setting are obvious. We start with the following definitions, which generalize those of fundamental lattices and fundamental frames to the modal setting in a natural way.

Definition 5.1 A fundamental modal lattice is a tuple $(L, \neg, \Box, \Diamond)$ such that $(L, \neg)$ is a fundamental lattice, and $\Box$ and $\Diamond$ are unary operations on L such that for any $a, b \in L$:

- $\Box(a \wedge b) = \Box a \wedge \Box b$, $\Box 1 = 1$;
- $\Diamond(a \vee b) = \Diamond a \vee \Diamond b$, $\Diamond 0 = 0$;
- $\Diamond\neg a \leq \neg\Box a$.

Definition 5.2 An *additive unified fundamental modal frame* (*AUFM frame* for short) is a tuple (X, R, M) such that (X, R) is fundamental frame and M is a relation on X satisfying the following conditions for any $x, y, z \in X$:

- $xMy\mathsf{Я}z \Rightarrow \exists x'\mathsf{Я}x\forall y'\mathsf{R}x'\exists z' : x'Mz'\mathsf{Я}z$;
- $xMy\mathsf{R}z \Rightarrow \exists x'\mathsf{R}x\forall y'\mathsf{Я}x'\exists z' : x'Mz'\mathsf{R}z$.

As shown in [10], given an additive unified fundamental modal frame (X, R, M), the unary operation $\Box_M : \mathcal{P}(X) \to \mathcal{P}(X)$ given by $\Box_M A = \{x \in X \mid \forall y \in X : xMy \Rightarrow y \in A\}$ restricts to a map from $\chi_{\mathsf{R}}(X)$ to $\chi_{\mathsf{R}}(X)$. Moreover, one can define a "pseudo-dual" $\Diamond_M$ to $\Box_M$, given by $\Diamond_M A = \neg_{\mathsf{R}}\Box_M\neg_{\mathsf{Я}}A$, which has the property that $\Diamond_M\neg_{\mathsf{R}}A \subseteq \neg_{\mathsf{R}}\Box_M A$ for any $A \in \chi_{\mathsf{R}}(X)$. Finally, any fundamental modal lattice $(L, \neg, \Box, \Diamond)$ modally embeds into the positive algebra of the expansion of its dual canonical frame $F(L)$ with a relation M given by xMx' iff $\Box a \in x_F$ implies $a \in x'_F$ and $\Diamond b \in x_I$ implies $b \in x'_I$, and this defines an $AUFM$ frame.

Let us now define the relevant notion of morphism between $AUFM$ frames.

Definition 5.3 Let (X, R, M) and (Y, S, N) be two $AUFM$ frames. An $AUFM$*-morphism* is a f-morphism $h : (X, \mathsf{R}) \to (Y, \mathsf{S})$ satisfying the following additional constraints for any $x, x' \in X$ and any $y \in Y$:

(i) $xMx' \Rightarrow h(x)Nh(x')$;

(ii) $h(x)Ny \Rightarrow \exists x' \in X : xMx'$ and $y \leq_{\mathsf{S}\text{Ƨ}} h(x')$,

where $\leq_{\mathsf{S}\text{Ƨ}} = \leq_{\mathsf{S}} \cap \leq_{\text{Ƨ}}$.

Lemma 5.4 *Let* $h : (X, \mathsf{R}, M) \to (Y, \mathsf{S}, N)$ *be an* $AUFM$*-morphism between two* $AUFM$ *frames. Then for any* $A \in \chi_{\mathsf{S}}(Y)$*,* $h^{-1}[\Box_N A] = \Box_M h^{-1}[A]$ *and* $h^{-1}[\Diamond_N A] = \Diamond_M h^{-1}[A]$.

Proof. Let us first show that $h^{-1}[\Box_N A] = \Box_M h^{-1}[A]$ for any $A \in \chi_{\mathsf{S}}(Y) \cup \chi_{\text{Ƨ}}(Y)$. Suppose first that $h(x) \in \Box_N A$, and let xMx'. Then by condition (i) on $AUFM$-morphisms, $h(x)Nh(x')$, so $x' \in h^{-1}[A]$. This shows the left-to-right inclusion. For the converse, assume that xMx' implies $h(x') \in A$, and let $y \in Y$ be such that $h(x)Ny$. By condition (ii) on $AUFM$-morphisms, there is $x' \in X$ such that xMx' and $y \leq_{\mathsf{S}\text{Ƨ}} h(x')$. But this means that $h(x') \in A$ and therefore also $y \in A$, since $A \in \chi_{\mathsf{S}}(Y) \cup \chi_{\text{Ƨ}}(Y)$ and we have that $y \leq_{\mathsf{S}} h(x')$ and $y \leq_{\text{Ƨ}} h(x')$.

Now, to show that h^{-1} also preserves the $\Diamond$ operation, recall that, for any $A \in \chi_{\mathsf{S}}(Y)$, $\Diamond_N A = \neg_{\mathsf{S}} \Box_N \neg_{\text{Ƨ}} A$. Since (Y, S, N) is additive we have that $\Box_N B \in \chi_{\text{Ƨ}}(Y)$ whenever $B \in \chi_{\text{Ƨ}}(Y)$. Hence for any $A \in \chi_{\mathsf{S}}(Y)$, we have the following chain of identities:

$$h^{-1}[\neg_{\mathsf{S}} \Box_N \neg_{\text{Ƨ}} A] = \neg_{\mathsf{R}} h^{-1}[\Box_N \neg_{\text{Ƨ}} A] = \neg_{\mathsf{R}} \Box_M h^{-1}[\neg_{\text{Ƨ}} A] = \neg_{\mathsf{R}} \Box_M \neg_{\text{Я}} h^{-1}[A],$$

where the first and third identities follow from the fact that h is an f-morphism. This completes the proof. □

Here again, our definition of an $AUFM$-morphism is general enough to capture all modal fundamental lattice homomorphisms.

Lemma 5.5 *Let* $(L, \neg, \Box_L, \Diamond_L)$ *and* $(N, \sim, \Box_N, \Diamond_N)$ *be fundamental modal lattices, and* $f : L \to N$ *a fundamental modal homomorphism. Then* $\text{ꟻ}(f) : \text{ꟻ}(N) \to \text{ꟻ}(L)$ *is an* $AUFM$*-morphism.*

Proof. By Lemma 2.12, we know already that $\text{ꟻ}(f)$, given by $\text{ꟻ}(f)(x) = (f^{-1}[x_F], f^{-1}[x_I])$ for any $x \in \text{ꟻ}(M)$, is a f-morphism. Hence we only need to check the extra two conditions in Definition 5.3. Let R_L and M_L be the following relations on $\text{ꟻ}(L)$:

- $x\mathsf{R}_L y$ iff $x_I \cap y_F = \emptyset$;
- $xM_L y$ iff $\{\Box_L a \mid a \in x_F\} \subseteq y_F$ and $\{\Diamond_L b \mid b \in x_I\} \subseteq y_I$,

and let R_N and M_N be defined similarly on $\text{ꟻ}(N)$. Suppose first that $xM_N y$, and let $\Box_L a \in (\text{ꟻ}(f)(x))_F$. Then $h(\Box_L a) = \Box_M h(a) \in x_F$, which means that $h(a) \in y_F$. Similarly, $\Diamond_L b \in (\text{ꟻ}(f)(x))_I$ implies that $h(b) \in y_I$. But this means that $\text{ꟻ}(f)(x) M_L \text{ꟻ}(f)(y)$.

To check that the second condition also holds, suppose now that $\mathcal{F}(f)(x)Ny$ for some $x \in \mathcal{F}(N), y \in \mathcal{F}(L)$. Let $x' = (F, I)$, where $F = \{a \in N \mid \Box_N a \in x_F\}$ and $I = \{b \in N \mid \Diamond_N b \in x_I\}$. To verify that $x' \in \mathcal{F}(N)$, we only need to check that $a \in F$ implies $\sim a \in I$. But if $a \in F$, we have that $\Box_N a \in x_F$, hence $\sim\Box_N a \in I$. Since $\Diamond_N {\sim} a \leq_N \sim\Box_N a$, it follows that $\Diamond_N {\sim} a \in x_I$, hence $\sim a \in I$. Moreover, we clearly have $xM_N x'$. Finally, for any $a, b \in L$, we have that:

$$a \in (\mathcal{F}(f)(x))_F \Leftrightarrow f(a) \in F \Leftrightarrow \Box_N f(a) \in x_F \Leftrightarrow \Box_L a \in h(x')_F \Rightarrow a \in y_F$$
$$b \in (\mathcal{F}(f)(x))_I \Leftrightarrow f(b) \in I \Leftrightarrow \Diamond_N f(a) \in x_I \Leftrightarrow \Diamond_L a \in h(x')_I \Rightarrow b \in y_I,$$

from which we conclude that $y \leq_{\lrcorner\mathfrak{R}_\llcorner} \mathcal{F}(f)(x')$. □

As a consequence, we may straightforwardly adapt our analysis of the duals of homomorphic images, subalgebras and products to the modal case.

Definition 5.6 Let (X, R, M) and (Y, S, N) be $AUFM$ frames. Then (X, R, M) is a *modal subframe* of (Y, S, N) if there is an embedding $h : (X, \mathsf{R}) \to (Y, \mathsf{S})$ which is also an $AUFM$-morphism, and it is a *dense modal image* of (Y, S, N) if there is a dense f-morphism $g : (Y, \mathsf{S}) \to (X, \mathsf{R})$ which is also an $AUFM$-morphism. Moreover, (X, R, M) is a *strong modal subframe* (resp. a *strongly dense modal image*) of (Y, S, N) if h is a strong embedding (resp. g is strongly dense).

Definition 5.7 The coproduct of a family $\{(X_i, \mathsf{R}_i, M_i)\}_{i \in I}$ of $AUFM$ frames is the frame (X_I, R_I, M_I), where (X_I, R_I) is the coproduct of the family of fundamental frames $\{(X_i, \mathsf{R}_i)\}_{i \in I}$, and M_I is the disjoint union of the relations $\{M_i\}_{i \in I}$.

It is routine to verify that the coproduct of a family of $AUFM$ frames is itself an $AUFM$ frame, and that its positive algebra is isomorphic to the product of the positive algebras of each frame in the family, so we will omit the proof. Similarly, we define the filter extension of an $AUFM$ frame in a standard way.

Definition 5.8 Let (X, R, M) be a $AUFM$ frame. Then the filter extension of (X, R, M) is the $AUFM$ frame $(Y, \mathsf{S}, N) = \mathcal{F}(\chi_{\mathsf{R}}(X), \neg_{\mathsf{R}}, \Box_M, \Diamond_M)$, where for any $x, y \in Y$, xNy iff ($\Box_M A \in x_F$ implies $A \in y_F$, and $\neg_{\mathfrak{R}} \Diamond_M A \in x_I$ implies $\neg_{\mathfrak{R}} A \in y_I$ for any $A \in \chi_{\mathsf{R}}(X)$).

Filter extensions correspond to canonical extensions in the modal case as well.

Lemma 5.9 *For any fundamental modal lattice* $(L, \neg, \Box, \Diamond)$, *the positive algebra of* $\mathcal{F}(L, \neg, \Box, \Diamond)$ *is the* π*-canonical extension of* $(L, \neg, \Box, \Diamond)$.

Proof. From Lemma 4.2, we already know that $(\chi_{\mathsf{R}}(\mathcal{F}(L, \neg)), \neg_{\mathfrak{R}})$ is the π-canonical extension of $(L, \neg)$. Hence we only need to verify that $\Box_M$ is the π-extension of $\Box$ and that $\Diamond_M$ is the π-extension of $\Diamond$. For the first one, we must show that for any $A \in \chi(\mathcal{F}(L))$,

$$\Box_M A = \bigcap \{\neg_{\mathsf{R}} \neg_{\mathfrak{R}} \bigcup_{a \in I} \widehat{\Box a} \mid I \subseteq L \text{ is an ideal and } A \subseteq \neg_{\mathsf{R}} \neg_{\mathfrak{R}} \bigcup_{a \in I} \widehat{a}\}.$$

First, we claim that for any ideal I on L, $A \subseteq \bigvee_{a\in I} \widehat{a}$ iff $x_F \cap I \neq \emptyset$ for all $x \in A$. For the left-to-right direction, if there is $x \in A$ such that $x_F \cap I = \emptyset$, then let $y = (\{1\}, I)$. We have that $y\mathsf{R}x$, but clearly, if $y\mathsf{R}z$, then $z \notin \bigcup_{a\in A} \widehat{a}$, hence $x \notin \bigvee_{a\in I} \widehat{a}$. For the right-to-left direction, if $x_F \cap I \neq \emptyset$ for all $x \in A$, then $A \subseteq \bigcup_{a\in I} \widehat{a} \subseteq \bigvee_{a\in I} \widehat{a}$. This completes the proof of the claim.

Now for the left-to-right direction of the equality, suppose that $x \in \Box_M A$, and let $y = (\{a \mid \Box a \in x_F\}, \{b \mid \Diamond b \in x_I\})$. Clearly, xMy, hence $y \in A$. But by the claim, this means that $y_F \cap I \neq \emptyset$. Hence $x \in \widehat{\Box a}$ for some $a \in I$. For the right-to-left direction, now suppose that $x \notin \Box_M A$. Then we have y, z such that $xMy\mathsf{R}z$ and $z \in \neg_{\mathsf{Я}} A$. Since $z \in \neg_{\mathsf{Я}} A$, this means that for any $w \in A$, $z_I \cap w_F \neq \emptyset$, and therefore, by the claim, $A \subseteq \bigvee_{a\in z_I} \widehat{a}$. Now it only remains to show that $x \notin \bigvee_{a\in z_I} \widehat{\Box a}$. Let $w = (\{1\}, I)$, where $I = \downarrow\{\Box b \mid b \in z_I\}$ (note that I is an ideal because $\Box a \vee \Box b \leq \Box(a \vee b)$ for any $a, b \in z_I$). Clearly, $w \in \neg\bigcup_{a\in z_I} \widehat{\Box a}$. But we also have $x\mathsf{R}w$: indeed, if $x_F \cap w_I \neq \emptyset$, then there is $b \in z_I$ such that $\Box b \in x_F$, and hence $b \in y_F$, contradicting our assumption. This shows that $x \notin \bigvee_{a\in z_I} \widehat{\Box a}$.

Finally, let us show that $\Diamond_M$ is the π-extension of $\Diamond$. For any $A \in \chi_{\mathsf{R}}(\mathcal{F}(L))$ and any $x \in \mathcal{F}(L)$, we need to show that $x \in \Diamond_M A$ iff $x \in \bigvee_{a\in I} \widehat{\Diamond a}$ for every ideal I such that $A \subseteq \bigvee_{a\in I} \widehat{a}$. For the left-to-right direction, suppose that $x \in \Diamond_M A$ and that I is an ideal such that $A \subseteq \bigvee_{a\in I} \widehat{a}$. Using the fact that $\Diamond_M = \neg_{\mathsf{R}}\Box_M\neg_{\mathsf{Я}}$, a simple computation shows that we need to show that $x \in \neg_{\mathsf{R}} \bigcap_{a\in I} \Box_M \breve{a}$. For this, it is enough to show that $\downarrow\{\Diamond a \mid a \in I\} \cap x_F \neq \emptyset$, since this will show that for any $y\mathsf{R}x$, $y \notin \Box_M \breve{a}$ for some $a \in I$. Suppose, towards a contradiction, that $\downarrow\{\Diamond a \mid a \in I\} \cap x_F = \emptyset$. Then there is $y\mathsf{R}x$ such that $y_I = \downarrow\{\Diamond a \mid a \in I\}$. But now we claim that $y \in \Box\neg_{\mathsf{Я}} A$. Indeed, if $yMz\mathsf{R}w$, then $w_F \cap I = \emptyset$ which, by the first claim above, means that $w \notin A$. Hence $x \notin \neg_{\mathsf{R}}\Box_M\neg_{\mathsf{Я}} A$, contradicting the assumption that $x \in \Diamond_M A$.

For the converse direction, suppose now that $x \notin \Diamond_M A = \neg_{\mathsf{R}}\Box_M\neg_{\mathsf{Я}} A$. Then we have $y \in \Box_M\neg_{\mathsf{Я}} A$ such that $y\mathsf{R}x$. Let $w = (\{a \mid \Box a \in y_F\}, \{b \mid \Diamond b \in y_I\})$. Then yMw, hence $w \in \neg_{\mathsf{Я}} A$. But this implies that $w_I \cap z_F = \emptyset$ for any $z \in A$, hence, by the first claim above, $A \subseteq \bigvee_{a\in w_I} \widehat{a}$. Now we claim that $x \notin \bigvee_{a\in w_I} \widehat{\Diamond a}$. To show this, it is enough to have that $x_F \cap \{\Diamond a \mid a \in w_I\} = \emptyset$. But the latter is clear, since $a \in w_I$ implies $\Diamond a \in y_I$, which in turns implies $\Diamond a \notin x_F$ by assumption on y. This completes the proof. $\Box$

As a direct consequence of Theorem 4.6, we obtain a Goldblatt-Thomason theorem for fundamental modal logic, as well as a characterization of classes of $AUFM$ frames of the form $Mod(L)$ for L a canonical superfundamental modal logic.

Theorem 5.10 *Let $\mathfrak{K}$ be a class of $AUFM$ frames closed under filter extensions. Then the following are equivalent:*

(i) *$\mathfrak{K}$ is axiomatic;*

(ii) *$\mathfrak{K}$ is closed under modal subframes, dense modals images and coproducts, and it reflects filter extensions;*

(iii) *$\mathfrak{K}$ is closed under strong modal subframes, strongly dense modal images and coproducts, and it reflects filter extensions.*

Corollary 5.11 *Let $\mathfrak{K}$ be a class of AUFM frames. The following are equivalent:*

(i) *$\mathfrak{K} = Mod(L)$, for L a canonical superfundamental modal logic;*

(ii) *$\mathfrak{K}$ is closed under filter extensions, modal subframes, dense modal images and coproducts, and it reflects filter extensions;*

(iii) *$\mathfrak{K}$ is closed under filter extensions, strong modal subframes, strongly dense modal images and coproducts, and it reflects filter extensions.*

The proofs of these two results completely mirror the non-modal case, and are therefore omitted.

Acknowledgements

I thank Wes Holliday for helpful discussions on the topic, as well as three anonymous referees for their comments which helped improve the clarity of the paper.

References

[1] Birkhoff, G., *On the structure of abstract algebras*, Mathematical proceedings of the Cambridge philosophical society **31** (1935), pp. 433–454.

[2] Conradie, W., A. Palmigiano and A. Tzimoulis, *Goldblatt-Thomason for le-logics* (2018).

[3] de Groot, J., *Goldblatt-Thomason theorems for modal intuitionistic logics* (2022).

[4] Dunn, J. M., M. Gehrke and A. Palmigiano, *Canonical extensions and relational completeness of some substructural logics*, The Journal of Symbolic Logic **70** (2005), pp. 713–740.

[5] Fitch, F. B., *Natural deduction rules for English*, Philosophical Studies **24** (1973), pp. 89–104.

[6] Goldblatt, R., *Axiomatic classes of intuitionistic models.*, J. Univers. Comput. Sci. **11** (2005), pp. 1945–1962.

[7] Goldblatt, R. I., *Semantic analysis of orthologic*, Journal of Philosophical logic (1974), pp. 19–35.

[8] Goldblatt, R. I. and S. K. Thomason, *Axiomatic classes in propositional modal logic*, in: *Algebra and Logic: Papers from the 1974 Summer Research Institute of the Australian Mathematical Society, Monash University, Australia*, Springer, 1975, pp. 163–173.

[9] Holliday, W. H., *A fundamental non-classical logic*, Logics **1** (2023), pp. 36–79.
URL `https://www.mdpi.com/2813-0405/1/1/4`

[10] Holliday, W. H., *Modal logic, fundamentally* (2024).
URL `https://arxiv.org/abs/2403.14043`

[11] Jónsson, B. and A. Tarski, *Boolean algebras with operators. part I*, American journal of mathematics **73** (1951), pp. 891–939.

[12] Jónsson, B. and A. Tarski, *Boolean algebras with operators. part II*, American journal of mathematics **74** (1952), pp. 127–162.

[13] Kripke, S. A., "Semantical Analysis of Intuitionistic Logic," North Holland Publishing Company, Amsterdam, 1965 pp. 92 – 130.

[14] Kunen, K., "Set theory: an introduction to independence proofs," Elsevier, 2014.

[15] Kurz, A. and J. Rosicky, *The Goldblatt-Thomason theorem for coalgebras*, in: *Algebra and Coalgebra in Computer Science: Second International Conference, CALCO 2007, Bergen, Norway, August 20-24, 2007. Proceedings 2*, Springer, 2007, pp. 342–355.

[16] Massas, G., *B-frame duality*, Annals of Pure and Applied Logic **174** (2023).
[17] Rodenburg, P., "Intuitionistic Correspondence Theory," Ph.D. thesis, University of Amsterdam (1986).

Coalgebraic Proof Translations for Non-Wellfounded Proofs

Borja Sierra Miranda [1] Thomas Studer [2] Lukas Zenger [3]

Institut für Informatik, Universität Bern
Neubrückstrasse 10, 3012 Bern, Switzerland

Abstract

Non-wellfounded proof theory results from allowing proofs of infinite height in proof theory. To guarantee that there is no vicious infinite reasoning, it is usual to add a constraint to the possible infinite paths appearing in a proof. Among these conditions, one of the simplest is enforcing that any infinite path goes through the premise of a rule infinitely often. Systems of this kind appear for modal logics with conversely well-founded frame conditions like GL or Grz.
In this paper, we provide a uniform method to define proof translations for such systems, guaranteeing that the condition on infinite paths is preserved. In addition, as particular instance of our method, we establish cut-elimination for a non-wellfounded system of the logic Grz. Our proof relies only on the categorical definition of corecursion via coalgebras, while an earlier proof by Savateev and Shamkanov uses ultrametric spaces and a corresponding fixed point theorem.

Keywords: Non-wellfounded Proof Theory, Coalgebra, Cut-Elimination.

1 Introduction

In proof theory proofs are traditionally viewed as finite trees that are labelled by sets of formulas (called sequents) according to some fixed set of rules. In recent decades a new notion of proof has gained prominence, namely the notion of a non-wellfounded proof. These are proofs in the traditional sense — i.e. labelled trees — that allow some branches to be infinitely long, whence the name non-wellfounded. Such proofs can be considered to be a formal counterpart of proofs by infinite descent [5]. In order to ensure soundness in the presence of infinite branches, a global correctness condition is imposed that distinguishes correct infinitary reasoning from vicious reasoning. One kind of global soundness condition is to enforce that every infinite branch must pass infinitely often

[1] borja.sierra@unibe.ch, supported by the Swiss National Science Foundation (SNSF) under grant agreement Nr. 200021_214820 (Non-wellfounded and cyclic proof theory)

[2] thomas.studer@unibe.ch, supported by the Swiss National Science Foundation (SNSF) under grant agreement Nr. 200021_214820 (Non-wellfounded and cyclic proof theory)

[3] lukas.zenger@unibe.ch, supported by the European Union's Horizon 2020 research and innovation program under the Marie Skłodowska-Curie grant agreement Nr. 101007627

through a specific premise of some rule. We call this a *local progress condition.* Non-wellfounded proof systems with a local progress condition have been developed for modal logics with conversely well-founded frame conditions like Gödel-Löb logic GL [16] or Grzegorczyk modal logic Grz [14]. It is this kind of soundness condition that is studied in this paper.

Apart from GL and Grz there are many other logical systems that allow for non-wellfounded proofs. In particular, non-wellfounded proof systems have been developed for modal fixed point logics, such as the modal mu-calculus (see e.g. [19,18,11]) as well as many of its fragments including the alternation free mu-calculus [10], common knowledge [6,12], temporal logics [9,3], or program logics such as PDL [8]. More recently non-wellfounded proof systems have been developed for intuitionistic modal fixed point logics [1,20] and also for Peano arithmetic [17]. However, proof systems for these logics usually require a stronger global constraint than the local progress condition, namely a so-called *trace condition.* As trace conditions are not considered in this paper, we will not give more details. The interested reader may consult one of the papers cited above.

Proof translations are commonly studied in structural proof theory, one example being cut-elimination, where arbitrary proofs are translated into cut-free proofs of the same sequents. For finite proofs, cut-elimination is established by showing how to permute instances of cuts upwards, and then how to eliminate instances of cuts at the leaves. It is then shown that applying such permutations recursively yields a cut-free proof. For non-wellfounded proofs such a translation is aggravated due to the fact that the local permutations must preserve the global soundness criterion. Neverthless, results about proof translations for non-wellfounded proofs are found in the literature. To name a few, Savateev and Shamkanov establish cut-elimination for non-wellfounded proofs for Grz [14] using ultrametric spaces and a fixed point theorem to guarantee that the limit of their construction is a (cut-free) proof. Baelde et al. [4] and Saurin [13] consider the problem of cut-elimination for non-wellfounded proofs for linear logic with fixed points, and Das and Pous [7] establish cut-elimination for non-wellfounded proofs for the equational theory for Kleene algebra. Recently, Afshari and Kloibhofer proved cut-elimination for a cyclic calculus for modal logic with an 'eventually' operator [2].

In this paper we study proof translations for non-wellfounded proofs based on a local progress condition. Our main contribution is the development of a uniform method to define such proof translations. Given two calculi $\mathcal{C}_0$ and $\mathcal{C}_1$ based on a local progress condition, we show how to define a map that translates proofs in $\mathcal{C}_0$ into proofs in $\mathcal{C}_1$. Depending on the considered translation, the map can be made to satisfy desirable properties, such as preserving the root sequent. This map is defined by first dissecting non-wellfounded proofs into finite fragments. It is then shown how the translation operates on a fragment locally, which we call a *translation step.* Finally, the entire translation is defined from translation steps by corecursion. As corecursion is essentially a coalgebraic technique, our method is formulated within the realm of coalgebra and draws essentially from its concepts. In particular, the corecursive definition of the translation map is given by the unique morphism from a coalgebra into the final coalgebra of a given endofunctor.

The developed method is uniform: it can be applied to all non-wellfounded proofs based on a local progress condition, independently of the specific logic, and to any kind of proof translation. We illustrate this by giving a concrete example: we establish cut-elimination for non-wellfounded proofs for Grz. In difference to the proof of Savateev and Shamkanov [14], we do not require to take a detour over ultrametric spaces. Instead it suffices to show how to push instances of cut out of the main fragment of a non-wellfounded proof. Our method then implies the existence of a proof translation mapping non-wellfounded proofs with cuts onto cut-free proofs. This also illustrates that proof translations within our framework can be considered to be close in spirit with proof translations for finite proofs. As for finite proofs, it requires to show how to translate a finite fragment of the proof tree to obtain the entire translation.

We assume that the reader is familiar with basic concepts of coalgebra, such as the definitions of a coalgebra, an endofunctor, a coalgebra morphism and a final coalgebra. An introduction to these concepts can be found in [21].

In Section 2 we introduce the mathematical machinery needed to develop our method. In particular we define the endofunctor that allows us to define proof translations by corecursion and so-called finite-fragmented trees, that formalize the intuition of dissecting proofs. In Section 3 we define proof translations and prove their correctness. Section 4 then consists of the cut-elimination proof for Grz.

2 Corecursion

2.1 Basic definitions

We will refer to (possibly empty) finite sequences of $\mathbb{N}$ as *words* and denote the collection of all words as $\mathbb{N}^*$. We will use w, v, u to denote words. The empty word is ϵ, $\mathbb{N}^+$ is the collection of non-empty words. Concatenation of words is denoted as juxtaposition, and so is adding a number to the start or the end of a word. The prefix order on words is denoted as $\sqsubseteq^w$. $w \perp v$ means that w, v are disjoint[4]. If $W \subseteq \mathbb{N}^*$ then we define $w \circ W = \{wv \mid v \in W\}$.

We will also work with finite sequences of $\mathbb{N}^+$, which are denoted by r, s. The empty sequence of words is denoted as nil, and concatenation is again simply juxtaposition. $w{:}r$ and $r{:}w$ denote the sequences obtained by adding the word w to the start of r or to the end of r respectively. The prefix order on these sequences is denoted as $\sqsubseteq^l$. The number of elements of r is denoted as $|r|$.

If f is a function and A is a subset of its domain, we will write $f{\restriction}A$ to denote the restriction of f to A. Notice that if w is a finite (infinite) sequent, then it is a function with domain a natural number (with domain $\mathbb{N}$). Then

[4] In words, that they do not have a common $\sqsubseteq^w$-upperbound.

$w{\restriction}n$ will just be the word consisting in the first n letters of w.

A *tree* is an ordered pair consisting of a (finite or infinite) collection of words (called *nodes*) closed by the prefix order and a labelling function. Trees are assumed to be finitely branching. Therefore each node w of a tree has an arity $k \in \mathbb{N}$ such that for any $i \in \mathbb{N}$, wi is a node iff $i < k$. The set of nodes of a tree τ is denoted by $\mathcal{N}^\tau$ and the set of *leaves* (maximal words by the prefix order belonging to the nodes) by $\mathcal{L}^\tau$. We fix an arbitrary set A whose elements will be used as labels as well as an object $*$ not belonging to A.

2.2 Endofunctor

One way to look at a non-wellfounded tree is as a collection of finite trees with instructions on how to glue the root of each of these finite trees (except one which is at the root) to the leaves of other finite trees in the collection. In order to specify to which leaves of a finite tree other finite trees can be glued to we introduce the following notion.

Definition 2.1 A *finite tree with non-wellfounded leaves* is a finite tree ι with labels in $A \cup \{*\}$ that satisfies the following two properties:

$$\text{(i) } \ell^\iota(\epsilon) \neq *, \qquad \text{(ii) } \ell^\iota(w) = * \text{ implies } w \in \mathcal{L}^\iota.$$

We denote finite trees with non-wellfounded leaves by ι and the collection of finite trees with non-wellfounded leaves by $\mathbb{T}^{\mathrm{nw}}$. Given $\iota \in \mathbb{T}^{\mathrm{nw}}$ we define the sets:

$$\mathcal{NW}^\iota = \{w \in \mathcal{N}^\iota \mid \ell^\iota(w) = *\}, \qquad \mathcal{PN}^\iota = \mathcal{N}^\iota \setminus \mathcal{NW}^\iota.$$

The elements of the first set are called *non-wellfounded leaves of* ι and the elements of the second set are called *proper nodes of* ι.

We define the endofunctor which we need to obtain the desired definition by corecursion.

Definition 2.2 We define an endofunctor $\mathcal{T}$ over **Set** (the category of sets) as:

(i) $\mathcal{T}(X) := \{(\iota, \mu) \mid \iota \in \mathbb{T}^{\mathrm{nw}} \text{ and } \mu : \mathcal{NW}^\iota \longrightarrow X\}$,

(ii) $\mathcal{T}(f : X \longrightarrow Y) = \Big((\iota, \mu) \mapsto (\iota, f \circ \mu)\Big)$.

From now on coalgebra will mean $\mathcal{T}$-coalgebra and coalgebra morphism will mean $\mathcal{T}$-coalgebra morphism.

Given a coalgebra (C, α) and an element $c \in C$ with $\alpha(c) = (\iota, \mu)$ we will write ι^α_c to refer to ι and μ^α_c to refer to μ. In case α can be deduced from context we will omit it.

Definition 2.3 Let (C, α) be a coalgebra, $c \in C$ and r be a finite sequence of $\mathbb{N}^+$. We say that r is an α *root-path of* c iff

nil is always a root-path of c,

$w{:}r$ is a root-path of c iff $w \in \mathcal{NW}^{\iota_c}$ and r root-path of $\mu^\alpha_c(w)$.

We will usually denote root-paths with r or s and the collection of α-root paths of c as $\mathcal{P}^{\alpha,c}$.

The notion of root-path is useful to traverse a coalgebra (C, α). Let us have a $c_0 \in C$ and $r \in \mathcal{P}^{\alpha,c}$. The first word w of r is a non-wellfounded leaf of ι_c^α. Using μ_c^α we can obtain a new element of the coalgebra $c_1 = \mu_{c_0}^\alpha(w)$. We can think of c_1 as the result of traversing c_0 using the w branch. We can iterate this procedure to exhaust the full root-path. This means that given an element of a coalgebra and one of its root-paths we can obtain a new element of the coalgebra. This concept is formalized in the next definition.

Definition 2.4 Let (C, α) be a coalgebra, $c \in C$ and $r \in \mathcal{P}^{\alpha,c}$. We define the *$\alpha$ subelement of c generated by r* by recursion in r as:

$$\mathsf{SE}^\alpha_{\mathsf{nil}}(c) = c, \qquad \mathsf{SE}^\alpha_{w:r}(c) = \mathsf{SE}^\alpha_r(\mu^\alpha_c(w)).$$

We can also talk about the tree with non-wellfounded leaves of a subelement, this is called a *fragment* of the original element of the coalgebra.

Definition 2.5 Let (C, α) be a coalgebra, $c \in C$ and $r \in \mathcal{P}^{\alpha,c}$. We define the *$\alpha$-fragment of c at r* by recursion in r as:

$$\mathsf{F}^\alpha_{\mathsf{nil}}(c) = \iota^\alpha_c, \qquad \mathsf{F}^\alpha_{w:r}(c) = \mathsf{F}^\alpha_r(\mu^\alpha_c(w)).$$

The fragment $\mathsf{F}^\alpha_{\mathsf{nil}}(c)$ is also called the *main α-fragment of c*.

The naming of these concepts: root-paths, fragments and subelements, are motivated from the final coalgebra of the endofunctor $\mathcal{T}$. The objects of the final coalgebra will be non-wellfounded trees expressed as a finite tree which has finite trees attached to some of its leaves, which may also have attached finite trees to some of their leaves, ... If we call roots the points where a finite tree start, a root-path is just a path of roots. A subelement will just be a subtree which do not break any of the finite trees in two and a fragment of a root-path is just a finite tree occurring at the end of the path of roots.

We state some basic properties of root-paths, fragments and subelements.

Lemma 2.6 *Let (C, α) be a coalgebra. We have that:*

(i) $r \in \mathcal{P}^{\alpha,c}$ *and* $s \in \mathcal{P}^{\alpha,\mathsf{SE}^\alpha_r(c)}$ *implies.* $rs \in \mathcal{P}^{\alpha,c}$, $\mathsf{F}^\alpha_s(\mathsf{SE}^\alpha_r(c)) = \mathsf{F}^\alpha_{rs}(c)$ *and* $\mathsf{SE}^\alpha_s(\mathsf{SE}^\alpha_r(c)) = \mathsf{SE}^\alpha_{rs}(c)$.

(ii) $r \in \mathcal{P}^{\alpha,c}$ *implies that for any* $w \in \mathbb{N}^+$, $r{:}w \in \mathcal{P}^{\alpha,c}$ *iff* $w \in \mathcal{NW}^{\mathsf{F}^\alpha_r(c)}$.

(iii) $s \sqsubseteq^l r$ *and* $r \in \mathcal{P}^{\alpha,c}$ *implies that* $s \in \mathcal{P}^{\alpha,c}$.

(iv) *If* $(C, \alpha), (D, \beta)$ *are coalgebras and* $\gamma : C \longrightarrow D$ *is a coalgebra morphism, we have that:* $r \in \mathcal{P}^{\alpha,c}$ *iff* $r \in \mathcal{P}^{\beta,\gamma(c)}$, $\mathsf{F}^\alpha_r(c) = \mathsf{F}^\beta_r(\gamma(c))$ *and* $\gamma(\mathsf{SE}^\alpha_r(c)) = \mathsf{SE}^\beta_r(\gamma(c))$.

Proof. All of these are simple inductions in r. □

2.3 Finite-fragmented trees

Finite-fragmented trees are non-wellfounded trees with a partition of the nodes. The idea is that each element of the partition gives a finite tree. This will imply that finite-fragmented trees with an appropiate function will form a final coalgebra of $\mathcal{T}$.

Definition 2.7 Let $F \subseteq \mathbb{N}^*$. We say that F fulfills the *fragmentation properties* iff the following conditions hold:

(i) (Finiteness) F is finite.

(ii) (Root) There is a $w \in F$ such that for any $v \in F$, $w \sqsubseteq^w v$.

(iii) (Convexity) For any $w, v \in F$ if $w \sqsubseteq^w u \sqsubseteq^w v$ then $u \in F$.

Given such a set F, we will denote by $\sqrt{F}$ its unique root, in other words its minimum element.

Definition 2.8 A finite-fragmented tree is a pair $\pi = (\tau, \mathcal{F})$ consisting of:

(i) A non-wellfounded tree τ.

(ii) A partition $\mathcal{F}$ of $\mathcal{N}^\tau$ such that any $F \in \mathcal{F}$ fulfills the fragmentation properties, this $\mathcal{F}$ is called *the fragmentation of* π.

We will usually denote finite-fragemented trees with π and the collection of all finite-fragmented trees as $\mathbb{T}^{\mathrm{ff}}$. Given $\pi = (\tau, \mathcal{F})$ we also denote $\mathcal{F}$ by $\mathcal{F}^\pi$ and τ by τ^π. Given $\pi \in \mathbb{T}^{\mathrm{ff}}$ we define the roots of π as the set $\mathcal{R}^\pi = \{\sqrt{F} \mid F \in \mathcal{F}^\pi\}$ and given a node w of π, ${}^\pi\sqrt{w}$ will be the minimum of the equivalence class w belongs to in π.

We will write $(\tau, \sim)$ for the finite-fragmented tree whose fragmentation is the partition given by the equivalence relation $\sim$.

The roots of a finite-fragmented tree mark the end of a finite tree and the start of a new one. We define an immediate successor relation for roots and also assign a measure to nodes depending on the roots below them.

Definition 2.9 For $\pi \in \mathbb{T}^{\mathrm{ff}}$ define the relation $\prec^\pi \subseteq \mathcal{R}^\pi \times \mathcal{R}^\pi$ as:

$$w \prec^\pi v \text{ iff } w \sqsubset^w v \text{ and there is no } u \in \mathcal{R}^\pi \text{ such that } w \sqsubset^w u \sqsubset^w v.$$

Observe that if $w \prec^\pi u, v \prec^\pi u$ then $w = v$. Given $w \in \mathcal{N}^\pi$ define its π-*fragmentation height* as

$$\mathrm{fhg}^\pi(w) = |\{v \sqsubset^w w \mid v \in \mathcal{R}^\pi\}|,$$

where $|X|$ means the cardinality of X.

Given a root w of π define the finite tree starting at w and the (finite-fragmented) subtree of π starting at w as follows.

Definition 2.10 Let $\pi \in \mathbb{T}^{\mathrm{ff}}$ and $w \in \mathcal{R}^\pi$. We define the *tree fragment in* π *given by* w, denoted $\mathsf{T}_w(\pi)$, as the finite tree with non-wellfounded leaves

$(\mathcal{N}, \ell)$, where:

$$\mathcal{N} = \{v \in \mathbb{N}^* \mid w \sim^\pi wv\} \cup \{u \in \mathbb{N}^+ \mid w \prec^\pi wu\},$$
$$\ell(v) = \begin{cases} \ell^\pi(wv) & \text{if } w \sim^\pi wv, \\ * & \text{if } w \prec^\pi wv. \end{cases}$$

Definition 2.11 Let $\pi \in \mathbb{T}^{\mathrm{ff}}$ and $w \in \mathcal{R}^\pi$. We define the *subtree of* π *generated by* w, denoted $\mathsf{ST}_w(\pi)$, as the finite-fragmented tree $(\mathcal{N}, \ell, \sim)$ where:

$$\mathcal{N} = \{v \in \mathbb{N}^* \mid wv \in \mathcal{N}^\pi\},$$
$$\ell(v) = \ell^\pi(wv),$$
$$v \sim u \text{ iff } wv \sim^\pi wu.$$

Roots and tree fragments characterize finite-fragmented trees:

Lemma 2.12 *Let* $\pi_0, \pi_1 \in \mathbb{T}^{\mathrm{ff}}$ *such that* $\mathcal{R}^{\pi_0} = \mathcal{R}^{\pi_1}$ *and for any* $w \in \mathcal{R}^{\pi_0} = \mathcal{R}^{\pi_1}$, $\mathsf{T}_w(\pi_0) = \mathsf{T}_w(\pi_1)$. *Then* $\pi_0 = \pi_1$.

Proof. Let $w \in \mathcal{N}^{\pi_0}$, define $v = \sqrt[\pi_0]{w}$ so $w = vu$ for some u and $u \in \mathcal{PN}^{\mathsf{T}_v(\pi_0)} = \mathcal{PN}^{\mathsf{T}_v(\pi_1)}$. By definition of tree fragment $w = vu \in \mathcal{N}^{\pi_1}$. So $\mathcal{N}^{\pi_0} \subseteq \mathcal{N}^{\pi_1}$, the other direction is analogous so $\mathcal{N}^{\pi_0} = \mathcal{N}^{\pi_1}$. Also, using the definition of tree fragment $\ell^{\pi_0}(w) = \ell^{\mathsf{T}_v(\pi_0)}(u) = \ell^{\mathsf{T}_v(\pi_1)}(u) = \ell^{\pi_1}(w)$. All left to show is that $w_0 \sim^{\pi_0} w_1$ iff $w_0 \sim^{\pi_1} w_1$, we show the left to right direction the other being analogous.

Let $w_0 \sim^{\pi_0} w_1$ and define $v = \sqrt[\pi_0]{w_0} = \sqrt[\pi_0]{w_1}$ so $w_0 = vu_0$ and $w_1 = vu_1$ for some u_0, u_1. By definition of tree fragment $u_0, u_1 \in \mathcal{PN}^{\mathsf{T}_v(\pi_0)} = \mathcal{PN}^{\mathsf{T}_v(\pi_1)}$ so $w_0 \sim^{\pi_1} v \sim^{\pi_1} w_1$, as desired. □

We state some basic properties of these constructions.

Lemma 2.13 *Let* $\pi \in \mathbb{T}^{\mathrm{ff}}$ *and* $w \in \mathcal{R}^\pi$. *We have that:*

(i) $v \in \mathcal{R}^{\mathsf{ST}_w(\pi)}$ *iff* $wv \in \mathcal{R}^\pi$.

(ii) $\mathsf{T}_v(\mathsf{ST}_w(\pi)) = \mathsf{T}_{wv}(\pi)$.

(iii) $\mathsf{ST}_v(\mathsf{ST}_w(\pi)) = \mathsf{ST}_{wv}(\pi)$.

Proof. To show (i) we just note that $[v]_{\sim^{\mathsf{ST}_w(\pi)}} = \{u \mid wu \in [wv]_{\sim^\pi}\}$ (1). It is clear then that if wv is the minimum of $[wv]_{\sim^\pi}$ then v is the minimum of $[v]_{\sim^{\mathsf{ST}_w(\pi)}}$, which shows right to left. For the other direction let $v \in \mathcal{R}^{\mathsf{ST}_w(\pi)}$, i.e. v is the minimum of $[v]_{\sim^{\mathsf{ST}_w(\pi)}}$ and let $u \in [wv]_{\sim^\pi}$. By the root property there must be a $u' = \sqrt[\pi]{[wv]_{\sim^\pi}}$ with $u' \sqsubseteq^w u$ and $u' \sqsubseteq^w wv$ (2). (2) implies that either $w \sqsubseteq^w u'$ or $u' \sqsubset^w w$, the second case being impossible since by convexity using that $u' \sqsubset^w w \sqsubseteq^w wv$ we get $w \in [u']_{\sim^\pi} = [wv]_{\sim^\pi}$ but w is a root so $w \sqsubseteq^w s$. Since $w \sqsubseteq^w u' \sqsubseteq^w u$ there is u_0 such that $wu_0 = u$ and by (1) $u_0 \in [v]_{\sim^{\mathsf{T}_w(\pi)}}$ so $v \sqsubseteq^w u_0$ and then $wv \sqsubseteq^w wu_0 = u$, as desired. The equalities (ii) and (iii) are just by unfolding definitions. □

With these constructions we can put a coalgebra structure onto $\mathbb{T}^{\mathrm{ff}}$.

Definition 2.14 We define the function destruct : $\mathbb{T}^{\mathrm{ff}} \longrightarrow \mathcal{T}(\mathbb{T}^{\mathrm{ff}})$ as:

$$\pi \mapsto (\mathsf{T}_\epsilon(\pi), w \mapsto \mathsf{ST}_w(\pi))$$

We will usually omit the destruct when we write $\mathcal{P}^{\mathrm{destruct},\pi}$, $\mathsf{F}_r^{\mathrm{destruct}}(\pi)$, $\mathsf{SE}_r^{\mathrm{destruct}}(\pi)$, and so on.

In words, given a finite fragmented tree we can destruct it into a finite fragment at its root, which is a tree with non-wellfounded leaves, and an attachment of finite-fragmented trees to the non-wellfounded leaves of the finite fragment.

Note that right now we have a duplication of concepts for finite-fragmented trees. We have roots, tree fragments and subtrees and from its coalgebraic structure with destruct we also have root-paths, fragments and subelements. The following definition and the subsequent lemma make clear that these concepts are in fact analogous (which justifies the naming of the coalgebraic concepts).

Definition 2.15 Let (C, α) be a coalgebra, $c \in C$ and $r \in \mathcal{P}^{\alpha,c}$. We define the *word of* r recursively in r as:

$$\mathsf{w}_{\mathsf{nil}} = \epsilon, \qquad \mathsf{w}_{w:r} = w\mathsf{w}_r.$$

We note that it is easy to show $\mathsf{w}_{r:w} = \mathsf{w}_r w$.

Given $\pi \in \mathbb{T}^{\mathrm{ff}}$ and $w \in \mathcal{R}^\pi$ it can be shown that there exists an unique sequence[5] $w_0, \dots, w_n \in \mathcal{R}^\pi$ such that $\epsilon = w_0 \prec^\pi \cdots \prec^\pi w_n = w$. We define $[w]_\pi = [w_1, w_2 - w_1, \dots, w_n - w_{n-1}]$.

Lemma 2.16 *Let* $\pi \in \mathbb{T}^{\mathrm{ff}}$, *we have that:*

(i) *If* $r \in \mathcal{P}^\pi$, *then:*
  (a) $\mathsf{w}_r \in \mathcal{R}^\pi$,
  (b) $[\mathsf{w}_r]_\pi = r$,
  (c) $\mathsf{F}_r(\pi) = \mathsf{T}_{\mathsf{w}_r}(\pi)$,
  (d) $\mathsf{SE}_r(\pi) = \mathsf{ST}_{\mathsf{w}_r}(\pi)$.

(ii) *If* $w \in \mathcal{R}^\pi$, *then:*
  (a) $[w]_\pi \in \mathcal{P}^\pi$,
  (b) $\mathsf{w}_{[w]_\pi} = w$,
  (c) $\mathsf{T}_w(\pi) = \mathsf{F}_{[w]_\pi}(\pi)$,
  (d) $\mathsf{ST}_w(\pi) = \mathsf{SE}_{[w]_\pi}(\pi)$.

Proof. Proof of (i). We show the four statements simultaneously by induction in $|r|$. The nil case is easy, so assume we have $r{:}w$. First, since $r{:}w \in \mathcal{P}\pi$ we have that $r \in \mathcal{P}\pi$ and then we know that $r{:}w \in \mathcal{P}\pi$ implies that $w \in \mathcal{NW}^{\mathsf{F}_r(\pi)} \stackrel{\text{I.H.}}{=} \mathcal{NW}^{\mathsf{T}_{\mathsf{w}_r}(\pi)}$. By definition of tree-fragment this means that $\mathsf{w}_r \prec^\pi \mathsf{w}_r w$ (1), so $\mathsf{w}_{r:w} = \mathsf{w}_r w \in \mathcal{R}^\pi$.

To show (b) we have to show that $[\mathsf{w}_{r:w}]_\pi = r{:}w$, so let $r = [w_0, \dots, w_{n-1}]$ i.e. $\mathsf{w}_r = w_0 \cdots w_{n-1}$. By I.H. we have that $[\mathsf{w}_r]_\pi = r$ which implies that $\epsilon \prec^\pi w_0 \prec^\pi w_0 w_1 \prec^\pi \cdots \prec^\pi w_0 \cdots w_{n-1}$. It suffices to show that $w_0 \cdots w_{n-1} = \mathsf{w}_r \prec^\pi \mathsf{w}_{r:w} = w_0 \cdots w_{n-1} w$. This is simply (1), showed above.

[5] For uniqueness just note that if $w \prec^\pi u$ and $v \prec^\pi u$ then $w = v$

To show (c) we just use the following reasoning:

$$\begin{aligned}\mathsf{F}_{r:w}(\pi) &\overset{(2)}{=} \mathsf{F}_{[w]}(\mathsf{SE}_r(\pi)) \overset{\text{I.H.}}{=} \mathsf{F}_{[w]}(\mathsf{ST}_{\mathsf{w}_r}(\pi)) \overset{(3)}{=} \mathsf{F}_{\mathsf{nil}}(\mathsf{ST}_w(\mathsf{ST}_{\mathsf{w}_r}(\pi))) \\ &\overset{(4)}{=} \mathsf{F}_{\mathsf{nil}}(\mathsf{ST}_{\mathsf{w}_r w}(\pi)) \overset{(3)}{=} \mathsf{T}_\epsilon(\mathsf{ST}_{\mathsf{w}_r w}(\pi)) \overset{(4)}{=} \mathsf{T}_{\mathsf{w}_r w}(\pi) = \mathsf{T}_{\mathsf{w}_{r:w}}(\pi),\end{aligned}$$

where (2) is thanks to Lemma 2.6, (3) are just unfolding of definitions and (4) are thanks to Lemma 2.13.

Finally (d) is analogous to (c):

$$\begin{aligned}\mathsf{SE}_{r:w}(\pi) &\overset{(5)}{=} \mathsf{SE}_{[w]}(\mathsf{SE}_r(\pi)) \overset{\text{I.H.}}{=} \mathsf{SE}_{[w]}(\mathsf{ST}_{\mathsf{w}_r}(\pi)) \overset{(6)}{=} \mathsf{ST}_w(\mathsf{ST}_{\mathsf{w}_r}(\pi)) \\ &\overset{(7)}{=} \mathsf{ST}_{\mathsf{w}_r w}(\pi) = \mathsf{ST}_{r:w}(\pi).\end{aligned}$$

where (5) is thanks to Lemma 2.6, (6) are just unfolding of definitions and (7) are thanks to Lemma 2.13.

Proof of (ii). We show the four statements simultaneously by induction in $\text{fhg}^\pi(w)$. If $\text{fhg}^\pi(w) = 0$ then $w = \epsilon$ and the proof is easy, let us assumme that $\text{fhg}^\pi(w) = n+1$, so there is $v \in \mathcal{R}^\pi$ such that $\text{fhg}^\pi(v) = n$ and $v \prec^\pi w$. Let u be such that $w = vu$, we note that $u \in \mathbb{N}^+$ since $v \sqsubset^w w$. By I.H. $[v]_\pi \in \mathcal{P}\pi$ and by definiton of fragment tree $u \in \mathsf{T}_v(\pi) \overset{\text{I.H}}{=} \mathsf{F}_{[v]_\pi}(\pi)$, so by Lemma 2.6 we get $[v]_\pi{:}u \in \mathcal{P}\pi$. But $[v]_\pi{:}u = [w]_\pi$ thanks to $v \prec^\pi w$.

To show (b) we remember that $[w]_\pi = [v]_\pi{:}u$ and $\mathsf{w}_{[v]_\pi} \overset{\text{I.H.}}{=} v$. Then $\mathsf{w}_{[w]_\pi} = \mathsf{w}_{[v]_\pi:u} = \mathsf{w}_{[v]_\pi}u = vu = w$, as desired. Finally, (c) and (d) are easy to derive using that we already shown (i), (ii)(a) and (ii)(b). □

2.4 Finite-fragmented trees as a final coalgebra

We establish the desired result that $(\mathbb{T}^{\text{ff}}, \text{destruct})$ is a final coalgebra of $\mathcal{T}$. This will allow us to define functions to $\mathbb{T}^{\text{ff}}$ by corecursion. First, we state two technical lemmas.

Lemma 2.17 *Let (C, α) be a coalgebra, $c \in C$, $r, s \in \mathcal{P}^{\alpha,c}$. Then:*

(i) *$|r| = |s|$ and $r \neq s$ implies $\mathsf{w}_r \perp \mathsf{w}_s$.*

(ii) *$|r| < |s|$ then either $\mathsf{w}_r \perp \mathsf{w}_s$ or $r \sqsubset^l s$.*

(iii) *Then $\mathsf{w}_r \sqsubseteq^w \mathsf{w}_s$ implies $r \sqsubseteq^l s$.*

Proof. (i) is proved by induction in $|r|$ using Lemma 2.6 and that two distinct leaves of the same finite tree must be disjoint. (ii) is proved by induction on the difference $|s| - |r|$ using (i). (iii) is a consequence of (i) and (ii) by cases in $|r| = |s|$, $|r| < |s|$ or $|s| < |r|$. □

Lemma 2.18 *Let (C, α) be a coalgebra and $c \in C$. For any $r, s \in \mathcal{P}^{\alpha,c}$ distinct, we have that: $(\mathsf{w}_r \circ \mathcal{PN}^{\mathsf{F}^\alpha_r(c)}) \cap (\mathsf{w}_s \circ \mathcal{PN}^{\mathsf{F}^\alpha_s(c)}) = \varnothing$.*

Proof. We can assume without loss of generaltity that $|r| \leq |s|$ and let v belong to the intersection, in particular $v = \mathsf{w}_r u_0 = \mathsf{w}_s u_1$ for some $u_0 \in \mathcal{PN}^{\mathsf{F}^\alpha_r(c)}$ and $u_1 \in \mathcal{PN}^{\mathsf{F}^\alpha_s(c)}$. By Lemma 2.17, either $\mathsf{w}_r \perp \mathsf{w}_s$ or $r \sqsubset^l s$. The first case is impossible since v is an upperbound of both, so we must have $r \sqsubset^l s$ and then $s = r[w, \ldots]$ with $w \sqsubseteq^w u_0$. Since s is a root-path an $r{:}w \sqsubseteq^l s$ we get that $r{:}w$ is a root-path which implies that $w \in \mathcal{NW}^{\mathsf{F}^\alpha_r(c)}$. Then $u_0 \in \mathcal{PN}^{\mathsf{F}^\alpha_r(c)}$ and $w \sqsubseteq^w u_0$ implies that $w \sqsubset^w u_0$, but u_0 is a node of $\mathsf{F}^\alpha_r(c)$ and w is a leaf of the same tree, so $w \sqsubset^w u_0$ is impossible. □

Finally, we can prove the main theorem of this section. [6]

Theorem 2.19 ($\mathbb{T}^{\mathit{ff}}$, destruct) *is a final coalgebra of* $\mathcal{T}$.

Proof. Existence. Let (C, α) be a coalgebra. We define $f : C \longrightarrow \mathbb{T}^{\mathrm{ff}}$ such that given $c \in C$ $f(c)$ is the finite-fragmented tree $\pi = (\mathcal{N}, \ell, \mathcal{F})$ where:

$$\mathcal{N} = \bigcup_{r \in \mathcal{P}^{\alpha,c}} \mathsf{w}_r \circ \mathcal{PN}^{\mathsf{F}^\alpha_r(c)},$$

$$\ell(w) = \ell^{\mathsf{F}^\alpha_r(c)}(u) \text{ where } r \text{ is the unique element of } \mathcal{P}^{\alpha,c} \text{ and}$$
$$u \text{ is the unique element of } \mathcal{PN}^{\mathsf{F}^\alpha_r(c)} \text{ such that } v = \mathsf{w}_r u,$$

$$\mathcal{F} = \{\mathsf{w}_r \circ \mathcal{PN}^{\mathsf{F}^\alpha_r(c)} \mid r \in \mathcal{P}^{\alpha,c}\}.$$

The labelling is well-defined (i.e. the things asserted to be unique are indeed unique) due to Lemma 2.18. Note that π is indeed a finite-fragmented tree.

We need to prove that f is a coalgebra morphism, i.e. we have to show that $\iota^\alpha_c = \mathsf{T}_\epsilon(\pi)$ and $f \circ \mu^\alpha_c = (w \in \mathsf{T}_\epsilon(\pi) \mapsto \mathsf{ST}_w(\pi))$.

Proof of $\iota^\alpha_c = \mathsf{T}_\epsilon(\pi)$. By definition the proper nodes of $\mathsf{T}_\epsilon(\pi)$ is the equivalence class of epsilon in π, which in this case is $\epsilon \circ \mathcal{PN}^{\iota^\alpha_c}$, with the same labels as in ι^α_c, The non-wellfounded leaves are the w such that $\epsilon \prec^\pi w$. To show the desired equality all we need to show is that for any $w \in \mathcal{N}^\pi$:

$$\epsilon \prec^\pi w \text{ if and only if } w \in \mathcal{NW}^{\iota^\alpha_c} = \mathcal{NW}^{\mathsf{F}^\alpha_{\mathsf{nil}}(c)}.$$

Left to right. Since $w \in \mathcal{N}^\pi$ there must be a $r \in \mathcal{P}^\alpha c$ such that $w = \mathsf{w}_r u$ where $u \in \mathcal{PN}^{\mathsf{F}^\alpha_r(c)}$. But $w \in \mathcal{R}^\pi$ means that it must be the minimum of its equivalecne class which is $\mathsf{w}_r \circ \mathcal{PN}^{\mathsf{F}^\alpha_r(c)}$, since $\epsilon \in \mathcal{PN}^{\mathsf{F}^\alpha_r(c)}$ we can conclude that $u = \epsilon$, i.e. $\mathsf{w}_r = w$ (which implies that $|r| > 0$ since $\epsilon \sqsubset^w w$). If $|r| > 1$ then $r = [v, \ldots]$ so $[v] \in \mathcal{P}^\alpha c$ and then $v \in \mathcal{R}^\pi$ and $\epsilon \sqsubset^w v \sqsubset^w w$, contradiction. So $|r| = 1$ and since $\mathsf{w}_r = w$ we must have $r = [w]$, which implies (Lemma 2.6) that $w \in \mathcal{NW}^{\mathsf{F}^\alpha_{\mathsf{nil}}(c)}$.

Right to left. Let $w \in \mathcal{NW}^{\mathsf{F}^\alpha_{\mathsf{nil}}(c)}$, since $\mathsf{nil} \in \mathcal{P}^\alpha c$ by Lemma 2.6 we get that $[w] \in \mathcal{P}^\alpha c$. It is easy see then that by definition of π, $\mathsf{w}_{[w]} = w \in \mathcal{R}^\pi$. We want to show that $\epsilon \prec^\pi w$, assume the contrary. Then there must be a

[6] Category-theory oriented readers may wonder why we take the burden to prove this. A more categorical approach would have proven the existence of a final coalgebra for $\mathcal{T}$ and define the finite-fragmented trees as elements of this coalgebra. However, for our purposes of providing translations between proof systems we need to transform (some) non-wellfounded trees to finite-fragmented trees and back. If we take this categorical approach providing this transformation will be equivalent to directly proving that our definition of finite-fragmented tree provides a final coalgebra of $\mathcal{T}$.

$v \in \mathcal{R}^\pi$ such that $\epsilon \sqsubset^w v \sqsubset^w w$, but then there must be (with a reasoning analogous to the one in the left to right direction to show that $w = \mathsf{w}_r$) a $s \in \mathcal{P}^\alpha c$ such that $\mathsf{w}_s = v$. Since $\mathsf{w}_s = v \sqsubset^w w = \mathsf{w}_{[w]}$ by Lemma 2.17 it must be the case that $s \sqsubset^l [w]$, which means that $s = \mathsf{nil}$. But this is absurd since $\epsilon = \mathsf{w}_{\mathsf{nil}} = \mathsf{w}_s = v > \epsilon$.

Let $w \in \mathcal{NW}^{\iota_c^\alpha} = \mathcal{NW}^{\mathsf{T}_\epsilon(\pi)}$, we have to show that $f(\mu_c^\alpha(w)) = \mathsf{ST}_w(\pi)$. We note that by definition we have that:

$$r \in \mathcal{P}^\alpha \mu_c^\alpha(w) \text{ iff } w{:}r \in \mathcal{P}^\alpha c,$$

$$\mathsf{F}_r^\alpha(\mu_c^\alpha(w)) = \mathsf{F}_{w:r}^\alpha(c).$$

Then

$$\mathcal{N}^{f(\mu_c^\alpha(w))} = \bigcup_{r \in \mathcal{P}^\alpha \mu_c^\alpha(w)} \mathsf{w}_r \circ \mathcal{PN}^{\mathsf{F}_r^\alpha(\mu_c^\alpha(w))} = \bigcup_{r \in \{r | w:r \in \mathcal{P}^\alpha c\}} \mathsf{w}_r \circ \mathcal{PN}^{\mathsf{F}_{w:r}^\alpha(c)}.$$

To show that $\mathsf{ST}_w(\pi)$ and $f(\mu_c^\alpha(w))$ have the same nodes it suffices to prove that:

$$\mathcal{N}^{\mathsf{ST}_w(\pi)} = \bigcup_{r \in \{r | w:r \in \mathcal{P}^\alpha c\}} \mathsf{w}_r \circ \mathcal{PN}^{\mathsf{F}_{w:r}^\alpha(c)}. \tag{i}$$

The right to left inclusion is trivial since it is clear that if v belongs to the RHS set then wv belongs to the nodes of π and then v belongs to the LHS set. Assume that $v \in \mathcal{N}^{\mathsf{ST}_w(\pi)}$, i.e. $wv \in \mathcal{N}^\pi$ so there is an $s \in \mathcal{P}^\alpha c$ and a $u \in \mathcal{PN}^{\mathsf{F}_s^\alpha(c)}$ such that $wv = \mathsf{w}_s u$. Since $w \in \mathcal{NW}^{\iota_c^\alpha}$ then $[w] \in \mathcal{P}^\alpha c$ and since $w \sqsubseteq^w wv = \mathsf{w}_s u$ we have that either $w \sqsubseteq^w \mathsf{w}_s$ or $\mathsf{w}_s \sqsubset^w w$. In the first case, by Lemma 2.17 we have that $[w] \sqsubseteq^l s$ which implies that $s = w{:}s'$ and then $\mathsf{w}_{s'} u = v$ is in the RHS set. If $\mathsf{w}_s \sqsubset^w w$ then $s \sqsubset^l [w]$ so $s = \mathsf{nil}$ which implies that $wv = \mathsf{w}_s u = \epsilon u = u$ and that $u \in \mathcal{PN}^{\mathsf{F}_s^\alpha(c)} = \mathcal{PN}^{\mathsf{F}_{\mathsf{nil}}^\alpha(c)} = \mathcal{PN}^{\iota_c^\alpha}$. This is a contradiction since $wv = u$ implies $w \sqsubseteq^w u$ but w is a non-wellfounded leaf of ι_c^α and u is not (so it cannot be equal and it cannot be bigger since w is a leaf).

To show that $f(\mu_c^\alpha(w))$ has the same labelling as $\mathsf{ST}_w(\pi)$ we let $v \in \mathcal{N}^{\mathsf{ST}_w(\pi)}$, then (by (i)) there are $w{:}r \in \mathcal{P}^\alpha c$ and $u \in \mathcal{PN}^{\mathsf{F}_{w:r}^\alpha(c)}$ such that $wv = \mathsf{w}_{w:r} u$. The following equalities straightforwardly follow from this:

$$\ell^{f(\mu_c^\alpha(w))}(v) = \ell^{\mathsf{F}_r^\alpha(\mu_c^\alpha(w))}(u) = \ell^{\mathsf{F}_{w:r}^\alpha(c)}(u) = \ell^\pi(wv) = \ell^{\mathsf{ST}_w(\pi)}(v).$$

Finally, we show that $f(\mu_c^\alpha(w))$ has the same fragmentation as $\mathsf{ST}_w(\pi)$. Note that $v \sim^{\mathsf{ST}_w(\pi)} u$ iff $wv \sim^\pi wu$ iff[7] there are r such that $w{:}r \in \mathcal{P}^\alpha c$ and $v_0, u_0 \in \mathcal{PN}^{\mathsf{F}_{w:r}^\alpha(c)}$ such that $\mathsf{w}_{w:r} v_0 = wv$ and $\mathsf{w}_{w:r} u_0 = wu$. This is equivalent to the existence of $r \in \mathcal{P}^\alpha \mu_c^\alpha(w)$ and $v_0, u_0 \in \mathcal{PN}^{\mathsf{F}_r^\alpha(\mu_c^\alpha(w))}$ such that $\mathsf{w}_r v_0 = v$ and $\mathsf{w}_r u_0 = u$, which is equivalent to $v \sim^{f(\mu_c^\alpha(w))} u$.

[7] In principle it would be equivalent to the existence of some $r \in \mathcal{P}^\alpha c$ and v_0, u_0 such that $\mathsf{w}_r v_0 = wv$ and $\mathsf{w}_r u_0 = wu$. This equalities with the fact that $w \in \mathcal{NW}^{\iota_c^\alpha}$ so $[w] \in \mathcal{P}^\alpha c$ allows us to derive the desired equivalence.

Uniqueness. Let (C, α) be a coalgebra and $\gamma_0, \gamma_1 : C \longrightarrow \mathbb{T}^{\mathrm{ff}}$ be coalgebra morphisms and let $\pi_0 = \gamma_0(c), \pi_1 = \gamma_1(c)$ with $c \in C$. By Lemma 2.12 it suffices to show that $\mathcal{R}^{\pi_0} = \mathcal{R}^{\pi_1}$ (i) and that for any $w \in \mathcal{R}^{\pi_0}$, $\mathsf{T}_w(\pi_0) = \mathsf{T}_w(\pi_1)$ (ii).

Proof of (i). We show $\subseteq$, the other inclusion being analogous. If $w \in \mathcal{R}^{\pi_0}$ then $[w]_\pi \in \mathcal{P}^{\pi_0}$ by Lemma 2.16. By Lemma 2.6 we have that

$$\mathcal{P}^{\pi_0} = \mathcal{P}^{\gamma_0(c)} = \mathcal{P}^{\alpha,c} = \mathcal{P}^{\gamma_1(c)} = \mathcal{P}^{\pi_1},$$

by virtue of γ_0, γ_1 being coalgebra morphisms. So $[w]_{\pi_0} \in \mathcal{P}^{\pi_1}$ and thanks to Lemma 2.16 we get $w = \mathsf{w}_{[w]_{\pi_0}} \in \mathcal{R}^{\pi_1}$.

Proof of (ii). Since we showed that $\mathcal{R}^{\pi_0} = \mathcal{R}^{\pi_1}$ we have that for any $w \in \mathcal{R}^{\pi_0}$, $[w]_{\pi_0} = [w]_{\pi_1}$. Fix $w \in \mathcal{R}^{\pi_0}$ and define $r = [w]_{\pi_0} = [w]_{\pi_1}$. Using Lemma 2.6 and Lemma 2.16 we get the desired

$$\mathsf{T}_w(\pi_0) = \mathsf{F}_r(\pi_0) = \mathsf{F}_r(\gamma_0(c)) = \mathsf{F}_r^\alpha(c) = \mathsf{F}_r(\gamma_1(c)) = \mathsf{F}_r(\pi_1) = \mathsf{T}_w(\pi_1).$$

□

Observe that since $(\mathbb{T}^{\mathrm{ff}}, \text{destruct})$ is a final coalgebra, destruct is a coalgebra isomorphism and thus has an inverse construct. It is easy to see that given a pair $(\iota, \mu : \mathcal{NW}^\iota \longrightarrow \mathbb{T}^{\mathrm{ff}})$, construct$(\iota, \mu)$ is just the result of "gluing" $\mu(w)$ at w in ι for each $w \in \mathcal{NW}^\iota$. Given a coalgebra (C, α) let us denote by $!_\alpha$ the unique coalgebra morphism from it to $(\mathbb{T}^{\mathrm{ff}}, \text{construct})$. We have that:

$$!_\alpha = \text{construct} \circ \mathcal{T}(!_\alpha) \circ \alpha.$$

This equation is what provides corecursion of its character. α can be considered as performing 1 step of the corecursion and $!_\alpha$ will be applying the whole corecursion. Then, the equation says that applying the whole corecursion is the same as applying 1 step obtaining a pair consisting of a finite tree and a function giving an element of the coalgebra for each leaf. Then apply the whole corecursion to the elements of the coalgebras obtaining finite-fragmented trees and glue these finite-fragmented trees to the finite tree given by α.

3 Proof translations

In this section we provide sufficient conditions to ensure that a function defined by corecursion transforms proofs in a local progress calculus to proofs in another local progress calculus. We start by defining exactly what we mean by local progress calculi. During this section we fix an arbitrary set $\mathbb{S}$ whose elements are called sequents.

Definition 3.1 A *rule instance* consists in a pair $P \Longrightarrow C$ where:

(i) P is a finite sequence of sequents, the *premises* of the instance.

(ii) C is a sequent, the *conclusion* of the instance.

A *rule* R is a non-empty set of rule instances. A *local-progress sequent calculus* is a pair $\mathcal{C} = (\mathcal{R}, L)$ where:

(i) $\mathcal{R}$ is a set of rules, called the *rules of* $\mathcal{C}$.

(ii) L is a function that given a rule $R \in \mathcal{R}$ and an instance of the rule $(S_0, \ldots, S_{k-1} \Longrightarrow S) \in R$ it returns a subset of $\{0, \ldots, k-1\}$. This subset is the set of premises of the rule instance that make progress. The function is called the *local progress function of* $\mathcal{C}$. We will denote the application of L to a rule R and rule instance $r \in R$ as $L_R(r)$.

We will use $\mathcal{C}$ to denote local progress sequent calculi.

From now on we assume that all trees considered are labelled with ordered pairs (S, R) where S is a sequent and R a rule. We denote the collection of non-wellfounded labelled trees by $\mathbb{T}^{\infty}$. Given a node w in a tree τ we will write τ-sequent(w) to mean the first component of its label and τ-rule(w) to mean the second component of its label. We define the usual notion of non-wellfounded proof in the following definition.

Definition 3.2 Let $\pi \in \mathbb{T}^{\infty}$. We say that π is a *pre-proof in* $\mathcal{C}$ iff for every node w of π we have that: $(S_0, \ldots, S_{k-1} \Longrightarrow S) \in R$, where w is k-ary in π, $S_i = \pi$-sequent(wi) for $i < k$, $S = \pi$-sequent(w) and $R = \pi$-rule(w). We denote the collection of pre-proofs of $\mathcal{C}$ as $\mathbb{P}_0^{\infty}(\mathcal{C})$.

Let $\tau \in \mathbb{P}_0^{\infty}(\mathcal{C})$ and $w \in \mathcal{N}^{\tau}$. We say that w *is progressing (in* τ *with respect to* $\mathcal{C}$*)* if there is $v \in \mathbb{N}^*, i \in \mathbb{N}$ such that $w = vi$ and $i \in L_R(S_0, \ldots, S_{k-1} \Longrightarrow S)$ where v is k-ary in τ, $S_j = \tau$-sequent(vj) for $j < k$, $S = \tau$-sequent(v) and $R = \tau$-rule(v).

We say that π is a *proof in* $\mathcal{C}$ iff it is a pre-proof and for any infinite branch b of π there are infinitely many i's such that $b \restriction i$ is progressing. The collection of proofs of $\mathcal{C}$ is denoted as $\mathbb{P}^{\infty}(\mathcal{C})$.

In the following two definitions, we define a notion of proof for finite-fragmented trees.

Definition 3.3 Let $\iota \in \mathbb{T}^{\mathrm{nw}}$ and $s : \mathcal{NW}^{\iota} \longrightarrow \mathbb{S}$. We say that (ι, s) is a proof fragment in $\mathcal{C}$ iff for any $w \in \mathcal{PN}^{\iota}$ of arity k in ι, if we define

$$S_i = \begin{cases} \iota\text{-sequent}\,(wi) & \text{for } wi \notin \mathcal{NW}^{\iota}, \\ s(wi) & \text{for } wi \in \mathcal{NW}^{\iota}, \end{cases} \quad \text{for } i < k$$

$$S = \iota\text{-sequent}\,(w)\,,$$
$$R = \iota\text{-rule}\,(w)\,,$$

then we have that:

(i) $(S_0, \ldots, S_{k-1} \Longrightarrow S) \in R$.

(ii) (Progress) For any $i < k$, $wi \in \mathcal{NW}^{\iota}$ iff $i \in L_R(S_0, \ldots, S_{k-1} \Longrightarrow S)$.

Definition 3.4 Let $\pi = (\tau, \sim) \in \mathbb{T}^{\mathrm{ff}}$. We say that π is a *finite-fragmented proof in* $\mathcal{C}$ iff for any $r \in \mathcal{P}^{\pi}$ [8] we have that the pair

$$(\mathsf{F}_r(\pi),\ v \in \mathcal{NW}^{\mathsf{F}_r(\pi)} \mapsto (\mathsf{F}_{r:v}(\pi))\text{-sequent}\,(\epsilon))$$

[8] Recall that we write $\mathcal{P}^{\pi}$ for $\mathcal{P}^{\mathrm{destruct},\pi}$.

is a proof fragment in $\mathcal{C}$. We will denote the set of finite-fragmented proofs in $\mathcal{C}$ as $\mathbb{P}^{\mathrm{ff}}(\mathcal{C})$.

We note that in the previous definition $(\mathsf{F}_{r:v}(\pi))$-sequent$(\epsilon) = (\mathsf{T}_{\mathsf{w}_r v}(\pi))$-sequent$(\epsilon) = \pi$-sequent$(\mathsf{w}_r v)$, so it is just the sequent corresponding to the node in the whole proof. Given a preproof of $\mathcal{C}$ we can define a partition as in the next definition.

Definition 3.5 Let $\tau \in \mathbb{P}_0^{\infty}(\mathcal{C})$, we define the relation $\rightsquigarrow^{\tau} \subseteq \mathcal{N}^{\tau} \times \mathcal{N}^{\tau}$ as:

$$w \rightsquigarrow^{\tau} v \text{ iff there is } i \in \mathbb{N} \text{ such that } v = wi \text{ and } v \text{ is not progressing.}$$

Let $\sim^{\tau}$ be the equivalence relation closure of $\rightsquigarrow^{\tau}$, in other words $w \sim^{\tau} v$ iff there is a sequence $w_0, \ldots, w_n$ such that $w = w_0, v = w_n$ and for any $i < n$, $w_i \rightsquigarrow^{\tau} w_{i+1}$ or $w_{i+1} \rightsquigarrow^{\tau} w_i$.

A relation between finite-fragmented proofs and non-wellfounded proofs is established and proven in the following lemma.

Lemma 3.6 *We have that*

(i) $\tau \in \mathbb{P}^{\infty}(\mathcal{C})$ *implies* $(\tau, \sim^{\tau}) \in \mathbb{P}^{ff}(\mathcal{C})$.[9]

(ii) $\pi = (\tau, \sim) \in \mathbb{P}^{ff}(\mathcal{C})$ *implies* $\tau \in \mathbb{P}^{\infty}(\mathcal{C})$.

Proof. To show that every node is an instance of a rule is easy in both implication, so we will focus in proving that the progressing condition is fulfilled in both implications.

Proof of (i). First we need to show that if τ is a proof then $\sim^{\tau}$ is a fragmentation. Convexity is shown by using that any path in a tree between w and $v \geq w$ must go through all the middle nodes, i.e. through $\{u \mid w \sqsubseteq^{w} u \sqsubseteq^{w} v\}$. The existence of root is shown by using that any path between w and v in a tree must go through $w \cap v$, i.e. the common part of w and v. Finiteness is shown using König's Lemma together with convexity and the existence of root for any equivalence class.

That the progress condition of proof-fragment is fulfilled follows from the definition of $\sim^{\tau}$.[10]

Proof of (ii). If we have an infinite branch in τ it must go through infinitely many fragments, since each fragment is finite. By the second condition of proof fragment and thanks to each fragment being a proof fragment this implies that the branch makes progress infinitely many times. □

With the notion of finite-fragmented proof at hand we are ready to define the sufficient conditions to obtain a proof translation by corecursion.

Definition 3.7 Let $\mathcal{C}_0$ and $\mathcal{C}_1$ be local progress sequent calculi and let $\alpha : \mathbb{T}^{\mathrm{ff}} \longrightarrow \mathcal{T}(\mathbb{T}^{\mathrm{ff}})$. We say that α *is a proof translation step from* $\mathcal{C}_0$ *to* $\mathcal{C}_1$ iff we have that for any $\pi \in \mathbb{P}^{\mathrm{ff}}(\mathcal{C}_0)$:

---

[9] In fact it can be shown that the fragmentation in a finite-fragmented proof is unique. This point implies that $\sim^{\tau}$ is that unique fragmentation.

[10] We also need to use that in a tree any path from w to wi must contain a transition (w, wi).

(i) $(\mathsf{F}^{\alpha}_{\mathsf{nil}}(\pi), w \mapsto \mathsf{F}^{\alpha}_{[w]}(\pi)\text{-sequent}\,(\epsilon))$ is a proof fragment in $\mathcal{C}_1$.

(ii) For any $w \in \mathcal{NW}^{\mathsf{F}^{\alpha}_{\mathsf{nil}}(\pi)}$ we have that $\mathsf{SE}^{\alpha}_{[w]}(\pi) \in \mathbb{P}^{\mathrm{ff}}(\mathcal{C}_0)$.

A function γ is a *finite-fragmented proof translation from* $\mathcal{C}_0$ *to* $\mathcal{C}_1$ iff $\gamma : \mathbb{P}^{\mathrm{ff}}(\mathcal{C}_0) \longrightarrow \mathbb{P}^{\mathrm{ff}}(\mathcal{C}_1)$. A function γ is a *non-wellfounded proof translation from* $\mathcal{C}_0$ *to* $\mathcal{C}_1$ iff $\gamma : \mathbb{P}^{\infty}(\mathcal{C}_0) \longrightarrow \mathbb{P}^{\infty}(\mathcal{C}_1)$.

Theorem 3.8 *If α is a proof translation step from $\mathcal{C}_0$ to $\mathcal{C}_1$ then*

(i) $!_{\alpha} \restriction \mathbb{P}^{ff}(\mathcal{C}_0)$, *i.e. the function defined by corecursion from α applied only to finite-fragmented proofs of $\mathcal{C}_0$, is a finite-fragmented proof translation from $\mathcal{C}_0$ to $\mathcal{C}_1$, and*

(ii) *The function $\tau \in \mathbb{P}^{\infty}(\mathcal{C}_0) \mapsto (\mathsf{p}_0 \circ !_{\alpha})(\tau, \sim^{\tau})$, where p_0 is the first projection from an ordered pair, is a non-wellfounded proof translation from $\mathcal{C}_0$ to $\mathcal{C}_1$.*

Proof. Proof of (i). Let $\pi_0 \in \mathbb{P}^{\mathrm{ff}}(\mathcal{C}_0)$ and $\pi_1 = !_{\alpha}(\pi_0)$. Thanks to the second condition in the definition of proof translation step, by induction in r we can show that if $\pi_0 \in \mathbb{P}^{\mathrm{ff}}(\mathcal{C}_0)$ then $\mathsf{SE}^{\alpha}_{r}(\pi_0) \in \mathbb{P}^{\mathrm{ff}}(\mathcal{C}_0)$ (1).

Let $r \in \mathcal{P}^{\pi_1}$, we have to show that $(\mathsf{F}_r(\pi_1), w \mapsto \mathsf{F}_{r:w}(\pi_1)\text{-sequent}\,(\epsilon))$ is a proof fragment in $\mathcal{C}_1$. By Lemma 2.6 and thanks to $!_{\alpha}$ being a coalgebra morphism, we know that $\mathcal{P}^{\pi_1} = \mathcal{P}^{\alpha,\pi_0}$ and $\mathsf{F}_r(\pi_1) = \mathsf{F}^{\alpha}_r(\pi_0) = \mathsf{F}^{\alpha}_{\mathsf{nil}}(\mathsf{SE}^{\alpha}_r(\pi_0))$. Given $w \in \mathcal{NW}^{\mathsf{F}_r(\pi_1)}$, with Lemma 2.6 again, we get $r{:}w \in \mathcal{P}^{\pi_1} = \mathcal{P}^{\alpha,\pi_0}$ and $\mathsf{F}_{r:w}(\pi_1) = \mathsf{F}^{\alpha}_{r:w}(\pi_0) = \mathsf{F}^{\alpha}_{[w]}(\mathsf{SE}^{\alpha}_r(\pi_0))$. So it suffices to show that $(\mathsf{F}^{\alpha}_{\mathsf{nil}}(\mathsf{SE}^{\alpha}_r(\pi_0)), w \mapsto \mathsf{F}^{\alpha}_{[w]}(\mathsf{SE}^{\alpha}_r(\pi_0))\text{-sequent}\,(\epsilon))$ is a proof fragment of $\mathcal{C}_1$. But this is the first condition in the definition of proof translation step, which is obtained thanks to $\mathsf{SE}^{\alpha}_r(\pi_0) \in \mathbb{P}^{\mathrm{ff}}(\mathcal{C}_0)$ by (1).

Proof of (ii). Note that if $\tau \in \mathbb{P}^{\infty}(\mathcal{C}_0)$ then $(\tau, \sim^{\tau}) \in \mathbb{P}^{\mathrm{ff}}(\mathcal{C}_0)$ by Lemma 3.6 and thanks to point (i) of this theorem $\pi_1 = !_{\alpha}(\tau, \sim^{\tau}) \in \mathbb{P}^{\mathrm{ff}}(\mathcal{C}_1)$. Applying Lemma 3.6 we conclude that $\mathsf{p}_0(\pi_1) \in \mathbb{P}^{\infty}(\mathcal{C}_1)$. □

Let us finish this section remarking that this result can be generalized. We can redefine the notion of proof translation step as starting in an arbitrary set, say that $\alpha : X \longrightarrow \mathcal{T}(X)$ *is a proof translation step from X to the calculus $\mathcal{C}$* iff for any $x \in X$, $(\mathsf{T}_{\mathsf{nil}}\,(\alpha)\,(x), w \mapsto \mathsf{T}_w\,(\alpha)\,(x)\text{-sequent}\,(\epsilon)$ is a proof fragment in $\mathcal{C}$. Then it is easy to adapt the proof of Theorem 3.8 to show that extending this kind of proof translation step via corecursion gives a function from X to proofs in $\mathcal{C}$. One possible use of this is to define corecursions that change their character. For example, imagine you want to define a function from X to proofs in $\mathcal{C}$, that first follows some proof translation step α and then it follows another proof translation step β. Then we can define a proof translation step

$\gamma : X \times \{0,1\} \longrightarrow \mathcal{C}$[11] as:

$$\gamma(x,i) = \begin{cases} (\iota^\alpha_x, (w \mapsto \mu^\alpha_x(w), 0)) & \text{if } i = 0 \text{ and the condition is not true,} \\ (\iota^\alpha_x, (w \mapsto \mu^\alpha_x(w), 1)) & \text{if } i = 0 \text{ and the condition is true,} \\ (\iota^\beta_x, (w \mapsto \mu^\beta_x(w), 1)) & \text{if } i = 1. \end{cases}$$

The desired translation of $x \in X$ would be to apply the translation obtained by this proof translation step γ to $(x, 0)$.

4 Cut-Elimination for Non-wellfounded Grz

In this section we illustrate the method of translations between local progress sequent systems developed in Section 3 by proving cut-elimination for a non-wellfounded calculus for Grz. The calculus was introduced by Savateev and Shamkanov [14], who also establish cut-elimination. We briefly introduce the language of Grz and the aforementioned non-wellfounded calculus. We then show how instances of cut can be pushed out of the main fragment of a proof, and apply our method to obtain the cut-elimination result.

The language of Grz is the language of modal logic with primitive symbols $\bot$, $\to$, and $\Box$. We will use p, q, r to denote propositional variables and ϕ, ψ, χ to denote formulas. Γ, Δ, Λ, Π will denote finite multisets of formulas. Sequents are ordered pairs of shape (Γ, Δ), usually denoted by S. Union of sequents is the componentwise union of multisets and is denoted by $S_0 \cup S_1$. We also define the notations:

$$S, \phi^\bullet \text{ to mean } S \cup (\{\phi\}, \varnothing), \qquad S, \phi^\circ \text{ to mean } S \cup (\varnothing, \{\phi\}),$$
$$S, \Gamma^\bullet \text{ to mean } S \cup (\Gamma, \varnothing), \qquad S, \Gamma^\circ \text{ to mean } S \cup (\varnothing, \Gamma).$$

where $\{\phi\}$ is the singleton multiset consisting of ϕ.

Definition 4.1 We define the following rules:

$$\frac{}{S, p^\bullet, p^\circ}\,\mathrm{Ax} \qquad \frac{}{S, \bot^\bullet}\,\bot^\bullet$$

$$\frac{S, \phi^\circ \qquad S, \psi^\bullet}{S, \phi \to \psi^\bullet}\,\to^\bullet \qquad \frac{S, \phi^\bullet, \psi^\circ}{S, \phi \to \psi^\circ}\,\to^\circ$$

$$\frac{S, \phi^\bullet, \Box\phi^\bullet}{S, \Box\phi^\bullet}\,\mathrm{Refl} \qquad \frac{S, \Box\Pi^\bullet, \phi^\circ \qquad \Box\Pi^\bullet, \phi^\circ}{S, \Box\Pi^\bullet, \Box\phi^\circ}\,\Box$$

In each rule the explicitly displayed formula in the conclusion is called *principal*. In the rule $\Box$ the sequent S is called the *weakening part*.

[11] Note that in order to γ being a proof translation step from X to the calculus $\mathcal{C}$ we also need a compatibility condition between α and β. The compatibility condition can be intuitively expressed as: if we find $(x, 0)$ and it meets the condition for changing the behaviour from α to β then $(\iota^\alpha_x, w \mapsto \iota^\beta_{\mu^\alpha_x(w)}\text{-sequent}(\epsilon))$ must be a proof fragment in $\mathcal{C}$.

Then Grz^∞ is the local progress sequent calculus with the rules above such that the only progress occurs in the right premise of the $\Box$ rule. We also define the rule:

$$\frac{S,\phi^\circ \qquad S,\phi^\bullet}{S}\,\mathsf{cut}$$

The system consisting of the rules of Grz^∞ plus cut and where the only progress occurs in the right premise of $\Box$ is denoted as $(\mathsf{Grz}+\mathsf{cut})^\infty$

Thanks to Section 3 we know that τ is an non-wellfounded proof of C iff $(\tau, \sim^\tau)$ ($\sim^\tau$ is defined in Definition 3.5) is a finite-fragmented proof of C. For that reason it will be common to identify both objects and for example talk about the main fragment of τ when we are strictly speaking of the main fragment of $(\tau, \sim^\tau)$. With this in mind, we define the notion of *local height* of a proof π as the height of its main fragment.

Definition 4.2 A function $f : \mathbb{P}^\infty(\mathsf{Grz}+\mathsf{cut})^\infty \longrightarrow \mathbb{P}^\infty(\mathsf{Grz}+\mathsf{cut})^\infty$ is said to:

(i) *Preserve local height* iff the local height of $f(\pi)$ is smaller or equal than the local height of π.

(ii) *Preserve freeness of cuts in the main fragment* iff whenever π is a proof with no cuts in its main fragment, so is $f(\pi)$.

Lemma 4.3 *We have functions* $\mathsf{wk}_{S'}$, $\mathsf{ctr}_{p^\bullet}$, ctr_{p°, $\mathsf{inv}_{\bot^\circ}$, $\mathsf{linv}_{\phi\to\psi^\bullet}$, $\mathsf{rinv}_{\phi\to\psi^\bullet}$, $\mathsf{inv}_{\phi\to\psi^\circ}$, $\mathsf{inv}_{\Box\phi^\circ} : \mathbb{P}^\infty(\mathsf{Grz}+\mathsf{cut})^\infty \longrightarrow \mathbb{P}^\infty(\mathsf{Grz}+\mathsf{cut})^\infty$ *such that:*

(i) *If* $\pi \vdash S$, *then* $\mathsf{wk}_{S'}(\pi) \vdash S, S'$.

(ii) *If* $\pi \vdash S, p^\bullet, p^\bullet$, *then* $\mathsf{ctr}_{p^\bullet}(\pi) \vdash S, p^\bullet$.

(iii) *If* $\pi \vdash S, p^\circ, p^\circ$, *then* $\mathsf{ctr}_{p^\circ}(\pi) \vdash S, p^\circ$.

(iv) *If* $\pi \vdash S, \bot^\circ$, *then* $\mathsf{inv}_{\bot^\circ}(\pi) \vdash S$.

(v) *If* $\pi \vdash S, \phi \to \psi^\bullet$, *then* $\mathsf{linv}_{\phi\to\psi^\bullet}(\pi) \vdash S, \phi^\circ$ *and* $\mathsf{rinv}_{\phi\to\psi^\bullet}(\pi) \vdash S, \psi^\bullet$.

(vi) *If* $\pi \vdash S, \phi \to \psi^\circ$, *then* $\mathsf{inv}_{\phi\to\psi^\circ}(\pi) \vdash S, \phi^\bullet, \psi^\circ$.

(vii) *If* $\pi \vdash S, \Box\phi^\circ$, *then* $\mathsf{inv}_{\Box\phi^\circ}(\pi) \vdash S, \phi^\circ$.

(viii) *All these functions preserve local height and freeness of cuts in the main fragment.*

Proof. All the items (i) to (vii) are provable by induction in the local height of π, (viii) is just an observation of the constructions during the induction. $\Box$

Lemma 4.4 *Let* $\pi \vdash_{(\mathsf{Grz}+\mathsf{cut})^\infty} S, \phi^\circ$ *and* $\tau \vdash_{(\mathsf{Grz}+\mathsf{cut})^\infty} S, \phi^\bullet$ *without any instance of* cut *in their main fragments. Then there is a proof of* $\vdash_{(\mathsf{Grz}+\mathsf{cut})^\infty} S$ *without any instance of* cut *in its main fragment.*

Proof. To any such pair of proofs (π, τ) we can assign an ordinal $n\omega+m$ where n is the rank of ϕ (the cut formula) and m is the sum of the local heights of π and τ. We proceed by strong induction in this ordinal and by cases in the shape of π and τ as follows.

Case 1: either π or τ consists of an initial sequent only. In this case it suffices to consider if S is still an initial sequent and in case it is not use $\mathsf{inv}_{\bot^\circ}$, $\mathsf{ctr}_{p^\bullet}$ or ctr_{p° from Lemma 4.3.

Case 2: the main formulas of π and τ are the cut-formula. There are two subcases, if both formulas are implication it must be that π ends in an application of $\to^\circ$ and τ in an application of $\to^\bullet$. Then, it is easy to see that the cut can be replaced using the I.H. in the premises since the rank of the cut-formula is smaller, applying wk from Lemma 4.3 when needed. The other possibility is that the main formulas are box-formulas. We have that π and τ have shape:

$$\dfrac{\overset{\pi_0}{S_0,\Box\Pi^\bullet,\phi_0^\circ} \quad \overset{\pi_1}{\Box\Pi^\bullet,\phi_0^\circ}}{S_0,\Box\Pi^\bullet,\Box\phi_0^\circ}\,\Box \qquad\qquad \dfrac{\overset{\tau_0}{S_0,\Box\Pi^\bullet,\phi_0^\bullet,\Box\phi_0^\bullet}}{S_0,\Box\Pi^\bullet,\Box\phi_0^\bullet}\,\mathsf{Refl}$$

where $S = S_0,\Box\Pi^\bullet$ and $\phi = \Box\phi_0$. By induction hypothesis applied to $\mathsf{wk}_{\phi_0^\bullet}(\pi),\tau_0$ (the local height of τ_0 is strictly smaller than the local height of τ and the local height of the other proof is smaller or equal that the local height of π by Lemma 4.3) we get a proof $\iota_0 \vdash S_0,\Box\Pi^\bullet,\phi_0^\bullet$ without cuts in its main fragment. We can apply the induction hypothesis to π_0,ι_0 (the rank of ϕ_0 is smaller than the rank of $\Box\phi_0$) and obtain $\iota_1 \vdash S_0,\Box\Pi^\bullet$ with no cuts in its main fragment. This is the desired proof.

Case 3: the principal formula of either π or τ is not the cut-formula. We have five subcases, three of them correspond to the principal formula not being the cut-formula being the result of $\to^\bullet,\to^\circ$ or refl. In these cases we just need to use linv and rinv,inv or wk (from Lemma 4.3), apply the I.H. and then apply $\to^\bullet,\to^\circ$ or refl respectively. We explicitely prove the other two subcases.

The last rule of τ (the case with π is analogous) is $\Box$ and the cut-formula is in the weakening part of the rule instance. π and τ have shape:

$$\overset{\pi}{S_0,\Box\Pi^\bullet,\Box\psi^\circ,\phi^\circ} \qquad\qquad \dfrac{\overset{\tau_0}{S_0,\Box\Pi^\bullet,\psi^\circ,\phi^\bullet} \quad \overset{\tau_1}{\Box\Pi^\bullet,\psi^\circ}}{S_0,\Box\Pi^\bullet,\Box\psi^\circ,\phi^\bullet}\,\Box$$

where $S = S_0,\Box\Pi^\bullet,\Box\psi^\circ$. Let $\iota_0 = \mathsf{inv}_{\Box\psi^\circ}(\pi) \vdash S_0,\Box\Pi^\bullet,\psi^\circ,\phi^\bullet$, we can apply the induction hypothesis to ι_0,τ_0 (the local height is reduced thanks to Lemma 4.3) to obtain $\iota_1 \vdash S_0,\Box\Pi^\bullet,\psi^\circ$ with no cuts in its main fragment. The desired proof is:

$$\dfrac{\overset{\iota_1}{S_0,\Box\Pi^\bullet,\psi^\circ} \quad \overset{\tau_1}{\Box\Pi^\bullet,\psi^\circ}}{S_0,\Box\Pi^\bullet,\Box\psi^\circ}\,\Box$$

The last rule of τ is $\Box$ and the cut-formula is not the principal formula but it is not in the weakening part of the rule instance (the case with π is impossible since in π the cut-formula must be on the right of the sequent, so if it is not the principal formula it appears in the weakening part). π and τ have the following shape:

$$\dfrac{\overset{\pi_0}{S_0,\Box\Pi^\bullet,\Box\psi^\circ,\phi_0^\circ} \quad \overset{\pi_1}{\Box\Pi^\bullet,\phi_0^\circ}}{S_0,\Box\Pi^\bullet,\Box\psi^\circ,\Box\phi_0^\circ}\Box \qquad \dfrac{\overset{\tau_0}{S_1,\Box\Lambda^\bullet,\psi^\circ,\Box\phi_0^\bullet} \quad \overset{\tau_1}{\Box\Lambda^\bullet,\psi^\circ,\Box\phi_0^\bullet}}{S_1,\Box\Lambda^\bullet,\Box\psi^\circ,\Box\phi_0^\bullet}\Box$$

where $S = S_0,\Box\Pi^\bullet,\Box\psi^\circ = S_1,\Box\Lambda^\bullet,\Box\psi^\circ$ (so $S_0,\Box\Pi^\bullet = S_1,\Box\Lambda^\bullet$ (i)) and $\phi = \Box\phi_0$. Note that $\Pi \cup (\Lambda \setminus \Pi) = \Lambda \cup (\Pi \setminus \Lambda)$ and define $S_2 = S_0 \cap S_1$, so $S_0 = S_2,\Box(\Lambda \setminus \Pi)^\bullet$ and $S_1 = S_2,\Box(\Pi \setminus \Lambda)^\bullet$. Let $\pi' = \mathsf{inv}_{\Box\psi^\circ}(\pi) \vdash S_0,\Box\Pi^\bullet,\psi^\circ,\Box\phi_0^\circ$. Due to (i) we can apply the induction hypothesis to π', τ_0 (the local height is reduced and there are no cuts in the main fragments thanks to Lemma 4.3 and because we are in the main fragment of τ still) and obtain a proof $\iota_0 \vdash S_0,\Box\Pi^\bullet,\psi^\circ = S_2,\Box\Pi^\bullet,\Box(\Lambda \setminus \Pi)^\bullet,\psi^\circ$ with no cuts in its main fragment. Let ρ be the proof:

$$\dfrac{\dfrac{\overset{\mathsf{wk}_{\Box(\Lambda\setminus\Pi)^\bullet,\psi^\circ}(\pi_1)}{\Box\Pi^\bullet,\Box(\Lambda \setminus \Pi)^\bullet,\psi^\circ,\phi_0^\circ} \quad \overset{\pi_1}{\Box\Pi^\bullet,\phi_0^\circ}}{\Box\Pi^\bullet,\Box(\Lambda \setminus \Pi)^\bullet,\psi^\circ,\Box\phi_0^\circ}\Box \qquad \overset{\mathsf{wk}_{\Box(\Pi\setminus\Lambda)^\bullet}(\tau_1)}{\Box\Pi^\bullet,\Box(\Lambda \setminus \Pi)^\bullet,\psi^\circ,\Box\phi_0^\bullet}}{\Box\Pi^\bullet,\Box(\Lambda \setminus \Pi)^\bullet,\psi^\circ}\mathsf{cut}$$

Then, the desired proof is

$$\dfrac{\overset{\iota_0}{S_2,\Box\Pi^\bullet,\Box(\Lambda \setminus \Pi)^\bullet,\psi^\circ} \quad \overset{\rho}{\Box\Pi^\bullet,\Box(\Lambda \setminus \Pi)^\bullet,\psi^\circ}}{S_2,\Box\Pi^\bullet,\Box(\Lambda \setminus \Pi)^\bullet,\Box\psi^\circ}\Box$$

We notice that the constructed proof does not have any cuts in its main fragment, since the right premise of $\Box$ is outside the main fragment of the proof. $\Box$

Lemma 4.5 *There is a function* $\mathsf{cuts\text{-}up} : \mathbb{P}^\infty(\mathsf{Grz}+\mathsf{cut})^\infty \longrightarrow \mathbb{P}^\infty(\mathsf{Grz}+\mathsf{cut})^\infty$ *such that if* $\pi \vdash S$ *then* $\mathsf{cuts\text{-}up}(\pi) \vdash S$ *and* $\mathsf{cuts\text{-}up}(\pi)$ *has no instances of* cut *in its main fragment.*

Proof. This is a simple induction in the number of cuts in the main fragment of π. For the inductive step just go to a top-most cut in the main fragment (i.e. a cut without any cuts that are above it in the main fragment) and apply Lemma 4.4 to get rid of that cut. $\Box$

Thanks to the tools of Section 3 the cut-elimination proof from Lemma 4.5 is straightforward.

Theorem 4.6 *If* $\vdash_{(\mathsf{Grz}+\mathsf{cut})^\infty} S$*, then* $\vdash_{\mathsf{Grz}^\infty} S$*.*

Proof. Define the function $\alpha : \mathbb{P}^{\mathrm{ff}}(\mathsf{Grz}+\mathsf{cut})^\infty \longrightarrow \mathcal{T}(\mathbb{P}^{\mathrm{ff}})$ that, given $\pi = (\tau, \sim)$, outputs $(\iota, \mu) = \mathrm{destruct}(\tau', \sim^{\tau'})$ where $\tau' = \mathsf{cuts\text{-}up}(\tau)$. Since ι is the main fragment of $\mathsf{cuts\text{-}up}(\tau)$, it is a proof-fragment in $(\mathsf{Grz}+\mathsf{cut})^\infty$. Since there are no instances of cut in ι by Lemma 4.5, and $(\mathsf{Grz}+\mathsf{cut})^\infty$ only progresses in the same rule instances as in Grz^∞, it is clear that ι is a proof fragment of Grz^∞. Since $\mu(w)$ is by definition a subtree of $\mathsf{cuts\text{-}up}$ which is a proof in $(\mathsf{Grz} + \mathsf{cut})^\infty$ we have that $\mu(w)$ is also a proof in $(\mathsf{Grz} + \mathsf{cut})^\infty$. We can conclude that the conditions to apply Theorem 3.8 are fulffiled. Hence we obtain a proof translation from $(\mathsf{Grz}+\mathsf{cut})^\infty$ to Grz^∞ that does not change the conclusion sequence (by definition of the α) so we have the desired result. □

We conclude that in any local progress sequent calculi where cuts can be pushed outside the main fragment, we can show cut-elimination. In particular, we remark that cut-elimination can be proven for weak Grzegorczyk modal logic wGrz (see [15] to see a non-wellfounded proof system for this logic and a proof of its cut-elimination using ultrametric spaces) by adapting the function $\mathsf{cuts\text{-}up}$ accordingly. This implies that cut-elimination for local progress sequent calculi is quite close to cut-elimination of finitary systems since the main fragment is always finite.

5 Final remarks and future work

We have introduced a uniform method to define proof translations for non-wellfounded proofs with a local progress condition. Our method is based on the observation that it suffices to translate a finite fragment of a proof and then define from such a translation step a map by corecursion. The translation method is uniform in the sense that the specific proof systems, between which a translation shall be defined, as well as the specific type of translation, does not matter. We have illustrated the method by proving cut-elimination for a non-wellfounded calculus for Grz.

A natural direction for further research is to study the relationship between our corecursive method for cut-elimination and the ultrametric approach by Shamkanov and Savateev. At the time of writing this paper it is unknown to us how exactly these two method relate and whether using ultrametric spaces is strictly more powerful than our approach. Another direction that we are currently pursuing is to explore which logics can be given a complete non-wellfounded local progress calculus. The standard examples are GL and Grz, however we suspect that many modal fixed point logics belonging to the alternation free fragment of the modal mu-calculus may also be given a local progress calculus. For example modal logic with the master modality admits such a calculus [12], leading to the question whether we can establish cut-elimination for this calculus using our corecursive method. Finally, another pressing open question is whether a similar method for proof translations can be defined for non-wellfounded calculi based on a trace condition. We expect this to be a difficult problem, since it is generally challenging to ensure that traces are preserved when translating proofs.

[12] Adding ordinal annotations, which may complicate the use of this method.

References

[1] Afshari, B., L. Grotenhuis, G. E. Leigh and L. Zenger, *Ill-founded proof systems for intuitionistic linear-time temporal logic*, in: R. Ramanayake and J. Urban, editors, *Automated Reasoning with Analytic Tableaux and Related Methods* (2023), pp. 223–241.

[2] Afshari, B. and J. Kloibhofer, *Cut elimination for cyclic proofs: A case study in temporal logic*, in: *Proceedings Twelfth International Workshop on Fixed Points in Computer Science* (to appear).

[3] Afshari, B., G. E. Leigh and G. Menéndez Turata, *A cyclic proof system for full computation tree logic*, in: *31st EACSL Annual Conference on Computer Science Logic (CSL 2023)*, Schloss Dagstuhl-Leibniz-Zentrum für Informatik, 2023.

[4] Baelde, D., A. Doumane and A. Saurin, *Infinitary Proof Theory: the Multiplicative Additive Case*, in: J.-M. Talbot and L. Regnier, editors, *25th EACSL Annual Conference on Computer Science Logic (CSL 2016)*, Leibniz International Proceedings in Informatics (LIPIcs) **62** (2016), pp. 42:1–42:17.
URL
`https://drops-dev.dagstuhl.de/entities/document/10.4230/LIPIcs.CSL.2016.42`

[5] Brotherston, J. and A. Simpson, *Complete sequent calculi for induction and infinite descent*, in: *Proceedings of LICS-22* (2007), pp. 51–60.

[6] Bucheli, S., R. Kuznets and T. Studer, *Two ways to common knowledge*, Electronic Notes in Theoretical Computer Science **262** (2010), pp. 83–98, proceedings of the 6th Workshop on Methods for Modalities (M4M-6 2009).
URL `https://www.sciencedirect.com/science/article/pii/S1571066110000290`

[7] Das, A. and D. Pous, *Non-Wellfounded Proof Theory For (Kleene+Action)(Algebras+Lattices)*, in: D. R. Ghica and A. Jung, editors, *27th EACSL Annual Conference on Computer Science Logic (CSL 2018)*, Leibniz International Proceedings in Informatics (LIPIcs) **119** (2018), pp. 19:1–19:18.
URL
`https://drops-dev.dagstuhl.de/entities/document/10.4230/LIPIcs.CSL.2018.19`

[8] Docherty, S. and R. N. S. Rowe, *A non-wellfounded, labelled proof system for propositional dynamic logic*, in: *Automated Reasoning with Analytic Tableaux and Related Methods: 28th International Conference, TABLEAUX 2019, London, UK, September 3-5, 2019, Proceedings* (2019), p. 335–352.
URL `https://doi.org/10.1007/978-3-030-29026-9_19`

[9] Kokkinis, I. and T. Studer, *Cyclic proofs for linear temporal logic*, in: D. Probst and P. Schuster, editors, *Concepts of Proof in Mathematics, Philosophy, and Computer Science*, De Gruyter, Berlin, Boston, 2016 pp. 171–192.
URL `https://doi.org/10.1515/9781501502620-011`

[10] Marti, J. and Y. Venema, *A focus system for the alternation-free μ-calculus*, in: *Automated Reasoning with Analytic Tableaux and Related Methods: 30th International Conference, TABLEAUX 2021, Birmingham, UK, September 6–9, 2021, Proceedings* (2021), p. 371–388.
URL `https://doi.org/10.1007/978-3-030-86059-2_22`

[11] Niwiński, D. and I. Walukiewicz, *Games for the μ-calculus*, Theoretical Computer Science **163** (1996), pp. 99–116.
URL `https://www.sciencedirect.com/science/article/pii/0304397595001360`

[12] Rooduijn, J. M. W. and L. Zenger, *An analytic proof system for common knowledge logic over s5*, in: *David Fernández-Duque, Alessandra Palmigiano and Sophie Pinchinat (eds.) Advances in Modal Logic*, 2022, pp. 659–680.

[13] Saurin, A., *A linear perspective on cut-elimination for non-wellfounded sequent calculi with least and greatest fixed-points*, in: R. Ramanayake and J. Urban, editors, *Automated Reasoning with Analytic Tableaux and Related Methods* (2023), pp. 203–222.

[14] Savateev, Y. and D. Shamkanov, *Non-well-founded proofs for the Grzegorczyk modal logic*, The Review of Symbolic Logic **14** (2018).

[15] Savateev, Y. and D. Shamkanov, *Cut elimination for the weak modal grzegorczyk logic via non-well-founded proofs*, in: R. Iemhoff, M. Moortgat and R. de Queiroz, editors, *Logic, Language, Information, and Computation* (2019), pp. 569–583.

[16] Shamkanov, D. S., *Circular proofs for the Gödel-Löb provability logic*, Mathematical Notes **96** (2014), pp. 575–585.
URL `http://dx.doi.org/10.1134/S0001434614090326`

[17] Simpson, A., *Cyclic arithmetic is equivalent to peano arithmetic*, in: J. Esparza and A. S. Murawski, editors, *Foundations of Software Science and Computation Structures* (2017), pp. 283–300.

[18] Stirling, C., *A proof system with names for modal mu-calculus*, Electronic Proceedings in Theoretical Computer Science **129** (2013).

[19] Studer, T., *On the proof theory of the modal mu-calculus*, Studia Logica **89** (2008), pp. 343–363.
URL `http://www.jstor.org/stable/40268983`

[20] Turata, G. M., *Cyclic proof systems for modal fixpoint logics*, ILLC Dissertation series (2024).

[21] Venema, Y., *Algebras and coalgebras*, in: P. Blackburn, J. Van Benthem and F. Wolter, editors, *Handbook of Modal Logic*, Studies in Logic and Practical Reasoning **3**, Elsevier, 2007 pp. 331–426.
URL `https://www.sciencedirect.com/science/article/pii/S1570246407800097`

Projectivity Meets Uniform Post-Interpolant: Classical and Intuitionistic Logic

Mojtaba Mojtahedi [1] Konstantinos Papafilippou [2]

Ghent University

Abstract

We examine the interplay between projectivity (in the sense that was introduced by S. Ghilardi) and uniform post-interpolant for the classical and intuitionistic propositional logic. More precisely, we explore whether a projective substitution of a formula is equivalent to its uniform post-interpolant, assuming the substitution leaves the variables of the interpolant unchanged. We show that in classical logic, this holds for all formulas. Although such a nice property is missing in intuitionistic logic, we provide Kripke semantical characterisation for propositions with this property.
As a main application of this, we show that the unification type of some extensions of intuitionistic logic are finitary. In the end, we study admissibility for intuitionistic logic, relative to some sets of formulae.
The first author of this paper recently considered a particular case of this relativised admissibility and found it useful in characterising the provability logic of Heyting Arithmetic.

Keywords: Unification, Projectivity, Admissible Rules, Intuitionistic Logic, Uniform Interpolation.

1 Introduction

The notion of projectivity for propositional logics was first introduced by S. Ghilardi [6,7,8]. Roughly speaking, a formula A is projective if there is a substitution θ which unifies A, i.e. $\vdash \theta(A)$, and $A \vdash \theta(B) \leftrightarrow B$ for every formula B. The importance of projective formulas is mainly due to the following observation: any projective formula, has a single unifier which is more general than all other unifiers. Then, by approximating formulas with projective ones, we may fully describe *all unifiers* of a given formula [7,8,4,1]. Having this strong tool in hand, we may then characterise admissible rules [3] [10,11,12,5,9,18] for a logic. The situation for classical logic is simple: Every satisfiable formula is

[1] Email: mojtahedy@gmail.com. This work is partially funded by FWO grant G0F8421N and BOF grant BOF.STG.2022.0042.01.

[2] Email: Konstantinos.Papafilippou@uGent.be. Funded by the FWO-FWF Lead Agency grant G030620N (FWO)/I4513N (FWF) and by the SNSF–FWO Lead Agency Grant 200021L_196176/G0E2121N.

[3] An inference rule A/B is admissible if every unifier of A is also a unifier of B.

also projective. Hence no approximation is required and also no non-derivable admissible rule exists (A/B is admissible iff $A \to B$ is derivable).

Additionally, we may consider propositional language with two sorts of atomics: variables and parameters (constants), annotated respectively by var and par. Variables are intended to be substituted, while parameters are considered as constants and are never substituted. Again one may wonder about the previous questions of unification [4] and admissibility [14,13] for this two-sorted language. But this time, even for classical logic, not all satisfiable formulas are projective. For example the formula $A := p \wedge x$ for atomic variable x and parameter p is satisfiable, while it is not unifiable. This will rule out A being projective only because of it not being unifiable. To mend this, we define the notion of E-projectivity by replacing condition $\vdash \theta(A)$ by $\vdash \theta(A) \leftrightarrow E$ in above definition. This time in our example, the formula A is p-projective. In this paper, we first observe that A can only be E-projective for a unique E (Lemma 7), which is annotated as A^*. Then we show (Theorem 3) that all A's are A^*-projective in classical logic. For intuitionistic logic, we find a Kripke semantical characterisation for those A's that are A^*-projective (Theorem 9).

On the other hand, we have the well-studied notion of the *uniform post-interpolant*. Given a formula A, the uniform post-interpolant of A with respect to par, is the strongest formula B in the variable-free language such that $A \to B$ holds. Of course, it might be the case that such a *strongest* formula does not exist at all. Fortunately, several logics indeed have uniform post-interpolants, namely intuitionistic logic [20] and all locally tabular (finite) logics like classical logic. Interestingly, it turns out (Theorem 7) that A^* is equal to the uniform post-interpolant of A. As an application, we show that extensions of intuitionistic logic with variable-free formulas have a finitary unification type. More precisely, in the mentioned logics, we show that every unifiable formula has a finite complete set of unifiers, i.e. a finite set of unifiers such that every unifier is less general than at least one of them. In the end, we observe that admissibility relative to the set of all variable-free formulas is trivial: $A \mathrel{|\!\!\sim}^{\mathsf{par}} B$[4] iff $\vdash A \to B$ for the intuitionistic logic.

Motivations for the study of relativised projectivity and admissibility come from their applications in the study of the provability logic of HA[5] [15]. More precisely, the first author studied projectivity and admissibility relative to the set NNIL[6] [16] and used it in a crucial way [15] to axiomatize the provability logic of HA. Interestingly, [3,2] have also found relative unification beneficiary for their study. Nevertheless, they consider the non-parametric language. The main aim of this paper is to look at this relativised notion of projectivity and admissibility, in some more general viewpoint. In this direction, Theorem 5 generalises the previous result in [16] from NNIL to all finite sets of formulas

---

[4] $A \mathrel{|\!\!\sim}^{\mathsf{par}} B$ is defined as follows: for every variable-free E and substitution θ such that $\vdash E \to \theta(A)$ we have $\vdash E \to \theta(B)$.

[5] Heyting Arithmetic

[6] NNIL is the set of formulas with No Nested Implications on the Left [21].

Γ. However, it appears that there is no straightforward generalisation for the characterisation of admissibility relative to NNIL.

2 Definitions and basic facts

Language

In the sequel, we assume that par and var are two sets of parameters and variables, respectively. Unless otherwise stated, we assume that both sets var and par are infinite. Then the propositional language $\mathcal{L}$, is the set of all Boolean combinations of atomic propositions $\mathsf{atom} := \mathsf{var} \cup \mathsf{par}$. Boolean connectives are $\bot$, $\wedge$, $\vee$ and $\rightarrow$. Note that all connectives are binary, except for falsity $\bot$ which is nullary, i.e. an operator without argument. Then $\top$, $\neg A$ and $A \leftrightarrow B$ are shorthand for $\bot \rightarrow \bot$, $A \rightarrow \bot$ and $(A \rightarrow B) \wedge (B \rightarrow A)$, respectively. Sets of formulas are indicated by Γ, Δ, Π, X and Y. We use $x, y, \ldots$, and $p, q, \ldots$, and $a, b, \ldots$, and $A, B, C, D, E, F, \ldots$ as meta-variables for variables, parameters, atomics and formulas, respectively. Optionally, we may also use subscripts for them. Let $\mathcal{L}(X)$ indicate all Boolean combinations of propositions in X. Hence, for example, $\mathcal{L}(\mathsf{var} \cup \mathsf{par})$ is equal to $\mathcal{L}$. Moreover, Γ^{par} indicates the set $\Gamma \cap \mathcal{L}(\mathsf{par})$. We use $\Gamma \vdash^{\mathsf{i}} A$ to indicate the intuitionistic derivability of A from the set Γ. We also use $\Gamma \vdash^{\mathsf{c}} A$ for the classical derivability. Whenever we state a definition for both intuitionistic and classical cases, we simply use $\vdash$ without superscript.

A par-*extension* of a logic is an extension of that logic by adding some $E \in \mathcal{L}(\mathsf{par})$ to its set of axioms. Note that this additional axiom is not considered as a schema.

Kripke models

A Kripke model is a tuple $\mathcal{K} = (W, \preccurlyeq, V)$, whose frame $(W, \preccurlyeq)$, is a reflexive transitive ordering. We let $w \prec u$ stand for $w \preccurlyeq u$ and $w \neq u$. Moreover, we use $\succcurlyeq$ and $\succ$ as inverses of the relations $\preccurlyeq$ and $\prec$, respectively. A root of $\mathcal{K}$ is a minimum element in $(W, \preccurlyeq)$ if it exists, and we say that $\mathcal{K}$ is rooted if it has a root. We say that $\mathcal{K}$ is finite if W is so. In this paper, it is assumed that all Kripke models are *finite* and *rooted.* We use the notation $\mathcal{K}, w \Vdash A$ for the validity/forcing of A at the node w in the model $\mathcal{K}$. $\mathcal{K} \Vdash A$ is defined as $\forall w \in W \ (\mathcal{K}, w \Vdash A)$. Given $w \in W$, we define $\mathcal{K}_w$ as the restriction of $\mathcal{K}$ to all nodes that are accessible from w. Furthermore, $\mathcal{K}(w)$ indicates the valuation of $\mathcal{K}$ at w, i.e. the set of all atomics a such that $w \ V \ a$.

Substitutions

Substitutions are functions $\theta : \mathcal{L} \longrightarrow \mathcal{L}$ with the following properties: (they are indicated by lowercase Greek letters $\theta, \tau, \lambda, \gamma, \ldots$)

- θ commutes with connectives. This means that for every binary connective $\circ$ we have $\theta(A \circ B) = \theta(A) \circ \theta(B)$ and for $\bot$ we have $\theta(\bot) = \bot$.
- $\theta(p) = p$ for every $p \in \mathsf{par}$.

Given two substitutions θ and γ, we define $\theta\gamma$ as the composition of θ and γ: $\theta\gamma(a) := \theta(\gamma(a))$. Given a substitution θ and the Kripke model $\mathcal{K} = (W, \preccurlyeq, V)$, define $\theta^*(\mathcal{K})$ as a Kripke model with the same frame of $\mathcal{K}$, and with valuation V' as follows: $w\ V'\ a$ iff $\mathcal{K}, w \Vdash \theta(a)$. Then it can be easily observed that $\mathcal{K}, w \Vdash \theta(A)$ iff $\theta^*(\mathcal{K}), w \Vdash A$, for every $w \in W$ and $A \in \mathcal{L}$. If no confusion is likely[7], we may simply use $\theta(\mathcal{K})$ instead of $\theta^*(\mathcal{K})$.

Given two substitutions θ and γ, we say that θ is more general than γ (γ less general than θ), annotated as $\gamma \leq \theta$, if there is a substitution λ such that $\vdash \gamma(x) \leftrightarrow \lambda\theta(x)$.

Relative projectivity

An *A-identity* is some θ such that

$$\forall a \in \mathsf{atom} \quad A \vdash \theta(a) \leftrightarrow a. \tag{1}$$

Let E be a proposition in the parametric language $\mathcal{L}(\mathsf{par})$. An E-fier of A is some θ such that $\vdash \theta(A) \leftrightarrow E$. An E-fier for some $E \in \Gamma^{\mathsf{par}} := \Gamma \cap \mathcal{L}(\mathsf{par})$ is also called a Γ-fier. Note that here Γ is not necessarily a subset of $\mathcal{L}(\mathsf{par})$. $\top$-fiers and $\mathcal{L}$-fiers are also called *unifiers* and *parametrifiers*, respectively. We say that A is *E-projective* (Γ-projective), if there is some θ which is an A-identity and E-fier (Γ-fier) of A. In this case we say that θ is a projective E-fier (Γ-fier) of A and E is *the* projection of A. As we will see in Lemma 7, the projection is unique, modulo provable equivalence. So we use the notation A^* for its unique projection, if it exists. Then $\mathcal{L}$-projectivity is also called par-projectivity. The well-known notion of *projectivity*, as introduced by S. Ghilardi [6,7], coincided with $\top$-projectivity in this general setting.

Given $A \in \mathcal{L}$ and a set Θ of substitutions, we say that Θ is a complete set of unifiers of A, if the following holds: (1) every $\theta \in \Theta$ is a unifier of A, (2) every unifier of A is less general than some $\theta \in \Theta$. We say that the unification type of a logic is:

- unitary, if every unifiable formula of the logic, has a one-element complete set of unifiers,
- finitary, if every unifiable formula of the logic, has a finite complete set of unifiers.

Note that in our definition, if the unification type of a logic is unitary, it is also finitary.

All the above definitions of projectivity rely on a *background logic* in the context. More precisely, we have two different notions of Γ-projectivity: one for classical logic and one for intuitionistic logic. To simplify the notation, we always hide the dependency on the logic. This will not cause much confusion, since it is quite clear from the context which logic we are talking about. To be more precise, in this section, we provide all statements for both classical and intuitionistic logic. Later in section 3, classical logic is our background logic, whereas for the rest of the paper, the background logic is intuitionistic logic.

---

[7] Confusions can be due to the fact that $(\theta\gamma)^* = \gamma^*\theta^*$.

Lemma 2.1 (Uniqueness of projections) *Projection is unique, modulo provable equivalence.*

Proof. Let $\vdash \theta_i(A) \leftrightarrow B_i$ and $A \vdash \theta_i(x) \leftrightarrow x$ and $B_i \in \mathcal{L}(\mathsf{par})$ for $i = 1, 2$ and every $x \in \mathsf{var}$. Then $A \vdash \theta_1(A) \leftrightarrow A$ and thus $\vdash A \to B_1$. Therefore, we have $\vdash \theta_2(A \to B_1)$ and since θ_2 is identity over the language $\mathcal{L}(\mathsf{par})$, we may deduce $\vdash \theta_2(A) \to B_1$, and thus $\vdash B_2 \to B_1$. By a similar argument we may prove $\vdash B_1 \to B_2$, and thus $\vdash B_1 \leftrightarrow B_2$. □

Uniform post-interpolant and upward approximations

Given a formula A, we say that B is its uniform post-interpolant with respect to par, if we have (1) $B \in \mathcal{L}(\mathsf{par})$, (2) $\vdash A \to B$ and (3) for every $C \in \mathcal{L}(\mathsf{par})$ s.t. $\vdash A \to C$ we have $\vdash B \to C$. It can be easily proved that such a B is unique (modulo provable equivalence), if it exists. Hence, we use the notation $\lceil A \rceil$ for the uniform post-interpolant of A with respect to par. It is a well-known fact that the uniform post-interpolant for classical and intuitionistic logic exists. In the following theorem, we will show that the unique projection is equal to the uniform post-interpolant with respect to the set par of atomics.

Theorem 2.2 *For every* par*-projective A, we have $A^* = \lceil A \rceil$.*

Proof. Let $A \vdash \theta(x) \leftrightarrow x$ for every $x \in \mathsf{var}$ and $\vdash \theta(A) \leftrightarrow A^*$ and $A^* \in \mathcal{L}(\mathsf{par})$. We will check that all 3 required conditions for it to be a uniform post-interpolant hold. (1) Trivially, we have $A^* \in \mathcal{L}(\mathsf{par})$. (2) By the argument provided in the proof of Lemma 7, we have $\vdash A \to A^*$. (3) Let $C \in \mathcal{L}(\mathsf{par})$ s.t. $\vdash A \to C$. Hence $\vdash \theta(A) \to \theta(C)$ and since $C \in \mathcal{L}(\mathsf{par})$ we get $\theta(C) = C$. Thus, $\vdash A^* \to C$. □

Given $A \in \mathcal{L}$, one might wonder if par-projectivity of A could be defined using the standard notion of projectivity. We will prove that par-projectivity of A is equivalent to projectivity of $A \leftrightarrow \lceil A \rceil$, for both classical and intuitionistic logic. One direction is easy and stated in the following lemma. However, for the other direction, we provide separate proofs for intuitionistic and classical logic. For intuitionistic logic, we take advantage of the semantic characterisation of par-projectivity (see Corollary 9). For classical logic, Lemma 3 and Lemma 3 imply that $A \leftrightarrow \lceil A \rceil$ is projective for every A!

Lemma 2.3 *Every projective unifier of $A \leftrightarrow \lceil A \rceil$ is also a projective parametrifier of A. Therefore, projectivity of $A \leftrightarrow \lceil A \rceil$ implies* par*-projectivity of A.*

Proof. Let θ be a projective unifier of $A \leftrightarrow \lceil A \rceil$. Hence $\vdash \theta(A \leftrightarrow \lceil A \rceil)$ and thus $\vdash \theta(A) \leftrightarrow \lceil A \rceil$ as $\lceil A \rceil$ contains only parameters. This shows that θ is a parametrifier of A. Also, for every variable x we have $A \leftrightarrow \lceil A \rceil \vdash \theta(x) \leftrightarrow x$. Since $\vdash A \to \lceil A \rceil$, we have $\lceil A \rceil \to A \vdash \theta(x) \leftrightarrow x$. Hence, a fortiori, we have $A \vdash \theta(x) \leftrightarrow x$. Thus, θ is a projective parametrifier of A. □

Theorem 2.4 *Every projective unifier of $A \leftrightarrow \lceil A \rceil$ is a most general $\lceil A \rceil$-fier of A.*

Proof. Let θ be a projective unifier of $A \leftrightarrow \lceil A \rceil$. It is obvious that θ is also a $\lceil A \rceil$-fier of A. To show that it is a most general $\lceil A \rceil$-fier of A, let γ be some

$\lceil A \rceil$-fier of A. Hence γ is also a unifier of $A \leftrightarrow \lceil A \rceil$. Since projective unifiers are most general unifiers, we may infer that γ is less general than θ. □

3 $\mathcal{L}$-Projectivity for Classical Logic

In this section we will prove that in Classical Logic, every $A \in \mathcal{L}$ is par-projective. First, let us see some definitions.

An interpretation is a function $I : \mathcal{L} \longrightarrow \{0,1\}$ (1 for truth and 0 for falsity) such that: $I(\bot) = 0$, $I(A \wedge B) = \min(I(A), I(B))$, $I(A \vee B) = \mathsf{max}(I(A), I(B))$ and $I(A \to B) = \mathsf{max}(1 - I(A), I(B))$. For an interpretation I we use the notation $I \models A$ for $I(A) = 1$. By soundness and completeness of classical propositional logic, we have $\vdash^{\mathsf{c}} A$ iff for every interpretation I we have $I \models A$.

We say that an atomic x is positive in A if for all interpretations I and J such that $I(x) \leq J(x)$ and $I(a) = J(a)$ for other atomics a, we have $I(A) \leq J(A)$. Furthermore, $A[x : B]$ denotes the substitution of B for x in A.

Lemma 3.1 *A variable x is positive in A iff $\vdash^{\mathsf{c}} A \to A[x : \top]$.*

Proof. Left-to-right: Let x be positive in A and $I \models A$ seeking to show that $I \models A[x : \top]$. Let J be the interpretation that is always equal to I except for $J(x) := 1$. Then obviously $I(x) \leq J(x)$ and thus $I(A) \leq J(A)$. Since $I \models A$ we get $J \models A$ and thus $I \models A[x : \top]$.
Right-to-left: Let $\vdash^{\mathsf{c}} A \to A[x : \top]$, $I(x) < J(x)$ (note that this implies $J \models x$) and $I(a) = J(a)$ elsewhere. Assume $I \models A$ to show $J \models A$. Since $I \models A \to A[x : \top]$ we get $I \models A[x : \top]$, which is just $J \models A$. □

Given $A \in \mathcal{L}$ and some $x \in \mathsf{var}$, define the substitution θ^A_x as follows:

$$\theta^A_x(y) := \begin{cases} y & : y \neq x \\ A \wedge y & : y = x \end{cases}$$

Lemma 3.2 *Given A, the variable x is positive in $\theta^A_x(A)$. Furthermore, if y is positive in A, then it is also positive in $\theta^A_x(A)$.*

Proof. Let $I \not\models x$ and $J \models x$ and $I(a) = J(a)$ for every atomic $a \neq x$. Moreover, assume that $I \models \theta^A_x(A)$, seeking to show $J \models \theta^A_x(A)$. If $I(A) = J(A)$, then obviously $I(\theta^A_x(A)) = J(\theta^A_x(A))$ and we are done. Hence, we have the following cases:

- $I \models A$ and $J \not\models A$. Then we have $I(\theta^A_x(x)) = J(\theta^A_x(x)) = I(x) = 0$ and thus $I(\theta^A_x(A)) = J(\theta^A_x(A)) = 1$.
- $I \not\models A$ and $J \models A$. Then we have $J(\theta^A_x(x)) = J(x)$ and thus $J(\theta^A_x(A)) = J(A)$. This means that $J \models \theta^A_x(A)$.

At the end, we must show that if y is positive in A, then it is also positive in $\theta^A_x(A)$. Let $I \not\models y$ and $J \models y$ and $I(a) = J(a)$ for every atomic $a \neq y$. Moreover, assume that $I \models \theta^A_x(A)$, seeking to show $J \models \theta^A_x(A)$. If $I(A) = J(A)$, then obviously $I(\theta^A_x(A)) = J(\theta^A_x(A))$ and we are done. So we have the following cases:

- $I \models A$ and $J \not\models A$. This case contradicts the positiveness of y in A.

- $I \not\models A$ and $J \models A$. Then we have $J(\theta_x^A(x)) = J(x)$ and thus $J(\theta_x^A(A)) = J(A)$. This means that $J \models \theta_x^A(A)$.

□

Lemma 3.3 *The following items are equivalent:*

(i) *A is unifiable.*

(ii) *A is projective.*

(iii) $\lceil A \rceil = \top$.

Proof. (i)⇒(ii): Let $\vdash \tau(A)$ and define the substitution ϵ_τ as follows:

$$\epsilon_\tau(x) := (A \wedge x) \vee (\neg A \wedge \tau(x)) \quad \text{for every variable } x.$$

It is obvious that ϵ_τ is A-identity. Moreover one may easily prove that both $A \vdash^{\mathsf{c}} \epsilon_\tau(A)$ and $\neg(A) \vdash^{\mathsf{c}} \epsilon_\tau(A)$ and thus $\vdash^{\mathsf{c}} \epsilon_\tau(A)$.

(ii)⇒(iii): This holds by Theorem 7.

(iii)⇒(i): We use induction on n_A, the number of non-positive variables occurring in A. First, observe that if $n_A = 0$, then for the substitution τ which replaces all variables with $\top$, by Lemma 3 we have $\vdash A \to \tau(A)$ and $\tau(A) \in \mathcal{L}(\mathsf{par})$. Then, since $\lceil A \rceil = \top$, we get $\vdash^{\mathsf{c}} \tau(A)$, and thus A is unifiable. As induction hypothesis, assume that every formula B with $n_B \leq n_A - 1$ and $\lceil B \rceil = \top$ is unifiable. Take some non-positive variable x in A and let $B := \theta_x^A(A)$. Since $A \vdash \theta_x^A(y) \leftrightarrow y$ for every y, we get $\vdash A \to B$ and thus $\lceil B \rceil = \top$. Moreover, Lemma 3 implies that $n_B \leq n_A - 1$ and hence by the induction hypothesis, there is a substitution γ which unifies B. Therefore, $\gamma \circ \theta_x^A$ unifies A, as desired. □

Lemma 3.4 $\lceil \lceil A \rceil \leftrightarrow A \rceil = \top$.

Proof. Since $\vdash^{\mathsf{c}} A \to \lceil A \rceil$, we have $\lceil \lceil A \rceil \leftrightarrow A \rceil = \lceil \lceil A \rceil \to A \rceil$. So it is enough to show $\lceil \lceil A \rceil \to A \rceil = \top$. Let $E \in \mathcal{L}(\mathsf{par})$ be such that $\vdash^{\mathsf{c}} (\lceil A \rceil \to A) \to E$. It suffices to show that $\vdash^{\mathsf{c}} E$. By assumption, we have $\vdash^{\mathsf{c}} A \to E$, and thus $\vdash^{\mathsf{c}} \lceil A \rceil \to E$. On the other hand, $\vdash^{\mathsf{c}} (\lceil A \rceil \to A) \to E$ also implies $\vdash^{\mathsf{c}} \neg \lceil A \rceil \to E$. Thus $\vdash^{\mathsf{c}} E$. □

Theorem 3.5 *All formulas are* par*-projective. Moreover, $\lceil A \rceil \leftrightarrow A$ for every A is projective.*

Proof. Given $A \in \mathcal{L}$, let $B := \lceil A \rceil \leftrightarrow A$. Lemma 3 implies that $\lceil B \rceil = \top$. Hence Lemma 3 implies that B is projective. Thus, Lemma 7 gives the desired result. □

Recall from section 2 that a par-extension of a logic is an extension by an $\mathcal{L}(\mathsf{par})$-formula.

Corollary 3.6 *The unification type of* par*-extensions of classical logic is unitary.*

Proof. Let $A \in \mathcal{L}$ and $E \in \mathcal{L}(\mathsf{par})$ and assume that $B := E \to A$ is unifiable, seeking to find a single unifier of B that is more general than all of its unifiers.

Since B is unifiable, there is some γ such that $\vdash^{\mathsf{c}} \gamma(B)$, and hence $\lceil B \rceil = \top$. Then Lemma 3 implies that B is projective. Let θ be its projective unifier. We claim that θ is also its most general unifier. Take any unifier τ of B. Since $B \vdash^{\mathsf{c}} \theta(x) \leftrightarrow x$, we get $\tau(B) \vdash^{\mathsf{c}} \tau\theta(x) \leftrightarrow \tau(x)$ and thus $\vdash^{\mathsf{c}} \tau\theta(x) \leftrightarrow \tau(x)$. This shows that τ is less general than θ. □

4 Γ-Projectivity for Intuitionistic Logic

As we have shown in previous section, classical logic is well-behaving for par-projectivity in the following sense: all formulas are classically par-projective (Theorem 3). Nevertheless, such a nice behaviour does not hold in intuitionistic logic. For example the formula $A := x \vee \neg x$, is not par-projective by the following argument. First, observe that $\lceil A \rceil = \top$.[8] Hence, a parametrifier of A should be a unifier. By the disjunction property for intuitionistic logic, A has only two unifiers: those who replace the variable x by truth, and by falsity respectively; namely $\theta_1(x) := \top$ and $\theta_2(x) := \bot$. This means that A does not have a most general unifier. On the other hand, if A were $\top$-projective, its projective unifier should be a most general unifier, a contradiction with our previous observation. Thus, A is not par-projective.

Given that not all formulas are par-projective, the question arises on which formulas are par-projective. In this section, we will characterise par-projectivity, via Kripke semantics. It is a variant of [7] in which it is proven that projectivity is equivalent to *extendability* for a given formula in the non-parametric language.[9] Let us first look at some definitions.

Given $X \subseteq \mathsf{atom}$, we say that $\mathcal{K}_1$ is a X-variant of $\mathcal{K}_2$, if (1) they share the same frame, (2) the evaluations at every node other than the root are the same, and (3) the evaluations at the root for all $a \in X$ are the same. An $\emptyset$-variant is simply called a variant. We say that A is weakly valid in $\mathcal{K}$, notation $\mathcal{K} \Vdash^{-} A$, if A is valid on every node other than the root.

Definition 4.1 We say that A is B-extendable, if $\vdash^{\mathsf{i}} A \to B$ and for every Kripke model $\mathcal{K}$ with $\mathcal{K} \Vdash^{-} A$ and $\mathcal{K} \Vdash B$, there is some par-variant $\mathcal{K}'$ of $\mathcal{K}$ such that $\mathcal{K}' \Vdash A$. Additionally, for a given set Γ of formulas, we say that A is Γ-extendable, if there is some $B \in \Gamma^{\mathsf{par}}$ such that A is B-extendable.

Given $A \in \mathcal{L}$ and a set of variables X, we define θ_A^X as follows:

$$\theta_A^X(x) := \begin{cases} A \to x & : x \in X \\ A \wedge x & : x \notin X \end{cases}$$

[8] One may show $\lceil A \rceil = \top$ by following semantical argument. For a given $E \in \mathcal{L}(\mathsf{par})$, if $\nvdash^{\mathsf{i}} E$ then there is a Kripke model $\mathcal{K} \nVdash E$. Since x does not appear in E, without loss of generality, we may also assume that $\mathcal{K} \Vdash x$ (or $\mathcal{K} \Vdash \neg x$). Hence $\mathcal{K} \nVdash (x \vee \neg x) \to E$. Thus $\nvdash^{\mathsf{i}} (x \vee \neg x) \to E$. This finishes showing that for every $E \in \mathcal{L}(\mathsf{par})$, if $\vdash^{\mathsf{i}} (x \vee \neg x) \to E$ then $\vdash^{\mathsf{i}} E$.

[9] Roughly speaking, an extendable formula is a formula whose Kripke models could be extended from below.

Now let $X := \{x_1, x_2, \ldots, x_m\}$ be the set of all variables occurring in A. Consider an ordering of the powerset of X, namely $P(X) = \{X_0, X_1, \ldots, X_k\}$ such that $X_i \subseteq X_j \Rightarrow i \geq j$. Then define $\theta_A := \theta_A^{X_0} \circ \theta_A^{X_1} \circ \ldots \circ \theta_A^{X_k}$.

Our construction of θ_A is by iterated compositions of the substitutions θ_A^X, and is the same as the one first introduced by S. Ghilardi in [7]. Since θ_A is the cumulative composition of A-identity substitutions, we may infer that θ_A is also an A-identity.

Lemma 4.2 *$\theta_A(\mathcal{K})$ forces A iff a* par*-variant of it forces A.*

Proof. We first make this observation that for any model $\mathcal{K}$, any set $Y \subseteq \mathsf{var}$ and any $x \in Y$, if $\mathcal{K} \nVdash A$ then $\theta_A^Y(\mathcal{K}) \Vdash x$ iff $x \in Y$, while if $\mathcal{K} \Vdash A$ then $\theta_A^Y(\mathcal{K}) \Vdash x$ iff $\mathcal{K} \Vdash x$.

We only prove the non-trivial direction, say right to left. For every $i \leq k$, we define $\theta_A^i = \theta_A^{X_0} \circ \ldots \circ \theta_A^{X_i}$ all of which are A-identities. By the observation above, it follows that if $\theta_A^i(\mathcal{K}) \Vdash A$ for some $i \leq k$, then $\theta_A^j(\mathcal{K}) \Vdash A$ for all $i \leq j \leq k$. Hence, $\theta_A(\mathcal{K}) \Vdash A$ iff there is some i such that $\theta_A^i(\mathcal{K}) \Vdash A$. We claim that if there is i such that $\mathcal{K}' \Vdash A$ is a par-variant of $\theta_A^i(\mathcal{K})$, then there is r such that $\theta_A^r(\mathcal{K}) \Vdash A$.

Let i be least such that $\theta_A^i(\mathcal{K}) \Vdash^- A$ and j be least such that there is a par-variant $\mathcal{K}' \Vdash A$ of $\theta_A^j(\mathcal{K})$. Observe that $i \leq j$. Since $\theta_A^i(\mathcal{K}) \Vdash^- A$, it follows that $\theta_A^i(\mathcal{K}_w) \Vdash x$ iff $\theta_A^j(\mathcal{K}_w) \Vdash x$ iff $\mathcal{K}'_w \Vdash x$ for every w other than the root. Thus $\theta_A^j(\mathcal{K})$ is a par-variant of $\theta_A^i(\mathcal{K})$ and so $i = j$.

Let Y be the valuation of $\mathcal{K}'$ at its root and let r be such that $Y \cap \mathsf{var} = X_r$. By the minimality of i, there is some w other than the root for which i is least with the property $\theta_A^i(\mathcal{K}_w) \Vdash A$. Therefore $\theta_A^i(\mathcal{K}_w) \Vdash x$ iff $x \in X_i$. Hence $X_r \subseteq X_i$ by the monotonicity of Kripke models and thus $r \geq i$. If $\theta_A^{r-1}(\mathcal{K}) \Vdash A$ (in the very specific case of $r = 0$, we define θ_A^{r-1} as identity substitution) then by our above observation we also have $\theta_A^r(\mathcal{K}) \Vdash A$ and we are done. Thus assume that $\theta_A^{r-1}(\mathcal{K}) \nVdash A$. By our above observation, $\theta_A^r(\mathcal{K}) \Vdash x$ iff $x \in X_r = Y \cap \mathsf{var}$. Therefore $\theta_A^r(\mathcal{K}) \Vdash A$ iff $\mathcal{K}' \Vdash A$, thus proving the claim. □

Theorem 4.3 *E-extendability and E-projectivity are equivalent notions for every $E \in \mathcal{L}(\mathsf{par})$.*

Proof. Right-to-left: Let θ be a projective E-fier of A, i.e. $A \vdash^{\mathsf{i}} \theta(x) \leftrightarrow x$ for every $x \in \mathsf{var}$ and $\vdash^{\mathsf{i}} \theta(A) \leftrightarrow E$. We will show that A is E-extendable. First observe that $A \vdash^{\mathsf{i}} \theta(A) \leftrightarrow A$ and hence $\vdash^{\mathsf{i}} A \to E$. Now let $\mathcal{K} \Vdash^- A$ with $\mathcal{K} \Vdash E$ and take $\mathcal{K}' := \theta(\mathcal{K})$. Then it is not difficult to observe that $\mathcal{K}'$ is indeed a par-variant of $\mathcal{K}$ and $\mathcal{K}' \Vdash A$.

Left-to-Right: Assume that A is E-extendable. Hence $\vdash^{\mathsf{i}} A \to E$ and for every $\mathcal{K}$ with $\mathcal{K} \Vdash E$ and $\mathcal{K} \Vdash^- A$, there is a par-variant $\mathcal{K}'$ of $\mathcal{K}$ such that $\mathcal{K}' \Vdash A$. We claim that θ_A, as defined just before Lemma 9, is a projective E-fier of A. We know that θ_A is an A-identity, so it only remains to show that θ_A is an E-fier of A. By induction on the height of the Kripke model $\mathcal{K}$, we prove $\theta_A(\mathcal{K}) \Vdash A \leftrightarrow E$. Note that since E does not have any variables, and substitutions are assumed to leave parameters unchanged, this implies $\mathcal{K} \Vdash \theta_A(A) \leftrightarrow E$, as desired.

Induction hypothesis, implies $\theta_A(\mathcal{K}_w) \Vdash A \leftrightarrow E$ for every w other than the root. If $\theta_A(\mathcal{K}) \nVdash E$, since $\vdash^{\mathsf{i}} A \to E$, we also have $\theta_A(\mathcal{K}) \nVdash A$ and hence $\theta_A(\mathcal{K}) \Vdash A \leftrightarrow E$. Next assume that $\theta_A(\mathcal{K}) \Vdash E$. Thus by induction hypothesis, $\theta_A(\mathcal{K}) \Vdash^{-} A$. Therefore, by E-extendability of A, there is a par-variant $\mathcal{K}'$ of $\theta_A(\mathcal{K})$ such that $\mathcal{K}' \Vdash A$. Then Lemma 9 implies that $\theta_A(\mathcal{K}) \Vdash A$ and so $\theta_A(\mathcal{K}) \Vdash A \leftrightarrow E$. □

Corollary 4.4 par-*projectivity of* A *is equivalent to projectivity of* $\lceil A \rceil \to A$.

Proof. One direction has already been proved in Lemma 7. For the other way around, assume that A is par-projective and, hence, $\lceil A \rceil$-projective. Then by Theorem 9, A is also $\lceil A \rceil$-extendable. Again, by Theorem 9, it is enough to show that $\lceil A \rceil \to A$ is $\top$-extendable. So let $\mathcal{K} \Vdash^{-} \lceil A \rceil \to A$, seeking to find some par-variant $\mathcal{K}'$ of $\mathcal{K}$ such that $\mathcal{K}' \Vdash \lceil A \rceil \to A$. If $\mathcal{K} \nVdash \lceil A \rceil$, take $\mathcal{K}' := \mathcal{K}$ and we are done. Otherwise, by $\lceil A \rceil$-extendability of A, there is some par-variant $\mathcal{K}'$ of $\mathcal{K}$ such that $\mathcal{K}' \Vdash A$, as desired. □

Question 1: With the above corollary, we have a nice characterisation of projectivity of $\lceil A \rceil \to A$. We may wonder about (characterisation of) its dual notion, namely projectivity of $A \to \lfloor A \rfloor$, in which $\lfloor A \rfloor$ indicates the uniform pre-interpolant of A.

5 Unification type of extensions of intuitionistic logic

Thanks to [7], we know that the unification type of intuitionistic logic is finitary. In this section, we show in Theorem 5 that the unification type of par-extensions of intuitionistic logic is also finitary.

Convention. All over this section, we assume that atom, the set of atomic formulas is finite.

We say that two Kripke models are par-equivalent ($\mathcal{K}_1 \equiv^{\mathsf{par}} \mathcal{K}_2$), if they share the same frame and for every node w and every $p \in \mathsf{par}$ we have $\mathcal{K}_1, w \Vdash p$ iff $\mathcal{K}_2, w \Vdash p$.

Given a class $\mathscr{M}$ of Kripke models, $\sum \mathscr{M}$ indicates the disjoint union of Kripke models in $\mathscr{M}$ with a fresh root below with empty valuation.

Definition 5.1 A class of Kripke models $\mathscr{K}$ is called B-extendable, if: $\mathscr{K} \Vdash B$ and for every finite (including empty) $\mathscr{M} \subseteq \mathscr{K}$, if a variant $\mathcal{K}$ of $\sum \mathscr{M}$ forces B, then there is a par-variant of $\mathcal{K}$ which belongs to $\mathscr{K}$. We say that $\mathscr{K}$ is Γ-extendable, if there is some $B \in \Gamma^{\mathsf{par}}$ such that $\mathscr{K}$ is B-extendable. Then $\top$-extendability is also called extendability. It is not difficult to observe that A is B-extendable (Definition 9) iff its class of models is so.

The *implicative degree* of a formula A, denoted as $c_{\to}(A)$, indicates the maximum number of nested implications in A. More precisely,

- $c_{\to}(A) = 0$ for $A \in \mathsf{atom} \cup \{\bot\}$.
- $c_{\to}(A \wedge B) = c_{\to}(A \vee B) = \mathsf{max}\{c_{\to}(A), c_{\to}(B)\}$.

- $c_{\to}(A \to B) := 1 + \max\{c_{\to}(A), c_{\to}(B)\}$.

In what follows, we will make use of the notions of bounded bisimilarity and bounded sub-bisimilarity. We define these notions inductively for given models $\mathcal{K}$ and $\mathcal{K}'$ with respective roots w_0 and w_0':

- $\mathcal{K} \rightleftharpoons_0 \mathcal{K}'$ iff $\mathcal{K}(w_0) = \mathcal{K}'(w_0')$.
- $\mathcal{K} \rightleftharpoons_{n+1} \mathcal{K}'$ iff
  - Forth: $\forall w \in \mathcal{K}\ \exists w' \in \mathcal{K}'\ \mathcal{K}_w \rightleftharpoons_n \mathcal{K}'_{w'}$ and
  - Back: $\forall w' \in \mathcal{K}'\ \exists w \in \mathcal{K}\ \mathcal{K}_w \rightleftharpoons_n \mathcal{K}'_{w'}$.
- $\mathcal{K} \leq_0 \mathcal{K}'$ iff $\mathcal{K}(w_0) \supseteq \mathcal{K}'(w_0')$.
- $\mathcal{K} \leq_{n+1} \mathcal{K}'$ iff
  - Forth: $\forall w \in \mathcal{K}\ \exists w' \in \mathcal{K}'\ \mathcal{K}_w \rightleftharpoons_n \mathcal{K}'_{w'}$.

Observe that $\rightleftharpoons_n$ is an equivalence relation and $\leq_n$ is a reflexive and transitive relation.

Lemma 5.2 *Let $\mathcal{K} \leq_n \mathcal{K}'$ and $c_{\to}(A) \leq n$. If $\mathcal{K}' \Vdash A$, then $\mathcal{K} \Vdash A$.*

Proof. See [7]. □

Then for a class of Kripke models $\mathscr{K}$, we define

$$\langle \mathscr{K} \rangle_n := \{\mathcal{K} : \exists \mathcal{K}' \in \mathscr{K}\ (\mathcal{K} \leq_n \mathcal{K}')\}.$$

We call a class $\mathscr{K}$ of Kripke models *stable*, iff for every $\mathcal{K} \in \mathscr{K}$ and every node w of $\mathcal{K}$, $\mathcal{K}_w \in \mathscr{K}$. Furthermore, we say that $\mathscr{K}$ is $\leq_n$-downward closed, if $\mathcal{K}' \leq_n \mathcal{K} \in \mathscr{K}$ implies $\mathcal{K}' \in \mathscr{K}$. Observe that since $\leq_n$ is transitive, $\langle \mathscr{K} \rangle_n$ is $\leq_n$-downward closed. Moreover, every $\leq_n$-downward closed class of Kripke models is also stable. Hence $\langle \mathscr{K} \rangle_n$ is stable.

Lemma 5.3 *Let $\mathscr{K}$ be a stable class of E-extendable Kripke models for some $E \in \mathcal{L}(\mathsf{par})$. Then for every $n > c_{\to}(E)$, $\langle \mathscr{K} \rangle_n$ is also stable and E-extendable.*

Proof. By our above observation, we only need to show that $\langle \mathscr{K} \rangle_n$ is E-extendable. Since $\mathscr{K} \Vdash E$ and $c_{\to}(E) < n$, we get from Lemma 5 that $\langle \mathscr{K} \rangle_n \Vdash E$. Let $\mathscr{M}$ be a finite subset of $\langle \mathscr{K} \rangle_n$; then due to the stability of $\mathscr{K}$, there is a minimal set $\mathscr{M}' \subseteq \mathscr{K}$ such that for every $\mathcal{M} \in \mathscr{M}$ and every $w \in \mathcal{M}$ there is a model $\mathcal{M}' \in \mathscr{M}'$ such that $\mathcal{M}_w \rightleftharpoons_{n-1} \mathcal{M}'$. Since $\mathscr{M}'$ is minimal, it will be finite as there can be at most finitely many non-$\rightleftharpoons_{n-1}$ Kripke models. Additionally, for every $\mathcal{M}' \in \mathscr{M}'$ there is some $\mathcal{M} \in \mathscr{M}$ and some $w \in \mathcal{M}$ such that $\mathcal{M}_w \rightleftharpoons_{n-1} \mathcal{M}'$ as otherwise the class $\mathscr{M}' \setminus \{\mathcal{M}'\}$ would contradict the minimality of $\mathscr{M}'$.

Now assume that there is a variant $\mathcal{K}$ of $\sum \mathscr{M}$ such that $\mathcal{K} \Vdash E$ and let $\mathcal{K}'$ be a variant of $\sum \mathscr{M}'$ with its root evaluated the same as $\mathcal{K}$. We show that $\mathcal{K} \rightleftharpoons_{n-1} \mathcal{K}'$.

- By our assumption for $\mathscr{M}'$ for every $w \in \mathcal{K}$ other than the root, there is some $\mathcal{M}' \in \mathscr{M}'$ such that $\mathcal{K}_w \rightleftharpoons_{n-1} \mathcal{M}'$ and thus $\mathcal{K}_w \rightleftharpoons_{n-2} \mathcal{M}'$.
- Similarly, for every $w' \in \mathcal{K}'$ other than the root, $\mathcal{K}'_{w'} = \mathcal{M}'_{w'}$ for some $\mathcal{M}' \in \mathscr{M}$. By the above, there is some $\mathcal{M} \in \mathscr{M}$ and $u \in \mathcal{M}$ such that

$\mathcal{M}_u \rightleftharpoons_{n-1} \mathcal{M}'$. Thus, by the definition of bounded bisimulation, there is some $w \in \mathcal{M}$ with $u \preccurlyeq w$ such that $\mathcal{M}_w \rightleftharpoons_{n-2} \mathcal{M}'_{w'}$.

- Finally, the roots are 0-bisimilar to each other and hence, by the above, $\mathcal{K} \rightleftharpoons_{n-1} \mathcal{K}'$.

Hence $\mathcal{K}' \Vdash E$, so there is a par-variant $\overline{\mathcal{K}'} \in \mathscr{K}$ of $\mathcal{K}'$. Consider the par-variant $\overline{\mathcal{K}}$ of $\mathcal{K}$ whose root is evaluated the same as in $\overline{\mathcal{K}'}$. One can then readily prove that $\overline{\mathcal{K}} \leq_n \overline{\mathcal{K}'}$ in a similar way as the proof of $\mathcal{K} \rightleftharpoons_{n-1} \mathcal{K}'$, only by omitting the backward direction. Therefore, $\overline{\mathcal{K}} \in \langle \mathscr{K} \rangle_n$. □

Corollary 5.4 *Let Γ be finite. Then there is m such that for every stable class $\mathscr{K}$ of Γ-extendable Kripke models and every $n > m$, $\langle \mathscr{K} \rangle_n$ is also stable and Γ-extendable.*

Proof. Let $m = \max\{c_{\rightarrow}(E) : E \in \Gamma^{\mathsf{par}}\}$ and apply Lemma 5. □

Lemma 5.5 *A class $\mathscr{K}$ of Kripke models is equal to the class of all models of some A with $c_{\rightarrow}(A) \leq n$ iff $\mathscr{K}$ is $\leq_n$-downward closed.*

Proof. See [7, Prop. 2]. □

Theorem 5.6 *Given a finite Γ and A, there is a finite set Π of Γ-projective formulas with the following properties:*

(i) $\vdash^{\mathsf{i}} \bigvee \Pi \rightarrow A$.

(ii) $A \mathrel{|\!\!\sim}_{\Gamma} \Pi$. *(see section 6 for the definition of $\mathrel{|\!\!\sim}_{\Gamma}$.)*

Proof. Let $m_0 := c_{\rightarrow}(A)$ and m_1 be the number m given from Corollary 5 and $n := \max\{m_0, m_1\}$. For a given $E \in \Gamma^{\mathsf{par}}$ and substitution θ with $\vdash^{\mathsf{i}} E \rightarrow \theta(A)$, we find some $B^{\theta}_{E} \in \mathcal{L}$ such that:

(i) $c_{\rightarrow}(B^{\theta}_{E}) \leq n$.

(ii) B^{θ}_{E} is E-projective.

(iii) $\vdash^{\mathsf{i}} B^{\theta}_{E} \rightarrow A$.

(iv) $\vdash^{\mathsf{i}} E \rightarrow \theta(B^{\theta}_{E})$.

Let us first see why this finishes the proof of this theorem. Define

$$\Pi := \{B^{\theta}_{E} : E \in \Gamma^{\mathsf{par}} \ \& \ \vdash^{\mathsf{i}} E \rightarrow \theta(A)\}.$$

Since atom is finite, by (i) we have the finiteness of Π. Γ-projectivity of elements of Π is guaranteed by (ii). Item (iii) implies $\vdash^{\mathsf{i}} \bigvee \Pi \rightarrow A$. Finally, by item (iv) we have $A \mathrel{|\!\!\sim}_{\Gamma} \Pi$.

So, let us go back to find B^{θ}_{E} with the listed properties. Define

$$\mathscr{K} := \{\theta(\mathcal{K}) : \mathcal{K} \Vdash E\}.$$

Observe that $\mathscr{K}$ is stable and since $E \in \mathcal{L}(\mathsf{par})$, we get from its definition that $\mathscr{K}$ is also E-extendable. Then Lemma 5 implies that $\langle \mathscr{K} \rangle_n$ is also stable and E-extendable. Lemma 5 implies that there is some formula B^{θ}_{E} with $c_{\rightarrow}(B^{\theta}_{E}) \leq n$ such that $\langle \mathscr{K} \rangle_n = \{\mathcal{K} : \mathcal{K} \Vdash B^{\theta}_{E}\}$. Hence (i) is satisfied. Since $\langle \mathscr{K} \rangle_n$ is extendable, Theorem 9 implies the validity of item (ii). To check the validity of item (iii), we argue semantically. Let $\mathcal{K} \Vdash B_E^\theta$. Therefore, $\mathcal{K} \leq_n \mathcal{K}'$ for some $\mathcal{K}' \in \mathscr{K}$. Since $\mathcal{K}' \in \mathscr{K}$, we have $\mathcal{K}' = \theta(\mathcal{K}'')$ for some $\mathcal{K}'' \Vdash E$. On the other hand, since $\vdash^{\mathsf{i}} E \to \theta(A)$, we have $\mathcal{K}'' \Vdash E \to \theta(A)$. Hence $\mathcal{K}'' \Vdash \theta(A)$ and thus $\theta(\mathcal{K}'') \Vdash A$. This means $\mathcal{K}' \Vdash A$ and since $c_{\to}(A) \leq n$ and $\mathcal{K} \leq_n \mathcal{K}'$, Lemma 5 implies $\mathcal{K} \Vdash A$. Thus, $\vdash^{\mathsf{i}} B_E^\theta \to A$. For the item (iv), let $\mathcal{K} \Vdash E$. Then $\theta(\mathcal{K}) \in \mathscr{K} \subseteq \langle \mathscr{K} \rangle_n$. Hence $\theta(\mathcal{K}) \Vdash B_E^\theta$ and thus $\mathcal{K} \Vdash \theta(B_E^\theta)$. □

Recall from section 2 that a par-extension of a logic is an extension by an $\mathcal{L}(\mathsf{par})$-formula.

Theorem 5.7 *The unification type of* par*-extensions of intuitionistic logic is finitary.*

Proof. Let $A \in \mathcal{L}$ and $E \in \mathcal{L}(\mathsf{par})$ be given, seeking a finite complete set of unifiers of A in the intuitionistic logic extended by E. Take $\Gamma := \{E\}$ and apply Theorem 5 to obtain a finite set Π with the properties mentioned. Then it can be easily verified that the set of projective E-fiers of elements of Π serves as a finite complete set of unifiers, as desired. □

6 Parametric Admissibility: Relative to Γ

Given a logic L, the (multi-conclusion) admissibility relation for L is defined as follows:

$$A \mathrel{|\!\!\sim}_{\mathsf{L}} \Delta \quad \text{iff} \quad \forall \theta\, (\mathsf{L} \vdash \theta(A) \;\Rightarrow\; \exists B \in \Delta \;\; \mathsf{L} \vdash \theta(B)).$$

We simply write $A \mathrel{|\!\!\sim}_{\mathsf{L}} B$ for $A \mathrel{|\!\!\sim}_{\mathsf{L}} \{B\}$. Given $E \in \mathcal{L}(\mathsf{par})$, we define $\mathrel{|\!\!\sim}_E$ as the admissibility relation for intuitionistic logic strengthened by an additional axiom E. Note that since $E \in \mathcal{L}(\mathsf{par})$, this logic is closed under substitutions, and hence it is a logic indeed. The admissibility for intuitionistic logic $\mathrel{|\!\!\sim}_{\mathsf{IPC}}$ is annotated as $\mathrel{|\!\!\sim}$ in the literature [10] and is known to be decidable [17]. One may easily observe that $A \mathrel{|\!\!\sim}_E B$ iff $(E \to A) \mathrel{|\!\!\sim} (E \to B)$ for every $E \in \mathcal{L}(\mathsf{par})$. Thus, by characterising the admissibility relation $\mathrel{|\!\!\sim}$, we also have a characterisation of $\mathrel{|\!\!\sim}_E$.

Then we define the Γ-admissibility relation $\mathrel{|\!\!\sim}_\Gamma$, as the intersection of all $\mathrel{|\!\!\sim}_E$'s with $E \in \Gamma^{\mathsf{par}} = \Gamma \cap \mathsf{par}$. In other words:

$$A \mathrel{|\!\!\sim}_\Gamma B \quad \text{iff} \quad \forall E \in \Gamma^{\mathsf{par}} \; A \mathrel{|\!\!\sim}_E B.$$

This time, the characterisation of $\mathrel{|\!\!\sim}_\Gamma$ is not that simple. For instance, the characterisation for $\Gamma = \mathsf{NNIL}$, the set of No Nested Implications on the Left, is the main result of [16]. This result has been used in an essential way in the proof for characterisation of intuitionistic provability logic [15]. In this paper, we will dig further and characterise one more salient case: $\Gamma = \mathcal{L}$, the full language. It turns out that $A \mathrel{|\!\!\sim}_{\mathcal{L}} B$ iff $\vdash^{\mathsf{i}} A \to B$ (Corollary 6).

There is yet another binary relation on $\mathcal{L}$, called preservativity [19] and annotated as $A \mathrel{|\!\!\approx}_\Gamma B$, which is defined as follows:

$$A \mathrel{|\!\!\approx}_\Gamma B \quad \text{iff} \quad \forall E \in \Gamma\, (\vdash^{\mathsf{i}} E \to A \Longrightarrow \vdash^{\mathsf{i}} E \to B).$$

Note the difference between the role of E in $\mathrel{|\!\sim}_\Gamma$ and $\mathrel{|\!\approx}_\Gamma$. In the first, we quantify E over the set Γ^{par}, a subset of Γ, whereas in the second, we quantify it over Γ itself. Since both definitions are universal over Γ, the following trivially holds:

Remark 6.1 If $\Gamma \subseteq \Delta$ then $\mathrel{|\!\sim}_\Gamma \supseteq \mathrel{|\!\sim}_\Delta$ and $\mathrel{|\!\approx}_\Gamma \supseteq \mathrel{|\!\approx}_\Delta$.

In the following lemma, we show that Γ-admissibility implies $\widehat{\Gamma}$-preservativity, where $\widehat{\Gamma}$ is the set of Γ-projective formulas.

Lemma 6.2 $\mathrel{|\!\sim}_\Gamma \subseteq \mathrel{|\!\approx}_{\widehat{\Gamma}}$.

Proof. Assume that $A \mathrel{|\!\sim}_\Gamma B$ and let $E \in \widehat{\Gamma}$ be such that $\vdash^{\mathsf{i}} E \to A$, seeking to show $\vdash^{\mathsf{i}} E \to B$. Since $E \in \widehat{\Gamma}$, we have $E^* \in \Gamma^{\mathsf{par}}$ and there is a projective E^*-fier θ of E. Therefore, by $\vdash^{\mathsf{i}} E \to A$, we have $\vdash^{\mathsf{i}} E^* \to \theta(A)$. Hence, by $A \mathrel{|\!\sim}_\Gamma B$ we have $\vdash^{\mathsf{i}} E^* \to \theta(B)$. Since by Theorem 7 $\vdash^{\mathsf{i}} E \to E^*$, we have $\vdash^{\mathsf{i}} E \to \theta(B)$, and thus by the E-projectivity of θ, we have $\vdash^{\mathsf{i}} E \to B$. □

A par-substitution is a substitution θ such that $\theta(x) \in \mathsf{par}$ for every $x \in \mathsf{var}$. Then we say that Γ is closed under par-substitutions if for every $A \in \Gamma$ and every par-substitution θ we have $\theta(A) \in \Gamma$.

Lemma 6.3 *Given Γ closed under* par*-substitutions and $A \in \Gamma$, the following are equivalent:*

(i) $\vdash^{\mathsf{i}} A \to B$.

(ii) $A \mathrel{|\!\sim}_\Gamma B$.

(iii) $A \mathrel{|\!\approx}_{\widehat{\Gamma}} B$.

Proof. The proof of (i)→(ii) is obvious. (ii)→(iii) holds by Lemma 6. We reason for (iii)→(i) as follows. Let $\nvdash^{\mathsf{i}} A \to B$. Also, assume that $X := \{x_1, \dots, x_n\}$ includes all variables occurring in A, and $\{p_1, \dots, p_n\}$ is a set of fresh parameters, i.e. parameters not appearing in A and B. Finally, define

$$E := A \wedge \bigwedge_{i=1}^{n} x_i \leftrightarrow p_i \quad \text{and} \quad \theta(x) := \begin{cases} p_i & : x \in X \text{ and } x = x_i \text{ for } 1 \leq i \leq n \\ x & : x \notin X \end{cases}$$

It can be easily verified that

- E is Γ-projective.
- $\vdash^{\mathsf{i}} E \to A$.
- $\nvdash^{\mathsf{i}} E \to B$.

This implies that $A \mathrel{|\!\not\approx}_{\widehat{\Gamma}} B$, as desired. □

Note that in the proof of the above lemma, we are taking advantage of the infiniteness of par.

As a result, if we take Γ as the full language $\mathcal{L}$, we have the following:

Corollary 6.4 *The following are equivalent:*

(i) $\vdash^{\mathsf{i}} A \to B$.

(ii) $A \mathrel{|\!\sim}_{\mathcal{L}} B$.

(iii) $A \mathrel{|\!\approx}_{\widehat{\mathcal{L}}} B$.

Given that $A \mathrel{|\!\sim}_{\mathcal{L}} B$ is equivalent to $\vdash^{\mathsf{i}} A \to B$, asking for par-projective approximations of a proposition A (Theorem 5), simplifies to the following question.

Question 2: Given $A \in \mathcal{L}$, is it possible to find a finite set Π_A of $\mathcal{L}$-projective propositions with

$$\vdash^{\mathsf{i}} A \leftrightarrow \bigvee \Pi_A.$$

By the arguments in this section, we already have a characterisation / decidability of $\mathrel{|\!\sim}_{\Gamma}$ for finite Γ and for $\Gamma = \mathcal{L}$. Moreover, [16] provides a characterisation for the case $\Gamma = \mathsf{NNIL}$, the set of No Nested Implications on the Left. [10]

We say that A is *weakly extendable* if every $\mathcal{K}$ with $\mathcal{K} \Vdash^{-} A$ has a variant $\mathcal{K}'$ such that $\mathcal{K}' \Vdash A$. Note that being weakly extendable not necessarily implies $\top$-extendable. For instance, the atomic formula $p \in \mathsf{par}$, is weakly extendable, while it is not $\top$-extendable. Otherwise, by Theorem 9 it would be $\top$-projective. Then by Theorem 7 we have $\top = \lceil p \rceil$, a contradiction. Also note that weak extendability, coincides [7, Theorem 5] with the standard notion of projectivity as defined in [7], by treating parameters as variables. In other words, A is weakly extendable iff there is a substitution θ which possibly replaces parameters, such that (1) $\vdash^{\mathsf{i}} \theta(A)$, (2) for every atomic a we have $A \vdash^{\mathsf{i}} \theta(a) \leftrightarrow a$.

Question 3: Axiomatize or provide decision algorithm for $\mathrel{|\!\sim}_{\Gamma}$ in the cases that Γ is the set of Weakly extendable formulae.

A formula A is *prime* if $\vdash^{\mathsf{i}} A \to (B \vee C)$ implies $\vdash^{\mathsf{i}} A \to B$ or $\vdash^{\mathsf{i}} A \to C$. Note that every weakly extendable formula is prime, but not necessarily vice versa. For example the formula $A := \neg p \to (q \vee r)$ is prime, while it is not weakly extendable. The reason that A is not weakly extendable, is that the Kripke model $\mathcal{K}$ with three worlds w_0, w_1, w_2 and with accessibility relations $w_0 \prec w_1$ and $w_0 \prec w_2$ and with $\mathcal{K}(w_0) = \emptyset$, $\mathcal{K}(w_1) := \{q\}$ and $\mathcal{K}(w_2) := \{r\}$. For this $\mathcal{K}$ we have $\mathcal{K} \Vdash^{-} A$ while no variant $\mathcal{K}'$ of $\mathcal{K}$ exists such that $\mathcal{K}' \Vdash A$. Also to show primeness of A, let $\nvdash^{\mathsf{i}} A \to B$ and $\nvdash^{\mathsf{i}} A \to C$ seeking to show $\nvdash^{\mathsf{i}} A \to (B \vee C)$. By $\nvdash^{\mathsf{i}} A \to B$ and $\nvdash^{\mathsf{i}} A \to C$ we have Kripke models $\mathcal{K}_B$ and $\mathcal{K}_C$ such that $\mathcal{K}_B \Vdash A$ and $\mathcal{K}_B \nVdash B$ and $\mathcal{K}_C \Vdash A$ and $\mathcal{K}_C \nVdash C$. Add a fresh root in beneath of disjoint union of $\mathcal{K}_B$ and $\mathcal{K}_C$ and let the valuation be empty at the new root. The resulted model is a counter-model of $A \to (B \vee C)$ and thus $\nvdash^{\mathsf{i}} A \to (B \vee C)$. Moreover, above argument obviously works for every weakly extendable A. This shows that every weakly extendable formula is also prime.

[10] Since its appearance in the literature [21], NNIL has shown itself to be of great importance in the study of intuitionistic logic.

Question 4: Axiomatize or provide decision algorithm for $\mathrel{|\!\!\sim}_\Gamma$ in the cases that Γ is the set of prime formulae.

One of the reasons for our interest in above-mentioned Γ's in Questions 3 and 4 is that in both cases we have the validity of the following disjunction rule for $\mathrel{|\!\!\sim}_\Gamma$:

$$\frac{B \mathrel{|\!\!\sim}_\Gamma A \qquad C \mathrel{|\!\!\sim}_\Gamma A}{B \vee C \mathrel{|\!\!\sim}_\Gamma A}.$$

References

[1] Alizadeh, M., M. Ardeshir, P. Balbiani and M. Mojtahedi, *Unification types in Euclidean modal logics*, Logic Journal of the IGPL **31** (2023), pp. 422–440.

[2] Baader, F. and B. Morawska, *Unification in the description logic EL*, Logical Methods in Computer Science **6** (2010).

[3] Balbiani, P. and Q. Gougeon, *Projective relative unification through duality*, Journal of Logic and Computation (2023), p. exad058.

[4] Balbiani, P. and M. Mojtahedi, *Unification with parameters in the implication fragment of classical propositional logic*, Logic Journal of the IGPL **30** (2021), pp. 454–464.

[5] Cintula, P. and G. Metcalfe, *Admissible rules in the implication–negation fragment of intuitionistic logic*, Annals of Pure and Applied Logic **162** (2010), pp. 162–171.

[6] Ghilardi, S., *Unification through projectivity*, J. Log. Comput. **7** (1997), pp. 733–752.

[7] Ghilardi, S., *Unification in Intuitionistic Logic*, Journal of Symbolic Logic **64** (1999), pp. 859–880.

[8] Ghilardi, S., *Best solving modal equations*, Annalas of Pure and Applied Logic **102** (2000), pp. 183–198.

[9] Goudsmit, J. P. and R. Iemhoff, *On unification and admissible rules in Gabbay–de Jongh logics*, Annals of Pure and Applied Logic **165** (2014), pp. 652–672.

[10] Iemhoff, R., *On the Admissible Rules of Intuitionistic Propositional Logic*, The Journal of Symbolic Logic **66** (2001), pp. 281–294.

[11] Jeřábek, E., *Admissible rules of modal logics*, Journal of Logic and Computation **15** (2005), pp. 411–431.

[12] Jeřábek, E., *Admissible rules of łukasiewicz logic*, Journal of Logic and Computation **20** (2010), pp. 425–447.

[13] Jeřábek, E., *Rules with parameters in modal logic II*, Annals of Pure and Applied Logic **171** (2020), p. 102829.

[14] Jeřábek, E., *Rules with parameters in modal logic I*, Annals of Pure and Applied Logic **166** (2015), pp. 881–933.

[15] Mojtahedi, M., *On Provability Logic of HA* (2022).

[16] Mojtahedi, M., *Relative unification in intuitionistic logic: towards provability logic of HA* (2022).

[17] Rybakov, V. V., *Bases of admissible rules of the modal system Grz and of intuitionistic logic*, Mathematics of the USSR-Sbornik **56** (1987), pp. 311–331.

[18] van der Giessen, I., *Admissible rules for six intuitionistic modal logics*, Annals of Pure and Applied Logic **174** (2023), p. 103233.

[19] Visser, A., *Substitutions of Σ_1^0 sentences: explorations between intuitionistic propositional logic and intuitionistic arithmetic*, Ann. Pure Appl. Logic **114** (2002), pp. 227–271, commemorative Symposium Dedicated to Anne S. Troelstra (Noordwijkerhout, 1999).

[20] Visser, A., "Uniform Interpolation and Layered Bisimulation," Lecture Notes in Logic, Cambridge University Press, 2017 p. 139–164.

[21] Visser, A., J. van Benthem, D. de Jongh and G. R. R. de Lavalette, NNIL, *a study in intuitionistic propositional logic*, in: *Modal logic and process algebra (Amsterdam, 1994)*, CSLI Lecture Notes **53**, CSLI Publ., Stanford, CA, 1995 pp. 289–326.

Natural Deduction, Normalization and Subformula Property for Kreisel-Putnam Logic

Valentin Müller [1]

78315 Radolfzell am Bodensee
Germany

Abstract

We introduce a new natural deduction system for Kreisel-Putnam logic. Our system is based on Schroeder-Heister's *calculus of higher-level rules*, an extension of ordinary natural deduction in which not only *formulas*, but also *rules* can act as assumptions that may be discharged in the course of a derivation. We will establish a *normalization theorem*, i.e., we will show that every deduction in our system can be converted into a deduction without 'detours'. The method used for this is new and might also be applied to other non-classical logics. As a consequence of this result, we will obtain an unrestricted subformula property and a separation theorem for our system.

Keywords: Kreisel-Putnam logic, intermediate logics, natural deduction, proof theory, normalization, subformula property, calculus of higher-level rules.

1 Introduction

The intermediate logic KP, known as *Kreisel-Putnam logic*, can be defined as the result of adding the scheme $(\neg\varphi \to (\psi \vee \chi)) \to ((\neg\varphi \to \psi) \vee (\neg\varphi \to \chi))$ to intuitionistic propositional logic (IPC). It was introduced in 1957 by Kreisel and Putnam [9] in order to disprove a conjecture by Łukasiewicz [11] according to which IPC is the only intermediate logic having the disjunction property.

In this paper, we provide a new natural deduction system for KP. Our system is based on the so-called *calculus of higher-level rules*, a generalized natural deduction formalism introduced by Schroeder-Heister [18,19,20], in which not only *formulas*, but also *rules* can be used as assumptions that may be discharged in the course of constructing a proof tree. We present an innovative normalization procedure which allows to convert every deduction in our system into a *normal deduction*, i.e., a deduction containing no 'detours'. The underlying method might also be used to obtain normalization results for other non-classical logics.[2] We will see that our system is fully *analytic*, in the sense that normal deductions always satisfy an unrestricted subformula property.

---

[1] valentin.johannes.mueller@gmail.com

[2] We already succeeded in finding similar natural deduction systems for propositional inquisitive logic and for a small number of intermediate logics different from KP (cf. [13]).

Over the years, proof theorists have developed various methods which allow to construct analytic proof systems for non-classical logics in a more or less uniform way [1,3,6,10,15]. The system KP occupies a somewhat special position here, as most of these methods do not seem to work when applied to it. A labelled sequent calculus for KP is presented in [7, Sect. 11.2], but this calculus internalizes the semantic characterization of KP given in [2, p. 55], so it is not purely syntactic in a strict sense. Simple natural deduction systems for KP and related logics are provided in [4,16], along with a computational interpretation in the style of Curry [5] and Howard [8]. These calculi, however, do not seem to satisfy a full subformula property (for a closely related system, this was pointed out in [14, pp. 35–36]). As far as we know, it is still an open question whether KP admits an analytic hypersequent calculus or nested sequent calculus.

The paper is structured as follows. In Section 2, we first recall some basic notions from the literature. In Sections 3–4, we introduce our natural deduction system for KP and demonstrate its completeness in a syntactic way. The proof of our normalization result is presented in Sections 5–7. In Section 8, we establish the subformula property and a separation theorem for our system.

2 Basic Notions

We start by defining some basic notions from the theory of superintuitionistic logics. For further details, we refer to [2, Sect. 4.1]. Throughout this paper, we assume a fixed set P of propositional atoms, denoted by the meta-variables p, q, r, etc. The *language of intuitionistic logic*, notation $\mathcal{L}_{\mathsf{I}}$, is defined as follows:

$$\varphi ::= p \mid \bot \mid \varphi \wedge \varphi \mid \varphi \vee \varphi \mid \varphi \to \varphi \qquad (p \in \mathsf{P}).$$

As usual, we write $\neg\varphi$ as an abbreviation for $\varphi \to \bot$. A formula $\varphi \in \mathcal{L}_{\mathsf{I}}$ is said to be a *prime formula*, if it is either an atom from the set P, or it is of the form $\varphi = \bot$. And φ is called *disjunction-free*, if it does not contain any occurrences of the symbol $\vee$. We also write IPC for the set of all formulas valid in *intuitionistic logic* and CPC for the set of all formulas valid in *classical logic*. By a *substitution*, we will mean a function $s : \mathsf{P} \to \mathcal{L}_{\mathsf{I}}$ which assigns, to each atom $p \in \mathsf{P}$, a formula $s(p) \in \mathcal{L}_{\mathsf{I}}$. Such a function s is extended to arbitrary formulas by putting $s(\bot) := \bot$ as well as $s(\varphi \odot \psi) := s(\varphi) \odot s(\psi)$ for all $\odot \in \{\wedge, \vee, \to\}$.

Definition 2.1 [Superintuitionistic Logic] A set of formulas $L \subseteq \mathcal{L}_{\mathsf{I}}$ is said to be a *superintuitionistic logic* (or an *si-logic* for short), if the following holds:

- $\mathsf{IPC} \subseteq L$,
- L is closed under modus ponens: if $\varphi \in L$ and $(\varphi \to \psi) \in L$, then $\psi \in L$,
- L is closed under substitution: for any substitution s, if $\varphi \in L$, then $s(\varphi) \in L$.

An si-logic L is *consistent*, if it satisfies the condition $\bot \notin L$. It is not difficult to show that every consistent si-logic L is *intermediate* between intuitionistic logic and classical logic, in the sense that $\mathsf{IPC} \subseteq L \subseteq \mathsf{CPC}$. A consistent si-logic is therefore also referred to as an *intermediate logic* [2, p. 109].

Clearly, for any set of formulas $\Gamma \subseteq \mathcal{L}_{\mathsf{I}}$, one can construct an si-logic by adding the formulas from Γ to IPC and by closing the resulting set of formu-

(A1) $\varphi \to (\psi \to \varphi)$	(A2) $(\varphi \to (\psi \to \chi)) \to ((\varphi \to \psi) \to (\varphi \to \chi))$
(A3) $(\varphi \wedge \psi) \to \varphi$, $(\varphi \wedge \psi) \to \psi$	(A4) $\varphi \to (\psi \to (\varphi \wedge \psi))$
(A5) $\varphi \to (\varphi \vee \psi)$, $\psi \to (\varphi \vee \psi)$	(A6) $(\varphi \to \chi) \to ((\psi \to \chi) \to ((\varphi \vee \psi) \to \chi))$
(A7) $\bot \to \varphi$	(KP) $(\neg\varphi \to (\psi \vee \chi)) \to ((\neg\varphi \to \psi) \vee (\neg\varphi \to \chi))$

Modus ponens: from $\Gamma \vdash_{\mathrm{H}} \varphi$ and $\Delta \vdash_{\mathrm{H}} \varphi \to \psi$, infer $\Gamma, \Delta \vdash_{\mathrm{H}} \psi$.

Fig. 1. The Hilbert-style system HKP.

las under modus ponens and substitution. The si-logic thus obtained will be denoted by $\mathsf{IPC} + \Gamma$. If Γ is of the form $\Gamma = \{\varphi_1, \dots, \varphi_n\}$, then we also write $\mathsf{IPC} + \varphi_1 + \dots + \varphi_n$ instead of $\mathsf{IPC} + \Gamma$. The intermediate logic KP, known as *Kreisel-Putnam logic*, was first considered in [9] and can be defined as follows:

$$\mathsf{KP} := \mathsf{IPC} + (\neg p \to (q \vee r)) \to ((\neg p \to q) \vee (\neg p \to r)).$$

The system KP has some historical significance, because it was the first counterexample to a conjecture by Łukasiewicz [11] according to which IPC is the only intermediate logic having the disjunction property. This conjecture was refuted by Kreisel and Putnam [9] who proved that the disjunction property also holds for KP: if we have $(\varphi \vee \psi) \in \mathsf{KP}$, then also $\varphi \in \mathsf{KP}$ or $\psi \in \mathsf{KP}$.

A sound and complete Hilbert-style system for KP is presented in Figure 1. The provability relation of this system is denoted by $\vdash_{\mathrm{H}}$ and defined in the usual way. That is, we write $\Gamma \vdash_{\mathrm{H}} \varphi$ and say that φ is *provable* from Γ in HKP, if one of the following holds: (1) φ is an element of Γ, or (2) φ is an axiom of HKP, or (3) there are subsets $\Delta, \Sigma \subseteq \Gamma$ and a formula ψ such that $\Gamma = \Delta \cup \Sigma$ as well as $\Delta \vdash_{\mathrm{H}} \psi$ and $\Sigma \vdash_{\mathrm{H}} \psi \to \varphi$. Using essentially the same argument as in [14, Theorem 1.5.4], it is easy to prove that HKP satisfies the *deduction theorem.*

Theorem 2.2 (Deduction Theorem) *We have* $\Gamma, \varphi \vdash_{\mathrm{H}} \psi$ *iff* $\Gamma \vdash_{\mathrm{H}} \varphi \to \psi$.

For later purposes, we also need to show that every negated formula of our language is intuitionistically equivalent to a disjunction-free formula. We first assign, to each $\varphi \in \mathcal{L}_{\mathsf{I}}$, a disjunction-free formula φ^{G} in the following way:

- $\bot^{\mathrm{G}} := \bot$ and $p^{\mathrm{G}} := \neg\neg p$ for all atoms $p \in \mathsf{P}$,
- $(\varphi \odot \psi)^{\mathrm{G}} := \varphi^{\mathrm{G}} \odot \psi^{\mathrm{G}}$ for $\odot \in \{\wedge, \to\}$, and $(\varphi \vee \psi)^{\mathrm{G}} := \neg(\neg\varphi^{\mathrm{G}} \wedge \neg\psi^{\mathrm{G}})$.

This translation is also known as the *Gödel-Gentzen negative translation* [21, p. 49]. We will write $\varphi \equiv_{\mathsf{IPC}} \psi$, if φ and ψ are equivalent in IPC, i.e., if we have $(\varphi \to \psi) \in \mathsf{IPC}$ and $(\psi \to \varphi) \in \mathsf{IPC}$. One can now establish the desired result: every negated formula $\neg\varphi$ is IPC-equivalent to the disjunction-free formula $\neg\varphi^{\mathrm{G}}$.

Lemma 2.3 *For any* $\varphi \in \mathcal{L}_{\mathsf{I}}$, *we have* $\neg\varphi \equiv_{\mathsf{IPC}} \neg\varphi^{\mathrm{G}}$.

Proof. Using induction on φ, it is easy to prove that $\neg\neg\varphi \equiv_{\mathsf{IPC}} \varphi^{\mathrm{G}}$. Moreover, we clearly have $\neg\varphi \equiv_{\mathsf{IPC}} \neg\neg\neg\varphi$. Hence, it follows $\neg\varphi \equiv_{\mathsf{IPC}} \neg\neg\neg\varphi \equiv_{\mathsf{IPC}} \neg\varphi^{\mathrm{G}}$. □

3 Higher-Level Rules and Natural Deduction for KP

We will now introduce an analytic natural deduction system for Kreisel-Putnam logic. As explained above, our system is based on Schroeder-Heister's *calculus of higher-level rules*, a generalization of ordinary natural deduction in which not only *formulas*, but also *rules* can serve as assumptions that may be discharged in the course of a derivation [18,19,20]. The basic idea is as follows: each proof tree in our system will be built up from two types of rules, which we will call *primitive rules* and *assumption rules.* The primitive rules are used to formalize the 'meaning' of the logical symbols in our language, so an application of such a rule will never count towards the open assumptions of a derivation. An application of an assumption rule, on the other hand, is treated as a *supposition* (rather than as a valid inference step), and it is considered to be *open* as long as it has not been discharged by another rule application in the proof tree.

First, we need to characterize the class of assumption rules that may occur in a proof tree of our system. An *assumption rule of level* 0 simply amounts to a formula $\varphi \in \mathcal{L}_\mathsf{I}$, so it behaves just like a *hypothesis* in ordinary natural deduction. By an *assumption rule of level* 1, we will mean a rule of the form

$$\frac{\varphi_1 \quad \varphi_2 \quad \cdots \quad \varphi_n}{\psi}$$

where $n \geq 1$. That is, a rule of level 1 simply allows to infer a formula ψ from a given set of premises $\varphi_1, \ldots, \varphi_n$, without discharging any other rules in the proof tree. In what follows, we will also use the linear notation '$\varphi_1, \ldots, \varphi_n \Rightarrow \psi$' as a name for the level-1 rule with premises $\varphi_1, \ldots, \varphi_n$ and conclusion ψ. It will sometimes be convenient to treat a level-0 rule φ (i.e., an ordinary hypothesis) as a level-1 rule with an empty set of premises. For this reason, we also write '$\emptyset \Rightarrow \varphi$' instead of '$\varphi$'. An *assumption rule of level* $\ell > 1$ is a rule of the form

$$u\,\frac{\begin{matrix}[\Gamma_1]^u \\ \mathcal{D}_1 \\ \varphi_1\end{matrix} \quad \begin{matrix}[\Gamma_2]^u \\ \mathcal{D}_2 \\ \varphi_2\end{matrix} \quad \cdots \quad \begin{matrix}[\Gamma_n]^u \\ \mathcal{D}_n \\ \varphi_n\end{matrix}}{\psi}$$

where $\Delta = \{\Gamma_1 \Rightarrow \varphi_1, \ldots, \Gamma_n \Rightarrow \varphi_n\}$ is a non-empty set of rules whose maximal level is $\ell - 1$.[3] In other words, a rule of this kind allows to infer a formula ψ from a set of premises $\varphi_1, \ldots, \varphi_n$, while discharging all open applications of the rules from Γ_i in the subtree ending with the premise φ_i. A rule of this form will also be denoted by '$(\Gamma_1 \Rightarrow \varphi_1, \ldots, \Gamma_n \Rightarrow \varphi_n) \Rightarrow \psi$'. Throughout this paper, the discharging of rules will be indicated by square brackets and by writing a suitable label (such as u, v, w) to the left of the corresponding inference line.

Definition 3.1 The class of *assumption rules* is defined in the following way:

- Every formula $\varphi \in \mathcal{L}_\mathsf{I}$ is an assumption rule of level 0.
- If $\varphi \in \mathcal{L}_\mathsf{I}$ is a formula and $\Gamma = \{R_1, \ldots, R_n\}$ with $n \geq 1$ is a set of rules whose maximal level is $\ell \geq 0$, then $\Gamma \Rightarrow \varphi$ is an assumption rule of level $\ell + 1$.

Note that, according to this definition, assumption rules are always *non-schematic*, in the sense that they are built up from *specific formulas*, rather than from syntactic meta-variables ranging over arbitrary formulas. Furthermore, we do not assign any order to the expressions occurring on the left-hand side of

[3] That is, each $\Gamma_i \Rightarrow \varphi_i$ is of level $< \ell$ and at least one $\Gamma_i \Rightarrow \varphi_i$ is of level exactly $\ell - 1$.

$$\frac{\varphi \quad \psi}{\varphi \wedge \psi}\wedge I \qquad \frac{\varphi_1 \wedge \varphi_2}{\varphi_k}\wedge E \quad (k=1,2) \qquad \frac{\varphi_k}{\varphi_1 \vee \varphi_2}\vee I \quad (k=1,2) \qquad \frac{\bot}{\varphi}\bot_{\mathsf{i}}$$

$$\begin{array}{c}[\varphi]^u\\ \mathcal{D}\\ u\dfrac{\psi}{\varphi \to \psi}\to I\end{array} \qquad \frac{\varphi \to \psi \quad \varphi}{\psi}\to E \qquad \begin{array}{ccc} [\Theta]^u & [\Theta \Rightarrow \varphi]^u & [\Theta \Rightarrow \psi]^u \\ \mathcal{D}_1 & \mathcal{D}_2 & \mathcal{D}_3 \end{array} \\ u\dfrac{\varphi \vee \psi \qquad \chi \qquad \chi}{\chi}\vee E$$

Side condition: In the rule $\vee E$, we require Θ to be a finite set of negative rules.

Fig. 2. Primitive rules of the natural deduction system NKP.

the symbol $\Rightarrow$ in an assumption rule. So, for example, $\varphi_1, (\varphi_2, \varphi_3 \Rightarrow \varphi_4) \Rightarrow \varphi_5$ and $(\varphi_3, \varphi_2 \Rightarrow \varphi_4), \varphi_1 \Rightarrow \varphi_5$ will be treated as *identical* assumption rules.

Definition 3.2 The *subrules* of an assumption rule are defined as follows:

- Every assumption rule is a subrule of itself.
- If $R = ((\Gamma_1 \Rightarrow \varphi_1, \ldots, \Gamma_n \Rightarrow \varphi_n) \Rightarrow \psi)$ is a rule of level $\ell \geq 2$, then every subrule of the rules contained in $\Gamma_1 \cup \ldots \cup \Gamma_n$ is also a subrule of R.

In other words, R is a subrule of R', just in case we either have $R = R'$, or R is one of the rules that are discharged by R' or by one of its other subrules.

Definition 3.3 [Prime Rule, Disjunction-Free Rule, Negative Rule] Consider an arbitrary assumption rule $R = (\Gamma \Rightarrow \varphi)$. We will say that:

- R is a *prime rule*, if every formula occurring in $\Gamma \cup \{\varphi\}$ is a prime formula.
- R is *disjunction-free*, if every formula occurring in $\Gamma \cup \{\varphi\}$ is disjunction-free.
- R is *negative*, if R is disjunction-free and of the form $R = (\Gamma \Rightarrow \bot)$ for $\Gamma \neq \emptyset$.

We are now ready to introduce a higher-level natural deduction system for Kreisel-Putnam logic.[4] Our system will be denoted by NKP and comprises the primitive rules depicted in Figure 2. As can be seen, the rules for conjunction, implication and falsum as well as the introduction rule for disjunction are simply the familiar ones from intuitionistic logic. In fact, the only new ingredient of our system is the elimination rule $\vee E$, which should be read as follows: let Θ be a finite set of *negative rules* and suppose that there exists a proof tree $\mathcal{D}_1$ for $\varphi \vee \psi$ such that each of the rules from Θ possibly has an undischarged application in $\mathcal{D}_1$. Furthermore, let $\mathcal{D}_2$ and $\mathcal{D}_3$ be two proof trees for χ such that $\Theta \Rightarrow \varphi$ possibly occurs among the open assumptions of $\mathcal{D}_2$, and $\Theta \Rightarrow \psi$ possibly occurs among the open assumptions of $\mathcal{D}_3$. Then, using the rule $\vee E$, we are allowed to infer χ and to discharge all open applications of the rules from Θ in $\mathcal{D}_1$, all open applications of $\Theta \Rightarrow \varphi$ in $\mathcal{D}_2$, and all open applications of $\Theta \Rightarrow \psi$ in $\mathcal{D}_3$. It is easy to see that the standard elimination rule for $\vee$ can be obtained as a special case of the rule $\vee E$, by taking Θ to be the *empty set*:

[4] A similar proof system for *inquisitive logic* was recently developed by the author in an unpublished manuscript [13]. An earlier version of this system was presented in [14, Ch. 2].

in this case, nothing is discharged in the subtree $\mathcal{D}_1$, and the additional rules $\Theta \Rightarrow \varphi$ and $\Theta \Rightarrow \psi$ simply amount to the level-0 rules φ and ψ, respectively.

Given an arbitrary application of $\vee E$, we will also use the following terminology: the negative rules from the set Θ are said to be the *auxiliary rules* of this application of $\vee E$. And the *side rules* are the two rules of the form $\Theta \Rightarrow \varphi$ and $\Theta \Rightarrow \psi$ which are discharged in the subtrees $\mathcal{D}_2$ and $\mathcal{D}_3$, respectively.

Definition 3.4 [The System NKP] We define NKP to be the higher-level natural deduction system having the primitive rules depicted in Figure 2.

Every proof tree built up from the primitive rules of our system and arbitrary (open or discharged) applications of assumption rules will be referred to as a *quasi-deduction*. By a *deduction* in NKP, we will mean a quasi-deduction in which all undischarged assumption rules are of level 0 (so these assumption rules are just ordinary *hypotheses*). The *derivability relation* of our system will be denoted by $\vdash_{\mathrm{N}}$. Thus, given an arbitrary set of assumption rules Γ, we will write $\Gamma \vdash_{\mathrm{N}} \varphi$ and say that φ is *derivable* from Γ, if there exists a quasi-deduction $\mathcal{D}$ with conclusion φ such that all open assumption rules in $\mathcal{D}$ are contained in Γ. In this case, $\mathcal{D}$ is also said to be a *quasi-deduction for* $\Gamma \vdash_{\mathrm{N}} \varphi$. If $\Gamma \vdash_{\mathrm{N}} \varphi$ holds for a set of *formulas* $\Gamma \subseteq \mathcal{L}_{\mathsf{I}}$, then φ is said to be *provable* from Γ.

Example 3.5 The following proof tree is a deduction in our system:

$$
z\,\dfrac{
\dfrac{\neg(p\wedge q)\to(\varphi\vee\psi) \qquad u\,\dfrac{\dfrac{\dfrac{[p\wedge q]^u}{p}\,\wedge E}{\bot}\,[p\Rightarrow\bot]^z}{\neg(p\wedge q)}\,{\to}I}{\varphi\vee\psi}\,{\to}E
\qquad
\dfrac{w\,\dfrac{v\,\dfrac{\dfrac{[\neg p]^w \quad [p]^v}{\bot}\,{\to}E}{\varphi}\,[(p\Rightarrow\bot)\Rightarrow\varphi]^z}{\neg p\to\varphi}\,{\to}I}{(\neg p\to\varphi)\vee(\neg p\to\psi)}\,\vee I
\qquad
\dfrac{y\,\dfrac{x\,\dfrac{\dfrac{[\neg p]^y \quad [p]^x}{\bot}\,{\to}E}{\psi}\,[(p\Rightarrow\bot)\Rightarrow\psi]^z}{\neg p\to\psi}\,{\to}I}{(\neg p\to\varphi)\vee(\neg p\to\psi)}\,\vee I
}{(\neg p\to\varphi)\vee(\neg p\to\psi)}\,\vee E
$$

In applications of primitive and non-primitive rules, we in general also permit *vacuous discharge of assumption rules*. However, one can show that, in applications of $\vee E$, vacuous discharge is always redundant. More precisely, every quasi-deduction $\mathcal{D}$ in our system can be rewritten in such a way that (1) each auxiliary rule and each side rule of any application of $\vee E$ in $\mathcal{D}$ is in fact discharged by this application, and (2) every proper subrule of an auxiliary rule (resp., side rule) of $\vee E$ in $\mathcal{D}$ is in fact discharged by at least one application of this auxiliary rule (resp., of a side rule). In what follows, we will tacitly assume that quasi-deductions always satisfy the conditions (1) and (2). In most cases, this does not make a big difference and the reader can easily transfer our arguments to the more general setting in which such a convention is not adopted.

4 Soundness and Completeness

We will now prove that our system is sound and complete with respect to the Hilbert-style system HKP (see Figure 1). First, we establish the completeness of our system: everything which is provable in HKP is also provable in NKP.

Lemma 4.1 *Let $\Gamma \subseteq \mathcal{L}_{\mathsf{I}}$ be a set of formulas. If $\Gamma \vdash_{\mathrm{H}} \varphi$ holds in the Hilbert-style system* HKP, *then $\Gamma \vdash_{\mathrm{N}} \varphi$ holds in the natural deduction system* NKP.

Proof. By induction on the structure of a Hilbert-style proof for $\Gamma \vdash_{\mathrm{H}} \varphi$. It suffices to show that all axioms of HKP are provable in NKP. For the schemes (A1)–(A7), this is straightforward. In order to show the provability of (KP), let φ be an arbitrary formula. By Lemma 2.3, we have $\neg\varphi \equiv_{\mathsf{IPC}} \neg\varphi^{\mathrm{G}}$, so there must be a deduction $\mathcal{D}_1$ for $\neg\varphi \vdash_{\mathrm{N}} \neg\varphi^{\mathrm{G}}$ and a deduction $\mathcal{D}_2$ for $\neg\varphi^{\mathrm{G}} \vdash_{\mathrm{N}} \neg\varphi$ in our system. Using these proof trees, we now construct the following deduction:

$$
a\,\dfrac{
z\,\dfrac{
\dfrac{[\neg\varphi \to (\psi \vee \chi)]^a \quad \begin{matrix} u\,\dfrac{\dfrac{[\varphi^{\mathrm{G}}]^u \quad [\varphi^{\mathrm{G}} \Rightarrow \bot]^z}{\bot}}{\neg\varphi^{\mathrm{G}}}\,{\to}I \\ \mathcal{D}_2 \\ \neg\varphi \end{matrix}}{\psi \vee \chi}\,{\to}E
\quad
\dfrac{w\,\dfrac{v\,\dfrac{\dfrac{\begin{matrix}[\neg\varphi]^w \\ \mathcal{D}_1 \\ \neg\varphi^{\mathrm{G}}\end{matrix} \quad [\varphi^{\mathrm{G}}]^v}{\bot}\,{\to}E \quad [(\varphi^{\mathrm{G}} \Rightarrow \bot) \Rightarrow \psi]^z}{\psi}}{\neg\varphi \to \psi}\,{\to}I}{(\neg\varphi \to \psi) \vee (\neg\varphi \to \chi)}\,{\vee}I
\quad
\dfrac{y\,\dfrac{x\,\dfrac{\dfrac{\begin{matrix}[\neg\varphi]^y \\ \mathcal{D}_1 \\ \neg\varphi^{\mathrm{G}}\end{matrix} \quad [\varphi^{\mathrm{G}}]^x}{\bot}\,{\to}E \quad [(\varphi^{\mathrm{G}} \Rightarrow \bot) \Rightarrow \chi]^z}{\chi}}{\neg\varphi \to \chi}\,{\to}I}{(\neg\varphi \to \psi) \vee (\neg\varphi \to \chi)}\,{\vee}I
}{(\neg\varphi \to \psi) \vee (\neg\varphi \to \chi)}\,{\vee}E
}{(\neg\varphi \to (\psi \vee \chi)) \to ((\neg\varphi \to \psi) \vee (\neg\varphi \to \chi))}\,{\to}I
$$

Note that, since φ^{G} is disjunction-free, $\varphi^{\mathrm{G}} \Rightarrow \bot$ is a *negative rule* (see Definition 3.3), so the application of $\vee E$ in this deduction is in fact correct. □

In order to establish the soundness of our system, we first assign, to each assumption rule R, a corresponding formula $F(R)$ in the following way:

- If $R = \varphi$ is of level 0, then we put $F(R) := \varphi$.
- If $R = (R_1, \ldots, R_n \Rightarrow \varphi)$ is of level $\ell \geq 1$, then $F(R) := (\bigwedge_{i=1}^{n} F(R_i)) \to \varphi$.

For any set of rules Γ, we also put $F(\Gamma) := \{F(R) \mid R \in \Gamma\}$. It is now easy to show that, if φ is derivable from a set of assumption rules Γ in NKP, then φ is provable from the set of formulas $F(\Gamma)$ in the Hilbert-style system HKP.

Lemma 4.2 *Let Γ be a set of assumption rules. If $\Gamma \vdash_{\mathrm{N}} \varphi$ holds in the natural deduction system* NKP*, then $F(\Gamma) \vdash_{\mathrm{H}} \varphi$ holds in the Hilbert-style system* HKP.

The proof proceeds by induction on the structure of a quasi-deduction for $\Gamma \vdash_{\mathrm{N}} \varphi$ in our system. A few representative cases are considered in Appendix A.

Theorem 4.3 (Soundness and Completeness) *Let Γ be a set of formulas. We have $\Gamma \vdash_{\mathrm{N}} \varphi$ in the system* NKP *iff $\Gamma \vdash_{\mathrm{H}} \varphi$ holds in the system* HKP.

Proof. The statement follows directly from Lemma 4.1 and Lemma 4.2. □

5 Detours in Deductions

In this and the next two sections, we will establish a *normalization theorem* for our system, so we will show that every deduction in NKP can be converted into a *normal deduction*. Intuitively, a deduction is said to be *normal*, if it does not contain any 'detours'. In order to make this precise, we first need to recall some standard terminology. Given an arbitrary application of a rule of inference, we will call the formula occurrences directly above the line the *premises* and the formula occurrence directly below the line the *conclusion* of the application. In applications of *elimination rules*, we will distinguish between two types of premises: the *major premise* is the premise which contains the 'eliminated' occurrence of the connective; all other premises (if any) are said to be *minor premises*. So, for example, the major premise of an application of $\to E$ is the premise of the form $\varphi \to \psi$, and the minor premise is the one of the form φ. The major premise of an application of $\vee E$, on the other hand, is the premise of the form $\varphi \vee \psi$, and the minor premises are the two premises of the form χ.

A detour in a deduction may now be conceived as a formula that is both the conclusion of an introduction rule and the major premise of an elimination rule. However, deductions may also contain *repetitions* of formulas, since the minor premises and the conclusion of an application of $\vee E$ are always of the same shape. Thus, rather than considering only single occurrences of formulas, we must consider sequences of consecutive occurrences of the same formula. Such a sequence is also referred to as a *segment*. The intuitive notion of a detour is now formalized by the concept of a *cut segment*, which is defined as follows.

Definition 5.1 [Cut Segment] A *cut segment of length* n in a proof tree $\mathcal{D}$ is a sequence $\xi_1, \ldots, \xi_n$ of n consecutive occurrences of a formula φ in $\mathcal{D}$ such that:

- ξ_1 is the conclusion of an introduction rule,
- each ξ_i with $1 \leq i < n$ is a minor premise of $\vee E$,
- ξ_n is the major premise of an elimination rule.

Example 5.2 The following deduction contains a cut segment of length 2:

$$\dfrac{\begin{array}{c}\vdots\\ \chi_1 \vee \chi_2\end{array} \qquad \dfrac{\begin{array}{cc}\mathcal{D}_1 & \mathcal{D}_2\\ \varphi & \psi\end{array}}{\varphi \wedge \psi}\wedge I \qquad \begin{array}{c}\vdots\\ \varphi \wedge \psi\end{array}}{\dfrac{\varphi \wedge \psi}{\varphi}\wedge E}\vee E$$

The unique formula occurring in a cut segment is said to be the *cut formula* of the segment. Our main goal is to show that cut segments can always be eliminated. However, this alone does not suffice to obtain a strong subformula property for our system, since there are also other types of detours that need to be considered. One problem arises from the intuitionistic absurdity rule $\bot_{\mathsf{i}}$, because this rule, too, can 'introduce' a formula which might then serve as the major premise of an elimination rule. This issue can easily be resolved by restricting $\bot_{\mathsf{i}}$ to instances in which the conclusion is atomic. Another, more severe, problem is related to the assumption rules involved in an application of $\vee E$. In order to understand this problem, consider the following deduction:

$$z\,\dfrac{\dfrac{\neg p \to (r_1 \vee r_2) \qquad u\,\dfrac{\dfrac{\dfrac{[p]^u \quad q}{p \wedge q}\wedge I}{\bot}[p \wedge q \Rightarrow \bot]^z}{\neg p}{\to}I}{r_1 \vee r_2}{\to}E \qquad \dfrac{w\,\dfrac{v\,\dfrac{\dfrac{[\neg p]^w \quad \dfrac{[p \wedge q]^v}{p}\wedge E}{\bot}{\to}E}{r_1}[(p \wedge q \Rightarrow \bot) \Rightarrow r_1]^z}{\neg p \to r_1}{\to}I}{(\neg p \to r_1) \vee (\neg p \to r_2)}\vee I \qquad \dfrac{y\,\dfrac{x\,\dfrac{\dfrac{[\neg p]^y \quad \dfrac{[p \wedge q]^x}{p}\wedge E}{\bot}{\to}E}{r_2}[(p \wedge q \Rightarrow \bot) \Rightarrow r_2]^z}{\neg p \to r_2}{\to}I}{(\neg p \to r_1) \vee (\neg p \to r_2)}\vee I}{(\neg p \to r_1) \vee (\neg p \to r_2)}\vee E$$

This deduction contains the formula $p \wedge q$ (acting as the premise of the auxiliary rule $p \wedge q \Rightarrow \bot$), but this formula is neither a subformula of an open hypothesis nor a subformula of the conclusion of the deduction. Note that, in a sense, $p \wedge q$ behaves just like a *cut formula* here: it is the conclusion of an introduction rule in the subtree on the left-hand side, and it is the major premise of an elimination rule in the two subtrees on the right-hand side. In order to avoid detours of this kind, we must not only eliminate cut segments, but we must also show that auxiliary rules can be restricted to *prime rules* (see Definition 3.3).

Definition 5.3 [Pure Quasi-Deduction] A quasi-deduction $\mathcal{D}$ is *pure*, if it satisfies the following two conditions: (1) all auxiliary rules of applications of $\vee E$ in $\mathcal{D}$ are prime rules, and (2) every conclusion of $\bot_\mathsf{i}$ in $\mathcal{D}$ is an atomic formula.

Definition 5.4 [Normal Quasi-Deduction] We will say that a quasi-deduction is *normal*, if it is a pure quasi-deduction and it also contains no cut segments.

Our normalization proof consists of two main steps: in Section 6, we will first see that every deduction can be turned into a *pure deduction*. In Section 7, we will then see that every pure deduction can be turned into a *normal deduction*.

6 Purification of Deductions

Let us start by showing that every deduction can be transformed into a pure deduction. We first note that, in every quasi-deduction of our system, $\bot_\mathsf{i}$ can be restricted to instances in which the conclusion is an atomic formula. The proof works in essentially the same way as in intuitionistic logic (see, e.g., [12, Proposition 4.2]). Thus, it only remains to show that auxiliary rules of $\vee E$ can be restricted to prime rules. To this end, we first assign, to each disjunction-free rule $R = (\Gamma \Rightarrow \varphi)$, a finite set of prime rules $\mathsf{pr}(R)$ in the following way:

- if φ is prime, then $\mathsf{pr}(\Gamma \Rightarrow \varphi) := \{\mathsf{pr}(\Gamma) \Rightarrow \varphi\}$ for $\mathsf{pr}(\Gamma) := \bigcup_{R' \in \Gamma} \mathsf{pr}(R')$,[5]
- $\mathsf{pr}(\Gamma \Rightarrow \psi \wedge \chi) := \mathsf{pr}(\Gamma \Rightarrow \psi) \cup \mathsf{pr}(\Gamma \Rightarrow \chi)$,
- $\mathsf{pr}(\Gamma \Rightarrow \psi \rightarrow \chi) := \mathsf{pr}(\Gamma, \psi \Rightarrow \chi)$.

Given any set of disjunction-free rules Γ, we also define $\mathsf{pr}(\Gamma) := \bigcup_{R \in \Gamma} \mathsf{pr}(R)$. Note that, if Γ is finite, then $\mathsf{pr}(\Gamma)$ is finite as well. The basic idea is to show that, if a formula φ is derivable using a disjunction-free rule R as an assumption, then it is also derivable using the prime rules from $\mathsf{pr}(R)$ as assumptions.

Definition 6.1 The *rank* of a formula φ, notation $rk(\varphi)$, is defined as follows:

- $rk(\bot) := 1$ and $rk(p) := 1$ for $p \in \mathsf{P}$,
- $rk(\psi \odot \chi) := rk(\psi) + rk(\chi) + 1$ for $\odot \in \{\wedge, \vee, \rightarrow\}$.

In other words, $rk(\varphi)$ simply amounts to the *length* of the string φ, provided that parentheses are ignored. By the *degree of a rule* $R = (\Gamma \Rightarrow \varphi)$, notation $dg(R)$, we will mean the sum of the ranks of all formulas occurring in $\Gamma \cup \{\varphi\}$. For any set of rules Γ, we also write $\Gamma \vdash_{\mathrm{P}} \varphi$, if φ is derivable from Γ by a *pure* quasi-deduction. We are now ready to prove the desired result: if a disjunction-

[5] If $\Gamma = \emptyset$, then we put $\mathsf{pr}(\Gamma) := \emptyset$. So, in this case, we have $\mathsf{pr}(\Gamma \Rightarrow \varphi) = \{\emptyset \Rightarrow \varphi\} = \{\varphi\}$.

free rule R occurs as (a component of) an open assumption in a quasi-deduction, then we can always replace R be the corresponding set of prime rules $\mathsf{pr}(R)$.

Lemma 6.2 *Let Γ and Δ be arbitrary sets of assumption rules and suppose that Δ is finite. Furthermore, let R be a disjunction-free rule. It holds:*

(i) *if $\Gamma, R \vdash_{\mathrm{P}} \varphi$, then $\Gamma, \mathsf{pr}(R) \vdash_{\mathrm{P}} \varphi$,*

(ii) *if $\Gamma, (\Delta, R \Rightarrow \psi) \vdash_{\mathrm{P}} \varphi$, then $\Gamma, (\Delta, \mathsf{pr}(R) \Rightarrow \psi) \vdash_{\mathrm{P}} \varphi$.*

Proof. Both statements are proved simultaneously, by induction on the degree of R. The base case is trivial. For the inductive step, let $dg(R) \geq 2$ and suppose that both of the claims hold for all disjunction-free rules of degree smaller than $dg(R)$. Without loss of generality, we may assume that $R = (\Pi \Rightarrow \chi)$, where χ is a disjunction-free formula and $\Pi = \{\Sigma_1 \Rightarrow \varphi_1, \ldots, \Sigma_n \Rightarrow \varphi_n\}$ is a possibly empty set of disjunction-free rules. There are the following possibilities.

Case 1: χ is a prime formula. In this case, one can easily derive (i) from the induction hypothesis for (ii), and (ii) from the induction hypothesis for (i).

Case 2: χ is not prime. In order to prove part (i), let $\mathcal{D}$ be a pure quasi-deduction for $\Gamma, R \vdash_{\mathrm{P}} \varphi$. We select a topmost open application of $R = (\Pi \Rightarrow \chi)$ in $\mathcal{D}$ and rewrite this application using one of the following two conversions:

$$\chi = \alpha\wedge\beta: \quad \begin{matrix} u\,\dfrac{\begin{matrix}[\Sigma_1]^u & & [\Sigma_n]^u\\ \mathcal{D}_1 & & \mathcal{D}_n\\ \varphi_1 & \cdots & \varphi_n\end{matrix}}{\alpha\wedge\beta}\,\Pi\Rightarrow\alpha\wedge\beta \\ \mathcal{D}_0\end{matrix} \quad\rightsquigarrow\quad \begin{matrix}\dfrac{u\,\dfrac{\begin{matrix}[\Sigma_1]^u & & [\Sigma_n]^u\\ \mathcal{D}_1 & & \mathcal{D}_n\\ \varphi_1 & \cdots & \varphi_n\end{matrix}}{\alpha}\,\Pi\Rightarrow\alpha \qquad v\,\dfrac{\begin{matrix}[\Sigma_1]^v & & [\Sigma_n]^v\\ \mathcal{D}_1 & & \mathcal{D}_n\\ \varphi_1 & \cdots & \varphi_n\end{matrix}}{\beta}\,\Pi\Rightarrow\beta}{\alpha\wedge\beta}\,\wedge I \\ \mathcal{D}_0\end{matrix}$$

$$\chi = \alpha\to\beta: \quad \begin{matrix} u\,\dfrac{\begin{matrix}[\Sigma_1]^u & & [\Sigma_n]^u\\ \mathcal{D}_1 & & \mathcal{D}_n\\ \varphi_1 & \cdots & \varphi_n\end{matrix}}{\alpha\to\beta}\,\Pi\Rightarrow\alpha\to\beta \\ \mathcal{D}_0\end{matrix} \quad\rightsquigarrow\quad \begin{matrix} v\,\dfrac{u\,\dfrac{\begin{matrix}[\Sigma_1]^u & & [\Sigma_n]^u & \\ \mathcal{D}_1 & & \mathcal{D}_n & \\ \varphi_1 & \cdots & \varphi_n & [\alpha]^v\end{matrix}}{\beta}\,\Pi,\alpha\Rightarrow\beta}{\alpha\to\beta}\,\to I \\ \mathcal{D}_0\end{matrix}$$

We only consider the case in which χ is of the form $\chi = \alpha \to \beta$, so we assume that the second of these conversions has been performed. As can be seen, the selected application of R is now replaced by an open application of $\Pi, \alpha \Rightarrow \beta$. Thus, by repeating the procedure for all other open applications of R, we finally obtain a pure quasi-deduction for $\Gamma, (\Pi, \alpha \Rightarrow \beta) \vdash_{\mathrm{P}} \varphi$. Now, observe that we have $dg(\Pi, \alpha \Rightarrow \beta) < dg(R)$. Hence, by the induction hypothesis for (i), we may conclude $\Gamma, \mathsf{pr}(\Pi, \alpha \Rightarrow \beta) \vdash_{\mathrm{P}} \varphi$ and therefore $\Gamma, \mathsf{pr}(R) \vdash_{\mathrm{P}} \varphi$, as desired.

In order to establish part (ii) of the lemma, let now $\mathcal{D}$ be a pure quasi-deduction for $\Gamma, (\Delta, R \Rightarrow \psi) \vdash_{\mathrm{P}} \varphi$. We select a topmost open application of $\Delta, R \Rightarrow \psi$ in $\mathcal{D}$ and rewrite this application using one of the following conversions (the vertical dots represent a collection of subtrees corresponding to the rules in Δ):

$$\chi = \alpha \wedge \beta: \quad u\dfrac{\begin{matrix}\vdots \\ \vdots\,\Delta \\ \vdots\end{matrix} \quad \begin{matrix}[\Pi]^u \\ \mathcal{D}_1 \\ \alpha \wedge \beta\end{matrix}}{\begin{matrix}\psi \\ \mathcal{D}_2\end{matrix}}\Delta, R \Rightarrow \psi \quad \rightsquigarrow \quad u\dfrac{\begin{matrix}\vdots \\ \vdots\,\Delta \\ \vdots\end{matrix} \quad \dfrac{\begin{matrix}[\Pi]^u \\ \mathcal{D}_1 \\ \alpha \wedge \beta\end{matrix}}{\alpha}\wedge E \quad \dfrac{\begin{matrix}[\Pi]^u \\ \mathcal{D}_1 \\ \alpha \wedge \beta\end{matrix}}{\beta}\wedge E}{\begin{matrix}\psi \\ \mathcal{D}_2\end{matrix}}\Delta, (\Pi \Rightarrow \alpha), (\Pi \Rightarrow \beta) \Rightarrow \psi$$

$$\chi = \alpha \to \beta: \quad u\dfrac{\begin{matrix}\vdots \\ \vdots\,\Delta \\ \vdots\end{matrix} \quad \begin{matrix}[\Pi]^u \\ \mathcal{D}_1 \\ \alpha \to \beta\end{matrix}}{\begin{matrix}\psi \\ \mathcal{D}_2\end{matrix}}\Delta, R \Rightarrow \psi \quad \rightsquigarrow \quad u\dfrac{\begin{matrix}\vdots \\ \vdots\,\Delta \\ \vdots\end{matrix} \quad \dfrac{\begin{matrix}[\Pi]^u \\ \mathcal{D}_1 \\ \alpha \to \beta\end{matrix} \quad [\alpha]^u}{\beta}\to E}{\begin{matrix}\psi \\ \mathcal{D}_2\end{matrix}}\Delta, (\Pi, \alpha \Rightarrow \beta) \Rightarrow \psi$$

Again, let us suppose that $\chi = \alpha \to \beta$, so the selected application of $\Delta, R \Rightarrow \psi$ is now replaced by an open application of $\Delta, R' \Rightarrow \psi$, where $R' := (\Pi, \alpha \Rightarrow \beta)$. By repeating the procedure, we thus obtain $\Gamma, (\Delta, R' \Rightarrow \psi) \vdash_{\mathrm{P}} \varphi$. Since we have $dg(R') < dg(R)$ and $\mathsf{pr}(R') = \mathsf{pr}(R)$, this implies $\Gamma, (\Delta, \mathsf{pr}(R') \Rightarrow \psi) \vdash_{\mathrm{P}} \varphi$ and therefore $\Gamma, (\Delta, \mathsf{pr}(R) \Rightarrow \psi) \vdash_{\mathrm{P}} \varphi$ by the induction hypothesis for part (ii). □

Proposition 6.3 *Let Γ be a set of formulas. If there exists a deduction for $\Gamma \vdash_{\mathrm{N}} \varphi$ in our system, then there also exists a pure deduction for $\Gamma \vdash_{\mathrm{N}} \varphi$.*

Proof. Suppose that there exists a deduction $\mathcal{D}$ for $\Gamma \vdash_{\mathrm{N}} \varphi$. Without loss of generality, we may assume that every conclusion of $\bot_{\mathsf{i}}$ in $\mathcal{D}$ is atomic. We now select an uppermost application of $\vee E$ in $\mathcal{D}$ such that not every auxiliary rule of this application is prime, and we then rewrite this application as follows:

$$u\dfrac{\begin{matrix}[\Theta]^u \\ \mathcal{D}_3 \\ \psi_1 \vee \psi_2\end{matrix} \quad \begin{matrix}[\Theta \Rightarrow \psi_1]^u \\ \mathcal{D}_1 \\ \chi\end{matrix} \quad \begin{matrix}[\Theta \Rightarrow \psi_2]^u \\ \mathcal{D}_2 \\ \chi\end{matrix}}{\begin{matrix}\chi \\ \mathcal{D}_4\end{matrix}}\vee E \quad \rightsquigarrow \quad u\dfrac{\begin{matrix}[\mathsf{pr}(\Theta)]^u \\ \mathcal{D}_3' \\ \psi_1 \vee \psi_2\end{matrix} \quad \begin{matrix}[\mathsf{pr}(\Theta) \Rightarrow \psi_1]^u \\ \mathcal{D}_1' \\ \chi\end{matrix} \quad \begin{matrix}[\mathsf{pr}(\Theta) \Rightarrow \psi_2]^u \\ \mathcal{D}_2' \\ \chi\end{matrix}}{\begin{matrix}\chi \\ \mathcal{D}_4\end{matrix}}\vee E$$

where each of the proof trees $\mathcal{D}_i'$ with $i = 1, 2, 3$ is obtained from $\mathcal{D}_i$ by applying Lemma 6.2 to the disjunction-free rules in Θ. Observe that $\mathsf{pr}(\Theta)$ is still a finite set of *negative* rules, so the new application of $\vee E$ is in fact correct. Therefore, by repeating the procedure, we finally obtain a pure deduction for $\Gamma \vdash_{\mathrm{N}} \varphi$. □

7 Normalization

Next, we will show that every pure deduction can be transformed into a normal deduction. For technical reasons, we first note the following fact: if $\mathcal{D}_1$ is a quasi-deduction containing some open occurrences of a rule $\Theta \Rightarrow \varphi$, and if $\mathcal{D}_2$ is another quasi-deduction with conclusion φ, then we can always construct a new quasi-deduction $\mathcal{D}_1\langle\mathcal{D}_2 : \Theta \Rightarrow \varphi\rangle$ by 'substituting' $\mathcal{D}_2$ for all undischarged applications of $\Theta \Rightarrow \varphi$ in $\mathcal{D}_1$. Moreover, if Θ contains only *prime rules*, then φ is the only new cut formula that may arise in the course of this construction.

Lemma 7.1 *Let Θ be a finite set of prime rules, let $\mathcal{D}_1$ be a pure quasi-deduction for $\Gamma, (\Theta \Rightarrow \varphi) \vdash_{\mathrm{N}} \psi$ and let $\mathcal{D}_2$ be a pure quasi-deduction for $\Delta, \Theta \vdash_{\mathrm{N}} \varphi$.*

Then there exists a pure quasi-deduction $\mathcal{D}_3$ for $\Gamma, \Delta \vdash_{\mathrm{N}} \psi$ such that every cut formula occurring in $\mathcal{D}_3$, except possibly φ, already occurs as a cut formula in $\mathcal{D}_1$ or $\mathcal{D}_2$. Furthermore, if the conclusion of $\mathcal{D}_3$ belongs to a segment that starts with the conclusion of an introduction rule, then so does the conclusion of $\mathcal{D}_1$, provided that $\varphi \neq \psi$. This quasi-deduction $\mathcal{D}_3$ is denoted by $\mathcal{D}_1\langle \mathcal{D}_2 : \Theta \Rightarrow \varphi\rangle$.

A proof is provided in Appendix B. By the *rank* of a cut segment σ, notation $rk(\sigma)$, we will mean the rank of the cut formula occurring in σ. A cut segment σ in a deduction $\mathcal{D}$ is said to be *maximal*, if it is a cut segment of highest rank in this deduction. And by the *cut rank* of a deduction $\mathcal{D}$, we will mean the pair of natural numbers $\textsc{cr}(\mathcal{D}) = (m, n)$ defined as follows: if $\mathcal{D}$ contains no cut segments, then we put $\textsc{cr}(\mathcal{D}) := (0, 0)$. Otherwise, we put $\textsc{cr}(\mathcal{D}) := (m, n)$, where m is the rank of a maximal cut segment in $\mathcal{D}$, and n is the number of maximal cut segments in $\mathcal{D}$. As usual, we will assume that cut ranks are ordered *lexicographically*. That is, given two deductions $\mathcal{D}$ and $\mathcal{D}'$ with $\textsc{cr}(\mathcal{D}) = (m, n)$ and $\textsc{cr}(\mathcal{D}') = (m', n')$, we will write $\textsc{cr}(\mathcal{D}) < \textsc{cr}(\mathcal{D}')$ and say that the cut rank of $\mathcal{D}$ is *smaller* than the cut rank of $\mathcal{D}'$, if we either have $m < m'$, or we have both $m = m'$ and $n < n'$. We first show that cut ranks of non-normal deductions can always be decreased: every pure deduction containing at least one cut segment can be transformed into a pure deduction of smaller cut rank.

Lemma 7.2 *Let $\mathcal{D}$ be a pure deduction for $\Gamma \vdash_{\mathrm{N}} \varphi$ such that $\textsc{cr}(\mathcal{D}) > (0, 0)$. Then there exists a pure deduction $\mathcal{D}'$ for $\Gamma \vdash_{\mathrm{N}} \varphi$ with $\textsc{cr}(\mathcal{D}') < \textsc{cr}(\mathcal{D})$.*

Proof. Let $\mathcal{D}$ be a pure deduction for $\Gamma \vdash_{\mathrm{N}} \varphi$ such that $\textsc{cr}(\mathcal{D}) > (0, 0)$. We select an *uppermost* maximal cut segment σ in $\mathcal{D}$, so σ is a cut segment of highest rank in $\mathcal{D}$, and it holds $rk(\pi) < rk(\sigma)$ for every cut segment π occurring above the conclusion of the elimination rule at the bottom of σ. Let ψ be the cut formula occurring in σ. While σ is of length greater than 1, we first permute the major premise at the end of σ over the minor premises of $\vee E$ in the 'middle part' of σ. This is achieved by applying the following *permutation conversion*, where (E) stands for an instance of an elimination rule with major premise ψ:

$$\begin{matrix}\dfrac{u\,\dfrac{\begin{matrix}[\Theta]^u\\ \mathcal{D}_3\\ \varphi_1\vee\varphi_2\end{matrix}\quad\begin{matrix}[\Theta\Rightarrow\varphi_1]^u\\ \mathcal{D}_1\\ \psi\end{matrix}\quad\begin{matrix}[\Theta\Rightarrow\varphi_2]^u\\ \mathcal{D}_2\\ \psi\end{matrix}}{\psi}\,\vee E\qquad \mathcal{D}_4}{\chi}\,(\mathrm{E})\\ \mathcal{D}_5\end{matrix}
\quad\rightsquigarrow\quad
\begin{matrix}\dfrac{u\,\begin{matrix}[\Theta]^u\\ \mathcal{D}_3\\ \varphi_1\vee\varphi_2\end{matrix}\quad\dfrac{\begin{matrix}[\Theta\Rightarrow\varphi_1]^u\\ \mathcal{D}_1\\ \psi\end{matrix}\quad\mathcal{D}_4}{\chi}\,(\mathrm{E})\quad\dfrac{\begin{matrix}[\Theta\Rightarrow\varphi_2]^u\\ \mathcal{D}_2\\ \psi\end{matrix}\quad\mathcal{D}_4}{\chi}\,(\mathrm{E})}{\chi}\,\vee E\\ \mathcal{D}_5\end{matrix}$$

A small difficulty arises from the fact that (E) itself could be an instance of $\vee E$ and, in this case, it might also discharge some open applications of auxiliary rules occurring in the subtree $\mathcal{D}_3$. Applying the permutation conversion in such a situation would turn these auxiliary rules into *open* assumptions. However, this problem has an easy solution, since it is always possible to 'shift' the corresponding auxiliary rules from the lower application of $\vee E$ to the upper application of $\vee E$. An explanation of the procedure is provided in Appendix C.

Clearly, after a finite number of permutations, we obtain a pure deduction in which σ has been turned into a cut segment of length 1. We now eliminate the cut formula ψ using one of the following *detour conversions* ($i = 1, 2$):

$$\psi = \varphi_1 \wedge \varphi_2: \quad \dfrac{\dfrac{\begin{matrix}\mathcal{D}_1 \\ \varphi_1\end{matrix} \quad \begin{matrix}\mathcal{D}_2 \\ \varphi_2\end{matrix}}{\varphi_1 \wedge \varphi_2}\wedge I}{\begin{matrix}\varphi_i \\ \mathcal{D}_3\end{matrix}}\wedge E \quad \rightsquigarrow \quad \begin{matrix}\mathcal{D}_i \\ \varphi_i \\ \mathcal{D}_3\end{matrix}$$

$$\psi = \varphi_1 \to \varphi_2: \quad \dfrac{u\dfrac{\begin{matrix}[\varphi_1]^u \\ \mathcal{D}_1 \\ \varphi_2\end{matrix}}{\varphi_1 \to \varphi_2}\to I \quad \begin{matrix}\mathcal{D}_2 \\ \varphi_1\end{matrix}}{\begin{matrix}\varphi_2 \\ \mathcal{D}_3\end{matrix}}\to E \quad \rightsquigarrow \quad \begin{matrix}\mathcal{D}_2 \\ \varphi_1 \\ \mathcal{D}_1 \\ \varphi_2 \\ \mathcal{D}_3\end{matrix}$$

$$\psi = \varphi_1 \vee \varphi_2: \quad u\dfrac{\dfrac{\begin{matrix}[\Theta]^u \\ \mathcal{D}_3 \\ \varphi_i\end{matrix}}{\varphi_1 \vee \varphi_2}\vee I \quad \begin{matrix}[\Theta \Rightarrow \varphi_1]^u \\ \mathcal{D}_1 \\ \chi\end{matrix} \quad \begin{matrix}[\Theta \Rightarrow \varphi_2]^u \\ \mathcal{D}_2 \\ \chi\end{matrix}}{\begin{matrix}\chi \\ \mathcal{D}_4\end{matrix}}\vee E \quad \rightsquigarrow \quad \begin{matrix}\mathcal{D}_i\langle \mathcal{D}_3 : \Theta \Rightarrow \varphi_i\rangle \\ \chi \\ \mathcal{D}_4\end{matrix}$$

where $\mathcal{D}_i\langle \mathcal{D}_3 : \Theta \Rightarrow \varphi_i\rangle$ is the proof tree obtained from Lemma 7.1. It is easy to see that every new cut formula possibly arising from the conversion is of smaller rank than ψ. Hence, the resulting deduction $\mathcal{D}'$ satisfies $\text{CR}(\mathcal{D}') < \text{CR}(\mathcal{D})$. □

Given an arbitrary deduction $\mathcal{D}$, we can now first apply Proposition 6.3 in order to convert $\mathcal{D}$ into a pure deduction $\mathcal{D}'$. Afterwards, we can repeatedly apply Lemma 7.2 in order to construct a sequence of pure deductions $\mathcal{D}_1$, $\mathcal{D}_2$, $\mathcal{D}_3$, etc., such that the associated cut ranks form a descending chain $\text{CR}(\mathcal{D}') > \text{CR}(\mathcal{D}_1) > \text{CR}(\mathcal{D}_2) > \ldots$ After a finite number of steps, this must yield a *normal deduction*, so we now obtain the desired normalization result for our system.

Theorem 7.3 (Normalization Theorem) *Let Γ be a set of formulas. Every deduction for $\Gamma \vdash_{\mathrm{N}} \varphi$ can be transformed into a normal deduction for $\Gamma \vdash_{\mathrm{N}} \varphi$.*

8 Properties of Normal Deductions

We will now establish an unrestricted *subformula property* for our system. First, we need to find a suitable notion of a *track* in a deduction. Intuitively, a track may be seen as a sequence of formula occurrences that are sufficiently 'related' to each other by their syntactic shape. In particular, any two consecutive formulas in a track (except possibly those involved in $\bot_{\mathrm{i}}$) should satisfy some kind of *local* subformula property, in the sense that one of these formulas is always a subformula of the other (cf. [21, Sect. 6.2]). Defining tracks for our system turns out to be somewhat complicated, because a deduction may now also contain premises of *assumption rules*, and it is not obvious what the successor of such a formula should be in a track. In order to address this difficulty, we first need to take a closer look at the relationship between the *auxiliary rules* and the *side rules* of an application of $\vee E$. We will say that φ is a P-*formula* (resp., a C-*formula*) of an assumption rule R, if φ is a premise (resp., the conclusion) of some subrule of R.[6] Given any instance of $\lor E$ with set of auxiliary rules Θ and major premise $\varphi \lor \psi$, it is now easy to verify the following facts:

- if a formula χ occurs as a P-formula in an auxiliary rule $R \in \Theta$, then χ also occurs as a C-formula in each of the two side rules $\Theta \Rightarrow \varphi$ and $\Theta \Rightarrow \psi$,
- if χ occurs as a P-formula in one of the side rules $\Theta \Rightarrow \varphi$ and $\Theta \Rightarrow \psi$, then χ also occurs as a C-formula in at least one auxiliary rule $R \in \Theta$.

In other words, for every P-formula of an auxiliary rule of $\lor E$, there exists a corresponding C-formula in each of the two side rules of this application of $\lor E$. And for every P-formula of a side rule, there exists a corresponding C-formula in one of the auxiliary rules. The notion of a track is now defined as follows.

Definition 8.1 [Track] A *track* in a deduction $\mathcal{D}$ is a sequence of pairwise distinct formula occurrences $\xi_1, \ldots, \xi_n$ in $\mathcal{D}$ such that the following holds:[7]

(i) ξ_1 is a hypothesis of $\mathcal{D}$, but not a C-formula of a side rule of $\lor E$.

(ii) Each ξ_i with $1 \leq i < n$ is not a minor premise of $\to E$ and it holds:
   (a) If ξ_i is the major premise $\varphi \lor \psi$ of an application of $\lor E$, then ξ_{i+1} is the conclusion of one of the side rules of this application.
   (b) If ξ_i is a P-formula of an auxiliary rule (resp., side rule) of $\lor E$, then ξ_{i+1} is an occurrence of the corresponding C-formula belonging to one of the side rules (resp., auxiliary rules) of this application of $\lor E$.
   (c) If ξ_i is neither a major premise of $\lor E$ nor a premise of an assumption rule, then ξ_{i+1} is the formula occurrence standing directly below ξ_i.

(iii) ξ_n is a minor premise of $\to E$ or the conclusion of $\mathcal{D}$.

A track in a deduction $\mathcal{D}$ is said to be a *main track*, if it ends with the conclusion of $\mathcal{D}$. For example, the normal deduction displayed below contains exactly six tracks: the main tracks of the deduction are the sequences $\tau_1 = (1, 5, 9, 10, 11, 18)$ and $\tau_2 = (1, 5, 15, 16, 17, 18)$. And the other four tracks are the sequences $\tau_3 = (6, 8, 3, 4)$, $\tau_4 = (12, 14, 3, 4)$, $\tau_5 = (2, 7)$ and $\tau_6 = (2, 13)$.

$$
\dfrac{
z\;\dfrac{(1)\ \neg p \to (\varphi \lor \psi) \qquad u\,\dfrac{\dfrac{(2)\ [p]^u}{(3)\ \bot}[p \Rightarrow \bot]^z}{(4)\ \neg p}{\to}I}{(5)\ \varphi \lor \psi}{\to}E
\qquad
\dfrac{w\,\dfrac{v\,\dfrac{\dfrac{(6)\ [\neg p]^w \quad (7)\ [p]^v}{(8)\ \bot}{\to}E}{(9)\ \varphi}[(p \Rightarrow \bot) \Rightarrow \varphi]^z}{(10)\ \neg p \to \varphi}{\to}I}{(11)\ (\neg p \to \varphi) \lor (\neg p \to \psi)}\lor I
\qquad
\dfrac{y\,\dfrac{x\,\dfrac{\dfrac{(12)\ [\neg p]^y \quad (13)\ [p]^x}{(14)\ \bot}{\to}E}{(15)\ \psi}[(p \Rightarrow \bot) \Rightarrow \psi]^z}{(16)\ \neg p \to \psi}{\to}I}{(17)\ (\neg p \to \varphi) \lor (\neg p \to \psi)}\lor I
}{(18)\ (\neg p \to \varphi) \lor (\neg p \to \psi)}\lor E
$$

Lemma 8.2 *Let $\mathcal{D}$ be a deduction. Every formula occurrence in $\mathcal{D}$ belongs to some track. Thus, in particular, $\mathcal{D}$ has at least one main track.*[8]

[6] In this definition, a level-0 rule ψ is treated as a zero-premise rule with conclusion ψ.

[7] We say that two formula occurrences are *distinct*, if they are *distinct occurrences* of formulas (but not necessarily occurrences of *distinct formulas*). Requiring $\xi_1, \ldots, \xi_n$ to be pairwise distinct is not essential here, but it simplifies our treatment a little: without this restriction, a deduction might also contain *infinitely many* tracks. Further details are provided in [13].

[8] Here, we are assuming the discharge convention introduced at the end of Section 3.

The proof idea is explained in Appendix D. By a *strictly positive subformula* (or an *sp-subformula*) of φ, we will mean (an occurrence of) a subformula of φ which is not included in the antecedent of an implication. A formula φ is said to be a *Harrop formula*, if it does not contain an sp-subformula with main connective $\vee$. Obviously, every track in a deduction may also be divided into a sequence of *segments*, rather than into a sequence of single occurrences of formulas. Here, a segment is understood to be any (maximal) subsequence of a track that consists of repeated occurrences of the same formula. So, in particular, a segment may now also contain repetitions of formulas arising from the 'switching' between P-formulas and C-formulas. We will say that a segment σ is the *conclusion* (resp., a *premise*) of a rule application, if the first (resp., last) formula occurrence in σ is the conclusion (resp., a premise) of this application.

Proposition 8.3 (Structure of Tracks) *Let $\mathcal{D}$ be a normal deduction, let τ be a track in $\mathcal{D}$, and let $\sigma_1, \ldots, \sigma_n$ be the sequence of segments in τ. There exists a segment σ_k in τ, called the* minimum segment *of the track, such that:*

(i) *Each σ_i with $1 \leq i < k$ is a major premise of an elimination rule and the formula occurring in σ_i is an sp-subformula of the one occurring in σ_1.*

(ii) *If $k \neq n$, then σ_k is a premise of an introduction rule or a premise of $\bot_\mathsf{i}$.*

(iii) *Each σ_i with $k < i < n$ is a premise of an introduction rule and the formula occurring in σ_i is an sp-subformula of the one occurring in σ_n.*

Proof. Let $\mathcal{D}$ be a normal deduction, let τ be a track in $\mathcal{D}$, and let $\sigma_1, \ldots, \sigma_n$ be the sequence of segments in τ. We first show that every major premise of an E-rule in τ must precede all premises of I-rules and all premises of $\bot_\mathsf{i}$ in τ. Towards a contradiction, suppose that there exists a premise of an I-rule or a premise of $\bot_\mathsf{i}$ that precedes a major premise of an E-rule in τ. Then, clearly, there must be a segment σ_j which is both the conclusion of an I-rule or the conclusion of $\bot_\mathsf{i}$, and the major premise of an E-rule. By the pureness of $\mathcal{D}$, we know that σ_j cannot be a conclusion of $\bot_\mathsf{i}$, and it also cannot contain any premises of assumption rules: otherwise, the formula occurring in σ_j would be prime, and such a formula can never be the major premise of an E-rule. Hence, σ_j is the conclusion of an I-rule and every formula occurrence in σ_j, except the last one, is a minor premise of $\vee E$. But this means that σ_j is a *cut segment*, which is a contradiction to the assumption that $\mathcal{D}$ is normal. Thus, every major premise of an E-rule precedes all premise of I-rules and all premises of $\bot_\mathsf{i}$ in τ. Using a similar argument, it is also easy to show that there can be at most one premise of $\bot_\mathsf{i}$ in τ, and this premise must precede all premises of I-rules in τ.

The desired minimum segment can now be determined as follows: if τ contains a premise of an I-rule or a premise of $\bot_\mathsf{i}$, then let σ_k be the segment that ends with the first such formula in τ. Otherwise, let σ_k be the end segment of τ. By what was said above, every segment preceding σ_k in τ is a major premise of an E-rule, and every segment succeeding σ_k in τ, except the last one, is a premise of an I-rule. Hence, σ_k does in fact have the desired properties.□

We are now ready to establish an unrestricted version of the subformula property for our system: every formula occurring in a normal deduction $\mathcal{D}$ is a subformula of an open hypothesis or a subformula of the conclusion of $\mathcal{D}$.

Theorem 8.4 (Subformula Property) *Every formula occurring in a normal deduction for* $\Gamma \vdash_{\overline{N}} \varphi$ *is a subformula of some formula in* $\Gamma \cup \{\varphi\}$.

Proof. Let Γ be a set of formulas and let $\mathcal{D}$ be a normal deduction for $\Gamma \vdash_{\overline{N}} \varphi$. Towards a contradiction, suppose that $\mathcal{D}$ contains a formula ψ which is *not* a subformula of some element of $\Gamma \cup \{\varphi\}$. Without loss of generality, we may assume that ψ is also a *largest* such formula in $\mathcal{D}$, i.e., every formula of rank greater than $rk(\psi)$ in $\mathcal{D}$ is a subformula of some element of $\Gamma \cup \{\varphi\}$. Let τ be a track to which ψ belongs and let $\xi_1, \ldots, \xi_n$ be the sequence of formulas in τ.

First, assume that ψ occurs after the minimum segment in τ. Then, by Proposition 8.3, ψ is a subformula of ξ_n, which is either the conclusion of $\mathcal{D}$ or a minor premise of $\to E$. If ξ_n would be the conclusion of $\mathcal{D}$, then ψ would be a subformula of φ, which is a contradiction to our assumption. Thus, ξ_n must be a minor premise of $\to E$. But then, clearly, ψ is a proper subformula of the associated major premise $\xi_n \to \chi$, which is a formula of rank greater than $rk(\psi)$ and therefore a subformula of some element of $\Gamma \cup \{\varphi\}$. As a consequence, ψ must also be such a subformula, so we arrive again at a contradiction.

Next, assume that ψ occurs in the minimum segment or before the minimum segment in τ. Then, by Proposition 8.3, ψ is a subformula of the hypothesis ξ_1 occurring at the top of τ. Clearly, ξ_1 cannot be discharged by an assumption rule: in this case, it would be a C-formula of an auxiliary rule of $\vee E$, so there would be a corresponding P-formula (belonging to a side rule of this application of $\vee E$) which would precede ξ_1 in τ. Hence, ξ_1 is either an open hypothesis, or it is discharged by $\to I$. If ξ_1 is an open hypothesis, then ψ is a subformula of some element of Γ, which is a contradiction to our assumption about ψ. And if ξ_1 is discharged by an application of $\to I$, then ψ must be a proper subformula of the conclusion $\xi_1 \to \chi$ of this application, which is a formula of rank greater than $rk(\psi)$ and therefore a subformula of some element of $\Gamma \cup \{\varphi\}$. Hence, ψ must also be such a subformula, so we obtain again a contradiction. □

As an immediate consequence of this result, we now also obtain a *separation theorem* for our system. Furthermore, using Proposition 8.3, it is easy to establish a strong form of the *disjunction property* for KP in a proof-theoretical manner. The argument is essentially the same as in [17, pp. 55–56].

Theorem 8.5 (Separation) *The only primitive rules applied in a normal deduction for* $\Gamma \vdash_{\overline{N}} \varphi$ *are the rules for the logical symbols occurring in* $\Gamma \cup \{\varphi\}$.

Theorem 8.6 (Disjunction Property) *Let* Γ *be a set of Harrop formulas. If we have* $\Gamma \vdash_{\overline{N}} \varphi \vee \psi$ *in our system, then also* $\Gamma \vdash_{\overline{N}} \varphi$ *or* $\Gamma \vdash_{\overline{N}} \psi$.

9 Conclusion

We introduced a higher-level natural deduction system for KP, presented a new normalization procedure and established an unrestricted subformula property for our system. Our results might be of interest not only with regard to KP itself, but also with regard to the proof theory of non-classical logics in general. In this area of research, one focuses primarily on *sequent-style* proof systems (such as, e.g., hypersequents, nested sequent calculi or labelled sequent calculi), but natural deduction systems often play only a minor role. Using the expressive power of higher-level rules, one might be able to overcome this limitation and to find analytic natural deduction systems for logics that do not admit a full subformula property within the framework of ordinary natural deduction.

For future work, we plan to adapt our normalization strategy to other non-classical logics, which might include not only intermediate logics different from KP, but also various modal logics. This is not entirely trivial, because some of our arguments are tailored to cases in which certain assumption rules are required to be *disjunction-free* (this applies primarily to the proof of Lemma 6.2). Apart from considering specific examples of non-classical logics, we also intend to develop a more general procedure which allows to construct higher-level natural deduction systems for modal and intermediate logics in a systematic way. This might be done by converting axiom schemes directly into 'equivalent' natural deduction rules and by establishing a generic normalization result for the proof systems thus obtained. First attempts in this direction have already led to various promising ideas that we plan to address in a subsequent study.

Acknowledgements. I am grateful to the anonymous reviewers for their insightful comments and suggestions. I would also like to thank Marianna Girlando and Peter Schroeder-Heister for their feedback on another paper of mine [13], whose content is closely related to the work presented here.

Appendix

A Proof of Lemma 4.2

Let Γ be a set of assumption rules and suppose that there exists a quasi-deduction $\mathcal{D}$ for $\Gamma \vdash_{\mathrm{N}} \varphi$ in the system NKP. Using induction on the structure of $\mathcal{D}$, we show that there also exists a Hilbert-style proof for $F(\Gamma) \vdash_{\mathrm{H}} \varphi$ in HKP.

For the base case, assume that $\mathcal{D} = \varphi$ is a one-element quasi-deduction. Then, since $\varphi \in \Gamma$ and $F(\varphi) = \varphi$, we trivially also have $F(\Gamma) \vdash_{\mathrm{H}} \varphi$ in HKP.

For the inductive step, suppose that $\mathcal{D}$ ends with a primitive rule of NKP or with an assumption rule of level $\ell \geq 1$. We only consider the following cases.

Case 1: $\mathcal{D}$ ends with an application of $\vee E$. In this case, $\mathcal{D}$ is of the form

$$\dfrac{\begin{matrix}[\Theta]^u \\ \mathcal{D}_3 \\ \psi_1 \vee \psi_2\end{matrix} \qquad \begin{matrix}[\Theta \Rightarrow \psi_1]^u \\ \mathcal{D}_1 \\ \varphi\end{matrix} \qquad \begin{matrix}[\Theta \Rightarrow \psi_2]^u \\ \mathcal{D}_2 \\ \varphi\end{matrix}}{\varphi}\,{\vee}E\ (u)$$

where $\Theta = \{\Pi_1 \Rightarrow \bot, \dots, \Pi_n \Rightarrow \bot\}$ is a finite set of negative rules, each $\mathcal{D}_i$ with $i = 1, 2$ is a quasi-deduction for $\Gamma, (\Theta \Rightarrow \psi_i) \vdash_{\mathrm{N}} \varphi$, and $\mathcal{D}_3$ is a quasi-deduction for $\Gamma, \Theta \vdash_{\mathrm{N}} \psi_1 \vee \psi_2$. Using the induction hypothesis, we first obtain

$$F(\Gamma), F(\Theta \Rightarrow \psi_i) \vdash_{\mathrm{H}} \varphi \quad \text{for } i = 1, 2, \tag{A.1}$$

$$F(\Gamma), F(\Theta) \vdash_{\mathrm{H}} \psi_1 \vee \psi_2 \tag{A.2}$$

in the system HKP. For simplicity, let us assume that $\Theta \neq \emptyset$. In this case, we have $F(\Theta \Rightarrow \psi_i) = \bigwedge F(\Theta) \to \psi_i$ for each $i = 1, 2$. Thus, from (A.1), it follows

$$F(\Gamma) \vdash_{\mathrm{H}} (\bigwedge F(\Theta) \to \psi_i) \to \varphi \quad \text{for } i = 1, 2 \tag{A.3}$$

by the deduction theorem for HKP (Theorem 2.2). For every k with $1 \leq k \leq n$, let now π_k be the formula given by $\pi_k := \bigwedge F(\Pi_k)$. Then, clearly, we have $F(\Pi_k \Rightarrow \bot) = \neg\pi_k$ for all $1 \leq k \leq n$, so it follows $F(\Theta) = \{\neg\pi_1, \dots, \neg\pi_n\}$. Consequently, from (A.2), we obtain $F(\Gamma), \neg\pi_1, \dots, \neg\pi_n \vdash_{\mathrm{H}} \psi_1 \vee \psi_2$. Using the deduction theorem and a subsequent application of axiom scheme (KP), this implies $F(\Gamma), \neg\pi_2, \dots, \neg\pi_n \vdash_{\mathrm{H}} (\neg\pi_1 \to \psi_1) \vee (\neg\pi_1 \to \psi_2)$. By repeating the argument and by using intuitionistic reasoning, it is easy to derive

$$F(\Gamma) \vdash_{\mathrm{H}} (\bigwedge F(\Theta) \to \psi_1) \vee (\bigwedge F(\Theta) \to \psi_2). \tag{A.4}$$

The desired Hilbert-style proof for $F(\Gamma) \vdash_{\mathrm{H}} \varphi$ can now be obtained from (A.3) and (A.4) using axiom scheme (A6) and three applications of modus ponens.

Case 2: $\mathcal{D}$ ends with an application of an assumption rule R of level $\ell \geq 1$. Clearly, this application must be open, so $\mathcal{D}$ is a quasi-deduction for $\Delta, R \vdash_{\mathrm{H}} \varphi$, where Δ is a set of rules satisfying $\Gamma = \Delta \cup \{R\}$. By assumption, R is of the form $R = (\Pi \Rightarrow \varphi)$, where $\Pi = \{\Sigma_1 \Rightarrow \psi_1, \dots, \Sigma_n \Rightarrow \psi_n\}$ is a set of rules whose maximal level is $\ell - 1$. As a consequence, $\mathcal{D}$ must be of the form

$$\dfrac{\begin{matrix}[\Sigma_1]^u \\ \mathcal{D}_1 \\ \psi_1\end{matrix} \qquad \cdots \qquad \begin{matrix}[\Sigma_n]^u \\ \mathcal{D}_n \\ \psi_n\end{matrix}}{\varphi}\,R \quad (u)$$

where each $\mathcal{D}_i$ is a quasi-deduction for $\Delta, \Sigma_i \vdash_{\mathrm{N}} \psi_i$. Let now i with $1 \leq i \leq n$ be arbitrary and let $R_i := (\Sigma_i \Rightarrow \psi_i)$. By induction hypothesis, we have $F(\Delta), F(\Sigma_i) \vdash_{\mathrm{H}} \psi_i$ in the system HKP. We first show that this yields $F(\Delta) \vdash_{\mathrm{H}} F(R_i)$. For simplicity, let us assume that $\Sigma_i \neq \emptyset$. From $F(\Delta), F(\Sigma_i) \vdash_{\mathrm{H}} \psi_i$, it follows $F(\Delta) \vdash_{\mathrm{H}} \bigwedge F(\Sigma_i) \to \psi_i$ by Theorem 2.2 and intuitionistic reasoning. But then, since $F(R_i) = \bigwedge F(\Sigma_i) \to \psi_i$, we also have $F(\Delta) \vdash_{\mathrm{H}} F(R_i)$, as desired. Now, as we have seen, it holds $F(\Delta) \vdash_{\mathrm{H}} F(R_i)$ for all $1 \leq i \leq n$. Hence, using axiom (A4) and modus ponens, we may conclude $F(\Delta) \vdash_{\mathrm{H}} \bigwedge F(\Pi)$. Moreover, since $F(R) = \bigwedge F(\Pi) \to \varphi$, it also holds $F(R) \vdash_{\mathrm{H}} \bigwedge F(\Pi) \to \varphi$. Together with $F(\Delta) \vdash_{\mathrm{H}} \bigwedge F(\Pi)$, this implies $F(\Delta), F(R) \vdash_{\mathrm{H}} \varphi$ by modus ponens. □

B Proof of Lemma 7.1

Let Θ be a finite set of prime rules, let $\mathcal{D}_1$ be a pure quasi-deduction for $\Gamma, (\Theta \Rightarrow \varphi) \vdash_{\mathrm{N}} \psi$ and let $\mathcal{D}_2$ be a pure quasi-deduction for $\Delta, \Theta \vdash_{\mathrm{N}} \varphi$. We establish the existence of $\mathcal{D}_1\langle\mathcal{D}_2 : \Theta \Rightarrow \varphi\rangle$ by induction on the level of $R := (\Theta \Rightarrow \varphi)$.

For the base case, assume that R is of level $\ell = 0$, so Θ is empty and $R = \varphi$. In this case, we define $\mathcal{D}_1\langle\mathcal{D}_2 : \Theta \Rightarrow \varphi\rangle$ to be the pure quasi-deduction obtained by writing the tree $\mathcal{D}_2$ above each open occurrence of the hypothesis $R = \varphi$ in $\mathcal{D}_1$. Clearly, φ is the only new cut formula that may arise in the course of this construction, so $\mathcal{D}_1\langle\mathcal{D}_2 : \Theta \Rightarrow \varphi\rangle$ does in fact have the desired properties.

For the inductive step, suppose that R is of level $\ell \geq 1$, so $\Theta = \{\Pi_1 \Rightarrow \chi_1, \ldots, \Pi_n \Rightarrow \chi_n\} \neq \emptyset$, where each $\Pi_i \Rightarrow \chi_i$ is a prime rule of level $< \ell$. We select a topmost open application of $\Theta \Rightarrow \varphi$ in $\mathcal{D}_1$, i.e., we write $\mathcal{D}_1$ in the form

$$\dfrac{\begin{matrix}[\Pi_1]^u \\ \mathcal{D}_1^* \\ \chi_1\end{matrix} \quad \cdots \quad \begin{matrix}[\Pi_n]^u \\ \mathcal{D}_n^* \\ \chi_n\end{matrix}}{\begin{matrix}\varphi \\ \mathcal{D}_0 \\ \psi\end{matrix}} {\scriptstyle u}\ \Theta \Rightarrow \varphi$$

where none of $\mathcal{D}_1^*, \ldots, \mathcal{D}_n^*$ contains another open application of $\Theta \Rightarrow \varphi$. For each i with $1 \leq i \leq n$, let now Σ_i be the smallest set of rules such that $\mathcal{D}_i^*$ is a quasi-deduction for $\Sigma_i, \Pi_i \vdash_{\mathrm{N}} \chi_i$. By assumption, $\mathcal{D}_2$ is a quasi-deduction for $\Delta, (\Pi_1 \Rightarrow \chi_1), \ldots, (\Pi_n \Rightarrow \chi_n) \vdash_{\mathrm{N}} \varphi$, and each $\Pi_i \Rightarrow \chi_i$ is of level $< \ell$. Hence, using the induction hypothesis n times, we obtain the pure quasi-deduction

$$\mathcal{D}^* := \mathcal{D}_2 \langle \mathcal{D}_1^* : \Pi_1 \Rightarrow \chi_1 \rangle \cdots \langle \mathcal{D}_n^* : \Pi_n \Rightarrow \chi_n \rangle$$

for $\Delta, \Sigma_1, \ldots, \Sigma_n \vdash_{\mathrm{N}} \varphi$ such that all cut formulas occurring in $\mathcal{D}^*$, except possibly $\chi_1, \ldots, \chi_n$, already occur in $\mathcal{D}_2$ or in one of the trees $\mathcal{D}_1^*, \ldots, \mathcal{D}_n^*$ (and therefore in $\mathcal{D}_1$). But note that, by assumption, each χ_i is prime, and such a formula can never be a cut formula. Thus, every cut formula in $\mathcal{D}^*$ already occurs as a cut formula in $\mathcal{D}_1$ or $\mathcal{D}_2$. We now construct the pure quasi-deduction

$$\begin{matrix}\mathcal{D}^* \\ \varphi \\ \mathcal{D}_0 \\ \psi\end{matrix}$$

Clearly, φ is the only new cut formula possibly arising from this construction. Hence, by repeating the procedure for the other open applications of $\Theta \Rightarrow \varphi$, we finally obtain the desired quasi-deduction $\mathcal{D}_1 \langle \mathcal{D}_2 : \Theta \Rightarrow \varphi \rangle$ for $\Gamma, \Delta \vdash_{\mathrm{N}} \psi$. □

C Supplement to the Proof of Lemma 7.2

We need to show that the permutation conversion can also be performed, if the elimination rule (E) at the bottom of a cut segment is itself an instance of $\vee E$. In order to understand the problem, let us consider a deduction $\mathcal{D}$ of the form

$$\dfrac{\dfrac{\begin{matrix}[\Theta]^u\ [\Omega]^v \\ \mathcal{D}_3 \\ \varphi_1 \vee \varphi_2\end{matrix} \quad \begin{matrix}[\Theta \Rightarrow \varphi_1]^u\ [\Omega]^v \\ \mathcal{D}_1 \\ \psi_1 \vee \psi_2\end{matrix} \quad \begin{matrix}[\Theta \Rightarrow \varphi_2]^u\ [\Omega]^v \\ \mathcal{D}_2 \\ \psi_1 \vee \psi_2\end{matrix}}{\psi_1 \vee \psi_2} {\scriptstyle u}\ \vee E \quad \begin{matrix}[\Omega \Rightarrow \psi_1]^v \\ \mathcal{D}_4 \\ \chi\end{matrix} \quad \begin{matrix}[\Omega \Rightarrow \psi_2]^v \\ \mathcal{D}_5 \\ \chi\end{matrix}}{\begin{matrix}\chi \\ \mathcal{D}_6\end{matrix}} {\scriptstyle v}\ \vee E$$

The upper application of $\vee E$ in this proof tree has the set of auxiliary rules Θ, and the lower application has the set of auxiliary rules Ω. As indicated by the label v, the lower application of $\vee E$ also discharges some of the rules from Ω in the subtree $\mathcal{D}_3$, so we cannot apply the desired permutation conversion directly, as this would turn the respective rules into *open* assumptions.

In order to resolve this issue, we will now demonstrate that one can 'shift' each of the auxiliary rules from Ω in $\mathcal{D}_3$ to the upper application of $\vee E$. For this purpose, let us say that a quasi-deduction is *structural*, if it contains only applications of assumption rules, but no applications of primitive rules. The following fact can be verified by an easy induction on the level of a rule $\Gamma \Rightarrow \varphi$.

Fact C.1 (Modus Ponens for Assumption Rules) *For every assumption rule* $\Gamma \Rightarrow \varphi$*, there exists a structural quasi-deduction for* $(\Gamma \Rightarrow \varphi), \Gamma \mid\!\!\frac{\;}{\text{N}}\, \varphi$.

Suppose now that, in the deduction above, Θ and Ω are the sets of negative rules given by $\Theta = \{\Pi_1 \Rightarrow \bot, \ldots, \Pi_n \Rightarrow \bot\}$ and $\Omega = \{\Sigma_1 \Rightarrow \bot, \ldots, \Sigma_m \Rightarrow \bot\}$. By applying Fact C.1 to the rules in Ω, we may obtain, for each $1 \leq i \leq m$, a structural quasi-deduction $\mathcal{D}_i^*$ for $(\Sigma_i \Rightarrow \bot), \Sigma_i \mid\!\!\frac{\;}{\text{N}}\, \bot$ and thus for $\Omega, \Sigma_i \mid\!\!\frac{\;}{\text{N}}\, \bot$. Let now $k = 1, 2$ be arbitrary and let $\mathcal{D}_k^+$ be the quasi-deduction obtained from the subtree $\mathcal{D}_k$ by rewriting every open application of $\Theta \Rightarrow \varphi_k$ as follows:

$$\dfrac{\begin{matrix}[\Pi_1]^w \\ \mathcal{D}_1' \\ \bot\end{matrix} \quad \cdots \quad \begin{matrix}[\Pi_n]^w \\ \mathcal{D}_n' \\ \bot\end{matrix}}{\begin{matrix}\varphi_k \\ \mathcal{D}_0\end{matrix}}\,w\;\Theta \Rightarrow \varphi_k \quad \rightsquigarrow \quad \dfrac{\begin{matrix}[\Pi_1]^w \\ \mathcal{D}_1' \\ \bot\end{matrix} \quad \cdots \quad \begin{matrix}[\Pi_n]^w \\ \mathcal{D}_n' \\ \bot\end{matrix} \quad \begin{matrix}\Omega\;[\Sigma_1]^w \\ \mathcal{D}_1^* \\ \bot\end{matrix} \quad \cdots \quad \begin{matrix}\Omega\;[\Sigma_m]^w \\ \mathcal{D}_m^* \\ \bot\end{matrix}}{\begin{matrix}\varphi_k \\ \mathcal{D}_0\end{matrix}}\,w\;\Theta, \Omega \Rightarrow \varphi_k$$

In the course of this conversion, additional open occurrences of the rules from Ω are added to the proof tree, and each open application of $\Theta \Rightarrow \varphi_k$ is replaced by an open application of $\Theta, \Omega \Rightarrow \varphi_k$. Note that, since every $\mathcal{D}_i^*$ is structural, the conversion does not introduce any new cut formulas. Using the proof trees $\mathcal{D}_1^+$ and $\mathcal{D}_2^+$ thus obtained, we now transform the whole deduction $\mathcal{D}$ into

$$\dfrac{\dfrac{\begin{matrix}[\Theta,\Omega]^u \\ \mathcal{D}_3 \\ \varphi_1 \vee \varphi_2\end{matrix} \quad \begin{matrix}[\Theta,\Omega \Rightarrow \varphi_1]^u\;[\Omega]^v \\ \mathcal{D}_1^+ \\ \psi_1 \vee \psi_2\end{matrix} \quad \begin{matrix}[\Theta,\Omega \Rightarrow \varphi_2]^u\;[\Omega]^v \\ \mathcal{D}_2^+ \\ \psi_1 \vee \psi_2\end{matrix}}{\psi_1 \vee \psi_2}\,u\;\vee E \quad \begin{matrix}[\Omega \Rightarrow \psi_1]^v \\ \mathcal{D}_4 \\ \chi\end{matrix} \quad \begin{matrix}[\Omega \Rightarrow \psi_2]^v \\ \mathcal{D}_5 \\ \chi\end{matrix}}{\begin{matrix}\chi \\ \mathcal{D}_6\end{matrix}}\,v\;\vee E$$

Clearly, this deduction has the same cut rank as the original one. However, the lower application of $\vee E$ does not discharge any assumption rules in the subtree $\mathcal{D}_3$ anymore, so one can now safely apply the desired permutation conversion.

D Proof of Lemma 8.2

We first define the notion of a *pre-track* in a quasi-deduction in exactly the same way as the notion of a track, except that (1) a pre-track may also start with an arbitrary C-formula belonging to an open application of an assumption rule, and (2) a pre-track always stops as soon as a P-formula of an open application of an assumption rule is encountered. Using induction on the structure of an arbitrary quasi-deduction $\mathcal{D}$, it is now possible to show that every formula occurrence in $\mathcal{D}$ belongs to some pre-track. The desired statement now follows from the fact that, if $\mathcal{D}$ is a *deduction* (i.e., if each assumption rule of level greater than 0 is discharged in $\mathcal{D}$), then every pre-track in $\mathcal{D}$ is also a track. □

References

[1] Brünnler, K., *Deep Sequent Systems for Modal Logic*, in: G. Governatori, I. Hodkinson and Y. Venema, editors, *Advances in Modal Logic (AiML), vol. 6*, College Publications, London, 2006 pp. 107–119.

[2] Chagrov, A. and M. Zakharyaschev, "Modal Logic," Clarendon Press, Oxford, 1997.

[3] Ciabattoni, A., N. Galatos and K. Terui, *From Axioms to Analytic Rules in Nonclassical Logics*, in: *LICS '08: Proceedings of the 2008 23rd Annual IEEE Symposium on Logic in Computer Science*, IEEE Computer Society, 2008 pp. 229–240.

[4] Condoluci, A. and M. Manighetti, *Admissible Tools in the Kitchen of Intuitionistic Logic*, in: S. Berardi and A. Miquel, editors, *Proceedings of the 7th International Workshop on Classical Logic and Computation*, Electronic Proceedings in Theoretical Computer Science, vol. 281, 2018 pp. 10–23, arXiv:1810.07372 [cs.LO].
URL `https://doi.org/10.48550/arXiv.1810.07372`

[5] Curry, H. B., *Functionality in Combinatory Logic*, Proceedings of the National Academy of Sciences of the United States of America **20** (1934), pp. 584–590.

[6] Dyckhoff, R. and S. Negri, *Proof Analysis in Intermediate Logics*, Archive for Mathematical Logic **51** (2012), pp. 71–92.

[7] Dyckhoff, R. and S. Negri, *Geometrisation of First-Order Logic*, The Bulletin of Symbolic Logic **21** (2015), pp. 123–163.

[8] Howard, W. A., *The Formulae-As-Types Notion of Construction*, in: J. P. Seldin and J. R. Hindley, editors, *To H. B. Curry: Essays on Combinatory Logic, Lambda Calculus and Formalism*, Academic Press, London, 1980 pp. 479–490.

[9] Kreisel, G. and H. Putnam, *Eine Unableitbarkeitsbeweismethode für den Intuitionistischen Aussagenkalkül*, Archiv für mathematische Logik und Grundlagenforschung **3** (1957), pp. 74–78.

[10] Lahav, O., *From Frame Properties to Hypersequent Rules in Modal Logics*, in: *LICS '13: Proceedings of the 2013 28th Annual ACM/IEEE Symposium on Logic in Computer Science*, IEEE Computer Society, 2013 pp. 408–417.

[11] Łukasiewicz, J., *On the Intuitionistic Theory of Deduction*, Indagationes Mathematicae **14** (1952), pp. 202–212.

[12] Mancosu, P., S. Galvan and R. Zach, "An Introduction to Proof Theory," Oxford University Press, Oxford, 2021.

[13] Müller, V., *Higher-Level Natural Deduction and Normalization for Inquisitive Logic* (2023), unpublished manuscript.

[14] Müller, V., *On the Proof Theory of Inquisitive Logic* (2023), MSc thesis. Institute for Logic, Language and Computation, University of Amsterdam.
URL `https://eprints.illc.uva.nl/id/eprint/2278/1/MoL-2023-26.text.pdf`

[15] Negri, S., *Proof Analysis in Modal Logic*, Journal of Philosophical Logic **34** (2005), pp. 507–544.

[16] Pezlar, I., *Constructive Validity of a Generalized Kreisel-Putnam Rule* (2023), preprint. arXiv:2311.15376 [cs.LO].
URL `https://doi.org/10.48550/arXiv.2311.15376`

[17] Prawitz, D., "Natural Deduction: A Proof-Theoretical Study," Almqvist & Wiksell, Stockholm, Göteborg and Uppsala, 1965.

[18] Schroeder-Heister, P., *Untersuchungen zur regellogischen Deutung von Aussagenverknüpfungen* (1981), PhD thesis. University of Bonn.
URL `http://dx.doi.org/10.15496/publikation-68454`

[19] Schroeder-Heister, P., *A Natural Extension of Natural Deduction*, The Journal of Symbolic Logic **49** (1984), pp. 1284–1300.

[20] Schroeder-Heister, P., *The Calculus of Higher-Level Rules, Propositional Quantification, and the Foundational Approach to Proof-Theoretic Harmony*, Studia Logica **102** (2014), pp. 1185–1216.

[21] Troelstra, A. S. and H. Schwichtenberg, "Basic Proof Theory," Cambridge University Press, Cambridge, 1996.

Positive Modal Logic Over Finite MV-Chains

Wolfgang Poiger [1]

University of Luxembourg

Abstract

In this paper, we introduce and study the positive (*i.e.*, the negation-free and implication-free) fragment of many-valued modal logic based on a finite MV-chain as algebra of truth-degrees. Besides the usual Kripke semantics, we introduce richer relational semantics based on partially ordered sets with local constraints on admissible valuations. We study this logic algebraically via the varieties of modal n-valued positive MV-algebras. Utilizing prior work on natural dualities for n-valued positive MV-algebras and the close relationship between these algebras and their distributive skeletons, we prove an algebraic completeness theorem. Furthermore, by means of an example regarding canonicity, we illustrate how the richer relational semantics presented in this paper are 'better-behaved' with respect to the positive many-valued modal logic than Kripke (or other intermediate) semantics are.

Keywords: Positive modal logic, many-valued modal logic, finite MV-chains, positive MV-algebras, algebraic completeness, distributive skeleton.

1 Introduction

Positive modal logic, introduced by Dunn [10], is the negation-free (and implication-free) fragment of classical modal logic. In this paper, we introduce and study the positive fragment of the *many-valued* modal logics which use some *finite* MV-*chain* $\mathbf{Ł}_n$ as algebra of truth-degrees (see, *e.g.*, [4,13]).

Algebraically speaking, moving from classical modal logic to positive modal logic amounts to replacing Boolean algebras with operators by *modal distributive lattices* (Definition 2.1). Similarly, on the propositional level we shift from MV_n-algebras [12] to certain *positive* MV-*algebras* [2]. More specifically, we work over the variety of PMV_n-*algebras* generated by the *finite positive* MV-*chain* $\mathbf{PŁ}_n$ (Definition 2.8) for this purpose. While this shift clearly leads to *less structure* on the algebraic side, at the same time we consider relational semantics which carry *more structure* than usual Kripke frames (throughout this paper, we refer to Kripke frames as Set-*frames*).

The semantics of positive modal logic on Set-frames originally considered by Dunn [10] have the shortcoming that there are consequence pairs which, in

---

[1] I want to thank the three anonymous referees for their helpful comments.
The author is supported by the Luxembourg National Research Fund under the project PRIDE17/12246620/GPS.

the presence of negation, define the same class of Set-frames and are canonical, but which are not inter-derivable in positive modal logic. To remedy this shortcoming, Celani and Jansana [5,6] introduced relational semantics on *ordered* frames (which we refer to as Pos-*frames* in this paper, see Definition 2.2). Similarly, Hansoul and Teheux [13,23] studied richer semantics for $\mathbf{Ł}_n$-valued modal logic on *frames with local constraints* (which we refer to as Set_n-*frames*, see Definition 2.5). Taking both of these into account, we introduce semantics for $\mathbf{PŁ}_n$-valued modal logic on Pos_n-*frames*, that is, frames which carry a partial order *and* local constraints on valuations (Definition 3.2).

The methodology used in this paper is primarily algebraic. We introduce the varieties mPMV_n of *modal* PMV_n-*algebras* (Definition 4.1), and prove a truth lemma (Lemma 4.5) which leads to an algebraic completeness theorem (Theorem 4.8). Furthermore, we show that the *canonical frames* arising from modal PMV_n-algebras always are Pos_n-frames (Proposition 4.4) and that mPMV_n is generated by the *complex algebras* of Pos_n-frames (Theorem 4.7). The proofs of these main results rely on the author's prior work [21], where topological (natural) dualities for the varieties PMV_n were established. In particular, the relationship between PMV_n-algebras and their *distributive skeletons* (Definition 2.9) studied therein [21, Subsection 4.2] provide a useful tool here.

While the distinction between Set-frames and Pos_n-frames does not play a role for the *minimal* modal logic, we show that it does make a difference when it comes to axiomatic extensions thereof. Indeed, towards the end of this paper we study an example of two consequence pairs which are canonical in $\mathbf{Ł}_n$-valued modal logic and define the same Set_n-frames, but which are not inter-derivable in $\mathbf{PŁ}_n$-valued modal logic. In our semantics, this is reflected by the fact that these consequence pairs define different classes of Pos_n-frames (Section 5). A more exhaustive study of this topic is one of many directions for future research we propose in the conclusion of this paper.

2 Preliminaries

In this section, we set up the scene by recalling some important definitions and prior results from positive modal logic (Subsection 2.1), modal logic over finite MV-chains (Subsection 2.2) and positive MV_n-algebras (Subsection 2.3).

2.1 Positive modal logic

Positive modal logic (with truth-constants), introduced by Dunn in [10], is the $\{\wedge, \vee, 0, 1, \Box, \Diamond\}$-fragment of classical modal logic. Algebraically speaking, removing the negation from the language amounts to replacing the underlying variety BA of Boolean algebras by the variety DL of bounded distributive lattices. Since $\Box$ and $\Diamond$ are then no longer inter-definable, their interplay is instead described by Dunn's *positivity-axioms*, that is, the axioms (P1)-(P2) in the following.

Definition 2.1 [Modal distributive lattice] A *modal distributive lattice* is an algebra $\langle \mathbf{D}, \Box, \Diamond \rangle$, where $\mathbf{D} \in \mathsf{DL}$ is a bounded distributive lattice and $\Box, \Diamond \colon D \to D$ are unary operations satisfying the equations

$$\begin{array}{ll} \text{(B1)}\ \Box 1 = 1, & \text{(D1)}\ \Diamond 0 = 0, \\ \text{(B2)}\ \Box(x \wedge y) = \Box x \wedge \Box y, & \text{(D2)}\ \Diamond(x \vee y) = \Diamond x \vee \Diamond y, \\ \text{(P1)}\ \Box(x \vee y) \leq \Box x \vee \Diamond y, & \text{(P2)}\ \Box x \wedge \Diamond y \leq \Diamond(x \wedge y). \end{array}$$

We denote the variety of modal distributive lattices by mDL.

The exact relationship between the variety mDL and positive modal logic was described by Jansana in [14].

While Dunn [10] only considered the usual relational semantics on Set-frames (throughout this paper, we use the term Set-*frame* instead of *Kripke frame* to avoid confusion), Celani and Jansana [5] proposed some 'better-behaved' semantics based on the category of quasi-orders, which we now recall. For a slight increase in simplicity, in this paper we work with the category Pos of posets instead of quasi-orders.

Given a poset $(X, \leq)$, we use $\leq_{\mathrm{EM}}$ to denote the *Egli-Milner-lifting* of $\leq$ to the powerset $\mathcal{P}(X)$, defined for $A, B \subseteq X$ by

$$A \leq_{\mathrm{EM}} B \Leftrightarrow \begin{cases} \forall a \in A \colon \exists b \in B \colon a \leq b \text{ and} \\ \forall b \in B \colon \exists a \in A \colon a \leq b. \end{cases}$$

With this convention, we may paraphrase [5, Definition 4.1] as follows.

Definition 2.2 [Pos-frame & Pos-model] A Pos-*frame* is a structure $(X, \leq, R)$, where $(X, \leq)$ is a poset and $R \subseteq X^2$ is a binary relation which satisfies that $R[x] := \{x' \in X \mid xRx'\}$ is always convex and for all $x, y \in X$ it holds that

$$x \leq y \Rightarrow R[x] \leq_{\mathrm{EM}} R[y].$$

Fixing a countable set Prop of propositional variables, a Pos-*model* is a structure $(X, \leq, R, \mathsf{Val})$ consisting of a Pos-frame together with a Pos-*valuation* $\mathsf{Val}\colon X \times \mathsf{Prop} \to \{0, 1\}$, which satisfies

$$x \leq y \Rightarrow \mathsf{Val}(x, p) \leq \mathsf{Val}(y, p)$$

for all $x, y \in X$ and $p \in \mathsf{Prop}$.

The condition on Pos-frames (regarding the Egli-Milner order) ensures that the extension of Val to all formulas φ built from Prop and connectives in $\{\wedge, \vee, 0, 1, \Box, \Diamond\}$ also has the property that $\mathsf{Val}(-, \varphi)$ is always order-preserving.

As pointed out in [5], one considerable advantage these Pos based semantics have over Set-based semantics comes to light in the study of canonicity. For example, the two consequence pairs

$$\Box p \vdash p \text{ and } p \vdash \Diamond p$$

both define the class of reflexive Set-frames but are not mutually inter-derivable in positive modal logic. In the richer semantics, however, both these formulas are canonical and correspond to the classes of Pos-frames $(X, \leq, R)$ where

$$R_\Box := R \circ \leq \text{ and } R_\Diamond := R \circ \leq^{-1}$$

are reflexive, respectively [5, Sections 5 & 7].

The advantages of Pos-based semantics over Set-based ones also become apparent in the study of positive modal logic via duality theory [6] and coalgebras [20,3,9]. In particular, the latter fully embraces this view and proposes positive coalgebraic logic as the *logic of ordered transition systems.*

2.2 Modal logic over finite MV-chains

Recall that the *standard* MV-*algebra* is the algebra

$$\text{Ł} = \langle [0,1], \odot, \oplus, \wedge, \vee, \neg, 0, 1 \rangle$$

based on the real unit interval with its usual bounded lattice structure $\wedge, \vee, 0, 1$ and additional operations defined by

$$x \odot y = \max\{0, x+y-1\}, \qquad x \oplus y = \min\{1, x+y\}, \qquad \neg x = 1-x.$$

A detailed overview of MV-algebras and their relationship to Łukasiewicz logic may be found in the books [7,18]. In this paper, we focus on (positive subreducts of) *finite* MV-*chains*, that is, the following finite subalgebras of the standard MV-algebra.

Definition 2.3 [Finite MV-chain] Let $n \geq 1$ be a natural number. The $(n+1)$-*element* MV-*chain* is given by

$$\text{Ł}_n = \langle \{0, \tfrac{1}{n}, \ldots \tfrac{n-1}{n}, 1\}, \odot, \oplus, \wedge, \vee, \neg, 0, 1 \rangle,$$

considered as a subalgebra of the standard MV-algebra. As usual, we use MV_n to denote the variety $\mathbb{HSP}(\text{Ł}_n)$.

The varieties MV_n, axiomatized by Grigolia [12], provide algebraic counterparts to Łukasiewicz *finitely-valued logics.* Many-valued modal logic over finite MV-chains was investigated by Hansoul and Teheux [13] (also see Bou *et al.* [4]) as follows.

The *modal* MV-*language* $\mathcal{L}^{\mathrm{m}}_{\mathsf{MV}} = \{\odot, \oplus, \wedge, \vee, \neg, 0, 1, \Box\}$ consists of the signature of MV together with a unary operation symbol $\Box$. The collection of *modal* MV *formulas* $\mathsf{Form}^{\mathrm{m}}_{\mathsf{MV}}$ is defined inductively from a countable collection of propositional variables Prop and the connectives of $\mathcal{L}^{\mathrm{m}}_{\mathsf{MV}}$ as usual.

Similarly to how Pos-frames provide richer semantics for positive modal logic, relational frames based on the following category were introduced by Hansoul and Teheux [13] as richer semantics for Ł_n-valued modal logic.

Definition 2.4 [The category Set_n] The category Set_n has objects (X, sub), where X is a set and $\mathsf{sub}\colon X \to \mathbb{S}(\text{Ł}_n)$ associates with every point of X a subalgebra of Ł_n.

A Set_n-*morphism* $(X, \mathsf{sub}) \to (X', \mathsf{sub}')$ is a map $f\colon X \to X'$ which satisfies $\mathsf{sub}'(f(x)) \subseteq \mathsf{sub}(x)$ for all $x \in X$.

The duality-theoretical reason behind this choice of base category is that Set_n is the 'discrete counterpart' of the category Stone_n of structured Stone spaces dually equivalent to MV_n [15,8,17]. From the logical perspective, we think of the additional structure on Set_n as *putting constraints on the admissible valuations*. This becomes apparent in the following definition of Set_n-based relational frames and models [13, Definition 7.2].

Definition 2.5 [Set_n-frame & Set_n-model] A Set_n-*frame* is a structure of the form (X, sub, R), where (X, sub) is a member of Set_n and $R \subseteq X^2$ is a binary relation such that the condition

$$xRy \Rightarrow \mathsf{sub}(y) \subseteq \mathsf{sub}(x)$$

holds for all $x, y \in X$.

A Set_n-*model* is given by $(X, \mathsf{sub}, R, \mathsf{Val})$, consisting of a Set_n-frame and a Set_n-*valuation* $\mathsf{Val}\colon X \times \mathsf{Prop} \to \mathbf{Ł}_n$, which needs to satisfy $\mathsf{Val}(x, p) \in \mathsf{sub}(x)$ for all $x \in X$ and $p \in \mathsf{Prop}$.

The condition on the accessibility relation of Set_n-frames is precisely what is needed to ensure that the extension of valuations to the set of all formulas in $\mathsf{Form}^{\mathrm{m}}_{\mathsf{MV}}$ still satisfies the constraints on valuations. More specifically, if $(X, \mathsf{sub}, R, \mathsf{Val})$ is a Set_n-model, we inductively extend Val to a map $\mathsf{Val}\colon X \times \mathsf{Form}^{\mathrm{m}}_{\mathsf{MV}} \to \mathbf{Ł}_n$, which is defined in the obvious way for the MV-connectives and for formulas of the form $\Box\varphi$ by

$$\mathsf{Val}(x, \Box\varphi) = \bigwedge\{\mathsf{Val}(x', \varphi) \mid xRx'\}.$$

The fact that $\mathsf{sub}(x)$ is a subalgebra of $\mathbf{Ł}_n$ and the condition on Set_n-frames ensure that $\mathsf{Val}(x, \varphi) \in \mathsf{sub}(x)$ holds for all $x \in X$ and $\varphi \in \mathsf{Form}^{\mathrm{m}}_{\mathsf{MV}}$.

We say that a formula $\varphi \in \mathsf{Form}^{\mathrm{m}}_{\mathsf{MV}}$ is *satisfied* at $x \in X$ in a Set_n-model $(X, \mathsf{sub}, R, \mathsf{Val})$ if $\mathsf{Val}(x, \varphi) = 1$ and Set_n-frame *validity* is defined analogous to classical modal logic. We denote by $\Lambda^{\mathrm{m}}_{\mathsf{MV}_n}$ the set of all modal MV-formulas which are valid in every Set_n-frame.

Every Set-frame (X, R) can be identified with a Set_n-frame $(X, \mathsf{sub}^{\mathrm{tr}}, R)$ which has the 'trivial constraint' $\mathsf{sub}^{\mathrm{tr}}(x) = \mathbf{Ł}_n$ at all $x \in X$ and it is easy to see that $\Lambda^{\mathrm{m}}_{\mathsf{MV}_n}$ coincides with the set of modal MV-formulas valid on all Set-frames in this sense. The additional 'richness' of semantics over Set_n-frames again becomes visible when one considers canonicity [13, Sections 7-9] and frame-definability [23]. For example, the formulas

$$\Box(x \oplus x) \to \Box x \text{ and } \Diamond(x \oplus x) \to \Diamond x$$

both define the class of Set_n-frames (X, sub, R) which satisfy

$$\forall x\colon \forall x' : xRx' \to \mathsf{sub}(x') = \mathbf{Ł}_1,$$

that is, frames where all states which are successors are crisp states (identifying $\mathbf{Ł}_1$ with the two-element Boolean algebra). Since the former formula is canonical [13, Example 8.28], the latter one is derivable from the axiomatic extension it yields. In Section 5, we study a similar situation in the positive fragment of the logic described here.

The algebraic counterpart to many-valued modal logic over $\mathbf{Ł}_n$ is given by the following variety of modal MV_n-algebras.

Definition 2.6 [Modal MV_n-algebra] A *modal* MV_n*-algebra* or mMV_n-algebra is an algebra $\langle \mathbf{A}, \Box \rangle$, where $\mathbf{A} \in \mathsf{MV}_n$ is a MV_n-algebra and $\Box\colon A \to A$ is a unary operation satisfying the equations

$$\text{(B1)}\ \Box 1 = 1, \qquad \text{(B2)}\ \Box(x \wedge y) = \Box x \wedge \Box y,$$
$$\text{(B3)}\ \Box(x \oplus x) = \Box x \oplus \Box x, \qquad \text{(B4)}\ \Box(x \odot x) = \Box x \odot \Box x.$$

We denote the variety of modal MV_n-algebras by mMV_n.

The following is a combined result of [13] and [4, Subsection 5.2].

Theorem 2.7 ([13,4]) *Let φ be a modal* MV*-formula. Then*

$$\varphi \in \Lambda^{\mathrm{m}}_{\mathsf{MV}_n} \text{ if and only if } \mathsf{mMV}_n \models \varphi \approx 1.$$

One of the main results of this paper is a similar theorem for the corresponding *positive* modal logic (see Theorem 4.8). On the underlying propositional level we consider positive MV_n-algebras, introduced in the next subsection.

2.3 Finite positive MV-algebras

The quasi-variety PMV of *positive* MV*-algebras*, that is, negation-free (and implication-free) subreducts of MV-algebras, was recently introduced and finitely axiomatized in [2]. Here, we focus on finite positive MV-chains defined as follows.

Definition 2.8 [Finite positive MV-chain] Let $n \geq 1$ be a natural number. The $(n+1)$*-element positive* MV*-chain* is given by

$$\mathbf{PŁ}_n = \langle \{0, \tfrac{1}{n}, \ldots \tfrac{n-1}{n}, 1\}, \odot, \oplus, \wedge, \vee, 0, 1 \rangle,$$

understood as a reduct of $\mathbf{Ł}_n$. We write PMV_n for the (quasi-)variety $\mathbb{ISP}(\mathbf{PŁ}_n) = \mathbb{HSP}(\mathbf{PŁ}_n)$ generated by $\mathbf{PŁ}_n$, and we refer to members of PMV_n as *positive* MV_n*-algebras* or PMV_n*-algebras*.

An axiomatization of PMV_n (involving a quasi-equation) can easily be obtained as combination of the axiomatization of PMV from [2] with the one of MV_n from [12], noting that the additional axioms which characterize MV_n inside MV do not involve negation or implication (for a purely equational axiomatization, also see the recent [1]).

Later on, we make frequent use of the fact that, for each $\ell \in \mathbf{PŁ}_n$, the unary map $\tau_\ell\colon \{0, \frac{1}{n}, \ldots, \frac{n-1}{n}, 1\} \to \{0, \frac{1}{n}, \ldots, \frac{n-1}{n}, 1\}$ given by

$$\tau_\ell(x) = \begin{cases} 1 & \text{if } \ell \leq x, \\ 0 & \text{otherwise} \end{cases}$$

is term-definable in $\mathbf{PŁ}_n$, since it can be term-defined in $\mathbf{Ł}_n$ as combination of expressions of the form $x \odot x$ and $x \oplus x$ only [19, pp. 344-345].

Another useful tool later on is a close relation between a PMV_n-algebra and its distributive skeleton, defined similarly to the Boolean skeleton of a MV_n-algebra.

Definition 2.9 [Distributive skeleton] Let $\mathbf{A} \in \mathsf{PMV}_n$ be a positive MV_n-algebra. The *distributive skeleton* of $\mathbf{A}$ is the bounded distributive lattice $\mathfrak{S}(\mathbf{A})$ with carrier set $\mathfrak{S}(A) = \{a \in \mathbf{A} \mid a \oplus a = a\}$ and the bounded lattice structure inherited from $\mathbf{A}$.

The following fact about distributive skeletons is crucial for the purposes of this paper. In practice, it means that prime filters of $\mathfrak{S}(\mathbf{A})$ can be uniquely extended to PMV_n-homomorphisms $\mathbf{A} \to \mathbf{P}\mathbf{Ł}_n$.

Theorem 2.10 ([21, Theorem 4.4]) *For every $\mathbf{A} \in \mathsf{PMV}_n$, there is a bijection between the sets of homomorphisms*

$$\mathsf{PMV}_n(\mathbf{A}, \mathbf{P}\mathbf{Ł}_n) \cong \mathsf{DL}(\mathfrak{S}(\mathbf{A}), \mathbf{2})$$

given by restriction $u \mapsto u|_{\mathfrak{S}(\mathbf{A})}$.

In Section 4, we consider modal extensions of PMV_n-algebras as algebraic counterpart of the $\mathbf{P}\mathbf{Ł}_n$-valued modal logic introduced in the next section.

3 Positive modal logic over finite MV-chains

We work in the language $\mathcal{L}^{\mathrm{m}}_{\mathsf{PMV}} = \{\odot, \oplus, \wedge, \vee, 0, 1, \Box, \Diamond\}$, that is, the signature of PMV together with two unary operation symbols $\Box$ and $\Diamond$. We define the set $\mathsf{Form}^{\mathrm{m}}_{\mathsf{PMV}}$ of modal PMV-formulas inductively from a countable collection Prop of propositional variables and the connectives in $\mathcal{L}^{\mathrm{m}}_{\mathsf{PMV}}$.

Similarly to what is discussed in Subsections 2.1 and 2.2, we consider relational semantics on frames with additional structure. We choose the following base-category, which entails both the categories Pos and Set_n.

Definition 3.1 [The category Pos_n] We define the category Pos_n as follows. The objects of Pos_n are structures of the form $(X, \leq, \mathsf{sub})$, such that $(X, \leq)$ is a poset and $(X, \mathsf{sub}) \in \mathsf{Set}_n$, that is, $\mathsf{sub}\colon X \to \mathbb{S}(\mathbf{P}\mathbf{Ł}_n)$.

A morphism $f\colon (X, \leq, \mathsf{sub}) \to (X', \leq', \mathsf{sub}')$ in Pos_n is an order-preserving map $f\colon (X, \leq) \to (X', \leq')$ which is also a Set_n-morphism, that is, $\mathsf{sub}'(f(x)) \subseteq \mathsf{sub}(x)$ holds for all $x \in X$.

Note that we can really consider the reduct (X, sub) of an object of Pos_n as a member of Set_n, since the subuniverses of $\mathbf{P}\mathbf{Ł}_n$ coincide with the ones of $\mathbf{Ł}_n$, as shown in [21, Proposition 3.2]. We also think of $\mathbf{P}\mathbf{Ł}_n$ itself as a member of Pos_n which we denote by

$$\underline{\mathbf{P}\mathbf{Ł}}_n = \left(\{0, \tfrac{1}{n}, \ldots, \tfrac{n-1}{n}, 1\}, \leq, \mathsf{sub}^{\mathrm{gen}}\right).$$

Here, $\leq$ is the the usual chain-order and $\mathsf{sub}^{\mathrm{gen}}(\ell) = \langle \ell \rangle$ is the subalgebra of $\mathbf{P}\mathbf{Ł}_n$ generated by $\{\ell\}$.

Members of Pos_n endowed with a compatible accessibility relation constitute our semantics for positive modal logic over finite MV-chains.

Definition 3.2 [Pos_n-frame] A Pos_n-*frame* is a structure of the form

$$\mathfrak{F} = (X, \leq, \mathsf{sub}, R),$$

where $(X, \leq, \mathsf{sub}) \in \mathsf{Pos}_n$ and $R \subseteq X^2$ is a binary relation such that $R[x]$ is always convex and the following two *compatibility conditions* are satisfied:

(i) For all $x, y \in X$ it holds that

$$x \leq y \Rightarrow R[x] \leq_{\mathrm{EM}} R[y].$$

(ii) Whenever $x, y \in X$ satisfy $y \in R[x]$, there exist $y', y'' \in R[x]$ with
- $y' \leq y \leq y''$ and
- $\mathsf{sub}(y'), \mathsf{sub}(y'') \subseteq \mathsf{sub}(x)$.

Here, in condition (i) the subscript EM denotes the Egli-Milner lifting of $\leq$ to $\mathcal{P}(X)$ as defined in the paragraph before Definition 2.2. The significance of the compatibility conditions (i) and (ii) should become clear after Proposition 3.5.

Example 3.3 The following are special examples of Pos_n-frames.

(1) Every Pos-frame $(X, \leq, R)$ as in Definition 2.2 can be identified with the Pos_n-frame $(X, \leq, \mathsf{sub}^{\mathrm{tr}}, R)$, where $\mathsf{sub}^{\mathrm{tr}}$ are the 'trivial constraints' $\mathsf{sub}^{\mathrm{tr}}(x) = \mathbf{P}\mathbf{Ł}_n$ for all $x \in X$.

(2) Every Set_n-frame (X, sub, R) as in Definition 2.5 can be identified with the Pos_n-frame $(X, \leq^{\mathrm{dis}}, \mathsf{sub}, R)$, where $\leq^{\mathrm{dis}}$ is the discrete order on X.

(3) In particular, combining (1) and (2), every Set-frame (X, R) can be identified with the Pos_n-frame $(X, \leq^{\mathrm{dis}}, \mathsf{sub}^{\mathrm{tr}}, R)$.

Not surprisingly, the corresponding Pos_n-models are also defined as a blend of Pos-models and Set_n-models.

Definition 3.4 [Pos_n-model] A Pos_n-*model* is a structure of the form

$$\mathfrak{M} = (X, \leq, \mathsf{sub}, R, \mathsf{Val}),$$

where $(X, \leq, \mathsf{sub}, R)$ is a Pos_n-frame and the Pos_n-*valuation* $\mathsf{Val}\colon X \times \mathsf{Prop} \to \mathbf{P}\mathbf{Ł}_n$ satisfies the following two conditions:

(i) If $x \leq y$, then $\mathsf{Val}(x, p) \leq \mathsf{Val}(y, p)$ for all $p \in \mathsf{Prop}$.

(ii) $\mathsf{Val}(x, p) \in \mathsf{sub}(x)$ for all $x \in X$ and $p \in \mathsf{Prop}$.

For example, every Pos-model, Set_n-model or Set-model can be seen as a Pos_n-model analogous to Example 3.3.

Given a Pos_n-model $(X, \leq, \mathsf{sub}, R, \mathsf{Val})$, we inductively extend its Pos_n-valuation to a map

$$\mathsf{Val}\colon X \times \mathsf{Form}^{\mathrm{m}}_{\mathsf{PMV}} \to \mathbf{P}\mathbf{Ł}_n,$$

which is defined in the obvious way for the PMV-connectives and for formulas of shape $\Box\varphi$ or $\Diamond\varphi$ via

$$\mathsf{Val}(x, \Box\varphi) = \bigwedge\{\mathsf{Val}(x', \varphi) \mid xRx'\} \text{ and } \mathsf{Val}(x, \Diamond\varphi) = \bigvee\{\mathsf{Val}(x', \varphi) \mid xRx'\}.$$

Conditions (i) and (ii) of Definition 3.4 are equivalent to the fact that $\mathsf{Val}(-,p)\colon (X,\leq,\mathsf{sub}) \to \underset{\sim}{\mathbf{PŁ}_n}$ is always a morphism in Pos_n (where we identify $\mathbf{PŁ}_n$ with a member of Pos_n as described after Definition 3.1). In the following, we show that the compatibility conditions required of Pos_n-frames (Definition 3.2) assure that this property extends to $\mathsf{Val}(-,\varphi)$ for every $\varphi \in \mathsf{Form}^{\mathrm{m}}_{\mathsf{PMV}}$.

Proposition 3.5 *Let* $(X,\leq,\mathsf{sub},R,\mathsf{Val})$ *be a* Pos_n*-model. Then, for every modal* PMV*-formula* $\varphi \in \mathsf{Form}^{\mathrm{m}}_{\mathsf{PMV}}$*, the following two conditions hold.*

(i) *If* $x \leq y$ *then* $\mathsf{Val}(x,\varphi) \leq \mathsf{Val}(y,\varphi)$ *for all* $x,y \in X$.

(ii) $\mathsf{Val}(x,\varphi) \in \mathsf{sub}(x)$ *for all* $x \in X$.

Proof. We proceed by induction on the formula φ. The case $\varphi = p \in \mathsf{Prop}$ is covered by Definition 3.4. If $\varphi = \psi_1 * \psi_2$ with $* \in \{\odot,\oplus,\wedge,\vee\}$ being a PMV-connective, condition (i) holds because all of these connectives are order-preserving and (ii) holds because $\mathsf{sub}(x)$ always is a subalgebra of $\mathbf{PŁ}_n$.

This leaves us with the case $\varphi = \Box\psi$ (the case $\varphi = \Diamond\psi$ being analogous). To prove condition (i), suppose that $\mathsf{Val}(y,\varphi) < 1$. Then there exists some $y' \in R[y]$ with $\mathsf{Val}(y,\varphi) = \mathsf{Val}(y',\psi)$. Since $R[x] \leq_{\mathrm{EM}} R[y]$, there exists some $x' \in R[x]$ with $x' \leq y'$ and therefore, by the inductive hypothesis, $\mathsf{Val}(x',\psi) \leq \mathsf{Val}(y',\psi)$. This yields

$$\mathsf{Val}(x,\Box\psi) = \bigwedge\{\mathsf{Val}(\tilde{x},\psi) \mid xR\tilde{x}\} \leq \mathsf{Val}(x',\psi) \leq \mathsf{Val}(y',\psi) = \mathsf{Val}(y,\Box\psi)$$

as desired.

To prove condition (ii), suppose towards contradiction that $\mathsf{Val}(x,\Box\psi) \notin \mathsf{sub}(x)$. Since $1 \in \mathsf{sub}(x)$, there needs to exist some $y \in R[x]$ with $\mathsf{Val}(x,\varphi) = \mathsf{Val}(y,\psi)$. By Definition 3.2(ii), there is some $y' \in R[x]$ with $y' \leq y$ and $\mathsf{sub}(y') \subseteq \mathsf{sub}(x)$. By the inductive hypothesis we have $\mathsf{Val}(y',\psi) \neq \mathsf{Val}(y,\psi)$ (since otherwise $\mathsf{Val}(y',\psi) \notin \mathsf{sub}(y') \subseteq \mathsf{sub}(x)$). This yields

$$\mathsf{Val}(y',\psi) < \mathsf{Val}(y,\psi) = \mathsf{Val}(x,\Box\psi) = \bigwedge\{\mathsf{Val}(\tilde{y},\psi) \mid xR\tilde{y}\},$$

a contradiction to xRy'. □

We say that a modal PMV-formula $\varphi \in \mathsf{Form}^{\mathrm{m}}_{\mathsf{PMV}}$ is *satisfied* at $x \in X$ in a Pos_n-model $(X,\leq,\mathsf{sub},R,\mathsf{Val})$ if $\mathsf{Val}(x,\varphi) = 1$. As usual, we say that φ is *valid* in a Pos_n-frame if it is satisfied at every state in every model based on that frame. We denote by $\Lambda^{\mathrm{m}}_{\mathsf{PMV}_n}$ the set of all modal PMV-formulas which are valid in every Pos_n-frame. More generally, similar to [10], we consider *consequence pairs* $\psi \vdash \varphi$ and say that such a consequence pair is *valid in a* Pos_n*-frame* if $\mathsf{Val}(x,\psi) = 1$ implies $\mathsf{Val}(x,\varphi) = 1$ in every Pos_n-model based on that frame. We say the consequence pair is *valid* if it is valid in all Pos_n-frames. In this case, we write $\psi \models_n \varphi$. The valid formulas $\varphi \in \Lambda^{\mathrm{m}}_{\mathsf{PMV}_n}$ are thus precisely the ones for which $1 \models_n \varphi$ holds.

As we show in the following, regarding the *minimal* modal logic over $\mathbf{PŁ}_n$, the distinction between Pos_n-based semantics and Set-based semantics is 'redundant'.

Proposition 3.6 *A consequence pair $\psi \vdash \varphi$ is valid (in all Pos_n-frames) if and only if it is valid in all* Set*-frames.*

Proof. If the consequence pair is valid in all Pos_n-frames, then it is valid in all Set-frames, since the latter can be identified with Pos_n-frames as in Example 3.3(3).

Conversely, if the consequence pair is not valid in all Pos_n-frames, then there is a Pos_n-model $(X, \leq, \mathsf{sub}, R, \mathsf{Val})$ witnessing this. The underlying Set-model (X, R, Val) then witnesses that the consequence pair is not valid in the Set-frame (X, R). □

Nevertheless, as in the case of classical positive modal logic (Subsection 2.1) and modal logic over $\mathbf{Ł}_n$ (Subsection 2.2), the additional structure of Pos_n-frames becomes interesting when considering questions about *modal definability* or *canonicity.* We begin to touch upon these questions (with a specific example) in Section 5. Before that, we study the logic introduced here algebraically in the following section.

4 Algebraic framework

In this section, we study the logic introduced in the previous section by algebraic means. The corresponding variety of modal algebras arises as a combination of the axioms of modal distributive lattices (see Definition 2.1) and modal MV_n-algebras (see Definition 2.6) as follows.

Definition 4.1 [Modal PMV_n-algebra] A *modal* PMV_n*-algebra* or mPMV_n-algebra is an algebra $\langle \mathbf{A}, \Box, \Diamond \rangle$, where $\mathbf{A} \in \mathsf{PMV}_n$ is a positive MV_n-algebra and $\Box, \Diamond \colon A \to A$ are unary operations satisfying the equations

$$
\begin{array}{ll}
\text{(B1)}\ \Box 1 = 1, & \text{(D1)}\ \Diamond 0 = 0, \\
\text{(B2)}\ \Box(x \wedge y) = \Box x \wedge \Box y, & \text{(D2)}\ \Diamond(x \vee y) = \Diamond x \vee \Diamond y, \\
\text{(B3)}\ \Box(x \oplus x) = \Box x \oplus \Box x, & \text{(D3)}\ \Diamond(x \oplus x) = \Diamond x \oplus \Diamond x, \\
\text{(B4)}\ \Box(x \odot x) = \Box x \odot \Box x, & \text{(D4)}\ \Diamond(x \odot x) = \Diamond x \odot \Diamond x, \\
\text{(P1)}\ \Box(x \vee y) \leq \Box x \vee \Diamond y, & \text{(P2)}\ \Box x \wedge \Diamond y \leq \Diamond(x \wedge y).
\end{array}
$$

We denote the variety of modal PMV_n-algebras by mPMV_n.

In practice, the following equivalent axiomatization of mPMV_n, using the unary terms τ_ℓ defined in Subsection 2.1, will usually be more convenient to work with.

Lemma 4.2 *An algebra $\langle \mathbf{A}, \Box, \Diamond \rangle$ with $\mathbf{A} \in \mathsf{PMV}_n$ is a modal PMV_n-algebra if and only if it satisfies* (B1)-(B2), (D1)-(D2), (P1)-(P2) *and*

$$
\begin{array}{l}
(\mathrm{B}\tau)\ \ \Box\tau_\ell(x) = \tau_\ell(\Box x)\ \textit{for all}\ \ell \in \mathbf{PŁ}_n \backslash \{0\}, \\
(\mathrm{D}\tau)\ \ \Diamond\tau_\ell(x) = \tau_\ell(\Diamond x)\ \textit{for all}\ \ell \in \mathbf{PŁ}_n \backslash \{0\}.
\end{array}
$$

Proof. It is shown in [19, pp. 344-345] that every $\tau_\ell(x)$ can be obtained exclusively as a combination of terms $x \oplus x$ and $x \odot x$. Therefore, the axioms

(B3)-(B4) imply (Bτ) and (D3)-(D4) imply (Dτ). Conversely, note that the equations

$$\tau_\ell(x \oplus x) = \tau_{\lceil \frac{\ell}{2} \rceil}(x) \text{ and } \tau_\ell(x \odot x) = \tau_{\lceil \frac{\ell+1}{2} \rceil}(x)$$

are satisfied in $\mathbf{P}\mathbf{Ł}_n$, where for $q \in \mathbb{Q}$ we define $\lceil q \rceil$ to be the smallest element of $\mathbf{P}\mathbf{Ł}_n$ which is above q. Therefore, using (Bτ), for every $\ell \in \mathbf{P}\mathbf{Ł}_n \backslash \{0\}$ we can compute

$$\tau_\ell(\Box x \oplus \Box x) = \tau_{\lceil \frac{\ell}{2} \rceil}(\Box x) = \Box \tau_{\lceil \frac{\ell}{2} \rceil}(x) = \Box \tau_\ell(x \oplus x) = \tau_\ell(\Box(x \oplus x)).$$

But this implies $\Box x \oplus \Box x = \Box(x \oplus x)$ since $\mathbf{P}\mathbf{Ł}_n$ satisfies the quasi-equation

$$\bigwedge_{\ell \neq 0} (\tau_\ell(x) \approx \tau_\ell(y)) \rightarrow x \approx y.$$

The case for $\Box(x \odot x)$, $\Diamond(x \oplus x)$ and $\Diamond(x \odot x)$ is analogous. □

Next, we explain how to obtain Pos_n-frames from mPMV_n-algebras.

Definition 4.3 [Canonical frame of a mPMV_n-algebra] Let $\langle \mathbf{A}, \Box, \Diamond \rangle$ be a modal PMV_n-algebra. The *canonical* Pos_n*-frame of* $\langle \mathbf{A}, \Box, \Diamond \rangle$ is the Pos_n-frame

$$\langle \mathbf{A}, \Box, \Diamond \rangle^+ := \big(\mathsf{PMV}_n(\mathbf{A}, \mathbf{P}\mathbf{Ł}_n), \leq^{\mathrm{pw}}, \mathsf{im}, R^{\mathrm{m}}\big),$$

where $\leq^{\mathrm{pw}}$ is the point-wise order, im takes a homomorphism u to its image $\mathsf{im}(u) = u(\mathbf{A}) \subseteq \mathbf{P}\mathbf{Ł}_n$ and the binary relation R^{m} is defined via

$$uR^{\mathrm{m}}u' \iff \forall a \in \mathbf{A} : u(\Box a) \leq u'(a) \leq u(\Diamond a).$$

This definition is justified, since in the following we show that canonical frames of mPMV_n-algebras really are Pos_n-frames.

Proposition 4.4 *Let* $\langle \mathbf{A}, \Box, \Diamond \rangle$ *be a modal* PMV_n*-algebra. Then its canonical frame* $\langle \mathbf{A}, \Box, \Diamond \rangle^+$ *is a* Pos_n*-frame.*

Proof. It is clear by definition that $(\mathsf{PMV}_n(\mathbf{A}, \mathbf{P}\mathbf{Ł}_n), \leq^{\mathrm{pw}}, \mathsf{im})$ is an object of the category Pos_n. We need to show that R^{m} satisfies the compatibility conditions (i) and (ii) from Definition 3.2.

To verify (i), first suppose that $u, w \in \mathsf{PMV}_n(\mathbf{A}, \mathbf{P}\mathbf{Ł}_n)$ are homomorphisms for which $u \leq^{\mathrm{pw}} w$ and $wR^{\mathrm{m}}w'$ hold. Then we need to find u' which satisfies $uR^{\mathrm{m}}u' \leq w'$. We define a filter $F \subseteq \mathfrak{S}(\mathbf{A})$ on the distributive skeleton of $\mathbf{A}$ (recall Definition 2.9) by

$$F = \{b \in \mathfrak{S}(\mathbf{A}) \mid u(\Box b) = 1\}$$

and an ideal $J \subseteq \mathfrak{S}(\mathbf{A})$ generated by the set

$$\{b \in \mathfrak{S}(\mathbf{A}) \mid w'(b) = 0 \text{ or } u(\Diamond b) = 0\}.$$

If $b \in J$, there are $b', b'' \in \mathfrak{S}(\mathbf{A})$ with $b \leq b' \vee b''$, $w'(b') = 0$ and $u(\Diamond b'') = 0$. We then compute

$$\begin{aligned} u(\Box b) &\leq u(\Box(b' \vee b'')) \\ &\leq u(\Box b') \vee u(\Diamond b'') && \text{(P1)} \\ &\leq w(\Box b') \vee u(\Diamond b'') && (u \leq^{\mathrm{pw}} w) \\ &\leq w'(b') \vee u(\Diamond b'') = 0. && (wR^{\mathrm{m}}w') \end{aligned}$$

Therefore, $b \notin F$ and we showed that F and J are disjoint. There is a prime filter U' extending F with $U' \cap J = \varnothing$, and by Theorem 2.10 there is a homomorphism $u' \colon \mathbf{A} \to \mathbf{P}\text{Ł}_n$ which extends (the characteristic function of) U'.

We show that u' has the desired properties. To see that $u'(a) \leq w'(a)$, in case $w'(a) \neq 1$ we can use $\ell = w'(a) + \frac{1}{n}$ and find $\tau_\ell(a) \in J$, which yields

$$\tau_\ell(u'(a)) = u'(\tau_\ell(a)) = 0,$$

that is, $u'(a) < \ell$ as desired. The argument for $u'(a) \leq u(\Diamond a)$ is similar, using (Dτ), and for $u(\Box a) \leq u'(a)$ using the construction of F and (Bτ).

For the second half of condition (i) we proceed similarly. Assuming that $u \leq^{\mathrm{pw}} w$ and $uR^{\mathrm{m}}u'$ hold, we want to find w' with $wR^{\mathrm{m}}w'$ and $u' \leq^{\mathrm{pw}} w'$. This time, let $F \subseteq \mathfrak{S}(\mathbf{A})$ be the filter generated by

$$\{b \in \mathfrak{S}(\mathbf{A}) \mid u'(b) = 1 \text{ or } w(\Box b) = 1\},$$

and let $J \subseteq \mathfrak{S}(\mathbf{A})$ be the ideal

$$J = \{b \in \mathfrak{S}(\mathbf{A}) \mid w(\Diamond b) = 0\}.$$

Given $b \in F$, we have $b \geq b' \wedge b''$ for some $b', b'' \in \mathfrak{S}(\mathbf{A})$ which satisfy $u'(b') = w(\Box b'') = 1$. We compute

$$\begin{aligned} w(\Diamond b) &\geq w(\Diamond(b' \wedge b'')) \\ &\geq w(\Diamond b') \wedge w(\Box b'') && \text{(P2)} \\ &\geq u(\Diamond b') \wedge w(\Box b'') && (u \leq^{\mathrm{pw}} w) \\ &\geq u'(b') \wedge w(\Box b'') = 1. && (uR^{\mathrm{m}}u') \end{aligned}$$

Therefore, $b \notin J$ and we showed that J and F are disjoint. The filter F can be extended to a prime filter $W' \subseteq \mathfrak{S}(\mathbf{A})$ which is disjoint from J, and by Theorem 2.10, there is a homomorphism $w' \colon \mathbf{A} \to \mathbf{P}\text{Ł}_n$ which restricts to (the characteristic function of) W' on $\mathfrak{S}(\mathbf{A})$.

To see that w' has the desired properties,

$$u'(a) \leq w'(a)$$

follows from $\tau_\ell(a) \in F$ with $\ell = u'(a)$ and similarly we get $w(\Box a) \leq w'(a)$. To see that $w'(a) \leq w(\Diamond a)$ holds, suppose $w(\Diamond a) \neq 1$ and set $\ell = w(\Diamond a) + \frac{1}{n}$.

Then $\tau_\ell(a) \in J$ implies $0 = w'(\tau_\ell(a)) = \tau_\ell(w'(a))$, which means $w'(a) \leq w(\Diamond a)$ by our choice of ℓ.

We now verify condition (ii) of Definiton 3.2, using an argument similar to [13, Lemma 7.4]. For the first part of condition (ii), let $u \in \mathsf{PMV}_n(\mathbf{A}, \mathbf{PŁ}_n)$ with $\mathsf{im}(u) = \mathbf{PŁ}_k$ and let $uR^{\mathrm{m}}w$. Suppose towards contradiction that there is no $w' \leq^{\mathrm{pw}} w$ with $uR^{\mathrm{m}}w'$ and $\mathsf{im}(w') \leq \mathbf{PŁ}_k$. Choose $\tilde{w} \in R^{\mathrm{m}}[u]$ and $a \in \mathbf{A}$ such that $\tilde{w} \leq^{\mathrm{pw}} w$ and $\tilde{w}(a) = \frac{1}{m} \in \mathbf{PŁ}_n \backslash \mathbf{PŁ}_k$ is the minimal value obtained by a member of $R^{\mathrm{m}}[u]$ below w. By the duality for PMV_n established in [21], we can think of $(\mathsf{PMV}_n(\mathbf{A}, \mathbf{PŁ}_n), \leq^{\mathrm{pw}})$ as Priestley space in which the subsets $\mathsf{PMV}_n(\mathbf{A}, \mathbf{PŁ}_k)$ and $R^{\mathrm{m}}[u]$ are closed. The algebra $\mathbf{A}$ can be identified with the collection of structure-preserving continuous maps $\mathsf{PMV}_n(\mathbf{A}, \mathbf{PŁ}_n) \to \underset{\sim}{\mathbf{PŁ}}_n$ via the isomorphism $a \mapsto \mathsf{ev}_a$. Now consider the closed sets

$$F = R^{\mathrm{m}}[u] \cap \mathsf{PMV}_n(\mathbf{A}, \mathbf{PŁ}_k) \text{ and } G = R^{\mathrm{m}}[u] \cap \tilde{w}{\downarrow}.$$

We have $F{\uparrow} \cap G{\downarrow} = \varnothing$, since otherwise there would be $w' \in R^{\mathrm{m}}[u] \cap \mathsf{PMV}_n(\mathbf{A}, \mathbf{PŁ}_k)$ with $w' \leq^{\mathrm{pw}} \tilde{w} \leq^{\mathrm{pw}} w$, contradicting our initial assumption. Therefore, there exists a clopen down-set Ω which contains G and is disjoint from F. Now we define the element

$$a' = a|_{\Omega} \cup 1|_{\mathsf{PMV}_n(\mathbf{A}, \mathbf{PŁ}_n) \backslash \Omega},$$

which is a well-defined member of $\mathbf{A}$ since Ω is downwards closed. Now by Lemma 4.5 below we find

$$u(\Box a') = \bigwedge\{v(a') \mid v \in R^{\mathrm{m}}[u]\} = \bigwedge\{v(a) \mid v \in R^{\mathrm{m}}[u] \cap \Omega\} = \tilde{w}(a),$$

which yields a contradiction to $\mathsf{im}(u) = \mathbf{PŁ}_k$, since $\tilde{w}(a) \notin \mathbf{PŁ}_k$.

For the proof of the second part of condition (ii) we proceed similarly. Take $u \in \mathsf{PMV}_n(\mathbf{A}, \mathbf{PŁ}_n)$ with $\mathsf{im}(u) = \mathbf{PŁ}_k$ and $uR^{\mathrm{m}}w$ and suppose towards contradiction that there is no $w \leq^{\mathrm{pw}} w''$ with $uR^{\mathrm{m}}w''$ and $\mathsf{im}(w'') \subseteq \mathbf{PŁ}_k$. Choose $\tilde{w} \in R^{\mathrm{m}}[u] \cap w{\uparrow}$ and $a \in \mathbf{A}$ such that $\tilde{w}(a) = \frac{m-1}{m}$ is the maximal value obtained by a member of $R[u]$ above w. Separate the closed sets

$$F = R^{\mathrm{m}}[u] \cap \tilde{w}{\uparrow} \text{ and } G = R^{\mathrm{m}}[u] \cap \mathsf{PMV}_n(\mathbf{A}, \mathbf{PŁ}_k),$$

by a clopen upset $\Omega \supseteq F$ and take

$$a' = a|_{\Omega} \cup 0|_{\mathsf{PMV}_n(\mathbf{A}, \mathbf{PŁ}_n) \backslash \Omega},$$

which is a well-defined member of $\mathbf{A}$ since Ω is upwards closed. Then, using Lemma 4.5, we compute

$$u(\Diamond a') = \bigvee\{v(a') \mid v \in R^{\mathrm{m}}[u]\} = \bigvee\{v(a) \mid v \in R^{\mathrm{m}}[u] \cap \Omega\} = \tilde{w}(a),$$

which yields $u(\Diamond a') = \frac{m-1}{m} \notin \mathbf{PŁ}_k$, which is a contradiction to our assumption $\mathsf{im}(u) = \mathbf{PŁ}_k$. □

As usual, the following truth lemma is the most significant ingredient to obtain an algebraic completeness result.

Lemma 4.5 (Truth Lemma) *Let $\langle \mathbf{A}, \Box, \Diamond \rangle$ be a mPMV_n-algebra. Then, in the canonical frame $\langle \mathbf{A}, \Box, \Diamond \rangle^+$, we have*

$$u(\Box a) = \bigwedge \{v(a) \mid uR^{\mathrm{m}}v\} \text{ and } u(\Diamond a) = \bigvee \{v(a) \mid uR^{\mathrm{m}}v\}$$

for all $u \in \mathsf{PMV}_n(\mathbf{A}, \mathbf{P Ł}_n)$ and $a \in \mathbf{A}$.

Proof. We start with the first equation. It is clear by definition that $u(\Box a) \leq \bigwedge\{v(a) \mid uR^{\mathrm{m}}v\}$ holds. Suppose towards contradiction that this inequality is strict, say

$$u(\Box a) < \bigwedge \{v(a) \mid uR^{\mathrm{m}}v\} =: d$$

holds for some $a \in \mathbf{A}$. We define a filter F and an ideal J on the distributive skeleton $\mathfrak{S}(\mathbf{A})$ as follows. Set

$$F = \{b \in \mathfrak{S}(\mathbf{A}) \mid u(\Box b) = 1\},$$

which is a filter due to axioms (B1) and (B2) and let $J \subseteq \mathfrak{S}(\mathbf{A})$ be the ideal generated by the set

$$\{\tau_d(a) \vee b' \in \mathfrak{S}(\mathbf{A}) \mid u(\Diamond b') = 0\}.$$

If $b \in J$, then $b \leq \tau_d(a) \vee b'_1 \vee \cdots \vee b'_n$ for some $b'_1, \ldots b'_n$ with $u(\Diamond b'_i) = 0$ for all $i = 1, \ldots, n$. Then we use the fact that u is a homomorphism and the properties of modal PMV_n-algebras to find

$$\begin{aligned} u(\Box b) &\leq u(\Box(\tau_d(a) \vee (b'_1 \vee \cdots \vee b'_n))) && \text{(B2)} \\ &\leq u(\Box \tau_d(a) \vee \Diamond(b'_1 \vee \cdots \vee b'_n)) && \text{(P1)} \\ &= u(\Box \tau_d(a)) \vee u(\Diamond b'_1 \vee \cdots \vee \Diamond b'_n) && \text{(D2)} \\ &= \tau_d(u(\Box a)) \vee u(\Diamond b'_1) \vee \cdots \vee u(\Diamond b'_n) = 0. && \text{(B}\mathcal{T}\text{)} \end{aligned}$$

Therefore, $u(\Box b) = 0$, which implies $b \notin F$. So we showed that F and J are disjoint. By the prime ideal theorem for distributive lattices, we can find a prime filter $W \subseteq \mathfrak{S}(\mathbf{A})$ with $F \subseteq W$ and $W \cap J = \varnothing$. By Theorem 2.10, there is a homomorphism $w \colon \mathbf{A} \to \mathbf{P Ł}_n$ which restricts to (the characteristic function of) W on $\mathfrak{S}(\mathbf{A})$.

We now show that $uR^{\mathrm{m}}w$ holds. Indeed, take $a' \in \mathbf{A}$ and set $u(\Box a') = \ell$. Then

$$\tau_\ell(w(a')) = w(\tau_\ell(a')) = 1$$

because $u(\Box \tau_\ell(a')) = \tau_\ell(u(\Box a')) = 1$ implies $\tau_\ell(a') \in F$. Therefore, we have

$$u(\Box a') \leq w(a') \text{ for all } a' \in \mathbf{A}.$$

Similarly, if $u(\Diamond a') \neq 1$, set $\ell = u(\Diamond a') + \frac{1}{n}$ and find

$$\tau_\ell(w(a')) = w(\tau_\ell(a')) = 0$$

because $\tau_\ell(a') \leq \tau_d(a) \vee \tau_\ell(a') \in J$ implies $\tau_\ell(a') \in J$. Therefore, we have $w(a') < u(\Diamond a') + \frac{1}{n}$, which implies

$$w(a') \leq u(\Diamond a') \text{ for all } a' \in \mathbf{A}$$

as desired.

So we showed that $uR^{\mathrm{m}}w$ holds. However, note that $\tau_d(a) \in J$ implies that

$$\tau_d(w(a)) = w(\tau_d(a)) = 0,$$

which by our choice of d means $w(a) < \bigwedge\{v(a) \mid uR^{\mathrm{m}}v\}$, a contradiction to $uR^{\mathrm{m}}w$. This finishes the proof of the first equation.

To prove the second equation, suppose towards contradiction that

$$\bigvee\{v(a) \mid uR^{\mathrm{m}}v\} < u(\Diamond a)$$

holds for some $a \in \mathbf{A}$. Set $d := \bigvee\{v(a) \mid uR^{\mathrm{m}}v\} + \frac{1}{n}$. Define a filter F and an ideal J of $\mathfrak{S}(\mathbf{A})$ as follows. Let

$$J = \{b \in \mathfrak{S}(\mathbf{A}) \mid u(\Diamond b) = 0\},$$

which is an ideal due to axioms (D1)-(D2) and let F be the filter generated by the set

$$\{\tau_d(a) \wedge b' \mid u(\Box b') = 1\}.$$

If $b \in F$, then $\tau_d(a) \wedge b'_1 \wedge \cdots \wedge b'_n \leq b$ for some $b'_1, \ldots b'_n$ with $u(\Box b'_i) = 1$ for all $i = 1, \ldots, n$. We compute

$$\begin{aligned} u(\Diamond b) &\geq u\big(\Diamond(\tau_d(a) \wedge (b'_1 \wedge \cdots \wedge b'_n))\big) && \text{(D2)}\\ &\geq u(\Diamond \tau_d(a) \wedge \Box(b'_1 \wedge \cdots \wedge b'_n)) && \text{(P2)}\\ &= u(\Diamond \tau_d(a)) \wedge u(\Box b'_1 \wedge \cdots \wedge \Box b'_n) && \text{(B2)}\\ &= \tau_d(u(\Diamond a)) \wedge u(\Box b'_1) \wedge \cdots \wedge u(\Box b'_n) = 1. && \text{(D}\tau\text{)} \end{aligned}$$

Therefore, $u(\Diamond b) = 1$, which implies $b \notin J$ and we showed that F and J are disjoint. As before, we can find a homomorphism $w \colon \mathbf{A} \to \mathbf{PL}_n$ with $w(F) = \{1\}$ and $w(J) = \{0\}$.

We show that $uR^{\mathrm{m}}w$ holds. Given $a' \in \mathbf{A}$, let $\ell = u(\Box a')$. Then $\tau_\ell(a') \geq \tau_d(a) \wedge \tau_\ell(a') \in F$ implies $w(\tau_\ell(a')) = 1$ and thus we showed

$$u(\Box a') \leq w(a') \text{ for all } a \in \mathbf{A}.$$

Conversely, if $u(\Diamond a') \neq 1$, take $\ell = u(\Diamond a') + \frac{1}{n}$. Then $u(\Diamond \tau_\ell(a')) = 0$ yields $\tau_\ell(a') \in J$ and thus $w(\tau_\ell(a')) = 0$. This means $w(a') < u(\Diamond a) + \frac{1}{n}$. Thus we showed

$$w(a') \leq u(\Diamond a') \text{ for all } a' \in \mathbf{A},$$

and altogether that $uR^{\mathrm{m}}w$ holds.

However, since $\tau_d(a) \in F$ holds, we have $\tau_d(w(a)) = w(\tau_d(a)) = 1$, which by our choice means $\bigvee\{v(a) \mid uR^{\mathrm{m}}v\} < w(a)$, a contradiction. $\square$

In the converse direction, we now explain how to obtain mPMV_n-algebras from Pos_n-frames.

Definition 4.6 [Complex algebra of a Pos_n-frame] Let $\mathfrak{F} = (X, \leq, \mathsf{sub}, R)$ be a Pos_n-frame. The *complex* mPMV_n*-algebra* of $\mathfrak{F}$ is the mPMV_n-algebra

$$\mathfrak{F}_+ := \langle \underset{\sim}{\mathbf{P\L}}_n^{\mathfrak{F}}, \Box_R, \Diamond_R \rangle,$$

where $\underset{\sim}{\mathbf{P\L}}_n^{\mathfrak{F}}$ consists of all Pos_n-morphisms $(X, \leq, \mathsf{sub}) \to \underset{\sim}{\mathbf{P\L}}_n$ with point-wise PMV_n-operations and

$$(\Box_R \alpha)(x) = \bigwedge_{xRx'} \alpha(x') \text{ and } (\Diamond_R \alpha)(x) = \bigvee_{xRx'} \alpha(x').$$

It is easy to check that this really is a mPMV_n-algebra.

Every mPMV_n-algebra can be embedded in the complex algebra of its canonical frame and vice versa. The proof of the following theorem is mostly straightforward.

Theorem 4.7 (Representation Theorem) *Let* $\mathbf{A}$ *be a* mPMV_n*-algebra and let* $\mathfrak{F}$ *be a* Pos_n*-frame.*

(i) *There is an embedding* $\mathbf{A} \hookrightarrow (\mathbf{A}^+)_+$ *via evaluations* $a \mapsto \mathsf{ev}_a$.

(ii) *There is an embedding* $\mathfrak{F} \hookrightarrow (\mathfrak{F}_+)^+$ *via evaluations* $x \mapsto \mathsf{ev}_x$.

Proof. The 'non-modal parts' of (i) and (ii) are both consequences of the dualities established in [21]. For the 'modal part' of (i), we need to show that $\Box_{R^{\mathrm{m}}} \mathsf{ev}_a = \mathsf{ev}_{\Box a}$ and $\Diamond_{R^{\mathrm{m}}} \mathsf{ev}_a = \mathsf{ev}_{\Diamond a}$. This follows from a direct computation

$$\Box_{R^{\mathrm{m}}} \mathsf{ev}_a(u) = \bigwedge_{uR^{\mathrm{m}}v} \mathsf{ev}_a(v) = \bigwedge_{uR^{\mathrm{m}}v} v(a) = u(\Box a) = \mathsf{ev}_{\Box a},$$

where we used Lemma 4.5. The other equation is shown completely analogous.

For the 'modal part' of (ii), we need to show

$$xRx' \Leftrightarrow \mathsf{ev}_x R^{\mathrm{m}} \mathsf{ev}_{x'}.$$

The direction '$\Rightarrow$' is immediate by definition since

$$\mathsf{ev}_x R^{\mathrm{m}} \mathsf{ev}_{x'} \Leftrightarrow \mathsf{ev}_x(\Box_R \alpha) \leq \mathsf{ev}_{x'}(\alpha) \leq \mathsf{ev}_x(\Diamond_R \alpha) \Leftrightarrow \bigwedge_{xRy} \alpha(y) \leq \alpha(x') \leq \bigvee_{xRy} \alpha(y)$$

and the latter clearly holds for all α if xRx'.

For the direction '$\Leftarrow$', suppose $x' \notin R[x]$. Then $R[x] \cap x'\!\uparrow = \varnothing$ or $R[x] \cap x'\!\downarrow = \varnothing$ needs to hold (since otherwise xRx' holds by convexity of $R[x]$). In the former case, define $\alpha\colon \mathfrak{F} \to \underset{\sim}{\mathbf{P\L}}_n$ by $\alpha(y) = 0$ if $y \in R[x]\!\downarrow$ and $\alpha(y) = 1$ otherwise. Then

$$\mathsf{ev}_{x'}(\alpha) = \alpha(x') = 1 \text{ but } \mathsf{ev}_x(\Diamond_R \alpha) = \bigvee_{xRy} \alpha(y) = 0$$

shows $\mathsf{ev}_{x'} \notin R^{\mathrm{m}}[\mathsf{ev}_x]$. In the other case a similar argument works with $\alpha(y) = 1$ if $y \in R[x]\!\uparrow$ and $\alpha(y) = 0$ otherwise. □

We are now ready to state the main result of this section. In the presence of the prior results of this section (in particular Lemma 4.5), the proof is fairly routine.

Theorem 4.8 (Algebraic Completeness) *Let* $\psi, \varphi \in \mathsf{Form}^{\mathrm{m}}_{\mathsf{PMV}}$.

(i) $\psi \models_n \varphi$ *if and only if* $\mathsf{mPMV}_n \models \psi \leq \varphi$.

(ii) *In particular,* $\varphi \in \Lambda^{\mathrm{m}}_{\mathsf{PMV}_n}$ *if and only if* $\mathsf{mPMV}_n \models \varphi \approx 1$.

Proof. Let $\langle \mathbf{F}, \Box, \Diamond \rangle = \mathsf{Free}_{\mathsf{mPMV}_n}(\mathsf{Prop})$ be the free mPMV_n-algebra generated by the countable set of propositional variables Prop. The *canonical* Pos_n-*model* is based on its canonical frame $\langle \mathbf{F}, \Box, \Diamond \rangle^+$ together with the *canonical* Pos_n-*valuation* defined by

$$\mathsf{Val}^{\mathrm{can}}(u, p) = u([p]),$$

where $[p]$ denotes the equivalence class of p. By Proposition 4.4, we know that $\langle \mathbf{F}, \Box, \Diamond \rangle^+$ really is a Pos_n-frame. Furthermore, we can easily check that the canonical Pos_n-model really is a Pos_n-model (Definition 3.4) since

(i) If $u \leq^{\mathrm{pw}} v$, then $\mathsf{Val}^{\mathrm{can}}(u, p) = u([p]) \leq v([p]) = \mathsf{Val}^{\mathrm{can}}(v, p)$.

(ii) $\mathsf{Val}^{\mathrm{can}}(u, p) = u([p]) \in \mathsf{im}(u)$.

Using the Truth Lemma (Lemma 4.5), it is easy to verify that the property

$$\mathsf{Val}^{\mathrm{can}}(u, \varphi) = u([\varphi])$$

extends to all modal PMV-formulas $\varphi \in \mathsf{Form}^{\mathrm{m}}_{\mathsf{PMV}}$.

Now suppose that $\mathsf{mPMV}_n \not\models \psi \leq \varphi$. Then in particular $[\psi] \leq [\varphi]$ does not hold in $\mathbf{F}$. Thus, there exists a homomorphism $u \colon \mathbf{F} \to \mathbf{P Ł}_n$ with $u([\psi]) = 1$ and $u([\varphi]) = 0$. This means that $\mathsf{Val}^{\mathrm{can}}(u, [\psi]) = 1$ and $\mathsf{Val}^{\mathrm{can}}(u, [\varphi]) = 0$, witnessing that $\psi \not\models_n \varphi$ in the canonical Pos_n-model. □

For the minimal $\mathbf{P Ł}_n$-valued modal logic, the additional structure of Pos_n-frames is 'redundant' (recall Proposition 3.6). However, in the next section we study a consequence pair which is only canonical with respect to the semantics over Pos_n.

5 A case study in canonicity

In this section, we give an example of a consequence pair which is canonical with respect to the semantics over Pos_n-frames but isn't with respect to either of the semantics over Pos-frames or Set_n-frames.

From now on, for simplicity we assume that n is a prime number, which implies $\mathbb{S}(\mathbf{P Ł}_n) = \{\mathbf{P Ł}_n, \mathbf{P Ł}_1\}$. We consider the consequence pairs

$$\Box(x \oplus x) \vdash \Box x \text{ and } \Diamond(x \oplus x) \vdash \Diamond x.$$

As noted in Subsection 2.2, over MV_n (*i.e.*, with negation) the formula $\Box(x \oplus x) \to \Box x$ is canonical [13, Example 8.28] and defines the class of Set_n-frames in which 'all successors are crisp'. In particular, this means that the formula $\Diamond(x \oplus x) \to \Diamond x$ which defines the same class of Set_n-frames is derivable from the corresponding axiomatic extension of modal logic over MV_n. In this section, we show that in modal logic over PMV_n this is not the case anymore. In the following, we identify the classes of Pos_n-frames which are defined by the above consequence pairs. As it turns out, the former consequence pair defines Pos_n-frames in which 'every successor has a crisp successor of the same element below' while the latter one defines the ones in which 'every successor has a crisp successor of the same element above'. Note that a Pos_n-frame $(X, \leq, \mathsf{sub}, R)$ can be identified with a first-order structure where sub is described by unary relations for all $\mathbf{PŁ}_k \in \mathbb{S}(\mathbf{PŁ}_n)$.

Proposition 5.1 *Let $\mathfrak{F} = (X, \leq, \mathsf{sub}, R)$ be a Pos_n-frame.*

(i) *The consequence pair $\Box(p \oplus p) \vdash \Box p$ is valid in $\mathfrak{F}$ if and only if $\mathfrak{F}$ satisfies*

$$\forall x \forall y\colon \big(xRy \to \exists y' : (xRy' \land y' \leq y \land \mathsf{sub}(y') = \mathbf{PŁ}_1)\big).$$

(ii) *The consequence pair $\Diamond(p \oplus p) \vdash \Diamond p$ is valid in $\mathfrak{F}$ if and only if $\mathfrak{F}$ satisfies*

$$\forall x \forall y\colon \big(xRy \to \exists y'' : (xRy'' \land y \leq y'' \land \mathsf{sub}(y'') = \mathbf{PŁ}_1)\big).$$

Proof. We first prove part (i) of the proposition. Assume that $\mathfrak{F}$ satisfies the first-order condition in the statement and let $(X, \leq, \mathsf{sub}, R, \mathsf{Val})$ be an arbitrary Pos_n-model based on $\mathfrak{F}$ which satisfies $\mathsf{Val}(x, \Box(p \oplus p)) = 1$ and $R[x] \neq \varnothing$ (otherwise, the consequence pair is trivially satisfied at x). If $y \in R[x]$, there is some $y' \in R[x] \cap y{\downarrow}$ with $\mathsf{sub}(y') = \mathbf{PŁ}_1$. Then it must hold that $\mathsf{Val}(y', p) = 1$ (since otherwise $\mathsf{Val}(y', p) = 0$ implies $\mathsf{Val}(x, \Box(p \oplus p)) = 0$). Since $y' \leq y$, this now implies $\mathsf{Val}(y, p) = 1$ as well. Since y was an arbitrary successor of x, we get that $\mathsf{Val}(x, \Box p) = 1$ as desired.

For the converse, assume that $\mathfrak{F}$ does not satisfy the first-order formula of the statement. In other words, there are xRy such that no xRy' with $y' \leq y$ satisfies $\mathsf{sub}(y') = \mathbf{PŁ}_1$. Then choose some $\ell \in \mathbf{PŁ}_n$ with $\frac{1}{2} \leq \ell < 1$ and consider any Pos_n-valuation which satisfies

$$\mathsf{Val}(w, p) = \begin{cases} \ell & \text{if } xRw \text{ and } w \leq y \\ 1 & \text{if } xRw \text{ and } w \not\leq y. \end{cases}$$

The Pos_n-model thus arising witnesses $\mathsf{Val}(x, \Box(p \oplus p)) = 1$ but $\mathsf{Val}(x, \Box p) = \ell$, which shows that the consequence pair is not valid in $\mathfrak{F}$.

To similarly prove part (ii) of the proposition, first suppose that $\mathfrak{F}$ satisfies the first-order formula in the statement, and suppose $(X, \leq, \mathsf{sub}, R, \mathsf{Val})$ is a Pos_n-model based on $\mathfrak{F}$ with $\mathsf{Val}(x, \Diamond(p \oplus p)) = 1$. Then there exists some successor $y \in R[x]$ with $\mathsf{Val}(y, p) \geq \frac{1}{2}$. By our assumption, there exists another successor $y'' \in R[x]$ with $y \leq y''$ and $\mathsf{sub}(y'') = \mathbf{PŁ}_1$. Since $\mathsf{Val}(y, p) \neq 0$ and y'' is crisp, we necessarily have $\mathsf{Val}(y'', p) = 1$, which implies $\mathsf{Val}(x, \Diamond p) = 1$ as desired.

For the converse, assume that the $\mathfrak{F}$ does not satisfy the first-order condition in the statement. Then there exist some xRy such that all $w \in R[x]$ with $y \leq w$ satisfy $\mathsf{sub}(w) = \mathbf{PŁ}_n$. Now pick some $\ell \in \mathbf{PŁ}_n$ with $\frac{1}{2} \leq \ell < 1$ and consider any valuation which satisfies

$$\mathsf{Val}(w, p) = \begin{cases} \ell & \text{if } xRw \text{ and } y \leq w \\ 0 & \text{if } xRw \text{ and } y \not\leq w. \end{cases}$$

The Pos_n-model thus arising witnesses $\mathsf{Val}(x, \Diamond(p \oplus p)) = 1$ but $\mathsf{Val}(x, \Diamond p) = \ell$, which shows that the consequence pair is not valid in $\mathfrak{F}$. □

To conclude this section, we note that the consequence pair $\Box(p \oplus p) \vdash \Box p$ is canonical with respect to Pos_n-frames. Together with Section 4, this is a consequence of the following.

Proposition 5.2 *Let $\langle \mathbf{A}, \Box, \Diamond \rangle$ be a modal PMV_n-algebra which satisfies the equation $\Box(x \oplus x) \leq \Box x$. Then the canonical frame $\langle \mathbf{A}, \Box, \Diamond \rangle^+$ satisfies the first-order condition from Proposition 5.1(i).*

Proof. The idea is similar to the proof of part (ii) of Proposition 4.4. Suppose towards contradiction that there are some $u, v \in \mathsf{PMV}_n(\mathbf{A}, \mathbf{P}\mathbf{Ł}_n)$ with $uR^{\mathrm{m}}v$ and $R^{\mathrm{m}}[u] \cap \mathsf{PMV}_n(\mathbf{A}, \mathbf{P}\mathbf{Ł}_1) \cap v{\downarrow} = \varnothing$. Consider the closed sets

$$F = R^{\mathrm{m}}[u] \cap \mathsf{PMV}_n(\mathbf{A}, \mathbf{P}\mathbf{Ł}_1) \text{ and } G = R^{\mathrm{m}}[u] \cap v{\downarrow}.$$

As in the proof of Proposition 4.4, we can separate these sets by a clopen downset $\Omega \supseteq G$ with $\Omega \cap F = \varnothing$. Thus, from $a \in \mathbf{A}$ with $v(a) = \frac{1}{n}$ (which exists because $\mathsf{im}(v) = \mathbf{P}\mathbf{Ł}_n$), we can construct a witness for $u(\Box a') = v(a) = \frac{1}{n}$, which yields $u(\Box(a' \oplus a')) > u(\Box a')$, contradicting the initial assumption that $\Box(a' \oplus a') \leq \Box a'$ holds in $\mathbf{A}$. □

Together with Proposition 5.1 and the results of Section 4, this implies that the consequence pair $\Diamond(p \oplus p) \vdash \Diamond p$ is not derivable in the corresponding axiomatic extension. While the semantics over Pos_n reflect this situation adequately, the semantics restricted to Set, Set_n or Pos all fail to do so.

6 Conclusion and further research

We introduced the positive fragment of $\mathbf{Ł}_n$-valued modal logic and studied it via its algebraic counterpart, modal PMV_n-algebras. We introduced relational semantics on Pos_n-frames for this logic and showed that canonical frames of modal PMV_n-algebras are Pos_n-frames. In an example, we illustrated how these richer semantics are 'well-behaved' with respect to canonicity. In the following, we collect some potential directions for future research along similar lines.

(1) Establish a 'Jónsson-Tarski-style' *topological duality* for the variety mPMV_n similar to the topological duality for mDL from [6] (which should correspond to the special case $n = 1$). Similarly to how the duality for mDL may be seen as extending Priestley duality, a duality for the variety mPMV_n may be be obtained as extension of the natural duality for PMV_n from [21].

(2) Investigate *canonicity and frame-definability* over Pos_n-frames more rigorously, for example by providing an analogue of the Sahlqvist Theorem or the Goldblatt-Thomason Theorem. This will likely be similar to what is found in [6] for classical positive modal logic and in [13,23] for $\mathbf{Ł}_n$-valued modal logic.

(3) Study positive modal logic over $\mathbf{P}\mathbf{Ł}_n$ *coalgebraically*, similarly to [3,9] for classical positive modal logic. Similarly to [16], one could also attempt

to study the relationship between classical positive modal logic and $\mathbf{PŁ}_n$-valued positive modal logic coalgebraically.

(4) Find axiomatizations and 'richer' semantics for broader classes of positive many-valued modal logics. For example, consider positive versions of some of the common infinitely-valued modal *fuzzy* logics [13,22,24]. Furthermore, the author agrees with one of the anonymous referees, who deemed it interesting to which extent the results of this paper can be extended to arbitrary finite lattice-based algebras of truth degrees (for which natural dualities always exist by [8, Corollary 3.3.9]).

In general, we hope this paper inspires future discussions of relational semantics *beyond* Set, in particular in many-valued modal logic where it still is customary to exclusively consider Set-based semantics (see, *e.g.*, [11,4,22]).

References

[1] Abbadini, M., P. Aglianò and S. Fioravanti, *Varieties of MV-monoids and positive MV-algebras* (2024), preprint available at `https://arxiv.org/abs/2405.08471`.

[2] Abbadini, M., P. Jipsen, T. Kroupa and S. Vannucci, *A finite axiomatization of positive MV-algebras*, Algebra universalis **83** (2022), pp. 1134–1164.

[3] Balan, A., A. Kurz and J. Velebil, *Positive fragments of coalgebraic logics*, in: R. Heckel and S. Milius, editors, *Algebra and Coalgebra in Computer Science* (2013), pp. 51–65.

[4] Bou, F., F. Esteva, L. Godo and R. O. Rodríguez, *On the minimum many-valued modal logic over a finite residuated lattice*, Journal of Logic and Computation **21** (2011), pp. 739–790.

[5] Celani, S. and R. Jansana, *A new semantics for positive modal logic*, Notre Dame Journal of Formal Logic **38** (1997), pp. 1–18.

[6] Celani, S. and R. Jansana, *Priestley duality, a Sahlqvist theorem and a Goldblatt-Thomason theorem for positive modal logic*, Logic Journal of the IGPL **7** (1999), pp. 683–715.

[7] Cignoli, R. L. O., I. M. L. D'Ottaviano and D. Mundici, "Algebraic Foundations of Many-Valued Reasoning," Trends in Logic **7**, Springer, 2000.

[8] Clark, D. M. and B. A. Davey, "Natural Dualities for the Working Algebraist," Cambridge studies in advanced mathematics **57**, Cambridge University Press, 1998.

[9] Dahlqvist, F. and A. Kurz, *The Positivication of Coalgebraic Logics*, in: F. Bonchi and B. König, editors, *7th Conference on Algebra and Coalgebra in Computer Science (CALCO 2017)*, Leibniz International Proceedings in Informatics (LIPIcs) **72** (2017), pp. 9:1–9:15.

[10] Dunn, J. M., *Positive modal logic*, Studia Logica **55** (1995), pp. 301–317.

[11] Fitting, M. C., *Many-valued modal logics*, Fundamenta Informaticae **15** (1991), pp. 235–254.

[12] Grigolia, R., *Algebraic analysis of Lukasiewicz-Tarski's n-valued logical systems*, in: R. Wójcicki and G. Malinowski, editors, *Selected papers on Łukasiewicz sentential calculi*, 1977, pp. 81–92.

[13] Hansoul, G. and B. Teheux, *Extending Łukasiewicz logics with a modality: Algebraic approach to relational semantics*, Studia Logica **101** (2013), pp. 505–545.

[14] Jansana, R., *Full models for positive modal logic*, Mathematical Logic Quarterly **48** (2002), pp. 427–445.

[15] Keimel, K. and H. Werner, *Stone duality for varieties generated by quasi-primal algebras*, Memoirs of the American Mathematical Society **148** (1974), pp. 59–85.

[16] Kurz, A., W. Poiger and B. Teheux, *Many-valued coalgebraic logic over semi-primal varieties* (2024), accepted at Logical Methods in Computer Science. Preprint available at `https://arxiv.org/abs/2308.14581`.

[17] Kurz, A., W. Poiger and B. Teheux, *New perspectives on semi-primal varieties*, Journal of Pure and Applied Algebra **228** (2024), p. 107525.

[18] Mundici, D., "Advanced Łukasiewicz calculus and MV-algebras," Trends in Logic **35**, Springer, 2011.

[19] Ostermann, P., *Many-valued modal propositional calculi*, Mathematical Logic Quarterly **34** (1988), pp. 343–354.

[20] Palmigiano, A., *A coalgebraic view on positive modal logic*, Theoretical Computer Science **327** (2004), pp. 175–195, selected Papers of CMCS '03.

[21] Poiger, W., *Natural dualities for varieties generated by finite positive MV-chains* (2023), preprint available at `https://arxiv.org/abs/2309.16998`.

[22] Rodriguez, R. O. and A. Vidal, *Axiomatization of crisp Gödel modal logic*, Studia Logica **109** (2021), pp. 367–395.

[23] Teheux, B., *Modal definability based on Łukasiewicz validity relations*, Studia Logica **104** (2016), pp. 343—-363.

[24] Vidal, A., F. Esteva and L. Godo, *On modal extensions of product fuzzy logic*, Journal of Logic and Computation **27** (2017), pp. 299–336.

Informative Presupposition in Inquisitive Logic

Vít Punčochář,

Institute of Philosophy, The Czech Academy of Sciences
Jilská 1, 11000 Prague, The Czech Republic

Ivo Pezlar [1]

Institute of Philosophy, The Czech Academy of Sciences
Jilská 1, 11000 Prague, The Czech Republic

Abstract

In this paper, we explore a logic of the modality of informative presupposition in the context of propositional intuitionistic inquisitive logic. We present a natural deduction calculus for this logic and show that it is sound and complete with respect to a Kripke semantics based on inquisitive Kripke models and an algebraic semantics based on the notion of inquisitive nucleus. We argue that our setting sheds some light on the relation between different approaches to intuitionistic inquisitive logic, namely, the approach developed by Wesley Holliday, on one side, and approaches based on the tensor disjunction, on the other side. Furthermore, we assess the connections of the presupposition modality to other related notions such as lax modality.

Keywords: Inquisitive logic, Lax logic, Information-based semantics, presupposition, informative content, nucleus.

---

1 Introduction

Inquisitive semantics and logic [6,7] is a framework for a logical analysis of questions. Standard inquisitive logic, as developed in [2,9], is based on classical logic for statements. In intuitionistic inquisitive logic the background logic of statements is intuitionistic. Various versions of propositional intuitionistic inquisitive logic have been considered in the literature.

The most simple version, that we call InqIL, replaces the intuitionistic disjunction with inquisitive disjunction (and thus in this version the background logic of statements is not the full intuitionistic logic but only its disjunction-free fragment). This logic was considered already in [15], though not in the context of modern inquisitive semantics. A generalization of the basic inquisitive semantics suitable for InqIL was introduced in [16]; an algebraic semantics

[1] This paper is an outcome of the project Logical Structure of Information Channels, no. 21-23610M, supported by the Czech Science Foundation and carried out at the Institute of Philosophy of the Czech Academy of Sciences.

for InqIL is developed in [18]; [21] proves that InqIL is the logic determined by proof-theoretic semantics; [11] shows that there is an intimate connection between InqIL and the notion of structural completeness; it follows from the main result of [12] that the schematic fragment of InqIL is the Medvedev logic of finite problems (a result that was stated without proof already in [15]).

In [8,17], the language of InqIL was expanded with an additional declarative disjunction (so that the background logic of statements is the full propositional intuitionistic logic). This disjunction was characterized by a semantic clause that is commonly used in the context of team semantics for dependence logic where the connective is called *tensor disjunction* (see, e.g., [23]). We call the resulting logic $\mathsf{InqIL}^{\otimes}$. A different treatment of declarative disjunction is developed by Holliday in [13] where inquisitive logic is construed in the framework of Beth semantics and declarative disjunction is captured through the notion of Beth nucleus. We denote Holliday's logic as HInqIL. Further work on intuitionistic inquisitive logic includes [19,20].

In this paper, we will add to InqIL a modality of informative presupposition $\circ$, by which we obtain the logic InqIL°. We will formulate a system of natural deduction for this logic, prove completeness, compactness, decidablity and several other results. Moreover, we will show that the presence of the presupposition modality leads to a treatment of declarative disjunction that will help us to clarify the relationship between Holliday's approach (related to HInqIL) and the approaches based on the tensor disjunction (related to $\mathsf{InqIL}^{\otimes}$).

2 Inquisitive Logics

The basic language of propositional inquisitive logic L is just like the language of intuitionistic logic with an unusual symbol for disjunction ($\vee\!\!\!\vee$). So, L is generated from a fixed set of atomic formulas with the following logical symbols: $\bot, \rightarrow, \wedge, \vee\!\!\!\vee$. Negation ($\neg$) and equivalence ($\leftrightarrow$) will be regarded as defined symbols: $\neg\varphi =_{def} \varphi \rightarrow \bot$, $\varphi \leftrightarrow \psi =_{def} (\varphi \rightarrow \psi) \wedge (\psi \rightarrow \varphi)$. In the context of L, as well as in the contexts of other languages employed in this paper, we will be distinguishing two categories of formulas: *declarative* formulas and *arbitrary* formulas. Declarative formulas, which form a particular subset of arbitrary formulas, correspond to statements. Arbitrary formulas encompass statements as well as questions. In the context of any of the considered languages, we will be using systematically the variables $\varphi, \psi, \chi, \vartheta$ for arbitrary formulas and α, β, γ for declarative formulas. In the language L, declarative formulas are defined as the $\vee\!\!\!\vee$-free fragment of L.

Let us introduce the usual intuitionistic Kripke semantics for L. A *Kripke frame* is a partially ordered set $\mathcal{F} = \langle S, \leq \rangle$. The elements of S will be called *states* in $\mathcal{F}$. A set $P \subseteq S$ is *upward closed* (in $\mathcal{F}$) if it satisfies the following condition: if $w \in P$ and $w \leq v$ then $v \in P$. The upward closed sets will be called *propositions*.

A *Kripke model* is a Kripke frame $\mathcal{F}$ equipped with a valuation V defined as a function that assigns to each atomic formula a proposition in $\mathcal{F}$. Given any Kripke model $\mathcal{M} = \langle S, \leq, V \rangle$, the relation $\Vdash$ between states of the model and formulas is defined in the usual recursive way:

- $w \Vdash p$ iff $w \in V(p)$; $w \nVdash \bot$; $w \Vdash \varphi \rightarrow \psi$ iff for any $v \geq w$, if $v \Vdash \varphi$, $v \Vdash \psi$;
- $w \Vdash \varphi \wedge \psi$ iff $w \Vdash \varphi$ and $w \Vdash \psi$; $w \Vdash \varphi \vee\!\!\!\vee \psi$ iff $w \Vdash \varphi$ or $w \Vdash \psi$.

We define $||\varphi||_{\mathcal{M}} = \{w \in S \mid w \Vdash \varphi \text{ in } \mathcal{M}\}$. $||\varphi||_{\mathcal{M}}$ is called *the proposition expressed by* φ *in* $\mathcal{M}$. This is justified by the fact that the relation $\Vdash$ is persistent: if $w \Vdash \varphi$ and $w \leq v$ then $v \Vdash \varphi$, and thus $||\varphi||_{\mathcal{M}}$ is upward closed. We will be omitting the subscript $\mathcal{M}$ if clear from the context.

We say that φ is *valid in* $\mathcal{M}$ if $w \Vdash \varphi$ holds in $\mathcal{M}$ for every $w \in S$, i.e. if $||\varphi|| = S$. It is well-known that a formula is intuitionistically valid if and only if it is valid in all Kripke models [14].

In accordance with [16], classical and intuitionistic *inquisitive* logics are defined as the logics determined by particular classes of Kripke models. Let $\mathcal{M} = \langle S, \leq, V\rangle$ be a Kripke model. The *inquisitive variant of* $\mathcal{M}$ is defined as the Kripke model $inq(\mathcal{M}) = \langle Up_{\emptyset}S, \supseteq, V^*\rangle$, where $Up_{\emptyset}S$ denotes the set of all *nonempty* upward closed sets of states in $\mathcal{M}$, $\supseteq$ is the superset relation and $V^*(p) = \{s \in Up_{\emptyset}S \mid s \subseteq V(p)\}$, for every atomic formula p.

The idea behind the construction of $inq(\mathcal{M})$ is that the states in $inq(\mathcal{M})$ correspond to consistent propositions or consistent pieces of information in $\mathcal{M}$. Hence, in the model $inq(\mathcal{M})$, formulas are evaluated relative to propositions of $\mathcal{M}$ rather than relative to single states of $\mathcal{M}$. If s is such a proposition (i.e. an upward closed set in $\mathcal{M}$) and α is a declarative formula then $s \Vdash \alpha$ in $inq(\mathcal{M})$ can be interpreted as stating that the information s supports the statement α. This interpretation aligns with the following crucial feature of this construction (for a proof, see [16]).

Claim 2.1 *Let* $\mathcal{M}$ *be a Kripke model,* s *a state in* $inq(\mathcal{M})$ *and* α *an* L*-formula. Then* $s \Vdash \alpha$ *in* $inq(\mathcal{M})$ *iff for all* $w \in s$, $w \Vdash \alpha$ *in* $\mathcal{M}$. *As a consequence,* α *is valid in* $inq(\mathcal{M})$ *if and only if* α *is valid in* $\mathcal{M}$.

This claim cannot be expanded to the full language L. The disjunction $\vee\!\!\!\vee$ becomes inquisitive in $inq(\mathcal{M})$. For instance, the formula $p \vee\!\!\!\vee q$ does not express the statement *that* p *or* q in $inq(\mathcal{M})$ but rather the question *whether* p *or* q. If s is a consistent piece of information (a nonempty upward closed set) in $\mathcal{M}$ then $s \Vdash p \vee\!\!\!\vee q$ in $inq(\mathcal{M})$ means that s either supports p, or it supports q, and thus provides an answer to the question whether p or q.

A Kripke model $\mathcal{N}$ is called *inquisitive* if it is the inquisitive variant of a Kripke model, i.e. $\mathcal{N} = inq(\mathcal{M})$, for some Kripke model $\mathcal{M}$. An inquisitive Kripke model is called *classical* if it is the inquisitive variant of a Kripke model where the ordering is identity. The states in such an underlying Kripke model $\mathcal{M} = \langle W, =, V\rangle$ can be viewed as *possible worlds* and the states in $\mathcal{N} = inq(\mathcal{M})$ are nonempty sets of possible worlds. So, classical inquisitive Kripke models are exactly those of the shape $\langle \mathcal{P}(W)\backslash\{\emptyset\}, \supseteq, V^*\rangle$, where for each p there is $U \subseteq W$ such that $V^*(p) = \mathcal{P}(U)\backslash\{\emptyset\}$.

Let us illustrate the semantics with a simple classical inquisitive Kripke model. Let $\mathcal{M} = \langle W, =, V\rangle$, where $W = \{w_1, w_2, w_3, w_4\}$, $V(p) = \{w_1, w_2\}$,

Table 1
The system $S(\mathsf{InqIL})$

ex falso	$\bot/\varphi$	
implication	$[\varphi/\psi]/\varphi \to \psi$	$\varphi \to \psi, \varphi/\psi$
conjunction	$\varphi, \psi/\varphi \wedge \psi$	(i) $\varphi \wedge \psi/\varphi$ (ii) $\varphi \wedge \psi/\psi$
disjunction	(i) $\varphi/\varphi \vee\!\!\!\vee \psi$ (ii) $\psi/\varphi \vee\!\!\!\vee \psi$	$\varphi \vee\!\!\!\vee \psi, [\varphi/\chi], [\psi/\chi]/\chi$
split	$\alpha \to (\varphi \vee\!\!\!\vee \psi)/(\alpha \to \varphi) \vee\!\!\!\vee (\alpha \to \psi)$	

and $V(q) = \{w_1, w_3\}$. The propositions in $\mathcal{M}$ are just subsets of W. For example, $||p|| = \{w_1, w_2\}$ in $\mathcal{M}$. In contrast, the propositions in $inq(\mathcal{M})$ are downward closed sets of nonempty subsets of W. Nevertheless, Claim 2.1 guarantees that for declarative formulas these sets of subsets are generated by the corresponding propositions in $\mathcal{M}$. For a simple example, we have $||p|| = \{\{w_1, w_2\}, \{w_1\}, \{w_2\}\}$ in $inq(\mathcal{M})$. In general, if inquisitive disjunction is involved, the corresponding proposition in $inq(\mathcal{M})$ is not generated by a single proposition in $\mathcal{M}$ but rather by a set of such propositions (representing the possible answers to a question). For example, $||p \vee\!\!\!\vee q|| = \{\{w_1, w_2\}, \{w_1, w_3\}, \{w_1\}, \{w_2\}, \{w_3\}\}$ in $inq(\mathcal{M})$.

In general, we say that a proposition P in $inq(\mathcal{M})$ is *declarative* if it is closed under arbitrary unions, i.e. whenever P contains some sets s_i, $i \in I$, then it contains also the set $\bigcup_{i \in I} s_i$. This means that declarative propositions are those that are generated by a single proposition in $\mathcal{M}$. That is, an upward closed set of states P in $inq(\mathcal{M})$ is declarative iff there is an upward closed set Q in $\mathcal{M}$ such that P is the set of all nonempty upward closed subsets of Q. We say that P is *inquisitive* if it is not declarative.

Note that according to our definition, the empty set is always declarative. Furthermore, if $inq(\mathcal{M})$ is classical then every declarative proposition in $inq(\mathcal{M})$ is the set of nonempty subsets of a set of possible worlds from $\mathcal{M}$.

The intuitionistic inquisitive logic InqIL is defined as the set of L-formulas valid in all inquisitive Kripke models. The classical (or basic) inquisitive logic InqB is the set of L-formulas valid in all classical inquisitive Kripke models.

It follows from Claim 2.1 that in the declarative fragment of L, InqIL coincides with the (disjunction-free fragment of) intuitionistic logic and InqB with classical logic. However, the presence of inquisitive disjunction makes a difference. The logic InqIL can be characterized by a system of natural deduction consisting of the rules for intuitionistic logic plus one additional rule called *split*: $\alpha \to (\varphi \vee\!\!\!\vee \psi)/(\alpha \to \varphi) \vee\!\!\!\vee (\alpha \to \psi)$, where the antecedent α is restricted to declarative formulas (see [16]). Using a nonstandard but clear and compact notation we summarize the whole system, which we call $S(\mathsf{InqIL})$, in Table 1. The logic is not closed under uniform substitution. This is related to the fact that it is determined by a class of Kripke models (not frames) with specific valuations.

A system for the logic InqB, which is also not closed under uniform substitution, is obtained by adding to the system $S(\mathsf{InqIL})$ the double negation law restricted to declarative formulas: $\neg\neg\alpha/\alpha$. Both these logics, InqIL and InqB, have the disjunction property and the finite model property.

3 The modality of informative presupposition

Consider a Kripke model $\mathcal{M} = \langle W, =, V\rangle$, where $W = \{w_1, \ldots, w_6\}$, $V(p) = \{w_1, w_2\}$, $V(q) = \{w_1, w_3\}$, and $V(r) = \{w_3, w_4\}$. The formula $p \vee q \vee r$ represents the question *whether p or q or r*. This question presupposes the information that at least one of the three alternatives p, q, r is true and thus excludes the worlds w_5 and w_6. We want to capture such an informative presupposition as a modality. This modality takes a possibly inquisitive proposition P and returns the corresponding declarative proposition iP respecting the boundaries of P but destroying its internal structure. We can depict the situation as follows:

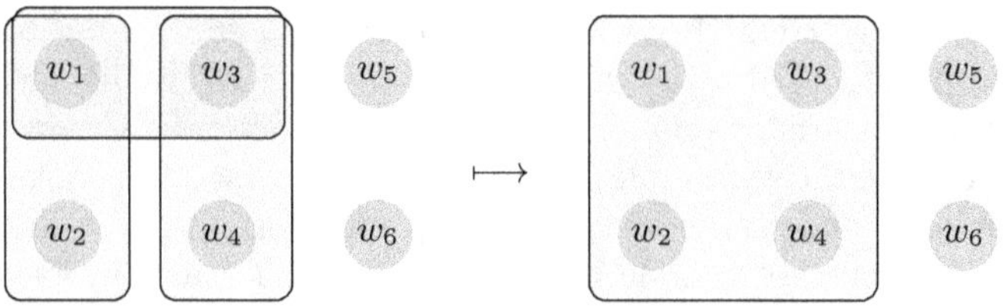

Recall that P and iP are closed under nonempty subsets and so only the maximal sets are depicted (so that the operation does not coincide with simple union). In general, we are concerned with the following operation. If $\mathcal{M} = \langle S, \leq, V\rangle$ is a Kripke model and P is a proposition in $inq(\mathcal{M})$ then

$$iP = \{s \in Up_\emptyset S \mid s \subseteq \bigcup P\}.$$

Note that the application of this operation always returns a proposition in $inq(\mathcal{M})$. In particular, iP is the strongest declarative proposition implied by P. This operation, also known as "non-inquisitive closure", has been considered and used in the context of InqB and its semantics based on classical inquisitive models. It is well-known that in this context the operation coincides with double negation and the exclamation mark has been used as a defined symbol for its syntactic representation: $!\varphi =_{def} \neg\neg\varphi$.[2]

However, in general, the presupposition modality is not identical with double negation. Consider the following Kripke model $\mathcal{M}$:

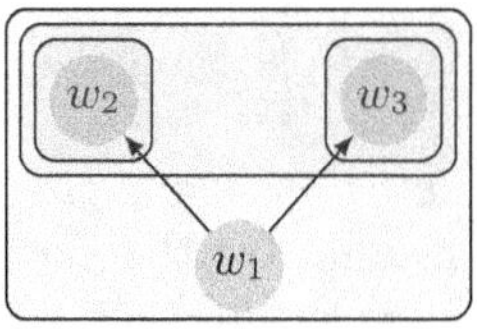

[2] Instead of the exclamation mark, we will use the symbol $\circ$ to indicate the connection to the lax modality [10].

Let $||p|| = \{w_2\}$ and $||q|| = \{w_3\}$ in $\mathcal{M}$. Then in $inq(\mathcal{M})$ we obtain: $||p \vee q|| = \{\{w_2\}, \{w_3\}\}$, $i||p \vee q|| = \{\{w_2, w_3\}, \{w_2\}, \{w_3\}\}$, $||\neg\neg(p \vee q)|| = \{\{w_1, w_2, w_3\}, \{w_2, w_3\}, \{w_2\}, \{w_3\}\}$.

Our main goal is to characterize the logic of informative presupposition in the context of intuitionistic inquisitive logic. Let us first make the following observation that shows that i is an operation of an important kind, namely a dense nucleus on the algebra of propositions.

Claim 3.1 *Let $\mathcal{M}$ be a Kripke model and P, Q propositions in $inq(\mathcal{M})$. Then: (a) $P \subseteq iP$, (b) $iiP = iP$, (c) $i(P \cap Q) = iP \cap iQ$, (d) $i\emptyset = \emptyset$.*

For a proof (of this and other results in this paper), see the appendix. Claim 3.1 suggests that there is a connection between the operation i and the lax modality [10] which is a logical counterpart of the notion of nucleus. We will study this connection in more detail in the next section. Inspired by the relation to the lax logic, we will be using the symbol $\circ$ for the presupposition modality. We denote the language L enriched with this modality as L°. Since $\circ\varphi$ expresses the information presupposed by φ, we can define the declarative disjunction as the presupposition of the corresponding inquisitive disjunction: $\varphi \vee \psi =_{def} \circ(\varphi \vee \psi)$.

The set of declarative formulas in L° is the smallest set of L°-formulas which is closed under $\to$ and $\wedge$ and contains $\bot$, all atomic formulas and all formulas of the form $\circ\varphi$. In contrast to L, declarative formulas in L° may involve inquisitive disjunction, but only in the scope of $\circ$. This definition of declarative formulas aligns with the general definition of declaratives in inquisitive modal logic [4].

If $\langle S, \leq \rangle$ is a partial order and $w \in S$ then $\uparrow w$ denotes the upward closed set generated by w, i.e. $\uparrow w = \{v \in S \mid w \leq v\}$. Relative to any inquisitive model $inq(\mathcal{M})$ and state s in $inq(\mathcal{M})$ we define the following support condition for $\circ$:

- $s \Vdash \circ\varphi$ iff for all $w \in s$, $\uparrow w \Vdash \varphi$.

With this support condition we obtain that $||\circ\varphi|| = i||\varphi||$. Now we would like to expand Claim 2.1 to the language L° but the semantic clause for $\circ$ is defined only for inquisitive models and so the formulation of Claim 2.1 would not make sense for declarative formulas of the shape $\circ\varphi$. Nevertheless, we can reformulate the claim in the following meaningful way that applies to L°.

Claim 3.2 *Let $\mathcal{M}$ be a Kripke model, s a state in $inq(\mathcal{M})$ and α a declarative L°-formula. Then $s \Vdash \alpha$ in $inq(\mathcal{M})$ iff for all $w \in s$, $\uparrow w \Vdash \alpha$ in $inq(\mathcal{M})$.*

The set of L°-formulas that are valid in all inquisitive Kripke models will be denoted as InqIL°. We also define the corresponding consequence relation. For any set of L°-formulas $\Delta \cup \{\varphi\}$ we define $\Delta \models_{(\mathsf{InqIL}^\circ)} \varphi$ as meaning that for every state s of any inquisitive Kripke model, if s supports all formulas from Δ then s supports φ.

Claim 3.2 can be reformulated as saying that $\alpha \leftrightarrow \circ\alpha \in \mathsf{InqIL}^\circ$, for every declarative α. The intuitive meaning of this fact is that the information presupposed by a statement is equivalent to that statement, which makes perfect sense because statements are (as is usual in logical semantics) reduced to their informative content.

Table 2
The system $S(\mathsf{InqIL}^\circ)$ consists of the rules of $S(\mathsf{InqIL})$ plus the following rules

$\circ$-*intro*: $\varphi/\circ\varphi$	$\circ$-*elim*: $\circ\varphi, [\varphi/\alpha]/\alpha$

The following observation demonstrates that in the intuitionistic inquisitive logic the presupposition modality is not expressible in the language L and so its addition to the language increases the expressive power.

Claim 3.3 *There is no L-formula χ such that $\circ(p \mathbin{\backslash\!\!\!\vee} q) \leftrightarrow \chi \in \mathsf{InqIL}^\circ$.*

An important feature of the logic InqIL° that can be easily established is the disjunction property for $\mathbin{\backslash\!\!\!\vee}$.

Claim 3.4 *If $\varphi \mathbin{\backslash\!\!\!\vee} \psi \in \mathsf{InqIL}^\circ$ then $\varphi \in \mathsf{InqIL}^\circ$ or $\psi \in \mathsf{InqIL}^\circ$.*

We will prove that the modality $\circ$ can be completely characterized by the introduction and elimination rules presented in Table 2. In $\circ$-*elim* we assume, as usual, that α is declarative. Note that in the context of the other rules *split* could be equivalently formulated purely schematically:

$$split^*: \quad \circ\chi \to (\varphi \mathbin{\backslash\!\!\!\vee} \psi)/(\circ\chi \to \varphi) \mathbin{\backslash\!\!\!\vee} (\circ\chi \to \psi)$$

In this reformulation, the only rule responsible for the failure of substitution would be $\circ$-*elim*. We will write $\vdash_{S(\mathsf{InqIL}^\circ)} \varphi$ to express that φ is derivable in the system $S(\mathsf{InqIL}^\circ)$ from Table 2. This will be identified with $\emptyset \vdash_{S(\mathsf{InqIL}^\circ)} \varphi$. We further write $\Delta \vdash_{S(\mathsf{InqIL}^\circ)} \varphi$ if φ is derivable in $S(\mathsf{InqIL}^\circ)$ from (a finite subset of) formulas in Δ.

The rules $\circ$-*intro* and $\circ$-*elim* illustrate that $\circ\varphi$ can be viewed as a degenerate disjunction with only one disjunct. From this perspective, $\circ$-*intro* is like disjunction introduction (the disjunction $\circ\varphi$ can be derived from every, i.e. the only disjunct φ) and $\circ$-*elim* resembles disjunction elimination (if we have the disjunction $\circ\varphi$ and are able to derive α from all its disjuncts, i.e. from φ, then we can infer α).

One can observe that, in the context of the other rules, $\circ$-*intro* and $\circ$-*elim* are respectively interderivable with the following rules:

$$\circ\text{-}intro^*: \ \circ\varphi \to \alpha/\varphi \to \alpha \qquad\qquad \circ\text{-}elim^*: \ \varphi \to \alpha/\circ\varphi \to \alpha$$

If we put these two facts together we obtain that the introduction and elimination rules for $\circ$ could be replaced with a single axiom: $(\varphi \to \alpha) \leftrightarrow (\circ\varphi \to \alpha)$. The following lemma then amounts to soundness of the system with respect to inquisitive semantics.

Claim 3.5 $(\varphi \to \alpha) \leftrightarrow (\circ\varphi \to \alpha) \in \mathsf{InqIL}^\circ$, *for all φ, ψ and all declarative α.*

In order to prove completeness let us define the sublanguage of L° generated from atomic formulas by $\bot, \to, \wedge, \vee$. This is indeed a sublanguage of L° because $\vee$ is a symbol defined in L°. The formulas of this sublanguage will be called *simple.* Note that every simple formula is declarative. Simple formulas simulate the declarative language of intuitionistic logic. Importantly, we can simulate the usual introduction and elimination rules for $\vee$ in $S(\mathsf{InqIL}^\circ)$. The introduction rules amount to $\alpha/\circ(\alpha \mathbin{\vee\!\!\!\vee} \beta)$ and $\beta/\circ(\alpha \mathbin{\vee\!\!\!\vee} \beta)$, which can be obtained as follows:

$$\dfrac{\dfrac{\alpha}{\alpha \mathbin{\vee\!\!\!\vee} \beta}\ \mathbin{\vee\!\!\!\vee}\text{-}intro}{\circ(\alpha \mathbin{\vee\!\!\!\vee} \beta)}\ \circ\text{-}intro \qquad \dfrac{\dfrac{\beta}{\alpha \mathbin{\vee\!\!\!\vee} \beta}\ \mathbin{\vee\!\!\!\vee}\text{-}intro}{\circ(\alpha \mathbin{\vee\!\!\!\vee} \beta)}\ \circ\text{-}intro$$

The elimination rule amounts to $\circ(\alpha \vee \beta), [\alpha/\gamma], [\beta/\gamma]/\gamma$, which is obtained in this way:

$$\dfrac{\circ(\alpha \vee \beta) \qquad \dfrac{[\alpha \mathbin{\vee\!\!\!\vee} \beta]^{(1)} \qquad \dfrac{[\alpha]^{(2)}}{\gamma} \qquad \dfrac{[\beta]^{(3)}}{\gamma}}{\gamma}\ (2),(3),\ \mathbin{\vee\!\!\!\vee}\text{-}elim}{\gamma}\ (1),\ \circ\text{-}elim$$

As a consequence, the deductive system for InqIL° has all the resources of full intuitionistic logic on the level of simple formulas. That is, when we restrict ourselves to simple formulas all the intuitionistic introduction and elimination rules for $\bot, \to, \wedge, \vee$ are either directly contained or can be simulated in $S(\mathsf{InqIL}^\circ)$.

Claim 3.6 *If α is an intuitionistically valid simple formula (with $\vee$ in the role of intuitionistic disjunction) then $\vdash_{S(\mathsf{InqIL}^\circ)} \alpha$.*

This observation allows us to prove strong completeness of the system for InqIL° in a way that is standard in inquisitive logic. Before we formulate this result we will need one more characteristic property of propositional inquisitive logic adapted to our setting. We define recursively a function $\mathcal{R}$ assigning to every formula φ a finite set of simple formulas, called the resolutions of φ:

$\mathcal{R}(p) = \{p\}$, $\mathcal{R}(\bot) = \{\bot\}$,
$\mathcal{R}(\varphi \to \psi) = \{\bigwedge_{\alpha \in \mathcal{R}(\varphi)}(\alpha \to f(\alpha)) \mid f : \mathcal{R}(\varphi) \to \mathcal{R}(\psi)\}$,
$\mathcal{R}(\varphi \wedge \psi) = \{\alpha \wedge \beta \mid \alpha \in \mathcal{R}(\varphi), \beta \in \mathcal{R}(\psi)\}$,
$\mathcal{R}(\varphi \mathbin{\vee\!\!\!\vee} \psi) = \mathcal{R}(\varphi) \cup \mathcal{R}(\psi)$,
$\mathcal{R}(\circ\varphi) = \{\bigvee \mathcal{R}(\varphi)\}$.

We obtain the inquisitive disjunctive normal form and consequently completeness, compactness, finite model property and decidability of InqIL°.

Claim 3.7 *If $\mathcal{R}(\varphi) = \{\alpha_1, \ldots, \alpha_n\}$, then $\varphi \leftrightarrow (\alpha_1 \mathbin{\vee\!\!\!\vee} \ldots \mathbin{\vee\!\!\!\vee} \alpha_n) \in \mathsf{InqIL}^\circ$.*

Theorem 3.8 *$\Delta \vdash_{S(\mathsf{InqIL}^\circ)} \varphi$ iff $\Delta \vDash_{(\mathsf{InqIL}^\circ)} \varphi$.*

Corollary 3.9 *$\Delta \vDash_{(\mathsf{InqIL}^\circ)} \varphi$ iff $\Delta' \vDash_{(\mathsf{InqIL}^\circ)} \varphi$, for some finite $\Delta' \subseteq \Delta$.*

Theorem 3.10 *If $\varphi \notin \mathsf{InqIL}^\circ$ then there is a finite Kripke model $\mathcal{M}$ such that φ is not valid in inq($\mathcal{M}$).*

Corollary 3.11 *InqIL° is decidable.*

4 Propositional lax logic and algebraic semantics

The previous section suggests that there is a close connection between InqIL° and propositional lax logic [10]. In this section, we focus on this connection.

Table 3
The system $S(\mathsf{PLL})$ consists of the rules of intuitionistic logic (in the language L°) plus the following rules

L1: $\varphi/\circ\varphi$	L2: $\circ\varphi, [\varphi/\circ\psi]/\circ\psi$

A system $S(\mathsf{PLL})$ of natural deduction for propositional lax logic formulated in the language L° is presented in Table 3. The system $S(\mathsf{DPLL})$ for dense propositional lax logic is obtained from $S(\mathsf{PLL})$ by adding the rule $\circ\bot/\bot$. Now, we introduce the system $S^{*}(\mathsf{InqIL}^{\circ})$ as an extension of $S(\mathsf{DPLL})$ with the rules of schematic *split* (*split**) and the rule of declarative atoms (*da*):

$$\begin{array}{ll} \textit{split}^{*}: & \circ\chi \to (\varphi \mathbin{\vee\!\!\!\vee} \psi)/(\circ\chi \to \varphi) \mathbin{\vee\!\!\!\vee} (\circ\chi \to \psi) \\ \textit{da}: & \circ p/p, \text{ for every atomic formula } p \end{array}$$

Since $S^{*}(\mathsf{InqIL}^{\circ})$ directly extends a system for propositional lax logic, the following result clarifies the connection of propositional lax logic and InqIL°.

Theorem 4.1 *$\Delta \vdash_{S(\mathsf{InqIL}^{\circ})} \varphi$ iff $\Delta \vdash_{S^{*}(\mathsf{InqIL}^{\circ})} \varphi$*

So, the system $S^{*}(\mathsf{InqIL}^{\circ})$ gives us an alternative characterization of InqIL°. This characterization is closely connected to a common alternative characterization of basic inquisitive logic InqB in which *split* is replaced by the Kreisel-Putnam rule $\neg\chi \to (\varphi \mathbin{\vee\!\!\!\vee} \psi)/(\neg\chi \to \varphi) \mathbin{\vee\!\!\!\vee} (\neg\chi \to \psi)$ and the double negation law is stated just for atomic formulas. This presentation reflects the original Hilbert style axiomatization of InqB in [2]. It is instructive to observe that we can replace $\neg\chi$ with $\neg\neg\chi$ in the schematic formulation of the Kreisel-Putnam rule and the result, as a schema, is in the context of intuitionistic logic equivalent to the original formulation. So, a system of natural deduction for InqB can be equivalently presented as the system for intuitionistic logic enriched with the modified Kreisel-Putnam rule (*mkp*) and double negation restricted to atomic formulas (*dna*):

$$\begin{array}{ll} \textit{mkp}: & \neg\neg\chi \to (\varphi \mathbin{\vee\!\!\!\vee} \psi)/(\neg\neg\chi \to \varphi) \mathbin{\vee\!\!\!\vee} (\neg\neg\chi \to \psi) \\ \textit{dna}: & \neg\neg p/p, \text{ for every atomic formula } p \end{array}$$

There is an obvious connection between this system for InqB and the system $S^{*}(\mathsf{InqIL}^{\circ})$: the rules *mkp* and *dna* are respectively obtained from *split** and *da* by replacing $\circ$ with double negation. Note that in $S(\mathsf{DPLL})$ one can derive $(\neg\neg\varphi \to \varphi) \to (\varphi \leftrightarrow \circ\varphi)$. Hence, adding the rule of full double negation $(\neg\neg\varphi/\varphi)$ to $S(\mathsf{DPLL})$ leads to the trivialization of $\circ$ (in the sense that $\varphi \leftrightarrow \circ\varphi$ becomes derivable). In $S^{*}(\mathsf{InqIL}^{\circ})$ we can moreover derive $(\neg\neg\alpha \to \alpha) \leftrightarrow (\neg\neg\alpha \leftrightarrow \circ\alpha)$, for every declarative α. This observation is related to the fact that adding *dna* to $S^{*}(\mathsf{InqIL}^{\circ})$ has the same effect as (is equivalent to) adding the fully schematic axiom identifying $\circ$ with double negation: $\neg\neg\varphi \leftrightarrow \circ\varphi$.

The connection of InqIL° to propositional lax logic also suggests a straightforward algebraic semantics for InqIL° that we briefly outline. A *nucleus* on a Heyting algebra $\mathcal{H} = \langle H, \sqcap, \sqcup, \Rightarrow, 0\rangle$ is a function $j : H \to H$ such that for each $x, y \in H$: (a) $x \leq j(x)$; (b) $j(j(x)) = j(x)$; (c) $j(x \sqcap y) = j(x) \sqcap j(y)$. A nucleus is *dense* if $j(0) = 0$. We say that a nucleus is *inquisitive* if it is dense and satisfies $j(x) \Rightarrow (y \sqcup z) = (j(x) \Rightarrow y) \sqcup (j(x) \Rightarrow z)$. A *nuclear algebra* is a Heyting algebra equipped with a nucleus. Note that every nucleus is a closure operator (but not the other way around). Moreover, for every nuclear algebra $\langle \mathcal{H}, j\rangle$, we can define the algebra $j\mathcal{H} = \langle jH, \sqcap_j, \sqcup_j, \Rightarrow_j, j(0)\rangle$, where jH is the set of j-fixed points in H; $\sqcap_j$ and $\Rightarrow_j$ coincide with $\sqcap$ and $\Rightarrow$ restricted to jH (using the fact that jH is always closed under $\sqcap$ and $\Rightarrow$); and $x \sqcup_j y = j(x \sqcup y)$. A crucial feature of nuclear algebras is that $j\mathcal{H}$ is always a Heyting algebra.

A *valuation* in a nuclear algebra $\langle \mathcal{H}, j\rangle$ is a function assigning to every atomic formula an element of H. A nuclear algebra equipped with a valuation will be called a *nuclear model*. We say that a valuation v in $\langle \mathcal{H}, j\rangle$ is *declarative* if $v(p) \in jH$, for every atom p. A nuclear model $\langle \mathcal{H}, j, v\rangle$ will be called *inquisitive* if j is an inquisitive nucleus and v is a declarative valuation.

Given a nuclear model $\langle \mathcal{H}, j, v\rangle$ we can assign to each L°-formula φ an element of H that will be denoted as $[\varphi]$. This assignment is defined recursively: $[p] = v(p)$, $[\bot] = 0$, $[\varphi \to \psi] = [\varphi] \Rightarrow [\psi]$, $[\varphi \wedge \psi] = [\varphi] \sqcap [\psi]$, $[\varphi \mathbin{\rotatebox[origin=c]{-90}{$\geqslant$}} \psi] = [\varphi] \sqcup [\psi]$, $[\circ\varphi] = j([\varphi])$. We say that φ is *valid* in the nuclear model if $[\varphi]$ is the top element of $\mathcal{H}$.

It is well-known that an L°-formula is valid in (dense) propositional lax logic if and only if it is valid in all (dense) nuclear models. With respect to InqIL° we obtain the following algebraic completeness.

Theorem 4.2 *$\varphi \in \mathsf{InqIL}^{\circ}$ iff φ is valid in all inquisitive nuclear models.*

Let us conclude this section with a suggestion for future research. The operation of adding *dna* to superintuitionistic intuitionistic logics was studied in depth in [1]. The logic obtained by adding *dna* to a logic L is called the negative variant of L. This study was motivated mainly by the fact that InqB can be obtained as the negative variant of several interesting superintuitionistic logics, e.g. the Maksimova logic ND, Kreisel-Putnam logic KP, or the Medvedev logic of finite problems ML. [3]

The observations that we have made in this section suggest that a natural generalization of this research would be the study of the operation of adding the rule of declarative atoms *da* to the extensions of PLL. We expect that many results of [1] would generalize to that setting.

5 Declarative disjunction and tensor

We have shown that in the language L°, we can define declarative disjunction $\varphi \vee \psi$ as $\circ(\varphi \mathbin{\rotatebox[origin=c]{-90}{$\geqslant$}} \psi)$. We will observe that one could alternatively take $\vee$ as primitive and define $\circ\varphi$ as $\varphi \vee \varphi$.

The disjunction $\vee$ differs from the tensor disjunction $\otimes$ used in the context of dependence logic. To make this difference clear we will consider the following languages. The language $L^{\vee}$ is built up from atomic formulas with the following logical symbols: $\bot, \to \wedge, \mathbin{\rotatebox[origin=c]{-90}{$\geqslant$}}, \vee$. The language $L^{\otimes}$ is like $L^{\vee}$ but

[3] It was proved in [9] that the negative variant of a superintuitionistic logic L coincides with InqB if and only if L is in the interval between ND and ML.

the connective $\vee$ is replaced with $\otimes$. Finally, the language $L^{\vee\otimes}$ contains both $\vee$ and $\otimes$ as primitive connectives. *Declarative formulas* in $L^{\vee\otimes}$ is the smallest set of formulas that includes $\bot$, all atomic formulas, all formulas of the form $\varphi \vee \psi$, and is closed under $\to$, $\wedge$ and $\otimes$. The declarative formulas of $L^{\vee}$ are the declarative $\otimes$-free formulas of $L^{\vee\otimes}$ and the declarative formulas of $L^{\otimes}$ are the declarative $\vee$-free formulas of $L^{\vee\otimes}$ (i.e. $\mathbin{\vee\!\!\!\vee}$-free formulas of $L^{\otimes}$).

Semantically, the connectives $\vee$ and $\otimes$ are characterized by the following clauses defined for any state s of any inquisitive model $inq(\mathcal{M})$ and for any $L^{\vee\otimes}$-formulas φ, ψ:

- $s \Vdash \varphi \vee \psi$ iff for all $w \in s$, $\uparrow w \Vdash \varphi$ or $\uparrow w \Vdash \psi$,
- $s \Vdash \varphi \otimes \psi$ iff $s \Vdash \varphi$ or $s \Vdash \psi$ or there are states t, u in $inq(\mathcal{M})$ such that $t \Vdash \varphi$, $u \Vdash \psi$, and $t \cup u = s$.

The semantic clause for $\vee$ is in accordance with our previous interpretation according to which the declarative disjunction is the presupposition of the corresponding inquisitive disjunction. The clause for $\otimes$ is equivalent to the one used in team semantics [23].[4] In the presence of both these conditions persistence of $\Vdash$ is guaranteed.

Note that $||\varphi \vee \psi||$ is always a declarative proposition (and thus every declarative formula of $L^{\vee\otimes}$ expresses a declarative proposition). This does not hold for $\otimes$. We can illustrate the difference with the following example. Consider the classical Kripke model $\mathcal{M}$ consisting of three worlds w_p, w_q, w_r, and a valuation V such that $V(p) = \{w_p\}$ and $V(q) = \{w_q\}$ and $V(r) = \{w_r\}$. Then, we obtain the following difference:

- $||(p \mathbin{\vee\!\!\!\vee} q) \vee r|| = ||p \vee q \vee r|| = \uparrow\{w_p, w_q, w_r\}$,
- $||(p \mathbin{\vee\!\!\!\vee} q) \otimes r|| = ||(p \otimes r) \mathbin{\vee\!\!\!\vee} (q \otimes r)|| = \uparrow\{w_p, w_r\} \cup \uparrow\{w_q, w_r\}$.

However, the connectives $\vee$, $\otimes$ have the same effect when applied to declarative formulas.

Claim 5.1 $||\alpha \vee \beta|| = ||\alpha \otimes \beta||$, *for any declarative $L^{\vee\otimes}$-formulas α, β and in any inquisitive model.*

The logic $\mathsf{InqIL}^{\vee}$ ($\mathsf{InqIL}^{\otimes}$, $\mathsf{InqIL}^{\vee\otimes}$) is the set of $L^{\vee}$-formulas ($L^{\otimes}$-formulas, $L^{\vee\otimes}$-formulas) valid in all inquisitive models. The corresponding consequence relations $\vDash_{(\mathsf{InqIL}^{\vee})}$, $\vDash_{(\mathsf{InqIL}^{\otimes})}$, $\vDash_{(\mathsf{InqIL}^{\vee\otimes})}$ are defined by preservation of support in inquisitive models. We will show that $\mathsf{InqIL}^{\vee}$ and $\mathsf{InqIL}^{\otimes}$ can be characterized by the systems $S(\mathsf{InqIL}^{\vee})$ and $S(\mathsf{InqIL}^{\otimes})$ presented in Tables 4 and 5. The system

[4] The usual clause for tensor is slightly different and a bit simpler: $s \Vdash \varphi \otimes \psi$ iff there are states t, u such that $t \Vdash \varphi$, $u \Vdash \psi$, and $t \cup u = s$. This is because tensor is usually considered in the context of team semantics or information state-based semantics for inquisitive logic where also the empty set (empty team or empty information state) is considered. However, in this paper, we formulate the framework in the context of Kripke semantics for intuitionistic logic where the empty set is not considered as a separate state. In the presence of the empty state our clause for $\otimes$ is equivalent to the usual one and so the disjuncts "$s \Vdash \varphi$ or $s \Vdash \psi$" can be omitted. Without the empty state we have to include them to guarantee persistence of $\Vdash$ and the soundness of $\varphi/\varphi \otimes \psi$ and $\psi/\varphi \otimes \psi$.

Table 4
The system $S(\mathsf{InqIL}^{\vee})$ consists of the rules of $S(\mathsf{InqIL})$ plus the following rules

$\vee$-*intro*: (i) $\varphi/\varphi\vee\psi$, (ii) $\psi/\varphi\vee\psi$	$\vee$-*elim*: $\varphi\vee\psi, [\varphi/\alpha], [\psi/\alpha]/\alpha$

Table 5
The system $S(\mathsf{InqIL}^{\otimes})$ consists of the rules of $S(\mathsf{InqIL})$ plus the following rules

T1: $\alpha\otimes\alpha/\alpha$	T2: $\varphi\otimes\psi/\psi\otimes\varphi$	T3: $\varphi/\varphi\otimes\psi$
T4: $\varphi\otimes\psi, [\varphi/\chi], [\psi/\vartheta]/\chi\otimes\vartheta$	T5: $\varphi\otimes(\psi\mathbin{\vee\!\!\!\vee}\chi)/(\varphi\otimes\psi)\mathbin{\vee\!\!\!\vee}(\varphi\otimes\chi)$	

$S(\mathsf{InqIL}^{\vee\otimes})$ is obtained by combining the rules of $S(\mathsf{InqIL}^{\vee})$ and $S(\mathsf{InqIL}^{\otimes})$.

Theorem 5.2 *(a)* $\Delta \vdash_{S(\mathsf{InqIL}^{\vee})} \varphi$ *iff* $\Delta \vDash_{(\mathsf{InqIL}^{\vee})} \varphi$, *(b)* $\Delta \vdash_{S(\mathsf{InqIL}^{\otimes})} \varphi$ *iff* $\Delta \vDash_{(\mathsf{InqIL}^{\otimes})} \varphi$, *(c)* $\Delta \vdash_{S(\mathsf{InqIL}^{\vee\otimes})} \varphi$ *iff* $\Delta \vDash_{(\mathsf{InqIL}^{\vee\otimes})} \varphi$.

Theorem 5.3 *All the logics* $\mathsf{InqIL}^{\vee}$, $\mathsf{InqIL}^{\otimes}$, $\mathsf{InqIL}^{\vee\otimes}$ *are compact, have the disjunction property, the finite model property and are decidable.*

The logic $\mathsf{InqIL}^{\otimes}$ coincides with the intuitionistic inquisitive logic from [8,17] (though there are some slight differences in the formulation of the corresponding deductive systems). The logic $\mathsf{InqIL}^{\vee}$ is closely connected to the logic InqIL°. The connection can be described as follows: We can introduce translation k from the language L° to the language $L^{\vee}$ and l from the language L° to the language $L^{\vee}$.

The translation k is defined as follows: $k\bot = \bot$, $kp = p$, for every atom p, and for every L°-formulas φ, ψ, $k(\varphi * \psi) = k\varphi * k\psi$, for each $* \in \{\to, \wedge, \mathbin{\vee\!\!\!\vee}\}$, and $k(\circ\varphi) = k\varphi \vee k\varphi$. The translation l is defined as follows: $l\bot = \bot$, $lp = p$, for every atom p, and for every $L^{\vee}$-formulas φ, ψ, $l(\varphi * \psi) = l\varphi * l\psi$, for each $* \in \{\to, \wedge, \mathbin{\vee\!\!\!\vee}\}$, and $l(\varphi \vee \psi) = \circ(l\varphi \mathbin{\vee\!\!\!\vee} l\psi)$.

The mutual definability of $\circ$ and $\vee$ then leads to the following connection between InqIL° and $\mathsf{InqIL}^{\vee}$.

Theorem 5.4 *Let* φ *be an* L°*-formula and* ψ *an* $L^{\vee}$*-formula. Then (a)* $\varphi \in \mathsf{InqIL}^{\circ}$ *iff* $k\varphi \in \mathsf{InqIL}^{\vee}$, *(b)* $\psi \in \mathsf{InqIL}^{\vee}$ *iff* $l\psi \in \mathsf{InqIL}^{\circ}$, *(c)* $\varphi \leftrightarrow lk\varphi \in \mathsf{InqIL}^{\circ}$, *(d)* $\psi \leftrightarrow kl\psi \in \mathsf{InqIL}^{\vee}$.

To make a comparison between the logics $\mathsf{InqIL}^{\vee}$ and $\mathsf{InqIL}^{\otimes}$ we note that $\mathsf{InqIL}^{\vee}$ could be equivalently characterized by the system consisting of the rules of $S(\mathsf{InqIL})$ plus the following four rules:

D1: $\alpha\vee\alpha/\alpha$ D2: $\varphi\vee\psi/\psi\vee\varphi$ D3: $\varphi/\varphi\vee\psi$ D4: $\varphi\vee\psi, [\varphi/\chi], [\psi/\vartheta]/\chi\vee\vartheta$

These rules are mirroring the rules T1–T4 from Table 5. Nevertheless, that does not mean that, if formulated in the same language, the logic $\mathsf{InqIL}^{\vee}$ would be weaker than $\mathsf{InqIL}^{\otimes}$. The rule D1 is actually much stronger than T1 because the languages $L^{\vee}$ and $L^{\otimes}$ have different definitions of declarative formulas ($\varphi\vee\psi$ is always declarative but $\varphi\otimes\psi$ is not). In fact, the logic $\mathsf{InqIL}^{\vee}$ is neither weaker nor stronger than $\mathsf{InqIL}^{\otimes}$.[5]

[5] E.g. $(p\vee(q\vee r)) \to (p\vee(q\mathbin{\vee\!\!\!\vee} r))$ is valid in $\mathsf{InqIL}^{\vee}$ but $(p\otimes(q\otimes r)) \to (p\otimes(q\mathbin{\vee\!\!\!\vee} r))$ is not valid in

Table 6
The system $S(\mathsf{HInqIL})$ consists of the rules for intuitionistic logic (with $\vee\!\!\!\vee$ in the role of intuitionistic disjunction) plus the following rules for $\vee$

H1: $\alpha \vee \alpha / \alpha$, for $\vee\!\!\!\vee$-free α	H2: $\varphi / \varphi \vee \varphi$	H3: $(\varphi \vee \varphi) \vee (\varphi \vee \varphi) / \varphi \vee \varphi$
H4: $(\varphi \wedge \psi) \vee (\varphi \wedge \psi) // (\varphi \vee \varphi) \wedge (\psi \vee \psi)$		H5: $(\varphi \vee\!\!\!\vee \psi) \vee (\varphi \vee\!\!\!\vee \psi) // \varphi \vee \psi$

6 Holliday's approach

In this section, we compare our approach to Holliday's inquisitive intuitionistic logic. Holliday's semantics is based on Beth semantics for intuitionistic logic which we will now briefly present.

A *Beth frame* is, just like Kripke frame, a partially ordered set $\mathcal{F} = \langle S, \leq \rangle$. A *chain* in $\mathcal{F}$ is a subset $C \subseteq S$ such that for all $s, t \in C$, $s \leq t$ or $t \leq s$. A *path* in $\mathcal{F}$ is a chain C that is closed under upper bounds, i.e. if $s \leq t$, for all $s \in C$, then $t \in C$. If C is a path and $s \in C$ then we say that C *is a path through* s.

The *Beth nucleus* is a particular nucleus j_B on the Heyting algebra of upward closed sets in $\mathcal{F}$. For every upward closed set X in $\mathcal{F}$ we define:

$$j_B(X) = \{s \in S \mid \text{ every path through } s \text{ intersects } X\}.$$

An upward closed set X in $\mathcal{F}$ is called *Beth closed* if $j_B(X) = X$. A *Beth valuation* in $\mathcal{F}$ is a function assigning to each atomic formula a Beth closed set. A *Beth model* is a Beth frame equipped with a Beth valuation. Relative to each Beth model $\langle \mathcal{F}, V \rangle$, Holliday defines a forcing relation between states in the model and $L^{\vee}$-formulas by the following clauses:

- $s \Vdash p$ iff $s \in V(p)$; not $s \Vdash \bot$; $s \Vdash \varphi \to \psi$ iff for any $t \geq s$, if $t \Vdash \varphi$, $t \Vdash \psi$;
- $s \Vdash \varphi \wedge \psi$ iff $s \Vdash \varphi$ and $s \Vdash \psi$; $s \Vdash \varphi \vee\!\!\!\vee \psi$ iff $s \Vdash \varphi$ or $s \Vdash \psi$;
- $s \Vdash \varphi \vee \psi$ iff every path through s goes through some t s.t. $t \Vdash \varphi$ or $t \Vdash \psi$.

Let $|\varphi| = \{s \in S \mid s \Vdash \varphi\}$. With the clause for $\vee$ we obtain $|\varphi \vee \psi| = j_B(|\varphi| \cup |\psi|)$. Moreover, $|\alpha|$ is Beth closed, for every declarative α, but union of Beth closed sets is not generally Beth closed and thus, in analogy to our semantics for inquisitive logic, inquisitive disjunction $\vee\!\!\!\vee$ generates propositions that are not Beth closed.

We say that φ is *valid* in a Beth model if φ is forced by every state of that model. Holliday's intuitionistic inquisitive logic, called $\mathsf{Inq(IPC)}$ in [13] and HInqIL in this paper, is defined as the set of $L^{\vee}$-formulas valid in all Beth models. Holliday's system for HInqIL (or rather its reformulation in natural deduction) is presented in Table 6. The double slash // in the formulation of H4 and H5 means that these are bi-directional rules allowing us to infer also the formula on the left side of // from the formula on the right side. The system directly reflects the nuclear character of $\vee$ and all the rules from Table 6 would remain to be sound with respect to the obvious generalized semantics with

$\mathsf{InqIL}^{\otimes}$ and $(p \otimes (q \vee\!\!\!\vee r)) \to ((p \otimes q) \vee\!\!\!\vee (p \otimes r))$ is valid in $\mathsf{InqIL}^{\otimes}$ but $(p \vee (q \vee\!\!\!\vee r)) \to ((p \vee q) \vee\!\!\!\vee (p \vee r))$ is not valid in $\mathsf{InqIL}^{\vee}$.

Table 7
Relations among the nuclei i, j_B, j_{DN} in inquisitive models

	finite	infinite
classical	$i = j_B = j_{DN}$	$i = j_{DN}, j_B \neq j_{DN}$
non-classical	$i \neq j_{DN}, j_B = j_{DN}$	$i \neq j_{DN}, j_B \neq j_{DN}, i \neq j_B$

j_B replaced by an arbitrary nucleus. The main result of [13] is the following completeness theorem.

Theorem 6.1 *$\varphi \in \mathsf{HInqIL}$ if and only if φ is derivable in $S(\mathsf{HInqIL})$.*

There are some strong analogies between Holliday's approach determining the logic HInqIL and the approach based on inquisitive models which determines the logic $\mathsf{InqIL}^{\vee}$. Both these approaches employ some specific kinds of nuclei and use them in order to define declarative disjunction as the nuclear closure of the corresponding inquisitive disjunction. However, the nuclei they employ are very different. Only in very specific cases the nucleus i that captures our notion of informative presupposition coincides with the nucleus j_B on $\langle Up_\emptyset S, \supseteq\rangle$. In every finite partial order, j_B corresponds to the double negation operation and we showed in Section 3 a simple counterexample of a finite inquisitive model in which i is different from double negation. Nevertheless, we can combine the fact that in finite models j_B coincides with double negation with the fact that in classical inquisitive models i coincides with double negation. This clarifies in which sense Holliday's approach generalizes the usual semantics of InqB.

Theorem 6.2 *In every finite classical inquisitive model the nucleus i is identical with the nucleus j_B. Hence, with a restriction to the language L, the logic of all finite classical inquisitive models in Beth semantics is InqB. Moreover, in these models $\varphi \vee \psi$ is according to the Beth semantics equivalent to $\neg\neg(\varphi \mathbin{\mathbb{V}} \psi)$ and thus also to $\neg(\neg\varphi \wedge \neg\psi)$.*

Nevertheless, Beth semantics diverges from the usual inquisitive semantics even in regards to classical models. Consider the following Kripke model $\mathcal{M} = \langle W, =, V\rangle$, where $W = \{w_n \mid n \in \mathbb{N}\}$, $V(p) = \{w_n \in W \mid n \text{ is even}\}$, and $V(q) = \{w_n \in W \mid n \text{ is odd}\}$. Take the proposition $P = ||p \mathbin{\mathbb{V}} q||$ in the classical inquisitive model $inq(\mathcal{M})$. It holds that $W \in iP$ but $W \notin j_B P$. To see the latter, consider the path $C = \{s_n \mid n \in \mathbb{N}\}$ where $s_n = \{w_m \mid m \geq n\}$. Then C is a path through W which does not intersect P. Table 7 summarizes the relationships between the nuclei i, j_B, and j_{DN} (standing for double negation) in finite and infinite, classical and non-classical inquisitive models.

Another crucial difference between Holliday's framework and the usual inquisitive semantics is that HInqIL does not validate *split*. As a Beth counterexample to *split* consider the following Beth model:

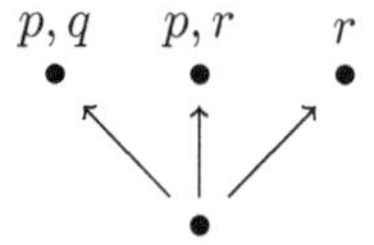

One can easily check that $|p|$, $|q|$, $|r|$ are all Beth closed (but note that without the state where only r holds the root state would have to be included into $|p|$ and the model would not work as a counterexample to *split*). In this model the root state forces $p \to (q \vee r)$ but not $(p \to q) \vee (p \to r)$.

In the light of these differences between Holliday's approach and the usual inquisitive semantics it is perhaps surprising that on the logical level there is a tight connection between Holliday's rules from $S(\mathsf{HInqIL})$ and the rules $\vee$-*intro* and $\vee$-*elim* of our system $S(\mathsf{InqIL}^{\vee})$. In fact, over intuitionistic logic $\vee$-*intro* and $\vee$-*elim* are interderivable with the collection of rules H1–H5. This insight leads to a more perspicuous reformulation of Holliday's logic and an intriguing observation that Holliday's disjunction can be completely characterized by the usual introduction and elimination rules where just a syntactic restriction on the conclusion is imposed in the elimination rule. Let $S^*(\mathsf{HInqIL})$ be the system $S(\mathsf{InqIL}^{\vee})$ without *split*. Then we obtain:

Theorem 6.3 $\Delta \vdash_{S(\mathsf{HInqIL})} \varphi$ *iff* $\Delta \vdash_{S^*(\mathsf{HInqIL})} \varphi$.

In the conclusion of his paper, Holliday formulates an open problem of characterizing *split* in the context of Beth semantics. Let us conclude this section with a related but a bit more specific open problem. We say that a path C_1 in a Beth frame *joins* a path C_2 if there is $s \in C_1 \cap C_2$ such that for all $t \geq s$, $t \in C_1$ iff $t \in C_2$. Consider the following constraint on Beth frames:

(C) If $s \leq t$ and $s \leq u$ then there is v such that $s \leq v$, $v \leq t$ and $v \leq u$, and every path through v joins a path that goes through t or through u.

Note that all the finite classical inquisitive Kripke frames have this property.

Claim 6.4 *If a Beth frame $\mathcal{F}$ satisfies (C) then* split *is valid in every Beth model on $\mathcal{F}$.*

So, $S(\mathsf{InqIL}^{\vee})$ is Beth sound with respect to the class of frames satisfying (C). We leave as an open problem for future research whether the system is also complete with respect to this class.

7 Final remarks

We conclude the paper by outlining a broader plan for future research. The modality of informative presupposition $\circ$ that we studied in this paper bears a striking similarity to propositional truncation modality $\triangle$. Propositional truncation is a crucial concept in many type theories utilizing the propositions-as-types correspondence, most notably in homotopy type theory [22]. It allows us to turn an arbitrary type A into a "mere proposition" $\triangle A$, i.e. a type that has at most one proof. More specifically, $\triangle A$ makes all proofs of a type A equal, thus suppressing all information contained in them other than their existence and erasing the computational content of A in the process. Interestingly, we can observe that this behavior of truncation is in many respects similar to the behavior of informative presupposition $\circ$. While presupposition turns arbitrary propositions into declarative propositions by hiding their inquisitive structure, truncation turns arbitrary types into mere propositions by hiding their computational structure.

This link between presupposition and truncation can also be observed in their respective inference rules. The truncation modality $\triangle$ is often specified by the following introduction and elimination rules and a single axiom (for simplicity, we omit the proof terms and focus only on the propositional side):

$\triangle$-*intro*: $A/\triangle A$ $\quad$ $\triangle$-*elim*: $A \to B/\triangle A \to B$ $\quad$ $\triangle$-*ax*: $\triangle A$ is a mere proposition

with the side condition that B is a mere proposition. We can observe that $\triangle$-*intro* and $\triangle$-*elim* correspond directly to our rules $\circ$-*intro* and $\circ$-*elim**, while $\triangle$-ax corresponds to our stipulation that $\circ\varphi$ is a declarative formula.

These initial observations seem to suggest that there is a link not only between the notions of presupposition and truncation but more generally between the ideas of inquisitive and computational content. This seems to be further strengthened by the fact that inquisitive proofs are known to have computational content in the sense of the propositions-as-types principle (see [3] and Section 4.2 in [6]). The connection to the truncation modality also seems to match the interpretation of non-inquisitive closure when we regard inquisitive propositions as information types as in [5]. We plan to focus on these intriguing connections in our future research.

Appendix

Proof of Claim 3.1: We will show just the least obvious part of this claim, namely that $iP \cap iQ \subseteq i(P \cap Q)$. Assume that $s \in iP \cap iQ$. Then $s \subseteq \bigcup P$ and $s \subseteq \bigcup Q$. So, for every $w \in s$ there is $t \in P$ such that $w \in t$ and there is $u \in Q$ such that $w \in u$. Then $w \in t \cap u$ and since P and Q are closed under nonempty subsets, $t \cap u \in P \cap Q$. So, $w \in \bigcup(P \cap Q)$. Since this holds for every $w \in s$, we obtain $s \subseteq \bigcup(P \cap Q)$, i.e. $s \in i(P \cap Q)$. Hence, $iP \cap iQ \subseteq i(P \cap Q)$.

Proof of Claim 3.3: Consider the Kripke model $\mathcal{M}$ depicted below on the left. The algebra of propositions in $inq(\mathcal{M})$ is depicted in the middle:

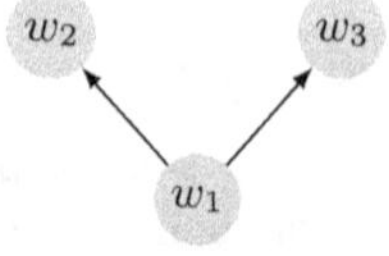

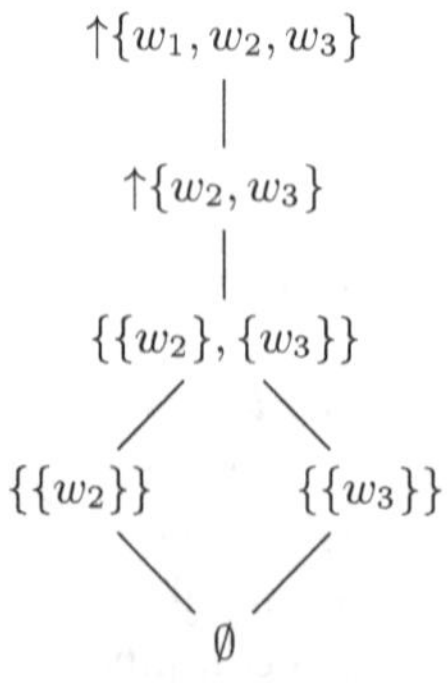

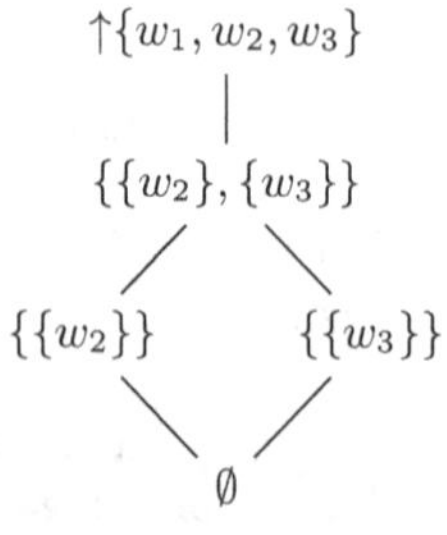

Assume that the valuation V in $\mathcal{M}$ is defined as follows: $V(p) = \{w_2\}$, $V(q) = \{w_3\}$, and $V(r) = \emptyset$, for every other atom r. Now we can observe that the algebra generated from the atomic propositions in $inq(\mathcal{M})$ by the logical operators from L is the one depicted on the right. In other words, no L-formula expresses in $inq(\mathcal{M})$ the proposition generated by $\{w_2, w_3\}$. But that is the proposition expressed by $\circ(p \mathbin{\vee\!\!\!\vee} q)$. Hence, no formula in L is equivalent to $\circ(p \mathbin{\vee\!\!\!\vee} q)$ over inquisitive Kripke models.

Proof of Claim 3.4: Assume that a model $inq(\mathcal{M}_1)$ contains a state s_1 that does not support φ and $inq(\mathcal{M}_2)$ a state s_2 that does not support ψ. Then $inq(\mathcal{M}_1 \uplus \mathcal{M}_2)$, where $\mathcal{M}_1 \uplus \mathcal{M}_2$ is the disjoint union of $\mathcal{M}_1$ and $\mathcal{M}_2$, contains the state $s'_1 \cup s'_2$, where s'_1 is the copy of s_1 and s'_2 is the copy of s_2. The state $s'_1 \cup s'_2$ does not support $\varphi \mathbin{\vee\!\!\!\vee} \psi$.

Proof of Claim 3.5: The right-to-left implication is equivalent to $\circ$-intro which holds by persistence. Let us show that the left-to-right implication also holds. Assume that s is a state of an inquisitive model $inq(\mathcal{M})$ such that $s \Vdash \varphi \to \alpha$. Take any nonempty $t \subseteq s$ such that $t \Vdash \circ\varphi$. Then for every $w \in t$, $\uparrow w \Vdash \varphi$. Since $\uparrow w \subseteq s$, we obtain $\uparrow w \Vdash \alpha$. Since this holds for every $w \in t$ we obtain $t \Vdash \alpha$, by Claim 3.2 and the assumption that α is declarative. So, $s \Vdash \circ\varphi \to \alpha$.

Proof of Claim 3.7: First, we need to make sure that equivalent formulas are interchangeable in $S(\mathsf{InqIL}^\circ)$. That is, if $\vdash_{S(\mathsf{InqIL}^\circ)} \varphi \leftrightarrow \psi$ and χ' is obtained from χ by replacing an occurrence of a subformula φ in χ with ψ then $\vdash_{S(\mathsf{InqIL}^\circ)} \chi' \leftrightarrow \chi$. To establish this claim we need to show that logical equivalence is preserved by application of all logical connectives. We show just the case of $\circ$. We need to show that if $\vdash_{S(\mathsf{InqIL}^\circ)} \varphi \leftrightarrow \psi$ then $\vdash_{S(\mathsf{InqIL}^\circ)} \circ\varphi \leftrightarrow \circ\psi$. This follows from the fact that if $\vdash_{S(\mathsf{InqIL}^\circ)} \varphi \to \psi$ then $\vdash_{S(\mathsf{InqIL}^\circ)} \circ\varphi \to \circ\psi$ which can be proved as follows:

$$\dfrac{\dfrac{\dfrac{\dfrac{\varphi \to \psi \qquad [\varphi]^{(1)}}{\psi}\,\to\text{-}elim}{\circ\psi}\,\circ\text{-}intro}{\varphi \to \circ\psi}\,(1),\ \to\text{-}intro}{\circ\varphi \to \circ\psi}\,\circ\text{-}elim^*$$

The rest is established by induction. All the steps are standard. See [17] for the inductive steps for $\to, \wedge, \mathbin{\vee\!\!\!\vee}$ (*split* is needed in the step for implication). Note that the inductive step for $\circ$ is a direct application of interchangeability of equivalent formulas.

Proof of Theorem 3.8: To prove this result we will need two auxiliary claims.

Lemma A.1 *For every set of L°-formulas $\Delta \cup \{\varphi\}$, if $\Delta \nvdash_{S(\mathsf{InqIL}^\circ)} \varphi$ then there is a set of simple formulas Δ^φ such that $\Delta^\varphi \nvdash_{S(\mathsf{InqIL}^\circ)} \varphi$ and for every $\psi \in \Delta$ there is $\beta \in \Delta^\varphi$ such that $\beta \in \mathcal{R}(\psi)$.*

Proof. If Δ is finite then the proof is straightforward. We will assume that Δ is infinite. Let $\psi_1, \psi_2, \ldots$ be all the formulas in Δ. We define a sequence of sets of formulas $\Delta_1, \Delta_2, \ldots$ such that $\Delta_1 = \Delta$ and $\Delta_{n+1} = \Delta_n \setminus \{\psi_n\}$. Then we define by induction a function f assigning to each $\psi \in \Delta$ a formula $\beta \in \mathcal{R}(\psi)$. Since $\{\psi_1\} \cup \Delta_2 \nvdash_{S(\mathsf{InqIL}^\circ)} \varphi$ there is $\beta_1 \in \mathcal{R}(\psi_1)$ such that $\{\beta_1\} \cup \Delta_2 \nvdash_{S(\mathsf{InqIL}^\circ)} \varphi$. Then we fix $f(\psi_1) = \beta_1$. Now assume that $f(\psi_1), \ldots, f(\psi_n)$ are already defined and $\{f(\psi_1), \ldots, f(\psi_n), \psi_{n+1}\} \cup \Delta_{n+2} \nvdash_{S(\mathsf{InqIL}^\circ)} \varphi$. Then there is $\beta_{n+1} \in \mathcal{R}(\psi_{n+1})$ such that $\{f(\psi_1), \ldots, f(\psi_n), \beta_{n+1}\} \cup \Delta_{n+2} \nvdash_{S(\mathsf{InqIL}^\circ)} \varphi$ and we fix $f(\psi_{n+1}) = \beta_{n+1}$. In this way the function f is defined for every $\psi \in \Delta$. Fix $\Delta^\varphi = \{f(\psi) \mid \psi \in \Delta\}$. Then $\Delta^\varphi \nvdash_{S(\mathsf{InqIL}^\circ)} \varphi$ and for every $\psi \in \Delta$ there is $\beta \in \Delta^\varphi$ such that $\beta \in \mathcal{R}(\psi)$. □

We say that a set of simple L°-formulas Δ is a *theory* if (a) Δ is nonempty; (b) $\bot \notin \Delta$; (c) Δ is closed under the rulés of intuitionistic logic. We say that a theory Δ is *prime* if $\varphi \vee \psi \in \Delta$ only if $\varphi \in \Delta$ or $\psi \in \Delta$. The set of all prime theories will be denoted as $Prime(\mathsf{IL})$.

Now, let us consider the usual canonical model of propositional intuitionistic logic $\mathcal{M}^{\mathsf{IL}} = \langle Prime(\mathsf{IL}), \subseteq, V \rangle$, where $V(p) = \{\Gamma \in Prime(\mathsf{IL}) \mid p \in \Gamma\}$, for every atomic formula p. For this construction, we can prove the following form of a truth lemma.

Lemma A.2 *Let α be a simple formula and X a nonempty upward closed set of prime theories, i.e. a state in $inq(\mathcal{M}^{\mathsf{IL}})$. Then*

$$X \Vdash \alpha \textit{ in } inq(\mathcal{M}^{\mathsf{IL}}) \textit{ iff } \alpha \in \bigcap X.$$

Proof. We proceed by induction. We show just the inductive step for $\vee$. Assume that the claim holds for α and β. Then the following claims are equivalent:

- $X \Vdash \alpha \vee \beta$, i.e. $X \Vdash \circ(\alpha \mathbin{\vee\!\!\!\vee} \beta)$,
- for all $\Gamma \in X$, $\uparrow\Gamma \Vdash \alpha$ or $\uparrow\Gamma \Vdash \beta$,
- for all $\Gamma \in X$, $\alpha \in \bigcap\uparrow\Gamma$ or $\beta \in \bigcap\uparrow\Gamma$,
- for all $\Gamma \in X$, $\alpha \in \Gamma$ or $\beta \in \Gamma$,
- for all $\Gamma \in X$, $\alpha \vee \beta \in \Gamma$,
- $\alpha \vee \beta \in \bigcap X$.

□

Using these lemmas, Theorem 3.8 can be established as follows. The left-to-right implication follows from Claim 3.5. We prove the right-to-left implication. Assume $\Delta \nvdash_{S(\mathsf{InqIL}^\circ)} \varphi$. Then, by Lemma A.1, there is a set of simple L°-formulas Δ^φ such that $\Delta^\varphi \nvdash_{S(\mathsf{InqIL}^\circ)} \varphi$ and it holds that for every $\psi \in \Delta$ there is $\beta \in \Delta^\varphi$ such that $\beta \in \mathcal{R}(\psi)$.

Take any $\alpha \in \mathcal{R}(\varphi)$. Then we have $\Delta^\varphi \nvdash_{S(\mathsf{InqIL}^\circ)} \alpha$ and by the standard construction, we can build a prime theory Γ^α such that $\Delta^\varphi \subseteq \Gamma^\alpha$ and $\alpha \notin \Gamma^\alpha$. Now we can define the following nonempty upward closed set of prime theories, i.e a state in $inq(\mathcal{M}^{\mathsf{IL}})$:

$$X^\Delta_\varphi = \{\Gamma \in Prime(\mathsf{IL}) \mid \Gamma^\alpha \subseteq \Gamma, \text{ for some } \alpha \in \mathcal{R}(\varphi)\}.$$

For each $\alpha \in \mathcal{R}(\varphi)$, $\alpha \notin \bigcap X^{\Delta}_{\varphi}$, and thus, by Lemma A.2, $X^{\Delta}_{\varphi} \nVdash \alpha$. Hence $X^{\Delta}_{\varphi} \nVdash \varphi$. Moreover, for every $\psi \in \Delta$ there is $\beta \in \mathcal{R}(\psi)$ such that $\beta \in \bigcap X^{\Delta}_{\varphi}$ and thus $X^{\Delta}_{\varphi} \Vdash \beta$. So, for every $\psi \in \Delta$, $X^{\Delta}_{\varphi} \Vdash \psi$, which yields the desired result that $\Delta \nvDash_{(\mathsf{InqIL}^{\circ})} \varphi$.

Proof of Theorem 3.10: We can proceed as in the proof of Theorem 3.8, but using the fact that intuitionistic logic has the finite model property. Assume $\varphi \notin \mathsf{InqIL}^{\circ}$. Then $\alpha \notin \mathsf{InqIL}^{\circ}$, for every $\alpha \in \mathcal{R}(\varphi)$. By the finite model property of intuitionistic logic, for each $\alpha \in \mathcal{R}(\varphi)$ there is a finite Kripke model refuting α. We can take the disjoint union $\mathcal{M}$ of these models. Then, for each $\alpha \in \mathcal{R}(\varphi)$, there is a state in $inq(\mathcal{M})$ that does not support α. The union of these states does not support φ.

Proof of Theorem 4.1: The rule L1 coincides with $\circ$-*intro* and it is clear that all the rules L2, $\circ\bot/\bot$, *split** and *da* can be simulated in $S(\mathsf{InqIL}^{\circ})$. For instance, *da* is obtained as follows:

$$\dfrac{\circ p \qquad [p]^{(1)}}{p}\ (1),\ \circ\text{-}elim$$

Hence, if $\Delta \vdash_{S^*(\mathsf{InqIL}^{\circ})} \varphi$ then $\Delta \vdash_{S(\mathsf{InqIL}^{\circ})} \varphi$. In order to show also the converse implication we will simulate the rule $\circ$-*elim** in $S^*(\mathsf{InqIL}^{\circ})$. First, we will show by induction on declarative α that $\circ\alpha \vdash_{S^*(\mathsf{InqIL}^{\circ})} \alpha$. The cases of $\alpha = \bot$ and $\alpha = \circ\varphi$ are by DPLL. The case of atomic formulas is by *da*. As an inductive assumption (IndA) we assume that we already have for some declarative formulas β, γ that $\circ\beta \vdash_{S^*(\mathsf{InqIL}^{\circ})} \beta$ and $\circ\gamma \vdash_{S^*(\mathsf{InqIL}^{\circ})} \gamma$. We just sketch the derivations for $\alpha = \beta \wedge \gamma$ and $\alpha = \beta \to \gamma$:

$$\dfrac{\dfrac{\dfrac{\dfrac{\circ(\beta\wedge\gamma)}{\circ\beta\wedge\circ\gamma}\ \mathsf{PLL}}{\circ\beta}\ \wedge\text{-}elim}{\beta}\ \text{IndA} \qquad \dfrac{\dfrac{\dfrac{\circ(\beta\wedge\gamma)}{\circ\beta\wedge\circ\gamma}\ \mathsf{PLL}}{\circ\gamma}\ \wedge\text{-}elim}{\gamma}\ \text{IndA}}{\beta\wedge\gamma}\ \wedge\text{-}intro \qquad\qquad \dfrac{\dfrac{\dfrac{\circ(\beta\to\gamma)}{\circ\beta\to\circ\gamma}\ \mathsf{PLL}}{\circ\beta\to\gamma}\ \mathsf{IL},\ \text{IndA}}{\beta\to\gamma}\ \mathsf{PLL}$$

Now, $\varphi \to \alpha \vdash_{S^*(\mathsf{InqIL}^{\circ})} \circ\varphi \to \alpha$ can be derived as follows:

$$\dfrac{\dfrac{\dfrac{[\circ\varphi]^{(1)} \qquad \dfrac{\dfrac{[\varphi]^{(2)} \qquad \varphi\to\alpha}{\alpha}\ \to\text{-}elim}{\circ\alpha}\ \text{L1}}{\circ\alpha}\ (2),\ \text{L2}}{\alpha}}{\circ\varphi\to\alpha}\ (1),\ \to\text{-}intro$$

Proof of Theorem 4.2: The proof of soundness is straightforward. To prove completeness we can either directly construct the Lindenbaum-Tarski algebra of InqIL°, or use our completeness result with respect to inquisitive Kripke models and observe that the algebras of propositions in these models are inquisitive nuclear algebras.

Proofs of Theorems 5.2 and 5.3: We can proceed as in the proofs of Theorems 3.8 and 3.10. Let us focus on the logic $\mathsf{InqIL}^{\vee}$. The role of simple formulas in L° can be played by $\vee\!\!\!\vee$-free $L^{\vee}$-formulas. We have to show (a) that we can obtain a result analogous to Claim 3.6, and (b) that we obtain a suitable disjunctive normal form result. As regards (a) the introduction and elimination rules for $\Vee$-free $L^{\vee}$-formulas are obtained directly from $\vee$-*intro* and $\vee$-*elim*. In order to prove the disjunctive normal form we need to extend the definition of resolution for $\vee$. We define:

- $\mathcal{R}(\varphi \vee \psi) = \{\bigvee(\mathcal{R}(\varphi) \cup \mathcal{R}(\psi))\}$.

Then we need to prove interchangeability of equivalent formulas for $\vee$ which follows from the fact that we can derive $(\varphi \vee \chi) \to (\psi \vee \chi)$ from $\varphi \to \psi$ in the following way (using the fact that $\psi \vee \chi$ is declarative):

$$\dfrac{\dfrac{[\varphi \vee \chi]^{(1)} \qquad \dfrac{\dfrac{[\varphi]^{(2)} \qquad \varphi \to \psi}{\psi}\,{\to}\text{-}elim}{\psi \vee \chi}\,\vee\text{-}intro \qquad \dfrac{[\chi]^{(3)}}{\psi \vee \chi}\,\vee\text{-}intro}{\psi \vee \chi}\,(2),(3),\vee\text{-}elim}{(\varphi \vee \chi) \to (\psi \vee \chi)}\,(1),{\to}\text{-}intro$$

In the inductive step for $\vee$ in the proof of disjunctive normal form we have to show that

$$\vdash_{S(\mathsf{InqIL}^{\vee})} (\alpha_1 \Vee \ldots \Vee \alpha_n) \vee (\beta_1 \Vee \ldots \Vee \beta_m) \leftrightarrow (\alpha_1 \vee \ldots \vee \alpha_n \vee \beta_1 \vee \ldots \vee \beta_m)$$

which is obtained by application of the introduction and elimination rules for $\vee$ and $\Vee$ using the fact that the formulas on both sides of the equivalence are declarative. In the case of $\mathsf{InqIL}^{\otimes}$ we need to employ a different resolution condition:

- $\mathcal{R}(\varphi \otimes \psi) = \{\alpha \otimes \beta \mid \alpha \in \mathcal{R}(\varphi), \beta \in \mathcal{R}(\psi)\}$.

The rule T5 is used in the inductive step for $\otimes$ in the proof of the disjunctive normal form result. In the case of $\mathsf{InqIL}^{\vee\otimes}$ we also use the fact that $(\alpha \vee \beta) \leftrightarrow (\alpha \otimes \beta)$ is provable in $S(\mathsf{InqIL}^{\vee\otimes})$ (using $\vee$-*intro*, $\vee$-*elim* and T1–T4).

Proof of Theorem 5.4: (a)–(d) follow from the mutual definability of $\vee$ and $\circ$. (c) is based on the fact that $\circ\varphi$ is equivalent to $\circ(\varphi \Vee \varphi)$ in InqIL°, and (d) is based on the fact that $\varphi \vee \psi$ is equivalent to $(\varphi \Vee \psi) \vee (\varphi \Vee \psi)$ in $\mathsf{InqIL}^{\vee}$.

Proof of Theorem 6.3: We can first observe that $\vee$-*intro* and $\vee$-*elim* are sound with respect to Beth semantics and thus can be derived in $S(\mathsf{HInqIL})$ because of its completeness. The soundness of $\vee$-*intro* is immediate. In order to show the soundness of $\vee$-*elim* we can take any state s of any Beth model and assume that $\varphi \vee \psi$, $\varphi \to \alpha$ and $\psi \to \alpha$ all hold at s. It follows from these assumptions that every path through s intersects $|\alpha|$. Since α is declarative we obtain $s \in |\alpha|$, i.e. $s \Vdash \alpha$, as desired.

In the other direction, we can observe that all instances of H1–H5 are derivable from $\vee$-*intro* and $\vee$-*elim*. For an illustration, we show how to derive both directions of H5. The left-to-right direction of H5 is obtained as follows:

$$
\dfrac{(\varphi \mathbin{\vee\!\!\!\vee} \psi) \vee (\varphi \mathbin{\vee\!\!\!\vee} \psi) \qquad \dfrac{[\varphi \mathbin{\vee\!\!\!\vee} \psi]^{(1)} \qquad \dfrac{[\varphi]^{(2)}}{\varphi \vee \psi}\,\vee\text{-}intro \qquad \dfrac{[\psi]^{(3)}}{\varphi \vee \psi}\,\vee\text{-}intro}{\varphi \vee \psi}\,(2),(3),\mathbin{\vee\!\!\!\vee}\text{-}elim}{\varphi \vee \psi}\,(1),\vee\text{-}elim
$$

The right-to-left direction of H5 is derived in the following way:

$$
\dfrac{\varphi \vee \psi \qquad \dfrac{\dfrac{[\varphi]^{(1)}}{\varphi \mathbin{\vee\!\!\!\vee} \psi}\,\mathbin{\vee\!\!\!\vee}\text{-}intro}{(\varphi \mathbin{\vee\!\!\!\vee} \psi) \vee (\varphi \mathbin{\vee\!\!\!\vee} \psi)}\,\vee\text{-}intro \qquad \dfrac{\dfrac{[\psi]^{(2)}}{\varphi \mathbin{\vee\!\!\!\vee} \psi}\,\mathbin{\vee\!\!\!\vee}\text{-}intro}{(\varphi \mathbin{\vee\!\!\!\vee} \psi) \vee (\varphi \mathbin{\vee\!\!\!\vee} \psi)}\,\vee\text{-}intro}{(\varphi \mathbin{\vee\!\!\!\vee} \psi) \vee (\varphi \mathbin{\vee\!\!\!\vee} \psi)}\,(1),(2),\vee\text{-}elim
$$

Proof of Claim 6.4: Take any state s of any Beth model based on a Beth frame satisfying the condition (C) and assume that in this model s does not force $(\alpha \rightarrow \varphi) \mathbin{\vee\!\!\!\vee} (\alpha \rightarrow \psi)$. Then there are $t, u \geq s$ such that $t \Vdash \alpha$ but not $t \Vdash \varphi$ and $u \Vdash \alpha$ but not $u \Vdash \psi$. Since the model satisfies (C) there is $v \geq u$ such that $v \leq t$ and $v \leq u$ and every path through v joins a path that goes through t or through u. It follows that every path through v intersects $|\alpha|$. Since α is declarative, $|\alpha|$ is Beth closed and thus $v \Vdash \alpha$. But, because of persistence, v does not force $\varphi \mathbin{\vee\!\!\!\vee} \psi$. It follows that s does not force $\alpha \rightarrow (\varphi \mathbin{\vee\!\!\!\vee} \psi)$.

References

[1] Bezhanishvili, N., G. Grilletti and D. E. Quadrellaro, *An algebraic approach to inquisitive and DNA-logics*, The Review of Symbolic Logic **15** (2022), pp. 950–990.

[2] Ciardelli, I., "Inquisitive Semantics and Intermediate Logics," Master's thesis, University of Amsterdam, Amsterdam (2009).

[3] Ciardelli, I., *Interrogative dependencies and the constructive content of inquisitive proofs*, in: *Logic, Language, Information, and Computation, WoLLIC* (2014), pp. 109–123.

[4] Ciardelli, I., *Modalities in the realm of questions: axiomatizing inquisitive epistemic logic*, in: *Proceedings of Advances in Modal Logic, AiML* (2014), pp. 94–113.

[5] Ciardelli, I., *Questions as information types*, Synthese **195** (2018), pp. 321–365.

[6] Ciardelli, I., "Inquisitive Logic: Consequence and Inference in the Realm of Questions," Springer, 2023.

[7] Ciardelli, I., J. Groenendijk and F. Roelofsen, "Inquisitive Semantics," Oxford University Press, 2019.

[8] Ciardelli, I., R. Iemhoff and F. Yang, *Questions and dependency in intuitionistic logic*, Notre Dame Journal of Formal Logic **61** (2020), pp. 75–115.

[9] Ciardelli, I. and F. Roelofsen, *Inquisitive logic*, Journal of Philosophical Logic **40** (2011), pp. 55–94.

[10] Fairtlough, M. and M. Mendler, *Propositional lax logic*, Information and Computation **137** (1997), pp. 1–33.

[11] Ferguson, T. and V. Punčochář, *Structural completeness and superintuitionistic inquisitive logics*, in: *Proceedings of the Workshop on Logic, Language, Information, and Computation, WoLLIC* (2023), p. 194–210.

[12] Grilletti, G., *Medvedev logic is the logic of finite distributive lattices without top element*, in: *Proceedings of Advances in Modal Logic, AiML* (2022), pp. 451–466.

[13] Holliday, W. H., *Inquisitive intuitionistic logic*, in: *Proceedings of Advances in Modal Logic, AiML* (2022), pp. 329–348.

[14] Kripke, S. A., *Semantical analysis of intuitionistic logic I*, in: M. A. E. Dummett and J. N. Crossley, editors, *Studies in Logic and the Foundations of Mathematics*, Elsevier, 1965 pp. 92–130.

[15] Miglioli, P., U. Moscato, M. Ornaghi, S. Quazza and G. Usberti, *Some results on intermediate constructive logics*, Notre Dame Journal of Formal Logic **30** (1989), pp. 543–562.

[16] Punčochář, V., *A generalization of inquisitive semantics*, Journal of Philosophical Logic **45** (2016), pp. 399–428.

[17] Punčochář, V., *Algebras of information states*, Journal of Logic and Computation **27** (2017), pp. 1643–1675.

[18] Punčochář, V., *Inquisitive Heyting algebras*, Studia Logica **109** (2021), pp. 995–1017.

[19] Quadrellaro, D. E., *On intermediate inquisitive and dependence logics: An algebraic study*, Annals of Pure and Applied Logic **173** (2022), article 103143.

[20] Sano, K., *Goldblatt-Thomason-style characterization for intuitionistic inquisitive logic*, in: *Proceedings of Advances in Modal Logic, AiML* (2020), pp. 541–560.

[21] Stafford, W., *Proof-theoretic semantics and inquisitive logic*, Journal of Philosophical Logic **50** (2021), pp. 1199–1229.

[22] The Univalent Foundations Program, "Homotopy type theory: Univalent foundations of mathematics," Institute for Advanced Study, Princeton, 2013.

[23] Yang, F. and J. Väänänen, *Propositional logics of dependence*, Annals of Pure and Applied Logic **167** (2016), pp. 557–589.

LEGO-Like Small Model Constructions for Åqvist's Logics

Dmitry Rozplokhas [1]

TU Wien, Austria

Abstract

Åqvist's logics (**E**, **F**, **F+(CM)**, and **G**) are among the best-known systems in the long tradition of preference-based approaches for modeling conditional obligation. While the general semantics of preference models align well with philosophical intuitions, more constructive characterizations are needed to assess computational complexity and facilitate automated deduction. Existing small model constructions from conditional logics (due to Friedman and Halpern) are applicable only to **F+(CM)** and **G**, while recently developed proof-theoretic characterizations leave unresolved the exact complexity of theoremhood in logic **F**. In this paper, we introduce alternative small model constructions assembled from elementary building blocks, applicable uniformly to all four Åqvist's logics. Our constructions propose alternative semantical characterizations and imply co-NP-completeness of theoremhood. Furthermore, they can be naturally encoded in classical propositional logic for automated deduction.

Keywords: deontic logic, conditional logic, preference models, small model property

1 Introduction

In deontic logic, the analysis of various normative scenarios and deontic paradoxes led to the formalization of obligations as conditionals, i.e. as dyadic modalities $\bigcirc(\gamma \mid \alpha)$ read as "γ is obligatory if α holds". Traditionally these modalities are formalized using preference-based logics, inspired by the rational choice theory and introduced in the deontic context by Hansson [11]. This approach considers preference models — a kind of relational models with a "relative goodness" relation between worlds; a conditional obligation $\bigcirc(\gamma \mid \alpha)$ is satisfied when γ is true in the "best" worlds satisfying α. Åqvist [1] formalized these ideas using the language of modal logic and his framework now serves as one of the standard implementations of the preference-based approach in deontic logic. Initially, the framework comprised three logics of increasing deductive strength: the logic **E** that places no restrictions on the preference relation, the logic **F** that considers *limited* preference relations to rule out contradictory obligations, and the logic **G** that assumes the preference relation to be a total and limited preorder. A later addition to Åqvist's family is the

[1] dmitry@logic.at

logic **F+(CM)** [16] that drops the totality assumption and considers *smooth* preorders, and it is axiomatized by extending **F** with the cautious monotony principle, well-known in non-monotonic reasoning [8]. These four logics provide a useful scale against which various deontic scenarios can be evaluated.

At the same time, preference models were applied in the neighboring field of conditional reasoning. Notable examples of conditional frameworks defined in terms of preference models include Lewis's family of logic for counterfactuals [15], Burgess's preferential conditional logic **PCL** and its extensions [3], and KLM logic for non-monotonic reasoning [13]. These frameworks consider similar models, so there is an intersection with the Åqvist family: **F+(CM)** coincides with Burgess's logic **PCA**, and **G** coincides with Lewis's logic **VTA** (using the terminology of [9]), while the flat fragment of both these logics coincides with KLM logic **P**. However, it is not common in conditional logics to consider relaxed notions of preference relations, since assuming transitivity and smoothness is necessary for a well-behaved consequence relation in the logic [18]. On the other hand, when the preference relation is treated as comparative goodness, the adequacy of these assumptions becomes controversial (see, e.g., [18] and [12, Sec. 2.3] for an overview of counterexamples). For this reason, the weaker logics **E** and **F** play an important role in the normative reasoning context.

In recent decades, significant progress has been made in exploring variations of preference-model characterization for Åqvist's logics and their corresponding axiomatizations, surveyed in [17]. Now, there's a growing focus on the computational properties of these logics, which is the main motivation for this paper too. In [17] the decidability of theoremhood for all four logics is proven through alternative semantics based on selection functions, and embedding of the weakest logic **E** into Higher-Order Logic (HOL) from [2] is suggested as a potential approach for automated deduction. These approaches however are not suitable for assessment of the exact complexity of logics, which requires more constructive characterizations. One such characterization came from the proof-theoretic side in the form of cut-free hypersequent calculi, developed recently for all four Åqvist's logics [4,5,6]. For **E**, **F+(CM)** and **G** the proof search in the calculi has optimal co-NP complexity, and polynomial-size preference countermodels can be reconstructed from failed derivations [4,9]. At the same time, the limitedness condition of **F** seems difficult to handle both model-theoretically and proof-theoretically. The calculus for **F** [5], which is an even more complicated variation of the calculi for logic **GL** [19], gives only a co-NEXP upper bound for theoremhood (which is the best estimation so far) and no countermodel construction.

Another powerful approach for establishing computational complexity of conditional logics is *small model constructions* proposed by Friedman and Halpern [7] for Burgess' logic **PCL** and its extensions, which transforms any satisfying model into a satisfying model of bounded size. Their approach covers in particular extensions **PCA** and **VTA** (i.e. Åqvist's logics **F+(CM)** and **G**), and establishes co-NP-completeness of theoremhood for them. However, this approach significantly relies on the smoothness and transitivity of the preference relation and therefore is not applicable for weaker logics **E** and **F** (see Remark 3.22 for details).

In this paper, we propose alternative small model constructions to handle all four Åqvist's logics uniformly. We compose a model of polynomial size by assembling elementary building blocks (chains, antichains, and cliques of worlds selected from any given model) like LEGO. We provide sufficient conditions for such construction to be a countermodel and define a suitable construction for each Åqvist's logic. There are two main applications for our constructions, obtained uniformly for all logics.

Alternative semantical characterizations of theoremhood. Our results imply that theoremhood can be characterized by finite models. Moreover, for finite models the complicated properties of limitedness and smoothness (which are not frame properties) can be replaced by natural frame properties: acyclicity and transitivity of the preference relation, respectively.

Complexity and automated deduction. The polynomial size of models together with easily checkable frame properties immediately imply co-NP-completeness of theoremhood (including logic **F**, for which it was an open problem) and allow for natural encodings in classical propositional logics, which can be utilized for efficient automated deduction using SAT-solvers.

2 Preliminaries

The syntax of Åqvist's logics extends the usual propositional language with two modalities: unary $\Box$ for necessity and binary $\bigcirc(\cdot\,|\,\cdot)$ for conditional obligation. We define the formulas over the set *Var* of propositional variables.

$$\mathcal{F} ::= x \in \mathit{Var} \mid \neg\mathcal{F} \mid \mathcal{F}\wedge\mathcal{F} \mid \Box\mathcal{F} \mid \bigcirc(\mathcal{F}|\mathcal{F})$$

We will use small Greek letters to denote formulas. $|\varphi|$ will denote size of the formula (number of symbols), $\mathit{Sub}\mathcal{F}(\varphi)$ will denote the set of all subformulas of φ (including φ), and $\mathrm{Cond}(\varphi) = \{\alpha \mid \bigcirc(\gamma\,|\,\alpha) \in \mathit{Sub}\mathcal{F}(\varphi)\}$.

Definition 2.1 A *preference model* is a triple $\langle W, \succeq, \mathbb{V}\rangle$ where W is a (non-empty) set of worlds, $\succeq$ is a binary relation on W, and $\mathbb{V}\colon \mathit{Var} \to 2^W$ is a valuation function. We denote by $W(M)$ the set of worlds of a given model.

The semantics of obligation is based on the notion of "best" worlds in the preference model. There are different definitions of bestness appearing in the literature (see [10,16] for the comparison of different definitions), we will use the most common one — maximality: a world is a best world when there are no worlds that are *strictly* more preferable. As usual we denote by $\succ$ a strict version of $\succeq$ ($w_1 \succ w_2$ when $w_1 \succeq w_2$ and $w_2 \not\succeq w_1$). We will use the notation $\mathit{Bet}_{\succ}(v) = \{w \in W \mid w \succ v\}$ for a set of worlds strictly preferable to (better than) a given one.

Definition 2.2 For a preference model $M = \langle W, \succeq, \mathbb{V}\rangle$ and $U \subseteq W$ we define $\max(U) = \{v \in U \mid \nexists u \in U\colon u \succ v\}$.

Logic	Limit conditions		Properties of $\succeq$	
	limited	smooth	transitive	total
E				
F	✓			
F+(CM)		✓	✓	
G		✓	✓	✓

Fig. 1. Preference-semantical characterizations for Åqvist's logics [17, Tab. 1 and 2] (with maximality as the notion of bestness).

Satisfaction of $\bigcirc(\gamma\,|\,\alpha)$ is defined using this notion of bestness: $\bigcirc(\gamma\,|\,\alpha)$ is true when γ is true in all maximal worlds satisfying α (we will call such worlds α-*maximal*). And $\Box\beta$ is true when β is true in all worlds (so we treat $\Box$ as the universal **S5** modality).

Definition 2.3 (Satisfaction) For a preference model $M = \langle W, \succeq, \mathbb{V}\rangle$ the *truth set* $||\varphi||^M$ of a formula φ is defined inductively:

- $w \in ||x||^M$ for $x \in Var$ when $w \in \mathbb{V}(x)$,
- $w \in ||\neg\psi||^M$ when $w \notin ||\psi||^M$,
- $w \in ||\psi_1 \wedge \psi_2||^M$ when $w \in ||\psi_1||^M$ and $w \in ||\psi_2||^M$,
- $w \in ||\Box\beta||^M$ when $||\beta||^M = W$,
- $w \in ||\bigcirc(\gamma\,|\,\alpha)||^M$ when $\max(||\alpha||^M) \subseteq ||\gamma||^M$.

We say that w satisfies φ in M (denoted $M, w \models \varphi$) when $w \in ||\varphi||^M$, and that M validates φ (denoted $M \models \varphi$) when $||\varphi||^M = W$. For $U \subseteq W$ we denote a set of formulas satisfiable in U as $Sat\mathcal{F}_M(U) = \{\psi \mid \exists u \in U : M, u \models \psi\}$.

Notice that the satisfaction of both $\Box\beta$ and $\bigcirc(\gamma\,|\,\alpha)$ does not depend on the world of evaluation.

Different Åqvist's logics are defined by different classes of preference models. Some of these classes are defined using the properties of preference relation $\succeq$ in the model, we will use two properties: transitivity ($\succeq$ is transitive when $w_1 \succeq w_2$ and $w_2 \succeq w_3$ imply $w_1 \succeq w_3$) and totalness ($\succeq$ is total when for any $w_1, w_2 \in W$ either $w_1 \succeq w_2$ or $w_2 \succeq w_1$). Another property used for the characterization of deontic logic is what Lewis called "limit assumption", which ensures the existence of best worlds. The are different formal definitions of this assumption in the literature, we will use two versions from [17]: *limitedness* and *smoothness*.

Definition 2.4 (Limit conditions) Let $M = (W, \succeq, \mathbb{V}) \in \mathcal{M}$. M is *limited* when for any formula α if $||\alpha||^M \neq \emptyset$ then $\max(||\alpha||^M) \neq \emptyset$. M is *smooth* when for any formula α and any world $w \in ||\alpha||^M$ there exists $u \in \max(||\alpha||^M)$ such that either $u = w$ or $u \succ w$.

We rely on the semantical characterizations of the four Åqvist logics in Fig. 1, which are presented (among various other characterizations) in [17].

Definition 2.5 Formula φ is a theorem of Åqvist's logic $\mathcal{L}$ iff $M \models \varphi$ for any preference model M that satisfies model conditions for logic $\mathcal{L}$ in Fig. 1.

We will call a preference model M a *countermodel for a formula* φ if $M \not\models \varphi$ and we will further call it an $\mathcal{L}$-*countermodel* if it belongs to a class of models corresponding to a logic $\mathcal{L}$ from Fig. 1.

3 Small Model Constructions

This section contains the main technical result of the paper: for every logic $\mathcal{L}$ from the Åqvist family, we will show how an arbitrary $\mathcal{L}$-countermodel M for a formula φ can be transformed into an $\mathcal{L}$-countermodel with the number of worlds bounded polynomially w.r.t. $|\varphi|$. We will achieve this by selecting a finite number of worlds from M, adding copies for some of them, and defining a new preference relation on the selected worlds without changing the valuation. We call such transformation a *rearrangement* of a model.

Definition 3.1 We say that a model $M' = \langle W', \succeq', \mathbb{V}' \rangle$ *rearranges* the model $M = \langle W, \succeq, \mathbb{V} \rangle$ when there exists a *prototype function* $prot\colon W' \to W$ such that $w' \in \mathbb{V}'(x)$ is equivalent to $prot(w) \in \mathbb{V}(x)$ for all $x \in Var$.

Our main goal for the rearranged model is to have each of its worlds satisfying the same subformulas of φ as its prototype does. Evaluation of a formula in a world involves other worlds only in the cases of $\Box$ and $\bigcirc(\cdot \mid \cdot)$ modalities. Therefore, we only need to ensure that the rearranged model validates the same modalities among subformulas of φ as the original model does, while the satisfaction (and non-satisfaction) of other subformulas will be preserved in the rearranged model automatically.

We will examine the cases of validated and non-validated modalities separately. Let us denote by $Box^+(\varphi, M)$ (resp. $Ob^+(\varphi, M)$) the set of subformulas of φ of the form $\Box\beta$ (resp. $\bigcirc(\gamma \mid \alpha)$) that are validated by M, and by $Box^-(\varphi, M)$ and $Ob^-(\varphi, M)$ the sets of subformulas of φ of the corresponding form that are not validated by M. To falsify $\Box\beta \in Box^-(\varphi, M)$ and $\bigcirc(\gamma \mid \alpha) \in Ob^-(\varphi, M)$ we need to take in M' some worlds that were falsifying these modalities in M. While the evaluation of $\Box\beta$ modalities relies only on the presence of the worlds satisfying β in the model, special care is needed to ensure that the evaluation of $\bigcirc(\gamma \mid \alpha)$ is the same. Namely, if a world w was made not α-maximal in M by some world $u \in ||\alpha||^M$ such that $u \succ w$ we need to preserve this violation of maximality in M'. Conversely, we need to ensure that we are not violating α-maximality in M' for the world falsifying $\bigcirc(\gamma \mid \alpha)$. This reasoning leads to the following four conditions sufficient to ensure that a rearranged model M' is a countermodel for φ.

Theorem 3.2 *Suppose* $M = \langle W, \succeq, \mathbb{V} \rangle$ *is a countermodel for* φ *and* $M' = \langle W', \succeq', \mathbb{V}' \rangle$ *rearranges* M *with the prototype function* $prot\colon W' \to W$. *Then the following conditions are sufficient for* $M' \not\models \varphi$.

(*i*) *There exists* $v' \in W'$ *such that* $M, prot(v') \not\models \varphi$.

(*ii*) *For any* $\Box\beta \in Box^-(\varphi, M)$ *there exists* $v' \in W'$ *such that* $M, prot(v') \not\models \beta$.

(*iii*) *For any* $\bigcirc(\gamma \mid \alpha) \in Ob^-(\varphi, M)$ *there exists* $v' \in W'$ *such that* $prot(v') \in \max(||\alpha||^M) \setminus ||\gamma||^M$ *and for all* $u' \succ' v'$ *holds* $prot(u') \succ prot(v')$.

(*iv*) *For any* $w' \in W'$*, for all* $\bigcirc(\gamma \mid \alpha) \in Ob^+(\varphi, M)$ *if there exists* $u \succ prot(w')$ *such that* $M, u \models \alpha$ *then there exists* $s' \succ' w'$ *such that* $M, prot(s') \models \alpha$.

Proof. We will prove a generalized statement: for any $w' \in W'$ and any $\psi \in Sub\mathcal{F}(\varphi)$ holds $M', w' \models \psi$ iff $M, prot(w') \models \psi$. Then $M' \not\models \varphi$ follows from condition (*i*). The proof is by induction on ψ with case analysis on ψ belonging to $Box^+(\varphi, M)$ or $Box^-(\varphi, M)$ for $\psi = \Box\beta$ and on ψ belonging to $Ob^+(\varphi, M)$ or $Ob^-(\varphi, M)$ for $\psi = \bigcirc(\gamma \mid \alpha)$. Conditions (*ii*), (*iii*), and (*iv*) directly cover cases $\psi \in Box^-(\varphi, M)$, $\psi \in Ob^-(\varphi, M)$, and $\psi \in Ob^+(\varphi, M)$ respectively (see appendix A for details). □

Ensuring conditions (*i*) and (*ii*) is simple: we need to take arbitrary worlds from $(W \setminus ||\varphi||^M)$ and from $(W \setminus ||\beta||^M)$ for each $\Box\beta \in Box^-(\varphi, M)$. For this, we will use a *representative function rep*: $(2^W \setminus \{\emptyset\}) \to W$ that chooses an element $rep(S) \in S$ from any given non-empty subset S of W (thus, we use the axiom of choice explicitly in our construction). We will also need representatives of $(\max(||\alpha||^M) \setminus ||\gamma||^M)$ for every $\bigcirc(\gamma \mid \alpha) \in Ob^-(\varphi, M)$ for condition (*iii*). Let us denote the set of all such falsifying worlds $Fal(\varphi, M)$.

Definition 3.3 (Falsifying worlds) For a model $M = \langle W, \succeq, \mathbb{V}\rangle$ such that $M \not\models \varphi$, $Fal(\varphi, M) = rep(W \setminus ||\varphi||^M) \cup Fal^{\Box}(\varphi, M) \cup Fal^{\bigcirc}(\varphi, M)$, where $Fal^{\Box}(\varphi, M) = \{rep(W \setminus ||\beta||^M) \mid \Box\beta \in Box^-(\varphi, M)\}$, $Fal^{\bigcirc}(\varphi, M) = \{rep(\max(||\alpha||^M) \setminus ||\gamma||^M) \mid \bigcirc(\gamma \mid \alpha) \in Ob^-(\varphi, M)\}$.

The rest of the rearranged model will be chosen to ensure the satisfaction of conditions (*iii*) and (*iv*). We will represent our small model constructions as *composite models*, assembled from *blocks*. A *block B* is a finite selection of worlds from M with some new preference relation on them (in our cases it will be either an empty relation, a strict linear order, or a universal relation).

Definition 3.4 (Block) A *block* on M is a tuple $\langle U, \succeq_U\rangle$ where $U \subseteq W(M)$ and $\succeq_U$ is a binary relation on U. We will use $W(B)$ to refer to the set of worlds in B. For a given M and $U \subseteq W(M)$ we will consider the blocks of the following forms:

- *antichain*$(U) = \langle U, \succeq^a\rangle$, where $\succeq^a$ is an empty relation;
- *chain*$(S) = \langle \{w_i\}_{i=1}^n, \succeq^{ch}\rangle$ if $S = [w_1, \ldots, w_n]$ is a finite ordered sequence of worlds and $w_i \succeq^{ch} w_j$ iff $i \leq j$;
- *clique*$(U) = \langle U, \succeq^{cl}\rangle$ where $u_1 \succeq_{cl} u_2$ for all $u_1, u_2 \in U$.

A composite construction consists of the number of blocks with an additional preference relation on them. Each composite construction generates a model rearranging M, in which the new preference relation is given by combining the relation between blocks and the relations inside blocks. To allow multiple occurrences of the same block in the construction we define the composite construction using labels and a labeling function.

Definition 3.5 (Composite construction) A *composite construction on* M is a tuple $\langle L, \succeq_L, \mathcal{B} \rangle$ where L is a set of labels, $\succeq_L$ is a binary relation on L, and $\mathcal{B}$ is a labeling function that maps every label from L into a block on M. Each composite construction $\mathfrak{C} = \langle L, \succeq_L, \mathcal{B} \rangle$ on $M = \langle W, \succeq, \mathbb{V} \rangle$ *generates* a model $gen(\mathfrak{C}) = \langle W^{gen}, \succeq^{gen}, \mathbb{V}^{gen} \rangle$, where

- $W^{gen} = \{(l, w) \mid l \in L, w \in W(\mathcal{B}(l))\}$;
- $(l_1, w_1) \succeq^{gen} (l_2, w_2)$ iff either $l_1 \succeq_L l_2$ or both $l_1 = l_2$ and $w_1 \succeq_U w_2$ for $\mathcal{B}(l_1) = \langle U, \succeq_U \rangle$;
- $(l, w) \in \mathbb{V}^{gen}(x)$ iff $w \in \mathbb{V}(x)$.

We can now simplify the conditions of Th. 3.2 for models generated by composite constructions. For conditions (i)-(iii) it is enough to have every world v from $Fal(\varphi, M)$ in some block in the construction, such that this block does not have any $\succ_U$-preferable worlds inside and all blocks $\succ_L$-prefferable to it have only worlds from $Bet_{\succeq}(v)$. Also, we can ensure (iv) separately for each block by either ensuring it inside this block or having another block $\succ_L$-preferred to it that has all worlds required in (iv). We express these condition using the following notions of block compatibility.

Definition 3.6 (Block compatibility properties). For a given model M and a formula φ we define the following properties of blocks on M:

- $\langle U, \succeq_U \rangle$ is *flat* when $w_1 \not\succ_U w_2$ for any $w_1, w_2 \in U$.
- B' is (iii)*-suitable for* B when $W(B') \subseteq Bet_{\succ}(w)$ for any $w \in W(B)$.
- $\langle U, \succeq_U \rangle$ is (iv)*-safe* when for every $w \in U$ and every $\alpha \in Sat\mathcal{F}_M(Bet_{\succ}(w)) \cap \mathrm{Cond}(\varphi)$ there is $w' \succ_U w$ such that $M, w' \models \alpha$.
- B' (iv)*-covers* B when $Sat\mathcal{F}_M(Bet_{\succ}(w)) \cap \mathrm{Cond}(\varphi) \subseteq Sat\mathcal{F}_M(W(B'))$ for any $w \in W(B)$.

Using these notions we can reformulate (the weaker version of) Th. 3.2 for composite constructions as follows.

Theorem 3.7 *Let* $M \not\models \varphi$ *and* $\mathfrak{C} = \langle L, \succeq_L, \mathcal{B} \rangle$ *be a composite construction on* M. *For* $gen(\mathfrak{C}) \not\models \varphi$ *it is sufficient that:*

(a) For every falsifying world $v \in Fal(\varphi, M)$ *there is a label* $b_v \in L$ *such that* $v \in W(\mathcal{B}(b_v))$ *and* $\mathcal{B}(b_v)$ *is flat and* $\mathcal{B}(b')$ *is* (iii)*-suitable for* $\mathcal{B}(b_v)$ *for every* $b' \succ_L b_v$.

(b) For every $b \in L$ *the block* $\mathcal{B}(b)$ *is either* (iv)*-safe or* (iv)*-covered by* $\mathcal{B}(b')$ *for some* $b' \succ_L b$.

Proof. $gen(\mathfrak{C})$ rearranges M (with $prot((l, w)) = w$), so we can apply Th. 3.2. (a) ensures conditions (i)-(iii) and (b) ensures condition (iv). □

We now define composite constructions for each Åqvist's logic satisfying the conditions from Th. 3.7 and the model conditions for the logic from Fig. 1.

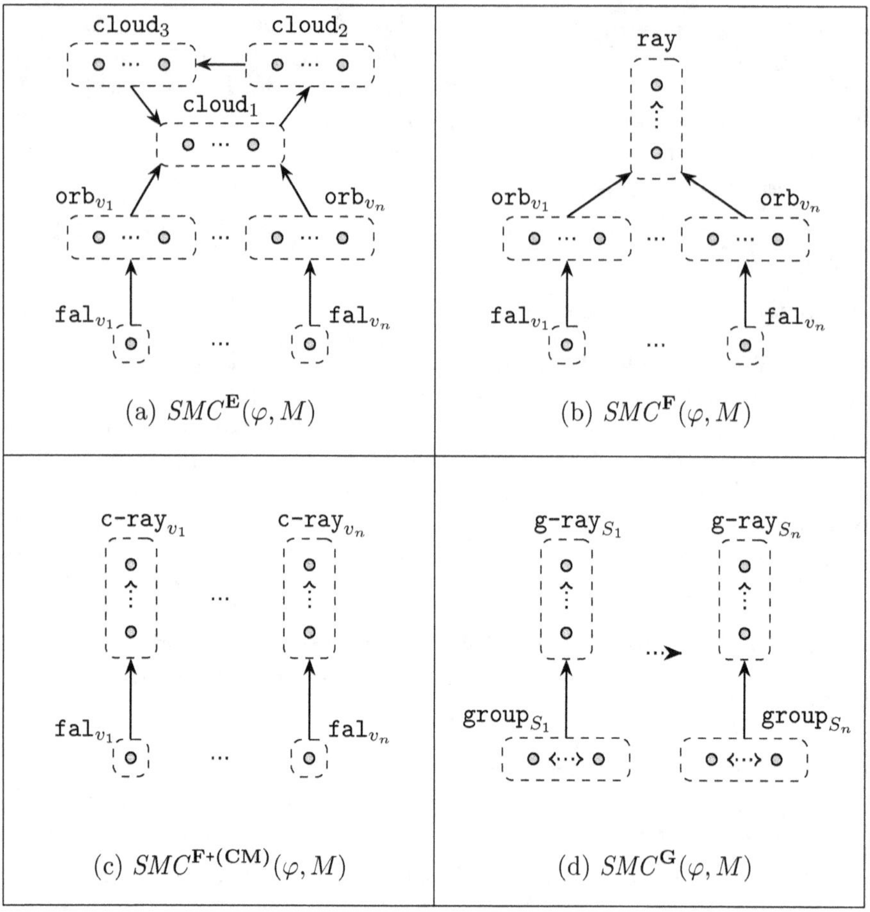

(a) $SMC^{\mathbf{E}}(\varphi, M)$ (b) $SMC^{\mathbf{F}}(\varphi, M)$ (c) $SMC^{\mathbf{F+(CM)}}(\varphi, M)$ (d) $SMC^{\mathbf{G}}(\varphi, M)$

Fig. 2. Small model constructions for Åqvist's logics. Gray circles represent worlds, dashed rectangles represent blocks. Symbol ⋯ inside a block indicates an antichain, $\hat{\vdots}$ indicates a chain, and ‹⋯› indicates a clique. Solid arrows represent the preference relation $\succeq_L$ between blocks: an arrow from a block l_1 to a block l_2 means $l_2 \succeq_L l_1$. The arrow ⋯➤ between blocks in construction $SMC^{\mathbf{G}}(\varphi, M)$ means that there is a linear order on blocks. Note that the preference relation in constructions $SMC^{\mathbf{E}}(\varphi, M)$ and $SMC^{\mathbf{F}}(\varphi, M)$ is not transitive.

3.1 Small Model Construction for Logic E

In the case of logic **E** there are no model conditions we need to satisfy in our countermodel, so we can use a preference relation that is non-transitive and contains cycles. In this simple case, all blocks of the countermodel construction will be antichains.

We start our composite construction with a dedicated one-world block *antichain*($\{v\}$) labeled $\texttt{fal}_v$ for each world $v \in \mathit{Fal}(\varphi, M)$. The simplest way to (*iv*)-cover such block with a (*iii*)-suitable block containing linearly many (w.r.t. $|\varphi|$) worlds is to go through formulas from Cond(φ) satisfied by some world in $Bet_{\succ}(v)$ and select one representative for each. Below such selection is defined more generally, for an arbitrary set of formulas $\mathcal{A}$ and an arbitrary set of worlds U to select from.

Definition 3.8 (Selection) For a set $U \subseteq W(M)$ and a set of formulas $\mathcal{A}$, $Sel_M(U, \mathcal{A}) = \{rep(||\alpha||^M \cap U) \mid \alpha \in \mathcal{A}, ||\alpha||^M \cap U \neq \emptyset\}$.

We can show that such a selection can (*iv*)-cover not only single-world blocks like $\mathtt{fal}_v$, but any block B as long as U contains all worlds $\succ$-preferable to some world in B.

Lemma 3.9 *If* $Bet_{\succ}(w) \subseteq U$ *for all* $w \in W(B)$ *in some block* B *then* $antichain(Sel_M(U, Cond(\varphi)))$ *(iv)-covers* B.

Proof. If $\alpha \in$ Cond(φ) is satisfiable in $Bet_{\succ}(w)$ for some $w \in W(B)$ then there will be a representative satisfying α in $Sel_M(U, \text{Cond}(\varphi))$. □

A block to (*iv*)-cover $\mathtt{fal}_v$, which we will call *orbit* and label $\mathtt{orb}_v$, can be defined as $Orbit(M, \varphi, v) = antichain(Sel_M(Bet_{\succ}(v), \text{Cond}(\varphi)))$. To (*iv*)-cover orbits themselves, we can make another selection, this time from the whole $W(M)$, as this block does not need to be (*iii*)-suitable. So the block $Cloud(M, \varphi) = antichain(Sel_M(W(M), \text{Cond}(\varphi)))$, which we will label $\mathtt{cloud}_1$, can be used to (*iv*)-cover all orbits. Finally, to (*iv*)-cover $\mathtt{cloud}_1$ we can add two more copies of $Cloud(M, \varphi)$ (labeled $\mathtt{cloud}_2$ and $\mathtt{cloud}_3$) and have a non-transitive loop on these three copies, which will (*iv*)-cover each other circularly. This leads to the following small model construction for **E**.

Definition 3.10 (Small Model Construction for E)
If M is an **E**-countermodel for φ, $SMC^{\mathbf{E}}(\varphi, M) = \langle L, \succeq_L, \mathcal{B}\rangle$ where $L = \{\mathtt{fal}_v, \mathtt{orb}_v \mid v \in Fal(\varphi, M)\} \cup \{\mathtt{cloud}_i \mid i \in \{1, 2, 3\}\}$, $\mathcal{B}(\mathtt{fal}_v) = antichain(\{v\})$, $\mathcal{B}(\mathtt{orb}_v) = Orbit(M, \varphi, v)$, $\mathcal{B}(\mathtt{cloud}_i) = Cloud(M, \varphi)$ and the preference relation $\succeq_L$ on blocks is demonstrated on Fig. 2a.

Theorem 3.11 *If* M *is a* **E***-countermodel for* φ *then* $gen(SMC^{\mathbf{E}}(\varphi, M))$ *is a* **E***-countermodel for* φ *and* $|W(gen(SMC^{\mathbf{E}}(\varphi, M)))| = \mathcal{O}(|\varphi|^2)$. [2]

Proof. $SMC^{\mathbf{E}}(\varphi, M)$ is a countermodel for φ by Th. 3.7, because $\mathcal{B}(\mathtt{orb}_v)$ is (*iii*)-suitable for $\mathcal{B}(\mathtt{fal}_v)$ and all blocks are (*iv*)-covered by Lem. 3.9. $|W(gen(SMC^{\mathbf{E}}(\varphi, M)))| = \mathcal{O}(|\varphi|^2)$ since $SMC^{\mathbf{E}}(\varphi, M)$ contains $(2 \cdot |Fal(\varphi, M)| + 3)$ blocks with at most $|\text{Cond}(\varphi)|$ worlds each. □

3.2 Small Model Construction for Logic F

For logic **F**, we will utilize the limitedness of the countermodel M to construct a small countermodel with an acyclic $\succ$, which will automatically make it limited (and thus an **F**-countermodel) too.

---

[2] As usual, the notation $f(\varphi, M) = \mathcal{O}(g(\varphi, M))$ for integer-valued functions f and g means that there exists a constant C such that $f(\varphi, M) \leq C \cdot g(\varphi, M)$ for all φ and M.

Lemma 3.12 *Model $\langle W, \succeq, \mathbb{V}\rangle$ is limited if W is finite and $\succ$ is acyclic.*

Proof. If there is some $w_0 \in ||\alpha||^M$, consider (any) longest path $w_0 \prec w_1 \prec w_2 \prec \ldots$ with worlds from $||\alpha||^M$ staring from w_0. Since W is finite and there can be no repetitions on the path (due to acyclicity of $\succ$), the path is finite and there is the last world w_m for which there is no $u \in ||\alpha||^M$ such that $u \succ w_m$, and so $w_m \in \max(||\alpha||^M)$ by definition. □

For acyclicity, we will modify our construction $SMC^{\mathbf{E}}(\varphi, M)$ by replacing a non-transitive cycle on blocks $\mathtt{cloud}_1, \mathtt{cloud}_2, \mathtt{cloud}_3$ with one finite chain. Our goal is to select a chain of polynomial size that is (iv)-safe and satisfies any $\alpha \in \mathrm{Cond}(\varphi)$ that is satisfiable in M (which will allow us to use the chain to (iv)-cover any block). We construct such a chain through an iterative process, that selects maximal worlds for disjunctions of conditions. At the beginning of the process, we have $\mathcal{A}_0 = \mathrm{Cond}(\alpha)$ as the set of conditions for which we need satisfying worlds. If at least one of conditions in $\mathcal{A}_0$ is satisfied by some world in M, then $||\bigvee_{\alpha\in\mathcal{A}_0} \alpha||^M \neq \emptyset$, then by limitedness there exists some $z_0 \in \max(||\bigvee_{\alpha\in\mathcal{A}_0} \alpha||^M)$. We can safely take z_0 as the first (i.e. most preferable) world in the chain, since there are no worlds $u \succ z_0$ in M satisfying conditions from $\mathcal{A}_0$. z_0 satisfies some conditions from $\mathcal{A}_0$ (since $M, z_0 \models \bigvee_{\alpha\in\mathcal{A}_0} \alpha$), therefore we can move on to the next step with a strictly smaller set $\mathcal{A}_1$ of conditions for which we still need satisfying worlds. We can safely repeat this process by taking worlds from $z_i \in \max(||\bigvee_{\alpha\in\mathcal{A}_i} \alpha||^M)$ at every iteration: z_i has no $\succ$-prefferable α-worlds for all remaining conditions $\alpha \in \mathcal{A}_i$, while for all already removed formulas there is a satisfying world somewhere earlier (i.e. preferrable to z_i) in the chain, thus condition (iv) will be satisfied for this world. After a linear number of iterations, the chain will contain satisfying worlds for all formulas from $\mathcal{A}$ satisfiable in M.

Below is the formal definition of the described chain of maximal worlds. We give a generalized version that selects maximal worlds from any given subset of worlds U and any given set of formulas $\mathcal{A}_i$, the same way as we did for $Sel_M(U, \mathcal{A})$. We will need this generalized version for logics **F+(CM)** and **G**. To define linear order in chains formally we will use the notation of lists: [ ] will denote an empty list, and $a :: S$ will denote the list in which element a is appended to the beginning of the list S.

Definition 3.13 For any $U \subseteq W(M)$ and a finite set of formulas $\mathcal{A}_i$,

$$MaxSeq_M(U, \mathcal{A}_i) = \begin{cases} [\,], & \text{if } \mathcal{A}_i = \emptyset \\ [\,], & \text{if } \mathcal{D}(U, \mathcal{A}_i) = \emptyset \\ d(U, \mathcal{A}_i) :: MaxSeq_M(U, \mathcal{A}_{i+1}), & \text{otherwise} \end{cases}$$

where $\mathcal{D}(U, \mathcal{A}) = U \cap \max(||\bigvee_{\alpha\in\mathcal{A}_i} \alpha||^M)$, $d(U, \mathcal{A}) = rep(\mathcal{D}(U, \mathcal{A}))$ and $\mathcal{A}_{i+1} = \{\alpha \in \mathcal{A}_i \mid M, d(U, \mathcal{A}) \not\models \alpha\}$.

Notice that for a finite $\mathcal{A}_0$ this sequence is well-defined (representative $d(U, \mathcal{A}_i)$ is always taken from a non-empty set and $|\mathcal{A}_i|$ decreases) and always has length at most $|\mathcal{A}_0|$. The reasoning above that shows (iv)-safeness of the chain built from this sequence works in the general case with arbitrary U and does not even require the limitedness of M.

Lemma 3.14 *$chain(MaxSeq_M(U, Cond(\varphi)))$ is (iv)-safe for any $U \subseteq W(M)$.*

Proof. Let $z_k \in W(chain(MaxSeq_M(U, \mathrm{Cond}(\varphi))))$ and $\bigcirc(\gamma \mid \alpha) \in Ob^+(\varphi, M)$. $z_k \in \max(||\bigvee_{\alpha \in \mathcal{A}_k} \alpha||^M)$ for some step k and set $\mathcal{A}_k$ of remaining conditions. If there is $u \succ z_k$ such that $M, u \models \alpha$, then $\alpha \notin \mathcal{A}_k$ due to maximality of z_k, which means that α was removed at some previous step, therefore there is z_j with $j < k$ such that $M, z_j \models \alpha$. □

For logic **F**, we select worlds in the chain from the whole $W(M)$: for a limited model M we define block $Ray(M, \varphi) = chain(MaxSeq_M(\mathrm{Cond}(\varphi), W(M)))$, which we will label `ray`. $Ray(M, \varphi)$ contains satisfying worlds for all conditions from $\mathrm{Cond}(\varphi)$ satisfiable in M so it (*iv*)-covers any block on M.

Lemma 3.15 *For a limited M, $MaxChain(M, \varphi)$ (iv)-covers any block.*

Proof. If a condition $\alpha \in \mathrm{Cond}(\varphi)$ is satisfiable in M then it can not be among the remaining conditions when the chain is built (otherwise $\mathcal{A}_m \neq \emptyset$ and $\mathcal{D}(W(M), \mathcal{A}_m) \neq \emptyset$ due to limitedness of M), therefore for some world z in the chain $M, z \models \alpha$. □

Replacement of non-transitive triangle in $SMC^E(\varphi, M)$ with $Ray(M, \varphi)$ gives us the small model construction $SMC^F(\varphi, M)$ with an acyclic strict version of preference relation.

Definition 3.16 (Small Model Construction for F) If M is an **F**-countermodel for φ, $SMC^{\mathbf{F}}(\varphi, M) = \langle L, \succeq_L, \mathcal{B} \rangle$ where $L = \{\mathtt{fal}_v, \mathtt{orb}_v \mid v \in Fal(\varphi, M)\} \cup \{\mathtt{ray}\}$, $\mathcal{B}(\mathtt{fal}_v) = antichain(\{v\})$, $\mathcal{B}(\mathtt{orb}_v) = Orbit(M, \varphi, Bet_{\succ}(v))$, $\mathcal{B}(\mathtt{ray}) = Ray(M, \varphi)$ and a preference relation $\succeq_L$ on blocks is demonstrated on Fig. 2b.

Theorem 3.17 *If M is a **F**-countermodel for φ then $gen(SMC^{\mathbf{F}}(\varphi, M))$ is a **F**-countermodel for φ and $|W(gen(SMC^{\mathbf{F}}(\varphi, M)))| = \mathcal{O}(|\varphi|^2)$.*

Proof. $gen(SMC^{\mathbf{F}}(\varphi, M))$ is a countermodel for φ by Th. 3.7, because $\mathcal{B}(\mathtt{orb}_v)$ is (*iii*)-suitable for $\mathcal{B}(\mathtt{fal}_v)$, $\mathcal{B}(\mathtt{ray})$ is (*iv*)-safe by Lem. 3.14 and all other blocks are (*iv*)-covered by Lem. 3.15 and Lem. 3.9. $gen(SMC^{\mathbf{F}}(\varphi, M))$ is an **F**-countermodel by Lem. 3.12. $|W(gen(SMC^{\mathbf{F}}(\varphi, M)))| = \mathcal{O}(|\varphi|^2)$ since $SMC^{\mathbf{F}}(\varphi, M)$ contains $(2 \cdot |Fal(\varphi, M)| + 1)$ blocks with at most $|\mathrm{Cond}(\varphi)|$ worlds each. □

3.3 Small Model Construction for Logic F+(CM)

For logic **F+(CM)** we need to ensure the transitivity of the preference relation in $gen(SMC^{\mathbf{F+(CM)}}(\varphi, M))$. It is enough to obtain an **F+(CM)**-countermodel since for finite models transitivity implies smoothness.

Lemma 3.18 *$M = \langle W, \succeq, \mathbb{V}\rangle$ is smooth if W is finite and $\succeq$ is transitive.*

Proof. First, notice that transitivity of $\succeq$ implies transitivity of $\succ$. Indeed, if $w_1 \succ w_2$ and $w_2 \succ w_3$ then $w_1 \succeq w_3$ by transitivity of $\succeq$ and $w_3 \not\succeq w_1$ since otherwise there would be a transitive triangle on these three worlds and none of them could be strictly preferable to another. Now, for an arbitrary $w_0 \in ||\alpha||^M$, consider (any) longest path $w_0 \prec w_1 \prec w_2 \prec \ldots$ with worlds from $||\alpha||^M$ staring from w_0. Since W is finite and there can be no repetitions on the path due to transitivity of $\succ$, the path is finite and there is the last world w_m for which there is no $u \in ||\alpha||^M$ such that $u \succ w_m$, so $w_m \in \max(||\alpha||^M)$ and either $w_m = w_0$ or $w_m \succ w_0$, so M is smooth. □

In $SMC^{\mathbf{F}}(\varphi, M)$ non-transitivity was essential: we can not put $\texttt{ray} \succ_L \texttt{fal}_v$ since we selected worlds in the maximal chain from the whole initial model, so $\mathcal{B}(\texttt{ray})$ can be not (*iii*)-suitable for $\mathcal{B}(\texttt{fal}_v)$. However, smoothness allows us to select a maximal chain only among worlds in $Bet_{\succ}(v)$. Specifiaclly, for every falsifying world v we introduce individual *chain-orbit* $RayOrb(M, \varphi, v) = chain(MaxSeq_M(Bet_{\succ}(v), \mathrm{Cond}(\varphi)))$. We already know that this block is (*iv*)-safe by Lem. 3.14, and we can show that for an **F+(CM)**-model M it covers block $\mathcal{B}(\texttt{fal}_v)$.

Lemma 3.19 *For a transitive and smooth M, $RayOrb(M, \varphi, v)$ (iv)-covers $antichain(\{v\})$.*

Proof. Suppose that (1) there is some $u \succ v$ in M such that $M, u \models \alpha$ for some $\alpha \in \mathrm{Cond}(\varphi)$, we need to show that there is a world z_k in the chain such that $M, z_k \models \alpha$. Similarly to Lem. 3.15, we show it by proving that in this case α is removed from the set of conditions $\mathcal{A}_i$ at some point. And to show this, it is enough to prove that for smooth models (*) $\alpha \in \mathcal{A}_i$ implies $\mathcal{D}(Bet_{\succ}(v), \mathcal{A}_i) \neq \emptyset$ from the definition of *MaxSeq* (then the sequence of maximal worlds cannot end while α belongs to $\mathcal{A}_i$).

Let us prove (*). Suppose that $\alpha \in \mathcal{A}_i$. From this and (1) follows $u \in ||\bigvee_{\alpha \in \mathcal{A}_i} \alpha||^M$. Due to smoothness of M it implies that (2) there is $u' \in \max(||\bigvee_{\alpha \in \mathcal{A}_i} \alpha||^M)$ such that either $u' = u$ or $u' \succ u$. In either case $u' \succ v$ (since $u \succ v$ by (1) and transitivity of $\succeq$ implies transitivity of $\succ$). So, we have (3) $u' \in Bet_{\succ}(v)$. (2) and (3) together imply $u' \in \mathcal{D}(Bet_{\succ}(v), \mathcal{A}_i)$, concluding the proof of (*). □

So we can obtain a small model construction for **F+(CM)** by replacing each orbit $\texttt{orb}_v$ with an individual maximal chain $RayOrb(M, \varphi, v)$ (which we will label $\texttt{c-ray}_v$). The common chain $\texttt{ray}$ from $SMC^{\mathbf{F}}(\varphi, M)$ is not needed anymore.

Definition 3.20 (Small Model Construction for F+(CM))
If M is an **F+(CM)**-countermodel for φ, $SMC^{\mathbf{F+(CM)}}(\varphi, M) = \langle L, \succeq_L, \mathcal{B}\rangle$ where $L = \{\texttt{fal}_v, \texttt{c-ray}_v \mid v \in Fal(\varphi, M)\}$, $\mathcal{B}(\texttt{fal}_v) = antichain(\{v\})$, $\mathcal{B}(\texttt{c-ray}_v) = RayOrb(M, \varphi, v)$ and a preference relation $\succeq_L$ on blocks is demonstrated on Fig. 2c.

Theorem 3.21 *If* M *is a* ***F+(CM)****-countermodel for* φ *then* $gen(SMC^{\mathbf{F+(CM)}}(\varphi, M))$ *is a* ***F+(CM)****-countermodel for* φ *and* $|W(gen(SMC^{\mathbf{F+(CM)}}(\varphi, M)))| = \mathcal{O}(|\varphi|^2)$.

Proof. $SMC^{\mathbf{F+(CM)}}(\varphi, M)$ is a countermodel for φ by Th. 3.7, because $\mathcal{B}(\texttt{c-ray}_v)$ is (*iii*)-suitable for $\mathcal{B}(\texttt{fal}_v)$, each $\mathcal{B}(\texttt{c-ray}_v)$ is (*iv*)-safe by Lem. 3.14 and each $\mathcal{B}(\texttt{fal}_v)$ is (*iv*)-covered by $\mathcal{B}(\texttt{c-ray}_v)$ by Lem. 3.19. $SMC^{\mathbf{F+(CM)}}(\varphi, M)$ is an **F+(CM)**-countermodel by Lem. 3.18. $|W(gen(SMC^{\mathbf{F+(CM)}}(\varphi, M)))| = \mathcal{O}(|\varphi|^2)$ since $SMC^{\mathbf{F+(CM)}}(\varphi, M)$ contains $(2 \cdot |Fal(\varphi, M)|)$ blocks with at most $|\mathrm{Cond}(\varphi)|$ worlds each. □

Remark 3.22 The form of the countermodel that we obtain — a union of incomparable finite chains — is the same as a Friedman-Halpern countermodel for logic **PCA** (i.e. **F+(CM)**) [7]. However, we have achieved it by using different methods: they use a finite-model property of **PCL** extensions (shown in [3]) and extend the preference relation to a linear order, then construct chains by selecting the greatest world w.r.t. extended order independently for each conditional in $Ob^+(\varphi, M)$, while we do it using an iterative procedure. The possibility of their selection fully relies on finitedness and transitivity, which due to Lem. 3.18 is only possible in smooth models, so it cannot be applied to the weaker logics **E** and **F**. Furthermore, the Horn fragment [3] of **PCA** was studied extensively in the area of non-monotonic reasoning, where it is known as the KLM logic **P** [13] of preferential reasoning. A small model construction for **P** has been introduced in [14] and consists of a single chain of polynomial size (by essentially the same method as Friedman-Halpern). Notice, that both Friedman-Halpern and our constructions turn into a single chain when restricted to Horn formulas.

3.4 Small Model Construction for Logic G

For logic **G**, we also need to ensure the totalness of the transformed model by leveraging the fact that the falsifying worlds in $Fal(\varphi, M)$ are ordered in the initial model by $\succeq$ which in a **G**-model is a total preorder.

Let us consider first a simple case where $\succeq$ in the given **G**-countermodel is asymmetric (and therefore a strict linear order). Then there exists an ordering $v_1 \prec \cdots \prec v_n$ of worlds from $Fal(\varphi, M)$. Then we can linearly order blocks of $SMC^{\mathbf{F+(CM)}}(\varphi, M)$ with the following order: $\texttt{fal}_{v_1} \prec_L \texttt{c-ray}_{v_1} \prec_L \ldots \prec_L \texttt{fal}_{v_n} \prec_L \texttt{c-ray}_{v_n}$. The (*iii*)-suitability will still be satisfied with such ordering, because for every $j \geq i$ we have $W(\texttt{c-ray}_{v_j}) \subseteq Bet_{\succ}(v_j)$ and $Bet_{\succ}(v_j) \subseteq Bet_{\succ}(v_i)$ due to transitivity of $\succeq$.

In general, $\succeq$ is not necessarily asymmetric, but we can generalize the same idea by grouping together $\succeq$-equivalent worlds as in the following definition.

Definition 3.23 (Stratification) For $M = \langle W, \succeq, \mathbb{V} \rangle$ and a finite set $U \subseteq W$, a sequence $[S_1, \ldots, S_n]$ of non-empty subsets of W is called a *stratification of U*

---

[3] Conditional Horn formula is a formula of a form $\bigcirc(\gamma_1 \,|\, \alpha_1) \wedge \cdots \wedge \bigcirc(\gamma_n \,|\, \alpha_n) \rightarrow \bigcirc(\gamma_0 \,|\, \alpha_0)$.

when U is the disjoint union of subsets $\{S_i\}_{i=1}^n$ and for every $u_i \in S_i, u_j \in S_j$ we have $s_i \succeq s_j$ iff $i \geq j$.

For total preorders the unique stratification of any finite set is given by its factorization w.r.t. $\succeq$-equivalence.

Lemma 3.24 *If $\succeq$ is transitive and total, there exists a unique stratification of any finite subset U.*

Proof. Consider an equivalence relation $\approx$ on U where $u_1 \approx u_2$ means that both $u_1 \succeq u_2$ and $u_2 \succeq u_1$. Consider further a relation $\succeq_S$ on the set of equivalence classes of U w.r.t. $\approx$ where $S_i \succeq_S S_j$ when there exist $u_i \in S_i$ and $u_j \in S_j$ such that $u_i \succeq u_j$. Notice that for a transitive and total $\succeq$ the relation $\succeq_S$ is a linear order: it is antisymmetric due to definitions of $\approx$ and $\succeq_S$, and it is transitive and total (and hense reflexive) due to the transitivity and totalness of $\succeq$. This linear ordering gives a stratification by definition. Notice also that it is the only stratification: every element of a stratification should be an equivalence class w.r.t. $\approx$ and their order in the list should be aligned with $\succeq_S$ (i.e. $i \geq j$ implies $S_i \succeq_S S_j$) by definition. □

Therefore, we can take the stratification $[S_1, \ldots, S_n]$ of $Fal(\varphi, M)$ w.r.t. $\succeq$ and create a block $clique(S_i)$ for every group S_i.

Notice that for $u_1, u_2 \in S_i$ both $u_1 \succeq u_2$ and $u_2 \succeq u_1$ so $Bet_{\succ}(u_1) = Bet_{\succ}(u_2)$ (due to transitivity). Therefore, to (iv)-cover block $clique(S_i)$ we can take orbit-chain $RayOrb(M, \varphi, rep(S_i))$ with arbitrary representative of S_i.

Lemma 3.25 *For a transitive and smooth $M = \langle W, \succeq, \mathbb{V}\rangle$ and $S \subseteq W(M)$, if $u \succeq u'$ for all $u, u' \in S$ then $RayOrb(M, \varphi, rep(S))$ (iv)-covers $clique(S)$.*

Proof. For any $u \in S_i$ holds $Bet_{\succ}(u) = Bet_{\succ}(rep(S))$ (since $\succeq$ is transitive), so any formula from $\mathrm{Cond}(\varphi)$ satisfied in some world from $Bet_{\succ}(v)$ is also satisfied by some world in $RayOrb(M, \varphi, rep(V))$ by Lem. 3.19. □

Thus we can take as the construction $SMC^{\mathbf{G}}(\varphi, M)$ a linearly-oredered sequence of blocks in which cliques $clique(S_i)$, labeled $\texttt{group}_{S_i}$, are interleaved with chain-orbits $RayOrb(M, \varphi, rep(S_i))$, labeled $\texttt{g-ray}_{S_i}$.

Definition 3.26 (Small Model Construction for G) For a **G**-countermodel $M = \langle W, \succeq, \mathbb{V}\rangle$ for a formula φ, let $[S_1, \ldots, S_n]$ be the unique stratification of $Fal(\varphi, M)$ w.r.t. $\succeq$. Then $SMC^{\mathbf{G}}(\varphi, M) = \langle L, \succeq_L, \mathcal{B}\rangle$ where $L = \bigcup_{i=1}^n \{\texttt{group}_{S_i}, \texttt{g-ray}_{S_i}\}$, $\mathcal{B}(\texttt{group}_{S_i}) = clique(V_i)$, $\mathcal{B}(\texttt{g-ray}_{S_i}) = chain(RayOrb(M, \varphi, rep(S_i)))$ and the blocks are ordered linearly as follows: $\texttt{group}_{S_1} \prec_L \texttt{g-ray}_{S_1} \prec_L \ldots \prec_L \texttt{group}_{S_n} \prec_L \texttt{g-ray}_{S_n}$.

Theorem 3.27 *If M is an **G**-countermodel for φ then $gen(SMC^{\mathbf{G}}(\varphi, M))$ is a **G**-countermodel for φ and $|W(gen(SMC^{\mathbf{G}}(\varphi, M)))| = \mathcal{O}(|\varphi|^2)$.*

Proof. $SMC^{\mathbf{G}}(\varphi, M)$ is a countermodel for φ by Th. 3.7: $\mathcal{B}(\texttt{group}_{S_i})$ is flat and all block $\succ_L$-preferable to it are (iii)-suitable for it (since $\succeq$ in M is transitive), each $\mathcal{B}(\texttt{g-ray}_{S_i})$ is (iv)-safe by Lem. 3.14 and each $\mathcal{B}(\texttt{group}_{S_i})$ is (iv)-covered by $\mathcal{B}(\texttt{g-ray}_{S_i})$ due to Lem. 3.25. $SMC^{\mathbf{G}}(\varphi, M)$ is a **G**-countermodel since its preference relation is transitive and total, and also smooth by Lem. 3.18. $|W(gen(SMC^{\mathbf{G}}(\varphi, M)))| = \mathcal{O}(|\varphi|^2)$ since $SMC^{\mathbf{G}}(\varphi, M)$ contains at most $(2 \cdot |Fal(\varphi, M)|)$ blocks with at most $|\text{Cond}(\varphi)|$ worlds each. □

Remark 3.28 Friedman and Halpern also provide a counter-model for logic **VTA** (i.e. **G**) [7]. They use an ad hoc approach, different from the one they use for the other extensions of **PCL**. For **VTA** for each conditional they simply take one world from the original model without changing the preference relation, resulting in a model of linear size. Although Th. 3.2 can be also used to establish the adequacy of their constriction, we provided a different construction (with a new explicitly defined preference relation and with potentially a quadratic number of worlds) for uniformity with the constructions for the other three Åqvist's logics.

4 Applications

In this section, we describe two applications of our small model constructions: alternative semantical characterizations, complexity and encodings in the classical propositional logic.

4.1 Alternative semantical characterizations

We will call a class $\mathfrak{M}$ of preference models a semantical characterization for theoremhood in logic $\mathcal{L}$ when any φ is a theorem of $\mathcal{L}$ iff φ is valid in all models from $\mathfrak{M}$. New semantical characterizations for theoremhood can be extracted from the specific form of our small model constructions. Namely, any model property satisfied by $SMC^{\mathcal{L}}(\varphi, M)$ that is stronger than some existing characterization for $\mathcal{L}$ (e.g. from Fig. 1) can be used as an alternative characterization.

Lemma 4.1 *Let $\mathfrak{M}$ be a class of models characterizing theoremhood in $\mathcal{L}$. If $\mathfrak{M}' \subseteq \mathfrak{M}$ and $SMC^{\mathcal{L}}(\varphi, M) \in \mathfrak{M}'$ for every φ and M, then $\mathfrak{M}'$ also characterizes theoremhood in $\mathcal{L}$.*

Proof. If φ is a theorem of $\mathcal{L}$, it is valid in all models in $\mathfrak{M}$, so it is also valid in all models of $\mathfrak{M}'$. If φ is not a theorem of $\mathcal{L}$, it is not valid in $SMC^{\mathcal{L}}(\varphi, M)$ that belongs to $\mathfrak{M}'$. □

We can use this method to characterize theoremhood in Åqvist logics with *frame properties*, i.e. properties of the preference relation. Notice that the limit conditions (limitedness and smoothness) used for the characterization of **F**, **F+(CM)**, and **G** are not frame properties: they impose conditions only on truth sets of the model. This choice plays a vital role in establishing correspondence between semantics and known axiomatizations of Åqvist's logics, but it makes it hard to work with these models since you need to distinguish which subsets of worlds can be a truth set. However, we can notice that our small model constructions satisfy the corresponding limit conditions for all subsets of worlds, therefore limitedness/smoothness on the level of frames can be used to characterize theoremhood also.

Logic	Cardinality of W	Properties of $\succeq$		
		acyclic	transitive	total
E	finite			
F	finite	✓		
F+(CM)	finite		✓	
G	finite		✓	✓

Fig. 3. Finite-model characterizations of theoremhood in Åqvist's logics (with maximality as the notion of bestness).

More importantly, our constructions satisfy some stronger frame properties. We already used these properties to prove that $SMC^{\mathcal{L}}(\varphi, M)$ generates an $\mathcal{L}$-countermodel. If we consider only finite models limit conditions can be replaced with natural conditions on preference relations: limitedness can be replaced with acyclicity, and smoothness can be dropped in presence of transitivity.

Theorem 4.2 *Formula φ is a theorem of Åqvist logic $\mathcal{L}$ iff $M \models \varphi$ for all finite models M satisfying the frame properties for logic $\mathcal{L}$ from Fig. 3.*

Proof. Using Lem. 4.1. Finite models with transitive preference realtion are smooth by Lem. 3.18. Finite models with acyclic preference relation are limited by Lem. 3.12 (acyclicity of $\succeq$ implies acyclicity of $\succ$), and although the preference relation in $SMC^{\mathbf{F}}(\varphi, M)$ has cycles in the form of reflexive loops inside the chain, these loops can be removed without affecting the satisfaction in the model, so any non-theorem has a finite countermodel with acyclic preference relation. □

In addition, our models for **E**, **F**, **F+(CM)** satisfy antisymmetry so this property can be added to finite-model characterization for these logics from Fig. 3, but not to the characterization of **G**.[4] Either reflexivity or irreflexivity can also be added since it is trivial to force them in any model without changing the satisfaction relation. Thus, **F+(CM)** is characterized by finite models where $\succeq$ is a partial order (strict or non-strict). At the same time, finite models where $\succeq$ is a linear order give some logic that is stronger than **G**. Even more specialized properties can be extracted from our construction via Lem. 4.1, e.g. **F+(CM)** can be characterized by models that are unions of non-comparable finite chains.

Remark 4.3 Note that here we are only concerned with semantical characterizations of theoremhood. Our results can not be extended to characterizing entailments $\Gamma \vdash \varphi$ if Γ is infinite (to provide a *strongly complete* characterization of the logics). This is a natural limitation for finite-model characterizations since entailments from infinite sets of premises can not be characterized using

[4] E.g., the principle of conditional excluded middle $\bigcirc(\gamma \mid \alpha) \vee \bigcirc(\neg\gamma \mid \alpha)$ is valid in all models with total and antisymmetric preference relation (since they have at most one α-best world), but it can be easily falsified in **G** by a model with two α-worlds preferable to each other (and thus both allowed to be α-best simultaniously).

only finite models for **S5** already (see a counterexample in Appendix B) and therefore for Åqvist's logics that extend **S5** too.

4.2 Complexity and automated deduction

Our small model constructions show that for any non-valid formula there exists a countermodel with at most $N(\varphi)$ worlds, where $N(\varphi)$ is a certain upper bound polynomial w.r.t. $|\varphi|$. Plus, the stronger frame properties from Fig. 3 can be easily checked in polynomial time w.r.t. the model size. This immediately implies co-NP-completeness of theoremhood.

Theorem 4.4 *Theoremhood is co-NP-complete for every Åqvist's logic.*

Proof. Non-theoremhood can be checked non-deterministically in polynomial time by guessing a countermodel M of size at most $N(\varphi)$ (i.e. guessing preference relation and valuation for all variables occurring in φ) and then checking $M \not\models \varphi$ and the required properties from Fig. 3. co-NP-hardness follows from co-NP-completeness of theoremhood in classical logic (since a propositional formula is a classical tautology iff it is a theorem of an Åqvist logic). □

Moreover, with simpler finite-model characterization from Fig. 3 a countermodel definition can be naturally encoded with a propositional formula of a polynomial size (see Appendix C for the full encodings). This propositional formula can be given to any SAT-solver for efficient theoremhood checking and countermodels can be reconstructed from classical models found by the solver.

Concluding remark

In this paper, we provide small model constructions for Åqvist's logics, which can be used to understand theoretical properties of these logics (such as finite-model semantical characterizations and complexity) and to generate countermodels for non-valid formulas using SAT-solvers. Ideally, this should be complemented by analytic calculi which provide transparent derivations for valid formulas. We plan to explore the relationship between our constructions and hypersequent calculi, aiming for simpler proof-theoretic characterizations, particularly for the challenging logic **F**.

Acknowledgements

I want to thank Agata Ciabattoni and Dominik Pichler for the helpful discussions and comments on early versions of this paper, Xavier Parent for the explanations regarding the history of preference-based approach in deontic logic, Roman Kuznets for the discussion about the relation between the strong completeness and the finite model property, and the anonymous reviewers for their useful remarks and suggestions.

This work was funded by the European Union's Horizon 2020 research and innovation programme under grant agreement No 101034440.

References

[1] Åqvist, L., *Deontic logic*, in: D. M. Gabbay and F. Guenthner, editors, *Handbook of Philosophical Logic: Volume 8*, Springer Netherlands, Dordrecht, 2002 pp. 147–264.

[2] Benzmüller, C., A. Farjami and X. Parent, *Åqvist's dyadic deontic logic E in HOL*, FLAP **6** (2019), pp. 733–754.

[3] Burgess, J. P., *Quick completeness proofs for some logics of conditionals*, Notre Dame Journal of Formal Logic **22** (1981), pp. 76–84.

[4] Ciabattoni, A., N. Olivetti and X. Parent, *Dyadic obligations: Proofs and countermodels via hypersequents*, in: R. Aydogan, N. Criado, J. Lang, V. Sánchez-Anguix and M. Serramia, editors, *PRIMA 2022: Principles and Practice of Multi-Agent Systems - 24th International Conference, Valencia, Spain, November 16-18, 2022, Proceedings*, Lecture Notes in Computer Science **13753** (2022), pp. 54–71.

[5] Ciabattoni, A., N. Olivetti, X. Parent, R. Ramanayake and D. Rozplokhas, *Analytic proof theory for Åqvist's system F*, in: J. Maranhão, C. Peterson, C. Straßer and L. van der Torre, editors, *Deontic Logic and Normative Systems - 16th International Conference, DEON 2023, Trois-Rivières, QC, Canada, July 5-7, 2023* (2023), pp. 79–98.

[6] Ciabattoni, A. and M. Tesi, *Sequents vs hypersequents for Åqvist systems*, in: *The proceedings of International Joint Conference on Automated Reasoning (IJCAR)*, 2024.

[7] Friedman, N. and J. Y. Halpern, *On the complexity of conditional logics*, in: J. Doyle, E. Sandewall and P. Torasso, editors, *Proceedings of the 4th International Conference on Principles of Knowledge Representation and Reasoning (KR'94). Bonn, Germany, May 24-27, 1994* (1994), pp. 202–213.

[8] Gabbay, D. M., *Theoretical foundations for non-monotonic reasoning in expert systems*, in: K. R. Apt, editor, *Logics and Models of Concurrent Systems* (1985), pp. 439–457.

[9] Girlando, M., "On the Proof Theory of Conditional Logics," Theses, Aix-Marseille Universite ; Helsinki University (2019).

[10] Grossi, D., W. van der Hoek and L. B. Kuijer, *Reasoning about general preference relations*, Artif. Intell. **313** (2022).

[11] Hansson, B., *An analysis of some deontic logics*, Noûs **3** (1969), pp. 373–398.

[12] Hansson, S. O. and T. Grüne-Yanoff, *Preferences*, in: E. N. Zalta, editor, *The Stanford Encyclopedia of Philosophy*, Metaphysics Research Lab, Stanford University, 2022, Spring 2022 edition .

[13] Kraus, S., D. Lehmann and M. Magidor, *Nonmonotonic reasoning, preferential models and cumulative logics*, Artif. Intell. **44** (1990), pp. 167–207.

[14] Lehmann, D. and M. Magidor, *What does a conditional knowledge base entail?*, Artif. Intell. **55** (1992), pp. 1–60.

[15] Lewis, D. K., "Counterfactuals," Blackwell, Malden, Mass., 1973.

[16] Parent, X., *Maximality vs. optimality in dyadic deontic logic*, J. Philos. Log. **43** (2014), pp. 1101–1128.

[17] Parent, X., *Preference semantics for Hansson-type dyadic deontic logic: a survey of results*, in: D. Gabbay, J. Horty, X. Parent, L. van der Torre and R. van der Meyden, editors, *Handbook of Deontic Logic and Normative Systems (vol. 2)*, College Publications, London, 2021 pp. 7–70.

[18] Putte, F. V. D. and C. Straßer, *Preferential semantics using non-smooth preference relations*, Journal of Philosophical Logic **43** (2014), pp. 903–942.

[19] Sambin, G. and S. Valentini, *The modal logic of provability. the sequential approach*, Journal of Philosophical Logic **11** (1982), pp. 311–342.

Appendix

A Detailed proof of Th. 3.2

Theorem A.1 *Let φ be a formula and $M = \langle W, \succeq, \mathbb{V}\rangle \in \mathcal{M}$ such that $M \not\models \varphi$. If a model $M' = \langle W', \succeq', \mathbb{V}'\rangle \in \mathcal{M}$ rearranges M with the prototype function $prot\colon W' \to W$ then the following four conditions are sufficient for $M' \not\models \varphi$.*

(*i*) *There exists $v' \in W'$ such that $M, prot(v') \not\models \varphi$.*

(*ii*) *For any $\Box\beta \in Box^-(\varphi, M)$ there exists $v' \in W'$ such that $M, prot(v') \not\models \beta$.*

(*iii*) *For any $\bigcirc(\gamma\,|\,\alpha) \in Ob^-(\varphi, M)$ there exists $v' \in W'$ such that $prot(v') \in \max(||\alpha||^M) \setminus ||\gamma||^M$ and for all $u' \succ' v'$ holds $prot(u') \succ prot(v')$.*

(*iv*) *For any $w' \in W'$, for all $\bigcirc(\gamma\,|\,\alpha) \in Ob^+(\varphi, M)$ if there exists $u \succ prot(w')$ such that $M, u \models \alpha$ then there exists $u' \succ' w'$ such that $M, prot(u') \models \alpha$.*

Proof. We will prove that for any $w' \in W'$ and any $\psi \in Sub\mathcal{F}(\varphi)$ holds $M', w' \models \psi$ iff $M, prot(w') \models \psi$. Then $M' \not\models \varphi$ follows by the condition (*i*). The proof is by induction on ψ (we use the abbreviation IH(s) to refer to the inductive hypothesis(-es)).

- $\psi = x \in Var$. $w \in \mathbb{V}(x)$ iff $prot(w) \in \mathbb{V}'(x)$ by the definition of the prototype function.
- $\psi = \neg\psi'$. Directly from IH for ψ'.
- $\psi = \psi_1 \wedge \psi_2$. Directly from IHs for ψ_1 and ψ_2.
- $\psi = \Box\beta$ and $\psi \in Box^+(\varphi, M)$. $M \models \Box\beta$, so for all $w' \in W'$ holds $M, prot(w') \models \beta$, so by IH for all $w' \in W'$ holds $M', w' \models \beta$, so $M' \models \Box\beta$.
- $\psi = \Box\beta$ and $\psi \in Box^-(\varphi, M)$. By (*ii*) there is $v' \in W'$ such that $M, prot(v') \not\models \beta$, so by IH holds $M', v' \not\models \beta$, so $M' \not\models \Box\beta$.
- $\psi = \bigcirc(\gamma\,|\,\alpha)$ and $\psi \in Ob^+(\varphi, M)$. Take any $w' \in W'$ such that $w' \in \max(||\alpha||^{M'})$. Then (1) $prot(w') \in \max(||\alpha||^M)$: $prot(w') \in ||\alpha||^M$ by IH, and there can be no $s \succ prot(w')$ such that $s \in ||\alpha||^M$ (otherwise there would be $u' \succ' w'$ such that $M', u' \models \alpha$ by (*iv*) and IH). Since $M \models \bigcirc(\gamma\,|\,\alpha)$, (1) implies $M, prot(w') \models \gamma$, which implies $M', w' \models \gamma$ by IH. Thus $M' \models \bigcirc(\gamma\,|\,\alpha)$.
- $\psi = \bigcirc(\gamma\,|\,\alpha)$ and $\psi \in Ob^-(\varphi, M)$. For the corresponding world $v' \in W'$ from (*iii*) we have $v' \notin ||\gamma||^{M'}$ (by IH and the choice of v') and $v' \in \max(||\alpha||^{M'})$ (since $v' \in ||\alpha||^{M'}$ by IH and the choice of v', and for all $u' \succ' v'$ we have $u' \notin ||\alpha||^{M'}$ by (*iii*) and IH), so $M' \not\models \bigcirc(\gamma\,|\,\alpha)$.

□

B Strong completeness vs finite-modal characterization

In this appendix, we show that the entailment from the infinite set of premises in **S5** can not be characterized with a class of models that contains only finite models. Consider the infinite set of propositional variables $\{x_i\}_{i=1}^{\infty}$ and an infinite sequence of formulas $\{\varepsilon_n\}_{n=1}^{\infty}$ defined as

$$\varepsilon_n = x_n \wedge \bigwedge_{i=1}^{n-1} \neg x_i$$

Clearly, two formulas ε_n and ε_m for $n \neq m$ can not be both satisfied in one world. Therefore, the formulas from the set $\Gamma^{\diamond} = \{\diamond\varepsilon_n\}_{n=1}^{\infty}$ can not all be simultaneously valid in any finite model. So entailment $\Gamma^{\diamond} \models \bot$ holds in all finite Kripke models, but does not hold in **S5**, since an infinite Kripke model satisfying all formulas from $\Gamma^{\diamond}$ simultaneously can be easily constructed.

C Propositional encoding for Åqvist's logics

This appendix provides an embedding of every Åqvist's logic into the classical propositional logic. Specifically, for any given modal formula φ in some Åqvist's logic $\mathcal{L}$ we define a propositional formula $F_{\mathcal{L}}(\varphi)$, such that there is a one-to-one correspondence between classical countermodels for $F_{\mathcal{L}}(\varphi)$ and preference countermodels for φ with $N(\varphi)$ worlds (where $N(\varphi)$ is a size bound given by our small model construction) satisfying model conditions for $\mathcal{L}$ in Fig. 3. As a result, φ is valid in $\mathcal{L}$ iff $F_{\mathcal{L}}(\varphi)$ is classically valid.

To encode a countermodel M for a formula φ with words $\{w_1, \ldots, w_{N(\varphi)}\}$ we will use the following variables:

- $p_{i,j}$ for $1 \leq i, j \leq N(\varphi)$ to encode the fact $w_i \succeq w_j$
- v_i^{ψ} for $1 \leq i \leq N(\varphi)$ and $\psi \in Sub\mathcal{F}(\varphi)$ to encode the fact $M, w_i \models \psi$

If ψ is not a propositional variable, v_i^{ψ} is determined by v-variables for the immediate subformulas of ψ and this can be straightforwardly encoded by definition by a set of propositional equivalences of polynomial size:

$$\begin{array}{rllll}
C^{ev}(\varphi) = & \{ & v_i^{\neg\psi} \Leftrightarrow (\neg v_i^{\psi}) & \mid (\neg\psi) \in Sub\mathcal{F}(\varphi) & \}_{1\leq i\leq N(\varphi)} \\
\cup & \{ & v_i^{\psi_1\wedge\psi_2} \Leftrightarrow (v_i^{\psi_1} \wedge v_i^{\psi_2}) & \mid (\psi_1 \wedge \psi_2) \in Sub\mathcal{F}(\varphi) & \}_{1\leq i\leq N(\varphi)} \\
\cup & \{ & v_i^{\Box\beta} \Leftrightarrow (\bigwedge_{j=1}^{N(\varphi)} v_j^{\beta}) & \mid (\Box\beta) \in Sub\mathcal{F}(\varphi) & \}_{1\leq i\leq N(\varphi)} \\
\cup & \{ & v_i^{\bigcirc(\gamma|\alpha)} \Leftrightarrow (\bigwedge_{j=1}^{N(\varphi)} (v_j^{\gamma} \vee \neg v_j^{\alpha} \vee (\bigvee_{t=1}^{N(\varphi)} (p_{t,j} \wedge \neg p_{j,t} \wedge v_t^{\alpha})))) & & \\
& & & \mid \bigcirc(\gamma|\alpha) \in Sub\mathcal{F}(\varphi) & \}_{1\leq i\leq N(\varphi)}
\end{array}$$

Transitivity and totality of $\succeq$ can be encoded straightforwardly by definition too:

$$\begin{array}{rcll}
C^{trans}(\varphi) & = & \{ \ (p_{i,j} \wedge p_{j,k}) \Rightarrow p_{i,k} & \}_{1\leq i,j,k\leq N(\varphi)} \\
C^{total}(\varphi) & = & \{ \quad p_{i,j} \vee p_{j,i} & \}_{1\leq i,j\leq N(\varphi)}
\end{array}$$

To encode acyclicity of $\succeq$ we can reformulate it equivalently as follows: there exists a relation $\succeq^t$ that is transitive, irreflexive and contains $\succeq$ (such relation exists iff the positive transitive closure of $\succeq$ is irreflexive, i.e. when $\succeq$ is acyclic). Introducing additional variables $t_{i,j}$ for $1 \leq i, j \leq N$ to encode the fact $w_i \succeq^t w_j$, we can then encode acyclicity with the following set of formulas:

$$
\begin{array}{rll}
C^{acyclic}(\varphi) = & \{\ (t_{i,j} \wedge t_{j,k}) \Rightarrow t_{i,k}\ \}_{1\leq i,j,k\leq N(\varphi)} \\
\cup \{ & \neg t_{i,i} & \}_{1\leq i\leq N(\varphi)} \\
\cup \{ & p_{i,j} \Rightarrow t_{i,j} & \}_{1\leq i,j\leq N(\varphi)}
\end{array}
$$

Putting everything together, we get the following encodings of countermodels (falsifying φ in the world w_1) in Åqvist's logics as formulas of polynomial size.

$$
\begin{array}{ll}
F_{\mathbf{E}}(\varphi) & = \neg v_1^{\varphi} \wedge \bigwedge C^{ev}(\varphi) \\
F_{\mathbf{F}}(\varphi) & = \neg v_1^{\varphi} \wedge \bigwedge C^{ev}(\varphi) \wedge \bigwedge C^{acyclic}(\varphi) \\
F_{\mathbf{F+(CM)}}(\varphi) & = \neg v_1^{\varphi} \wedge \bigwedge C^{ev}(\varphi) \wedge \bigwedge C^{trans}(\varphi) \\
F_{\mathbf{G}}(\varphi) & = \neg v_1^{\varphi} \wedge \bigwedge C^{ev}(\varphi) \wedge \bigwedge C^{trans}(\varphi) \wedge \bigwedge C^{total}(\varphi)
\end{array}
$$

On the System of Positive Slices in the Structure of Superintuitionistic Predicate Logics

Mikhail Rybakov

Higher School of Modern Mathematics MIPT and HSE University, Moscow, Russia

Dmitry Shkatov

School of Computer Science and Applied Mathematics, University of the Witwatersrand, Johannesburg, South Africa

Dmitrij Skvortsov

Federal Research Center "Computer Science and Control", Russian Academy of Sciences, Moscow, Russia

Abstract

We study the system of classes of superintuitionistic predicate logics induced by the equivalence relation identifying logics with the same positive fragment. We call such classes positive slices. We state a condition guaranteeing that logics determined by classes of Kripke frames or Kripke sheaves share a positive fragment, and so belong to the same positive slice. We then use this condition to prove that some well-known superintuitionistic predicate logics have the same positive fragment. We also present an example of a continuum of logics whose positive slices are singletons.

Keywords: superintuitionistic predicate logic, system of slices, positive fragment.

1 Introduction

Studying classes of logics, defined either syntactically or semantically, rather than particular logical systems, has proved a fruitful approach to the study of superintuitionistic and modal *propositional* logics; see, e.g., [1,24] for systematic surveys. A similar approach has also been fruitful in the study of superintuitionistic and modal *predicate* logics; see, e.g., [5,15,4,11]. Often, in particular in [5,15,4,11], the unit of study is then a class of logics lying between two particular system; such classes are often called *intervals* or, as in this paper, *segments*. For example, [11] proves that all superintuitionistic predicate logics between the intuitionistic predicate logic **QH** and the predicate counterpart **QKC** of the propositional logic **KC** of the weak law of the excluded middle are undecidable in languages with two individual variables and a single monadic predicate letter.

Recently, Skvortsov [20], developing ideas of Hosoi [6,7] for propositional logics, proposed to classify superintuitionistic predicate logics into systems of *slices*, which are convex classes induced by an equivalence relations on the class of all logics (we note that segments are convex, and so are subsumed by this approach). This slice-based approach to the study of predicate logics appears to be more powerful than the segment-based one: segments are usually chosen so that all their logics share a common property; thus, by considering slices of all logics sharing a property, rather than particular segments, we might obtain more general, in the sense of the set of the logics covered, results. Obviously, different equivalence relations, and therefore different systems of slices, turn out to be useful for different purposes.

The present paper is meant to be a first sketch of a study of the system of slices induced by the equivalence relation identifying logics with the same positive fragment. This equivalence relation proves useful in, among others, the study of the computational properties of superintuitionistic predicate logics [11,12]. For a study of another system of slices, see [18,20].

In this paper, we obtain (Proposition 5.8) a continuum of singleton positive slices, thus proving the existence of a continuum of logics with a unique positive fragment. On the other hand, we identify, in Theorem 5.12 (Main Theorem), conditions ensuring that logics determined by classes of Kripke frames or Kripke sheaves have the same positive fragment; this allows us to obtain examples of non-trivial positive slices of superintuitionistic predicate logics. Theorem 5.12 generalizes an observation by Yankov [25, Theorem] that the intuitionistic propositional logic **H** has the same positive fragment as the propositional logic $\mathbf{KC} = \mathbf{H} + \neg p \vee \neg\neg p$, as well as a similar observation about corresponding predicate logics [13, Proposition 10.2]. Using Theorem 5.12, we, in particular, prove (Propositions 5.14, 5.16, and 5.17) that some well-known superintuitionistic predicate logics have identical positive fragments. We note that, in this context, the use of Kripke sheaf semantics leads to stronger results than the use of the more familiar Kripke frame semantics (see Remark 5.21). We mention important classes of posets to which Theorem 5.12 does not apply; the study of positive fragments of the logics of those posets requires techniques other than those used in this paper.

The paper is structured as follows. Section 2 contains preliminaries on superintuitionistic predicate logics and formulas. In Section 3, we define the system of positive slices in the lattice of superintuitionistic predicate logics and show that the lattice of slices is isomorphic to the lattice of positively axiomatizable logics. In Section 4, we recall Kripke sheaf and Kripke frame semantics for superintuitionistic predicate logics. In Section 5, we present our main results on positive slices. Subsection 5.1 contains an example of a continuum of degenerate (singleton) positive slices. Subsection 5.2 contains our Main Theorem stating conditions guaranteeing that logics determined by classes of Kripke frames or Kripke sheaves have the same positive fragment; it also contains consequences of Main Theorem concerning well-known superintuitionistic predicate logics. Lastly, in Subsection 5.3, we mention important logics to which Main Theorem does not apply. Section 6 outlines directions for future work.

2 Preliminaries on logics and formulas

We consider logics in a pure (i.e., without individual constants, function symbols, or equality) predicate language $\mathcal{L}$ containing the following symbols: countably many individual variables; for every $n \geqslant 0$, countably many n-ary predicate letters (nullary letters are identified with proposition letters); the propositional constant $\bot$; the binary connectives $\wedge$, $\vee$, and $\rightarrow$; the quantifier symbols $\exists$ and $\forall$. The definition of $\mathcal{L}$-formulas (or, simply, formulas) is standard. We use the standard abbreviations $\neg A = A \rightarrow \bot$ and $A \leftrightarrow B = (A \rightarrow B) \wedge (B \rightarrow A)$, and adopt the usual conventions about omitting parentheses. In what follows, the language $\mathcal{L}$ is identified with the set of its formulas. A formula is *propositional* if it contains no individual variables, and hence no non-nullary predicate letters and no quantifier symbols. The free variables of a formula are its *parameters*. The universal closure of a formula A, which may be assumed to be unique up to the enumeration of A's parameters, is denoted by $\bar{\forall} A$.

2.1 The lattice of superintuitionistic predicate logics

A *superintuitionistic predicate logic*, or simply *logic*, is a set of formulas including the intuitionistic predicate logic **QH** and closed under Predicate Substitution, Modus Ponens, and Generalisation; thus, **QH** is the smallest superintuitionistic predicate logic. If $\mathbf{L}$ is a logic and A a formula, then $\mathbf{L} \vdash A$ means the same as $A \in \mathbf{L}$. The smallest logic including a logic $\mathbf{L}$ and a set Γ of formulas is denoted by $\mathbf{L} + \Gamma$; if A is a formula, we write $\mathbf{L} + A$ instead of $\mathbf{L} + \{A\}$. The *logical sum* of a family $\{\mathbf{L}_\theta : \theta \in \Theta\}$, where Θ is an index set, of logics is the smallest logic including $\bigcup\{\mathbf{L}_\theta : \theta \in \Theta\}$; notice that the logical sum of the empty family of logics is **QH** and that the logical sum of L_1 and L_2 is the logic $L_1 + L_2$. If A and B are formulas and $\mathbf{L}$ a logic, then we say that

- A *implies* B *in* $\mathbf{L}$, and write $A \Rightarrow_{\mathbf{L}} B$, if $\mathbf{L} \vdash A \rightarrow B$;
- B is *derivable* from A, and write $A \vdash B$, if $\mathbf{QH} + A \vdash B$;
- A and B are *equivalent in* $\mathbf{L}$ if $\mathbf{L} \vdash A \leftrightarrow B$;
- A and B are *deductively equivalent* if they are mutually derivable, i.e., if $A \vdash B$ and $B \vdash A$, or, equivalently, if $\mathbf{QH} + A = \mathbf{QH} + B$.

It should be clear that, if A and B are deductively equivalent, then $A \in \mathbf{L}$ if, and only if, $B \in \mathbf{L}$, for every logic $\mathbf{L}$.

It is well known that the set of all logics forms a lattice with respect to the set-theoretic inclusion; we denote this lattice by $\mathfrak{L}$; algebraically, the meet of $\mathbf{L}_1, \mathbf{L}_2 \in \mathfrak{L}$ is the logic $\mathbf{L}_1 \cap \mathbf{L}_2$, and the join of $\mathbf{L}_1, \mathbf{L}_2 \in \mathfrak{L}$ is the logic $\mathbf{L}_1 + \mathbf{L}_2$. The least element of $\mathfrak{L}$ is the logic **QH**; the greatest, the (absolutely) inconsistent logic $\mathcal{L}$ (the set of all formulas), which, obviously, coincides with both $\mathbf{QH} + p$ and $\mathbf{QH} + \bot$. The lattice $\mathfrak{L}$ is complete since intersections and logical sums of arbitrary (finite or infinite) families of logics are themselves logics.

2.2 Delta-operation on formulas

In Section 5.3, we use the δ-operation on formulas introduced by Hosoi [7]. Here, we recall its definition and the syntactic properties used later on; for more background, consult [4, Section 1.16].

If A is a formula and p a proposition letter not occurring in A, then $\delta A = q \vee (q \to A)$. The following is well known (see, e.g., [4, Lemma 1.16.6]) and easy to check:

Fact 2.1 *For every formulas A and B,*

(1) $\mathbf{QH} \vdash A \to \delta A$.

(2) $\mathbf{QH} \vdash \delta(A \to B) \to (\delta A \to \delta B)$.

(3) $\mathbf{QH} \vdash \delta(A \wedge B) \leftrightarrow (\delta A \wedge \delta B)$.

2.3 Some important formulas

We shall consider the following standard formulas (here, $h < \omega$ and $n < \omega$):

$$
\begin{array}{lcl}
J & = & \neg p \vee \neg\neg p; \\
Z & = & (p \to q) \vee (q \to p); \\
CD & = & \forall x\,(P(x) \vee q) \to \forall x\, P(x) \vee q; \\
K & = & \forall x\, \neg\neg P(x) \to \neg\neg \forall x\, P(x); \\
E & = & \neg\neg \exists x\, P(x) \to \exists x\, \neg\neg P(x); \\
JE & = & \neg \exists x\, P(x) \vee \exists x\, \neg\neg P(x); \\
U & = & \forall x \forall y\, (P(x) \to P(y)); \\
U' & = & \exists x\, P(x) \to \forall x\, P(x); \\
P_0 & = & \bot; \\
P_{h+1} & = & \delta P_h \; = \; q_h \vee (q_h \to P_h); \\
P_0^+ & = & \bot; \\
P_{h+1}^+ & = & \forall x\, (Q_h(x) \vee (Q_h(x) \to P_h^+)); \\
Wid_n & = & \bigvee\limits_{i=0}^{n} (p_i \to \bigvee\limits_{j \neq i} p_j).
\end{array}
$$

Thus, P_h is a propositional formula with proposition letters $q_0, \ldots, q_{h-1}$, and P_h^+ is a formula with monadic predicate letters $Q_0, \ldots, Q_{h-1}$. We note that both P_h and P_h^+ are deductively equivalent to positive formulas obtained by replacing occurrences of $\bot$ in them with fresh proposition letters. The following is well known and easy to check:

Fact 2.2

(1) $\mathbf{QH} \vdash U \leftrightarrow U'$;

(2) $\mathbf{QH} + JE = \mathbf{QH} + J \wedge E$;

(3) $\mathbf{QH} + Z \vdash J$.

Lemma 2.3 $\mathbf{QH} + CD + J \vdash E$.

Proof It is not hard to see that

$$\mathbf{QH} + J \vdash \forall x\,(\neg P(x) \vee \exists x\,\neg\neg P(x)).$$

Hence,

$$\mathbf{QH} + CD + J \vdash \forall x\,\neg P(x) \vee \exists x\,\neg\neg P(x).$$

Since $\mathbf{QH} \vdash \forall x\,\neg P(x) \leftrightarrow \neg\exists x\,P(x)$, it follows that

$$\mathbf{QH} + CD + J \vdash JE.$$

Thus, by Fact 2.2 (2), $\mathbf{QH} + CD + J \vdash E$. □

Corollary 2.4 $\mathbf{QH} + CD + Z \vdash E$.

Proof Immediate from Lemma 2.3 and Fact 2.2 (3). □

Remark 2.5 Without CD, the formula E is not derivable from either J or Z, i.e., neither $\mathbf{QH} + J \vdash E$ nor $\mathbf{QH} + Z \vdash E$; see Lemma 5.20.

Lemma 2.6 *$\mathbf{QH} + J + P_2 \vdash Z$; moreover, $\mathbf{QH} \vdash J \wedge P_2 \to Z$.*

Proof Observe that

$$\begin{aligned}(\neg q \vee \neg\neg q) \wedge (p \vee (p \to q \vee \neg q)) &\Rightarrow_{\mathbf{QH}} \neg q \vee p \vee (p \to q)\\ &\Rightarrow_{\mathbf{QH}} (q \to p) \vee (p \to q).\end{aligned}$$

□

2.4 Positive formulas and positively axiomatizable logics

An $\mathcal{L}$-formula is *positive* if it does not contain occurrences of $\bot$. The set of positive formulas is denoted by $\mathcal{L}^+$. A logic $\mathbf{L}$ is *positively axiomatizable* if it is axiomatizable over $\mathbf{QH}$ only by positive formulas; thus, positively axiomatizable logics are those representable as $\mathbf{QH} + \Gamma$, with $\Gamma \subseteq \mathcal{L}^+$. The set of all positively axiomatizable logics shall be denoted by $\mathfrak{L}_{pos}$.

Proposition 2.7 *$\mathfrak{L}_{pos}$ is a sublattice of $\mathfrak{L}$.*

Proof Obviously, $\mathfrak{L}_{pos}$ is closed under arbitrary (finite and infinite) logical sums. We next show that it is closed under intersections. Let $\mathbf{L}_1$ and $\mathbf{L}_2$ be positively axiomatizable. Then, there exist $\Gamma_1, \Gamma_2 \subseteq \mathcal{L}^+$ such that $\mathbf{L}_i = \mathbf{QH} + \Gamma_i$, for $i \in \{1, 2\}$. By [4, Proposition 2.10.1 (1)],

$$\mathbf{L}_1 \cap \mathbf{L}_2 = \mathbf{QH} + \{\bar{\forall} A_1^m \vee \bar{\forall} A_2^m : A_1 \in \Gamma_1,\, A_2 \in \Gamma_2,\, m \geqslant 0\},$$

where $\bar{\forall} A^m$ denotes the universal closure of an m-shift [1] of the formula A. Since m-shifts of positive formulas are positive, it follows that $\mathbf{L}_1 \cap \mathbf{L}_2$ is positively axiomatizable. □

Proposition 2.8 *The lattice $\mathfrak{L}_{pos}$ is complete.*

[1] An m-shift of a formula A is obtained from A by a substitution of a special form intended to increase arities of predicate letters of A by m, using a fixed list of m fresh variables; for details, see [4, Section 2.5].

Proof As we have seen, $\mathfrak{L}_{pos}$ is a sublattice of $\mathfrak{L}$ with respect to arbitrary logical sums; hence, each family $\{\mathbf{L}_\theta : \theta \in \Theta\}$ of elements of $\mathfrak{L}_{pos}$ has a supremum, its logical sum; therefore, it also has an infimum, which is the logical sum of $\{\mathbf{L} : \mathbf{L} \subseteq \mathbf{L}_\theta, \text{for all } \theta \in \Theta\}$. □

We do not know if the infimum of an infinite family of elements of $\mathfrak{L}_{pos}$ coincides with its intersection:

Problem 2.9 *Is the lattice $\mathfrak{L}_{pos}$ closed under arbitrary intersections?*

We give some examples of logics that are not positively axiomatizable in Subsection 5.4.

3 Positive slices

3.1 Convex sets in the lattice of logics

We say that a set S of the elements of the lattice $\mathfrak{L}$ of logics is *convex* if

$$\forall \mathbf{L}_1, \mathbf{L}_2 \in S\ \forall \mathbf{L}_0 \in \mathfrak{L}\ (\mathbf{L}_1 \subseteq \mathbf{L}_0 \subseteq \mathbf{L}_2 \Rightarrow \mathbf{L}_0 \in S).$$

We shall only be interested in non-empty convex subsets of $\mathfrak{L}$ (the empty set is trivially convex). Special types of convex subsets of $\mathfrak{L}$ are *segments* and *intervals*: if $\mathbf{L}_1, \mathbf{L}_2 \in \mathfrak{L}$ and $\mathbf{L}_1 \subseteq \mathbf{L}_2$, then a segment in $\mathfrak{L}$ is a set

$$[\mathbf{L}_1, \mathbf{L}_2] = \{\mathbf{L} : \mathbf{L}_1 \subseteq \mathbf{L} \subseteq \mathbf{L}_2\},$$

and an interval in $\mathfrak{L}$ is a (possibly empty) set

$$(\mathbf{L}_1, \mathbf{L}_2) = \{\mathbf{L} : \mathbf{L}_1 \subset \mathbf{L} \subset \mathbf{L}_2\}.$$

Notice that for the interval $(\mathbf{L}_1, \mathbf{L}_2)$ to be non-empty it is necessary, but not sufficient, that $\mathbf{L}_1 \subset \mathbf{L}_2$.

It should be clear that a convex set is a segment if, and only if, it contains a least and a greatest element. On the other hand, a set S of the elements of $\mathfrak{L}$ is convex if, and only if, S contains the segment $[\mathbf{L}_1, \mathbf{L}_2]$ whenever $\mathbf{L}_1, \mathbf{L}_2 \in S$.

3.2 Systems of slices in the lattice of logics

A *system* $(S_\theta : \theta \in \Theta)$ *of slices in* $\mathfrak{L}$, where Θ is an index set, is a partition of $\mathfrak{L}$ into convex subsets, called the *slices* of the system. Thus, slices are non-empty, mutually disjoint convex subsets of $\mathfrak{L}$ whose union coincides with $\mathfrak{L}$. It is well known that every such partition is induced by an equivalence on $\mathfrak{L}$.

Of course, not every system of slices is worth studying—either the slices themselves or the corresponding equivalence relation should be meaningful and interesting. In this paper, we consider one such example, a system of positive slices.

3.3 Positive slices and partial ordering on positive slices

3.3.1 Positive fragments and positive slices

The *positive fragment* of a logic $\mathbf{L}$ is the set $\mathbf{L}^+ = \mathbf{L} \cap \mathcal{L}^+$ of its positive formulas.

Lemma 3.1 *If $\mathbf{L}$ is a logic, then $\mathbf{L}^+ = (\mathbf{QH} + \mathbf{L}^+)^+$.*

Proof Clearly, $\mathbf{QH}+\mathbf{L}^+ \subseteq \mathbf{L}$; hence, $(\mathbf{QH}+\mathbf{L}^+)^+ \subseteq \mathbf{L}^+$. On the other hand, since $\mathbf{L}^+ \subseteq \mathbf{QH}+\mathbf{L}^+$, it follows that $\mathbf{L}^+ \subseteq (\mathbf{QH}+\mathbf{L}^+) \cap \mathcal{L}^+ = (\mathbf{QH}+\mathbf{L}^+)^+$.□

We next introduce a preorder on logics with respect to the inclusion of their positive fragments; we also introduce the equivalence induced by this preorder: for every $\mathbf{L}_1, \mathbf{L}_2 \in \mathfrak{L}$, put

$$\mathbf{L}_1 \leqslant_{pos} \mathbf{L}_2 \rightleftharpoons \mathbf{L}_1^+ \subseteq \mathbf{L}_2^+;$$
$$\mathbf{L}_1 \equiv_{pos} \mathbf{L}_2 \rightleftharpoons \mathbf{L}_1 \leqslant_{pos} \mathbf{L}_2 \text{ and } \mathbf{L}_2 \leqslant_{pos} \mathbf{L}_1.$$

Thus, $\mathbf{L}_1 \equiv_{pos} \mathbf{L}_2$ means that $\mathbf{L}_1^+ = \mathbf{L}_2^+$. We call the relation $\equiv_{pos}$ the *positive equivalence* on $\mathfrak{L}$ and denote the equivalence class of a logic $\mathbf{L}$ under $\equiv_{pos}$ by $[\mathbf{L}]_{pos}$.

It should be clear that sets of the form $[\mathbf{L}]_{pos}$ are convex subsets of $\mathfrak{L}$. Hence, the relation $\equiv_{pos}$ induces a system of slices on $\mathfrak{L}$, which we call the *system of positive slices* on $\mathfrak{L}$ and denote by $\mathfrak{L}/{\equiv_{pos}}$; we call elements of $\mathfrak{L}/{\equiv_{pos}}$ *positive slices*. The relation $\leqslant_{pos}$ naturally induces a partial order $\preccurlyeq_{pos}$ on $\mathfrak{L}/{\equiv_{pos}}$: if $\mathcal{S}_1, \mathcal{S}_2 \in \mathfrak{L}/{\equiv_{pos}}$, then

$$\mathcal{S}_1 \preccurlyeq_{pos} \mathcal{S}_2 \rightleftharpoons \mathbf{L}_1 \leqslant_{pos} \mathbf{L}_2 \text{ whenever } \mathbf{L}_1 \in \mathcal{S}_1 \text{ and } \mathbf{L}_2 \in \mathcal{S}_2.$$

The relation $\preccurlyeq_{pos}$ is well defined since whether $\mathbf{L}_1 \leqslant_{pos} \mathbf{L}_2$ is independent of the choice of $\mathbf{L}_1 \in \mathcal{S}_1$ and $\mathbf{L}_2 \in \mathcal{S}_2$.

3.3.2 Least logics of positive slices

It should be clear that every positive slice is closed under all (finite and infinite) non-empty intersections, and hence contains a least logic. We next characterize least logics of positive slices:

Proposition 3.2 *For every logic* $\mathbf{L}$*, the following conditions are equivalent:*

- *(i)* $\mathbf{L}$ *is the least logic of a positive slice;*
- *(ii)* $\mathbf{L} \subseteq \mathbf{QH} + \mathbf{L}^+$*;*
- *(iii)* $\mathbf{L} = \mathbf{QH} + \mathbf{L}^+$*;*
- *(iv)* $\mathbf{L}$ *is a positively axiomatizable logic.*

Proof *(i)* ⇒ *(ii)*: Let $\mathbf{L}$ be the least logic of a positive slice $\mathcal{S}$. By Lemma 3.1, $(\mathbf{QH}+\mathbf{L}^+) \in \mathcal{S}$. Since $\mathbf{L}$ is the least in $\mathcal{S}$, it follows that $\mathbf{L} \subseteq \mathbf{QH}+\mathbf{L}^+$.

(ii) ⇒ *(iii)*: Since $\mathbf{QH} \subseteq \mathbf{L}$ and $\mathbf{L}^+ \subseteq \mathbf{L}$, surely $\mathbf{QH}+\mathbf{L}^+ \subseteq \mathbf{L}$.

(iii) ⇒ *(iv)*: Immediate from the definition of positive axiomatizability.

(iv) ⇒ *(i)*: Suppose that $\mathbf{L} = \mathbf{QH}+\Gamma$, for some $\Gamma \subseteq \mathcal{L}^+$. Then, $\Gamma \subseteq \mathbf{L}^+$. Let $\mathbf{L} \equiv_{pos} \mathbf{L}_0$, i.e., $\mathbf{L}^+ = \mathbf{L}_0^+$. Then, $\Gamma \subseteq \mathbf{L}_0^+ \subseteq \mathbf{L}_0$. Hence, $\mathbf{QH}+\Gamma \subseteq \mathbf{L}_0$, i.e., $\mathbf{L} \subseteq \mathbf{L}_0$. □

Lemma 3.3 *If logics* $\mathbf{L}_1$ *and* $\mathbf{L}_2$ *are positively axiomatizable, then the following conditions are equivalent:*

- *(i)* $\mathbf{L}_1 \leqslant_{pos} \mathbf{L}_2$*;*
- *(ii)* $\mathbf{L}_1 \subseteq \mathbf{L}_2$.

Proof $(i) \Rightarrow (ii)$: Let $\mathbf{L}_1 \leqslant_{pos} \mathbf{L}_2$, i.e., $\mathbf{L}_1^+ \subseteq \mathbf{L}_2^+$. Then, $\mathbf{QH}+\mathbf{L}_1^+ \subseteq \mathbf{QH}+\mathbf{L}_2^+$. By Proposition 3.2, $\mathbf{L}_1 = \mathbf{QH} + \mathbf{L}_1^+$ and $\mathbf{L}_2 = \mathbf{QH} + \mathbf{L}_2^+$. Thus, $\mathbf{L}_1 \subseteq \mathbf{L}_2$.

$(ii) \Rightarrow (i)$: If $\mathbf{L}_1 \subseteq \mathbf{L}_2$, then $\mathbf{L}_1^+ \subseteq \mathbf{L}_2^+$, i.e., $\mathbf{L}_1 \leqslant_{pos} \mathbf{L}_2$. □

We denote the least logic of a positive slice $\mathcal{S}$ by $\mathbf{L}^{\mathcal{S}}_{pos}$. Proposition 3.2 immediately gives us the following:

Lemma 3.4 *Let $\mathcal{S}$ be a positive slice. Then,*

(1) $\mathbf{L}^{\mathcal{S}}_{pos} = \mathbf{QH} + \mathbf{L}^+$, *for every* $\mathbf{L} \in \mathcal{S}$;

(2) $\mathbf{L}^{\mathcal{S}}_{pos}$ *is the unique positively axiomatizable logic in* $\mathcal{S}$.

Proposition 3.5 *The system $\mathfrak{L}/{\equiv_{pos}}$, partially ordered by $\preccurlyeq_{pos}$, is a complete lattice isomorphic to $\mathfrak{L}_{pos}$.*

Proof By Lemma 3.4 (2), there exists a bijection $f\colon \mathcal{S} \mapsto \mathbf{L}^{\mathcal{S}}_{pos}$ between $\mathfrak{L}/{\equiv_{pos}}$ and $\mathfrak{L}_{pos}$. By definition of $\preccurlyeq_{pos}$,

$$\mathcal{S} \preccurlyeq_{pos} \mathcal{S}' \iff \mathbf{L}^{\mathcal{S}}_{pos} \leqslant_{pos} \mathbf{L}^{\mathcal{S}'}_{pos}.$$

Moreover, by Lemma 3.3,

$$\mathbf{L}^{\mathcal{S}}_{pos} \leqslant_{pos} \mathbf{L}^{\mathcal{S}'}_{pos} \iff \mathbf{L}^{\mathcal{S}}_{pos} \subseteq \mathbf{L}^{\mathcal{S}'}_{pos}.$$

Hence, the lattices $\mathfrak{L}/{\equiv_{pos}}$ and $\mathfrak{L}_{pos}$ are isomorphic. □

Proposition 3.5 shall be relied on in a more detailed study of positive slices that we hope to undertake in the near future.

3.3.3 Maximal logics of positive slices

The logical sum of a family of logics with the same positive fragment might itself have a positive fragment larger than the positive fragment of the summands. Hence, positive slices are not guaranteed to be closed under logical sums, and so are not guaranteed to have largest logics. It should, however, be obvious that, if $\mathcal{C}$ is a chain of logics with the same positive fragment, say P, then $(\bigcup \mathcal{C})^+ = P$. Hence, by Zorn's lemma, every logic of a positive slice $\mathcal{S}$ is included in a logic maximal in $\mathcal{S}$. Thus, every positive slice $\mathcal{S}$ is bounded below by the least logic $\mathbf{L}^{\mathcal{S}}_{pos}$ and above by the antichain $\mathfrak{M}^{\mathcal{S}}_{pos}$ of its maximal logics. In other words, every positive slice $\mathcal{S}$ is representable as follows:

$$\{\mathbf{L} : \exists \mathbf{L}' \in \mathfrak{M}^{\mathcal{S}}_{pos}\ (\mathbf{L}^{\mathcal{S}}_{pos} \subseteq \mathbf{L} \subseteq \mathbf{L}')\}.$$

If $\mathbf{L}_0 \in \mathfrak{L}$ and $\mathfrak{A}$ is an antichain of logics such that $\mathbf{L}_0 \subseteq \mathbf{L}$ whenever $\mathbf{L} \in \mathfrak{A}$, then we say that the set $\{\mathbf{L} : \exists \mathbf{L}' \in \mathfrak{A}\ (\mathbf{L}_0 \subseteq \mathbf{L} \subseteq \mathbf{L}')\}$ is a *tulip* in $\mathfrak{L}$. Thus, every positive slice $\mathcal{S}$ is a tulip contained between $\mathbf{L}^{\mathcal{S}}_{pos}$ and $\mathfrak{M}^{\mathcal{S}}_{pos}$.

Notice that segments are just tulips whose upper antichains are singletons.

4 Kripke semantics

In this section, we recall Kripke sheaf and Kripke frame semantics for superintuitionistic predicate logics and define some important logics characterized using these types of semantics.

4.1 Kripke sheaves and frames

A *Kripke sheaf* [4, Section 3.6] is a triple $F = \langle W, D, \rho \rangle$ where

- W is a non-empty poset with the partial order $\leqslant$; elements of W are called *points*, or *worlds*;
- $D = \{D_u : u \in W\}$ is a system of non-empty domains;
- $\rho = \{\rho_{uv} : u \leqslant v\}$ is a system of *transition maps* $\rho_{uv} : D_u \to D_v$, subject to the following conditions:
  - ρ_{uu} is the identity on D_u, for all $u \in W$;
  - $\rho_{uw} = \rho_{vw} \circ \rho_{uv}$, for all $u, v, w \in W$, i.e.,

$$\forall u, v, w \in W \ [u \leqslant v \leqslant w \Rightarrow \forall a \in D_u \, \rho_{uw}(a) = \rho_{vw}(\rho_{uv}(a))].$$

We say that a Kripke sheaf $F = \langle W, D, \rho \rangle$ is a *Kripke sheaf over the poset* W. If $\boldsymbol{a} = \langle a_1, \ldots, a_n \rangle \in D_u^n$, then $\rho_{uv}(\boldsymbol{a})$ denotes the tuple $\langle \rho_{uv}(a_1), \ldots, \rho_{uv}(a_n) \rangle$.

A *Kripke frame* is a Kripke sheaf satisfying the following conditions:

- *expanding domains condition*: if $u \leqslant v$, then $D_u \subseteq D_v$;
- *the identity condition*: if $u \leqslant v$ and $a \in D_u$, then $\rho_{uv}(a) = a$.

Since in frames transition maps are uniquely determined by the identity condition, to simplify notation, we omit the mention of transition maps when talking about Kripke frames, presenting them simply as pairs $\langle W, D \rangle$. We say that a Kripke frame has a *constant domain* if $D_u = D_v$ whenever $u, v \in W$. The corresponding Kripke sheaves are obtained by requiring all maps ρ_{uv} to be surjective; we are not aware of special category-theoretic terminology for such Kripke sheaves; here we call them *surjective*.

A *valuation* on a sheaf $F = \langle W, D, \rho \rangle$ is a map ζ sending an n-ary predicate letter P and a world $u \in W$ to a subset $\zeta(u, P^n)$ of D_u^n; the map ζ is required to satisfy the following *heredity condition*: for all $u, w \in W$ and $\boldsymbol{a} \in D_u^n$,

$$u \leqslant v \;\&\; \boldsymbol{a} \in \zeta(u, P^n) \implies \rho_{uv}(\boldsymbol{a}) \in \zeta(v, P^n).$$

If F is a Kripke sheaf and ζ a valuation on F, then a tuple $M = \langle F, \zeta \rangle$ is called a *Kripke sheaf model*; if, additionally, F is a Kripke frame, then M is called simply a *Kripke model*. If $u \in W$, then a D_u*-sentence* is an expression obtained from a formula B by substituting (constants corresponding to) elements of D_u for all occurrences of parameters in B. Note that sentences, i.e., formulas without parameters, are just D_u-sentences without any occurrences of constants from D_u. We occasionally write $A(a_1, \ldots, a_n)$ to mean that A is a D_u-sentence containing no constants from D_u beside $a_1, \ldots, a_n$.

Truth of a D_u-sentence $A(\boldsymbol{a})$, with $\boldsymbol{a} \in D_u^n$, for some n, in a Kripke sheaf model M at a point u is defined by recursion (the clauses for $\bot$, $\vee$, and $\wedge$ are as in propositional logic):

- $M, u \Vdash P(\boldsymbol{a})$ if $\boldsymbol{a} \in \zeta(u, P)$;
- $M, u \Vdash (A_1 \to A_2)(\boldsymbol{a})$ if
$$\forall u' \geqslant u \ [M, u' \Vdash A_1(\rho_{uu'}(\boldsymbol{a})) \Rightarrow M, u' \Vdash A_2(\rho_{uu'}(\boldsymbol{a}))];$$

- $M, u \Vdash (\exists x\, A_1)(\boldsymbol{a})$ if $\exists b \in D_u\ [M, u \Vdash A_1(b, \boldsymbol{a})]$;
- $M, u \Vdash (\forall x\, A_1)(\boldsymbol{a})$ if $\forall u' \geqslant u\, \forall b \in D_{u'}\ [M, u' \Vdash A_1(b, \rho_{uu'}(\boldsymbol{a}))]$.

We say that a formula A is

- *true at a point* u of a model M, and write $M, u \Vdash A$, if the D_u-sentence $\overline{\forall} A$ is true in M at u;
- *valid on a Kripke sheaf* F, and write $F \Vdash A$, if $M, u \Vdash A$ holds for every point u of F and every model M over F;
- *valid on a class of Kripke sheaves* if it is valid on every sheaf from the class.

The set of formulas valid on a class $\mathcal{F}$ of Kripke sheaves is denoted by $\mathbf{L}\mathcal{F}$; if $\mathcal{F} = \{F\}$, we write $\mathbf{L}F$ instead of $\mathbf{L}\{F\}$. It is well known that, if F is a Kripke sheaf, then $\mathbf{L}F$ is a logic, called the *logic of* F. Consequently, if $\mathcal{F}$ is a class of Kripke sheaves, then $\mathbf{L}\mathcal{F} = \bigcap\{\mathbf{L}F : F \in \mathcal{F}\}$ is a logic, called the *logic of* $\mathcal{F}$. Logics representable as $\mathbf{L}\mathcal{F}$, for some class $\mathcal{F}$ of Kripke sheaves, are called *Kripke sheaf complete.* Logics representable as $\mathbf{L}\mathcal{F}$, for some class $\mathcal{F}$ of Kripke frames, are called *Kripke complete.*

It is well known [4, Lemma 3.6.20] that the logic of all Kripke sheaves over a class $\mathcal{W}$ of posets coincides with the logic of all Kripke frames over $\mathcal{W}$; hence, we denote this logic by $\mathbf{L}\mathcal{W}$. The same is true for constant domains: the logic of all surjective Kripke sheaves over a class $\mathcal{W}$ of posets coincides with the logic of all Kripke frames with constant domains over $\mathcal{W}$; hence, we denote this logic by $\mathbf{L}_c\mathcal{W}$.

4.2 Subsheaves and rooted sheaves

If W is a poset with a partial order $\leqslant$ and $w_0 \in W$, then a *poset generated by* w_0, denoted by $W{\uparrow}w_0$, is a substructure of W with the set of points $\{w \in W : w_0 \leqslant w\}$. If $F = \langle W, D, \rho\rangle$ is a Kripke sheaf and $w_0 \in W$, then the *subsheaf of* F *generated by* w_0 is the Kripke sheaf $F{\uparrow}w_0 = \langle W{\uparrow}w_0, D', \rho'\rangle$ where D' and ρ' are restrictions to $W{\uparrow}w_0$ of, respectively, D and ρ. A world $w_0 \in W$ is a *root* of F if $F{\uparrow}w_0 = F$. A sheaf is *rooted* if it has a root.

In Section 5.3, we use the following fact [4, Lemma 1.16.3] about formulas of the form δA (see Section 2.2):

Proposition 4.1 *Let* $F = \langle W, R, D\rangle$ *be a Kripke sheaf with root* w_0. *Then,*

$$F \Vdash \delta A \iff \forall w \in W \setminus \{w_0\}\ F{\uparrow}w \Vdash A.$$

4.3 Some families of posets and their logics

For the purposes of this paper, the following classes of posets shall be of interest:

- $\mathcal{W}_{po}$, the class of all posets;
- $\mathcal{W}_g$, the class of all posets with a greatest element;
- $\mathcal{HEI}_h$, where $h < \omega$, the class of posets of height at most h;
- $\mathcal{WID}_n$, where $n < \omega$, the class of posets of width at most n;

- $\mathcal{HEI}_\omega = \bigcup\{\mathcal{HEI}_h : h < \omega\}$, the class of posets of finite height;
- $\mathcal{WID}_\omega = \bigcup\{\mathcal{WID}_n : n < \omega\}$, the class of posets of finite width;
- $\mathcal{CH}_h = \mathcal{WID}_1 \cap \mathcal{HEI}_h$, where $h < \omega$, the class of chains of height at most h;
- $\mathcal{CH}_\omega = \bigcup\{\mathcal{C}_h : h < \omega\} = \mathcal{WID}_1 \cap \mathcal{HEI}_\omega$, the class of all finite chains;
- $\mathcal{FIN} = \mathcal{HEI}_\omega \cap \mathcal{WID}_\omega$, the class of all finite posets.

It should be clear that

$$\begin{array}{llll} \mathbf{L}\mathcal{HEI}_\omega & = \bigcap\limits_{h<\omega} \mathbf{L}\mathcal{HEI}_h; & \mathbf{L}_c\mathcal{HEI}_\omega & = \bigcap\limits_{h<\omega} \mathbf{L}_c\mathcal{HEI}_h; \\ \mathbf{L}\mathcal{WID}_\omega & = \bigcap\limits_{n<\omega} \mathbf{L}\mathcal{WID}_n; & \mathbf{L}_c\mathcal{WID}_\omega & = \bigcap\limits_{n<\omega} \mathbf{L}_c\mathcal{WID}_n. \end{array}$$

We next recall known facts about logics of these classes of posets and about some closely related logics:

(4.1) As shown in [8],

$$\begin{array}{ll} \mathbf{L}\mathcal{W}_{po} & = \mathbf{QH}; \\ \mathbf{L}_c\mathcal{W}_{po} & = \mathbf{QH} + CD. \end{array}$$

(4.2) By [3, Theorems 5.8 and 5.12],

$$\begin{array}{ll} \mathbf{L}\mathcal{W}_g & = \mathbf{QH} + K + J; \\ \mathbf{L}_c\mathcal{W}_g & = \mathbf{QH} + K + J + CD. \end{array}$$

(4.3) As shown in [26],

$$\mathbf{L}\mathcal{HEI}_h = \mathbf{QH} + P_h^+.$$

(4.4) Even though formulas P_h are propositionally complete ($\mathbf{H} + P_h$ is the propositional logic of $\mathcal{HEI}_h$), if $h > 1$, then the logic $\mathbf{QH} + P_h$ is Kripke, and hence Kripke sheaf, incomplete: H. Ono proved that $\mathbf{QH} + P_h \nvdash K$ whenever $h > 1$; on the other hand, for every such h, the class $\mathcal{HEI}_h$ validates K. The logic $\mathbf{QH} + P_1$ is Kripke complete since $\mathbf{QH} + P_1 = \mathbf{L}\mathcal{HEI}_1 = \mathbf{QC}$ (here, $\mathbf{QC}$ is the classical predicate logic).

(4.5) As follows from [10, Theorem 11],

$$\mathbf{L}_c\mathcal{HEI}_h = \mathbf{QH} + P_h \wedge CD = \mathbf{QH} + P_h^+ \wedge CD.$$

(4.6) As follows from [14, Theorem 3.9], for every $n < \omega$,

$$\mathbf{L}_c\mathcal{WID}_n = \mathbf{QH} + Wid_n \wedge CD.$$

(4.7) As shown in [2,17],

$$\mathbf{L}\mathcal{WID}_1 = \mathbf{L}\mathbb{Q} = \mathbf{L}\mathbb{R} = \mathbf{QH} + Z.$$

(4.8) As shown in [21],

$$\mathbf{L}_c\mathcal{WID}_1 = \mathbf{L}_c\mathbb{Q} = \mathbf{QH} + Z \wedge CD.$$

(4.9) As shown in [2, p. 334],

$$\begin{aligned}\mathbf{L}(\mathcal{WID}_1 \cap \mathcal{W}_g) &= \mathbf{QH} + K + Z;\\ \mathbf{L}_c(\mathcal{WID}_1 \cap \mathcal{W}_g) &= \mathbf{QH} + K + Z + CD.\end{aligned}$$

(4.10) $\mathbf{L}_c\mathbb{R}$ is a finitely axiomatizable proper extension of $\mathbf{L}_c\mathbb{Q}$ [21].

(4.11) $\mathbf{L}\mathcal{HEI}_\omega$ and $\mathbf{L}_c\mathcal{HEI}_\omega$ are Π^0_1-hard [15, Corollary 1.2] and hence not recursively axiomatizable.

(4.12) As shown in [10], $\mathbf{L}\mathcal{CH}_h = \mathbf{QH} + P_h^+ + Z$; on the other hand, if $h > 1$, then $\mathbf{QH} + P_h + Z$ is Kripke sheaf incomplete.

(4.13) $\mathbf{L}_c\mathcal{CH}_h = \mathbf{QH} + P_h^+ + Z + CD = \mathbf{QH} + P_h + Z + CD$.

(4.14) As shown in [15], $\mathbf{L}\mathcal{CH}_\omega$ and $\mathbf{L}_c\mathcal{CH}_\omega$ are both Π^0_1-hard and are both in Π^0_2.

(4.15) $\mathbf{L}\mathcal{FIN}$ and $\mathbf{L}_c\mathcal{FIN}$ are both Π^0_1-hard [15, Corollary 2.1] and are both in Π^0_2.

Remark 4.2 It is not known whether logics $\mathcal{WID}_n$, with $n > 1$, are recursively axiomatizable; we conjecture that the answer is negative; if our conjecture is true, then logics $\mathbf{QH}+Wid_n$ are Kripke incomplete, i.e., are proper sublogics of $\mathbf{L}\mathcal{WID}_n$.[2]

5 Main results

In this section, we present our results on positive slices obtained so far.

5.1 A continuum of degenerate positive slices

5.1.1 Degenerate slices

We say that a positive slice $\mathcal{S}$ is *degenerate* if it is a singleton. By Proposition 3.2, every degenerate slice $\mathcal{S}$ has the form $\{\mathbf{L}^{\mathcal{S}}_{pos}\}$. Moreover, the following is true:

Lemma 5.1 *A slice of a logic* $\mathbf{L}$ *is degenerate if, and only if, the following conditions simultaneously hold:*

(1) $\mathbf{L}$ *is a positively axiomatizable logic;*

(2) *for every logic* $\mathbf{L}_0$*, if* $\mathbf{L} \subset \mathbf{L}_0$*, then* $\mathbf{L}^+ \subset \mathbf{L}_0^+$.

Proof Suppose that (1) and (2) hold, and let $\mathcal{S}$ be the positive slice of $\mathbf{L}$. By (1) and Proposition 3.2, $\mathbf{L}$ is the least logic of $\mathcal{S}$. Due to (2), $\mathbf{L}$ is also a maximal logic of $\mathcal{S}$. Hence, $\mathcal{S} = \{\mathbf{L}\}$.

Conversely, suppose that $\mathcal{S}$ is a positive slice such that $\mathcal{S} = \{\mathbf{L}\}$. Then, $\mathbf{L}$ is the least logic of $\mathcal{S}$; hence, by Proposition 3.2, $\mathbf{L}$ is positively axiomatizable, i.e., (1) holds. Since $\mathcal{S}$ contains no logics beside $\mathbf{L}$, no logic other than $\mathbf{L}$ has the same positive fragment as $\mathbf{L}$; hence, (2) holds, as well. □

[2] The third author had established Kripke incompleteness of $\mathbf{QH} + Wid_2 \wedge P_3$; he believes that the Kripke completion of this logic is not recursively enumerable; however, a proof, as well as a proof of incompleteness for $n \geqslant 2$ and $h \geqslant 3$, is likely to be quite complicated.

We say that a logic $\mathbf{L}$ is *hereditarily positively axiomatizable* if both $\mathbf{L}$ and all its proper extensions are positively axiomatizable.

Corollary 5.2 *The slice of a hereditarily positively axiomatizable logic is degenerate.*

Proof Let $\mathbf{L}$ be a hereditarily positively axiomatizable logic. Due to Lemma 5.1, it suffices to show that $\mathbf{L} \subset \mathbf{L}_0$ implies $\mathbf{L}^+ \subset \mathbf{L}_0^+$, for every $\mathbf{L}_0$.

Suppose that $\mathbf{L} \subset \mathbf{L}_0$. Then, $\mathbf{L}^+ \subseteq \mathbf{L}_0^+$ and $\mathbf{L}_0$ is positively axiomatizable. Suppose, for contradiction, that $\mathbf{L}^+ = \mathbf{L}_0^+$. Since $\mathbf{L}$ and $\mathbf{L}_0$ are positively axiomatizable, it follows, by Proposition 3.2, that $\mathbf{L} = \mathbf{QH}+\mathbf{L}^+$ and $\mathbf{L}_0 = \mathbf{QH}+\mathbf{L}_0^+$, and so $\mathbf{L} = \mathbf{L}_0$, contrary to the assumption. Hence, $\mathbf{L}^+ \subset \mathbf{L}_0^+$. □

5.1.2 Superclassical logics

We shall consider the lattice $\mathfrak{L}_{\mathbf{QC}} = \{\mathbf{L} \in \mathcal{L} : \mathbf{QC} \subseteq \mathbf{L}\}$ of *superclassical logics*, i.e., superintuitionistic predicate logics extending the classical predicate logic $\mathbf{QC}$. We begin with the following decreasing chain of all Kripke complete extensions of $\mathbf{QC}$ (notice that all these logics are, indeed, superintuitionistic predicate logics):

- $\mathbf{QC}_m$, where $0 < m < \omega$, is the logic of Kripke frames over singleton posets with m-element domains; thus, $\mathbf{QC}_m$ is the set of formulas classically valid over m-element domains;
- $\mathbf{QC}_\omega = \bigcap\limits_{m=1}^{\infty} \mathbf{QC}_m$ is the logic of Kripke frames over singleton posets with finite domains; thus, $\mathbf{QC}_\omega$ is the set of formulas classically valid over finite domains;
- $\mathbf{QC}_0$ is the inconsistent logic $\mathcal{L}$.[3]

It is well known that both $\mathbf{QC}_0$ and $\mathbf{QC}$ are finitely positively axiomatizable over $\mathbf{QH}$ by, respectively, p and Pierce's law $((p \to q) \to p) \to p$. By Trakhtenbrot's theorem [22], $\mathbf{QC}_\omega$ is Π_1^0-complete, and so is not finitely axiomatizable.

The lattice $\mathfrak{L}_{\mathbf{QC}}$ includes the infinite segment $[\mathbf{QC}, \mathbf{QC}_\omega]$ of logics between $\mathbf{QC}$ and $\mathbf{QC}_\omega$. The following is known [18, Section 0.6]:

Fact 5.3 *The lattice* $\mathfrak{L}_{\mathbf{QC}_\omega} = \{\mathbf{L} \in \mathcal{L} : \mathbf{QC}_\omega \subseteq \mathbf{L}\}$ *of all extensions of* $\mathbf{QC}_\omega$ *is just the decreasing* $(\omega+1)$*-chain* $\{\mathbf{QC}_m : 0 \leqslant m \leqslant \omega\}$.

Fact 5.4 *The lattice* $\mathfrak{L}_{\mathbf{QC}}$ *is just* $\mathfrak{L}_{\mathbf{QC}_\omega} \cup [\mathbf{QC}, \mathbf{QC}_\omega]$*; in other words,* $\mathfrak{L}_{\mathbf{QC}}$ *does not contain logics incomparable with* $\mathbf{QC}_\omega$.

Wajsberg [23] had shown that the cardinality of the lattice $\mathfrak{L}_{\mathbf{QC}}$, and hence of the segment $[\mathbf{QC}, \mathbf{QC}_\omega]$, is continuum.

---

[3] By analogy with the case when $m > 0$, one might think of $\mathbf{QC}_0$ as the logic of Kripke frames that are singleton posets with 0-element domains: such Kripke frames do not exist, hence $\mathbf{QC}_0$ is the logic of the empty class of Kripke frames.

5.1.3 A continuum of logics with degenerate slices

To obtain a continuum of logics with degenerate slices, we show that the classical predicate logic **QC** is hereditarily positively axiomatizable:

Proposition 5.5 *Every superclassical logic is positively axiomatizable.*

Proof For every formula A, we choose a nullary letter q not in A and define positive formulas A' and A'' as follows:

$$A' = [q/\bot]A, \qquad A'' = (q \to A') \to A'.$$

We shall prove that A and A'' are deductively equivalent in **QC**.

We start by proving that

$$\mathbf{QH} \vdash (q \vee \neg q) \wedge A \to A''. \tag{5.1}$$

We use reasoning by cases. Case q: It should be clear that $\mathbf{QH} \vdash q \to A''$; hence, $\mathbf{QH} \vdash q \wedge A \to A''$. Case $\neg q$: Since $\mathbf{QH} \vdash \neg q \to (\bot \leftrightarrow q)$, it follows, by induction on A, using the equivalence replacement rule, that $\mathbf{QH} \vdash \neg q \to (A \leftrightarrow A')$. Since $\mathbf{QH} \vdash A' \to ((q \to A') \to A')$, we obtain $\mathbf{QH} \vdash \neg q \to (A \to A'')$. This proves (5.1).

Now, as $\mathbf{QC} \vdash q \vee \neg q$, it follows, by (5.1), that

$$\mathbf{QC} \vdash A \to A''. \tag{5.2}$$

Second, we show that

$$\mathbf{QH} + A'' \vdash A. \tag{5.3}$$

Substituting $\bot$ for q in A'', we obtain $(\bot \to A) \to A$. Since the latter formula is equivalent in **QH** to A, this gives us (5.3).

Thus, A and A'' are deductively equivalent in **QC**, i.e.

$$\mathbf{QC} + A = \mathbf{QC} + A''. \tag{5.4}$$

Now, let **L** be a superclassical logic. By (5.4), $\mathbf{L} + A = \mathbf{L} + A''$, and so

$$\mathbf{L} = \mathbf{QH} + \mathbf{L} = \mathbf{QH} + \{A'' : A \in \mathbf{L}\}. \tag{5.5}$$

Since formulas of the form A'' are positive, (5.5) immediately implies the statement of the proposition. [4] □

The proof of Proposition 5.5 also gives us the following:

Corollary 5.6

(1) *A superclassical logic is finitely positively axiomatizable if, and only if, it is finitely axiomatizable.*

(2) *A superclassical logic is recursively positively axiomatizable if, and only if, it is recursively axiomatizable.*

Since extensions of a superclassical logic are themselves superclassical, Proposition 5.5 immediately gives us the following:

---

[4] Clearly, in (5.5), i.e., in a positive axiomatization of **L**, it suffices to use only closed formulas from **L**.

Proposition 5.7 *Every superclassical logic is hereditarily positively axiomatizable.*

Finally, by Proposition 5.7 and Corollary 5.2, we obtain the following:

Proposition 5.8 *The positive slice of every superclassical logic is degenerate.*

5.1.4 Finite positive axiomatizations of $\mathbf{QC}_m$

We conclude this section by presenting explicit finite positive axiomatizations for superclassical logics of finite domains, i.e., of logics $\mathbf{QC}_m$, with $0 < m < \omega$. We rely on the following fact, first observed by Skvortsov [18, Section 0.6]:

Fact 5.9 *For every predicate formula A,*

$$\mathbf{QC} + A = \mathbf{QC}_m \iff A \in \mathbf{QC}_m \setminus \mathbf{QC}_{m+1}.$$

Now, for every m with $0 < m < \omega$, we define the formula

$$DOM_m^* = \bigwedge_{i=0}^{m} \exists x\, P_i(x) \to \bigvee_{i \neq j} \exists x\, (P_i(x) \wedge P_j(x)).$$

It is not hard to see that $DOM_m^* \in \mathbf{QC}_m \backslash \mathbf{QC}_{m+1}$, i.e., DOM_m^* is classically valid on domains with m elements, but not on domains with $m+1$ elements. This observation, together with Fact 5.9, immediately gives us the following: [5]

Proposition 5.10 $\mathbf{QC} + DOM_m^* = \mathbf{QC}_m$.

5.2 Non-degenerate slices: main theorem and its corollaries

In this section, we show that positive slices of many well-known Kripke complete and Kripke sheaf complete logics are non-degenerate.

Define the *g-extension* of a poset W, denoted by W^g, to be the poset obtained by adding to W the greatest element; by default, the greatest element of a poset will be denoted by g.

Define the *g-extension* of a Kripke frame $F = \langle W, D \rangle$ by letting

- $D_u^g = D_u$, for every $u \in W$;
- $D_g^g = \bigcup\{D_u : u \in W\}$ (thus, the domain of the greatest point g is the union of all domains from F);
- $F^g = \langle W^g, D^g \rangle$.

Similarly, define the *g°-extension* of a Kripke sheaf $F = \langle W, D, \rho \rangle$ by letting

- $D_u^\circ = D_u$, for every $u \in W$;
- D_g° to be a singleton domain $\{t\}$;
- $\rho_{ug}^\circ(a) = t$, for every $u \in W$ and $a \in D_u$ (thus, t is a common inheritor of all individuals from F);
- $F^\circ = (W^g, D^\circ, \rho^\circ)$.

[5] Axioms DOM_m^* seem to be both simple and natural; we are not, however, aware of their use in the literature.

We note that Kripke sheaf semantics, unlike Kripke frame semantics, allows us to glue together all the individuals of the greatest world, resulting in a compact, convenient, and effective construction.

If $\mathcal{F}$ is a class of Kripke frames, we define $\mathcal{F}^g = \{F^g : F \in \mathcal{F}\}$. We say that a class $\mathcal{F}$ of Kripke frames is *g-closed* if $\mathcal{F}^g \subseteq \mathcal{F}$. Similarly, if $\mathcal{F}$ is a class of Kripke sheaves, we define $\mathcal{F}^\circ = \{F^\circ : F \in \mathcal{F}\}$. We say that a class $\mathcal{F}$ of Kripke sheaves is *g°-closed* if $\mathcal{F}^\circ \subseteq \mathcal{F}$.

We now obtain a sufficient condition for inclusion of positive fragments of Kripke complete and Kripke sheaf complete logics:

Lemma 5.11 (Main lemma)

(1) *If $\mathcal{F}$ is a class of Kripke frames, then $(\mathbf{L}\mathcal{F}^g)^+ \subseteq (\mathbf{L}\mathcal{F})^+$, i.e., $\mathbf{L}\mathcal{F}^g \leqslant_{pos} \mathbf{L}\mathcal{F}$.*

(2) *If $\mathcal{F}$ is a class of Kripke sheaves, then $(\mathbf{L}\mathcal{F}^\circ)^+ \subseteq (\mathbf{L}\mathcal{F})^+$, i.e., $\mathbf{L}\mathcal{F}^\circ \leqslant_{pos} \mathbf{L}\mathcal{F}$.*

Proof (1): Let A be a positive formula not in $(\mathbf{L}\mathcal{F})^+$ (we may assume that A is closed). Then, there exists a Kripke model $M = \langle F, \zeta \rangle$ over a Kripke frame $F = \langle W, D \rangle$ from $\mathcal{F}$ and a world $u_0 \in W$ such that $M, u_0 \not\Vdash A$. Expand the valuation ζ to the valuation ζ^g over the Kripke frame F^g as follows: let $\zeta^g \upharpoonright W = \zeta$ and, for every n-ary predicate letter P, let $\zeta^g(g, P)$ be the set of all n-tuples of elements from D^g_g. Then, all atoms are true at g in M^g. Put $M^g = \langle F^g, \zeta^g \rangle$. It should be clear that M^g satisfies the heredity condition; hence, M^g is a Kripke model. A straightforward induction on D^g_g-sentences shows that

$$M^g, g \Vdash B, \text{ for every positive } D^g_g\text{-sentence } B. \tag{5.6}$$

Now, straightforward induction, using (5.6), shows that, for every $u \in W$ and every positive D_u-sentence B,

$$M, u \Vdash B \iff M^g, u \Vdash B;$$

in other words, the values of positive D_u-sentences are preserved at all non-greatest worlds of M^g. Hence, $M^g, u_0 \not\Vdash A$, and so $A \notin (\mathbf{L}\mathcal{F}^g)^+$.

(2): The argument here is similar. We define a Kripke sheaf model over a Kripke sheaf F° analogously to the definition of the Kripke model M^g from the proof of (1). Since the definition of M^g from (1) did not require to distinguish values of atoms on different individuals from the domain of world g, gluing these individuals together does not affect the truth of formulas: thus, we make all atoms true on the unique individual of the domain of g. The remainder of the argument is essentially identical. □

We next obtain a sufficient condition for the equality of positive fragments of Kripke complete and Kripke sheaf complete logics (this will give us examples of extensive non-degenerate positive slices):

Theorem 5.12 (Main theorem)

(1) *If $\mathcal{F}$ is a g-closed class of Kripke frames, then* $\mathbf{L}\mathcal{F} \equiv_{pos} \mathbf{L}\mathcal{F}^g$.

(2) *If $\mathcal{F}$ is a g°-closed class of Kripke sheaves, then* $\mathbf{L}\mathcal{F} \equiv_{pos} \mathbf{L}\mathcal{F}^\circ$.

Proof (1) Since $\mathcal{F}$ is g-closed, $\mathcal{F}^g \subseteq \mathcal{F}$. Hence $\mathbf{L}\mathcal{F} \subseteq \mathbf{L}\mathcal{F}^g$, and so $(\mathbf{L}\mathcal{F})^+ \subseteq (\mathbf{L}\mathcal{F}^g)^+$, i.e., $\mathbf{L}\mathcal{F} \leqslant_{pos} \mathbf{L}\mathcal{F}^g$. The converse follows by Lemma 5.11.

(2) The argument here is similar to (1). □

Corollary 5.13

(1) $\mathbf{L}\mathcal{W}^g \leqslant_{pos} \mathbf{L}\mathcal{W}$ *and* $\mathbf{L}_c\mathcal{W}^g \leqslant_{pos} \mathbf{L}_c\mathcal{W}$, *for every class $\mathcal{W}$ of posets.*

(2) $\mathbf{L}\mathcal{W}^g \equiv_{pos} \mathbf{L}\mathcal{W}$ *and* $\mathbf{L}_c\mathcal{W}^g \equiv_{pos} \mathbf{L}_c\mathcal{W}$, *for every g-closed class $\mathcal{W}$ of posets.*

Proof Immediate from Lemma 5.11 and Theorem 5.12. □

We next apply Main Theorem to some well-known logics.

Proposition 5.14

(1) $\mathbf{QH} \equiv_{pos} \mathbf{QH} + J + K$;

(2) $\mathbf{QH} + \mathit{CD} \equiv_{pos} \mathbf{QH} + \mathit{CD} + J + K$;

(3) $\mathbf{QH} + Z \equiv_{pos} \mathbf{QH} + Z + K$;

(4) $\mathbf{QH} + \mathit{CD} + Z \equiv_{pos} \mathbf{QH} + \mathit{CD} + Z + K$.

Proof We use facts from Section 4.3.

(1): Since the class $\mathcal{W}_{po}$ of all posets is g-closed, it follows, by Corollary 5.13 (2), that $\mathbf{L}\mathcal{W}_{po} \equiv_{pos} \mathbf{L}\mathcal{W}^g_{po}$. By (4.1), $\mathbf{L}\mathcal{W}_{po} = \mathbf{QH}$. Since $\mathcal{W}^g_{po} = \mathcal{W}_g$, it follows that $\mathbf{L}\mathcal{W}^g_{po} \vdash J \wedge K$. Hence, $\mathbf{QH} \equiv_{pos} \mathbf{QH} + J + K$.

(2)–(4): The argumentation here is similar; use (4.1), (4.7), and (4.8). □

The scope of Proposition 5.14 will become clearer if the reader consults Fact 2.2 (3), Lemma 2.3, and Remark 2.5.

Due to (4.2), and (4.9), the results of Proposition 5.14 are the maximal ones that Theorem 5.12 (1) enables us to obtain. However, using Kripke sheaf semantics, we shall next obtain stronger (see Remark 5.21) results.

Lemma 5.15 *Let F be a Kripke sheaf. Then,* $F^\circ \Vdash J \wedge \neg\neg U$.

Proof First, $F^\circ \Vdash J$ since F° is a sheaf over a poset with the greatest element g. Second, $F^\circ \Vdash \neg\neg U$ since the domain of g is a singleton and hence $F^\circ \uparrow g \Vdash U$. □

Proposition 5.16 *Let $\mathcal{C}$ be one of the following classes of posets:* $\mathcal{W}_{po}$, $\mathcal{WID}_n$, *for some n such that* $1 < n < \omega$, $\mathcal{WID}_\omega$, $\mathcal{HEI}_\omega$, *and* $\mathcal{FIN}$; *let also* $\mathbf{L} \in \{\mathbf{L}\mathcal{C}, \mathbf{L}_c\mathcal{C}\}$. *Then,* $\mathbf{L} \equiv_{pos} \mathbf{L} + J + \neg\neg U$.

Proof By Lemma 5.15, formulas J and $\neg\neg U$ are valid on every Kripke sheaf of the form F°. Every class of posets mentioned in the proposition is g-closed; thus, the corresponding classes of Kripke sheaves and of surjective Kripke sheaves are g°-closed. Hence, the statement follows by Theorem 5.12 (2). □

Proposition 5.17 *If* $\mathbf{L} \in \{\mathbf{QH} + Z, \mathbf{QH} + Z + CD\}$, *then* $\mathbf{L} \equiv_{pos} \mathbf{L} + \neg\neg U$.

Proof Similar to the proof of Proposition 5.16 (recall that, by Fact 2.2 (3), $\mathbf{QH} + Z \vdash J$). □

Remark 5.18 Proposition 5.17 can be transferred to the logics of the class $\mathcal{CH}_\omega$ of all finite chains.

Remark 5.19 Notice that $\neg\neg U'$, which, by Fact 2.2 (1), is equivalent in $\mathbf{QH}$ to $\neg\neg U$, implies both K and E: namely, in the presence of $\neg\neg U'$, i.e., in the logic $\mathbf{L} = \mathbf{QH} + \neg\neg U = \mathbf{QH} + \neg\neg U'$, we obtain

$$\forall x \neg\neg P(x) \Rightarrow_{\mathbf{L}} \exists x \neg\neg P(x) \Rightarrow_{\mathbf{L}} \neg\neg\exists x\, P(x) \Rightarrow_{\mathbf{L}} \neg\neg\forall x P(x)$$

and

$$\neg\neg\exists x P(x) \Rightarrow_{\mathbf{L}} \neg\neg\forall x P(x) \Rightarrow_{\mathbf{L}} \forall x \neg\neg P(x) \Rightarrow_{\mathbf{L}} \exists x \neg\neg P(x).$$

We next show that Propositions 5.16 and 5.17 are stronger than Proposition 5.14. To that end, we need the following lemma:

Lemma 5.20 $\mathbf{QH} + Z + P_2^+ \nvdash E$.

Proof Recall from Section 4.3 that formulas Z and P_2^+ are valid, respectively, on chains and on posets of height at most 2. Define a Kripke model $M = \langle W, D, \zeta \rangle$ so that $W = \{u, v\}$ is a two-element poset where $u < v$, $D(u) = \{a\}$, $D(v) = \{a, b\}$, $\zeta(u, P) = \varnothing$, and $\zeta(v, P) = \{b\}$. Since W is a chain of height 2, surely $\langle W, D \rangle \Vdash \{J, P_2^+\}$. On the other hand, it is straightforward to check that $M, u \nVdash E$. □

Remark 5.21 To compare Propositions 5.16 and 5.14, observe that, since $\mathbf{QH} + \neg\neg U \vdash E$, Proposition 5.16 implies that

$$\mathbf{QH} \equiv_{pos} \mathbf{QH} + E. \tag{5.7}$$

On the other hand, Proposition 5.14 implies that $\mathbf{QH} \equiv_{pos} \mathbf{L}$ only if $\mathbf{L} \subseteq \mathbf{QH} + J + K$; this statement is weaker than (5.7) since, as we next show, E does not belong to the latter logic: indeed, since $\mathbf{QH} + Z \vdash J$ and $\mathbf{QH} + P_2^+ \vdash K$, it follows, by Lemma 5.20, that $\mathbf{QH} + J + K \nvdash E$.

Similar observations about logics with the axiom Z apply to Propositions 5.17 and 5.14.

Proposition 5.22 $\mathbf{L}_c\mathbb{R} \equiv_{pos} \mathbf{L}_c\mathbb{R} + \neg\neg U$.

Proof We show that $\mathbf{L}_c\mathbb{R} + \neg\neg U \leqslant_{pos} \mathbf{L}_c\mathbb{R}$, i.e., that

$$(\mathbf{L}_c\mathbb{R} + \neg\neg U)^+ \subseteq (\mathbf{L}_c\mathbb{R})^+. \tag{5.8}$$

Denote by $\mathcal{F}$ and $\mathcal{F}^*$ the classes of surjective Kripke sheaves over, respectively, $\mathbb{R}$ and $\mathbb{R}^g$, ordered by the usual $\leqslant$ relation. Then, $\mathbf{L}\mathcal{F} = \mathbf{L}_c\mathbb{R}$ and

$\mathbf{L}\mathcal{F}^* = \mathbf{L}_c\mathbb{R}^g$. Takano proved [21, Theorem (2°) and Proposition 5.1 (2°)] that $\mathbf{L}_c\mathbb{R}^g = \mathbf{L}_c\mathbb{R} + K$.[6] Hence, $\mathbf{L}_c\mathbb{R} \subseteq \mathbf{L}_c\mathbb{R}^g$.[7]

Now, let $\mathcal{F}^\circ = \{F^\circ : F \in \mathcal{F}\}$. Since every sheaf of the form F°, with $F \in \mathcal{F}$, is obviously surjective, surely $\mathcal{F}^\circ \subseteq \mathcal{F}^*$. Hence, $\mathbf{L}\mathcal{F}^* \subseteq \mathbf{L}\mathcal{F}^\circ$, and so

$$\mathbf{L}_c\mathbb{R} \subseteq \mathbf{L}_c\mathbb{R}^g = \mathbf{L}\mathcal{F}^* \subseteq \mathbf{L}\mathcal{F}^\circ.$$

By Lemma 5.15, $\neg\neg U \in \mathbf{L}\mathcal{F}^\circ$. Hence,

$$\mathbf{L}_c\mathbb{R} + \neg\neg U \subseteq \mathbf{L}\mathcal{F}^\circ. \tag{5.9}$$

By Lemma 5.11 (2), $(\mathbf{L}\mathcal{F}^\circ)^+ \subseteq (\mathbf{L}\mathcal{F})^+$, i.e.

$$(\mathbf{L}\mathcal{F}^\circ)^+ \subseteq (\mathbf{L}_c\mathbb{R})^+. \tag{5.10}$$

Lastly, (5.9) and (5.10) immediately imply (5.8). □

We note that Proposition 5.22 can be obtained from Main Theorem since the g-closure of $\mathbb{R}$ validates Takano's axioms for $\mathbf{L}_c\mathbb{R}$ [21, Proposition 5.2]; we, however, believe that our proof of Proposition 5.22 is simpler and more immediate than an appeal to Main Theorem.

5.3 Some restrictions on the application of the main theorem

In this section, we give examples of some interesting classes of posets to which Theorem 5.12 does not apply; the study of their positive fragments shall require techniques that differ from those used here.

Since no class $\mathcal{HEI}_h$, where $h < \omega$, is g-closed, the following is not unexpected (recall from (4.3), (4.4), and (4.5) that $\mathbf{QH}+P_h \subset \mathbf{QH}+P_h^+ = \mathbf{L}\mathcal{HEI}_h$ and $\mathbf{QH}+CD+P_h = \mathbf{QH}+CD+P_h^+ = \mathbf{L}_c\mathcal{HEI}_h$, for every h with $1 < h < \omega$):

Proposition 5.23 *Let* $\mathbf{L}$ *be one of the logics* $\mathbf{QH} + P_h$, $\mathbf{QH} + P_h^+$, *or* $\mathbf{QH} + CD + P_h$, *where* $1 < h < \omega$. *Then,* $\mathbf{L} \not\equiv_{pos} \mathbf{L} + J$.

Proof We first consider the case when $h = 2$. Denote by W_3 a rooted 3-element poset with two maximal elements ('fork'); this is a tree of height 2. It should be clear that $\mathbf{L}_c\mathcal{HEI}_2 \subseteq \mathbf{L}_cW_3$, but $W_3 \not\Vdash Z$; hence, $\mathbf{L}_c\mathcal{HEI}_2 \nvdash Z$. On the other hand, by Lemma 2.6, $\mathbf{QH} + J + P_2 \vdash Z$. Hence, $Z \in (\mathbf{L} + J) \setminus \mathbf{L}$, and so $\mathbf{L} \not\equiv_{pos} \mathbf{L} + J$, for every logic $\mathbf{L}$ from the segment $[\mathbf{QH} + P_2, \mathbf{L}_c\mathcal{HEI}_2]$.

The case when $h > 2$ is similar, using the δ-operation on formulas (see Section 2.2). Recall that that $P_h = \delta^{h-2}P_2$. Due to Lemma 2.6, $\mathbf{QH}+J \vdash P_2 \to Z$. Hence, by Fact 2.1 (2), $\mathbf{QH}+J \vdash \delta^{h-2}P_2 \to \delta^{h-2}Z$, i.e., $\mathbf{QH}+J \vdash P_h \to \delta^{h-2}Z$. On the other hand, since, as we have seen, $\mathbf{L}_c\mathcal{HEI}_2 \nvdash Z$, it follows, by Proposition 4.1, that $\mathbf{L}_c\mathcal{HEI}_h \nvdash \delta^{h-2}Z$. Therefore, $\delta^{h-2}Z \in (\mathbf{L} + J) \setminus \mathbf{L}$, and so $\mathbf{L} \not\equiv_{pos} (\mathbf{L} + J)$, for every logic $\mathbf{L}$ from the segment $[\mathbf{QH} + P_h, \mathbf{L}_c\mathcal{HEI}_h]$. □

---

[6] Even though Takano characterized these logics using Kripke frames, the same logics can, as we have seen at the end of Section 4.1, be characterized using surjective Kripke sheaves.

[7] By the way, this inclusion, besides Takano's completeness results, also follows from the existence of a p-morphism of $\mathbb{R}$ onto $\mathbb{R}^g$; for information on p-morphisms of Kripke frames, consult [4, Section 3.3].

Remark 5.24 Due to Lemma 2.6, $\mathbf{QH} \vdash J \wedge P_2 \to Z$. Hence, by Fact 2.1 (2) and Fact 2.1 (3), $\mathbf{QH} \vdash \delta^{h-2} J \wedge \delta^{h-2} P_2 \to \delta^{h-2} Z$, for every $h > 2$.

Hence, if $h > 2$, then, for every logic $\mathbf{L}$ from Proposition 5.23 parameterized by the said h,

$$\mathbf{L} \not\equiv_{pos} (\mathbf{L} + \delta^{h-2} J).$$

We note that, by Fact 2.1 (1), $\mathbf{L} + \delta^{h-2} J \subset \mathbf{L} + J$, for such logics $\mathbf{L}$.

Proposition 5.25 *Let* $\mathbf{L}$ *be one of the following logics:*

- *an extension of* $\mathbf{QH} + P_2$ *not containing* Z*;*
- $\mathbf{QH} + P_h + Wid_n$ *and* $\mathbf{QH} + P_h^+ + Wid_n$*, with* $1 < h < \omega$ *and* $1 < n < \omega$*;*
- $\mathbf{QH} + CD + P_h + Wid_n$ *(i.e.,* $\mathbf{QH} + CD + P_h^+ + Wid_n$*), with* $1 < h < \omega$ *and* $1 < n < \omega$*.*

Then $\mathbf{L} \not\equiv_{pos} \mathbf{L} + J$*.*

Proof Similar to the proof of Proposition 5.23. □

Remark 5.26 Observe that logics mentioned in the first item of Proposition 5.25 include the logics $\mathbf{QH} + P_2^+ + Wid_n$ of n-branching trees of height 2, as well as Kripke incomplete logics $\mathbf{QH} + P_2 + Wid_n$, with $1 < n < \omega$. The logics of constant domains $\mathbf{QH} + CD + P_2 + Wid_n$ (i.e., $\mathbf{QH} + CD + P_2^+ + Wid_n$) are included, as well.

5.4 On logics that are not positively axiomatizable

Our results have immediate corollaries concerning lack of positive axiomatizability for classes of logics. We give only one example (it is not hard to extend it to similar cases):

Proposition 5.27 *Neither* $\mathbf{QH} + J$*, nor any logic in the interval* $(\mathbf{QH}, \mathbf{QH} + J)$ *is positively axiomatizable.*

Proof Immediate from Proposition 3.2 and Proposition 5.14 (1). □

We next give an example independent from Proposition 3.2 (once again, it is not hard to produce similar examples). Recall that we denote by W_3 a 3-element rooted poset with two maximal elements ('fork'); we also denote by W_4 a 4-element poset where a root sees a two-element anti-chain whose elements see the greatest element ('rhombus').

Proposition 5.28 *No logic in the segment* $[\mathbf{L}W_4, \mathbf{L}_c W_4]$ *is positively axiomatizable.*

Proof Clearly, $W_4 = W_3^g$. Hence, by Corollary 5.13 (1), $\mathbf{L}_c W_4 \leqslant_{pos} \mathbf{L}_c W_3$, i.e., $(\mathbf{L}_c W_4)^+ \subseteq (\mathbf{L}_c W_3)^+$. On the other hand, $J \in \mathbf{L}_c W_4 \setminus \mathbf{L}_c W_3$. Hence, if $\mathbf{L} \in [\mathbf{L}W_4, \mathbf{L}_c W_4]$, then $J \in \mathbf{L} \setminus (\mathbf{QH} + \mathbf{L}^+)$, which implies the statement of the proposition. □

We do not know how to explicitly axiomatize the least logic in the positive slice of $\mathbf{L}W_4$. Logics $\mathbf{L}W_3$ and $\mathbf{L}_c W_3$, as well as logics mentioned in Remark 5.26, are positively axiomatizable.

Remark 5.29 It is well known that propositional extensions of the logic $\mathbf{H}+Z$, i.e., propositional logics of (finite) chains, as well as propositional extensions of $\mathbf{H}+P_2$, i.e., propositional logics of (finite) trees of height at most 2 (and so, in particular, proper extensions of the propositional logic of W_4, i.e., the logic of rhombus), are all finitely positively axiomatizable. It remains unknown if the analogous facts hold for predicate logics.

Problem 5.30 Are $\mathbf{QH} + CD + Z$ (the logic of all chains with constant domains) and $\mathbf{QH} + CD + P_2^+$ (the logic of all trees of height at most 2 with constant domains) hereditarily positively axiomatizable?

Note that the class of extensions of $\mathbf{QH} + CD + Z$ includes the infinite family of the predicate logics of ordinals. The constant domain logics of ordinals have been studied by Minari, Takano, and Ono [9]. It is known [15] that all these logics except logics of finite chains, as well as their expanding domains counterparts, are Π^1_1-hard, and so are not arithmetical.

6 Directions for future work

The present paper is but a first sketch of the study of the system of positive slices in the lattice of superintuitionistic predicate logics.

As already mentioned, our Main Theorem generalises an observation made by Yankov [25] about superintuitionistic propositional logics. In fact, Yankov proved that the propositional logic $\mathbf{H}+J$ of the weak law of the excluded middle is the greatest propositional logic whose positive fragment coincides with that of the intuitionistic propositional logic $\mathbf{H}$ (i.e., in our terminology, the positive slice of $\mathbf{H}$ is the segment $[\mathbf{H}, \mathbf{H} + J]$). Our Main Theorem does not imply an analogous statement for predicate logics. Moreover, we do not know if any of the non-degenerate positive slices of predicate logics are segments. These are questions for future study.

Acknowledgements

We are grateful to anonymous reviewers for helping to improve the presentation of the paper. The first author has been supported by the HSE Academic Fund Program, Project 23-00-022.

References

[1] Chagrov, A. and M. Zakharyaschev, "Modal Logic," Oxford University Press, 1997.

[2] Corsi, G., *Completeness theorem for Dummett's LC quantified and some of its extensions*, Studia Logica **51** (1992), pp. 317–335.

[3] Corsi, G. and S. Ghilardi, *Directed frames*, Archive for Mathematical Logic **29** (1989), pp. 53–67.

[4] Gabbay, D., V. Shehtman and D. Skvortsov, "Quantification in Nonclassical Logic, Volume 1," Studies in Logic and the Foundations of Mathematics **153**, Elsevier, 2009.

[5] Ghilardi, S., *Incompleteness results in Kripke semantics*, Journal of Symbolic Logic **56** (1991), pp. 517–538.

[6] Hosoi, T., *On intermediate logics, I*, Journal of the Faculty of Science of the University of Tokyo, Section I **14** (1967), pp. 293–312.

[7] Hosoi, T., *On intermediate logics, II*, Journal of the Faculty of Science of the University of Tokyo, Section I **16** (1969), pp. 1–12.
[8] Kripke, S., *Semantical analysis of intuitionistic logic I*, in: J. Crossley and M. Dummett, editors, *Formal Systems and Recursive Functions*, Studies in Logic and the Foundations of Mathematics **40**, Elsevier, 1965 pp. 92–130.
[9] Minari, P., M. Takano and H Ono. *Intermediate Predicate Logics Determined by Ordinals*. Journal of Symbolic Logic **55**, pp. 1099–1124
[10] Ono, H., *On finite linear intermediate predicate logics*, Studia Logica **47** (1988), pp. 391–399.
[11] Rybakov, M. and D. Shkatov, *Undecidability of first-order modal and intuitionistic logics with two variables and one monadic predicate letter*, Studia Logica **107** (2019), pp. 695–717. Corrected version available at `https://arxiv.org/abs/1706.05060`.
[12] Rybakov, M. and D. Shkatov, *Algorithmic properties of first-order superintuitionistic logics of finite Kripke frames in restricted languages*, Journal of Logic and Computation **31** (2021), pp. 494–522.
[13] Rybakov, M. and D. Shkatov, *Variations on the Kripke trick*, To appear in Studia Logica, `https://doi.org/10.1007/s11225-023-10093-y`.
[14] Shimura, T., *Kripke completeness of some intermediate predicate logics with the axiom of constant domain and a variant of canonical formulas*, Studia Logica **52** (1993), pp. 23–40.
[15] Skvortsov, D., *On non-axiomatizability of superintuitionistic predicate logics of some classes of well-founded and dually well-founded Kripke frames*, Journal of Logic and Computation **16** (2006), pp. 685–695.
[16] Skvortsov, D., *A remark on superintuitionistic predicate logics of Kripke frames with constant and with nested domains*, Journal of Logic and Computation **21** (2009), pp. 697–713.
[17] Skvortsov, D., *On an axiomatization of superintuitionistic predicate logic of Kripke frames with nested domains over the set of reals*, Topology, Algerbra and Categories in Logic, TACL 2009, ILLC, Amsterdam, 2009.
[18] Skvortsov, D., *On finite domains based slices in the structure of superintuitionistic predicate logics, preview*, Logical Investigations **29** (2023), pp. 101–113.
[19] Skvortsov, D., *On systems of slices in the structure of superintuitionistic predicate (or propositional) logics I*, Proceedings of the 20th Russian Conference on Artificial Intelligence, Vol. 1, Moscow, 2022, pp. 311–321 (in Russian).
[20] Skvortsov, D., *On systems of slices in the structure of superintuitionistic predicate (or propositional) logics I*, Pattern Recognition and Image Analysis **33** (2023), pp. 511–516 (English translation of [19]).
[21] Takano, M., *Ordered sets R and Q as bases of Kripke models*, Studia Logica (1987), pp. 137–148.
[22] Trakhtenbrot, B. A., *Impossibility of an algorithm for the decision problem in finite classes*, American Mathematica Society Translations **23** (1963), pp. 1–5.
[23] Wajsberg, M., *Beitrag zur Metamathematik*, Mathematische Annalen **109**, pp. 200- 229. (English translation: *A contribution to metamathematics*, in: M. Wajsberg, "Logical Works", edited by S. J. Surma, Ossolineum 1977, 62-88.)
[24] Wolter, F. and M. Zakharyaschev, *Modal decision problems*, in: P. Blackburn, J. V. Benthem and F. Wolter, editors, *Handbook of Modal Logic*, Elsevier, 2007 pp. 427–489.
[25] Yankov, V. A., *The calculus of the weak "Law of excluded middle"*, Mathematics of the USSR–Izvestiya **32** (1968), pp. 1044-1051.
[26] Yokota, S., *Axiomatization of the first-order intermediate logics of bounded Kripkean heights I*, Mathematical Logic Quarterly **35** (1989), pp. 415–421.

A Tree Rewriting System for the Reflection Calculus

Sofía Santiago-Fernández [1] Joost J. Joosten [2]
David Fernández-Duque [3]

Universitat de Barcelona
Barcelona, Spain

Abstract

The *Reflection Calculus* (**RC**, c.f. [6], [8]) is the fragment of the polymodal logic **GLP** in the language $\mathcal{L}^+$ whose formulas are built up from $\top$ and propositional variables using conjunction and diamond modalities. **RC** is complete with respect to the arithmetical interpretation that associates modalities with reflection principles and has various applications in proof theory, specifically ordinal analysis.
We present TRC, a tree rewriting system (c.f. [2]) that is adequate and complete with respect to **RC**, designed to simulate **RC** derivations. TRC is based on a given correspondence between formulas of $\mathcal{L}^+$ and modal trees $\mathsf{Tree}^\diamond$. Modal trees are presented as an inductive type (c.f. [9], [16]) allowing precise positioning and transformations which give rise to the formal definition of rewriting rules and facilitates formalization in proof assistants. Furthermore, we provide a rewrite normalization theorem for systematic rule application. The normalization of the rewriting process enhances proof search efficiency and facilitates implementation (c.f. [20], [12], [15]).
By providing TRC as an efficient provability tool for **RC**, we aim to help on the study of various aspects of the logic such as the subformula property and rule admissibility.

Keywords: Strictly positive logics, Reflection Calculus, tree rewriting systems, abstract rewriting systems, polymodal logics, normalized rewriting, proof normalization, proof search, proof theory, type theory.

1 Introduction

Modal logics provide an attractive alternative to first or higher order logic for computational applications, largely due to the fact that they often enjoy a decidable consequence relation while remaining expressive enough to describe intricate processes. However, decidability alone does not suffice for practical implementation when complexity comes at a hefty price tag; even propositional

[1] sofia.santiago@ub.edu
[2] jjoosten@ub.edu
[3] fernandez-duque@ub.edu

logic is NP-complete, which quickly becomes intractable as formula size and especially the number of variables is large.

This is no longer an issue when working in *strictly positive* fragments (see e.g. [14]), which in contrast enjoy a polynomially decidable consequence relation. Strictly positive formulas do not contain negation and instead are built from atoms and $\top$ using conjunction and $\Diamond$ (or, more generally, a family of modalities $\langle i \rangle$ indexed by i in some set I). Strictly positive formulas tend to be contingent, so validity and satisfiability are no longer the most central problems, but the consequence relation is indeed useful for example for reasoning about ontologies and is the basis for some description logics [1].

One remarkable success story for strictly positive logics comes from the *reflection calculus* (**RC**) [8,7]. Beklemishev has shown how Japaridze's polymodal provability logic **GLP** [13] can be used to perform a proof-theoretic analysis of Peano aritmetic and its fragments [5]; however, the logic **GLP** is notoriously difficult to work with, especially as it is not Kripke-complete. In contrast, its strictly positive fragment is rather tame from both a theoretical and computational point of view, yet suffices for the intended proof-theoretic applications.

The current work is inspired by two distinct ideas that have arisen in the study of strictly positive logics. The first is the tree representation of formulas, which yield a way to decide strictly positive implications. This was developed by Kikot et al. [14] in a general setting and by Beklemishev [7] in the context of **RC**. In both cases, one assigns to each strictly positive formula φ a finite, tree-like Kripke model $T(\varphi)$ with the crucial property that $\varphi \to \psi$ is valid if and only if $T(\varphi) \models \psi$. Thus the study of strictly positive fragments can be reduced to the study of their tree-like Kripke models.

The second is the connection of strictly positive calculi to term rewrite systems. Strictly positive formulas and, particularly, those built exclusively from $\top$ and the modalities $\langle i \rangle$, may be regarded as *words* (or 'worms'). This has prompted Beklemishev [4] to view strictly positive fragments as term-rewriting systems [2], but connections between such systems and modal logic are not new and can be traced back to Foret [10].

Term rewriting is a discipline that integrates elements of logic, universal algebra, automated theorem proving, and functional programming. It has applications in algebra (e.g. Boolean algebra), recursion theory (computability of rewriting rules), software engineering and programming languages (especially functional and logic programming [19]), with the λ-calculus perhaps being the most familiar example [3]. Of particular interest to us, tree rewriting systems [11] are term rewriting systems such that terms are trees.

When terms represent formulas, rewrite rules are similar to deep inference rules, i.e. rules which may be applied to strict subformulas of the displayed formulas. This is the approach taken by Shamkanov [17] for developing a cut-free calculus for **GLP**. As is the case for other technical differences between **GLP** and the reflection calculus, our tree rewrite system makes up for the loss in expressive power with increased simplicity and transparent Kripke semantics.

Our approach is to recast **RC** as an abstract rewriting system in which terms are trees. In the parlance of rewrite systems, cut-elimination can be viewed as a normalization procedure for derivations. In our setting we do not have an analogue of the cut rule, but we do obtain a rewriting normalization theorem which states that the rewriting process can be consistently and efficiently executed by grouping rewriting rules by their kinds and applying them in a designated sequence. By doing so, it enhances our comprehension of the dynamics of the tree rewriting system, offering valuable insights into the nature of the rewriting process and the interplay among rules. Moreover, it furnishes an efficient framework for proof search methodologies. Thanks to the normalization theorem, the need for exhaustive exploration is minimized by focusing on normalized rewriting sequences, which mitigates the risk of redundancy in rewriting. Consequently, when searching for a proof of a certain result, we only need to consider the normalized derivations, thereby reducing the proof search space and improving computational efficiency [12]. Furthermore, it serves as a practical guide for implementing the rewriting process in proof assistants [15].

In our presentation, we make use of the inductive type of lists within the framework of type theory (cf. [16], [9]) to define the trees in our tree rewriting system. The use of lists allows to define inductive structures with an order, facilitating the specification of internal positions and transformations for rewriting systems, and its formalization in proof assistants.

Since lists play such a central role in our work, we conclude this introduction by establishing some notation. A list of elements of type $\mathcal{A}$ is either the empty list $\varnothing$ or $[x] \frown L$ for x an element of type $\mathcal{A}$, a list L of elements of type $\mathcal{A}$ and $\frown$ the operator of concatenation of lists. We write $x \frown L$ and $L \frown x$ to denote $[x] \frown L$ and $L \frown [x]$, respectively. The length of a list L of elements of type $\mathcal{A}$ is denoted by $|L|$.

2 From $\mathbf{K}^+$ to RC

In this section we present the basic sequent-style system $\mathbf{K}^+$ for the language of strictly positive formulae, concluding by an introduction to the Reflection Calculus (**RC**) as an extension of $\mathbf{K}^+$.

We consider the language of strictly positive formulae $\mathcal{L}^+$ composed from propositional variables p,q,..., in Prop, the constant $\top$, and connectives $\wedge$ for conjunction and $\langle\alpha\rangle$ for diamond modalities for each ordinal $\alpha \in \omega$. Formally, the strictly positive formulae φ of $\mathcal{L}^+$ are generated by the following grammar:

$$\varphi ::= \top \mid p \mid \langle\alpha\rangle\varphi \mid (\varphi \wedge \varphi), \quad \alpha \in \omega \text{ and } p \in \mathsf{Prop}.$$

The *modal depth* of φ, denoted by $\mathsf{md}(\varphi)$, is recursively defined as $\mathsf{md}(\top) := 0$, $\mathsf{md}(p) := 0$ for $p \in \mathsf{Prop}$, $\mathsf{md}(\langle\alpha\rangle\varphi) := \mathsf{md}(\varphi) + 1$ and $\mathsf{md}(\varphi \wedge \psi) := \max\{\mathsf{md}(\varphi), \mathsf{md}(\psi)\}$.

Sequents are expressions of the form $\varphi \vdash \psi$ for $\varphi, \psi \in \mathcal{L}^+$. If L is a logic, we write $\varphi \vdash_{\mathsf{L}} \psi$ for the statement that $\varphi \vdash \psi$ is provable in L. We write $\varphi \equiv_{\mathsf{L}} \psi$ to denote $\varphi \vdash_{\mathsf{L}} \psi$ and $\psi \vdash_{\mathsf{L}} \varphi$.

Polymodal **K** can be readily adapted to its strictly positive variant, where most notably the necessitation rule is replaced by distribution for each diamond modality.

Definition 2.1 ($\mathbf{K}^+$, [4]) The basic sequent-style system $\mathbf{K}^+$ is given by the following axioms and rules:

$\varphi \vdash_{\mathbf{K}^+} \varphi$; $\varphi \vdash_{\mathbf{K}^+} \top$;

if $\varphi \vdash_{\mathbf{K}^+} \psi$ and $\psi \vdash_{\mathbf{K}^+} \phi$ then $\varphi \vdash_{\mathbf{K}^+} \phi$ (cut);

$\varphi \wedge \psi \vdash_{\mathbf{K}^+} \varphi$ and $\varphi \wedge \psi \vdash_{\mathbf{K}^+} \psi$ (elimination of conjunction);

if $\varphi \vdash_{\mathbf{K}^+} \psi$ and $\varphi \vdash_{\mathbf{K}^+} \phi$ then $\varphi \vdash_{\mathbf{K}^+} \psi \wedge \phi$ (introduction to conjunction);

if $\varphi \vdash_{\mathbf{K}^+} \psi$ then $\langle\alpha\rangle\varphi \vdash_{\mathbf{K}^+} \langle\alpha\rangle\psi$ (distribution).

For Π a finite list of strictly positive formulae, $\bigwedge\Pi$ is defined by $\top$ for $\Pi = \varnothing$ and $\varphi \wedge \bigwedge\hat{\Pi}$ for $\Pi = \varphi \frown \hat{\Pi}$. Note that $\bigwedge(\Pi_1 \frown \Pi_2) \equiv_{\mathbf{K}^+} \bigwedge\Pi_1 \wedge \bigwedge\Pi_2$ for Π_1 and Π_2 finite lists of strictly positive formulae.

A diamond modality can be distributed over a conjunction of formulas for $\mathbf{K}^+$ as follows.

Lemma 2.2 $\langle\alpha\rangle(\varphi_1 \wedge ... \wedge \varphi_n) \vdash_{\mathbf{K}^+} \langle\alpha\rangle\varphi_1 \wedge ... \wedge \langle\alpha\rangle\varphi_n$.

Proof. By an easy induction on n. □

We aim to define a tree rewriting system adequate and complete w.r.t. the Reflection Calculus, the strictly positive fragment of Japaridze's polymodal logic formulated as an extension of $\mathbf{K}^+$.

Definition 2.3 (**RC**, [6], [8]) The *Reflection Calculus* (**RC**) is the strictly positive modal logic extending $\mathbf{K}^+$ by the following axioms:

$\langle\alpha\rangle\langle\alpha\rangle\varphi \vdash_{\mathbf{RC}} \langle\alpha\rangle\varphi$ (transitivity);

$\langle\alpha\rangle\varphi \vdash_{\mathbf{RC}} \langle\beta\rangle\varphi$, $\alpha > \beta$ (monotonicity);

$\langle\alpha\rangle\varphi \wedge \langle\beta\rangle\psi \vdash_{\mathbf{RC}} \langle\alpha\rangle(\varphi \wedge \langle\beta\rangle\psi)$, $\alpha > \beta$ (J).

3 Modal trees

In this section we present modal trees, a concrete set of inductively defined trees on which our rewriting system is based, and the corresponding framework for their manipulation. Modal trees are finite labeled trees with nodes labeled with lists of propositional variables and edges labeled with ordinals less than ω. Specifically, modal trees can be regarded as tree-like Kripke models of the form $(\mathcal{W}, \{R_\alpha\}_{\alpha\in\omega}, v)$ such that an ordinal α labels an edge if R_α relates the corresponding nodes, and a list of propositional variables labels a node if its elements are the only propositional variables being true under the valuation v in that node. However, for technical convenience, both in presenting the rewrite system and in formalizing our results in a proof assistant, it will be convenient to present modal trees as inductively defined structures. In particular, the children of a node of a modal tree are given by lists instead of sets, providing a default ordering on its children useful for unambiguously determining positions in the tree.

Definition 3.1 ($\mathsf{Tree}^\diamond$) The set of modal trees $\mathsf{Tree}^\diamond$ is defined recursively to be the set of pairs $\langle \Delta; \Gamma \rangle$ where Δ is a finite list of propositional variables and Γ is a finite list of pairs $(\alpha, \mathtt{T})$, with $\alpha < \omega$ and $\mathtt{T} \in \mathsf{Tree}^\diamond$.

Elements of $\mathsf{Tree}^\diamond$ will be denoted by $\mathtt{T}$ and $\mathtt{S}$. Note that we employ distinct notations to enhance clarity on wether a pair is a modal tree: $\langle \cdot ; \cdot \rangle$ is used for a pair representing a modal tree, while $(\cdot, \cdot)$ denotes a pair comprising an ordinal and a modal tree. The *root* of a modal tree $\langle \Delta; \Gamma \rangle$ is Δ and its *children* is the list $[\mathtt{S} | (\alpha, \mathtt{S}) \in \Gamma]$. Note that, in general we write $[f(\alpha, \mathtt{S}) | (\alpha, \mathtt{S}) \in \Gamma]$ to denote the list $[f(\alpha_1, \mathtt{S}_1), ..., f(\alpha_n, \mathtt{S}_n)]$ for $\Gamma = [(\alpha_1, \mathtt{S}_1), ..., (\alpha_n, \mathtt{S}_n)]$, $n \geq 0$ and f a function of domain $\mathsf{Ord}^{<\omega} \times \mathsf{Tree}^\diamond$. For the sake of readability we write $\gamma \in \Gamma$ and $\mathtt{T} \in \Gamma$ to denote $\gamma \in [\alpha | (\alpha, \mathtt{S}) \in \Gamma]$ and $\mathtt{T} \in [\mathtt{S} | (\alpha, \mathtt{S}) \in \Gamma]$ respectively, since the context permits a clear distinction. A modal tree is called a *leaf* if it has an empty list of children. The *height* of a modal tree $\mathtt{T}$, denoted by $\mathsf{h}(\mathtt{T})$, is inductively defined as $\mathsf{h}(\langle \Delta; \varnothing \rangle) := 0$ and $\mathsf{h}(\langle \Delta; \Gamma \rangle) := \mathsf{max}[\mathsf{h}(\mathtt{S}) | \mathtt{S} \in \Gamma] + 1$.

The sum of modal trees is the tree obtained by concatenating their roots and children.

Definition 3.2 The *sum* of modal trees $\mathtt{T}_1 = \langle \Delta_1; \Gamma_1 \rangle$ and $\mathtt{T}_2 = \langle \Delta_2; \Gamma_2 \rangle$ is defined as $\mathtt{T}_1 + \mathtt{T}_2 := \langle \Delta_1 \frown \Delta_2; \Gamma_1 \frown \Gamma_2 \rangle$.

More generally, for Λ a finite list of modal trees, $\sum \Lambda$ is defined as $\langle \varnothing; \varnothing \rangle$ if $\Lambda = \varnothing$ and $\mathtt{T} + \sum \hat{\Lambda}$ if $\Lambda = \mathtt{T} \frown \hat{\Lambda}$. Note that $\mathsf{h}(\mathtt{T}_1 + \mathtt{T}_2) = \mathsf{max}\{\mathsf{h}(\mathtt{T}_1), \mathsf{h}(\mathtt{T}_2)\}$ for $\mathtt{T}_1, \mathtt{T}_2 \in \mathsf{Tree}^\diamond$.

A standard numbering of the nodes of the tree by strings of positive integers allows us to refer to positions in a tree. Specifically, the set of positions of a tree includes the root position, defined as the empty string, and the positions from its children which are obtained by appending the order of each child to its positions.

Definition 3.3 (Set of positions) The set of positions of a modal tree $\mathtt{T} = \langle \Delta; \Gamma \rangle$, denoted by $\mathsf{Pos}(\mathtt{T}) \in \mathcal{P}(\mathbb{N}^*)$, is inductively defined as

- $\mathsf{Pos}(\langle \Delta; \varnothing \rangle) := \{\epsilon\}$ for $\epsilon \in \mathbb{N}^*$ the empty string,
- $\mathsf{Pos}(\langle \Delta; \Gamma \rangle) := \{\epsilon\} \cup \bigcup_{i=1}^{n} \{i\mathbf{k} | \mathbf{k} \in \mathsf{Pos}(\mathtt{S}_i)\}$ for $\Gamma = [(\alpha_1, \mathtt{S}_1), ..., (\alpha_n, \mathtt{S}_n)]$.

Using the precise position apparatus we can define derived notions like, for example, that of subtree.

Definition 3.4 (Subtree) The subtree of $\mathtt{T} \in \mathsf{Tree}^\diamond$ at a position $\mathbf{k} \in \mathsf{Pos}(\mathtt{T})$, denoted by $\mathtt{T}|_{\mathbf{k}}$, is inductively defined over the length of $\mathbf{k}$ as

- $\mathtt{T}|_\epsilon := \mathtt{T}$,
- $\mathtt{T}|_{i\mathbf{r}} := \mathtt{S}_i|_{\mathbf{r}}$ for $1 \leq i \leq n$ such that $\mathtt{T} = \langle \Delta; [(\alpha_1, \mathtt{S}_1), ..., (\alpha_n, \mathtt{S}_n)] \rangle$.

We can now define subtree replacement based on the precise positioning provided.

Definition 3.5 (Replacement) Let $\mathtt{T}, \mathtt{S} \in \mathsf{Tree}^\diamond$ and $\mathbf{k} \in \mathsf{Pos}(\mathtt{T})$. The *tree obtained from* $\mathtt{T}$ *by replacing the subtree at position* $\mathbf{k}$ *by* $\mathtt{S}$, denoted by $\mathtt{T}[\mathtt{S}]_{\mathbf{k}}$, is inductively defined over the length of $\mathbf{k}$ as

- $\mathtt{T}[\mathtt{S}]_\epsilon := \mathtt{S}$,
- $\mathtt{T}[\mathtt{S}]_{i\mathbf{r}} := \langle \Delta; [(\alpha_1, \mathtt{S}_1), ..., (\alpha_i, \mathtt{S}_i[\mathtt{S}]_\mathbf{r}), ..., (\alpha_n, \mathtt{S}_n)] \rangle$ for $1 \leq i \leq n$ and $\mathtt{T} = \langle \Delta; [(\alpha_1, \mathtt{S}_1), ..., (\alpha_n, \mathtt{S}_n)] \rangle$.

Here below we present useful results on positioning and replacement in a modal tree.

Lemma 3.6 *Let* $\mathtt{T}, \mathtt{S}, \hat{\mathtt{S}} \in \mathsf{Tree}^\diamond$ *be modal trees. Then, for* $\mathbf{k}$ *and* $\mathbf{r}$ *belonging to the adequate position sets we have*

(i) $(\mathtt{T}|_\mathbf{k})|_\mathbf{r} = \mathtt{T}|_{\mathbf{kr}}$;

(ii) $\mathtt{T}[\mathtt{T}|_\mathbf{k}]_\mathbf{k} = \mathtt{T}$;

(iii) $(\mathtt{T}[\mathtt{S}]_\mathbf{k})|_\mathbf{k} = \mathtt{S}$;

(iv) $(\mathtt{T}[\hat{\mathtt{S}}]_\mathbf{k})[\mathtt{S}]_\mathbf{k} = \mathtt{T}[\mathtt{S}]_\mathbf{k}$ *(transitivity of replacement);*

(v) $(\mathtt{T}[\hat{\mathtt{S}}]_\mathbf{k})[\mathtt{S}]_{\mathbf{kr}} = \mathtt{T}[\hat{\mathtt{S}}[\mathtt{S}]_\mathbf{r}]_\mathbf{k}$.

Proof. We proceed by induction on the tree structure of $\mathtt{T}$ for each statement. If $\mathtt{T}$ is a leaf, the results follow easily since $\mathbf{k} = \epsilon$. Otherwise, we continue by cases on the length of $\mathbf{k}$. For ϵ the statements trivially hold. Finally consider $i\hat{\mathbf{k}}$ for $1 \leq i \leq n$ and $\hat{\mathbf{k}} \in \mathsf{Pos}(\mathtt{S}_i)$ such that $\mathtt{T} = \langle \Delta; [(\alpha_1, \mathtt{S}_1), ..., (\alpha_n, \mathtt{S}_n)] \rangle$. Then, by definition and each statement's inductive hypothesis for $\mathtt{S}_i$, we conclude

1. $(\mathtt{T}|_{i\hat{\mathbf{k}}})|_\mathbf{r} = (\mathtt{S}_i|_{\hat{\mathbf{k}}})|_\mathbf{r} = \mathtt{S}_i|_{\hat{\mathbf{k}}\mathbf{r}} = \mathtt{T}|_{i\hat{\mathbf{k}}\mathbf{r}}$;
2. $\mathtt{T}[\mathtt{T}|_{i\hat{\mathbf{k}}}]_{i\hat{\mathbf{k}}} = \mathtt{T}[\mathtt{S}_i|_{\hat{\mathbf{k}}}]_{i\hat{\mathbf{k}}} = \langle \Delta; [(\alpha_1, \mathtt{S}_1), ..., (\alpha_i, \mathtt{S}_i[\mathtt{S}_i|_{\hat{\mathbf{k}}}]_{\hat{\mathbf{k}}}), ..., (\alpha_n, \mathtt{S}_n)] \rangle = \mathtt{T}$;
3. $(\mathtt{T}[\mathtt{S}]_{i\hat{\mathbf{k}}})|_{i\hat{\mathbf{k}}} = (\langle \Delta; [(\alpha_1, \mathtt{S}_1), ..., (\alpha_i, \mathtt{S}_i[\mathtt{S}]_{\hat{\mathbf{k}}}), ..., (\alpha_n, \mathtt{S}_n)] \rangle)|_{i\hat{\mathbf{k}}} = (\mathtt{S}_i[\mathtt{S}]_{\hat{\mathbf{k}}})|_{\hat{\mathbf{k}}} = \mathtt{S}$;
4. $$\begin{aligned}(\mathtt{T}[\hat{\mathtt{S}}]_{i\hat{\mathbf{k}}})[\mathtt{S}]_{i\hat{\mathbf{k}}} &= (\langle \Delta; [(\alpha_1, \mathtt{S}_1), ..., (\alpha_i, \mathtt{S}_i[\hat{\mathtt{S}}]_{\hat{\mathbf{k}}}), ..., (\alpha_n, \mathtt{S}_n)] \rangle)[\mathtt{S}]_{i\hat{\mathbf{k}}} \\ &= \langle \Delta; [(\alpha_1, \mathtt{S}_1), ..., (\alpha_i, (\mathtt{S}_i[\hat{\mathtt{S}}]_{\hat{\mathbf{k}}})[\mathtt{S}]_{\hat{\mathbf{k}}}), ..., (\alpha_n, \mathtt{S}_n)] \rangle \\ &= \langle \Delta; [(\alpha_1, \mathtt{S}_1), ..., (\alpha_i, \mathtt{S}_i[\mathtt{S}]_{\hat{\mathbf{k}}}), ..., (\alpha_n, \mathtt{S}_n)] \rangle = \mathtt{T}[\mathtt{S}]_{i\hat{\mathbf{k}}};\end{aligned}$$
5. $$\begin{aligned}(\mathtt{T}[\hat{\mathtt{S}}]_{i\hat{\mathbf{k}}})[\mathtt{S}]_{i\hat{\mathbf{k}}\mathbf{r}} &= (\langle \Delta; [(\alpha_1, \mathtt{S}_1), ..., (\alpha_i, \mathtt{S}_i[\hat{\mathtt{S}}]_{\hat{\mathbf{k}}}), ..., (\alpha_n, \mathtt{S}_n)] \rangle)[\mathtt{S}]_{i\hat{\mathbf{k}}\mathbf{r}} \\ &= \langle \Delta; [(\alpha_1, \mathtt{S}_1), ..., (\alpha_i, (\mathtt{S}_i[\hat{\mathtt{S}}]_{\hat{\mathbf{k}}})[\mathtt{S}]_{\hat{\mathbf{k}}\mathbf{r}}), ..., (\alpha_n, \mathtt{S}_n)] \rangle \\ &= \langle \Delta; [(\alpha_1, \mathtt{S}_1), ..., (\alpha_i, \mathtt{S}_i[\hat{\mathtt{S}}[\mathtt{S}]_\mathbf{r}]_{\hat{\mathbf{k}}}), ..., (\alpha_n, \mathtt{S}_n)] \rangle = \mathtt{T}[\hat{\mathtt{S}}[\mathtt{S}]_\mathbf{r}]_{i\hat{\mathbf{k}}}.\end{aligned}$$ □

4 Relating formulas and modal trees

Our tree rewriting system is based on a correspondence between the language of $\mathcal{L}^+$ and $\mathsf{Tree}^\diamond$. Thereby we can ensure that transformations within the structure of modal trees accurately simulate derivations in a considered proof system. For this purpose, we introduce the tree embedding operator $\mathcal{T}$ inductively defined over the set of strictly positive formulas mapping them to modal trees. This definition is inspired by the canonical tree representation of strictly positive formulae presented by Beklemishev (see [7]) as a combinatiorial tool for proving the polytime decidability of **RC**. Additionally, we define $\mathcal{F}$ mapping modal trees to formulas. Ultimately, we prove that composition $\mathcal{T} \circ \mathcal{F}$ serves as the identity over $\mathsf{Tree}^\diamond$, while $\mathcal{F} \circ \mathcal{T}$ acts as the identity on $\mathcal{L}^+$ modulo equality for **K+**.

Definition 4.1 ($\mathcal{T}$) The modal tree embedding is the function $\mathcal{T} : \mathcal{L}^+ \longrightarrow \mathsf{Tree}^\diamond$ inductively defined over the structure of strictly positive formulae as

- $\mathcal{T}(\top) := \langle \varnothing; \varnothing \rangle$,
- $\mathcal{T}(p) := \langle [p]; \varnothing \rangle$ for $p \in \mathsf{Prop}$,
- $\mathcal{T}(\langle \alpha \rangle \varphi) := \langle \varnothing; [(\alpha, \mathcal{T}(\varphi))] \rangle$ for $\varphi \in \mathcal{L}^+$,
- $\mathcal{T}(\varphi \wedge \psi) := \mathcal{T}(\varphi) + \mathcal{T}(\psi)$ for $\varphi, \psi \in \mathcal{L}^+$.

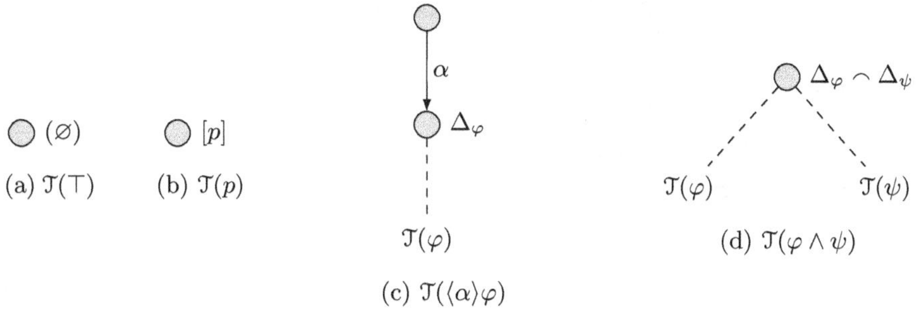

Fig. 1. Modal tree embedding

The modal depth of a formula coincides with the height of the modal tree which it is mapped to.

Lemma 4.2 $\mathsf{h}(\mathcal{T}(\varphi)) = \mathsf{md}(\varphi)$ *for* $\varphi \in \mathcal{L}^+$.

Proof. By an easy induction on the structure of φ. □

We also introduce a corresponding embedding in the opposite direction.

Definition 4.3 ($\mathcal{F}$) The strictly positive formulae embedding is the function $\mathcal{F} : \mathsf{Tree}^\diamond \longrightarrow \mathcal{L}^+$ defined as

$$\mathcal{F}(\langle \Delta; \Gamma \rangle) := \bigwedge \Delta \wedge \bigwedge [\langle \alpha \rangle \mathcal{F}(\mathtt{S}) | (\alpha, \mathtt{S}) \in \Gamma].$$

For the sake of readability, we will simply write $\Diamond\Gamma$ to denote $[\langle \alpha \rangle \mathcal{F}(\mathtt{S}) | (\alpha, \mathtt{S}) \in \Gamma]$.

We conclude this section by providing a relation among strictly positive formulas and modal trees through the composition of the embeddings.

Proposition 4.4 (Embedding composition) $\mathcal{T} \circ \mathcal{F} = id_{\mathsf{Tree}^\diamond}$ *and* $\mathcal{F} \circ \mathcal{T} = id_{\mathcal{L}^+/\equiv_{\mathbf{K}+}}$.

Proof. We firstly prove $\mathcal{T} \circ \mathcal{F}(\mathtt{T}) = \mathtt{T}$ for $\mathtt{T} \in \mathsf{Tree}^{\diamond}$ by induction on the modal tree structure. The leaf case follows by definition since $\mathcal{T}(\mathcal{F}(\langle\Delta;\varnothing\rangle)) = \mathcal{T}(\bigwedge\Delta \wedge \top) = \langle\Delta;\varnothing\rangle$. Otherwise, assuming $\mathcal{T}(\mathcal{F}(\mathtt{S})) = \mathtt{S}$ for every $\mathtt{S} \in \Gamma$, we conclude

$$\begin{aligned}\mathcal{T}(\mathcal{F}(\langle\Delta;\Gamma\rangle)) &= \mathcal{T}(\bigwedge\Delta \wedge \bigwedge\diamond\Gamma) = \mathcal{T}(\bigwedge\Delta) + \mathcal{T}(\bigwedge\diamond\Gamma)\\ &= \langle\Delta;\varnothing\rangle + \sum[\mathcal{T}(\langle\alpha\rangle\mathcal{F}(\mathtt{S}))|(\alpha,\mathtt{S}) \in \Gamma]\\ &= \langle\Delta;\varnothing\rangle + \sum[\langle\varnothing;[(\alpha,\mathcal{T}(\mathcal{F}(\mathtt{S})))]\rangle|(\alpha,\mathtt{S}) \in \Gamma]\\ &= \langle\Delta;\varnothing\rangle + \sum[\langle\varnothing;[(\alpha,\mathtt{S})]\rangle|(\alpha,\mathtt{S}) \in \Gamma] = \langle\Delta;\Gamma\rangle.\end{aligned}$$

Finally, we prove $\mathcal{F} \circ \mathcal{T}(\varphi) \equiv_{\mathbf{K}^+} \varphi$ for $\varphi \in \mathcal{L}^+$ by induction on the structure of φ.

- $\mathcal{F} \circ \mathcal{T}(\top) = \top \wedge \top \equiv_{\mathbf{K}^+} \top$; $\mathcal{F} \circ \mathcal{T}(p) = (p \wedge \top) \wedge \top \equiv_{\mathbf{K}^+} p$.
- Let us assume $\mathcal{F} \circ \mathcal{T}(\psi) \equiv_{\mathbf{K}^+} \psi$. Then,

$$\mathcal{F} \circ \mathcal{T}(\langle\alpha\rangle\psi) = \mathcal{F}(\langle\varnothing;[(\alpha,\mathcal{T}(\psi))]\rangle) = \top \wedge \langle\alpha\rangle\mathcal{F} \circ \mathcal{T}(\psi) \wedge \top \equiv_{\mathbf{K}^+} \langle\alpha\rangle\psi.$$

- Let $\mathcal{T}(\psi) = \langle\Delta_\psi;\Gamma_\psi\rangle$ and $\mathcal{T}(\phi) = \langle\Delta_\phi;\Gamma_\phi\rangle$. Assuming $\mathcal{F} \circ \mathcal{T}(\psi) \equiv_{\mathbf{K}^+} \psi$ and $\mathcal{F} \circ \mathcal{T}(\phi) \equiv_{\mathbf{K}^+} \phi$,

$$\mathcal{F} \circ \mathcal{T}(\psi \wedge \phi) \equiv_{\mathbf{K}^+} \bigwedge\Delta_\psi \wedge \bigwedge\diamond\Gamma_\psi \wedge \bigwedge\Delta_\phi \wedge \bigwedge\diamond\Gamma_\phi \equiv_{\mathbf{K}^+} \psi \wedge \phi.$$

□

5 The tree rewriting system for RC

We introduce the tree rewriting system for **RC**, an abstract rewriting system for $\mathsf{Tree}^{\diamond}$ which will be proven adequate and complete w.r.t. **RC** in the next section. Additionally, we present useful results for the rewriting, along with the Inside Rewriting Property which involves transforming a subtree while preserving the remaining parts invariant.

An *abstract rewriting system* is a pair $(A, \{\hookrightarrow^\mu\}_{\mu\in I})$ consisting of a set A and a sequence of binary relations $\hookrightarrow^\mu$ on A, also called rewriting rules (r.r.). Instead of $(a,b) \in \hookrightarrow^\mu$ we write $a \hookrightarrow^\mu b$ and call b *is obtained by applying* μ *to* a or b *is obtained by performing one step* μ *to* a. The composition of rewriting rules μ_1 and μ_2 is written $a \hookrightarrow^{\mu_1} \circ \hookrightarrow^{\mu_2} b$ and denotes that there is $\hat{a} \in A$ such that $a \hookrightarrow^{\mu_1} \hat{a} \hookrightarrow^{\mu_2} b$.

In particular, the rewriting rules of the tree rewriting system for **RC** transform a tree by replacing a subtree with a predetermined tree. The rules are classified into five kinds according to the performed transformation: atomic, structural, replicative, decreasing, and modal rewriting rules. Atomic rules duplicate or eliminate propositional variables in the lists labeling the nodes; the structural rule permutes the order of a node's children; the replicative rule duplicates a child of a node; decreasing rules either eliminate a child or remove a node and its children under certain conditions; and modal rules either decrease the label of an edge or apply a transformation simulating the J axiom of **RC**.

Atomic r.r.	$\mathtt{T} \hookrightarrow^{\rho^+} \mathtt{T}[\langle \Delta^{+i}; \Gamma \rangle]_{\mathbf{k}}$	ρ^+-rule
$0 < i \leq \lvert\Delta\rvert$	$\mathtt{T} \hookrightarrow^{\rho^-} \mathtt{T}[\langle \Delta^{-i}; \Gamma \rangle]_{\mathbf{k}}$	ρ^--rule
Structural r.r. $0 < i, j \leq \lvert\Gamma\rvert$, $i \neq j$	$\mathtt{T} \hookrightarrow^{\sigma} \mathtt{T}[\langle \Delta; \Gamma^{i \leftrightarrow j} \rangle]_{\mathbf{k}}$	σ-rule
Replicative r.r. $0 < i \leq \lvert\Gamma\rvert$	$\mathtt{T} \hookrightarrow^{\pi^+} \mathtt{T}[\langle \Delta; \Gamma^{+i} \rangle]_{\mathbf{k}}$	π^+-rule
Decreasing r.r.	$\mathtt{T} \hookrightarrow^{\pi^-} \mathtt{T}[\langle \Delta; \Gamma^{-i} \rangle]_{\mathbf{k}}$	π^--rule
$0 < i \leq \lvert\Gamma\rvert$ $0 < j \leq \lvert\tilde{\Gamma}\rvert$	$\mathtt{T} \hookrightarrow^{4} \mathtt{T}[\langle \Delta; \Gamma[(\beta, \mathtt{S})]_i \rangle]_{\mathbf{k}}$ for $\#_i\Gamma = (\beta, \langle \tilde{\Delta}; \tilde{\Gamma} \rangle)$ and $\#_j\tilde{\Gamma} = (\beta, \mathtt{S})$	4-rule
Modal r.r. $\alpha > \beta$	$\mathtt{T} \hookrightarrow^{\lambda} \mathtt{T}[\langle \Delta; \Gamma[(\beta, \mathtt{S})]_i \rangle]_{\mathbf{k}}$ for $\#_i\Gamma = (\alpha, \mathtt{S})$	λ-rule
$0 < i, j \leq \lvert\Gamma\rvert$ $i \neq j$	$\mathtt{T} \hookrightarrow^{\mathsf{J}} \mathtt{T}[\langle \Delta; (\Gamma[(\alpha, \langle \tilde{\Delta}; \tilde{\Gamma} \frown (\beta, \mathtt{S}) \rangle)]_i)^{-j} \rangle]_{\mathbf{k}}$ for $\#_i\Gamma = (\alpha, \langle \tilde{\Delta}; \tilde{\Gamma} \rangle)$ and $\#_j\Gamma = (\beta, \mathtt{S})$	J-rule

Table 1
Rewriting rules of $\mathcal{R}$ for $\mathtt{T} \in \mathsf{Tree}^{\diamond}$, $\mathbf{k} \in \mathsf{Pos}(\mathtt{T})$ and $\mathtt{T}|_{\mathbf{k}} = \langle \Delta; \Gamma \rangle$.

To define the rewriting rules, we introduce the following notation. Let $\mathtt{T} = \langle \Delta; \Gamma \rangle$ be a modal tree, $0 < i, j \leq \lvert\Gamma\rvert$ and $n \leq \lvert\Delta\rvert$ such that $\Gamma = [(\alpha_1, \mathtt{S}_1), ..., (\alpha_m, \mathtt{S}_m)]$. The ith element of Γ is denoted by $\#_i\Gamma$. The list obtained by erasing the ith element of Γ, i.e. $[(\alpha_1, \mathtt{S}_1), ..., (\alpha_{i-1}, \mathtt{S}_{i-1}), (\alpha_{i+1}, \mathtt{S}_{i+1}), ..., (\alpha_m, \mathtt{S}_m)]$, is denoted by Γ^{-i}. The list obtained by duplicating the ith element of Γ and placing it at the beginning, i.e. $(\alpha_i, \mathtt{S}_i) \frown \Gamma$, is denoted by Γ^{+i}. Analogously, the list obtained by erasing the nth element of Δ and the list obtained by duplicating the nth element of Δ and placing it at the beginning are denoted by Δ^{-n} and Δ^{+n}, respectively. The *list obtained from* Γ *by replacing its ith element* by $(\alpha, \mathtt{S})$ is defined by $\Gamma[(\alpha, \mathtt{S})]_i := [(\alpha_1, \mathtt{S}_1), ..., (\alpha_{i-1}, \mathtt{S}_{i-1}), (\alpha, \mathtt{S}), (\alpha_{i+1}, \mathtt{S}_{i+1}), ..., (\alpha_m, \mathtt{S}_m)]$. Note that we use the same notation for replacement in a list of pairs and replacement in a modal tree since the context allows for a clear distinction. Finally, the list obtained by interchanging the ith element with the jth element, i.e. $(\Gamma[\#_i\Gamma]_j)[\#_j\Gamma]_i$, is denoted by $\Gamma^{i \leftrightarrow j}$.

We can now present the tree rewriting system for Reflection Calculus.

Definition 5.1 (TRC) The *tree rewriting system for* **RC**, denoted by TRC, is the abstract rewriting system $(\mathsf{Tree}^{\diamond}, \{\hookrightarrow^{\mu}\}_{\mu \in \mathcal{R}})$ for $\mathcal{R} = \{\rho^+, \rho^-, \sigma, \pi^+, \pi^-, 4, \lambda, \mathsf{J}\}$. The rewriting rules of $\mathcal{R}$ are defined in Table 1.

Due to the transformations they induce, the rules are named as follows:

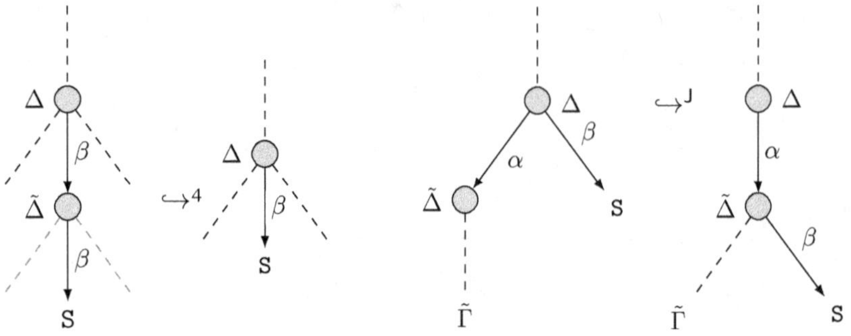

Fig. 2. 4-rule and J-rule.

the ρ^+-rule is called *atom duplication*, the ρ^--rule is called *atom elimination*, the σ-rule is called *child permutation*, the π^+-rule is called as *child duplication*, the π^--rule is called *child elimination*, the 4-rule is called *transitivity*, and the λ-rule is called *monotonicity*.

More generally, the tree rewriting relation is the union of the rewriting rules.

Definition 5.2 ($\hookrightarrow$) The *tree rewriting relation* $\hookrightarrow$ is defined as

$$\hookrightarrow := \hookrightarrow^{\rho^+} \cup \hookrightarrow^{\rho^-} \cup \hookrightarrow^{\sigma} \cup \hookrightarrow^{\pi^+} \cup \hookrightarrow^{\pi^-} \cup \hookrightarrow^{4} \cup \hookrightarrow^{\lambda} \cup \hookrightarrow^{\mathsf{J}} .$$

We say that *the step in* $\mathtt{T} \hookrightarrow \mathtt{T}'$ *has been performed at position* $\mathbf{k}$ if the applied rule replaces the subtree at position $\mathbf{k} \in \mathsf{Pos}(\mathtt{T})$. We say that $\mathtt{T}$ *rewrites to* $\mathtt{T}'$, denoted by $\mathtt{T} \hookrightarrow^* \mathtt{T}'$, if $\mathtt{T}'$ is the result of applying zero, one or several rewriting rules of $\mathcal{R}$ to $\mathtt{T}$. In other words, $\hookrightarrow^*$ denotes the reflexive transitive closure of $\hookrightarrow$. The trees $\mathtt{T}$ *and* $\mathtt{T}'$ *are* TRC*-equivalent*, denoted by $\mathtt{T} \overset{*}{\leftrightarrow} \mathtt{T}'$, if $\mathtt{T} \hookrightarrow^* \mathtt{T}'$ and $\mathtt{T}' \hookrightarrow^* \mathtt{T}$. If $\mathtt{T}$ rewrites to $\mathtt{T}'$ by applying the rewriting rule μ zero, one or several times, we write $\mathtt{T} \hookrightarrow^{\mu *} \mathtt{T}'$. For Ω a list of rewriting rules, we define $\mathtt{T} \hookrightarrow^{\Omega} \mathtt{S}$ inductively as $\mathtt{T} \hookrightarrow^* \mathtt{T}$ by applying no rewriting rules if $\Omega = \varnothing$, and $\mathtt{T} \hookrightarrow^{\mu} \circ \hookrightarrow^{\hat{\Omega}} \mathtt{S}$ for $\Omega = \mu \frown \hat{\Omega}$. Likewise, for Ω_1 and Ω_2 lists of rewriting rules, $\mathtt{T} \hookrightarrow^{\Omega_1} \circ \hookrightarrow^{\Omega_2} \mathtt{S}$ denotes that there is $\hat{\mathtt{S}}$ such that $\mathtt{T} \hookrightarrow^{\Omega_1} \hat{\mathtt{S}} \hookrightarrow^{\Omega_2} \mathtt{S}$.

Modal trees with permuted lists labeling the nodes are TRC-equivalent.

Lemma 5.3 $\langle \Delta_1 \frown \Delta_2; \Gamma \rangle \overset{*}{\leftrightarrow} \langle \Delta_2 \frown \Delta_1; \Gamma \rangle$.

Proof. It suffices to show $\langle \Delta_1 \frown \Delta_2; \Gamma \rangle \hookrightarrow^* \langle \Delta_2 \frown \Delta_1; \Gamma \rangle$ by induction on the length of Δ_2. If Δ_2 is empty, the result trivially holds. Assuming $\langle \Delta \frown \Delta_2; \Gamma \rangle \hookrightarrow^* \langle \Delta_2 \frown \Delta; \Gamma \rangle$ for any list of propositional variables Δ, using atom duplication and atom elimination rewriting rules we conclude

$$\begin{aligned} \langle \Delta_1 \frown (p \frown \Delta_2); \Gamma \rangle = \langle (\Delta_1 \frown p) \frown \Delta_2; \Gamma \rangle \hookrightarrow^* \langle \Delta_2 \frown (\Delta_1 \frown p); \Gamma \rangle \\ \hookrightarrow^{\rho^+} \langle p \frown \Delta_2 \frown \Delta_1 \frown p; \Gamma \rangle \hookrightarrow^{\rho^-} \langle (p \frown \Delta_2) \frown \Delta_1; \Gamma \rangle. \end{aligned}$$

□

Here are some useful results on rewriting a sum of modal trees.

Lemma 5.4 *Let* $\mathtt{T}_1, \mathtt{T}_2, \mathtt{T}_3, \mathtt{S}_1, \mathtt{S}_2 \in \mathsf{Tree}^\diamond$. *Then,*

(i) $\mathtt{T}_1 \overset{*}{\leftrightarrow} \mathtt{T}_1 + \mathtt{T}_1$;

(ii) $\mathtt{T}_1 + \mathtt{T}_2 \overset{*}{\leftrightarrow} \mathtt{T}_2 + \mathtt{T}_1$;

(iii) $\mathtt{T}_1 + \mathtt{T}_2 \hookrightarrow^* \mathtt{T}_1$ *and* $\mathtt{T}_1 + \mathtt{T}_2 \hookrightarrow^* \mathtt{T}_2$;

(iv) *If* $\mathtt{T}_1 \hookrightarrow^* \mathtt{S}_1$, *then* $\mathtt{T}_1 + \mathtt{T}_2 \hookrightarrow^* \mathtt{S}_1 + \mathtt{T}_2$;

(v) *If* $\mathtt{T}_1 \hookrightarrow^* \mathtt{T}_2$ *and* $\mathtt{T}_1 \hookrightarrow^* \mathtt{T}_3$ *then* $\mathtt{T}_1 \hookrightarrow^* \mathtt{T}_2 + \mathtt{T}_3$;

(vi) *If* $\mathtt{S}_1 \hookrightarrow^* \mathtt{T}_1$ *and* $\mathtt{S}_2 \hookrightarrow^* \mathtt{T}_2$, *then* $\mathtt{S}_1 + \mathtt{S}_2 \hookrightarrow^* \mathtt{T}_1 + \mathtt{T}_2$.

Proof. The implication from left to right of the first statement holds by atom and child duplication, and the inverse implication by atom and child elimination. The second result holds by Lemma 5.3 and child permutation. The third follows by atom and child elimination. The fourth result holds by induction on the number of rewriting steps performed in $\mathtt{T}_1 \hookrightarrow^* \mathtt{S}_1$ and by cases on the rewriting rules. The fifth statement holds by the fourth result using the statements one and two:

$$\mathtt{T}_1 \hookrightarrow^* \mathtt{T}_1 + \mathtt{T}_1 \hookrightarrow^* \mathtt{T}_2 + \mathtt{T}_1 \hookrightarrow^* \mathtt{T}_1 + \mathtt{T}_2 \hookrightarrow^* \mathtt{T}_3 + \mathtt{T}_2 \hookrightarrow^* \mathtt{T}_2 + \mathtt{T}_3.$$

Finally, by the fifth statement it suffices to show $\mathtt{S}_1 + \mathtt{S}_2 \hookrightarrow^* \mathtt{T}_1$ and $\mathtt{S}_1 + \mathtt{S}_2 \hookrightarrow^* \mathtt{T}_2$ to prove the sixth result, which follow by the third statement and the hypotheses. □

The following results state that transformations in subtrees can be extended to the entire tree, allowing for systematic and consistent modifications throughout the tree. In consequence, we can effectively manipulate and modify complex tree structures while leaving the other parts untouched.

Proposition 5.5 (Inside Rewriting Property) *If* $\mathtt{S} \hookrightarrow^* \mathtt{S}'$, *then* $\mathtt{T}[\mathtt{S}]_{\mathbf{k}} \hookrightarrow^* \mathtt{T}[\mathtt{S}']_{\mathbf{k}}$.

Proof. By induction on the number of rewriting steps that are performed in $\mathtt{S} \hookrightarrow^* \mathtt{S}'$. If no rewriting step is performed, the result is trivially satisfied. Now assume $\mathtt{T}[\mathtt{S}]_{\mathbf{k}} \hookrightarrow^* \mathtt{T}[\hat{\mathtt{S}}]_{\mathbf{k}}$ for $\mathtt{S} \hookrightarrow^* \hat{\mathtt{S}}$ by performing n rewriting steps. Moreover, let $\mathtt{S} \hookrightarrow^* \hat{\mathtt{S}} \hookrightarrow^\mu \hat{\mathtt{S}}[\tilde{\mathtt{S}}]_{\mathbf{r}}$ for $\tilde{\mathtt{S}} \in \mathsf{Tree}^\diamond$ and $\mathbf{r} \in \mathsf{Pos}(\hat{\mathtt{S}})$ according to the rewriting rule μ. Since $\mathtt{T}[\mathtt{S}]_{\mathbf{k}} \hookrightarrow^* \mathtt{T}[\hat{\mathtt{S}}]_{\mathbf{k}}$ by the inductive hypothesis, it suffices to show $\mathtt{T}[\hat{\mathtt{S}}]_{\mathbf{k}} \hookrightarrow^\mu \mathtt{T}[\hat{\mathtt{S}}[\tilde{\mathtt{S}}]_{\mathbf{r}}]_{\mathbf{k}} = (\mathtt{T}[\hat{\mathtt{S}}]_{\mathbf{k}})[\tilde{\mathtt{S}}]_{\mathbf{kr}}$ by Lemma 3.6. We proceed by induction on the tree structure of $\mathtt{T}$. For $\mathtt{T}$ a leaf, by the hypothesis,

$$\mathtt{T}[\hat{\mathtt{S}}]_\epsilon = \hat{\mathtt{S}} \hookrightarrow^\mu \hat{\mathtt{S}}[\tilde{\mathtt{S}}]_{\mathbf{r}} = (\mathtt{T}[\hat{\mathtt{S}}]_\epsilon)[\tilde{\mathtt{S}}]_{\epsilon\mathbf{r}}.$$

Let $\mathtt{T} = \langle \Delta; [(\alpha_1, \mathtt{S}_1), ..., (\alpha_m, \mathtt{S}_m)] \rangle$ such that $\mathtt{S}_i[\hat{\mathtt{S}}]_{\mathbf{l}} \hookrightarrow^\mu (\mathtt{S}_i[\hat{\mathtt{S}}]_{\mathbf{l}})[\tilde{\mathtt{S}}]_{\mathbf{lr}}$ for $\mathbf{l} \in \mathsf{Pos}(\mathtt{S}_i)$ and $1 \leq i \leq m$. We continue by cases on the length of $\mathbf{k}$. For $\mathbf{k} = \epsilon$, the result is trivially satisfied. For $\mathbf{k} = i\hat{\mathbf{k}}$ such that $\hat{\mathbf{k}} \in \mathsf{Pos}(\mathtt{S}_i)$ and $1 \leq i \leq n$, we conclude by performing μ at position $\mathbf{kr}$ using the inductive hypothesis for $\mathtt{S}_i$ as follows,

$$\begin{aligned} \mathtt{T}[\hat{\mathtt{S}}]_{i\hat{\mathbf{k}}} &= \langle \Delta; [(\alpha_1, \mathtt{S}_1), ..., (\alpha_i, \mathtt{S}_i[\hat{\mathtt{S}}]_{\hat{\mathbf{k}}}), ..., (\alpha_m, \mathtt{S}_m)] \rangle \\ &\hookrightarrow^\mu \langle \Delta; [(\alpha_1, \mathtt{S}_1), ..., (\alpha_i, (\mathtt{S}_i[\hat{\mathtt{S}}]_{\hat{\mathbf{k}}})[\tilde{\mathtt{S}}]_{\hat{\mathbf{k}}\mathbf{r}}), ..., (\alpha_m, \mathtt{S}_m)] \rangle = (\mathtt{T}[\hat{\mathtt{S}}]_{i\hat{\mathbf{k}}})[\tilde{\mathtt{S}}]_{i\hat{\mathbf{k}}\mathbf{r}}. \end{aligned}$$

□

Corollary 5.6 *If* $\mathtt{T}|_{\mathbf{k}} \hookrightarrow^* \mathtt{S}$, *then* $\mathtt{T} \hookrightarrow^* \mathtt{T}[\mathtt{S}]_{\mathbf{k}}$ *for* $\mathbf{k} \in \mathsf{Pos}(\mathtt{T})$.

Proof. By Proposition 5.5 since $\mathtt{T} = \mathtt{T}[\mathtt{T}|_{\mathbf{k}}]_{\mathbf{k}}$ (Lemma 3.6). □

6 Adequacy and completeness

We aim to show how the tree rewriting system TRC faithfully simulates logical derivations in **RC** thanks to the embeddings defined in Section 4. Thereby, adequacy and completeness theorems are presented as key results that underscore the efficacy of tree rewriting systems in relating logical inference and structural transformation.

Firstly, we show adequacy by proving that if a rewriting step is performed, the sequent of formulas which the trees are mapped to is provable in **RC**.

Proposition 6.1 *If* $\mathtt{T} \hookrightarrow \mathtt{T}'$, *then* $\mathcal{F}(\mathtt{T}) \vdash_{\mathbf{RC}} \mathcal{F}(\mathtt{T}')$ *for* $\mathtt{T}, \mathtt{T}' \in \mathsf{Tree}^{\diamond}$.

Proof. By induction on the tree structure of $\mathtt{T}$. For $\mathtt{T}$ a leaf the result follows easily by cases on the performed rewriting rule. Now consider $\mathtt{T} = \langle \Delta; \Gamma \rangle$ for $\Gamma = [(\alpha_1, \mathtt{S}_1), ..., (\alpha_m, \mathtt{S}_m)]$ such that $\mathcal{F}(\mathtt{S}_i) \vdash_{\mathbf{RC}} \mathcal{F}(\mathtt{S}')$ if $\mathtt{S}_i \hookrightarrow \mathtt{S}'$ for $1 \leq i \leq m$. Assuming $\langle \Delta; \Gamma \rangle \hookrightarrow^{\mu} \mathtt{T}'$ for μ a rewriting rule, we show $\mathcal{F}(\mathtt{T}) \vdash_{\mathbf{RC}} \mathcal{F}(\mathtt{T}')$ by cases on the length of the position at which μ is performed.

First consider a rewriting at a position $i\mathbf{k} \in \mathsf{Pos}(\mathtt{T})$. By definition, $\mathtt{T}'$ is of the form $\langle \Delta; [(\alpha_1, \mathtt{S}_1), ..., (\alpha_i, \mathtt{S}'), ..., (\alpha_m, \mathtt{S}_m)] \rangle$ for $\mathtt{S}_i \hookrightarrow^{\mu} \mathtt{S}'$ by rewriting at position $\mathbf{k}$. Hence, by the inductive hypothesis, $\mathcal{F}(\mathtt{S}_i) \vdash_{\mathbf{RC}} \mathcal{F}(\mathtt{S}')$. Thus,

$$\begin{aligned}\mathcal{F}(\mathtt{T}) &\equiv_{\mathbf{K}^+} \bigwedge \Delta \wedge \langle \alpha_i \rangle \mathcal{F}(\mathtt{S}_i) \wedge \bigwedge \diamond(\Gamma^{-i}) \\ &\vdash_{\mathbf{RC}} \bigwedge \Delta \wedge \langle \alpha_i \rangle \mathcal{F}(\mathtt{S}') \wedge \bigwedge \diamond(\Gamma^{-i}) \equiv_{\mathbf{K}^+} \mathcal{F}(\mathtt{T}').\end{aligned}$$

For rewriting performed at position ϵ, the proof concludes easily by cases on μ. Let see in some detail the cases of transitivity and J rewriting rules.

4-rule: Consider $\mathtt{T} \hookrightarrow^4 \langle \Delta; \Gamma[(\beta, \mathtt{S})]_i \rangle$ such that $\#_i \Gamma = (\beta, \langle \tilde{\Delta}; \tilde{\Gamma} \rangle)$ for $0 < i \leq |\Gamma|$ and $\#_j \tilde{\Gamma} = (\beta, \mathtt{S})$ for $0 < j \leq |\tilde{\Gamma}|$. By Lemma 2.2 and transitivity rule for **RC** we conclude,

$$\begin{aligned}\mathcal{F}(\mathtt{T}) &\equiv_{\mathbf{K}^+} \bigwedge \Delta \wedge \langle \beta \rangle \mathcal{F}(\langle \tilde{\Delta}; \tilde{\Gamma} \rangle) \wedge \bigwedge \diamond(\Gamma^{-i}) \\ &\equiv_{\mathbf{K}^+} \bigwedge \Delta \wedge \langle \beta \rangle (\bigwedge \tilde{\Delta} \wedge \langle \beta \rangle \mathcal{F}(\mathtt{S}) \wedge \bigwedge \diamond(\tilde{\Gamma}^{-j})) \wedge \bigwedge \diamond(\Gamma^{-i}) \\ &\vdash_{\mathbf{K}^+} \bigwedge \Delta \wedge \langle \beta \rangle \langle \beta \rangle \mathcal{F}(\mathtt{S}) \wedge \bigwedge \diamond(\Gamma^{-i}) \\ &\vdash_{\mathbf{RC}} \bigwedge \Delta \wedge \langle \beta \rangle \mathcal{F}(\mathtt{S}) \wedge \bigwedge \diamond(\Gamma^{-i}) \\ &\equiv_{\mathbf{K}^+} \mathcal{F}(\langle \Delta; \Gamma[(\beta, \mathtt{S})]_i \rangle).\end{aligned}$$

J-rule: Consider $\mathtt{T} \hookrightarrow^{\mathsf{J}} \langle \Delta; (\Gamma[(\alpha, \langle \tilde{\Delta}; \tilde{\Gamma} \frown (\beta, \mathtt{S}) \rangle)]_i)^{-j} \rangle$ such that $\#_i \Gamma = (\alpha, \langle \tilde{\Delta}; \tilde{\Gamma} \rangle)$ and $\#_j \Gamma = (\beta, \mathtt{S})$ for $0 < i, j, \leq |\Gamma|$ satisfying $i \neq j$. Let $i < j$

without loss of generality. By the J rule for **RC** we conclude,

$$\begin{aligned}\mathcal{F}(\mathtt{T}) &\equiv_{\mathbf{K}^+} \bigwedge \Delta \wedge \langle\alpha\rangle\mathcal{F}(\langle\tilde{\Delta};\tilde{\Gamma}\rangle) \wedge \langle\beta\rangle\mathcal{F}(\mathtt{S}) \wedge \bigwedge \Diamond((\Gamma^{-j})^{-i})\\ &\vdash_{\mathbf{RC}} \bigwedge \Delta \wedge \langle\alpha\rangle(\mathcal{F}(\langle\tilde{\Delta};\tilde{\Gamma}\rangle) \wedge \langle\beta\rangle\mathcal{F}(\mathtt{S})) \wedge \bigwedge \Diamond((\Gamma^{-j})^{-i})\\ &\equiv_{\mathbf{K}^+} \bigwedge \Delta \wedge \langle\alpha\rangle(\mathcal{F}(\langle\tilde{\Delta};\tilde{\Gamma} \frown (\beta,\mathtt{S})\rangle)) \wedge \bigwedge \Diamond((\Gamma^{-j})^{-i})\\ &\equiv_{\mathbf{K}^+} \mathcal{F}(\langle\Delta;(\Gamma[(\alpha,\langle\tilde{\Delta};\tilde{\Gamma} \frown (\beta,\mathtt{S})\rangle)]_i)^{-j}\rangle).\end{aligned}$$

□

Theorem 6.2 (Adequacy) *If* $\mathtt{T} \hookrightarrow^* \mathtt{T}'$, *then* $\mathcal{F}(\mathtt{T}) \vdash_{\mathbf{RC}} \mathcal{F}(\mathtt{T}')$ *for* $\mathtt{T},\mathtt{T}' \in \mathsf{Tree}^{\Diamond}$.

Proof. By an easy induction on the number of rewriting steps that are performed using Proposition 6.1. □

We conclude this section by showing that TRC is complete with respect to **RC**.

Theorem 6.3 (Completeness) *If* $\varphi \vdash_{\mathbf{RC}} \psi$, *then* $\mathcal{T}(\varphi) \hookrightarrow^* \mathcal{T}(\psi)$ *for* $\varphi,\psi \in \mathcal{L}^+$.

Proof. By induction on the length of the **RC** derivation.

If $\varphi \vdash_{\mathbf{RC}} \varphi$, then $\mathcal{T}(\varphi) \hookrightarrow^* \mathcal{T}(\varphi)$ by applying no rewriting rule; if $\varphi \vdash_{\mathbf{RC}} \top$, then $\mathcal{T}(\varphi) \hookrightarrow^* \mathcal{T}(\top)$ by atom and child elimination.

(Cut): If $\varphi \vdash_{\mathbf{RC}} \psi$ and $\psi \vdash_{\mathbf{RC}} \phi$ such that $\mathcal{T}(\varphi) \hookrightarrow^* \mathcal{T}(\psi)$ and $\mathcal{T}(\psi) \hookrightarrow^* \mathcal{T}(\phi)$, then $\mathcal{T}(\varphi) \hookrightarrow^* \mathcal{T}(\phi) \hookrightarrow^* \mathcal{T}(\phi)$.

(Elimination of conjunction): If $\varphi\wedge\psi \vdash_{\mathbf{RC}} \varphi$, by Lemma 5.4 follows $\mathcal{T}(\varphi\wedge\psi) = \mathcal{T}(\varphi) + \mathcal{T}(\psi) \hookrightarrow^* \mathcal{T}(\varphi)$. Analogously, if $\varphi \wedge \psi \vdash_{\mathbf{RC}} \psi$ then $\mathcal{T}(\varphi \wedge \psi) \hookrightarrow^* \mathcal{T}(\psi)$ by Lemma 5.4.

(Introduction to conjunction): If $\varphi \vdash_{\mathbf{RC}} \psi$ and $\varphi \vdash_{\mathbf{RC}} \phi$ such that $\mathcal{T}(\varphi) \hookrightarrow^* \mathcal{T}(\psi)$ and $\mathcal{T}(\varphi) \hookrightarrow^* \mathcal{T}(\phi)$, by Lemma 5.4 we conclude $\mathcal{T}(\varphi) \hookrightarrow^* \mathcal{T}(\psi)+\mathcal{T}(\phi) = \mathcal{T}(\psi \wedge \phi)$.

(Distribution): If $\varphi \vdash_{\mathbf{RC}} \psi$ such that $\mathcal{T}(\varphi) \hookrightarrow^* \mathcal{T}(\psi)$, then $\mathcal{T}(\langle\alpha\rangle\varphi) \hookrightarrow^* \mathcal{T}(\langle\alpha\rangle\psi)$ by the inside rewriting property (Corollary 5.6).

(Transitivity): If $\langle\alpha\rangle\langle\alpha\rangle\varphi \vdash_{\mathbf{RC}} \langle\alpha\rangle\varphi$, by the 4-rule,

$$\mathcal{T}(\langle\alpha\rangle\langle\alpha\rangle\varphi) = \langle\varnothing;[(\alpha,\langle\varnothing;[(\alpha,\mathcal{T}(\varphi))]\rangle)]\rangle \hookrightarrow^4 \langle\varnothing;[(\alpha,\mathcal{T}(\varphi))]\rangle = \mathcal{T}(\langle\alpha\rangle\varphi).$$

(Monotonicity): If $\langle\alpha\rangle\varphi \vdash_{\mathbf{RC}} \langle\beta\rangle\varphi$ for $\alpha > \beta$, by the λ-rule,

$$\mathcal{T}(\langle\alpha\rangle\varphi) = \langle\varnothing;[(\alpha,\mathcal{T}(\varphi))]\rangle \hookrightarrow^\lambda \langle\varnothing;[(\beta,\mathcal{T}(\varphi))]\rangle = \mathcal{T}(\langle\beta\rangle\varphi).$$

(J): Consider $\alpha > \beta$ and $\mathcal{T}(\varphi) = \langle\Delta;\Gamma\rangle$. If $\langle\alpha\rangle\varphi \wedge \langle\beta\rangle\psi \vdash_{\mathbf{RC}} \langle\alpha\rangle(\varphi \wedge \langle\beta\rangle\psi)$, by the J-rule,

$$\begin{aligned}\mathcal{T}(\langle\alpha\rangle\varphi \wedge \langle\beta\rangle\psi) = \langle\varnothing;[(\alpha,\langle\Delta;\Gamma\rangle),(\beta,\mathcal{T}(\psi))]\rangle &\hookrightarrow^{\mathsf{J}} \langle\varnothing;[(\alpha,\langle\Delta;\Gamma \frown (\beta,\mathcal{T}(\psi))\rangle)]\rangle\\ &= \langle\varnothing;[(\alpha,\mathcal{T}(\varphi)+\mathcal{T}(\langle\beta\rangle\psi))]\rangle = \mathcal{T}(\langle\alpha\rangle(\varphi \wedge \langle\beta\rangle\psi)).\end{aligned}$$

□

This concluding corollary states that a sequence can be shown in **RC** by identifying transformations within the trees in which the involved formulas are embedded into. Likewise, a rewriting can be proven by showing the corresponding **RC** sequence for the formulas embedding the corresponding trees.

Corollary 6.4 *$\mathcal{T}(\varphi) \hookrightarrow^* \mathcal{T}(\psi) \implies \varphi \vdash_{\mathbf{RC}} \psi$ and $\mathcal{F}(\mathtt{T}) \vdash_{\mathbf{RC}} \mathcal{F}(\mathtt{T}') \implies \mathtt{T} \hookrightarrow^* \mathtt{T}'$.*

Proof. By adequacy and completeness (Theorems 6.2 and 6.3, respectively) using the embedding composition property (Proposition 4.4). □

7 The Rewrite Normalization Theorem

In this section we present our rewriting normalization theorem, which allows us to perform rewriting in a designated sequence of rewriting rules according to their kinds. Specifically, we can always perform the rewrite process using normal rewriting sequences.

Definition 7.1 (Normal rewrite) A list of rewriting rules Ω is a *normal rewriting sequence* if it is of the form

$$\Omega_{\pi^+} \frown \Omega_\diamond \frown \Omega_\delta \frown \Omega_\rho \frown \Omega_\sigma$$

for $\Omega_{\pi^+}, \Omega_\diamond, \Omega_\delta, \Omega_\rho$ and Ω_σ lists of replicative, modal, decreasing, atomic and structural rewriting rules, respectively.

The order of the presented normal rewriting sequence adheres to the following principles. Firstly, performing any kind of rewriting rule before a replicative one cannot be equivalently reversed. Similarly, the order of modal and decreasing rewriting rules cannot be interchanged (for example, it may be necessary to decrease the label of an edge in order to apply 4-rule). Finally, atomic and structural rewriting rules, which pertain to node labels and child permutation, are placed last because nodes and children may be removed during the rewriting process.

In the following, we disclose the results needed to establish the normal rewriting theorem. Their proofs delve into technical intricacies, requiring meticulous attention to the rewritten positions and the arguments determining the application of the concerned rewriting rules. If two rewriting rules are applied in different positions of the tree such that their outcome transformation is disjoint, their permutation is proven straightforward. However, if their application intersects, additional transformations may be required. To illustrate these scenarios, we provide simplified examples alongside the proofs.

Lemma 7.2 *If $\mathtt{T} \hookrightarrow^\sigma \circ \hookrightarrow^\mu \mathtt{S}$, then $\mathtt{T} \hookrightarrow^\mu \circ \hookrightarrow^{\sigma*} \mathtt{S}$ for $\mu \in \mathcal{R} \setminus \{\sigma\}$.*

Proof. By cases on μ. Each case is proven by induction on the length of the position of σ and by cases on the position of μ. □

Corollary 7.3 *If $\mathtt{T} \hookrightarrow^{\sigma*} \circ \hookrightarrow^\Omega \mathtt{S}$, then $\mathtt{T} \hookrightarrow^\Omega \circ \hookrightarrow^{\sigma*} \mathtt{S}$ for Ω a list of atomic, replicative, decreasing and modal rewriting rules.*

Proof. By induction on the length of Ω and by induction on the number of σ-steps performed using Lemma 7.2. □

Lemma 7.4 *Let ρ be an atomic rule. If* $\mathtt{T} \hookrightarrow^{\rho} \circ \hookrightarrow^{\mu} \mathtt{S}$, *then* $\mathtt{T} \hookrightarrow^{\mu} \circ \hookrightarrow^{\rho *} \mathtt{S}$ *for* $\mu \in \{\pi^+, \pi^-, 4, \lambda, \mathsf{J}\}$.

Proof. By cases on μ. Each case is shown by induction on the length of the position of ρ and by cases on the position of μ. If μ is π^+, we prove that either $\mathtt{T} \hookrightarrow^{\pi^+} \circ \hookrightarrow^{\rho} \mathtt{S}$ or $\mathtt{T} \hookrightarrow^{\pi^+} \circ \hookrightarrow^{\rho} \circ \hookrightarrow^{\rho} \mathtt{S}$ holds (see Figure 3). If μ is a decreasing rewriting rule, then either $\mathtt{T} \hookrightarrow^{\mu} \mathtt{S}$ or $\mathtt{T} \hookrightarrow^{\mu} \circ \hookrightarrow^{\rho} \mathtt{S}$. Otherwise we show $\mathtt{T} \hookrightarrow^{\mu} \circ \hookrightarrow^{\rho} \mathtt{S}$. □

Corollary 7.5 *Let Ω_ρ be a list of atomic rewriting rules. If* $\mathtt{T} \hookrightarrow^{\Omega_\rho} \circ \hookrightarrow^{\mu} \mathtt{S}$, *then* $\mathtt{T} \hookrightarrow^{\mu} \circ \hookrightarrow^{\Omega'_\rho} \mathtt{S}$ *for* $\mu \in \{\pi^+, \pi^-, 4, \lambda, \mathsf{J}\}$ *and* Ω'_ρ *a list of atomic rewriting rules.*

Proof. By an easy induction on the length of Ω_ρ using Lemma 7.4. □

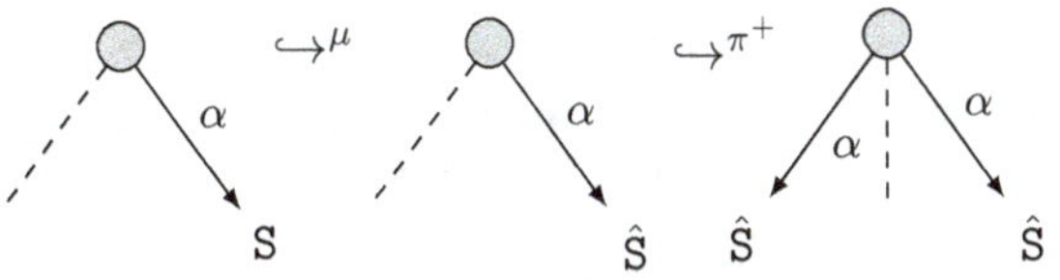

(a) $\mathtt{T} \hookrightarrow^{\mu} \circ \hookrightarrow^{\pi^+} \mathtt{S}$ (for $\mathtt{S} \hookrightarrow^{\mu} \hat{\mathtt{S}}$).

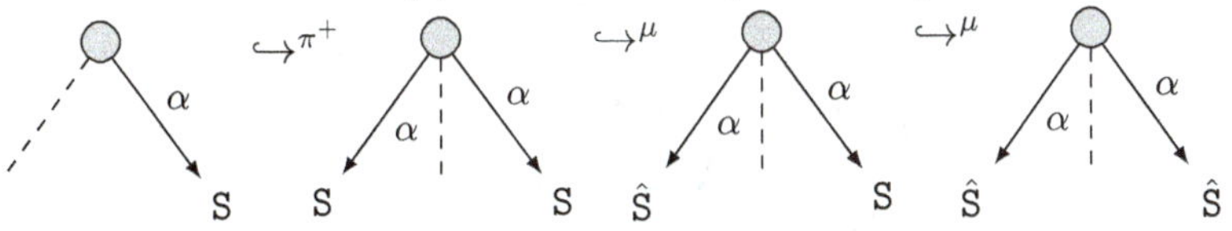

(b) $\mathtt{T} \hookrightarrow^{\pi^+} \circ \hookrightarrow^{\mu} \circ \hookrightarrow^{\mu} \mathtt{S}$.

Fig. 3. $\mathtt{T} \hookrightarrow^{\mu} \circ \hookrightarrow^{\pi^+} \mathtt{S} \Longrightarrow \mathtt{T} \hookrightarrow^{\pi^+} \circ \hookrightarrow^{\mu} \circ \hookrightarrow^{\mu} \mathtt{S}, \mu \in \mathcal{R} \setminus \{\pi^+\}$.

Proposition 7.6 *Let δ be a decreasing rule. If* $\mathtt{T} \hookrightarrow^{\delta} \circ \hookrightarrow^{\mu} \mathtt{S}$, *then* $\mathtt{T} \hookrightarrow^{\mu *} \circ \hookrightarrow^{\delta *} \mathtt{S}$ *for* $\mu \in \{\pi^+, \lambda, \mathsf{J}\}$.

Proof. By induction on the length of the position of δ and by cases on the length of the position of μ.

Let δ be π^-. If μ is π^+, we show that either $\mathtt{T} \hookrightarrow^{\pi^+} \circ \hookrightarrow^{\pi^-} \mathtt{S}$ or $\mathtt{T} \hookrightarrow^{\pi^+} \circ \hookrightarrow^{\pi^-} \circ \hookrightarrow^{\pi^-} \mathtt{S}$ holds (see Figure 3). Otherwise we prove $\mathtt{T} \hookrightarrow^{\mu} \circ \hookrightarrow^{\pi^-} \mathtt{S}$.

Let δ be 4. If μ is π^+ we show that either $\mathtt{T} \hookrightarrow^{\pi^+} \circ \hookrightarrow^{4} \mathtt{S}$ or $\mathtt{T} \hookrightarrow^{\pi^+} \circ \hookrightarrow^{4} \circ \hookrightarrow^{4} \mathtt{S}$ (see Figure 3). For μ being λ we prove that either $\mathtt{T} \hookrightarrow^{\lambda} \circ \hookrightarrow^{4} \mathtt{S}$ or $\mathtt{T} \hookrightarrow^{\lambda} \circ \hookrightarrow^{\lambda} \circ \hookrightarrow^{4} \mathtt{S}$ (see Figure 4). Finally, for μ being J we show that either $\mathtt{T} \hookrightarrow^{J} \circ \hookrightarrow^{4} \mathtt{S}$ or $\mathtt{T} \hookrightarrow^{J} \circ \hookrightarrow^{J} \circ \hookrightarrow^{4} \mathtt{S}$ (see Figure 5). □

Corollary 7.7 *Let δ and θ be decreasing and modal rewriting rules respectively, and Ω_δ be a list of decreasing rewriting rules.*

(i) *If* $\mathtt{T} \hookrightarrow^{\Omega_\delta} \circ \hookrightarrow^{\pi^+} \mathtt{S}$, *then* $\mathtt{T} \hookrightarrow^{\pi^+} \circ \hookrightarrow^{\Omega'_\delta} \mathtt{S}$ *for* Ω'_δ *a list of decreasing rules.*

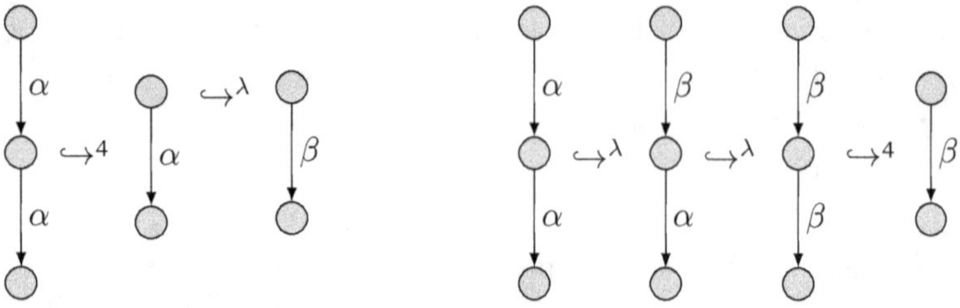

(a) $\mathsf{T} \hookrightarrow^4 \circ \hookrightarrow^\lambda \mathsf{S}$. (b) $\mathsf{T} \hookrightarrow^\lambda \circ \hookrightarrow^\lambda \circ \hookrightarrow^4 \mathsf{S}$.

Fig. 4. $\mathsf{T} \hookrightarrow^4 \circ \hookrightarrow^\lambda \mathsf{S} \Longrightarrow \mathsf{T} \hookrightarrow^\lambda \circ \hookrightarrow^\lambda \circ \hookrightarrow^4 \mathsf{S}$.

(ii) *If* $\mathsf{T} \hookrightarrow^\delta \circ \hookrightarrow^{\theta*} \mathsf{S}$, *then* $\mathsf{T} \hookrightarrow^{\theta*} \circ \hookrightarrow^\delta \mathsf{S}$.

(iii) *If* $\mathsf{T} \hookrightarrow^{\Omega_\delta} \circ \hookrightarrow^{\mu*} \mathsf{S}$ *for* $\mu \in \{\pi^+, \lambda, \mathsf{J}\}$, *then* $\mathsf{T} \hookrightarrow^{\mu*} \circ \hookrightarrow^{\Omega'_\delta} \mathsf{S}$ *for* Ω'_δ *a list of decreasing rewriting rules.*

Proof. The first statement follows by induction on the length of Ω_δ and Proposition 7.6. The second one is shown by induction on the number of modal rewriting rules applied and Proposition 7.6. The third result follows by cases on μ. For $\mu \in \{\lambda, \mathsf{J}\}$, we conclude by induction on the length of Ω_δ and the second statement. Otherwise, we conclude by induction on the number of μ-rules applied and the first statement. □

Proposition 7.8 *Let* θ *be a modal rewriting rule. If* $\mathsf{T} \hookrightarrow^\theta \circ \hookrightarrow^{\pi^+} \mathsf{S}$, *then* $\mathsf{T} \hookrightarrow^{\pi^+*} \circ \hookrightarrow^{\theta*} \circ \hookrightarrow^{\sigma*} \mathsf{S}$.

Proof. By induction on the length of the position of θ and by cases on the length of the position of π^+. If θ is λ, we show that either $\mathsf{T} \hookrightarrow^{\pi^+} \circ \hookrightarrow^\lambda \mathsf{S}$ or $\mathsf{T} \hookrightarrow^{\pi^+} \circ \hookrightarrow^\lambda \circ \hookrightarrow^\lambda \mathsf{S}$ holds (see Figure 3). For θ being J, we prove that either $\mathsf{T} \hookrightarrow^{\pi^+} \circ \hookrightarrow^{\mathsf{J}} \mathsf{S}$ or $\mathsf{T} \hookrightarrow^{\pi^+} \circ \hookrightarrow^{\mathsf{J}} \circ \hookrightarrow^{\mathsf{J}} \mathsf{S}$ (see Figure 3) or $\mathsf{T} \hookrightarrow^{\pi^+} \circ \hookrightarrow^{\pi^+} \circ \hookrightarrow^{\mathsf{J}} \circ \hookrightarrow^{\mathsf{J}} \mathsf{S}$ (see Figure 6) or $\mathsf{T} \hookrightarrow^{\pi^+} \circ \hookrightarrow^{\mathsf{J}} \circ \hookrightarrow^{\mathsf{J}} \circ \hookrightarrow^\sigma \mathsf{S}$ (see Figure 7). □

We now want to show permutation of the application of multiple π^+-rules after multiple modal rewriting rules. To show permutability of π^+-rules following a J-rule, we need to reorganize duplications applied at positions in increasing depth in the tree. Furthermore, we require flexibility to choose which branch to duplicate first.

For readability, we write $\mathsf{T} \hookrightarrow^{\pi^+(\mathbf{k},i)} \mathsf{S}$ to denote that S is obtained by applying the π^+-rule to duplicate the ith child at position $\mathbf{k} \in \mathsf{Pos}(\mathsf{T})$. Similarly, $\mathsf{T} \hookrightarrow^{\sigma(\mathbf{k},i,j)} \mathsf{S}$ denotes the application of the σ-rule to permute the ith and jth children at position $\mathbf{k} \in \mathsf{Pos}(\mathsf{T})$. Given this notation, the following result aids in reorganizing duplications at positions in increasing depth.

Lemma 7.9 (Depth-level permutability of π^+)

(i) *If* $\mathsf{T} \hookrightarrow^{\pi^+(l\mathbf{k},i)} \circ \hookrightarrow^{\pi^+(\epsilon,j)} \mathsf{S}$ *for* $j \neq l$, *then* $\mathsf{T} \hookrightarrow^{\pi^+(\epsilon,j)} \circ \hookrightarrow^{\pi^+((l+1)\mathbf{k},i)} \mathsf{S}$.

(ii) *If* $\mathsf{T} \hookrightarrow^{\pi^+(j\mathbf{k},i)} \circ \hookrightarrow^{\pi^+(\epsilon,j)} \mathsf{S}$, *then* $\mathsf{T} \hookrightarrow^{\pi^+(\epsilon,j)} \circ \hookrightarrow^{\pi^+((j+1)\mathbf{k},i)} \circ \hookrightarrow^{\pi^+(1\mathbf{k},i)} \mathsf{S}$.

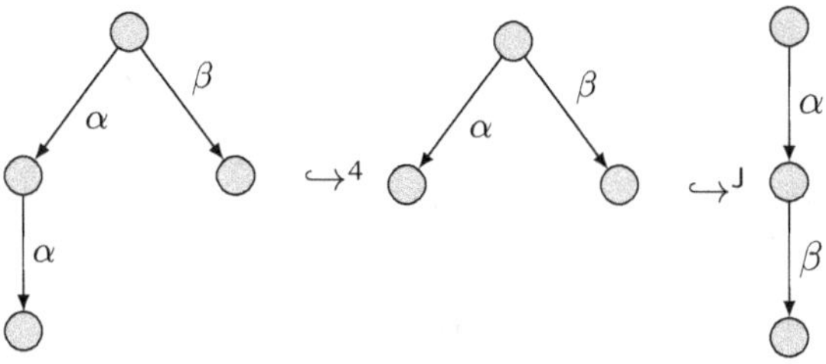

(a) $\mathtt{T} \hookrightarrow^4 \circ \hookrightarrow^{\mathsf{J}} \mathtt{S}$.

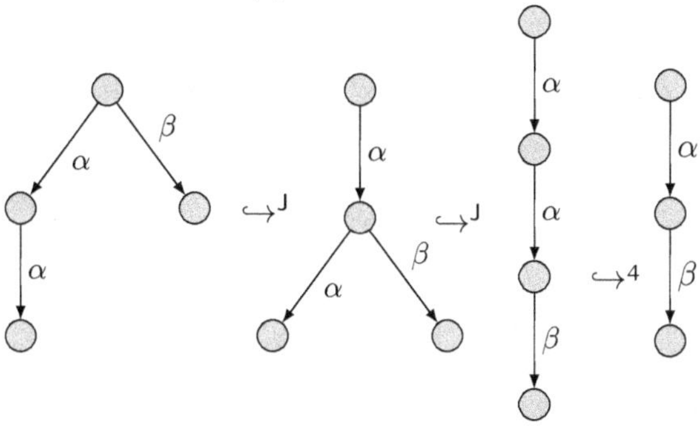

(b) $\mathtt{T} \hookrightarrow^{\mathsf{J}} \circ \hookrightarrow^{\mathsf{J}} \circ \hookrightarrow^4 \mathtt{S}$.

Fig. 5. $\mathtt{T} \hookrightarrow^4 \circ \hookrightarrow^{\mathsf{J}} \mathtt{S} \Longrightarrow \mathtt{T} \hookrightarrow^{\mathsf{J}} \circ \hookrightarrow^{\mathsf{J}} \circ \hookrightarrow^4 \mathtt{S}$.

(iii) *If* $\mathtt{T} \hookrightarrow^{\pi^+(l\mathbf{k}_l,i)} \circ \hookrightarrow^{\pi^+(n\mathbf{k}_n,j)} \mathtt{S}$ *for* $|n\mathbf{k}_n| < |l\mathbf{k}_l|$, *then* $\mathtt{T} \hookrightarrow^{\pi^+(n\mathbf{k}_n,j)} \circ \hookrightarrow^{\pi^+(l\mathbf{k}_l,i)} \mathtt{S}$.

Proof. The first two statements are trivially satisfied by definition. The third follows by induction on the tree structure of T. □

Likewise, the lemma below permits to choose which branch to duplicate first.

Remark 7.10 (Width-level permutability of π^+)

(i) If $\mathtt{T} \hookrightarrow^{\pi^+(\epsilon,i)} \circ \hookrightarrow^{\pi^+(\epsilon,j)} \mathtt{S}$ for $1 < j$ and $j \neq i+1$, then $\mathtt{T} \hookrightarrow^{\pi^+(\epsilon,j-1)} \circ \hookrightarrow^{\pi^+(\epsilon,i+1)} \hookrightarrow^{\sigma(\epsilon,1,2)} \mathtt{S}$.

(ii) If $\mathtt{T} \hookrightarrow^{\pi^+(n\mathbf{k}_n,i)} \circ \hookrightarrow^{\pi^+(l\mathbf{k}_l,j)} \mathtt{S}$ for $n \neq l$, then $\mathtt{T} \hookrightarrow^{\pi^+(l\mathbf{k}_l,j)} \circ \hookrightarrow^{\pi^+(n\mathbf{k}_n,i)} \mathtt{S}$.

Therefore, we can show permutation of the application of arbitrary π^+-rules after a J-rule.

Proposition 7.11 *If* $\mathtt{T} \hookrightarrow^{\mathsf{J}} \circ \hookrightarrow^{\pi^+*} \mathtt{S}$, *then* $\mathtt{T} \hookrightarrow^{\pi^+*} \circ \hookrightarrow^{\mathsf{J}*} \circ \hookrightarrow^{\sigma*} \mathtt{S}$.

Proof. If the J-rule and the π^+-rules are applied in different positions of the tree such that the performed transformations are disjoint, their permutation is proven straightforward as shown in Proposition 7.8. Therefore, we will use Lemma 7.9 and Remark 7.10 to reorganise the application of the π^+-rules to easily permute the disjoint cases. Otherwise, in cases where the subtree

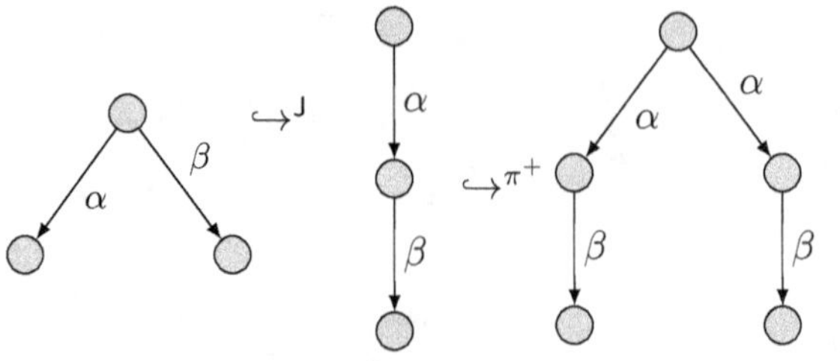

(a) $\mathtt{T} \hookrightarrow^{\mathsf{J}} \circ \hookrightarrow^{\pi^+} \mathtt{S}$.

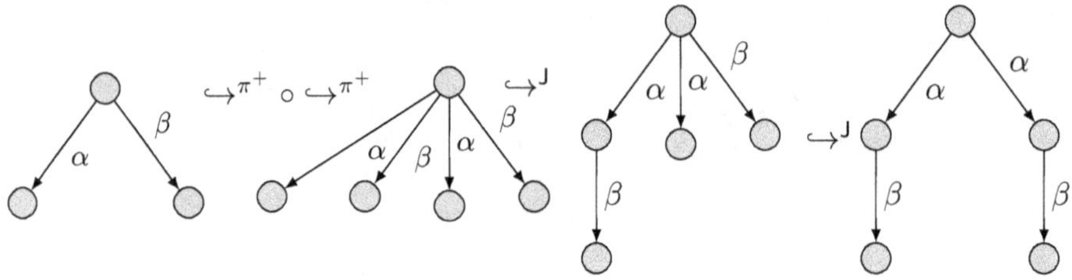

(b) $\mathtt{T} \hookrightarrow^{\pi^+} \circ \hookrightarrow^{\pi^+} \circ \hookrightarrow^{\mathsf{J}} \circ \hookrightarrow^{\mathsf{J}} \mathtt{S}$.

Fig. 6. $\mathtt{T} \hookrightarrow^{\mathsf{J}} \circ \hookrightarrow^{\pi^+} \mathtt{S} \Longrightarrow \mathtt{T} \hookrightarrow^{\pi^+} \circ \hookrightarrow^{\pi^+} \circ \hookrightarrow^{\mathsf{J}} \circ \hookrightarrow^{\mathsf{J}} \mathtt{S}$.

affected by the J-rule is duplicated, we use a unique labeling of the nodes to track the number of times those subtrees are duplicated. Specifically, we label the nodes with their positions. For $\mathtt{T} \hookrightarrow^{\mathsf{J}} \hookrightarrow^{*} \mathtt{S}$ such that J has been performed at position $n\mathbf{k} \in \mathsf{Pos}(\mathtt{T})$, we call the J-*label* to the label n. Moreover, if J has been performed at the empty position relating the ith and the jth children according to the notation of its definition, we call the *upper* J-*label* to the label i and the *lower* J-*label* to the label j.

Therefore, the proof proceeds by induction on the length of the position in which the J-rule has been applied. For the J-rule applied at the empty position, we proceed by induction on the occurrences of the upper J-label. The inductive step is shown by Proposition 7.8 and Corollary 7.3, using Lemma 7.9 and Remark 7.10 to get a suitable reorganisation of the applications of the π^+-rules. The base case in which the upper J-label occurs once in S is shown by induction on the occurrences of the lower J-label using the same strategy.

Lastly, if the J-rule is applied at a non-empty position, we proceed by induction on the occurrences of the J-label. The base case follows by Proposition 7.8 using Lemma 7.9 and Remark 7.10 together with the inductive hypothesis on the position in which J has been performed. The inductive step is similarly proven by Proposition 7.8 and Corollary 7.3 using Lemma 7.9 and Remark 7.10. □

Corollary 7.12 *Let $\Omega_\diamond$ be a list of modal rewriting rules.*

(i) *If* $\mathtt{T} \hookrightarrow^{\lambda} \circ \hookrightarrow^{\pi^+ *} \mathtt{S}$, *then* $\mathtt{T} \hookrightarrow^{\pi^+ *} \circ \hookrightarrow^{\lambda *} \mathtt{S}$ *by applying the same number of π^+-rules.*

(ii) *If* $\mathtt{T} \hookrightarrow^{\Omega_\diamond} \circ \hookrightarrow^{\pi^+ *} \mathtt{S}$, *then* $\mathtt{T} \hookrightarrow^{\pi^+ *} \circ \hookrightarrow^{\Omega'_\diamond} \circ \hookrightarrow^{\sigma *} \mathtt{S}$ *for $\Omega'_\diamond$ a list of modal*

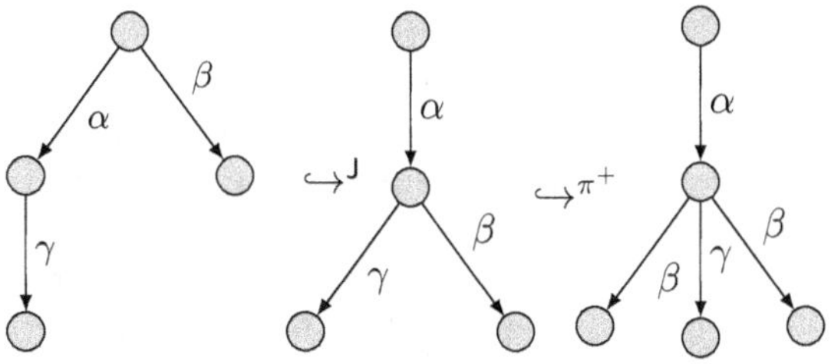

(a) $\mathtt{T} \hookrightarrow^{\mathsf{J}} \circ \hookrightarrow^{\pi^+} \mathtt{S}$.

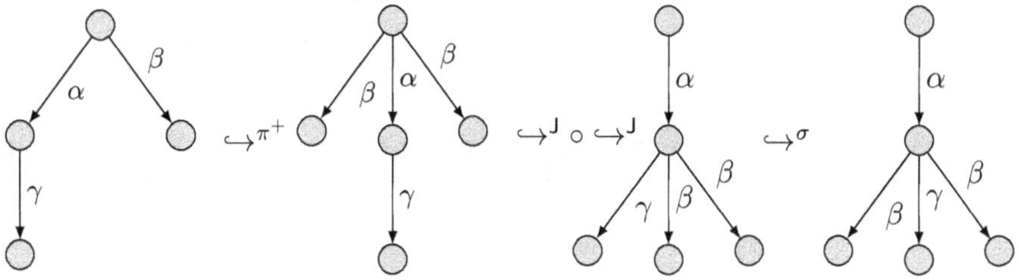

(b) $\mathtt{T} \hookrightarrow^{\pi^+} \circ \hookrightarrow^{\mathsf{J}} \circ \hookrightarrow^{\mathsf{J}} \circ \hookrightarrow^{\sigma} \mathtt{S}$.

Fig. 7. $\mathtt{T} \hookrightarrow^{\mathsf{J}} \circ \hookrightarrow^{\pi^+} \mathtt{S} \Longrightarrow \mathtt{T} \hookrightarrow^{\pi^+} \circ \hookrightarrow^{\mathsf{J}} \circ \hookrightarrow^{\mathsf{J}} \circ \hookrightarrow^{\sigma} \mathtt{S}$.

rewriting rules.

Proof. The first statement follows by induction on the number of π^+-rules applied and Proposition 7.8. The second statement follows by induction on the length of $\Omega_\diamond$. The inductive step is shown by cases on the considered modal rewriting rule: the λ-rule case follows by the first statement and the J-rule case is shown by Proposition 7.11 and Corollary 7.2. □

Finally, we present the main theorem of the section.

Theorem 7.13 (Rewrite Normalization Theorem) *If* $\mathtt{T} \hookrightarrow^* \mathtt{S}$, *then* $\mathtt{T} \hookrightarrow^{\Omega} \mathtt{S}$ *for* Ω *a normal rewriting sequence.*

Proof. By induction on the number of rewriting steps that are performed using Corollaries 7.3, 7.5, 7.7 and 7.12. □

8 Conclusions and future work

We have provided a method for designing tree rewriting systems for different strictly positive logics and in particular re-cast the reflection calculus in this framework. Although not explicitly stated, it easily follows from the presented techniques that a rewriting system for $\mathbf{K}^+$ can be defined by only considering atomic, structural, extensional and π-rules, and we are confident that our framework will find applications in positive fragments for other (poly)modal logics. The use of abstract rewriting systems aids in the analysis of further properties like subformula property and admissibility of rules and provides a foundation for computationally efficient implementation.

Our work is based on a type-theoretic presentation of trees, which has the dual benefit of allowing for rules to be described in a precise and succinct manner and being particularly amenable to formalization in proof assistants. Many of our results have already been implemented in the proof assistant Coq and our goal is to fully formalize our work. Since rewriting normalization theorem and proof-theoretic investigations in general require checking multiple distinct cases in detail, the benefit of formalization is particularly clear in this type of proof, and indeed the community has been gravitating towards formalized proofs (see e.g. [18]). In the framework of **RC**, this has the added advantage that results often need not only to be *true*, but provable in suitable systems of arithmetic, and the latter can be made no more transparent than via formalization. Last but not least, proofs implemented in Coq can be automatically extracted into fully verified algorithms, paving the road to reasoners with the highest degree of reliability attainable by current technology.

References

[1] Baader, F., D. Calvanese, D. L. McGuinness, D. Nardi and P. F. Patel-Schneider, editors, "The Description Logic Handbook," Cambridge University Press, Cambridge, UK, 2007, 2 edition.

[2] Baader, F. and T. Nipkow, "Term rewriting and all that," Cambridge university press, 1998.

[3] Barendregt, H. P., "The lambda calculus - its syntax and semantics," Studies in logic and the foundations of mathematics **103**, North-Holland, 1985.

[4] Beklemishev, L., *A note on strictly positive logics and word rewriting systems*, in: S. Odintsov, editor, *Larisa Maksimova on Implication, Interpolation, and Definability*, Springer Verlag, 2018 .

[5] Beklemishev, L. D., *Provability algebras and proof-theoretic ordinals, I*, Annals of Pure and Applied Logic **128** (2004), pp. 103–124.

[6] Beklemishev, L. D., *Calibrating provability logic*, , **9**, College Publications, London, 2012 pp. 89–94.

[7] Beklemishev, L. D., *Positive provability logic for uniform reflection principles*, Annals of Pure and Applied Logic **165** (2014), pp. 82–105.

[8] Dashkov, E. V., *On the positive fragment of the polymodal provability logic GLP*, Mathematical Notes **91** (2012), pp. 318–333.

[9] Fitting, M., "Types, tableaus, and Gödel's god," Springer Science & Business Media, 2002.

[10] Foret, A., *Rewrite rule systems for modal propositional logic*, The Journal of Logic Programming **12** (1992), pp. 281–298.

[11] Genest, B., A. Muscholl, O. Serre and M. Zeitoun, *Tree pattern rewriting systems*, in: S. S. Cha, J.-Y. Choi, M. Kim, I. Lee and M. Viswanathan, editors, *Automated Technology for Verification and Analysis* (2008), pp. 332–346.

[12] Goubault-Larrecq, J. and I. Mackie, "Proof theory and automated deduction," Springer Science & Business Media, 2001.

[13] Japaridze, G., *The polymodal provability logic*, in: *Intensional logics and logical structure of theories: material from the Fourth Soviet-Finnish Symposium on Logic*, Metsniereba, Telavi, 1988 In Russian.

[14] Kikot, S., A. Kurucz, Y. Tanaka, F. Wolter and M. Zakharyaschev, *Kripke completeness of strictly positive modal logics over meet-semilattices with operators*, The Journal of Symbolic Logic **84** (2019), pp. 533–588.

[15] Newborn, M., "Automated theorem proving: theory and practice," Springer Science & Business Media, 2000.

[16] Pierce, B. C., "Types and programming languages," MIT press, 2002.
[17] Shamkanov, D., *Nested sequents for provability logic GLP*, Logic Journal of the IGPL **23** (2015), pp. 789–815.
[18] Shillito, I. and R. Goré, *Direct elimination of additive-cuts in GL4ip: verified and extracted*, in: D. Fernández-Duque, A. Palmigiano and S. Pinchinat, editors, *Advances in Modal Logic, AiML 2022, Rennes, France, August 22-25, 2022* (2022), pp. 429–449.
[19] Terese, "Term Rewriting Systems," Cambridge Tracts in Theoretical Computer Science **55**, Cambridge University Press, 2003.
[20] Wos, L., R. Overbeek, E. Lusk and J. Boyle, "Automated reasoning introduction and applications," McGraw-Hill, Inc., 1992.

Point-Set Neighborhood Logic

Yanjing Wang [1]

Department of Philosophy, Peking University

Junhua Yu [2]

Department of Philosophy, Tsinghua University

Abstract

In this paper, we propose Point-set Neighborhood Logic (PSNL) to reason about neighborhood structures. The bimodal language of PSNL is defined via a mutual induction of point-formulas and set-formulas. We show that this simple language is equally expressive as the language of Instantial Neighborhood Logic (INL). As the main results, we first give two complete proof systems, one in Hilbert-style and one in Gentzen sequent-style, each featuring two intertwined K-like systems. The proof of strong completeness of the Hilbert-style system is based on a direct canonical model construction without relying on a normal form. Based on the sequent calculus, we establish constructively the uniform interpolation property of PSNL, from which that of INL follows.

Keywords: instantial neighborhood logic, normal modal logic, uniform interpolation

---

1 Introduction

Neighborhood structures are generalizations of relational models and topological models for modal logic, which can facilitate weaker logics than normal modal logic [3]. Beyond being technical tools for syntactically driven logics, neighborhood structures can also be viewed as genuine models for reasoning about evidence [21], games [14], knowledge and belief [5], hypergraphs [4] and many more [13]. It is natural to ask whether the standard modal language is strong enough to describe neighborhood structures. This motivates the more expressive framework of instantial neighborhood logic (INL) proposed in [20]. Since then, INL has been studied in both theoretical perspectives like duality theory [1], Sahlqvist theory [24], proof theory [22,23], and applicational perspectives like logical dynamics [19], multi-agency [7].

In INL, the basic modal formulas are of the form $\Box(\alpha_1, \ldots, \alpha_n; \beta)$ (where $n \in \mathbb{N}$) expressing that there is a neighborhood (a set X) of the current possible world (a point) s.t. $\alpha_1, \ldots, \alpha_n$ each holds somewhere in X and β holds

---

[1] email: y.wang@pku.edu.cn, ORCID: 0000-0002-9499-416X.

[2] email: junhua.yu.5036@outlook.com, ORCID: 0000-0002-0714-1709.

everywhere in X. The language of INL extends the standard modal language over the monotonic neighborhood semantics, leaving $\Box\beta$ in the latter as an abbreviation of the instantial formula when $n = 0$. Although this extension looks natural, it brings technical challenges to axiomatization and establishment of meta-properties. For example, existing completeness proofs of the Hilbert-style axiomatization of INL are indirect via a normal form theorem [20,18]. Moreover, since INL is not locally tabular, from its constructively established Craig-Lyndon interpolation property [23] one cannot directly derive its uniform interpolation property, although the latter is supported suggestively by coalgebraic methods and general results [16,12].

These technical difficulties are closely related to the highly integrated nature of the syntax of INL formulas. Perhaps too much information is packed into a single operator, making it difficult to manipulate in a proof system.

In this paper, we take one step back and rethink whether we can keep the expressivity of INL but make the syntax easier to handle. The semantics of $\Box(\alpha_1, \ldots, \alpha_n; \beta)$ suggests a *bimodal* perspective as already mentioned in [20]: there is a unary modality $\Diamond_1$ to pick a neighborhood first, and only then unary $\Diamond_2\alpha_i$ and $\Box_2\beta$ are evaluated within the given neighborhood. Similar ideas also appeared in [8] for standard neighborhood semantics. However, differing from the standard bimodal logic, these two types of modalities cannot be taken as unrestricted constructors: $\Diamond_1$ picks a set given a point, and $\Diamond_2$ picks a point given a set. For example, $\Diamond_1 p$ does not make much sense for p is not evaluated over a set. Nested modalities $\Diamond_2 \Diamond_2 \alpha$ raise a similar confusion.

One way to go around is to transform neighborhood models into multi-sorted Kripke models, viewing neighborhoods (sets of points) as another sort of points [11]. Similar ideas also appeared in many-sorted coalgebraic modal logic [10] where a powerful logic was axiomatized. On the other hand, we can keep the models as they are and restrict the syntax by introducing two types of formulas (via a mutual induction) evaluated respectively on points and sets, as briefly discussed in [24] on the Sahlqvist correspondence of INL. In this paper, we follow this idea by using a bimodal language featuring both *point-formulas* and *set-formulas* and go much further by studying the proof systems using this language to obtain results that can be transferred back to the original INL.

The main results in this paper are summarized below:

- It is shown that the bimodal point-set neighborhood logic (PSNL) is equally expressive as INL.
- A simple Hilbert-style axiomatization of PSNL featuring two intertwined K systems is given. Its strong completeness is proved by directly constructing a canonical model.
- A complete sequent calculus of PSNL is given, based on which PSNL is shown to have uniform interpolation property via a constructive proof-theoretic method from [2]. The uniform interpolation property of INL follows.

PSNL can talk about neighborhoods directly via set-formulas that can be manipulated by proof rules. This, in turn, helps us to reason about point-formulas with more flexibility, enabling standard methods to be applied.

The rest of this paper is structured as follows. In Section 2 we lay out the basics of the point-set neighborhood logic, and show that this language has an equally expressive power as that of INL. In Section 3, we introduce a Hilbert-style axiomatization, whose strong-completeness proof also gives us a canonical model construction to INL. In Section 4, we introduce a cut-free sequent calculus that supports mechanical proof search as well as our uniform interpolant construction in Section 5.

2 Point-set Neighborhood Language

2.1 Language and semantics

The following language is a variant of the bimodal language introduced by [24] under minor modifications.

Definition 2.1 (Language $\mathcal{L}^{\mathrm{ps}}(\boxdot, \boxtimes)$) *Given the set of propositional letters P, the language $\mathcal{L}^{\mathrm{ps}}(\boxdot, \boxtimes)$ of point-formulas α is defined by the following mutual induction with the set-formulas φ:*

$$\alpha ::= \bot \mid p \mid (\alpha \to \alpha) \mid \boxdot\varphi$$
$$\varphi ::= \neg\varphi \mid (\varphi \to \varphi) \mid \boxtimes\alpha$$

where $p \in P$. [3]

We also simply call the point-formulas in $\mathcal{L}^{\mathrm{ps}}(\boxdot, \boxtimes)$, the *formulas* of $\mathcal{L}^{\mathrm{ps}}(\boxdot, \boxtimes)$. p, q, r are used for propositional letters and P, Q for sets of them; $\alpha, \beta, \gamma, \delta, \theta$ are used for point-formulas and $\Gamma, \Delta, \Theta, \Omega, \Upsilon$ for sets/multi-sets of them; $\varphi, \psi, \pi, \sigma, \xi$ are used for set-formulas and $\Phi, \Psi, \Pi, \Sigma, \Xi$ for sets/multi-sets of them. Point-/set-formulas should be carefully kept distinct, e.g., $\boxtimes p$ is not a point-formula but $\boxdot \boxtimes p$ is. It is not hard to see that no one can be both a point-formula and a set-formula. Clearly, $\mathcal{L}^{\mathrm{ps}}(\boxdot, \boxtimes)$ is *not* closed under the usual notion of taking subformulas. Also, note that φ can be viewed as a Boolean combination of "atomic" formulas in the shape of $\boxtimes\alpha$. In particular, propositional letters are point-formulas but not set-formulas.

As abbreviations we define $\vee$, $\wedge$, $\leftrightarrow$, $\bigwedge$, $\bigvee$ classically for both point-formulas and set-formulas, and define $\neg\alpha$, the negation of point-formula α, as $\alpha \to \bot$. Moreover, we abbreviate $\neg\bot$ as $\top$, $\neg \boxdot \neg\varphi$ as $\Diamond\varphi$, and $\neg \boxtimes \neg\alpha$ as $\otimes\alpha$.

Definition 2.2 (Models) *A (neighborhood) model is a tuple $\langle W, N, V\rangle$ where:*

- *W is a non-empty set of possible worlds (i.e., points),*
- *$N : W \to \mathcal{P}(\mathcal{P}(W))$ is a neighborhood function,*
- *$V : P \to \mathcal{P}(W)$ is a valuation function.*

Given a model M, we refer to its components as W^M, N^M, V^M. When $w \in W^M$, the tuple $\langle M, w\rangle$ is called a pointed model.

[3] Precisely speaking, one should take distinct symbols for point- and set-implications, as we did for boxes. Yet we decided to step a bit back in view of keeping notations less cumbersome.

Definition 2.3 (Semantics) *Given a model M, the satisfaction relation $\vDash$ between a world and a point-formula α is defined mutually with the relation $\Vdash$ between a set X of worlds and a set-formula φ:*

$$
\begin{array}{rcl}
M, w \vDash \bot & \Leftrightarrow & \textit{never} \\
M, w \vDash p & \Leftrightarrow & w \in V(p) \\
M, w \vDash (\alpha \to \beta) & \Leftrightarrow & M, w \nvDash \alpha \textit{ or } M, w \vDash \beta \\
M, w \vDash \boxdot\varphi & \Leftrightarrow & \textit{for all } X \in N(w)\textit{: } M, X \Vdash \varphi \\
M, X \Vdash \neg\varphi & \Leftrightarrow & M, X \nVdash \varphi \\
M, X \Vdash (\varphi \to \psi) & \Leftrightarrow & M, X \nVdash \varphi \textit{ or } M, X \Vdash \psi \\
M, X \Vdash \boxtimes\alpha & \Leftrightarrow & \textit{for all } v \in X\textit{: } M, v \vDash \alpha
\end{array}
$$

For a set Γ (resp. Φ) of point-(set-)formulas, we write $M, w \vDash \Gamma$ ($M, X \vDash \Phi$) when $M, w \vDash \gamma$ ($M, X \vDash \varphi$) holds for each $\gamma \in \Gamma$ ($\varphi \in \Phi$). Γ (resp. Φ) is said to be satisfiable, if $M, w \vDash \Gamma$ ($M, X \vDash \Phi$) holds for some model M and $w \in W^M$ ($X \subseteq W^M$). Γ (resp. Φ) is said to be finitely satisfiable, if each of its finite subsets is satisfiable. A formula is said to be satisfiable, if the singleton with that formula as the sole element is satisfiable. When $\alpha \to \bot$ (resp. $\neg\varphi$) is not satisfiable, α (φ) is said to be valid, in notation: $\vDash \alpha$ ($\Vdash \varphi$). If $\vDash \alpha \leftrightarrow \beta$ (resp. $\Vdash \varphi \leftrightarrow \psi$), then we say α and β (φ and ψ) are equivalent.

It is not hard to derive the intended semantics for defined operators:

$$
\begin{array}{rcl}
M, w \vDash \neg\alpha & \Leftrightarrow & M, w \nvDash \alpha \\
M, w \vDash \alpha \wedge \beta & \Leftrightarrow & M, w \vDash \alpha \text{ and } M, w \vDash \beta \\
M, w \vDash \alpha \vee \beta & \Leftrightarrow & M, w \vDash \alpha \text{ or } M, w \vDash \beta \\
M, w \vDash \diamondsuit\varphi & \Leftrightarrow & \text{for some } X \in N(w)\text{: } M, X \Vdash \varphi \\
\hline
M, X \Vdash \varphi \wedge \psi & \Leftrightarrow & M, X \Vdash \varphi \text{ and } M, X \Vdash \psi \\
M, X \Vdash \varphi \vee \psi & \Leftrightarrow & M, X \Vdash \varphi \text{ or } M, X \Vdash \psi \\
M, X \Vdash \otimes\alpha & \Leftrightarrow & \text{for some } v \in X\text{: } M, v \vDash \alpha
\end{array}
$$

If we view sets of worlds as worlds of a different sort, as in the two-sorted first-order interpretation of neighborhood structures [8], then it is clear that both $\boxtimes$ and $\boxdot$ are classical and normal. Thus, it is straightforward to see the following:

Proposition 2.4 *The following point-/set-formulas are valid:*

$$
\begin{array}{ll}
\boxdot\varphi \leftrightarrow \neg \diamondsuit \neg\varphi & \boxtimes\alpha \leftrightarrow \neg \otimes \neg\alpha \\
\boxdot(\varphi \wedge \psi) \leftrightarrow \boxdot\varphi \wedge \boxdot\psi & \boxtimes(\alpha \wedge \beta) \leftrightarrow \boxtimes\alpha \wedge \boxtimes\beta \\
\diamondsuit(\varphi \vee \psi) \leftrightarrow \diamondsuit\varphi \vee \diamondsuit\psi & \otimes(\alpha \vee \beta) \leftrightarrow \otimes\alpha \vee \otimes\beta
\end{array}
$$

In our setting, validity can be transferred from one type to the other.

Proposition 2.5 *The following rules hold:*

$$
\frac{\Vdash \varphi \leftrightarrow \psi}{\vDash \boxdot\varphi \leftrightarrow \boxdot\psi} \qquad
\frac{\Vdash \varphi \leftrightarrow \psi}{\vDash \diamondsuit\varphi \leftrightarrow \diamondsuit\psi} \qquad
\frac{\vDash \alpha \leftrightarrow \beta}{\Vdash \boxtimes\alpha \leftrightarrow \boxtimes\beta} \qquad
\frac{\vDash \alpha \leftrightarrow \beta}{\Vdash \otimes\alpha \leftrightarrow \otimes\beta}
$$

In the following, we prove that formulas in $\mathcal{L}^{\mathrm{ps}}(\boxdot, \boxtimes)$ have a normal form.

Lemma 2.6 (Normal form) *Each point-formula* $\gamma \in \mathcal{L}^{\text{ps}}(\boxdot, \boxtimes)$ *is equivalent to a Boolean combination of point-formulas with shape* $\diamonddot(\otimes\alpha_1 \wedge \cdots \wedge \otimes\alpha_n \wedge \boxtimes\beta)$. *That Boolean combination shares the same set of propositional letters with* γ.

Proof As each $\boxdot\varphi$ is equivalent to $\neg \diamonddot \neg\varphi$, we can just consider $\diamonddot\varphi$ formulas and try to massage it into an equivalent Boolean combination of formulas in the shape of $\diamonddot(\otimes\alpha_1 \wedge \cdots \wedge \otimes\alpha_n \wedge \boxtimes\beta)$.

As the semantics of set-formulas is classical for all Boolean connectives, we can rewrite $\diamonddot\varphi$ as $\diamonddot\psi$, where ψ is a *disjunctive normal form* in the shape of $\psi^1 \vee \cdots \vee \psi^n$ ($n > 0$) s.t. each ψ^i is a conjunction of a bunch of $\boxtimes\beta$ and $\otimes\alpha$.

Now we rewrite $\diamonddot(\psi^1 \vee \cdots \vee \psi^n)$ into the equivalent $\diamonddot\psi^1 \vee \cdots \vee \diamonddot\psi^n$ according to the semantics. Then we focus on a particular $\diamonddot\psi^i$, and suppose it is

$$\diamonddot(\otimes\alpha_1 \wedge \cdots \wedge \otimes\alpha_m \wedge \boxtimes\beta_1 \wedge \cdots \wedge \boxtimes\beta_k)$$

where $m + k > 0$. Then it is easy to see that $\diamonddot\psi^i$ is equivalent to

$$\diamonddot(\otimes\alpha_1 \wedge \cdots \wedge \otimes\alpha_m \wedge \boxtimes(\beta_1 \wedge \cdots \wedge \beta_k))$$

based on Propositions 2.4 and 2.5.

In sum, we can rewrite any $\mathcal{L}^{\text{ps}}(\boxdot, \boxtimes)$ point-formula into a Boolean combination of point-formulas with shape $\diamonddot(\otimes\alpha_1 \wedge \cdots \wedge \otimes\alpha_n \wedge \boxtimes\beta)$. □

In the following, we will use this normal form to show that $\mathcal{L}^{\text{ps}}(\boxdot, \boxtimes)$ is equally expressive as the language of instantial neighborhood logic [20].

2.2 Connections with INL

Recall the language and semantics of instantial neighborhood logic [20]:

Definition 2.7 (Language and semantics of $\mathcal{L}^{\text{inl}}(\Box)$) *Formulas in* $\mathcal{L}^{\text{inl}}(\Box)$ *are defined via*

$$\alpha ::= \bot \mid p \mid (\alpha \to \alpha) \mid \Box(\alpha, \ldots, \alpha; \alpha)$$

and $\neg, \wedge, \vee$ *are defined as abbreviations as usual. The semantics is also based on neighborhood structures:*

$M, w \Vdash \Box(\alpha_1, \ldots, \alpha_n; \beta)$	$\Leftrightarrow$	*there is* $X \in N^M(w)$ *that satisfies:* *for each* $1 \le i \le n$*, there is* $v_i \in X$ *s.t.* $M, v_i \Vdash \alpha_i$*;* *and for all* $v \in X$*:* $M, v \Vdash \beta$.

Theorem 2.8 $\mathcal{L}^{\text{ps}}(\boxdot, \boxtimes)$ *and* $\mathcal{L}^{\text{inl}}(\Box)$ *are equally expressive.*

Proof Define the following translation $t : \mathcal{L}^{\text{inl}}(\Box) \to \mathcal{L}^{\text{ps}}(\boxdot, \boxtimes)$ as

$$\begin{aligned} t[\bot] &:= \bot \\ t[p] &:= p \\ t[\alpha_1 \to \alpha_2] &:= t[\alpha_1] \to t[\alpha_2] \\ t[\Box(\alpha_1, \ldots, \alpha_n; \beta)] &:= \diamonddot(\otimes t[\alpha_1] \wedge \cdots \wedge \otimes t[\alpha_n] \wedge \boxtimes t[\beta]) \end{aligned}$$

We can check by induction that given $\alpha \in \mathcal{L}^{\mathtt{inl}}(\Box)$ the image $t[\alpha] \in \mathcal{L}^{\mathtt{ps}}(\boxdot, \boxtimes)$ is indeed well-formed. It is then not hard to show that for every pointed model M, w and $\alpha \in \mathcal{L}^{\mathtt{inl}}(\Box)$: $M, w \Vdash \alpha$ iff $M, w \vDash t[\alpha]$.

For the other direction, given $\gamma \in \mathcal{L}^{\mathtt{ps}}(\boxdot, \boxtimes)$, using Lemma 2.6 repeatedly we get an equivalent formula $\breve{\gamma}$ in the following fragment of $\mathcal{L}^{\mathtt{ps}}(\boxdot, \boxtimes)$:

$$\alpha ::= \bot \mid p \mid (\alpha \to \alpha) \mid \diamonddot(\otimes\alpha \wedge \cdots \wedge \otimes\alpha \wedge \boxtimes\alpha)$$

Define the translation t' from that fragment into $\mathcal{L}^{\mathtt{inl}}(\Box)$ as

$$\begin{aligned} t'[\bot] &:= \bot \\ t'[p] &:= p \\ t'[\alpha_1 \to \alpha_2] &:= t'[\alpha_1] \to t'[\alpha_2] \\ t'[\diamonddot(\otimes\alpha_1 \wedge \cdots \wedge \otimes\alpha_n \wedge \boxtimes\beta)] &:= \Box(t'[\alpha_1], \ldots, t'[\alpha_n]; t'[\beta]) \end{aligned}$$

Then an induction shows that for each pointed model M, w and $\gamma \in \mathcal{L}^{\mathtt{ps}}(\boxdot, \boxtimes)$: $M, w \vDash \gamma$ iff $M, w \Vdash t'[\breve{\gamma}]$. $\Box$

As consequences, the bisimulation notion for $\mathcal{L}^{\mathtt{inl}}(\Box)$ [20] also works for $\mathcal{L}^{\mathtt{ps}}(\boxdot, \boxtimes)$, and the satisfiability problem of $\mathcal{L}^{\mathtt{ps}}(\boxdot, \boxtimes)$ is decidable.

2.3 Connection with cover modality and coalgebraic logic

In [20], to draw the connections between INL and coalgebraic logics, a cover modality ∇ is defined by: $\mathcal{M}, w \vDash \nabla\{\Gamma_1, \ldots, \Gamma_k\}$ if and only if:

(forth) for every $Z \in N(w)$ there is some Γ_i such that:
for every $u \in Z$ there is some $\alpha \in \Gamma_i$ with $u \vDash \alpha$, and
for every $\alpha \in \Gamma_i$ there is some $u \in Z$ with $u \vDash \alpha$;

(back) for every Γ_i there is some $Z \in N(w)$ such that:
for every $u \in Z$ there is some $\alpha \in \Gamma_i$ with $u \vDash \alpha$, and
for every $\alpha \in \Gamma_i$ there is some $u \in Z$ with $u \vDash \alpha$.

Denote the language with ∇ to be $\mathcal{L}(\nabla)$. Now, given the language of PSNL and Theorem 2.8, we can try to compare the expressivity of $\mathcal{L}(\nabla)$ and $\mathcal{L}^{\mathtt{inl}}(\Box)$ via $\mathcal{L}^{\mathtt{ps}}(\boxdot, \boxtimes)$.

First of all, let $\Gamma = \{\Gamma_1, \ldots \Gamma_k\}$, it is not hard to show that:

$$\vDash \nabla\{\Gamma_1, \ldots, \Gamma_k\} \leftrightarrow \boxdot \bigvee_{\Gamma_i \in \Gamma} (\boxtimes \bigvee_{\alpha \in \Gamma_i} \alpha \wedge \bigwedge_{\alpha \in \Gamma_i} \otimes\alpha) \wedge \bigwedge_{\Gamma_i \in \Gamma} \diamonddot(\boxtimes \bigvee_{\alpha \in \Gamma_i} \alpha \wedge \bigwedge_{\alpha \in \Gamma_i} \otimes\alpha) \quad (\star)$$

Note that ∇ is equivalent to the nesting of two usual cover modalities, one for $\boxdot$ and the other for $\boxtimes$, which justifies the chosen symbol of ∇. As it is well-known, these two cover modalities can define $\boxdot$ and $\boxtimes$, respectively. Therefore, the difference between $\mathcal{L}^{\mathtt{ps}}(\boxdot, \boxtimes)$ and $\mathcal{L}(\nabla)$ lies in that the latter bundles the two cover modalities together.

By $(\star)$ we can conclude that $\mathcal{L}(\nabla)$ is not more expressive than $\mathcal{L}^{\mathtt{inl}}(\Box)$ and $\mathcal{L}^{\mathtt{ps}}(\boxdot, \boxtimes)$. The remaining question is to see whether $\mathcal{L}(\nabla)$ is *equally* expressive as the other two. In [20], it is shown that each INL-formula is equivalent to a disjunction of "balanced" formulas $\Box(\alpha_1, ..., \alpha_n; \bigvee_{i=1}^{n} \alpha_i)$. Over models without empty neighborhoods, i.e., $\varnothing \notin N(w)$ for all w, such an balanced formula can be expressed by $\mathbb{W}\{\{\top\},\{\alpha_1,\dots,\alpha_n\}\}$. Without assuming the absence of empty neighborhoods, it is not so clear whether each $\mathcal{L}^{\mathtt{inl}}(\Box)$-formula can be expressed by $\mathcal{L}(\mathbb{W})$. A seemingly possible candidate would be $\mathbb{W}\{\varnothing,\{\top\},\{\alpha_1,\dots,\alpha_n\}\}$, as in the equivalent PSNL-formula ($\star$), the first conjunct is equivalent to $\top$. However, in the set prefixed by $\mathbb{W}$, elements $\varnothing$ and $\{\top\}$ require an empty and a non-empty neighborhood, hence, this candidate is not a desired one.

Nevertheless, *if* $\mathcal{L}(\mathbb{W})$ and $\mathcal{L}^{\mathtt{inl}}(\Box)$ are proven to be equally expressive, then it will facilitate applications of known techniques in coalgebraic logic with cover modalities developed in [12], which can show that $\mathcal{L}(\mathbb{W})$ has the uniform interpolation property semantically. In that case, the same property follows immediately for both INL and PSNL. Independent from the coalgebraic method, in Section 5 we will use a constructive proof-theoretical method to show PSNL (and thus INL) has the uniform interpolation property.

3 Axiomatization and completeness

3.1 A Hilbert-style proof system

In this section, we propose a system making use of two provability symbols $\vdash_{\mathtt{p}}$ and $\vdash_{\mathtt{s}}$ for point-formulas and set-formulas respectively.

Definition 3.1 (System $\mathsf{HK}^{\boxdot}_{\boxtimes}$) *Axiom schemes and rules are shown below:*

Axiom schemas

TAUTP	$\vdash_{\mathtt{p}}$ CPL$_{\mathtt{p}}$
DIST⊡	$\vdash_{\mathtt{p}} \boxdot(\varphi\to\psi)\to(\boxdot\varphi\to\boxdot\psi)$
TAUTS	$\vdash_{\mathtt{s}}$ CPL$_{\mathtt{s}}$
DIST⊠	$\vdash_{\mathtt{s}} \boxtimes(\alpha\to\beta)\to(\boxtimes\alpha\to\boxtimes\beta)$

where CPL$_{\mathtt{p}}$/CPL$_{\mathtt{s}}$ *stands for classical propositional axiom schemas for point-/set-formulas.*

Rules

MPP	$\dfrac{\vdash_{\mathtt{p}}\alpha \quad \vdash_{\mathtt{p}}\alpha\to\beta}{\vdash_{\mathtt{p}}\beta}$
MPS	$\dfrac{\vdash_{\mathtt{s}}\varphi \quad \vdash_{\mathtt{s}}\varphi\to\psi}{\vdash_{\mathtt{s}}\psi}$
NEC⊡	$\dfrac{\vdash_{\mathtt{s}}\varphi}{\vdash_{\mathtt{p}}\boxdot\varphi}$
NEC⊠	$\dfrac{\vdash_{\mathtt{p}}\alpha}{\vdash_{\mathtt{s}}\boxtimes\alpha}$

A proof of $\vdash_{\mathtt{p}}\alpha$ ($\vdash_{\mathtt{s}}\varphi$) is a finite sequence of $\vdash_{\mathtt{p}}$ and $\vdash_{\mathtt{s}}$ statements ending with $\vdash_{\mathtt{p}}\alpha$ ($\vdash_{\mathtt{s}}\varphi$) s.t. each item in the sequence is either an instance of an axiom or the result of applying a rule to previous items. For any set of point-formulas $\Gamma\cup\{\alpha\}$, we write $\Gamma\vdash_{\mathtt{p}}\alpha$ iff there are finitely many $\beta_1,\dots,\beta_n\in\Gamma$ s.t. $\vdash_{\mathtt{p}}\beta_1\wedge\dots\wedge\beta_n\to\alpha$ is provable. Similarly, we can define $\Sigma\vdash_{\mathtt{s}}\varphi$ for any set of set-formulas $\Sigma\cup\{\varphi\}$.[4]

Remark 3.2 The above system can be viewed as a mixture of the two normal K-systems with NEC⊡ and NEC⊠ as *bridging rules* between the two types of formulas. Thus a proof of $\vdash_{\mathtt{p}}\alpha$ in $\mathsf{HK}^{\boxdot}_{\boxtimes}$ may involve using $\vdash_{\mathtt{s}}\varphi$. Therefore it is not the *fusion* (or *product*) of the two K-systems [20].

By using DIST⊡, DIST⊠ and NEC⊡, NEC⊠, it is easy to show:

[4] The subscripts of $\vdash$ might be omitted as the shape of a formula itself can tell whether it is a point-formula or a set-formula. We include them for conceptual clarity and convenience.

Proposition 3.3 *The following monotonicity rules are admissible in* $\mathsf{HK}^{\boxdot}_{\boxtimes}$.

$$\frac{\vdash_s \varphi \to \psi}{\vdash_p \boxdot\varphi \to \boxdot\psi} \qquad \frac{\vdash_s \varphi \to \psi}{\vdash_p \diamonddot\varphi \to \diamonddot\psi} \qquad \frac{\vdash_p \alpha \to \beta}{\vdash_s \boxtimes\alpha \to \boxtimes\beta} \qquad \frac{\vdash_p \alpha \to \beta}{\vdash_s \otimes\alpha \to \otimes\beta}$$

We can show by inductions the replacement rules are admissible.

Proposition 3.4 *The following rules are admissible in* $\mathsf{HK}^{\boxdot}_{\boxtimes}$.

$$\frac{\vdash_s \varphi \leftrightarrow \psi}{\vdash_p \alpha \leftrightarrow \alpha[\varphi/\psi]} \qquad \frac{\vdash_p \alpha \leftrightarrow \beta}{\vdash_s \varphi \leftrightarrow \varphi[\alpha/\beta]} \qquad \frac{\vdash_p \alpha \leftrightarrow \beta}{\vdash_p \gamma \leftrightarrow \gamma[\alpha/\beta]} \qquad \frac{\vdash_s \varphi \leftrightarrow \psi}{\vdash_s \xi \leftrightarrow \xi[\varphi/\psi]}$$

By mimicking the standard proofs in Hilbert system K, we can show:

Proposition 3.5 *The following are provable in* $\mathsf{HK}^{\boxdot}_{\boxtimes}$.

$$\begin{array}{ll} \vdash_p \boxdot\varphi \leftrightarrow \neg\diamonddot\neg\varphi & \vdash_s \boxtimes\alpha \leftrightarrow \neg\otimes\neg\alpha \\ \vdash_p \boxdot(\varphi\wedge\psi) \leftrightarrow (\boxdot\varphi\wedge\boxdot\psi) & \vdash_s \boxtimes(\alpha\wedge\beta) \leftrightarrow (\boxtimes\alpha\wedge\boxtimes\beta) \\ \vdash_p \diamonddot(\varphi\vee\psi) \leftrightarrow (\diamonddot\varphi\vee\diamonddot\psi) & \vdash_s \otimes(\alpha\vee\beta) \leftrightarrow (\otimes\alpha\vee\otimes\beta) \end{array}$$

As an example, we show how to prove one crucial axiom schema `Case`

$$\Box(\alpha_1,...,\alpha_n;\beta) \to \Box(\alpha_1,...,\alpha_n,\gamma;\beta) \vee \Box(\alpha_1,...,\alpha_n;\beta\wedge\neg\gamma)$$

in the axiomatization of INL [20].

Example 3.6 We prove the corresponding formula of the `Case` schema:

$$\begin{aligned} \vdash_p\ & \diamonddot(\otimes\alpha_1\wedge\cdots\wedge\otimes\alpha_n\wedge\boxtimes\beta) \to \\ & \diamonddot(\otimes\alpha_1\wedge\cdots\wedge\otimes\alpha_n\wedge\otimes\gamma\wedge\boxtimes\beta) \vee \diamonddot(\otimes\alpha_1\wedge\cdots\wedge\otimes\alpha_n\wedge\boxtimes(\beta\wedge\neg\gamma)) \end{aligned}$$

To illustrate the idea, we only prove the following simplified ($n = 1$) schema:

$$\vdash_p \diamonddot(\otimes\alpha\wedge\boxtimes\beta) \to \diamonddot(\otimes\alpha\wedge\otimes\gamma\wedge\boxtimes\beta) \vee \diamonddot(\otimes\alpha\wedge\boxtimes(\beta\wedge\neg\gamma))$$

Proof Using `TAUTS` and other modal transformations, we have:

$$\vdash_s \otimes\alpha\wedge\boxtimes\beta \to (\otimes\alpha\wedge\otimes\gamma\wedge\boxtimes\beta) \vee (\otimes\alpha\wedge\neg\otimes\gamma\wedge\boxtimes\beta),$$

$$\vdash_s \otimes\alpha\wedge\boxtimes\beta \to (\otimes\alpha\wedge\otimes\gamma\wedge\boxtimes\beta) \vee (\otimes\alpha\wedge\boxtimes(\neg\gamma\wedge\beta)).$$

By the admissible monotonicity rule,

$$\vdash_p \diamonddot(\otimes\alpha\wedge\boxtimes\beta) \to \diamonddot((\otimes\alpha\wedge\otimes\gamma\wedge\boxtimes\beta) \vee (\otimes\alpha\wedge\boxtimes(\beta\wedge\neg\gamma))),$$

and $\vdash_p \diamonddot(\otimes\alpha\wedge\boxtimes\beta) \to \diamonddot(\otimes\alpha\wedge\otimes\gamma\wedge\boxtimes\beta) \vee \diamonddot(\otimes\alpha\wedge\boxtimes(\beta\wedge\neg\gamma))$ then follows. □

Theorem 3.7 (Soundness) *For any point-formulas* α, *if* $\vdash_p \alpha$ *then* $\vDash \alpha$. *For any set-formulas* φ, *if* $\vdash_s \varphi$ *then* $\Vdash \varphi$.

Proof By definition, a proof for $\vdash_p \alpha$ is a finite sequence of $\vdash_p$ and $\vdash_s$ statements in which each statement is either an axiom or the result of applying a rule. Via a mutual induction on proof length, we can show that each axiom is valid w.r.t. the corresponding $\vDash$ or $\Vdash$, and that all the rules preserve the corresponding $\vDash$ or $\Vdash$ validity. □

3.2 Completeness proof

Theorem 3.8 (Strong completeness) *For every set* $\Gamma \cup \{\alpha\}$ *of point-formulas and every set* $\Sigma \cup \{\varphi\}$ *of set-formulas, we have:*

$$\Gamma \vdash_{\mathsf{p}} \alpha \textit{ iff } \Gamma \vDash \alpha \quad \textit{and} \quad \Sigma \vdash_{\mathsf{s}} \varphi \textit{ iff } \Sigma \Vdash \varphi.$$

Proof We show that each $\vdash_{\mathsf{p}}$-consistent set of point-formulas has a pointed model, and each $\vdash_{\mathsf{s}}$-consistent set of set-formulas has a model w.r.t. some set X of worlds. Actually, we just need to build a single canonical model $M^c = \langle W^c, N^c, V^c \rangle$ where

- W^c is the set of all maximal $\vdash_{\mathsf{p}}$-consistent sets of *point-formulas*,
- $N^c(\Delta) = \{X \subseteq W^c \mid \Delta^\flat \subseteq \text{supp}(X)\}$ for each $\Delta \in W^c$,
- $V^c(p) = \{\Delta \in W^c \mid p \in \Delta\}$;

where $\Delta^\flat := \{\varphi \mid \Box\varphi \in \Delta\}$ and $\text{supp}(X)$ is the collection of all set-formulas *supported* by $X \subseteq W^c$ in the following sense:

$$\begin{array}{rl} X \text{ supports } \boxtimes\alpha & \text{iff } \alpha \in \Theta \text{ for all } \Theta \in X \\ X \text{ supports } \neg\varphi & \text{iff } X \text{ does not support } \varphi \\ X \text{ supports } \varphi \to \psi & \text{iff } X \text{ does not support } \varphi \text{ or } X \text{ supports } \psi. \end{array}$$

Before we proceed to the truth lemma, here is a claim at first:

Claim (#) Every $\vdash_{\mathsf{s}}$-consistent set of set-formulas is supported by an $X \subseteq W^c$.
Proof of Claim (#): Given a $\vdash_{\mathsf{s}}$-consistent set Σ of set-formulas, we extend it to a maximal $\vdash_{\mathsf{s}}$-consistent set Σ^+ of set-formulas. We want to build a set X of $\vdash_{\mathsf{p}}$-MCSs s.t. elements of Σ^+ are all supported by X. Due to the definition of support, we only need to consider "literals" in Σ^+. Group them in $\Sigma_1^+ = \{\boxtimes\beta_1, \boxtimes\beta_2, \dots\}$ and $\Sigma_0^+ = \{\neg\boxtimes\gamma_1, \neg\boxtimes\gamma_2, \dots\}$. We need to build a set X of $\vdash_{\mathsf{p}}$-MCSs s.t. (a) each $\neg\boxtimes\gamma_i$ is witnessed (for each γ_i there is a $\vdash_{\mathsf{p}}$-MCS in X containing $\neg\gamma_i$), and (b) each $\boxtimes\beta_j$ is supported (all $\vdash_{\mathsf{p}}$-MCSs in X should contain all the β_j formulas). Now we just need to show for each i, $(\Sigma_1^+)^\flat \cup \{\neg\gamma_i\}$ is $\vdash_{\mathsf{p}}$-consistent, where $(\Sigma_1^+)^\flat := \{\beta \mid \boxtimes\beta \in \Sigma_1^+\} = \{\beta_1, \beta_2 \dots\}$. If so, we can extend it to a $\vdash_{\mathsf{p}}$-MCS and add it into our X. By collecting all these witnesses, we will have a desired X that supports Σ.

Suppose towards a contradiction that for some i, $(\Sigma_1^+)^\flat \cup \{\neg\gamma_i\}$ is not $\vdash_{\mathsf{p}}$-consistent. Then there are $\beta_{k_1}, \dots, \beta_{k_m} \in (\Sigma_1^+)^\flat$ s.t. $\vdash_{\mathsf{p}} \beta_{k_1} \wedge \dots \wedge \beta_{k_m} \to \gamma_i$. By $\mathtt{NEC}\boxtimes$ and $\mathtt{DIST}\boxtimes$, $\vdash_{\mathsf{s}} \boxtimes(\beta_{k_1} \wedge \dots \wedge \beta_{k_m}) \to \boxtimes\gamma_i$ whence $\vdash_{\mathsf{s}} \boxtimes\beta_{k_1} \wedge \dots \wedge \boxtimes\beta_{k_m} \to \boxtimes\gamma_i$. As Σ^+ is maximally $\vdash_{\mathsf{s}}$-consistent and $\boxtimes\beta_{k_1}, \dots, \boxtimes\beta_{k_m} \in \Sigma^+$, we have $\boxtimes\gamma_i \in \Sigma^+$. But then Σ^+ is no longer $\vdash_{\mathsf{s}}$-consistent as $\neg\boxtimes\gamma_i \in \Sigma_0^+ \subseteq \Sigma^+$, contradiction! □

Now we show the following **truth lemma** to complete the proof. For every point-formula α, set-formula φ, $\Delta \in W^c$ and $X \subseteq W^c$:

$$\alpha \in \Delta \text{ iff } M^c, \Delta \vDash \alpha \quad \text{and} \quad \varphi \in \text{supp}(X) \text{ iff } M^c, X \Vdash \varphi$$

In an induction, Boolean cases are trivial, and here we show:

$$(1)\, \Box\varphi \in \Delta \text{ iff } M^c, \Delta \vDash \Box\varphi \quad \text{and} \quad (2)\, \boxtimes\alpha \in \text{supp}(X) \text{ iff } M^c, X \Vdash \boxtimes\alpha$$

From left to right: (1) Suppose $\boxdot\varphi \in \Delta$, then $\varphi \in \Delta^\flat$ and hence $\varphi \in \mathrm{supp}(X)$ for each $X \in N^c(\Delta)$. By IH, $M^c, X \Vdash \varphi$, and arbitrariness of X then entails $M^c, \Delta \vDash \boxdot\varphi$.

(2) Suppose $\boxtimes\alpha \in \mathrm{supp}(X)$ then $\alpha \in \Theta$ for every $\Theta \in X$. By IH, $M^c, \Theta \vDash \alpha$, and arbitrariness of Θ then entails $M^c, X \Vdash \boxtimes\alpha$.

From right to left: (1) For the contrapositive: suppose $\boxdot\varphi \notin \Delta$ then $\diamondminus\neg\varphi \in \Delta$, we need to show $M^c, \Delta \vDash \diamondminus\neg\varphi$. By the semantics, we need to find an $X \in N^c(\Delta)$ s.t. $M^c, X \Vdash \neg\varphi$, i.e., $M^c, X \nVdash \varphi$. Using the IH, this amounts to finding an $X \in N^c(\Delta)$ s.t. $\neg\varphi \in \mathrm{supp}(X)$. In the following we construct such an X. We first collect $\Sigma = \{\neg\varphi\} \cup \Delta^\flat$ and claim its $\vdash_{\mathsf{s}}$-consistency.

Suppose towards a contradiction that Σ is not $\vdash_{\mathsf{s}}$-consistent, then there are $\psi_1, \ldots, \psi_n \in \Delta^\flat$ s.t. $\vdash_{\mathsf{s}} \psi_1 \wedge \cdots \wedge \psi_n \to \varphi$ and $\boxdot\psi_1, \ldots, \boxdot\psi_n \in \Delta$. Then we have $\vdash_{\mathsf{p}} \boxdot(\psi_1 \wedge \cdots \wedge \psi_n \to \varphi)$ by $\mathtt{NEC}\boxdot$, $\vdash_{\mathsf{p}} \boxdot(\psi_1 \wedge \cdots \wedge \psi_n) \to \boxdot\varphi$ by $\mathtt{DIST}\boxdot$, and finally $\vdash_{\mathsf{p}} \boxdot\psi_1 \wedge \cdots \wedge \boxdot\psi_n \to \boxdot\varphi$. However, as $\boxdot\psi_1, \ldots, \boxdot\psi_n \in \Delta$, we see $\boxdot\varphi \in \Delta$, contradiction.

With $\vdash_{\mathsf{s}}$-consistency of Σ verified, Claim (#) above ensures an $X \subseteq W^c$ supporting $\Sigma \ni \neg\varphi$. By the definition of Σ, we see $X \in N^c(\Delta)$ as desired.

(2) Suppose $M^c, X \Vdash \boxtimes\alpha$ then $M^c, \Theta \vDash \alpha$ for all $\Theta \in X$. By IH, $\alpha \in \Theta$ for all $\Theta \in X$. Then by definition, $\boxtimes\alpha \in \mathrm{supp}(X)$.

Based on the truth lemma, every $\vdash_{\mathsf{p}}$-consistent set Γ of point-formulas is satisfied at some world $\Delta \supseteq \Gamma$ in M^c, and each $\vdash_{\mathsf{s}}$-consistent set of set-formulas is satisfiable for some $X \subseteq W^c$ due to Claim (#). □

The canonical model construction can be transformed into an equivalent one in the setting of INL (cf. also [18]). As a usual consequence of strong completeness, we can see PSNL is compact, and thus INL is compact as they are equally expressive. Therefore, the axiomatization in [20] is also strongly complete.

4 A sequent calculus $\mathsf{G3K}^{\boxdot}_{\boxtimes}$ for PSNL

From now on, denote the set of all point-(set-)formulas by $\mathcal{F}_{\mathsf{p}}$ (and $\mathcal{F}_{\mathsf{s}}$) for brevity. A point-/set-*sequent* is an ordered pair of finite multi-sets of point-/set-formulas, written as $\Gamma \Rightarrow \Delta$ and $\Pi \Rrightarrow \Sigma$. A point-/set-sequent is *empty*, if both multi-sets in the pair are empty. Adopt usual conventions when writing multi-sets in sequents, e.g. (where $\cup$ is multi-set union) $\alpha, \Gamma \Rightarrow \Delta, \boxdot\xi$ means $\{\alpha\} \cup \Gamma \Rightarrow \Delta \cup \{\boxdot\xi\}$, and $\Rrightarrow \varphi, \varphi$ means $\varnothing \Rrightarrow \{\varphi, \varphi\}$. Also, let $\boxdot\Pi := \{\boxdot\pi \mid \pi \in \Pi\}$ and $\boxtimes\Gamma := \{\boxtimes\gamma \mid \gamma \in \Gamma\}$ be multi-sets of point-/set-formulas. The *degree* of a point-/set-formula is defined as the size of its parsing tree -1, which is also the number of primitive operator (except $\bot$) occurrences in that point-/set-formula. Formulas with degree 0 (those in P and $\bot$) are said to be *prime*. The *degree of a sequent* is defined as the sum of degree of its formulas (count all copies in multi-sets).

Definition 4.1 ($\mathsf{G3K}^{\boxdot}_{\boxtimes}$) *The calculus* $\mathsf{G3K}^{\boxdot}_{\boxtimes}$ *has following rules (where* $p \in P$*):*

$$\frac{}{p, \Gamma \Rightarrow \Delta, p}\,(\mathrm{p}Ax) \qquad \frac{}{\bot, \Gamma \Rightarrow \Delta}\,(\mathrm{p}L\bot)$$

$$\frac{\Pi \Rrightarrow \Sigma, \varphi}{\neg\varphi, \Pi \Rrightarrow \Sigma}\,(\mathrm{s}L\neg) \qquad \frac{\varphi, \Pi \Rrightarrow \Sigma}{\Pi \Rrightarrow \Sigma, \neg\varphi}\,(\mathrm{s}R\neg)$$

$$\frac{\Gamma \Rightarrow \Delta, \alpha \quad \beta, \Gamma \Rightarrow \Delta}{\alpha \to \beta, \Gamma \Rightarrow \Delta}\,(\mathrm{p}L\to) \qquad \frac{\alpha, \Gamma \Rightarrow \Delta, \beta}{\Gamma \Rightarrow \Delta, \alpha \to \beta}\,(\mathrm{p}R\to)$$

$$\frac{\Pi \Rrightarrow \Sigma, \varphi \quad \psi, \Pi \Rrightarrow \Sigma}{\varphi \to \psi, \Pi \Rrightarrow \Sigma}\,(\mathrm{s}L\to) \qquad \frac{\varphi, \Pi \Rrightarrow \Sigma, \psi}{\Pi \Rrightarrow \Sigma, \varphi \to \psi}\,(\mathrm{s}R\to)$$

$$\frac{\Pi \Rrightarrow \sigma}{\boxdot\Pi, \Gamma \Rightarrow \Delta, \boxdot\sigma}\,(\mathtt{p}\boxdot) \qquad \frac{\Gamma \Rightarrow \delta}{\boxtimes\Gamma, \Pi \Rrightarrow \Sigma, \boxtimes\delta}\,(\mathtt{s}\boxtimes)$$

For each rule scheme and its appliction, each formula-occurrence in the conclusion has a unique role (weak/side/principal) defined as usual (cf. [17]). Each rule respects polarities (positive/negative) of occurrences (also cf. [17]). A $\mathsf{G3K}^{\boxdot}_{\boxtimes}$-*proof* (or simply a *proof* when the context is clear) is a finite directed tree with sequent-occurrences as nodes, and each node is the conclusion of a rule scheme applied to all its child-nodes in order as premises. Each proof is called a proof of its root (sequent). By $\vdash^n_{\mathsf{G3K}^{\boxdot}_{\boxtimes}} \Gamma \Rightarrow \Delta$, we mean $\Gamma \Rightarrow \Delta$ has a $\mathsf{G3K}^{\boxdot}_{\boxtimes}$-proof that (as a directed tree) has height[5] $\leq n$, and such a proof is said to *exemplify* the statement $\vdash^n_{\mathsf{G3K}^{\boxdot}_{\boxtimes}} \Gamma \Rightarrow \Delta$ of *n-provability.* A similar notation is employed for provability of set-sequents. When height n is inessential, we may write $\vdash_{\mathsf{G3K}^{\boxdot}_{\boxtimes}}$ for simplicity. Note that for each rule, the number of premises is finite, and the degree of a premise is always strictly smaller than that of the conclusion. Therefore, we see the following:

Proposition 4.2 $\mathsf{G3K}^{\boxdot}_{\boxtimes}$ *enjoys the sub-formula property. Provability of sequents in* $\mathsf{G3K}^{\boxdot}_{\boxtimes}$ *is decidable, and* $\mathsf{G3K}^{\boxdot}_{\boxtimes}$ *supports mechanical proof-search.*

It is well-known that sequent calculus and tableau system of purely classical propositional logic can be seen as dual versions of each other, and that correspondence generalizes to many modal logics including K [6]. In the perspective of proof-search, $\mathsf{G3K}^{\boxdot}_{\boxtimes}$ can also be viewed as the dual version of a tableau system. In $\mathsf{G3K}^{\boxdot}_{\boxtimes}$ appearances of "built-in" weakenings (weak occurrences in $(\mathtt{p}\boxdot)$ and $(\mathtt{s}\boxtimes)$) indicate the "destructive" nature of such tableau systems.

Definition 4.3 Consider an arbitrary rule scheme (X) with finitely-many premises. If for all instances of (X), provability of all its premises entails provability of its conclusion, then we say (X) is *admissible.* If for all instances of (X) and all $n \in \mathbb{N}$, n-provability of all its premises entails n-provability of its conclusion, then we say (X) is *height-preserved-admissible.*

The following admissibility of Weakening is trivial.

[5] Height is defined as the distance between endpoints of the longest directed path. For instance, a tree with only one node has height 0.

Lemma 4.4 *The following schemes are height-preserved-admissible in* $\mathsf{G3K}^{\boxdot}_{\boxtimes}$*:*

$$\frac{\Gamma \Rightarrow \Delta}{\theta, \Gamma \Rightarrow \Delta}(\mathtt{p}LW) \quad \frac{\Gamma \Rightarrow \Delta}{\Gamma \Rightarrow \Delta, \theta}(\mathtt{p}RW) \quad \frac{\Pi \Rrightarrow \Sigma}{\xi, \Pi \Rrightarrow \Sigma}(\mathtt{s}LW) \quad \frac{\Pi \Rrightarrow \Sigma}{\Pi \Rrightarrow \Sigma, \xi}(\mathtt{s}RW)$$

From now on, let $(\mathtt{p}W)$ (resp. $(\mathtt{s}W)$) denote a finite (possibly empty) list of consecutively applied $(\mathtt{p}LW)$ and $(\mathtt{p}RW)$ (resp. $(\mathtt{s}LW)$ and $(\mathtt{s}RW)$).

By a mutual induction on α and φ, we have the following lemma to justify our routine choice of allowing only propositional letters as principal formula-occurrences of $(\mathtt{p}Ax)$.

Lemma 4.5 $\vdash_{\mathsf{G3K}^{\boxdot}_{\boxtimes}} \alpha \Rightarrow \alpha$ *and* $\vdash_{\mathsf{G3K}^{\boxdot}_{\boxtimes}} \varphi \Rrightarrow \varphi$*, for each* $\alpha \in \mathcal{F}_{\mathsf{p}}$ *and* $\varphi \in \mathcal{F}_{\mathsf{s}}$.

This also leads to admissibilities of rule schemes for defined operators.

Lemma 4.6 *All of the following schemes are admissible in* $\mathsf{G3K}^{\boxdot}_{\boxtimes}$*:*

$$\frac{\Gamma \Rightarrow \Delta, \alpha}{\neg\alpha, \Gamma \Rightarrow \Delta}(\mathtt{p}L\neg) \qquad \frac{\alpha, \Gamma \Rightarrow \Delta}{\Gamma \Rightarrow \Delta, \neg\alpha}(\mathtt{p}R\neg)$$

$$\frac{\alpha, \beta, \Gamma \Rightarrow \Delta}{\alpha \wedge \beta, \Gamma \Rightarrow \Delta}(\mathtt{p}L\wedge) \qquad \frac{\Gamma \Rightarrow \Delta, \alpha \quad \Gamma \Rightarrow \Delta, \beta}{\Gamma \Rightarrow \Delta, \alpha \wedge \beta}(\mathtt{p}R\wedge)$$

$$\frac{\varphi, \psi, \Pi \Rrightarrow \Sigma}{\varphi \wedge \psi, \Pi \Rrightarrow \Sigma}(\mathtt{s}L\wedge) \qquad \frac{\Pi \Rrightarrow \Sigma, \varphi \quad \Pi \Rrightarrow \Sigma, \psi}{\Pi \Rrightarrow \Sigma, \varphi \wedge \psi}(\mathtt{s}R\wedge)$$

$$\frac{\alpha, \Gamma \Rightarrow \Delta \quad \beta, \Gamma \Rightarrow \Delta}{\alpha \vee \beta, \Gamma \Rightarrow \Delta}(\mathtt{p}L\vee) \qquad \frac{\Gamma \Rightarrow \Delta, \alpha, \beta}{\Gamma \Rightarrow \Delta, \alpha \vee \beta}(\mathtt{p}R\vee)$$

$$\frac{\varphi, \Pi \Rrightarrow \Sigma \quad \psi, \Pi \Rrightarrow \Sigma}{\varphi \vee \psi, \Pi \Rrightarrow \Sigma}(\mathtt{s}L\vee) \qquad \frac{\Pi \Rrightarrow \Sigma, \varphi, \psi}{\Pi \Rrightarrow \Sigma, \varphi \vee \psi}(\mathtt{s}R\vee)$$

Consider an arbitrary rule scheme (X) with a positive but finite number, say n, of premises. For $i \in \{1, 2, ..., n\}$, denote by $(X)_i^{-1}$ the rule scheme with (X)'s conclusion as its sole premise, and (X)'s i-th premise as its conclusion. If $(X)_i^{-1}$ is admissible/height-preserved-admissible for all $i \in \{1, 2, ..., n\}$, then (X) is said to be *invertible/height-preserved-invertible.* As routine, we have:

Lemma 4.7 *All of* $(\mathtt{s}L\neg)$, $(\mathtt{s}R\neg)$, $(\mathtt{p}L\rightarrow)$, $(\mathtt{p}R\rightarrow)$, $(\mathtt{s}L\rightarrow)$, $(\mathtt{s}R\rightarrow)$ *are height-preserved-invertible.*

Then follows height-preserved-admissibility of Contraction.

Lemma 4.8 *The following schemes are height-preserved-admissible in* $\mathsf{G3K}^{\boxdot}_{\boxtimes}$*:*

$$\frac{\theta, \theta, \Gamma \Rightarrow \Delta}{\theta, \Gamma \Rightarrow \Delta}(\mathtt{p}LC) \qquad \frac{\Gamma \Rightarrow \Delta, \theta, \theta}{\Gamma \Rightarrow \Delta, \theta}(\mathtt{p}RC)$$

$$\frac{\xi, \xi, \Pi \Rrightarrow \Sigma}{\xi, \Pi \Rrightarrow \Sigma}(\mathtt{s}LC) \qquad \frac{\Pi \Rrightarrow \Sigma, \xi, \xi}{\Pi \Rrightarrow \Sigma, \xi}(\mathtt{s}RC)$$

Proof A routine mutual induction on proof-height for all the four. Discuss by cases roles of the two displayed θ's or ξ's. The case when one of them is weak is trivial. The case when both of them are side (the last rule-applied must be a negation or implication rule) is covered by IH's. The case when one of them is side and the other is principle (the last rule-applied must be a negation or implication rules) is covered by Lemma 4.7 together with IH's. The case when both of them are principle (possible only when encountering (pLC) or (sLC) with last rule-applied being (p⊡) or respectively (s⊠)) is covered by IH from (sLC) or respectively (pLC). □

From now on, we denote by (pC) (resp. (sC)) a finite (possibly empty) list of consecutively applied (pLC) and (pRC) (resp. (sLC) and (sRC)).

We are now ready to consider the following (context-sharing) Cut rules:

$$\frac{\Gamma \Rightarrow \Delta, \theta \quad \theta, \Gamma \Rightarrow \Delta}{\Gamma \Rightarrow \Delta}(\mathtt{p}Cut) \qquad \frac{\Pi \Rightarrow \Sigma, \xi \quad \xi, \Pi \Rightarrow \Sigma}{\Pi \Rightarrow \Sigma}(\mathtt{s}Cut)$$

Note here that height-preserved admissibilities of Weakening, inverted Boolean rule schemes, and Contraction (Lemmas 4.4,4.7,4.8) *also hold* when (pCut) and (sCut) of any restricted cut-rank (the degree of the "cut-formula" θ or ξ) are included into $\mathsf{G3K}^{\boxdot}_{\boxtimes}$. We are ready to establish admissibility of Cut.

Theorem 4.9 *Both* (pCut) *and* (sCut) *are admissible in* $\mathsf{G3K}^{\boxdot}_{\boxtimes}$.

Proof We perform an induction on "cut-rank" (the degree of the "cut-formula" θ or ξ) with a sub-induction on "cut-level" (sum of heights of premises of the Cut-application), always eliminating a top-most application of maximal-rank Cut. The only thing not routine is, we need mutual inductions on both sorts (point/set) of Cuts.

Case 1, the cut-formula is weak at one of the two premises, we take the same subproof without that weak occurrence of the cut-formula.

Case 2, the cut-formula is side at one of the two premises, then in the subproof rooted at that premise, the last rule-applied must be a negation or implication rule, or a Cut with lower rank. We use IH as in the proof of cut-admissibility of purely classical propositional calculus (cf. [17]).

Case 3, the cut-formula is principal at both premises. We present here the only subcases not involved in purely classical propositional calculus - when cut-formula is ⊡-(or ⊠-)prefixed. Consider a subproof ending with a top-most application of maximal-rank Cut. Assume that is an application of (sCut) (the (pCut) subcase is quite similar), i.e.:

$$\frac{\dfrac{\Gamma \Rightarrow \alpha}{\boxtimes\Gamma, \Pi \Rightarrow \Sigma, \boxtimes\alpha}(\mathtt{s}\boxtimes) \quad \dfrac{\alpha, \Theta \Rightarrow \beta}{\boxtimes\alpha, \boxtimes\Theta, \Phi \Rightarrow \Psi, \boxtimes\beta}(\mathtt{s}\boxtimes)}{\boxtimes\Gamma, \Pi \Rightarrow \Psi, \boxtimes\beta}(\mathtt{s}Cut)$$

where as multi-set unions $(\boxtimes\Gamma) \cup \Pi = (\boxtimes\Theta) \cup \Phi$, and $\Sigma = \Psi \cup \{\boxtimes\beta\}$. Replacing that subproof by

$$\dfrac{\dfrac{\dfrac{\dfrac{\Gamma \Rightarrow \alpha}{\Gamma, \Theta \Rightarrow \alpha, \beta}\,(\mathrm{p}W) \quad \dfrac{\alpha, \Theta \Rightarrow \beta}{\alpha, \Gamma, \Theta \Rightarrow \beta}\,(\mathrm{p}W)}{\Gamma, \Theta \Rightarrow \beta}\,(\mathrm{p}Cut)}{\boxtimes\Gamma, \boxtimes\Theta, \Pi, \Phi \Rrightarrow \Psi, \boxtimes\beta}\,(\mathrm{s}\boxtimes)}{\boxtimes\Gamma, \Pi \Rrightarrow \Psi, \boxtimes\beta}\,(\mathrm{s}C)$$

where the displayed $(\mathrm{p}Cut)$-application has a strictly smaller cut-rank. □

With the help of admissible Weakening and Contraction, as a direct corollary we have the admissibility of context-splitting Cuts.

Corollary 4.10 *The following schemes are admissible in* $\mathsf{G3K}^{\boxdot}_{\boxtimes}$*:*

$$\dfrac{\Gamma_0 \Rightarrow \Delta_0, \theta \quad \theta, \Gamma_1 \Rightarrow \Delta_1}{\Gamma_0, \Gamma_1 \Rightarrow \Delta_0, \Delta_1}\,(\mathrm{p}Cut') \qquad \dfrac{\Pi_0 \Rrightarrow \Sigma_0, \xi \quad \xi, \Pi_1 \Rrightarrow \Sigma_1}{\Pi_0, \Pi_1 \Rrightarrow \Sigma_0, \Sigma_1}\,(\mathrm{s}Cut')$$

We now verify that $\mathsf{G3K}^{\boxdot}_{\boxtimes}$ is indeed a sequent calculus for the (semantically defined) logic PSNL.

Theorem 4.11 *(1) For each* $\alpha \in \mathcal{F}_{\mathrm{p}}$*:* α *is valid iff* $\vdash_{\mathsf{G3K}^{\boxdot}_{\boxtimes}} \Rightarrow \alpha$*;*
(2) For each $\varphi \in \mathcal{F}_{\mathrm{s}}$*:* φ *is valid iff* $\vdash_{\mathsf{G3K}^{\boxdot}_{\boxtimes}} \Rrightarrow \varphi$*.*

Proof For soundness, by induction on $n \in \mathbb{N}$ we show that if (1) $\vdash^n_{\mathsf{G3K}^{\boxdot}_{\boxtimes}} \Gamma \Rightarrow \Delta$ or (2) $\vdash^n_{\mathsf{G3K}^{\boxdot}_{\boxtimes}} \Pi \Rrightarrow \Sigma$, then (1) $\vdash_{\mathsf{HK}^{\boxdot}_{\boxtimes}(\mathrm{p})} \bigwedge\Gamma \to \bigvee\Delta$ or resp. (2) $\vdash_{\mathsf{HK}^{\boxdot}_{\boxtimes}(\mathrm{s})} \bigwedge\Pi \to \bigvee\Sigma$. Cases for prime and Boolean rules are routine. For

$$\dfrac{\Pi \Rrightarrow \sigma}{\boxdot\Pi, \Gamma \Rightarrow \Delta, \boxdot\sigma}\,(\mathrm{p}\boxdot)$$

by IH(2), $\vdash_{\mathsf{HK}^{\boxdot}_{\boxtimes}(\mathrm{s})} \bigwedge\Pi \to \sigma$. By Proposition 3.3, $\vdash_{\mathsf{HK}^{\boxdot}_{\boxtimes}(\mathrm{p})} \boxdot\bigwedge\Pi \to \boxdot\sigma$, and then by Proposition 3.5, $\vdash_{\mathsf{HK}^{\boxdot}_{\boxtimes}(\mathrm{p})} \bigwedge\boxdot\Pi \to \boxdot\sigma$, hence $\vdash_{\mathsf{HK}^{\boxdot}_{\boxtimes}(\mathrm{p})} \bigwedge\Gamma \wedge \bigwedge\boxdot\Pi \to \boxdot\sigma \vee \bigvee\Delta$. The $(\mathrm{s}\boxtimes)$ case can be treated similarly. Now suppose $\vdash_{\mathsf{G3K}^{\boxdot}_{\boxtimes}} \Rightarrow \alpha$ or $\vdash_{\mathsf{G3K}^{\boxdot}_{\boxtimes}} \Rrightarrow \varphi$, we then have $\vdash_{\mathsf{HK}^{\boxdot}_{\boxtimes}(\mathrm{p})} \alpha$ or resp. $\vdash_{\mathsf{HK}^{\boxdot}_{\boxtimes}(\mathrm{s})} \varphi$, which entails validity of α or resp. φ by soundness of $\mathsf{HK}^{\boxdot}_{\boxtimes}$.

For completeness, assume validity of α or φ. By completeness of $\mathsf{HK}^{\boxdot}_{\boxtimes}$, we see $\vdash_{\mathsf{HK}^{\boxdot}_{\boxtimes}(\mathrm{p})} \alpha$ or $\vdash_{\mathsf{HK}^{\boxdot}_{\boxtimes}(\mathrm{s})} \varphi$. By induction on $\mathsf{HK}^{\boxdot}_{\boxtimes}$-proofs, it can be shown that $\vdash_{\mathsf{G3K}^{\boxdot}_{\boxtimes}} \Rightarrow \alpha$ and $\vdash_{\mathsf{G3K}^{\boxdot}_{\boxtimes}} \Rrightarrow \varphi$ for every point-/set-formula α or φ provable in $\mathsf{HK}^{\boxdot}_{\boxtimes}$. Specifically, $\vdash_{\mathsf{G3K}^{\boxdot}_{\boxtimes}} \Rightarrow \boxdot(\varphi \to \psi) \to (\boxdot\varphi \to \boxdot\psi)$ follows from Lemmas 4.5,4.4, and rules $(\mathrm{s}L{\to})$, $(\mathrm{p}\boxdot)$, $(\mathrm{p}R{\to})$; $\vdash_{\mathsf{G3K}^{\boxdot}_{\boxtimes}} \Rrightarrow \boxtimes(\alpha \to \beta) \to (\boxtimes\alpha \to \boxtimes\beta)$ follows from Lemmas 4.5,4.4, and rules $(\mathrm{p}L{\to})$, $(\mathrm{s}\boxtimes)$, $(\mathrm{s}R{\to})$. Rules `MPP` and `MPS` are covered by IH's, Lemma 4.7 and Corollary 4.10. Rules `NEC`$\boxdot$ and `NEC`$\boxtimes$ are covered by IH's, rule $(\mathrm{p}\boxdot)$ and rule $(\mathrm{s}\boxtimes)$. □

In a tableau system for modal logic K, rules for modalities can be restricted to apply only on "Boolean exhausted" branches. We reformulate this idea in $\mathsf{G3K}^{\boxdot}_{\boxtimes}$ as "Boolean-clean" provability.

Definition 4.12 A $\mathsf{G3K}^{\boxdot}_{\boxtimes}$-proof is said to be *Boolean-clean*, iff in that proof:

(a) in each application of $(\mathrm{p}Ax)$ and $(\mathrm{p}L\bot)$, weak occurrences[6] are all prime or $\boxdot$-prefixed;

(b) in each applicaiton of $(\mathrm{p}\boxdot)$, negative weak occurrences are all prime, and positive weak occurrences are either prime or $\boxdot$-prefixed;

(c) in each applicaiton of $(\mathrm{s}\boxtimes)$, there are no negative weak occurrences, and positive weak occurrences are all $\boxtimes$-prefixed.

Lemma 4.13 *Each* $\mathsf{G3K}^{\boxdot}_{\boxtimes}$*-provable point-/set-sequent has a Boolean-clean proof.*

Proof We offer a procedure which, when given a $\mathsf{G3K}^{\boxdot}_{\boxtimes}$-proof (a finite tree), can transform it via a root-to-leaf reduction, to a Boolean-clean $\mathsf{G3K}^{\boxdot}_{\boxtimes}$-proof of the same root sequent. For a better presentation, we temporarily define the "error degree" of a sequent to be the (finite) number of occurrences of set-$\neg$, point-$\rightarrow$, and set-$\rightarrow$ that are not in the scope of any $\boxdot$-/$\boxtimes$-occurrence.[7]

Suppose the procedure encounters an application of $(\mathrm{p}Ax)$, whose conclusion $\Gamma \Rightarrow \Delta$ has error degree > 0. In such a point-sequent, only $\rightarrow$ can occur out of the scope of $\boxdot$-occurrences, hence the error degree is solely due to $\rightarrow$-occurrences. Assume $\Gamma \Rightarrow \Delta$ has the form of $\alpha \rightarrow \beta, \Gamma_0 \Rightarrow \Delta$. Note that $\alpha \rightarrow \beta$ can only be weak in $(\mathrm{p}Ax)$, and $(\mathrm{p}Ax)$ has no restriction on form or number of weak occurrences. This allows the procedure to replace the $(\mathrm{p}Ax)$-application encountered by sub-proof

$$\dfrac{\dfrac{}{\Gamma_0 \Rightarrow \Delta, \alpha}\,(\mathrm{p}Ax) \qquad \dfrac{}{\beta, \Gamma_0 \Rightarrow \Delta}\,(\mathrm{p}Ax)}{\alpha \rightarrow \beta, \Gamma_0 \Rightarrow \Delta}\,(\mathrm{p}L\rightarrow)$$

that has two $(\mathrm{p}Ax)$-applications each with a strictly smaller error degree. The procedure repeats what above on each of these "less-errored" $(\mathrm{p}Ax)$-applications, round by round, and eventually terminates when error degrees of all $(\mathrm{p}Ax)$-applications remain go down to 0. The other subcase, $\Gamma \Rightarrow \Delta$ has the form of $\Gamma \Rightarrow \Delta_0, \alpha \rightarrow \beta$, can be treated similarly.

When encountering an application of $(\mathrm{p}L\bot)$, apply a similar trick.

When the procedure encounters an application of $(\mathrm{p}\boxdot)$, since its conclusion is a point-sequent, only $\rightarrow$ can contribute to the error degree. By the form of $(\mathrm{p}\boxdot)$, it is clear that implications are all weak, and $(\mathrm{p}\boxdot)$ has no restriction on form or number of weak occurrences. This allows the procedure to employ the trick used in the $(\mathrm{p}Ax)$ case, insert $(\mathrm{p}L\rightarrow)/(\mathrm{p}R\rightarrow)$-applications between premise and conclusion of the $(\mathrm{p}\boxdot)$-application, and break it into finitely many $(\mathrm{p}\boxdot)$-applications with error degree 0. It is now sufficient to take care of nega-

---

[6] Recall that in $\mathsf{G3K}^{\boxdot}_{\boxtimes}$, only $(\mathrm{p}Ax), (\mathrm{p}L\bot), (\mathrm{p}\boxdot), (\mathrm{s}\boxtimes)$ permit weak occurrences in their conclusion. These include (as in Definition 4.1), displayed Γ and Δ in $(\mathrm{p}Ax), (\mathrm{p}L\bot), (\mathrm{p}\boxdot)$, as well as displayed Π and Σ in $(\mathrm{s}\boxtimes)$. Occurrences to the left of $\Rightarrow/\Rrightarrow$ are said to be negative, and those to the right are said to be positive.

[7] For instance, $\neg(\neg q \rightarrow \boxdot \boxtimes (p \rightarrow q)) \Rrightarrow \neg p, \neg p$ has error degree 5.

tive weak occurrences that are $\boxdot$-prefixed.[8] This is done by re-assigning them all as negative principal occurrences, as allowed by the form of (p$\boxdot$). Namely,

$$\frac{\Pi \Rightarrow \sigma}{\boxdot\Pi, \boxdot\Phi, \Gamma_0 \Rightarrow \Delta_0, \boxdot\sigma}\,(\mathrm{p}\boxdot) \quad\rightsquigarrow\quad \frac{\dfrac{\Pi \Rightarrow \sigma}{\Pi, \Phi \Rightarrow \sigma}\,(\mathrm{s}W)}{\boxdot\Pi, \boxdot\Phi, \Gamma_0 \Rightarrow \Delta_0, \boxdot\sigma}\,(\mathrm{p}\boxdot)$$

where all occurrences in Γ_0 are prime and all occurrences in Δ_0 are prime or $\boxdot$-prefixed. Termination of the procedure is respected, since $(\mathrm{s}W)$ is height-preserved-admissible (Lemma 4.4).

When the procedure encounters an application of (s$\boxtimes$), since its conclusion is a set-sequent, only $\neg$ and $\rightarrow$ can contribute to the error degree. The procedure works similar as in the (p$\boxdot$) case. Since there is no prime set-formula, after the procedure works through, there will be no negative weak occurrences, and all positive weak occurrences are $\boxtimes$-prefixed. □

5 Uniform interpolation

For each $\alpha \in \mathcal{F}_\mathrm{p}$, let V_α be the set of propositional letters that occur in α. Similarly, we have V_φ for each $\varphi \in \mathcal{F}_\mathrm{s}$, and this notation also generalizes to multi-sets of formulas. As usual, $A - B$ means the difference of sets A and B. For multi-sets, $A \setminus\!\!\setminus B$ means the resulting multi-set of removing from A all copies of elements that have at least one copy included in B.[9]

Uniform *pre-interpolants* (cf. [9]) of a point-/set-formula w.r.t. a finite set of propositional letters can be defined as follows:

Definition 5.1 (Uniform interpolants) *Let $\beta \in \mathcal{F}_\mathrm{p}$, $\psi \in \mathcal{F}_\mathrm{s}$, and $Q \subseteq P$ be finite.*

(1) By a pre-interpolant *of $\langle\beta, Q\rangle$, we mean $\theta \in \mathcal{F}_\mathrm{p}$ that meets:*
(a) $V_\theta \subseteq V_\beta - Q$; (b) $\theta \rightarrow \beta$ is valid; and (c) for each $\alpha \in \mathcal{F}_\mathrm{p}$ s.t. $V_\alpha \cap Q = \varnothing$ and $\alpha \rightarrow \beta$ is valid, $\alpha \rightarrow \theta$ is also valid;

(2) By a pre-interpolant *of $\langle\psi, Q\rangle$, we mean $\xi \in \mathcal{F}_\mathrm{s}$ that meets:*
(a) $V_\xi \subseteq V_\psi - Q$; (b) $\xi \rightarrow \psi$ is valid; and (c) for each $\varphi \in \mathcal{F}_\mathrm{s}$ s.t. $V_\varphi \cap Q = \varnothing$ and $\varphi \rightarrow \psi$ is valid, $\varphi \rightarrow \xi$ is also valid.

It follows from the definition that, for each $\beta \in \mathcal{F}_\mathrm{p}$ (resp. $\psi \in \mathcal{F}_\mathrm{s}$) and $Q \subseteq P$, the pre-interpolant (if exists) of $\langle\beta, Q\rangle$ ($\langle\psi, Q\rangle$) is unique up to equivalence. One may also define *post-interpolants* in a similar way. This is not necessary for our purpose, since PSNL and its calculi are equipped with classical negations,[10] which ensures that, say, if θ is a pre-interpolant of $\langle\neg\beta, Q\rangle$, then $\neg\theta$ is a post-interpolant of $\langle\beta, Q\rangle$. Therefore, if pre-interpolant exists for each $\beta \in \mathcal{F}_\mathrm{p}$ and $Q \subseteq P$, then post-interpolant also exists for each β and Q.

---

[8] Some of them are not independent occurrences in the original (p$\boxdot$)-application, but are contents discharged when the procedure breaks implications. Our procedure takes care of $\boxdot$-prefixed negative weak occurrences only after implications.

[9] For instance, $\{2,1\} - \{3,1\} = \{2\}$ and $\{2,1,1,2,1\} \setminus\!\!\setminus \{3,1,1\} = \{2,2\}$.

[10] Both the primitive set-$\neg$ and the defined point-$\neg$ are classical.

In order to establish uniform interpolation property of PSNL, we construct uniform pre-interpolants in a generalized sense for sequents. What we proceed is actually a mutual version of Marta Bílková's method [2], which in turn was adopted from [15] authored by Andrew Pitts.

Given a non-empty point-/set-sequent s: (1) by $\mathsf{c}(s)$ we mean the closure of s under inverted rule schemes listed in Lemma 4.7; (2) s is said to be *critical*, if s is non-empty and $\mathsf{c}(s) = \{s\}$;[11] and (3) let $\mathsf{cl}(s) := \{x \in \mathsf{c}(s) \mid x \text{ is critical}\}$. It is obvious by forms of inverted rules that: $\mathsf{c}(s)$ *is always finite*, and if s is a point-/set-sequent then so are all sequents in $\mathsf{c}(s)$. For a multi-set $\Theta \subseteq \mathcal{F}_{\mathsf{p}}$, let $\Theta^0 := \{\theta \in \Theta \mid \theta \text{ is prime}\}$ and $\Theta^\natural := \{\theta \in \Theta \mid \theta \text{ is } \boxdot\text{-prefixed}\}$ be multi-sets of point-formulas, and $\Theta^\flat := \{\xi \mid \boxdot\xi \in \Theta\}$ be a multi-set of set-formulas. Likewise, for a multi-set $\Xi \subseteq \mathcal{F}_{\mathsf{s}}$, let $\Xi^\natural := \{\xi \in \Xi \mid \xi \text{ is } \boxtimes\text{-prefixed}\}$ be a multi-set of set-formulas, and $\Xi^\flat := \{\theta \mid \boxtimes\theta \in \Xi^\natural\}$ be a multi-set of point-formulas.

Theorem 5.2 *Assume $Q \subseteq P$ is finite.*

(1) For each point-sequent $\Gamma \Rightarrow \Delta$, there is $\theta^Q_{\Gamma\Delta} \in \mathcal{F}_{\mathsf{p}}$ s.t.: (a) $V_{\theta^Q_{\Gamma\Delta}} \subseteq V_{\Gamma\cup\Delta} - Q$; (b) $\vdash_{\mathsf{G3K}^{\boxdot}_{\boxtimes}} \Gamma, \theta^Q_{\Gamma\Delta} \Rightarrow \Delta$; and (c) $\vdash_{\mathsf{G3K}^{\boxdot}_{\boxtimes}} \Omega \Rightarrow \theta^Q_{\Gamma\Delta}, \Upsilon$ for every point-sequent $\Omega \Rightarrow \Upsilon$ s.t. $V_{\Omega\cup\Upsilon} \cap Q = \varnothing$ and $\vdash_{\mathsf{G3K}^{\boxdot}_{\boxtimes}} \Omega, \Gamma \Rightarrow \Delta, \Upsilon$.

(2) For each set-sequent $\Pi \Rrightarrow \Sigma$, there is $\xi^Q_{\Pi\Sigma} \in \mathcal{F}_{\mathsf{s}}$ s.t.: (a) $V_{\xi^Q_{\Pi\Sigma}} \subseteq V_{\Pi\cup\Sigma} - Q$; (b) $\vdash_{\mathsf{G3K}^{\boxdot}_{\boxtimes}} \Pi, \xi^Q_{\Pi\Sigma} \Rrightarrow \Sigma$; and (c) $\vdash_{\mathsf{G3K}^{\boxdot}_{\boxtimes}} \Phi \Rrightarrow \xi^Q_{\Pi\Sigma}, \Psi$ for every set-sequent $\Phi \Rrightarrow \Psi$ s.t. $V_{\Phi\cup\Psi} \cap Q = \varnothing$ and $\vdash_{\mathsf{G3K}^{\boxdot}_{\boxtimes}} \Phi, \Pi \Rrightarrow \Sigma, \Psi$.

Proof Construct $\theta^Q_{\Gamma\Delta}$ and $\xi^Q_{\Pi\Sigma}$ by a mutual induction on degree of sequents:

$$\theta^Q_{\Gamma\Delta} := \begin{cases} \bot & \text{if } \Gamma = \Delta = \varnothing \\ \top & \text{else if } \Gamma \Rightarrow \Delta \text{ is critical and } Q \cap \Gamma^0 \cap \Delta^0 \neq \varnothing \\ \diamondsuit\xi^Q_{\Gamma^\flat\varnothing} \vee \bigvee_{\sigma\in\Delta^\flat} \boxdot\xi^Q_{\Gamma^\flat\{\sigma\}} \vee \bigvee\neg(\Gamma^0 \backslash\backslash Q) \vee \bigvee(\Delta^0 \backslash\backslash Q) & \\ & \text{else if } \Gamma \Rightarrow \Delta \text{ is critical and } Q \cap \Gamma^0 \cap \Delta^0 = \varnothing \\ \bigwedge_{i\in I} \theta^Q_{\Gamma_i\Delta_i} & \text{else, where } \mathsf{cl}(\Gamma \Rightarrow \Delta) = \{\Gamma_i \Rightarrow \Delta_i\}_{i\in I} \end{cases}$$

$$\xi^Q_{\Pi\Sigma} := \begin{cases} \diamondtimes\bot & \text{if } \Pi = \Sigma = \varnothing \\ \diamondtimes\theta^Q_{\Pi^\flat\varnothing} \vee \bigvee_{\delta\in\Sigma^\flat} \boxtimes\theta^Q_{\Pi^\flat\{\delta\}} & \text{else if } \Pi \Rrightarrow \Sigma \text{ is critical} \\ \bigwedge_{i\in I} \xi^Q_{\Pi_i\Sigma_i} & \text{else, where } \mathsf{cl}(\Pi \Rrightarrow \Sigma) = \{\Pi_i \Rrightarrow \Sigma_i\}_{i\in I} \end{cases}$$

It remains to show the correctness of $\theta^Q_{\Gamma\Delta}$ and $\xi^Q_{\Pi\Sigma}$ as constructed.

Both **(1a)** and **(2a)** are obvious.

For **(1b)** and **(2b)**, we perform a mutual induction (on degree of point-/set-sequents). (1b) $\vdash_{\mathsf{G3K}^{\boxdot}_{\boxtimes}} \Gamma, \theta^Q_{\Gamma\Delta} \Rightarrow \Delta$, there are four cases. Case 1, $\Gamma = \Delta = \varnothing$, which is trival by $(\mathrm{p}L\bot)$. Case 2, $\Gamma \Rightarrow \Delta$ is critical and $Q \cap \Gamma^0 \cap \Delta^0 \neq \varnothing$, which is trivial by $(\mathrm{p}Ax)$. Case 3, $\Gamma \Rightarrow \Delta$ is critical and $Q \cap \Gamma^0 \cap \Delta^0 = \varnothing$. By admissible $(\mathrm{p}L\vee)$, it is sufficient to show $\vdash_{\mathsf{G3K}^{\boxdot}_{\boxtimes}} \Gamma, \eta \Rightarrow \Delta$ for every disjunct η in $\theta^Q_{\Gamma\Delta}$. For

[11] That is, nothing listed in Lemma 4.7 is applicable to s.

$\neg p$ where $p \in \Gamma^0 \setminus\!\setminus Q$ and $p \in \Delta^0 \setminus\!\setminus Q$, the desired provability is direct. For $\diamond\xi^Q_{\Gamma^\flat\varnothing}$, IH(2b) gives $\vdash_{\mathsf{G3K}^{\boxdot}_{\boxtimes}} \Gamma^\flat, \xi^Q_{\Gamma^\flat\varnothing} \Rightarrow \varnothing$, and then we proceed like

$$\dfrac{\dfrac{\dfrac{\Gamma^\flat, \xi^Q_{\Gamma^\flat\varnothing} \Rightarrow \varnothing}{\Gamma^\flat \Rightarrow \neg\xi^Q_{\Gamma^\flat\varnothing}}(\mathsf{s}R\neg)}{\Gamma \Rightarrow \boxdot\neg\xi^Q_{\Gamma^\flat\varnothing}, \Delta}(\mathsf{p}\boxdot)}{\Gamma, \diamond\xi^Q_{\Gamma^\flat\varnothing} \Rightarrow \Delta}(\mathsf{p}L\neg)$$

For $\boxdot\xi^Q_{\Gamma^\flat\{\sigma\}}$ where $\sigma \in \Delta^\flat$, IH(2b) gives $\vdash_{\mathsf{G3K}^{\boxdot}_{\boxtimes}} \Gamma^\flat, \xi^Q_{\Gamma^\flat\{\sigma\}} \Rightarrow \sigma$, which entails the desired $\vdash_{\mathsf{G3K}^{\boxdot}_{\boxtimes}} \Gamma, \boxdot\xi^Q_{\Gamma^\flat\{\sigma\}} \Rightarrow \Delta$ by a further application of $(\mathsf{p}\boxdot)$. Case 4, $\Gamma \Rightarrow \Delta$ is non-critical and $\mathsf{cl}(\Gamma \Rightarrow \Delta) = \{\Gamma_i \Rightarrow \Delta_i\}_{i\in I}$. By admissible $(\mathsf{p}L\wedge)$, it is sufficient to show $\vdash_{\mathsf{G3K}^{\boxdot}_{\boxtimes}} \Gamma, \{\theta^Q_{\Gamma_i\Delta_i}\}_{i\in I} \Rightarrow \Delta$. By IH(1b), $\vdash_{\mathsf{G3K}^{\boxdot}_{\boxtimes}} \Gamma_i, \theta^Q_{\Gamma_i\Delta_i} \Rightarrow \Delta_i$ for each $i \in I$. Then by Lemma 4.4, $\vdash_{\mathsf{G3K}^{\boxdot}_{\boxtimes}} \Gamma_i, \{\theta^Q_{\Gamma_i\Delta_i}\}_{i\in I} \Rightarrow \Delta_i$ for each $i \in I$, which entails the desired $\vdash_{\mathsf{G3K}^{\boxdot}_{\boxtimes}} \Gamma, \{\theta^Q_{\Gamma_i\Delta_i}\}_{i\in I} \Rightarrow \Delta$ via definition of cl. For (2b) $\vdash_{\mathsf{G3K}^{\boxdot}_{\boxtimes}} \Pi, \xi^Q_{\Pi\Sigma} \Rightarrow \Sigma$, a verification is quite similar and a bit simpler (as there are no prime formulas in critical cases).

For **(1c)** and **(2c)**, by Lemma 4.13, statements $\vdash_{\mathsf{G3K}^{\boxdot}_{\boxtimes}} \Omega, \Gamma \Rightarrow \Delta, \Upsilon$ and $\vdash_{\mathsf{G3K}^{\boxdot}_{\boxtimes}} \Phi, \Pi \Rightarrow \Sigma, \Psi$ are exemplified by Boolean-clean proofs. We proceed by a mutual induction on proof heights, and focus on (1c) at first.

Case 1, $\Omega, \Gamma \Rightarrow \Delta, \Upsilon$ is the conclusion of a $(\mathsf{p}Ax)$-application. By Boolean-cleanness, $\Omega, \Gamma \Rightarrow \Delta, \Upsilon$ is critical, and then so is $\Gamma \Rightarrow \Delta$. Denote the principal occurrence (which is a propositional letter) by q. Subcase 1a, $q \in Q$. Since $V_{\Omega\cup\Upsilon} \cap Q = \varnothing$, we see $q \notin V_{\Omega\cup\Upsilon}$ and then $q \in \Gamma^0 \cap \Delta^0$. Thus, $\theta^Q_{\Gamma\Delta}$ is simply $\top$, and $\vdash_{\mathsf{G3K}^{\boxdot}_{\boxtimes}} \Omega \Rightarrow \theta^Q_{\Gamma\Delta}, \Upsilon$ is obvious. Subcase 1b, $q \in (\Gamma^0 \cap \Delta^0) \setminus\!\setminus Q$. Then $\theta^Q_{\Gamma\Delta}$ either is $\top$ or has both q and $\neg q$ as its disjuncts, and hence $\vdash_{\mathsf{G3K}^{\boxdot}_{\boxtimes}} \Omega \Rightarrow \theta^Q_{\Gamma\Delta}, \Upsilon$. Subcase 1c, $q \in (\Gamma^0 \cap \Upsilon) \setminus\!\setminus Q$. Then $\theta^Q_{\Gamma\Delta}$ either is $\top$ or has $\neg q$ as its disjunct. Since $q \in \Upsilon$, we see $\vdash_{\mathsf{G3K}^{\boxdot}_{\boxtimes}} \Omega \Rightarrow \theta^Q_{\Gamma\Delta}, \Upsilon$. Subcase 1d, $q \in (\Omega \cap \Delta^0) \setminus\!\setminus Q$. This is similar to subcase 1c. Subcase 1e, $q \in (\Omega \cap \Upsilon) \setminus\!\setminus Q$. Trivial via a $(\mathsf{p}Ax)$-application.

Case 2, $\Omega, \Gamma \Rightarrow \Delta, \Upsilon$ is the conclusion of a $(\mathsf{p}L\bot)$-application, so $\bot \in \Omega \cup \Gamma$. Similar to case 1, $\Gamma \Rightarrow \Delta$ is critical. Subcase 2a, $\bot \in \Omega$. Trivial. Subcase 2b, $\bot \in \Gamma$. Then $\theta^Q_{\Gamma\Delta}$ either is $\top$ or has $\neg\bot$ as its disjunct, hence $\vdash_{\mathsf{G3K}^{\boxdot}_{\boxtimes}} \Omega \Rightarrow \theta^Q_{\Gamma\Delta}, \Upsilon$.

Case 3, $\Omega, \Gamma \Rightarrow \Delta, \Upsilon$ is the conclusion of a $(\mathsf{p}L\rightarrow)/(\mathsf{p}R\rightarrow)$-application. In this case, $\Omega, \Gamma \Rightarrow \Delta, \Upsilon$ is not critical. Let $\mathsf{cl}(\Gamma \Rightarrow \Delta) = \{\Gamma_i \Rightarrow \Delta_i\}_{i\in I}$ and $\mathsf{cl}(\Omega \Rightarrow \Upsilon) = \{\Omega_j \Rightarrow \Upsilon_j\}_{j\in J}$. By Boolean-cleanness, for each $i \in I$ and $j \in J$, $\vdash_{\mathsf{G3K}^{\boxdot}_{\boxtimes}} \Omega_j, \Gamma_i \Rightarrow \Delta_i, \Upsilon_j$ is exemplified by a Boolean-clean proper sub-proof. Note that $\Omega_j, \Gamma_i \Rightarrow \Delta_i, \Upsilon_j$ is critical and $V_{\Omega_j\cup\Upsilon_j} \cap Q \subseteq V_{\Omega\cup\Upsilon} \cap Q = \varnothing$. By IH(1c), $\vdash_{\mathsf{G3K}^{\boxdot}_{\boxtimes}} \Omega_j \Rightarrow \theta^Q_{\Gamma_i\Delta_i}, \Upsilon_j$. Since $\theta^Q_{\Gamma\Delta} = \bigwedge_{i\in I} \theta^Q_{\Gamma_i\Delta_i}$ (this holds no matter whether $\Gamma \Rightarrow \Delta$ is critical or not), admissible $(\mathsf{p}R\wedge)$ gives $\vdash_{\mathsf{G3K}^{\boxdot}_{\boxtimes}} \Omega_j \Rightarrow \theta^Q_{\Gamma\Delta}, \Upsilon_j$ for each $j \in J$. By $(\mathsf{p}L\rightarrow)/(\mathsf{p}R\rightarrow)$-applications, $\vdash_{\mathsf{G3K}^{\boxdot}_{\boxtimes}} \Omega \Rightarrow \theta^Q_{\Gamma\Delta}, \Upsilon$ as desired.

Case 4, $\Omega, \Gamma \Rightarrow \Delta, \Upsilon$ is the conclusion of a $(\mathsf{p}\boxdot)$-application. In this case,

$\Omega, \Gamma \Rightarrow \Delta, \Upsilon$ is critical. What we need to show is trivial when $\theta^Q_{\Gamma\Delta}$ is $\top$, and hence in what follows, we assume that it is not the case, i.e., $\theta^Q_{\Gamma\Delta}$ is

$$\Diamond\xi^Q_{\Gamma^\flat\varnothing} \vee \bigvee_{\sigma\in\Delta^\flat} \boxdot\xi^Q_{\Gamma^\flat\{\sigma\}} \vee \bigvee\neg(\Gamma^0\backslash\!\backslash Q) \vee \bigvee(\Delta^0\backslash\!\backslash Q).$$

By Boolean-cleanness, $\Omega = \Omega^0 \cup \Omega^\natural$ (a disjoint union), and the similar holds for Γ, Δ, Υ. By the form of (p$\boxdot$), there is a unique positive principal occurrence (denote it by $\boxdot\sigma$), and that (p$\boxdot$)-application has the form of

$$\frac{\Omega^\flat, \Gamma^\flat \Rightarrow \sigma}{\Omega, \Gamma \Rightarrow \Delta, \Upsilon}\ (\mathrm{p}\boxdot)$$

in which $\boxdot\sigma \in \Delta \cup \Upsilon$. Subcase 4a, $\boxdot\sigma \in \Upsilon$. Since $V_{\Omega^\flat\cup\{\sigma\}} \cap Q \subseteq V_{\Omega\cup\Upsilon} \cap Q = \varnothing$, by IH(2c), $\vdash_{\mathsf{G3K}^{\boxdot}_{\boxtimes}} \Omega^\flat \Rightarrow \xi^Q_{\Gamma^\flat\varnothing}, \sigma$. Then we have

$$\cfrac{\cfrac{\cfrac{\cfrac{\Omega^\flat \Rightarrow \xi^Q_{\Gamma^\flat\varnothing}, \sigma}{\Omega^\flat, \neg\xi^Q_{\Gamma^\flat\varnothing} \Rightarrow \sigma}\ (\mathrm{s}L\neg)}{\Omega, \boxdot\neg\xi^Q_{\Gamma^\flat\varnothing} \Rightarrow \Upsilon}\ (\mathrm{p}\boxdot)}{\Omega \Rightarrow \Diamond\xi^Q_{\Gamma^\flat\varnothing}, \Upsilon}\ (\mathrm{p}R\neg)}{\Omega \Rightarrow \theta^Q_{\Gamma\Delta}, \Upsilon}\ (\mathrm{p}W + \mathrm{p}R\vee)$$

where in the final step, since $\Gamma \Rightarrow \Delta$ is critical (recall that $\Gamma = \Gamma^0 \cup \Gamma^\natural$ and $\Delta = \Delta^0 \cup \Delta^\natural$), $\Diamond\xi^Q_{\Gamma^\flat\varnothing}$ is a disjunct of $\theta^Q_{\Gamma\Delta}$. Subcase 4b, $\boxdot\sigma \in \Delta$ (so $\sigma \in \Delta^\flat$). Since $V_{\Omega^\flat\cup\varnothing} \cap Q \subseteq V_{\Omega\cup\Upsilon} \cap Q = \varnothing$, by IH(2c), $\vdash_{\mathsf{G3K}^{\boxdot}_{\boxtimes}} \Omega^\flat \Rightarrow \xi^Q_{\Gamma^\flat\{\sigma\}}$. Then we have

$$\cfrac{\cfrac{\Omega^\flat \Rightarrow \xi^Q_{\Gamma^\flat\{\sigma\}}}{\Omega \Rightarrow \boxdot\xi^Q_{\Gamma^\flat\{\sigma\}}, \Upsilon}\ (\mathrm{p}\boxdot)}{\Omega \Rightarrow \theta^Q_{\Gamma\Delta}, \Upsilon}\ (\mathrm{p}W + \mathrm{p}R\vee)$$

where in the final step, since $\Gamma \Rightarrow \Delta$ is critical and $\sigma \in \Delta^\flat$, $\boxdot\xi^Q_{\Gamma^\flat\{\sigma\}}$ is a disjunct of $\theta^Q_{\Gamma\Delta}$.

This finishes our proof for (1c), and (2c) can be treated similarly. A set-sequent may be derived by an application of (s$L\neg$), (s$R\neg$), (s$L\rightarrow$), (s$R\rightarrow$), or (s$\boxtimes$). Each of these five is similar to implication or modality cases in (1c).□

As a corollary, we now have *uniform interpolation propery* of PSNL.

Corollary 5.3 *If $Q \subseteq P$ is finite, $\beta \in \mathcal{F}_\mathrm{p}$ and $\psi \in \mathcal{F}_\mathrm{s}$, then pre-interpolant and post-interpolant of $\langle\beta, Q\rangle$ as well as that of $\langle\psi, Q\rangle$ always exist.*

Proof It is sufficient to offer pre-interpolants. We only present the case for point-formulas, as those of set-formulas is similar.

By Theorem 5.2(1) in the sense that $\Gamma = \varnothing$ and $\Delta = \{\beta\}$, there is $\theta \in \mathcal{F}_{\mathsf{p}}$ that meets: $V_\theta \subseteq V_\beta - Q$, $\vdash_{\mathsf{G3K}^{\boxdot}_{\boxtimes}} \theta \Rightarrow \beta$, and $\vdash_{\mathsf{G3K}^{\boxdot}_{\boxtimes}} \alpha \Rightarrow \theta$ for every $\alpha \in \mathcal{F}_{\mathsf{p}}$ s.t. $V_\alpha \cap Q = \varnothing$ and $\vdash_{\mathsf{G3K}^{\boxdot}_{\boxtimes}} \alpha \Rightarrow \beta$. By Lemma 4.7 and Theorem 4.11, we know that $\vdash_{\mathsf{G3K}^{\boxdot}_{\boxtimes}} \alpha \Rightarrow \beta$ iff $\vdash_{\mathsf{G3K}^{\boxdot}_{\boxtimes}} \Rightarrow \alpha \to \beta$ iff $\alpha \to \beta$ is valid, and similarly for $\alpha \to \theta$ and $\theta \to \beta$. Therefore, θ is a pre-interpolant of $\langle \beta, Q \rangle$. □

By inter-language translations t and t' in the proof of Theorem 2.8, we get:

Corollary 5.4 *INL has uniform interpolation property.*

Proof Since INL is equipped with classical negation, it is sufficient to offer pre-interpolants for each $\eta \in \mathcal{L}^{\mathtt{inl}}(\Box)$ and finite $Q \subseteq P$. By Corollary 5.3, $\langle t[\eta], Q \rangle$ has a pre-interpolant $\theta \in \mathcal{F}_{\mathsf{p}}$. Then $t'[\breve{\theta}] \in \mathcal{L}^{\mathtt{inl}}(\Box)$ is a pre-interpolant of $\langle \eta, Q \rangle$.

To see this, note that both $t[\cdot]$ and $t'[\breve{\cdot}]$ respect propositional letters, and hence language conditions set by Q are automatically met. INL-validity of $t'[\breve{\theta}] \to \eta$ follows from PSNL-validity of $\theta \to t[\eta]$. For each $\zeta \in \mathcal{L}^{\mathtt{inl}}(\Box)$ s.t. $V_\zeta \cap Q = \varnothing$ and $\zeta \to \eta$ is INL-valid, we see $V_{t[\zeta]} \cap Q = \varnothing$ and $t[\zeta] \to t[\eta]$ is PSNL-valid. From the fact that θ is a pre-interpolant of $\langle t[\eta], Q \rangle$, PSNL-validity of $t[\zeta] \to \theta$ follows, which then entails INL-validity of $\zeta \to t'[\breve{\theta}]$. □

6 Conclusions

In this paper, we study the framework of PSNL, which is equally expressive as INL. By breaking down the "bundle" of the original modality in INL, we work with a bimodal language that is intuitive to use and technically easy to handle, as demonstrated by the completeness proof and the proof of uniform interpolation property. The definitions and proofs are mostly based on mutual inductions of two types of formulas, although set-formulas can be viewed as auxiliary machinery to define point-formulas. The ability to manipulate set-formulas by axioms and rules offers great flexibility, and enables us to (indirectly) show properties of INL with much easier proofs. Compared to the usual multi-modal modal logic, we do not base our work on Kripke models with two sorts of worlds and keep the neighborhoods as sets, which can facilitate stronger logics over models that impose set-theoretical closure properties of neighborhoods. Compared to the methods of coalgebraic modal logic for related axiomatizations and interpolation results, our proofs are mostly elementary.

Acknowledgements

The authors would like to thank the three anonymous reviewers of AiML2024 for their helpful and detailed comments, and especially pointers to closely related work in coalgebraic logics. The authors are in debt to Johannes Marti for his detailed explanation of the cover modality in the coalgebraic setting and how uniform interpolation property follows based on the techniques in his work [12]. The authors also thank Nick Bezhanishvili and the audience at Tsinghua University and Sun Yat-sen University where earlier versions of this work were presented, for their helpful comments and pointers. Yanjing Wang is supported by the grant 19BZX135 from National Social Science Fund of China. Junhua

Yu is supported by Beijing Municipal Social Sciences Foundation 18ZXC013 and Tsinghua University Humanities Development Program.

References

[1] Bezhanishvili, N., S. Enqvist and J. de Groot, *Duality for Instantial Neighbourhood Logic via Coalgebra*, in: *Coalgebraic Methods in Computer Science - LNCS 12094* (2020), pp. 32–54.
[2] Bílková, M., *Uniform Interpolation and Propositional Quantifiers in Modal Logics*, Studia Logica **85** (2007), pp. 1–31.
[3] Chellas, B. F., "Modal Logic: An Introduction," Cambridge University Press, 1980.
[4] Ding, Y., J. Liu and Y. Wang, *Someone Knows That Local Reasoning on Hypergraphs is a Weakly Aggregative Modal Logic*, Synthese **201** (2023), pp. 1–27.
[5] Fagin, R. and J. Y. Halpern, *Belief, Awareness, and Limited Reasoning*, Artif. Intell. **34** (1987), pp. 39–76.
[6] Fitting, M., *Modal Proof Theory*, in: *Handbook of Modal Logic* (2006).
[7] Goranko, V. and S. Enqvist, *Socially Friendly and Group Protecting Coalition Logics*, in: *Proceedings of AAMAS '18* (2018), pp. 372–380.
[8] Hansen, H. H., "Monotonic Modal Logics," Master's thesis, University of Amsterdam (2003).
[9] Henkin, L., *An Extension of the Craig–Lyndon Interpolation Theorem*, Journal of Symbolic Logic **28** (1963), pp. 201–216.
[10] Jacobs, B., *Many-sorted coalgebraic modal logic : a model-theoretic study*, RAIRO - Theoretical Informatics and Applications - Informatique Théorique et Applications **35** (2001), pp. 31–59.
[11] Leustean, I., N. Moanga and T. Serbanuta, *A Many-sorted Polyadic Modal Logic*, Fundam. Informaticae **173** (2020), pp. 191–215.
[12] Marti, J., "Relation Liftings in Coalgebraic Modal Logic," Master's thesis, University of Amsterdam (2011).
[13] Pacuit, E., "Neighborhood Semantics for Modal Logic," Springer, 2017.
[14] Pauly, M., *A Modal Logic for Coalitional Power in Games*, J. Log. Comput. **12** (2002), pp. 149–166.
[15] Pitts, A., *On an Interpretation of Second Order Quantification in First Order Intuitionistic Propositional Logic*, Journal of Symbolic Logic **57** (1992), pp. 33–52.
[16] Seifan, F., L. Schröder and D. Pattinson, *Uniform Interpolation in Coalgebraic Modal Logic*, in: F. Bonchi and B. König, editors, *Proceedings of CALCO 2017*, Leibniz International Proceedings in Informatics (LIPIcs) **72** (2017), pp. 21:1–21:16.
[17] Troelstra, A. S. and H. Schwichtenberg, "Basic Proof Theory, 2nd ed." Cambridge University Press, 2000.
[18] Tuyt, O., "Canonical Rules on Neighbourhood Frames," Master's thesis, University of Amsterdam (2016).
[19] van Benthem, J., N. Bezhanishvili and S. Enqvist, *A Propositional Dynamic Logic for Instantial Neighborhood Semantics*, Studia Logica **107** (2019), pp. 719–751.
[20] van Benthem, J., N. Bezhanishvili, S. Enqvist and J. Yu, *Instantial Neighbourhood Logic*, Rev. Symb. Log. **10** (2017), pp. 116–144.
[21] van Benthem, J., D. Fernández-Duque and E. Pacuit, *Evidence and Plausibility in Neighborhood Structures*, Ann. Pure Appl. Log. **165** (2014), pp. 106–133.
[22] Yu, J., *A Tableau System for Instantial Neighborhood Logic*, in: *Logical Foundations of Computer Science - LNCS 10703* (2018), pp. 337–353.
[23] Yu, J., *Lyndon Interpolation Theorem of Instantial Neighborhood Logic - Constructively via a Sequent Calculus*, Ann. Pure Appl. Log. **171** (2020).
[24] Zhao, Z., *Sahlqvist Correspondence for Instantial Neighbourhood Logic*, Studies in Logic **14** (2021), pp. 1–23.

A Modal Logic for Reasoning in Contexts

Xuefeng Wen [1]

Sun Yat-sen University
135 Xingang Road West, Guangzhou, 510275, China

Abstract

We propose a new semantics for indicatives and epistemic modals as well as a new notion of validity, according to which an inference is not valid or invalid per se, but valid or invalid under a set of assumptions. In the new semantics, a formula is not evaluated at a pointed model, but at a pointed model together with a context, provided either by the assumptive premises or by the antecedents of indicatives. Armed with the new semantic framework, we differentiate three forms of inferences, two notions of equivalence, and two notions of inconsistency, which can be used to solve some famous puzzles concerning indicatives and epistemic modals in a unified way. Our semantics may shed some light on natural language reasoning in contexts.

Keywords: indicatives, epistemic modality, validity, modus ponens, import-export, Moore sentences, Wittgenstein sentences, strict-tolerant logic

1 Three Puzzles

1.1 The barbershop paradox

In 1894, Lewis Carroll proposed a famous logical paradox [5], whose reformulation amounts to a counterexample to Modus Tollens. Though Modus Tollens seems equally valid as Modus Ponens, unlike Modus Ponens, counterexamples to Modus Tollens have been frequently discovered in the literature. Some involve nested conditionals [33], like Carroll's example. Others involve deontic, epistemic, or probabilistic operators [16,37].

Carroll's logical paradox invites us to consider the following scenario. Three barbers Allen, Brown, and Carr runs a barbershop. The barbershop is always open, which means at least one of the barbers must be in. Allen has been very nervous after a fever so that he never leaves the shop unless Brown goes out with him. Can we conclude that Carr is in the shop? Obviously not, for when Brown is in and Carr is out, both the constraints about the barbershop can be satisfied. On the other hand, we also have the following inference.

(1) A counterexample to Modus Tollens.

[1] This paper was supported by National Social Science Foundation of China (Grant No. 22BZX129). We thank the three anonymous referees for their detailed and suggestive comments.

(a) If Carr is out then if Allen is out then Brown is in.
(b) It is not the case that if Allen is out then Brown is in.
(c) ? Therefore, it is not the case that Carr is out.

The first premise (1 a) rings true, because at least one barber must be in. The second premise (1 b) is also true, because when Allen is out Brown must be out with him. Using Modus Tollens, we obtain (1 c), which is false. [2]

There are basically three approaches to solving the paradox. One is denying that both premises are true. The other is denying that the inference is an instance of Modus Tollens. The third is denying that Modus Tollens is universally valid. The first two approaches seem ad hoc. The third approach demands a new semantics and logic for indicative conditionals, which invalidates Modus Tollens but can also explain why in most cases Modus Tollens is innocuous.

1.2 Might

George Moore noticed that sentences like (2) sounds strange.

(2) It's raining and I don't know it's raining.

On one hand, it can be true. On the other hand, it seems that I cannot consistently assert it. One solution to dispel the tension is appealing to pragmatics. According to this view, the sentence is consistent in semantics but inconsistent in pragmatics. One evidence supporting this solution is that if we replace 'I' by my name in the sentence and the modified sentence is asserted by others, there will be no inconsistency. Another evidence supporting the solution is that if we put the sentence into an assumption rather than asserting it, there will be no inconsistency either. For example, the following sentences sound felicitous.

(3) (a) Suppose it's raining and I don't know it.
(b) If it's raining and I don't know it, then...

Wittgenstein noticed that sentences like (4) is also weird.

(4)# It's raining outside and it might not be raining outside.

On one hand, the sentence which can be formulated by $p \wedge \Diamond\neg p$ is consistent in standard semantics of modal logic. On the other hand, there seems to be some inconsistency in the sentence too. However, we cannot use the same strategy to dispel the tension. Even if the sentence is assert by someone else, there is still inconsistency. And we cannot put it into an assumption or use it as the antecedent of a conditional. The following sentences are still weird, which had been noticed by Yacin [36].

(5) (a) #Suppose it's raining outside and it might not be raining outside.
(b) #If it's raining outside and it might not be raining outside, then...

[2] Carroll's original formulation of the puzzle used Reduction to Absurdity rather than Modus Tollens to obtain (1 c). The argument roughly goes as follows. Suppose Carr is out (C). Then if Allen is out then Brown is in (A), since at least one must be in. On the other hand, we also have if Allen is out then Brown is out (B), which contradicts A. Therefore, $\neg C$. To validate the argument in formal logic, however, we still need some other assumptions to make A and B contradictory. Our formulation avoids the trouble by simply taking A to be false.

Sentences like (4) are sometimes called Wittgenstein sentences, distinguished from Moore sentences like (2). The above comparison indicates that unlike Moore sentences, the inconsistency in Wittgenstein sentences lies in semantics rather than in pragmatics, which demands a new semantics for modality.

1.3 Three barbers × Might

The third puzzle involves both indicatives and epistemic modality. Consider the same scenario as the barbershop paradox and the following inference.

(6) A counterexample to Hypothetical Syllogism.
- (a) If Allen is out then Brown is out.
- (b) If Brown is out, Allen might be in.
- (c) ?Therefore, if Allen is out he might be in.

Both (6 a) and (6 b) are true in the scenario. And the inference is an instance of Hypothetical Syllogism, which is valid in classical logic. But the conclusion rings false.

This example appears to provide a counterexample to Hypothetical Syllogism, which had already been challenged by Lewis [17] for inferences involving counterfactuals. For indicatives, however, whether Hypothetical Syllogism is valid is controversial. If it is not valid, then we need a logical theory that can cope with both indicatives and epistemic modals.

1.4 Main idea

This paper aims to solve the above puzzles, among others. To solve the puzzles, we provide a new semantics for indicatives and epistemic modals, together with a new notion of validity.

According to the new notion of validity, an inference is not valid or invalid per se, but valid or invalid under a set of assumptions. An inference is thus tripart: a set of actual premises, which are true at the actual world, a set of assumptive premises, which are assumed to be true at all epistemically possible worlds, and a conclusion. An inference rule now has three different forms: the standard form, in which all premises are supposed to be actual, the assumptive form, in which all premises are assumed to be true in all epistemically possible worlds, and the hybrid form, in which some premises are supposed to be actual, some are assumptive. The same inference in different forms may have different results of validity.

In the new semantics, a formula is not evaluated at a pointed model, but at a pointed model together with a context, namely a set of worlds in the model. Contexts are provided either by the assumptive premises or by the antecedents of indicatives.

By differentiating three forms of inferences, the new semantics together with the new notion of validity predicts our intuitions of the validity and invalidity of the inference rules involving indicatives and epistemic modals. Not only the three puzzles above are resolved, but also some related puzzles can be solved in a unified way, which may shed some light on contextual reasoning, as well as the fundamental concepts of inferences and validity in logic.

The remaining part of the paper is organized as follows. Section 2 proposes our new semantics for indicatives and epistemic modals, with some properties of the semantics prepared. Section 3 contends the new notion of validity, distinguishing three forms of inferences, two notions of equivalence, and two notions of inconsistency. Section 4 applies the new semantic framework to the three puzzles above, among others, showing how linguistic data are predicted by our theory. Section 5 concludes the paper and suggests some future work. All proofs are put in the Appendix.

2 A Semantics for Indicatives and Epistemic Modals

Our formal language $\mathcal{L}$ is generated from a given set At of atomic formulas by the usual Boolean connectives $\neg, \wedge, \vee$ together with a binary connective $>$ for indicatives and a unary operator $\Diamond$ for epistemic modality. The modal operator $\Box$ is defined as the dual of $\Diamond$, as usual. Formulas without $>$ and $\Diamond$ are called Boolean. We abbreviate $p_0 \wedge \neg p_0$ by $\bot$ for the first atom p_0 in At. We stipulate that $\wedge$ and $\vee$ have stronger power of combination than $>$, so that $p \wedge q > r$ means $(p \wedge q) > r$.

We use selection models for our semantics. But the same idea can also be realized in other models for conditionals, including Lewis' sphere systems [17]. To avoid using an absurd world as in Stalnaker's semantics for conditionals [28], we treat selection functions as partial.

Definition 2.1 (Models) *A partial selection model is a triple $\mathfrak{M} = (W, f, V)$, where*

- $W \neq \emptyset$, *is a set of possible worlds;*
- $f : W \times \wp(W) \to W$, *is a partial selection function, satisfying the following conditions:*
  (id) *if $f(w, X)$ is defined then $f(w, X) \in X$,*
  (cs) *if $w \in X$ then $f(w, X) = w$;*
- $V : At \to \wp(W)$, *is a valuation function, mapping each atom to the set of worlds in which it is true.*

Unlike most semantics in modal logic, we evaluate a formula not simply at a pointed model but at a pointed model together with a set of worlds in the model, which is intended to represent a context. As selection functions are partial, a formula can be neither true nor false. Hence we propose both truth and falsity conditions. An atomic p is true (false, respectively) at w in X iff p is true (false, respectively) at the closest world in X to w. Boolean connectives are defined as usual. An indicative $\varphi > \psi$ is true (false, respectively) at w in X iff ψ is true (false, respectively) at w in the context updated by φ, which is the intersection of X and the truth set of φ in the context X. An epistemic modal $\Diamond\varphi$ is true at w in X iff φ is true at some world in X. More precisely, we have the following definition.

Definition 2.2 (Truth and falsity conditions) *Given a partial selection model $\mathfrak{M} = (W, f, V)$, that φ is true (false, respectively) at (w, X) in $\mathfrak{M}$,* *denoted* $\mathfrak{M}, w, X \Vdash \varphi$ $(\mathfrak{M}, w, X \dashv\!\mid \varphi$, *respectively), is inductively defined as follows.*

- $\mathfrak{M}, w, X \Vdash p$ *iff* $f(w, X)$ *is defined and* $f(w, X) \in V(p)$;
- $\mathfrak{M}, w, X \dashv\!\mid p$ *iff* $f(w, X)$ *is defined and* $f(w, X) \notin V(p)$;
- $\mathfrak{M}, w, X \Vdash \neg\varphi$ *iff* $\mathfrak{M}, w, X \dashv\!\mid \varphi$;
- $\mathfrak{M}, w, X \dashv\!\mid \neg\varphi$ *iff* $\mathfrak{M}, w, X \Vdash \varphi$;
- $\mathfrak{M}, w, X \Vdash \varphi \wedge \psi$ *iff* $\mathfrak{M}, w, X \Vdash \varphi$ *and* $\mathfrak{M}, w, X \Vdash \psi$;
- $\mathfrak{M}, w, X \dashv\!\mid \varphi \wedge \psi$ *iff* $\mathfrak{M}, w, X \dashv\!\mid \varphi$ *or* $\mathfrak{M}, w, X \dashv\!\mid \psi$;
- $\mathfrak{M}, w, X \Vdash \varphi \vee \psi$ *iff* $\mathfrak{M}, w, X \Vdash \varphi$ *or* $\mathfrak{M}, w, X \Vdash \psi$;
- $\mathfrak{M}, w, X \dashv\!\mid \varphi \vee \psi$ *iff* $\mathfrak{M}, w, X \dashv\!\mid \varphi$ *and* $\mathfrak{M}, w, X \dashv\!\mid \psi$;
- $\mathfrak{M}, w, X \Vdash \Diamond\varphi$ *iff* $\mathfrak{M}, u, X \Vdash \varphi$ *for some* $u \in X$;
- $\mathfrak{M}, w, X \dashv\!\mid \Diamond\varphi$ *iff* $\mathfrak{M}, u, X \dashv\!\mid \varphi$ *for all* $u \in X$;
- $\mathfrak{M}, w, X \Vdash \varphi > \psi$ *iff* $\mathfrak{M}, w, X \cap [\![\varphi]\!]_+^X \Vdash \psi$, *where*

$$[\![\varphi]\!]_+^X = \{u \in W \mid \mathfrak{M}, u, X \Vdash \varphi\};$$

- $\mathfrak{M}, w, X \dashv\!\mid \varphi > \psi$ *iff* $\mathfrak{M}, w, X \cap [\![\varphi]\!]_+^X \dashv\!\mid \psi$.

We write $\mathfrak{M}, w, X \Vdash \Gamma$, if $\mathfrak{M}, w, X \Vdash \varphi$ for all $\varphi \in \Gamma$. Given a partial selection model $\mathfrak{M} = (W, f, V)$, we denote by $[\![\varphi]\!]_-^X$ the set $\{u \in W \mid \mathfrak{M}, u, X \dashv\!\mid \varphi\}$. For brevity, we write $[\![\Delta]\!]_+^X$ for $\bigcap_{\delta\in\Delta}[\![\delta]\!]_+^X$. We often omit the brackets $\{\ \}$ to denote a set $\{\varphi_1, \ldots, \varphi_n\}$.

The following lemma says that a formula cannot be both true and false, which is straightforward by induction.

Lemma 2.3 *For any partial selection model* $\mathfrak{M}$ *and* X *in* $\mathfrak{M}$, *for any* $\varphi \in \mathcal{L}$, *we have* $[\![\varphi]\!]_+^X \cap [\![\varphi]\!]_-^X = \emptyset$.

The following lemma indicates that Boolean formulas have the same truth and falsity conditions as atomic formulas.

Lemma 2.4 *If* φ *is Boolean, then*

(i) $\mathfrak{M}, w, X \Vdash \varphi$ *iff* $\mathfrak{M}, f(w, X), W \Vdash \varphi$

(ii) $\mathfrak{M}, w, X \dashv\!\mid \varphi$ *iff* $\mathfrak{M}, f(w, X), W \dashv\!\mid \varphi$

The following lemma says that a context X updated first by φ and then by ψ is the same as that updated by $\varphi \wedge \psi$, provided that ψ is Boolean.

Lemma 2.5 *Let* $\mathfrak{M} = (W, f, V)$ *be a normal model. If* ψ *is Boolean, then*

$$X \cap [\![\varphi]\!]_+^X \cap [\![\psi]\!]_+^{X\cap[\![\varphi]\!]_+^X} = X \cap [\![\varphi \wedge \psi]\!]_+^X.$$

The following lemma says that if w is in the context, then Boolean formulas at w are context insensitive.

Lemma 2.6 *Let* φ *be Boolean. If* $w \in X \cap Y$, *then*

(i) $\mathfrak{M}, w, X \Vdash \varphi$ *iff* $\mathfrak{M}, w, Y \Vdash \varphi$;

(ii) $\mathfrak{M}, w, X \dashv\!\mid \varphi$ *iff* $\mathfrak{M}, w, Y \dashv\!\mid \varphi$.

3 A New Notion of Validity

3.1 Three forms of inferences

The following are typical inferences in natural language.

(i) Either Carr is in or out. He is not in. Therefore, he is out.

(ii) Suppose Carr is out. Then Allen or Brown is in. (So, if Carr is out then Allen or Brown is in.)

(iii) Suppose Carr is out. Then Allen or Brown is in, since at least one is in.

The three inferences have the following forms, respectively.

Standard Form: A. Therefore, C. (Or, C, since A.)

Assumptive Form: Suppose A. Then C.

Hybrid Form: Suppose A. Then C, since B.

The difference between the standard form and the assumptive form is that the premise A in the former is a fact whereas it is an assumption in the latter. The assumptive form is usually followed by the conditional "if A then C", cancelling the assumption, which is called Conditional Proof. In formal logic, the standard form and the assumptive form are usually not distinguished. Both are formalized by $A \models C$. In particular, the conditional proof is formalized by inferring $\models A > C$ from $A \models C$. For the same reason, neither are the two premises A and B distinguished in the hybrid form, which is formalized by $A, B \models C$.

In natural language reasoning, however, factual premises and assumptive ones are different. The logical effect of the former is only in the actual world, whereas the logical effect of the latter is in all epistemically possible worlds. In mathematical reasoning, the difference is unimportant, as a mathematical fact is a fact true in all epistemically possible worlds. This is why the three forms are not distinguished in standard formal logic. Nevertheless, if we want to characterize reasoning in natural language better, we should distinguish the three forms in our formalization.

3.2 Formalizing the three forms of inferences

To formalize the three forms of inferences, we define a new notion of validity, according to which an inference is not valid or invalid per se, but valid or invalid under a set of assumptions.

Definition 3.1 (Validity) *The inference from* Γ *to* φ *under* Δ *is valid, denoted* $\Gamma \models^{\Delta} \varphi$, *if there is no partial selection model* $\mathfrak{M} = (W, f, V)$ *and* $w \in W$ *such that* $\mathfrak{M}, w, [\![\Delta]\!]^W_+ \Vdash \Gamma$ *and* $\mathfrak{M}, w, [\![\Delta]\!]^W_+ \dashv\!\mid \varphi$.

According to the definition, factual premises and conclusions are evaluated in the contexts specified by assumptive premises. Apart from generalizing logical consequence from a binary relation to a ternary one, we adopt the *st*-validity proposed in Strict-Tolerant logic [6,26] rather than the standard notion of truth preserving. The *st*-validity can be regarded as a formalization of Strawson entailment [32, pp. 176–177], according to which the inference from A to B is valid if and only if B is true whenever A is true and all presuppositions involved are satisfied. If we assume that a statement is neither true nor false if its presuppositions are not satisfied, and a statement cannot be both true and false, then it turns out that Strawson entailment and *st*-validity coincide. One merit of *st*-validity as shown by Strict-Tolerant logics is that when restricted to Boolean formulas, the semantics yields the same logical consequence as classical logic, unlike most three-valued logics, which are sublogics of classical logic.

In terms of the new notion of validity, the three forms of inferences above can be formulated as follows.

Standard Form: $A \models^{\emptyset} C$

Assumptive Form: $\emptyset \models^{A} C$

Hybrid Form: $B \models^{A} C$

For brevity, we will write $\Gamma \models \varphi$ for $\Gamma \models^{\emptyset} \varphi$ and $\models^{\Delta} \varphi$ for $\emptyset \models^{\Delta} \varphi$, respectively. Within the new semantic framework, we can define two notions of equivalence as follows.

Definition 3.2 (Equivalence) *We say that φ and ψ are*

(i) *semantically equivalent, denoted $\varphi \equiv \psi$, if for all partial selection models $\mathfrak{M}$ and X in $\mathfrak{M}$, we have $[\![\varphi]\!]^X_+ = [\![\psi]\!]^X_+$ and $[\![\varphi]\!]^X_- = [\![\psi]\!]^X_-$;*

(ii) *hypothetically equivalent, denoted $\varphi \asymp \psi$, if $\models (\varphi > \psi) \wedge (\psi > \varphi)$, or equivalently, $\models^{\varphi} \psi$ and $\models^{\psi} \varphi$.*

We can also define two notions of inconsistency as follows.

Definition 3.3 (Inconsistency) *We say that φ is*

(i) *inconsistent, if $\varphi \models \bot$;*

(ii) *self-refuting, if $\models^{\varphi} \neg\varphi$.*

It is desirable that if an inference only involves Boolean formulas, it should be characterized by classical logic. After all, most challenges to classical logic are centering around conditionals and modals rather than Boolean operators. The following theorem says that confined to Boolean formulas, there is no difference between the three forms of inferences in our framework, which are indeed all equivalent to the inferences in classical logic.

Theorem 3.4 *If all the formulas in $\Gamma \cup \{\varphi\}$ are Boolean, then $\Sigma \models^{\Delta} \varphi$ iff $\Gamma \models_{cl} \varphi$, where $\Sigma \cup \Delta = \Gamma$ and $\models_{cl}$ is the logical consequence of classical logic.*

3.3 Import-Export

The following lemma shows that a restricted version of Import-Export, which had been proposed by Mandelkern [21], is valid in our semantics.

Lemma 3.5 *If ψ is Boolean then $\varphi > (\psi > \chi) \equiv \varphi \wedge \psi > \chi$.*

Validating Import-Export, however, is not without price. As Gibbard [10] noted, there is a tension between Import-Export and Modus Ponens. Together with the innocuous rule $\varphi \vDash \psi \Longrightarrow \vDash \varphi > \psi$, validating both Modus Ponens and Import-Export leads to the collapse of conditionals, i.e., conditionals are logically equivalent to material implication, which is undesirable.

Noting that Modus Ponens does have counterexamples, McGee [22] proposed a semantics for conditionals that validates Import-Export and invalidates Modus Ponens. One most famous counterexample given by McGee invites us to consider the 1980 US presidential election. Opinion polls showed that the republican Ronald Reagan was decisively ahead of the democrat Jimmy Carter, who was in turn decisively ahead of the other republican John Anderson. Then the premises (7 a) and (7 b) below ring true, while the conclusion (7 c) does not, for if Reagan does not win, Carter will.

(7) A counterexample to Modus Ponens.
   (a) If a republican wins the election, then if it's not Reagan who wins it will be Anderson.
   (b) A republican will win the election.
   (c) Therefore, if it's not Reagan who wins, it will be Anderson.

Many don't think that this is a counterexample to Modus Ponens. Before discussing the (in)validity of Modus Ponens, we'd like to point out why Import-Export is only restrictedly valid.

According to Mandelkern [19,21], when evaluating a nested conditional $\varphi > (\psi > \chi)$, the nested antecedent ψ should not be evaluated only in the same context as that of the whole conditional $\varphi > (\psi > \chi)$, namely the global context, but should also be evaluated in the context of φ, which is the local context of ψ. When ψ is Boolean, φ will not have any effect on ψ. So we can ignore the local context provided by φ. But if ψ itself is a conditional, then it will be affected by φ. This means that when ψ is a conditional, $\varphi \wedge \psi > \chi$ is not equivalent to $\varphi > (\psi > \chi)$, as ψ is evaluated independently of φ in the former, whereas ψ is affected by the local context φ in the latter. Thus, both Importation and Exportation are invalid when ψ contains conditionals, as instantiated by Manderlkern [21] below.

Consider the same scenario as in McGee's counterexample, except that we only know that Reagan is well ahead of Carter and Anderson without knowing the ordering between Carter and Anderson.

(8) A counterexample to Exportation.
   (a) If a Republican will win the election, and Anderson will win if Reagan doesn't win, then both Republicans are currently in a stronger position to win than Carter.
   (b) If a Republican will win the election, then if Anderson will win if Reagan doesn't, then both Republicans are currently in a stronger position to win than Carter.

Intuitively, (8 a) is true whereas (8 b) is false. In formulas, we have $\varphi \wedge \psi > \chi$ but not $\varphi > (\psi > \chi)$.

The following counterexample is based on the same scenario as McGee's.

(9) A counterexample to Importation.
   (a) If a Republican will win the election, then if Anderson will win if Reagan doesn't, then the polling data we've just received are correct.
   (b) If a Republican will win the election, and Anderson will win if Reagan doesn't, then the polling data we've just received are correct.

Intuitively, (9 a) is true whereas (9 b) is false. In formulas, we have $\varphi > (\psi > \chi)$ but not $\varphi \wedge \psi > \chi$. Our semantics predicts Manderkern's counterexamples.

Proposition 3.6 *$p > ((q > r) > s)$ and $p \wedge (q > r) > s$ are not equivalent: either can be true with the other being false.*

4 Solving the Puzzles

Before solving the three puzzles in Section 1, we first discuss the (in)validity of Modus Ponens.

4.1 Modus Ponens

Proposition 4.1 *Modus Ponens is (in)valid in the following forms.*

(i) $p, p > (q > r) \not\models q > r$

(ii) $p, p > \Box q \not\models \Box q$

(iii) $\models^{p, p>(q>r)} q > r$

(iv) $\models^{p, p>\Box q} \Box q$

(v) $\varphi, \varphi > \psi \models \psi$, *where ψ is Boolean*

(vi) $\models^{\varphi, \varphi>\psi} \psi$, *where ψ is Boolean*

(vii) $\varphi > \psi \models^{\varphi} \psi$, *where φ is Boolean*

In our semantics, the standard form of Modus Ponens is invalid, if it involves nested conditionals, as in McGee's semantics. Our semantics also predicts that if epistemic modals are involved, Modus Ponens is also invalid, as exemplified by the following example, assuming that in the 1980 US election, there were only two candidates, Reagan and Carter.

(10) A counterexample to Modus Ponens involving 'must'.
   (a) If a republican wins the election, then the winner must be Reagan.
   (b) A republican will win the election.
   (c) Therefore, the winner must be Reagan.

Over [25], among others, objected to McGee's counterexamples. He argued that if the premises of Modus Ponens have been certainly assumed, the conclusion cannot be false. This intuition can be formulated by the assumptive form of Modus Ponens in our semantics. Though it is not generally valid, its restricted form for the two examples above are valid, predicted by (iii) and (iv) of Proposition 4.1.

Now in the literature concerning the validity of Modus Ponens, often two notions of validity are distinguished (cf. [31]). Mandelkern [20] categorized them as truth-preserving validity, which is the standard one, and informational validity, which was defended in [2], based on the semantics initiated by Veltman [33] and developed by Gillies [11] and Yalcin [36], among others. The two notions of validity yield two forms of Modus Ponens. The truth-preserving form reads: if φ and $\varphi > \psi$ are true, then ψ is true. The informational form reads: if φ and $\varphi > \psi$ are fully accepted, then rationality requires ψ to be fully accepted.

Bledin [3] argues that, though truth-preserving Modus Ponens may not be valid, informational Modus Ponens is still valid. Consider McGee's example again. If we fully accept that a republican will win, and that if a republican wins then if it is not Reagan who wins it will be Anderson, then we must accept that if Reagan does not win, then Anderson will win. The intuition is similar to Over's, which could also be formulated by the assumptive Modus Ponens.

But to dispel the counterexample does not require us to fully accept both premises. Fully accepting that a republican will win is enough for us to infer from "if a republican wins then if it is not Reagan who wins it will be Anderson" to its consequent, which is predicted by (vii) of Proposition 4.1. Note that fully accepting the other premise only does not make the inference valid.

4.2 Modus Tollens

Proposition 4.2 *Modus Tollens is (in)valid in the following forms.*

(i) $p > (q > r), \neg(q > r) \not\models \neg p$

(ii) $p > \Box q, \neg\Box q \not\models \neg p$

(iii) $\not\models^{p>(q>r),\neg(q>r)} \neg p$

(iv) $\not\models^{p>\Box q,\neg\Box q} \neg p$

(v) $\varphi > \psi, \neg\psi \models \neg\varphi$, *where ψ is Boolean*

(vi) $\models^{\varphi>\psi,\neg\psi} \neg\varphi$, *where φ and ψ are Boolean*

According to our semantics, the inference in the barbershop puzzle is indeed invalid. Unlike McGee's counterexamples to Modus Ponens, even if the premises of the inference in the barbershop puzzle are assumed, the inference is still invalid, which is predicted by (iii) of Proposition 4.2. This may explain why counterexamples to Modus Tollens were found much earlier and more often than those to Modus Ponens. Though Modus Ponens and Modus Tollens seem two symmetric rules, the latter is not as valid as the former.

4.3 Hypothetical Syllogism

Proposition 4.3 *Hypothetical syllogism is (in)valid in the following forms.*

(i) $p > q, q > r \not\models p > r$

(ii) $q > r \not\models^{p>q} p > r$

(iii) $\models^{\varphi>\psi,\psi>\chi} \varphi > \chi$, *where φ, ψ, χ are Boolean*

(iv) $\varphi > \psi \models^{\psi>\chi} \varphi > \chi$, *where ψ and χ are Boolean*

(v) $\not\models^{p>(q>r),(q>r)>s} p > s$

(vi) $p > (q > r) \not\models^{(q>r)>s} p > s$

(vii) $\not\models^{p>q, q>\Diamond r} p > \Diamond r$

(viii) $p > q \not\models^{q>\Diamond r} p > \Diamond r$

It has been well known since Lewis [17] that Hypothetical Syllogism is invalid for counterfactuals. Alleged counterexamples for indicatives were also found. The following is a typical one.

(11) An alleged counterexample to Hypothetical Syllogism.
   (a) If Ann drops out, then Bill will win.
   (b) If Bill wins, Ann will finish second.
   (c) Therefore, if Ann drops out, she will finish second.

Intuitively, we can think of a scenario in which both (11 a) and (11 b) are true, whereas (11 c) is definitely false. Like Stalnaker-Lewisian conditional logics and McGee's semantics, our semantics predicts (11). Note that even if (11 a) is assumed, the inference is still invalid, which is predicted by (iv) and (v) of Proposition 4.3.

Why do some argue that 11 is not a counterexample to Hypothetical Syllogism? If (11 b) is assumed, then it excludes the possibility that Ann drops out. So under this assumption, the antecedent of the conclusion (11 c) cannot be true, and thus the whole conditional cannot be false. Indeed, whenever the second premise of Hypothetical Syllogism is assumed, the inference is valid, provided that the formulas are Boolean, which is predicted by (iii) and (iv) of of Proposition 4.3.

Hypothetical Syllogism is not valid even in its assumptive form, if it involves nested conditionals. The following example is adapted from Mandelkern's counterexample to Import-Export, using the same scenario as McGee's.

(12) A counterexample to Hypothetical Syllogism involving nested conditionals.
   (a) If a republican wins the election, then if it's not Reagan who wins it will be Anderson.
   (b) If Anderson will win if Reagan doesn't, then the polling data we've just received are incorrect.
   (c) ?Therefore, if a republican wins the election, then the polling data we've just received are incorrect.

According to our semantics, the third puzzle in Section 4.3 does provide a counterexample to Hypothetical Syllogism, as even if the second premise in Hypothetical Syllogism is assumed, it is still invalid. The following is another counterexample in the same form.

(13) A counterexample to Hypothetical Syllogism involving epistemic modals.
   (a) If the flower is red, then it is not blue.
   (b) If the flower is not blue, it might be purple.
   (c) ?Therefore, if the flower is red, it might be purple.

Both (13 a) and (13 b) are true, whereas (13 c) rings false, even if (13 b) is assumed, or both (13 a) and (13 b) are assumed.

The counterexamples 12 and 13 are predicted by (v)–(viii) of Proposition 4.3. [3]

4.4 Simplification of Disjunctive Antecedents

The inference from $\varphi \vee \psi > \chi$ to $(\varphi > \chi) \wedge (\psi > \chi)$ is called Simplification of Disjunctive Antecedents (SDA, henceforth). It sounds a valid inference. If I'll be happy if Ann or Bill comes to my party, then I'll be happy if Ann comes to my party and I'll be happy if Bill comes to my party. However, the standard form of SDA together with Left Logical Equivalents (LLE), namely, equating $\varphi > \chi$ with $\psi > \chi$ from the equivalence of φ and ψ, which seems innocuous, will recover the standard form of Antecedent Strengthening, which is undesirable.

To circumvent the dilemma, one solution is to accept SDA and abandon LLE by constructing hyperintensional logics (e.g. [24,8]). Another solution is to reject SDA and account for its intuitive validity by translation lore (e.g. [18]). Both approaches appear ad hoc. In our framework, SDA is invalid in its standard form but valid in its assumptive form, if the formulas involved are Boolean.

Proposition 4.4 *SDA is (in)valid in the following forms.*

(i) $p \vee q > r \not\models (p > r) \wedge (q > r)$;

(ii) $\models^{\varphi \vee \psi > \chi} (\varphi > \chi) \wedge (\psi > \chi)$, *where* φ, ψ, χ *are Boolean;*

(iii) $\not\models^{p \vee q > \Diamond r} (p > \Diamond r) \wedge (q > \Diamond r)$.

On the other hand, SDA does have counterexamples, even in its assumptive form, if the consequent involves epistemic modals, as exemplified below.

(14) A counterexample to SDA involving 'might'.
  (a) If Ann or Bill goes to the party, there might be a woman in the party.
  (b)? Therefore, if Bill goes to the party, there might be a woman in the party.

Even if (14 a) is assumed, (14 b) can still be false. The counterexample is predicted by (iii) of Proposition 4.4.

Counterexamples to SDA without involving modality were also found in the literature (e.g. [4]), as given below.

(15) An alledged counterexample to SDA.
  (a) If Ivan is playing tennis or playing baseball, then he is playing baseball.
  (b) Therefore, if Ivan is playing tennis, he is playing baseball.

If Ivan plays baseball and nearly does not play tennis then (15 a) rings true but (15 b) does not. According to our analysis, (15) is only a counterexample to the standard form of SDA. It is not to the assumptive form. If we have

[3] The same analysis can be applied to Antecedent Strengthening, the rule inferring $\varphi \wedge \psi > \chi$ from $\varphi > \chi$, and Contraposition, the rule inferring $\neg\psi > \neg\varphi$ from $\varphi > \psi$. For lack of space, we omit it here.

assumed that (15 a), then whenever Ivan is playing one of the two games, he is playing baseball. It follows that Ivan never plays tennis (if we assume that one cannot play the two games at the same time). Hence, the conclusion is not false (though not true either). Our intuition of rejecting (15 b) is because we presuppose the possibility of its antecedent. But the possibility is already excluded by assuming (15 a). This analysis is similar to that given by Fine [8], except that his analysis is about the counterexample for counterfactuals given in [23].

4.5 Or-to-If

The inference from $\varphi \vee \psi$ to $\neg\varphi > \psi$ is called Or-to-If (a.k.a., the Direct Argument), which is invalid in Stalnaker-Lewisian conditional logics. This has to be so, as their logics validate Modus Ponens, together with which, Or-to-If leads to collapse: $\varphi > \psi$ will be equivalent to $\neg\varphi \vee \psi$. Or-to-If is not valid in our semantics either. Otherwise, the restricted Modus Ponens (which is valid in our semantics) also leads to collapse.

Nevertheless, Or-to-If seems pervasive and innocuous in natural language reasoning. If there are only two cases, then ruling out one surly leads us to the other. To explain the plausibility of this inference, Stalnaker [29] distinguished reasonable inferences from valid ones. An inference is reasonable if in every context in which the premises could appropriately be asserted or supposed, it is impossible for anyone to accept the premises without accepting the conclusion. This idea is in line with informational validity mentioned in Section 4.1, which can be formulated in the assumptive form of Or-to-If. Moreover, it seems that we can not only infer $\neg\varphi > \psi$ from $\varphi \vee \psi$, but also $\neg\varphi > \Box\psi$. The following is a common example.

(16) A valid inference of modal Or-to-If.
   (a) Either the butler or the gardener did it.
   (b) Therefore, if the butler didn't do it, it must be the gardener.

We call it modal Or-to-If. Besides, Or-to-If also has a conditional reading: If $\varphi \vee \psi$, then if $\neg\varphi$ then ψ. All the forms of Or-to-If can be formulated in our framework.

Proposition 4.5 *Or-to-If is (in)valid in the following forms, where φ and ψ are Boolean.*

(i) $p \vee q \nvDash \neg p > q$
(ii) $\vDash^{\varphi \vee \psi} \neg\varphi > \psi$
(iii) $\vDash^{\varphi \vee \psi} \neg\varphi > \Box\psi$
(iv) $\vDash (\varphi \vee \psi) > (\neg\varphi > \psi)$

Clause (ii) can be regarded as a formalization of Stalnaker's reasonable inference. Clause (iii) predicts (16) above. Note that Clause (iv) does not hold in Stalnaker-Lewisian conditional logics, since by Modus Ponens it leads to collapse.

Some authors advocate that indicative implication is no more than material implication in semantics (e.g. [14,12]). Our semantics inherits some basic ideas of Stalnaker's theory of conditionals. Can we also incorporate the theory of material implication for conditionals? Yes. According to our semantics, $\varphi > \psi$ and $\neg\varphi \vee \psi$ are indeed 'equivalent'. The equivalence, however, is not in the sense that if one is *true* so is the other. It is in the sense that if one is *assumed* then the other cannot be false. In our terminology, they are not semantically equivalent, but hypothetically equivalent, as given by the following theorem.

Theorem 4.6 *If φ and ψ are Boolean, then $\varphi > \psi \asymp \neg\varphi \vee \psi$, i.e., $\varphi > \psi$ and $\neg\varphi \vee \psi$ are hypothetically equivalent.*

4.6 Wittgenstein Sentences

Armed with two notions of inconsistency, the puzzle about Wittgenstein sentences can be resolved.

Proposition 4.7 *Let $\omega = p \wedge \Diamond\neg p$ or $\omega = \neg p \wedge \Diamond p$. Then*

(i) $\omega \not\models \bot$

(ii) $\models^{\omega} \neg\omega$

Clause 1 says that the Wittgenstein sentence ω is consistent, which explains why it can be true. Clause 2 says that ω is self-refuting, which explains why it cannot be consistently asserted or assumed. By assuming $p \wedge \Diamond\neg p$, we exclude all worlds in which p is false in the context. Then $\Diamond\neg p$ can no longer be true in the context and thus $p \wedge \Diamond\neg p$ can no longer be true either. The existence of such self-refuting formulas also makes the semantic equivalence between $\Diamond\varphi$ and $\neg(\varphi > \neg\varphi)$, which holds in most conditional logics, no longer hold. It can be verified that $\Diamond\omega$ can be true, whereas $\neg(\omega > \neg\omega)$ cannot be true, since $\omega > \neg\omega$ is valid and hence cannot be false.

Moreover, our semantics also predicts the order sensitivity between p and $\Diamond p$, which cannot be explained by Yalcin's semantics. As noted by Veltman [33], there are differences between the following two sequences of statements.

(17) Order sensitivity involving 'might'.
   (a) Someone is knocking at the door. Maybe it's Bill (and thus not Ann). It's Ann.
   (b) #Someone is knocking at the door. It's Ann (and thus not Bill). Maybe it's Bill.

(17 a) is felicitous, whereas (17 b) sounds odd. The moral is that after asserting $\Diamond p$ you can still assert $\neg p$. But after asserting $\neg p$ you can no longer assert $\Diamond p$. The difference is predicted by the following result.

Proposition 4.8 *For any $p \in At$,*

(i) $\neg p \not\models^{\Diamond p} \bot$

(ii) $\Diamond p \models^{\neg p} \bot$

Clause 1 says that $\neg p$ is consistent under $\Diamond p$. Assuming or asserting $\Diamond p$

does not rule out the possibility of $\neg p$, and thus $\neg p$ can be true and asserted. Clause 2 says that $\Diamond p$ is inconsistent under $\neg p$. Assuming or asserting $\neg p$ has ruled out the possibility of p, and thus $\Diamond p$ cannot be true or asserted.

4.7 Conditional Excluded Middle

Conditional Excluded Middle (CEM, henceforth) refers to the scheme $(\varphi > \psi) \vee (\varphi > \neg\psi)$, which is valid in Stalnaker's conditional logic. Though admitting the plausibility of CEM, Lewis [17] gave an If-Might argument against CEM, which was further developed by Bennett [1]. The argument roughly goes as follows. First, there is strong intuition that $\varphi > \neg\psi$ is inconsistent with $\varphi > \Diamond\psi$, which can be formulated as follows. [4]

If-Might Contradiction (IMC) $(\varphi > \neg\psi) \wedge (\varphi > \Diamond\psi) \vDash \bot$

From IMC, using reduction to absurdity, it follows that $\varphi > \Diamond\psi \vDash \neg(\varphi > \neg\psi)$. By CEM, using disjunctive syllogism, we have $\neg(\varphi > \neg\psi) \vDash \varphi > \psi$. Then by the transitivity of $\vDash$, it follows that $\varphi > \Diamond\psi \vDash \varphi > \psi$. Analogously, we obtain $\varphi > \psi \vDash \varphi > \Diamond\psi$. But it is absurd that $\varphi > \Diamond\psi$ and $\varphi > \psi$ are logically equivalent. So if $\vDash$ is a conservative extension of classical logic, CEM is incompatible with IMC.

We argue that the If-Might argument against CEM is flawed. The problem of the If-Might argument lies in the formulation of the If-Might contradiction. The inconsistency between $\varphi > \neg\psi$ and $\varphi > \Diamond\psi$ is not that they cannot both be true. It is that they cannot both be asserted or assumed, as the following examples shows.

(18) #If it rains, the game will not be held. But even if it rains, the game might be held.

(19) #Suppose that if it rains, the game will not be held, and that if it rains, the game might be held.

Analogous to the analysis of Wittgenstein sentences, in our terminology, $(\varphi > \neg\psi) \wedge (\varphi > \Diamond\psi)$ is consistent but self-refuting, demonstrated below.

Proposition 4.9 *Let* $\lambda = (p > \neg q) \wedge (p > \Diamond q)$. *Then*

(i) $\lambda \nvDash \bot$

(ii) $\vDash^{\lambda} \neg\lambda$

In fact, according to our semantics, $(p > \neg q) \wedge (p > \Diamond q)$ is equivalent to a conditional Wittgenstein sentence, namely, $p > (\neg q \wedge \Diamond q)$, which can be regarded as the Wittgenstein sentence $\neg q \wedge \Diamond q$ in the context of p. If we agree that $\neg q \wedge \Diamond q$ is consistent (in the sense of possibly being true), then it should also be consistent in the context of p. If we think that $\neg q \wedge \Diamond q$ is inconsistent (in the sense of impossibly being assumed), then so is it in the context of p.

Since the correct If-Might contradiction should be formulated by Clause

---

[4] Lewis used a single might-conditional $\Diamond\!\!\rightarrow$ rather than combining two logical operators as we formulate here. Our formulation follows the tradition in formal semantics about indicatives and epistemic modals. See, e.g., [27].

(ii) rather than (i) of Proposition 4.9 above, the reasoning from IMC to the equivalence of $\varphi > \psi$ and $\varphi > \Diamond\psi$ is unjustified. In fact, more and more authors are defending CEM now (e.g., [30,7,35,15,27]). It is also valid in our semantics.

Proposition 4.10 $\models (\varphi > \psi) \vee (\varphi > \neg\psi)$.

5 Conclusion

We define a new semantics for indicatives and epistemic modals, together with a ternary notion of validity, by which an inference is not (in)valid per se, but (in)valid under a set of assumptions. Thus, an inference has three different forms: standard form (no premises are assumed), assumptive form (all premises are assumed), and hybrid form (some premises are assumed), which are not equivalent to each other. We also define two kinds of equivalence and two kinds of inconsistency. Armed with these differentiations, we give a unified solution to several puzzles concerning indicatives and epistemic modals, by predicting linguistic data in natural language reasoning. This is a preliminary attempt to unify two kinds of validity proposed in the literature. Future work includes a full comparison to existing works, including Fitting's work on ternary consequence for modal logic [9] [5] and other logics for conditionals and modals (e.g., [34]). Our logic also demands a proof theory suitable for its ternary notion of validity, so that the logic can be better understood and compared to other logics.

Note that our semantics in the current form cannot deal with Wittgenstein disjunctions (disjunction of two Wittgenstein sentences), which were argued by Mandelkern and Holliday in [19,13] to be as inconsistent as Wittgenstein sentences. In fact, they take Wittgenstein sentences to be semantically equivalent to a contradiction. They would also take $p > \neg q \wedge \Diamond q$ to be a contradiction, which we only regard as self-refuting. We argue that both $\neg q \wedge \Diamond q$ and $p > \neg q \wedge \Diamond q$ can be true. Suppose I will take a flight (p). In the most plausible worlds for me, there will not be an accident of my flight ($\neg q$). So I believe $\neg q$. Meanwhile, I do not exclude the possibility of an accident. That's why I will take out an insurance for my flight. So under the assumption of p, I believe both $\neg q$ and $\Diamond q$. On the other hand, our semantic framework leaves room for treating Wittgenstein sentences as real contradictions. One strategy is to apply local contexts also to conjunctions and disjunctions, as in [19]. We leave the investigation of whether this constitutes a better semantics for future work.

[5] Using Fitting's logical consequence and our notation, $\Gamma \models^{\Delta} \varphi$ iff for all models $\mathfrak{M}$ in which all formulas in Δ are globally true (i.e., true at all worlds in $\mathfrak{M}$), if all formulas in Γ are (locally) true at some world in $\mathfrak{M}$ then φ is also (locally) true at the world. The formulas in Δ are called global assumptions, and those in Γ are called local assumptions, paralleling our differentiation between assumptive premises and actual premises. Nevertheless, since a Wittgenstein sentence can never be globally true, in Fitting's logical consequence, any inference from Γ to φ under a Wittgenstein sentence is (vacuously) valid , which does not hold in our semantics. The set Δ in Fitting's logical consequence is used to choose models for Γ, whereas in ours it is used to modify contexts.

Appendix: Proofs

Proof of Lemma 2.4. By induction on φ.

Case 1: $\varphi = p$. Then $\mathfrak{M}, w, X \Vdash \varphi$ iff $f(w, X)$ is defined and $f(w, X) \in V(p)$ iff $f(w, X)$ is defined and $f(f(w, X), W) \in V(p)$ iff $\mathfrak{M}, f(w, X), W \Vdash \varphi$, where the second 'iff' is by (cs). Analogously, we have $\mathfrak{M}, w, X \dashv\!\mid \varphi$ iff $\mathfrak{M}, f(w, X), W \dashv\!\mid \varphi$.

Case 2: $\varphi = \neg\psi, \psi \wedge \chi, \psi \vee \chi$. Immediate from the inductive hypothesis.

Proof of Lemma 2.5. By Lemma 2.4 and (cs), we have

$$\begin{aligned} X \cap [\![\varphi]\!]_+^X \cap [\![\psi]\!]^{X \cap [\![\varphi]\!]_+^X} &= X \cap [\![\varphi]\!]_+^X \cap \{w \in W \mid \mathfrak{M}, f(w, X \cap [\![\varphi]\!]_+^X), W \Vdash \psi\} \\ &= \{w \in X \cap [\![\varphi]\!]_+^X \mid \mathfrak{M}, f(w, X \cap [\![\varphi]\!]_+^X), W \Vdash \psi\} \\ &= \{w \in X \cap [\![\varphi]\!]_+^X \mid \mathfrak{M}, w, W \Vdash \psi\} \\ &= [\![\varphi]\!]_+^X \cap \{w \in X \mid \mathfrak{M}, w, W \Vdash \psi\} \\ &= [\![\varphi]\!]_+^X \cap \{w \in X \mid \mathfrak{M}, f(w, X), W \Vdash \psi\} \\ &= [\![\varphi]\!]_+^X \cap \{w \in X \mid \mathfrak{M}, w, X \Vdash \psi\} \\ &= X \cap [\![\varphi]\!]_+^X \cap \{w \in W \mid \mathfrak{M}, w, X \Vdash \psi\} \\ &= X \cap [\![\varphi]\!]_+^X \cap [\![\psi]\!]_+^X \\ &= X \cap [\![\varphi \wedge \psi]\!]_+^X \end{aligned}$$

Proof of Lemma 2.6. For (i), we have $\mathfrak{M}, w, X \Vdash \varphi$ iff $f(w, X)$ is defined and $\mathfrak{M}, f(w, X), W \Vdash \varphi$ iff $\mathfrak{M}, w, W \Vdash \varphi$, where the first 'iff' is by Lemma 2.4, and the second 'iff' is by (cs), since $w \in X$. Similarly, we have $\mathfrak{M}, w, Y \Vdash \varphi$ iff $\mathfrak{M}, w, W \Vdash \varphi$. Hence, $\mathfrak{M}, w, X \Vdash \varphi$ iff $\mathfrak{M}, w, Y \Vdash \varphi$. (ii) can be proved analogously.

Proof of Theorem 3.4. We prove the contrapositive. Suppose $\Gamma \not\models_{cl} \varphi$. Then there is a valuation $V : At \to \{1, 0\}$ such that $V(\Gamma) = 1$ and $V(\varphi) = 0$. Let $\mathfrak{M} = (\{w\}, f, V')$, where $f(w, \{w\}) = w$, $V'(p) = \{w\}$ iff $V(p) = 1$ for all $p \in At$. It can be verified by induction that $\mathfrak{M}, w, \{w\} \Vdash \psi$ iff $V(\psi) = 1$ and $\mathfrak{M}, w, \{w\} \dashv\!\mid \psi$ iff $V(\psi) = 0$ for all Boolean ψ. Hence, $\mathfrak{M}, w, [\![\Delta]\!]_+^{\{w\}} \Vdash \Sigma$ and $\mathfrak{M}, w, [\![\Delta]\!]_+^{\{w\}} \dashv\!\mid \varphi$, whence $\Sigma \not\models^{\Delta} \varphi$.

For the other direction, suppose $\Sigma \not\models^{\Delta} \varphi$. Then there is a contextually pointed model $(\mathfrak{M}, w, X)$ with $\mathfrak{M} = (W, f, V)$ such that $\mathfrak{M}, w, [\![\Delta]\!]_+^W \Vdash \Sigma$ and $\mathfrak{M}, w, [\![\Delta]\!]_+^W \dashv\!\mid \varphi$. Define $V' : At \to \{1, 0, u\}$ such that for all $p \in At$, $V'(p) = 1$ iff $f(w, [\![\Delta]\!]_+^W) \in V(p)$. By Lemma 2.4, V' is well-defined. It can be verified by induction that $\mathfrak{M}, w, [\![\Delta]\!]_+^W \Vdash \psi$ iff $V'(\psi) = 1$ and $\mathfrak{M}, w, [\![\Delta]\!]_+^W \dashv\!\mid \psi$ iff $V'(\psi) = 0$. Moreover, since $\mathfrak{M}, w, [\![\Delta]\!]_+^W \dashv\!\mid \varphi$, by Lemma 2.4, $f(w, [\![\Delta]\!]_+^W)$ is defined. Then by (id), we have $\mathfrak{M}, f(w, [\![\Delta]\!]_+^W), W \Vdash \Delta$. By Lemma 2.4 again, we have $\mathfrak{M}, w, [\![\Delta]\!]_+^W \Vdash \Delta$. Together with $\mathfrak{M}, w, [\![\Delta]\!]_+^W \Vdash \Sigma$, it follows that $\mathfrak{M}, w, [\![\Delta]\!]_+^W \Vdash \Gamma$. Hence, $V'(\psi) = 1$ for all $\psi \in \Gamma$. Note that we also have $V'(\varphi) = 0$. It follows that $\Gamma \not\models_{cl} \varphi$.

Proof of Lemma 3.5. Given any partial selection model $\mathfrak{M} = (W, f, V)$, we have

$$\begin{aligned}
[\![\varphi > (\psi > \chi)]\!]_+^X &= [\![\psi > \chi]\!]_+^{X \cap [\![\varphi]\!]_+^X} \\
&= [\![\chi]\!]_+^{X \cap [\![\varphi]\!]_+^X \cap [\![\psi]\!]_+^{X \cap [\![\varphi]\!]_+^X}} \\
&= [\![\chi]\!]^{X \cap [\![\varphi \wedge \psi]\!]_+^X} \\
&= [\![\varphi \wedge \psi > \chi]\!]_+^X,
\end{aligned}$$

where the third '=' is by Lemma 2.5. Analogously, we have $[\![\varphi > (\psi > \chi)]\!]_-^X = [\![\varphi \wedge \psi > \chi]\!]_-^X$.

Proof of Proposition 3.6. Let $r = p$. Let $\mathfrak{M} = (W, f, V)$, where $W = \{w, u, v\}$, $f(w, \{u, v\}) = \{u\}$, $f(u, \{v\}) = \{v\}$, $V(p) = \{u, v\}$, $V(q) = \{v\}$, $V(s) = \{v\}$. Then it can be verified that $\mathfrak{M}, w, W \Vdash p > ((q > r) > s)$ but $\mathfrak{M}, w, W \dashV\!\!\vdash p \wedge (q > r) > s$.

Let $\mathfrak{M}' = (W, f, V')$ be the same as $\mathfrak{M}$, except that $V'(s) = W - V(s)$. Then it can be verified that $\mathfrak{M}', w, W \Vdash p \wedge (q > r) > s$ but $\mathfrak{M}', w, W \dashV p > ((q > r) > s)$.

Proof of Proposition 4.1. For (i), let $\mathfrak{M} = (W, f, V)$ where $W = \{w, u\}$, $V(p) = \{w\}$, $V(q) = \{w, u\}$, $V(r) = \{w\}$. Then $\mathfrak{M}, w, W \Vdash p$ and $\mathfrak{M}, w, W \Vdash p > (q > r)$ but $\mathfrak{M}, w, W \dashV q > r$.

For (ii), let $\mathfrak{M} = (W, f, V)$, where $W = \{w, u\}$, $V(p) = \{w\}$, $V(q) = \{w\}$. Then $\mathfrak{M}, w, W \Vdash p$ and $\mathfrak{M}, w, W \Vdash p > \Box q$ but $\mathfrak{M}, w, W \dashV \Box q$.

For (iii), suppose otherwise $\mathfrak{M}, w, [\![p, p > (q > r)]\!]_+^W \dashV q > r$. Then $\mathfrak{M}, w, [\![p, p > (q > r)]\!]_+^W \cap [\![q]\!]_+^X \dashV r$, where $X = [\![p, p > (q > r)]\!]_+^W$. Let $Y = [\![p, p > (q > r)]\!]_+^W \cap [\![q]\!]_+^X$. By Lemma 2.4, we have $\mathfrak{M}, f(w, Y), W \dashV r$. Let $u = f(w, Y)$. By (id), we have $\mathfrak{M}, u, W \Vdash p$, $\mathfrak{M}, u, W \Vdash p > (q > r)$, and $\mathfrak{M}, u, X \Vdash q$. By $\mathfrak{M}, u, W \Vdash p > (q > r)$ and Lemma 3.5, we have $\mathfrak{M}, u, W \Vdash p \wedge q > r$. Note that $u \in X$ and $u \in Y$. By Lemma 2.6, we have $\mathfrak{M}, u, W \Vdash q$. Then $\mathfrak{M}, u, W \Vdash p \wedge q$. It follows by (iii) that $\mathfrak{M}, u, W \not\dashV r$, contradicting $\mathfrak{M}, f(w, Z), W \dashV r$.

For (iv), suppose otherwise $\mathfrak{M}, w, [\![p, p > \Box q]\!]_+^W \dashV \Box q$. Then there exists $u \in [\![p, p > \Box q]\!]_+^W \dashV q$ such that $\mathfrak{M}, u, [\![p, p > \Box q]\!]_+^W \dashV q$. By Lemma 2.4, there exists $v \in W$ such that $f(u, [\![p, p > \Box q]\!]_+^W) = v$ and $\mathfrak{M}, v, W \dashV q$. On the other hand, by (id), we have $\mathfrak{M}, v, W \Vdash p$ and $\mathfrak{M}, v, W \Vdash p > \Box q$. By the latter, we have $\mathfrak{M}, v, [\![p]\!]_+^W \Vdash \Box q$. Note that $v \in [\![p]\!]_+^W$. It follows that $\mathfrak{M}, v, [\![p]\!]_+^W \Vdash q$. Then by Lemma 2.6, we have $\mathfrak{M}, v, W \Vdash q$, contradiction.

For (v), suppose $\mathfrak{M}, w, W \Vdash \varphi$ and $\mathfrak{M}, w, W \Vdash \varphi > \psi$. By the latter, we have $\mathfrak{M}, w, [\![\varphi]\!]_+^W \Vdash \psi$. By the former, we have $w \in [\![\varphi]\!]_+^W$. Then by Lemma 2.6, we have $\mathfrak{M}, w, W \Vdash \psi$. Hence, $\mathfrak{M}, w, W \not\dashV \psi$.

For (vi), suppose otherwise $\mathfrak{M}, w, [\![\varphi, \varphi > \psi]\!]_+^W \dashV \psi$. Then there exists $u \in W$ such that $f(w, [\![\varphi, \varphi > \psi]\!]_+^W) = u$ and $\mathfrak{M}, u, W \dashV \psi$. By (id), $\mathfrak{M}, u, W \Vdash \varphi$ and $\mathfrak{M}, u, W \Vdash \varphi > \psi$. It follows by (iii) that $\mathfrak{M}, u, W \not\dashV \psi$, contradiction.

For (vii), suppose $\mathfrak{M}, w, [\![\varphi]\!]_+^W \Vdash \varphi > \psi$. Then $\mathfrak{M}, w, [\![\varphi]\!]_+^W \cap [\![\varphi]\!]^{[\![\varphi]\!]_+^W} \Vdash$

ψ. Since φ is Boolean, by Lemma 2.6, we have $\mathfrak{M}, w, [\![\varphi \wedge \varphi]\!]_+^W \Vdash \psi$, i.e., $\mathfrak{M}, w, [\![\varphi]\!]_+^W \Vdash \psi$. By Lemma 2.3, we have $\mathfrak{M}, w, [\![\varphi]\!]_+^W \not\dashv\!\!| \psi$, as required.

Proof of Proposition 4.2. (i) and (iii) can be proved using the same counter-models as in the proof of (i) of Proposition 4.1. (ii) and (iv) can be proved using the same counter-models as in the proofs of (ii) of Proposition 4.1.

For (v), suppose otherwise there exist $\mathfrak{M} = (W, f, V)$ and $w \in W$ such that $\mathfrak{M}, w, W \Vdash \varphi > \psi$, $\mathfrak{M}, w, W \Vdash \neg\psi$, and $\mathfrak{M}, w, W \dashv\!\!| \neg\varphi$. Then $\mathfrak{M}, w, [\![\varphi]\!]_+^W \Vdash \psi$ and $\mathfrak{M}, w, W \Vdash \varphi$. It follows by Lemma 2.6 that $\mathfrak{M}, w, W \Vdash \psi$, contradicting $\mathfrak{M}, w, W \Vdash \neg\psi$.

For (vi), suppose otherwise there exists $\mathfrak{M} = (W, f, V)$ and $w \in W$ such that $\mathfrak{M}, w, [\![\varphi > \psi, \neg\psi]\!]_+^W \dashv\!\!| \neg\varphi$, i.e., $\mathfrak{M}, w, [\![\varphi > \psi, \neg\psi]\!]_+^W \Vdash \varphi$. Since φ is Boolean, there exits $u \in W$ such that $f(w, [\![\varphi > \psi, \neg\psi]\!]_+^W) = u$ and $\mathfrak{M}, u, W \Vdash \varphi$. By (id), we have $\mathfrak{M}, u, W \Vdash \varphi > \psi$ and $\mathfrak{M}, u, W \Vdash \neg\psi$. By (v), we have $\mathfrak{M}, u, W \Vdash \neg\varphi$, contradiction.

Proof of Proposition 4.3. For (i), consider $\mathfrak{M} = (W, f, V)$, where $W = \{w, u, v\}$, $f(w, \{u\}) = u$, $f(w, \{u, v\}) = v$, $V(p) = \{u\}$, $V(q) = \{u, v\}$, $V(r) = \{v\}$. Then $\mathfrak{M}, w, W \Vdash p > q$ and $\mathfrak{M}, w, W \Vdash q > r$ but $\mathfrak{M}, w, W \dashv\!\!| p > r$.

For (ii), consider $\mathfrak{M} = (W, f, V)$, where $W = \{w, u, v\}$, $f(w, \{u, v\}) = u$, $f(w, \{v\}) = f(u, \{v\}) = f(v, \{v\}) = v$, $V(p) = \{v\}$, $V(q) = \{u, v\}$, $V(r) = \{u\}$. Then $[\![p > q, q]\!]_+^W = \{w, u, v\} \cap \{u, v\} = \{u, v\}$, $[\![p > q, p]\!]_+^W = \{w, u, v\} \cap \{v\} = \{v\}$. It follows that $\mathfrak{M}, w, [\![p > q]\!]_+^W \Vdash q > r$ and $\mathfrak{M}, w, [\![p > q]\!]_+^W \dashv\!\!| p > r$. Hence, $q > r \not\models^{p>q} p > r$.

For (iii), suppose $\mathfrak{M}, w, [\![\psi > \chi]\!]_+^W \Vdash \varphi > \psi$. Then $\mathfrak{M}, w, [\![\psi > \chi]\!]_+^W \cap [\![\varphi]\!]^X \Vdash \psi$, where $X = [\![\psi > \chi]\!]_+^X$. Let $Y = [\![\psi > \chi]\!]_+^W \cap [\![\varphi]\!]^X$. By Lemma 2.4, there exists $u \in W$ such that $f(w, Y) = u$ and $\mathfrak{M}, u, W \Vdash \psi$. By (id), we have $\mathfrak{M}, u, W \Vdash \psi > \chi$. It follows by (v) of Proposition 4.1 that $\mathfrak{M}, u, W \not\dashv\!\!| \chi$. By Lemma 2.4 again, we have $\mathfrak{M}, w, Y \not\dashv\!\!| \chi$. Hence, $\mathfrak{M}, w, [\![\psi > \chi]\!]_+^W \not\dashv\!\!| \varphi > \chi$, as required.

For (iv), suppose otherwise there exist $\mathfrak{M} = (W, f, V)$ and $w \in W$ such that $\mathfrak{M}, w, [\![\varphi > \psi, \psi > \chi]\!]_+^W \dashv\!\!| \varphi > \chi$. Then $\mathfrak{M}, w, [\![\varphi > \psi, \psi > \chi]\!]_+^W \cap [\![\varphi]\!]_+^X \dashv\!\!| \chi$, where $X = [\![\varphi > \psi, \psi > \chi]\!]_+^W$. By Lemma 2.5, we have $\mathfrak{M}, w, [\![\varphi > \psi, \psi > \chi, \varphi]\!]_+^W \dashv\!\!| \chi$. Let $Y = [\![\varphi > \psi, \psi > \chi, \varphi]\!]_+^W$. By Lemma 2.4, there exists $u \in W$ such that $f(w, Y) = u$ and $\mathfrak{M}, u, W \dashv\!\!| \chi$. On the other hand, by the proof of (v) of Proposition 4.1 and (id), we have $\mathfrak{M}, u, W \Vdash \chi$, contradiction.

For (v), consider $\mathfrak{M} = (W, f, V)$, where $W = \{w, u, v\}$, $f(w, \{u, v\}) = u$, $f(w, \{u\}) = u$, $V(p) = W$, $V(q) = \{u, v\}$, $V(r) = \{u\}$. Then it can be verified that $\mathfrak{M}, w, [\![(p > (q > r), (q > r) > s]\!]_+^W \dashv\!\!| p > s$.

(vi) can be proved using the same counter-model as in (v).

For (vii), consider $\mathfrak{M} = (W, f, V)$, where $W = \{w, u\}$, $f(w, \{w\}) = f(w, \{w, u\}) = w$, $f(u, \{w\}) = w$, $f(u, \{u\}) = f(u, \{w, u\}) = u$, $V(p) = \{w\}$, $V(q) = \{w, u\}$, $V(r) = \{u\}$. Then it can be verified that $\mathfrak{M}, w, [\![p > q, q > \Diamond r]\!]_+^W \dashv\!\!| p > \Diamond r$.

(viii) can be proves using the same counter-model as in (vii).

Proof of Proposition 4.4. For (i), consider $\mathfrak{M} = (W, f, V)$, where $W = \{w, u\}$, $f(w, \{w, u\}) = w$, $f(w, \{u\}) = u$, $V(p) = V(r) = \{w\}$, $V(q) = \{u\}$. Then $\mathfrak{M}, w, W \Vdash p \vee q > r$ but $\mathfrak{M}, w, W \dashV q > r$.

For (ii), suppose otherwise there exist $\mathfrak{M} = (W, f, V)$ and $w \in W$ such that $\mathfrak{M}, w, [\![\varphi \vee \psi > \chi]\!]^W_+ \dashV (\varphi > \chi) \wedge (\psi > \chi)$. Then $\mathfrak{M}, w, [\![\varphi \vee \psi > \chi]\!]^W_+ \dashV \varphi > \chi$ or $\mathfrak{M}, w, [\![\varphi \vee \psi > \chi]\!]^W_+ \dashV \psi > \chi$. Let $X = [\![\varphi \vee \psi > \chi]\!]^W_+$. Then $\mathfrak{M}, w, X \cap [\![\varphi]\!]^X \dashV \chi$ and $\mathfrak{M}, w, X \cap [\![\psi]\!]^X \dashV \chi$. By Lemma 2.5, it follows that $\mathfrak{M}, w, Y \dashV \chi$ or $\mathfrak{M}, w, Z \dashV \chi$, where $Y = [\![\varphi \vee \psi > \chi, \varphi]\!]^W_+$, $Z = [\![\varphi \vee \psi > \chi, \psi]\!]^W_+$. Suppose $\mathfrak{M}, w, Y \dashV \chi$. By Lemma 2.5, there exists $u \in W$ such that $f(w, Y) = u$ and $\mathfrak{M}, u, W \dashV \chi$. By (id), we have $\mathfrak{M}, u, W \Vdash \varphi \vee \psi > \chi$ and $\mathfrak{M}, u, W \Vdash \psi$. By the latter, we have $\mathfrak{M}, u, W \Vdash \varphi \vee \psi$. Then by (v) of Proposition 4.1, we have $\mathfrak{M}, u, W \not\dashV \chi$, contradiction. A similar contradiction can be derived if $\mathfrak{M}, w, Z \dashV \chi$. Hence, $\mathfrak{M}, w, [\![\varphi \vee \psi > \chi]\!]^W_+ \not\dashV (\varphi > \chi) \wedge (\psi > \chi)$, as required.

For (iii), consider $\mathfrak{M} = (W, f, V)$, where $W = \{w, u\}$, $f(w, \{w, u\}) = f(w, \{w\}) = w$, $f(u, \{w, u\}) = u$, $f(u, \{w\}) = w$, $V(p) = \{w\}$, $V(q) = V(r) = \{u\}$. Then it can be verified that $\mathfrak{M}, w, [\![p \vee q > \Diamond r]\!]^W_+ \dashV p > \Diamond r$.

Proof of Proposition 4.5. For (i), consider $\mathfrak{M} = (W, f, V)$, where $W = \{w, u\}$, $f(w, \{u\}) = u$, $V(p) = V(q) = \{w\}$. Then it is easily verified that $\mathfrak{M}, w, W \Vdash p \vee q$ but $\mathfrak{M}, w, W \dashV \neg p > q$.

For (ii), suppose otherwise there exist $\mathfrak{M} = (W, f, V)$ and $w \in W$ such that $\mathfrak{M}, w, [\![\varphi \vee \psi]\!]^W_+ \dashV \neg\varphi > \psi$. Then $\mathfrak{M}, w, [\![\varphi \vee \psi]\!]^W_+ \cap [\![\neg\varphi]\!]^{[\![\varphi \vee \psi]\!]^W_+} \dashV \psi$. By Lemma 2.5, it follows that $\mathfrak{M}, w, [\![(\varphi \vee \psi) \wedge \neg\varphi]\!]^W_+ \dashV \psi$. By Lemma 2.4, there exists $u \in W$ such that $f(w, [\![(\varphi \vee \psi) \wedge \neg\varphi]\!]^W_+) = u$ and $\mathfrak{M}, u, W \dashV \psi$. On the other hand, we have $[\![(\varphi \vee \psi) \wedge \neg\varphi]\!]^W_+ \subseteq [\![\psi]\!]^W_+$. Then by (id), it follows that $\mathfrak{M}, u, W \Vdash \psi$, contradiction.

(iii) is straightforward by (ii), and (iv) is a reformulation of (ii).

Proof of Theorem 4.6. By (iv) of Proposition 4.5, it suffices to show that $\vDash (\neg\varphi > \psi) > (\varphi \vee \psi)$. Suppose otherwise there exist $\mathfrak{M} = (W, f, V)$ and $w \in W$ such that $\mathfrak{M}, w, W \dashV (\neg\varphi > \psi) > (\varphi \vee \psi)$. Then $\mathfrak{M}, w, [\![\neg\varphi > \psi]\!]^W_+ \dashV \varphi \vee \psi$, whence $\mathfrak{M}, w, [\![\neg\varphi > \psi]\!]^W_+ \dashV \varphi$ and $\mathfrak{M}, w, [\![\neg\varphi > \psi]\!]^W_+ \dashV \psi$. By Lemma 2.4, there exists $u \in W$ such that $f(w, [\![\neg\varphi > \psi]\!]^W_+ = u$ and $\mathfrak{M}, u, W \dashV \varphi$ and $\mathfrak{M}, u, W \dashV \psi$. By the former, we have $\mathfrak{M}, u, W \Vdash \neg\varphi$. By (id), we also have $\mathfrak{M}, u, W \Vdash \neg\varphi > \psi$. It follows from (v) of Proposition 4.1 that $\mathfrak{M}, u, W \not\dashV \psi$, contradiction.

Proof of Proposition 4.7. We prove the case for $\omega = p \wedge \Diamond\neg p$. The other case can be prove analogously.

For (i), it is easily seen that ω can be true in a model with two worlds, one verifying p and the other falsifying p.

For (ii), suppose otherwise there exist $\mathfrak{M} = (W, f, V)$ and $w \in W$ such that $\mathfrak{M}, w, [\![p \wedge \Diamond\neg p]\!]^W_+ \dashV \neg(p \wedge \Diamond\neg p)$. Then $\mathfrak{M}, w, [\![p \wedge \Diamond\neg p]\!]^W_+ \Vdash \Diamond\neg p$. Thus there exists $u \in [\![p \wedge \Diamond\neg p]\!]^W_+$ such that $\mathfrak{M}, u, [\![p \wedge \Diamond\neg p]\!]^W_+ \Vdash \neg p$. Then $f(u, [\![p \wedge \Diamond\neg p]\!]^W_+)$ is defined and $f(u, [\![p \wedge \Diamond\neg p]\!]^W_+) \notin V(p)$. On the other hand, by (id), we have $f(u, [\![p \wedge \Diamond\neg p]\!]^W_+) \in [\![p]\!]^W_+ = V(p)$, contradiction.

Proof of Proposition 4.8. For (i), consider $\mathfrak{M} = (W, f, V)$, where $W = \{w, u\}$, $f(w, W) = w$, $V(p) = \{u\}$. Then $\mathfrak{M}, w, [\![\Diamond p]\!]^W_+ \Vdash \neg p$, since $f(w, [\![\Diamond p]\!]^W_+) = f(w, W) = w$. And $\mathfrak{M}, w, [\![\Diamond p]\!]^W_+ \dashv\Vdash \bot$. Hence, $\neg p \nvDash^{\Diamond p} \bot$.

For (ii), suppose there exist $\mathfrak{M} = (W, f, V)$ and $w \in W$ such that $\mathfrak{M}, w, [\![\neg p]\!]^W_+ \Vdash \Diamond p$. Then $\mathfrak{M}, u, [\![\neg p]\!]^W_+ \Vdash p$ for some $u \in [\![\neg p]\!]^W_+$. By Lemma 2.6, we have $\mathfrak{M}, u, W \Vdash p$. On the other hand, by (id), we have $\mathfrak{M}, u, W \dashv\Vdash p$, contradiction.

Proof of Proposition 4.9. For (i), consider $\mathfrak{M} = (W, f, V)$, where $W = \{w, u\}$, $f(w, W) = \{w\}$, $f(u, W) = \{u\}$, $V(p) = W$, $V(q) = \{u\}$. Then $\mathfrak{M}, w, W \Vdash p > \neg q$ and $\mathfrak{M}, u, [\![p]\!]^W_+ \Vdash q$ and hence $\mathfrak{M}, w, W \Vdash p > \Diamond q$.

For (ii), suppose otherwise there exist $\mathfrak{M} = (W, f, V)$ and $w \in W$ such that $\mathfrak{M}, w, [\![p > \neg q, p > \Diamond q]\!]^W_+ \dashv\Vdash \neg((p > \neg q) \wedge (p > \Diamond q))$. Then $\mathfrak{M}, w, X \Vdash p > \Diamond q$, where $X = [\![p > \neg q, p > \Diamond q]\!]^W_+$. It follows that $\mathfrak{M}, w, X \cap [\![p]\!]^X_+ \Vdash \Diamond q$. Thus there exists $u \in X \cap [\![p]\!]^X_+$ such that $\mathfrak{M}, u, X \cap [\![p]\!]^X_+ \Vdash q$. By Lemma 2.6, we have $\mathfrak{M}, u, W \Vdash q$. On the other hand, by (id), we have $\mathfrak{M}, u, W \Vdash p > \neg q$ and $\mathfrak{M}, u, X \Vdash p$. Since $u \in X$, by Lemma 2.6 again, we have $\mathfrak{M}, u, W \Vdash p$. Then by (v) of Proposition 4.1, we have $\mathfrak{M}, u, W \nVdash q$, contradiction.

Proof of Proposition 4.10. Suppose otherwise there exist $\mathfrak{M} = (W, f, V)$ and $w \in W$ such that $\mathfrak{M}, w, W \dashv\Vdash (\varphi > \psi) \vee (\varphi > \neg\psi)$. Then $\mathfrak{M}, w, [\![\varphi]\!]^W_+ \dashv\Vdash \psi$ and $\mathfrak{M}, w, [\![\varphi]\!]^W_+ \dashv\Vdash \neg\psi$. From the latter we have $\mathfrak{M}, w, [\![\varphi]\!]^W_+ \Vdash \psi$, contradicting the former by Lemma 2.3.

References

[1] Bennett, J., "A Philosophical Guide to Condtionals," Oxford University Press, 2003.

[2] Bledin, J., *Logic informed*, Mind **123** (2014), pp. 277–316.

[3] Bledin, J., *Modus ponens defended*, Journal of Philosophy **112** (2015), pp. 57–83.

[4] Carlstrom, I. F. and C. S. Hill, *The Logic of Conditionals by Ernest W. Adams*, Philosophy of Science **45** (1978), pp. 155–158.

[5] Carroll, L., *A Logical Paradox*, Mind **3** (1894), pp. 436–438.

[6] Cobreros, P., P. Egré, D. Ripley and R. van Rooij, *Tolerant, classical, strict*, Journal of Philosophical Logic **41** (2012), pp. 347–385.

[7] Cross, C. B., *Jonathan Bennett on 'even If'*, Linguistics and Philosophy **8** (1985), pp. 353–357.

[8] Fine, K., *Counterfactuals without possible worlds*, The Journal of Philosophy **109** (2012), pp. 221–246.

[9] Fitting, M., "Proof Methods for Modal and Intuitionistic Logics," Springer, 1983.

[10] Gibbard, A., *Two recent theories of conditionals*, in: W. L. Harper, R. Stalnaker and G. Pearce, editors, *Ifs: Conditionals, Belief, Decision, Chance, and Time*, Reidel, Dordrecht, 1981 pp. 211–248.

[11] Gillies, AS., *Epistemic Conditionals and Conditional Epistemics*, Noûs **4** (2004), pp. 585–616.

[12] Grice, H. Paul., "Studies in the Way of Words," Harvard University Press, 1989.

[13] Holliday, W. H. and M. Mandelkern, *The Orthologic of Epistemic Modals*, Journal of Philosophical Logic (2024).

[14] Jackson, F., *On Assertion and Indicative Conditionals*, The Philosophical Review **88** (1979), pp. 565–589.

[15] Klinedinst, N., *Quantified Conditionals and Conditional Excluded Middle*, Journal of Semantics **28** (2011), pp. 149–170.

[16] Kolodny, N. and J. MacFarlane, *Ifs and Oughts*, Journal of Philosophy **107** (2010), pp. 115–143.
[17] Lewis, D., "Counterfactuals," Harvard University Press, 1973.
[18] Lewis, D., *Possible-World Semantics for Counterfactual Logics: A Rejoinder*, Journal of Philosophical Logic **6** (1977), pp. 359–363.
[19] Mandelkern, M., *Bounded Modality*, Philosophical Review **128** (2019), pp. 1–61.
[20] Mandelkern, M., *A Counterexample to Modus Ponenses*, Journal of Philosophy **117** (2020), pp. 315–331.
[21] Mandelkern, M., *Import-Export and 'And'*, Philosophy and Phenomenological Research **100** (2020), pp. 118–135.
[22] McGee, V., *A counterexample to modus ponens*, Journal of Philosophy **82** (1985), pp. 462–471.
[23] McKay, T. and P. van Inwagen, *Counterfactuals with Disjunctive Antecedents*, Philosophical Studies **31** (1977), pp. 353–356.
[24] Nute, D., *Simplification and Substitution of Counterfactual Antecedents*, Philosophia **2** (1978), pp. 317–325.
[25] Over, D., *Assumptions and the Supposed Counterexamples to Modus Ponens*, Analysis **47** (1987), pp. 142–146.
[26] Ripley, D., *Conservatively Extending Classical Logic with Transparent Truth*, The Review of Symbolic Logic **5** (2012), pp. 354–378.
[27] Santorio, P., *Path Semantics for Indicative Conditionals*, Mind **131** (2022), pp. 59–98.
[28] Stalnaker, R., *A Theory of Conditionals*, in: N. Rescher, editor, *Studies in Logical Theory*, Basil Blackwell Publishers, 1968 pp. 98–112.
[29] Stalnaker, R., *Indicative Conditionals*, Philosophia **5** (1975), pp. 269–286.
[30] Stalnaker, R. C., *A Defense of Conditional Excluded Middle*, in: W. L. Harper, R. Stalnaker and G. Pearce, editors, *IFS: Conditionals, Belief, Decision, Chance and Time*, The University of Western Ontario Series in Philosophy of Science, Springer Netherlands, Dordrecht, 1981 pp. 87–104.
[31] Stern, R. and S. Hartmann, *Two sides of modus ponens*, Journal of Philosophy **115** (2018), pp. 605–621.
[32] Strawson, P. F., "Introduction to Logical Theory," Introduction to Logical Theory, Wiley, Oxford, England, 1952, x, 266 pp.
[33] Veltman, F., *Defaults in Update Semantics*, Journal of Philosophical Logic **25** (1996), pp. 221–261.
[34] Weiss, Y., *Semantics for Counterpossibles*, The Australasian Journal of Logic **14** (2017), pp. 383–407.
[35] Williams, J. R. G., *Defending Conditional Excluded Middle*, Noûs **44** (2010), pp. 650–668.
[36] Yalcin, S., *Epistemic Modals*, Mind **116** (2007), pp. 983–1026.
[37] Yalcin, S., *A counterexample to modus tollens*, Journal of Philosophical Logic **41** (2012), pp. 1001–1024.

A First-order Modal Logic of Strict Implication on Varying-Domain Models

Yuanzhe Yang [1]

Department of Philosophy and Religious Studies
Peking University

Abstract

In studies of first-order modal logic (FOML), in order to give a semantics on varying-domain models, we usually use free logic. However, we can also keep the classical first-order logic, but revise the semantics of the modal operator instead, as shown in works on the so-called "common sense predicate modal logic". In this paper, following the latter approach, we introduce a binary strict implication operator $\phi \Rrightarrow \psi$, which is more expressive than the common sense $\Box$-operator, and is able to characterize a lot of our reasoning about necessity (or time, knowledge, etc.) in natural language that involves existence and non-existence of objects. We offer complete axiomatizations of the FOML of $\Rrightarrow$ on the class of all varying-domain models, the class of symmetric models, as well as the class of transitive models with a special property we call the "continual existence property".

Keywords: First-order modal logic, varying-domain semantics, strict implication, completeness.

1 Introduction

In natural language, it is very common to say something like "I could have not been born", "The sun will no longer exist after ten billion years", or "He is still alive, but she does not know it" - in other words, when we are talking and reasoning about necessity, time, knowledge, etc., it is very natural to involve the existence and non-existence of objects, either explicitly or implicitly.

This motivates us to consider *first-order modal logic* (FOML for short) on *varying-domain models*. But this turns out to be a tough subject to deal with - even defining a reasonable semantics is not easy. We are facing a dilemma here: on the one hand, classical first-order logic only allows us to talk about *existent objects*; on the other hand, according to the standard semantics of modal logic, when assessing the truth value of a modal formula, we need to go from one world to its successors, where existent objects might become *non-existent* since we are working on varying-domain models - but this forces us to talk about non-existent objects. As a result, in a varying-domain semantics for FOML,

[1] 1900014924@pku.edu.cn

it seems impossible to have both classical first-order logic and the standard semantics for the modal operator at the same time.

In present studies of FOML, the most popular approach to deal with varying-domain models is to keep the standard semantics of $\Box\phi$, and employ *free logic* for the first-order part, which allows all talk about non-existent objects.[2] However, this produces conceptual absurdities: for example, at the *present* moment, it sounds strange to say either "Socrates *is* white" or "Socrates *is not* white" - one would naturally feel hesitant to give them any truth value, but according to the semantics of free logic, one of them must be true, while the other one is false. On the technical level, we would also lose classical first-order logic as a result.[3]

Is there a way to avoid using free logic, then? The answer is affirmative: we can also keep the standard semantics of first-order logic, but change the semantics of the modal operator instead. For example, consider the following alternative semantics for $\Box\phi$:

$$\mathcal{M}, w, \sigma \vDash \Box\phi \iff \begin{array}{l}\text{For all } v \in W, \text{ if } wRv \text{ and } \sigma[FV(\phi)] \subseteq D_v, \\ \text{then } \mathcal{M}, v, \sigma \vDash \phi\end{array}$$

Such a semantics was introduced by Johan van Benthem[4], and is further studied by Jeremy Seligman[5] and also in [4] and [5] under the name "*common sense modal predicate logic*". Conceptually, such a semantics indeed makes sense in many cases: for example, when we say "Socrates would necessarily become a philosopher", we mean that in every possible world *where Socrates existed*, he would become a philosopher, and such a meaning can be precisely expressed by $\Box\text{Phil}(x)$.

So far, in the literature, the common sense semantics is the only alternative semantics for the modal operator on varying-domain models. But there are also natural expressions that cannot be formalized using the common sense $\Box$. For example, consider the sentence "I could have not been born." It is true iff there is some possible world where I am not born, but such a meaning cannot be captured by the common sense modal formula $\neg\Box\text{Born}(x)$, which says that there is a world *where I exist* but am not born. Thus, it is natural to consider also the following semantics for the modal operator:

$$\mathcal{M}, w, \sigma \vDash \boxdot\phi \iff \begin{array}{l}\text{For all } v \in W, \text{ if } wRv, \\ \text{then } \sigma[FV(\phi)] \subseteq D_v \text{ and } \mathcal{M}, v, \sigma \vDash \phi\end{array}$$

---

[2] Other approaches to deal with varying-domain models, e.g. further confining the objects we can talk about on a point or employing three-valued logic, are also discussed in books like [1] and [3]. But note that they both eventually arrive at the conclusion that we should use free logic. A more philosophical discussion is offered in [2].

[3] Of course, it is not the case that classical first-order logic is unconditionally more preferable than free logic. But nor is it easy to offer a conceptual justification for the use of free logic in FOML that is completely *non-ad hoc*.

[4] According to van Benthem, he started using this definition in his modal logic course in the 1970s. But in published works, it first appeared in [7] (and also in [6]).

[5] Unfortunately, we have not found any published work from Seligman on this subject.

With $\boxdot$, then, we can formalize the above sentence as $\neg\boxdot\text{Born}(x)$.

But even when we have both $\Box$ and $\boxdot$ in our language, the expressive power of the language still seems too weak. For example, consider the sentence "I will not live longer than you." Intuitively, it says that in every possible future where I am still alive (existing), you are also alive (existing); but neither the common sense formula $\Box(\text{Alive}(x) \to \text{Alive}(y))$ nor $\boxdot(\text{Alive}(x) \to \text{Alive}(y))$ can express such a meaning.

This motivates us to consider a more expressive modal operator. Instead of unary operators like $\Box$ and $\boxdot$, we now introduce a *binary* operator $\phi \Rrightarrow \psi$, which revives the notion of *strict implication*. Roughly, the semantics of $\phi \Rrightarrow \psi$ is that, in every possible world, if the objects mentioned by ϕ exist and ϕ is true, then the objects mentioned by ψ exist and ψ is true (we will define its semantics strictly in Section 2). While in the propositional case, strict implication can be defined by $\Box(\phi \to \psi)$, the kind of strict implication we are considering here cannot be defined by $\Box(\phi \to \psi)$ or $\boxdot(\phi \to \psi)$ - it goes beyond the expressivity of $\Box$ and $\boxdot$. Such a strict implication operator has very strong expressive power: not only can we express both $\Box$ and $\boxdot$ using $\Rrightarrow$ (we will show this in section 2), but sentences like the one we discussed in the previous paragraph can also be easily formalized. Thus, we believe the FOML of $\Rrightarrow$ is an interesting subject to study, since it offers us a powerful formal tool to reason about existence and non-existence in the modal context without resorting to free logic.[6]

The main results of this paper are a number of strongly complete axiomatizations of the FOML of $\Rrightarrow$, as summarized in the following table:

Frame properties	—	Seriality	Reflexivity
—	$\mathbf{QK}^{\Rrightarrow}$	$\mathbf{QD}^{\Rrightarrow}$	$\mathbf{QT}^{\Rrightarrow}$
Symmetry	$\mathbf{QKB}^{\Rrightarrow}$	$\mathbf{QDB}^{\Rrightarrow}$	$\mathbf{QTB}^{\Rrightarrow}$
Transitivity (+ continual existence)	$\mathbf{QK4}^{\Rrightarrow}_{\text{CE}}$	$\mathbf{QD4}^{\Rrightarrow}_{\text{CE}}$	$\mathbf{QS4}^{\Rrightarrow}_{\text{CE}}$

The structure of the paper is as follows: In section 2, we formally introduce the language of $\Rrightarrow$ and its semantics. Then, in section 3, we offer the minimal logic of $\Rrightarrow$ and prove the corresponding completeness theorem; in section 4, we consider axiomatizations on symmetric models; and in section 5, we consider axiomatizations on transitive models with a special property we call the "continual existence property". Finally, in section 6, we conclude the results of the paper, and consider directions for future work.

[6] It should be noted, however, that our work is essentially a work on first-order modal logic, rather than a work on *conditional logic*. We are interested in the strict implication operator $\Rrightarrow$ mainly because it has very interesting expressive power on varying-domain models, and we can axiomatize the FOML of $\Rrightarrow$ in an elegant way, but *not* because we think it constitutes a reasonable semantics for conditionals or counterfactuals - as a candidate for the semantics of conditionals, $\Rrightarrow$ is basically as problematic as standard strict implication.

2 Language and semantics

We fix an at most countable set $\mathcal{P}$ of predicates and a countable set $\mathbf{X}$ of variables. In this paper, we restrict ourselves to the *countable* case.[7]

Definition 2.1 ($\mathcal{L}^{\Rrightarrow}$) $\mathcal{L}^{\Rrightarrow}$-formulas are defined recursively as follows:

$$\phi ::= Px_1, ..., x_n \mid \mathsf{E}x \mid \neg\phi \mid (\phi \wedge \phi) \mid \forall x\phi \mid (\phi \Rrightarrow \phi)$$

where $x, x_1, ..., x_n \in \mathbf{X}$, and $P \in \mathcal{P}$ is n-ary.

Other Boolean connectives and $\exists x\phi$ are defined in the usual way.[8]

We may abbreviate a sequence of variables $x_1, ..., x_n$ as $\vec{x}$, and a sequence of quantifiers $\forall x_1 \cdots \forall x_n$ (or $\exists x_1 \cdots \exists x_n$) as $\forall\vec{x}$ (or $\exists\vec{x}$).

$\phi[y/x]$ denotes the formula obtained by replacing every free occurrences of x in ϕ with a y. We say $\phi[y/x]$ is *admissible* if in ϕ, x does not occur free within the scope of $\exists y$.

$FV(\phi)$ denotes the free variables of ϕ.

Remark 2.2 Note that the *existential predicate* E is also introduced to our language. But this will *not* lead us to free logic, since, as we will soon see, $\mathsf{E}x$ is always trivially true in our logic (in fact, we can even define $\mathsf{E}x$ as the abbreviation of $(Px \vee \neg Px)$ for some random unary predicate P). Its behavior only becomes interesting within the scope of $\Rrightarrow$. This is also an interesting feature of our logic: we can do non-trivial reasoning using a trivial existential predicate.

Now, we give the semantics of the language. First, first-order Kripke models are defined as follows:

Definition 2.3 (First-order Kripke model) A *first-order Kripke model* is a 5-tuple $\mathcal{M} = (W, D, \delta, R, \rho)$, where

- $W \neq \emptyset$ is the set of possible worlds, and $D \neq \emptyset$ is the set of objects;
- $\delta : W \to \wp(D)$ assigns each $w \in W$ a non-empty set of objects, called the local domain on w (we will often denote $\delta(w)$ as D_w);
- $R \subseteq W \times W$ is the accessibility relation on worlds;
- $\rho : \mathcal{P} \times W \to \wp(D_w^{<\omega})$ is the interpretation of the model.

As noted in the introduction, we do not want to allow all talk about non-existent objects, as is the case of free logic. Thus, on the formal level, the truth value of a formula on a world is defined only when the corresponding assignment is a *relevant assignment* in the following sense:

Definition 2.4 (Relevant assignment) Given a model $\mathcal{M} = (W, D, \delta, R, \rho)$, a $w \in W$ and $\phi \in \mathcal{L}^{\Rrightarrow}$, a *relevant assignment* σ w.r.t. $\mathcal{M}, w$ and ϕ is a function from $\mathbf{X}$ to D s.t. for all $x \in FV(\phi)$, $\sigma(x) \in D_w$ (i.e. $\sigma[FV(\phi)] \subseteq D_w$).

[7] In this paper, by "countable", we always mean *countably infinite*.

[8] We do not include the identity relation in our language; nor do we include any constant symbol or function symbol. It seems better to start considering them in our language after we become more familiar with the behavior of the $\Rrightarrow$-operator.

For a set Δ of $\mathcal{L}^{\Rrightarrow}$-formulas, We also say σ is a relevant assignment w.r.t. $\mathcal{M}, w$ and Δ if for all $\phi \in \Delta$, σ is a relevant assignment w.r.t $\mathcal{M}, w$ and ϕ.

Then, we define the following semantics with the help of the above notions:

Definition 2.5 ($\vDash$) Given $\mathcal{M} = (W, D, \delta, R, \rho)$, $w \in W$, $\phi \in \mathcal{L}^{\Rrightarrow}$ and an assignment σ relevant to $\mathcal{M}, w$ and ϕ, the satisfaction relation $\vDash$ is defined recursively as follows: [9]

$\mathcal{M}, w, \sigma \vDash Px_1, ..., x_n$	$\iff$	$(\sigma(x_1), ..., \sigma(x_n)) \in \rho(P, w)$
$\mathcal{M}, w, \sigma \vDash \mathsf{E}x$	$\iff$	$\sigma(x) \in D_w$
$\mathcal{M}, w, \sigma \vDash \neg\phi$	$\iff$	$\mathcal{M}, w, \sigma \nvDash \phi$
$\mathcal{M}, w, \sigma \vDash \phi \wedge \psi$	$\iff$	$\mathcal{M}, w, \sigma \vDash \phi$ and $\mathcal{M}, w, \sigma \vDash \psi$
$\mathcal{M}, w, \sigma \vDash \forall x\phi$	$\iff$	For all $a \in D_w$, $\mathcal{M}, w, \sigma[x \mapsto a] \vDash \phi$
$\mathcal{M}, w, \sigma \vDash \phi \Rrightarrow \psi$	$\iff$	For all $v \in W$ s.t. wRv, if $\sigma[FV(\phi)] \subseteq D_v$ and $\mathcal{M}, v, \sigma \vDash \phi$, then $\sigma[FV(\psi)] \subseteq D_v$ and $\mathcal{M}, v, \sigma \vDash \psi$

For the *semantic consequence* relation, we write $\Delta \vDash \phi$ if for all pointed model $\mathcal{M}, w$ and all assignment σ which is relevant to $\mathcal{M}, w$ and $\Delta \cup \{\phi\}$, if $\mathcal{M}, w \vDash \Delta$, then $\mathcal{M}, w \vDash \phi$. As usual, ϕ is *valid* if $\emptyset \vDash \phi$.

Note that we consider only relevant assignments in our semantics - in other words, if σ is not a relevant assignment w.r.t. $\mathcal{M}, w$ and ϕ, then the satisfaction relation between $\mathcal{M}, w, \sigma$ and ϕ is left undefined. As a result, $\mathsf{E}x$ is true so long as its truth value is defined.

Also note that even when σ is relevant w.r.t. $\mathcal{M}, w$ and ϕ, it does not follow that for all successor v of w, σ is also relevant w.r.t. $\mathcal{M}, v$ and ϕ - In fact, this is exactly what makes the semantics for FOML on varying-domain models difficult to define. As a response, our semantics for $\Rrightarrow$ guarantees that the truth value of $\phi \Rrightarrow \psi$ is defined for $\mathcal{M}, w, \sigma$ as long as σ is relevant w.r.t. $\mathcal{M}, w$ and ϕ, ψ, since whether the truth values of ϕ and ψ are defined on a successor of w also constitutes a part of the truth condition of $\phi \Rrightarrow \psi$ on w. This should be viewed as a generalization of the idea behind the common sense $\Box$.

As we have noted in the introduction, according to the semantics of $\Rrightarrow$, we can easily formalize the sentence we discussed in the introduction, "I will not live longer than you", with a formula of the form Alive(x) $\Rrightarrow$ Alive(y) (or simply $\mathsf{E}x \Rrightarrow \mathsf{E}y$). Moreover, as noted in the introduction, using $\Rrightarrow$, we can express both the common sense $\Box$ and the operator $\boxdot$. Their definitions in $\mathcal{L}^{\Rrightarrow}$ are as follows: [10]

$$\Box\phi := \neg\phi \Rrightarrow \bot$$
$$\boxdot\phi := \top \Rrightarrow \phi$$

[9] $\sigma[x \mapsto a]$ is the assignment which is identical with σ except that $\sigma[x \mapsto a](x) = a$. It is not hard to check that the semantics for $\forall x\phi$ and $\phi \Rrightarrow \psi$ are indeed well-defined: though their semantics involves changes in assignments or points, in the end, we are still dealing with assignments relevant to the point and formula in question.

[10] $\bot$ and $\top$ stand for logical contradiction and tautology with no free variable, respectively. For example, we may define $\top$ as the abbreviation for $\exists x \mathsf{E}x$, and $\bot$ as $\neg\top$.

It is easy to check that the above definitions work in the expected way. In the following discussions, we take these definitions to be our "official" definitions of $\Box\phi$ and $\boxdot\phi$ in the language $\mathcal{L}^{\Rrightarrow}$ (accordingly, $\Diamond\phi$ is viewed as the abbreviation for $\neg(\phi \Rrightarrow \bot)$).[11]

Conversely, there clearly exist formulas in $\mathcal{L}^{\Rrightarrow}$ that cannot be expressed even when we are allowed to use both $\Box$ and $\boxdot$. For an example, see Example A.1 in the appendix. Thus, $\mathcal{L}^{\Rrightarrow}$ is strictly more expressive than the first-order modal language with only $\Box$ and $\boxdot$ (we will not offer a precise characterization of the expressivity of $\mathcal{L}^{\Rrightarrow}$ in this paper, though).

For convenience, we also introduce the following abbreviation:

$$\Diamond^{-\psi}\phi := \neg(\phi \Rrightarrow \psi)$$

Intuitively, $\Diamond^{-\psi}\phi$ says that the present point can see a successor where ϕ is defined and true, and ψ is either undefined or false. To put it more strictly:

$$\mathcal{M}, w, \sigma \vDash \Diamond^{-\psi}\phi \iff \begin{array}{l}\text{There is } v \in W \text{ s.t. } wRv,\ \sigma[FV(\phi)] \subseteq D_v,\\ \mathcal{M}, v, \sigma \vDash \phi, \text{ and either } \sigma[FV(\psi)] \nsubseteq D_v \text{ or}\\ \sigma[FV(\psi)] \subseteq D_v \text{ but } \mathcal{M}, v, \sigma \nvDash \psi\end{array}$$

In particular, when the formula ψ in $\Diamond^{-\psi}\phi$ is of the form $\mathsf{E}x$, $\Diamond^{-\mathsf{E}x}\phi$ says that the present point can see a successor where ϕ is defined and true, and the object denoted by x does not exist (since $\mathsf{E}x$ cannot be false when its truth value is defined, as we have noted above). Thus, we may further abbreviate $\Diamond^{-\mathsf{E}x}\phi$ as $\Diamond^{-x}\phi$. Formally, the semantics of $\Diamond^{-x}\phi$ is the following one:

$$\mathcal{M}, w, \sigma \vDash \Diamond^{-x}\phi \iff \begin{array}{l}\text{There is } v \in W \text{ s.t. } wRv,\ \sigma[FV(\phi)] \subseteq D_v,\\ \mathcal{M}, v, \sigma \vDash \phi, \text{ and } \sigma(x) \notin D_v\end{array}$$

In fact, we can also treat $\Diamond^{-x}\phi$ as a primitive operator instead, and add it to the language of common sense FOML. The resulting language have exactly the same expressivity as $\mathcal{L}^{\Rrightarrow}$, since $\phi \Rrightarrow \psi$ can be expressed by $\neg\Diamond^{-\psi}\phi$, and $\Diamond^{-\psi}\phi$ can be defined using $\Diamond^{-x}$ and the common sense $\Diamond$ as follows:

$$\Diamond^{-\psi}\phi := \Diamond(\phi \wedge \neg\psi) \vee \bigvee_{x \in FV(\psi)} \Diamond^{-x}\phi$$

In this sense, $\mathcal{L}^{\Rrightarrow}$ is more expressive than language of common sense FOML exactly because it has formulas of the form $\Diamond^{-x}\phi$, so it is essentially a language that extends common sense FOML by adding the $\Diamond^{-x}$ operator. That being said, we still choose to take the binary operator $\Rrightarrow$ (rather than $\Diamond$ and $\Diamond^{-x}$) as the primitive operator of our language, since the semantics of $\phi \Rrightarrow \psi$ is rather natural, and when $\Rrightarrow$ is the primitive operator of the language, the logics of $\Rrightarrow$ can be axiomatized in a very intuitive way, as we will soon see in the next section.

---

[11] Of course, a probably more straightforward way to define $\Box\phi$ is to use the $\mathcal{L}^{\Rrightarrow}$-formula $(\bigwedge_{x \in FV(\phi)} \mathsf{E}x) \Rrightarrow \phi$. But we choose to define $\Box\phi$ with $\neg\phi \Rrightarrow \bot$ because this definition seems more elegant, and it is also easier to prove some of the theorems involving $\Box\phi$ in the axiomatic systems when $\Box\phi$ is defined this way.

3 The minimal logic

3.1 The logic $\mathbf{QK}^{\Rrightarrow}$

We now introduce the minimal logic of $\Rrightarrow$, which corresponds to the class of all varying-domain first-order Kripke models.

The logic $\mathbf{QK}^{\Rrightarrow}$

Axioms		**Rules**	
TAUT	All propositional tautologies	MP	$\phi, \phi \to \psi \vdash \psi$
UI	$\forall x\phi \to \phi[y/x]$ ($\phi[y/x]$ is admissible)	UG	$\dfrac{\vdash \psi \to \phi}{\vdash \psi \to \forall x\phi}$ $(x \notin FV(\psi))$
E	$\mathsf{E}x$		
CH$^{\Rrightarrow}$	$((\phi \Rrightarrow \psi) \wedge (\psi \Rrightarrow \chi)) \to (\phi \Rrightarrow \chi)$		
AND$^{\Rrightarrow}$	$((\phi \Rrightarrow \psi) \wedge (\phi \Rrightarrow \chi)) \to (\phi \Rrightarrow (\psi \wedge \chi))$	NEC$^{\Rrightarrow}$	$\dfrac{\vdash \phi \to \psi}{\vdash \phi \Rrightarrow \psi}$ $(FV(\psi) \subseteq FV(\phi))$
OR$^{\Rrightarrow}$	$((\phi \Rrightarrow \chi) \wedge (\psi \Rrightarrow \chi)) \to ((\phi \vee \psi) \Rrightarrow \chi)$		
EXP$^{\Rrightarrow}$	$\bot \Rrightarrow \phi$		

Proposition 3.1 *(Soundness of* $\mathbf{QK}^{\Rrightarrow}$*)* $\mathbf{QK}^{\Rrightarrow}$ *is sound w.r.t. the class of all varying-domain first-order Kripke models.*

As we can see, within $\mathbf{QK}^{\Rrightarrow}$, we still have an intact classical first-order logic - unlike the case of free logic, we do not need the existential predicate when writing the first-order axioms and rules. In fact, the only axiom that mentions the existential predicate here is E, which says that $\mathsf{E}x$ always holds trivially.

As for the modal part of the logic, the axioms for $\Rrightarrow$ are just what one would naturally expect in a logic of strict implication; the rule NEC$^{\Rrightarrow}$, on the other hand, distinguishes $\Rrightarrow$ from standard strict implication, since it sets an requirement on the pattern of *free variables* of the formula to which it can be applied. Such a rule should remind us of Seligman's axiom $\mathtt{K}_{inv}$ used in the axiomatizations of common sense FOML (which can be found in [4] and [5]):

$$\mathtt{K}_{inv} \quad \Box(\phi \to \psi) \to (\Box\phi \to \Box\psi) \quad (FV(\phi) \subseteq FV(\psi))$$

And $\mathtt{K}_{inv}$ can indeed be deduced as a theorem in $\mathbf{QK}^{\Rrightarrow}$, when $\Box\phi$ is viewed as the abbreviation of $\neg\phi \Rrightarrow \bot$.

It is also worth noting that no interaction axiom or rule between the first-order part and the modal part is needed in this minimal logic. This is not always true for axiomatizations of FOML under varying-domain semantics.[12]

Below is a list of theorems and derivable rules of $\mathbf{QK}^{\Rrightarrow}$. The way to deduce them is shown in the proof of Proposition A.2 in the appendix.

---

[12] For example, we sometimes need rules of the following form in axiomatizations of varying-domain FOML, which involves both quantifier and modal operator (see, for example, Chapter 16 of [1]):

$$\frac{\vdash \psi_0 \to \Box(\psi_1 \to \Box(\psi_2 \to \cdots \Box(\psi_n \to \Box\phi)\cdots))}{\vdash \psi_0 \to \Box(\psi_1 \to \Box(\psi_2 \to \cdots \Box(\psi_n \to \Box\forall x\phi)\cdots))} \quad (x \notin \bigcup_{i \leq n} FV(\psi_i))$$

ID$^{\Rrightarrow}$	$\phi \Rrightarrow \phi$
ANT$^{\Rrightarrow}$	$(\phi \Rrightarrow \psi) \to (\phi \Rrightarrow (\phi \wedge \psi))$
AS$^{\Rrightarrow}$	$(\phi \Rrightarrow \psi) \to ((\phi \wedge \chi) \Rrightarrow \psi)$
MONO$^{\Rrightarrow}$	$\dfrac{\vdash \phi \to \psi}{\vdash (\psi \Rrightarrow \chi) \to (\phi \Rrightarrow \chi)}$ $(FV(\psi) \subseteq FV(\phi))$
SUB$^{\Rrightarrow}$	$\dfrac{\vdash \phi \leftrightarrow \psi}{\vdash (\phi \Rrightarrow \chi) \leftrightarrow (\psi \Rrightarrow \chi)}$ $(FV(\psi) = FV(\phi))$
DIV$^{\Rrightarrow}$	$\Diamond^{-\psi}\exists\vec{x}\phi \to (\Diamond^{-\psi}\exists\vec{x}(\phi \wedge \chi) \vee \Diamond^{-\psi}\exists\vec{x}(\phi \wedge \neg\chi))$ $(FV(\chi) \subseteq FV(\phi))$
K$^{\Rrightarrow}$	$(\Diamond^{-\psi}(\phi \wedge \alpha) \wedge (\alpha \Rrightarrow \beta)) \to \Diamond^{-\psi}(\phi \wedge \beta)$
K$_{inv}$	$\Box(\phi \to \psi) \to (\Box\phi \to \Box\psi)$ $(FV(\phi) \subseteq FV(\psi))$
$\Box$-Alt	$\Box\phi \leftrightarrow ((\bigwedge_{x \in FV(\phi)} \mathsf{E}x) \Rrightarrow \phi)$
$\Rrightarrow$to$\Box$	$(\phi \Rrightarrow \psi) \to \Box(\phi \to \psi)$
$\Rrightarrow$-Rdc	$\Diamond^{-\psi}\phi \leftrightarrow (\Diamond(\phi \wedge \neg\psi) \vee \bigvee_{x \in FV(\psi)} \Diamond^{-x}\phi)$

3.2 Completeness

Now, we show that the logic $\mathbf{QK}^{\Rrightarrow}$ is strongly complete w.r.t. the class of all variable-domain models. We will construct a kind of varying-domain canonical model in accordance to the special semantics of the $\Rrightarrow$-operator.

In order to define the canonical model, there are some preparations that should be done in advance. First, we introduce a larger set of variables $\mathbf{X}^+$ satisfying that $\mathbf{X} \subseteq \mathbf{X}^+$ and $\mathbf{X}^+ \setminus \mathbf{X}$ is countable; we use $\mathcal{L}_+^{\Rrightarrow}$ to denote the $\Rrightarrow$-language which takes $\mathbf{X}^+$ as its set of variables. $\mathbf{X}^+$ would provide us with enough witnesses for the existential formulas.

We also need the following notions: for a set Δ of $\mathcal{L}_+^{\Rrightarrow}$-formulas, we say that Δ has the *FV-$\exists$-property*, iff for all $x \in \mathbf{X}^+$ and $\phi \in \mathcal{L}_+^{\Rrightarrow}$ s.t. $FV(\exists x\phi) \subseteq FV(\Delta)$, there is $y \in \mathbf{X}^+$ s.t. $\exists x\phi \to \phi[y/x] \in \Delta$ (where $\phi[y/x]$ is admissible); and given a logic $\mathbf{L}$, we say that Δ is an *$\mathbf{L}$-FV-maximal consistent set* (*$\mathbf{L}$-FV-MCS* for short), iff Δ is $\mathbf{L}$-consistent, and for all $\phi \in \mathcal{L}_+^{\Rrightarrow}$ s.t. $FV(\phi) \subseteq FV(\Delta)$, either $\phi \in \Delta$ or $\neg\phi \in \Delta$.[13]

Unlike standard MCSs, different FV-MCSs can have different free variables. Thus, we can use them to construct a *varying-domain* canonical model in a natural way (by taking the sets of free variables as the local domains).

It is routine to prove the following lemma:

Lemma 3.2 *Let $\mathbf{L}$ be a logic at least as strong as $\mathbf{QK}^{\Rrightarrow}$, and let Δ_0 be an $\mathbf{L}$-consistent set of $\mathcal{L}_+^{\Rrightarrow}$-formulas. Then, the following hold:*

- *If $\mathbf{Y} \subseteq \mathbf{X}^+ \setminus FV(\Delta_0)$ is countable, then there is $\Delta \subseteq \mathcal{L}_+^{\Rrightarrow}$ s.t. $\Delta_0 \subseteq \Delta$, $FV(\Delta) = FV(\Delta_0) \cup \mathbf{Y}$, Δ is $\mathbf{L}$-consistent and has the FV-$\exists$-property;*
- *There is $\Delta \subseteq \mathcal{L}_+^{\Rrightarrow}$ s.t. $\Delta_0 \subseteq \Delta$, $FV(\Delta) = FV(\Delta_0)$, and Δ is an $\mathbf{L}$-FV-MCS.*

Now, we are ready to offer our definition of the canonical model.

[13] It is not hard to check that FV-MCSs still have properties that resemble standard MCSs (e.g. if Δ is an FV-MCS, $\Delta \vdash \phi$ and $FV(\phi) \subseteq FV(\Delta)$, then $\phi \in \Delta$).
In fact, the notions of MCS and witness property introduced in [4] are very similar to the ones we introduced here.

Definition 3.3 (Canonical model) Given a logic $\mathbf{L}$, the corresponding *canonical model* $\mathcal{M}_{\mathbf{L}}^c = (W_{\mathbf{L}}^c, D_{\mathbf{L}}^c, \delta_{\mathbf{L}}^c, R_{\mathbf{L}}^c, \rho_{\mathbf{L}}^c)$ is defined as follows:

- $W_{\mathbf{L}}^c = \{\Delta \subseteq \mathcal{L}_+^{\Rrightarrow} \mid \Delta$ is an $\mathbf{L}$-FV-MCS with the FV-$\exists$-property and $\mathbf{X}^+ \setminus FV(\Delta)$ is countable$\}$, and $D_{\mathbf{L}}^c = \mathbf{X}^+$;
- $\delta_{\mathbf{L}}^c(\Delta) = FV(\Delta)$ for all $\Delta \in W_{\mathbf{L}}^c$;
- $R_{\mathbf{L}}^c$ satisfies that for all $\Delta, \Theta \in W^c$, $\Delta R_{\mathbf{L}}^c \Theta$ iff for all $\phi, \psi \in \mathcal{L}_+^{\Rrightarrow}$, if $\phi \Rrightarrow \psi \in \Delta$ and $\phi \in \Theta$, then $\psi \in \Theta$;
- $\rho_{\mathbf{L}}^c$ satisfies that for all $\Delta \in W^c$, $P \in \mathcal{P}$ and $\vec{x} \in (FV(\Delta))^{<\omega}$, $\vec{x} \in \rho_{\mathbf{L}}^c(P, \Delta)$ iff $P\vec{x} \in \Delta$.

The canonical assignment σ^c is the assignment s.t. $\sigma^c(x) = x$ for any $x \in \mathbf{X}^+$.

For simplicity, we now fix a logic $\mathbf{L}$ no weaker than $\mathbf{QK}^{\Rrightarrow}$, and omit all the prefix and subscript "$\mathbf{L}$" in the following proofs.

Note that for any $\Delta \in W^c$, we have $\exists x E x \in \Delta$, so $FV(\Delta) \neq \emptyset$ by the FV-$\exists$-property of Δ. Hence, the local domains in the canonical model are all non-empty.

We then prove the most non-trivial step in the completeness proof: the *existence lemma*.

Lemma 3.4 *(Existence) For any $\Delta \in W^c$ and $\phi, \psi \in \mathcal{L}_+^{\Rrightarrow}$, if $\Diamond^{-\psi}\phi \in \Delta$ and $\mathbf{Y} \subseteq \mathbf{X}^+ \setminus FV(\Delta)$ is countable, then there is some $\Theta \in W^c$ s.t. $FV(\Theta) \subseteq FV(\Delta) \cup \mathbf{Y}$, $\Delta R^c \Theta$, $\phi \in \Theta$ and $\psi \notin \Theta$.*[14]

Proof Let $\Delta \in W^c$ and $\phi, \psi \in \mathcal{L}_+^{\Rrightarrow}$ be arbitrary, and assume that $\Diamond^{-\psi}\phi \in \Delta$. Also let $\mathbf{Y}$ be an arbitrary countable subset of $\mathbf{X}^+ \setminus FV(\Delta)$.

We first construct a countable sequence $(\Phi_k)_{k \in \omega}$ of finite formula sets s.t. for all $k \in \omega$, $\Diamond^{-\psi} \bigwedge \Phi_k \in \Delta$ (for simplicity, we use ϕ_k to denote $\bigwedge \Phi_k$). The union of all such Φ_k's will form a "basis" of the successor Θ we want to construct, which guarantees that $\Delta R^c \Theta$.

As preparation, we first enumerate all formulas in Δ of the form $\alpha \Rrightarrow \beta$ as $\alpha_0 \Rrightarrow \beta_0$, $\alpha_1 \Rrightarrow \beta_1$, $\alpha_2 \Rrightarrow \beta_2$, ...

As the starting point, let $\Phi_0 = \{\phi\}$. Clearly we have $\Diamond^{-\psi}\phi \in \Delta$ by assumption.

Then, given Φ_k s.t. $\Diamond^{-\psi}\phi_k \in \Delta$, we construct Φ_{k+1} in the following way. Let l be the least natural number s.t. $FV(\alpha_l) \subseteq FV(\Phi_k)$, but $\neg\alpha_l \notin \Phi_k$ and $\beta_l \notin \Phi_k$. (If there does not exist such an l, simply let $\Phi_{k+1} = \Phi_k$.) Then, since $\Diamond^{-\psi}\phi_k \in \Delta$ and $FV(\alpha_l) \subseteq FV(\phi_k)$, by $\mathtt{DIV}^{\Rrightarrow}$, we have $\Diamond^{-\psi}(\phi_k \wedge \alpha_l) \in \Delta$ or $\Diamond^{-\psi}(\phi_k \wedge \neg\alpha_l) \in \Delta$. If the former holds, then since we also have $\alpha_l \Rrightarrow \beta_l \in \Delta$, by $\mathtt{K}^{\Rrightarrow}$, $\Diamond^{-\psi}(\phi_k \wedge \beta_l) \in \Delta$; so in this case, let $\Phi_{k+1} = \Phi_k \cup \{\beta_l\}$. Otherwise, we have $\Diamond^{-\psi}(\phi_k \wedge \neg\alpha_l) \in \Delta$, so we simply let $\Phi_{k+1} = \Phi_k \cup \{\neg\alpha_l\}$. In both cases, we clearly have $\Diamond^{-\psi}\phi_{k+1} \in \Delta$.

[14] Note that the lemma allows us to choose the variable set $\mathbf{Y}$ which provides witnesses for the existential formulas. The power to make such a choice is necessary for the proof of Lemma 5.2 below.

Finally, let $\Theta_0 = \bigcup_{k\in\omega} \Phi_k$. Clearly Θ_0 is consistent: if not, then there is some $k \in \omega$ s.t. $\vdash \phi_k \to \bot$, so $\phi_k \Rrightarrow \bot \in \Delta$ by $\texttt{NEC}^{\Rrightarrow}$; then, by $\texttt{EXP}^{\Rrightarrow}$, we also have $\phi_k \Rrightarrow \psi \in \Delta$, contradicting that $\Diamond^{-\psi}\phi_k \in \Delta$.

Then, let $\mathbf{Z}$ be a random countable subset of $\mathbf{Y}$ s.t. $\mathbf{Y} \setminus \mathbf{Z}$ is countable. By Lemma 3.2, we can extend Θ_0 into an FV-MCS Θ with the FV-$\exists$-property s.t. $FV(\Theta) = FV(\Theta_0) \cup \mathbf{Z}$. Clearly, $\Theta \in W^c$ and $\phi \in \Theta$.

Then, we check that $\Delta R^c \Theta$. Assume that $\alpha \Rrightarrow \beta \in \Delta$, $\alpha \in \Theta$, and suppose (towards a contradiction) that $\beta \notin \Theta$. Say the index number of the pair $\alpha \Rrightarrow \beta$ is i, i.e. $\alpha \Rrightarrow \beta = \alpha_i \Rrightarrow \beta_i$. Then, since $FV(\Theta) = FV(\Theta_0) \cup \mathbf{Z}$ and $\mathbf{Z} \cap FV(\Delta) = \emptyset$, we have $FV(\alpha_i) \subseteq FV(\Theta_0)$, so there is some $k \in \omega$ s.t. $FV(\alpha_i) \subseteq FV(\Phi_k)$. Then, there must exist some $m \geq k$, for which i becomes the least natural number s.t. $FV(\alpha_i) \subseteq FV(\Phi_m)$, $\neg\alpha_i \notin \Phi_k$ and $\beta_i \notin \Phi_k$. Hence, $\Phi_{m+1} = \Phi_m \cup \{\beta_i\}$ or $\Phi_{m+1} = \Phi_m \cup \{\neg\alpha_i\}$; and in both cases, a contradiction can be reached. Therefore, $\beta_i \in \Theta$, i.e. $\beta \in \Theta$.

Finally, we check that $\psi \notin \Theta$. If $FV(\psi) \nsubseteq FV(\Theta_0)$, then clearly $FV(\psi) \nsubseteq FV(\Theta)$, so $\psi \notin \Theta$. Thus, we assume that $FV(\psi) \subseteq FV(\Theta_0)$. Then, there is some $k \in \omega$ s.t. $FV(\psi) \subseteq FV(\Phi_k)$. Again, since $\psi \Rrightarrow \psi \in \Delta$ (by $\texttt{ID}^{\Rrightarrow}$), there is some $m \geq k$ s.t. $\Phi_{m+1} = \Phi_m \cup \{\psi\}$ or $\Phi_{m+1} = \Phi_m \cup \{\neg\psi\}$; and since $\Diamond^{-\psi}\phi_{m+1} \in \Delta$, clearly Φ_{m+1} can only be $\Phi_m \cup \{\neg\psi\}$, so $\neg\psi \in \Phi_{m+1} \subseteq \Theta$. Thus, in this case, we also have $\psi \notin \Theta$. □

Note that for all $\Delta \in W^c$ and $\phi \in \mathcal{L}^{\Rrightarrow}$, the canonical assignment σ^c is relevant w.r.t. $\mathcal{M}^c, \Delta$ and ϕ iff $FV(\phi) \subseteq FV(\Delta)$. Thus, with the existence lemma in hand, it is now routine to prove the following *truth lemma*:

Lemma 3.5 *(Truth) For all $\Delta \in W^c$ and all $\phi \in \mathcal{L}^{\Rrightarrow}_{+}$ s.t. $FV(\phi) \subseteq FV(\Delta)$, we have*

$$\mathcal{M}^c, \Delta, \sigma^c \vDash \phi \iff \phi \in \Delta$$

Then, since any $\mathbf{L}$-consistent set Δ_0 of $\mathcal{L}^{\Rrightarrow}$-formulas can be extended into an $\mathbf{L}$-FV-MCS Δ with the FV-$\exists$-property (using Lemma 3.2) which exists in $W^c_{\mathbf{L}}$, Δ_0 can be satisfied in $\mathcal{M}^c_{\mathbf{L}}$ according to the truth lemma.

Therefore, when the logic $\mathbf{L}$ in question is just $\mathbf{QK}^{\Rrightarrow}$, we have the following completeness theorem:

Theorem 3.6 *(Completeness of $\mathbf{QK}^{\Rrightarrow}$) $\mathbf{QK}^{\Rrightarrow}$ is sound and strongly complete w.r.t. the class of all varying-domain first-order Kripke models.*

It is also routine to check that if $\texttt{D}^{\Box} : \Diamond\top$ is an axiom of $\mathbf{L}$, then $\mathcal{M}^c_{\mathbf{L}}$ is serial, and if $\texttt{T}^{\Box} : \Box\phi \to \phi$ is an axiom of $\mathbf{L}$, then $\mathcal{M}^c_{\mathbf{L}}$ is reflexive. Thus, the following completeness results are also direct:

Theorem 3.7 *(Completeness of $\mathbf{QD}^{\Rrightarrow}$ and $\mathbf{QT}^{\Rrightarrow}$)*

- $\mathbf{QD}^{\Rrightarrow} := \mathbf{QK}^{\Rrightarrow} \oplus \texttt{D}^{\Box}$ *is sound and strongly complete w.r.t. the class of all serial models;*
- $\mathbf{QT}^{\Rrightarrow} := \mathbf{QK}^{\Rrightarrow} \oplus \texttt{T}^{\Box}$ *is sound and strongly complete w.r.t. the class of all reflexive models.*

4 Symmetric models

In this section, we consider axiomatizations on symmetric models, which is more complex than the basic case.[15] The complexity mainly comes from the fact that on symmetric models, we can express a "conditional version" of the Barcan formula in $\mathcal{L}^{\Rrightarrow}$.

4.1 The axiom $\mathtt{Bf}^{\Rrightarrow}_{\mathtt{Cond}}$

In axiomaitzations on symmetric models, besides the axiom $\mathtt{B}^{\Box} : \phi \to \Box\Diamond\phi$, we also need the following Barcan-like axiom:

$$\mathtt{Bf}^{\Rrightarrow}_{\mathtt{Cond}} \quad \Diamond^{-\psi}\exists x(\phi \wedge (\chi \Rrightarrow \mathsf{E}x)) \to (\chi \to \exists x\Diamond^{-\psi}\phi) \quad (x \notin FV(\psi) \cup FV(\chi))$$

Proposition 4.1 $\mathtt{Bf}^{\Rrightarrow}_{\mathtt{Cond}}$ *is valid on all symmetric models.*

Similar to the standard Barcan formula $\Diamond\exists x\phi \to \exists x\Diamond\phi$[16], the axiom $\mathtt{Bf}^{\Rrightarrow}_{\mathtt{Cond}}$ also tells us for wRv, which object in D_v should also exist in D_w. But unlike the Barcan formula, which says that *every* object in D_v also exists in D_w unconditionally, $\mathtt{Bf}^{\Rrightarrow}_{\mathtt{Cond}}$ only says that an object a in D_v also exists in D_w when some χ is satisfied on w, and $\chi \Rrightarrow \mathsf{E}x$ is satisfied on v, with x taking value a - such a condition forces a to exist in D_w, since the models we are now considering are symmetric. (In contrast, we do not need rules like $\mathtt{Bf}^{\Rrightarrow}_{\mathtt{Cond}}$ in the minimal logic $\mathbf{QK}^{\Rrightarrow}$, because in that case there does not exist a kind of condition that forces an object in D_v to also exist in D_w for all wRv).

In fact, we can also view $\mathtt{Bf}^{\Rrightarrow}_{\mathtt{Cond}}$ as a kind of $\mathtt{B}$ axiom. We can view $\neg\phi \Rrightarrow \psi$ as a special kind of $\Box$-operator $\Box^{-\psi}\phi$ (just like we treat $\neg(\phi \Rrightarrow \psi)$ as a special kind of $\Diamond$-operator $\Diamond^{-\psi}\phi$), and write $\mathtt{Bf}^{\Rrightarrow}_{\mathtt{Cond}}$ in the following equivalent form:

$$\forall x\Box^{-\psi}\neg\chi \to (\phi \to \Box^{-\psi}\forall x(\chi \to \Diamond^{-x}\phi))$$

It is not hard to see its connection with $\mathtt{B}^{\Box} : \phi \to \Box\Diamond\phi$ - we may say that it is essentially the $\Diamond^{-x}$ version of the $\mathtt{B}$ axiom.

4.2 Completeness

Unfortunately, even for a logic which includes both $\mathtt{B}^{\Box}$ and $\mathtt{BF}^{\Rrightarrow}_{\mathtt{Cond}}$ and is no weaker than $\mathbf{QK}^{\Rrightarrow}$, the kind of canonical model we defined previously is not automatically symmetric. Thus, in order to prove the completeness theorem for the symmetric case, we introduce the following notion of symmetric canonical models, which are obtained by deleting all the asymmetric edge in the original canonical model.

Definition 4.2 (Symmetric canonical model) Given an logic $\mathbf{L}$, the corresponding *symmetric canonical model* is $\mathcal{M}^s_{\mathbf{L}} = (W^c_{\mathbf{L}}, D^c_{\mathbf{L}}, \delta^c_{\mathbf{L}}, R^s_{\mathbf{L}}, \rho^c_{\mathbf{L}})$, where $\mathbf{R}^s_{\mathbf{L}} = \{(\Delta, \Theta) \in W^c_{\mathbf{L}} \times W^c_{\mathbf{L}} \mid \Delta R^c_{\mathbf{L}}\Theta \text{ and } \Theta R^c_{\mathbf{L}}\Delta\}$.

---

[15] A model $\mathcal{M} = (W, D, \delta, R, \rho)$ is symmetric, if for all $w, v \subset W$, if wRv, then vRw.

[16] Note that when $\Diamond$ takes the common sense semantics, the Barcan formula still corresponds to the decreasing-domain property, as shown in [6].

For simplicity, we now fix a logic **L** which is at least as strong as $\mathbf{QK}^{\Rrightarrow}$ and contains both $\mathtt{B}^{\Box}$ and $\mathtt{BF}^{\Rrightarrow}_{\mathtt{Cond}}$, and omit all prefix and subscript "**L**" in the following discussion.

Since we now have fewer edges in the symmetric canonical model, we need to prove a stronger version of existence lemma. The proof of such a lemma is more complex than the one for Lemma 3.4, since we can no longer handle the modal part and the first-order part separately here.

Lemma 4.3 *(Symmetric existence) For any $\Delta \in W^c$ and $\phi, \psi \in \mathcal{L}^{\Rrightarrow}_{+}$, if $\Diamond^{-\psi}\phi \in \Delta$ and $\mathbf{Y} \subseteq \mathbf{X}^+ \setminus FV(\Delta)$ is countable, then there is some $\Theta \in W^c$ s.t. $FV(\Theta) \subseteq FV(\Delta) \cup \mathbf{Y}$, $\Delta R^s \Theta$, $\phi \in \Theta$ and $\psi \notin \Theta$.*

Proof Let $\Delta \in W^c$ and $\phi, \psi \in \mathcal{L}^{\Rrightarrow}_{+}$ be arbitrary, and assume that $\Diamond^{-\psi}\phi \in \Delta$. Also let $\mathbf{Y}$ be an arbitrary countable subset of $\mathbf{X}^+ \setminus FV(\Delta)$, and let $\mathbf{Z}$ be some countable subset of $\mathbf{Y}$ s.t. $\mathbf{Y} \setminus \mathbf{Z}$ is countable.

We now construct a countable sequence $(\Phi_k)_{k \in \omega}$ s.t. for all $k \in \omega$, $\Diamond^{-\psi}\exists\vec{y_k} \bigwedge \Phi_k \in \Delta$, where $\vec{y_k}$ collects exactly all free variables in Φ_k that is not free in Δ (again, for simplicity, we use ϕ_k to denote $\bigwedge \Phi_k$).

As preparation, we first make the following enumerations:

- Enumerate all formulas as $\chi_0, \chi_4, \chi_8, \ldots$
- Enumerate all formulas of the form $\exists x \xi$ as $\exists x_1 \xi_1, \exists x_5 \xi_5, \exists x_9 \xi_9, \ldots$
- Enumerate all formulas of the form $\alpha \Rrightarrow \beta$ as $\alpha_2 \Rrightarrow \beta_2, \alpha_6 \Rrightarrow \beta_6, \alpha_{10} \Rrightarrow \beta_{10}, \ldots$
- Enumerate all formulas of the form $\gamma \Rrightarrow \mathsf{E}z$ as $\gamma_3 \Rrightarrow \mathsf{E}z_3, \gamma_7 \Rrightarrow \mathsf{E}z_7, \gamma_{11} \Rrightarrow \mathsf{E}z_{11}, \ldots$

As the starting point, let $\Phi_0 = \{\phi\}$. Clearly $\Diamond^{-\psi}\phi_0 \in \Delta$ by assumption.

Then, given Φ_k satisfying $\Diamond^{-\psi}\exists\vec{y_k}\phi_k$, we show how to construct Φ_{k+1}. Let l be the least natural number satisfying one of the followings:

(i) $FV(\chi_l) \subseteq FV(\Phi_k)$, but $\chi_l \notin \Phi_k$ and $\neg\chi_l \notin \Phi_k$;

(ii) $FV(\exists x_l \xi_l) \subseteq FV(\Phi_k)$, but $\exists x_l \xi_l \to \xi_l[y/x] \notin \Phi_k$ for all $y \in \mathbf{X}^+$;

(iii) $\alpha_l \Rrightarrow \beta_l \in \Delta$, $FV(\alpha_l) \subseteq FV(\Phi_k)$, but $\neg\alpha_l \notin \Phi_k$ and $\beta_l \notin \Phi_k$;

(iv) $\gamma_l \in \Delta$, $z_l \in \mathbf{Z}$, and $\gamma_l \Rrightarrow \mathsf{E}z_l \in \Phi_k$.

Correspondingly, we have four kinds of operations:

- If l satisfies (i), then we add either χ_l or $\neg\chi_l$ to Φ_k;
- If l satisfies (ii), then we add a witness formula $\exists x_l \xi_l \to \xi_l[y/x]$ to Φ_k, where y is a fresh variable in $\mathbf{Z}$ that we have *never* used before; [17]
- If l satisfies (iii), then we add either $\neg\alpha_l$ or β_l to Φ_k;

[17] Note that the variable y we choose here is not only a variable that does not occur freely in Φ_k, but a variable we have *never* used in the previous construction. Since we may also *remove* variables in $\mathbf{Z}$ from Φ_k's during the construction, we need this constraint to guarantee that for any index l, we need to deal with l for at most once.

- If l satisfies (iv), then we replace all free occurrences of z_l in Φ_k with a variable in $FV(\Delta)$, using $\mathtt{Bf}^{\Rrightarrow}_{\mathtt{Cond}}$ and the FV-$\exists$-property of Δ.

The first operation can be easily carried out using $\mathtt{DIV}^{\Rrightarrow}$, the second operation is basically only a matter of relettering, and the third operation is already demonstrated in the proof of Lemma 3.4. Thus, we only write down the fourth operation in detail here.

Assume that l satisfies (iv); for simplicity, also assume that $\overrightarrow{y_k} = z_l\overrightarrow{y}$. Since $\Diamond^{-\psi}\exists z_l \exists \overrightarrow{y}\phi_k \in \Delta$ and $\gamma_l \Rrightarrow \mathsf{E}z_l \in \Phi_k$, by $\mathtt{SUB}^{\Rrightarrow}$, we have

$$\Diamond^{-\psi}\exists z_l \exists \overrightarrow{y}(\phi_k \wedge (\gamma_l \Rrightarrow \mathsf{E}z_l)) \in \Delta$$

Since $\gamma_l \in \Delta$, we have $FV(\gamma_l) \subseteq FV(\Delta)$, so $\exists\overrightarrow{y}$ does not bound γ_l. Hence, we can push the quantifier $\exists\overrightarrow{y}$ inside (using $\mathtt{SUB}^{\Rrightarrow}$) to obtain the following:

$$\Diamond^{-\psi}\exists z_l (\exists \overrightarrow{y}\phi_k \wedge (\gamma_l \Rrightarrow \mathsf{E}z_l)) \in \Delta$$

Then, since we also have $\gamma_l \in \Delta$ and $z_l \notin FV(\gamma_l)$, by $\mathtt{Bf}^{\Rrightarrow}_{\mathtt{Cond}}$, it follows that $\exists z_l \Diamond^{-\psi}\exists \overrightarrow{y}\phi_k \in \Delta$. Thus, by the FV-$\exists$-property of Δ, there is some admissible $x \in FV(\Delta)$ s.t. $\Diamond^{-\psi}\exists\overrightarrow{y}\phi_k[x/z_l] \in \Delta$. Hence, let $\Phi_{k+1} = \Phi_k[x/z_l]$.[18]

Then, based on $(\Phi_k)_{k\in\omega}$, we define the following formula set:

$$\Theta = \{\theta \mid \text{ there is some } m \in \omega \text{ s.t. } \theta \in \bigcap_{k\geq m} \Phi_k\}$$

It is not hard to check that Θ is an FV-MCS with the FV-$\exists$-property, and satisfies that $\phi \in \Theta$, $\psi \notin \Theta$ and $\Delta R^c \Theta$. We only show in detail that $\Theta R^c \Delta$. It suffices to show that for all $\gamma \in \Delta$ and $z \in \mathbf{Z} \cap FV(\Theta)$, $\Diamond\gamma \in \Theta$ and $\Diamond^{-z}\gamma \in \Theta$.

For $\Diamond\gamma$: If $\Diamond\gamma \notin \Theta$, then $\Box\neg\gamma \in \Theta$, so there is some $k \in \omega$ s.t. $\Box\neg\gamma \in \Phi_k$. But then, we have $\Diamond^{-\psi}\Box\neg\gamma \in \Delta$, so $\Diamond\Box\neg\gamma \in \Delta$, and thus $\neg\gamma \in \Delta$ by $\mathtt{B}^{\Box}$, causing a contradiction. Thus, $\Diamond\gamma \in \Theta$.

For $\Diamond^{-z}\gamma$: If $\Diamond^{-z}\gamma \notin \Theta$, then $\gamma \Rrightarrow \mathsf{E}z \in \Theta$, so there is some $m \in \omega$ s.t. $\gamma \Rrightarrow \mathsf{E}z \in \bigcap_{k\geq m}\Phi_k$. But then, there must be some $n \geq m$ s.t. $\Phi_{n+1} = \Phi_n[x/z]$ for some $x \in FV(\Delta)$, contradicting that $z \in FV(\Theta)$. Thus, $\Diamond^{-z}\gamma \in \Theta$. $\Box$

With the help of the above existence lemma, the truth lemma can be easily proved. Thus, we have the following completeness theorem:

Theorem 4.4 *(Completeness of* $\mathbf{QKB}^{\Rrightarrow}$, $\mathbf{QDB}^{\Rrightarrow}$ *and* $\mathbf{QTB}^{\Rrightarrow}$*)*

- $\mathbf{QKB}^{\Rrightarrow} := \mathbf{QK}^{\Rrightarrow} \oplus \{\mathtt{B}^{\Box}, \mathtt{Bf}^{\Rrightarrow}_{\mathtt{Cond}}\}$ *is sound and strongly complete w.r.t. the class of all symmetric models;*
- $\mathbf{QDB}^{\Rrightarrow} := \mathbf{QD}^{\Rrightarrow} \oplus \{\mathtt{B}^{\Box}, \mathtt{Bf}^{\Rrightarrow}_{\mathtt{Cond}}\}$ *is sound and strongly complete w.r.t. the class of all serial and symmetric models;*
- $\mathbf{QTB}^{\Rrightarrow} := \mathbf{QT}^{\Rrightarrow} \oplus \{\mathtt{B}^{\Box}, \mathtt{Bf}^{\Rrightarrow}_{\mathtt{Cond}}\}$ *is sound and strongly complete w.r.t. the class of all reflexive and symmetric models.*

[18] To avoid using the axiom of choice, we may assume that $\mathbf{X}^+$ is indexed by natural numbers, and let the x here be the variable in $\mathbf{X}^+$ with the least index number s.t. $\Diamond^{-\psi}\exists\overrightarrow{y}\phi_k[x/z_l] \in \Delta$.

5 Transitivity and continual existence

In this section, we consider axiomatizations on transitive models.[19] Unfortunately, we still do not know how to axiomatize the logic of $\Rrightarrow$ on the class of *all* transitive models. But we can offer a rather simple axiomatization for a special kind of transitive models - namely, transitive models with the *continual existence property.*

5.1 The continual existence property

The continual existence property is defined as follows:

Definition 5.1 (CEP) We say a model $\mathcal{M} = (W, D, \delta, R, \rho)$ has the *continual existence property* (CEP for short), iff for all $w, v, u \in W$ s.t. wRv and vRu, we have $\delta(w) \cap \delta(u) \subseteq \delta(v)$.

Such a property seems very natural when we interpret the accessibility relation in the models in a *temporal* sense: interpreted this way, the property says that the existence of an object should always be *continual*: if an object exists at some time t_0, and also exists at some later time t_1, then it should also exist at any time between t_0 and t_1.[20]

On the technical level, it is also easier to offer axiomatizations for transitive models with CEP. A transitive model with such a property validates the following two axioms:

$$\begin{array}{ll} 4^{\Rrightarrow}_{\Box} & (\phi \Rrightarrow \psi) \to \Box(\phi \Rrightarrow \psi) \\ 4^{\Rrightarrow}_{\Diamond} & (\phi \Rrightarrow \psi) \to (\Diamond\phi \Rrightarrow \Diamond\psi) \end{array}$$

In contrast, on a transitive model without CEP, while $4^{\Rrightarrow}_{\Box}$ is still valid, $4^{\Rrightarrow}_{\Diamond}$ might not hold. But our completeness proof below would not work without the latter axiom.

5.2 Completeness

Again, even when a logic no weaker than $\mathbf{QK}^{\Rrightarrow}$ has both $4^{\Rrightarrow}_{\Box}$ and $4^{\Rrightarrow}_{\Diamond}$ as its axioms, the corresponding canonical model is still non-transitive, nor does it has CEP. Thus, we will find *sub-models* within the canonical model with all the properties we need.

Now, we fix a logic $\mathbf{L}$ which is at least as strong as $\mathbf{QK}^{\Rrightarrow}$ with both $4^{\Rrightarrow}_{\Box}$ and $4^{\Rrightarrow}_{\Diamond}$, and omit all the prefix and subscript "$\mathbf{L}$" in the following proofs.

The following lemma shows how to construct the sub-models we need for a FV-MCSs in the canonical model. Such a construction gives us a better control on the local domains of the points in the model, so that we can secure transitivity and CEP for the model.

---

[19] A model $\mathcal{M} = (W, D, \delta, R, \rho)$ is *transitive* if for all $w, v, u \in W$, wRv and vRu implies wRu.

[20] We have to admit that when the accessibility relation in the models is interpreted in terms of knowledge, belief or necessity, it becomes more difficult to make sense of CEP. In particular, note that if the accessibility relation is *symmetric*, then CEP implies *constant domain.*

Lemma 5.2 *Let $\Delta \in W^c$. Then, there is a sub-model $\mathcal{M}_\Delta = (W, D, \delta, R, \rho)$ of $\mathcal{M}^c$, s.t. $\Delta \in W$, and the followings hold:*

- *$\mathcal{M}_\Delta$ is transitive and has CEP;*
- *For all $\Theta \in W$ and $\Diamond^{-\psi}\phi \in \Theta$, there is $\Gamma \in W$ s.t. $\Theta R \Gamma$, $\phi \in \Gamma$ and $\psi \notin \Gamma$.*

Proof In order to construct the model $\mathcal{M}_\Delta$, we construct a countable sequence $(\Delta_k, R_k)_{k\in\omega}$, s.t. for each $k \in \omega$, we have $\Delta_k \in W^c$, $\mathbf{X}^+ \setminus \bigcup_{i\leq k} FV(\Delta_i)$ is countable, $R_k \subseteq R^c$ is a transitive relation on $\{\Delta_i \mid i \leq k\}$, and for all $i, m, j \leq k$, $\Delta_i R_k \Delta_m R_k \Delta_j$ implies $FV(\Delta_i) \cap FV(\Delta_j) \subseteq FV(\Delta_m)$. At each step of the construction, we add a witness for a $\Diamond^{-\psi}\phi$ formula in some MCS that has already been introduced.

As preparation, we first enumerate all pairs of natural number and formulas in $\mathcal{L}_+^{\Rrightarrow}$ of the form $\Diamond^{-\psi}\phi$ as $(n_0, \Diamond^{-\psi_0}\phi_0)$, $(n_1, \Diamond^{-\psi_1}\phi_1)$, $(n_2, \Diamond^{-\psi_2}\phi_2)$, ...

As the starting point of the construction, let $(\Delta_0, R_0) = (\Delta, \emptyset)$. Clearly it satisfies all conditions we require.

Then, given (Δ_0, R_0), ..., $(\Delta_k R_k)$, we show how to construct (Δ_{k+1}, R_{k+1}). Let l be the least natural number s.t. $n_l \leq k$, $\Diamond^{-\psi_l}\phi_l \in \Delta_{n_l}$, but there is no $i \leq k$ s.t. $\Delta_{n_l} R_k \Delta_i$, $\phi \in \Delta_i$ and $\psi \notin \Delta_i$ (if there is no such l, just let $\Delta_{k+1} = \Delta_k$ and $R_{k+1} = R_k$). Since $\mathbf{X}^+ \setminus \bigcup_{i\leq k} FV(\Delta_i)$ is countable, let $\mathbf{Y}$ be a countable subset of $\mathbf{X}^+ \setminus \bigcup_{i\leq k} FV(\Delta_i)$ s.t. $\mathbf{X}^+ \setminus (\bigcup_{i\leq k} FV(\Delta_i) \cup \mathbf{Y})$ is still countable. Then Lemma 3.4 gives us a $\Theta \in W^c$ s.t. $FV(\Theta) \subseteq FV(\Delta_{n_l}) \cup \mathbf{Y}$, $\Delta_{n_l} R^c \Theta$, $\phi \in \Theta$ and $\psi \notin \Theta$. Also note that according to our construction in Lemma 3.4, $FV(\Theta) \cap \mathbf{Y} \neq \emptyset$, so $\Theta \notin \{\Delta_i \mid i \leq k\}$.

Then, let $\Delta_{k+1} = \Theta$, and define R_{k+1} as follows:

$$R_{k+1} = R_k \cup \{(\Delta_{n_l}, \Theta)\} \cup \{(\Delta_i, \Theta) \mid i \leq k \text{ and } \Delta_i R_k \Delta_{n_l}\}$$

Clearly $\Delta_{k+1} = \Theta \in W^c$, $\mathbf{X}^+ \setminus \bigcup_{i\leq k+1} FV(\Delta_i)$ is countable and R_{k+1} is transitive. The only non-trivial part is to show that (i) $R_{k+1} \subseteq R^c$ and (ii) for all $i, m, j \leq k+1$, $\Delta_i R_{k+1} \Delta_m R_{k+1} \Delta_j$ implies $FV(\Delta_i) \cap FV(\Delta_j) \subseteq FV(\Delta_m)$.

For (i): We only need to show that for all $i \leq k$, $\Delta_i R_k \Delta_{n_l}$ implies $\Delta_i R^c \Theta$. Assume that $\Delta_i R_k \Delta_{n_l}$. Let $\alpha \Rrightarrow \beta \in \Delta_i$ be arbitrary, and also assume that $\alpha \in \Theta$. Then, $FV(\alpha) \subseteq FV(\Delta_i) \cap FV(\Theta)$, which implies $FV(\alpha) \subseteq FV(\Delta_{n_l})$, according to our construction of Θ. Thus, $\Diamond\alpha \in \Delta_{n_l}$. Then, since $\alpha \Rrightarrow \beta \in \Delta_i$, by $4_\Diamond^{\Rrightarrow}$, $\Diamond\alpha \Rrightarrow \Diamond\beta \in \Delta_i$; so since $\Delta_i R_k \Delta_{n_l}$ (and thus $\Delta_i R^c \Delta_{n_l}$), $\Diamond\beta \in \Delta_{n_l}$. Thus, $FV(\alpha \Rrightarrow \beta) \subseteq FV(\Delta_{n_l})$. Then, by $4_\Box^{\Rrightarrow}$, we also have $\Box(\alpha \Rrightarrow \beta) \subset \Delta_i$, so $\alpha \Rrightarrow \beta \in \Delta_{n_l}$. Finally, since $\Delta_{n_l} R^c \Theta$, we have $\beta \in \Theta$. Thus, $\Delta_i R^c \Theta$.

For (ii): Since $\Theta \notin \{\Delta_i \mid i \leq k\}$, we only need to show that for all $i, j \leq k$, $\Delta_i R_k \Delta_j R_{k+1} \Theta$ implies $FV(\Delta_i) \cap FV(\Theta) \subseteq FV(\Delta_j)$. Assume that $\Delta_i R_k \Delta_j R_{k+1} \Theta$. As we have noted above, clearly $FV(\Delta_i) \cap FV(\Theta) \subseteq FV(\Delta_{n_l})$; and according to our construction of R_{k+1}, either $\Delta_j R_k \Delta_{n_l}$ or $\Delta_j = \Delta_{n_l}$. Thus, $FV(\Delta_i) \cap FV(\Theta) \subseteq FV(\Delta_i) \cap FV(\Delta_{n_l}) \subseteq FV(\Delta_j)$.

Finally, let $\mathcal{M}_\Delta = (W, D, \delta, R, \rho)$, where $W = \{\Delta_k \mid k \in \omega\}$, $D = D^c$, $\delta = \delta^c \restriction_W$, $R = \bigcup_{k\in\omega} R_k$, and $\rho = \rho^c \restriction_{\mathcal{P}\times W}$. It is not hard to check that Δ has all the properties we need. □

Remark 5.3 Also notice that when $\mathtt{T}^{\Box}$ is in $\mathbf{L}$, the reflexive closure of the kind of sub-model constructed above also has all the properties we need.

Then, we can also easily prove a version of the truth lemma for these sub-models, so the following completeness results are direct:

Theorem 5.4 *(Completeness of* $\mathbf{QK4}^{\Rrightarrow}_{CE}$*,* $\mathbf{QD4}^{\Rrightarrow}_{CE}$ *and* $\mathbf{QS4}^{\Rrightarrow}_{CE}$*)*

- $\mathbf{QK4}^{\Rrightarrow}_{CE} := \mathbf{QK}^{\Rrightarrow} \oplus \{4^{\Rrightarrow}_{\Box}, 4^{\Rrightarrow}_{\Diamond}\}$ *is sound and strongly complete w.r.t. the class of all transitive models with CEP;*
- $\mathbf{QD4}^{\Rrightarrow}_{CE} := \mathbf{QD}^{\Rrightarrow} \oplus \{4^{\Rrightarrow}_{\Box}, 4^{\Rrightarrow}_{\Diamond}\}$ *is sound and strongly complete w.r.t. the class of all serial and transitive models with CEP;*
- $\mathbf{QS4}^{\Rrightarrow}_{CE} := \mathbf{QT}^{\Rrightarrow} \oplus \{4^{\Rrightarrow}_{\Box}, 4^{\Rrightarrow}_{\Diamond}\}$ *is sound and strongly complete w.r.t. the class of all reflexive and transitive models with CEP.*

6 Conclusion

In this paper, we introduced a strict implication operator $\Rrightarrow$, which enables us to make reasoning involving the existence and non-existence of objects on varying-domain Kripke models, in a way that does not resort to free logic. We offered complete axiomatizations for the FOML of $\Rrightarrow$ on the class of all varying-domain models, as well as the class of symmetric models and the class of transitive models with the continual existence property.

Based on our work, there are a lot of interesting directions for future work.

For example, we may further study axiomatizations of the logics of $\Rrightarrow$. There are a lot of model classes that have not been touched upon in this paper: even the axiomatization on the class of transitive models without CEP is still unknown, not to mention the class of $S5$-models; and the techniques we used in this paper do not apply to these cases directly. We may also add constant symbols, functional symbols and the identity relation to our language, and consider how to characterize their behavior in the axiomatizations.

It also seems interesting to study the other aspects of the $\Rrightarrow$-operator, e.g. its proof theoretical properties or frame definability. In [4], a sequent calculi for common sense FOML is offered, and it is also shown that cut elimination holds for such a calculi; so it is natural to wonder whether similar results can also be shown for the FOML of $\Rrightarrow$. On the other hand, in [6] and [5], frame correspondences of the common sense $\Box$ is discussed; then, since $\Rrightarrow$ is more expressive than the common sense $\Box$, it also seems interesting to consider what frame classes can we define using $\Rrightarrow$.

We may also consider further generalizations of the $\Rrightarrow$-operator: for example, for each $n \in \omega$, we can introduce an $(n+1)$-ary operator $\phi \Rrightarrow (\psi_1, ..., \psi_n)$ with the following semantics:

$$\begin{array}{lcl} \mathcal{M}, w, \sigma \vDash \phi \Rrightarrow (\psi_1, ..., \psi_n) & \iff & \text{for all } v \in W \text{ s.t. } wRv, \\ & & \text{if } \sigma[FV(\phi)] \subseteq D_v \text{ and } \mathcal{M}, v, \sigma \vDash \phi, \\ & & \text{then there is } 1 \le i \le n \text{ s.t.} \\ & & \sigma[FV(\psi_i)] \subseteq D_v \text{ and } \mathcal{M}, v, \sigma \vDash \psi_i \end{array}$$

A language with such polyadic operators would clearly be more expressive than the language with only a binary $\Rrightarrow$-operator, but it should not be very difficult to generalize the axiomatizations presented in this paper to the polyadic case. It also seems interesting to consider whether the language with all such polyadic strict implications is in a sense "maximal" in terms of expressivity.

Acknowledgement

I would like to thank Johan van Benthem and Jeremy Seligman for offering me information about the common sense modal predicate logic. I would also like to thank Yanjing Wang for his advices on the writing of this paper. Finally, I would like to thank three anonymous reviewers from AiML 2024 for their critical advices and insightful comments on this work.

References

[1] Cresswell, M. J. and G. E. Hughes, "A New Introduction to Modal Logic," Routledge, New York, 1996.

[2] Fitting, M. and R. L. Mendelsohn, "First-Order Modal Logic," Kluwer Academic Publishers, Dordrecht, Netherland, 1998.

[3] Garson, J. W., *Quantification in modal logic*, in: D. M. Gabbay and F. Guenthner, editors, *Handbook of Philosophical Logic*, Springer Netherlands, Dordrecht, 2001 pp. 267–323.

[4] Sawasaki, T. and K. Sano, *Proof-theoretic results of common sense modal predicate calculi*, in: B. Liao and Y. N. Wáng, editors, *Context, Conflict and Reasoning* (2020), pp. 127–138.

[5] Sawasaki, T. and K. Sano, *Frame definability, canonicity and cut elimination in common sense modal predicate logics*, Journal of Logic and Computation **31** (2021), pp. 1933–1958.

[6] van Benthem, J., *Frame correspondences in modal predicate logic*, Learning and Motivation - LEARN MOTIV (2010).

[7] van Benthem, J., "Modal Logic for Open Minds," Center for the Study of Language and Information, Stanford, California, 2010.

A Appendix

Example A.1 Consider the sentence $\exists x \forall y(\mathsf{E}x \Rrightarrow \mathsf{E}y)$: we can find a pair of models that can be distinguished by this sentence, but cannot be distinguished in the first-order modal language with both $\Box$ and $\boxdot$ (and without identity).

Let $\mathcal{M} = (W, D, \delta, R, \rho)$ and $\mathcal{N} = (W', D, \delta', R', \rho')$, where

- $W = \{w_0, w_1, w_2, w_3\}$, $W' = \{v_0, v_1, v_2\}$, and $D = \{a, b\}$;
- $\delta = \{(w_0, \{a, b\}), (w_1, \{a, b\}), (w_2, \{a\}), (w_3, \{b\})\}$, and $\delta' = \{(v_0, \{a\}), (v_1, \{a\}), (v_2, \{b\})\}$;
- $R = \{(w_0, w_i) \mid i \in \{1, 2, 3\}\}$, and $R' = \{(v_0, w_i) \mid i \in \{1, 2\}\}$;
- For all $P \in \mathcal{P}$, ρ satisfies that $\rho(P, w) = \emptyset$ for all $w \in W$, and ρ' satisfies that $\rho(P, v) = \emptyset$ for all $v \in W'$.

Below is a simple diagram of $\mathcal{M}$ and $\mathcal{N}$:

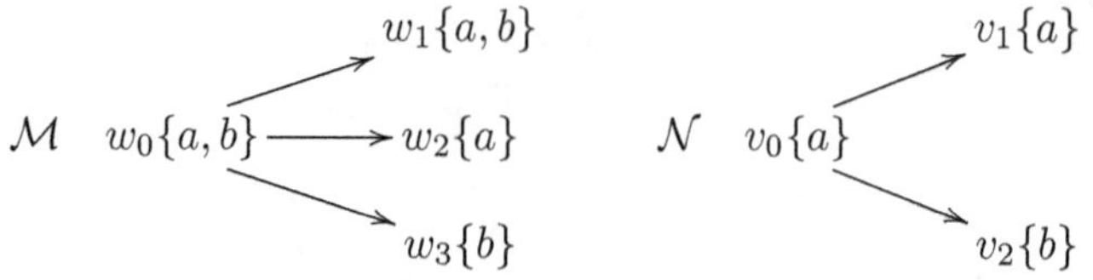

It is easy to check that $\mathcal{M}, w_0 \not\models \exists x\forall y(\mathsf{E}x \Rrightarrow \mathsf{E}y)$ and $\mathcal{N}, v_0 \models \exists x\forall y(\mathsf{E}x \Rrightarrow \mathsf{E}y)$. However, it is also not hard to check by induction that $\mathcal{M}$ and $\mathcal{N}$ cannot be distinguished in the first-order modal language with both $\Box$ and $\boxdot$ (and without identity).

Proposition A.2 ID$^{\Rrightarrow}$, ANT$^{\Rrightarrow}$, AS$^{\Rrightarrow}$, MONO$^{\Rrightarrow}$, SUB$^{\Rrightarrow}$, DIV$^{\Rrightarrow}$, K$^{\Rrightarrow}$, K_{inv}, $\Box$-Alt, $\Rrightarrow$to$\Box$ *and* $\Rrightarrow$-Rdc *are all theorems / derivable rules of* $\mathbf{QK}^{\Rrightarrow}$.

Proof (For simplicity, in the following proof, when we write $\vdash \phi$, we mean that ϕ is a theorem of $\mathbf{QK}^{\Rrightarrow}$.)

For ID$^{\Rrightarrow}$: Clearly we have $\vdash \phi \to \phi$. Then, since $FV(\phi) \subseteq FV(\phi)$, we can apply NEC$^{\Rrightarrow}$ to obtain $\vdash \phi \Rrightarrow \phi$.

For ANT$^{\Rrightarrow}$: By AND$^{\Rrightarrow}$, we have $\vdash ((\phi \Rrightarrow \phi) \wedge (\phi \Rrightarrow \psi)) \to (\phi \Rrightarrow (\phi \wedge \psi))$; and by ID$^{\Rrightarrow}$, we have $\vdash \phi \Rrightarrow \phi$. So we can combine them to obtain $\vdash (\phi \Rrightarrow \psi) \to (\phi \Rrightarrow (\phi \wedge \psi))$.

For AS$^{\Rrightarrow}$: Clearly, we have $\vdash (\phi \wedge \chi) \to \phi$. Then, since $FV(\phi) \subseteq FV(\phi \wedge \chi)$, we can apply NEC$^{\Rrightarrow}$ to obtain $\vdash (\phi \wedge \chi) \Rrightarrow \phi$. Then, by CH$^{\Rrightarrow}$, it follows that $\vdash (\phi \Rrightarrow \psi) \to ((\phi \wedge \chi) \Rrightarrow \psi)$.

For MONO$^{\Rrightarrow}$: Assume that $\vdash \phi \to \psi$ and $FV(\psi) \subseteq FV(\phi)$. Then, by NEC$^{\Rrightarrow}$, $\vdash \phi \Rrightarrow \psi$; so by CH$^{\Rrightarrow}$, it follows that $\vdash (\psi \Rrightarrow \chi) \to (\phi \Rrightarrow \chi)$. For SUB$^{\Rrightarrow}$, Just apply MONO$^{\Rrightarrow}$ twice.

For DIV$^{\Rrightarrow}$: Assume that $FV(\chi) \subseteq FV(\phi)$, we prove the following:

$$\vdash (\exists\vec{x}(\phi \wedge \chi) \Rrightarrow \psi) \wedge (\exists\vec{x}(\phi \wedge \neg\chi) \Rrightarrow \psi) \to (\exists\vec{x}\phi \Rrightarrow \psi)$$

Then, since $FV(\chi) \subseteq FV(\phi)$, by SUB$^{\Rrightarrow}$, we have

$$\vdash ((\exists\vec{x}(\phi \wedge \chi) \vee \exists\vec{x}(\phi \wedge \neg\chi)) \Rrightarrow \psi) \to (\exists\vec{x}\phi \Rrightarrow \psi)$$

Moreover, by OR$^{\Rrightarrow}$, we have

$$\vdash (\exists\vec{x}(\phi \wedge \chi) \Rrightarrow \psi) \wedge (\exists\vec{x}(\phi \wedge \neg\chi) \Rrightarrow \psi) \to ((\exists\vec{x}(\phi \wedge \chi) \vee \exists\vec{x}(\phi \wedge \neg\chi)) \Rrightarrow \psi)$$

Thus, the desired theorem can be obtained using MP.

For $\mathtt{K}^{\Rrightarrow}$: Note that $\mathtt{K}^{\Rrightarrow}$ is equivalent to the following formula, according to our definition of $\Diamond^{-\psi}$:

$$(\alpha \Rrightarrow \beta) \wedge ((\phi \wedge \beta) \Rrightarrow \psi) \to ((\phi \wedge \alpha) \Rrightarrow \psi)$$

Hence, we only show that this formula is a theorem of $\mathbf{QK}^{\Rrightarrow}$. First, by $\mathtt{AS}^{\Rrightarrow}$ (and also $\mathtt{SUB}^{\Rrightarrow}$), we have $\vdash (\alpha \Rrightarrow \beta) \to ((\phi \wedge \alpha) \Rrightarrow \beta)$. Then, by $\mathtt{ANT}^{\Rrightarrow}$, it follows that $\vdash (\alpha \Rrightarrow \beta) \to ((\phi \wedge \alpha) \Rrightarrow (\phi \wedge \alpha \wedge \beta))$. We can also easily show $\vdash (\phi \wedge \alpha \wedge \beta) \Rrightarrow (\phi \wedge \beta)$ using $\mathtt{NEC}^{\Rrightarrow}$, so we have $\vdash (\alpha \Rrightarrow \beta) \to ((\phi \wedge \alpha) \Rrightarrow (\phi \wedge \beta))$ by $\mathtt{CH}^{\Rrightarrow}$. Finally, by $\mathtt{CH}^{\Rrightarrow}$ again, we can obtain $\vdash (\alpha \Rrightarrow \beta) \wedge ((\phi \wedge \beta) \Rrightarrow \psi) \to ((\phi \wedge \alpha) \Rrightarrow \psi)$, as intended.

For $\mathtt{K}_{inv}$: Assume that $FV(\phi) \subseteq FV(\psi)$. We first show that in this case, we have $\vdash \Box(\phi \to \psi) \to (\neg\psi \Rrightarrow \neg\phi)$. Recall that $\Box(\phi \to \psi)$ is the abbreviation for $\neg(\phi \to \psi) \Rrightarrow \bot$, i.e. $(\phi \wedge \neg\psi) \Rrightarrow \bot$. Thus, by $\mathtt{EXP}^{\Rrightarrow}$ and $\mathtt{CH}^{\Rrightarrow}$, it is easy to see that $\vdash \Box(\phi \to \psi) \to (\phi \wedge \neg\psi) \Rrightarrow \neg\phi$. On the other hand, by $\mathtt{NEC}^{\Rrightarrow}$, we also have $\vdash (\neg\phi \wedge \neg\psi) \Rrightarrow \neg\phi$. Thus, by $\mathtt{OR}^{\Rrightarrow}$, it follows that

$$\vdash \Box(\phi \to \psi) \to (((\phi \wedge \neg\psi) \vee (\neg\phi \wedge \neg\psi)) \Rrightarrow \neg\phi)$$

Since $FV(\phi) \subseteq FV(\psi)$, by $\mathtt{SUB}^{\Rrightarrow}$, it follows that $\vdash \Box(\phi \to \psi) \to (\neg\psi \Rrightarrow \neg\phi)$. Finally, since $\Box\phi$ and $\Box\psi$ are abbreviations of $\neg\phi \Rrightarrow \bot$ and $\neg\psi \Rrightarrow \bot$ respectively, by $\mathtt{CH}^{\Rrightarrow}$, we have $\vdash \Box(\phi \to \psi) \to ((\neg\phi \Rrightarrow \bot) \to (\neg\psi \Rrightarrow \bot))$, i.e. $\vdash \Box(\phi \to \psi) \to (\Box\phi \to \Box\psi)$.

For $\Box$-$\mathtt{Alt}$: We first show that $\vdash \Box\phi \to ((\bigwedge_{x \in FV(\phi)} \mathsf{E}x) \Rrightarrow \phi)$. Since $\Box\phi$ is the abbreviation for $\neg\phi \Rrightarrow \bot$, by $\mathtt{EXP}^{\Rrightarrow}$ and $\mathtt{CH}^{\Rrightarrow}$, we have $\vdash \Box\phi \to (\neg\phi \Rrightarrow \phi)$. Thus, by $\mathtt{AS}^{\Rrightarrow}$, it follows that $\vdash \Box\phi \to ((\bigwedge_{x \in FV(\phi)} \mathsf{E}x \wedge \neg\phi) \Rrightarrow \phi)$. On the other hand, by $\mathtt{NEC}^{\Rrightarrow}$, we have $\vdash (\bigwedge_{x \in FV(\phi)} \mathsf{E}x \wedge \phi) \Rrightarrow \phi$. Thus, by $\mathtt{OR}^{\Rrightarrow}$, it follows that

$$\vdash \Box\phi \to (((\bigwedge_{x \in FV(\phi)} \mathsf{E}x \wedge \phi) \vee (\bigwedge_{x \in FV(\phi)} \mathsf{E}x \wedge \neg\phi)) \Rrightarrow \phi)$$

Thus, $\vdash \Box\phi \to ((\bigwedge_{x \in FV(\phi)} \mathsf{E}x) \Rrightarrow \phi)$ by $\mathtt{SUB}^{\Rrightarrow}$.

Next, we show that $\vdash ((\bigwedge_{x \in FV(\phi)} \mathsf{E}x) \Rrightarrow \phi) \to \Box\phi$. By $\mathtt{MONO}^{\Rrightarrow}$, we have $\vdash ((\bigwedge_{x \in FV(\phi)} \mathsf{E}x) \Rrightarrow \phi) \to (\neg\phi \Rrightarrow \phi)$; and by $\mathtt{ANT}^{\Rrightarrow}$, we also have $\vdash (\neg\phi \Rrightarrow \phi) \to (\neg\phi \Rrightarrow (\neg\phi \wedge \phi))$, so $\vdash (\neg\phi \Rrightarrow \phi) \to (\neg\phi \Rrightarrow \bot)$, i.e. $\vdash (\neg\phi \Rrightarrow \phi) \to \Box\phi$. Hence, by $\mathtt{MP}$, we obtain $\vdash ((\bigwedge_{x \in FV(\phi)} \mathsf{E}x) \Rrightarrow \phi) \to \Box\phi$.

For $\Rrightarrow\mathtt{to}\Box$: Recall that $\Box(\phi \to \psi)$ is the abbreviation for $\neg(\phi \to \psi) \Rrightarrow \bot$, which is equivalent to $(\phi \wedge \neg\psi) \Rrightarrow \bot$. Thus, we only need to show that $\vdash (\phi \Rrightarrow \psi) \to ((\phi \wedge \neg\psi) \Rrightarrow \bot)$. By $\mathtt{MONO}^{\Rrightarrow}$, we have $\vdash (\phi \Rrightarrow \psi) \to ((\phi \wedge \neg\psi) \Rrightarrow \psi)$; and by $\mathtt{NEC}^{\Rrightarrow}$, we also have $\vdash (\phi \wedge \neg\psi) \Rrightarrow \neg\psi$. Then, by $\mathtt{AND}^{\Rrightarrow}$, it follows that $\vdash (\phi \Rrightarrow \psi) \to ((\phi \wedge \neg\psi) \Rrightarrow (\psi \wedge \neg\psi))$, so $\vdash (\phi \Rrightarrow \psi) \to ((\phi \wedge \neg\psi) \Rrightarrow \bot)$. Hence, $\vdash (\phi \Rrightarrow \psi) \to \Box(\phi \to \psi)$.

For $\Rrightarrow$-RDC: We first show that $\vdash \Diamond^{-\psi}\phi \to (\Diamond(\phi \wedge \neg\psi) \vee \bigvee_{x\in FV(\psi)} \Diamond^{-x}\phi)$. We only need to show that we have the following theorem in $\mathbf{QK}^{\Rrightarrow}$:

$$(\Box(\phi \to \psi) \wedge \bigwedge_{x\in FV(\psi)} (\phi \Rrightarrow \mathsf{E}x)) \to (\phi \Rrightarrow \psi)$$

By $\texttt{AND}^{\Rrightarrow}$, we have $\vdash \bigwedge_{x\in FV(\psi)}(\phi \Rrightarrow \mathsf{E}x) \to (\phi \Rrightarrow \bigwedge_{x\in FV(\psi)} \mathsf{E}x)$. Moreover, by $\texttt{ANT}^{\Rrightarrow}$, we have $\vdash (\phi \Rrightarrow \bigwedge_{x\in FV(\psi)} \mathsf{E}x) \to (\phi \Rrightarrow (\phi \wedge \bigwedge_{x\in FV(\psi)} \mathsf{E}x))$; and by $\texttt{NEC}^{\Rrightarrow}$, we also have $\vdash (\phi \wedge \bigwedge_{x\in FV(\psi)} \mathsf{E}x) \Rrightarrow \bigwedge_{x\in FV(\phi)\cup FV(\psi)} \mathsf{E}x$. Thus, by $\texttt{CH}^{\Rrightarrow}$, we have

$$\vdash \bigwedge_{x\in FV(\psi)} (\phi \Rrightarrow \mathsf{E}x) \to (\phi \Rrightarrow \bigwedge_{x\in FV(\phi)\cup FV(\psi)} \mathsf{E}x)$$

Then, by $\Box$-$\texttt{Alt}$, $\vdash \Box(\phi \to \psi) \to ((\bigwedge_{x\in FV(\phi)\cup FV(\psi)} \mathsf{E}x) \Rrightarrow (\phi \to \psi))$. Thus, by $\texttt{CH}^{\Rrightarrow}$, we obtain the following:

$$\vdash (\Box(\phi \to \psi) \wedge \bigwedge_{x\in FV(\psi)} (\phi \Rrightarrow \mathsf{E}x)) \to (\phi \Rrightarrow (\phi \to \psi))$$

Thus, we only need to check that $\vdash (\phi \Rrightarrow (\phi \to \psi)) \to (\phi \Rrightarrow \psi)$. This is easy: by $\texttt{ANT}^{\Rrightarrow}$, we have $\vdash (\phi \Rrightarrow (\phi \to \psi)) \to (\phi \Rrightarrow (\phi \wedge (\phi \to \psi))$; and by $\texttt{NEC}^{\Rrightarrow}$, we have $\vdash (\phi \wedge (\phi \to \psi)) \Rrightarrow \psi$. Thus, $\vdash (\phi \Rrightarrow (\phi \to \psi)) \to (\phi \Rrightarrow \psi)$ follows by $\texttt{CH}^{\Rrightarrow}$, and we are done.

Then, we show that $\vdash (\Diamond(\phi \wedge \neg\psi) \vee \bigvee_{x\in FV(\psi)} \Diamond^{-x}\phi) \to \Diamond^{-\psi}\phi$. This is the relatively easy direction: we only need to show that $\vdash (\phi \Rrightarrow \psi) \to \Box(\phi \to \psi)$ and $\vdash (\phi \Rrightarrow \psi) \to (\phi \Rrightarrow \mathsf{E}x)$ for all $x \in FV(\psi)$. The former is just the theorem $\Rrightarrow\texttt{to}\Box$, and the latter follows directly from $\texttt{NEC}^{\Rrightarrow}$ and $\texttt{CH}^{\Rrightarrow}$. Thus, this direction of the implication is also proved. $\Box$

www.ingramcontent.com/pod-product-compliance
Lightning Source LLC
Chambersburg PA
CBHW072017240426
43667CB00043B/1457
* 9 7 8 1 8 4 8 9 0 4 6 7 5 *